TANDEM MASS SPECTROMETRY – APPLICATIONS AND PRINCIPLES

Edited by **Jeevan K. Prasain**

Tandem Mass Spectrometry – Applications and Principles
Edited by Jeevan K. Prasain

Published by InTech

Janeza Trdine 9, 51000 Rijeka, Croatia

Publishing Process Manager Martina Durovic

Technical Editor Teodora Smiljanic

Cover Designer InTech Design Team

Additional hard copies can be obtained from orders@intechopen.com

Tandem Mass Spectrometry – Applications and Principles,
Edited by Jeevan K. Prasain
p. cm.
ISBN 978-953-51-0141-3

Meet the editor

Dr Jeevan K. Prasain received his Ph.D. in 1998 from the Toyama Medical & Pharmaceutical University, Japan. He was a visiting fellow at the Biotechnology Research Institute, Montreal, Canada and then joined the Institute of Physical and Chemical Research (RIKEN), Japan in 2000 as a postdoctoral fellow where he worked on the use of tandem mass spectrometry in dereplication of natural products. He moved to the USA in 2001 and joined Dr. Stephen Barnes' group at the University of Alabama at Birmingham (UAB). Currently, he is an Assistant Professor at the UAB and his research interests include the use of tandem mass spectrometry in the study of metabolomics, lipidomics, bioavailability and pharmacokinetics of dietary natural products.

Contents

Preface

Over the last two decades tandem mass spectrometry has become one of the most versatile analytical techniques in the study of biochemicals and it is expected to expand further from its classical applications in chemical research to the study of chemical entities at the systems level. The main objective of this book is to provide a current review of the principles and various applications of tandem mass spectrometry. Although there are a number of books dealing with mass spectrometry, there is seemingly no book available in the market that can authoritatively describe the qualitative and quantitative analyses of small and macro molecules by mass spectrometry in a single text. This book comprises 30 chapters and is divided into five main sections. Attempts have been made to offer a detailed discussion on background, experimental methods and results with many illustrations in each chapter.

Thanks are expressed to the contributing authors who have attempted to provide comprehensive review from the basic principles to cutting edge research of mass spectrometry. Finally, I would like to thank Ms. Alida Lesnjakovic, Publishing Process Manager, and other InTech publication staffs for their wonderful support.

Jeevan K. Prasain
Department of Pharmacology & Toxicology
University of Alabama at Birmingham
USA

Part 1

Introduction

1

Applications of Tandem Mass Spectrometry: From Structural Analysis to Fundamental Studies

Paulo J. Amorim Madeira and M. Helena Florêncio
Centro de Química e Bioquímica, Departamento de Química e Bioquímica, Faculdade de Ciências da Universidade de Lisboa
Portugal

1. Introduction

1.1 Mass spectrometry brief history and recent developments

The first and most important question to be asked: What is mass spectrometry?

The basic principle of mass spectrometry (MS) is to generate ions from either inorganic or organic compounds by any suitable method, to separate these ions by their mass-to-charge ratio (m/z) and to detect them qualitatively and quantitatively by their respective m/z and abundance. The analyte may be ionized thermally, by electric fields or by impacting energetic electrons, ions or photons. The ... ions can be single ionized atoms, clusters, molecules or their fragments or associates. Ion separation is effected by static or dynamic electric or magnetic fields.(Gross 2004)

Although this definition dates back to 1968, when mass spectrometry was at its childhood, it is still valid. Nevertheless, two additions should be included. Firstly, besides electrons, (atomic) ions or photons, energetic neutral atoms and heavy cluster ions can also be used to ionize the analyte. Secondly, ion separation by *m/z* can be effected in field free regions, as effectively demonstrated by the time-of-flight analyser, provided the ions possess a well-defined kinetic energy at the entrance of the flight path.

From the 1950s to the present, mass spectrometry has evolved tremendously. The pioneering mass spectrometrist had a home-built naked instrument, typically a magnetic sector instrument with electron ionization. Nowadays, highly automated commercial systems, able to produce thousands of spectra per day, are now concealed in a "black box", a nicely designed and beautifully coloured unit resembling more an espresso machine or tumble dryer than a mass spectrometer.

Mass spectrometry (MS) is probably the most versatile and comprehensive analytical technique currently available in the chemists and biochemists arsenal. Mass spectrometry measures precisely the molecular masses of individual compounds by converting them into ions and analysing them in what is called a mass analyser. This is the simplest, but somewhat reductionist, definition of mass spectrometry. The days of the simple determination of the *m/z* ratio of an organic compound are over. Today, mass spectrometry can be used to determine molecular structures, to study reaction dynamics and ion chemistry, to provide thermochemical and physical properties such as ionization energy, appearance energy, reaction enthalpies, proton and ion affinities, gas-phase acidities, and so on.

Mass spectrometry is extremely versatile and several areas of physics, pharmaceutical sciences, archaeology, forensics and environmental sciences, just to mention a few, have benefited from the advances in this instrumental technique.

The history of mass spectrometry starts in 1898 with the work of Wien, who demonstrated that canal rays could be deflected by passing them through superimposed parallel electric and magnetic fields. Nevertheless, its birth can be credited to Sir J. J. Thomson, Cavendish Laboratory of University of Cambridge, through his work on the analysis of negatively and positively charged cathode rays with a parabola mass spectrograph, the great grand-father of the modern mass spectrometers.(Thomson 1897; Thomson 1907) In the next two decades, the developments of mass spectrometry continued in the hands of renowned physicists like Aston,(Aston 1919) Dempster,(Dempster 1918) Bainbridge,(Bainbridge 1932; Bainbridge and Jordan 1936) and Nier.(Nier 1940; Johnson and Nier 1953)

In the 1940s, chemists recognized the great potential of mass spectrometry as an analytical tool, and applied it to monitor petroleum refinement processes. The first commercial mass spectrometer became available in 1943 through the Consolidated Engineering Corporation. The principles of time-of-flight (TOF) and ion cyclotron resonance (ICR) were introduced in 1946 and 1949, respectively.(Sommer, Thomas et al. 1951; Wolff and Stephens 1953)

Applications to organic chemistry started to appear in the 1950s and exploded during the 1960s and 1970s. Double-focusing high-resolution mass spectrometers, which became available in the early 1950s, paved the way for accurate mass measurements. The quadrupole mass analyser and the ion traps were described by Wolfgang Paul and co-workers in 1953.(Paul 1990) Development of the GC/MS in the 1960s marked the beginning of the analysis of seemingly complex mixtures by mass spectrometry.(Ryhage 2002; Watson and Biemann 2002) The 1960s witnessed the development of tandem mass spectrometry and collision-induced decompositions,(Jennings 1968) being a high point in the field of structural analysis, in unambiguous identification by mass spectrometry, as well as in the development of soft ionization techniques such as chemical ionization.(Munson and Field 2002)

By the 1960s, mass spectrometry had become a standard analytical tool in the analysis of organic compounds. Its application to the biosciences, however, was lacking due to the inexistence of suitable methods to ionize fragile and non-volatile compounds of biological origin. During the 1980s the mass spectrometry range of applications increased "exponentially" with the development of softer ionization methods. These included fast atom bombardment (FAB) in 1981,(Barber, Bordoli et al. 1981) electrospray ionization (ESI) in 1984-1988,(Fenn, Mann et al. 1989) and matrix-assisted laser desorption/ionization (MALDI) in 1988.(Karas and Hillenkamp 2002) With the development of the last two methods, ESI and MALDI, the upper mass range was extended beyond 100kDa and had an enormous impact on the use of mass spectrometry in biology and life sciences. This impact was recognized in 2002 when John Fenn (for his work on ESI) and Koichi Tanaka (for demonstrating that high molecular mass proteins could be ionized using laser desorption) won the Nobel Prize in Chemistry.

Concurrent with the ionization methods development, several innovations in mass analyser technology, such as the introduction of high-field and superfast magnets, as well as the improvements in the TOF and Fourier transform ion cyclotron resonance (FTICR), enhanced the sensitivity and the upper mass range. The new millennium brought new advances such as two new types of ion traps, the orbitrap in 2000 by the hands of Alexander Makarov(Makarov 2000) and the linear quadrupole ion trap (LIT) in 2002 by James W. Hager. (Hager 2002)

The coupling of high-performance liquid chromatography (HPLC) with mass spectrometry was first demonstrated in the 1970s (Dass 2007). Nevertheless, it was with the development and commercialization of atmospheric pressure ionization sources (ESI, APCI) that for the first time the combination of liquid chromatography and mass spectrometry entered the realm of routine analysis.(Voyksner 1997; Covey, Huang et al. 2002; Whitehouse, Dreyer et al. 2002; Rodrigues, Taylor et al. 2007)

A full description of ionization sources and their underlying mechanisms is outside the scope of this chapter. The reader in then encouraged to seek more information in other sources. For example, there are several excellent reviews and books that cover nearly all aspects of electrospray ionization (ESI), (Pramanik, Ganguly et al. 2002; Gross 2004; Covey, Thomson et al. 2009; Kebarle and Verkerk 2009; Cole 2010; Crotti, Seraglia et al. 2011) matrix-assisted laser desorption/ionization (MALDI), (Gross 2004; Hillenkamp and Peter-Katalinic 2007; Cole 2010) atmospheric pressure photoionization (APPI) (Raffaelli and Saba 2003) and atmospheric pressure chemical ionization (APCI),(Byrdwell 2001; Covey, Thomson et al. 2009), among others.

1.2 Tandem mass spectrometry basics

I feel sure that there are many problems in Chemistry which could be solved with far greater ease by this than by any other method. The method is surprisingly sensitive – more so even than that of spectrum analysis – requires an infinitesimal amount of material and does not require this to be specially purified:...(Thomson 1921)

J. J. Thomson made this statement about mass spectrometry, but it may be even more apt in describing tandem mass spectrometry. As such, besides being the father of mass spectrometry, Thomson can also be considered the forefather of tandem mass spectrometry. In fact, to demonstrate experimentally the processes of neutralization and collisional ionization, Thomson built the first MS/MS instrument, which consisted in a serial arrangement of two magnets, with the field on one magnet oriented perpendicular to the other.(Busch, Glish et al. 1988)

Tandem mass spectrometry (MS/MS) can be considered as any general method involving at least two stages of mass analysis, either in conjunction with a dissociation process or in a chemical reaction that causes a change in the mass or charge of an ion.(Hoffmann and Stroobant 2007)

In the most common tandem mass spectrometry experiment a first analyser is used to isolate the precursor ion (m_p^+), which then undergoes fragmentation (this could be achieved either spontaneously or by making use of some activation technique) to yield product ions (m_f^+) and neutral fragments (m_n) which are then analysed by a second mass analyser. This reaction is depicted in equation 1.

$$m_p^+ \rightarrow m_f^+ + m_n \quad (1)$$

An activation barrier must be surmounted before the general reaction depicted in equation 1 can occur. The energy to overcome this barrier can come from one of two sources (Busch, Glish et al. 1988):

i. By the excess energy deposited onto the precursor ion by the ionization process. Nevertheless, this is valid only we dealing with electron ionization at high energies.
ii. By means of activation methods such as collision activated/induced dissociation (CAD and CID), infrared multiphoton dissociation (IRMPD), electron capture/electron

transfer dissociation (ECD and ETD) and surface induced dissociation (SID), for which the fundamental aspects will be discussed in section 2 of this chapter.

The principle of MS/MS is illustrated in Fig. 1. Tandem mass spectrometry can be conceived in two ways: in space by the coupling of two mass spectrometers, or in time by an appropriate sequence of events in an ion storage device. This consequently leads to two main categories of instruments that allow for tandem mass spectrometry experiments: tandem mass spectrometers in space or in time.(Hoffmann and Stroobant 2007)

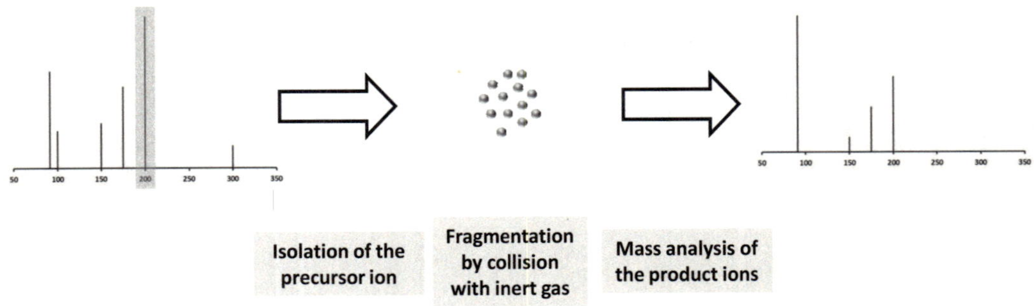

Fig. 1. Principle of MS/MS: a precursor ion is isolated by the first mass analyser, fragmented by collision with an inert gas and the products ions are analysed by the second mass spectrometer.

Tandem mass spectrometers in space are either double focusing mass spectrometers (Fig. 2) or instruments of the quadrupole type (Fig. 3).

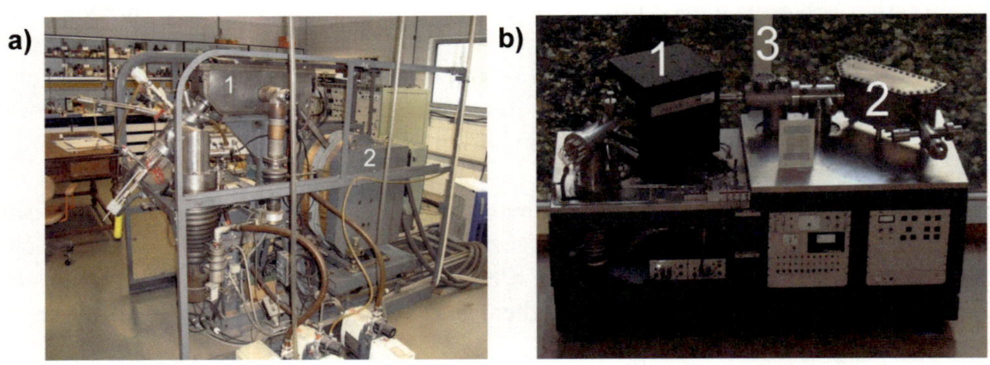

Fig. 2. A)Double focusing mass spectrometer MS9 (AEI) installed at the Forensics Biological and Environmental Mass Spectrometry Laboratory of the Faculdade de Ciências da Universidade de Lisboa, Portugal. It is a Nier-Johnson geometry mass spectrometer, electric sector-E- (1) is followed by the magnetic sector-B- (2) - EB - hence it is possible to study metastable ions through linked scans. B) Double focusing mass spectrometer ZAB 2F (VG Analytical) on display at the Faculdade de Ciências da Universidade de Lisboa, Portugal. It is a reverse geometry mass spectrometer, magnetic sector (1) followed by an electric sector (2), and it is equipped with a collision cell (3) making possible to perform CID experiments.

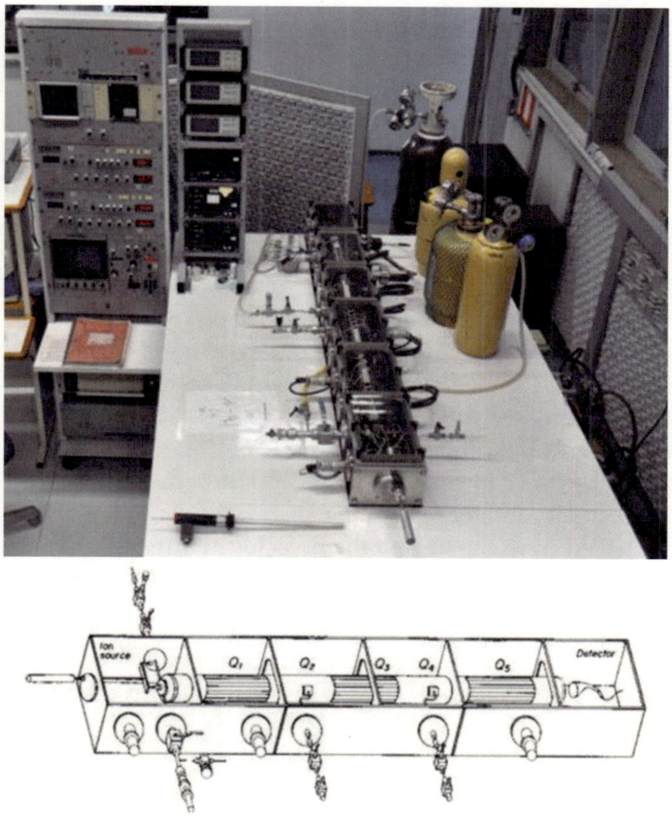

Fig. 3. Extrel Pentaquadrupole QqQqQ instaled at the ThoMSon Lab, Universidade Estadual de Campinas, Brazil. The photo was kindly supplied by Professor Marcos N. Eberlin.

Tandem mass spectrometers in time are ion traps (Fig. 4), Fourier transform ion cyclotron resonance mass spectrometers (Fig. 5) and orbitrap.
To note that the theory underlying each mass analyser will not be addressed in this chapter so the reader is encouraged to seek this information in other sources. There are several reviews on the principle of quadrupole mass spectrometry,(Douglas 2009) ion traps, (Louris, Cooks et al. 1987; Douglas, Frank et al. 2005; March 2009) FTICR, (Comisarow and Marshall 1974; Marshall, Comisarow et al. 1979; Marshall, Hendrickson et al. 1998; Barrow, Burkitt et al. 2005) time-of-flight (Price and Williams 1969; Guilhaus, Selby et al. 2000) and information on sector mass spectrometry that can easily be found in mass spectrometry textbooks.(Chapman 1985; Gross 2004; Dass 2007; Hoffmann and Stroobant 2007)
Common tandem mass spectrometers in space have, at least, two mass analysers. Quadrupoles are frequently used as mass analysers. For example, the QqQ configuration indicates an instrument with three quadrupoles, where the second one (indicated by the lower case q) is the reaction region working in RF-only mode (i.e. serving as a lens for all the ions). Other instruments combine electric and magnetic sectors (E and B) or electric and magnetic sectors and quadrupoles (E, B and qQ). Time-of-Flight (TOF) instruments equipped with a reflectron or a combination of quadrupoles with TOF instruments are also used to perform tandem mass spectrometry in space. (Hoffmann and Stroobant 2007)

Higher order MS^n spectra can be acquired by combining more analysers; nevertheless, this will certainly increase the complexity of the instrument and consequently its cost. Theoretically, any number of analysers can be sequentially combined, but since the fraction of ions transmitted in each step is low, the practical maximum, for the particular case of beam instruments, is three or four.(Hoffmann and Stroobant 2007) Nevertheless, there is always room to improve; researchers of the ThoMSon Lab (University of Campinas, Brazil) built a one-of-a-kind instrument, the pentaquadrupole (Fig. 3A), QqQqQ. In fact, the ThoMSon lab is best known as "The home of the pentaquadrupole". This instrument is still working and it is used mainly for the study of gas-phase ion-molecule reactions.
Tandem mass spectrometry experiments can also be performed through time separation with analysers such has ion traps, FTICR and orbitraps, programmed so that different steps are successively carried out in the same instrument. The maximum number of steps for these instruments is generally seven to eight (even though manufacturers claim to go as far as 10). In these instruments the proportion of ions transmitted is high, but, at each step, the mass of the fragments becomes lower.(Hoffmann and Stroobant 2007) Considering ion traps and FTICR, there are some differences among them. The most important is that in an ion trap the ions are expelled from the trap to the detector in order to get a signal (i.e. they can be observed only once at the end of the process), while in a FTICR the ions are detected in a non-destructive manner and can be measured at each step in the sequential fragmentation process.(Hoffmann and Stroobant 2007)

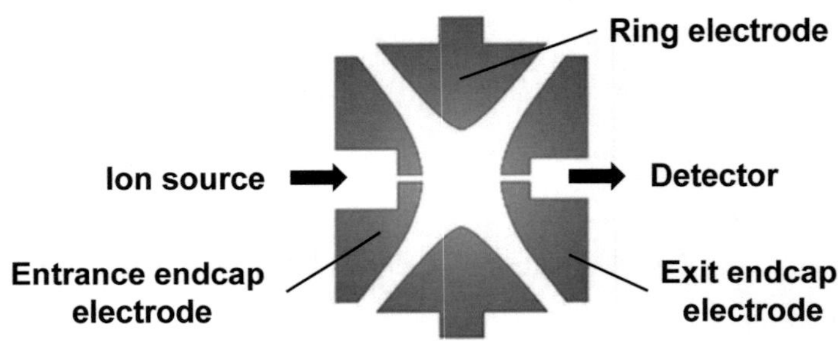

Fig. 4. Schematic representation of an ion trap mass analyser.

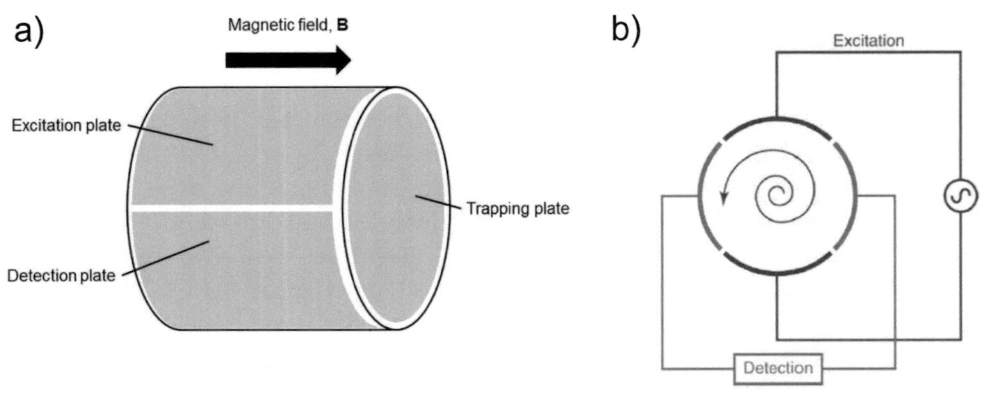

Fig. 5. Schematic representation of: a) an ion cyclotron resonance cell (to note that the cell is composed of 6 plates, 2 for excitation, 2 for detection and 2 for trapping of the ions), b) excitation of an ion packet to higher orbit radius to allow detection.

The four main scan modes available using MS/MS, product ion scan, precursor ion scan, neutral loss scan and selected reaction monitoring - SRM - (also known as multiple reaction monitoring, MRM) are depicted in Fig. 6. A symbolism proposed by Cooks and co-workers (Schwartz, Wade et al. 1990) to easily describe the various scan modes is also depicted in Fig. 6.

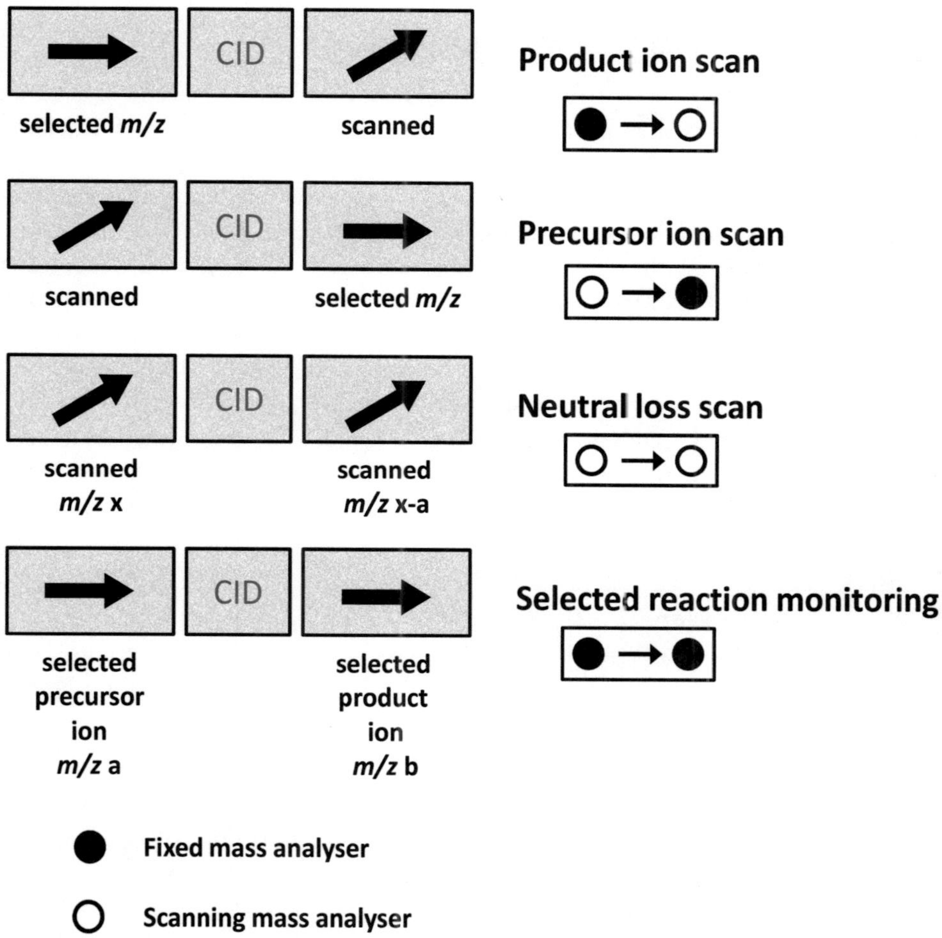

Fig. 6. Tandem mass spectrometry main scan modes, CID stands for Collision induced dissociation, and symbolism proposed by Cooks and co-workers (Schwartz, Wade et al. 1990) for the easy representation of the various scan modes.

Product ion scan, consists of selecting a precursor ion of a given *m/z* ratio and determining the product ions resulting from fragmentation.

Precursor ion scan consists of selecting a given product ion and determining the precursor ions. This scan mode cannot be performed with time-based mass spectrometers (ion traps and FTICRs).

Neutral loss scan consists of selecting a neutral fragment and detecting the fragmentations that lead to such loss. As in the case of precursor ion scan, this scan mode is not available in time-based mass spectrometers.

Selected reaction monitoring consists of selecting a fragmentation reaction. In this scan mode the first and second mass analysers are focused on selected *m/z* ratios thus increasing sensitivity.
Even though the precursor and neutral loss scans are not available in time-based instruments, these can perform quite easily MS^n experiments.

2. Fragmentation methods (Activation methods)

Early mass spectrometers were capable of only one stage of mass analysis and structure elucidation relied on the dissociation of the molecular ion in the course of its formation and within the ion source.
Soft ionization techniques (FAB, ESI and MALDI), that extended the range of application of mass spectrometry to polar, thermally labile compounds yield primarily protonated (or deprotonated) molecules with little or no fragmentation, limiting also the structural information to a single stage mass spectrum. To circumvent this limitation, tandem mass spectrometry emerged as an essential technique for structural analysis of a wide range of compounds with a clear focus on biologically relevant compounds (drugs, peptides and proteins, nucleic acids among others) over the past years.(Sleno and Volmer 2004)
Tandem mass spectrometry involves the activation of a precursor ion formed in the source and the mass analysis of its fragmentation products. The ion activation step is crucial and ultimately defines the type of product ions that are observed.(Sleno and Volmer 2004)

Method	Energy range	Instruments	Description
Collision Induced Dissociation (CID)	Low (1-100eV)	QqQ, IT, QqTOF, QqLIT, FTICR	Collision-induced dissociation by collision of precursor ions with inert target gas molecules.
	High (keV range)	Tandem TOF, Sector instruments	
Infrared Multiphoton Dissociation (IRMPD)	Low	IT, FTICR	Continuous-wave low-energy infrared laser activates the precursor ions by multiphoton absorption with consequent fragmentation
Electron Capture Dissociation (ECD)	Low	FTICR	Low-energy beam of electrons resulting in electron capture at protonation, or cationic site, with subsequent fragmentation following radical ion chemistry.
Surface Induced Dissociation (SID)	Low	Hybrid (BqQ), QqQ, IT, FTICR	Collisions between precursor ions and a solid target surface with or without a self-assembled monolayer causing fragmentation as well as other side reactions.
	High	Tandem TOF, RETOF	
IT- ion trap; FTICR – Fourier transform ion cyclotron resonance; TOF – time-of-fligth; QqQ – triple quadrupole			

Table 1. General description of the activation methods to be presented in this chapter.

2.1 Collision Activated and Collision Induced Dissociations (CAD/CID)

Collision-Induced Dissociation (CID also called Collision Activated Dissociation - CAD) remains the most common ion activation method available in present day instruments.(Jennings 2000; Shukla and Futrell 2000) Collisions between the precursor ion and a neutral gas target are accompanied by an increase in internal energy. This increase in internal energy induces decomposition with a higher fragmentation probability than, for example, metastable unimolecular dissociations.(Sleno and Volmer 2004)

The inelastic collision of an ion (with high kinetic energy) with a neutral, results on the conversion of part of the kinetic energy into internal energy leading to the decomposition of the ion.(Levsen 1978)

The conversion of kinetic energy into internal energy can be explained by the laws of physics involving a mobile species (ion) and a static target (gas). In order to simplify the description it is best to work in the center-of-mass (cm) framework instead of in the laboratory reference frame, simply because the center-of-mass momentum is always zero.(Sleno and Volmer 2004) The energy transfer in collisional activation and the kinetic energy release during ion dissociation are deduced by applying, to the process, the conservation of momentum and conservation of energy. The entire system is treated as a whole and the ion and neutral velocities are stated as velocities relative to each other.(Busch, Glish et al. 1988; Sleno and Volmer 2004)

The total energy available for conversion from kinetic to internal energy (E_{cm}) can be determined by equation 2

$$E_{cm}=[N/(m_p+N)]\times E_{Lab} \tag{2}$$

Where N is the mass of the neutral, m_p is the mass of the precursor ion and E_{lab} is the ion's kinetic energy. Equation 2 also tells us that the CID process is highly dependent on the relative masses of the two species. For example, E_{cm} increases with the target's mass, allowing more of the ion's kinetic energy to be converted into internal energy. It decreases, however, with $1/m_p$, meaning that larger precursor ions have less internal energy available for fragmentation through collision processes.(Busch, Glish et al. 1988)

Collisional activation mechanisms have been extensively studied for diatomic ions with neutral target atoms; but they are not well defined for polyatomic ions.(Sleno and Volmer 2004) The subsequent dissociation of activated ions is adequately described by the Rice, Ramsperger Kassel and Marcus (RRKM) theory (Marcus 1952) and less well by the quasi-equilibrium theory (QET).(Rosenstock, Wallenstein et al. 1952)

The overall CID process, which is depicted in equation 3, is assumed to occur by a two-step process encompassing the excitation of the precursor ions and subsequent fragmentation.

$$m_p^+ + N \rightarrow m_p^{+\neq} \rightarrow m_f^+ + m_N \tag{3}$$

Since the activation of fast-moving ions is much faster than their dissociation, it is reasonable to consider that the two processes are separate in time.

Based primarily on the kinetic energy of the precursor ion it is possible to separate the CID processes into two categories (Table 1): low-energy CID and high-energy CID.

Low-energy CID occurs in the 1-100eV range and is common in ion traps, quadrupoles and FTICR instruments. High-energy CID occurs in the kiloelectronvolt range and can be performed in sector and TOF/TOF instruments. There are important differences in the resulting CID spectra under low and high collision energy some of which will be briefly described in this chapter. In order to get more information the reader is encouraged to read an excellent review on the subject published in the Journal of Mass Spectrometry.(Sleno and Volmer 2004)

Low-energy CID is mostly performed in quadrupole instruments (such as triple quadrupoles) and trapping instruments (such as ion traps and Fourier transform ion cyclotron resonance instruments).

Taking as example a triple quadrupole, QqQ, the collision cell is the second quadrupole (denoted q) and it is only operated in r.f.- mode (allowing the ions to be focused). This collision cell is filled with a neutral inert gas (usually N_2 or Ar) and ion activation is achieved by multiple collisions. Both the nature of the collision gas and its pressure inside the collision cell are important factors. At higher collision gas pressures the number of ions undergoing collisions and the probability for an individual ion to collide increases. In fact, increasing the collision gas pressure is a convenient means of increasing the degree of dissociation of higher mass ions (for which E_{cm} is low) and ions that are particularly stable.(Sleno and Volmer 2004)

In an ion trap, the precursor ions are isolated and accelerated by on-resonance excitation causing collisions to occur and product ions are detected by subsequent ejection from the trap. With on-resonance excitation, the isolated precursor ion is excited by applying a small (tickle) a.c. potential across the end-caps. Ion activation times of the order of tens of milliseconds can be used without significant ion losses, hence multiple collisions can occur during the excitation period. Because of this relatively long time-scale, this excitation technique falls into the category of the slow-heating processes.(McLuckey and Goeringer 1997) Nevertheless, for a slow-heating process, excitation in an ion trap is still fairly fast, due to the high pressure of gas (normally He at $\approx$ 1 mTorr) present in the trap. There are other slow-heating methods with much longer excitation times, such as sustained off-resonance irradiation (SORI) which will also be described in this chapter.

Much like ion traps, in FTICR instruments the isolation and excitation phenomena take place in the same confined space where ions are trapped for a specific time, in a combined magnetic and electrostatic field. On-resonance excitation can be achieved by using a short (hundreds of μs) high-amplitude a.c. signal at the natural cyclotron frequency of the precursor ion. This excites the ion rapidly via multiple collisions and deposits a large amount of energy in the ion. The short irradiation time is necessary to minimize precursor ion losses.

Off-resonance irradiation, the SORI technique, is usually applied for collisional activation of precursor ions in FTICR instruments. In SORI, the precursor ion is excited at a frequency slightly higher than the natural cyclotron frequency. Ions undergo multiple acceleration/deceleration cycles as they repeatedly increase and decrease their orbital radii in the FTICR cell before dissociation take place. The ion translational energy is small compared with on-resonance excitation and much longer activation times (from hundreds of milliseconds to seconds) are used without ion loss.(McLuckey and Goeringer 1997) Consequently, a large number of collisions take place and the ion sequentially absorbs more and more collision energy until the collision threshold is reached.

The mass of the neutral target has a more important role for low-energy CID. With heavier targets more energy is transferred. Even though the average energy deposited per collision is lower than in high-energy CID, product ion yields are very high due to the occurrence of multiple collisions which are allowed by the collision gas pressure employed and the length of the collision cell (in triple quadrupoles) or the time allotted for CID (in ion traps).

Collisional activation in the keV range occurs in sector and TOF instruments, where the precursor ions have very high translational energies. In both types of instruments, a collision cell is placed between the two mass analysers. The precursor ion beam, with a kinetic energy of a few keV, can enter the collision cell, usually causing single collisions before mass analysis of the product ions. High-energy CID usually employs He as collision gas, nevertheless, the collision yield can be increased by using a heavier gas (such as Ar or Xe).
At high kinetic energies, ion excitation is mainly electronic, (Hitoshi, Pham et al. 1969) however, vibrational and rotational energies can also play an important role in excitation.(Franchetti, Freiser et al. 1978) The conversion of kinetic into internal energy is most efficient when the interaction time coincides with the period of the internal mode which is undergoing excitation.(Beynon, Boyd et al. 1986)
The two possible CID regimes often yield different products and this can be used to our advantage since structurally important fragment ions can be formed in both techniques. For unimolecular dissociations, direct bond cleavages require energies higher than rearrangement reactions do.(Levsen 1978)

2.2 Infrared multiphoton dissociation (IRMPD)

Gaseous ions can be excited and subsequently fragmented by the absorption of one or more photons. This type of activation has been performed with a wide range of photon energies by using lasers of different wavelengths. In the early days, UV and visible lasers were used whereas in recent years there has been an increase in the use of IR lasers. When using an UV laser the absorption of only one photon provided enough energy to initiate the dissociation of an isolated precursor ion. The same does not apply for the particular case of IR lasers. These are of energies lower than the UV lasers hence, multiphoton processes are needed to sufficiently excite ions for efficient fragmentation to occur. With the rise in popularity of trapping instruments (such as ion traps and FTICR) there has been an increase in the number of applications for IRMPD as an activation technique in tandem mass spectrometry. Typically, the stored ions are activated by a low power (< 100W) continuous-wave CO_2 (10.6 μm) laser during a selected irradiance time (usually in the order of tens to hundreds of milliseconds), followed by the detection of the resulting product ions.
Generally, photodissociation can be viewed by the mechanism depicted in equation 4.

$$m_p^+ + N \xrightarrow{nh\upsilon} m_p^{+\neq} \xrightarrow{k_{dis}} m_f^+ + m_N \quad (4)$$

Where n is the number of absorbed photons, $h\upsilon$ is the photon energy and k_{dis} is the rate constant for photodissociation.
The activation mechanism is assumed to occur through absorption of IR radiation by IR active modes present in the ion. After photon absorption, rapid energy redistribution over all vibrational degrees of freedom occurs, resulting in a statistical internal energy distribution, similar to CID. The dissociation occurs by low energy pathways, often the lowest possible.
There are several requirements for photodissociation to occur: 1) the precursor ion must be able to absorb energy in the form of photons producing excited species above the dissociation threshold for the ion of interest; 2) the energy gained by photon absorption

must overcome the energy lost by photon emission from the excited ions and deactivation by collisions.

Ion activation by absorption of photons is rather non-selective, i.e. product ions derived from the precursor ion can be further excited into dissociative states. Nevertheless, this outcome can be disadvantageous if too many ions are formed and the resulting spectrum becomes a complex collection of peaks.(Sleno and Volmer 2004)

The advantages of IRMPD are numerous. The amount of energy available is well defined, e.g. for a 10.6 µm CO_2 laser, the absorption of one photon corresponds to 0.117 eV of energy.(Sleno and Volmer 2004) If, for ion activation, enough time is given, the dissociation efficiency is good. In the case of FTICR, there is no need to add gas to the cell to promote activation and dissociation. Nevertheless, despite these advantages, it is a costly fragmentation technique.

2.3 Electron Capture and Electron Transfer Dissociations (ECD/ETD)

Electron Capture Dissociation (ECD) was developed by McLafferty's group in 1998.(Zubarev, Kelleher et al. 1998) Briefly, it involves the capture of low-energy electrons by multiply charged ions, with charge state reduction and subsequent fragmentation. ECD is the result of several important observations from ion-electron reactions which are all described in an excellent review by Zubarev et al.(Zubarev, Haelmann et al. 2002) The study of dissociative recombination (DR) correlates well with ECD and involves the fragmentation of gaseous positive ions following electron capture. The excited neutral then dissociates into two neutral (one radical and one even-electron species). Stabilization of the captured electron is faster than electron emission. Therefore, bond dissociation occurs faster than a typical bond vibration.

The main advantage of ECD is its ability to cause dissociation of very large biomolecules (≈40 kDa) at many sites, where other fragmentation methods are less effective.

ECD employs a tungsten filament (or a indirectly heated cathode) to emit a beam of very low-energy electrons (< 0.2 eV) for activation of the precursor ion. FTICR mass spectrometers are ideal for the application of ECD (McLafferty, Horn et al. 2001) and, accordingly, all the currently commercial FTICR instruments are equipped with this dissociation technique. The main reason for this is that the ICR cell traps the ions due to a combination of a strong magnetic field and a weak electrostatic field, without having any influence on the kinetic energy of the electrons during their interaction with the precursor ions.(Zubarev 2003) However, it remains difficult to perform ECD in ion traps, where the ions are trapped by strong RF potentials applied to the ring-electrode, thus affecting the movement of electrons inside the trap. To solve this problem, electron transfer dissociation (ETD) was developed by Hunt's group at the University of Virginia (Syka, Coon et al. 2004). This will be briefly described at the end of this section.

Charge neutralization occurs when a singly charged ion undergoes electron capture and the resulting neutral cannot, obviously, be detected by the mass spectrometer. As such, ECD is only applicable to multiply charged cationic species. The dissociation mechanism involves the fragmentation of an odd-electron ion resulting from electron capture by an even-electron species. The dissociation of the ion is restricted to specific protonation (or cationic) sites where the electron is captured. As a result, radical ion chemistry governs the bond cleavages.(McLafferty and Turecek 1993). Nevertheless, the electron is not necessarily captured directly at the charge site. The electron can land far from the cationic site with subsequent electron transfer to the highest charge density site. The intramolecular potential

difference, which causes this secondary electron transfer, is especially apparent in multiply charged metal ion complexes. Metal ions, such as Fe^{3+} and Zn^{2+}, serve as an electron sink and less fragmentation of the regions close to the metal is observed.(Zubarev, Horn et al. 2000) The landing of an electron on a multiply charged precursor ion creates an hypervalent (hydrogen-excess) unstable species. Groups with high affinity for the hot hydrogen atom, such as carbonyl or disulphide bonds, are principal sites for capture. The exothermicity of charge reduction and H• capture leads to bond cleavage initiated at the radical site.
ECD has been described as a non-ergodic process,(Turecek and McLafferty 1984) where the energy is not redistributed over the whole molecule and weakest bonds are not preferentially broken following ion activation. ECD assumed to occur much faster than other methods (CID and IRMPD) permitting the occurrence of direct bond cleavages only. As a result, it has been reported that the strong backbone N-C_α bonds of peptides are cleaved, affording c and z-type ions (see Fig. 7). Disulfide bonds, which are fairly stable under other activation methods, are also preferentially broken.

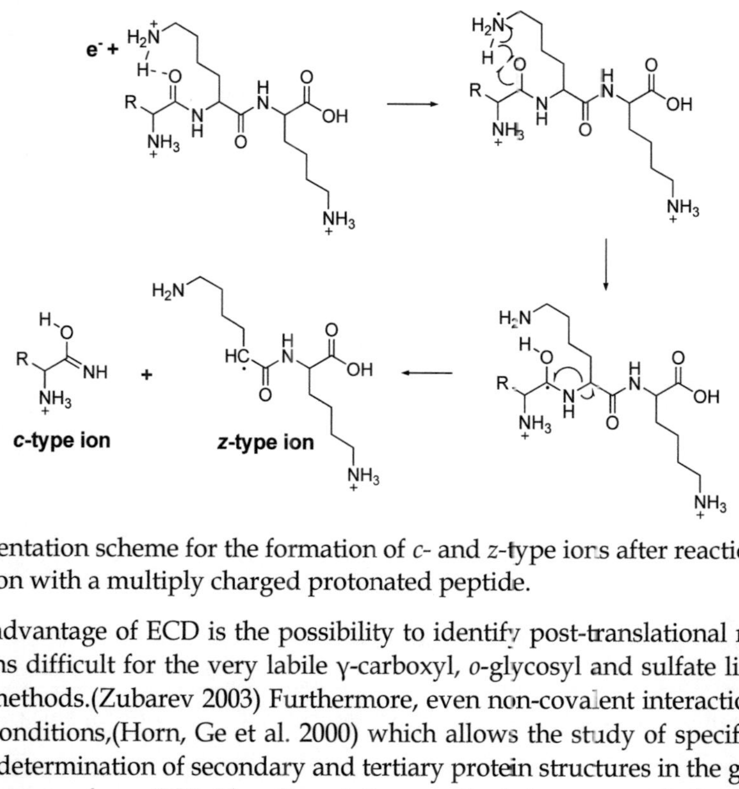

Fig. 7. Fragmentation scheme for the formation of *c*- and *z*-type ions after reaction of a low-energy electron with a multiply charged protonated peptide.

An obvious advantage of ECD is the possibility to identify post-translational modifications which remains difficult for the very labile γ-carboxyl, *o*-glycosyl and sulfate linkages when using other methods.(Zubarev 2003) Furthermore, even non-covalent interactions are stable under ECD conditions,(Horn, Ge et al. 2000) which allows the study of specific interaction sites and the determination of secondary and tertiary protein structures in the gas-phase.
The development of an ECD-like dissociation method for use with low-cost, widely accessible mass spectrometers, such as ion traps and linear ion traps, would be of obvious utility for protein sequence analysis. As stated above, storage of thermal electrons in a RF ion-containment field is, at best, problematic. To circumvent this, Hunt and co-workers (Syka, Coon et al. 2004) investigated the possibility of using anions as vehicles for delivering electrons to multiply charged peptide cations. They relied on their experience on negative ion chemical ionization (Hunt, Stafford et al. 1976; Hunt and Crow 1978) and concluded that anions with sufficiently low electron affinities could function as suitable one-electron

donors. Electron transfer to protonated peptides should be exothermic by 4-5.5 eV, trigger the release of a hydrogen radical and initiate fragmentation via the same non-ergodic pathways accessed in ECD.(Syka, Coon et al. 2004)

2.4 Surface Induced Dissociation (SID)

Surface Induced Dissociation (SID) is an activation method that is similar to CID, except that a solid surface is used as collision target instead of an inert gas. It was first developed by Cooks and co-workers in the 1970s and has been studied by several research groups, including those of McLafferty (Chorush, Little et al. 1995), Wysocki (Dongré, Somogyi et al. 1996; Nair, Somogyi et al. 1996; Schaaff, Qu et al. 1998) and Futrell.(Laskin, Denisov et al. 2002; Laskin, Bailey et al. 2003; Laskin, Beck et al. 2003).

The main idea behind this activation method is that energy transfer in CID is limited by the energy available in the center-of-mass reference frame (E_{cm}) which depends on the mass of the target gas. By increasing the mass of the target, E_{cm} becomes larger and energy transfer can be improved. Assuming that collisions occur with the entire surface, instead of with individual surface molecules, the mass of the target is effectively infinite. In theory, energy conversion should be more efficient in SID.(Laskin and Futrell 2003)

Cooks and co-workers classified the collision events at surfaces according to the collision energy (velocity) of the impacting projectile (neutral molecules, ions, atoms).(Verena, Jianwei et al. 2001) SID reactions fall in the 'hyperthermal' regime with energies ranging from 1eV to 100eV.

Hyperthermal ions can scatter on a surface in an elastic, inelastic or chemically reactive fashion (Table 2).

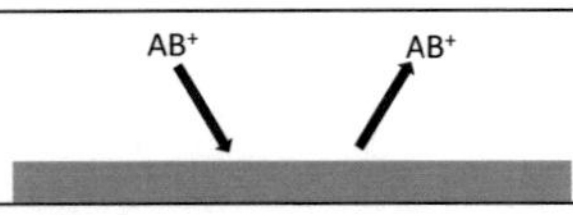	Elastic scattering	$AB^+ + SX \rightarrow AB^+ + SX$
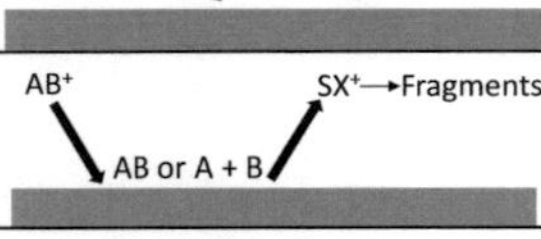	Charge transfer	$AB^+ + SX \rightarrow SA + SB + SX^+$
	Chemical sputtering	$AB^+ + SX \rightarrow AB + SX^+ \rightarrow S^+/X^+$
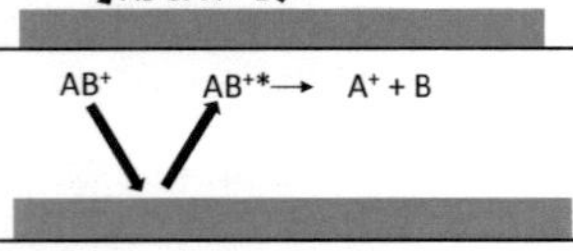	Surface-induced dissociation	$AB^+ + SX \rightarrow AB^{+*} + SX \rightarrow A^+ + B + SX$
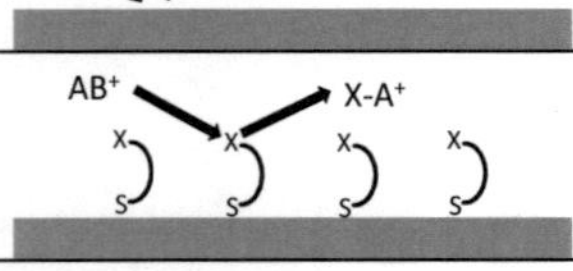	Ion-surface reactions	$AB^+ + SX \rightarrow AX^+ + SB$

Table 2. Elementary processes in ion-surface interactions (SX^+ stands for surface ions).

Inelastic and reactive scattering events are the most common processes observed at the collision energies normally used in SID and reactive scattering reactions are undesired side reactions that lead to reduced efficiency and sensitivity.

Given that in general SID and CID afford similar products, it is reasonable to rationalize the activation mechanism for SID as a two-step process. Initially, the incident ion collides (inelastically) with the solid surface, forming an internally excited ion, which then undergoes unimolecular dissociations. The interaction time with the surface is of the order of 10^{-12} s, a short period compared with the dissociation time for polyatomic ions.(Sleno and Volmer 2004)

It is believed that the SID experiment involves collisional activation at the surface, followed by delayed gas-phase dissociation of the scattered projectile ion. Nevertheless, at higher collision energies, shattering of the projectile ion can occur at the surface, (Laskin and Futrell 2003) resulting in fragment ions with the same kinetic energy, since they all originate at the surface.

The internal energy distributions of excited ions for SID are relatively narrow (Wysocki, Ding et al. 1992), which offers an advantage over CID; where the internal energy of ions is not as narrow. In general, all incident ions collide directly with the surface, whereas most collisions in CID are glancing. Hence, the energy transferred can easily be varied, simply by changing the impact energy.

The center-of-mass collision energy for SID is difficult to determine, and it has been shown that several experimental factors[1] influence the amount of kinetic energy converted into internal energy of the ion.(Merouch and Hase 2002) Due to the relatively narrow range of internal energies, SID has a potential application in isomer distinction.(Schaaff, Qu et al. 1998)

Several different instruments have been employed for SID of polyatomic ions however, several important mass spectrometer modifications are necessary. Nevertheless, for FTICR in particular, SID would be especially advantageous, given that to perform CID in these instruments it is necessary to introduce a gas in the ICR cell, which has a negative effect on mass resolutions. In fact, SID in an FTICR has been reported to be effective for the dissociation of large peptides.(Williams, Henry et al. 1990)

This activation method remains a very promising technique especially for high mass ions with relatively high dissociation thresholds, e.g. peptides and proteins. Nevertheless, it still awaits incorporation into a commercial instrument.

3. Applications

In the next sections we will present some examples where tandem mass spectrometry together with collision induced dissociations were helpful, namely for the identification of isomers (Section 3.1) and for the determination of thermochemical quantities, proton affinities and gas-phase acidities, in particular (Section 3.2).

3.1 Structure elucidation

We will present some examples of the application of tandem mass spectrometry together with collision induced dissociation to study the gas-phase behaviour of some aniline derivatives (Madeira 2010), α,β-unsaturated γ-lactones fused to sugars(Madeira, Rosa et al. 2010) and isoflavone aglycones.(Madeira, Borges et al.)

Gas-phase behaviour of aniline derivatives

The structures of the aniline derivatives studied are depicted in Fig.8.

The electrospray ionization MS^2 spectra of the protonated haloanilines studied are depicted in Fig. 9.

---

[1] These are surface composition, projectile structure, collision energy and incidence angle, just to name a few.

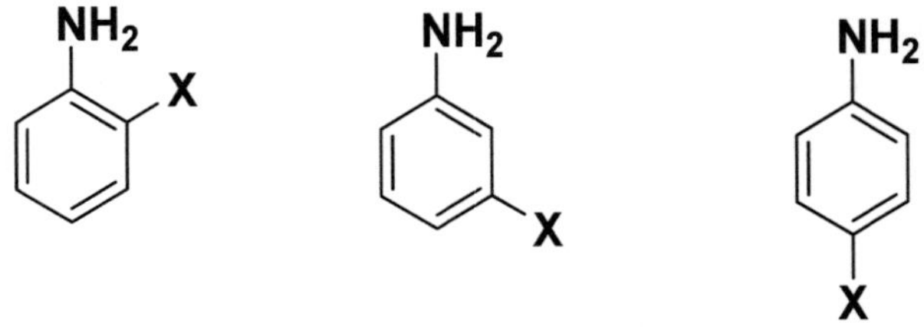

Fig. 8. Structures of the aniline derivatives studied. (X=F, Cl, Br, I and NO_2).

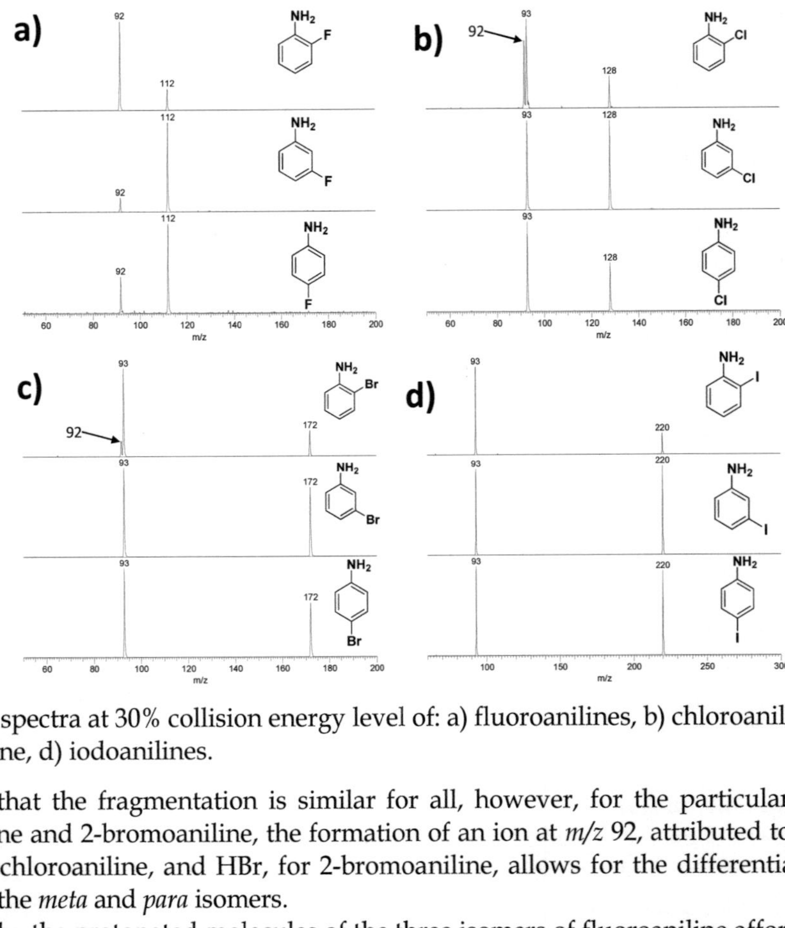

Fig. 9. MS^2 spectra at 30% collision energy level of: a) fluoroanilines, b) chloroanilines, c) bromoaniline, d) iodoanilines.

It is clear that the fragmentation is similar for all, however, for the particular case of 2-chloroaniline and 2-bromoaniline, the formation of an ion at *m/z* 92, attributed to the loss of HCl, for 2-chloroaniline, and HBr, for 2-bromoaniline, allows for the differentiation of the *ortho* from the *meta* and *para* isomers.

Interestingly, the protonated molecules of the three isomers of fluoroaniline afforded the ion at *m/z* 92, attributed to the loss of HF, which was already reported for fluoranilines under electron ionization conditions.(Tajima, Ueki et al. 1996) This loss was proposed to occur via a "ring-walk" mechanism, where the fluorine atom migrates to the *ortho* position. A similar behaviour has also been reported for 4-fluorotoluene and 4-chlorotoluene.(Parry, Fernandez et al. 1992) Taking into account that the typical ion residence time within the ion trap is a few hundred milliseconds (the maximum ejection time for these MS^2 experiments was set to 200 ms) and that this phenomenon was detected in a sector mass spectrometer,(Tajima, Ueki et al. 1996) for which the typical ion residence time within the ion source is ca. 10^{-6} s, it is reasonable to assume that this ring walk is possible in ion trap mass spectrometry conditions. With this in mind, it is understandable that the *m/z* 92 for 2-fluoroaniline has the highest relative abundance of the three. The proposed fragmentation pathways for the haloanilines are presented in Fig. 10.

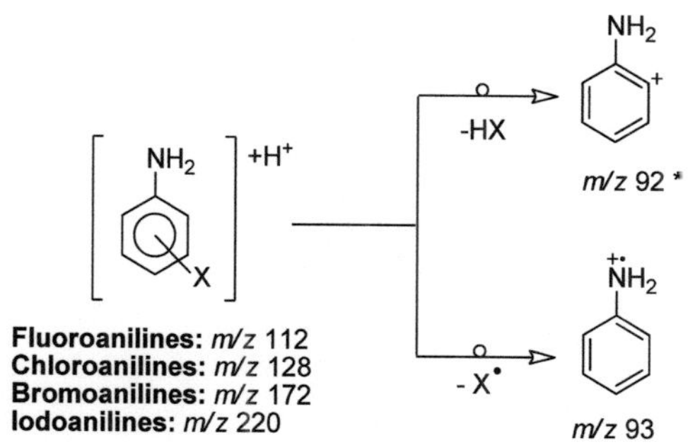

Fig. 10. Proposed fragmentation pathways for the protonated molecules of the haloanilines (*- the *m/z* ion was only detected for the fluoroanilines, 2-chloroaniline and 3-bromoaniline).

The electrospray ionization MS^2 spectra of the nitroanilines studied are depicted in Fig. 11. It is quite clear that the three isomeric forms can easily be distinguished from the fragmentation pattern alone.

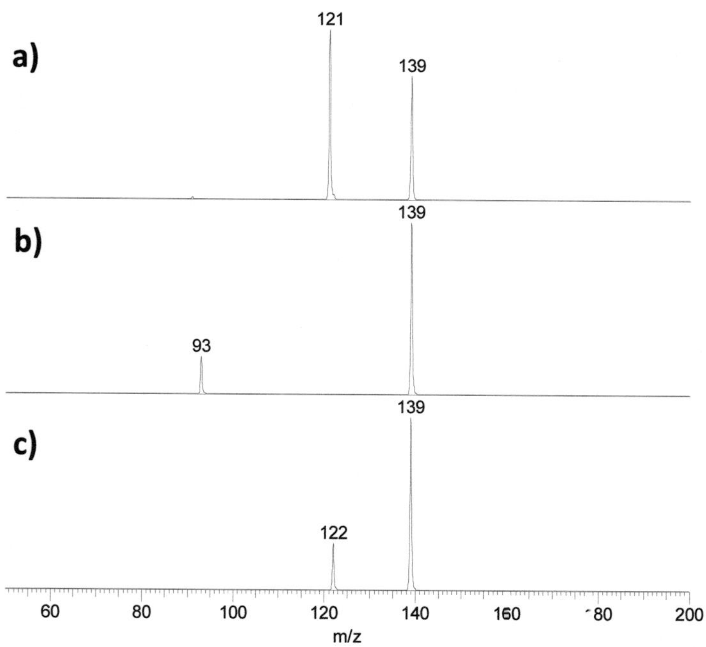

Fig. 11. MS^2 spectra at 30% collision energy level of: a) 2-nitroaniline, b) 3-nitroaniline, c) 4-nitroaniline.

The protonated molecule of 2-nitroaniline, Fig.11a, loses 18 Da, attributed to H_2O, affording an ion at *m/z* 121. Since this loss is not detected for the other two nitroanilines, it can be attributed to an *ortho* effect. This effect is well documented in the literature for electron ionization,(Schwarz 1978; Bobyleva, Kulikov et al. 1989; Attygalle, Ruzicka et al. 2006; Jariwala, Figus et al. 2008), but it was also observed when using electrospray ionization.(Holman, Wright et al. 2008) A possible pathway for this fragmentation is depicted in Fig. 12 and involves the formation of a six-membered ring intermediary.

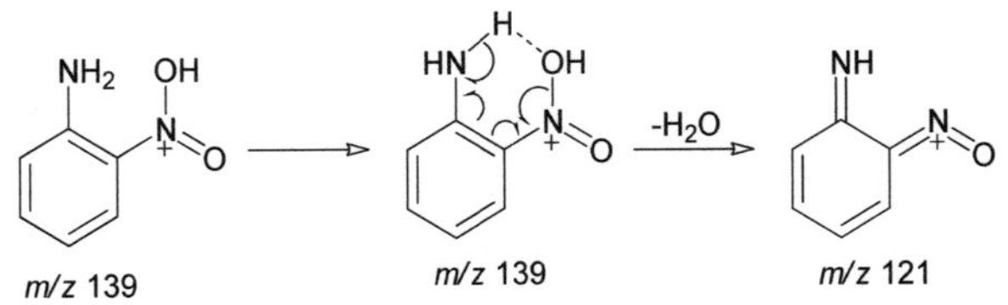

Fig. 12. Possible fragmentation pathway to afford the *m/z* 121 ion from the protonated molecule of 2-nitroaniline (*m/z* 139).

Gas-phase behaviour of α,β-unsaturated γ-lactones fused to sugars

In this study (Madeira, Rosa et al. 2010) we came across an interesting behaviour of two of the compounds studied (structures depicted in Fig. 13), that are isomers. The MS^2 spectra of the deprotonated molecules of these compounds are depicted in Fig. 14.

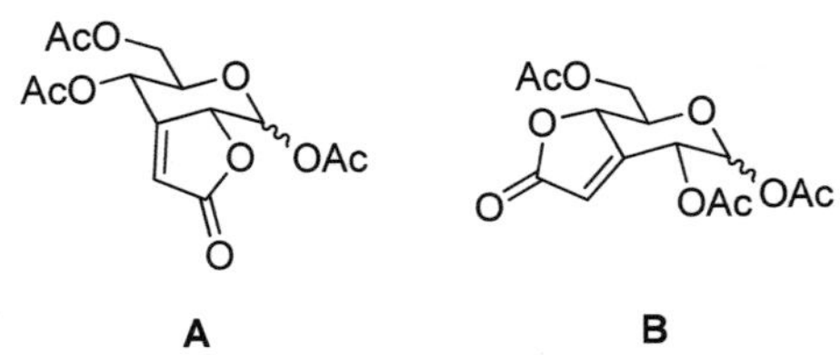

Fig. 13. Structures of the two isomers included in the electrospray ionization study.

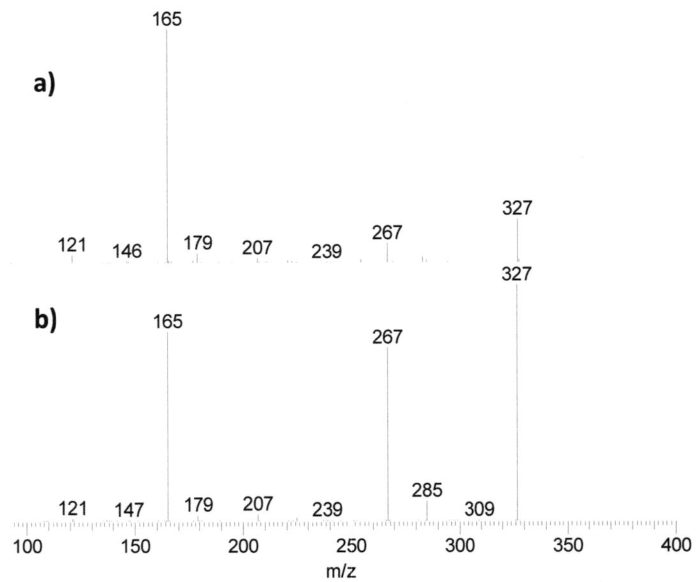

Fig. 14. Negative ion mode ESI MS^2 spectra at normalized collision energy of 15% of the deprotonated molecules of: a) compound **A**; b) compound **B**. (Reprinted with kind permission of John Wiley & Sons)

These compounds are isomers that differ in the positions of the lactone moiety and of one of the acetyl groups within the sugar ring. Regarding their fragmentation, the first significant difference is the loss of 42 Da, ketene (H_2CCO), from the deprotonated molecule to afford the ion at *m/z* 185. This loss is more pronounced for compound **B** than for **A**, and this behaviour might be related with the different substitution patterns of these two compounds. Both compounds lose 60 Da, CH_3COOH, affording the product ions at *m/z* 267 and, again, this loss is more pronounced for compound **B** than for **A**. This behaviour was attributed to the fact that compound **B** possesses an acetyl group at C^2 which is more labile than the other substituents. In this particular case the MS^2 experiments allowed the distinction between both isomers.

Gas-phase behaviour of isoflavone aglycones

Regarding the gas-phase behaviour of isoflavones (Madeira, Borges et al. 2010) it was also possible to differentiate two isomers (prunetin and biochanin A) since their fragmentation afforded different diagnostic ions for each. The ESI-MS^2 spectra of the protonated molecules of formononetin, prunetin and biochanin A are depicted in Fig. 15.

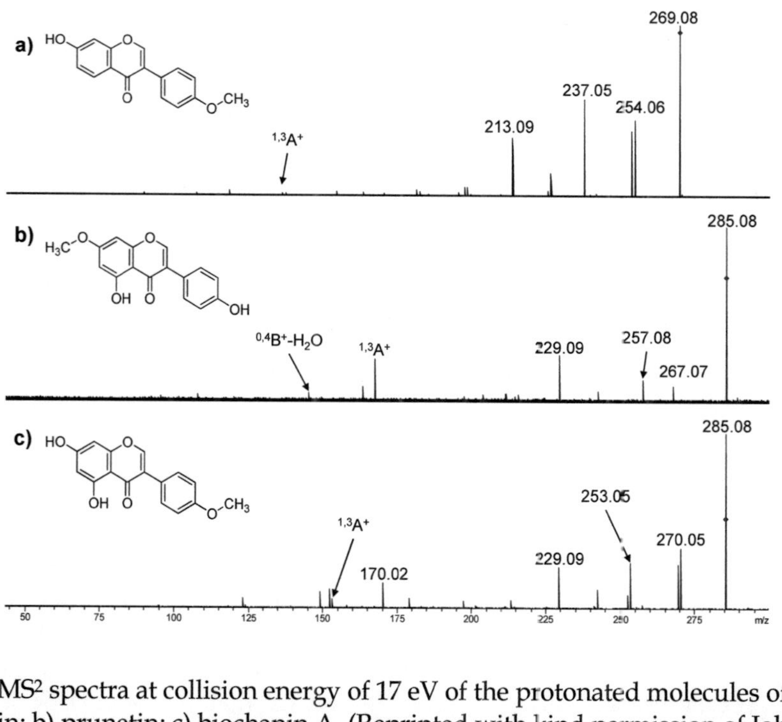

Fig. 15. ESI-MS^2 spectra at collision energy of 17 eV of the protonated molecules of: a) formononetin; b) prunetin; c) biochanin A. (Reprinted with kind permission of John Wiley & Sons)

The position of the methoxy group also influences the fragmentation pattern. For instance, the losses of $^{\bullet}CH_3$ and CH_3OH were only detected for formononetin and biochanin A,which have the methoxy group attached to the B-ring. The ions afforded by these fragmentations can also serve to access the B-ring substitution.

3.2 Gas-phase ion thermochemistry

Tandem mass spectrometry has other uses than the frequently and currently used structural analysis or the identification of proteins. MS/MS techniques can, for example, also be used to determine gas-phase thermochemical properties of several classes of molecules (biomolecules included).

The kinetic method was developed in order to determine thermochemical properties whenever the equilibrium method was either not applicable or the instrumentation was not available, and was based on the rates of competitive dissociation of mass-selected cluster ions.(Cooks and Wong 1998) Over the years, the kinetic method has been subjected to a great deal of discussion, for example see the Special Feature issue of volume 34 of Journal of Mass Spectrometry.(McLuckey, Cameron et al. 1981; Armentrout 1999; Capriolli 1999; Cooks, Koskinen et al. 1999; Drahos and Vékey 1999).

Proton affinities and Gas-phase basicities of aniline derivatives

We will start with a simple example using only compounds with known thermochemical properties. In this example we are going to deal only with proton affinities (PA) and gas-phase basicities (GB) which can be determined in the positive ion mode.(Madeira 2010) The kinetic method can, however, be used to determine other thermochemical properties. For example, the gas-phase acidity is determined in the negative ion mode.

The thermochemical properties of several aniline derivatives are well known (Table 3) and we will use these to illustrate how to apply the kinetic method in its simplest form.

	PA (kJ mol^{-1})	GB (kJ mol^{-1})
4-Nitroaniline	866.0	834.2
3-Fluoroaniline	867.3	835.5
3-Chloroaniline	868.1	836.3
4-Fluoroaniline	871.5	839.7
3-Bromoaniline	873.2	841.4
4-Chloroaniline	873.8	842.0

Table 3. Proton affinities (PA) and gas-phase basicities (GB) of haloanilines and 4-nitroaniline. (Hunter and Lias) These values were assumed to have an error of 8.4 kJ mol-1.

For this example we will determine the proton affinity and gas-phase basicity of 4-nitroaniline by pairing it with the other reference compounds presented on Table 3. The mass spectrum of a mixture of 4-nitroaniline and 4-fluoroaniline is presented in Fig. 16a and the MS^2 spectra at different normalized collision energies are depicted in Fig. 16b-e.

In the MS^2 spectra presented earlier (Fig. 15b-e) it is clear that the ion at *m/z* 112 (protonated 4-fluoroaniline) has a higher abundance than the ion at *m/z* 139 (protonated 4-nitroaniline). It is therefore reasonable to say that 4-fluoroaniline has a higher proton affinity (and gas-phase basicity) than 4-nitroaniline. The abundance ratio between the reference ion (4-fluoroaniline) and the unknown ion (4-nitroaniline) is a key element for the application of the kinetic method. Repeating the procedure for the other references in Table 3 we obtain the data presented in Table 4.

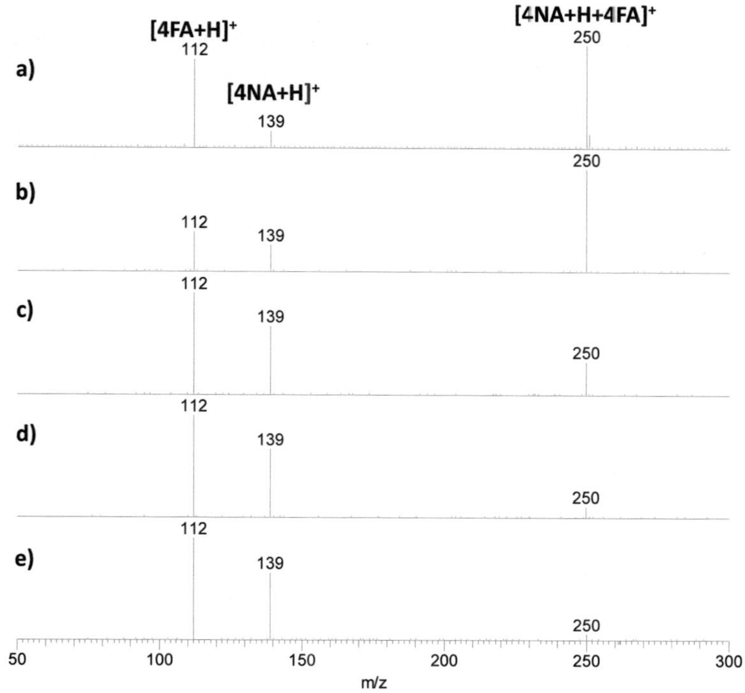

Fig. 16. Full scan mass spectrum of a mixture of 4-nitroaniline (4NA) and 4-fluoroaniline (4FA); b) MS^2 spectrum of the heterodimer $[4NA+H+4FA]^+$ (*m/z* 250) at a normalized collision energy (NCE) of 5%; c) MS^2 spectrum of the heterodimer at NCE 7%; d) MS^2 spectrum of the heterodimer at NCE 10%; e) MS^2 spectrum of the heterodimer at NCE 15%.

	PA (kJ mol^{-1})	GB (kJ mol^{-1})	Ab(AH$^+$)/Ab(BH$^+$)	Ln [Ab(AH$^+$)/Ab(BH$^+$)]
3-fluoroaniline	867.3	835.5	2.3	0.8
3-chloroaniline	868.1	836.3	2.7	1.0
4-fluoroaniline	871.5	839.7	0.7	-0.4
3-bromoaniline	*873.2*	*841.4*	2.6	*1.0*
4-chloroaniline	*873.8*	*842.0*	2.3	*0.8*

Table 4. Values of Ab(AH+)/Ab(BH+) and Ln[Ab(AH+)/Ab(BH+)]. (AH+ reference ion, BH+ unknown ion). 3-bromoaniline and 4-chloroaniline are presented in italic because they were not considered when constructing the ln(abundance ratio) vs GB or PA graphical representation. See explanation in the text.

3-Bromo and 4-chloroanilines were removed from the data, since their behaviour was somewhat unexpected (Fig. 17a and b). Indeed, according to the gas-phase thermochemical data presented in Table 3, after fragmentation of the heterodimer the abundance of the protonated molecule of 4-nitroaniline should have been lower than that of 4-chloroaniline and 3-bromoaniline. The MS^2 presented in Fig. 17 show the opposite thus, these two systems were not considered when calculating the proton affinity and gas-phase basicity of 4-nitroaniline.

For these two systems, isomerization reactions could be responsible for this behaviour. For the particular case of 4-chloroaniline, the migration from the *para* to the *meta* position should be energetically favourable. This type of migration has already been reported by Parry et al.(Parry, Fernandez et al. 1992) for *para*-halotoluenes and it seems to be a reasonable explanation for this observation. In fact, the branching ratio found for 4-chloroaniline is comparable to the one found for 3-chloroaniline, which supports the assumption of isomerization.

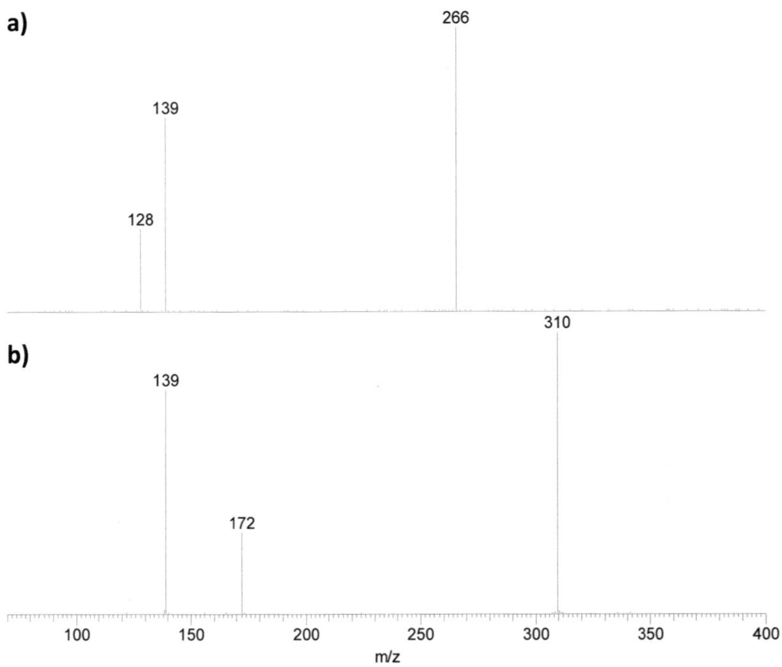

Fig. 17. ESI-MS^2 spectra at NCE 5% of a) 4-chloroaniline + 4-nitroaniline proton-bound heterodimer (*m/z* 266); b) 3-bromoaniline + 4-nitroaniline proton-bound heterodimer (*m/z* 310).

The data on Table 4 allows us to construct the graphical representation that will enable us to determine the proton affinity and gas-phase basicity of our unknown, 4-nitroaniline. This graphical representation is depicted in Fig 18.

The proton affinity/gas-phase basicity is the intercept of the least mean squares trendline with the x-axis. For 4-nitroaniline, the proton affinity was estimated to be 866.3±0.8 kJ mol^{-1} and the gas-phase acidity 834.5±0.8 kJ mol^{-1}. These values are in close agreement with the literature values (Table 3), which are 866.0 kJ mol^{-1} and 834.2 kJ mol^{-1} for the proton affinity and gas-phase basicity, respectively.

Gas-phase acidities of substituted phenols

Some antioxidant mechanisms displayed by several phenolic compounds relate with OH bond dissociation energy (DH^0(ArO-H). One way to determine it in the gas phase is to combine the gas-phase acidity, $\Delta_{ac}H^0$(ArOH), the electron affinity of the phenoxyl radical, E_{ea}(ArO$^\bullet$), and the ionization energy of the hydrogem atom, E_i(H) (equation 5).

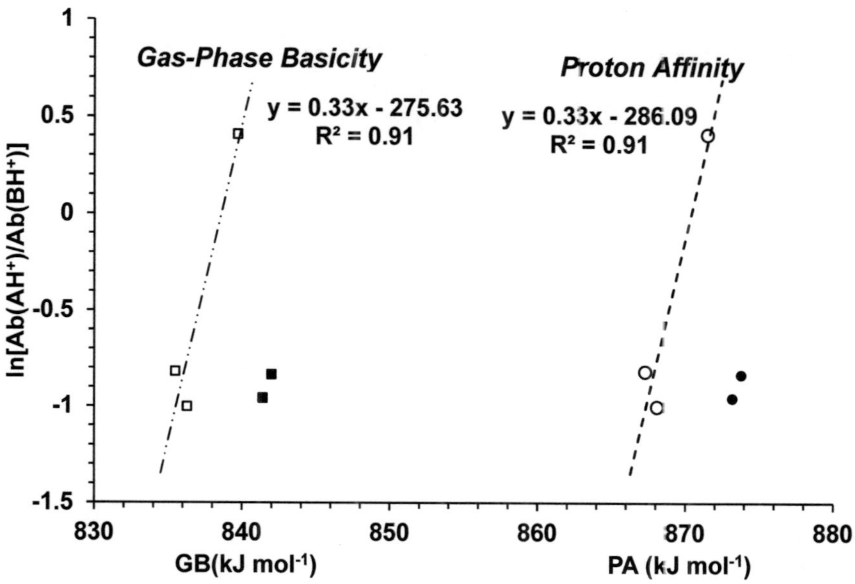

Fig. 18. ln[Ab(AH+)/Ab(BH+)] vs gas-phase basicity (kJ mol-1) and proton affinity (kJ mol-1). (AH+ - reference ion, BH+ - unknown ion).

$$DH^0(ArO\text{-}H)= \Delta_{ac}H^0(ArOH)+E_{ea}(ArO^\bullet)-E_i(H) \quad (5)$$

In the past years we have determined the gas-phase acidity for several substituted phenols, including dimethylphenols,(Madeira, Costa et al. 2008) dimethoxyphenols and chromanol(Madeira, Faddoul et al. 2011), with the purpose of establishing a bridge towards vitamin E.

Fig. 19a and b depicts the negative chemical ionization mass spectrum of a mixture of 3,5-dimethylphenol and 4-methylphenol and the MS^2 spectrum of the isolated heterodimer (*m/z* 229). To note that in the case of proton affinities and gas-phase basicities the work is done in the positive ion mode while for gas-phase acidities the mass spectra are acquired in the negative ion mode.

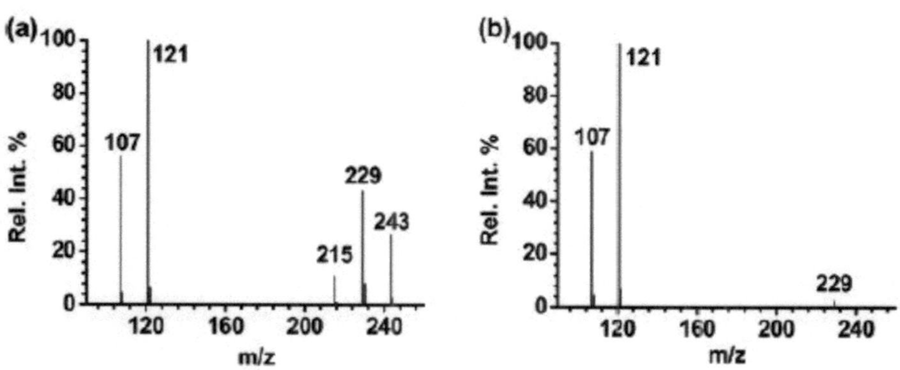

Fig. 19. (a) Chemical ionization mass spectrum of 3,5-dimethylphenol and 4-methylphenol mixture. (b) MS^2 spectrum of the isolated heterodimer (*m/z* 229, 6 ms of delay time). With kind permission from Springer Science+Business Media (Madeira, Costa et al. 2008)

Same as for the aniline derivatives presented earlier, if we pair the unknown compounds with other compounds with known gas-phase acidity we can apply the kinetic method formalism and determine its gas-phase acidity. The gas-phase acidity scales for dimethylphenols (Madeira, Costa et al. 2008) and for dimethoxyphenols and chromanol (Madeira, Faddoul et al. 2011) are graphically depicted in Fig. 20.

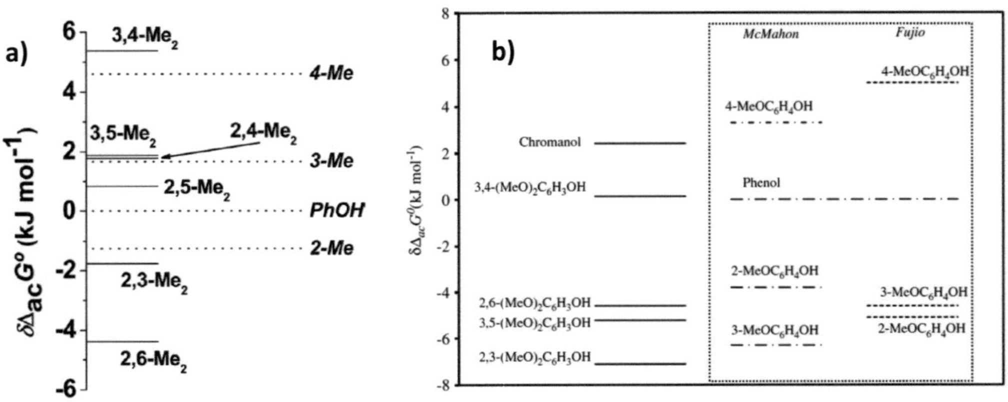

Fig. 20. Experimental gas-phase acidity scales relative to phenol for: a) dimethylphenols, with kind permission from Springer Science+Business Media (Madeira, Costa et al. 2008); b) dimethoxyphenols, with kind permission from John Wiley and Sons (Madeira, Faddoul et al. 2011).

4. Acknowledgment

The authors acknowledge the funding provided by Fundação para a Ciência e a Tecnologia (PEst-OE/QUI/UI0612/2011 and REDE/1501/REM/2005). The authors gratefully acknowledge the pentaquadrupole photos kindly supplied by Professor Marcos N. Eberlin, (ThoMSon Lab, Universidade Estadual de Campinas, Brazil).

5. References

Armentrout, P. B. (1999). *Is the kinetic method a thermodynamic method?* Journal of Mass Spectrometry 34(2): 74-78.

Aston, F. W. (1919). *LXXIV. A positive ray spectrograph.* Philosophical Magazine Series 6 38(228): 707 - 714.

Attygalle, A., J. Ruzicka, et al. (2006). *An unprecedented ortho effect in mass spectrometric fragmentation of even-electron negative ions from hydroxyphenyl carbaldehydes and ketones.* Tetrahedron Letters 47(27): 4601-4603.

Bainbridge, K. T. (1932). *The Isotopic Weight of* H_2. Physical Review 42(1): 1-10.

Bainbridge, K. T. and E. B. Jordan (1936). *Mass Spectrum Analysis 1. The Mass Spectrograph. 2. The Existence of Isobars of Adjacent Elements.* Physical Review 50(4): 282-296.

Barber, M., R. S. Bordoli, et al. (1981). *Fast atom bombardment of solids (F.A.B.): a new ion source for mass spectrometry.* Journal of the Chemical Society, Chemical Communications (7): 325-327.

Barrow, M. P., W. I. Burkitt, et al. (2005). *Principles of Fourier transform ion cyclotron resonance mass spectrometry and its application in structural biology.* Analyst 130(1): 18-28.

Beynon, J. H., R. K. Boyd, et al. (1986). *Charge permutation reactions.* Chischester, John Wiley & Sons.

Bobyleva, M. S., N. S. Kulikov, et al. (1989). *Mass spectrometry and structure of heterocyclic ions by collisional activation.* Chemistry of Heterocyclic Compounds 25(3): 274-279.

Busch, K. L., G. L. Glish, et al. (1988). *Mass spectrometry/mass spectrometry: techniques and applications of tandem mass spectrometry.* New York, VCH.

Byrdwell, W. (2001). *Atmospheric pressure chemical ionization mass spectrometry for analysis of lipids.* Lipids 36(4): 327-346.

Capriolli, R. (1999). *The kinetic method.* Journal of Mass Spectrometry 34(2): 73.

Chapman, J. R. (1985). *Practical organic mass spectrometry.* Chichester, John Wiley & Sons.

Chorush, R. A., D. P. Little, et al. (1995). *Surface-Induced Dissociation of Multiply-Protonated Proteins.* Analytical Chemistry 67(6): 1042-1046.

Cole, R. B., Ed. (2010). *Electrospray and MALDI Mass Spectrometry: Fundamentals, Instrumentation, Practicalities, and Biological Applications.* Hoboken, New Jersey, USA, John Wiley & Sons.

Comisarow, M. B. and A. G. Marshall (1974). *Fourier transform ion cyclotron resonance spectroscopy.* Chemical Physics Letters 25(2): 282-283.

Cooks, R. G., J. T. Koskinen, et al. (1999). *The kinetic method of making thermochemical determinations.* Journal of Mass Spectrometry 34(2): 85-92.

Cooks, R. G. and P. S. H. Wong (1998). *Kinetic Method of Making Thermochemical Determinations: Advances and Applications.* Accounts of Chemical Research 31(7): 379-386.

Covey, T. R., E. C. Huang, et al. (2002). *Structural characterization of protein tryptic peptides via liquid chromatography/mass spectrometry and collision-induced dissociation of their doubly charged molecular ions.* Analytical Chemistry 63(13): 1193-1200.

Covey, T. R., B. A. Thomson, et al. (2009). *Atmospheric pressure ion sources.* Mass Spectrometry Reviews 28(6): 870-897.

Crotti, S., R. Seraglia, et al. (2011). *Some thoughts on electrospray ionization mechanisms.* European Journal of Mass Spectrometry 17(2): 85-100.

Dass, C. (2007). *Fundamentals of Contemporary Mass Spectrometry.* Hoboken, New Jersey, John Wiley & Sons.

Dempster, A. J. (1918). *A new Method of Positive Ray Analysis.* Physical Review 11(4): 316-325.

Dongré, A. R., Á. Somogyi, et al. (1996). *Surface-induced Dissociation: An Effective Tool to Probe Structure, Energetics and Fragmentation Mechanisms of Protonated Peptides.* Journal of Mass Spectrometry 31(4): 339-350.

Douglas, D. J. (2009). *Linear quadrupoles in mass spectrometry.* Mass Spectrometry Reviews 28(6): 937-960.

Douglas, D. J., A. J. Frank, et al. (2005). *Linear ion traps in mass spectrometry.* Mass Spectrometry Reviews 24(1): 1-29.

Drahos, L. and K. Vékey (1999). *How closely related are the effective and the real temperature.* Journal of Mass Spectrometry 34(2): 79-84.

Fenn, J. B., M. Mann, et al. (1989). *Electrospray ionization for mass spectrometry of large biomolecules*. Science 246(4926): 64-71.

Franchetti, V., B. S. Freiser, et al. (1978). *Excitation in ion–molecule collisions. A study of protonated aromatic molecules by ion kinetic energy spectrometry*. Organic Mass Spectrometry 13(2): 106-110.

Gross, J. H. (2004). *Mass Spectrometry: A Textbook*. Berlin, Springer-Verlag.

Guilhaus, M., D. Selby, et al. (2000). *Orthogonal acceleration time-of-flight mass spectrometry*. Mass Spectrometry Reviews 19(2): 65-107.

Hager, J. W. (2002). *A new linear ion trap mass spectrometer*. Rapid Communications in Mass Spectrometry 16(6): 512-526.

Hillenkamp, F. and J. Peter-Katalinic, Eds. (2007). *MALDI MS: a practical guide to instrumentation, methods and applications*. Weinheim, Wiley-VCH.

Hitoshi, Y., D. Pham, et al. (1969). *Energetics of the Collision-Induced Dissociations* $C_2H_2^+ \rightarrow C_2H^+ + H$ *and* $C_2H_2^+ \rightarrow H^+ + C_2H$. The Journal of Chemical Physics 51(8): 3465-3476.

Hoffmann, E. d. and V. Stroobant (2007). *Mass spectrometry: Principles and Applications*. Chichester, John Wiley & Sons.

Holman, S. W., P. Wright, et al. (2008). *High-throughput approaches towards the definitive identification of pharmaceutical drug metabolites. 1. Evidence for an ortho effect on the fragmentation of 4-benzenesulfinyl-3-methylphenylamine using electrospray ionisation mass spectrometry*. Rapid Communications in Mass Spectrometry 22(15): 2355-2365.

Horn, D. M., Y. Ge, et al. (2000). *Activated Ion Electron Capture Dissociation for Mass Spectral Sequencing of Larger (42 kDa) Proteins*. Analytical Chemistry 72(20): 4778-4784.

Hunt, D. F. and F. W. Crow (1978). *Electron capture negative ion chemical ionization mass spectrometry*. Analytical Chemistry 50(13): 1781-1784.

Hunt, D. F., G. C. Stafford, et al. (1976). *Pulsed positive negative ion chemical ionization mass spectrometry*. Analytical Chemistry 48(14): 2098-2104.

Hunter, E. P. and S. G. Lias *Proton Affinity Evaluation*, In *NIST Chemistry WebBook, NIST Standard Reference Database Number 69*. P. J. Linstrom and W. G. Mallard. Gaithersburg, MD, National Institute of Standards and Technology.

Jariwala, F. B., M. Figus, et al. (2008). *Ortho Effect in Electron Ionization Mass Spectrometry of N-Acylanilines Bearing a Proximal Halo Substituent*. Journal of the American Society for Mass Spectrometry 19(8): 1114-1118.

Jennings, K. R. (1968). *Collision-induced decompositions of aromatic molecular ions*. International Journal of Mass Spectrometry and Ion Physics 1(3): 227-235.

Jennings, K. R. (2000). *The changing impact of the collision-induced decomposition of ions on mass spectrometry*. International Journal of Mass Spectrometry 200(1-3): 479-493.

Johnson, E. G. and A. O. Nier (1953). *Angular Aberrations in Sector Shaped Electromagnetic Lenses for Focusing Beams of Charged Particles*. Physical Review 91(1): 10-17.

Karas, M. and F. Hillenkamp (2002). *Laser desorption ionization of proteins with molecular masses exceeding 10,000 daltons*. Analytical Chemistry 60(20): 2299-2301.

Kebarle, P. and U. H. Verkerk (2009). *Electrospray: From ions in solution to ions in the gas phase, what we know now*. Mass Spectrometry Reviews 28(6): 898-917.

Laskin, J., T. H. Bailey, et al. (2003). *Shattering of Peptide Ions on Self-Assembled Monolayer Surfaces*. Journal of the American Chemical Society 125(6): 1625-1632.

Laskin, J., K. M. Beck, et al. (2003). *Surface-Induced Dissociation of Ions Produced by Matrix-Assisted Laser Desorption/Ionization in a Fourier Transform Ion Cyclotron Resonance Mass Spectrometer*. Analytical Chemistry 76(2): 351-356.

Laskin, J., E. V. Denisov, et al. (2002). *Surface-Induced Dissociation in a Fourier Transform Ion Cyclotron Resonance Mass Spectrometer: Instrument Design and Evaluation*. Analytical Chemistry 74(14): 3255-3261.

Laskin, J. and J. Futrell (2003). *Surface-induced dissociation of peptide ions: Kinetics and dynamics*. Journal of the American Society for Mass Spectrometry 14(12): 1340-1347.

Laskin, J. and J. H. Futrell (2003). *Collisional activation of peptide ions in FT-ICR mass spectrometry*. Mass Spectrometry Reviews 22(3): 158-181.

Levsen, K. (1978). *Fundamental aspects of organic mass spectrometry*. Weinheim, Verlag Chemie.

Louris, J. N., R. G. Cooks, et al. (1987). *Instrumentation, applications, and energy deposition in quadrupole ion-trap tandem mass spectrometry*. Analytical Chemistry 59(13): 1677-1685.

Madeira, P., P. Costa, et al. (2008). *Determination of gas-phase acidities of dimethylphenols: Combined experimental and theoretical study*. Journal of the American Society for Mass Spectrometry 19(11): 1590-1599.

Madeira, P. J. A. (2010). A mass spectrometry study of compounds with environmental and biological interest. Faculdade de Ciências. Lisboa, Universidade de Lisboa. PhD Thesis.

Madeira, P. J. A., C. M. Borges, et al. *Electrospray ionization Fourier transform ion cyclotron resonance mass spectrometric and semi-empirical calculations study of five isoflavone aglycones*. Rapid Communications in Mass Spectrometry 24(23): 3432-3440.

Madeira, P. J. A., C. M. Borges, et al. (2010). *Electrospray ionization Fourier transform ion cyclotron resonance mass spectrometric and semi-empirical calculations study of five isoflavone aglycones*. Rapid Communications in Mass Spectrometry 24(23): 3432-3440.

Madeira, P. J. A., M. Faddoul, et al. (2011). *On the way to understand antioxidants: chromanol and dimethoxyphenols gas-phase acidities*. Journal of Mass Spectrometry 46(7): 640-648.

Madeira, P. J. A., A. M. Rosa, et al. (2010). *Electrospray ionization mass spectrometric analysis of newly synthesized α,β-unsaturated γ-lactones fused to sugars*. Rapid Communications in Mass Spectrometry 24(7): 1049-1058.

Makarov, A. (2000). *Electrostatic Axially Harmonic Orbital Trapping: A High-Performance Technique of Mass Analysis*. Analytical Chemistry 72(6): 1156-1162.

March, R. E. (2009). *Quadrupole ion traps*. Mass Spectrometry Reviews 28(6): 961-989.

Marcus, R. A. (1952). *Unimolecular Dissociations and Free Radical Recombination Reactions*, AIP.

Marshall, A. G., M. B. Comisarow, et al. (1979). *Relaxation and spectral line shape in Fourier transform ion resonance spectroscopy*. Journal of Chemical Physics 71(11): 4434-4444.

Marshall, A. G., C. L. Hendrickson, et al. (1998). *Fourier transform ion cyclotron resonance mass spectrometry: A primer*. Mass Spectrometry Reviews 17(1): 1-35.

McLafferty, F. W., D. M. Horn, et al. (2001). *Electron capture dissociation of gaseous multiply charged ions by fourier-transform ion cyclotron resonance.* Journal of the American Society for Mass Spectrometry 12(3): 245-249.

McLafferty, F. W. and F. Turecek (1993). *Interpretation of mass spectra.* Sausalito, University Science Books.

McLuckey, S. A., D. Cameron, et al. (1981). *Proton affinities from dissociations of proton-bound dimers.* Journal of the American Chemical Society 103(6): 1313-1317.

McLuckey, S. A. and D. E. Goeringer (1997). *Slow Heating Methods in Tandem Mass Spectrometry.* Journal of Mass Spectrometry 32(5): 461-474.

Meroueh, O. and W. L. Hase (2002). *Dynamics of Energy Transfer in Peptide-Surface Collisions.* Journal of the American Chemical Society 124(7): 1524-1531.

Munson, M. S. B. and F. H. Field (2002). *Chemical Ionization Mass Spectrometry. I. General Introduction.* Journal of the American Chemical Society 88(12): 2621-2630.

Nair, H., Á. Somogyi, et al. (1996). *Effect of Alkyl Substitution at the Amide Nitrogen on Amide Bond Cleavage: Electrospray Ionization/Surface-induced Dissociation Fragmentation of SubstanceP and Two Alkylated Analogs.* Journal of Mass Spectrometry 31(10): 1141-1148.

Nier, A. O. (1940). *A Mass Spectrometer for Routine Isotope Abundance Measurements.* Review of Scientific Instruments 11(7): 212-216.

Parry, A., M. T. Fernandez, et al. (1992). *Thermodynamic studies of gas-phase proton transfer equilibria involving benzene: a reassessment of earlier data.* Journal of the Chemical Society, Faraday Transactions 88(22): 3331-3337.

Paul, W. (1990). *Electromagnetic traps for charged and neutral particles.* Reviews of Modern Physics 62(3): 531-540.

Pramanik, B. N., A. K. Ganguly, et al., Eds. (2002). *Applied electrospray mass spectrometry.* New York, Taylor & Francis.

Price, D. and J. E. Williams, Eds. (1969). *Time-of-flight mass spectrometry.* Oxford, Pergamon Press.

Raffaelli, A. and A. Saba (2003). *Atmospheric pressure photoionization mass spectrometry.* Mass Spectrometry Reviews 22(5): 318-331.

Rodrigues, J. A., A. M. Taylor, et al. (2007). *Mass Spectrometry of Carbohydrates: Newer Aspects,* In *Advances in Carbohydrate Chemistry and Biochemistry.* D. Horton. London, Academic Press. Volume 61: 59-141.

Rosenstock, H. M., M. B. Wallenstein, et al. (1952). *Absolute Rate Theory for Isolated Systems and the Mass Spectra of Polyatomic Molecules.* Proceedings of the National Academy of Sciences 38(8): 667-678.

Ryhage, R. (2002). *Use of a Mass Spectrometer as a Detector and Analyzer for Effluent Emerging from High Temperature Gas Liquid Chromatography Columns.* Analytical Chemistry 36(4): 759-764.

Schaaff, T. G., Y. Qu, et al. (1998). *Investigation of the trans effect in the fragmentation of dinuclear platinum complexes by electrospray ionization surface-induced dissociation tandem mass spectrometry.* Journal of Mass Spectrometry 33(5): 436-443.

Schwartz, J. C., A. P. Wade, et al. (1990). *Systematic delineation of scan modes in multidimensional mass spectrometry.* Analytical Chemistry 62(17): 1809-1818.

Schwarz, H. (1978). *Some newer aspects of mass spectrometric ortho effects*, In *Topics in Current Chemistry: Organic Chemistry*. Berlin, Springer-Verlag. 73: 231-263.

Shukla, A. K. and J. H. Futrell (2000). *Tandem mass spectrometry: dissociation of ions by collisional activation*. Journal of Mass Spectrometry 35(9): 1069-1090.

Sleno, L. and D. A. Volmer (2004). *Ion activation methods for tandem mass spectrometry*. Journal of Mass Spectrometry 39(10): 1091-1112.

Sommer, H., H. A. Thomas, et al. (1951). *The Measurement of e/M by Cyclotron Resonance*. Physical Review 82(5): 697-702.

Syka, J. E. P., J. J. Coon, et al. (2004). *Peptide and protein sequence analysis by electron transfer dissociation mass spectrometry*. Proceedings of the National Academy of Sciences of the United States of America 101(26): 9528-9533.

Tajima, S., M. Ueki, et al. (1996). *Unimolecular HF Loss from the Molecular Ions of Fluorophenols and Fluoroanilines. A 'Ring-walk' Mechanism of a Fluorine Atom*. Rapid Communications in Mass Spectrometry 10(9): 1076-1078.

Thomson, J. J. (1897). *Cathode Rays*. Philosophical Magazine Series 5 44(269): 293-316.

Thomson, J. J. (1907). *On Rays of Positive Electricity*. Philosophical Magazine Series 6 13(77): 561-575.

Thomson, J. J. (1921). *Rays of positive electricity and their application to chemical analyses*. London, Longmans, Green and Co.

Turecek, F. and F. W. McLafferty (1984). *Non-ergodic behavior in acetone-enol ion dissociations*. Journal of the American Chemical Society 106(9): 2525-2528.

Verena, G., S. Jianwei, et al. (2001). *Collisions of ions with surfaces at chemically relevant energies: Instrumentation and phenomena*, AIP.

Voyksner, R. D. (1997). *Combining Liquid Chromatography with Electrospray Mass Spectrometry*, In *Electrospray Ionization Mass Spectrometry: Fundamentals, Instrumentation, and Applications*. R. B. Cole. New York, John Wiley & Sons: 323-342.

Watson, J. T. and K. Biemann (2002). *Direct Recording of High Resolution Mass Spectra of Gas Chromatographic Effluents*. Analytical Chemistry 37(7): 844-851.

Whitehouse, C. M., R. N. Dreyer, et al. (2002). *Electrospray interface for liquid chromatographs and mass spectrometers*. Analytical Chemistry 57(3): 675-679.

Williams, E., K. Henry, et al. (1990). *Surface-induced dissociation of peptide ions in Fourier-transform mass spectrometry*. Journal of the American Society for Mass Spectrometry 1(5): 413-416.

Wolff, M. M. and W. E. Stephens (1953). *A Pulsed Mass Spectrometer with Time Dispersion*. Review of Scientific Instruments 24(8): 616-617.

Wysocki, V., J.-M. Ding, et al. (1992). *Surface-induced dissociation in tandem quadrupole mass spectrometers: A comparison of three designs*. Journal of the American Society for Mass Spectrometry 3(1): 27-32.

Zubarev, R. A. (2003). *Reactions of polypeptide ions with electrons in the gas phase*. Mass Spectrometry Reviews 22(1): 57-77.

Zubarev, R. A., K. F. Haelmann, et al. (2002). *Towards an understanding of the mechanism of electron- capture dissociation: a historical perspective and modern ideas*. European Journal of Mass Spectrometry 8(5): 337-349.

Zubarev, R. A., D. M. Horn, et al. (2000). *Electron Capture Dissociation for Structural Characterization of Multiply Charged Protein Cations*. Analytical Chemistry 72(3): 563-573.

Zubarev, R. A., N. L. Kelleher, et al. (1998). *Electron Capture Dissociation of Multiply Charged Protein Cations. A Nonergodic Process*. Journal of the American Chemical Society 120(13): 3265-3266.

Part 2

Proteomics/Macromolecular Analysis

2

Tandem Mass Spectrometry of Peptides

Renata Soares[1], Elisabete Pires[1], André M. Almeida[1,2], Romana Santos[1,3], Ricardo Gomes[1], Kamila Koči[1], Catarina Ferraz Franco[1] and Ana Varela Coelho[1]
[1]*ITQB/Universidade Nova de Lisboa*
[2]*Instituto de Investigação Cientifica Tropical& Centro Interdisciplinar de Investigação em Sanidade Animal*
[3]*Unidade de Investigação em Ciências Orais e Biomédicas, Faculdade de Medicina Dentária/Universidade de Lisboa*
Portugal

1. Introduction

Tandem MS is considered as a mass spectrum of a mass spectrum raised to the power *n*, where *n* is the number of MS/MS spectra, obtained per molecular ion. When a peptide is analyzed by mass spectrometry, the obtained MS spectrum provides the peptide mass. When tandem MS (or MS/MS) is performed, such peptide is fragmented into daughter ions, which provide information regarding the amino acid sequence of the peptide. Due to this particularity, peptide tandem MS can be used in several applications involving protein characterization and identification, such as the study of proteomes (Franco *et al.*, 2011a,b) and differential proteomics (Puerto *et al.*, 2011) of different organs, tissues, cells or biological fluids. In our laboratory we have also been using this approach for the characterization of proteins with specific function in adhesives (Santos *et al.*, 2009), the identification of protein adducts as potential biomarkers of toxicity (Antunes *et al.*, 2010), and protein glycation (Gomes *et al.*, 2008), it has also been useful in the identification of peptides with immunomodulation properties (Koči *et al.*, 2010). Two types of information obtained by mass spectrometry experiments are used for the study of proteins and peptides. They can be identified or characterized based on their peptide map and primary structure. The advantage of using tandem MS, is that it provides further data, and hence, confirm the assigned peptide identification, thus reducing the chance of obtaining wrongly assigned peptide/protein identifications. Additionally, the information obtained allows localizing of post-translational or chemical modifications at the amino acid residue level.

2. Peptide fragmentation

2.1 Proteins into amino acids

Nowadays, peptide fragmentation is the most commonly used MS information for protein identification in proteomic studies. In this type of studies, there are two approaches that can be used: a gel-based approach (Franco *et al.*, 2001a) or a gel-free approach (Washburn *et al.*, 2001), where the former involves an initial separation of a complex mixture of proteins in, for example, SDS- or two-dimensional-PAGE. Instead, in the latter, the separation of the complex mixture of previously digested proteins is accomplished using one or two different types of chromatographic separations. In this case, before the separation step, the proteins need to be cleaved into smaller compounds, namely peptides. This is usually accomplished through the use of specific proteases, most commonly trypsin. Different peptides are generated during protein digestion with different proteases, for example trypsin cleaves after lysine or arginine, and hence tryptic peptides will always end in one of these residues. This digestion is necessary as, for most mass spectrometers, proteins are too big to be analysed by tandem MS. This type of approach, where protein digestion is performed before mass spectrometry analysis, is called bottom-up proteomics. Alternatively, intact proteins can be directly analysed in particular mass spectrometers, for example FTICR and Orbitrap mass spectrometers, for protein identification and characterization without the need of proteolytic digestion. In this strategy, named top-down proteomics, proteins are introduced into the mass spectrometer, its mass measured and directly fragmented in the equipment (Reid & McLuckey, 2002). This procedure has several applications related to protein analysis, such as the characterization of post-translational modifications, protein confirmations, protein-ligand and protein-protein complexes, among others (see section 4.2 of this chapter). For more detailed information, we suggest the review by Cui and colleagues (2011).

Sample handling for tandem mass spectrometry analysis should always be done with special care due to the fact that contaminations and induced modifications during sample processing can occur. A common contamination that can occur when performing, in particular, protein digestions is with keratin, which can hinder the identification of proteins. The most common keratin contaminations occur from human hair and hands from the operator , indicating the crucial use of clean lab coat, cap and nitrile gloves. The use of an electrostatic eliminator can also be used to reduce this type of contamination, as demonstrated by Xu *et al.* (2011).

Standard protein digestion procedures are time consuming and involve multisteps. Automatization has been introduced, although it is only cost effective and efficient for a large number of samples. Alternatively simplified protocols have been developed. In a study by Ren and co-workers (2009) a quick digestion protocol was implemented in order to minimize the digestion-induced modifications on proteins, such as asparagine deamidation and N-terminal glutamine cyclization, found to be directly proportional to incubation time in reduction or alkylation and depending on digestion buffer composition. The authors tested this modified protocol on immunoglobulin gamma, which allowed reducing the total experimental procedure to a few hours, beside a protein coverage of 98.6% for IgG.

Particular attention must be driven to the amount of salts and detergents that protein digests have prior to MS or MS/MS analysis due to the fact that these can interfere with the ionization of peptides and can create strong interference on the mass spectrum signal. This can be by-passed by using a HPLC system prior to the MS analysis due to the fact that salts and detergents do not bind to the typical reverse phase columns used for peptide separation eluting in the first steps of the chromatographic run. If a HPLC system is not used, then alternative simpler methods should be used, such as micro columns packed with reverse -phase resins or graphite (Larsen *et al.*, 2007) and, when using MALDI-TOF-TOF, this desalting step can be done directly onto the MALDI target plate (Jia *et al.*, 2007).

2.2 The fragmentation process

The MS fragmentation process occurs in the mass spectrometer mass analyzer or in a collision cell through the action of collision energy on gas phase ions generated in the mass spectrometer ion source. Several parameters influence this fragmentation process, including amino acid composition, size of the peptide, excitation method, time scale of the instrument used, ion charge state, etc (Paiz & Suhai, 2005). Presently, there are several fragmentation processes available in commercial mass spectrometers, namely collision induced dissociation (CID), electron capture dissociation (ECD), electron transfer dissociation (ETD), etc.

The CID fragmentation method can be of two types: low- or high-energy, where the former uses up to 100 eV and the latter from hundreds eV up to several keV (Wells & McLuckey, 2005). The low-energy CID can be found in quadrupoles whereas the high-energy is used in, for example, tof-tof instruments.

The nomenclature of the daughter ions generated by CID was first established by Roepstorff & Fohlmann (1984) and later reviewed by Biemann (1988), where the *b*-series ions extend from the N-terminal and the *y*-series ions extend from the C-terminal. (see Fig. 1). The calculation of the mass difference between consecutive daughter ions belonging to the same ion series (for example *b* or *y* ion-series), allows the determination of peptide's primary sequence.

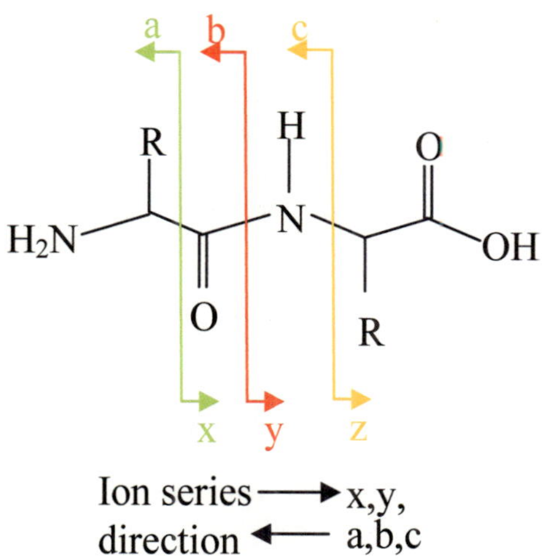

Fig. 1. Schematic diagram of daughter ion nomenclature, adapted from Roepstorff & Fohlmann (1984). A positively charged peptide (in black) is fragmented and the daughter ions are shown (*a, b, c, x, y, z*).

CID fragmentation of peptides uses an inert gas, such as helium, nitrogen or argon, to hit the peptides resulting in the excitation of the molecular ion leading to the polypeptide chain breaking, and to a lower extent to that of the amino acid side chains, generating daughter ions. The low-energy CID allows the rearrangement of the peptide after the loss of a fragment (Yague *et al.*, 2003) and it allows for multiple collisions (up to 100) in a time scale for dissociation of up to milliseconds whereas, the high-energy CID does not permit such process as it is too fast, breaking the peptide. Advantages of the high-energy CID are the differentiation of isobaric amino acids, such as leucine and isoleucine, as well as more reproducible fragmentation patterns, however only a few collisions can occur (usually no more than ten) due to the fast time scale for dissociation. The information obtained from both types of CID is different; the low-energy CID will generate predominantly *b*- and *y*-series ions, whereas with the high-energy CID *a*-, *x*- and immonium ions are also observed (Fig. 1).

There are several models of how peptides fragment using CID, as reviewed by Paizs & Suhai (2005), namely the *mobile proton* and the *pathways in competition* models. According with the *mobile proton model* a peptide can acquire a positive charge at several sites, named the protonation sites, and these include the terminal amino group, amide oxygens and nitrogens and side chain groups creating several isomers. So, as soon as the peptide becomes excited, a proton is added and this proton will move from protonation site to protonation site before fragmentation (Paiz & Suhai, 2005). The *mobile proton model* can be mainly used to interpret MSMS spectra in a qualitative manner. With the *pathways in competition model*, peptide fragmentation is seen as a competition between charge-remote and -directed in peptide fragmentation pathways, where generation of different peptides follows probability rules based on energetic and kinetic characterization of their major fragmentation pathways. For the charge–remote peptides selective cleavage can occur at the asparagines containing peptides and at oxidized methionines, leading to fragment ions containing information regarding the peptide's amino acid sequence, whereas for the charge-directed peptides fragment ions correspond to the loss of water, ammonia or other neutral losses (Paiz & Suhai, 2005).

A disadvantage of CID is that the side-chain of peptides can be lost. With the other types of fragmentation processes (ECD and ETD) this does not occur, which is particularly important for the identification and characterization of some labile peptide modifications (Zubarev, 2004). While with CID the CO-N bonds are broken along the peptide backbone, with ECD and ETD the N-Cα bonds are broken creating *c*-ions from the N-terminal and *z*-ions from the C-terminal (Fig. 1). Both processes allow extensive peptide backbone fragmentation, while preserving labile side chains (Bakhtiar & Guan, 2006). In ECD there is a reaction

$$[M+nH]^{n+} + e^{-} \rightarrow \left[[M+nH]^{(n-1)+}\right]^{*} \rightarrow fragments$$

where electrons attach to protonated peptides hence creating peptide cations with an additional electron. After this, the peptide undergoes a rearrangement leading to dissociation (Mikesh *et al.*, 2006). In most cases the information obtained with ECD is complementary to the information obtained with CID (Zubarev, 2004). ECD is mainly used in FTICR mass spectrometers, however it has been developed a similar type of collision energy, ETD, for ion-trap quadrupoles. The ETD advantage is the ability to analyze larger, non-tryptic peptides, allowing the detection of multiple PTMs. ETD dissociates peptides in the same bonds as ECD creating the same ion series. However, ETD does not use free electrons but employs radical anions where A is the anion.

$$[M+nH]^{n+} + A^{-} \rightarrow \left[[M+nH]^{(n-1)+}\right]^{*} + A \rightarrow fragments$$

This radical anion transfers an electron to the protonated peptide leading to its fragmentation (Mikesh *et al.*, 2006). ETD cleaves randomly along the peptide backbone while side chains and modifications, such as phosphorylation, are left intact. The technique only works well for higher charge state ions (z>2), however comparing to CID, ETD is advantageous for the fragmentation of longer peptides or even entire proteins, making this technique important for top-down proteomics.

ECD and ETD produce a complete or almost complete fragmentation spectrum of peptides leading to more information regarding the peptide sequence (Mikesh *et al.*, 2006). However, these types of fragmentation are usually associated with expensive mass spectrometers. Despite the fact that CID is the routine method for fragmentation of peptides, ETD has been described as a preferred method for peptides carrying labile PTMs.

A new method for fragmentation of peptides has been developed by Budnik and co-workers (2001) named electron detachment dissociation (EDD).

$$[\mathbf{M}-n\mathbf{H}]^{n-} + e^-_{fast} \rightarrow [\mathbf{M}-n\mathbf{H}]^{(n-1)-\bullet} + 2e^-$$

Like ECD and ETD, in EDD, fragmentation of peptides occurs in the N-Cα and provides information regarding the primary structure of peptides. This method is particularly useful for the analysis of acidic proteins (Ganisl *et al.*, 2011) as they ionize better using electrospray and are detected in negative ion mode. A drawback of EDD is the common loss of small molecules (such as CO_2) from the amino acid side-chains.

3. Analysis of tandem mass spectrometry data

After the mass spectrometer analysis a file is created containing a list of masses observed for the peptides, which could be used as precursor ions To each peptide selected for fragmentation there is, in this experimental file, associated peptide fragments and all this data can be analysed in order to obtain information regarding the amino acid sequence of the peptide and further used for protein identification or characterization (Cottrell, 2011).

There are basically two ways to analyse this type of data: manually or submitting the data to search engines where they are compared to selected protein sequence databases. The manual analysis is done by looking at the mass difference between peaks of a tandem mass spectrum and determining if this mass difference corresponds to the mass of a particular amino acid (see section 3.2 for *de novo* spectrum analysis). However, this is quite a difficult and time-consuming task with several cons, as both ion series are observed in the same spectrum and each series might not be completed, as there can be missing fragments, which produce gaps in the analyzed amino acid sequence. Additionally other fragments, mainly arising from neutral losses and ions from other ion series are usually detected. As a first option fragment patterns interpretation using automatic search engines is usually tried. Although when homologous proteins are not available in protein sequence databases, this approach is not successful. Alternatively, *de novo* interpretation softwares can be used and the obtained results should be carefully checked, usually involving a manual case by case inspection. This situation arises frequently for proteins isolated from organisms with non-sequenced genomes (section 4.1) or if the protein is strongly modified (section 4.3).

3.1 Search engines for protein identification

There are several softwares available for this purpose, based on mathematical algorithms that can help on the interpretation of this type of data. There are some free-ware softwares,

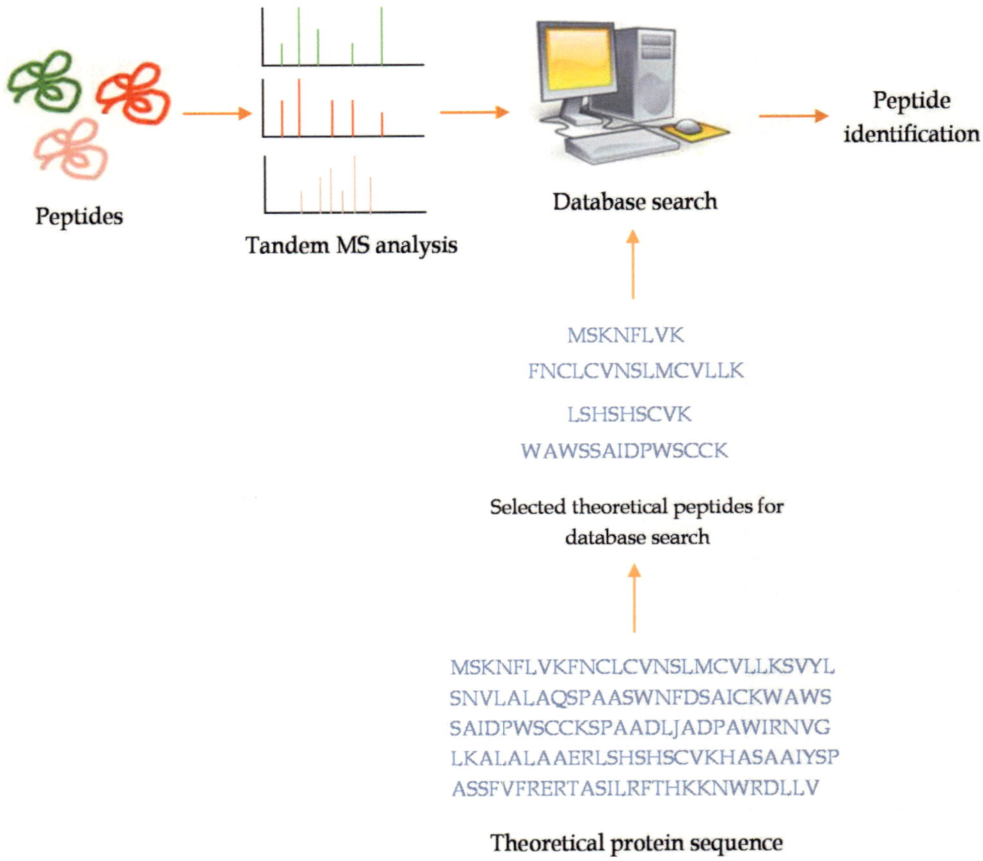

Fig. 2. Schematic diagram of the common workflow of database search for peptide identification.

for example MASCOT (Perkins *et al.*, 1999) from Matrix Science (Mowse algorithm), Crux (Park *et al.*, 2008) from University of Washington (Sequest algorithm), and commercially available softwares, for example Protein Pilot (Shilov, *et al.*, 2007) from AB Sciex (which uses Mowse and Paragon algorithms), Bioworks from Thermo Electron (Sequest algorithm) or Peaks from Bioinformatic Solutions. But how do these softwares perform the analysis of the peptide mass spectrometry data? Basically, they make a theoretical digestion, or *in silico* digestion, of all proteins present in a database and generate theoretical fragments of these peptides. After, it is done a comparison between the virtual and experimental data obtained in the mass spectrometer, attributing a score to the peptide or protein identified, where the highest the score the higher match is achieved and, consequently, more confident is the identification generated (Fig. 2). The search process is governed by specified parameters. In a recent review by Eng and co-workers (2011), it is presented a description of these parameters and criteria common in available search engines, as well as their impact on the identification results. Briefly, the mass tolerance for peptides and their fragments should be defined (which varies depending on the resolution and mass accuracy of the mass spectrometer used), the modifications introduced during sample treatment (for example, the alkylation with iodoacetamide) as well as known modifications of the protein under study, the protease used for digestion and the maximum number of miss cleavages allowed, as well as the database and taxonomy restriction to apply. The major limitation of database searching is that only the peptides that are present in the database can be identified in the search and sometimes by having one different amino acid, the peptide may not be found. The most popular search algorithms include Sequest (Eng *et al.*, 1994), Mowse and X!Tandem.

A recent methodology for peptide identification from tandem MS data, consensus-based method, implies the use of several search engines, hence different algorithms for peptide identification, merging results files and rescoring of identified peptides using platforms such as Scaffold (from Proteome Software Inc.) (Dagda *et al.*, 2010). With this type of strategy it is possible to increase the accuracy, sensitivity and specificity comparing to the use of individual search engines due to different mathematical and algorithmic strategies considered. An advantage of this approach is that it minimizes false positive identifications by using different types of search engines, the disadvantage of one is overcome by the other. The usual search engines used together for this type of strategy are Mowse, Sequest and X!tandem. The use of a fourth search algorithm does not appear to improve the results to a further extend (Dagda *et al.*, 2010).

3.2 *De novo* analysis

De novo sequencing of peptides by MS/MS is the process used for the determination of peptide primary structure not based on the information available in databases. It is ideal for the identification of proteins from organisms without sequenced genome or for the characterization of novel proteins, their isoforms, biological or induced post-translational modifications or peptides with non-proteic amino acids. An interesting review on the subject was published in early 2010 (Seidler *et al.*, 2010). Here we will present further details on manual sequencing and available automatic algorithms, as well as some examples on the characterization of peptides containing modified amino acids.

De novo sequencing analysis of MS/MS fragmentation data can be done manually or using specific softwares. As previously pointed in this section, manual analysis of peptide tandem MS spectra raises usually some difficulties, due to their complexity. Typically a sequence of steps is followed during this exercise, although some have to be performed interactively: a) assignment of major ion series, *b*- and *y*- or *c*- and *z*-; b) calculation of mass differences between peaks attributed to the same ion-series, in order to determine the amino acid residues present in the sequence; c) checking the presence of characteristic patterns for the type of fragmentation process used, namely relative intensity of fragment peaks along the *m/z* range, detection of neutral losses, immonium ions. In order to simplify the fragmentation pattern, several peptide derivatization procedures have been developed that allow the reduction of the peptide fragments to only one ion series, since the second loses its ionization capacities (An *et al.*, 2010, Franck *et al.*, 2010, Hennrich *et al.*, 2010, Miyashita *et al.*, 2011, Nakajima *et al.*, 2011). This strategy presents some drawbacks, namely due to interference from the excess reagent and corresponding side products and low yield of derivatization products. In our laboratory we have successfully used this approach for the identification and N-terminal characterization of a cutinase purified from *Colletotrichum kahawae*, a causal agent of the coffee berry disease (Zhenjia *et al.*, 2007). Although having four tryptic peptides common to *C. gloeosporioides* cutinase, the 21-kDa protein from *C. kahawae*

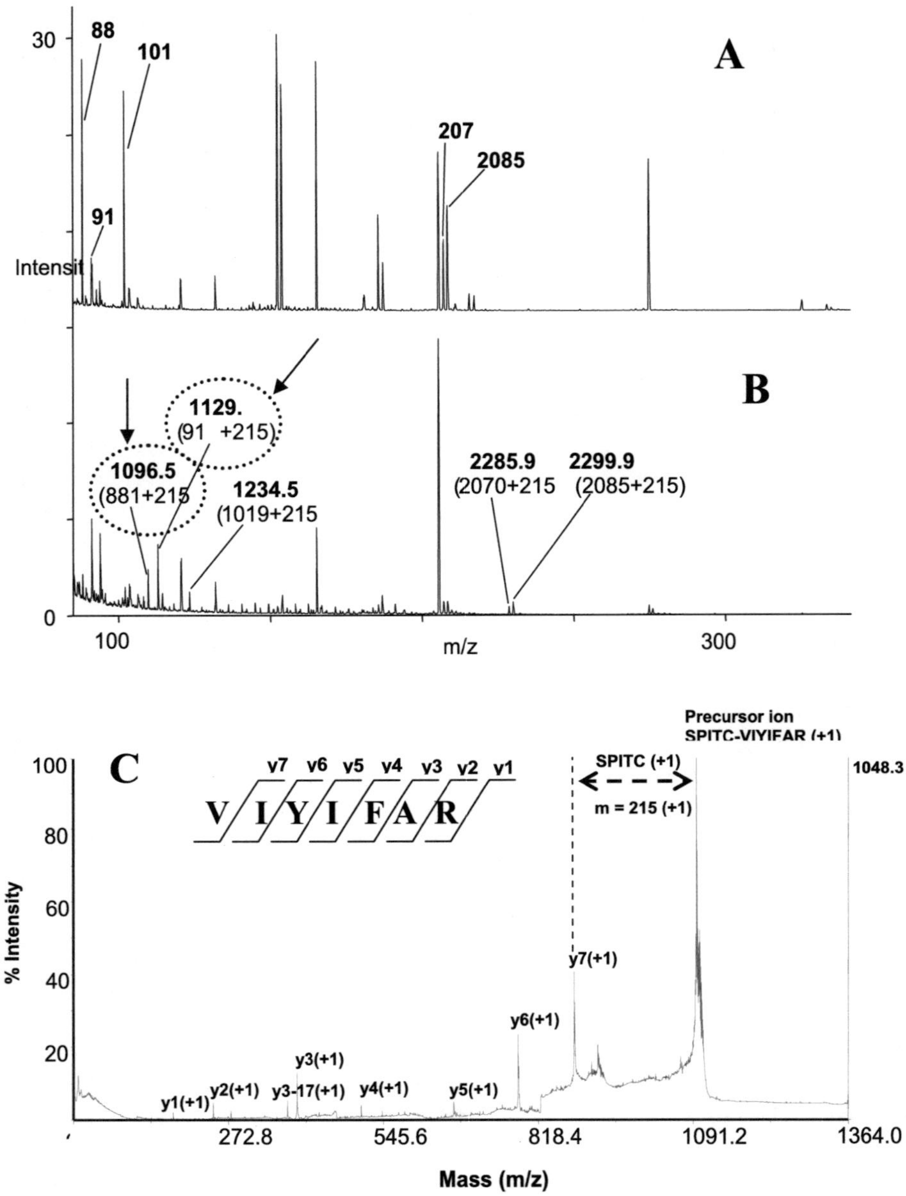

Fig. 3. Mass spectra from *C. kahawae* after trypsin cleavage. A) MALDI-TOF MS spectra of the tryptic peptides before derivatization with SPITC (peptides that were derivatized are labelled). B) MALDI-TOF MS spectra of the tryptic peptides after derivatization with SPITC. Peptides derivatized with SPITC (increase of 215 Da in *m/z* value) were subjected to PSD analysis to obtain sequence information. Arrowheads indicate the peptides that were subjected to further sequencing. C) Peptide sequence from peptide *m/z* 1.096. The resulting *y* ions from *m/z* 1.096 fragmentation were identified in the database as the sequence VIYIFAR (MASCOT, MS/MS ions search). This peptide was found in cutinase precursor form *C. capsici*. The score of 63 for this identification is higher than the minimum of 38, which indicates identity or extensive homology (p<0.05). Figure from Zhenjia *et al*, 2007.

could not be identified with a significant score. To enhance fragmentation towards the peptide bond, as described by Wang *et al.* (2004), a sulphonic acid group was introduced at the N-terminus of the tryptic peptides using 4-sulphophenyl isothiocyanate reagent. Two of the peptides derivatized were sequenced (Fig. 3). After a search in the database for identification of the MS/MS ions generated, one of the peptides revealed 100% homology to a highly conserved peptide from a cutinase precursor from *C. gloeosporioides*. Additionally, in the peptide mass map of the 21-kDa protein, a peptide with a monoisotopic mass of 1.653 ($[M+H]^+$) is present, suggesting the blockage of the N-terminus with a glucuronamide residue.

Several software packages for *de novo* sequencing interpretation of MS/MS spectra are available, either proprietary, publicly available academic interfaces or commercial programs. Bringans *et al.* (2008) published a comparative study of the accuracy of three commonly used programs DeNovo Explorer™ (Applied Biosystems), Peaks Studio 4.5 (Bioinformatics Solutions, Inc) and PepNovo (Pitzer *et al.*, 2007). Combining the number of correct residues assigned for Q TRAP and MALDI-TOF/TOF data, Pep-Novo (64.9%) and PEAKS (64.5%) presented a higher accuracy than De Novo ExplorerTM (38.2%). The obtained results indicated that the middle of the peptide was more accurately sequenced and the optimal peptide length is 10-12 residues. As expected, the percentage accuracies were generally related to the quality of the data. PEAKS and PepNovo although presenting very similar accuracy results, often made different choices, consistent with the difference in their algorithm design. For increased confidence in the accuracy of a derived sequence, a combination of algorithms is advised. More recently several other programs for *de novo* analysis have been developed (Chi *et al.*, 2010, He & Ma, 2010, Tessier *et al.*, 2010). The determined amino acid sequences can be used for further functional analysis by homology search of protein sequence databases using adequate programs, namely BLAST-related algorithms (Gaeta, 1998), as "bait for gene fishing" or as a probe for antibody production.

The above-described workflow (MS and MS/MS spectra acquisition, *de novo* sequencing and Blast search) was used for the identification of proteins present in sea urchin (*Paracentrotus lividus*) tube foot adhesive (Santos *et al.*, 2009). Beside the six proteins identified by homology–database search, other MALDI-MS/MS spectra did not allowed direct protein identification and were further used to automatically generate *de novo* sequences in an attempt to obtain more information on these proteins. Five *de novo*-generated peptide sequences were found that were not present in the available protein databases, suggesting that they might belong to novel or modified proteins.

4. Aplications of peptide tandem mass spectrometry

4.1 Protein Identification in non-sequenced organisms

As previously mentioned, the inherent characteristics of tandem MS-based protein identification provides a high-throughput and efficient methodology of utmost importance in the context of research in a wide range of research subjects: Plant, Marine, Agronomical, Human and Veterinary Medicine, Physiology, Parasitology, etc. Despite the efficiency and reliability frequently associated to tandem MS, identifications obtained are nevertheless hindered and highly dependent on how further the proteins of a specific study organism have been described and entered in public databases. In this section we will address such issue and the possible solutions that might overcame when a researcher is faced with the need of conducting proteomics studies using non-model organisms.

The number of protein entries per organism in public databases is extremely varied. In fact, if the organism is either sequenced or a model organism, chances are that the number of entries in public databases are representative enough to lead to a robust and publishable tandem MS-based proteomics study. On the contrary, non-sequenced or non-model organisms are however frequently poorly represented in databases, which may strongly condition the success of any tandem MS-based proteomics study. A quick analysis on the number of entries per organism in a public database, such as NCBI (http://www.ncbi.nlm.nih.gov/pubmed/), is presented in table 1 and is highly illustrative of the above mentioned.

When working with a poorly studied or poorly described organism, protein searches have to rely on protein entries that show a closer level of similarity with those of the protein of interest to the studied biological issue. Such searches are termed homology searches as they are dependent on the existence of homologies between the protein of interest and information available on the internet. An illustrative example of a homology search conducted by our research group, was the identification of tannin binding proteins in the saliva of sacred baboon (*Papio hamadryas*) using MS/MS data (Mau *et al.*, 2011). Sacred baboons are very poorly represented in public databases and, concerning NCBI, a mere total of 641 entries exist (September 2011). As a consequence, all proteins submitted for identification were identified in primate species other than *Papio hamadryas*, nevertheless with higher levels of representation in databases: Macaque, Colobus and Pan species (respectively 70.000, 1.057 and 43.890). Another striking example was the study involving Mediterranean mussel (*Mytilus galloprovincialis*) and the Vietnamese clam (*Corbicula fluminea*) when exposed to *Cylindrospermopsis raciborskii* cells (Puerto *et al.*, 2011). Again, both studied species are very poorly represented in public databases (1575 for *M. galloprovincialis* and 107 for *C. fluminea*). Consequently, all 26 identifications were obtained with higher level of confidence in other species, with special relevance to the blue mussel (*Mytilus edulis*), a species with 3.154 entries in NCBI.

Interestingly, public generalist databases may have remarkable differences regarding the number of entries for a specific organism. Such difference will have necessary consequences on the success rate of the identifications achieved. For instance, our research laboratory has conducted a study on the proteome of cattle pathogen *Ehrlichia ruminantium*, the agent of the tick-borne disease heartwater or cowdriosis (results unpublished). Identifications were extremely low when conducting the search using the UNIPROT KB curated database (2.369 entries for *Ehrlichia ruminantium*). On the contrary, when the search was conducted resorting to NCBI database, success rate of the identifications was substantially increased, as NCBI more than triples the number of entries for *E. ruminantium*. Such difference is probably the result of the fact that the number of laboratories conducting research on those specific bacteria is extremely low. Consequently, the choice on the public database to deposit protein sequences has necessarily significant implications. Additionally, curated databases only allow entries for verified protein sequences. Such rule limits strongly the access to entries based on, for instance, theoretical sequences obtained from genome-based information, often the most relevant sources of information.

Resorting to public databases may itself strongly limit the probability of obtaining robust and reliable identifications. In fact, the number of proteins entered in public databases is ultimately dependent on the availability of the researcher and host institutions to enter the

Sequenced or Model Organisms		Non-Sequenced or Non-Model Organisms	
Organism	**Number of Entries**	**Organism**	**Number of Entries**
Arabidopsis thaliana	552.550	Red cabbage (*Brassica oleracea*)	2.290
Barrel Medic (*Medicago truncatulla*)	10.223	Lucerne (*Medicago sativa*)	2.222
Rat (*Rattus norvegicus*)	113.981	Black rat (*Rattus rattus*)	937
Mouse (*Mus musculus*)	266.662	Ryukyu mouse (*Mus caroli*)	205
Cattle (*Bos taurus*)	108.129	Water Buffalo (*Bubalus bubalis*)	2.089
Rabbit (*Oryctolagus cuniculus*)	40.707	European hare (*Lepus europaeus*)	478
Human (*Homo sapiens*)	1.963.679	Gorilla (*Gorilla gorilla*)	2.100
Chimpanzee (*Pan troglodytes*)	43.890	Pigmy Chimpanzee (*Pan paniscus*)	1.117
Rhesus macaque (*Macaca mulatta*)	47.029	Celebes macaque (*Macaca nigra*)	73
Saccharomyces cerevisiae	225.898	*Saccharomyces bayanus*	541
Search performed on the 20th September 2011			

Table 1. Protein entries in NCBI database – sequenced and model organisms vs. non-sequenced and closely related non-model organisms.

information obtained in the database. Consequently, and imagining a putative consortium or laboratory would be sequencing a given organism, if the information generated would not be included in public databases, all such information would not be accessible to the general research community, particularly what concerns protein identification using mass spectrometry. Such information is, however, frequently compiled in the form of non-publicly disclosed dedicated databases that are organized and updated by the same consortium that generated it. Access to such databases may be granted in the form of collaboration or by ceding it to interested researchers. A pertinent example of the above-mentioned is a study we have conducted in our research group related to somatic embryogenesis in *Medicago truncatula* (results not published). In this study, protein identifications were conducted first using generalist databases, namely NCBI, with rather poor identification rates of around 50%. However, when the same information was submitted to a dedicated database, generated, kept and kindly ceded by the Samuel Roberts Noble Foundation (Ardmore, OK, USA), identifications rates increased significantly to rates over 85% of the searched proteins.

An interesting solution to the lack of information available on a specific organism may rely on the use of a dedicated database on a specific organism to which the studied organism is closely related and to which it would be expected that proteins would share a high level of homology. Such a strategy would associate the specificity of a dedicated database to high homologies of a very close species. It is likely that such strategy would only be surpassed, in successful protein identification rates, by the use of dedicated databases specific to the studied genus or species. Our research group has conducted one of such strategies to the proteome characterization of coelomocytes in sea star (*Marthasterias glacialis*) (Franco *et al.*,

2011b). *Marthasterias glacialis* sea stars are poorly represented in public databases with only 90 and 26 entries in NCBI and UniprotKB, respectively. Our research group has therefore opted to conduct protein identification using a dedicated database on *Strongylocentrotus purpuratus* (purple sea urchin), a sequenced organism, interestingly with over 1.500 entries in Uniprot and over 6.000 in NCBI. With this highly successful strategy and using both nanoLC-MALDI-TOF/TOF MS and two-dimensional electrophoresis with MALDI-TOF/TOF, we were able to identify over 350 proteins of a highly complex proteome.

Conducting research in non-sequenced or non-model organisms poses interesting challenges to proteomics researchers resorting to a uniquely reliable method for protein identification, such as tandem MS. To overcome such difficulties, the use of either homology searches or dedicated specific databases or even the combination of both may be a useful and interesting strategy. Alternatively, the use of *de novo* sequencing may also be an interesting and effective strategy that was the object of a previous section (section 3.2) in this chapter.

4.2 Top-down proteomics

So far, top-down mass spectrometry has been a less common application in tandem mass spectrometry. This global-view technique has been of great interest as a first approach on protein characterization, without the need of prior knowledge, for the determination of protein's exact molecular mass, post-translational modifications, etc. This examination is made without the need of protein digestion, intact proteins are directly fragmented in the mass spectrometer preventing artificial modifications that can occur during sample handling and preserving some PTM information.

In a recent work by Zhang and co-workers (2011), top-down mass spectrometry has been used for the analysis of human and mouse cardiac Troponin T. In this work, top-down mass spectrometry and fragmentation techniques such as CID and ECD were used to identify and characterize the modification sites of phosphorylation, acetylation, proteolysis, and spliced isoforms of the highly acidic N-terminal of cardiac Troponin T. In a preliminary approach, there has also been an attempt to employ this technology to clinical assays, as it was described by Théberge and co-workers (2011) where the software, BUPID (Boston University Protein Identifier) was developed in order to determine variant and/or modified protein sequencing. This analysis was applied to patient samples for the identification of transthyretin and haemoglobin in an automated and fast method, although it was concluded that at this stage it works mainly for small and abundant clinically relevant proteins.

A new technology has now been applied to top-down mass spectrometry, namely travelling wave ion mobility. A recent work by Halgand and colleagues (2011) involved the study of the heterogeneity of the recombinant phosphoprotein domain of the measles virus expressed in *Escherichia coli*. The use of this technique was particularly useful regarding the reduction of top-down MSMS spectra and the elucidation of the origin of subtle sample heterogeneity, regardless of the mechanisms responsible for the amino acid substitutions.

4.3 Post-translational modifications of proteins and peptides containing adducts

Another application of tandem mass spectrometry with biological and clinical relevance is the identification and characterization of post-translational modification (PTMs) of proteins, where mass spectrometry is the method of choice for this purpose. The modification of proteins post-translationally is of great importance in protein activity and cellular metabolism regulation. The protein's amino acids may be modified after translation, altering the protein's molecular mass, and therefore it cannot be predicted from genome information.

There are over 400 post-translational modifications of proteins described (Creasy & Cottrell, 2004). The most common and naturally occurring PTMs include phosphorylation, glycosylation, cleavage, formylation, methionine oxidation and ubiquitination (see table 2, as reviewed by Farley & Link (2009). We will focus mainly on applications of the most common and also labile PTMs, phosphorylation, glycosylation, and also on glycation.

Phoshporylation and glycosylation are considered labile PTMs. In these cases, the site of modification is lost, including during the fragmentation process, and more precisely with CID. In a work performed by Carapito and co-workers (2009) a method to detect modified peptides in a complex peptide sample and establish the nature of the modification was developed based on the alternation between MS spectra acquisition using different collision conditions. This experiment led to the cleavage of the substituents and hence to the detection of modified peptides based on their specific fragmentation and on the detection of low mass reporter ions. The described approach allows the detection of multiple modifications without prior knowledge on its type.

Phosphorylation is a PTM that is particularly challenging, as it is present in low-stoichiometry amounts in samples, therefore enrichment techniques are usually required prior to mass spectrometry analysis, as reviewed by Leitner *et al.* (2011). Presently, there are several techniques effective and available for this purpose, namely IMAC, titanium dioxide, among others, and recently a highly efficient new enrichment technique using lathanum ions for the precipitation of this type of modified peptides was developed (Pink *et al.*, 2011). The advantage of the lanthanum enrichment is that it is based on a single step precipitation and phosphoproteins can be isolated from frozen tissue or cells, after cell lysis, using several buffer systems compatible with tandem mass spectrometry analysis.

Phosphorylated peptides can be analyzed using tandem mass spectrometry with CID. However, this type of modified peptides can undergo neutral loss, particularly if phosphorylated in serine or threonine amino acid residues. This means that the MSMS spectrum would not contain much information regarding the site of modification and peptide sequence as the energy for peptide fragmentation is directed towards the phosphoric acid. To overcome this disadvantage, neutral loss LC-MS experiments can be done, where MS^n experiments are established. An advantage of this type of methods is that the peptide after the loss of phosphoric acid can be fragmented again and information regarding the modification site and amino acid composition can be obtained. A disadvantage however is that the duty cycle of the equipment is long and the information obtained is more complex for bioinformatic analysis (Leitner *et al.*, 2011). The development of ETD, prevents this type of problem, as the PTM is preserved during fragmentation, allowing more informative MSMS spectra. Top-down has also been extensively used in the identification and characterization of phosphorylation events, where in this case exact mass of the phosphoprotein is obtained followed by direct characterization of the site of modification, without the need for proteolytic digestion.

The other most common PTM is glycosylation. This PTM is important in protein function and cell fate differentiation. Glycosylation, like phosphorylation, is regulated by enzymes and it involves the binding of glycans to proteins being attached in linked or branched chains with various glycan composition and length (Lazar *et al.*, 2011), hence having an impact on charge, conformation and stability of proteins. Glycosylation has been shown to have an important role in several diseases, including cancer, inflammatory diseases and congenital disorders, etc. The study of glycosylation of proteins has been useful for the development of novel vaccines and biomarkers discovery for diagnosis (Lazar *et al.*, 2011).

PTM	Nominal mass shift (Da)	Stability	Proposed biological function
Phosphorylation			
pSer, pThr	+ 80	Very labile	Cellular signaling processes, enzyme activity, intermolecular interactions
pTyr	+ 80	Moderately labile	
Glycosylation			
O-linked	203, >800	Moderately labile	Regulatory elements, *O*-GlcNAc
N-linked	>800	Moderately labile	Protein secretion, signaling
Proteinaceous			
Ubiquitination	>1000	Stable	Protein degradation signal
Sumoylation	>1000	Stable	Protein stability
Nitrosative			
Nitration, nTyr	+ 45	Stable	Oxidative damage
Nitrosylation, nSer, nCys	+ 29	Stable	Cell signaling
Methylation	+ 14	Stable	Gene expression
Acetylation	+ 42	Stable	Histone regulation, protein stability
Sulfation, sTyr	+ 80	Very labile	Intermolecular interactions
Deamidation	+ 1	Stable	Intermolecular interactions, sample handling artifact
Acylation			
Farnesyl	+ 204	Stable	Membrane tethering, intermolecular interactions, cell localization signals
Myristoyl	+ 210	Stable	
Palmitoyl	+ 238	Moderately labile	
Disulfide bond	− 2	Moderately labile	Protein structure and stability
Alkylation, aCys	+ 57	Stable	Sample handling
Oxidation, oMet	+ 16	Stable	Sample handling

Table 2. Common PTMs of proteins, showing the mass difference of the modified protein as well as its stability and biological function (Farley & Link, 2009).

Due to the recent advances in technology, particularly in high performance liquid chromatography and mass spectrometry, these methods have been the preferred choice for the identification and characterization of glycosylated proteins. Just like with phosphorylation, glycosylation modified protein/peptides need to be enriched prior to mass spectrometry analysis. A common methodology for this purpose includes lectin enrichment, which can selectively separate complex carbohydrate structures, as reviewed by Lazar *et al.* (2011).

In mass spectrometry, the analysis of glycosylated peptides is particularly challenging, due to the heterogeneity of oligossacharide moieties and to proteolysis difficulties, since glycan motifs can prevent access of site-specific endoproteases, reducing protein coverage and identification. A solution to prevent this is to remove the glycans chemically or enzymatically, for example with PNGase F, prior to endoproteinase digestion. Another problem with this type of PTM is the fact that there is an increase of peptide mass due to the glycan moiety, which can lead to outside the adequate resolution range of most mass spectrometers. To overcome this, a combination of broad specificity proteases can be used in order to increase protein coverage and identification confidence, as well as characterization of glycosylation events. The ionization of these modified peptides is also challenging due to the fact that they can have low ionization efficiency, which implies lower sensitivity in peptide detection. The ions generated can be due to the sequential loss of the sugar residues, however, the ions from the fragmentation of the peptide backbone can be in low abundance or even absent (Lazar *et al.,* 2011).

Glycation is a type of chemical modification that can occur *in vivo*, but it is not controlled enzymatically. Glycation is involved in alterations of structure and stability of proteins, hence affecting protein function. This PTM has been shown to be involved in several diseases, such as diabetes and amyloidotic neuropathies. In a study performed in collaboration with our group, the effects of glycation *in vivo* and *in vitro* were studied on yeast enolase. Peptide mass data were used to determine methylglyoxal-derived advanced glycation end-product (MAGE) nature and location. Since only lysine and arginine residues are modified, tryptic digestion of glycated proteins will produce peptides with at least one miss cleavage associated to a defined mass increase corresponding to a specific MAGE. Using this approach for MS data interpretation important differences were observed between *in vivo* and *in vitro* glycation, that is the same residues were consistently modified *in vivo* suggesting that it is a specific process, whereas *in vitro* this is not the case, that is several residues are modified with different glycation end products (Gomes *et al.,* 2008). Another recent work by Oliveira and co-workers (2011) studied the mechanism of insulin fibril formation in the presence of methylglyoxal, the most significant glycation agent *in vivo*. To unequivocally identify glycated peptides and amino acid residues, non-glycated and glycated insulin were digested using chymotrypsin followed by MS and MS/MS analysis. A modified glycated peptide should be exclusively present in the MS spectrum of glycated insulin with a mass value corresponding to the insulin peptide plus the specific mass increment characteristic of a MAGE modification. This information was used to construct an inclusion list of modified peptides to be fragmented by an additional MS/MS experiment using the MALDI-TOF/TOF instrument. The sequence information thus obtained allowed the unequivocal identification of MAGE-modified peptides and also assignment of specific modified amino acids. This study showed that glycation by methylglyoxal agent stabilizes soluble aggregates that retain native-like structures of insulin.

Another application of tandem mass spectrometry of peptides is the study of protein adducts which is of great importance in toxicology studies. In a recent review by Rappaport and co-workers (2011) it is described a strategy for the identification of adducts in human blood, more precisely in haemoglobin and human serum albumin. In this review, a recent mass spectrometry methodology is described, named fixed-step selected reaction monitoring (FS-SRM), that involves a list of theoretical parent and product ions for the detection of all modifications of targeted nucleophile within a range of adduct masses. Another approach for this type of tandem MS application is described by Switzar and colleagues (2011), where the protein digestion protocol conditions were optimized (pH, temperature and time) in order to increase protein coverage and signal intensity of the modified peptides.

This tandem mass spectrometry application has also been employed in the study of the toxicity effect of anti-viral drugs, more precisely nevirapine (NVP) employed against human immunodeficiency virus type-1 (HIV-1) (Antunes *et al.*, 2010). MALDI-TOF-TOF-MS of tryptic digests was used to identify which human serum albumin and human hemoglobin amino acid residues were bound to NVP upon incubation with the synthetic model electrophile 12-mesyloxy-NVP, used as a surrogate for the Phase II metabolite 12-sulfoxy-NVP. The adopted strategy consisted of (i) comparison of the MS spectra of unmodified and NVP-modified HSA and Hb digests. The presence of new m/z peaks in the latter was presumed to correspond to potential NVP-amino acid adducts. (ii) The m/z values observed exclusively in the MS spectra of the tryptic digests of the modified proteins were compared to the theoretical tryptic peptide mass list for each protein, taking into account the mass increase characteristic of NVP modification. (iii) This information was used to construct an inclusion list of possible NVP-modified peptides to be fragmented by an additional MS/MS experiment using the MALDI-TOF-TOF instrument. The amino acid sequence information thus obtained allowed the unequivocal identification of NVP-modified peptides and the assignment of the specific NVP-modified amino acids. This study prompts to the identification of multiple modification sites suggesting several possible biomarkers of nevirapine toxicity that can be useful for monitoring the toxicity of this drug in patients.

Reference should also be made to the characterization of disulfide bonds using tandem mass spectrometry as these covalent bonds are important in protein folding and aggregation and hence the detection of this type of PTM and identification of the involved cysteine residues is particularly important for monitoring recombinant protein production, namely in the biopharmaceutical industry. The wrong establishment of disulfide bonds may have a remarkable importance in protein 3-dimensional structures and, consequently in their functional activity, thus affecting protein drug selection, assay development and drug testing. The conventional methodology for studying this PTM is, with HPLC or mass spectrometry, to compare protein enzymatic digests using the target protein in their reduced and native forms, where the chromatographic peaks or masses obtained are compared and the differences obtained in the native form are considered due to possible disulfide bonded peptides. In a recent study by Janecki & Nemeth (2011) an efficient method using MALDI-TOF-TOF and high-energy CID was applied to identify disulfide-bonded peptides in proteins with well-documented disulfide bond networks (namely bovine insulin and human serum albumin) and on recombinant proteins where disulfide bonds were not defined, without previous separation of the protein native digests. This method was based on the fact that a number of fragmentation processes happen around the S-S bond leading to a "triplet peak" signature in the spectrum (Fig. 4), which results from the symmetric cleavage of the S-

S bond originating a cysteine fragment (middle peak of the "triplet peak signature") and asymmetric cleavage of the S-S bond originating dehydroalanine and thiocysteine fragments with a mass difference from the middle peak of -34 and +32 Da, respectively (Fig. 4; Janecki & Nemeth, 2011). Sometimes a smaller peak identified as the dehydrocysteine peak can occur around the cysteine *m/z* peak. Hence this "triplet peak" signature can be used for identifying peptides containing one or more inter-disulfide bonds.

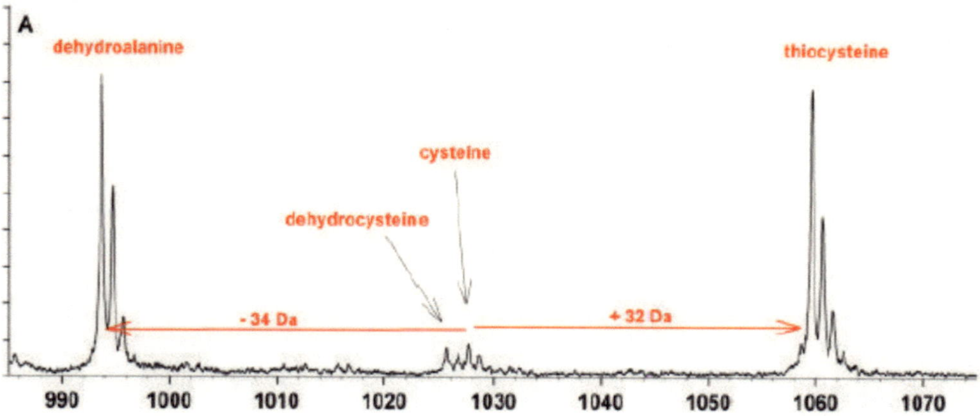

Fig. 4. Zoomed region of the tandem mass spectrum of the *m/z* peak 2071.01 of the native peptic digest of bovine insulin showing the "triplet peak" signature and the characteristic S-S bond fragments (from Janecki & Nemeth, 2011).

Top-down mass spectrometry has also been applied for the identification and characterization of disulfide bonds, however this technique is still limited. In a recent study by Chen and co-workers (2010), native chicken lysozyme was used to evaluate CID fragmentation of protein disulfide bonds by LTQ Orbitrap in positive ion mode. They identified fragments for low-charged protein precursor ions that correspond to the breakage of disulfide-bonds and of protein backbone. These related disulfide-bond fragments resulted from the addition or subtraction of a hydrogen atom or sulfhydryl group with mass changes of -32, -2, +2 and +32 Da for –SH, -H, +H, and +SH, respectively, similar to the patterns obtained by Janecki & Nemeth (2011) for tryptic peptides.

5. Concluding remarks

In this chapter we have reviewed the tandem mass spectrometry approach and its relevance as a technique for the analysis of proteins. To date there are several methodological developments namely, at the levels of sample preparation and experimental procedure and on instrumental and software innovation. Additionally, the number of applications regarding peptide identification, characterization of post-translational modifications, peptide/protein adduct identification, *de novo* studies and proteome studies (including non-sequenced organisms) are still being improved and diversified.

6. Acknowledgments

We would like to acknowledge *Fundação para a Ciência e a Tecnologia* (Lisbon, Portugal) for financial support in the form of PhD grant SFRH/BD/29799/2006 (Catarina Franco) and Post-Doc grants SFRH/BPD/41037/2007 (Ricardo Gomes) and REDE-1504 (Kamila Koči), as well as research contract by Ciência 2008 (Renata Soares and Romana Santos) and Ciência 2007 (André Almeida) programs; project grants PTDC/MAR/104058/2008 and PTDC/QUI/70610/2006 and equipment acquisition within the framework of the National Re-equipment Program – National Network Mass Spectrometry (REDE/1504/REM/2005).

7. References

An, M., Dai, J., Wang, Q., Tong, Y. & Ji, J. (2010). Efficient and clean charge derivatization of peptides for analysis by mass spectrometry. *Rapid Communications in Mass Spectrometry*, Vol.24, No.13 (July 2010), pp.1869-1874, ISSN 1097-0231

Antunes, A.M.M., Godinho, A.L.A., Martins, I.L., Oliveira, M.C., Gomes, R.A., Coelho, A.V., Beland, F.A. & Marques, M.M. (2010). Proteins Adducts as Prospective Biomarkers of Nevirapine Toxicity. *Chemical Research in Toxicology*, Vol.23, No.11, (November 2010), pp.1714-1725

Bai, L., Romanova, E.V. & Sweedler, J.V. (2011). Distinguishing Endogenous D-Amino Acid-Containing Neuropeptides in Individual Neurons Using Tandem Mass Spectrometry. *Analytical Chemistry*, Vol.83, No.7, (March 2011), pp.2794-2800, ISSN 1520-6882

Bakhtiar, R. & Guan Z. (2006). Electron Capture Dissociation Mass Spectrometry in Characterization of Peptides and Proteins. *Biotechnology Letters*, Vol.28, No.14, (July 2006), pp.1047-1059, ISSN 1573-6776

Budnik, B.A., Haselmann, K.F. & Zubarev, R.A. (2001). Electron Detachment Dissociation of Peptide Di-Anions: An Electron-Hole Recombination Phenomenon. *Chemical Physics Letters*, Vol.342, No.3-4, (July 2001), pp.299-302, ISSN 0009-2614

Carapito, C, Klemm, C., Aebersold, R. & Domon B. (2009). Systematic LC-MS Analysis of Labile Post-Translational Modifications in Complex Mixtures. *Journal of Proteome Research*, Vol.8, No.5, (March 2009), pp.2608-2614, ISSN 1535-3907

Chen, J., Shiyanov, P., Zhang, L., Schlager, J.J. & Green-Church, K.B. (2010). Top-Down Characterization of a Native Highly Intralinked Protein: Concurrent Cleavages of Disulfide and Protein Backbone Bonds. *Analytical Chemistry*, Vol.82, No.14, (July 2010), pp.6079-6089, ISSN 1520-6882

Chi, H., Sun, R.X., Yang, B., Song, C.Q., Wang, L.H., Liu, C., Fu, Y., Yuan, Z.F., Wang, H.P., He S.M. & Dong, M.Q. (2010). pNovo: *de novo* peptide sequencing and identification using HCD spectra. *Journal of Proteome Research*, Vol.9, No.5, (May 2010), pp.2713-2724, ISSN 1535-3907

Cottrell, J.S. (2011). Protein Identification using MS/MS data. *Journal of Proteomics*, Vol.74, No.10, (September 2011), pp.1842-1851, ISSN 1874-3919

Creasy, D.M. & Cottrell, J.S. (2004). Unimod: Protein Modifications for Mass Spectrometry. *Proteomics*, Vol.4, No.6, (June 2004), pp.1534-1536, ISSN 1615-9861

Cui, W., Rohrs, H.W. & Gross, M.L. (2011). Top-Down Mass Spectrometry: Recent Developments, Applications and Perspectives. *Analyst*, Vol.136, No.19, (August 2011), pp.3854-3864

Dagda, R.K., Sultana, T. & Lyons-Weiler, J. (2010). Evaluation of the Consensus of Four Peptide Identification Algorithms for Tandem Mass Spectrometry Based Proteomics. *Journal of Proteomics and Bioinformatics*, Vol.5, No.3, (February 2010), pp.39-47

Eng, J.K., McCormack, A.L. & Yates III, J.R. (1994). An Approach to Correlate Tandem Mass Spectral Data of Peptides with Amino Acid Sequences in a Protein Database. *Journal of the American Society for Mass Spectrometry*, Vol.5, No.11, (November 1994), pp.976-989, ISSN 1879-1123

Eng, J.K., Searle, B.C., Clauser, K.R. & Tabb, D.L. (2011). A Face in the Crowd: Recognizing Peptides through Database Search. *Molecular and Cellular Proteomics*, in press

Farley, A.R. & Link, A.J. (2009). Identification and Quantification of Protein Post-Translational Modifications, In: *Methods in Enzymology*, Burgess, R.R. & Deutscher, M.P., pp.725-763, Elsevier, ISBN 978-0-12-374536-1

Franck, J., Ayed, M.E., Wisztorski, M., Salzet, M. & Fournier, I. (2010). On tissue protein identification improvement by N-terminal peptide derivatization. *Methods in Molecular Biology*, Vol.656, No. 3, pp.323-338, ISSN 1064-3745

Franco, C.F., Santos, R. & Coelho, A.V. (2011a). Exploring the Proteome of an Echinoderm Nervous System: 2DE of the Sea Star Radial Nerve Cord and the Synaptosomal Membranes Subproteome. *Proteomics*, Vol.11, No.7, (April 2011), pp.1359-1364, ISSN 1615-9861

Franco, C.F., Santos, R. & Coelho, A.V. (2011b). Proteome Characterization of the Sea Star Coelomocytes - the Innate Immune Effector Cells of Echinoderms. *Proteomics*, Vol.11, No.17, (September 2011), pp.3587-3592, ISSN 1615-9861

Gaeta, B.A. (1998). Database searching using short peptide sequences. In: *ABRF News*, September 2011, Available from: http://www.abrf.org/ABRFNews/1997/December1997/dec97tips.html

Ganisl, B., Valovka, T., Hartl, M., Taucher, M., Bister, K. & Breuker K. (2011). Electron Detachment Dissociation for Top-Down Mass Spectrometry of Acidic Proteins. *Chemistry – A European Journal*, Vol.17, No.16, (April 2011), pp.4460-4469

Gomes, R.A., Oliveira, L.M., Silva, M.C., Ascenso, C., Quintas, A., Costa, G., Coelho, A.V., Sousa Silva, M., Ferreira, A.E., Ponces Freire, A. & Cordeiro, C. (2008). Protein Glycation *in vivo*: Functional and Structural Effects on Yeast Enolase. *Biochemical Journal*, Vol.416, No.3, (December 2008), pp.317-326

Halgand, F., Habchi, J., Cravello, L., Martinho, M., Guigliarelli, B. & Longhi S. (2011). Dividing to Unveil Protein Microheterogeneities: Travelling Wave Ion Mobility Study. *Analytical Chemistry*, e-pub, (July 2011), ISSN 1520-6882

He, L. & Ma, B. (2010). ADEPTS: advanced peptide *de novo* sequencing with a pair of tandem mass spectra. *Journal of Bioinformatics and Computational Biology*, Vol.8, No.6, (December 2010), pp.981-994, ISSN 1757-6334

Hennrich, M.L., Mohammed, S., Altelaar, A.F. & Heck, A.J. (2010). Dimethyl isotope labeling assisted de novo peptide sequencing. *Journal of American Society for Mass Spectrometry*, Vol.21, No.12, (December 2010), pp. 1957-1965, ISSN 1044-0305

Janecki, D.J. & Nemeth, J.F. (2011). Application of MALDI TOF/TOF Mass Spectrometry and Collision-Induced Dissociation for the Identification of Disulfide-Bonded Peptides. *Journal of Mass Spectrometry*, Vol.46, No.9, (September 2011), pp.677-688, ISSN 1096-9888

Jia, W., Wu, H., Lu, H., Li, N., Zhang, Y., Cai, R. & Yang, P. (2007). Rapid and Automatic On-Plate Desalting Protocol for MALDI-MS: Using Imprinted Hydrophobic Polymer Template. *Proteomics*, Vol.7, No.15, (August 2007), pp.2497-2506, ISSN 1615-9861

Koči, K., Binetti, A., Soares, R., Haag, E., Vinderola, G., Costa, M.C., Reinheimer, J. & Coelho, A.V. (2010). Insights in the Peptidomics of a Functional of a Functional Food Prototype - The Buttermilk Fermentation Product by *Lactobacillus delbrueckii* subsp. *lactis*, presented at 4th Portuguese Mass Spectrometry Meeting, Lisbon, Portugal, December, 2010

Larsen, M.R. (2007). Preparation of Reversed-Phase Microcolumns. *Cold Spring Harbor Protocols*, Vol.11, (November 2007), ISSN 1559-6095

Lazar, I.M., Lazar, A.C., Cortes, D.F. & Kabulski, J.L. (2011). Recent Advances in the MS Analysis of Glycoproteins: Theoretical Considerations. *Electrophoresis*, Vol.32, No.1, (January 2011), pp.3-13, ISSN 1522-2683

Mau, M., Almeida, A.M., Coelho, A.V. & Südekum, K.H. (2011). First Identification of Tannin-Binding Proteins in Saliva of *Papio hamadryas* using MS/MS Mass Spectrometry. *American Journal of Primatology*, Vol.73, No.9, (September 2011), pp.896-902, ISSN 1098-2345

Mikesh, L.M., Ueberheide, B., Chi, A., Coon, J.J., Syka, J.E.P., Shabanowitz, J. & Hunt, D.F. (2006). The Utility of ETD Mass Spectrometry in Proteomic Analysis. *Biochimica et Biophysica Acta*, Vol.1764, No.12, (December 2006), pp.1811-1822

Miyashita, M., Hanai, Y., Awane, H., Yoshikawa, T. & Miyagawa, H. (2011). Improving peptide fragmentation by N-terminal derivatization with high proton affinity. Rapid Communications of Mass Spectrometry, Vol.25, No.9, (May 2011), pp.1130-1140, ISSN 1097-0231

Nakajima, C., Kuyama, H., Nakazawa, T., Nishimura, O. & Tsunasawa, S. (2011).A method for N-terminal de novo sequencing of Nα-blocked proteins by mass spectrometry. *Analyst*,Vol.136, No.1, (January 2011), pp.113-119

Oliveira, L.M.A., Lages, A., Gomes, R.A., Neves, H., Família, C., Coelho, A.V. & Quintas, A. (2011). Insulin Glycation by Methylglyoxal Results in Native-like Aggregation and Inhibition of Fibril Formation. *BioMed Central Biochemistry*, Vol.12, No.41, (August 2011), pp.1-13, ISSN 1471-2091

Paizs, B. & Suhai, S. (2005). Fragmentation Pathways of Protonated Peptides. *Mass Spectrometry Reviews*, Vol.24, No.4, (July/August 2005), pp.508-548

Park, C.Y., Klammer , A.A., Käll, L., MacCoss, M.J. & Stafford Noble, W. (2008). Rapid and accurate peptide identification from tandem mass spectra. *Journal of Proteome* Research, Vol.7, No.7, (May 2008), pp.3022-3027, ISSN 1535-3907

Perkins, D.N., Pappin, D.J., Creasy, D.M. & Cottrell, J.S. (1999). Probability-Based Protein Identification by Searching Sequence Databases using Mass Spectrometry Data. *Electrophoresis*, Vol. 20, No.18, (December 1999), pp.3551-3567, ISSN 1522-2683

Pink, M., Verma, N., Polato, F., Bonn, G.K., Baba, H.A., Rettenmeier, A.W. & Schmitz-Spanke, S. (2011). Precipitation by Lanthanum Ions: A Straightforward Approach to Isolating Phosphoproteins. *Journal of Proteomics*, in press, (August 2011), ISSN 1874-3919

Pitzer, E., Masselot, A. & Colinge, J. (2007). Assessing peptide *de novo* sequencing algorithms performance on large and diverse data sets. *Proteomics*, Vol.7, No.17, (September 2007), pp. 3051-3054, ISSN 1615-9861

Puerto, M., Campos, A., Prieto, A., Cameán, A., de Almeida, A.M., Coelho, A.V. & Vasconcelos, V. (2011). Differential Protein Expression in Two Bivalve Species; *Mytilus galloprovincialis* and *Corbicula fluminea;* Exposed to *Cylindrospermopsis raciborskii* cells. *Aquatic Toxicology*, Vol.101, No.1, (January 2011), pp.109-116, ISSN 0166-445X

Rappaport, S.M., Li, H., Grigoryan, H., Funk, W.E. & Williams, E.R. (2011). Adductomics: Characterizing Exposures to Reactive Electrophiles. *Toxicology Letters*, in press, ISSN 0378-4274

Reid, G.E. & McLuckey, S.A. (2002). "Top-Down" Protein Characterization via Tandem Mass Spectrometry. *Journal of Mass Spectrometry*, Vol.37, No.7, (July 2002), pp.663-675, ISSN 1096-9888

Ren, D., Pipes, G.D., Liu, D., Shih, L.Y., Nichols, A.C., Treuheit, M.J., Brems, D.N. & Bondarenko, P.V. (2009). An Improved Trypsin Digestion Method Minimizes Digestion-Induced Modifications on Proteins. *Analytical Biochemistry*, Vol.392, No.1, (September 2009), pp. 12-21, ISSN 0003-2697

Santos, R., da Costa, G., Franco, C., Alves, P., Flammang, P. & Coelho, A.V. (2009). First Insights into the Biochemistry of Tube Foot Adhesive from the Sea Urchin *Paracentrotus lividus* (Echinoidea, Echinodermata). *Marine Biotechnology,* Vol.11, No.6, (December 2009), pp.686-698, ISSN 1436-2236

Seidler, J., Zinn, N., Boehm, M.E. & Lehmann, W.D. (2010). *De novo* sequencing of peptides by MS/MS. *Proteomics*, Vol.10, No. 4, (February 2010), pp.634–649, ISSN 1615-9861

Shilov, I.V., Seymour, S.L., Patel, A.A., Loboda, A., Tang, W.H.,Keating, S.P., Hunter, C.L., Nuwaysir, L.M. & Schaeffer, D.A. (2007). The Paragon Algorithm, a Next Generation Search Engine That Uses Sequence Temperature Values and Feature Probabilities to Identify Peptides from Tandem Mass Spectra. *Molecular and Cellular Proteomics*, Vol.6, No.9, (May 2007), pp.1638-1655, ISSN 1535-9484

Switzar, L., Giera, M., Lingeman, H., Irth H. & Niessen, W.M.A. (2011). Protein Digestion Optimization for Characterization of Drug-Protein Adducts using Response Surface Modelling. *Journal of Chromatography A,* Vol.1218, No.13, (April, 2011), pp.1715-1723, ISSN 0021-9673

Tessier, D., Yclon, P., Jacquemin, I., Larré, C. & Rogniaux, H. (2010). OVNIp: an open source application facilitating the interpretation, the validation and the edition of proteomics data generated by MS analyses and *de novo* sequencing. *Proteomics*, Vol.10, No.9, (May 2010), pp.1794-1801, ISSN 1615-9861

Théberge, R., Infusini, G., Tong, W., McComb, M.E. & Costello, C.E. (2011). Top-Down Analysis of Small Plasma Proteins using an LTQ-Orbitrap. Potential for Mass Spectrometry-Based Clinical Assays for Transthyretin and Hemoglobin. *International Journal of Mass Spectrometry*, Vol.300, No.2-3, (March 2011), pp.130-142, ISSN 1387-3806

Wang, D., Kalb, S. & Cotter, R. (2004). Improved procedures for N-terminal sulfonation of peptides for matrix-assisted laser desorption/ionization post-source decay peptide sequencing. *Rapid Communications in Mass Spectrometry*, Vol.18, No.1, (January 2004), pp.96–102, ISSN 1097-0231

Washburn, M.P., Wolters, D. & Yates, J.R.3rd. (2001). Large-scale analysis of the Yeast Proteome by Multidimentional Protein Identification Technology. *Nature Biotechnology*, Vol.19, No.3, (March 2001), pp.242-247, ISSN 1087-0156

Wells, J.M. & McLuckey, S.A. (2005). Collision-Induced Dissociation (CID) of Peptides and Proteins, In: *Methods in Enzymology*, Burlingame A.L. (Ed.), pp.148-185, ISBN 978-0-12-182807-3

Xu, B., Zhang, Y., Zhao, Z., Yoshida, Y., Magdeldin, S., Fujinaka, H., Ismail, T.A., Yaoita, E. & Yamamoto, T. (2011). Usage of Electrostatic Eliminator Reduces Human Keratin Contamination Significantly in Gel-Based Proteomics Anaysis. *Journal of Proteomics*, Vol.74, No.7, (June 2011), pp. 1022-1029, ISSN 1874-3919

Yague, J., Paradela, A., Ramos, M., Ogueta, S., Marina, A., Barahona, F., Lopez de Castro, J.A. & Vasquez, J. (2003). Peptide Arrangement during Quadrupole Ion Trap Fragmentation: Added Complexity to MS/MS Spectra. *Analytical Chemistry*, Vol.75, No.6, (March 2003), pp.1524-1535, ISSN 0003-2700

Zhang, J., Zhang, H., Ayaz-Guner, S., Chen, Y.C., Dong, X. & Xu, Q. (2011). Phosphorylation, but Not Alternative Splicing or Proteolytic Degradation, is Conserved in Human and Mouse Cardiac Troponin T. *Biochemistry*, Vol.50, No.27, (June 2011), pp.6081-6092, ISSN 1520-4995

Zhenjia, C., Franco, C.F., Baptista, R.P., Cabral, J.M.S., Coelho A.V., Rodrigues, C.J. & Melo, E.P. (2007).Purification and identification of cutinases from *C. kahawae* and *C. Gloesporioides*. *Applied Microbiology and Biotechnology*, Vol.73, No.6 (January 2007), pp.1306-1313, ISSN 1432-0614

Zubarev, R.A. (2004). Electron-Capture Dissociation Tandem Mass Spectrometry. *Current Opinion in Biotechnology*, Vol.15, No.1, (February 2004), pp.12-16

3

Evolutionary Proteomics: Empowering Tandem Mass Spectrometry and Bioinformatics Tools for the Study of Evolution

Irving E. Vega[1], Dan Rittschof[2], Gary H. Dickinson[3] and Ian Musgrave[4]
[1]*Department of Biology and Protein Mass Spectrometry Core Facility, College of Natural Science, University of Puerto Rico – Río Piedras Campus, Puerto Rico,*
[2]*MSC Division, Nicholas School and Biology Department Duke University,*
[3]*Department of Oral Biology, University of Pittsburgh School of Dental Medicine,*
[4]*Discipline of Pharmacology, School of Medical Sciences, University of Adelaide, Adelaide,*
[1,2,3]*USA*
[4]*Australia*

1. Introduction

Darwin's and Wallace's evolution hypotheses had as their basis the survival of the fittest. The morphological characteristics that an organism displays are acted upon by selective forces, which control the likelihood that those particular characteristics will be transferred to the next generation. Morphological characteristics observed by scientists in the 19th century were understood by their successors in the 20th century as a phenotypic representation of the complex expression pattern of an organism's genome. Today, we recognize this fundamental concept as the interaction between genome and environment (Qui & Cho, 2008; Dick, 2011; Tzeng et al., 2011). Phenotype is environmentally influenced through molecular changes that begin with gene transcription (Kappeler & Meaney, 2010). Genome complexity, organized and established during embryonic development, is the stage where divergence between species begins (Wang et al., 2010).

Embryology is the intersection between evolutionary and developmental biology (Goodman & Coughlin, 2000). Developmental biology focuses on questions of how genetics control cell growth, differentiation and, ultimately, morphogenesis. For developmental biologists, the timing and regulation of genome expression is crucial to achieving differentiation. In contrast, how biological diversity is transformed through time is a key question for evolutionary biologists. Nevertheless, both disciplines base their theoretical concepts on the expression patterns of the genome.

Genomics as an approach ushered in the beginning of a new scientific era; a logical progression is using genomic approaches to study diversity at the organismal level. Significantly, shortly after completion of the human genome project, the scientific community discovered that only 2% of the human genome is composed of functional genes (Human Genome Project [HGP], 2008). It is fascinating that ~80% of the 20,000 to 25,000 genes in human DNA can be found in invertebrate genomes (Prachumwat & Li, 2008), suggesting that just a small set of genes is responsible for observed species diversity. Thus, a series of questions arise in the genomics era: what makes us different from other organisms? If all organisms have a largely common set of genes, which genes are responsible for, or susceptible to molecular evolution? What molecular mechanisms maintain biological diversity among species?

Insights from the genomics era illustrated that the genome is like an orchestra. The orchestra uses the same instruments (genes) to generate different songs (organisms) that are distinct and unique. But a song can be modified from one music genre to another just by changes in tempo (phenotype) without losing its complete identity. Thus, the human's song comes in hundreds of different tempos without losing its identity, while the human song is very different from the mouse's song. Some scientists believe that epigenetics, factors causing differential gene expression, is the molecular mechanism that explains these variations in "tempo". After all, it has been demonstrated that epigenetic changes induce differential expression of genetic material and contributes to biological diversity (for examples see Levin & Moran, 2011; Day & Sweatt, 2011). However, changes in gene expression alone cannot fully explain diversity among species when most genes are shared. Thus, changes at the protein level are likely to play a role in inducing or maintaining species level biological diversity. Proteome dynamics in response to environmental cues or cellular insults can further contribute to biological diversity. The same protein, differentially modified or localized, could perform diverse functions in a specific cellular state. Although proteome dynamics could serve as the basis to understand species level biological diversity, the technology employed to identify and characterize protein changes require a theoretical context in order to inform molecular evolution.

Here, we put forth the concept that understanding biological commonalities through evolution could lead to the conceptualization of what contributes to the origin and maintenance of biological diversity. Proteomics approaches need not be limited by indexed and sequenced genomes from model organisms. The fact that most genes are conserved among species suggests that we could design experiments to search for evolutionary relationships from conserved proteins among members of genus, family, order, class or phylum. In this chapter, we discuss how tandem mass spectrometry and improved bioinformatics approaches could serve as powerful tools in the quest to uncover the molecular enigmas of evolution.

2. Tandem mass spectrometry: a historical perspective

Tandem mass spectrometry is an invaluable tool for identifying and sequencing proteins, and assessing their modifications. Protein sequencing methods provide insight into cellular and molecular mechanisms. Protein analysis has progressed continuously since the first reports by Dr. Edman, which illustrated that proteins are a sequential arrangement of amino acids (Edman, 1950). These key observations were soon followed by Dr. Sanger's sequencing of the first protein, insulin (Stretton, 2002). The technological revolution of the 1980's and 90's included crucial developments in mass spectrometry instrumentation by Dr. Fenn and Mr. Tanaka, which enabled the use of mass spectrometry for the study of biological macromolecules (Tanaka et al., 1988; Fenn, 2002). These developments, for which Dr. Fenn and Mr. Tanaka received the 2002 Nobel Prize in Chemistry, started a technical revolution that moved protein analysis from sequencing purified proteins one at a time in isolation, to the identification and quantification of multiple proteins in cellular extracts. These technological innovations, combined with a growing body of genomic knowledge and data, brought about the birth of *proteomics* (Patterson & Aebersold, 2003).

The monumental task of sequencing the genome of different species is an essential component for the identification of proteins and determination of their functions (Marshall, 2011). Genomic insight has made clear that protein diversity and biological function cannot be explained using just gene expression (Adami et al., 2000). Biological function will ultimately be determined by protein expression levels, in combination with each protein's structure, location and its interactions with other proteins and the environment. Posttranslational modifications (e.g. phosphorylation, glycosylation, hydroxylation) and processing (e.g. proteolytic cleavage, degradation), which alter protein sequence, structure and interactions, will further refine biological function. Proteomics as a field is essentially interested in characterizing each of these levels of complexity, for each protein, at each developmental stage. This type of approach would be impossible using traditional protein isolation, purification, and sequencing techniques.

With optimized sample preparation, tandem mass spectrometry instrumentation, and database selection and search algorithms (as described later in this chapter), multiple proteins within a complex mixture can be rapidly and definitively identified, and even quantified. The sub-field of *comparative proteomics* quantitatively compares the diversity, identity, and expression levels of proteins among samples, generally among individuals of a specific species exhibiting different phenotypes. Comparative proteomic approaches have great potential in medical diagnosis and treatment, as pathological conditions can be compared at the protein level to healthy individuals (see Sanchez et al., 2004 for detailed examples). A logical extension of comparative proteomics, *evolutionary proteomics*, assesses the diversity, identity, and expression levels of proteins among samples obtained from different species, in order to elucidate evolutionary patters and highly conserved mechanisms (for example see Budovskaya et al. 2005; Heyl et al. 2007). A unique example of evolutionary proteomics applications is the analysis of reproductive (seminal) fluids, since these fluids are directly acted upon by selective forces and show rapid diversification (Ramm et al., 2009; Marshall et al., 2011; reviewed in Findlay & Swanson 2010). Ramm et al. (2008) analyzed the array of proteins found in rodent seminal fluids among 18 muriod species. The study found significant variation in molecular mass of the same seminal proteins among species, suggestive of amino acid divergence, and showed evidence for sperm competition as a selective force.

To illustrate the potential of evolutionary proteomics in understanding the origin and maintenance of biological diversity, we present examples of highly conserved proteins that withstand evolution forces through multiple phyla. These proteins and biochemical mechanisms have been conserved through evolution because they serve a crucial function in an organism's survival and reproduction and can be adapted to serve multiple functions.

2.1 Highly conserved proteins

The integration of genomics and proteomics led to important discoveries including the identification of biomarkers and identification of conserved proteins from bacteria to humans. In the 1990's, the orthologous proteins of mammalian actin and tubulin were identified in bacteria, indicating that cytoskeletal structures are crucial components that preceded multicellular organisms (Bi & Lutkenhaus, 1991; Desai & Mitchison, 1998; Graumann, 2004). Therefore, other molecular changes through evolution, such as protein sequence and structure divergence, may explain the cytoskeletal differences between prokaryotes and eukaryotes. Understanding these types of changes through a proteomics approach will provide insight on molecular level changes that promote and/or enable species divergence. Prerequisite knowledge of highly conserved proteins and biochemical mechanisms will guide hypothesis testing and enable narrow database searches. As in the case of actin and tubulin, proteins and biochemical mechanisms that are highly conserved through evolution are essential to the fitness of an organism (i.e. its likelihood to survive and reproduce). We present two examples of highly conserved proteins: collagen, and the proteins involved in blood coagulation. These proteins have been the focus of recent evolutionary proteomics-based reports and will be further detailed in the technical aspects of tandem mass spectrometry section of this chapter (section 3).

Collagen is found in all animals and is the most abundant protein in the majority of vertebrates. It serves as the key structural protein in vertebrates, composing the vast majority of the extracellular matrix, including bone matrix, skin and tendons. Collagen has a unique hierarchical helical structure that provides the mechanical strength and stability necessary to serve as a structural protein. At the amino acid level, collagen has a highly repetitive sequence of Xaa-Yaa-Glycine, where Xaa is most often proline and Yaa is most often hydroxyproline, although other amino acids can fill these positions. This repetitive sequence enables formation of a helical structure, composed of three polypeptides, stabilized by hydrogen bonds (i.e. tropocollagen molecules). Tropocollagen molecules pack together in a consistent staggered array producing fibers, which are cross-linked together, increasing mechanical strength. Collagen is highly adaptable, due to variations at the genetic, post-translational, and processing levels, enabling it to serve a wide range of structural functions. There are at least 28 different types of vertebrate collagen, which vary in function and/or distribution (Shoulders & Raines, 2009). Collagen variants likely evolved through a process of gene duplication and drift within duplicated genes (Boot-Handford & Tuckwell 2003).

The presence and structural importance of collagen is equally important in invertebrates. Collagen is present in all metazoans, although fibrillar forms of collagen appear to be lacking in arthropods and nematodes (Garrone, 1999; Boot-Handford & Tuckwell 2003). For example in sponges, the most primitive metazoans, spongin (short-chained collagens) are involved in both adhesion to the substrate and as a skeletal matrix (Garrone, 1999). Collagen also serves a key role in adhesion of marine mussels. Marine mussels adhere using a bundle of byssal threads, which extend from the foot of the organism to the substrate, where they display adhesive pads. Byssal threads are composed of a large collagen domain flanked by two elastin domains, which provide the byssus with both mechanical strength and flexibility (Coyne et al., 1997).

A second salient example of highly conserved proteins is those involved in blood coagulation, a process that stems the loss of blood during injury and without which an animal could not survive. In vertebrates, blood clot formation is brought about by two closely interrelated and converging proteolytic cascades, first described by Davie and Rantoff (1964) and MacFarlane (1964). Fibrinogen is proteolytically activated by this cascade, producing fibrin monomers, which form a network and are covalently cross-linked by a transglutaminase (factor XIII). In the blood coagulation cascade, inactive proteases (zymogens) are converted to their active form by limited proteolysis and then in turn activate the next protease in the cascade. All of the major coagulation cascade enzymes are trypsin-like serine proteases, which cleave on the carboxyl side of arginine residues (Neurath, 1984, 1986; Davie, 2003). Within this conserved system, variability in the non-proteolytic domains of these enzymes allow for substrate and cofactor binding specificity (Patthy, 1993; Neurath, 1999).

A similar process has been shown to be involved in invertebrate blood coagulation. For example, the blood of horseshoe crabs coagulates through two converging cascades of trypsin-like serine proteases, contained within blood cells as inactive proteases, and released upon exposure to minute quantities of pathogens (reviewed in Muta & Iwanaga, 1996; Sritunyalucksana & Soderhall, 2000; Osaki & Kawabata, 2004; Theopold et al., 2004). Blood coagulation in crustaceans occurs though the action of a Ca^{2+} activated transglutaminase (Fuller & Doolittle 1971; Lorand, 1972; Kopacek et al., 1993), which is homologous to vertebrate factor XIIIa (Wang et al., 2001). Involvement of trypsin-like serine proteases in the crustacean blood coagulation process has also been shown (Durliat & Vranckx 1981; Madaras et al., 1981; Soderhall; 1981).
The activity of proteolytic enzymes, such as those involved in vertebrate and invertebrate blood coagulation, is widespread in biological systems (Neurath & Walsh, 1976; Neurath, 1986; Krem & Di Cera, 2002). These highly adaptable enzymes serve simple digestive function in primitive organisms, yet have evolved to regulate complex physiological control in higher organisms (Neurath, 1984; Krem & Di Cera, 2002). Apart from blood coagulation, proteolytic cascades of serine proteases comprise a variety of systems including the complement reaction, fibrinolysis, and dorsal-ventral patterning in drosophila (Neurath 1984; Krem & Di Cera, 2002). In each case, the proteolytic cascade enables amplification of a small stimulus into a physiological response (Neurath & Walsh 1976; Neurath, 1986). Proteolytic cascades are highly conserved because they work well, are adaptable, and can be regulated with inhibitors, cofactors and specific feedback mechanisms. Amino acid sequence analyses supports the hypothesis that the proteolytic cascades of vertebrate blood coagulation, horseshoe crab blood coagulation, and drosophila dorsal-ventral patterning, all evolved from a common ancestral cascade (Krem & Di Cera, 2002).
Variability on a common theme or conserved mechanisms at the molecular level may enable species divergence. For example at the transcription level, variation in Hox genes, a highly conserved group of genes that regulate body plan and structure through development, has been shown to drive morphological evolution (Heffer et al., 2010). Post-translation modifications, such as phosphorylation (Boekhorst et al., 2008) are highly conserved, and variations in the extent or specific site of phosphorylation parallel phylogenetic divergence. The identification and understanding of commonalities among distant species could provide the basis to dissect protein and, ultimately, specie divergence. Thus, with an appreciation of evolutionary proteomics approaches and highly evolutionarily conserved proteins, we move on to discuss technical aspects of tandems mass spectrometry, which will ultimately determine the success or failure of tandem mass spectrometry-based research.

3. Tandem mass spectrometry: technical aspects

Tandem mass spectrometry is defined as the sequential analysis of ions and their respective fragmentation patterns (Hunt et al., 1986; Mann & Kelleher, 2008). In the case of protein mass spectrometry analysis, this approach is also known as "bottom-up" proteomics. "Bottom-up" proteomics is based on the fragmentation pattern of an ionized peptide, which is unique to the corresponding sequence of that specific peptide. The uniqueness of those product ions facilitates the identification of the peptide and, consequently, the corresponding protein. This powerful discovery tool provides the opportunity to analyze protein samples from different organisms, extinct or living, in the quest to uncover the secrets of molecular evolution. To do so, there are several important technical aspects to be considered in tandem mass spectrometry: 1) sample preparation, 2) ionization, 3) mass spectrometer capabilities and, 4) validation.

3.1 Sample preparation

The most important aspect in mass spectrometry analysis is sample preparation. The increase in sensitivity, below attomole (10^{-18} M) levels, brought about by advancements in technology makes sample preparation crucial, and inarguably the most important step in mass spectrometry analysis (reviewed in Patterson & Aebersol, 2003). The sensitivity of the current generation of mass spectrometers leaves no room for sample preparation error. In the quest to identify proteins across different species, the integrity, quality and purity of the samples are crucial. Different sample preparation paradigms have been recommended for tandem mass spectrometry analysis. These paradigms are not exempt from potential pitfalls and each project may require customization. The most important aspect is to prevent decomposition, adducts and contamination. The first step after homogenization of tissue or cells is the isolation of proteins from the membranous fraction. Lipids, from membrane or detergents, are a contaminant that affects the analysis of proteins through tandem mass spectrometry (see Table 2, Section 3.2). But, the most common contaminant in tandem mass spectrometry analysis is keratin. Keratin is the principle structural component of the human epidermis and is found in hair and nails. Thus, the first obstacle that a user of tandem mass spectrometry technology needs to overcome is avoiding the identification of his or her own keratin. Contamination with keratin results from touching sample tubes with bare hands, skin peels falling into the sample due to dandruff, hair contamination, or simply air flow in a crowded laboratory environment. Table 1 lists keratin sources and possible solutions to prevent this common contaminant. However, keratin contamination is just the first hurdle to overcome in the application of tandem mass spectrometry to molecular evolution studies.

Keratin Source	Possible Solution
Skin, hair, nails	Wash sample tubes with ethanol three times, dry on speed vacuum Always clean the surface area of benches and instruments to be used with ethanol Work under a laminar flow hood Always use gloves and lab coat with cuffed long sleeves Always wear disposable hair cover
Aerosol	Work under a laminar flow hood Use filter-pipette tips Separate a set of pipettes that are kept in a clean area Avoid crowed laboratory areas Maintain proper ventilation
Tissue Sample	Avoid contamination with skin or hair from the tissue sample source Rinse the tissue with buffer before homogenization Avoid preparing samples close to areas used for euthanization, surgery or dissection of animals

Table 1. Sources of keratin and possible solutions.

When sample preparation is ideal, the integration of tandem mass spectrometry analysis in molecular evolution studies could provide significant advances in the understanding of molecular conservation and divergence. The concept that most genetic information is conserved between species could serve to develop hypothesis-based approaches that uncover similarities between modern organisms and those extinct millions of years ago, and raises the question of what is the actual unit of evolution. The identification of similarities could provide valuable insights about conserved mechanisms and molecular divergence between distant species such as dinosaurs and birds or arthropods and humans.

However, the use of tandem mass spectrometry in molecular evolution studies remains controversial. This is due to the lack of available protein databases for extinct or living organisms that would enable comparison between taxa and the potential for contamination with "modern" proteins during sample preparation. These concerns were the basis of a debate over a report published in *Science* in 2007. Asara et al. (2007) targeted collagen as a highly conserved and resilient protein in vertebrates, proposing that based on its molecular characteristics it could be extracted from fossil bone. Asara et al. (2007) published the identification of collagen α1t1 from two extinct animals, mastodon (*Mammut americanun*) and T. rex dinosaur (*Tyrannosaurus rex*). The authors went further to indicate that the identified *T. rex's* collagen peptides were more similar to birds' (chicken) collagen than to other species, while mastodon collagen peptides were similar to mammals (dog, bovine, human and elephant). In the same *Science* issue, Schweitzer et al. (2007) provided evidence of collagen I in *T. rex's* cortical and medullary bone, using multiple techniques, atomic force microscopy (AFM), *in situ* immunohistochemisty, and TOF-SIMS, validating the results obtained by Asara et al. (2007). Both articles were subjected to an intense scrutiny and public debate within the scientific community (Buckley et al. 2008; Pevzner et al. 2008). Part of the scrutiny is valuable as it pushes researchers to ensure the integrity of their sample preparation and analysis, while other part hinders progress, with obstacles difficult to overcome. Interestingly, and not debated, the authors acknowledge the identification of human keratin as a contaminant. Although unlikely due to fast rate decomposition of soft tissue in comparison to bone, no one questioned if the keratin was actually a conserved peptide from the well preserved dinosaur's sample.

There were three major concerns in accepting the tandem mass spectrometry data published by Asara et al. (2007), namely: sample preparation, database search, and validation of the results (Buckley et al. 2008; Pevzner et al. 2008). The bioinformatics component in tandem mass spectrometry is a bottleneck and the most quietly accepted limitation. Thus, an entire section of this chapter will be dedicated to a discussion of bioinformatics with regard to its use in molecular evolution studies. Validation is an important issue in mass spectrometry and it is discussed within this section.

Concerns about sample preparation are essential, especially when the tissue under study is a fossil that has been exposed to nature, decomposition and other organisms (big and small) for millions of years. Thus questions of the purity of the *T. rex* samples are reasonable. We are fascinated that only the results obtained from the *T. rex* samples were questioned and not the mastodon, which was exposed to the same sources of contamination though for lesser time.

Protein identification from a fossil, such as *T. rex's* bone, is similar to searching for water on Mars. As scientists, we are trained to use existing knowledge in the quest to generate new knowledge. The use of tandem mass spectrometry for the analysis of protein extracted from a fossil is based on the assumption that the amino acids that form "modern" proteins are the same as from ancient proteins and just the arrangement or amino acid order in the protein sequence has been subjected to evolution. This is similar to the assumption that the atmosphere and atoms prevalent on Mars are similar to those on Earth, and therefore there should be water on Mars. Yes, water in Mars and amino acids on proteins from extinct organisms may exist as we know them. But, it is reasonable to argue that some differences may exist. For example, one challenge in sample preparation for extinct organisms is the assumption that the chemistry and modifications of amino acids from millions of year old proteins is the same as those recognized today. This also brings up the importance of understanding the chemistry of amino acids, peptides, and proteins in the process of sample preparation. Extraction, isolation and purification of peptides and proteins depend on the chemical properties of amino acids and their interactions within the sequence. Sample preparation often includes treatment with compounds that induce modifications of the amino acids, such as hydroxylation, dehydration, oxidation and deamination (reviewed in Lubec & Afjehi-Sadat, 2007). These modifications change the molecular mass of the amino acids and, when not taken into consideration, affect the identification of peptides, due to deviations from the theoretical molecular mass registered in the database (see below). In reality, these considerations need to be taken into account for all samples not just for fossil samples.

Hypothesis based-approaches for molecular evolution studies of living organisms is a lesser challenge than for those of extinct animals, since issues of sample degradation and chemical changes over time are less important. An example of such a study was the identification of human coagulation factors in barnacle cement samples (Dickinson et al., 2009). The aggregation and cross-linking of barnacle cement proteins is crucial for the barnacle's survival since the cement anchors the organism to the surface, enabling feeding and reproduction. Based on the essential role of cement formation in the life cycle, it was hypothesized that the formation of this structure in an aqueous environment could be related to the molecular process involved in blood coagulation. As described in section 2.1, highly evolutionarily conserved proteins comprise the blood coagulation cascade. Blood coagulation is a life or death process that also involves the aggregation and cross-linking of soluble proteins.

Examination of the cement using AFM revealed a mesh of fibrous proteins, structurally similar to a fibrin blood clot. The first technical obstacle in analyzing the cement was sample preparation. Dickinson el al. (2009) developed a strategy to obtain proteins secreted by the barnacle during the process of cement release but before the secretion cured; the curing process renders most of the proteins insoluble. Complete proteins can only be collected prior to curing which involves at least one and probably other types of cross-linking (Dickinson et al., 2009). AFM of cement collected in this manner indicated the formation of fibrous structures upon polymerization, indistinguishable from those made by the barnacle *in situ*, validating the collection technique.

Next, tandem mass spectrometry was used to uncover molecular similarities in proteins between barnacles and humans (Dickinson et al., 2009). The barnacle cement's extracted proteins were resolved in one dimensional SDS-PAGE and subjected to trypsin digestion. The purified peptides were then analyzed by tandem mass spectrometry and the obtained spectrum subjected to sequence analysis against the human database. This approach led to the identification of two conserved peptides that correspond to the protein Transglutaminase, factor XIIIA, which plays a crucial role in the process of blood coagulation (Dickinson et al., 2009). This result suggests that coagulation is a conserved molecular mechanism crucial for the survival of all organisms and that has been preserved through evolution. Interestingly, the peptides identified correspond to only one factor in a complex and multifactorial molecular mechanism. Thus, the result obtained serves as foundation to uncover the similarities between blood coagulation and cement formation, but importantly the information gathered could be used to delineate the divergence of this process in organisms from two very distantly related taxa.

3.2 Ionization

The popular phrase "garbage in, garbage out" describes the importance of sample preparation in tandem mass spectrometry. However, there is one more important component that needs to be considered before the analytes reach the mass analyzer of a mass spectrometer: ionization efficiency. The process of sample ionization is carried out by two principal ionization sources used for tandem mass spectrometry analysis, namely Electrospray ionization (ESI) and Matrix-assisted laser desorption ionization (MALDI). These two sources of soft ionization were crucial in the integration of mass spectrometry for the analysis of biomolecules (Tanaka et al., 1988; Fenn, 2002). After all, only ions that are generated by an ionization method can be analyzed by a mass spectrometer.

Chemical	Source	Effect
Lipid	Cellular membrane, organelles	Affect resolution when liquid chromatography is used Produce prominent and persistent ions that could suppress or mask other ions
Detergent	Sample preparation buffers, surfactants	Affect resolution when liquid chromatography is used Produce prominent and persistent ions that could suppress or mask other ions Induce the formation of adducts
Salt	Sample preparation buffers	Induce the formation of adducts Ion suppression Ion overloading
Trifluoroacetic Acid (TFA)	Sample preparation solvent, reverse chromatography	Ionization suppression

Table 2. Chemicals and biomolecules that affect ionization and ion detection.

The efficiency of ESI to generate ions depends on three important factors: temperature, flow, and voltage. However, other factors may suppress ion formation as described in Table 2. A temperature of 200°C is constantly used at the ion transfer tube, which serves as the entrance to the mass analyzer. This high temperature allows the evaporation of highly volatile solvents in which analytes are dissolved (Table 3). The evaporation of the sprayed solvent containing the analytes generates the separation of one drop into smaller drops in a physical concept known as "Coloumbic explosion." This process continues until the analytes are completely dry and attracted to the mass analyzer where the ions are detected. Evaporation of the volatile solvent is a prerequisite for ion detection and, therefore, directly related to sensitivity.

Based on this principle, flow rate is inversely related to sensitivity. The lower the flow rate (200-300 nL/min) the faster the solvent is evaporated promoting more ions to be dried out and detected by the mass analyzer. Thus, lower flow rate implies higher sensitivity. However, ions are also formed if voltage is applied to the sample. In ESI, voltages applied to the sample range from 1.0 to 2.5 kV. The combination of temperature, flow and voltage generates the spray of ionized analytes detected by the mass analyzer.

MALDI requires use of a chemical matrix in which the analytes are embedded (Lubec & Afjehi-Sadat, 2007). Table 3 shows matrices frequently used. Since laser energy is absorbed by the matrix and transferred to the analyte, the solubility of the analytes in the matrix affects the ionization efficiency. In practice, the three most common matrices are used to determine the efficiency of ionization at the level of ion detection. Thus, it is recommended to dissolve the same sample in each of the three matrices and evaluate results. Energy from the laser is equivalent to voltage used in ESI, and therefore requires fine tuning and optimization. The energy promotes ionization and transition to the gas phase. As for ESI, the MALDI matrices are volatile so that the ions formed can be detected by the mass analyzer (see below).

Ionization source	Solvent
ESI	(10-80%) Acetonitrile/ (90-20%) Water/ (0.1-0.2%) Formic Acid (10-80%) Methanol/ (90-20%) Water/ (0.1-1.0%) Acetic Acid
MALDI	2,5-Dihydroxybenzoic acid (DHB) 3,5-Dimethoxy-4-hydroxycinnamic acid (sinapinic acid) α-Cyano-4-hydroxycinnamic acid (CHCA) 3-amino-4-hydroxybenzoic acid

Table 3. Ionization source and common solvents.

ESI and MALDI both have their advantages and disadvantages (Lubec & Afjehi-Sadat, 2007). ESI provides the advantage of ionizing the analytes as they are eluted from a column. Liquid chromatography online to the ionization source provides a better resolution of the analytes, avoiding clustering and exceeding the dynamic range (i.e. a measure of the detection range of a detector; ratio of the largest to smallest detectable signal) of the instrument. In MALDI, the chromatography needs to be carried out off-line and the fractions analyzed individually. Off-line chromatography allows the analysis of only the desired fractions, while in on-line chromatography all elution steps are analyzed. Older instrumentation for on-line chromatography did not allow the recovery of fractions at a specific retention time for re-analysis. At present, there are instruments, such as NanoMate TriVersa (Advion), that collect fractions through out the chromatography gradient without interrupting the ionization. Although controversial, MALDI is recognized as a better ionization source for its capacity of assisted-energy transfer to the analyte, which reduces the fragmentation or decomposition of the analyte at the ionization source. MALDI ionizes certain peptides better than ESI, and vice versa. Thus, in molecular evolutionary studies is important, when possible, to take in consideration and use both ionization sources for tandem mass spectrometry analysis of a particular sample.

In summary, ionization efficiency plays a crucial role in tandem mass spectrometry. It is important to understand that the most abundant peptide in the prepared sample is not necessarily the most abundant ion in the mass spectrum, because each peptide that reaches the ionization source will have different ionization efficiency. Samples derived from preserved fossils or live organisms need to be prepared in accordance with the requirements of the ionization source used. Good sample preparation will increase the chances of protein identification or characterization. However, the lack of fine tuning of the ionization source could induce fragmentation, clustering and ionic suppression, converting sample preparation efforts in a futile exercise. Once the conditions are conducive to increasing the ionization efficiency, the mass spectrometer itself is the next component that requires optimization.

3.3 Mass spectrometers

Mass analyzers have progressed at a fast and steady pace (Patterson & Aebersold, 2003). Sir Joseph John Thomson would be delighted to see how the measurement of mass-to-charge (m/z) ratio is done today. Crookes tubes used by Dr. JJ Thomson are pieces of history in a science museum. From instruments with very low resolution and dynamic range in the 1940's (e.g. MS-2) to today's high resolution instruments, such as the 14-Tesla Fourier transform ion cyclotron resonance (FT-ICR) instrument, mass spectrometers continue pushing forward our capacity to discover and understand molecular processes. Increased mass accuracy, high resolving power, scanning speed and affordability are some of the attributes of modern instruments (Mann & Kelleher, 2008). Given the technological progress in mass spectrometry, this can become a reliable tool in the understanding of molecular evolution. Nonetheless, the efficiency of a mass spectrometer depends on both of the parameters discussed above, sample preparation and ionization efficiency.

The controversy surrounding the *T. rex's* peptides sequences discussed above provide the basis to understand why ion mass accuracy is an important criterion in the selection of a mass spectrometer for a specific application (Mann & Kelleher, 2008). Why is ion mass accuracy important? The identification of a peptide depends on the bioinformatics analysis of the mass spectra recorded from a specific sample. The mass spectrometer detects ions, which we know are from a digested protein(s) or from "chemical noise" in the sample. The requirement for ion mass accuracy is directly related to the bioinformatics tools available for data mining that depends on the mass of the precursor and product ions in a tandem mass spectrometry analysis. The debate on the confidence level in the correlation of the MS and MS/MS data with the database starts at the level of the MS spectrum (Pevzner et al., 2008). In the *T. rex* example, the authors used an instrument that is considered a low-resolution instrument, LTQ (linear ion trap), which has a lower mass accuracy than other available instruments (Asara et al., 2007). Although mass accuracy is intrinsically related to the resolving power of the mass spectrometer, it does not mean that a low-resolution mass spectrometer generates poor quality data. The main point here is that the user needs to consider the instrument's capabilities in regard to the research application(s) and goal(s).

The resolving power of a mass spectrometer is defined, in simple terms, as the mass analyzer's capability to set apart two ions with similar m/z ratios. At the mass spectrum level, the instrument resolving power can be seen as the *"valley"* or distance between two ions with similar m/z (Mann & Kelleher, 2008). Examples of low- and high-resolution instruments are listed in Table 4. The main difference among these instruments is the type of mass analyzer used. Low-resolution mass spectrometers are more affordable and widely used for proteomics approaches in life sciences and biomedical research. Most of these instruments are very sensitive, have high scanning speed and robust performance. Linear ion trap mass analyzers are an example of a low-resolution mass spectrometer, but they compensate with a high scanning speed and dynamic range. On the other hand, high-resolution instruments generate accurate mass data, but with lower sensitivity and robustness. Although it is not always feasible, ideally selection between low- and high-resolution mass spectrometer is not just an issue of affordability (i.e. funds available), but also of application.

Resolution	Mass Analyzer	Instruments (examples)
Low	Quadrupole	Xevo TQ, Waters TSQ, ThermoElectron
	Ion trap	LCQ and LTQ, ThermoElectron 240-MS, Agilent amaZon, Bruker
	Time-of-flight	Autoflex, Bruker TripleTOF 5600, AB Sciex
High	Hybrid	LTQ Orbitrap, ThermoElectron Synapt G2-S, Waters
	Fourier Transform Ion Cyclotron Resonance	Apex Qh 9.4TFT-ICR and solarixFTICR, Bruker LTQ-FTICR, ThermoElectron

Table 4. Low- and high-resolution mass spectrometers.

For example, generation of data from complex protein samples could be achieved using an instrument with high scanning speed and sensitivity. The scanning speed is how fast the instrument can target an ion for fragmentation. In complex protein samples many peptides will elute from a column (LC-ESI) or ionize (MALDI) at the same time. Data acquisition from such a sample will depend on how many ions were selected and subject to fragmentation. In this case, a low-resolution mass spectrometer could extract the most information from the sample. However, for applications that require precise determination of m/z, such as quantification studies and identification of posttranslational modifications, high resolution mass spectrometers (e.g. Fourier Transform Ion Cyclotron Resonance, FT-ICR) are more appropriate. Based on scientific needs and marketing considerations, the industry produces hybrid mass spectrometers, which provide the best of both worlds. These instruments use a low-resolution mass analyzer (e.g. linear ion trap) to obtain information from complex samples, taking advantage of high scanning speed and sensitivity. Then, after the first data acquisition, ions can be transferred to FT mass analyzer where highly accurate mass is obtained from the selected ions. Thus, hybrid mass spectrometers are versatile instruments that should be considered as the most suitable tool in evolutionary proteomics approaches, since they are able to analyze complex mixtures of proteins as well as enable quantitative comparisons and high mass accuracy.

A relevant issue raised above is the identification of proteins from fossil samples or distant taxa using the information and knowledge on "modern" proteins or genomes with limited known sequences. The identification of proteins depends on detection of the precursor ion and its fragmentation pattern in the mass analyzer. The MS and MS/MS spectra are then compared with a database to extract the information that will lead to the identification of a peptide and, consequently, the corresponding protein. But, why cannot all peptides produced from a digested protein be identified? In addition to differential ionization efficiency between peptides, sensitivity, dynamic range and mass accuracy are crucial components that contribute to sequence coverage. For example, mass accuracy contributes to differentiating between isobaric amino acids, such as lysine (128.095) and glutamine (128.059). Another isobaric pair is leucine (131.094) and isoleucine (131.094), which are differentiated by fragmentation of their side chain. Amino acid modifications can also induce isobaric pairs, for example oxidized methionine (147.035) and phenylalanine (147.068). It is feasible to hypothesize that, in preserved fossil tissue or living organisms from different taxa, amino acids may be differentially modified or the peptides may contain unidentified amino acids. Thus, the question is: how many peptides are needed to correctly identify a protein present in a sample using tandem mass spectrometry? The rule of thumb is that two peptides are needed for a positive identification of a protein. Needless to say, more is always better, but when databases from living organisms are used to identify proteins extracted from fossil tissue this rule of thumb shortens the abysmal age differential between the extinct and the living. If we do identify matching peptides, how can we be certain that the identified protein is definitely in the sample? The answer is validation, validation, validation.

3.4 Validation, validation, validation...

Optimization and understanding the different components that are required for tandem mass spectrometry contributes to the generation of high quality and quantifiable data. The attention to detail during sample preparation contributes to avoiding contamination, undesirable modifications and decomposition of proteins. Calibration and fine tuning of the ionization source and mass analyzer increase the sensitivity and maximize the capabilities of the mass spectrometer. Despite all these considerations, the data generated by tandem mass spectrometry depend on the power of the computational tools and data mining capabilities available to the user. There is a phrase commonly used in the laboratory about the interpretation of MS data; "*if you did not detect a peptide, it does not mean that it is not there, but if a protein was identified you better validate it...*"

Validation of the results obtained by tandem mass spectrometry analysis is a requirement in today's proteomics approach. This fact recognizes that false positives could be detected despite the selection of the most stringent filter parameters and the requirement of powerful computational tools. However, the need for validation does not minimize the importance of tandem mass spectrometry as a discovery tool. It is just another example of how the scientific community finds ways to improve the quality and validity of their results. After all, a result could trigger a hypothesis, and the hypothesis can only be accepted as theory if it persists in scientific scrutiny over time.

Returning to the Asara et al. (2007) example and the controversy surrounding the identification of collagen peptides conserved from dinosaurs to chickens, the authors used different validation strategies to certify their tandem mass spectrometry analysis results, which were based on a predetermined search for collagen peptides. The authors search for collagen because it is a structurally strong protein that could withstand exposure to harsh environmental conditions and be readily detected in bone tissue samples. First, the authors generated a "database" that allowed them to search for the predicted collagen peptides in the raw MS/MS-MS spectra obtained in the tandem mass spectrometry analysis of proteins extracted from the bone tissue. They generated a "database" using conserved sequences among different taxa and point-assisted mutation matrices to take into consideration amino acid divergence throughout evolution. The collection of peptides in the database was used to identify peptides in the protein sample from *T. rex* bone tissue. This hypothesis-based approach allowed the identification of 33% sequence coverage for collagen α1t1 and 16% collagen α2t1 (Asara et al., 2007). Validation of these results is necessary.

For validation of the identified collagen peptides from *T. rex*, Asara et al. (2007) used synthetic peptides. The synthesized collagen peptides were subjected to tandem mass spectrometry and the resulting fragmentation patterns (i.e. MS/MS) were compared to the identified peptide from the *T. rex* sample. This strategy demonstrated that product ions from the *T. rex* sample were similar to the corresponding synthetic peptide. Although this is a valid strategy, this validation depends on the same technique that was used to generate the data under scrutiny. The best validation strategies are described in a complementary report by Schweitzer et al. (2007), published back-to-back with Asara et al (2007) in the same *Science* issue. Schweitzer et al. (2007) used immunochemistry to locate collagen protein in the tissue using an antibody against avian collagen. The antibody clearly showed immunoreactivity in both cortical and medullary bone tissue from *T. rex*. In order to detected immunoreactivity, the fixed tissue would need to have an epitope recognized by the anti-avian collagen protein, suggesting that this protein may be present in the bone tissue. Taken together with AFM and TOF-SIMS analyses, the immunochemical analysis of *T. rex* tissue confirmed the tandem mass spectrometry data, indicating that at least, peptide sequences homologous to collagen were present in the dinosaurs sample.

The second example described earlier in this chapter was the identification of coagulation factors conserved from barnacle to human. Dickinson et al. (2009) demonstrated that the polymerization of barnacle cement is similar to blood coagulation. The authors validated the identification of Factor XIIIA (transglutaminase), using western blot and functional analyses. Immunoreactivity to human factor XIIIA antibody was found in barnacle cement at approximately 75 kDa, remarkably close to the 83 kDa mass of human factor XIIIA despite roughly a billion years of evolutionary divergence. Dickinson et al. (2009) went on to conduct two functional analyses: quantification of transglutaminase activity level and an amino acid composition analysis procedure that specifically probed the end-product of transglutaminase cross-linking (as described by Pisano et al., 1968, 1969). Both functional analyses confirmed the presence and activity of a transglutaminase in barnacle cement.

The report by Dickinson et al. (2009) further expanded the evolutionary hypotheses established by testing for other components of the blood coagulation cascade, trypsin-like serine proteases, not specifically identified by tandem mass spectrometry. Evidence of trypsin activity was shown through SDS-PAGE, i.e. a pattern of pro-forms of proteins, active proteins and proteolytic clips. Activity of trypsin in barnacle cement was demonstrated by trypsin activity quantification (using an arginine ester substrate), and the reliance of the polymerization process on proteolytic activity was shown through trypsin inhibition assays. Furthermore, western blot analysis showed immunoreactivity to bovine trypsin antibodies, indicating shared epitopes between barnacle and vertebrate trypsin. Variations in non-proteolytic domains of these enzymes are likely to enable substrate specificity. The results demonstrated that coagulation is a conserved biological process relevant to the fitness and survival of organisms as distant, in evolutionary terms, as barnacles and humans. This on-going work is then directed toward using these similarities as a basis to dissect the molecular requirements for barnacle adhesion in order to identify functional divergence from the mammalian blood coagulation process.

Validation of tandem mass spectrometry results is based on the application or source of the sample used in the analysis. Protein identification could be validated by immunological techniques such as western blot, immunohistochemistry and immunocytochemistry. Also, immunoprecipitation or chromatography purification, followed by western blot analysis could be used to validate the identification of a novel protein as part of a known protein complex. Vega et al. (2008) demonstrated, using tandem mass spectrometry analysis of co-immunoprecipitated proteins, the association between the novel mouse protein EFhd2 and the human microtubule-associated protein tau (Tau) expressed in the brain of the tauopathy mouse model JNPL3 (Vega et al., 2008). The identified mouse's EFhd2 protein is 93% identical to its human counterpart. Thus, immunoprecipitation of Tau from human brain samples followed by western blot analysis validated its association with the human EFhd2 (Vega et al., 2008). This is not only an example of validation, but also how conserved proteins from different living organisms and distant taxa (i.e. mouse to human) could be used to identify proteins that were conserved through evolution. Thus, validation of results obtained by tandem mass spectrometry analysis is a way to reinforce an already powerful discovery tool.

4. Bioinformatics

As we have discussed in the preceding sections, two issues come to the forefront when performing evolutionary proteomics; is our protein of interest *accurately* identified and can we *reliably* infer the phylogeny of the organism in question? These questions become more acute when considering fossil material, where we will only have fragments of proteins to work with. We have already considered some aspects of this issue earlier in section 3.4. In this section we discuss bioinformatics approaches and how they can address issues of reliability.

4.1 Identifying proteins, databases and search strategies

4.1.1 Sensitivity: In both recovering sequences from fossil material (either bone or preserved tissue like Thylacine tissues) or from biological materials where we hope to have mechanistic insights through homology (e.g. barnacle glues), the issue of sensitivity is important. For fossil material, degradation means that peptides of interest will be at low abundance. For other materials, the proteins of interest may be of low abundance. Peptide MS can achieve exquisite sensitivity, in the attomole range or sometimes even below. This brings fossil material into the realm of the possible. However, while sensitivity is important, another issue is dynamic range. High-abundance proteins limit what we can observe with the given dynamic range of detection. For detection of peptides in complex mixtures, such as we see with a sample such as barnacle cement, the dynamic range is currently in the range of 10^3 to 10^4 (Mann & Kelleher 2008). This dynamic range issue may be why serine proteases were not detected in MS/MS scans of barnacle cement. For fossil material, exogenous contaminants such as bacterial proteins and keratin can be of high abundance compared to the desired protein (Edwards, 2011), making detection of low abundance material difficult.

4.1.2 Search Programs: The most common approach to identifying proteins in modern MS/MS data is to search un-interpreted MS/MS data directly. There are a large number of programs, both commercial and free, which can implement these searches. Commercial programs include SEQUEST, the original MS/MS database searching program, Phenyx, ProteinLynx and many others. A selection of popular free programs is given in Table 5.

Program	URL
InSPect (web based, requires registration)	http://proteomics.ucsd.edu/LiveSearch/
Mascot (free web version and licensed version)	http://www.matrixscience.com/
OMSSA (web based)	http://pubchem.ncbi.nlm.nih.gov/omssa/
Sonar (web based)	http://hs2.proteome.ca/prowl/knexus.html
X!-tandem (computer based -Linux and Windows)	http://www.thegpm.org/tandem/

Table 5. Freely available *de novo* MS/MS peptide identification software.

Each program uses a variety of strategies to determine sequence identity. SEQUEST uses correlation factors to match peptides while the Open Mass Spectrometry Search Algorithm (OMSSA), Mascot and X!-tandem use a probabilistic approach. As well, there are a bewildering variety of options available; all allow a choice of digest conditions, with varying options. For example OMSSA has a menu of 20 different digest conditions, including no digestion, while Sonar has only seven (and does not have a "no digestion" option). Similarly, there are a number of options for modifications, both fixed and variable. InSpect has the fewest options available of the free programs. All have a number of database options to search, which will be discussed in the next section.

4.1.3 Data formats: There are a wide variety of data formats produced by MS/MS instruments, many of which are proprietary and not readable by many of the programs mentioned here. These formats will need to be converted into a common format. Usually the vendor of the instrument will provide some data export capability, but there are also some open source tools at Proteowizard (http://proteowizard.sourceforge.net/) and the Trans Proteomic Pipeline (http://tools.proteomecenter.org/software.php). Conversion of these files may cause some loss of information (e.g. loss of metadata when converting to DTA format), which can potentially impact on the identification results. File formats include DTA, a simple text file format with no metadata, PKL, MGF (Mascot Generic Format) and a variety of formats to encode metadata using XML (mzXML, mzData and mzML). This impacts the programs you can use to search for matches, SEQUEST uses the DTA format, X! Tandem is set up to use DTA, PKL or MGF files, and OMMSA can handle DTA, XML encapsulated DTA and PKL or MGF files. The PRIDE database of MS spectra (http://www.ebi.ac.uk/pride/init.do) , which can be helpful to validate search strategies, uses the mzData format.

Given the variety of programs and formats available, which do you choose? Head-to-Head comparisons have been performed on a limited number of these programs using validated MS spectra; Boutilier et al (2005) compared SEAQUEST, Mascot, Sonar and Pepsea on an LCQ and a high resolution mass spectrometer. They found limited overlap between the programs, and this overlap was different with each instrument used. One approach to overcome this limitation is to use multiple search engines. This comes with a significant computational cost, but if differing programs identify the same peptides then there is greater confidence in the result (Boutilier et al. 2005). This approach was used by Asara et al., (2007) where both Mascot and SEQUEST scores were used as part of the process to validate *T. rex* collagen sequences.

4.1.4 Databases: Database choice is a trade-off between sensitivity and the time it takes to search the database. Smaller, selective databases will take a short time to search but may miss important peptides, while larger databases take much longer to search and may produce results of lower statistical significance.

Database	URL
RefSeq Protein	http://www.ncbi.nlm.nih.gov/RefSeq/
UniProt/Swiss Prot	http://www.uniprot.org/
Human Proteomics Database	http://www.hprd.org/
UniRef	http://www.ebi.ac.uk/uniref/
International Protein Index (closes Sept 2011)	http://www.ebi.ac.uk/IPI/IPIhelp.html

Table 6. Databases suitable for MS/MS peptide identification in fossil and phylogenetic material.

The National Centre for Biotechnology Information (http://www.ncbi.nlm.nih.gov/RefSeq/) and the European Bioinformatics Institute (http://www.ebi.ac.uk/) house two major curated sequence databases, along with search engines and other bioinformatics tools. These can be searched online through the MS search program, or the databases downloaded locally for faster searching if the MS/MS predictor program allows searching local databases (Table 6). For meaningful searches, it is imperative that high quality, curated databases be used. Within even these narrow limits the choice of database is, as mentioned above, a trade-off between speed of searching and low false positive rate versus more exclusive searches (Duncan, 2010; Edwards, 2011; Cottrell, 2011).

For fossil material, smaller databases may be appropriate. For example, the SwissProt section of the UniProtKB database is a non-redundant database. This means that it contains a consensuses sequence for each protein, the known variants being encapsulated in a single entry. Searching SwissProt will give fast results, at the risk of missing proteins with a low representation. In the case of fossil bone, where we are typically trying to recover collagen sequences, the speed and low error rate of such databases may be acceptable. More comprehensive databases, where all the known sequences are represented (such as UniRef100 or NCBInr) may be more appropriate with material from contemporary samples where low abundance proteins are important.

Within a database, it may be worthwhile to restrict the search taxonomically to organisms with the highest potential match. Most of the programs mentioned above have some capacity to do this. Some (e.g. SEQUEST and X!-tandem), can run custom databases if the peptides are likely to be poorly represented in larger databases. In follow up to the Dickinson et al. (2009) report, the barnacle cement sequences were compared to a database of custom non-redundant sequences from *Pacifastacus leniusculus*, a crustacean in which coagulation has been well studied (Dickinson, Vega, Rittschof, Musgrave, *unpublished*).

Anyone wishing to undertake identification of fossil proteins should make use of the PRIDE database of MS spectra (http://www.ebi.ac.uk/pride/init.do) which contains MS spectra for six extinct species, including *T. Rex* and Mammoth, to validate their search strategies against these fossil spectra before proceeding.

4.1.5 When matches are found: With programs like SEQUEST, the final scores are given as cross correlation. This does not typically provide enough information to decide if the protein identifications are valid. Usually the SEQUEST output is passed through a series of validation stages, typically the program Scaffold (http://www.proteomesoftware.com/Proteome_software_prod_Scaffold.html) to validate the peptide identities, and ProteinProphet (http://proteinprophet.sourceforge.net/: Nesvizhskiiet al., 2003) to convert the cross correlation scores to probabilities. Programs like Mascot and X!-tandem return probabilities that the matches are wrong (an empirical expectation value in the case of X!-tandem and a theoretical P-value in the case of Mascot). While X!-tandem and Mascot do not explicitly need post-processing, conversion of their statistics to the probability of a true match with ProteinProphet can be helpful, especially when comparing outputs. As stated above, when matches are found the rule of thumb is validation, validation, validation.

4.1.6 Contaminants: Searching your results against a contaminants database is very important, especially with fossil samples where contamination is a perennial problem (see section 3.1 above for methods to reduce contamination; see also Edwards, 2011 and Cottrell, 2011). There are two collections of contaminants databases available for researchers to download. The Max Planck Institute of Biochemistry, Martinsried, maintains a file of proteins selected from the International Protein Index (http://www.biochem.mpg.de/en/rd/maxquant/110606_backup/Downloads/Downloads.html). The Global Proteome Machine Organization maintains a common Repository of Adventitious Proteins (cRAP) which contains proteins selected from UniProt (http://www.thegpm.org /crap/index.html). Additionally, most algorithms (e.g. BioWorks, Proteome Discover) include filters for the most prevalent peptides produced from the major proteases used in tandem mass spectrometry analysis. For example, by enabling this feature, peptides from trypsin (or even keratin) could be filtered from the data. On the other hand, an exclusion list with the m/z ratios of the most prominent ions from expected contaminants in the sample preparation or solvent used could be created as part of the method. The mass analyzer will not waste time fragmenting ions with the m/z ratios listed. Although this pre-analysis exclusion strategy will generate "*cleaner*" data, it has the disadvantage that in low-resolution instruments peptides with similar m/z ratios of the one excluded will not be analyzed.

4.1.7 False discovery rates: When running peptides through very large databases, there will inevitably be false positive matches (above and beyond the contaminant issue above). Estimates of the False Discovery Rate (FDR) are now used alongside the statistical measures for indentifying peptides (Cottrell, 2011; Edwards, 2011). The FDR must be computed for each sequencing run, and can be computed by running the spectra against a decoy. The decoy run must be with identical search parameters to the real run, and the number of matches from the decoy database gives an estimate of the number of false positives in the run against the genuine database. Decoys can be sequences from the genuine database which have been either reversed or shuffled. Some search engines can run the genuine and decoy databases simultaneously. Cutoff for the FDR can be between 1% and 10%, however, when running fossil sequences it is best to be as conservative as possible (Ramos-Fernández et al., 2008).

4.2 Finding phylogenies

With your peptide sequences in hand, you can now approach building phylogenies. This is no trivial task, for any group of organisms, there exist an enormous number of possible phylogenetic trees which much be searched to find the optimal tree. As well, homoplasy (convergent evolution, parallel evolution and site reversal) can confuse the analysis even more. With MS/MS derived sequences, there is an additional problem if we do not have the full protein sequence since some potentially useful phylogenetic sequence information will be missing. Nonetheless, with careful attention useful phylogenies can be obtained.

Program	URL
BLAST (standalone or web-based versions)	http://blast.ncbi.nlm.nih.gov/Blast.cgi
COMPASS (web based multiple Sequence alignment)	http://prodata.swmed.edu/compass/compass.php
MAFFT (web-based multiple Sequence alignment and phylogeny)	http://mafft.cbrc.jp/alignment/server/index.html
MUSCLE (web-based multiple Sequence alignment and phylogeny)	http://www.drive5.com/muscle/
EMBL-EBI ClustalW (web-based phylogeny)	http://www.ebi.ac.uk/Tools/phylogeny/clustalw2_phylogeny/
Mobile@Pasture (web-based Multiple phylogenetic methods)	http://mobyle.pasteur.fr/cgi-bin/portal.py?#welcome

Table 7. Phylogenetics programs suitable for evolutionary MS/MS.

The first step in producing phylogenies is to produce multiple alignments with the database of sequences with which you wish to form a phylogeny. This aspect of the process is often given less attention than the others, but the accuracy of the alignments can have a significant impact on the phylogeny derived from them (Ogden & Rosenberg, 2006). The problem is particularly acute with fossil material, as coverage of the sequence will be incomplete (e.g. only 32% of mammoth α1t1 collagen sequence was recovered; Asara et al., 2007), and alignment matching may have significant mis-matches. Programs like MUSCLE, COMPASS or MAFFT which use iterative alignment gaps can have better selectivity than simple CLUSTAL-W alignments. MAFFT also has the advantage of being tolerant of large gaps, such as found in the MS/MS peptides recovered.

Once an alignment is produced then a phylogeny can be derived. There are many different approaches to creating the phylogeny. Methods include Maximum Likelihood, Maximum Parsimony, Bayesian, Neighbor Joining and the Unweighted Pair Group Method with Arithmetic Mean (UPGMA). As with the choice of databases, the choice of a phylogenetic algorithm is a trade-off between speed and accuracy. In head to head comparisons, Maximum Likelihood and Bayesian approaches outperformed Maximum Parsimony and Neighbor Joining (Ogden & Rosenberg, 2006) and Bayesian and UPGMA outperform Maximum Likelihood (Douady et al 2003, Kahn et al., 2008). In general, with the large gaps in MS/MS sequence data, Neighbor Joining does poorly. Unfortunately, at present, Maximum Likelihood approaches can be prohibitively slow.

A possible approach is to use multiple tree-making programs to determine the consensus. This approach was used by Organ et al., (2008) who used two independent implementations of Maximum Likelihood and one Neighborhood Joining scheme to determine the taxonomic status of peptides retrieved from mammoth and *T. Rex*. All retrieved the *T. Rex* sequences as being closer to chicken than to mammals, amphibians and fish; however, Neighborhood Joining misplaced the alligator and mammoth sequences (Organ et al., 2008). UPGMA does a much better job of developing relationships, but places mammoth in *Canis*.

Choosing the sequences to enter into your program can be critical; the tree constructed from just aligning the top sequences picked up with BLAST and PHI-BLAST places *T. Rex* next to *Rattus*. Excessive mammalian sequences can bias the result. A balance of taxonomic forms is needed as well as a moderate sized-search space (Liska & Shevenko, 2003). A problem with studying the other example described earlier, barnacle cement, is that that representative proteins from related organisms are quite limited.

As for peptide prediction, there are a large number of programs available to perform phylogenies, with many web-based systems (see the Table 7 for a short list of common programs). Mobile@Pasture is a one stop shop, where a variety of different phylogenetic methods are available (Maximum Likelihood, Bayesian and UPGMA) often with more than one variation. The web versions of MUSCLE and MAFFT both come with inbuilt phylogeny tests, but they are quite limited (Neighbor Joining and UPGMA). Again, anyone contemplating constructing fossil phylogenies should test their programs against the FASTA dataset in the supplemental data from Organ et al (2008).

5. Future directions

All aspects discussed above are important to be considered by the novice and contemplated by the expert user of tandem mass spectrometry. They provide the basis to suggest future directions that can enhance the use of tandem mass spectrometry in molecular evolution studies. Special attention is directed to the subjects discussed below, with the understanding that they are just some of the important areas to be considered in the quest to move from hypothesis-based to discovery-based approaches in evolutionary proteomics.

5.1 Protein identification by *de novo* sequencing

The identification of proteins from tandem mass spectrometry data heavily relies on databases generated from sequenced genomes (Colinge & Bennet, 2007; Kumar & Mann, 2009). As described above, this dependence hinders the capacity of tandem mass spectrometry as a discovery tool. Additionally, many of the MS/MS data generated evade the process of identification due to amino acid modifications not taken in consideration, algorithms and filters used, among others. Analyzing each spectrum generated by tandem mass spectrometry eliminates the high throughput capabilities of the instrument. Discovery-based approaches are hindered by the limitation of not being able to detect non-conserved peptides in a sample from extinct or living organisms using available databases. These facts call to automatize the search for novel sequences in MS/MS data through a process called *de novo* sequencing.

Today, identification of protein sequences by tandem mass spectrometry depends mostly on database search algorithms. As discussed above, the ions detected in a MS spectrum are selected for induced fragmentation, generating product ions in a second MS spectrum or MS/MS. For example, figure 1 shows a MS spectrum (insert) where an ion corresponding to

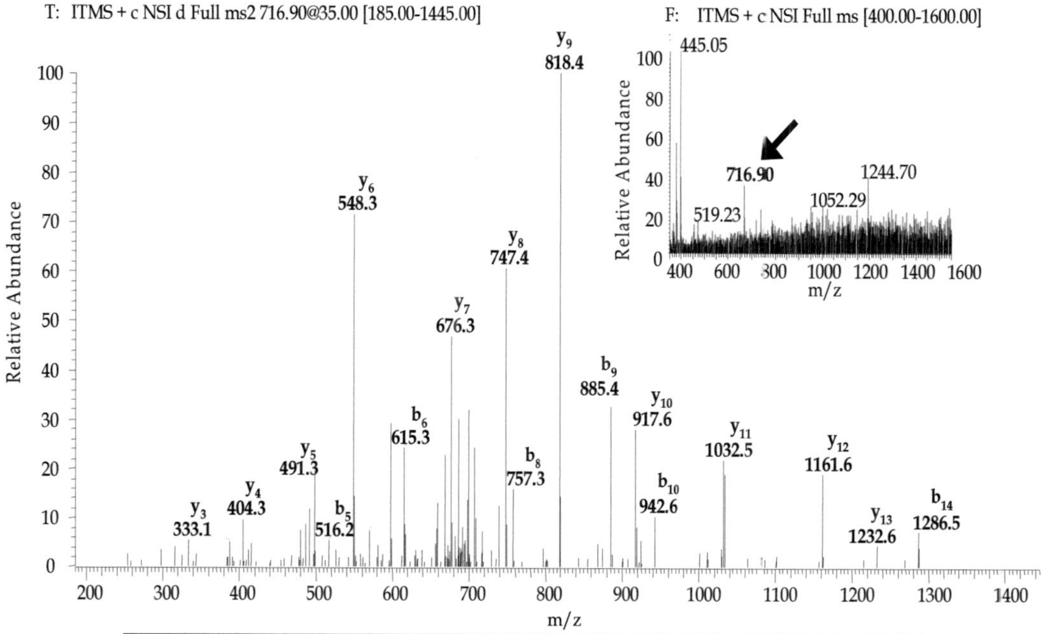

b ion		Amino Acid		y ion
-----	1	E	15	-----
201.1	2	A	14	1303.6
272.1	3	A	13	1232.6
401.2	4	E	12	1161.5
516.2	5	D	11	1032.5
615.3	6	V	10	917.5
686.3	7	A	9	818.4
757.3	8	A	8	747.4
885.4	9	Q	7	676.3
942.4	10	G	6	548.3
1029.4	11	S	5	491.2
1100.5	12	A	4	404.2
1157.5	13	G	3	333.2
1286.5	14	E	2	276.2
-----	15	K	1	147.1

Fig. 1. MS and MS/MS spectra of a specific peptide. The figure illustrates the MS spectrum (insert). The arrow indicates the m/z ratio of the ion corresponding to the 2+ charge state of the peptide (black arrow). The MS/MS spectrum illustrates the product ions of fragmentation. The specific peptide fragments identified by database search analysis are indicated. These correspond to the expected product ions as shown in the table.

the 2+ charge state (black arrow) of the peptide listed on the table was detected. Fragmentation of a peptide is produced by different techniques (see above). In this example, collision-induced dissociation (CID) was used. CID is based on the movement of helium atoms inside the linear-ion trap that collide against the selected ion (peptide). The peptide bond is rigid based on its partial double-bond character. The instrument is tuned to induced fracture at the peptide bond. An analogy that explains this concept well is an accident that has probably happened to all of us at one time in our life. Remember when you threw a baseball against a glass window? Yes, it broke! However, if the baseball had been thrown, at the same speed, to the glove of a friend, the glove did not break. Well, the glove is flexible while the glass is rigid. The collision of helium atoms against the peptide induces its fragmentation. A fragmentation event at the peptide bond generates two product ions, known as b and y ions. The b ion is the fragment containing the carboxyl terminal of the peptide bond and the y ion contains the amino terminal of the peptide bond. Thus, if a peptide of 15 amino acids (as the one shown in figure 1, table) is fragmented at the fifth amino acid of the sequence, b5 ion and y10 ion are generated. The products of the fragmentation events are shown in the MS/MS spectrum and analyzed against the database. The database search algorithms identified the corresponding sequence based on a comparison with those in the database. Dependence of the comparison against known sequences hinders the development of discovery-based approaches.

De Novo sequencing allows the subtraction of sequence information directly from MS/MS spectra without the need to confront the data against a pre-established database (Dancik et al., 1999). Programs developed for *de novo* sequencing started in the late 1980s. At the time several computational problems needed to be overcome. First, the algorithms needed to overcome the limitations of being instrument or method specific. Ion fragmentation is conducted by different techniques. Collision-induced dissociation (CID), electron capture dissociation (ECD), infrared multiphoton dissociation (IRMPD) and the combination of these are some of the techniques used for the fragmentation of ions to extract sequencing data. The difference among them is not only in the method used, but also in the fragmentation efficiency of a particular sequence. This brings the second computational problem. How do you assign sequence information to incomplete fragmentation products? Not a trivial question to address even with today's advances in computational biology. Thirdly, the scoring system used to determine levels of confiability to select the correct sequence from all possible alternatives needs to be robust and verifiable. Needless to say, new algorithms need to overcome these three major problems.

A new generation of *de novo* sequencing algorithms has become available. DeNoS in 2005 claimed to be the first algorithm to sequence peptides with >95% reliability (Savitski et al. 2005). Other more commonly known algorithms are PEAKS and Lutefisk, which can be used to analyze data obtained from both CID and ECD fragmentation (B. Ma et al., 2003; Johnson & Taylor, 2002). Similarly, newly developed algorithms such as ADEPTS and ScanRanker provide a platform not only to extract unassigned MS/MS spectra but also to increase the confidence of peptide sequence alignments (He & B. Ma, 2010; Z.Q. Ma et al., 2011). Ultimately, "shotgun protein sequencing" takes advantages of sequence alignment to group unidentified MS/MS spectra and identify modified and unmodified peptide variants, generating a new method to study unknown proteins from tandem mass spectrometry approaches (Bandeira, 2011). Undoubtedly, these approaches should serve to strengthen discovery-based approaches to identifying conserved and novel proteins from extinct and living organisms from different taxa.

5.2 Identification of biologically active peptides

As with most truly powerful scientific approaches, the immediate future of tandem mass spectrometry is studies in which the sophisticated user applies his/her knowledge and expertise to cross-disciplinary investigations of regulatory and physiological processes that are now in reach. Routinely these studies provide insight as to the nature of evolutionarily conserved as well as evolutionarily variable biochemical processes. The advancement of the specific field of enquiry as well as linking to molecular and biochemical approaches provides insight into how evolution works. A relatively simple example of this kind of problem is the processing of pro-forms of proteins into their active protein and signaling peptides. As described in sections 2.1 and 3.4, proteolytic activation of inactive proteins is the basis for a wide range of physiological processes including blood coagulation, the complement reaction, fibrinolysis, dorsal-ventral patterning in drosophila, and adhesion in barnacles. At one level these are some of the best understood processes in physiology and biology. At the evolutionary level they are a frontier.

The following example of the potential cross disciplinary use of Mass Spectrometry is signaling from marine chemical ecology. The signals are based upon the action of trypsin like serine proteases, evolutionarily ancient and highly conserved enzymes, which are important in sequencing proteins in Tandem Mass spectrometry. The signal molecules are pheromones or resource cues that are found in at least 4 phyla of marine invertebrates and that have strong similarity if not homology to the vertebrate complement cascade (Petts et al., 1993; Rittschof & Cohen, 2004). These signaling peptides are all peptides whose carboxyl terminal sequences end in arginine or lysine. The peptides coordinate a wide variety of processes such as prey location by gastropods, hermit crab shell location, larval release in decapod crustaceans, and induction of larval settlement in barnacles, and have been called keystone molecules by some ecologists (Zimmer & Butman, 2000).

Although these peptides have been known to exist for over 30 years, none have been sequenced. Moreover, only in the case of barnacle settlement pheromones have the precursors to the peptides, in this case, settlement inducing protein complex (SIPC; Dreanno et al., 2006a; 2006b; 2007) been identified. These peptides which are active at nM (10^{-9}) to pM (10^{-12}) should be low hanging fruit for modern techniques which have aM (10^{-18}) sensitivity. Knowing sequences and identifying precursors would enable the development of theory of the evolution of peptide signaling and shed light on the evolution of activation of the pro-forms of proteins, including which processes are highly conserved and which can tolerate change and enable diversity to evolve. Cross-disciplinary interest would include proteomics, enzymology, biochemistry, physiology, neurophysiology, behavior and ecology.

5.3 Identification of posttranslational modifications: A case for substituted amino sugars

Another area with high potential for the application of Tandem Mass Spectrometry is sequencing and determining substituted structures of complex amino sugar polymers associated with glucose amino glycans, such as heparin, chondroitan sulfate, related polymers and their products. These polymers from vertebrates are a hot topic in materials science (Ornitz, 2000). The related structures in estuarine and marine fish and invertebrates are the source of substituted amino sugar signal molecules (Forward & Rittschof, 2000; Rahmen et al., 2000). Again, in marine systems the actual structure of the native signal molecules has not been determined. Rather, indirect enzymatic methods and purified molecules from higher vertebrates, have been used in conjunction with bioassays to provide insight as to the structure of active signals.

We see an interesting parallel between studying complex amino sugar polymers and peptide sequencing. Sequencing of complex amino sugars is more difficult than peptide sequencing because there are usually at least two kinds of common linkages that can be degraded by enzymes and the sugars themselves can be substituted at a variety of sites with sulfates or acetates (Rittschof & Cohen, 2004). The first step in characterizing the sugar back bone is to strip off the substituted groups. This is frustrating as these substitutions confer biological activity. If techniques were developed using enzymes to sequence the polymers with their substituents, one could begin to discern the evolutionary relationships between the classes of substituted sugars and their conserved and variable regions.

Glucose amino glycans (GAG's) are a very important group of structural polymers which have evolved signaling functions. They are the ultimate in post synthesis modification of proteins and promise to entertain researchers for decades. Understanding their structures, their conserved and variable components and their relations throughout biology is a formidable challenge that will require coordinated use of techniques and approaches from a wide range of disciplines, from medicine and agriculture to materials science and chemical engineering. Mass spectrometry would inform the enquiry.

5.4 Funding issues

The use of tandem mass spectrometry in molecular evolution studies relies on a multidisciplinary and interdisciplinary approach. As it is obvious from all the issues discussed above, tandem mass spectrometry transcends many disciplines that range from computational biology and bioinformatics to chemistry and biochemistry. The integration of all of this knowledge is required to fulfill all of the capabilities that this technology puts in our hands. Therefore, the complex nature of tandem mass spectrometry as a tool is a cohesive force among scientists from diverse disciplines, whose professional goals and motivation is discovery and pushing forward the boundaries of knowledge. However, there is always a force in the opposite direction that pulls scientists away from each other, FUNDING.

Why should funding agencies consider investing in the improvement of tools for molecular evolution studies? First, the discovery of conserved mechanisms could provide valuable information to dissect the pathways that are crucial for cellular function. This information is essential for understand the pathophysiology of diseases that leads to cell death. Additionally, drug development could benefit from this information since highly conserved proteins could be "hubs" in a protein network involved in a variety of cellular processes (Zhu et al., 2007). Second, the identification of conserved proteins paves the way to dissect differences among organisms of different taxa, contributing to the understanding of divergence and speciation. The dissected differences may contribute to the understanding of molecular processes in a specific organism. Third, identification of conserved proteins and, consequently, molecular mechanisms may uncover new experimental tools and applications that benefit human kind. From the Dickinson et al. (2009) example, which showed similarities between barnacle cement and human blood coagulation, elucidating the proteome involved in barnacle adhesion may allow this organism to be used as an invertebrate model to discover and validate anti-coagulant drugs and to develop biomedical glues that could be later used in humans. This knowledge can also be employed in *preventing* adhesion of barnacles; the settlement of barnacles on ships and other man-made objects in the sea (i.e. biofouling) cost maritime industries billions of dollars annually. Anticoagulants that have been well-developed by the pharmaceutical industry may prove to effectively prevent barnacle adhesion when employed in a ship paint, which would greatly improve fuel efficiency (Rittschof et al., 2011).

This is not a request without precedent. NIH invested $30 million in a project to unravel the human brain's connections. This effort brought together neuroscientists, microscopists, chemists, biochemists, neurologists, radiologists, and others in order to construct the connection map that directs our mind. A similar effort to unravel the mysteries of molecular evolution could bring together evolutionists, developmental biologists, mass spectrometrists, bioinformaticians, computational biologists, statisticians, biochemists, cellular biologists, chemists and neuroscientists. The benefits of such an effort go beyond our inclination to understand the past, transcending to better understand our present.

6. Conclusions

In this chapter, the use of tandem mass spectrometry in molecular evolution studies has been discussed from a user perspective. Important considerations, from sample preparation to data mining, were explained to bear in the mind of those that want to venture on in these studies and to entice the creativity of the expert to push the boundaries forward. The examples used are recent debates and discoveries using hypothesis-based approaches to understand the past and conserved molecular mechanisms. Although controversial, the identification of conserved proteins from extinct organisms led us to think that, as stated above, our *song* has been modified by changes in *tempo* to adjust in space and time. Essential mechanisms that render the organism more fit to survive or respond to its environment open our understanding on how conservation is an integral component of evolution. In the future, scientists will continue joining forces in the quest to develop more instrumental and bioinformatics tools to uncover the mysteries of the past for the benefit of the present and future.

7. Acknowledgements

IEV acknowledge the support of NIH-COBRE grant 5P20RR016439 and UPR Institutional funds that contributed to the establishment of the Protein Mass Spectrometry Facility, and NIH-NINDS funds (SC1NS066988) that support, in part, this work. DR and GHD acknowledge support of the U.S. Office of Naval Research, Coatings and Biofouling Program, for continued support of interdisciplinary research.

8. References

Adami, C., Ofria, C., & Collier, T.C. (2000) Evolution of biological complexity. *Proceedings of the National Academy of Sciences of the United States of America,* 97, pp. 4463-4468

Asara, J.M., Schweitzer, M.H., Freimark, L.M., Phillips, M., & Cantley, L.C. (2007) Protein sequences from mastodon and *Tyrannosaurus rex* revealed by mass spectrometry. *Science,* 316, pp. 280-285

Bandeira, N. (2011) Protein Identification by spectral networks analysis. *Methds Mol. Biol.* 694, pp. 151-168

Bi, E., & Lutkenhaus, J. (1991) FtsZ ring structure associated with division in *Escherichia coli. Nature,* 354, pp. 161-164

Boekhorst, J., van Breukelen, B., Heck, A.J.R., & Snel., B. (2008) Comparative phosphoproteomics reveals evolutionary and functional conservation of phosphorylation across eukaryotes. *Genome Biology,* 9, pp. R144.

Boot-Handford, R.P. & Tuckwell, D.S. (2003). Fibrillar collagen: the key to vertebrate evolution? A tale of molecular incest. *BioEssays,* 25 pp. 142-151

Boutilier, K., Ross, M., Podtelejnikov, A.V., Orsi, C., Taylor, R., Taylor, P., & Figeys, D. (2005) Comparison of different search engines using validated MS/MS test datasets. *Analytica Chimica Acta,* 534, pp. 11-20

Budovskaya, Y.V., Stephan, J.S., Deminoff, S.J. & Herman, P.K. (2005) An evolutionary proteomics approach identifies substrates of the cAMP-dependent protein kinase. *Proceedings of the National Academy of Sciences of the United States of America,* 102, pp. 13933-139398

Buckley, M., Walker, A., Ho, S. Y. W., Yang, Y., Smith, C., Ashton, P., Thomas Oates,J., Cappellini, E., Koon, H., Penkman, K., Elsworth, B., Ashford, D., Solazzo, C., Andrews, P., Strahler, J., Shapiro, B., Ostrom, P., Gandhi, H., Miller, W., Raney, B., Zylber, M.I., Gilbert, M.T.P., Prigodich, R.V., Ryan, M., Rijsdijk, K.F., Janoo, A., & Collins, M.J. (2008) Comment on "Protein sequence from Mastodon and *Tyrannosaurus rex* revealed by mass spectrometry. *Science,* 319, pp. 33

Colinge, J., & Bennett, K.L. (2007) Introduction to Computational Proteomics. *PLoS Computational Biology,* 3, pp. 1151-1160

Cottrell, J.S. (2011) Protein identification using MS/MS data. *Journal of Proteomics,* 74, pp. 1842-1851

Coyne, K.J., Xiao-Xia, Q., & Waite, J.H. (1997). Collagen in mussel byssus: a natural block copolymer. *Science,* 277, pp. 1830-1832

Dancik, V., Addona, T.A., Clauser, KR., Vath, J.E., & Pevzner, P.A. (1999) De Novo sequencing via tandem mass spectrometry. *J. Computational Biol.,* 6, pp. 327-342

Davie, E. (2003) JBC Centennial 1905-2005: 100 years of biochemistry and molecular biology. A brief historical review of the waterfall/cascade of blood coagulation. *Journal of Biological Chemistry,* 278, pp. 50819-50832

Davie, E. W., & Rantoff, O. D. (1964) Waterfall sequence for intrinsic blood clotting. *Science,* 145, pp. 1310-1312

Day, J.J., & Sweatt, J.D. (2011) Epigenetic mechanisms in cognition. *Neuron,* 70, pp. 813-829

Desai, A., & Mitchison T.J. (1998) Tubulin and FtsZ structures: functional and therapeutic implications. *Bioessays,* 20, pp. 523-527

Dick, D.M. (2011) Gene-environment interaction in psychological traits and disorders. *Annual Review of Clinical Psychology.* 7, pp. 383-409

Dickinson, G.H., Vega, I.E., Wahl, K.J., Orihuela, B., Beyley, V., Rodriguez, E.N., Everett, R.K., Bonaventura, J., Rittschoff, D. (2009) Barnacle cement: a polymerization model based on evolutionary concepts. *Journal of Experimental Biology,* 212, pp. 3499-3510

Douady, C.J., Delsuc, F., Boucher, Y., Doolittle, W.F., & Douzery, E.J. (2003) Comparison of Bayesian and maximum likelihood bootstrap measures of phylogenetic reliability. *Molecular Biology and Evolution,* 20, pp. 248-254

Duncan, M.W., Aebersold, R., & Caprioli, R.M. (2010) The pros and cons of peptide-centric proteomics. *Nature Biotechnology,* 28, pp. 659-664

Dreanno, C., Matsumura, K., Dohmae, N., Takio, K., Hirota, H., Kirby, R.R. & Clare, A.S. (2006a) An alpha(2)-macroglobulin-like protein is the cue to gregarious settlement of the barnacle *Balanus amphitrite. Proceedings of the National Academy of Sciences of the United States of America,* 103, pp. 14396-14401

Dreanno, C., Kirby, R.R., & Clare, A.S. (2006b) Locating the barnacle settlement pheromone: spatial and ontogenetic expression of the settlement-inducing protein complex of

Balanus amphitrite. *Proceedings of the Royal Society B-Biological Sciences*, 273, pp. 2721-2728

Dreanno, C., Kirby, R.R., & Clare, A.S. (2007) Involvement of the barnacle settlement-inducing protein complex (SIPC) in species recognition at settlement. *Journal of Experimental Marine Biology and Ecology*, 351, pp. 276-282

Durliat, M., & Vranckx, R. (1981) Action of various anticoagulants on hemolymphs of lobsters and spiny lobsters. *Biological Bulletin* 160, pp. 55-68

Edman, P. (1950) Method for determination of the amino acid sequence in peptides. *Acta Chemica Scandinavica*, 4, pp. 283-293

Edwards, N.J. (2011) Protein identification from tandem mass spectra by database searching. *Methods in molecular biology*, 694, pp. 119-138

Findlay, G.D., & Swanson, W.J. (2010) Proteomics enhances evolutionary and functional analysis of reproductive proteins. *Bioessays*, 32, pp. 26-36

Forward, R.B., & Rittschof, D. (2000) Alteration of photoresponses involved in diel vertical migration of a crab larva by fish mucus and degradation products of mucopolysaccharides. *Journal of Experimental Marine Biology and Ecology*, 245, pp. 277-292

Fuller, G. M., & Doolittle, R. F. (1971) Studies of invertebrate fibrinogen. II. Transformation of lobster fibrinogen into fibrin. *Biochemistry*, 10, pp. 1311-1315

Garrone, R. (1999). Collagen, a common thread in extracellular matrix evolution. *Proceedings of the Indian Academy of Sciences: Chemical Sciences*, 111, pp. 51-56

Goodman, C.S., & Coughlin, B.C. (2000) The evolution of evo-devo biology. *Proceedings of the National Academy of Sciences of the United States of America*, 97, pp. 4424-4425

Graumann, P.L. (2004) Cytoskeletal elements in bacteria. *Current Opinion in Microbiology*, 7, pp. 565-571

He, L., & Ma, B. (2010) ADEPTS: advanced peptide de novo sequencing with a pair of tandem mass spectra. *J. Bioinform Comput. Biol.*, 8, pp. 981-994

Heffer, A., Shultz, J.W., & Pick, L. (2010) Surprising flexibility in a conserved Hox transcription factor over 550 million years of evolution. *Proceedings of the National Academy of Sciences of the United States of America*, 107, pp. 18040-18045

Human Genome Project (March 26, 2008) The science behind the human genome project, In: *From the Genome to the Proteome*, August 25, 2011, Available from: http://www.ornl.gov/sci/techresources/Human_Genome/project/info.shtml

Hunt, D.F., Yates III, J.R., Schabanowitz, J., Winston, S., & Hauer C.R. (1986) Protein sequencing by tandem mass spectrometry. *Proceedings of the National Academy of Sciences of the United States of America* 83, pp. 6233-6237

Johnson, R.S., & Taylor, J.A. (2002) Searching sequence databases via de novo peptide sequencing by tandem mass spectrometry. *Molecular Biotechnology*, 22, pp. 301-315

Kappeler, L., & Meaney, M.J. (2010) Epigenetics and parental effects. *Bioessays*, 32, pp. 818-827

Khan, H.A., Arif, I.A., Bahkali, A.H., Al Farhan, A.H., & Al Homaidan, A.A. (2008) Bayesian, maximum parsimony and UPGMA models for inferring the phylogenies of antelopes using mitochondrial markers. *Evolutionary Bioinformatics Online*, 4, pp. 263-270

Kopacek, P., Hall, M., & Soderhall, K. (1993) Characterization of a clotting protein, isolated from plasma of the fresh water crayfish *Pacifastacus leniusculus. European Journal of Biochemistry,* 213, pp. 591-597

Kumar, C., & Mann, M. (2009) Bioinformatics analysis of mass spectrometry-based proteomics data sets, *FEBS Letters,* 583, pp. 1703-1712

Levin, H.L., & Moran, J.V. (2011) Dynamic interactions between transposable elements and their hosts. *Nature Reviews Genetics,* 12, pp. 615.627

Liska, A.J., & Shevchenko, A. (2003) Expanding the organismal scope of proteomics: cross-species protein identification by mass spectrometry and its implications. *Proteomics,* 3, pp. 19-28

Lorand, L. (1972) Fibrinoligase: the fibrin-stabilizing factor system of blood plasma. *Annals of the New York Academy of Sciences,* 202, pp. 6-30

Lubec, G., & Afjehi-Sadat L. (2007) Limitations and pitfalls in protein identification by mass spectrometry, *Chemical Reviews,* 107, pp. 3568-3584

Ma, B., Zhang, K., Hendrie, C., Liang, C., Li, M., Doherty-Kirby, A., & Lajoie, G. (2003) PEAKS: powerful software for peptidede novo sequencing by tandem mass spectrometry. *Rapid Communications in Mass Spectrometry,* 17, pp. 2337–2342

Ma, Z.Q., Chambers, M.C. Ham, A.J., Cheek, K.L., Whitwell, C.W., Aerni, H.R., Schilling, B., Miller, A.W., Caprioli, R.M., & Tabb, D.L. (2011) Scan Ranker: Quality assessment of tandem mass spectra via sequence tagging. *Journal of Proteome Research,* 10, pp. 2896-2904

MacFarlane, R. G. (1964) An enzyme cascade in the blood clotting mechanism, and its function as a biochemical amplifier. *Nature,* 202, pp. 498-499

Madaras, F., Chew, M. Y., & Parkin, J. D. (1981) Purification and characterization of the sand crab (*Ovalipes bipustulatus*) coagulogen (fibrinogen). *Thrombosis and Haemostasis,* 45, pp. 77-81

Mann, M., & Kelleher, N.L. (2008) Precision proteomics: the case for high resolution and high mass accuracy. *Proceedings of the National Academy of Sciences of the United States of America,* 105, pp. 18132-18138

Marshall, E. (2011) Waiting for the Revolution. *Science,* 331, pp. 526-529

Marshall, J.L., Huestis, D.L., Garcia, C., Hiromasa, Y., Wheeler, S., Noh, S., Tomich, J.M., Howard, D.J. (2011) Comparative proteomics uncovers the signature of natural selection acting on the ejaculate proteomes of two cricket species isolated by postmating, prezygotic phenotypes. *Molecular Biology and Evolution,* 28, 423-435

Muta, T., & Iwanaga, S. (1996) The role of hemolymph coagulation in innate immunity. *Current Opinion in Immunology,* 8, pp. 41-47

Nesvizhskii, A.I., Keller, A., Kolker, E., & Aebersold, R. (2003) A statistical model for identifying proteins by tandem mass spectrometry. *Analytical chemistry,* 75, pp. 4646-4658

Neurath, H. (1984) Evolution of proteolytic enzymes. *Science,* 224, pp. 350-357

Neurath, H. (1986) The versatility of proteolytic enzymes. *Journal of Cellular Biochemistry,* 32, pp. 35-49

Neurath, H. (1999) Proteolytic enzymes, past and future. *Proc. Natl. Acad. Sci.,* 96, pp. 10962-10963

Neurath, H., & Walsh, K. A. (1976) Role of proteolytic enzymes in biological regulation (a review). *Proceedings of the National Academy of Sciences of the United States of America*, 73, pp. 3825-3832

Ogden, T.H., & Rosenberg, M.S. (2006) Multiple sequence alignment accuracy and phylogenetic inference. *Systematic Biology*, 55, pp. 314-328

Osaki, T., & Kawabata, S. (2004) Structure and function of coagulogen, a clottable protein in horseshoe crabs. *Cellular and Molecular Life Sciences*, 61, pp. 1257-1265

Organ, C.L., Schweitzer, M.H., Zheng. W., Freimark, L.M., Cantley, L.C., & Asara, J.M. (2008) Molecular phylogenetics of mastodon and *Tyrannosaurus rex*. *Science*, 320, pp. 499

Ornitz, D.M. (2000) FGFs, heparin sulfate and FGFRs: complex interactions essential for development. *Bioessays*, 22, pp. 108-112

Patternson, S.D., & Aerbersold, R.H. (2003) Proteomics: the first decade and beyond. *Nat. Genet.*, 33 Suppl., pp. 311-323

Patthy, L. (1993) Modular design of proteases of coagulation, fibrinolysis, and complement activation: implications for protein engineering and structure - function studies. Pages 10-22 in L. Lorand and K. G. Mann, editors. Methods in Enzymology. Academic Press, Inc., San Diego.

Pettis, R.J., Erickson, B.W., Forward, R.B. & Rittschof, D. (1993) Superpotent synthetic tripeptide mimics of the mud-crab pumping pheromone. *International Journal of Peptide and Protein Research*, 42, pp. 312-319

Pevzner, P.A., Kim, S., & Ng, J. (2008) Comment on "Protein sequence from Mastodon and Tyrannosaurus rex revealed by mass spectrometry. *Science*, 321, pp. 1040

Pisano, J. J., Finlayso, J. S., & Peyton, M. P. (1968). Cross-link in fibrin polymerized by factor XIII epsilon-(gamma-glutamyl)lysine. *Science* 160, pp. 892-89

Pisano, J. J., Finlayso, J. S. & Peyton, M. P. (1969). Chemical and enzymatic detection of protein cross-links. Measurement of epsilon-(gamma-glutamyl)lysine in fibrin polymerized by factor XIII. *Biochemistry* 8, pp. 871-876

Prachumwat, A., & Wen-Hsiung, L. (2008) Gene number expansion and contraction in vertebrate genomes with respect to invertebrate genomes. *Genome Research*, 18, pp. 221-232

Qi, L., & Cho, Y.A. (2008) Gene-environment interaction and obesity. *Nutrition Reviews*, 66, pp. 684-694

Rahman, Y.J., Forward, R.B., & Rittschof, D. (2000) Responses of mud snails and periwinkles to environmental odors and disaccharide mimics of fish odor. *Journal of Chemical Ecology*, 26, pp. 679-696

Ramos-Fernandez, A., Paradela, A., Navajas, R., & Albar, J.P. (2008) Generalized method for probability-based peptide and protein identification from tandem mass spectrometry data and sequence database searching. *Molecular and Cellular Proteomics*, 7, pp. 1748-1754

Ramm, S.A., McDonald, L., Hurst, J.L., Beynon, R.J., & Stockley, P. (2009) Comparative proteomics reveals evidence for evolutionary diversification of rodent seminal fluid and its functional significance in sperm competition. *Molecular Biology and Evolution*, 26, pp. 189-198

Rittschof, D., Cohen, J.H. (2004) Crustacean peptide and peptide-like pheromones and kairomones. *Peptides*, 25, pp. 1503-1516

Rittschof, D., Dickinson, G.H., Orihuela, B., & Holm, E.R. (2011) Anticoagulants as antifouling agents. US Pub. No. 2011/0041725 A1.

Sadreyev, R., & Grishin, N. (2003) COMPASS: a tool for comparison of multiple protein alignments with assessment of statistical significance. *Journal of Molecular Biology*, 326, pp. 317-336

Sanchez, J.C., Corthals, G.L., & Hochstrasser, D.F., editors (2004) Biomedical applications of proteomics. Weinheim: Wiley-VCH. 435 p.

Savitski, M.M., Nielsen, M. L., Kjeldsen, F., & Zubarev, R. A. (2005) Proteomics-Grade de Novo Sequencing Approach. *Journal of Proteome Research*, 4, pp. 2348–2354

Schweitzer, M.H., Suo, Z., Avci, R., Asara, J.M., Allen, M.A., Arce, F.T. & Homer J.R. (2007) Analyses of soft tissue from *Tyrannosaurus rex* suggest the presence of protein. *Science*, 316, pp. 277-280

Shoulders, M.D., & Raines, R.T. (2009). Collagen structure and stability. *Annual Review of Biochemistry*, 78, pp. 929-958

Soderhall, K. (1981) Fungal cell-wall beta-1,3-glucans induce clotting and phenoloxidase attachment to foreign surfaces of crayfish hemocyte lysate. *Developmental and Comparative Immunology*, 5, pp. 565-573

Stretton, A.O.W. (2002) The first sequence: Fred Sanger and Insulin. *Genetics*, 162, pp. 527-532

Sritunyalucksana, K., & Soderhall, K. (2000) The proPO and clotting system in crustaceans. *Aquaculture*, 191, pp. 53-69

Tanaka, K., Hiroaki, W., Yutaka, I., Satoshi, A., Yoshikazu, Y., Tamio, Y., & Matsuo, T. (1988) Protein and polymer analyses up to *m/z* 100 000 by laser ionization time-of-flight mass spectrometry. *Rapid Communications in Mass Spectrometry*, 2, pp. 151-153

Theopold, U., Schmidt, O., Soderhall, K., & Dushay, M. S. (2004) Coagulation in arthropods: defense, wound closure and healing. *Trends in Immunology*, 25, pp. 289-294

Tzeng, J.Y., Zhang, D., Pongpanich, M., Smith, C., McCarthy, M.I., Sale, M.M., Worrall, B.B., Hsu, F.C., Thomas, D.C. & Sullivan, P.F. (2011) Studying gene and gene-environment effects of uncommon and common variants on continuous traits: a marker-set approach using gene-trait similarity regression. *American Journal of Human Genetics*, 89, pp. 277-288

Wang, R., Liang, Z., Hall, M. & Soderhall, K. (2001) A transglutaminase involved in the coagulation system of the freshwater crayfish, *Pacifastacus leniusculus*. Tissue localisation and cDNA cloning. *Fish and Shellfish Immunology*, 11, pp. 623-637

Zhu, X., Gerstein, M., & Snyder M. (2007) Getting connected: analysis and principles of biological networks. *Genes and Development*, 21, pp. 1010-1024

Zimmer, R.K., & Butman, C.A. (2000) Chemical signaling processes in the marine environment. *Biological Bulletin*, 198, pp. 168–87

4

Proteomic Strategies for Comprehensive Identification of Post-Translational Modifications of Cellular Proteins Including Low Abundant and Novel Modifications

Jaeho Jeong, Jae-Jin Lee and Kong-Joo Lee
The Center for Cell Signaling & Drug Discovery Research, College of Pharmacy, Division of Life & Pharmaceutical Sciences, Department of Bioinspired Science, Ewha Womans University, Republic of Korea

1. Introduction

Post-translational modifications (PTMs) enable proteins carryout multiple biological functions. PTMs facilitate and regulate various protein functions by changing the chemical and physical characteristics of a protein, which affect its stability, cellular localization, and its interaction with other proteins as well as non-protein molecules which endow the protein-protein interactions, cellular localization and with new or altered biological activities. Understanding how and to what extent cellular proteins are post-translationally modified, at which sites on the protein, how the modifications affect the stoichiometry of the protein sequence, and what the functional consequences of each of the modifications are, therefore becomes indispensible for understanding cellular function and regulation of a protein. Recent advances that facilitated combining affinity-based enrichment of low abundant and novel PTMs with peptide sequencing with tandem mass spectrometric analysis have increased its speed and sensitivity of analytical approaches. Although it has been demonstrated that mass spectrometry can be an ideal tool for both qualitative and quantitative analysis of protein modifications, comprehensive identification of PTMs using mass spectrometry (MS) in a high-throughput manner remains a highly challenging task because of the diversity, dynamic complexities, low abundance and heterogeneity of PTMs. In addition, difficulties in interpreting tandem MS spectra for peptide sequencing, poor peptide fragmentation, and surprising appearance of totally unexpected modifications, among others, have compounded this challenge and limited the application of mass spectrometry to the identification of a few types of PTMs.

We will describe in this chapter, a strategy we have been employing successfully for rapid, efficient and sensitive identification of diverse and novel PTMs occurring in mammalian biological systems *in vivo*. Our strategy includes: separation of unmodified and modified proteins on 2-dimensional gel electrophoresis (2D-PAGE); and detection of low abundant peptide modifications employing a selectively excluded mass screening analysis (SEMSA) of unmodified peptides during LC-ESI-q-TOF MS/MS through replicated runs (Seo *et al.*, 2008), in conjunction with a searching algorithm MOD^i (Kim *et al.*, 2006; Na *et al.*, 2008) and

DBond (Choi *et al.*, 2010), which we developed. We will also provide examples of novel some low abundant cysteine modifications which we identified in cellular proteins employing our strategy (Lee *et al.*, 2009; Hwang *et al.*, 2009; Jeong *et al.*, 2011).
Since it is difficult in many cases, to predict the type and position of modifications from autotranslated DNA sequences, it is highly important in the postgenomic era, to experimentally obtain this information about PTMs. For, the type and position of PTMs differ, depending on the organism, cell type, signaling process and its kinetics. Proteomics allows deciphering the global picture of protein-related processes in the cell. But proteomics cannot be successfully employed without defining the dynamic PTM maps, which in turn requires the availability of a fast, reliable, and sensitive procedure for PTM characterization. We believe that the highly sensitive and efficient strategies for unrestrictive blind PTM identification that will be described in this chapter will help develop such PTM maps.

2. Diversity of post-translational modifications and their identification using mass spectrometry

Structural changes in proteins can occur from various events including genomic mutations, alternative splicing, proteolytic cleavage, chemical and enzymatic modifications in amino acid side chains. These changes occurring inside cells and tissues cannot be recognized in genomic level, other than by proteomic analysis which makes it possible to identify these alternations. Since these modifications are highly diverse, many proteomic approaches are required to identify them. This review will describe the diverse characteristics of PTMs in proteins and the analytical approaches including mass spectrometry employed to identify them.

2.1 Cellular protein modifications

Modifications in the side chain of amino acids of proteins can alter the protein's charge, polarity and spatial features, and induce conformational changes which in turn cause changes in a variety of protein characteristics. These modifications can occur reversibly or irreversibly, and be mediated by enzymes or non-enzymatic way. In addition, many changes at the genomic level such as point mutation, deletion and insertion, alternative splicing etc. induce alterations in amino acid sequences. Identification of these changes may require different analytical MS methodologies, depending on the type of the modification.

2.1.1 Amino acid substitution

Amino acid substitution, which can occur by point mutation in the genome, or chemical conversion of one amino acid to another, is involved in 346 of the listed 900 modifications in the database of protein modifications for mass spectrometry, UniMod (www.unimod.org). The Human Gene Mutation Database (www.hgmd.org), as of August 2011, lists more than 113,000 mutations of 4,122 human genes, which include missense/nonsense mutation, splicing, small and gross deletion/insertion, complex rearrangement and repeat variation. Although a single amino acid substitution usually does not seem affect a protein's structure in most cases, several proteins are known in which a change in single amino acid change caused significant changes in the protein structure and stability. An example is a single mutation in hemoglobin causes sickle cell anemia, one of several monogenic diseases.

Another way to occur amino acid substitution is by chemical conversion of one amino acid to another, for example, Asn and Gln can be converted to Asp and Glu by deamination. Cys residues are converted to Ser during oxidative stress (Jeong *et al.*, 2011). Mass spectrometry has been extremely useful in identifying unexpected amino acid substitutions and deletion and insertion of amino acids (Fig. 1). However, finding unknown substitutions with routine peptide sequencing in conjunction with tandem MS is not to easy, because MS/MS spectra searching algorithm matches only known sequences. In order to find these unexpected amino acid changes, in MS/MS spectra, it is necessary to carry out *de novo* partial sequencing of a peptide from an MS/MS spectrum. To find unknown amino acid changes, we suggest using the searching algorithm MOD^{i} combined with peptide sequencing using tandem MS.

2.1.2 Alternative splicing

Insertion and deletion of large amino acid sequences can markedly affect the protein structure, in contrast to changes in single amino acid residues which mostly do not. Alternative splicing can occur in a number of ways including exon skipping, exon insertion, alternative 5′ initiation, 3′ termination and even intron inclusions. About 74% of multi-exon genes produce alternatively spliced protein variants in humans. These PTMs result in proteins differing in binding affinity, enzymatic activity, localization within the cell, and stability. Alternative splicing therefore, is a common phenomenon in biology. Existence of the alternatively spliced variants can be detected only at the protein level. However, it is not easy to find spliced variants without separating each variant population. 2D-PAGE is an useful tool to separate the different variants because of their differences in their charges (isoelectric points) and molecular weights (Fig. 1). Separation of each population of protein variants is a prerequisite for employing mass spectrometry to identify the sequence charges. By comparing the chromatogram of each population, it is possible to find peptide products generated from alternative splicing, and then *de novo* sequencing the differential peptides.

2.1.3 Proteolytic cleavage

Proteolytic cleavage is obviously an irreversible protein modification. Proteins can be cleaved by various proteolytic enzymes, which are classified into serine proteases, cysteine proteases, aspartic acid proteases and metalloprotease, depending on their active sites of action. Employing genome sequence analysis, 553 proteases were identified in the human genome: 203 Ser proteases, 143 Cys proteases, 21 Asp proteases and 186 metalloproteases. Since the substrate specificity of each protease is not known in many cases, the proteolytic products can be further identified only at protein level. Unexpected cleavage products are not easy to detect without mass spectrometry (Fig. 1). Full peptide sequencing with tandem MS is the only way to characterize the cleaved products using the same strategy employed to detect the alternative splicing variants.

2.1.4 Enzymatic modifications

Cellular protein modifications are designed by nature to initiate and regulate essential cellular processes. The mechanisms for PTM regulation are not fully understood because of their complexity. More than 200 kinds of enzymes (>5% of total proteins) have been shown to be involved in catalyzing the various chemical modifications of protein side chains. In the human genome, it has been shown that more than 500 proteases, more than 500 protein

Wild type	ABCDE FGHIJ KLMNO PQRST UVWXY Z
AA substitution	ABCDE XGHIJ KLMNO PQRST UVWXY Z
AA deletion	ABCDE -GHIJ KLMNO PQRST UVWXY Z
AA insertion	ABCDE FXGHIJ KLMNO PQRST UVWXY Z
Alternative splicing	FGHIJ KLMNO PQRST UVWXY Z
	ABCDE -------- KLMNO PQRST UVWXY Z
Proteolytic cleavage	HIJ KLMNO PQRST UVWXY Z
	ST UVWXY Z

Fig. 1. Protein amino acid sequence changes with various ways generate unexpected peptide fragments in MS analysis.

kinases, more than 150 protein phosphatases, 5 class methyltransferases, a series of acetyltransferase and deacetylase, oxidoreductases, E1, E2, E3 for ubiquitination, sumoylation and neddylation, operate among others. Since most enzyme induced protein modifications are reversible, they are readily removed during the biological processes after they function as signalling molecules.

Enzymatic modification of amino acid side chains occurs in various ways depending on the species of amino acids, as shown in Table 1. For example, phosphorylation at -OH in Ser, Thr and Tyr residues, can be promoted by various protein kinases and one third of total human proteome is estimated to comprise substrates of various proteins kinases, predicted to number more than 500. Phosphorylated proteins serve as substrates of kinases. They are readily dephosphorylated by phosphatases. Tyrosine kinase receptors (e.g. PDGF receptor, VEGF receptor) are auto-phosphorylated by ligand binding and transduce the signalling as turn 'on' switch, and later dephosphorylated by phosphatases, as turn 'off' switch. These reversible modifications act as on/off switches in various signalling pathways by controlling phosphorylation and dephosphorylation. Thus it becomes important to identify, which residues are phosphorylated, which kinase acts on them, how long signalling is turned on and which phosphatases are involved in the turn off.

In addition to kinases and phosphatases which signal phosphorylation/dephosphorylation, other enzyme pairs are involved in reversible modifications; e.g. acetyltransferase and deacetylase, acyltransferase and deacylase, methyltransferase and demethylase, ubiquitinating and deubiquitinating enzymes, chemical oxidation and peroxidase and reductase, glycosylase and deglycosylase, among others. Irreversible modifications are also possible, for example, crosslinkiing of the proteins by transglutaminase. Although many modifications have been identified thus far, an enormous number of unknown protein modifications are waiting to be discovered. Mass spectrometry offers an useful analytical approach to identifying and quantifying the unknown novel modifications, modification sites, and the accompanying structural changes. The relevant methodologies are detailed in section 2.2.

2.1.5 Non-enzymatic chemical modifications

Protein modifications can occur inside cells by chemical reactions rather than by enzymatic action. Oxidation of Cys, Met, Pro, Trp residues is promoted by reactive oxygen species (ROS) resulting in loss of protein activity or alteration of the protein's biological function by modifying its cellular localization and interactions with other proteins. Nitrosylation, formation of metabolic protein adducts e.g. 4-hydoxynonenal (HNE) adducts involving His, Cys, Lys, chemical adducts with biotin, lipoic acid and phosphopantetheine are well defined modifications (Table 1). Many metabolite adducts of proteins will be further investigated using *de novo* sequencing using tandem MS and searching algorithm.

PTM type	Modified amino acid residue	Monoisotopic mass change	Remarks
Phosphorylation	S T Y H D	+79.966331	Appearance of ΔM = -97.976896 Da peak for dehydration in the phosphorylation sites of Ser and Thr. Use phosphatase inhibitor in sample
Cysteine oxidation	C		Careful for artifactual oxidation
Dehydroalanine		-33.987721	
Cys to Ser		-15.977156	
Disulfide		-2.0145	
Sulfenic acid		+15.994915	
Sulfinic acid		+31.98983	
Sulfonic acid		+47.984745	
Thiosulfonic acid		+63.961901	
Selenylation		+79.916520	Cys-S-SeH
C-nitrosylation		+28.990164	
Glutathionylation		+305.068156	
Acetylation	K N-term	+42.010565	Change (+) charge to neutral Use deacetylase inhibitors in sample
Acylation			Increase hydrophobicity of protein to be anchored into membrane
Octanoylation	S T	+126.104465	
Farnesylation	C	+204.187801	
Palmitoylation	C K N-term	+238.229666	
Geranyl-geranylation	C K N-term	+272.250401	
Retinal	K	+266.203451	
Diacylglycrol	C	+576.511761	
Ubiquitination	K	+114.042927	Ubiquitin C-term Gly-Gly adduct Use proteasome inhibitor in sample
Glycosylation			
Fucose	S T	+146.057909	
Hexoseamine	N T	+161.068808	
Hexose	N T	+162.052824	
O-GlcNAcylation	N S T	+203.0743	N-acetylhexoseamine
Methylation			Sterically bulky without charge change
Mono-methylation	K R	+14.01565	
Di-methylation	K R	+28.0313	
Tri-methylation	K R	+42.04695	
Sulfonation	S T Y C	+79.956815	
Deamidation	N Q	+0.984016	Amino acid substitution to D E
Chemical adduct			
Allysine	K	-1.031634	Lys oxid to aminoadipic semialdehyde
Oxidation	P D K N P Y R	+15.994915	Oxidation or hydroxylation
Hypusine	K	+87.068414	
4-hydroxynonenal	C H K	+156.115030	HNE
Lipoyl	K	+188.032956	
N-term			
Met-loss	M	-131.040485	N-term Met removal by aminopeptidase
Met-loss + Acetyl	M	-89.029920	Acetylation in new N-term after Met loss
Artifactual adduct			
Propionamide	C K N-term	+71.037114	Acrylamide adduct in PAGE
Carbamidomethyl	C K N-term H D	+57.021464	Iodoacetamide derivative

Table 1. The list of identified post-translational modifications.

2.2 Identification of post-translational modifications by mass spectrometry

Protein modifications play key role in protein structure, stability, its interactions and cellular localizations. MS is an ideal analytical tool for analyzing protein sequences, identifying many unknown post-translational modifications, sites modified, and amino acid sequences inserted, deleted or repalced. Studies employing MS along with other techniques, have recorded more than 900 PTMs in UniMod database (www.unimod.org). Number of PTM list in this database is rapidly increased. This informations on the nature of PTMs should facilitate a fuller understanding protein structure and function. A list of reported PTMs and the corresponding sites of amino acids, monoisotopic mass change by PTM, and sample preparation for identifying each PTM is shown in Table 1. These proteomic studies for PTM analysis can be divided into two groups. One group consists of large scale analysis of one kind PTM after enriching the modified proteins and identifying PTMs, necessitated by low abundance of modified proteins. The second group comprises comprehensive analyses of multiple modifications in one protein because the diversity of modifications in the chosen protein (Fig. 2).

2.2.1 Large scale analysis of same type of PTMs in many proteins

Large scale identification of PTMs is not easy problem to be solved because each modification has its distinct and unique chemical property including chemical affinity, solubility, charge and hydrophobicity. In order to identify one kind of PTM in many proteins with high throughput analysis, the enrichment of modified proteins or peptides was combined with peptide sequencing with tandem MS. Affinity-based enrichment of post-translationally modified proteins and peptides can provide to increase the relative abundance of a selected PTM in the sample.

Several studies have reported the systemic identification of same PTM to understand wide signaling cascade (Table 2). Phosphorylation, the most extensively studied PTM, has been reported in more than 10,000 proteins, variously concerned with specific signaling, in cell cycle, cancer, receptor function, and stress responses (Olsen *et al.*, 2010; Dephoure *et al.*, 2008; Rikova *et al.*, 2007; Kim *et al.*, 2002, Kim *et al.*, 2007a and 2007b). Intriguingly, although phosphorylation seems to play a pivotal role in various signal cascades, phosphorylated peptides constitute minor fraction of the total protein mileau. Due to the low abundance of phosphopeptides and low degrees of phosphorylation, enrichment phosphopeptides is essential prior to MS analysis (Fig. 2A). There are several strategies for enrichment of phosphoproteins, including immobilized metal ion affinity chromatography (IMAC), titanium dioxide (TiO_2) chromatography, and immunoaffinity chromatography with anti-pY antibodies (Thingholm *et al.*, 2009). These enrichment techniques themselves have advantages and disadvantages, related to different specificities of the proteins under study. Immunoaffinity chromatography is a traditional biochemical technique, based on the binding of proteins or peptides to phospho-specific antibodies. Highly selective phospho-specific antibodies are available that can be utilized for the enrichment of phosphorylated proteins prior to analysis by MS. IMAC takes advantage of the affinity of chelated (Fe^{3+}, Al^{3+}, Ga^{3+}, or Co^{2+}) ions towards the negative phosphate group of phosphopeptides. Titanium dioxide chromatography utilizes the affinity for phosphate ions in aqueous solutions. Titanium dioxide chromatography sometimes can be combined with IMAC in which mono- and multiply phosphorylated peptides are efficiently enriched from highly complex samples. When combined with other enrichment methods, the method offers an efficient enrichment strategy for phosphoproteomic studies. In another approach, cellular proteins with or without treatment, were separated on 2D-PAGE and phosphorylated proteins on Tyr residues were detected by anti-pY antibody, and each spot was analyzed with tandem MS (Kim *et al.*, 2007a and 2007b). Separation of modified protein from abundant unmodified proteins on 2D-PAGE is an informative way to obtain only modified proteins.

However, attempts to enrich other PTMs have met with limited success. Acetylation of Lys residue is another abundant PTM with fundamentally important regulatory function. Acetylation occurs as a co-translational and post-translational modification of proteins, such as histones, p53 and tubulins. The acetyl group can become attached to either the α-amino group at the protein N-terminus or the ε-amino groups of Lys residues and eliminate the positive charge of the amino group to make it uncharged. Thus, Lys acetylation, a reversible PTM, regulates protein interaction with negatively charged DNA, thereby playing a key regulatory role in gene expression. For example, acetylation of histone or p53 inhibits the DNA binding and renders DNA more relaxed, and deacetylation reverses this process. However, enrichment methods, except for immunoaffinity purification, have not been developed thus far. There remains a need for robust methods that can address the complexity and dynamic range of the cellular proteome.

In contrast to phosphorylation and acetylation as small chemical modifications, ubiquitination, a PTM involving the covalent attachment of ubiquitin, a 76-residue polypeptide, to Lys residue, induce bulk change of protein. Depending on the nature and site of linkage, ubiquitination regulates protein degradation, signal transduction, intracellular localization and DNA repair. Ubiquitination is readily detected by peptide sequencing with tandem MS, because Gly-Gly adduct of ubiquitin C-terminal (+114 Da increase) can easily be detected at Lys ubiquitinated residues in tryptic peptides. Most common enrichment methodology for ubiquinated proteins is immunoaffinity purification employing exogenously tagged ubiquitin (Danielsen *et al.*, 2011). Recent studies have employed endogenous ubiquitin enrichment using ubiquitin binding motifs such as ubiquitin interacting motif (UIM) or ubiquitin associated (UBA) domain, rather than Ub-antibody (Manzano *et al.*, 2008). About 200 ubiquitinated proteins binding to UBA domain of p62 were identified in Arabidopsis.

Glycosylated proteins exist as heterogenous populations characterized by a range of molecular weight, and play important role in membrane surface localization and as receptors by raising the hydrophilicity and changing the surface charge. The O-linked β-N-acetylglucosamine (O-GlcNAcylation) also plays a vital role modifying Ser and Thr residues in many cellular processes, including signal transduction, protein degradation, and regulation of gene expression. Since the lectin, Wheat Germ Agglutinin (WGA), has affinity for terminal N-acetylglucosamine (GlcNAc) and sialic acid residues, lectin immobilized affinity chromatography has been recently used for enrichment O-GlcNAc modified peptides (Vosseller *et al.*, 2006). Biotinylated O-GlcNac peptides are captured by avidin chromatography (Wang *et al.*, 2010).

Oxidative modification has emerged as a major PTM involved in oxidative stress. 4-Hydroxy-2-nonenal (HNE), generated during lipid peroxidation, modifies proteins. 4-HNE adducts are commonly enriched by immunoaffinity chromatography or solid-phase hydrazide enrichment strategy (Mendez *et al.*, 2010; Roe *et al.*, 2007). Other oxidation adducts of reactive Cys residues play key roles in cellular regulations. Recently dimedone, and chemicals that specifically label sulfenic acid, have been used to enrich sulfenic acid using biotin tagged dimedone, and to identify ROS sensitive Cys residues (Giron *et al.*, 2011,

Seo YH *et al.,* 2009, Leonard *et al.,* 2009). Large scale proteomic identifications of proteins containing the same kind of PTM, e.g., phosphorylation or acetylation, can be interesting if the relationship between a specific modification and biological function can be established.

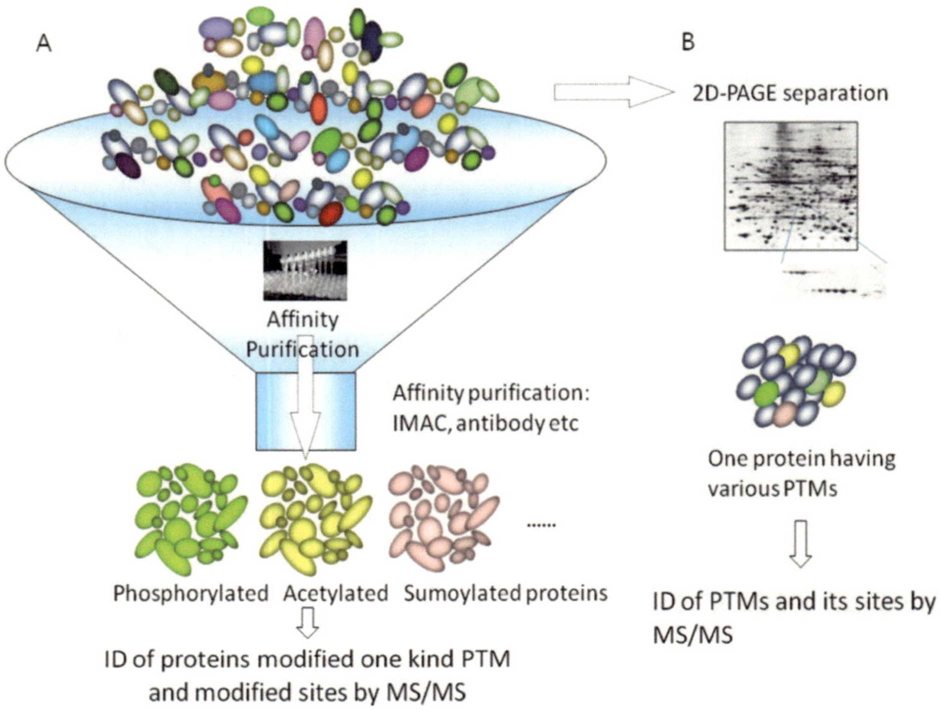

Fig. 2. Schematic diagram to identify one type modification in cellular proteins (A) and comprehensive modifications in one protein (B).

2.2.2 Comprehensive identification of multiple PTMs in the same protein by MS

If the same protein has multiple and diverse functions, one can hazard the assumption that it exists in several forms and contains different PTMs. Comprehensive identification of PTMs in a single protein can therefore help understand how the protein exerts multiple biological functions of multiply modified proteins as shown in Fig. 2B. However, it is not an easy task to clearly identify the PTMs in the same protein, because biological samples of proteins are mixtures of unmodified and modified populations, with unmodified molecules abundantly predominating and the much less abundant modified molecules. A 100% peptide coverage with MS/MS is required for identifying all modifications.

The low abundant PTMs can only be identified after adequately enriching their populations. When complex enrichment methods are needed, use of 2D-PAGE based separation in combination with mass spectrometry is beneficial. 2D-PAGE separates proteins based on their isoelectric points and molecular mass, and makes it possible to separate various modified proteins, spliced variants and proteolytic cleaved fragments. Proteomics has been traditionally exploited power of 2D-PAGE, to separate proteins, especially coupling it with MALDI-TOF MS, for the qualitative and quantitative analysis of proteins in complex extracts. However, the limitations of this approach in terms of throughput analysis of protein mixes have required the development of other proteomics approaches, based on separation of peptides rather than of proteins, or on direct protein identification and selection on dedicated arrays (protein chips). However, 2D-PAGE based separation is adequate in the comprehensive identification of PTMs because it can show the range of complex PTMs.

PTM type (Mass Shift)	**Enrichment methods**	**# of sites identified**	**Biological Association of the protein**	**References**
Phosphorylation (79.9663)	IMAC, and TiO_2	20,443 sites on 6027 proteins	cell cycle	Olsen *et al.*, 2010
		1400 sites on >1000 proteins	Mitosis	Dephoure *et al.*, 2008
	IMAC	654 sites	cAMP/vasopressin-dependent signaling pathways	Gunaratne *et al.*,2010
	Immuno-affinitiy purification with	4551 sites on 2550 proteins	Lung cancer	Rikova *et al.*, 2007
	Anti-pY Ab	628 sites	Cancer cells	Rush *et al.*, 2005
	2D-PAGE, Anti-pY Ab	81 proteins	Heat shock	Kim *et al.*, 2002
		95 proteins	Endothelial cells VEGF-, ROS-signaling	Kim *et al.*, 2007a
		66 proteins	Angiopoietin-1 signaling	Kim *et al.*, 2007b
Acetylation (42.0105)	Immuno-affinitiy purification with	3600 sites on 1750 proteins	Human acute myeloid leukemia cell line	Choudhary *et al.*, 2009
	Anti-acetyl Lys Ab	1981 sites on 1013 proteins	Drosophila	Weinert *et al.*, 2011
		1000 sites	Human liver tissue	Guan *et al.*, 2010
Ubiquitination (114.0429)	Tagged ubiquitin	753 site 471 proteins	Human osteosarcoma cells	Danielsen *et al.*, 2011
		110 sites 72 proteins		Peng, J. *et al.*, 2003
Sumoylation	Tagged SUMO	339 putative polySUMO	polySUMO conjugates	Bruderer *et al.*, 2011
		17 sites on 205 proteins	HEK293 cells	Galisson *et al.*, 2010
		14 sites on 12 proteins		Blomster *et al.*, 2010
O-GlcNAcylation (203.0793)	Lectin-affinity purification	141 sites on 64 proteins	Cytokinesis	Wang *et al.*, 2010
		65 sites	Postsynaptic density preparation.	Vosseller *et al.*, 2006
4-HNE (156.1150)	Immunoaffinity	12 proteins	Human RBCs	Mendez *et al.*, 2010
	Solid-phase hydrazide enrichment	125 sites on 67 proteins		Roe *et al.*, 2007
	2D-PAGE, MS		Early Alzheimer's disease	Redd *et al.*, 2009
Sulfenic acid (15.9949)	Biotin-dimedone purification	39 proteins	HeLa cells	Leonard *et al.*, 2009
	2D-PAGE	35 proteins	Hypertensive Rat	Tyther *et al.*, 2010
biotinylated 15d-PGJ2 (316.2038)	Streptoavidin affinity	12 proteins	Neuroblastoma	Aldini *et al.*, 2007

Table 2. Large scale identification of each PTM in proteins from divergent sources and pathways.

For example, phosphorylated, acetylated, glycosylated and oxidized proteins move in acidic direction, ubiqutinated and sumoylated proteins move upward by increasing molecular weight and disulfide bonded proteins move either upward with intermolecular disulfide or downward with intramolecular disulfide bond. This information can allow the prediction of the type of PTM and overcomes the limitations due to the complexity of PTMs.

Following the separation of the heterogeneous populations of modified proteins on 2D-PAGE, we try to obtain comprehensive PTM information via replicate nanoLC-ESI MS/MS analysis by raising the modified peptide coverage (Seo *et al.*, 2008). To facilitate the characterization of PTMs as much as possible, we devised the strategy of selective exclusion acquisition in replicate run analysis. In data dependent acquisition (DDA) mode, most intense precursor ions are redundantly acquired in nanoLC-ESI MS/MS run. If the exclusion list is not used, identification of low-intensity ions in the presence of high-intensity ions would be far less successful in randomly repeated runs. The number of obtainable MS/MS spectra is limited in a single run analysis, because we select MS/MS spectra having appropriate quality and quantity by optimizing the experimental procedure including sensitivity, scan time, number of ion channel, time for return to MS scan from MS/MS scan, elution time in LC etc. This is the reason for selectively excluding unwanted high-intensity MS/MS data generation. Exclusion methodology is a way to separate wanted peptides from unwanted ones. The overall scheme of **S**electively **E**xcluded **M**ass **S**creening **A**nalysis (SEMSA) is shown in Fig. 3. An exclusive implementation using this unmodified peptide library, resulted in efficient identification of low abundant PTMs (Fig. 4). As the SEMSA progressed, exclusion list is cumulated and then separation of unwanted peptide can ameliorate the quality of MS/MS spectra. The LC-MS procedure is repeated three times to obtain more MS/MS data. The MS data of the first run are then processed by ProteinLynx v2.1 for peak deconvolution and peak list generation. The resulting MS/MS spectra are then generated and submitted to Mascot and ProteinLynx database searches to obtain peptide identifications. Only unmodified peptides now serve as candidates for a precursor exclusion list, in terms of m/z and LC run times in the subsequent run. After the peak list of the second separation is generated, the peaks are matched to peptides previously identified and included in the exclusion list are automatically blocked from the peak selection prior to MS/MS acquisition. The ranges of tolerance windows of excluding peak are typically determined by mass accuracy and resolution in the MS scan and peak widths in the chromatogram. This PTM specific exclusion strategy enables less intense PTM peptides to be identified, thereby enhancing confidence level of PTM identifications.

To determine whether the GAPDH spots on 2D-PAGE have differential modifications, we exhaustively examined the PTMs in each population using SEMSA (Seo *et al.*, 2008). Diverse PTM populations were identified in various peptides (Fig. 5). Especially, multiple modifications of Cys residue in the peptide containing active site (^{152}CTTNC^{156}) were clearly demonstrated: intra-disulfide between 152C and 156C, oxidation to cysteic acid (152C), and transformation of Cys to Ser (152C). Simultaneously, 247C was shown to transform to sulfinic acid, mainly dehydroalanine and cysteic acid. This indicates that reactive Cys residues in GAPDH can be oxidized to various oxidation states depending upon tertiary structural environments. Results of these studies clearly established the exact oxidation sites, oxidation species, and the levels of oxidation states. This strategy was applied for finding many low abundant modifications including phosphorylation, acetylation, glutathionylation and some novel modifications (Hwang *et al.*, 2009; Lee *et al.*, 2010; Jeong *et al.*, 2011).

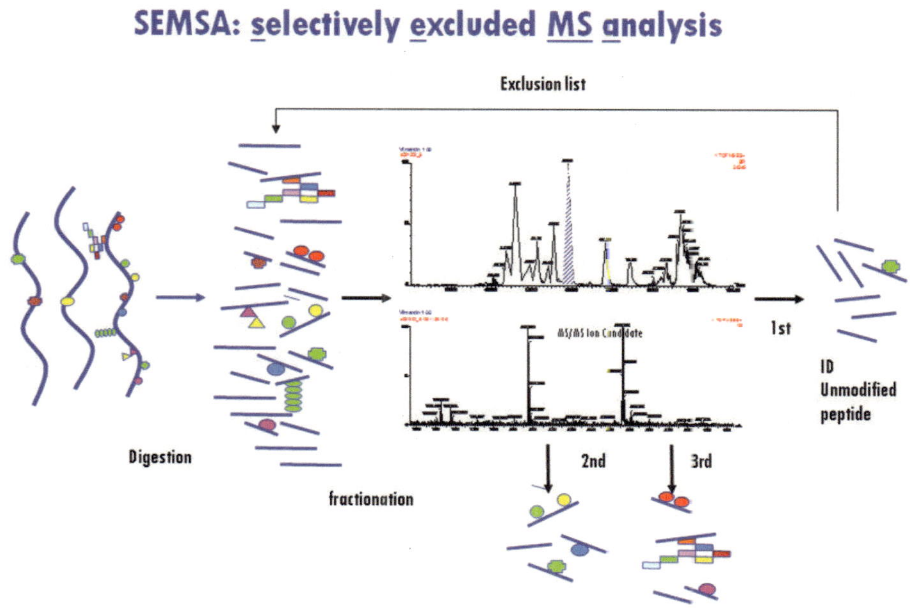

Fig. 3. Schematic diagram of SEMSA to enrich the low abundant modified peptides by excluding abundant unmodified peptides obtained in the first identification in the 2nd and 3rd runs.

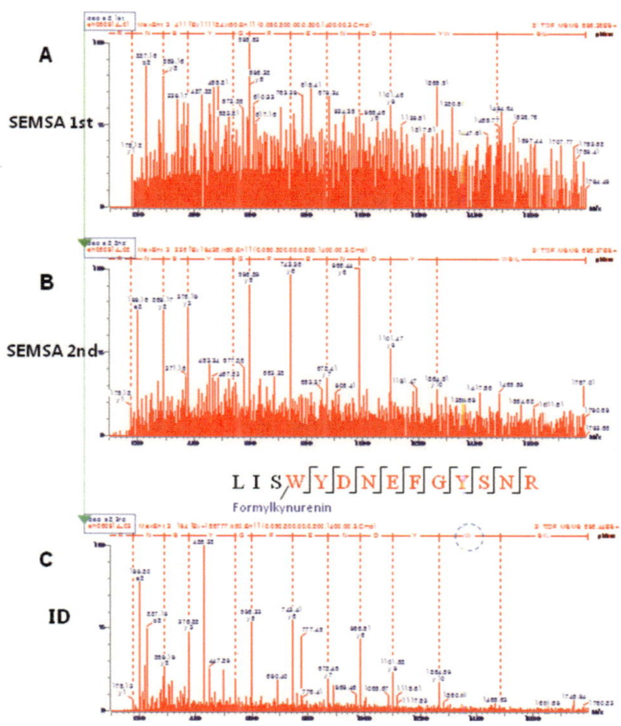

Fig. 4. Comparison of MS/MS spectra of chosen low-density peptides under without (A) and with (B,C) SEMSA strategy (cited from Seo *et al.*, 2008).

Combination of 2D-PAGE for separating modified populations and MS/MS analysis using SEMSA, makes it possible to identify low abundant modified peptides and to raise the identified peptide coverage nearly over 90%.

2.2.3 Bioinformatic tools for identifying multiple and novel modifications

The types and sites of PTMs in a protein vary widely. Although MS allows rapid identification of many types of PTMs, data analysis and interpretation of MS/MS spectra for identification of PTMs remain a major challenge. Most of the available search tools accept only a few types of PTMs as input. We have developed interpretative tools called MOD^i (Na *et al.*, 2008) and DBond (Choi *et al.*, 2010) for rapidly interpreting tandem mass spectra of peptides with all known types of PTMs simultaneously without limiting a multitude of modified sites.

Early approaches to PTM identification using MS/MS involved exhaustive searches of all possible combinations of PTMs for each peptide from a protein database (Eng *et al.*, 1994; Perkins *et al.*, 1999). Because the search space grows exponentially as the number of PTMs increases, these early approaches performed a restrictive search that took into account only a few types of PTMs during data analysis, ignoring all others. Investigators were obliged to guess the PTMs expected to exist in a sample prior to a search, and many potentially important PTMs may have been overlooked. A few tools were recently introduced for blind PTM search. MS-Alignment (Tsur *et al.*, 2005) predicts PTMs expected in a sample by spectral alignment between a database peptide and a spectrum followed by InsPecT search (Tanner *et al.*, 2005). ModifiComb (Savitski *et al.*, 2006) introduced a ΔM histogram between unassigned spectra and base peptides found in a database. These blind approaches predict PTMs based on the frequency of mass shifts (indicating potential PTMs) in a sample. Thus, they all have the intrinsic weakness of missing rare or infrequently observed PTMs that might provide important clues to understanding the function of a protein. Although many approaches have been developed to take into account several types of PTMs, most of them assume that there will be a single variable PTM per peptide and ignore peptides with multiple modifications. MOD^i (pronounced "mod eye") is essentially a sequence tag approach (Mann *et al.*, 1994; Tabb *et al.*, 2003) (Fig. 6). It constructs a partial sequence of a peptide from an MS/MS spectrum using *de novo* sequencing. MOD^i differs from previous approaches in that it simultaneously uses multiple sequence tags derived from a spectrum by introducing a notion of a *tag chain*, a combination structure of multiple sequence tags. A tag chain offers an effective localization of modified regions within a spectrum and thus allows rapid identification of multiple PTMs in a peptide, obviating search space explosion by inspecting PTMs only in the modified regions of a peptide.

The tag chain algorithm resists *de novo* sequencing errors, whereas most tag-based approaches depend critically on good *de novo* interpretations. This approach is scalable and performs well even when more than 900 types of modification are considered and the number of potential PTMs in a peptide increases. Compared with established tools, MOD^i reliably identifies a greater variety of modification types in multiply modified peptides and even detects modifications of low abundance.

Another new algorithm called "DBond" analyzes disulfide linked peptides based on specific features of disulfide bonds (Choi *et al.*, 2010). Identifying the sites of disulfide bonds in a protein is essential for thorough understanding of a protein's tertiary and quaternary structures and its biological functions. Disulfide linked peptides are usually identified indirectly by labeling free sulfhydryl groups with alkylating agents, followed by chemical

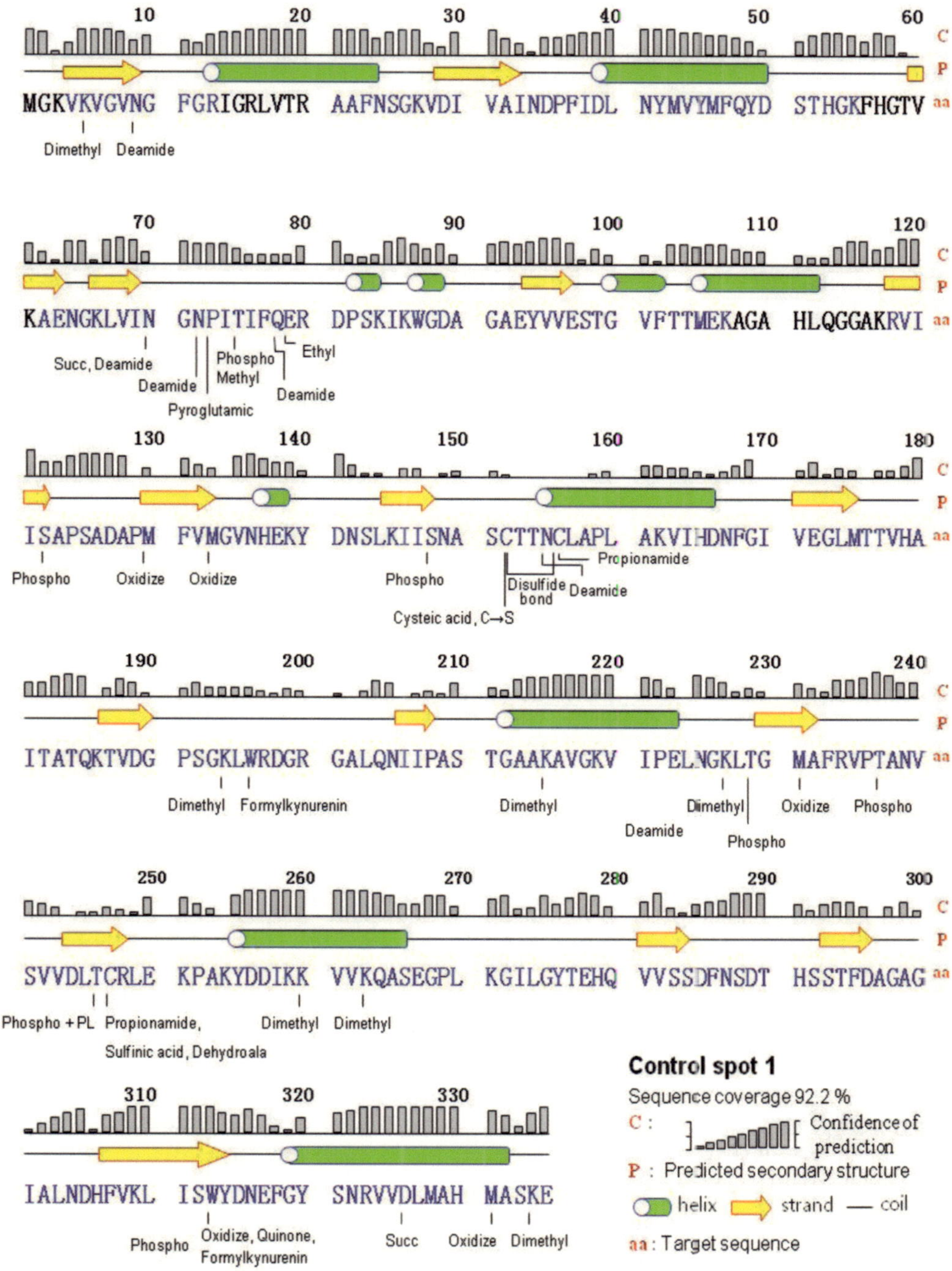

Fig. 5. One example of comprehensive identifications of PTMs in cellular GAPDH (cited from Seo *et al.*, 2008).

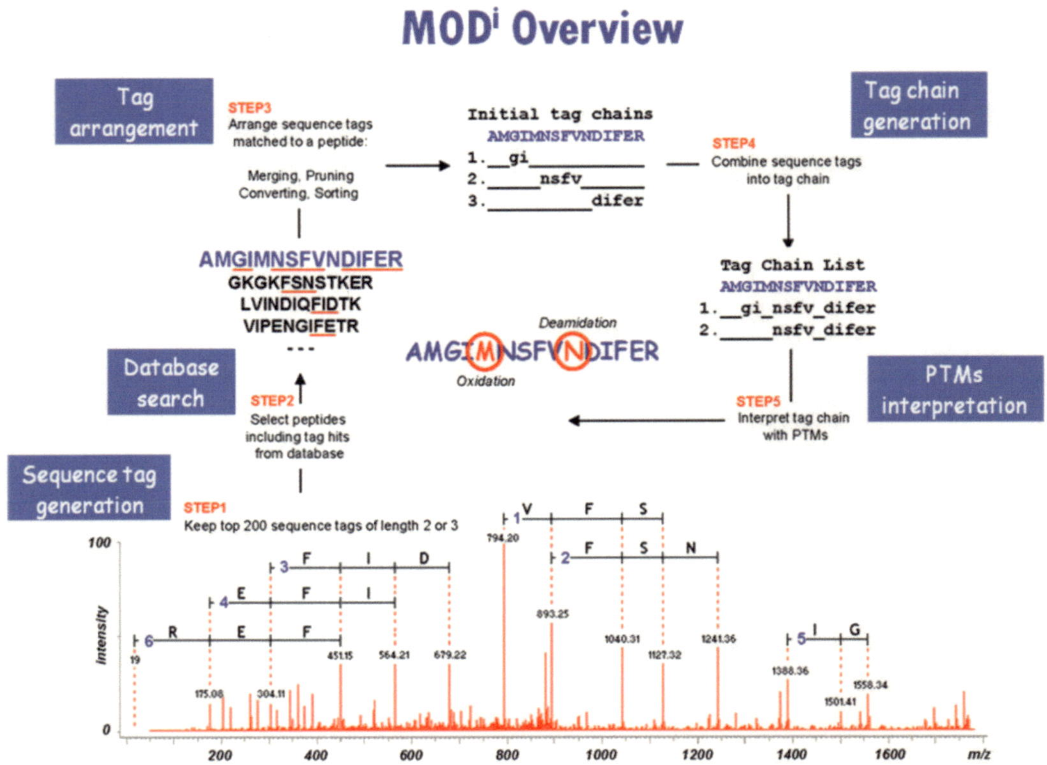

Fig. 6. Overview of MODi algorithm (cited from Na *et al.*, 2008).

reduction and mass spectral comparison or by detecting the expected masses of disulfide linked peptides on mass scan level. However, these approaches for determination of disulfide bonds become ambiguous when the protein is highly bridged and modified. For accurate identification of disulfide linked peptides, we developed an algorithmic solution for the analysis of MS/MS spectra of disulfide bonded peptides under non-reducing condition. To determine disulfide linked sites, DBond takes into account fragmentation patterns of disulfide linked peptides in nucleoside diphosphate kinase (NDPK) as a model protein, considering fragment ions including cysteine, cysteine thioaldehyde (-2 Da), cysteine persulfide (+32 Da) and dehydroalanine (-34 Da). Using this algorithm, we successfully identified about a dozen novel disulfide bonds in a hexa EF-hand calcium binding protein secretagogin and in methionine sulfoxide reductase. We believe that DBond, which takes into account disulfide bond fragmentation characteristics and post-translational modifications, offers a novel approach for automatic identification of unknown disulfide bonds as well as their sites in proteins from MS/MS spectra (Choi *et al.*, 2010).

2.2.4 Examples of multiply modified proteins

Employing SEMSA, a sensitive mass spectrometric method for detecting low abundant protein modifications, and MODi and DBond algorithm for searching for unknown modifications of separated protein on 2D-PAGE, we characterized the nature as well as the relative abundances of these hitherto unknown Cys modifications in cellular GAPDH purified on 2D-PAGE (Jeong *et al.*, 2011). We found unexpected mass shifts at active site Cys residue (ΔM = -16, -34 and +64 Da) in addition to those of previously known oxidation products including sulfinic and sulfonic acids, and disulfide bonds. Similar changes were also found in other ROS-sensitive proteins including NDPK A, PRX6 and mitochondrial proteins. Mass differences of -16, -34 and +64 Da are presumed to reflect the conversion of Cys to Ser, DHA and Cys-SO_2-SH respectively. The plausible pathways leading to their formation from Cys were deduced from the distribution of the disulfide bonds and were confirmed in model systems by analyzing three dimensional protein structures and by employing model chemical reactions. Also sulfenic and sulfinic acids were detected as acrylamide adducts (ΔM = +87 and +103 Da) in samples on SDS-PAGE. These findings suggest that diverse modifications of redox-active Cys can be generated by ROS. These findings of unknown modifications are due to the sensitive mass spectrometric method, SEMSA, and unrestricted PTM search tool, MOD^i and DBond. These strategies should lead to identification of many unknown modifications which doubtless occur in various signaling pathways and in health and disease.

Combining 2D-PAGE for separating heterogenous population of one protein, SEMSA for analyzing low abundant modified peptide with MS/MS and MOD^i for searching algorithm to identify novel and multiple modifications, makes it possible to identify multiple and low abundant modifications of protein comprehensively.

3. Conclusion

Many protein modifications including amino acid substitutions, alternative splicing and post-translational modifications regulate the various biological functions of proteins by altering the protein-protein interactions, protein localization and protein activity. Understanding the relationship between protein modifications and their physiological activities will help define the principles on biological regulation. To demonstrate the exact protein function and its regulation, comprehensive identifications of low abundant and unexpected modification is essential. For the comprehensive identification of PTMs, peptide sequencing using tandem MS can give the most effective results. However, identification of PTMs does not simple as protein identification using MS. The enrichment of low abundant PTMs is necessary. Because most enrichment approaches of many type of PTMs are insufficient, we suggest the use of a combination of approaches: first separating the heterogeneous populations of a protein with multiple modifications on 2D-PAGE, then comprehensively identifying the PTMs employing SEMSA of MS/MS for detecting low abundant modifications and raising peptide coverage up to 100%, and searching MS/MS spectra using searching algorithm MOD^i for multiple and unexpected modifications. With these combined powerful proteomic tools, it will be possible to discover hitherto unknown PTMs in proteins, that influence various signaling pathways. This will hold the key for fully understanding the roles of proteins in biological and pathological processes in molecular level.

4. Acknowledgment

This work was supported by National Research Foundation in Korea, through the Center for Cell Signaling Research and Drug Discovery Research (CCS & DDR, R15-2006-020) at Ewha Womans University from the National Core Research Center (NCRC) program, and by WCU project (R31-2008-000-10010-0). Lee JJ was supported by the Brain Korea 21 project.

5. References

Aldini G, Carini M, Vistoli G, Shibata T, Kusano Y, Gamberoni L, Dalle-Donne I, Milzani A, Uchida K. (2007) Identification of actin as a 15-deoxy-Delta12,14-prostaglandin J2 target in neuroblastoma cells: mass spectrometric, computational, and functional approaches to investigate the effect on cytoskeletal derangement. *Biochemistry* 46(10), 2707-2718.

Blomster HA, Imanishi SY, Siimes J, Kastu J, Morrice NA, Eriksson JE, Sistonen L. (2010) In vivo identification of sumoylation sites by a signature tag and cysteine-targeted affinity purification. *J Biol Chem* 285(25), 19324-19329.

Bruderer R, Tatham MH, Plechanovova A, Matic I, Garg AK, Hay RT. (2011) Purification and identification of endogenous polySUMO conjugates. *EMBO Rep* 12(2), 142-148.

Choi S, Jeong J, Na S, Lee HS, Kim HY, Lee KJ, Paek E. (2010) A new algorithm for the identification of intact disulfide linkages based on fragmentation characteristics in tandem mass spectra. *J Proteome Res* 9(1), 626-635.

Choudhary C, Kumar C, Gnad F, Nielsen ML, Rehman M, Walther TC, Olsen JV, Mann M. (2009) Lysine acetylation targets protein complexes and co-regulates major cellular functions. *Science* 325(5942), 834-840.

Danielsen JM, Sylvestersen KB, Bekker-Jensen S, Szklarczyk D, Poulsen JW, Horn H, Jensen LJ, Mailand N, Nielsen ML. (2011) Mass spectrometric analysis of lysine ubiquitylation reveals promiscuity at site level. *Mol Cell Proteomics* 10(3), M110.003590.

Dephoure N, Zhou C, Villén J, Beausoleil SA, Bakalarski CE, Elledge SJ, Gygi SP. (2008) A quantitative atlas of mitotic phosphorylation. *Proc Natl Acad Sci* 105(31), 10762-10767.

Eng JK, McCormack AL, Yates JR. (1994) An approach to correlate tandem mass spectral data of peptides with amino acid sequences in a protein database. *J Am Soc Mass Spectrom* 5, 976–989.

Galisson F, Mahrouche L, Courcelles M, Bonneil E, Meloche S, Chelbi-Alix MK, Thibault P. (2011) A novel proteomics approach to identify SUMOylated proteins and their modification sites in human cells. *Mol Cell Proteomics* 10(2), M110.004796.

Giron P, Dayon L, Sanchez JC. (2011) Cysteine tagging for MS-based proteomics. *Mass Spectrom Rev* 30(3), 366-395.

Gunaratne R, Braucht DW, Rinschen MM, Chou CL, Hoffert JD, Pisitkun T, Knepper MA. (2010) Quantitative phosphoproteomic analysis reveals cAMP/vasopressin-dependent signaling pathways in native renal thick ascending limb cells. *Proc Natl Acad Sci* 107(35), 15653-15658.

Guan KL, Yu W, Lin Y, Xiong Y, Zhao S. (2010) Generation of acetyllysine antibodies and affinity enrichment of acetylated peptides. *Nat Protoc* 5(9), 1583-1595.

Hwang NR, Yim SH, Kim YM, Jeong J, Song EJ, Lee Y, Lee JH, Choi S, Lee KJ. (2009) Oxidative Modifications of Glyceraldehyde-3-Phosphate Dehydrogenase Play a Key Role in Its Multiple Cellular Functions. *Biochem J* 423(2), 253-264.

Jamie D. Dunn, Gavin E. Reid, Merlin L. Bruening (2010) Techniques for phosphopeptide enrichment prior to analysis by mass spectrometry. *Mass Spectrometry Reviews* 29 (1), 29–54.

Jeong J, Jung Y, Na S, Jeong J, Lee E, Kim MS, Choi S, Shin DH, Paek E, Lee HY, Lee KJ. (2011) Novel oxidative modifications in redox-active cysteine residues. *Mol Cell Proteomics* 10(3), M110.000513.

Kim HJ, Song EJ, Lee KJ. (2002) Proteomic analysis of protein phosphorylations in heat shock response and thermotolerance. *J Biol Chem* 277(26), 23193-23207.

Kim S, Na S, Sim JW, Park H, Jeong J, Kim H, Seo Y, Seo J, Lee KJ, Paek E. (2006) MODi : a powerful and convenient web server for identifying multiple post-translational peptide modifications from tandem mass spectra. *Nucleic Acids Research* 34, W258–W263.

Kim SC, Sprung R, Chen Y, Xu Y, Ball H, Pei J, Cheng T, Kho Y, Xiao H, Xiao L, Grishin NV, White M, Yang XJ, Zhao Y (2006). Substrate and functional diversity of lysine acetylation revealed by a proteomics survey. *Mol Cell* 23(4), 607–618.

Kim YM, Song EJ, Seo J, Kim HJ, Lee KJ. (2007) Proteomic analysis of tyrosine phosphorylations in vascular endothelial growth factor- and reactive oxygen species-mediated signaling pathway. *J Proteome Res* 6(2), 593-601.

Kim YM, Seo J, Kim YH, Jeong J, Joo HJ, Lee DH, Koh GY, Lee KJ. (2007) Systemic analysis of tyrosine phosphorylated proteins in angiopoietin-1 induced signaling pathway of endothelial cells. *J Proteome Res* 6(8), 3278-3290.

Lee E, Jeong J, Kim SE, Song EJ, Kang SW, Lee KJ. (2009) Multiple Functions of Nm23-H1 are Regulated by Oxido-Reduction System. *PLoS ONE* 4(11), e7949.

Leonard SE, Reddie KG, Carroll KS. (2009) Mining the thiol proteome for sulfenic acid modifications reveals new targets for oxidation in cells. *ACS Chem Biol* 4(9), 783-799.

Mann M, Wilm M. (1994) Error-tolerant identification of peptides in sequence databases by peptide sequence tags. *Anal Chem* 66, 4390–4399.

Manzano C, Abraham Z, López-Torrejón G, Del Pozo JC. (2008) Identification of ubiquitinated proteins in Arabidopsis. *Plant Mol Biol* 68(1-2), 145-158.

Mendez D, Hernaez ML, Diez A, Puyet A, Bautista JM. (2010) Combined proteomic approaches for the identification of specific amino acid residues modified by 4-hydroxy-2-nonenal under physiological conditions. *J Proteome Res* 9(11), 5770-5781.

Na S, Jeong J, Park H, Lee KJ, Paek E. (2008) Unrestrictive identification of multiple post-translational modifications from tandem mass spectrometry using an error-tolerant algorithm based on an extended sequence tag approach. *Mol Cell Proteomics* 7(12), 2452-2463.

Olsen JV, Vermeulen M, Santamaria A, Kumar C, Miller ML, Jensen LJ, Gnad F, Cox J, Jensen TS, Nigg EA, Brunak S, Mann M. (2010) Quantitative phosphoproteomics reveals widespread full phosphorylation site occupancy during mitosis. *Sci Signal* 3(104), ra3.

Peng J, Schwartz D, Elias JE, Thoreen CC, Cheng D, Marsischky G, Roelofs J, Finley D, Gygi SP. (2003) A proteomics approach to understanding protein ubiquitination. *Nat Biotechnol* 21(8), 921-926.

Perkin DN, Pappin DJC, Creasy DM, Cottrell JS. (1999) Probability-based protein identification by searching sequence databases using mass spectrometry data. *Electrophoresis* 20(18), 3551–3567.

Reed TT, Pierce WM, Markesbery WR, Butterfield DA. (2009) Proteomic identification of HNE-bound proteins in early Alzheimer disease: Insights into the role of lipid peroxidation in the progression of AD. *Brain Res* 1274, 66-76.

Rikova K, Guo A, Zeng Q, Possemato A, Yu J, Haack H, Nardone J, Lee K, Reeves C, Li Y, Hu Y, Tan Z, Stokes M, Sullivan L, Mitchell J, Wetzel R, Macneill J, Ren JM, Yuan J, Bakalarski CE, Villen J, Kornhauser JM, Smith B, Li D, Zhou X, Gygi SP, Gu TL,

Polakiewicz RD, Rush J, Comb MJ. (2007) Global survey of phosphotyrosine signaling identifies oncogenic kinases in lung cancer. *Cell* 131(6), 1190-1203.

Roe MR, Xie H, Bandhakavi S, Griffin TJ. (2007) Proteomic mapping of 4-hydroxynonenal protein modification sites by solid-phase hydrazide chemistry and mass spectrometry. *Anal Chem* 79(10), 3747-3756.

Rush J, Moritz A, Lee KA, Guo A, Goss VL, Spek EJ, Zhang H, Zha XM, Polakiewicz RD, Comb MJ. (2005) Immunoaffinity profiling of tyrosine phosphorylation in cancer cells. *Nat Biotechnol* 23(1), 94-101.

Savitski MM, Nielsen ML, Zubarev RA. (2006) ModifiComb, a new proteomic tool for mapping substoichiometric post-translational modifications, finding novel types of modifications, and fingerprinting complex protein mixtures. *Mol Cell Proteomics* 5(12), 935–948.

Seo J, Jeong J, Kim YM, Hwang N, Paek E, Lee KJ. (2008) A strategy for Comprehensive Identification of Post-Translational Modifications in Cellular Proteins, Including Low Abundance Modifications: Application to Glyceraldehyde-3-phosphate Dehydrogenase. *J Proteome Res* 7(2), 587-602.

Seo YH, Carroll KS. (2009) Profiling protein thiol oxidation in tumor cells using sulfenic acid-specific antibodies. *Proc Natl Acad Sci* 106(38), 16163-16168.

Tabb DL, Saraf A, Yates JR. (2003) GutenTag: high-throughput sequence tagging via an empirically derived fragmentation model. *Anal Chem* 75(23), 6415–6421.

Tanner S, Shu H, Frank A, Wang LC, Zandi E, Mumby M, Pevzner PA, Bafna V. (2005) InsPecT: identification of posttranslationally modified peptides from tandem mass spectra. *Anal Chem* 77(14), 4626–4639.

Thingholm TE, Jensen ON, Larsen MR. (2009) Analytical strategies for phosphoproteomics. *Proteomics* 9(6), 1451–1468.

Tsur D, Tanner S, Zandi E, Bafna V, Pevzner PA. (2005) Identification of post-translational modifications by blind search of mass spectra. *Nat Biotechnol* 23(12), 1562–1567.

Tyther R, Ahmeda A, Johns E, McDonagh B, Sheehan D. (2010) Proteomic profiling of perturbed protein sulfenation in renal medulla of the spontaneously hypertensive rat. *J Proteome Res* 9(5), 2678-2687.

Vosseller K, Trinidad JC, Chalkley RJ, Specht CG, Thalhammer A, Lynn AJ, Snedecor JO, Guan S, Medzihradszky KF, Maltby DA, Schoepfer R, Burlingame AL. (2006) O-linked N-acetylglucosamine proteomics of postsynaptic density preparations using lectin weak affinity chromatography and mass spectrometry. *Mol Cell Proteomics* 5(5), 923-934.

Wang Z, Udeshi ND, Slawson C, Compton PD, Sakabe K, Cheung WD, Shabanowitz J, Hunt DF, Hart GW. (2010) Extensive crosstalk between O-GlcNAcylation and phosphorylation regulates cytokinesis. *Sci Signal* 3(104), ra2.

Weinert BT, Wagner SA, Horn H, Henriksen P, Liu WR, Olsen JV, Jensen LJ, Choudhary C. (2011) Proteome-wide mapping of the Drosophila acetylome demonstrates a high degree of conservation of lysine acetylation. *Sci Signal* 4(183), ra48.

5

Tandem Mass Spectrometry and Glycoproteins

Pavlina Dolashka
Institute of Organic Chemistry with Centre of Phytochemistry,
Bulgarian Academy of Sciences, Sofia,
Bulgaria

1. Introduction

Glycoproteins and glycopeptides are organic compounds that are composed of both polypeptide and carbohydrate chains bonded together. For many years glycoproteins and glycopeptides have been a subject of interest; however, recently they have aroused the interest of biochemists and biologists from a wide range of fields. This increased interest is partly due to the fact that glycoproteins were discovered to be abundant in living organisms and glycoproteins appear in nearly every biological process studied. Glycoproteins help many systems within the human body to function properly and optimally, and deficiencies can be responsible for a whole spectrum of diseases, conditions and ailments. Examples of glycoproteins include antibodies that interact directly with antigens. Major histocompatibility complex molecules that interact with the T-cells as part of an adaptive immune response and hemocyanins (Hcs) fudge molecules that transport oxygen in mollusks and arthropods.

Detailed knowledge of protein glycosylation at the proteome level is becoming an important aspect of post-genomic research. Moreover, glycobiology seeks to identify the molecular structure of glycopeptides and to further explore the function of such peptides in relation to other cells and molecules in the body. By determining how glycopeptides are structured and in order to better understand how they work, researchers working in the field of glycobiology may be able to produce treatments and therapies that improve health and may prolong life. Therefore, the development and application of different analysis techniques will increase the knowledge of their structure and function. A number of reviews have been published in the last few years on analytical methods, including chromatography, electrophoresis and MS, for the characterization of glycans and glycoproteins (Zaia, 2004; Morelle et al. 2005; Dell and Morris 2010), and on general MS-based proteome and peptide analysis methods (Medzihradszky, 2005; Domon et al. 2006; Froehlich et al. 2011).

Mass spectrometric (MS) techniques play a key role in glycoprotein and glycan analysis, to study protein glycosylation at the glycopeptide level. Therefore, MS is becoming an increasingly important aspect in proteomics. Current informatics tools are designed for large, high-throughput mass-spectrometry datasets.

One of the analytical instruments used in laboratories a tandem mass spectrometer. This instrument can analyze numerous compounds, such as those in body fluids and in the environment. Enrichment and separation techniques for glycoprotein and glycopeptide from complex (glyco-) protein mixtures and digests are summarized below. In addition to detection by mass spectrometry, the microarray platform has also become an essential tool to characterize glycan structures and to study glycosylation-related biological interactions; here one uses probes as a means to interrogate the spotted or captured glycosylated molecules on the arrays.

2. Glycoproteomics

Almost all secreted and membrane-associated proteins are glycosylated through post-translational modification. Protein glycosylation is one of the most important post-translational modifications in eukaryotes, but remains poorly investigated. N-linked glycans, with a common monosaccharide core and specific attachment motif, appear to be amenable to analysis, but significant difficulties remain.

The key structural issues of glycoproteomics are protein identification, glycosylation site determination, and glycan profiling at individual attachment sites (Tadjiri et al. 2005; Harvey, 2006). The structure of glycoproteins and glycopeptides is composed of a peptide chain with one or more carbohydrate moieties which constitute from less than 1% to more than 80 % of the total protein mass. Glycoproteins usually exist as complex mixtures of glycosylated variants (glycoforms). Glycosylation occurs in the endoplasmic reticulum (ER) and Golgi compartments of the cell and involves a complex series of reactions catalyzed by membrane-bound glycosyltransferases and glycosidases.

Based on the saccharide chains, referred to as glycans, two main classes of glycoprotein structures are known: N-glycosylation, in which the oligosaccharide is attached to an asparagine residue, and O-glycosylation, in which the oligosaccharide is attached to a serine or threonine residue. The O-linked glycans consist of N-acetylgalactosamine attached to the O-terminus of a threonine (Thr) or serine (Ser) residue (Fig. 1). The most common type of O-linked glycans contain an initial GalNAc residue (or Tn epitope); they are commonly referred to as mucin-type glycans. Other O-linked glycans include glucosamine, xylose, galactose, fucose, or mannose as the initial sugar bound to the Ser/Thr residues.

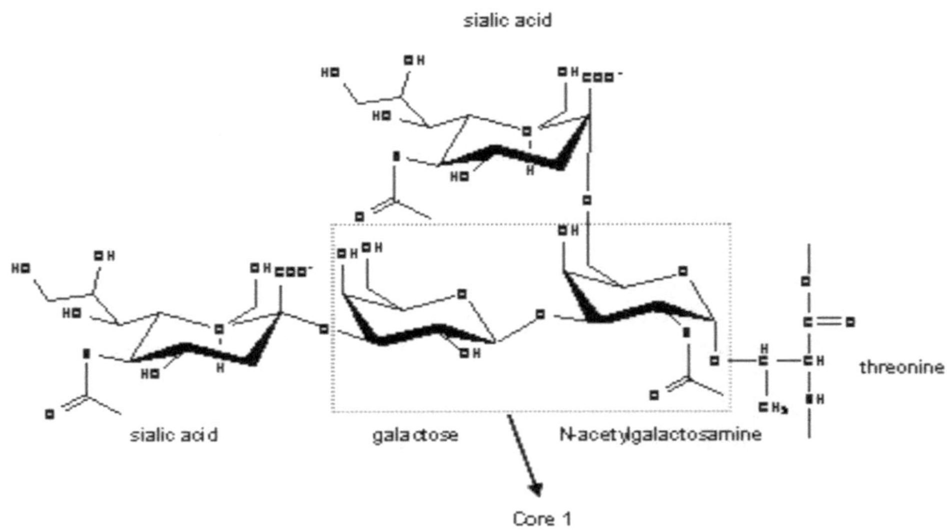

Fig. 1. O-linked glycan contain terminal N- acetyl neuraminic acid attached to a threonine residue.

The other class of glycoproteins are the N-linked glycans. These molecules consist of an N-acetylglucosamine bond to the amide nitrogen of an asparagine molecule, where X can be any amino acid (-X-Asn-X-Thr-) (Fig.2A,B).

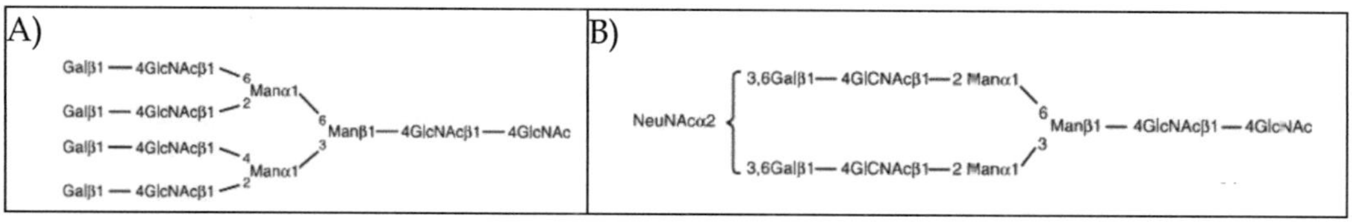

Fig. 2. A) Structures of a glycan at m/z 2393[M+Na]+ determined in matrix 2,5-dihydroxybenzoic acid (DHB) B). Structure of an acidic glycan at m/z 1932 [M-H]- determined in matrix 6-aza-2-thiothymine and 2,4,6-trihydroxyacetophenone (THAP).

Because carbohydrates and proteins by themselves serve in a vast number of biological functions, such as structure, enzymes, protection, carriers, immunological, defense, inhibitors, reproduction, etc. several techniques and analytical methods including chromatography, electrophoresis and MS-based proteome and peptide analysis are applied for the characterization of glycans and glycoproteins (Wada 2008; Sandra et al. 2004, 2007).

Another of these methods is nuclear magnetic resonance spectroscopy (NMR). The method is used to obtain information about the structure and dynamics of proteins and glycans. However, mass spectrometry has several advantages over NMR with respect to analysis of H/D exchange reactions: much less material is needed, the concentration of protein or glycan can be very low (as low as 0.1 uM), the size limit is much greater, and data can usually be collected and interpreted much more quickly.

A very important advantage is that mass spectrometry offers different techniques and approaches. Tandem mass spectrometry of the glycopeptides, isolated from tryptic digests of glycoproteins can be performed in a rapid and sensitive manner. Electrospray ionization (ESI) or matrix-assisted laser desorption/ionization (MALDI) mass spectrometry can be used. In the case of MALDI, profiling of oligosaccharides is achieved more appropriately by the linear time-of-flight (TOF) mode than in the reflection mode. Robust and reliable identification of proteins and the determination of the attachment sites often requires multiple-stage tandem mass spectrometry (Wada 2008).

Combining gel and capillary electrophoresis, nano-LC and mass spectrometry, the elucidation of post-translational modifications of *Trichoderma reesei* cellobiohydrolase I (Sandra et al. 2004 a) of hemocyanin isolated from marine snails *Rapana venosa* (Sandra et al.2007, Dolashka-Angelova et al. 2004) and of superoxide dismutase (Cu/Zn-SOD) from *K. marxianus* NBIMCC 1984 yeast (Dolashka et al. 2010) were analysed.

The strategy used for the characterisation of RvH1 N-glycosylation combines two main approaches (Fig. 3), where the intact RvH1 was subjected to PNGase F digestion and/or tryptic digestion. The first approach includes reduction and alkylation of RvH1 followed by PNGase F digestion of oligosaccharires from the protein. The N-glycans were analysed by MALDI-TOF and CE-MS and followed by 8-aminopyrene-1,3,6-trisulphonate (APTS) and 3-aminopyrazole (3-AP) labelling. Derivatisation of the oligosaccharides with APTS permits high-resolution CE, allows the simultaneous detection of uncharged and charged glycans and provides easily interpretable spectra.

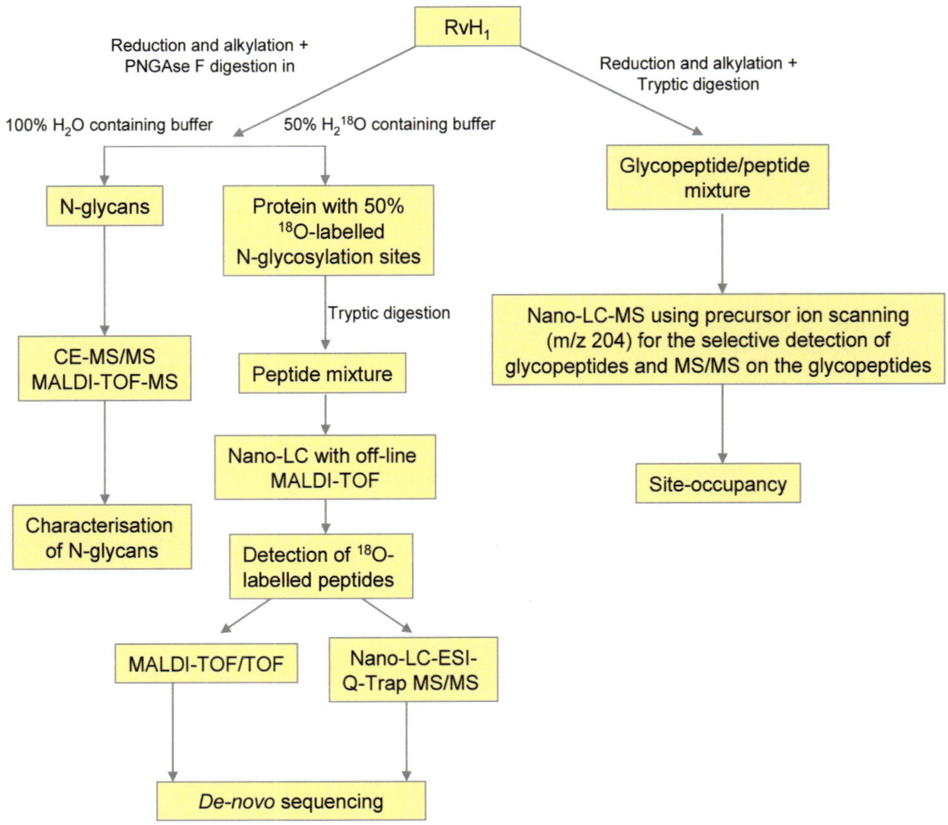

Fig. 3. Strategy used for the characterisation of the oligosaccharide structure of RvH_1.

The second approach includs reduction and alkylation followed by tryptic digestion of the glycoprotein. The N-glycopeptides were analysed by MALDI-TOF and nano-LC-MS or nano-LC-ESI- techniques. Since the protein sequence of RvH_1 is currently unknown, de novo MS sequencing had to be performed on the glycopeptides. Proteomic techniques, such as HPLC coupled to tandem mass spectrometry (LC-MS/MS), have proven to be useful for the identification of specific glycosylation sites of glycoproteins (glycoproteomics).

However, glycosylation sites of glycopeptides produced by trypsinization of complex glycoprotein mixtures are particularly difficult to identify because glycopeptides are usually present in relatively low abundance (2% to 5%) in peptide mixtures compared to the non-glycosylated peptides, and because the sugar fragments in the MS spectrum often dominate the peptide fragments due to glycosidic bonds being more labile than peptide bonds. Some approaches have been developed to overcome this problem, mainly based on multistage MS (Bateman et al. 1998, Demelbauer et al. 2004), electron capture dissociation (ECD) (Håkansson et al. 2001, 2003) or electron transfer dissociation (ETD) (Hogan et al. 2005).

3. Tandem mass spectrometry technique for glycoprotein and glycopeptide analysis

An important application using tandem mass spectrometry is in protein identification. The simplest form of the technique combines two mass spectrometers. One of these instruments is a Matrix-Assisted-Laser-Desorption/Ionization-Time-of-Flight-Mass-Spectrometer (MALDI-TOF/TOF-MS), where the first mass spectrometer is used to select a single (precursor) mass from the MS spectrum of a mixture of the precursor. After collisional activation (CA) or collision-induced dissociation (CID), the second mass spectrometer is used to separate the fragments ions according to their masses. The resulting "MS/MS" spectrum consists only of production ions from the selected precursor (Fig.4).

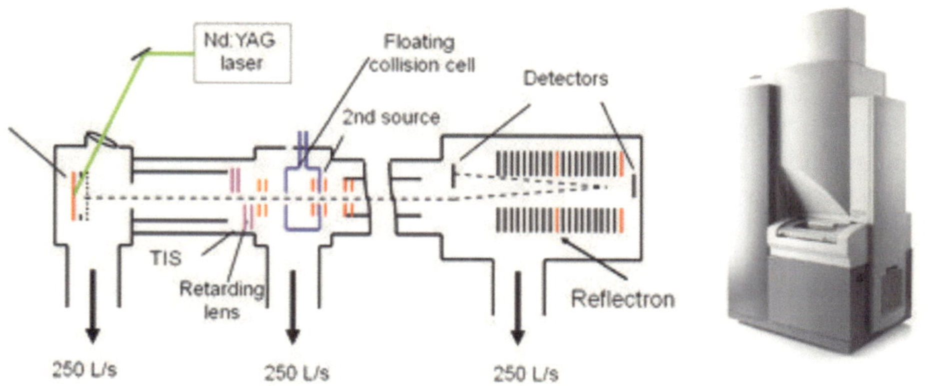

Fig. 4. Matrix-Assisted-Laser-Desorption/Ionization-Time-Of-Flight-Mass-Spectrometer (MALDI-TOF/TOF-MS).

There are various methods for fragmenting molecules by tandem MS, including:

- collision- induced dissociation (CID),
- electron capture dissociation (ECD),
- electron transfer dissociation (ETD),
- infrared multiphoton dissociation (IRMPD),
- blackbody infrared radiative dissociation (BIRD),
- electron-detachment dissociation (EDD) and
- surface-induced dissociation (SID).

Tandem mass spectrometry of glycopeptides is known as one of the most important tools in structural glycoproteomics. Various tandem MS (MS/MS) techniques for the analysis of glycopeptides as MALDI, MALDI-TOF/TOF-MS or MALDI-quadrupole-TOF were applied and compared with respect to the information they provide on peptide sequence, glycan attachment site and glycan structure. Glycopeptide ionization was performed mainly by CID or ETD.

3.1 Collision-induced dissociation

Early experiments with ESI and collision-induced dissociation (CID) on a triple-quadrupole mass spectrometer have already established several of the key features of CID of glycopeptides (Huddleston et al. 1993; Medzihradszky et. al. 1996). On the basis of this pioneering work, ESI with CID of glycopeptides has become a key tool in glycoproteomics. The potential of nano-ESI with a quadrupole-TOF mass analyzer for the characterization of N- and O-glycopeptides has been shown to be a sensitive tool that provides information on glycan structure, glycan attachment site, and peptide sequence. This method has been successfully applied to the characterization of O-glycosylated peptides carrying the Tn-antigen (GalNAcβ1-), the T-antigen disaccharide, or other slightly more elongated O-glycans based on β-linked GalNAc, attached to serine or threonine residues (Chalabi et al. 2006). Another type of glycosylations, analyzed by nano-ESI–quadrupole–TOF are O-fucosylation (Macek et al. 2001), O-linked N-acetylglucosamine (Vosseller et al. 2006), as well as C-mannosylation (Gonzales de Peredo et al. 2002). The C-linked mannose appeared to be very stable in CID, in contrast to O-glycans and N-glycans.

3.2 Electron-transfer dissociation (ETD)

Similar to the peptide structural information obtained from CD, electron-transfer dissociation has recently emerged as an MS/MS technique complementary to CID and RMPD. Peptide fragmentation is generated through gas-phase electron-transfer eactions from singly charged anions to multiply charged protonated peptides. Singly charged anions are used as vehicle for the electron delivery to the multiply protonated peptides.

Analogous to ECD, dissociation from electron transfer results in peptide backbone fragmentation into c- and z•-type ions. There fore this fragmentation is more useful for the analysis of posttranslational modification (PTM) such as phosphorylation (Syka et al. 2004) and glycosylation (Hogan et al. 2005). The feature makes this technique, together with ECD, a very attractive tool for the localization of the PTM attachment.

4. Sample preparation and characterization

The methods to study glycoproteins, glycopeptides or glycans in MS-based analyzes vary according to the specific research question. In many cases, these methods have been developed with the aim to selectively obtain the N-glycoproteome of a particular sample by MS analysis of the corresponding (deglycosylated) tryptic peptides to identify the underlying proteins. In these types of studies, N-glycosylation sites in tryptic glycopeptides usually are identified by conventional LC-MS/MS or MALDI-MS/MS analysis based on the conversion of Asn to Asp upon treatment with N-glycanase (PNGase), or on the localisation of the remaining GlcNAc-Asn tag upon treatment with endo-N-acetylglucosaminidases.

For the isolation of glycoproteins or glycopeptides by affinity chromatography various lectins have been used so far. Lectin chromatography using concanavalin A (Con A) has been reported for the enrichment of N-glycoproteins from diverse sources (Nasia et al. 2009). Enrichment techniques applied in combination with advanced MS/MS methods for the direct analysis of intact glycopeptides to obtain sequence information of both the glycan and the peptide moiety have been less commonly used. Now approaches using fragmentation techniques in glycopeptide and glycan analyses such as ESI, MALDI and LC/MS/MS–Q-Trap, are very popular to provide information on the peptide and glycan sequences, as well as on the attachment site.

4.1 Characterisation of glycoproteins and glycopeptides

Several methods and techniques have been applied to analyse the oligosaccharide structure of glycoproteins and glycopeptides. Applying both CID and ETD fragmentation techniques sequentially to protonated glycopeptides provides information on the glycan structure (CID) as well as the provide information on the peptide sequence and the glycan attachment site (ETD). The combination of these complementary data sets allows the detailed structural

characterization of glycopeptides species characterization of glycopeptides is performed with different instrumental configurations, such as:
- MALDI-MS/MS
- Liquid chromatography systems compatible with on-line MS of glycopeptides
- ESI-MS/MS
- Capillary electrophoresis with on-line MS

4.1.1 MALDI–MS and MS/MS analyses of glycopeptides

MALDI-MS/MS of glycopeptides has been performed using the following instrumental configurations:
- MALDI-TOF with post source decay (PSD),
- MALDI-TOF/TOF,
- MALDI-quadrupole-TOF, and
- MALDI-IT/TOF MS.

MALDI-TOF/TOF MS of glycopeptides in protonated form has been established using 2,5-dihydroxybenzoic acid (DHB) as a matrix (Uematsu et al. 2005; Wuhrer et al. 2004): the observed fragments result in the MS spectrum. Three different groups of fragment ion signals in MS are observed which provide information on both the peptide and glycan moiety of the glycopeptides. MALDI-TOF/TOF-MS of N-glycopeptides results in all the fragment ions retaining the peptide moiety, as well as in a set of cleavages at or near the innermost N-acetylglucosamine residue. The signals $[Mpep+H]^+$ and $[Mpep+H-17]^+$ usually arising from the cleavage of the side-chain amide bond of the glycosylated asparagine are observed. Fragmentation characteristics are very similar to those observed with MALDI-quadrupole /TOF-MS (Krokhin et al. 2005; Bykova et al. 2006) and MALDI-IT/TOF-MS (Demelbauer et al. 2004; Takemori et al. 2006).

4.1.1.1 MALDI-MS and MS/MS analyses of glycosylation sites and determination of the occupancy at a particular site

Determination of which sites in the glycoprotein are glycosylated and determination of the extent of occupation at each site are generally accomplished by performing trypsinolysis or other degradative reactions, generating a peptide/glycopeptide mixture. Glycopeptides can usually be selectively detected in such mixtures by precursor ion scanning, followed by MS/MS analysis of detected glycopeptides. This approach, however, is only straight forward for proteins with known amino acid sequence. This restriction, which originates from the dominance of the sugar fragments, can be overcome by performing MS/MS or MS^3 on the Y_1 ion (peptide + GlcNAc) generated via in-source fragmentation or MS/MS, respectively (Sandra et al. 2007).

Since the protein sequence of one subunit of *Rapana venosa* hemocyanin (RvH_1) is currently unknown, this approach was not applicable. Therefore, de novo MS sequencing was performed on the glycopeptides, obtained after tryptic digestion of RvH1 (Sandra et al. 2007). The glyco-moiety was removed from the glycopeptide because the sugar fragments often dominate the peptide fragments due to glycosidic bonds being more labile than peptide bonds. This is a phenomenon typically observed when using CID or PSD as fragmentation techniques. To solve this problem electron capture dissociation or electron transfer dissociation were applied. A method of labelling the N-glycosylation sites was used by performing a PNGase F digestion in a buffer containing 50% $H_2{}^{18}O$ (Sandra et al. 2004a). The ^{18}O-labelling step can be performed either prior or subsequent to trypsinolysis. When choosing this approach, it is important to remove the remaining $H_2{}^{18}O$ to prevent unspecific incorporation of the label into all tryptic peptides. Also, trypsin needs to be removed from the hydrolysate. Since $H_2{}^{18}O$ can easily be removed by membrane filtration, labelling the protein prior to the tryptic digestion was preferred. The complexity of the tryptic peptide mixture is reflected in the nano-LC-UV chromatogram. By way of example the MALDI-TOF spectrum corresponding to fraction 54, and an expanded view of the region containing the ^{18}O-labelling peptide ions at 2406.3296, are presented in Fig. 5. Sequence information can be obtained by performing higher-order MS. The peptides containing the glycosylation sites were detected via the 2 Da spacing between the unlabelled and labelled ions using MS (Fig.5).

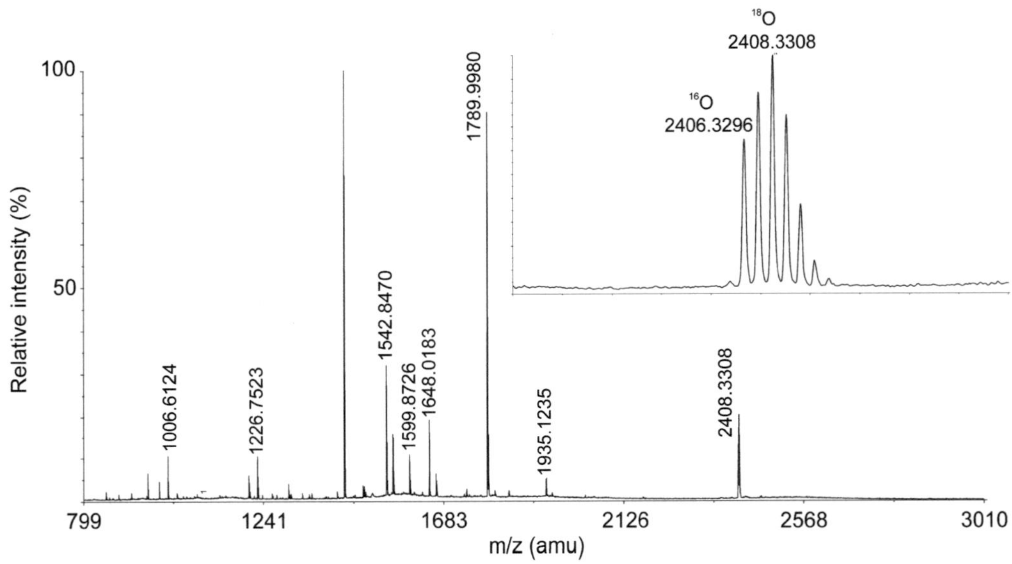

Fig. 5. The nano-LC-MALDI-TOF spectra corresponding to fraction 54 (b) and an expanded view of the regions containing the ^{18}O-labelling peptide ions at m/z 2406.3296 (inset).

Using this approach, 6 glycopeptides were identified from the enormous complexity of the RvH1 tryptic digest. Nano-LC was used as a preceding separation step, and fractions were directly collected onto a MALDI-target.

4.1.2 Characterisation of glycopeptides using MALDI–MS/MS and Q-Trap MS/MS analyses

The structure $Man_5GlcNAc_2$ of the glycan with mass of 1257.4 Da is represented with a specific fragmentation nomenclature in Q-Trap MS/MS spectrum. The MS/MS spectrum of the underivatised singly charged sodium-adduct is shown in Figure 6. The most dominant ions are Y and B that arise from glycosidic cleavages. C and Z ions are also observed. They can be differentiated from the Y and B ions without derivatisation, because of the asymmetrical nature of the molecule. At this stage it is impossible to demonstrate the existence of a branched structure. Many interesting cross-ring cleavage ions are also present as $_{0,2}A_5$, $_{2,4}A_4$, $_{0,3}A_3$, and $_{3,5}A_3$ ions rendue. They are very informative because they indicate a linkage of a trihexose at carbon 6 of a hexose. When the structure is not methylated, minor information can be extracted on the branching from the cross-ring cleavages.

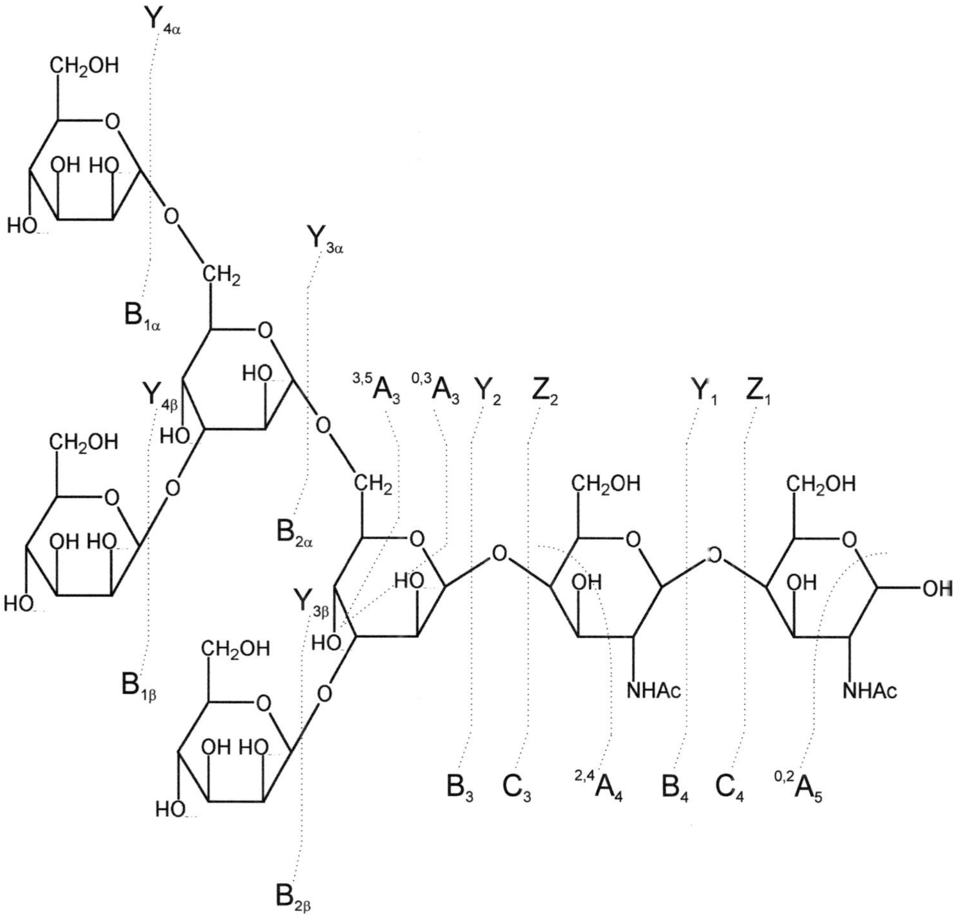

Fig. 6. Structure of the RNase B. Man $_5$GlcNAc$_2$ N-glycan and fragmentation nomenclature.

MALDI–MS/MS and Q-Trap MS/MS have been performed to analyse the carbohydrate structure in the polypeptide chain of Cu/Zn-SOD from *K. marxianus* NBIMCC 1984 yeast (Dolashka-Angelova et al. 2010). The obtained fractions after treatment of the protein with trypsin were separated by HPLC and their amino acid sequences were determined by MALDI-TOF-TOF. One putative linkage site was observed in the sequence [EVWN(I/L)TGNSPNA(I/L)R] of the peptide with a mass of 1773.51 Da. The orcinol/H_2SO_4 test was effectively positive, confirming that one glycopeptide is present in a fraction eluted at 19 min by HPLC. Structure of this glycopeptide was analysed by MALDI–MS/MS and Q-Trap MS/MS.

The amino acid sequence of the peptide chain EVWN(I/L)TGNSPNA(I/L)R was determined by MALDI-MS/MS, based on the singly-charged ions (Fig. 7). The difference between two single ions at m/z 1571.01 and m/z 1773.51 is corresponding to one GlcNAc connected to the peptide, which also demonstrates that the peptide of 1773.51 Da is glycosylated. The on at 1359.23 is the same peptide as y11 [at m/z 1156.74, N(I/L)TGNSPNA(I/L)R], with one GlcNAc being connected to linkage site Asp-Ile/Leu-Thr. Evidently the linkage site (-Asn-Leu/Ile-Thr-) at position 33-35 is the glycosylated one. This linkage site is conserved in several SODs.

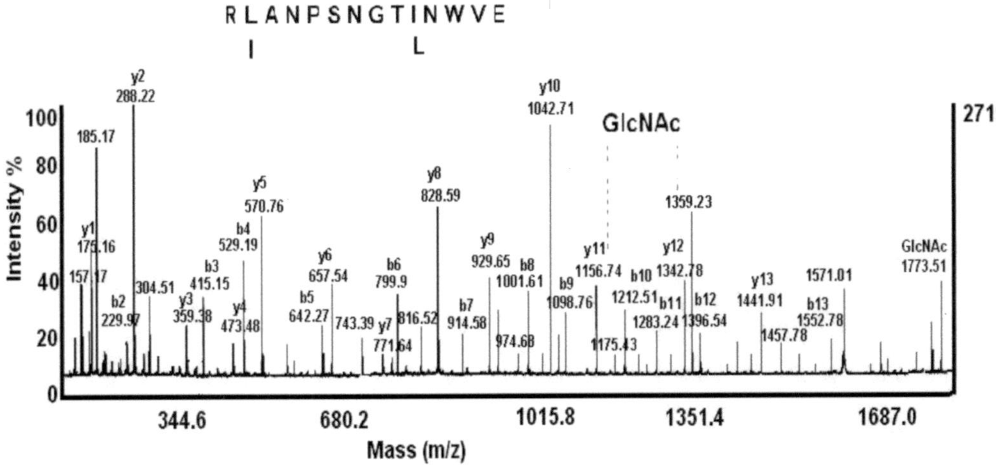

Fig. 7. MALDI-TOF-MS spectra of the N-glycan isolated from Cu/Zn-SOD from *K. marxianus* NBIMCC 1984. 1 µl of a 1:1 sugar-matrix mixture was applied onto the MALDI target. A matrix DHB (10 mg/ml dihydroxybenzoic acid solution in 50% AcN) was used.

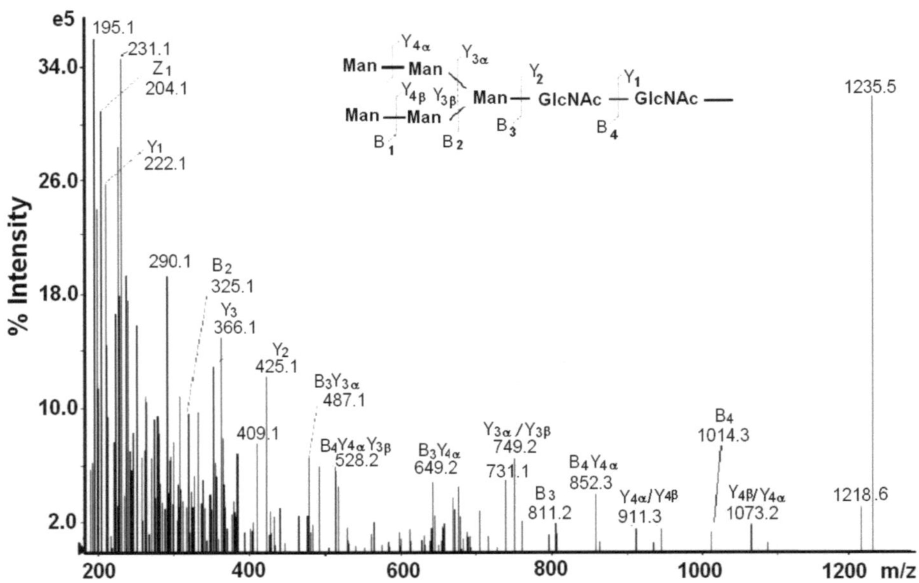

Fig. 8. MS/MS spectra on a Q-Trap mass spectrometer and structure with fragmentation nomenclature of the single charged $[M+Na]^+$ of the glycan at m/z 1235.52, isolated from Cu/Zn-SOD from *K.marxianus* NBIMCC 1984.

After treatment of the peptide with PNGase F a single peak at 1257.3 $[M+Na]^+$ was detected which suggests a uniform oligosaccharide chain. The structure of this carbohydrate chain was determined by Q-Trap MS/MS of the singly-charged ion with a mass of 1235.52 Da $(M+H)^+$ as shown in Figure 8. The sequence can easily be read when considering the Y ions and the combination of B and Y ions (m/z 222.1, 325.1, 425.1, 811.2, 1014.3). The structure, given as inset in the figure is a classical high mannose type of sugar (GlcNAc2, Man5), with a calculated mass of 1234.4 Da.

4.1.3 Electrospray ionization–mass spectrometry (ESI-MS) of glycopeptides

Electrospray is a soft ionization technique and can be performed on solid or liquid samples. Therefore, ESI is typically used to determine the molecular weights of proteins, peptides, and other biological macromolecules. Soft ionization is a useful technique when considering biological molecules of large molecular mass, as it turns the macromolecule being ionized into small fragments.

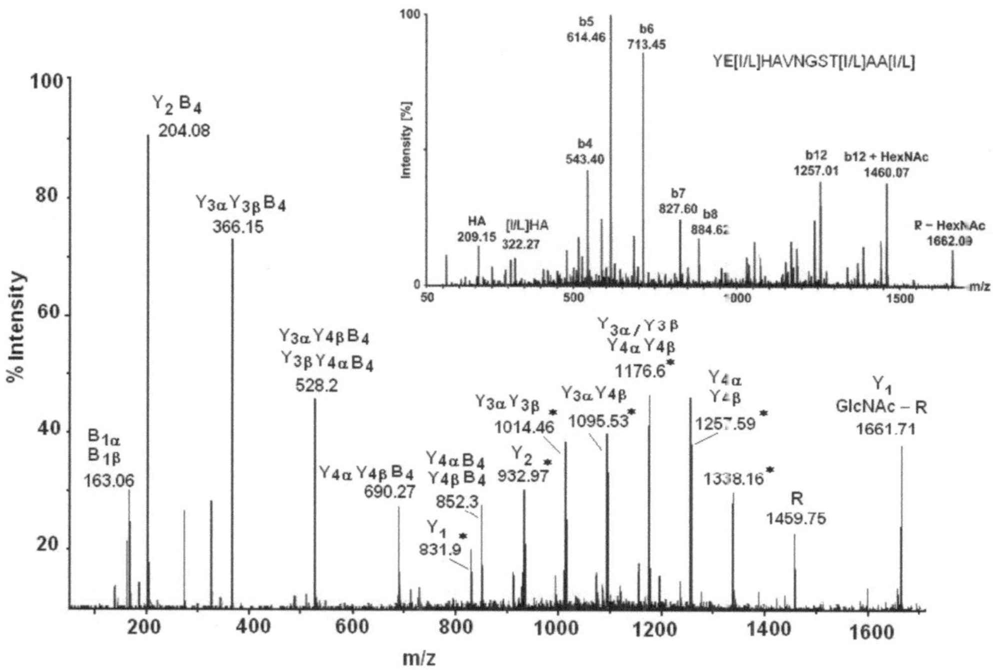

Fig. 9. Fragment spectra of the glycopeptide G1 with determined m/z 1338.16 $[M+2H]^{2+}$. A) MS/MS analysis of the sugar moiety using a collision energy of 37 eV. Annotation of sugar fragments. The insert shows the complete carbohydrate structure and the cleavage points leading to the respective fragments. * - corresponds to the doubly charged ions B) Pseudo-MS/MS/MS experiment (in-source and collision-induced fragmentation) of the peptide moiety (YEXHAVN*GSTXAAX) still carrying one HexNAc (m/z 1662.09, $[M - H]^{+}$), using a cone voltage of 95 V and a collision energy of 80 eV. R-corresponds to the peptide.

ESI and collision-induced dissociation (CID) on a triple-quadrupole mass spectrometer have already established several of the key features of CID of glycopeptides. ESI-MS was applied to analyse the glycopeptides obtained after treatment of Rapana hemocyanin with trypsin and being separated by HPLC (Beck et al. 2007; Sandra et al. 2007).

Six glycopeptides were characterized in detail by analyzing the corresponding HPLC fraction by ESI-MS/MS. The MS/MS spectra of one peptide appearing as a doubly protonated molecule of m/z 1338.16 $[M+2H]^{2+}$ in the MS spectrum itself (not shown) are displayed in Figure 9.

The MS/MS spectrum of the fragment ion at m/z 1661.71, corresponding to the peptide (designated R-, m/z 1459.75), which is N-glycosylated, with a single residual GlcNAc residue, is shon on Figure 9. The spectrum is further dominated by glycan fragments which correspond to a classical mannose type of oligosaccharide, consisting of two N-acetyl-hexosamines (HexNAc) monomers as well as five hexose (Hex) residues. In a pseudo-MS_3 experiment, the sugar side chain was removed by in-source fragmentation and the peptide with one remaining HexNAc residue was further fragmented in the hexapole collision cell. This allowed the sequence analysis of the peptide and the determination of the site to which the sugar chain was attached (Fig. 9, insert).

The amino acid sequence of the peptide was revealed to be YEXHAVN*GSTXAAX. The glycosylation site N* was identified to be part of the typical N-glycosylation motif N-aa-S/T, where aa can be any amino acid except proline. X represents a leucine or isoleucine residue which could not be distinguished by the applied MS methods.

4.1.4 Characterisation of glycopeptides using a Q-Trap LC/MS/MS system

As was described in 4.1.2 and 4.1.3, MS analysis of glycopeptides may be performed after extensive purification, using HPLC fractions analyzed by flow injection analysis on a Q-TOF or ESI-mass spectrometer. In practice, characterisation of glycopeptides is very difficult, because they are often analyzed from complex peptide/glycopeptide mixtures. When such samples are subjected to LC-ESI-MS/MS analysis with collision-induced fragmentation, data evaluation methods which highlight the relevant glycopeptide MS data within the complex overall data set are required. Strategies applied to achieve this goal comprise the generation of diagnostic fragment ions in the MS-mode (without precursor selection) and/or MS/MS-mode (with precursor selection).

The Q-Trap system with its capabilities to perform typical triple quadrupole scans was additionally used and several glycopeptides from RvH1 were identified and characterized (Sandra et al. 2007). Glycopeptides, selectively detected in a proteolytic mixture by the appearance of collisionan induced marker oxonium ions, such as m/z 163 (Hex^+), 204 ($HexNAc^+$) or 366 ($HexHexNAc^+$),, were sequenced. The insert of Figure 9A shows the LC/MS/MS total ion current (TIC) chromatogram of the precursor ion scan (monitoring m/z 204) of the HPLC fraction at time 31.24 min.

The Enhanced Resolution scan (not shown) showed that the glycopeptide with mass 2511.91 eluting at this moment was triply charged at m/z 837.97 $[M+3H]_{3+}$. The precursor ion scan at time 31.24 min is presented in Figure 10 (insert). The MS/MS spectrum is dominated by glycan fragmentation series of Y- and B-ions, according to the Domon/Costello nomenclature.

However, peptide fragmentation (Roepstorff/Biemann cleavages) became more dominant when the collision energy was increased, allowing to deduce the peptide sequence MGQYGD(I/L)STNNTR from the series of y- and b-ions (Fig. 10). The ion b_{13} (m/z 1439.6) or y13 (m/z 1457.5) correspond to the peptide which contains two potential linkage sites - D(L/I)S- and -NNT-.

Normally, D(L/I)S is not expected to be a linkage site, but, in the unlikely event that deglycosylation of a glycan linked asparagine had occurred during sample preparation, we thoroughly investigated this option. The ion y7 at m/z 806.4 corresponds to the C-terminal fragment of the peptide -(I/L)STNNTR, and the ion at m/z 1009.6 represents the same fragment still containing one GlcNAc moiety. This suggests that only the linkage site -NNT- is glycosylated, most likely via a high mannose like structure.

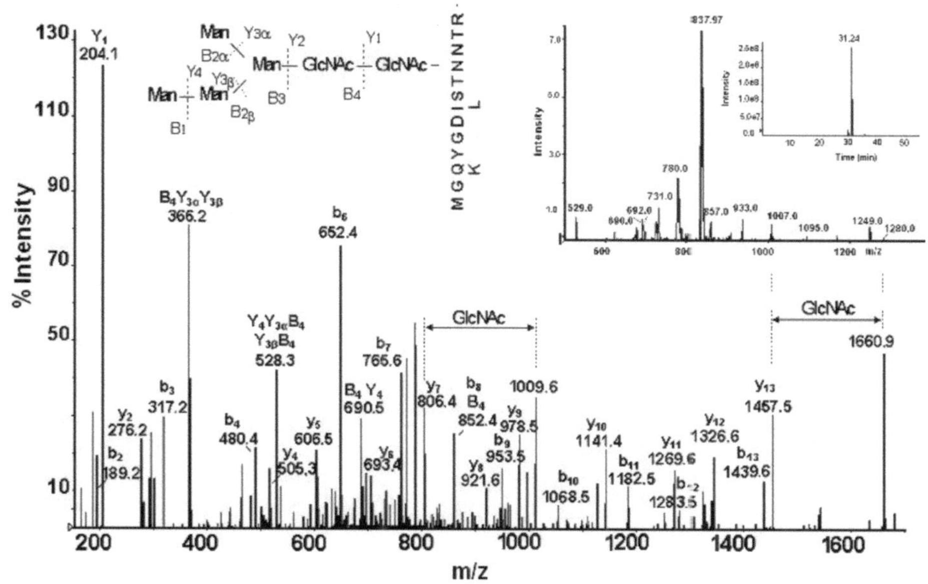

Fig. 10. Precursor ion scan (insert) at time 31,24 min of the chromatographic separation, and enhanced product ion (EPI) scan of the ion (B) at m/z 837.97.

The ion at m/z 1660.9 corresponds to the intact peptide, represented as ion y13 (m/z 1457.5), which is N-glycosylated with a single GlcNAc residue. The glycan structure of this peptide was determined following the typical ions as Y and B in MS/MS spectrum at m/z 204.1 (GlcNAc), m/z 366.2 (HexGlcNAc), m/z 528.3 (Hex2GlcNAc), 690.5 (Hex3GlcNAc), and 852.4 (Hex4GlcNAc) (Fig. 10).
Combining all data, the carbohydrate structure of the glycan with mass 1054.0 Da (Hex2Man3GlcNAc2) was suggested and the peptide sequence was determined MGQYGD(I/L)STNNTR.

4.2 Structural characterization of glycans by MS/MS analyses

While CID MS/MS techniques do routinely provide information on the glycan moieties of glycopeptides, they rarely reveal information on the peptide sequence and glycan attachment site(s). For a more detailed characterization of protein glycosylation, these techniques may be combined with other types of experiments. Glycopeptides may be treated with exoglycosidases revealing the nature and anomericity of terminal monosaccharide residues.
Alternatively, by treatment of glycopeptides with peptide-N-glycosidase F or A, N-glycans can be released and deglycosylated peptide moieties can be obtained. The peptides may then be subjected to mass spectrometric characterization.
Some glycans, obtained by us after digestion of proteins with PNGase F were identified by Q-Trap analysis. Analysis of the purified N-glycans by Q-Trap is a very sensitive and useful method and several glycans of the high mannose or complex types were identified following the sequence of B and Y ions in the MS/MS spectrum.
Two approaches were applied to analyse the isolated glycans after treatment of RvH1 with PNGase F (Dolashka-Angelova et al. 2010). The first approach included sequencing of the glycans by specific glycosidases and analysis of the fragments via MS before and after

treatment with the enzymes. This approach provided only preliminary results about the structures of the glycans. Therefore, in the second approach, tandem mass spectrometry was applied. The glycan structure was derived from the MS/MS spectra, obtained on a hybrid quadrupole-linear ion trap mass spectrometer.

The same method and techniques were applied and the configurations of 15 N-glycans released from *Haliotis tuberculata* hemocyanin (HtH1) after PNGase F-treatment were performed by Q-Trap tandem mass spectrometry (Velkova et al. 2011). Using both the MALDI-TOF-MS analysis after treatment of the glycans with the specific α1-6(>2,3,4) fucosidase and the Q-Trap-MS/MS analysis, core- structures containing Fuc(α1-6) GlcNAc, were detected.

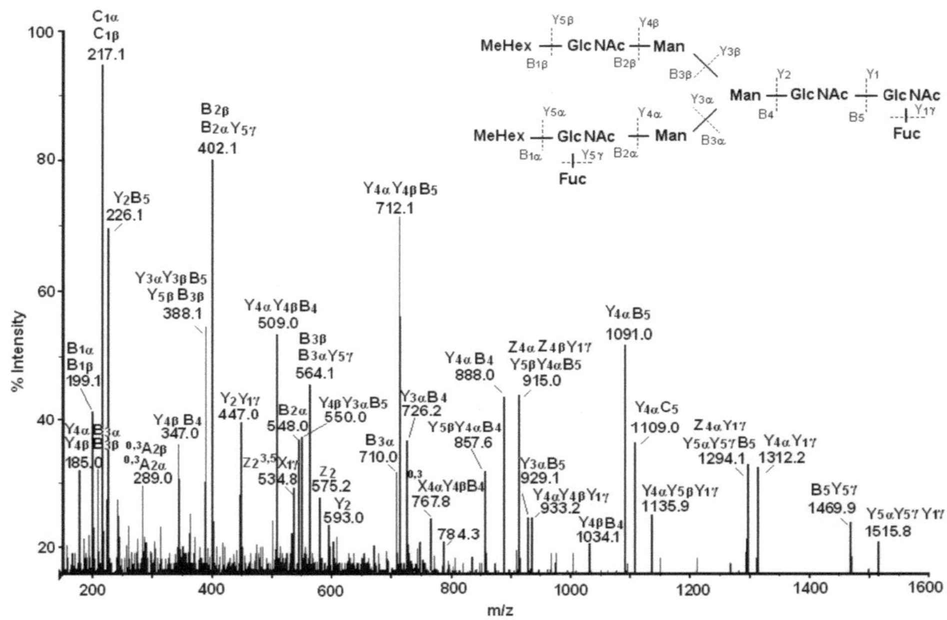

Fig. 11. MS/MS spectra and structures with fragmentation nomenclature of the double charged $[M+2Na]^{2+}$ of the glycan at m/z 1002.8 isolated from HtH1.

In the MS/MS spectrum the most dominant Y and B ions provided information on the sequence and the branching, but the positions of the monosaccharides were confirmed by the C, Z, X and A ions. The interesting structure of one glycan was determined after sequencing the doubly-charged ion ($[M+2Na]^{2+}$, m/z 1002.8, (Figure 11). As is shown in the spectrum, two deoxyhexose and two terminal methyl-Hex residues are linked to the internal GlcNAc of a molecule with the composition $MeHex_2HexMan_3GlcNAc_4Fuc_2$. Three ions, $Z_{2}{}^{3,5}X_{1\gamma}$ at m/z 534.8, Z_2 at m/z 575.2 and Y_2 at m/z 593.0, demonstrate the presence of a core-linked Fuc(α1-6) residue. Confirmation of additional Fuc(α1-3) branching to GlcNAc are the ions $B_{2\alpha}$ at m/z 548.0 and $B_{3\alpha}$ at m/z 710.0, as well as the ions $B_{3\beta}$ at m/z 564.1 and $B_5Y_{5\gamma}$ at m/z 1469.9, supporting the branching of two terminal MeHex residues. The observed cross-ring fragment ions $X_{4\beta}Y_{4\alpha}B_4$, at m/z 767.8, as well as ${}^{0,3}A_{2\alpha}$, are additional evidence of the suggested structure Fuc(α1-3)GlcNAc. The ion $Y_{4\beta}B_4$ at m/z 1034.1 corresponds to the composition $Man_3GlcNAcFucMeHex$ (carrying a (α1-3)-fucose terminal linkage at GlcNAc).

Consequently, cross-ring fragment ion $X_{4\alpha}Y_{4\beta}B_4$, at m/z 767.8, is resulting from the ion $Y_{4\beta}B_4$ at m/z 1034.1 without $_{0,3}A_{2\alpha}$ fragment, containing a MeHex moiety. The evidence of MeHex being linked to GlcNAc, as observed in most molluscan hemocyanins, is derived from the ions C1α and C1β at m/z 217.1, ions $B_{2\beta}$ and $B_{2\alpha}Y_{5\gamma}$ at m/z 402.1, as well as from the ions $B_{3\beta}$ and $B_{3\alpha}Y_{5\gamma}$ at m/z 564.1. Two alternatively interpretations of the signal at m/z 1294 as $Y_{5\alpha}Y_{5\gamma}B_5$ and $Z_{4\alpha}Y_{1\gamma}$ confirm the positions of two fucoses.

Based on these data the structures of the isolated N-glycans from HtH1 were proposed (Velkova et al. 2011). It was found that most of the glycans have a common structural feature, one α1-6 linked fucose being attached to the trimannosyl core. Some of the structures have being partially modified by the methyl group, and in a few glycans a second or a third fucosyl residue in the Fuc(α1-3)GlcNAc motif were identified.

4.3 Capillary electrophoresis and mass spectrometry

4.3.1 Characterization of glycopeptides by CE

Capillary electrophoresis (CE) and capillary gel electrophoresis have been widely used for complex carbohydrate separation (Sandra et al. 2004b) because of enhanced separation efficiency and shorter analysis times. CE-based carbohydrate analysis can be applied easily to determine the molar ratio, and the degree of polymerisation of oligosaccharides, and to detect changes in the extent or nature of the oligosaccharide distribution (fingerprinting).

Using capillary electrophoresis, MALDI-MS and ESI-MS in combination with glycosidase digestion we studied the N-terminal functional unit (FU) RvH1-a of the structural subunit RvH1 (Dolashka-Angelova et al. 2004)(Fig. 12).

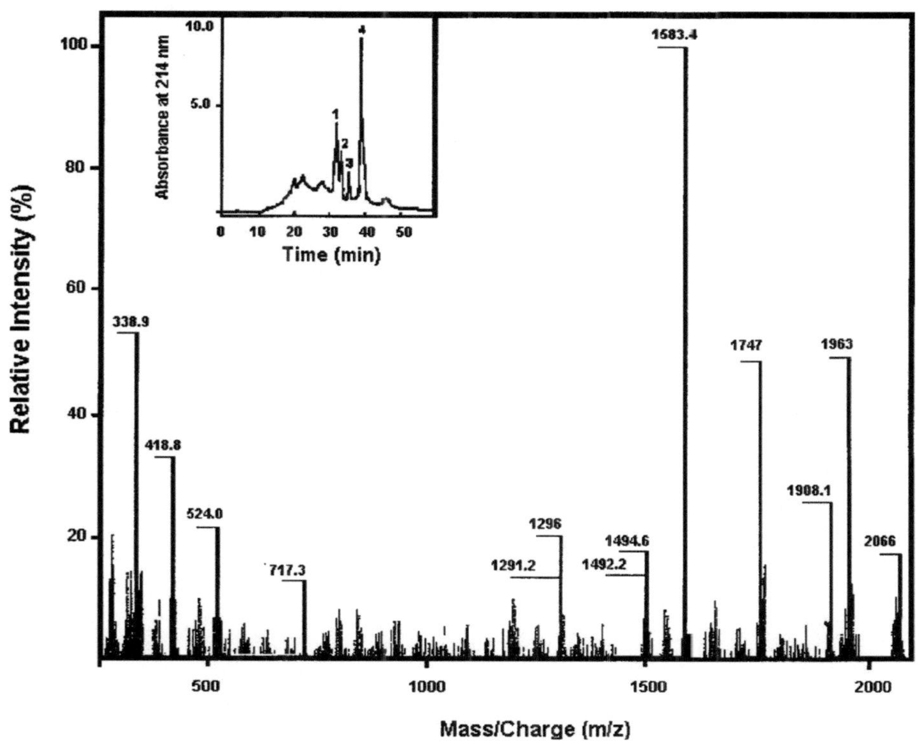

Fig. 12. ESI-MS and capillary electrophoresis of Glp1 following glycosidase digestion.

Oligosaccharide fragments were released from the glycoprotein by Smith degradation and separated by means of a Superdex 300 column. Glycopeptide fragments, giving a positive reaction for the orcinol/H_2SO_4 method, were separated by HPLC. The glycopeptide 1 with mass 2786 Da, was analyzed by MALDI-MS before and after treatment with PNGase-F. In the MALDI mass spectrum, acquired before PNGase-F treatment only one molecular ion at 2786 m/z was detected, generated by the intact Glp1 sodium adduct ion $[Glp1+Na]^+$. In contrast, three signals were observed after PNGas-F treatment. The signal at 2786 m/z is related to the intact Glp1, whereas the signal at m/z 1609 $[M+Na]^+$ is attributed to an oligosaccharide. The first signal (1177 m/z) corresponds to the deglycosylated peptide. To determine the oligosaccharide sequence, ESI-MS and capillary electrophoresis of Glp1 were used, following glycosidase digestion (Fig. 12).
Different glycolytic enzymes were added to glycopeptide 1: β1-2,3,4,6-GlcNAcase; α1-2,3-mannosidase, α1-2,3,6- mannosidase, β1-3,4,6-galactosidase. The mass of this carbohydrate moiety (1586 Da, Table 1) would account for an oligosaccharidic chain containing (SO_4) MeGal $GlcNAc_4$ Man_3 being connected to the peptide.

No	Enzymes	$[M+H]^+$ (m/z)	Structure of Glycopeptide 1
A	α1-2,3- and α1-2,3,6-mannosidase and β1-2,3,4,6-N-GlcNAcase	1747	← Man(β1-4) GlcNAc(β1-4) GlcNAc ¬P
B	α1-2,3-mannosidase and β1-2,3,4,6-N-GlcNAcase	1908	Man(α1-6) ↖ ← Man(β1-4) GlcNAc(β1-4) GlcNAc ¬P
C	α1-2,3,6-mannosidase and α1-2,3-mannosidase	1953	3MeGlcNAc(β1-2)←Man(β1-4)GlcNAc(β1-4)GlcNAc¬P
D	β1-2,3,4,6-N-GlcNAcase	1583	-GlcNAc(β1-4) GlcNAc¬P

Table 1. Proposed structures of the carbohydrate chains of Glp1. The carbohydrate chains were calculated on the basis of observed [M +H]+signals in the ESI mass spectra from Fig.12. P represents the peptide with sequence FANATSIDGPNA (1177 Da).

Depending on the specificity of the individual enzymes, the different linkages were identified by recording the pattern of molecular weights resulting from the digestion with the pool of glycosidases. After a 24 h incubation at 37°C, the sample was first analyzed by capillary electrophoresis. Four different peaks were detected, indicating that the treatment causes heterogeneity resulting from different cleavage sites (Fig. 12, insert). On the basis of these results and an amino acid sequence analysis it was concluded that the functional unit RvH_1-a contains 7% of oligosaccharides, that the glycans are N-glycosidically attached to Asn262 and Asn401, and that the structure is as follows:

Glp 1 SO_4Man(α1-6)
↖
3MeGlcNAc(β1-2) ← Man(β1-4) GlcNAc(β1-4) GlcNAc ¬
↙ FANATSIDGPNA
3MeGal(β1-3) GlcNAc(β1-2) Man(α1-3)

Using capillary electrophoresis, in combination with MALDI-MS, ESI-MS and Q-Trap-MS/MS (Fig. 13), the carbohydrate structure of several glycoproteins have been identified (Sandra et al. 2004 b; 2007).

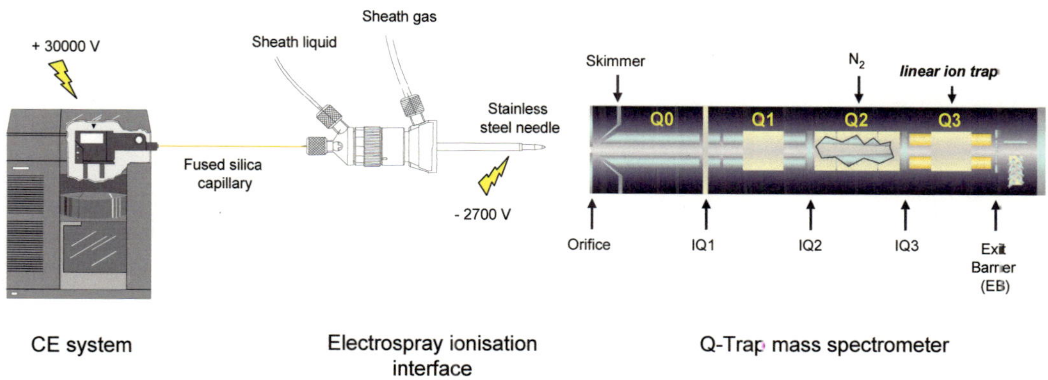

Fig. 13. Combination of CE, ESI and Q-Trap MS/MS system.

4.3.2 Characterization of glycans by CE

Labelling of the glycans has recently been proven to be of great value in the CE-MS/MS analysis of glycoproteins (Sandra et al. 2004 b).It permits high-resolution CE, allows the simultaneous detection of uncharged and charged glycans, and provides easily interpretable spectra. The latter label, representing a novel sugar tag, allows CE-MS detection in the positive ion mode and provides complementary information. The intact RvH1 was subjected to peptide-N-glycosidase F (PNGase F) digestion and the obtained glycans were labelled by 8-aminopyrene-1,3,6-trisulfonate (APTS) as well as by 3-aminopyrazole (3-AP). The N-glycans were analyzed by MALDI-TOF and CE-MS/MS.

MS/MS on APTS-derivatised sugars resulted in easily interpretable MS/MS spectra, since Y-fragments predominated. The CE-MS electropherogram of the APTS-labelled RvH_1 N-glycans is presented in Figure 14 (insert) and, by way of an example, the on-line MS spectrum of the compound migrating at time 10.0 min is presented. This glycan is detected as a doubly and triply negatively charged ion and the composition corresponds to a fucosylated $Man_3GlcNAc_2$ core structure. From the MS/MS data obtained on the $[M-2H]^{2-}$ ion at m/z 748, one can deduce that the fucose is attached to the proximal GlcNAc residue (Fig. 14).

The CE-MS(/MS) is a much more sensitive technique which allows identification of the presence of unusual structures. CE-MS/MS was performed to provide additional information of the oligosaccharide structure of RvH1. A new structure was suggested for the four times negatively charged ion at m/z 555.7 (Sandra et al. 2007). A sulphated oligosaccharide was observed before in RvH_1 [10], and for the ion at m/z 555.7 the following sulphated structures can indeed be proposed: $SO_4FucHex_4Man_3GlcNAc_2$ and $(SO_4)_2FucHexHexNAc_2Man_3GlcNAc_2$.

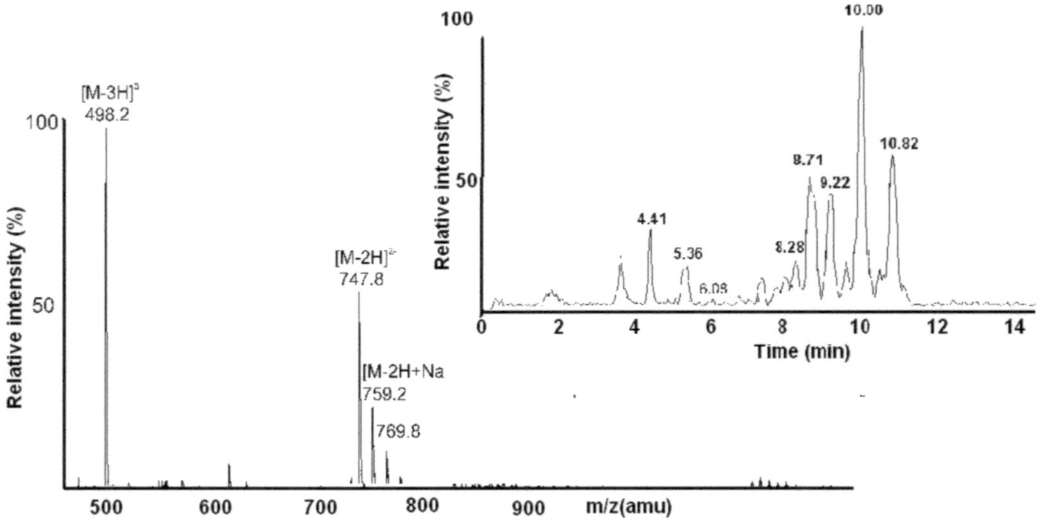

Fig. 14. On-line MS spectrum of the double charged ion at m/z 747.8 (insert). CE-MS base peak electropherogram of the APTS-labelled RvH1 N-glycans.

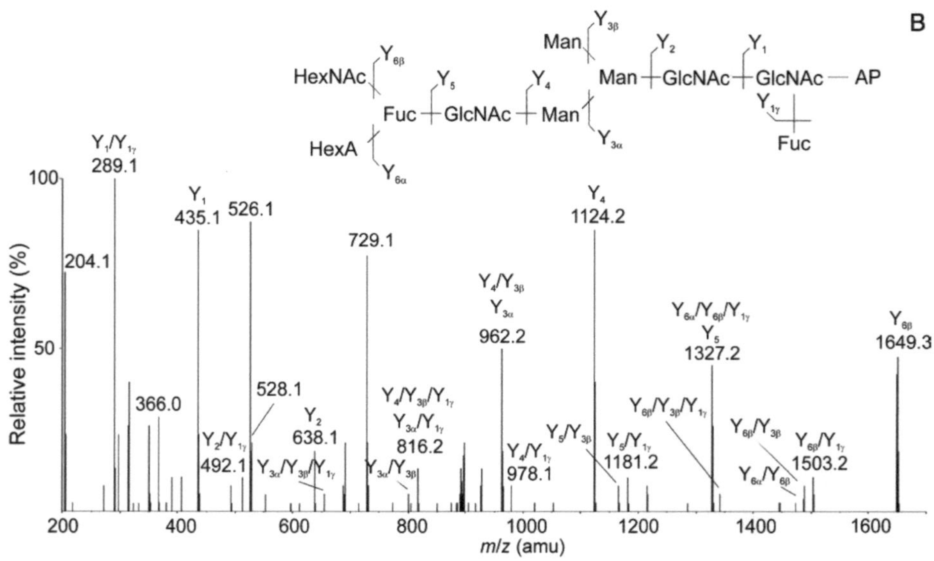

Fig. 15. CE-MS/MS electropherogram and tandem mass spectra of the compound at m/z 555.7 (P, precursor ion).

In order to distinguish these two possibilities, the sample was reinjected for an on-line MS/MS experiment, keeping Q1 fixed to select the ion at m/z 555.7 (Figure 15). Apparently, CE could separate two isomeric compounds (eluted after 10.5 and 10.9 min) with similar tandem mass spectra and, hence, the difference did not reside in the sequence. The spectra do not fit any of the sulfated glycans mentioned earlier.

Following MS/MS data, a structure with the HexAHexNAcFucGlcNAc oligosaccharide at both the α-1,3 and the α-1,6-arm could be suggested.

5. Conclusion

Protein glycosylation plays an important role in a multitude of biological processes such as cell-cell recognition, growth, differentiation, and cell death. It has been shown that specific glycosylation changes are key in disease progression and can have a diagnostic value for a variety of disease types such as cancer and inflammation. The complexity of carbohydrate structures and their derivatives makes their study a real challenge.

Mass spectrometric (MS) techniques play a key role in glycoprotein and glycan analysis, study protein glycosylation at the glycopeptide level. Tandem mass spectrometry provides both the separation of glycopeptides and the ability to determine the glycan composition and site-specific glycosylation. Current informatics tools are designed for large, high-throughput mass-spectrometry datasets and can analyze numerous compounds, such as those in body fluids. The methods and strategies being developed are compatible with the problems of microheterogeneity commonly found, allowing characterization of even very complex minor components. Therefore, MS is becoming an increasingly important aspect in proteomics of eukaryotic cells.

6. Acknowledgments

This work was supported by a research grant 451 VS.016.009N of FWO (Belgium) and the Bulgarian Academy of Sciences; the Bulgarian National Science Fund TK01-496/2009, HTC01-187 (Ukraine), DFG-01/2008 (Germany) and CNR (Italy). The author is indebted to Prof. Josef Van Beemen (Gent, Belgium) for critically reading and correcting the text.

7. References

Ball, L.E.; Berkaw, M.N.; Buse, M.G. (2006). Identification of the Major Site of *O*-Linked β-*N*-Acetylglucosamine Modification in the C Terminus of Insulin Receptor Substrate-, *Molecular and Cellular Proteomics,* Vol. 5, (February 2006), pp.313-323, ISSN 1535-9484.

Bateman, K.P.;White, R.L.;Yaguchi, M.;Thibault, P. (1998). Characterization of proteinglycoforms by capillary-zone electrophoresis-nanoelectrospray mass spectrometry. *Journal of Chromatography A*, Vol.794, No.1/2,(January 1998),pp. 327–344, ISSN 0021-9673.

Beck, A.; Hillen, N.; Dolashki, A.; Stevanovic, S.; Salvato, B.; Voelter, W.; Dolashka-Angelova, P. (2007). Oligosaccharide structure of a functional unit RvH1-b of *Rapana venosa* hemocyanin using HPLC/electrospray ionization mass spectrometry. *Biochimie*, Vol.89, No.8, (August 2007), pp.938-949, ISSN 0300-9084.

Bykova, N.V.; Rampitsch, C.; Krokhin, O.; Standing, K.G.; Ens, W. (2006) Determination and Characterization of Site-Specific N-Glycosylation Using MALDI-Qq-TOF Tandem Mass Spectrometry: Case Study with a Plant Protease. *Analytical Chemistry*, Vol.78, No. 4, (January 2006), pp.1093-1103, ISSN 1520-6882.

Chalabi, S.; Panico, M.; Sutton-Smith, M.; Haslam, S.M.; Patankar, M.S.; Lattanzio, F.A.; Morris, H.R.; Clark, G.F.; Dell, A. (2006) Differential O-Glycosylation of a Conserved Domain Expressed in Murine and Human ZP3. *Biochemistry*, Vol.45, No.2, (2006), pp.637-647, ISSN 1520-4995.

Dell, A.; Morris, H. R.(2001). Glycoprotein Structure Determination by Mass Spectrometry. *Science, Vol.251. No. 5512, (March 2001), pp. 2351-2356, ISSN* 1095-9203

Demelbauer, U.M.; Zehl, M.; Plematl, A.; Allmaier, G.; Rizzi, A. (2004) Determination of glycopeptide structures by multistage mass spectrometry with low-energy collision-induced dissociation: comparison of electrospray ionization quadrupole ion trap and matrix-assisted laser desorption/ionization quadrupole ion trap reflectron time-of-flight approaches. *Rapid Communications in Mass Spectrometry*, Vol.18, No. 14, (July 2004), pp. 1575-1582, ISSN 1097-0231.

Dolashka-Angelova, P.; Beck, A.; Dolashki, A.; Beltramini, M.; Stevanovic, S.; Salvato, B.;Hristova, R.; Velkova, L.; Voelter, W. (2004). Carbohydrate moieties of molluscan *Rapana venosa* hemocyanin. *Micron,* Vol. 35, No. 1-2, (January/February 2004), pp. 101–104, ISSN 0968-4328

Dolashka-Angelova, P.; Moshtanska, V; Kujumdzieva, A.; Atanasov, B.; Petrova, V.; Voelter, W.; Van Beeumen, J. (2010). Structure of glycosylated Cu/Zn-superoxide dismutase from kluyveromyces yeast NBIMCC 1984. *Journal of Molecular Structure,* Vol.980, No.1-3, (September 2010), pp.18-23, ISSN 0022-2860.

Domon, B.; Aebersold, R. Mass Spectrometry and Protein Analysis. *Science,* Vol. 312, No. 5771, (April 2006), pp. 212-217,ISSN 1095-9203

Froehlich, J.W.; Barboza, M.; Chu, C.; Lerno, L.A.; Clowers, B.H.; Zivkovic, A.M.; German, J.B. and Lebrill, C.B. (2011) Nano-LC-MS/MS of Glycopeptides Produced by Nonspecific Proteolysis Enables Rapid and Extensive Site-Specific Glycosylation Determination. *Analytical Chemistry,* Vol.83, No.14, (June 2011), pp.5541-5547, ISSN ISSN 1520-6882.

Gonzales de Peredo, A.; Klein, D.; Macek, B.; Hess, D.; Peter-Katalinic, J.; Hofsteenge, J. (2002). *C*-Mannosylation and -Fucosylation of hrombospondin Type 1 Repeats. *Molecular and Cellular Proteomics,* Vol.1, (2002), pp.11-18, ISSN 1535-9484.

Håkansson, K.; Chalmers, M.J.; Quinn, J.P.; McFarland, M.A.; Hendrickson, C.L.;Marshall, A.G. (2003). Combined electron capture and infrared multiphoton dissociation for multistage MS/MS in a Fourier transform ion cyclotron resonance mass spectrometer. *Analytical Chemistry.* Vol.75, No.13, (May 2003), pp. 3256–3262, ISSN 1520-6882.

Håkansson, K.; Cooper, H.J.; Emmett, M.R.; Costello, C.E.; Marschall, A.G.; Nilsson, C.L. (2001). Electron capture dissociation and infrared multiphoton dissociation MS/MS of an N-glycosylated tryptic peptide to yield complementary sequence information. *Analytical Chemistry,* Vol. 73, No.18,(August 2001), pp.4530–4536, ISSN 1520-6882.

Harvey, D.J. (2006). Analysis of carbohydrates and glycoconjugates by matrixassisted laser desorption/ionization mass spectrometry: An update covering the period 1999-2000. *Mass Spectrometry Reviews,* Vol.25, No. 4, (July/August 2006), pp.595–662, ISSN 1098-2787.

Hogan, J.M.; Pitteri, S.J.; Chrisman, P.A.; McLuckey, S.A. (2005). Complementary Structural Information from a Tryptic *N*-Linked Glycopeptide via Electron Transfer Ion/Ion Reactions and Collision-Induced Dissociation. *Journal of Proteome Research,* Vol.4, 2, (April 2005), pp.628-632, ISSN 1535-3907.

Huddleston, M.J.; Bean, M.F.; Carr, S.A. (1993) Collisional fragmentation of glycopeptides by electrospray ionization LC/MS and LC/MS/MS: methods for selective detection of glycopeptides in protein digests. *Analytical Chemistry*, Vol. 65, No. 7, (April 1993), pp. 877-884, ISSN 1520-6882.

Krokhin, O.V.; Ens, W.; Standing, K.G. (2005) MALDI Q-TOF MS Combined with Off-line HPLC for Characterization of Protein Primary Structure and Post-Translational Modifications. *Journal of Biomolecular Techniques*, Vol. 16, No.4, (December 2005), pp.429-440, ISSN 1524-0215

Macek, B.; Hofsteenge, J.; Peter-Katalinic, J. (2001) Direct determination of glycosylation sites in O-fucosylated glycopeptides using nano-electrospray quadrupole time-of-flight mass spectrometry. *Rapid Communications in Mass Spectrometry*, Vol. 15, No. 10, (May 2001), pp. 771-777, ISSN 1097-0231.

Medzihradszky, K.F.; Gillece-Castro, B.L.; Townsend, R.R.; Burlingame, A.L.; Hardy, M.R. (1996) Structural elucidation of *o*-linked glycopeptides by high energy collision-induced dissociation. *Journal of the American Society for Mass Spectrometry*,Vol. 7, No. 4, (April 1996), pp. 319-328, ISSN 1044-0305.

Medzihradszky, K.F. (2005). Peptide sequence analysis. *Methods Enzymology*, Vol.402, (2005), pp. 209-244, ISSN 0076-6879.

Morelle, W.; Michalski, J.C. (2005). The Mass Spectrometric Analysis of Glycoproteins and Their Glycan Structures. *Current Analytical Chemistry*,Vol. 1, No.1, (January 2005), pp. 29-57, ISSN 1573-4110.

Nasia, A.; Picariellob, G.; Ferrantia, P. (2009). Proteomic approaches to study structure, functions and toxicity of legume seeds lectins. Perspectives for the assessment of food quality and safety. *Journal of Proteomics*, Vol. 72, No.3, (April 2009), pp. 527-538, ISSN 1874-3919.

Patwa, T.; Li, Chen; Simeone D. M., Lubman, D. M. (2010). Clycoprotein analysis using protein microarrays and mass spectrometry. *Mass Spectrometry Reviwes*, Vol.29, No.5, (September/October 2010), pp.830-844, ISSN 1098-2787.

Sandra, K.; Dolashka-Angelova, P.; Deversee, B. and Beeumen, J.V. (2007). New insights in *Rapana venosa* hemocyanin N-glycosylation resulting from on-line mass spectrometric analyses. *Glycobiology*, Vol.17, No.2, (2007), pp. 141-156, ISSN 1460-2423.

Sandra, K.; Stals, I.; Sandra, P.; Claeyssens, M.; Beeumen, J.V and Devreese, B. (2004a). Combining gel and capillary electrophoresis, nano-LC and mass spectrometry for the elucidation of post-translational modifications of *Trichoderma reesei* cellobiohydrolase. *Journal of Chromatography A*, Vol. 1058, No.1-2, (November 2004), pp. 263-272, ISSN 0021-9673.

Sandra, K.; Van Beeumen, J.; Stals, I.; Sandra, P.; Claeyssens, M.; Devreese, B. (2004b). Characterization of cellobiohydrolase I N-glycans and differentiation of their phosphorylated isomers by capillary electrophoresis-Q-Trap mass spectrometry. *Analytical Chemistry*, Vol.76, No.19, (August 2004), pp.5878–5886, ISSN 1520-6882.

Syka, J.E.P.; Coon, J.J.; Schroeder, M.J.; Shabanowitz, J.; Hunt, D.F. (2004) Peptide and protein sequence analysis by electron transfer dissociation mass spectrometry. *Proc. Natl. Acad. Sci. U.S.A.*, Vol. 101 (2004) 9528-9533.

Tadjiri, M.; Yoshida, S.; Wada, Y. (2005) Differential analysis of site-specific glycans on plasma and cellular fibronectins: application of a hydrophilic affinity method for glycopeptide enrichment.*Glycobiology*, Vol.15, No.12,(July 2005), pp. 1332-1340, ISSN 1460-2423.

Takemori, N.; Komori, N.; Matsumoto, H. (2006). Highly sensitive multistage mass spectrometry enables small-scale analysis of protein glycosylation from two-dimensional polyacrylamide gels. *Electrophoresis*, Vol. 27, No.7, (Aprl 2006), pp. 1394-1406, ISSN 1522-2683.

Uematsu, R.; Furukawa, J.; Nakagawa, H.; Shinohara, Y.; Deguchi, K.; Monde, K.; Nishimura, S. (2005) High Throughput Quantitative Glycomics and Glycoform-focused Proteomics of Murine Dermis and Epidermis. *Molecular and Cellular Proteomics*, Vol.4 (December 2005), pp.1977-1989, ISSN 1535-9484.

Velkova, L.; Dolashka, P.; Lieb, B.; Dolashki, A.; Voelter, W.; Beeumen, J. V. and Devreese, B. (2011). Glycan structures of the structural subunit (HtH1) of *Haliotis tuberculata* hemocyanin. *Glycoconjugate Journal*, Vol. 28, No. 6,(August 2011), pp. 385-395, ISSN 1573-4986.

Vosseller, K.; Trinidad, J.C.; Chalkley, R.J.; Specht, C.G.; Thalhammer, A.; Lynn, A.J.; Snedecor, J.H.; Guan, S.; Medzihradszky, K.F.; Maltby, D.A.; Schoepfer, R.; Burlingame, A.L. (2006) *O*-Linked *N*-Acetylglucosamine Proteomics of Postsynaptic Density Preparations Using Lectin Weak Affinity Chromatography and Mass Spectrometry. *Molecular and Cellular Proteomics*, Vol.5, (May 2006), pp.923-934, ISSN 1535-9484.

Wada, Y. (2008) Mass Spectrometry of Glycopeptides. *Experimental Glycoscience: Glycochemistry*, Springer 2008, Part 1, Section III, EISBN: 978-4-431-77924-7.

Wuhrer, M.; Hokke, C.H.; Deelder, A.M. (2004) Glycopeptide analysis by matrix-assisted laser desorption/ionization tandem time-of-flight mass spectrometry reveals novel features of horseradish peroxidase glycosylation. *Rapid Communcations in Mass Spectrometry*, Vol.18, No. 15, (August 2004), pp.1741-1748, ISSN 1097-0231.

Zaia, J. (2004). Mass spectrometry of oligosaccharides. *Mass Spectrometry Reviews*, Vol.23, No.3, (May/June 2004), pp.161-227, ISSN 1098-2787.

Proteomics Methodology Applied to the Analysis of Filamentous Fungi - New Trends for an Impressive Diverse Group of Organisms

Carlos Barreiro, Carlos García-Estrada and Juan F. Martín
Instituto de Biotecnología de León (INBIOTEC). Proteomics Unit. Parque Científico de León. León, Spain

1. Introduction

Life comprises three domains: the Bacteria, the Archaea and the Eukaryota. Within the last one, the Fungi kingdom forms a monophyletic group of the eukaryotic crown group, which includes the largest group of organisms, where more than 500,000 species have been described and more than one million might exist. The Fungi kingdom consists of a heterogeneous group that includes yeasts, moulds and mushrooms characterized by their lack of photosynthetic pigment and their chitinous cell wall (Hawksworth, 1991). Fungi are one of the most important groups of organisms on the planet, for they play remarkable roles in several ways.

From the environmental point of view fungi, together with bacteria, are responsible for most of the organic matter decomposition process, thus recycling dead material and recalcitrant compounds such as lignin; thereby the organic material utilization by the microbial community is enhanced. This fungal recycling activity has a negative effect when it is exerted on food. Fungi can be responsible for large losses of stored food, particularly food which contains any moisture. This may be a major problem where large quantities of food are being produced seasonally and then require storage until they are needed.

Biocontrol, an alternative to chemical pesticides, is another important role played by fungi. Some species of this kingdom parasitize insects, which can be extremely useful for controlling insect pests of crops, such as the Colorado potato beetles, spittlebugs, leaf hoppers and citrus rust mites. In addition, mycorrhizal associations between plants and fungi are necessary for optimal growth of most plants, such as crops. However, some fungi can also cause negative effects on crops, since many plants are susceptible to fungal attack.

A wide range of fungi also live on and in humans, other animals or plants, but most coexist harmlessly. Nevertheless, some fungal species are involved in disease interactions, either by means of their direct action as disease agents or through the production of secondary metabolites (eg: host-specific toxins). In fact, filamentous fungi produce a diverse array of secondary metabolites, which have a tremendous impact on society because they are exploited for their ability to produce antibiotic (penicillins, cephalosporins, etc) or pharmaceuticals (cyclosporin and other immunosuppressants). Indeed, fungi are the second group, after Actinobacteria, of secondary metabolites producers with industrial application.

Among secondary metabolites, antibiotics (and more precisely β-lactam antibiotics) are especially relevant. The discovery of β-lactam antibiotics is one of the most significant milestones of the human history and entailed a revolution in modern chemoteraphy. Members of this family of antibiotics are commonly prescribed world widely due to their high activity and low toxicity and have helped medicine to drastically reduce the mortality rate.

Finally, we should not forget that some fungi are important for food industry. Many mushrooms (basidiomycetes) are edible and different species are cultivated for sale worldwide. Fungi are also widely used in the production of many foods and drinks, such as cheese, beer, wine, bread, juices, cakes, some soya bean products, due to their ability to produce useful enzyme cocktails for white biotechnology.

Several new fungal species are discovered every year, which indicates that a huge number of unrecognized and unidentified fungal species (more than 90%), could be found in association with plants, insects, animals, as lichen-forming fungi or in undisturbed areas (Hawksworth 2001). The three main fungal phyla are: Ascomycota (comprises the majority of all described fungi), Zygomycota and Basidiomycota, which theoretically diverged from the Chytridiomycota approximately 550 million years ago. The split between Ascomycota and Basidiomycota happened 400 million years ago, which was an evolutionary step previous to the land invasion of the plants. Intriguingly, the antiquity of the group strongly contrasts with the current low number of complete fungal genomes deposited in the databases, which is usually the main key that opens the door to the Proteomics application. Since 1996, when the first completed fungal genome of the ascomycetous yeast *Saccharomyces cerevisiae* was released and until 2011, sequencing of around 20 complete fungal genomes had been completed. These sequencing data support the incredible increment of Proteomics applied manuscripts related to filamentous fungi published just in the last decade. It was not until 2004 when the analysis of the secreted proteins of *Aspergillus flavus* was described, or 2005 when the secretome analysis of the edible fungus *Pleurotus sapidus* was published. These data support the interest of a methodological review of filamentous fungal Proteomics. This approach can be done using the three Proteomics levels analysed until our days: i) intracellular, ii) extracellular and iii) membrane and organelle proteomes.

2. Proteomics, a useful tool for the analysis of fungi

Proteomics, as the analysis of the protein components of organisms, is a well-known tool providing a global perspective (whole-organism approach) of cellular physiology that allows understanding the cellular protein expression alterations in response to various biotic and abiotic stresses. Due to its capacity to yield definitive information on protein identity, localization, posttranslational modification and the accuracy of *in silico* gene model prediction in fungi, Proteomics has become an integral component of all large-scale "omic" and systems approaches to understanding the rich complexity of fungal biochemistry (Doyle, 2011). The increase of plant-pathogen interaction analysis, focused on the plant-fungus association is a topic of rising interest in the last five years; even when the limitation in sequence availability in public databases is also challenging (Mehta et al., 2008). These studies are concerning to the understanding of the pathogen interaction with their respective hosts to combat the crop diseases, inductor effect of the susceptible host root plant extracts over pathogenic fungi or novel sources of host resistance and the design of increased disease resistance in crop plants. In addition to these modern Proteomics analyses, traditionally the research about fungal diseases has been applied to humans (eg: invasive pulmonary aspergillosis), plants (eg: grapevine decline) or other animals (eg: entomopathogenic fungus). Two main approaches are used for Proteomics: i) individual protein identification by MALDI-TOF/TOF or ii) tandem LC-MS. These techniques not only allow the elaboration of proteome reference maps, but also the comparative analysis between different experimental conditions, strains or mutants.

Due to the role that ascomycetes play in causing animal diseases and to their ability to be used as cell factories for biotechnology, proteomic studies focused on ascomycetes exceed those carried out using basidiomycetes. All fungal data sets include either "predicted proteins" or "hypothetical proteins" (Martin et al., 2008; Ferreira de Oliveira et al., 2010) and this information should be replaced by "proteins of unknown function" since it is rather clear that if these proteins have been identified, they do exist and are no longer "hypothetical". Providing a function to those "proteins of unknown function" represents one of the major challenges in fungal Proteomics.

3. Analysis of the intracellular proteome

The intracellular proteome of a eukaryotic organism mainly consists of those proteins that are present in the cytosol and inside organelles. Due to the heterogeneity of the fungi kingdom, the most basic procedures, such as good quality protein extraction for bidimensional analysis, should be updated for each species. Thus, a general overview of sample preparation, as well as proteomics applications to fungal analysis are developed along the next sections.

3.1 Sample preparation

Since most fungi possess a robust cell wall, the protein extraction protocol is a crucial step for fungal intracellular proteins analysis. A bibliographic search shows different procedures for protein extraction, precipitation or protein solubilisation that should be adapted for each fungus. These extraction procedures for intracellular proteomes, in the case of free living microorganisms, have some common steps, such as the mechanical breaking, the use of protease inhibitors, reducing agents and protein precipitation.

The process to obtain cytoplasmic proteins should include an initial step to discard the culture media, unbroken mycelia, cell wall and membrane contaminants in order to obtain the intracellular proteome as clean as possible. Thus, when cultures reach the desired conditions, the most common mycelia collection process is filtration, since centrifugation of mycelial microorganisms does not generate compact pellets that make the subsequent washing steps difficult. This procedure can be done by filtering through a Whatman 3MM paper (Whatman, Maidstone, England) or a nylon cloth filters (Nytal Maissa, Barcelona, Spain) (Lim et al., 2001; Fernández Acero et al., 2006; Coumans et al., 2010; Jami et al., 2010a; Yildirim et al., 2011). Hence, mycelia is collected and washed and the media can be stored for further analysis of secreted proteins. Washing steps, which allow the media elimination that can interfere in the protein purification, are usually done at 4°C in order to diminish the protein lysis by intracellular proteases. The most commonly employed washing solutions are: i) distilled water (Kniemeyer et al., 2006; Shimizu et al., 2009; Coumans et al., 2010); ii) phosphate buffered saline (PBS) (Lu et al., 2010) or iii) combination of 0.9 % sodium chloride and water (Jami et al., 2010a). In order to prevent protein degradation, samples collected at different time-points or conditions are washed, paper-dried and immediately stored at -20ºC to -80ºC for several months.

Two common steps included in all the protocols that are crucial for intracellular protein extraction are keeping the temperature as low as possible and adding protease inhibitors. Thus, the use of liquid nitrogen to decrease warming of the disruption systems is a widely used method. The main cell-breaking system is the traditional pre-chilled mortar grinding due to its efficiency against the fungal cell wall (Cobos et al., 2010; Lu et al., 2010; Vödisch et al., 2011), although waring blender machines (Lim et al., 2001), or glass bead beating systems, either combined with a 10 mM Tris-HCl buffer (Oh et al., 2010) or with a phenol buffer (Coumans et al., 2010; Vödisch et al., 2011), have been successfully applied.

After the breaking step, the protein solubilisation buffer always includes a protease inhibitor [e.g.: protease inhibition cocktail for fungi and yeast (Sigma), COMPLETE (Roche)] (Fernández-Acero et al., 2006, Jami et al., 2010a), in addition to a reducing agent [e.g.: 2 mercaptoethanol (2ME), dithiothreitol (DTT)] that reduces the disulfide linkages between two cysteines (Fernández Acero et al., 2006, Kniemeyer et al., 2006). The reducing agent is frequently also added to the precipitation solution (0.093% 2ME, 0.14% DTT). This precipitation step (previous to resuspension in sample solution) is used to selectively purify proteins from contaminants such as salts, nucleic acids, detergents or lipids, which interfere with the final bidimensional analysis. Although the combination of trichloroacetic acid (TCA) and acetone is usually more effective than either TCA or acetone alone (Bhadauria et al., 2007), other mixtures, such as 100 mM ammonium acetate in methanol, have been described for fungi (Vödisch et al., 2011). Optionally, the sample can be treated, previously to precipitation, with a nuclease mix [0.5 mg mL^{-1} DNase, 0.25 mg mL^{-1} RNase and 50 mmol L^{-1} $MgCl_2$; or commercially available e.g.: Benzonase (Merck)] (Lu et al., 2010; Barreiro et al., 2005). When the precipitation step is omitted, direct solubilisation in homogenization buffer is done, which includes fungal DNase/RNase as described by Oh and co-workers (2010).

Proteins may be difficult to resolubilize and may not completely resolubilize after TCA precipitation ["2-D Electrophoresis. Principles and Methods" GE Healthcare)]. Thus, residual TCA must be intensively removed by cold acetone washing steps, e.g.: i) 2x acetone plus 0.07% DTT and 1x 80% acetone (Jami et al., 2010a); ii) 2x acetone containing 1% 2ME (Kniemeyer et al., 2006; Shimizu et al., 2009); iii) 3x acetone/0.3% DTT (wt/vol) (Vödisch et al., 2011). A helpful tip after TCA precipitation and washing is to avoid long drying periods of time in order to improve protein resuspension (1-2 minutes is enough).

The main components of the final sample buffer or homogenization buffer described in fungal bibliography have urea (7-9 M), thiourea (0-2 M), CHAPS (1-4% w/v) and ampholyte (0,5-2%), which are combined as indicated in Table 1.

The protocol described by Kniemeyer and co-workers (2006) for protein extraction of *Aspergillus fumigatus* summarized part of the steps described above, since after the mycelia homogenization in a pre-cooled (in liquid nitrogen) mortar they did the extraction, cleaning and precipitation steps together, using the Clean Up kit (GE Healthcare). This procedure produced quite good results when it was applied to *P. chrysogenum* showing a good protein yield (Jami et al., 2010a). However, some minor problems, such as the poor representation of large proteins, stripping in the basic region and the improper cleaning of the middle sized protein region (50-100 kDa), were observed. Nonetheless, this procedure is highly recommendable for fungal protein extraction.

Based on the methods described by Fernandez-Acero and co-workers (2006) for *Botrytis cynerea,* the mortar grinding approach combined with a phosphate buffer extraction was used for *P. chrysogenum* (Jami et al., 2010a). This method improved the final bidimensional gel results, avoiding the problems observed in the direct Clean Up kit method. In addition, this procedure yielded large protein amounts, which allowed Blue silver Coomassie colloidal staining (Candiano et al., 2004), thus simplifying the 2D gels handling. In fact, Coomassie staining of proteins is still the visualization method of choice for the 2D analysis of fungal proteomes.

HOMOGENIZATION BUFFER	REFERENCE
8 M urea, 2% (w/v) CHAPS, 20 mM DTT, 0.5% (v/v) biolytes	Fernández-Acero et al., 2006
7 M urea, 2 M thiourea, 4% (w/v) CHAPS, 0.8% (v/v) ampholytes, 20 mM Tris, 20 mM DTT	Kniemeyer et al., 2006
7 M urea, 2 M thiourea, 4% (w/v) CHAPS, 20 mM DTT, 1.0% (v/v) IPG buffer	Shimizu et al., 2009
7 M urea, 2 M thiourea, 4% (w/v) CHAPS, 1% (w/v) DTT, 20 mM Tris, 1% (v/v) Pharmalyte pH 3-10	Lu et al., 2010
8 M urea, 4% (w/v) CHAPS, 40 mM Tris-pH 7.4, 100 mM DTT, 0.2% (w/v) ampholite	Oh et al., 2010
7 M urea, 2 M thiourea, 2% (w/v) CHAPS, 1% (w/v) Zwittergent 3-10, 20 mM Tris	Vödisch et al., 2011
8 M urea, 2 M thiourea, 1% (w/v) CHAPS, 20 mm DTT, 0,5% (v/v) ampholyte 3-10	Yildirim et al., 2011

Table 1. Different solubilisation buffers used for the Proteomics analysis of fungi

3.2 Topics analysed using the proteomics approach on intracellular proteins

Two main technologies are used for the identification of proteins: i) individual protein identification by MALDI-TOF/TOF following SDS-PAGE or 2D-PAGE fractionation and ii) "shotgun" Proteomics, where total protein digests of fungal origin are analysed by tandem LC-MS to generate constituent protein data sets.

3.2.1 Intracellular reference maps

Proteome reference maps provide an impression of the kind and quantity of proteins detectable by either of these two approaches. Furthermore, proteomic data has the potential to improve gene annotation and to discover alternative splice variants of transcripts and isoforms of proteins (Wright et al., 2009; Chang et al., 2010).

The availability of multiple fungal genome sequences during the last five years (over 20 fungal genomes have been completed) has been the key of the large progress of fungal Proteomics research. As a consequence, the number of proteins successfully identified from one specific fungus has drastically increased from some few to several hundreds. Now, the major challenge in modern fungal biology is to understand the expression, function and regulation of the entire set of proteins encoded by fungal genomes.

The combination of 2D-PAGE and mass spectrometry for the identification of proteins from filamentous fungi was first used by Grinyer and co-workers (2004), although reports on 2D-PAGE visualization of fungal proteins had been previously carried out (Hernández-Macedo et al., 2002). Combination of 2D-PAGE with both MALDI-TOF (Kratos Analytical AXIMA CFR plus) and LC-MS/MS [LCQ Deca IT MS (Thermo Finnigan)] techniques, allowed the initial identification of 25 proteins (out of 96 attempted) from the hundreds of proteins resolved in the initial proteome map of *Trichoderma harzianum*, which was performed using whole-cell protein extracts.

Aspergillus sp. is a model organism for filamentous fungi and genome sequences from several species of this genus have been released in the last years (Galagan et al., 2005; Machida et al., 2005; Nierman et al., 2005; Payne et al., 2006; Pel et al., 2007). However, the number of global protein expression studies on *Aspergillus* is still relatively low. The majority of the *Aspergillus* proteome research is still represented by quantitative 2D studies (Carberry and Doyle, 2007; Kim et al., 2008; Kniemeyer, 2011) and less than 10% of the predicted whole proteome of *Aspergillus* sp. has been identified and quantified until now. The mycellial intracellular proteome of the pathogenic fungus *A. fumigatus* was characterized in two studies; the first one identified 54 proteins (Carberry et al., 2006) by 2D-PAGE and Ettan MALDI-TOF (Amersham Biosciences) and more recently, the second one identified 381 spots representing 334 proteins (Vödisch et al., 2009) using the same approach and an Ultraflex MALDI-TOF/TOF (Bruker Daltonics). The majority of those proteins were involved in cellular metabolism, protein synthesis, transport processes and cell cycle. The intracellular proteome of dormant conidia (asexually produced spores) of *A. fumigatus* was carried out by Teutschbein et al. (2010), who used 2D-PAGE and MALDI-TOF techniques to find that these structures contained in particular high amounts of proteins, which are required for stress tolerance and rapid reactivation of metabolic processes. The cytosolic proteome reference map of the well-known producer of primary metabolites and extracellular proteins *Aspergillus niger* was also characterized. 2D electrophoresis combined with a mass spectrometry analysis [Ultraflex MALDI-TOF/TOF (Bruker Daltonics) and ESI-QqTOF MS/MS (Micromass)] allowed the identification of around 100 proteins (Lu et al., 2010). In all studies, proteins primarily involved in translation, energy metabolism, transport processes and the stress response were most abundant.

Penicillium chrysogenum, the microorganism industrially used for the production of penicillin, is another example of filamentous fungus whose genome sequence has been recently released (van den Berg et al., 2008). In this specific case, a first study was reported about the intracellular proteome reference map (Jami et al., 2010a). This map was carried out using 2D-PAGE (http://isa.uniovi.es/P_chrysogenum_proteome/) and further individual protein identification by means of a 4800 MALDI-TOF/TOF Analyzer (Applied Biosystems), which allowed the identification of almost 1000 proteins and isoforms, the majority of them being involved in energy metabolism and the stress responses. Although this is an adequate number of proteins for 2D-PAGE analysis, it represents less than 10% of the predicted whole proteome of *P. chrysogenum*. Therefore, proteome research on filamentous fungi, such as *Aspergillus* sp. or *P. chrysogenum*, clearly lag behind the deep proteome analyses of *S. cerevisiae* and other microbial model organisms with identification of around two-third of the predicted open reading frames (de Godoy et al., 2008; Picotti et al., 2009). However, it is likely that in the more complex filamentous fungi only a fraction of the total genome gives rise to proteins under the laboratory culture conditions. Other sets of proteins may be expressed under different nutritional and environmental conditions in nature.

The ascomycete *B. cinerea* is a phytopathogenic fungus important for being the causative agent of disease in a number of important crops. A first approach for the characterization of the *B. cinerea* proteome, detected over 400 spots in 2D-PAGE, although only 22 protein spots were identified by means of an Ultraflex MALDI-TOF/TOF (Bruker Daltonics) mass spectrometer or an ESI IT MS/MS [Esquire HCT IT (Bruker-Daltoniks)]. Some of them corresponded to isoforms of malate dehydrogenase and glyceraldehyde-3-phosphate dehydrogenase (Fernández-Acero et al., 2006).

In summary, the 2D-PAGE reference maps for filamentous fungi revealed that especially high-abundant proteins involved in metabolic processes and the general stress response are easily accessible to gel-based proteomic approaches. However, this technique is not appropriate for hydrophobic, membrane and low-abundant proteins (see Section 4 of this chapter).

3.2.2 Comparative proteomics

Besides the elaboration of proteome reference maps, the dynamic nature of fungal proteomes in response to different conditions (comparative Proteomics) has been also investigated. For example, the discovery of novel sources of host resistance, the evolution of strains due to the industrial improvement or the effects of carbon sources, antifungal drugs and gene deletion have been analysed in detail (Fernández-Acero et al., 2010; Jami et al., 2010a; Cagas et al., 2011b).

One of the earliest intracellular filamentous fungal proteomic studies was performed on the wood-degrading fungi *Phanerochaete chrysosporium* and *Lentinula edodes* (Hernández-Macedo et al., 2002). These authors used 2D-PAGE to compare cytoplasmic protein expression patterns in the presence or absence of iron and although they visualized 21 proteins related to iron uptake, the identification of such proteins was deficient. A more recent report on *P. chrysosporium* analysed total soluble proteome alterations of this fungus in response to lead exposure using 2D in combination with a 4700 MALDI-TOF Proteome-Analyzer (Applied Biosystems) (Yildirim et al., 2011). This study highlighted the particular role of the elements of DNA repair, post-transcriptional regulation and heterotrimeric G protein signalling in response to different doses of lead.

The proteome response of *Aspergillus nidulans* to the presence of the macrolide antibiotic concanamycin A was also analysed by 2D electrophoresis coupled to an ESI Qq-TOF mass spectrometer (Micromass) (Melin et al. 2002). In addition to glyceraldehydes-3-phosphate dehydrogenase, three proteins were found down-regulated under these conditions; a homolog to a cadmium-induced protein, a homolog to LovC (involved in the biosynthesis of the secondary metabolite lovastatin) and a homolog to a protein of unknown function. Osmoadaptation was also studied in *A. nidulans* by 2D-PAGE and MALDI-TOF [Autoflex series MALDI-TOF (Bruker Daltonics)] (Kim et al. 2007), revealing that glyceraldehyde-3-phosphate dehydrogenase, aldehyde dehydrogenase and a hypothetical protein with a domain of unknown function DUF1349 were overrepresented, whereas enolase and TCA cycle enzymes were less abundant in osmoadapted cells. These results suggest a shift from energy metabolism to the biosynthesis of glycerol, most likely important for control of cell turgor under osmoadptation. The major proteome differences found between *A. nidulans* grown under hypoxia or under normal aeration were also analysed by 2D electrophoresis combined with AXIMA MALDI-TOF (Shimadzu) protein identification (Shimizu et al. 2009). Differences in metabolic enzymes and enzymes for energy production together with an overrepresentation of proteins involved in nucleotide salvage under hypoxic conditions, suggested that activation of nucleotide salvage is a fungal mechanism of adaptation to oxygen deprival. In another study, long-term exposure to menadione was investigated by LC/MSMS Proteomics [3D IT LCQ Fleet (Thermo Fisher)] and transcriptomics in *A. nidulans* vegetative cells. Enzymes in the vitamin B2 and B6 biosynthetic pathways were repressed concomitantly with the repression of some protein folding chaperones and nuclear transport elements. Under long-term oxidative stress, the peroxide-detoxifying peroxiredoxins and cytochrome C peroxidase were replaced by thioredoxin reductase, a nitroreductase and a flavohemoprotein, and protein degradation became predominant to eliminate damaged proteins (Pusztahelyi et al., 2011). Those proteins involved in early phase of conidia germination were also analysed through Proteomics using 2D in conjunction with a Voyger-DE STR MALDI-TOF (PerSeptive Biosystems) mass spectrometer (Oh et al., 2010). During the early phase of the germination stage, levels of proteins involved in metabolism, protein synthesis and transcription highly increased, confirming the importance of metabolic activation and new protein synthesis for the germination process.

The effect of the deletion of a specific gene has been also investigated in *A. nidulans* by comparative Proteomics of the mutant versus the wild-type strain. Proteomic analysis of a strain deleted in the glutathione reductase gene (*glrA*) by 2D-PAGE revealed that 13 proteins were overrepresented, whereas 7 proteins reduced their expression in the *A. nidulans* mutant compared to the wild-type (Sato et al., 2009). The deleted mutant shifted to a temperature sensitive phenotype with decreased intracellular glutathione and reduced resistance to oxidative stress. Analysis of the upregulated proteins by means of an AXIMA MALDI-TOF (Shimadzu) mass spectrometer identified a thioredoxin reductase, cytochrome c peroxidase and catalase B, in addition to a number of peroxiredoxins. It was concluded that increased levels of those enzymes in the glrA-deletion strain revealed interplay between the glutathione system and both the thioredoxin system and hydrogen peroxide defence mechanisms. Moreover, upregulation of an elongation factor 1β (ElfA; 2.5-fold) and a glutathione s-transferase (GstB; 2.6-fold) was also observed in *A. nidulans ΔglrA*. Orthologues of these two proteins were also reported to be upregulated in response to oxidative stress in *A. fumigatus* (Burns et al., 2005; Carberry et al., 2006). In addition, increased levels of catalase B and cytochrome c peroxidase may be responsible for the depleted H_2O_2 levels observed in the deleted mutant. Comparative Proteomics was also applied to another strain of *A. nidulans* deleted in the *hapC* gene, which encodes a component of the transcriptional regulator AnCF controlling redox balancing and coordinating the oxidative stress response. Upregulation of several redox-active proteins, such as thioredoxin, peroxiredoxin A and glutathione was obseved by 2D-PAGE combined to mass spectrometry using an Ultraflex I MALDI-TOF/TOF (Bruker Daltonics) mass spectrometer in the mutant compared to the wild-type strain (Thön et al., 2010).

The dynamic nature of the *A. fumigatus* proteome in response to different conditions has been also subject of study. Comparison of the 2D-PAGE maps from *A. fumigatus* grown on either glucose or ethanol as sole carbon sources was carried out and the main enzymes involved in alcohol metabolism were identified by means of an Ultraflex I MALDI-TOF/TOF (Bruker Daltonics) mass spectrometer (Kniemeyer et al., 2006). Ethanol led to upregulation, among others, of an alcohol dehydrogenase and a particular aldehyde dehydrogenase. The latter was suggested to play a role as acetaldehyde dehydrogenase for the production of acetyl-CoA.

Comparison of developmental stages was also carried out in *A. fumigatus*. Analysis of the proteome of germinating conidia with hyphae revealed that a CipC-like protein was one of the major hyphal-specific proteins. Although the biological function of this cytosolic protein has not been elucidated yet, a putative role during invasive growth in the host has been suggested (Bauer et al., 2010). Four stages of early development were evaluated using the gel free system of isobaric tagging for relative and absolute quantitation (iTRAQ) to determine the full proteomic profile of the pathogen *A. fumigatus*. A total of 461 proteins were identified with a 4800 Plus MALDI-TOF/TOF Analyzer (Applied Biosystems), at 0, 4, 8, and 16 hours and several fold changes for each were established. Ten proteins including the hydrophobin rodlet protein RodA and a protein involved in melanin synthesis Abr2 were found to decrease relative to conidia (Cagas et al., 2011a).

The proteins differentially synthesized in *A. fumigatus* cultivated with the antifungal drug amphotericin B were also investigated in order to identify resistance mechanisms to this antifungal drug (Gautam et al., 2008). Differential expression levels for 85 proteins (76 upregulated and 9 downregulated) were detected by 2D electrophoresis and a 4800 Plus MALDI-TOF/TOF Analyzer (Applied Biosystems). Forty-eight of them were identified with high confidence and included ergosterol pathway proteins (a key amphotericin B target), cell stress proteins, cell wall proteins and transport proteins. Expressions of three genes, a Rho-GDP dissociation inhibitor, a secretory-pathway GDI and Mn SOD, were detectable at both microarray and proteomic levels. Matching between Proteomics and Transcriptomics pointed out the genes encoding those potential target proteins for the development of new antifungal drugs. In a more recent work, the same group followed a similar proteomic-transcriptomic approach to assess the *A. fumigatus* response to artemisinin, an antimalarial drug reported to have antifungal activity against some fungi (Gautam et al., 2011). Proteomic profiles of *A. fumigatus* treated with artemisinin showed modulation of 175 proteins (66 upregulated and 109 downregulated) and peptide mass fingerprinting led to the identification of 85 proteins (29 upregulated and 56 downregulated), 65 of which were unique proteins. The differentially expressed proteins belonged to carbohydrate metabolism, cell stress, amino acid metabolism, translation, ubiquitin-dependent protein degradation, transcription, cytoskeletal proteins, cell wall and associated proteins and others including hypothetical proteins.

Proteome analysis of the enzymatic reactive oxygen intermediates detoxifying system was studied in *A. fumigatus* challenged by H_2O_2 together with the comparative proteome analysis of a mutant deleted in the transcription factor Afyap1, which is involved in the oxidative stress response (Lessing et al., 2007). Differential gel electrophoresis (DIGE) analysis, followed by mass spectrometry using an Ultraflex I MALDI-TOF/TOF (Bruker Daltonics) mass spectrometer, identified 27 protein spots that displayed an increase and 17 protein spots that displayed a decrease in intensity following *A. fumigatus* exposure to H_2O_2. Examples of upregulated proteins were the Allergen AspF3 (putative thioredoxin peroxidase), a mitochondrial peroxredoxin Prx1 (which likely has thioredoxin peroxidase activity), Cu, Zn superoxide dismutase or spermidine synthase. Proteome analysis of the ΔAfyap1 mutant strain challenged with 2 mM H_2O_2 indicated that 29 proteins are controlled directly or indirectly by AfYap1, including catalase 2.

The modification of *P. chrysogenum* from a Proteomics point of view during the industrial strain improvement process was analysed in detail using high-resolution 2D-PAGE coupled to MALDI-TOF/TOF (Jami et al. 2010a). Three strains were compared in this study; the wild type *P. chrysogenum*, a strain with a small improvement for penicillin biosynthesis that is used as a reference in the laboratory and a penicillin high-producing strain. The used experimental conditions allowed for estimating the main proteome changes between different stages of the industrial strain improvement program. Cysteine (a penicillin precursor) biosynthetic enzymes, enzymes of the pentose phosphate pathway and stress response proteins were overrepresented in the high producer strains, whereas proteins for virulence and biosynthesis of other secondary metabolites different from penicillin (pigments and isoflavonoids) were down-regulated. This interesting study concluded that the increased penicillin titers reached by the high-producing strains is a consequence of rebalancing among different metabolic pathways.

Combination of proteomic analysis with metabolic measurements is a very interesting approach for the study of fungal metabolic processes. Matsuzaki and co-workers (2008) analysed the intracellular processes of the white-rot basidiomycete *P. chrysosporium* involved in the metabolism of benzoic acid at the proteome and metabolome levels by high-resolution 2D electrophoresis coupled to LC-MS/MS. Aryl-alcohol dehydrogenase, aryl-aldehyde dehydrogenase, and cytochrome P450s were upregulated after the addition of exogenous benzoic acid. Intracellular metabolic shifts from the short-cut TCA/glyoxylate bicycle system to the TCA cycle and an increased flux in the TCA cycle indicated activation of the heme biosynthetic pathway and the production of NAD(P)H. In addition, combined analyses of proteome and metabolome clearly showed the role of trehalose as a storage disaccharide and that the mannitol cycle plays a role in an alternative energy-producing pathway. Conclusions could be drawn for the process of lignin degradation based on enzyme expression values (assessed by Proteomics) and the amount of chemical species related to these enzymes (assessed by Metabolomics).

In another example of combination of Proteomics with metabolomics the analysis of the effect of lactate and starch on fumonisin B2 biosynthesis (a mycotoxin) in *A. niger* was carried out using 2D-PAGE and a 4800 MALDI-TOF/TOF Analyzer (Applied Biosystems) (Sørensen et al., 2009). These authors reported that there is a specific relation between the increase in fumonisin B2 and the enzymes affecting the intracellular levels of acetyl-CoA.

4. Proteome analysis of fungal secretomes

Greenbaum and co-workers (2001) defined the secretome as the population of gene products that are secreted from the cell. In 2010, Makridakis and Vlahou specified the definition as the rich complex set of molecules secreted from living cells, which in a less strict definition includes molecules shed from the surface of living cells. Thus, the secretome or extracellular proteome consists of the totality of the proteins secreted from a cell, organism or entity. These proteins can be subdivided into freely released proteins and proteins associated to the outer cell wall. Even the cellular machinery involved in the protein secretion was proposed as part of secretome (Tjalsma et al., 2000).

Many fungi secrete a huge number of proteins to facilitate their saprotrophic lifestyle, which justifies the secretome-related studies to understand this filamentous fungi way of life. Thus, Bennett (2006) wrote that "animals eat their food and then digest it; fungi digest their food and then eat it", which illustrates the vast number of extracellular enzymes necessary to assimilate the whole potential substrates. Studies of the secreted proteins from filamentous fungi show that they are not only natural toolboxes for white biotechnology (degradation process, industrial application, etc.), but also that they are the most important indoor allergen generators. Besides, some of these proteins are important as pathogenicity biomarkers in humans and plants, while others have demonstrable potential for the control of insect pests.

The considerable number of available publications on the analysis of human pathogenic bacteria and yeasts contrasts with the limited number of secretome studies on filamentous asco- and basidiomycetes. Since Medina and co-workers (2004) described the extracellular proteins involved in degradation of the flavonoid rutin by *A. flavus*, a few publications in the last years have described the fungal secretomes applied to different fields as described in the next sections.

4.1 Fungal allergen identification by 2D analysis

Naturally and synthetically produced compounds are responsible for the increase of human health conditions by allergies. Successful diagnosis and therapy are conditioned by the identification of the allergenic compounds, which mainly come from food and inhalation. Thus, pollens of grass and trees are the most important outdoor airborne allergens. In addition, other important allergenic source is airborne mould (Benndorf et al., 2008). A significant association between asthma and sensitization to *Alternaria* and *Cladosporium* (up to 20%) has been described (Zureik et al., 2002). These fungal species occur typically outdoors with a maximum level of spore concentration in summer and autumn, whereas species such as *Aspergillus* and *Penicillium*, especially *Aspergillus versicolor, Penicillium expansum* and *P. chrysogenum*, dominate indoors in winter in humid and cold climates (Cruz et al., 1997; Benndorf et al., 2008). Commercialized test systems to diagnose allergic reactions to this mould species are not properly developed. The allergens studied by means of Proteomics combining the resolution of 2D electrophoresis, the specificity of the immunoblotting and the mass spectrometry-based protein identification allow exploring the so-called immunoproteome or "immunome" (Doyle, 2011). This approach, which is usually performed using immunoglobulin E (IgE) from sera of sensitive patients, is an emerging strategy for the identification of immunoreactive fungal antigens (Benndorf et al., 2008; Ishibashi et al., 2009). In fact, this allergen screening system is only possible by means of the bidimesional electrophoresis, since it is the proper way to combine several sensitive sera over the same antigens. This procedure has been used for allergen identification of atopic dermatitis caused by *Malassezia globosa* [using an oMALDI Qq-TOF (Applied Biosystems) and N-terminal sequence analysis in a Procise 494 protein sequencer (Applied Biosystems) (Ishibashi et al., 2009)], enolase antigen from *Penicillium citrinum* and *A. fumigatus* [Edman degradation (Lai et al., 2002)], spore allergens from the indoor mould *A. versicolor* [by means of a LC/MSD TRAP XCT system (Agilent Technologies) (Benndorf et al., 2008)], or antigen of *A. fumigatus* from allergic broncho-pulmonary aspergillosis patients [using a Q-TOF Ultima Global mass spectrometer (Micromass) (Singh et al., 2010)]. Figure 1 shows the antigen detection system by 2D electrophoresis. As first step, the sera from sensitive patients are tested in order to find common allergens by monodimensional electrophoresis. Those common reactive sera can be used over the same blotted 2D gel (stripping for reprobing) to locate the antigenic protein, which is identified by mass spectrometry or N-terminal sequencing (Edman degradation). Thus, the peptides obtained are used as the template for degenerated primer design. These primers combined with the rapid amplification of cDNA ends method (RACE) allow the gene cloning from cDNA.

The characterization of secreted proteins has also other medical applications since surface-associated and secreted proteins represent primarily exposed components of fungi during host infection. Thus, several secreted proteins contribute to pathogenicity since its role in defence mechanisms or immune evasion is known. These studies lead to the identification of possible biomarkers for the verification of diseases caused by *Aspergillus* species as *A. fumigatus* combining 2D electrophoresis and mass spectrometry identification [Ultraflex I MALDI-TOF/TOF (Bruker Daltonics)] (Wartenberg et al., 2011).

4.2 The environment and fungal secretomes

Over 8,000 species of fungi cause more plant diseases than any other group of plant pests. As a result, some of the world's great famines, human suffering and largest human migrations on the planet have been due to plant-pathogenic fungi. The destruction of wheat

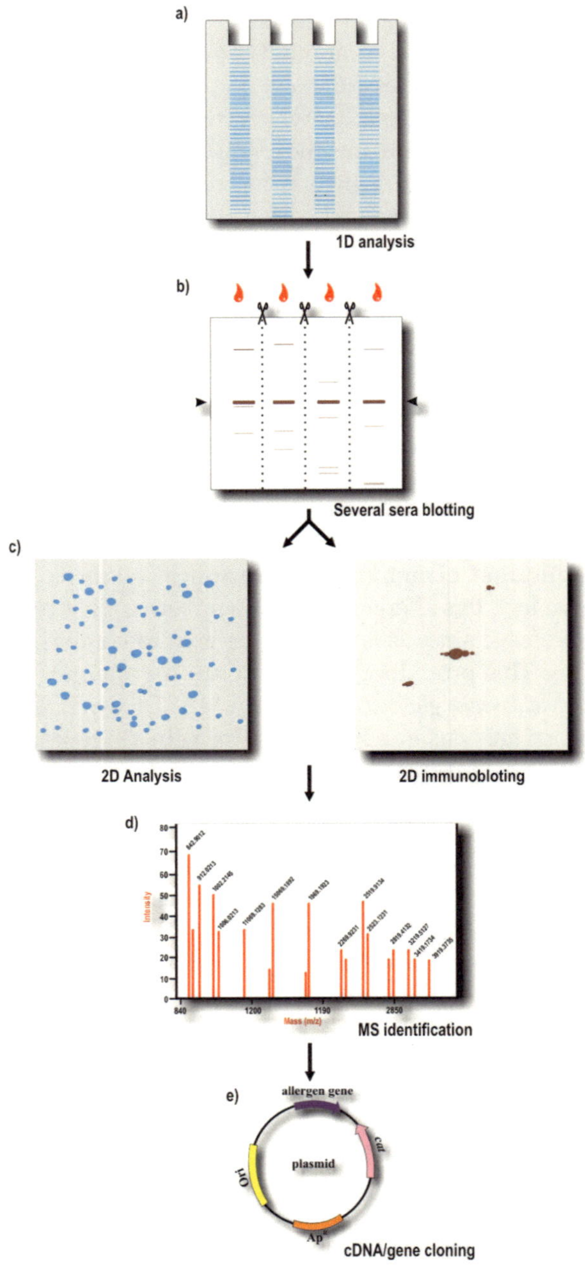

Fig. 1. Proteomics approach to allergen identification in fungi. The identification of fungal antigens needs several sera from different sensitive patients. After protein extraction, a monodimensional gel allows protein separation (a). The immunoblotting using different sera allows allergen localization and western condition optimization (b). 2D electrophoresis is used for individual antigen identification. The Coomassie stained gel (control) is used for protein isolation (c). Mass spectrometry is carried out for peptide identification (d). As final step, degenerated primers are designed and the gene encoding the allergen is cloned by combination of these primers and RACE methodology (e).

crops in the Middle Ages caused by the fungus called bunt or stinking smut (*Tilletia* spp.), the potato blight in Ireland and northern Europe caused by *Phytophthora infestans* or the strike in the grape vineyards of central Europe due to the downy mildew (*Plasmopara viticola*) in the 1870s are clear examples of the strong influence of the fungi over mankind (Ellis et al., 2008). Therefore, the interaction fungi-plant is in one the most interesting clues to be faced up for Proteomics due to its global economic implications.

The recognition plant-pathogen is either host specific or almost generalist. Thus, the genus *Botrytis*, which presents some species that seem to have host specificity, also includes *B. cinerea*. This fungus is able to infect more than 200 different plant species, at any plant stage or tissue, including fruits during storage and distribution. As a result, it inflicts high economic damage at various levels in the food industry (Fernández-Acero et al., 2010). The ability for *B. cinerea* to infect is due to its wide variety of infection strategies, which are facilitated by a set of pathogenicity/virulence factors. The study of the secreted proteins has been faced up by means of different carbon sources from fruit (tomato, strawberry, kiwifruit, glucose, carboxymethyl cellulose, starch, pectin or tomato cell walls) and using solid and liquid cultures (Shah et al., 2008; Espino et al., 2010; Fernández-Acero et al., 2010). The identified proteins by 2D electrophoresis and MALDI TOF-TOF [Autoflex MALDI-TOF or UltraflexIII MALDI TOF/TOF (Bruker-Daltonics)] or LC-MS/MS [LTQ linear IT (Thermo Fisher)] were involved in the degradation of plant defensive barriers, proteases, pathogenicity factors and virulence factors (e. g.: pectin methyl esterases, xylanases and proteases). This comparative approach is a promising strategy for discovering new pathogenicity factors and dissecting infection mechanisms.

The ultrafast genome sequencing is a boosted fashion trend, which is generating thousands of gene sequences every day. In spite of this fact, several fungi which are producers of severe infections in fruit trees (fruit blight) and grapevines, like *Diplodia seriata*, keep fully unsequenced. Despite this limitation, 2D electrophoresis application combined with *de novo* sequencing and BLAST similarity search by a 4800 MALDI-TOF/TOF Analyzer (Applied Biosystems) led to the protein identification of virulence factors described in other fungal strains when they grew in medium supplemented with carboxymethyl cellulose (Cobos et al., 2010). This fact shows the wide range of Proteomics applications.

The use of fungi as effective biocontrol agents is a promising field. Thus, *Trichoderma* spp., which are mycoparasites of several soil-borne fungal plant pathogens, have been analysed using the Proteomics approach. Suárez and co-workers (2005) visualized on 2D gels more than 250 extracellular proteins when chitin served as a carbon source, but comprehensive protein identification by mass spectrometry was not performed. Recently, Monteiro and co-workers (2010) studied the overgrow, mycelia degradation and protein secreted by *T. harzianum* on different plant pathogens (*Macrophomina phaseolina*, *Fusarium* sp. or *Rhizoctonia solani*) by means of 2D electrophoresis and mass spectrometry [MALDI-TOF Reflex IV (Bruker Daltonics)]. Intriguingly, hyphae of *T. harzianum* did not present coiling around *Fusarium* sp. supporting the idea that mycoparasitism may be different among plant pathogens. These data were concordant with the differently observed secretion patterns, which are related to phytopathogen cell-wall composition.

Tan spots are associated with the development of extensive chlorotic and/or necrotic lesions on the leaves of susceptible host genotypes caused by the fungus *Pyrenophora tritici-repentis*. It is an economically important foliar disease of wheat with a worldwide distribution (Cao et al., 2009). These authors compared the secretomes of avirulent *versus* virulent race isolates of the pathogen by means of 2D electrophoresis and later ESI-q-TOF MS/MS (Micromass) identification. Therefore, various metabolic enzymes in addition of µ-mannosidase, exo-b-1,3-glucanase, heat-shock and BiP proteins were found as up-regulated in the virulent race. Such differences could reflect an adaptation to a saprophytic habit in avirulent isolates of the fungus and it may suppose the application of avirulent races as biological control system to displace the virulent ones.

An environmentally friendly alternative to chemical insecticides for agricultural pest control are those fungi involved in insect-fungus interactions. These can result in biological insecticides deduced from entomopathogenic fungus. Thus, *Metarhizium* spp. are a good model system for studying these interactions and a resource of genes for biotechnology. Previously to the genome sequence publication of *Metarhizium anisopliae* and *Metarhizium acridum* (Gao et al., 2011), the intercellular proteome from conidia and mycelia of *M. acridum* was analysed by 2D electrophoresis and mass spectrometry [MALDI-TOF/TOF-MS (Axima Performace) or MALDI MS/MS (Waters)] (Barros et al., 2010) observing clear differences in the expression of identical proteins and isoform occurrence between conidia and mycelia. The secretome analysis of *M. anisopliae* combining 2D electrophoresis and mass spectrometry [UltraFlex II MALDI-TOF/TOF (Bruker Daltonics)] showed the presence of proteases, reductases and acetyltransferase enzymes in presence of the exoskeleton of *Callosobruchus maculatus* (main pest of cowpea) (Murad et al., 2006). In addition, the genome sequencing of fungal pathogens of trees that live as beetle symbionts such as *Grosmannia clavigera* (DiGuistini et al., 2011) will facilitate the analysis of the secretome evolution and its applicability to pest control. This analysis of secreted proteins of entomopathogenic fungi is a poorly explored field, which supposes one of the next and promising Proteomics challenges.

4.3 Proteomics of industrially interesting fungi

Filamentous fungi have an extraordinary ability to secrete proteins, secondary metabolites and organic acids to the culture medium. Thus, in comparison to other eukaryotic expression systems, such as yeast, algae or insect cells, filamentous fungi possess the exceptional advantage of an unbeatable secretion capacity. This is the reason why filamentous fungi have been used for food maturation and beverage processes for more than 1,500 years. Therefore, this traditional use has been transformed in commercial applications for food and beverage industries (Peberdy, 1994; Punt et al., 2002; Cardoza et al., 2003). Specially interesting is the list of commercial enzymes produced by Aspergilli that includes amylases, chymosin, glucose oxidases, catalases, cellulases, pectinases, lipases, proteases, phytases and xylanases in food, detergent, textile, pulp and paper industries (see review: Fleißner & Dersch, 2010). As an example, the production of fermented foods like miso, sake or soy sauce is traditionally performed by the ascomycete *Aspergillus oryzae*, which presents high protein secretion levels. This fungus is also an interesting host for enzyme production (homologous and heterologous) (Bouws et al., 2008). The extracellular proteome analysis of *A. oryzae* at different times in submerged and solid-state cultures containing wheat bran showed an impressive number of enzymes related to cell wall degradation (glucoamylase A, xylanase G2, α-glucosidase A, cellulase B, α-amylase, β-glucosidase and xylanase F3) (Oda et al., 2006). Thus, *A. oryzae* is also used in modern biotechnology due to this array of secreted proteins (Machida et al., 2005).

Even when the analysis of the extracellular proteome does not suggest a clear relation with the antibiotics production in *P. chrysogenum*, it can offer new clues either for the general understanding of the fungal metabolism or for possible industrial applications of the secreted proteins. Thus, some secreted proteins identified in the culture medium of *P. chrysogenum* (isoamyl alcohol oxidase, sulfydryl oxidase, dihydroxyacid dehydratase, polygalacturonases, pectate lyases, ferulic acid esterases) by 2D electrophoresis and mass spectrometry [4800 MALDI-TOF/TOF Analyzer (Applied Biosystems)] are especially relevant because of their interest for food industry as sake and other alcoholic beverages production, burnt flavor removal from sterilized milk, preparation of juices or health product preparation (Jami et al., 2010b).

The public recognition that environmental pollution is a worldwide threat to public health has given rise to a new massive industry for environmental restoration. Hence, paper and pulp industries, as well as food and feed, beverage, textile and several other industrial production processes are linked by the importance of the degradation of plant cell wall polysaccharides. The most efficient degraders of lignocelluloses are saprophytic asco- and basidiomycetes, since they are robust organisms that have a high tolerance to toxic environments. These wood degraders are typically divided in: i) white rot fungi (degradation of hemicelluloses and lignin by basidiomycetes and ascomycetes) and ii) brown rot fungi (attack of cellulose and hemicelluloses, exclusively, accomplished by basidiomycetous) (Bouws et al., 2008). *Postia placenta* is a brown-rot fungi common inhabitant of forest and also largely responsible for the destructive decay of wooden structures. Its genome, transcriptome and secretome have been recently studied revealing unique extracellular enzyme systems, which include an unusual repertoire of extracellular glycoside hydrolases (Martínez et al., 2009). Interestingly, the comparisons with the closely related white-rot fungus *P. chrysosporium* sustain an evolutionary shift from white-rot to brown-rot showing the loss of the capacity for efficient depolymerization of lignin.

Biopulping is an industrial biotechnology application of natural fungi to convert wood chips to paper pulp in an ecosafety process diminishing the chemical and mechanical problems of paper production. Thus fungal pre-treatment of wood led to a significant increase in pulp yield and a better bleachability of the pulp, which can be explained by the production of specific lignocellulose-degrading enzymes through the Proteomics analysis combining 2D electrophoresis and MALDI-QTOF MS (Applied Biosystems)] (Ravalason et al., 2008).

4.4 Guaranteeing the secretome quality: how to distinguish secretome from 'degradome'?

The previous sections have shown the relevance of the extracellular proteome analysis, but the crucial step for the secreted protein analysis is to distinguish these proteins really secreted from those "contaminant" proteins resulting from cell lyses events. This capital process can be sequentially tackled from three different points of view: i) sample collection; ii) post-identification protein analysis; iii) analysis of "moonlighting proteins" and unconventional secreted proteins.

Some tips in the collection of secreted protein are shown in the figure 2 following the method described for *P. chrysogenum* (Jami et al., 2010b), such as the careful mycelia elimination process (by the use of nylon filters, rigorous centrifugation and filtration through 0.45 µm filters) or the use of low temperatures to avoid mycelia degradation. Sample collection is also crucial to discard those non-natural secreted protein. The scientific literature describes how the presence of extracellular proteins in the culture medium is directly correlated to the growth phase (Peberdy, 1994). Jami and co-workers (2010b) showed the linear correlation existing between the presence of proteins in the culture medium and the culture growth (biomass formation). Interestingly, the protein levels that were present in the broth increased even after the culture reached the stationary phase. These data strongly suggest that some of those proteins arrived to the culture medium as a result of cell lysis. Consequently, taking samples at an early time point is recommended in concordance with the optimal moment of extracellular protein secretion. For example, *P. chrysogenum* does not show significant amounts of secreted proteins before 24 hours.

Regarding the post-identification protein analysis, Jami and co-workers (2010b) did the comparison between secreted proteins at 40 h *versus* those secreted at 68 h in *P. chrysogenum*. They demonstrated that the expression differences observed during the time course were mainly due to the presence at 40 h of isoforms from those lately expressed proteins or proteins expressed only at late stages. As a conclusion, the amount of "possible" contaminant intracellular proteins in the secretome was reported to be very low (6.09%) in *P. chrysogenum* at 68 h (Jami et al., 2010b). Thus, the knowledge of the protein abundance at the intracellular proteome helps to clarify possible cell lysis events. As an example, malate dehydrogenase and glyceraldehyde-3-phosphate dehydrogenase in *B. cinerea* or flavohemoglobin and manganese superoxide dismutase in *P. chrysogenum* should be highly detected in the case of lysis, since these are the most abundant intracellular proteins (Shah et al., 2009; Jami et al., 2010a). However, sometimes it is difficult to ascertain whether a protein is truly secreted or it is present in the culture medium as a consequence of cell lysis. Recently, new bioinformatics tools have been developed to characterize the secreted proteins either in a classical or non-classical way by means of prediction of secretion signal motifs. Therefore, SignalP (for classical secretion signal motifs) and SecretomeP (for non-classical signal motifs) are useful prediction software applications (Bendtsen et al., 2004 a, b) that can be used for real secretion understanding.

Even when the above-mentioned information is taken into account and the researcher has the certainty of the protein sample quality, some unconventional proteins, dubbed "moonlighting proteins" (Jeffery, 1999), can be detected. These proteins present unconventional protein secretion systems (Nickel & Rabouille, 2009) and then, it is difficult to define their authenticity as secreted proteins. One meaning of moonlighting is "to do paid work, usually at night, in addition to one's regular employment". Therefore, "moonlighting proteins" is a proper designation for this group of multifunctional proteins, which are widespread among organisms ranging from bacteria to mammals (Gancedo & Flores, 2008). In spite of this ubiquity, more attention has been traditionally given to the "moonlighting proteins" found in higher eukaryotes, but recently the yeasts have caught the attention for studying these proteins in lower eukaryotes. Thus, in many cases, the fusion of two genes that initially encoded proteins with single functions has been the origin of that duality. On the other hand, a significant number of proteins can perform dissimilar functions (see reviews: Gancedo & Flores, 2008; Flores & Gancedo, 2011).

An interesting phenomenon that has been observed in the *P. chrysogenum* secretome is the fact that some extracellular proteins were also previously identified in the microbody matrix of this fungus (Kiel et al., 2009). Therefore, an alternative explanation given to the presence of intracellular proteins in the culture broth was the selective autophagic degradation of peroxisomes (pexophagy), which can give a new sense to the strange "moonlighting proteins" as it was explained in *P. chrysogenum* (Jami et al., 2010b; Martín et al., 2010). It is well known that peroxisome abundance can be rapidly decreased through autophagic pathways, which selectively degrade peroxisomes by fusion to lysosomes or vacuoles (Oku and Sakai, 2010). Thus, integration of peroxisomes into vacuoles may lead to secretion of the proteins located in the peroxisomal matrix by exocytosis, a mechanism that has been discussed as an alternative route for the release of penicillin from peroxisomes to the culture medium (Martín et al., 2010).

Even when the mycelia lysis is an undesirable event, it can suppose a quite informative source. Thus, the analysis of the secretome of the pathogenic fungus *Fusarium graminearum in vitro* and *in planta* demonstrates the presence of thirteen non-secreted proteins only under *in planta* conditions combining SDS-PAGE protein separation and LTQ/FT MS (ThermoElectron) identification (Paper et al., 2007). These proteins, some of which are potent immunogens secreted by animal pathogenic fungi, indicate a significant fungal lysis during plant pathogenesis, which shows the lysis as a useful event for the infection even if it involves partial degradation of the fungal population.

4.5 Bioinformatics tools available for secreted protein predictions

Around 90% of human secreted proteins and almost 90% of the *A. niger* identified extracellular proteins by mass spectrometry contain classical N-terminal signal peptides. These signal peptides, which redirect the ribosomes to the rough endoplasmic reticulum, are typically 15–30 amino acids long and consist of 15–20 hydrophobic amino acid residues cleaved off during translocation across the membrane. Besides, there are also examples of non-classically secreted proteins in fungi, but generally the main group of secreted fungal proteins is processed by the classical secretory pathway (Lum & Min, 2011).

Bendtsen and co-workers (2004a) developed the software SignalP 3.0 (http://www.cbs.dtu.dk/services/SignalP/) for the identification of classical secretion signal motifs presents in proteins. This software is widely extended and commonly integrated in different bioinformatics platforms for secreted proteins identification. On the other hand, Phobius software (Kall et al., 2004) is used to identify signal peptides and to discriminate putative transmembrane domains (http://phobius.cgb.ki.se/index.html). The same developers of SignalP 3.0 software created a prediction program of secreted proteins by means of non-classical signal motifs (Bendtsen et al., 2004b). It was called SecretomeP (http://www.cbs.dtu.dk/services/SecretomeP/). Even when this program was developed for mammalian applications and later improved for bacteria, it has shown good results in the fungal protein analysis (Shah et al., 2009; Jami et al., 2010b).

The availability of several fungal genomes in databases together with the use of those prediction programs for secretory proteins have allowed the recent development of platforms for the annotation of fungal secretomes, such as the Fungal Secretome Database [FSD (Choi et al., 2010)] or the Fungal Secretome KnowledgeBase [FunSecKB (Lum & Min, 2011)], which allow the proper identification of secreted proteins by the compilation of different software. Thus, the Fungal Secretome Database (http://fsd.snu.ac.kr/) is a hierarchical identification pipeline based on nine prediction programs (SignalP 3.0, SigCleave, SigPred, RPSP, TMHMM 2.0c, TargetP 1.1b, PSort II, SecretomeP 1.0f and predictNLS), which predict the presence of signal peptides or nuclear localization signals, trans-membrane helixes, protein probable location and secretion by non-classical pathways. In contrast, Lum and Min (2011), who suppose that the FSD platform significantly over-estimate the number of secreted proteins in fungi, developed FunSecKB (http://proteomics.ysu.edu/secretomes/fungi.php) platform. This system offers a more conservative prediction tool. Nevertheless, the FunSecKB and the FSD databases could complement each other as different data sources, prediction tools and data access utilities.

The connection between the bioinformatics predictions and the really secreted proteins has been analysed in the white-rot basidiomycete *P. chrysosporium* (Wymelenberg et al., 2005;

2006) and in the filamentous ascomycete *A. niger* (Tsang et al., 2009) under different culture conditions. They demonstrated that SignalP results in higher specificity than Phobius, but the combined results from both yielded higher specificity of prediction than separately (Tsang et al., 2009). In addition, the computational secretome is clearly incomplete, but the predicted proteins include many interesting sequences that provide a framework for future investigations (Wymelenberg et al., 2006).

A reference map is a useful tool for the location of secreted proteins in addition to provide a good chance to identify those proteins. Thus, the first secretome reference maps just presented a few dozens of spots as that of *P. sapidus*, a white-rot fungus able to attack lignified biopolymers (Zorn et al., 2005). But the protein purification systems have evolved and several hundreds can be easily observed, as in case of *T. harzianum* (Suárez et al., 2005) or *P. chrysogenum* (Jami et al., 2010b) reference maps. Nowadays, there are several webpages that allow the public deposit of 2D reference maps [e.g.: World-2DPAGE Repository (http://world-2dpage.expasy.org/repository/)], where the proteins are easily identified clicking on the spot. However, these reference maps are usually from intracellular proteomes. The last step of this useful tool is the development of specific webpages by the research groups. Thus, the yeast *S. cerevisiae* has its own intracellular reference map website [Yeast Protein Map (http://www.ibgc.u-bordeaux2.fr/YPM/)]. The extracellular proteome of *P. chrysrogenum* that has been fully completed by Jami and co-workers (2010b) and up to our knowledge it has been the first secretome reference map on-line generated by the users. It is available on-line through the website: http://isa.uniovi.es/P_chrysogenum_secretome/

4.6 Extraction methods and buffer compositions of filamentous fungi secretomes

Table 2 summarizes some of the most recent methodological updates for the extracellular protein analysis of filamentous fungi.

Fungus	Source/Precipitation	Resuspension condition	Ref.
A. flavus	**Liquid:** Whatman filtration, lyophilization O/N. Resuspension in 5mL of ddH_2O, store at -20ºC. TCA precipitation (200 g/L) at -20ºC, 1 h. centrifugation 5 min, 15000g at 4ºC.	8M urea, 2% CHAPS, 50mM DTT, 0.2% (w/v) ampholytes 3/10, trace of bromophenol blue.	Medina et al., 2004
B. cinerea	**Solid:** Cellophane membranes containing germinated fungal spores were floated onto 37.5 mM sodium acetate buffer (pH 4.4) in the dark for 10 days. 1 ml was frozen / lyophilized to 30 μL	Mix 30 μL of sample plus 10 μL of SDS sample and buffer (Invitrogen). Boil 10 min, cool to R/T. Load 20 μL of mixture onto a NuPAGE 12% Bis-Tris precast gel.	Shah et al., 2008
	Liquid: Centrifugate (15000xg, 4ºC, 10 min) the medium. Filtrate supernatant through several layers of filter paper (Whatman No. 1). Froze at -20ºC O/N (allow polysaccharide precipitation). Filtrate through filter paper. Add TCA [6% (v/v) final concentration] and precipitate 1 h on ice. Centrifugate (15000xg, 4ºC, 10 min). Washed 3 times 96% ethanol. Resuspend in 8M urea. Centrifugate at 15000xg, 5 min to eliminate the insoluble material. Precipitated again with methanol-chloroform. Stored dry at -20ºC. Dissolve dry protein in ReadyPrep sequential extraction Reagent 2 to 0.04 mg/mL.	ReadyPrep sequential extraction Reagent 2: 8 M urea, 4% CHAPS, 40 mM Tris, 0.2% Bio-Lyte 3/10 ampholyte.	Espino et al., 2010

M. anisopliae	**Liquid:** 0.3 mm Millipore filtration, ddH_2O dialysis with a 1.0 kDa cutoff at 4ºC. Lyophilized. Store -80ºC. Resuspend 2000 mg in 0.005 M Tris/EDTA buffer (1 mM PMSF and 1 mM E-64 protease inhibitors). Precipitate by 2D Clean-Up kit (GE HealthCare).	To 500 mg of protein add 250 ml solution: 2% CHAPS, 8 M urea, 7 mg DTT ml^{-1} and 2% IPG buffer.	Murad et al., 2006
L. maculans/ L. bicolor	**Liquid:** **i)** Ultrafiltrate (Amicon Ultra-15) 15 mL of dialyzed secretome by centrifugation 45 min (4ºC, 5000xg) until 500 mL. -TCA precipitation: add 1.5 mL of 10% w/v TCA/0.007% v/v 2-ME in cold acetone to 500 mL of sample. -Phenol extraction: add 500 mL 2M sucrose buffer to 500 mL of the ultrafiltered sample, incubate on ice 15 min. Add 1 mL of Tris-saturated phenol (pH 8). The dried pellet was resolubilized in 500 mL of solution R. **ii)** TCA/acetone precipitation: mix 10 mL of sample and 40 mL 10% w/v TCA/0.007% v/v 2ME in cold acetone (-20ºC). Centrifugate: 30 min, -20ºC, 18000xg, discard supernatant. Repeat 3 times adding 10 ml of secretome every time. Wash in 0.002% v/v 2ME/ acetone. Dry pelet 2 h at 200 mbars and resolubilized in 500 mL of solution R. **iii)** Phenol/ether extraction: mix 10 mL of sample and 15 ml of phenol. Vortex 20 s; centrifugate 12000xg, 5 min. Discard suppernatant. Add 2 volumes of ether. Vortex 20 s; centrifugate 12000, 5 min. Rerpeat this step. Dry lower aquous phase under vaccum. Resolubilizate in 500 mL of solution R. **iii)** Lyophilization: freeze-dry 15 mL of sample and resolubilizate in 500 mL of solution R.	Solution R: 7M urea, 2M thiourea, 2% w/v CHAPS, 1% w/v DTT, 0.5% v/v proteinase inhibitor mix (GE Healthcare), 0.5% v/v pH 4–7 ampholites and 0.5% v/v pH 3–10 ampholites. 50 mL of solution R per microgram of pellet.	Vincent et al., 2009
P. tritici-repentis	**Liquid:** Filtrate medium through a 0.22 μm cellulose nitrate filter. Dialyze (cut-off: 1000 Da) against ddH_2O at 4ºC. Lyophilize and resuspend in 1.2 mL rehydration buffer containing 2 mM tributylphosphine (BioRad). Prior to 2D electrophoreis, samples were desalted using a ReadyPrepTM 2-D Cleanup kit (BioRad).	8 M urea, 2% CHAPS, 50 mM DTT, 0.2% Bio-Lyte 3/10 ampholyte, 0.001% bromophenol blue	Cao et al., 2009
T. harzianum	**Liquid:** Dialyzed sample against ddH_2O. Concentrate by ultrafiltration. Precipitate 400 μg of proteins with 2-D clean-up kit (GE Healthcare). Dissolve in rehydration solution.	8 M urea, 0.5% CHAPS, 0.2% IPG buffer, 15 mM DTT and trace amounts of bromophenol blue	Monteiro et al., 2010
A. niger	**Liquid:** Clarify supernatant by 0.2 μm filtration. Precipitate with TCA (10% w/v) O/N at 4 °C. Wash with ice cold acetone. Resuspend pelet in 0.1% SDS	0.1% SDS buffer	Adav et al., 2010

Table 2. Summary of methodologies used for secretome analyses of filamentous fungi. Microorganism, protein extraction method, buffer composition and reference are indicated. 2ME: 2-Mercaptoethanol (β-mercaptoethanol); TCA: Trichloroacetic acid.

5. Proteome analysis of membranes and organelles

Kim and co-workers (2007) propose the term 'subproteomics' to describe proteomic analysis of a defined subset of an organism's protein complement, primarily specific organelles based on the previous review of Cordwell and co-workers (2000). In the case of bacteria: cytosolic, membrane, cell surface-associated and extracellular proteins (secretome) are considered as subproteomes (Hecker et al., 2010). Thus, in eukaryotes the organelle proteomes are a clear example of a subproteome.

The connection of filamentous fungi between the local environment and inner media is provided by the embedded proteins present in the cell wall, plasma and organelle membranes. Thus, the cell wall, which is the largest organelle of a filamentous fungus, represents an essential dynamic structure fulfilling many vital functions, such as physical protection, osmotic stability, selective permeability barrier, immobilized enzyme support, cell-cell interactions and morphogenesis (Pitarch et al., 2008).

To date, most MS-based proteomic analyses of filamentous fungi have had three targets: the whole cell (mycelial extract), the cytosolic proteins and the secretome. Nevertheless, the membranes of organelles, like mitochondria, which includes proteins of the transport systems (e.g.: porins), as well as proteins involved in fusion, fission, morphology and the inheritance of the organelle, remains unknown. Thus, organelle proteome mapping of different cell compartments is an interesting way to reveal various aspects of fungal metabolite production, e.g. penicillin production in the microbodies of *P. chrysogenum* (Ferreira de Oliveira & de Graaff, 2011; Kiel et al. 2009).

5.1 Organelle isolation and enrichment for proteome analysis

A crucial step in the organelle Proteomics is organelle enrichment, since downstream analysis is dependent on its good-quality purification. The filamentous fungi idiosyncrasy results in tricky organelle isolation due to their protease production, the polarized growth substantiated by microtubules (increase the organelle separation) and their compact cell wall that difficult cell disruption. Thus, Ferreira de Oliveira and de Graaff (2011) propose two capital steps for organelle enrichment: i) cell disruption; ii) crude organelle separation and enrichment by additional separation techniques. An example of these steps is summarized in figure 2. Firstly, cell disruption should be as homogeneous and reproducible as possible, thus automatic grinders or French pressure cell are recommended. Anyway, bead beating or manual mortar handle can be used. An interesting and popular method to obtain intact organelles is the enzymatic degradation for protoplast formation followed by gentle lysis. In spite of its popularity, the time consumption (2-3 hours) of the process and batch enzymes variability are cons of this method. Secondly, different methods can be used for debris elimination as filtration; low-speed centrifugation; differential detergent fractionation; centrifugal elutriation; ultracentrifugation (linear or density gradients); immunomagnetic separation, which has been used for organelle isolation (mitochondria vacuoles, microbodies, endosomes, vesicles) in other eukaryotes; or miniaturized free-flow electrophoresis (FFE) (Kohlheyer et al. 2008, Ferreira de Oliveira & de Graaff, 2011).

A large number of diseases and developmental abnormalities result from mitochondrial functions. This fact has caught the attention over the mitochondrial biology, since our understanding of mitochondrial molecular biology has been obtained from in *Saccharomyces* and *Neurospora*. The combination of different electrophoretic techniques such as blue native,

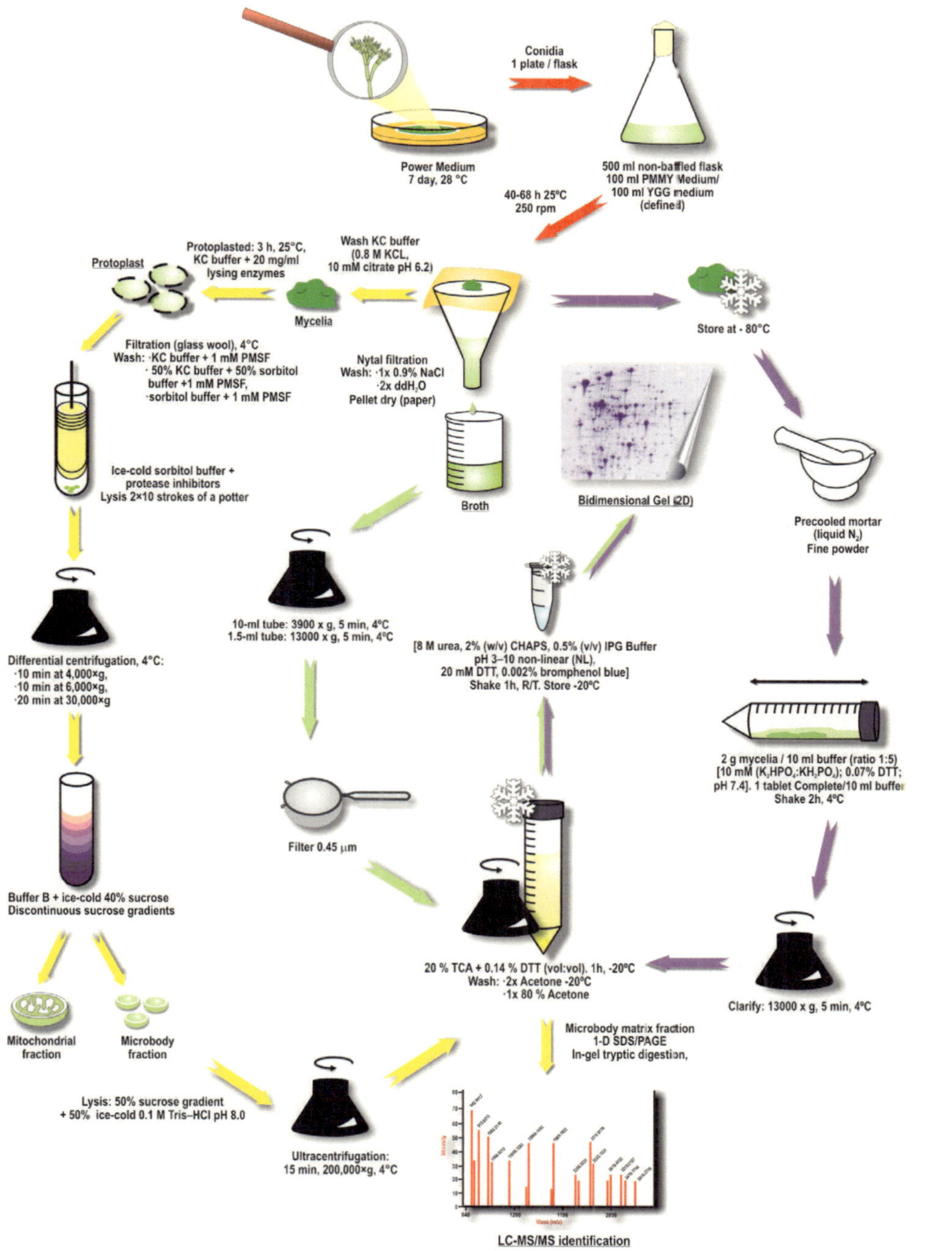

Fig. 2. Schematic representation of the optimized method for mycelia, secreted and microbody proteins collection of *P. chrysogenum*. Red arrows show the common steps for all the proteome extractions. Green/purple arrows represent common steps for intra- and extracelullar proteomes. Green arrows represent the specific paces for secreted proteins isolation. Purple arrows show those specific steps for intracellular proteome isolation. Yellow arrows present those particular steps of microbody proteins extraction (Based on: Jami et al., 2010a, 2010b; Kiel et al., 2009).

2D geles or 2D tricine on *Neurospora* has allowed the identification of 260 mitochondrial proteins, which include 55 previously predicted or hypothetical annotated and 101 proteins not previously identified in mass spectrometry studies [AutoFlex or UltraFlex II instruments (Bruker Daltonics)](Keeping et al., 2011).

An increase in the number of microbodies enhances antibiotic production by *P. chrysogenum*. This finding has focused the attention on the analysis of microbody matrix proteins (figure 2), since a part of the penicillin biosynthesis pathway is located in the peroxisomal lumen. Kiel and co-workers (2009) by means of LC-MS/MS analysis using a 4800 MALDI-TOF/TOF analyzer (Applied Biosystems) identified 89 microbody proteins, 79 with a putative microbody targeting signal.

5.2 Membrane and cell wall proteomes: an almost unexplored field

The hydrophobicity of the membrane or the cell wall embedded proteins cause a difficult isolation and analysis of those proteins, which is consistent with the low number of articles of membrane proteome in filamentous fungi. Initially, Bowman and co-workers (1981) described a protocol for plasma membrane extraction with high ATPase activity of *Neurospora*. Later, Mezence & Boiron (1995) developed a protocol for outer membrane protein extraction for *Madurella mycetomatis*. Both methods were used by Hernández-Macedo and co-worker (2002) for plasma membrane and cell wall protein extraction of *P. chrysosporium* and *L. edodes*, although the proteins were only visualized in one-dimensional SDS–PAGE rather than 2D electrophoresis.

Microbiologists, plant pathologists and commercial companies have paid a considerable attention to the production and application of biological control agents. Thus, *Trichoderma* genus has generated a special attention as active agent for biological control of plant pathogenic fungi. Proteins associated with the cell envelope of *Trichoderma reesei* were analysed by 2D electrophoresis allowing the location of 220 proteins and the identification of 32 spots by nanoelectrospray tandem mass spectrometry [Q-TOF MS (Micromass)] and amino acid sequence (Lim et al., 2001). The most abundant protein was HEX1, the major protein in Woronin body, suggesting that this structure unique to filamentous fungi is linked to the cell envelope.

Glycosylphosphatidylinositol-linked (GPI) proteins anchored to the plasma membrane are known to play a role in fungal morphogenesis (filamentation, mating, floculation, or adhesion to the external matrix) and an active role in cell wall organization. In order to identify GPI-anchored proteins involved in cell wall biogenesis, a proteomic analysis in *A. fumigatus*, a mould causing most of the invasive fungal lung infections in immunocompromised patients, was carried out. The GPI-anchored proteins were released from a membrane preparation [3 min in a CO_2-cooled MSK homogenizer with glass beads (1 mm-diameter) and centrifuged at 100000x*g* for 1 h at 4ºC] by an endogenous GPI-phospholipase C, sequentially purified by liquid chromatography, separated by 2D electrophoresis and characterized by MALDI-TOF [Voyager DE-STR MS (PerSeptive Biosystems)] and by internal amino acid sequencing [Finnigan TSQ 7000 (ThermoQuest)]. Thus, nine GPI-anchored proteins were identified in *A. fumigatus* (Bruneau et al., 2001). Following this analysis, de Groot and co-workers (2009) found ten predicted GPI-anchored proteins in cell wall fractions of *A. nidulans*, which consist of six proteins identified as cell wall carbohydrate-active enzymes including three orthologous proteins similar to those described before in *A. fumigatus* by Bruneau and co-workers (2001).

Fungus	Proteome Type	Proteomics approach	Application	Reference
A. fumigatus	Intracellular	2DE / MALDI-TOF	Reference map / Human pathogenicity	Carberry et al., 2006
		2DE / MALDI-TOF/TOF	Human pathogenicity / Alcohol metabolism	Kniemeyer et al., 2006
			Oxidative stress response	Lessing et al., 2007
			Antibiotic resistance	Gautam et al., 2008
			Reference map / Human pathogenicity	Vödisch et al., 2009
		2DE / MALDI-TOF	Dormant conidia / Pathogenicity	Teutschbein et al., 2010
		iTRAQ / 2D LC-MALDI-TOF/TOF	Human pathogenicity / Development	Cagas et al., 2011a
		2DE / MALDI-TOF-MS/MS	Antibiotic resistance	Gautam et al., 2011
A. nidulans	Intracellular	2DE / ESI-QqTOF	Bacterial-fungal interaction	Melin et al., 2002
		2DE / MALDI-TOF	Osmoadaptation	Kim et al., 2007
			Hypoxic responses	Shimizu et al., 2009
			Oxidative stress response	Sato et al., 2009
			Conidia germination	Oh et al., 2010
		2DE / MALDI-TOF/TOF	Oxidative stress response	Thön et al., 2010
				Thön et al., 2010
		2DE / Nano HPCL MS/MS		Pusztahelyi et al., 2011
A. niger	Intracellular	2DE / MALDI-TOF/TOF	Biotechnological processes	Sørensen et al., 2009
		2DE / MALDI-TOF/TOF / ESI-QqTOF	Reference map / Biotechnological processes	Lu et al., 2010
B. cinerea	Intracellular	2DE / MALDI-TOF/ ESI IT	Reference map / Vegetal pathogenicity	Fernández-Acero et al., 2006
P. chrysogenum	Intracellular	2DE / MALDI-TOF/TOF	Reference map / Penicillin production	Jami et al., 2010a
P. chrysosporium	Intracellular	2DE / LC-MS/MS	Environmental pollution	Matsuzaki et al., 2008
		2DE / MALDI-TOF		Yildirim et al., 2011
T. harzianum	Intracellular	2DE / MALDI-TOF / LC-MS/MS	Reference map / Biocontrol agent	Grinyer et al., 2004
A. fumigatus	Extracellular	2DE / immunoblotting / Q-TOF	Immunome	Singh et al., 2010
		2DE / MALDI-TOF-MS/MS	Human pathogenicity	Wartenberg et al., 2011

Fungus	Proteome Type	Proteomics approach	Application	Reference
A. niger	Extracellular	1D / Nano HPCL MS/MS	Reference map / Biotechnological processes	Lu et al., 2010
A. oryzae	Extracellular	2DE / MALDI-TOF	Food production	Oda et al., 2006
		2DE / MALDI-TOF		Machida et al., 2005
A. versicolor	Extracellular	2DE / immunoblotting / LC-MS/MS	Immunome	Benndorf et al., 2008
B. cinerea	Extracellular	1DE / LC-MS/MS	Vegetal pathogenicity / virulence factors	Shah et al., 2008
		2DE / MALDI-TOF / LC-MS/MS		Espino et al., 2010
		2DE / MALDI-TOF/TOF		Fernández-Acero et al., 2010
D. seriata	Extracellular	2DE / MALDI-TOF/TOF	Vegetal pathogenicity / virulence factors	Cobos et al., 2010
M. acridum	Extracellular	2DE / MALDI-TOF	Biocontrol agent	Barros et al., 2010
M. anisopliae	Extracellular	2DE / MALDI-TOF/TOF	Biocontrol agent	Murad et al., 2006
M. globosa	Extracellular	2DE / immunoblotting / oMALDI Qq-ToF / N-terminal aa sequencing	Immunome	Ishibashi et al., 2009
P. citrinum / A. fumigatus	Extracellular	2DE / immunoblotting / N-terminal aa sequencing	Immunome	Lai et al., 2002
P. chrysogenum	Extracellular	2DE / MALDI-TOF/TOF	Food production	Jami et al., 2010b
P. placenta	Extracellular	1DE / LC-MS/MS	Environmental pollution	Martínez et al., 2009
P. tritici-repentis	Extracellular	2DE / ESI-q-TOF MS/MS	Biocontrol agent	Cao et al., 2009
T. harzianum	Extracellular	2DE	Biocontrol agent	Suárez et al., 2005
		2DE / MALDI-TOF		Monteiro et al., 2010
A. fumigatus	Plasma-membrane anchored proteins	2DE / MALDI-TOF	Biomarker	Bruneau et al., 2001
		1DE/ LC-MS/MS		Groot et al., 2009
		1DE / 2D LC-MS/MS		Ouyang et al., 2010
	Conidial surface	2DE / LC-MS/MS		Asif et al., 2005
	cell wall and plasma membrane	iTRAQ / LC-MALDI-TOF/TOF	Antibiotic resistance	Cagas et al., 2011b
Neurospora spp.	Organelle	2DE / 2DE-tricine / MALDI-TOF	Disease analysis	Keeping et al., 2011
P. chrysogenum	Organelle	1DE / LC-MALDI-TOF/TOF	Penicillin production	Kiel et al., 2009
P. chrysosporium / L. edodes	Plasma membrane / cell wall	1DE	Procedure optimization	Hernández-Macedo et al., 2002
T. reesei	Cell envelope	2DE / ESI-Q-TOF	Biocontrol agent	Lim et al., 2001

Table 3. Summary of the major proteomes identified from filamentous fungi using different proteomic approaches.

In other proteomic analysis, Asif and co-workers (2005) provided the first conidial surface subproteome map of *A. fumigatus* with the goal of finding potential therapeutic targets against this human pathogen. The intact viable *Aspergillus* conidia were extracted with a mild alkaline buffer (0.1 M Tris-HCl-buffer pH 8.5) in the presence of a 1,3-β-glucanase containing 1 mM 1,10 phenanthrolin. Thus, the combination of 2D electrophoresis and LC-MS/MS [Q-TOF Ultima Global MS (Waters)] allowed the identification of 26 different proteins of *A. fumigatus*, twelve of which contain a signal for secretion. Among the proteins without a secretory signal the well-known allergen AspF3 was identified.

Recently, Ouyang and co-workers (2010) in close connection with the previous paper of Bruneau and co-workers (2001) attempted to identify membrane proteins associated with cell wall biosynthesis and glycoconjugates of total membrane preparations from *A. fumigatus*. Thus, the combination of 1D gels and 2D LC-MS/MS allowed the identification of 530 proteins that include 9 membrane proteins involved in cell wall biosynthesis and 8 enzymes involved in sphingolipid synthesis. These can potentially be used as antifungal drug targets following the way designed by Asif and co-workers (2005).

The identification of resistance mechanisms to antifungal drugs is an interesting topic, which has been investigated in *A. fumigatus* by determining the fungal proteomic response to drug exposure (Cagas et al., 2011b). These authors analysed the proteomic response of *A. fumigatus* to caspofungin by 2D electrophoresis from 4 subcellular compartment fractions (secreted, cytoplasmic, microsomal and cell wall and plasma membrane) and later gel-free iTRAQ method [using a MALDI-TOF/TOF MS (Applied Biosystems)] and microarrays were performed from the secreted and cell wall and plasma membrane fractions. These analyses presented the high potential to identify biomarkers of the employed techniques that assess the efficacy of caspofungin drug therapy.

On the one hand, the cell wall proteins are often highly glycosylated (hampers MS analysis) and they are difficult to separate from cell wall polysaccharides. On the other hand, membrane proteins are hard to dissolve and difficult to digest by standard methods (hydrophobic parts lack tryptic cleavage sites) due to their hydrophobicity. Besides, they usually present a dramatic low abundance. Thus, despite the technical progresses, the proteomic analysis of cell wall and membrane proteins remains as a tough task in filamentous fungi supposing a trending Proteomics challenge (Kniemeyer, 2011).

6. Conclusion

Proteomics is a powerful tool, which has emerged as a result of summarize methodological and technical findings in mass spectrometry, protein visible and fluorescent staining, bidimensional electrophoresis, isobaric labelling, etc. The genome sequence facilities generated in the last years have supported this current trending topic that proteomes and subproteomes analyses represent. Along the present review the gradient of knowledge existing about filamentous fungi Proteomics is clearly presented, which is perfectly correlated with the protein isolation difficulty. Thus, several studies have been carried out over the intracellular proteome that helped to the development of a huge number of procedures for protein extraction and analysis. These analyses collect from technical updates to multiple strain comparison. As far as the protein isolation became more and more complex the numbers of published analyses diminished due to the complexity of the organelle isolation as well as hydrophobic membrane protein analysis. Anyway, the Proteomics approach applied to the filamentous fungi is opening a new world of strain improvement, drug resistance analysis, antigen detection, biomarker discovery and environment applications, which was difficult to predict only ten years ago. Thus, Proteomics application to filamentous fungi is a present reality (summarized in table 3) with a promising future that still has several challenges to be faced up.

7. Acknowledgment

This work was supported by grants from the European Union (Eurofung Grant QLRT-1999-00729 and Eurofungbase). C. Barreiro was supported by the European Union program ERA-IB [BioProChemBB (EIB.08.008)]. The technical support on Proteomics of M. D. Gutiérrez (Universidad Complutense de Madrid, Spain), and E. Calvo and L. E. Camafeita (CNIC, Madrid, Spain) are acknowledged. We recognize the technical assistance of the INBIOTEC staff: B. Martín, J. Merino, A. Casenave and A. Mulero.

8. References

Adav, S.S.; Li, A.A.; Manavalan, A.; Punt, P. & Sze, S.K. (2010). Quantitative iTRAQ secretome analysis of *Aspergillus niger* reveals novel hydrolytic enzymes. *J Proteome Res*, Vol. 9, No. 8, (August 2010), pp. 3932-3940, ISSN 1535-3893.

Asif, A.R.; Oellerich, M.; Amstrong, V.W.; Riemenschneider, B.; Monod, M. & Reichard, U. (2006). Proteome of conidial surface associated proteins of *Aspergillus fumigatus* reflecting potential vaccine candidates and allergens. *J Proteome Res*, Vol. 5, No.4, (April 2006), pp. 954-62, ISSN 1535-3893.

Bhadauria, V.; Zhao, W.S.; Wang, L.X.; Zhang, Y.; Liu, J.H., et al. (2007). Advances in fungal Proteomics. *Microbiol Res*, Vol. 162, No. 3, (May 2007), pp. 193-200, ISSN 0953-7562.

Barreiro, C.; González-Lavado, E.; Brand, S.; Tauch, A. & Martín, J.F. (2005). Heat shock proteome analysis of wild-type *Corynebacterium glutamicum* ATCC 13032 and a spontaneous mutant lacking GroEL1, a dispensable chaperone. *J Bacteriol*, Vol. 187, No. 3, (February 2005), pp. 884-889, ISSN 0021-9193.

Barros, B.H.; da Silva, S.H.; dos ReisMarques Edos, R.; Rosa, J.C.; Yatsuda, A.P., et al. (2010). A proteomic approach to identifying proteins differentially expressed in conidia and mycelium of the entomopathogenic fungus *Metarhizium acridum*. *Fungal Biol*, Vol. 114, No. 7, (July 2010), pp. 572-9. ISSN 1878-6146.

Bauer, B.; Schwienbacher, M., Broniszewska, M.; Israel, L.; Heesemann, J. & Ebel, F. (2010). Characterisation of the CipC-like protein AFUA_5G09330 of the opportunistic human pathogenic mould *Aspergillus fumigatus*. *Mycoses*, Vol. 53, No. 4, (July 2010), pp. 296-304, ISSN 1439-0507.

Bendtsen, J.D.; Nielsen, H.; von Heijne, G. & Brunakm, S. (2004a) Improved prediction of signal peptides: Signal P 3.0. *J. Mol. Biol*, Vol. 340, No. 4, (July 2004), pp. 783–795, 2004, ISSN 0022-2836.

Bendtsen, J.D.; Jensen, L.J.; Blom, N.; von Heijne, G. & Brunak, S. (2004b). Feature based prediction of non-classical and leaderless protein secretion. *Protein Eng. Des. Sel*, Vol. 17, No. 4, (April 2004), pp. 349 356, ISSN 1741-0126.

Benndorf, D.; Müller, A.; Bock, K.; Manuwald, O.; Herbarth, O. & von Bergen. M. (2008). Identification of spore allergens from the indoor mould *Aspergillus versicolor*. *Allergy*, Vol. 63, No. 4, (April 2008), pp. 454-60, ISSN 0105-4538.

Bennett, J.W. (2006) The molds of Katrina. In: The Sciences 2.0. The Molds of Katrina: Report from the Field. Dan Van Atta (ed.), New York Academy of Sciences, (March/April 2006).

Bouws, H.; Wattenberg, A. & Zorn, H. (2008). Fungal secretomes—nature's toolbox for white biotechnology. *Appl Microbiol Biotechnol*, Vol. 80. No. 3, (September 2008), pp. 381–388, ISSN 0175-7598.

Bowman, E.J.; Bowman, B.J. & Slayman, C.W. (1981). Isolation and characterization of plasma membranes from wild type *Neurospora crassa*. *J Biol Chem*, Vol. 256, No. 23 (December 1981), pp. 12336-42, ISSN 0021-9258.

Bruneau, J.M.; Magnin, T.; Tagat, E.; Legrand, R.; Bernard, M., et al. (2001). Proteome analysis of *Aspergillus fumigatus* identifies glycosylphosphatidylinositol-anchored proteins associated to the cell wall biosynthesis, *Electrophoresis*, Vol. 22, No. 13, (August 2001), pp. 2812–2823, ISSN 1522-2683.

Burns, C.; Geraghty, R.; Neville, C.; Murphy, A.; Kavanagh, K. & Doyle, S. (2005). Identification, cloning, and functional expression of three glutathione transferase genes from *Aspergillus fumigatus*. *Fungal Genet Biol*, Vol. 42, No. 4, (April 2005), pp. 319-327, ISSN 1087-1845.

Cagas, S.E.; Raja-Jain, M.; Li, H. & Perlin, D.S. (2011a). The proteomic signature of *Aspergillus fumigatus* during early development. *Mol Cell Proteomics*. In Press (August 2011), ISSN 1535-9476.

Cagas, S.E.; Jain, M.R.; Li, H. & Perlin, D.S. (2011b). Profiling the *Aspergillus fumigatus* proteome in response to caspofungin. *Antimicrob Agents Chemother*, Vol. 55, No. 1, (January 2011), pp. 146-154, ISSN 0066-4804.

Candiano, G.; Bruschi, M.; Musante, L.; Santucci, L.; Ghiggeri, G.M., et al. (2004). Blue silver: a very sensitive colloidal Coomassie G-250 staining for proteome analysis. *Electrophoresis*, Vol. 25, No. 9, (May 2004), pp. 1327-1333, ISSN 1522-2683.

Cao, T.; Kim, Y.M.; Kav, N.N.V. & Strelkov, S.E. (2009). A proteomic evaluation of *Pyrenophora tritici-repentis*, causal agent of tan spot of wheat, reveals major differences between virulent and avirulent isolates. *Proteomics*, Vol. 9, No. 5 (March 2009), pp. 1177–1196, ISSN 1615-9853.

Cardoza, R.E.; Gutiérrez, S.; Ortega, N.; Colina, A.; Casqueiro, J. & Martín, J.F. (2003). Expression of a synthetic copy of the bovine chymosin gene in *Aspergillus awamori* from constitutive and pH-regulated promoters and secretion using two different pre-pro sequences. *Biotechnol. Bioeng*, Vol. 83, No. 3, (August 2003), pp. 249–259, ISSN 0006-3592.

Carberry, S. & Doyle, S. (2007). Proteomic studies in biomedically and industrially relevant fungi. *Cytotechnology*, Vol. 53, No. 1-3, (April 2007), pp. 95-100, ISSN 0920-9069.

Carberry, S.; Neville, C.M.; Kavanagh, K.A. & Doyle, S. (2006). Analysis of major intracellular proteins of *Aspergillus fumigatus* by MALDI mass spectrometry: identification and characterisation of an elongation factor 1B protein with glutathione transferase activity. *Biochem Biophys Res Commun*, Vol. 341, No. 4, (March 2006), pp. 1096-1104, ISSN 0006-291X.

Chang, K.Y.; Georgianna, D.R.; Heber, S.; Payne, G.A. & Muddiman, D.C. (2010). Detection of alternative splice variants at the proteome level in *Aspergillus flavus*. *J Proteome Res*, Vol. 9, No. 3, (March 2010), pp. 1209-1217, ISSN 1535-3893.

Choi, J; Park, J.; Kim, D.; Jung, K.; Kang, S. & Lee, YH. (2010). Fungal Secretome Database: Integrated platform for annotation of fungal secretomes. *BMC Genomics*, Vol. 11, (February 2010), pp. 105, ISSN 1471-2164.

Cobos, R.; Barreiro, C.; Mateos, R.M. & Coque, J.J.R. (2010). Cytoplasmic- and extracellular-proteome analysis of *Diplodia seriata*: a phytopathogenic fungus involved in grapevine decline. *Proteome Sci*, Vol. 8, (September 2010), pp. 46, ISSN 1477-5956.

Cordwell, S.J.; Nouwens, A.S.; Verrills, N.M.; Basseal, D.J. & Walsh, B.J. (2000). Subproteomics based upon protein cellular location and relative solubilities in conjunction with composite two-dimensional electrophoresis gels. *Electrophoresis*, Vol. 21, No. 6, (April 2000), pp. 1094-1103, ISSN 1522-2683.

Coumans, J.V.; Moens, P.D.; Poljak, A.; Al-Jaaidi, S.; Pereg, L. & Raftery, M.J. (2010). Plant-extract-induced changes in the proteome of the soil-borne pathogenic fungus *Thielaviopsis basicola*. *Proteomics*, Vol. 10, No. 8, (April 2010), pp. 1573-1591, ISSN 1615-9853.

Cruz, A.; Saenz de Santamaría, M.; Martínez, J.; Martínez, A.; Guisantes, J. & Palacios, R. (1997). Fungal allergens from important allergenic fungi imperfecti. *Allergol Immunopathol (Madr)*, Vol. 2, No. 3 (May-June 1997), pp. 153-8, ISSN 0301-0546.

de Godoy, L.M.; Olsen, J.V.; Cox, J.; Nielsen, M.L; Hubner, N.C., et al. (2008). Comprehensive mass-spectrometry-based proteome quantification of haploid versus diploid yeast. *Nature*, Vol. 455, No. 7217, (October 2008), pp. 1251-1254, ISSN 0028-0836.

de Groot, P.W.; Brandt, B.W.; Horiuchi, H.; Ram, A.F.; de Koster, C.G. & Klis, F.M. (2009). Comprehensive genomic analysis of cell wall genes in *Aspergillus nidulans*. *Fungal Genet Biol*, Vol. 46, Suppl 1, (March 2009), pp. S72-81, ISSN 1087-1845.

DiGuistini, S.; Wang, Y.; Liao, N.Y.; Taylor, G.; Tanguay, P., et al. (2011). Genome and transcriptome analyses of the mountain pine beetle-fungal symbiont *Grosmannia clavigera*, a lodgepole pine pathogen. *Proc Natl Acad Sci U S A*. Vol. 108, No. 6, (Febrary 2011), 2504-2509.

Doyle, S. (2011). Fungal Proteomics: from identification to function. *FEMS Microbiol Lett*, Vol. 321, No. 1, (August 2011), pp. 1–9, ISSN 0378-1097.

Espino, J.J; Gutiérrez-Sánchez, G.; Brito, N.; Shah, P.; Orlando, R. & González, C. (2010). The *Botrytis cinerea* early secretome. *Proteomics*, Vol. 10, No. 16, (Auygust 2010), pp. 3020–3034, ISSN 1615-9853.

Ellis, S.D.; Boehm, M. J. & Mitchell, T. K. (2008). Fungal and Fungal-like Diseases of Plants. In: Fact sheet: Agriculture and Natural Resources. The Ohio State University. pp. 401.07.

Fernández-Acero, F.J.; Jorge, I.; Calvo, E.; Vallejo, I.; Carbú, M., et al. (2006). Two-dimensional electrophoresis protein profile of the phytopathogenic fungus *Botrytis cinerea*. *Proteomics*, Vol. 6, Suppl. 1, (April 2006), pp. S88-S96, ISSN 1615-9853.

Fernández-Acero, F.J; Colby, T.; Harzen, A.; Carbu, M; Wieneke, U., et al. (2010). 2-DE proteomic approach to the *Botrytis cinerea* secretome induced with different carbon sources and plant-based elicitors. *Proteomics*, Vol. 10, No. 12, (June 2010), pp. 2270–2280, ISSN 1615-9853.

Ferreira de Oliveira, J.M. & de Graaff, L.H. (2011). Proteomics of industrial fungi: trends and insights for biotechnology. *Appl Microbiol Biotechnol*, Vol. 89, No. 2, (January 2011), pp. 225-37, ISSN 0175-7598.

Ferreira de Oliveira, J.M.; van Passel, M.W.; Schaap, P.J. & de Graaff, L.H. (2010). Shotgun Proteomics of *Aspergillus niger* microsomes upon D-xylose induction. *Appl Environ Microbiol*, Vol. 76, No. 13, (July 2010), pp. 4421-4429, ISSN 0099-2240.

Fleissner, A. & Dersch, P. (2010) Expression and export: recombinant protein production systems for *Aspergillus*. *Appl Microbiol Biotechnol*, Vol. 87, No. 4, (July 2010), pp. 1255-1270, ISSN: 0175-7598.

Flores, C.L. & Gancedo, C. (2011). Unraveling moonlighting functions with yeasts, *IUBMB Life*, Vol. 63, No. 7, (July 2011), pp. 457-462, ISSN 1521-6551.

Galagan, J.E.; Calvo, S.E.; Cuomo, C.; Ma, L.J.; Wortman, J.R., et al. (2005). Sequencing of *Aspergillus nidulans* and comparative analysis with *A. fumigatus* and *A. oryzae*. *Nature*, Vol. 438, No. 7071, (December 2005), pp. 1105-1115, ISSN 0028-0836.

Gancedo, C. & Flores, C.L. (2008). Moonlighting proteins in yeasts, *Microbiol Mol Biol Rev*, Vol. 72, No. 1, (March 2008), pp.197-210, ISSN 0005-3678.

Gao, Q.; Jin, K.; Ying, S.H.; Zhang, Y.; Xiao, G., et al.. (2011). Genome sequencing and comparative transcriptomics of the model entomopathogenic fungi *Metarhizium anisopliae* and *M. acridum*. *PLoS Genet*, Vol. 7, No. 1 (January 2011), pp. e1001264, ISSN 1553-7404.

Gautam, P.; Shankar, J.; Madan, T.; Sirdeshmukh, R.; Sundaram, C.S., et al. (2008). Proteomic and transcriptomic analysis of *Aspergillus fumigatus* on exposure to amphotericin B. *Antimicrob Agents Chemother*, Vol. 52, No. 12, (December 2008), pp. 4220-4227, ISSN 0066-4804.

Gautam, P.; Upadhyay, S.K.; Hassan, W.; Madan, T.; Sirdeshmukh, R., et al. (2011). Transcriptomic and Proteomic Profile of *Aspergillus fumigatus* on Exposure to Artemisinin. *Mycopathologia*. In Press (July 2011), ISSN 0301-486X.

Greenbaum, D.; Luscombe, N.M.; Jansen, R.; Qian, J. & Gerstein, M. (2001). Interrelating different types of genomic data, from proteome to secretome: 'oming in on function. *Genome Res*, Vol. 11, No. 9, (September 2001), pp. 1463-1468.

Grinyer, J.; McKay, M.; Nevalainen, H. & Herbert, B.R. (2004). Fungal Proteomics: initial mapping of biological control strain *Trichoderma harzianum*. *Curr Genet*, Vol. 45, No. 3, (March 2004), pp. 163-169, ISSN 0172-8083.

Hawksworth, D.L. (1991). The fungal dimension of biodiversity: magnitude, significance, and conservation. *Mycol. Res.* Vol. 95, (June 1991), pp. 641–655, ISSN 0953-7562.

Hawksworth, D.L. (2001). The magnitude of fungal diversity: the 1.5 million species estimate revisited. *Myco. Res*, Vol. 105, No. 12, (December 2011), pp. 1422–1432, ISSN 0953-7562.

Hecker, M.; Becher, D.; Fuchs, S. & Engelmann, S. (2010). A proteomic view of cell physiology and virulence of *Staphylococcus aureus*. *Int J Med Microbiol*, Vol. 300, No. 2-3, (Februrary 2010), pp. 76-87, ISSN 1438-4221.

Hernández-Macedo, M.L.; Ferraz, A.; Rodríguez, J.; Ottoboni, L.M. & De Mello, M.P. (2002). Iron-regulated proteins in *Phanerochaete chrysosporium* and *Lentinula edodes*: differential analysis by sodium dodecyl sulfate polyacrylamide gel electrophoresis and two-dimensional polyacrylamide gel electrophoresis profiles. *Electrophoresis*, Vol. 23, No. 4 (February 2002), pp. 655-61, ISSN 1522-2683.

Ishibashi, Y.; Kato, H.; Asahi, Y.; Sugita, T. & Nishikawa, A. (2009). Identification of the major allergen of *Malassezia globosa* relevant for atopic dermatitis. *J Dermatol Sci*, Vol. 55, No. 3, (September 2009), pp. 185-192, ISSN 0923-1811.

Jami, M.S.; Barreiro, C.; García-Estrada, C. & Martín, J. F. (2010a). Proteome analysis of the penicillin producer *Penicillium chrysogenum*: characterization of protein changes during the industrial strain improvement. *Mol Cell Proteomics*, Vol. 9, No. 6, (June 2010), pp. 1182-1198, ISSN 1535-9476.

Jami, M.S.; García-Estrada, C.; Barreiro, C.; Cuadrado, A.A; Salehi-Najafabadi, Z & J. F. Martín. (2010b). The *Penicillium chrysogenum* extracellular proteome. Conversion from a food-rotting strain to a versatile cell factory for white biotechnology. *Mol Cell Proteomics*, Vol. 9, No. 12, (December 2010), pp. 2729-2744, ISSN 1535-9476.

Jeffery, C.J. (1999). Moonlighting proteins. *Trends Biochem. Sci.* Vol. 24, No. 1, (January 1999), pp. 8–11, ISSN 0968-0004.

Kall, L.; Krogh, A. & Sonnhammer, E.L. (2004). A combined transmembrane topology and signal peptide prediction method. *J. Mol. Biol*, Vol. 338, No. 5, (May 2004), pp. 1027–1036, ISSN 0022-2836.

Keeping, A.; Deabreu, D.; Dibernardo, M. & Collins, R.A. (2011). Gel-based mass spectrometric and computational approaches to the mitochondrial proteome of *Neurospora*. *Fungal Genet Biol*, Vol. 48, No. 5, (May 2011), pp. 526-36, ISSN: 1087-1845.

Kiel, J.A; van den Berg, M.A.; Fusetti, F.; Poolman, B.; Bovenberg, R.A., et al. (2009). Matching the proteome to the genome: the microbody of penicillin-producing *Penicillium chrysogenum* cells. *Funct Integr Genomics*, Vol. 9, No. 2, (May 2009), pp. 167-184, ISSN: 1438-793X.

Kim, Y.; Nandakumar, M.P. & Marten, M.R. (2007). Proteome map of *Aspergillus nidulans* during osmoadaptation. *Fungal Genet Biol*, Vol. 44, No. 9, (September 2007), pp. 886-895, ISSN 1087-1845.

Kim, Y.; Nandakumar, M.P. & Marten, M.R. (2008). The state of proteome profiling in the fungal genus *Aspergillus*. *Brief Funct Genomic Proteomic*, Vol. 7, No. 2, (March 2008), pp. 87-94, ISSN 1477-4062.

Kniemeyer, O. (2011). Proteomics of eukaryotic microorganisms: The medically and biotechnologically important fungal genus *Aspergillus*. *Proteomics*, Vol. 11, No. 15, (August 2011), pp. 3232-3243, ISSN 1615-9853.

Kniemeyer, O.; Lessing, F.; Scheibner, O.; Hertweck, C. & Brakhage, A.A. (2006). Optimisation of a 2-D gel electrophoresis protocol for the human-pathogenic fungus *Aspergillus fumigatus*. *Curr Genet*, Vol. 49, No. 3, (March 2006), pp. 178-189, ISSN 0172-8083.

Kohlheyer, D.; Eijkel, J.C.; van den Berg, A. & Schasfoort, R.B. (2008). Miniaturizing free-flow electrophoresis—a critical review. *Electrophoresis*, Vol. 29, No. 5, (March 2008), pp. 977–993, ISSN: 0173-0835.

Lai, H.Y.; Tam, M.F.; Tang, R.B.; Chou, H.; Chang, C.Y., et al. (2002). cDNA cloning and immunological characterization of a newly identified enolase allergen from *Penicillium citrinum* and *Aspergillus fumigatus*. *Int Arch Allergy Immunol*, Vol. 127, No. 3, (March 2002), pp. 181-190. ISSN 1018-2438.

Lessing, F.; Kniemeyer, O.; Wozniok, I.; Loeffler, J., Kurzai, O., et al. (2007). The *Aspergillus fumigatus* transcriptional regulator AfYap1 represents the major regulator for defense against reactive oxygen intermediates but is dispensable for pathogenicity in an intranasal mouse infection model. *Eukaryot Cell*, Vol. 6, No. 12, (December 2007), pp. 2290-2302, ISSN 1535-9778.

Lim, D.; Hains, P.; Walsh, B.; Bergquist, P. & Nevalainen, H. (2001). Proteins associated with the cell envelope of *Trichoderma reesei*: a proteomic approach, *Proteomics*, Vol. 1, No. 7, (July 2001), pp. 899-909, ISSN 1615-9853.

Lu, X.; Sun, J.; Nimtz, M.; Wissing, J.; Zeng, A.P. & Rinas, U. (2010). The intra- and extracellular proteome of *Aspergillus niger* growing on defined medium with xylose or maltose as carbon substrate. *Microb Cell Fact*, Vol. 20, No. 9, (April 2010), pp. 23, ISSN 1475-2859.

Lum, G. & Min, X.J. (2011). FunSecKB: the Fungal Secretome KnowledgeBase. *Database (Oxford)*, Vol. 2011, (February 2011), ISSN 1758-0463.

Machida, M.; Asai, K.; Sano, M.; Tanaka, T.; Kumagai, T.; et al. (2005). Genome sequencing and analysis of *Aspergillus oryzae*. *Nature*, Vol. 438, No. 7071 (December 2005), pp. 1157–1161, ISSN 0028-0836.

Makridakis, M. & Vlahou, A. (2010). Secretome Proteomics for discovery of cancer biomarkers. *J Proteomics*, Vol. 73, No. 12, (November 2010), pp. 2291-2305, ISSN 1874-3919.

Martin, F.; Aerts, A.; Ahrén, D.; Brun, A.; Danchin, E.G., et al. (2008). The genome of *Laccaria bicolor* provides insights into mycorrhizal symbiosis. *Nature*, Vol. 452, No. 7183, (March 2008), pp. 88–92, ISSN 0028-0836.

Martín, J.F.; Ullán, R.V. & García-Estrada, C. (2010). Regulation and compartmentalization of beta-lactam biosynthesis, *Microbial Biotechnol*, Vol. 3, No. 3, (May 2010), pp. 285-299, ISSN 1751-7915.

Martínez, D; Challacombe, J.; Morgenstern, I.; Hibbett, D. & Schmoll, M. (2009). Genome, transcriptome, and secretome analysis of wood decay fungus *Postia placenta* supports unique mechanisms of lignocellulose conversion. *PNAS*, Vol. 106, No. 6, (February 2009), pp. 1954–1959, ISSN 0027-8424.

Matsuzaki, F.; Shimizu, M. & Wariishi, H. (2008). Proteomic and metabolomic analyses of the white-rot fungus *Phanerochaete chrysosporium* exposed to exogenous benzoic acid. *J Proteome Res*, Vol. 7, No. 6, (June 2008), pp. 2342-2350, ISSN 1535-3893.

Medina, M.L.; Kiernan, U.A. & Francisco, W.A. (2004). Proteomic analysis of rutin-induced secreted proteins from *Aspergillus flavus*. *Fungal Genet Biol*. Vol. 41, No. 3, (March 2004), pp. 327-35. ISSN: 1087-1845.

Mehta, A.; Brasileiro, A.C.; Souza, D.S.; Romano, E.; Campos, M.A., et al. (2008). Plant-pathogen interactions: what is Proteomics telling us? *FEBS J*, Vol. 275, No. 15 (August 2008), pp. 3731-3746, ISSN 1742-4658.

Melin, P.; Schnürer, J. & Wagner, E.G. (2002). Proteome analysis of *Aspergillus nidulans* reveals proteins associated with the response to the antibiotic concanamycin A, produced by *Streptomyces species*. *Mol Genet Genomics*, Vol. 267, No. 6, (August 2002), pp. 695-702, ISSN 1617-4615.

Mezence, M.I. & Boiron, P. (1995). Studies on siderophore production and effect of iron deprivation on the outer membrane proteins of *Madurella mycetomatis*. *Curr Microbiol*, Vol. 31, No. 4, (October 1995), pp. 220-223, ISSN 0343-8651.

Monteiro, V.N.; do Nascimento Silva, R.; Steindorff, A.S.; Costa, F.T.; Noronha, E.F., et al. (2010). New Insights in *Trichoderma harzianum* Antagonism of Fungal Plant Pathogens by Secreted Protein Analysis. *Curr Microbiol*, Vol. 61, Vol. 4, (October 2010), pp. 298–305, ISSN 0343-8651.

Murad, A.M.; Laumann, R.A.; Lima Tde, A.; Sarmento, R.B.; Noronha, E.F., et al. (2006). Screening of entomopathogenic *Metarhizium anisopliae* isolates and proteomic analysis of secretion synthesized in response to cowpea weevil (*Callosobruchus maculatus*) exoskeleton. *Comp Biochem Physiol C Toxicol Pharmacol*. Vol. 142, No. 3-4, (March-April 2006), pp. 365-370, ISSN 1532-0456.

Nickel, W. & Rabouille, C. (2009). Mechanisms of regulated unconventional protein secretion. *Nat. Rev. Mol. Cell Biol*, Vol. 10, No. 2, (February 2009), pp. 148–155, ISSN 1471-0072.

Nierman, W.C.; Pain, A.; Anderson, M.J.; Wortman, J.R.; Kim, H.S.; et al. (2005). Genomic sequence of the pathogenic and allergenic filamentous fungus *Aspergillus fumigatus*. *Nature*, Vol. 438, No. 7071, (December 2005), pp. 1151-1156, ISSN 0028-0836.

Oda, K.; Kakizono, D.; Yamada, O.; Iefuji, H.; Akita, O. & Iwashita, K. (2006). Proteomic analysis of extracellular proteins from *Aspergillus oryzae* grown under submerged and solid-state culture conditions. *Appl Environ Microbiol*, Vol. 72, No. 5, (May 2006), pp. 3448–3457, ISSN: 0099-2240.

Oh, Y.T.; Ahn, C.S.; Kim, J.G.; Ro, H.S.; Lee, C.W. & Kim, J.W. (2010). Proteomic analysis of early phase of conidia germination in *Aspergillus nidulans*. *Fungal Genet Biol*, Vol. 47, No. 3, (March 2010), pp. 246-253, ISSN ISSN 1087-1845.

Oku, M. & Sakai, Y. (2010). Peroxisomes as dynamic organelles: autophagic degradation. *FEBS J*, Vol. 277, No. 16, (august 2010), pp. 3289-94, ISSN 1742-4658.

Ouyang, H.; Luo, Y.; Zhang, L.; Li, Y. & Jin, C. (2010). Proteome Analysis of *Aspergillus fumigatus* Total Membrane Proteins Identifies Proteins Associated with the Glycoconjugates and Cell Wall Biosynthesis Using 2D LC-MS/MS. *Molecular Biotechnology*, Vol. 44, No. 3, (March 2010), pp. 177-189, ISSN 1073-6085.

Paper, J.M.; Scott-Craig, J.S.; Adhikari, N.D., Cuomo, C.A. & Walton, J.D. (2007). Comparative Proteomics of extracellular proteins in vitro and in planta from the pathogenic fungus *Fusarium graminearum*. *Proteomics*, Vol. 7, No. 17, (September 2007), pp. 3171-3183, ISSN 1615-9853.

Payne, G.A.; Nierman, W.C.; Wortman, J.R.; Pritchard, B.L.; Brown, D.; et al. (2006). Whole genome comparison of *Aspergillus flavus* and *A. oryzae*. *Med Mycol*, Vol. 44, (June 2006), pp. S9–S11, ISSN 1369-3786.

Peberdy, J.F. (1994) Protein secretion in filamentous fungi—trying to understand a highly productive black box. *Trends Biotechnol*. Vol. 12, No. 2, (February 1994), pp. 50–57, ISSN: 0167-7799.

Pel, H.J.; de Winde, J.H.; Archer, D.B.; Dyer, P.S.; Hofmann, G.; et al. (2007). Genome sequencing and analysis of the versatile cell factory *Aspergillus niger* CBS 513.88. *Nat Biotechnol*, Vol. 25, No. 2, (February 2007), pp. 221-231, ISSN 1087-0156.

Picotti, P.; Bodenmiller, B.; Mueller, L.N.; Domon, B. & Aebersold, R. (2009). Full dynamic range proteome analysis of *S. cerevisiae* by targeted Proteomics. *Cell*, Vol. 138, No. 4, (August 2009), pp. 795-806, ISSN 0092-8674.

Pitarch, A.; Nombela, C. & Gil, C. (2008). Cell wall fractionation for yeast and fungal Proteomics. *Methods Mol Biol*, Vol. 425, pp. 217-239, ISSN 1064-3745.

Punt, P.J.; van Biezen, N.; Conesa, A.; Albers, A.; Mangnus, J. & van den Hondel, C. (2002) Filamentous fungi as cell factories for heterologous protein production. *Trends Biotechnol*, Vol. 20, No. 5, (May 2002), pp. 200–206, ISSN: 0167-7799.

Pusztahelyi, T.; Klement, E.; Szajli, E.; Klem, J.; Miskei, M., et al. (2011). Comparison of transcriptional and translational changes caused by long-term menadione exposure in *Aspergillus nidulans*. *Fungal Genet Biol*, Vol. 48, No. 2, (February 2011), pp. 92-103, ISSN 1087-1845.

Ravalason, H.; Jan, G.; Mollé, D.; Pasco, M.; Coutinho, P.M., et al. (2008). Secretome analysis of *Phanerochaete chrysosporium* strain CIRM-BRFM41 grown on softwood. *Appl Microbiol Biotechnol*, Vol. 80, No. 4 (September 2008), pp. 719-733. ISSN 0175-7598.

Sato, I.; Shimizu, M.; Hoshino, T. & Takaya, N. (2009). The glutathione system of *Aspergillus nidulans* involves a fungus-specific glutathione S-transferase. *J Biol Chem*, Vol. 284, No. 12, (March 2009), pp. 8042-8053, ISSN 0021-9258.

Shah, P.; Atwood, J.A.; Orlando, R.; El Mubarek, H.; Podila, G.K. & Davis, M.R. (2009). Comparative proteomic analysis of *Botrytis cinerea* secretome. *J. Proteome Res.* Vol. 8, No. 3, (March 2009), pp. 1123–1130, ISSN 1535-3907.

Shimizu, M.; Fujii, T.; Masuo, S.; Fujita, K. & Takaya, N. (2009). Proteomic analysis of *Aspergillus nidulans* cultured under hypoxic conditions. *Proteomics*, Vol. 9, No. 1, (January 2008), pp. 7-19, ISSN 1615-9853.

Singh, B.; Oellerich, M.; Kumar, R.; Kumar, M.; Bhadoria, D.P., et al. (2010). Immuno-reactive molecules identified from the secreted proteome of *Aspergillus fumigatus*. *J Proteome Res*, Vol. 9, No. 11, (November 2010), pp. 5517-5529, ISSN 1535-3893.

Sørensen, L.M.; Lametsch, R.; Andersen, M.R.; Nielsen, P.V. & Frisvad, J.C. (2009). Proteome analysis of *Aspergillus niger*: lactate added in starch-containing medium can increase production of the mycotoxin fumonisin B2 by modifying acetyl-CoA metabolism. *BMC Microbiol*, Vol. 9, (December 2009), pp. 255, ISSN 1471-2180.

Suárez, M.B.; Sanz, L.; Chamorro, M.I.; Rey, M.; González, F.J., et al. (2005) Proteomic analysis of secreted proteins from *Trichoderma harzianum*. Identification of a fungal cell wallinduced aspartic protease. *Fungal Genet Biol*, Vol. 42, No. 11, (November 2005), pp. 924-934, ISSN 1087-1845.

Teutschbein, J.; Albrecht, D.; Pötsch, M.; Guthke, R.; Aimanianda, V., et al. (2010). Proteome profiling and functional classification of intracellular proteins from conidia of the human-pathogenic mold *Aspergillus fumigatus*. *J Proteome Res*, Vol. 9, No. 7, (July 2010), pp. 3427-3442, ISSN 1535-3893.

Thön, M.; Al Abdallah, Q.; Hortschansky, P.; Scharf, D.H.; Eisendle, M., et al. (2010). The CCAAT-binding complex coordinates the oxidative stress response in eukaryotes. *Nucleic Acids Res*, Vol. 38, No. 4, (March 2010), pp. 1098-1113, ISSN 0305-1048.

Tjalsma, H.; Bolhuis, A.; Jongbloed, J.D.H.; Bron, S. & van Dijl, J.M. (2000). Signal peptide-dependent protein transport in *Bacillus subtilis*: a genome-based survey of the secretome. *Microbial Mol Biol Rev*, Vol. 64, No. 3, (September 2000), pp. 515–547, ISSN 1092-2172.

Tsang, A.; Butler, G.; Powlowski, J.; Panisko, E.A. & Baker, S.E. (2009). Analytical and computational approaches to define the *Aspergillus niger* secretome. *Fungal Genetics and Biology*, Vol. 46, (March 2009), pp. S153–S160, ISSN 1087-1845.

van den Berg, M.A.; Albang, R.; Albermann, K.; Badger, J.H.; Daran, J.M.; et al. (2008). Genome sequencing and analysis of the filamentous fungus *Penicillium chrysogenum*. *Nat Biotechnol*, Vol. 26, No. 10, (October 2008), pp. 1161-1168, ISSN 1087-0156.

Vincent, D.; Balesdent, M.H.; Gibon, J., Claverol, S.; Lapaillerie, D.; et al. (2009). Hunting down fungal secretomes using liquid-phase IEF prior to high resolution 2-DE. *Electrophoresis*, Vol 30, No. 23, (December 2009), pp. 4118-4136, ISSN 1522-2683.

Vödisch, M.; Albrecht, D., Lessing, F.; Schmidt, A.D.; Winkler, R., et al. (2009). Two-dimensional proteome reference maps for the human pathogenic filamentous fungus *Aspergillus fumigatus*. *Proteomics*, Vol. 9, No. 5, (March 2009), pp. 1407-1415, ISSN 1615-9853.

Vödisch, M.; Scherlach, K.; Winkler, R.; Hertweck, C.; Braun, H.P., et al. (2011). Analysis of the *Aspergillus fumigatus* Proteome Reveals Metabolic Changes and the Activation of the Pseurotin A Biosynthesis Gene Cluster in Response to Hypoxia. *J Proteome Res*, Vol. 10, No. 5, (May 2011), pp. 2508-2524, ISSN 1535-3893.

Wartenberg, D.; Lapp, K.; Jacobsen, I.D.; Dahse, H.M.; Kniemeyer, O., et al. (2011). Secretome analysis of *Aspergillus fumigatus* reveals Asp-hemolysin as a major secreted protein. *Int J Med Microbiol*. In Press, ISSN 1438-4221.

Wright, J.C.; Sugden, D.; Francis-McIntyre, S.; Riba-Garcia, I.; Gaskell, S.J., et al. (2009). Exploiting proteomic data for genome annotation and gene model validation in *Aspergillus niger*. *BMC Genomics*, Vol. 4, No. 10, (February 2009), pp. 61, ISSN 1471-2164.

Wymelenberg, A.V.; Minges, P.; Sabat, G.; Martinez, D.; Aerts, A., et al. (2006). Computational analysis of the *Phanerochaete chrysosporium* v2.0 genome database and mass spectrometry identiWcation of peptides in ligninolytic cultures reveal complex mixtures of secreted proteins. *Fungal Genetics and Biology*, Vol. 43, No. 5, (May 2006), pp. 343–356, ISSN 1087-1845.

Wymelenberg, A.V.; Sabat, G.; Martinez, D.; Rajangam, A.S.; Teeri, T.T., et al. (2005). The *Phanerochaete chrysosporium* secretome: Database predictions and initial mass spectrometry peptide identifications in cellulose-grown medium. *Journal of Biotechnology*, Vol. 118, No. 1, (July 2005), pp. 17–34, ISSN 0168-1656.

Yildirim, V.; Ozcan, S.; Becher, D.; Büttner, K.; Hecker, M. & Ozcengiz, G. (2011). Characterization of proteome alterations in *Phanerochaete chrysosporium* in response to lead exposure. *Proteome Sci*, Vol. 9, No. 9, (March 2011), pp. 12, ISSN 1477-5956.

Zorn, H.; Peters, T.; Nimtz, M. & Berger, R.G. (2005). The secretome of *Pleurotus sapidus*. *Proteomics*, Vol. 5, No. 18, (December 2005), pp. 4832–4838, ISSN 1615-9853.

Zureik, M.; Neukirch, C.; Leynaert, B.; Liard, R.; Bousquet, J. & Neukirch, F. (2002). Sensitisation to airborne moulds and severity of asthma: cross sectional study from European Community respiratory health survey. *BMJ*, Vol. 325, No. 7361, (August 2002), pp. 411–414, ISSN 0959-8138.

7

ETD and ECD Mass Spectrometry Fragmentation for the Characterization of Protein Post Translational Modifications

Lisa Elviri
Dipartimento di Chimica Generale ed Inorganica, Chimica Analitica, Chimica Fisica, Università degli Studi di Parma, Parma, Italy

1. Introduction

The introduction of electron capture dissociation (ECD) by McLafferty and co-workers, and further of electron transfer dissociation (ETD), mechanism allows gas-phase fragmentation of multiply charged protein and peptide ions upon capture of a low-energy (<1 eV) electron or electron transfer in a gas phase ion-ion chemistry. The odd-electron species then undergoes rearrangement with subsequent cleavage of $N-C_\alpha$ backbone. Peptide fragmentation can take place inducing the formation of c- and z- type fragment ions without loss of the information on the PTM localization. The key to the success of this approach is the selection of intact protein molecular ions and its profound potential for PTM characterization as alternative to vibrational excitation techniques. With ECD, that occurs within 10^{-12} s, modifications that are rapidly lost upon vibrational excitation, such as phosphorylation, N- and O-glycosylation, sulfation and γ-carboxylation, can be easily established.

In this Chapter, we will illustrate the ECD and ETD protein and peptides fragmentation mechanism, the MS instruments used and the parameters involved and we will provide an overview on the most recent applications of protein PTM characterization.

2. Overview

Transcription of DNA to RNA and translation of RNA into protein result in a broad variety of final products so called "proteome" much more complex than the encoding genome system [1]. During or after synthesis protein post-translational modifications (PTMs) can occur resulting in the chemical modification of the amino acid side chains and the amine and carboxy terminus of the protein. PTMs can occur at one or more sites extending the function of a protein at any step of its life cycle [2]. The regulation or the alteration of these processes determine the cellular activity toward protein localization, protein/receptor, protein/protein, protein/DNA, protein/small ligands interactions. More than 200 PTMs, such as phosphorylation, glycosylation, acylation, nitrosylation, etc. are known and they play a pivotal role in the understanding of normal cell functions as well of degenerative diseases, diabetes, hearth diseases or cancer [3]. Proteomic methods strives to gain insights in the complexity of proteins, protein network, they role and biological functions. Over the last decade the role of mass spectrometry (MS) techniques to study proteins, peptides and other biomolecules was proved of fundamental importance to elucidate physiological pathways [4, 5]. Among the different applications of mass spectrometry to protein analysis, a primary interest is related to the investigation of PTMs able to regulate a variety of biological mechanisms. The development of analytical instrumentations and strategies exhibiting sensitivity, selectivity and accuracy suitable to identify and quantify them is thus demanding [6]. Several publications illustrated the excellent results obtained by MS and tandem MS (MS/MS) experiments enabling the site-specific assignment of PTMs even at the resolution of individual amino acids in proteins [7, 8]. In addition, the miniaturization of liquid chromatography (LC) techniques together with the development of nano-electrospray sources allows in combination with MS/MS methods the analysis of very small volumes of complex protein and peptide mixtures [9]. However, while the modern soft ionization techniques, matrix assisted laser desorption ionization (MALDI) and electrospray (ESI), allow to obtain intact gas-phase protein ions with even a high degree of modification, PTM analysis still remains a challenge.

Usually, proteins are digested into smaller peptides with an enzyme (i.e. trypsin, Lys C, Glu C etc.), that are separated by ionic exchange and/or reversed phase chromatography before mass spectrometry analysis. These peptides ionize via ESI or MALDI to different charge states (generally +1, +2 or +3), which are suitable for sequencing by collision-activated dissociation (CAD) process. Low-energy collision induced dissociation (CID) is the most commonly method used to sequence peptide ions by MS/MS. Upon collision with a neutral gas, fragmentation usually occurs at the amide bond level in the peptide backbone leading to the formation of singly charged b- and y-type fragment ions (Figure 1) [10]. The CID process is highly effective for small (15-20 aa residues), low-charged peptides (+1, +2, +3 charge state), and most of the algorithms for data interpretation prefers singly charged fragment ions. The reliability to identify a peptide involves the generation of an extensive fragmentation of the amino acid backbone. On the other hand fragmentation depends on the protein sequence, the peptide length or the presence of PTMs. In addition, the presence of several basic residues can prevent random protonation along the peptide backbone inducing site specific dissociation and few sequence ions. CAD is a widely used technique able to provide several useful information, however is not suitable for fragmentation of intact proteins and could prevent the detection of PTMs. Several PTMs are labile (i.e. phosphorylation, nitrosylation, sulfonation) and could be difficult to identify the sites of the modification or the relationship between them by CAD. For example, in the case of phosphorylated serine (Ser) or threonine (Thr) residues, the phosphate group competes with the peptide backbone as the preferred site of cleavage. The activation of these peptides upon collision displaces phosphoric acid from the peptide, and frequently results in the peptide's backbone bonds remaining intact [11]. The resultant spectra contain minimal peptide sequence information and may not allow for unambiguous peptide sequence assignment.

While phosphorylation of serine and threonine residues is widely documented a labile chemical modifications, other PTMs such as glycosylation [7], sulfonation [7], and nitrosylation [12] belong to this category. Loss of these labile PTMs during CAD process results in product ion spectra that does not contain sufficient information on the peptide sequence. However, even with this limitation, the methodology of proteolytic digestion followed by CAD MS/MS has successfully provided sequence identifications (including PTM locations) of thousands of peptides with labile PTMs (13-15).

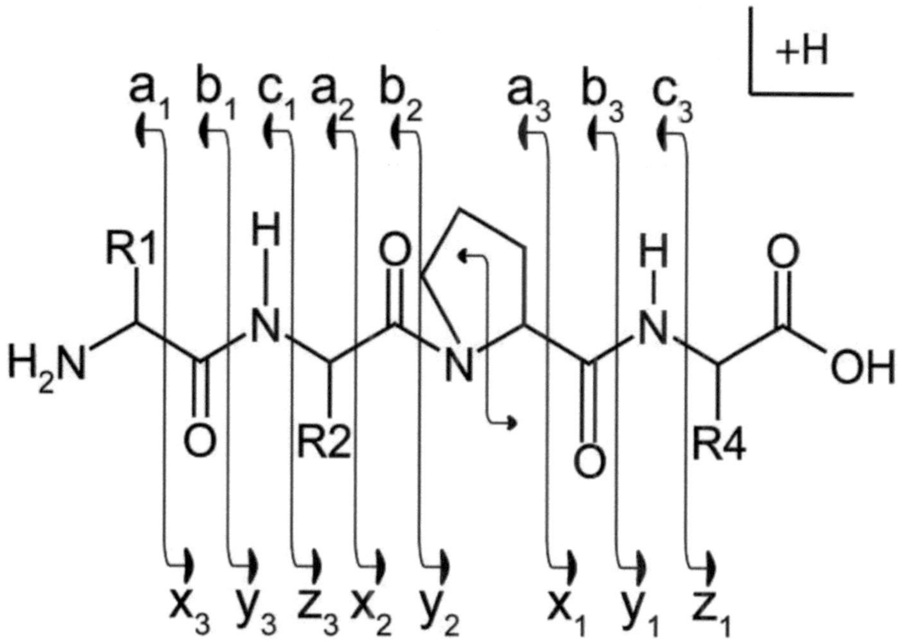

Fig. 1. Roepstorff Nomenclature Scheme. Illustration of fragment ions formed from the backbone cleavage of protonated peptides. Fragment ions retaining the positive charge on the amino terminus are termed a-, b-, or c-type ions. Fragment ions retaining the positive charge on the carboxy terminus are termed x-, y-, or z-type ions. [Ref 10, 54].

An alternative fragmentation technique, the electron capture dissociation (ECD), was introduced by McLafferty and co-workers, based on the gas-phase fragmentation of multiply charged protein and peptide ions upon capture of a low-energy (<1 eV) electron within a penning trap of an FT-ICR (Fourier Transform-Ion Cyclotron Resonance) MS [16]. The odd-electron species undergo rearrangement with subsequent cleavage of $N-C_\alpha$ backbone (Figure 2) [17]. Peptide fragmentation can take place following a lower energy pathway than CAD, inducing the formation of c- and z- type fragment ions without loss of the information on the PTM localization. The key to the success of this approach is the selection of intact protein molecular ions and its profound potential for PTM characterization as alternative to vibrational excitation techniques. With ECD, that occurs within 10^{-12} s, modifications that are rapidly lost upon vibrational excitation, such as phosphorylation, N- and O-glycosylation, sulfation and γ-carboxylation, can be easily established.

To go into the details, in the ECD process multiply charged peptides or proteins are isolated in the trap of a Fourier transform ion cyclotron resonance (FTICR) mass spectrometer and exposed to electrons with near-thermal energies [16]. The capture of a thermal electron by a protonated peptide is an exothermic process (6 eV) that does not involve intramolecular vibrational energy redistribution and is able to induce fragmentation of the peptide backbone by a nonergodic process. The initial step involves the formation of an odd-electron hypervalent species ($RNH_3\bullet$) that dissociates to produce RNH_2 and a hydrogen radical. As described by Zubarev et al. addition of H• to the carbonyl groups of the peptide backbone leads to a homologous series of complementary fragment ions of types c and z.

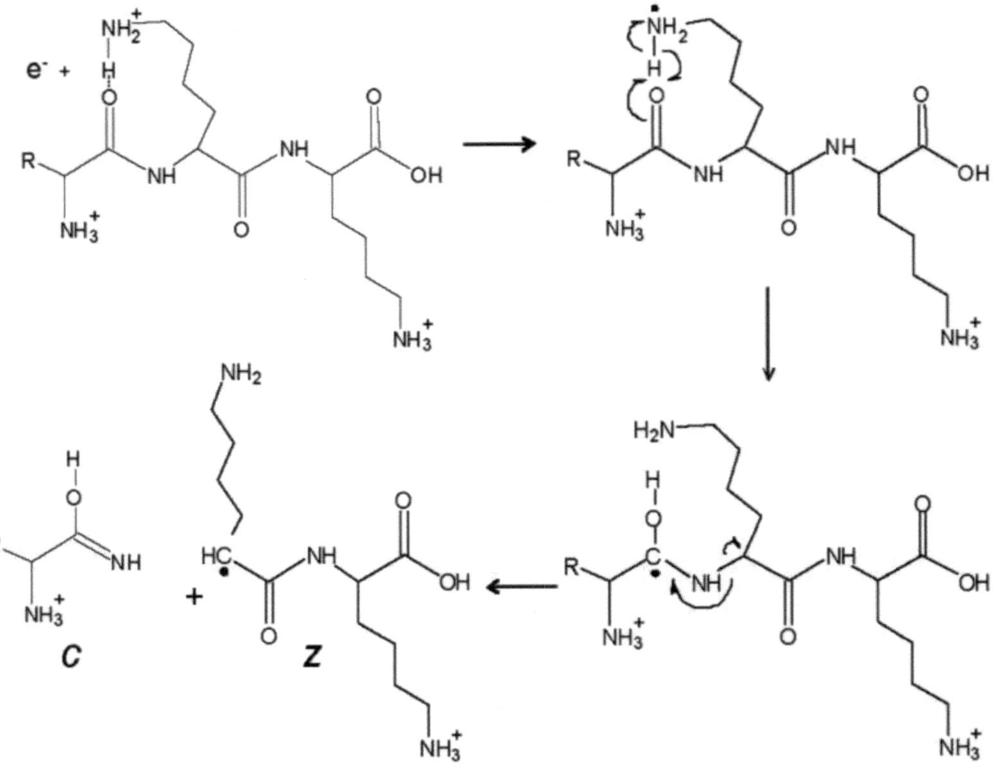

Fig. 2. ETD fragmentation scheme. Fragmentation scheme of a multiply protonated peptide after reaction with a low energy electron to produce c- and z-type ions [17].

ECD process occurs along the peptide backbone and is able to retain PTMs, for this reason was the technique of choice for the analysis of peptide and proteins with FTICR mass spectrometers [18-21]. Unfortunately, the efficiency of the ECD process requires that the precursor sample ions be immersed in a dense population of near-thermal electrons and can be performed only in the FTICR mass spectrometers. Thermal electrons introduced into the RF fields of RF 3D quadrupole ion trap (QIT), quadrupole time-of-flight, or RF linear 2D quadrupole ion trap (QLT) instruments maintain their thermal energy only for a fraction of a microsecond and are not trapped.

For this reason, QLT mass spectrometer were modified to enable ion_ion experiments and the capability to perform electron transfer dissociation (ETD) process.

ETD can be either utilized to fragment intact proteins by a top-down approach by direct introduction into the mass spectrometer analyzer. As reported in the literature, proteins or large peptides are isolated and reacted with the fluoranthene radical anion, generated by chemical ionization. McAlister et al. proposed in the 2008 the modification of the quadrupole linear ion trap-orbitap mass spectrometer to accommodate a negative chemical ionization source for the production of fluoranthene radical anions and the ETD process [22]. The resulting highly charged product ions are reacted with even electron anions of benzoic acid in order to reduce the charge states of the c and z-type fragment ions into predominantly singly charged species. This proton transfer (PTR) charge reduction along with ETD is needed to reduce the complexity of the spectra generated from these highly charged (z>10) species. ETD followed by proton transfer of larger peptides and proteins typically only generate the first 15–40 amino acids at both the N- and Ctermini. Nonetheless, information from current ETD mass spectrometers is more than sufficient to identify the protein. For this purpose, the accurate determination of the precursor charge state is fundamental for a suitable identification of the proteins by using the bioinformatics tools and the sequence data bases.

A complete protocol to isolate and characterize intact proteins by using liquid chromatography, MALDI-TOF-MS and ETD/PTR in a quadrupcle ion trap together with different applications were recently provided by Hart SR et al. [23].

To better explain, the ETD process involves two fundamental reactions: proton and electron transfer reaction. The ET reaction allows the formation of products that do not dissociate upon ion/ion-reactions as well as side chain losses from aa residues. Theoretically PTR from multiply protonated species to singly charged anions is a facile process for any negative ion [24]. As reported by Wiesner at al., the ET process can compete with PTR only if the species generating the negative ion has appropriate activation energy and the respective anion has favorable Frank-Condon factors associated with the transition from anion to neutral. Taking into account that the exothermicity for ET process for an anion can be determined by the recombination energy of the cationic species, low (60 kcal/mol) and high (62 kcal/mol) electron affinities hardly provide any probability for ET [25-27]. In case of ion/ion reactions, however, both kinds of reactions (ET and PT reactions) are exothermic, in particular if the peptide is multiprotonated, whereas for the majority of singly charged anions, PT from a multiply protonated peptide is thermodynamically favored over ET (Figure 3).

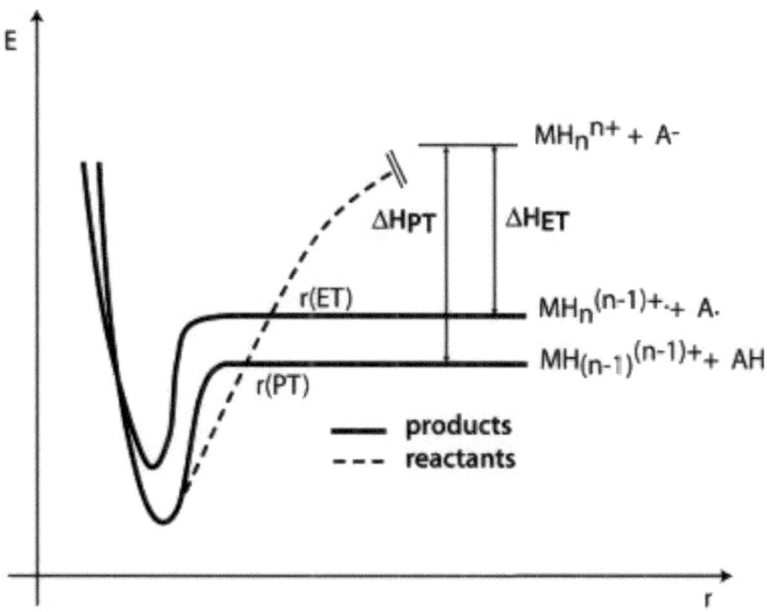

Fig. 3. Schematic potential energy curve of an ion/ion reaction between a peptide precursor [MHn]n1 and a reagent anion A2. As the entrance channel crosses the exit channels of either electron or PT, the reaction can occur. With regard to ET happening at greater distances, this reaction is more likely than PT for the typical anion/cation pairs. Slightly modified after [26].

As evident, the ion/ion reaction for ET might be reached before the point at which PT is likely and opens up the possibility for ET. Since fragmentation does not occur after PT, the charge-reduced species are produced. For these reasons, intact ET products should have higher masses than PT products that differ in mass by 1 Da. [28, 29].

The success of ETD experiments depends upon several parameters and Xia et al. provided an excellent overview of the features required by the precursor peptide ions in terms of the effects of cation charge-site identity and position on ETD of polypeptide cations.

Turecek et al. [30], showed the pivotal role played by hydrogen transfer from a neutralized charge site in providing high fragmentation efficiencies, thus peptide cations containing protonated lysine, arginine, and histidine provide a similar degree of ET which is significantly higher than for those peptides having fixed charge sites. However, the backbone fragmentation can still be observed suggesting the occurrence of direct ET to an amide linkage and the formation of some c- and z-type ions. In conclusion, even it is not a strict prerequisite, the presence of multiply charges facilitates the cleavages of bonds. As previously reported [30], among the different charge sites, protonated histidine reveals the highest degree of ET without dissociation while apparently no intact ET products are observed for peptides with protonated lysine and arginine. A possible explanation might be given by the stability of the imidazolium ion that is formed upon electron capture and can delocalize the radical along the aromatic ring of histidine.

Xia et al. demonstrated that all cation types show aa side chain losses with arginine containing peptides yielding the largest fraction and lysine peptides the smallest [28]. This feature fits with the ability to transfer a hydrogen atom from a neutralized charge site to the peptide backbone. Generally, electron attachment is favored at N-terminal lysine sites which reveal the lowest reaction energy value, thus decreased side-chain fragmentation is observed for these species. Apart from its identity, the amount of precursor charges also plays a role for the ETD outcome. ETD contributes a 20- fold higher confidence for the identification of triply charged peptides than doubly charged ones [31]. In addition, the product ratios depends also on the selected anion. Some anions function as strong bases and react exclusively as proton acceptors, whereas others participate in both proton and ET (*e.g.*, 1,3-dinitrobenzene results in a higher degree of PT than azobenzene) [32, 33].

As observed by Gunawardena et al., in general, a high degree of spectra similarities can be observed for different kinds of anions reacting with the same peptide [26].

Moreover, the MS instrument parameters, such as the filling of the IT space charge limit with anionic reagent, can affect ETD efficiency [34]. The space charge limit for anions in such instruments is higher than for cations, resulting in a critical cation current: too many will produce space-charge-associated mass shifts, whereas too few will result in low product ion intensities and the necessity of spectral averaging. The integrated automatic gain control (AGC) or ion charge control (ICC) regulate the anion and cation populations [35, 36].

As for instrumentation, a combination (compared with the FTICR MS) of ETD with a high-resolution mass spectrometer was recently introduced (Figure 4) [22]. As proposed for the LIT (LTQ XL) instruments with ETD, an negative CI was implemented at the back side of the instrument. In this way, positive peptide precursors are initially trapped within the first section of the LIT and subsequently, reagent anions are orientated in few ms through the ion optics at the rear side of the trap. After the ion/ion reactions, anions are removed and product ions are either detected within the LIT or with high-resolution within the Orbitrap mass analyzer.

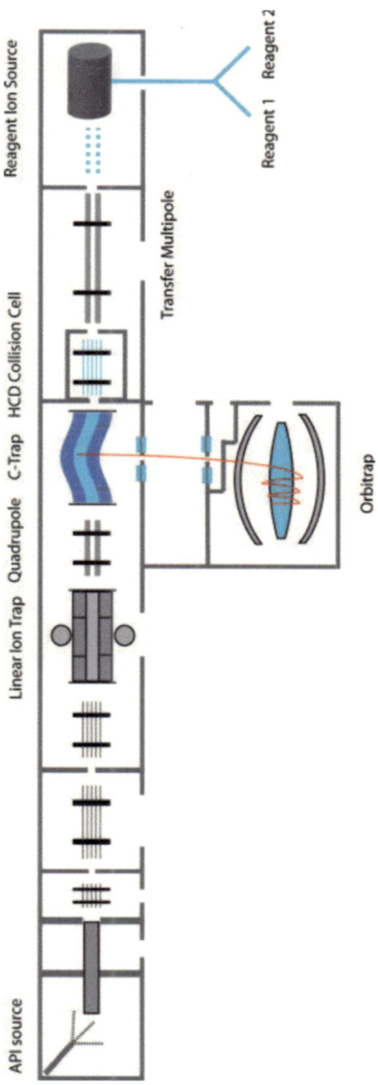

Fig. 4. Scheme of the newly developed LTQ-Orbitrap XL with ETD. The reagent ion source is implemented at the back side of the instrument. Slightly modified after [22].

Frese et al. very recently reported the results obtained by performing ETD experiments on a new generation Orbitrap [37]. A comprehensive experimental comparison of using ETD, ion-trap CID, and beam type CID (HCD) in combination with either linear ion trap or Orbitrap readout for the large-scale analysis of tryptic peptides was carried out. They demonstrated that the combination of fragmentation technique and mass analyzer is able to provide the best performance for the analysis of distinct peptide populations such as N-acetylated, phosphorylated, and tryptic peptides with up to two missed cleavages (Figure 5). They found that HCD provides more peptide identifications than CID and ETD for doubly charged peptides. In terms of Mascot score, ETD FT outperforms the other techniques for peptides with charge states higher than 2. They data shows that there is a trade-off between spectral quality and speed when using the Orbitrap for fragment ion

	mascot score	CID	HCD	ETD	CID FT	ETD FT	
2+ peptides	20-30	1283	1539	965	1152	725	A
	30-40	2057	2031	1828	1387	1365	
	40-50	1591	1852	1240	1266	1044	
	50-60	1109	1552	714	1214	730	
	60-70	787	1178	386	919	282	
	70-80	480	812	173	754	187	
	80-90	263	574	80	605	78	
	90-100	128	470	39	494	33	
	>100	77	861	20	990	17	
	total	7775	10869	5445	8781	4461	
3+ peptides	20-30	1074	895	342	996	191	B
	30-40	1084	1105	816	1007	351	
	40-50	613	1038	1084	889	516	
	50-60	300	998	1045	687	575	
	60-70	153	756	815	540	593	
	70-80	57	631	566	375	602	
	80-90	18	454	283	297	483	
	90-100	7	302	123	213	361	
	>100	3	519	89	484	383	
	total	3309	6698	5163	5488	4055	
4+ peptides	20-30	296	442	234	495	33	C
	30-40	173	477	517	459	74	
	40-50	69	470	476	349	117	
	50-60	17	337	302	259	142	
	60-70	10	224	165	183	194	
	70-80	3	183	93	103	193	
	80-90	5	134	20	72	222	
	90-100	0	91	6	53	251	
	>100	0	112	2	95	698	
	total	573	2470	1815	2068	1924	
phosphopeptides	20-30	430	470	223	331	193	D
	30-40	490	474	365	370	281	
	40-50	328	339	265	345	196	
	50-60	168	265	165	273	177	
	60-70	91	182	99	185	91	
	70-80	34	118	62	115	50	
	80-90	14	69	25	87	31	
	90-100	2	31	11	56	11	
	>100	1	34	7	60	21	
	total	1558	1982	1222	1822	1051	
N-acetylated peptides	20-30	121	97	79	113	70	E
	30-40	180	127	126	99	114	
	40-50	145	141	119	101	84	
	50-60	118	116	87	130	60	
	60-70	75	83	63	84	47	
	70-80	80	96	41	83	44	
	80-90	120	97	39	87	31	
	90-100	126	114	16	64	20	
	>100	53	234	13	305	30	
	total	1018	1105	583	1066	500	

Fig. 5. Histograms illustrating the Mascot ion score distribution of the doubly (A), triply (B), quadruply (C), phosphorylated (D), and N-acetylated (E) peptides that were identified from the analyses of SCX-fractionated tryptic peptides. Samples were analyzed consecutively for five times employing either CID, HCD, ETD, CID FT, or ETD FT. Data bars are normalized to the highest value within each population. [37]

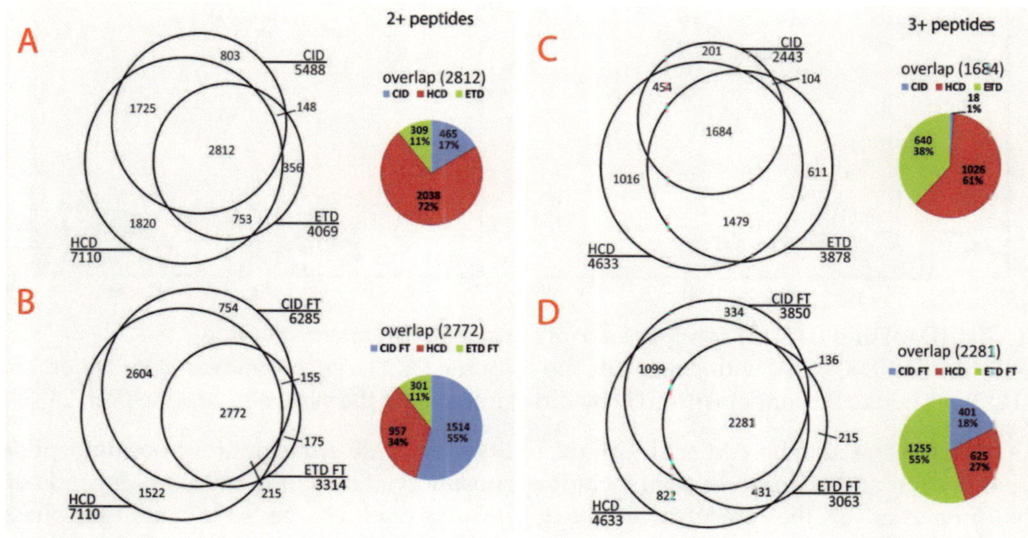

Fig. 6. Overlap of the doubly (A, B) and triply (C, D) charged peptides that were identified from the SCX-fractionated tryptic HeLa digest. Venn diagrams illustrate the overlap of the unique peptides identified between CID, HCD, and ETD (A, C) and between CID FT, HCD, and ETD FT (B, D). Pie charts show the number of unique peptides with the highest Mascot score per fragmentation mode within the peptides that were identified by all three fragmentation modes [37].

detection, concluding that a decision-tree regulated combination of higher-energy collisional dissociation (HCD) and ETD can improve the average Mascot score (Figure 6).

3. Applications of ECD/ETD process to PTM characterization

3.1 Phosphorylation

Phosphorylation consists in the addition of a phosphate (PO_4^{3-}) group to one or more amino acid side chain (Ser, Thr and Tyr) of eukaryotic proteins. Protein phosphorylation is a reversible reaction that regulates a wide range of cellular processes and plays a predominant role in biochemistry research. Liquid chromatography-tandem mass spectrometry techniques are commonly used in combination with selective sample enrichment strategies (i.e. TiO_2 cartridges or immobilized metal affinity chromatography, etc.) to identify phospho-peptides in complex mixtures. Many papers report about the use of CID/CAD strategies based on product-ion scan, neutral-loss scan or precursor ion scan acquisition modes for the identification of both phospho-peptides and phosphorylation sites. Usually, the CID fragmentation of a phospho-serine and -threonine results in the extremely favorite neutral loss of phosphoric acid (H_3PO_4; Δ98 Da) and the mass spectrum is characterized by a base peak corresponding to the loss of phosphoric acid from the parent mass and few low-intense fragments.

By using ETD the fragmentation process follows different rules, as reported above, and it is possible to observe in the mass spectrum the cleavage of the peptide backbone in to the c and z-type ions without loss of the phosphate group improving global protein profiling (Figure 7) [24].

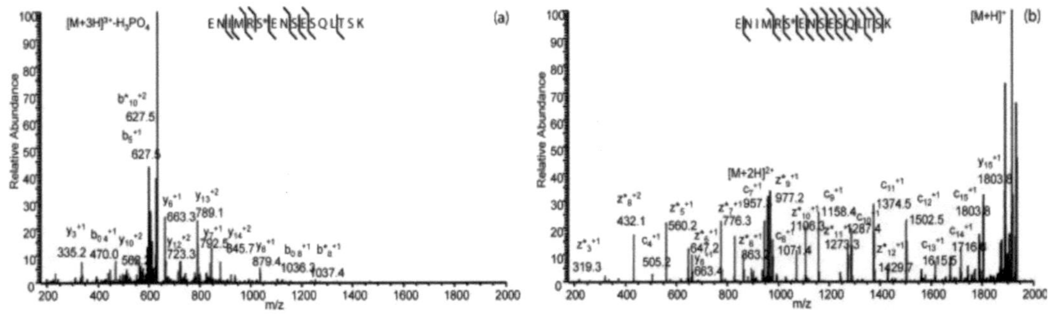

Fig. 7. CID (a) und ETD (b) spectra of a triply protonated synthetic peptide ENIMRS*ENSESQLTSK with phosphate modification (*). The spectrum was recorded on an LTQ XL (Thermo Scientific) with ETD by direct infusion of the synthetic peptide [Ref. 24].

As reported by Palumbo AM et al. [25] the utility of MS/MS strategies for phosphopeptide identification and characterization, including phosphorylation site localization depends on the properties of the precursor ion (e.g. polarity and charge state) and gas-phase mechanisms.

Very recently, Kim MS et al., reported the comparison of CID and ETD phosphopeptide fragmentation of a complex mixture [31]. They reported the identification of 2504, 491, 2584 and 3249 phospho-peptide spectrum from CID alone, ETD alone, decision tree-based CID/ETD, and altenating CID and ETD, respectively. They concluded that although alternating CID and ETD experiments for phosphopeptide identification are desirable for increasing the confidence of identifications, merging spectra prior to database search has to be carefully evaluated further in the context of the various algorithms before adopting it as a routing strategy.

3.2 O- and N-linked glycosylation

N- or O-glycosylation are a cell PTM dependent process of paramount importance in cell life, diseases and therapeutics. OGlcNAc was a labile modification not extensively studied by mass spectrometry until the last years mainly because of the analytical methodologies limitations. The introduction of ETD significantly improved the number of information available and thus facilitate the understanding of the role of this PTM.

Zhao P et al reported the use of a combination of HCD and EDT fragmentation techniques for the characterization of O-GlcNAc modified proteins. By enriching O-GlcNAc proteins from HEK293T cells, 83 modified sites were identified with high degree of confidence allowing to provide useful information about O-GlcNAc transferase activity (Figure 8) [38].

As for fragmentation process, a comparison between CAD/HCD and CAD/ETD performances for the characterization of glycopeptides were recently reported by Scott NE et al [39]. They developed a strategy of sample enrichment and LC-MS/MS analysis for the characterization of glycopeptides from gastrointestinal pathogen C.jejuni HB93-13. The results showed that CAD/HCD enabled the identification of glycan structure and peptide backbone allowing glycopeptide identification, whereas CAD/ETD enabled the elucidation of the glycosylation sites by retaining the glycan-peptide linkage. This paper reported the first use of HCD fragmentation for the identification of glycopeptide presenting intact glycan. At the end 130 glycopeptides, representing 75 glycosilation sites, were identified by LC-MS/MS using a zwitterionic HILIC (hydrophilic interaction chromatography). In details,

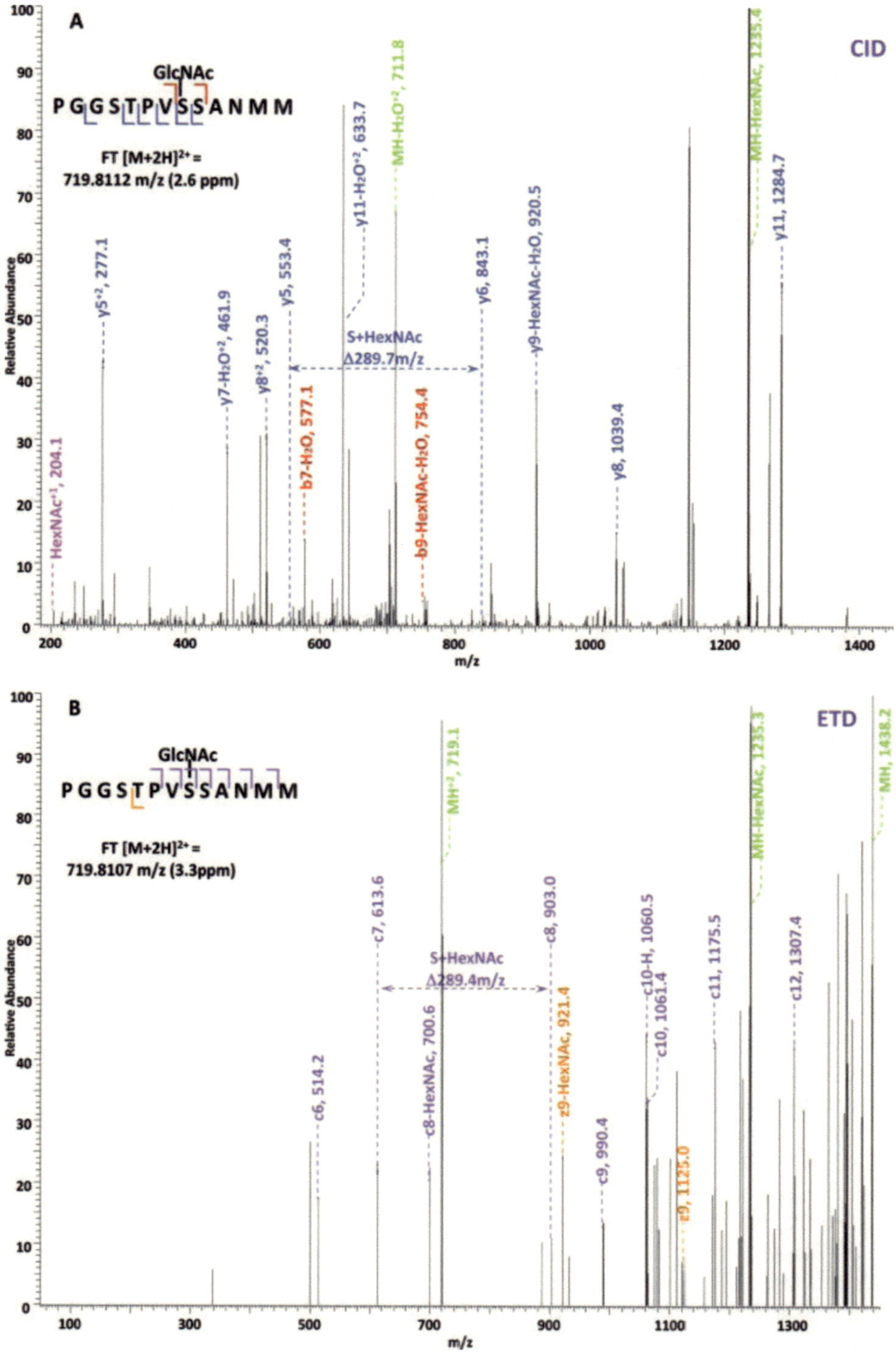

Fig. 8. Respective CID, ETD, and HCD spectra of standard O-GlcNAc modified peptides CKII and BPP. (A–B) CID and ETD spectra of O-GlcNAc modified CKII peptide, respectively [Ref.38].

CAD/HCD provided the majority of the identifications (73 sites) compared with ETD (26 sites) showing the capabilities of different approaches.

3.3 Nitrosylation

Among more than 200 different types of PTMs, S-nitrosylation, i.e the replacement of the hydrogen atom in the thiol group of cysteine (Cys) residues by a -NO moiety (SNO), has emerged as a key reaction in several metabolic processes occurring in animals and in plants [40, 41]. The study of NO signalling in animals or human endothelial cells resulted in the identification of more than a hundred modified proteins [42] even though the technical limitations in characterizing this modification due to its low stability caused a delay in appreciating its biological involvement. Different indirect methods were proposed for the identification of such PTM [43]. Among these, the Biotin Switch method, consisting in the selective reduction and substitution of the SNO groups with a biotin moiety, was successfully applied in mammalian cells and opened new insights on the S-nitrosylation of plant proteins [44]. However the sensitivity and specificity of this method are still a matter of debate. A modified version of the Biotin Switch called SNOSID (SNO Site Identification) has been proposed for high-throughput identification of SNO-Cys residues in complex protein mixture from tissues or cells [45-46]. More recently, a His-Tag switch method was introduced as a novel approach to identify proteins modified by nitric oxide allowing the unambiguous localization of the modified cysteine residues by MS [47]. While numerous published studies on protein S-nitrosylation are based on indirect labelling techniques, the direct MS measurement of nitrosylated cysteines is still poorly represented in the literature, due to the difficulty of preserving the integrity of the SNO groups during the analysis. The crucial point rely with the lability of the S-NO bond that readily undergoes fragmentation in the gas phase of a mass spectrometry source [48]. Most of the reported examples focus on single purified proteins, preferably *in vitro* overnitrosylated by chemical agents. ETD dissociation can represent an attractive alternative technique to retain the nitrosilation modification.

As an example, in the literature nitrosylated bovine insulin beta chain was analyzed by CAD and ETD as a model of this type of PTM [49]. The majority of the signal in the CAD spectrum of the $(M+5H)^{+5}$ of FVNQHLnCGSHLVEALYLVnCGERGFFYTPKA corresponded to the neutral loss of both NO groups on the cysteine residues $(M+5H\text{-}2NO)^{+5}$. Minimal peptide backbone fragmentation is obtained as only a few product ions are observed above 5% relative abundance (y_3, y_{13}–NO+2, b_{16}–NO^{+2}, b_{17}–NO^{+2}, and b_{24}–NO^{+2}). In the ETD spectrum of the same peptide, the following charge reduced (electron transfer without fragmentation) species with and without losses of NO were observed: $(M+4H\text{–}NO)^{+4}$ (may also be z_7), $(M+3H)^{+3}$, $(M+3H\text{–}NO)^{+3}$, $(M+3H\text{–}2NO)^{+3}$, $(M+2H)^{+2}$, $(M+2H\text{–}NO)^{+2}$, and $(M+2H\text{–}2NO)^{+2}$. The loss of NO from the charged reduced species may be acting as its own proton transfer reagent directing mostly charge reduction of the nitrosylated insulin as opposed to fragmentation. However, 6-7 low level c and z-type ion were observed (2% or less of the largest ion in the spectrum) demonstrating the retention of NO on the insulin product ions after ETD.

3.4 Disulfide linkage

The disulfide linkage of two cysteine residues plays an important role in the correct protein folding and thus its function. ETD fragmentation can be successfully used for the determination of disulfide linkages instead of CAD process.

Different examples are reported in the literature. In particular, Wu SL et al. reported the results of an LC-MS CAD (MS2), ETD (MS2) and CAD of isolated product ion derived from ETD (MS3) combined strategy for the characterization of disulfide linkages of recombinant human proteins [50]. In the case of human growth hormone (Nutropin), disulfide bond was not broken by using CAD process, whereas ETD fragmentation resulted in to two separated peptide ions along with typical fragmentation pattern of the backbone cleavages (c and z ions) with several high intensity ions consisting of charge-reduced species of the precursor ion ($[M+3H]^{2+\cdot\cdot}$, $[M\text{-}NH_3+3H]^{2+\cdot\cdot}$, $[M+3H]^{+\cdot\cdot}$, $[M\text{-}NH_3+3H]^{+\cdot\cdot}$, $[M\text{-}2NH_3+3H]^{+\cdot\cdot}$).

As previously reported, charge-reduced species rather than the disulfide-dissociated peptides are dominant product ions in the ETD spectrum for precursor ions with m/z>900 (Figure 4B and 3B) [51, 52]. The generation of several charge-reduced species with high intensities in the ETD spectrum allowed the determination of the charge state of the precursor ion (4+). Additionally, the Authors observed that the disulfide-dissociated peptides became the major ions in the MS3 step. Final suggestions are the use of combined fragmentation approach and high resolution mass spectrometers (Q-TOF or Orbitrap) for the characterization of multi- and close-disulfide bonds in proteins.

Wang Y et al. proposed the use of ETD process for the characterization and comparison of disulfide linkages and scrambling pattens in therapeutic monoclonal antibodies [53].

3.5 Sulfonation

Very few examples are reported in the literature about the use of ETD dissociation and characterization of O-sulfonation, a PTMs involved in protein assembly and signal transduction. CAD fragmentation of sulfonated peptides results in the neutral loss of SO_3 as predominant product ion. ETD process allows to retain the SO_3 group and to observe c and z-type ion series (Figure 9) and thus the aminoacid localization [54]. Mikesh et al, reported that the isobaric modifications, sulfonation from phosphorylation, differentiated by their different CAD neutral loss patterns (80 amu vs. 98 amu), can be detected by using a ETD modified LTQ for sequential CAD-ETD analysis.

A very interesting example comes from Cook et al. that investigated the dissociation behavior of phosphorylated and sulfonated peptide anions using metastable atom-activated dissociation mass spectrometry (MAD-MS) and CID. In this paper phosphorylated and sulfonated peptides in the 3- and 2- charge states were exposed to a beam of high kinetic energy helium metastable atoms. Unlike CID, where phosphate losses are dominant, the major dissociation channels observed using MAD were C(α)-C peptide backbone cleavages and neutral losses of CO_2, H_2O, and $[CO_2 + H_2O]$ from the charge reduced (oxidized) product ion. Regardless of charge state or modification, MAD provides ample backbone cleavages with little modification loss, which allows for unambiguous PTM site determination resulting a complementary dissociation technique to CID, ETD, ECD for peptide sequencing and modification identification. MAD offers the unique ability to analyze highly acidic peptides that contain few to no basic amino acids in either negative or positive ion mode [55].

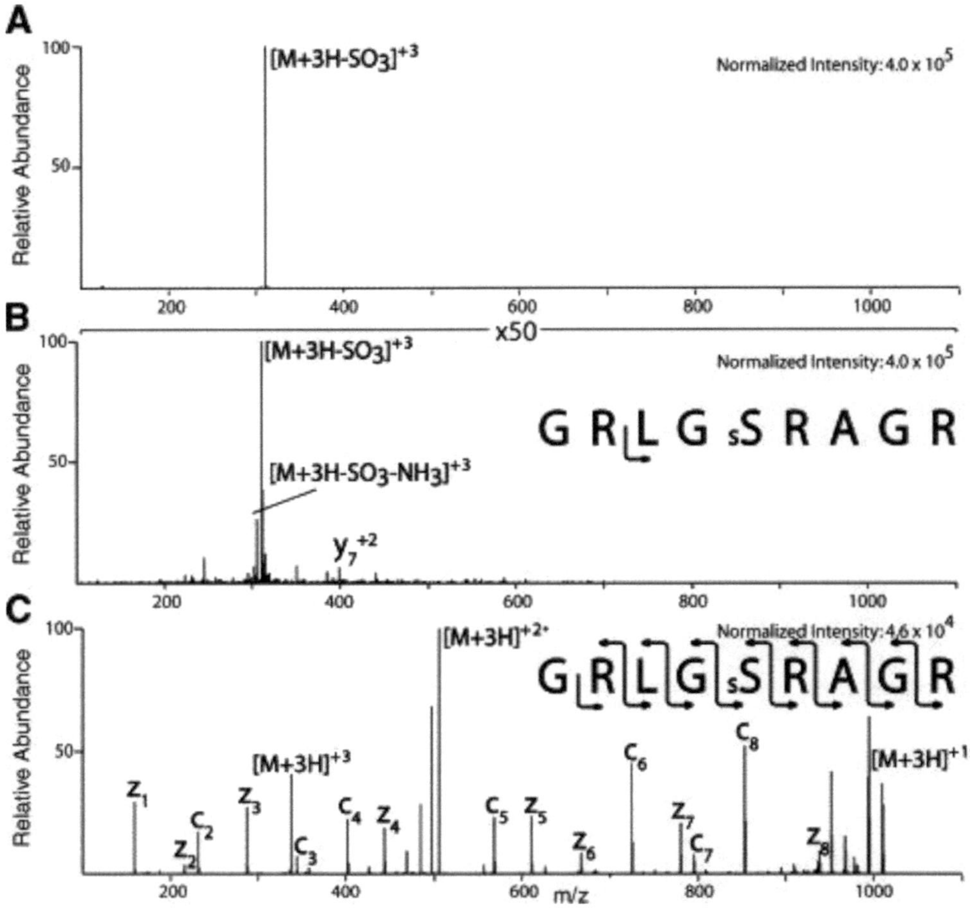

Fig. 9. Comparison of CAD vs. ETD spectrum of a sulfonated peptide. Sulfonation of peptides was achieved by reacting the peptide with 5% chlorosulfonic acid in neat trifluoroacetic acid (TFA) for 20 min at room temperature. The reaction was terminated by the addition of water and was purified by RP-HPLC. Mass spectrometry analysis before and after sulfonation confirmed reaction. Sulfonated peptides (1 pmol/µl) were infused at a flow rate of 60 nl/min into a ThermoElectron LTQ ion trap mass spectrometer modified to perform ETD. (A) The CAD spectrum of the illustrated sulfonated peptide contains one major ion corresponding to the neutral loss of SO_3 from the $(M + 3H)^{+3}$ precursor ion. (B) Magnification of the spectrum shown in A by 50×. (C) Fragmentation of the sulfonated peptide by ETD where a complete c and z-type ion series was observed with no detectable loss of SO_3 from the precursor ion or the peptide backbone. [Ref.54].

4. Conclusions

This chapter describes that ETD dissociation is able to generate product ions from peptides carrying PTMs, multiply charged peptides and intact proteins.

Whereas CID/CAD peptide dissociation usually induces significant loss of labile chemical groups, i.e. phosphoric acid, NO or SO_3, the ETD process allows to preserve these weak chemical bonds thus providing reliable and accurate AA site assignment.

Additionally, ETD has the capability to fragment intact proteins providing useful information on sequences, presence of disulfide bonds, protein isoforms.
Several application of ETD in proteomic research are reported in the literature involving many life-science areas, i.e. clinical diagnosis, biomarker discovery, cancer and degenerative diseases, cell signaling etc.. In these years we are assisting to a significant improvement in the mass spectrometry instruments that will turn in the increasing the ETD capabilities and applications.

5. References

[1] Renuse S, Chaerkady R, Pandey A, Proteogenomics, *Proteomics*, 11, 2011, 620-30.
[2] Krishna RG, Wold F, Post-traslational modification of proteins, *Adv. Enzymol. Relat. Areas Mol. Biol.*, 67, 1993, 265-298.
[3] Yang XJ, Multisite protein modification and intramolecular signaling, *Oncogene*, 24, 2005, 1653-1662.
[4] Aebersold R, Mann M, Mass spectrometry-based proteomics, *Nature*, 422, 2003, 198-207.
[5] Jates JR3rd, Gilchrist A, Howell KE, Bergeron JJ, Proteomics of organelles and large cellular structures, *Nature Rev. Mol. Cell Biol.*, 6, 2005, 702-714.
[6] Jensen ON, Modification-specific proteomics: characterization of post-translational modifications by mass spectrometry, *Curr. Opin. Chem. Biol.*, 8, 2004, 33-41.
[7] Mann M, Jensen ON, Proteomic analysis of post translational modifications, *Nature Biotechnol*, 21, 2003, 255-261.
[8] Cohen P, The regulation of protein function by multisite phosphorylation-a 25 year update. Trends Biochem. Sci., 25, 2000, 596-601.
[9] Mitulovic G, Mechtler K, HPLC techniques for proteomics analysis—a short overview of latest developments, 5, 2006, 249-260.
[10] Wysoki VH, Resing KA, Zhang QF, Cheng GL, Mass spectrometry of peptides and proteins, Methods, 35, 2005, 211-222.
[11] Larsen MR, Trelle B, Thinholm TE, Jensen ON, Analysis of posttranslational modifications of proteins by tandem mass spectrometry, Biotechniques, 40, 2006, 790-8.
[12] U.A. Mirza, B.T. Chait, H.M. Lander, Monitoring reactions of nitric oxide with peptides and proteins by electrospray ionization-mass spectrometry, J. Biol. Chem. 270 (1995) 17185–17188.
[13] Zhao Y, Jensen ON, Modification-specific proteomics: strategies for characterization of post translational modifications using enrichment techniques, *Proteomics*, 9, 2009, 4632-4641.
[14] Hoffman MD, Sniatynski M, Kast J, Current approaches for global post translational modification discovery and mass spectrometry analysis, *Anal. Chim. Acta*, 627, 2008, 50-61.
[15] Zubarev RA, Electron capture dissociation tandem mass spectrometry, *Curr. Opin. Biotechnol.*, 15, 2004, 12-16.
[16] Zubarev RA, Kelleher NL, McLafferty FW, ECD of multiply charged protein cations. A non-ergodic process, J. Am. Chem. Soc. 120 (1998) 3265–3266.
[17] Biemann K, Appendix 5. Nomenclature for peptide fragment ions (positive ions). *Methods Enzymol* 1990, 193, 886-887.

[18] Lei S, Shuying L, RA Zubarev, ECD Technology in biological mass spectrometry, Progress in Chemistry, 23, 2011, 1710-1718.

[19] Shi SD, Hemling ME, Carr SA, Horn DM, Lindh I, McLafferty FM, Phosphopeptide/phosphoprotein mapping by electron capture dissociation mass spectrometry, Anal. Chem. 73 (2001) 19–22.

[20] Mirgorodskaya E, Roepstorff P, Zubarev RA, Localization of Oglycosylation sites in peptides by electron capture dissociation in a Fourier transform mass spectrometer, Anal. Chem. 71 (1999) 4431–4436.

[21] Kelleher NL, Zubarev RA, Bush K, Furie B, Furie BC, McLafferty FW, Walsh CT, Localization of labile posttranslational modifications by electron capture dissociation: the case of gamma-carboxyglutamic acid, Anal. Chem. 71 (1999) 4250–4253.

[22] McAlister GC, Berggren WT, Griep-Raming J, Horning S. *et al.*, A Proteomics grade electron transfer dissociationenabled hybrid linear ion trap-orbitrap mass spectrometer. *J. Proteome Res.* 2008, *7*, 3127–3136.

[23. Hart SR, Sarah R, A protocol for top-down proteomics using HPLC and ETD/PTR-MS, Methods Mol. Biol. 2010, 658:339-53.

[24] Wiesner J, Premsler T, Sickman A, Application of electron transfer dissociation (ETD) for the analysis of posttranslational modifications, Proteomics, 8, 2008, 4466-4483.

[25] Palumbo AM, Tandem mass spectrometry strategies for phosphoproteome analysis, Mass Spectrom. Rev. 2001, 30(4): 600-25.

[26] Gunawardena HP, He M, Chrisman PA., Pitteri, S. J. *et al.*, Electron transfer versus proton transfer in gas-phase ion/ ion reactions of polyprotonated peptides. *J. Am. Chem. Soc.* 2005, *127*, 12627–12639.

[27] Wells JM., Chrisman PA, McLuckey SA., Formation and characterization of protein-protein complexes *in vacuo*. *J. Am. Chem. Soc.* 2003, *125*, 7238–7249.

[28] Xia Y, Gunawardena HP, Erickson DE, McLuckey SA, Effects of cation charge-site identity and position on electron- transfer dissociation of polypeptide cations. *J. Am. Chem. Soc.* 2007, *129*, 12232–12243.

[29. Chrisman PA, Pitteri SJ, Hogan JM., McLuckey SA., SO2-* electron transfer ion/ion reactions with disulfide linked polypeptide ions. *J. Am. Soc. Mass Spectrom.* 2005, *16*, 1020–1030.

[30] Turecek F, N-C(alpha) bond dissociation energies and kinetics in amide and peptide radicals. Is the dissociation a nonergodic process? *J. Am. Chem. Soc.* 2003, *125*, 5954–5963.

[31] Kim MS et al. Systematic evaluation of alternating CID and ETD fragmentation for phosphorylated peptides, Proteomics, 2011, 11 (12) 2568-72.

[32] Syka JE, Coon JJ, Schroeder M J, Shabanowitz J, Hunt DF, Peptide and protein sequence analysis by electron transfer dissociation mass spectrometry. *Proc. Natl. Acad. Sci. USA* 2004, *101*, 9528–9533.

[33] Swaney DL, McAlister GC.,Wirtala M., Schwartz JC. *et al.*, Supplemental activation method for high-efficiency electron-transfer dissociation of doubly protonated peptide precursors. *Anal. Chem.* 2007, *79*, 477–485.

[34] Good DM, Wirtala M, McAlister GC, Coon JJ, Performance characteristics of electron transfer dissociation mass spectrometry. *Mol. Cell. Proteomics* 2007, *6*, 1942– 1951.

[35] Stafford GC, Taylor DM, Bradshaw SC, Syka JEP, Enhanced sensitivity and dynamic range on an ion trap mass spectrometer with automatic gain control, *35th Annual Conference of the American Society of Mass Spectrometry*, Denver, CO 1987.

[36] Wang, Y., Franzen, J., *Int. J. Mass Spectrom. Ion Process.* 1992, *112*, 167–178.

[37] Frese CK, Maarten Altelaar AF, Hennrich ML, Nolting D, Zeller M, Griep-Raming J, Heck AJR, Mohammed S, Improved Peptide Identification by Targeted Fragmentation Using CID, HCD and ETD on an LTQ-Orbitrap Velos, *J. Proteome Res.*, 2011, *10* (5), 2377–2388.

[38] Zhao P, Viner R, Teo CF, Boons GJ, Horn D, Wells L, Combining High-Energy C-Trap Dissociation and Electron Transfer Dissociation for Protein O-GlcNAc Modification Site Assignment, J. Proteome Res. 2011, 10 (9): 4088-104

[39] Scott NE, Parker BL, Conolly AM, Paulech J, Edwards AVG, Falconer L, Kolarich D, Djordjevic SP, Hojrup P, Packer NH, Larsen MR, Cordwell SJ, Mol. Cell. Proteom. 10 (2011).

[40] Kanski J, Schoneich C, Protein nitration in biological aging: proteomic and tandem mass spectrometric characterization of nitrated sites, Methods Enzymol. 396 (2005) 160.

[41] Besson-Bard, A. Pugin, D. Wendehenne, New insights into nitric oxide signaling in plants Annu. Rev. Plant Biol. 59 (2008) 21.

[42] Martínez-Ruiz A, Lamas S, Proteomic identification of S-nitrosylated proteins in endothelial cells, *Methods Mol Biol.* 357 (2007) 215.

[43] Kaneko R, Wada Y, Decomposition of protein nitrosothiolsin matrix-assisted laser desorption/ionization and electrospray ionization mass spectrometry, *J. Mass Spectrom.* 38 (2003) 526.

[44] Torta F, Usuelli V, Malgaroli A, Bachi A, Proteomic analysis of S-nitrosylation, *Proteomics* 8 (2008) 4484.

[45] Jaffrey SR, Erdjument-Bromage H, Ferris CD, Tempst P, Snyder SH, Protein S-nitrosylation: a physiological signal for neuronal nitric oxide. *Nature Cell Biol.* 3 (2001) 193.

[46] Hao G, Derakhshan B, Shi L, Campagne F, Gross SS, SNOSID, a proteomic method for identification of cysteine S-nitrosylation sites in complex protein mixtures. *Proc. Natl. Acad. Sci.* 103 (2006) 1012.

[47] Torta F, Elviri L, Bachi A, Direct and indirect methods for the analysis of S-nitrosylate peptides and proteins, *Methods Enzymol*, 473, 2010, 265-80.

[48] Elviri L, Speroni F, Careri M, Mangia A, di Toppi LS, Zottini M, Identification of in vivo nitrosylated phytochelatins in Arabidopsis thaliana cells by liquid chromatography-direct electrospray-linear ion trap-mass spectrometry, *J Chromatogr* A, 1217, 2010, 4120-6.

[49] Cook SL, Jackson GP, Characterization of tyrosine nitration and cysteine nitrosylation modifications by metastable atom-activation dissociation mass spectrometry, *J Am Soc Mass Spectr.*, 22, 2011, 232.

[50] Wu SL, Jiang H, Lu Q, Dai S, Hancock WS, Karger BL, Mass spectrometric determination of disulfide linkages in recombinant therapeutic proteins using online LC-MS with electron transfer dissociation, *Anal. Chem.* 2009 8(1):112-22.

[51] Wu SL, Huhmer AF, Hao Z, Karger BL, On-line LC-MS approach combining collision-induced dissociation (CID), electron-transfer dissociation (ETD), and CID of an

isolated charge-reduced species for the trace level characterization of proteins with post-traslational modifications, J. Proteome Res. 2007 6:4230-44;
[52] Xia Y, Han H, McLuckey SA, Activation of intact electron-transfer products of polypeptides and proteins in cation transmission mode ion/ion reactions, Anal. Chem. 2008 80:1111-7
[53] Wang Y, Lu Q, Wu SL, Karger BL, Hancock WS, Characterization and comparison of disulfide linkages and scrambling pattens in therapeutic monoclonal antibodies: using LC-MS with electron transfer dissociation, Anal Chem 83, 2011, 3133-40).
[54] LM Mikesh, B Ueberheide, A Chi, J J. Coon, J. E.P. Syka, J. Shabanowitz , D. F. Hunt, The utility of ETD mass spectrometry in proteomic analysis, Biochimica et Biophysica Acta 1764 (2006) 1811–1822.
[55] SL Cook, GP Jackson, Metastable Atom-Activated Dissociation Mass Spectrometry of Phosphorylated and Sulfonated Peptides in Negative Ion Mode J. Am. Soc. Mass Spectrom. 22, 2011, 1088-1097.

8

Tandem Mass Spectrometry for Simultaneous Qualitative and Quantitative Analysis of Protein

Lay-Harn Gam
School of Pharmaceutical Sciences, Universiti Sains Malaysia
Malaysia

1. Introduction

Tandem mass spectrometry has long been recognized as a technique for qualitative analysis of proteins via amino acid sequencing. The use of tandem mass spectrometry in quantitative analysis for proteins is much limited. Nevertheless, application of chromatographic separation prior to mass spectrometry analysis can be used as a device for quantitative analysis of proteins. Following chromatographic separation, specific compound in a complex mixture can be determined with minimal interference; this is possible by monitoring only the selected m/z ratio of the compound, which is the characteristic of the compound of interest, an approach know as selected ion monitoring (SIM). The application of SIM using a combined GC-MS instrument was first demonstrated by Sweeley et al. in 1966.

On the other hand, quantitative data can also be obtained by repetitive scanning of compounds during elution of a sample from the gas chromatograph (GC-MS) or liquid chromatograph (LC-MS). The detection limits for such techniques are generally much poorer than those of SIM; this is because the instrument spends very little time at each m/z ratio during scanning. Such techniques can either scan the entire mass range of analysis or only scan a limited mass range with a greater sensitivity; subsequently the quantitative data can be calculated from the peak area or peak height displayed by the extracted chromatograms of the selected masses (Shoemaker & Elliott, 1991). The selected masses can be a molecular ion or a fragment ion provided it is sufficiently intense. Repetitive scanning together with automated data processing (library search of the recorded spectra) for biological samples has been used to measure a large number of biological compounds in complex samples. Such technique was found to be reliable, accurate and considerably more cost-effective than operator-mediated methods (Slivon et al., 1985).

Compared to a number of analytical techniques, tandem mass spectrometry is a technique which is able to provide more reliable data due to its high specificity and sensitivity. The major disadvantages of mass spectrometry are the high capital costs and the relatively low sample throughput. Therefore, the use of mass spectrometry in quantitative analysis is preferably dedicated to sensitivity and specificity rather than throughput. Since the original publication (Sweeley *et al.* 1966), technology and methodology on mass spectrometry have progressed and the applications of SIM for quantitative analysis, particularly in the analyses of small compounds have increased rapidly. Nevertheless, Bellar & Budde (1988) and Eichelberger, *et al.* (1983) cited that although improved sensitivity and precision is available with SIM, the loss of qualitative information is significant and usually unacceptable. This is especially true for quantitative analysis of minute protein in biological samples. Although coupled with effective sample cleanups, it is common for such mass spectrometry analysis to generate large amount of data caused by interferences from unrelated compounds derived from the complex sample matrix. In this condition, false identification of protein in SIM quantitative analysis will concomitantly increase.

Other quantitative analyses available are selected reaction monitoring (SRM) and multiple reaction monitoring (MRM). Both SRM and MRM were based on the similar principle, where the target precusor ion will be isolated and subsequently one of the fragment ions (SRM) or multiple fragment ions (MRM) will be monitored and quantified. The specificity of SRM and MRM is better than that of SIM. However, in terms of quantification of peptides and proteins, the methods' specificity vary depending on the nature of the peptide or the protein. This is especially true when the protein or peptide is belonged to a family of proteins, where the proteins in the family shared a great extend of similarity in their amino acid sequence. Furthermore, it is relatively common that the target peptide or protein be found in other forms than its intact form in biological matrix, where these other forms of proteins or peptides are the result of partial digestion of the protein or peptide through the activity of proteases in biological matrix. In these instances, the qualitative data of the target protein or peptide becomes important feature to discriminate the targeted peptides from these other forms of proteins or peptides. An illustration of the complexity of protein or peptide analysis will be demonstrated in the quantification analysis of human chorionic gonadotropin (hCG), a glycoprotein belongs to gonadotropin family. Besides the closely resemblance of hCG and LH, there are also the presence of nicked hCG in the biological matrix. In these circumstances, a complete qualitative data to indicate the identity of the target peptide will surely lead to higher confident in protein or peptides quantification.

Other method for quantitative analysis of protein is by ELISA, where the amount of the protein present in a solution of biological matrix is measured by the degree of antibody-antigen interaction that is expressed in the intensity of color developed by the antibody-tag enzymatic reaction with its substrate. ELISA is a device for partial quantitative analysis. Furthermore, ELISA method may lead to false positive result caused by the cross reaction of the antibody with other unrelated proteins or compounds as the assay depends solely on the specificity of the antibody used.

An approach for simultaneous qualitative and quantitative analysis of proteins was developed. In this method, qualitative and quantitative analysis of proteins can be conducted in a single tandem mass spectrometry analysis. The method not only provides unambiguous identification of protein via amino acid sequencing, at the same time, quantitative data can be generated from the same tandem mass spectrometry data. This method is recommended as it can achieve a very low limit of quantification.

2. The concept of simultaneous qualitative and quantitative analysis of proteins

The main problem encounters in the quantification of minute protein in a biological sample matrix is the isolation of the target proteins from the sample matrix. One of the most direct approaches for isolation of protein of interest is by using immunoaffinity extraction of the targeted protein through specific antibody-antigen interaction (Gam et al, 2006), this approach is possible if the antibody for the protein is available. Although Such technique can efficiently isolate target protein from the complex sample matrix, co-extraction of other unrelated proteins cannot be avoided. This may result from the non-specific interaction of the antibody with other proteins or may also result from the non-specific binding of other proteins with the antibody coating materials. Besides, it is also common that non-protein materials being extracted in such procedure. All these unrelated proteins or compounds will interfere with the subsequence analysis of the target protein using tandem mass spectrometry.

Mass spectrometer is an instrument with high sensitive but low selectivity, when analysis is carried out in the positive ion mode, the instrument will register all the positive charged ions that enter the detector. Therefore, quantification of minute protein extracted from complex biological matrix using mass spectrometer poses a great challenge. It is common that the target ion cannot be detected as it was suppressed or masked by other ions from the impurities in the sample. In this scenario, quantification of target ions cannot be achieved as it cannot fulfill the signal to noise ratio of greater than 3.

One way to solve such problem is to have a very clean sample, a demand that is hard to fulfill as the target protein is only present in trace quantity in the complex biological sample. The alternative way of solving this problem is to get rid of all the unrelated ions during mass spectrometry analysis making the spectrum to be very clean for quantification analysis. One way of getting rid of the unrelated ions is by filtering them out from the spectrum. Based on this understanding, we have developed a quantification method for minute protein in a complex mixture. The first issue to solve is how to filter the unrelated ions out from the target ion spectrum, surely we cannot place a mass filter at the inlet of sample nebulizer, where it will also filter off our target protein. The other filtering device that possible is by using the MS/MS scan, in this device, only the target ions that were exceeded the threshold programmed in MS scan will be isolated and excited to MS/MS scan. In a way, the target ion is being filtered from other unrelated ions.

In the analysis of protein using ion trap mass spectrometer, it is not advisable to analyze intact protein, where identification of intact protein is based solely on its deconvoluted molecular weight. Furthermore, analysis of intact glycoprotein by ion trap mass spectrometer possess additional problem where deconvolution of protein molecular weight may not be able to carry out successfully as the ionization of glycoprotein is inconsistent due to the presence of sugar components, which caused variation in the ionization of intact glycoprotein.

One of the advantages of using ion trap mass spectrometer is its ability to ionize peptide into multiple charged ions and to perform MS/MS scan, however it is not possible to perform MS/MS scan on most of the intact proteins, especially the high molecular weight proteins as the data generated cannot be interpreted. It is commonly understood that MS/MS scan is best performed on tryptic digested peptides, where the length of the peptides digested by trypsin is manageable by MS/MS scan of most types of mass spectrometer. Furthermore, the site of trypsin digestion is either Arginine or Lysine, the basic amino acids that will favor the formation of double charged ion to the peptides in positive ion mode scanning. The formation of double charged ion is an added advantage for collision induced dissociation (CID) in the MS/MS scan, according to the mobile proton hypothesis (McCormack et al, 1993), kinetic energy from collision induced dissociation will be converted to vibrational energy that releases through fragmentation reactions directed by the site of protonated amide bond. In this hypothesis, fragmentation of double charged peptide under the collision induced dissociation will result in detection of both y and b ions from the fragmentation, this is because each of the product ions carry a proton to themselves. When comparing to single charged ions, only one of the product ions resulted from the fragmentation will carry a proton and thus carry a charge for detection. The generation of comprehensive product ions by the double charged peptide ion gives a higher confident in determining its amino acid sequence, where the more y and b product ions identified, the more significance the identity of the peptide.

In the currently developed method, both the qualitative and quantitative analyses of protein are based on the data of MS/MS scan, which displays product ions of the selected peptide. In order to give significant qualitative value of a protein, the peptide marker of the target protein must be uniquely different from peptides of other proteins. In another word, the amino acid sequence of the peptide marker must be specific for the target protein. Furthermore, upon collision induced dissociation, such peptide must produce comprehensive spectrum of y and b product ions which give a definite identity to the target protein.

In terms of quantitative analysis of protein, the same peptide marker will be used. Similar with qualitative analysis, the peptide marker will be subjected to collision induced dissociation to generate y and b product ions, as mentioned earlier, these product ions are used to confirm the identity of target protein. Once the identity of the protein is confirmed, subsequence extracted product ions chromatogram will be performed. When performing extracted product ions chromatogram, two orders of mass filters were in place, the first mass filter is peptide marker mass filter that will single out the molecular ion for the peptide marker, while the second mass filter is quantitative ion (or selected product ions) mass filter that will single out the quantitative ion. In doing so, this method create a high selectivity to our currently developed quantitative method. This is because the first mass filter isolates the double charged m/z value of the peptide marker, subsequently the second mass filter isolates the selected quantitative ion/s that was generated from the peptide marker that has undergone the first mass filter. Quantitative ion/s selected must be the most stable and intense product ion/s (either y or b ions) from the fragmentation of the peptide marker. Quantitative ion can be a single product ion or can be a sum of a few product ions. In the event of the summation of a few product ions, limit of quantification (LOQ) of the method will concomitantly reduce, this is because the total peak areas of product ions will certainly enhance the signal to noise ratio. However, the usefulness of selected product ions to act as quantitative ions must be validated, where its peak area under extracted product ions chromatogram must be proportional to the concentration of the target protein. In the other word, the intensity of collision energy plays an important role in this currently developed method, where the collision energy used must be kept constant throughout the analysis.

The beauty of this method is it is able to simultaneously provide qualitative and quantitative data on trace amount of protein in a complex mixture. The two layers of filtering event will ensure only the correct ions being quantified. This method is very useful for quantification of trace amount of protein in a complex biological sample. It is not impossible that many compounds with similar masses can be filtered in by the first layer of mass filter, nevertheless, these similar masses from the impurities of the sample cannot generate the product ions spectrum similar to that of target peptide marker, as they do not contain the amino acid sequence of the peptide marker that is unique to the target protein. Therefore, these unrelated masses cannot pass the second layer of mass filter.

As a result of the selectivity of the two layers mass filter described above, the extracted product ion chromatogram produced is usually free from any back-ground noises, giving rise to a very low limit of quantification of the method and therefore trace protein in complex sample matrix can be quantified. Furthermore, the method gives a very high confident level in terms of qualitative information of the target protein, where false positive data are completely omitted.

In the course of my study, I was given a task to develop a quantitative method for human chorionic gonadotropin (hCG) in human urine. hCG is a glycoprotein misused by male athletes to induce endogenous secretion of testosterone. hCG is present as trace component in urine, a complex biological matrix. In order to concentrate and purify the glycoprotein from urine, I have implemented immunoaffinity purification technique to extract hCG from urine matrix. This purification technique did not produce pure hCG as expected, the numerous amount of contaminants can be visualized when subjecting the extracted hCG to mass spectrometry analysis.

3. Materials and methods

3.1 Immunoaffinity protein purification

Approximately 11 ml of urine was centrifuged at 1500 rpm for 5 minutes to precipitate any particulate matters. A volume of 10 ml of the centrifuged urine was transferred to a clean polypropylene tube for hCG extraction. The immunoaffinity column was first flushed with 6 ml distilled water to remove the storage buffer and then conditioned with 5 ml of 0.01 M PBS at pH 7.2. During this step, the column flow was adjusted to 7-9 drops per minute. A 2 ml volume of the centrifuged urine was loaded onto the immunoaffinity column (column volume was 2 ml). A 20 minutes incubation time was allowed for the antibody-antigen association to take place. The urine was then removed from the column by flushing with 2 ml of 0.01 M PBS pH 7.2. The column was then reloaded with another 2 ml aliquot of urine and incubated for 20 minutes. This process was repeated until all the 10 ml urine had passed through the column. Finally, the column was washed with 15 ml (7 bed volumes) of washing buffer (0.1% (v/v) Tween 20 in 0.1 M PBS, pH 7.2) followed by 2 ml of elution buffer (1 M citric acid adjusted to pH 2.2 with 10 M NaOH). After the first 1 ml of the elution buffer had entered the gel, the collection of the eluate began. When the 2 ml of the elution buffer had fully immersed in the gel, a 5 minutes equilibration time was allowed to enable complete antibody-antigen dissociation to take place. This was followed by 8 ml of elution buffer. A total of 9 ml eluate was collected. Eluate was concentrated and desalted using a protein concentrator column (Jones Chromatography).

3.2 Preparation of tryptic digestion product of hCG

3.2.1 Digestion of hCG using trypsin

Protein sample (hCG) was desalted using protein concentrator column (C_{18}, 2 cm x 4.6mm ID, Jones Chromatography). A syringe pump (Harward Apparatus) was used to pump the protein solution through the column at 1 ml/min. The column was then flushed with 25 ml of deionized distilled H_2O and the protein was recovered by eluting with 70% acetonitrile 0.1 % formic acid. The eluted protein was dried under N_2 at 37'C.

The dried protein was denatured using denaturing buffer (6M Guanidine HCl, 0.5M Tris, 2mM EDTA pH8.6). A volume of 10 µl of 1 M dithiothrietol was added to the mixture and

incubated at 37°C for 30 minutes. After which 25 µl of 1M iodoacetic acid in 1M NAOH was added and the mixture was further incubated for 30 minute at room temperature.
The excess reagents were removed from the protein sample by using the protein concentrator column (as mentioned above). The dried hCG was then reconstituted in 50 µl of 50 mM NH_4HCO_3. A volume of 2 µl (0.25 µg/µl) of trypsin solution was added and the mixture was incubated at room temperature for 20 hours. This was followed by another addition of the same amount of trypsin and the sample then further incubated for 4 hours at room temperature. The digested hCG was lyophilized and stored at -20°C.

3.3 HPLC separation

The tryptic digested hCG was first reconstituted in 25 µl of high purity distilled H_2O (Maxima, ELGA); 10 µl of the sample was then injected into the C-18 Vydac column (300 Å, 5 µm, 1 mm X 50 mm). Separation of the peptides was performed using a Hewlett Packard series 1100 HPLC. The flow rate was set at 1 ml/min and further split by a fused silica splitting device to 20 µl/min through-column flow rate. Mobile phase A was 0.05% TFA in H_2O and B was 0.05% TFA in ACN. The gradient used was 5-95% B for 20 minutes and held at 95% B constant for 5 minutes. The HPLC was interfaced to an ion trap mass spectrometer (LCQ, ThermoQuest).

3.4 Mass spectrometry

Mass spectrometric analysis was carried out using the ion trap mass spectrometer (LCQ, ThermoQuest). Data dependent experimental method was created for the analysis of tryptic peptides of hCG.

3.4.1 Creating a data dependent experimental method for qualitative analysis

The MS data was acquired at heating capillary temperature 200°C, sheath gas flow rate is 60arb, spray voltage at 4kV, tube lens offset is -60V and the capillary voltage is at 38V.
Data dependent experimental method was created for the analysis. The experimental method was consisted of 2 scan events. The first scan event was full scan MS, the second was MS/MS scan, which were dependent on the results of the full scan MS. This linkage is known as data dependent scan. The parameters of data dependent scan were default collision energy of 25, charge state of 2, minimum signal acquired was 1 x 10^5 counts, isolation width was 2 m/z.

3.4.2 Creating a data dependent experimental method for simultaneous qualitative and quantitative analysis of hCG

Doubly charged parent ion for peptide VLQGVLPALPQVVCNYR, $[964.7]^{2+}$ was programmed into the parent ion list in data dependent scan. The parameters set for data dependent scan (MS/MS scan) were default collision energy = 25, default charge state = 2, minimum signal acquired = 1 x 10^4 counts, and the isolation width = 2 m/z.

4. A demonstration of the method by using human chorionic gonadotropin hormone

Human charionic gonadotropin or hCG is a hormone misuses by male athlete to induce endogenous production of testosterone (Boer et al, 1991). It will be interesting to use hCG for the demonstration of this analysis method, this is because the hormone belongs to a family of gonadotropin. Other hormones in the same family are follicle stimulating hormone (FSH), lutropin hormone (LH) and thyroid stimulating hormone (TSH) (Canfield et al, 1976). All these hormones share a similar characteristic in molecular structure, they are glycoprotein hormones that made up of one alpha-subunit and one beta-subunit. The alpha-subunit of the hormones is identical (Vaitukaitis et al, 1976), hence, it cannot serve as marker for hCG. On the other hand, the beta-subunits are basically similar with only minor differences in certain amino acid residues. This is especially true between the beta-subunits of hCG and LH (Figure 1). Due to the high resemblance of the hormones, they share relatively similar electrophoretic mobility in gel electrophoresis separation. Therefore, qualitative and quantitative analysis of hCG possess extra challenges.

Human Chorionic gonadotropin (hCG) is synthesized by the trophoblast cells of the placenta (Canfield, *et al.*, 1971). The hormone is release in the first few weeks of pregnancy. Between the 7th to 12th week of pregnancy, the plasma level of hCG rises to extremely high levels where approximately 11,000 - 289,000 mIU/ml of hCG were released. This is followed by a decline during the last two trimester (Braunstein, *et al.*, 1978).The molecular weight of intact hCG, αhCG (α-subunit) and βhCG (β-subunit) are approximately 36.7, 14.5 and 22.2 kDa, respectively. It has been estimated that 30% of the total weight of hCG is contributed by the carbohydrate content (Canfield, *et al.*, 1976) and these carbohydrates account for the heterogeneity properties of hCG. Besides the intact hCG, αhCG and βhCG, the other commonly found hCG fragments are β-core fragment and nicked hCG fragment. These fragments are formed by proteolytic degradation of hCG in kidney. The β-core fragment composes of two polypeptides linked by a disulfide bond. The molecular weight of the purified β-core fragment is between 12-16kDa. Reduction of β-core fragment resulted in its dissociation into two fragments between 8-12kDa and 5-6kDa molecular weights (Endo, *et al.*, 1992).

S*K*EPLRP*R*-CRPINgA*T*LAVEK-EGCPVCITVNgTTICAGYC

PTM*T*R-VLQ*G*VLP*A*LPQVVC*N*YR-DVR-FESIR-LPGCPR-

GV*N*PVVS*YA*VALSC*Q*C*AL*CR-R-ST*T*DCGGPK-DHPLTCD

*D*P*R-FQDSSSSgK-APPPSgLPSPSgR-LPGPSgDTPILPQ*

Fig. 1. Amino acid sequence for hCG β-subunit, (-) indicates the site of trypsin digestion. The amino acid residues which are different from those in LH are in italic and red. The site of glycosylation is indicated by g.

In our method, hCG was subjected to reduction, alkylation and digestion using trypsin enzyme prior to tandem mass spectrometry analysis. The expected peptide fragments derived from the digestion were listed in Table 1. A total of 16 peptides and glycopeptides fragment were expected from the tryptic digestion of hCG, these peptides were (S *K* E P L R P *R*), (C R P I Ng A *T* L A V E K), (E G C P V C I T V Ng T T I C A G Y C P T M *T* R), (V L Q *G* V L P *A* L P Q V V C *N* Y R), (D V R), (F E S I R), (L P G C P R), (G V *N* P V V S *Y A* V A L S C *Q* C *A L* C R), (R), (S T *T* D C G G P K), (D H P L T C D *D* P *R*), (*F Q D S S S Sg K*), (*A P P P Sg L P S P Sg R*), (*L P G P Sg D T P I L P Q*). (note: amino acid residue in red is different from that of LH while *g* indicated the site of glycosylation). Amongst these peptides, only 10 peptides show at least one amino acid residue different from that of LH, the difference in amino acid residues can be used to distinguish hCG from LH in tandem mass spectrometry analysis.

Position no.	$[M + H]^+$	Selected data dependent parent ions	Sequence
1-2	234.1	-	SK
3-8	767.4	$[384.5]^{2+}$	EPLRPR
1-8	983.2	$[491.8]^{2+}$	SKEPLRPR
9-20	Glycopeptide	-	[a]CPINATLAVEK
21-43	Glycopeptide	-	[a]EGCPVCITVNTTICAGYTCPT MTR
44-60	1928.4	$[964.7]^{2+}$	VLQGVLPALPQVVCNYR
61-63	389.2	$[389.2]^{+}$	DVR
64-68	651.3	$[326.1]^{2+}$	FESIR
69-74	700.8	$[350.5]^{2+}$	LPGCPR
75-94	2228.6	$[743.5]^{3+}$	GVNPVVSYAVALSCQCALCR
95	175.2	-	R
96-104	924.0	$[462.5]^{2+}$	STTDCGGPK
105-114	1227.3	$[614.1]^{2+}$	DHPLTCDDPR
115-122	Glycopeptide	-	[b]FQDSSSSK
123-133	Glycopeptide	-	[b]APPPSLPSLSR
134-145	Glycopeptide		[b]LPGPSDTPILPQ

a shows N-link glycopeptide at the bold amino acid
b shows O-link glycopeptide at the bold amino acid/s

Table 1. Predicted tryptic digested βhCG fragments.

4.1 Selection of peptide marker

In the selection of suitable peptide marker for simultaneous qualitative and quantitative analysis of protein, three criteria are implemented:

a. The amino acid sequence of the peptide marker must be unique to the protein. This is important as the marker will be used in the qualitative analysis to differentiate the protein of interest from other unrelated proteins.
b. The length of the peptide marker must be suitably long. The long marker will lead to generation of a more comprehensive spectrum of product ions which will give higher confident level in the qualitative analysis. Moreover, the generation of greater number of product ions will also lead to greater choice of quantifying ions to be used in subsequent quantitative analysis.

c. The degree of ionization of the peptide marker. Only the peptide marker that can be easily ionized by the ionization mode of mass spectrometer to produce high abundant molecular ions will be selected as peptide marker. This is important in the analysis of protein from complex biological sample, where interferences from the matrix and chemical noise may mask the low abundance ions.

The β-subunit of hCG has an extension of 30 amino acid residues at its C-terminus compared to LH. This distinctive region of hCG is used as the antigenic epitope for raising of hCG-specific antibodies. This region is highly glycosylated and following trypsin digestion, a total of 3 glycopeptides, F Q D S S S Sg K, A P P P Sg L P S P Sg R, L P G P Sg D T P I L P Q are expected to be formed from this particular region. However, we did not detect any of these glycosylated peptides in MS scan, the possible reason is the presence of carbohydrate in the peptides suppressed the ions signals of glycopeptides and therefore reduced the detection of these glycopeptides.

On the other hand, the molecular ions of two other glycopeptides C R P I Ng A *T* L A V E K and E G C P V C I T V Ng T T I C A G Y C P T M *T* R derived from the β-subunit of hCG were detected in different glycoforms, which indicates more than one type of sugar moieties were present in each individual glycopeptide. However, when subjecting these molecular ions to MS/MS scan, there is no useful MS/MS data to indicate the identity of these glycopeptides. Once the sugar moieties were removed (deglycosylation), four out of the five deglycosylated glycopeptides derived from the β-subunit of hCG were detected (Gam *et al*, 2006). Nevertheless, the deglycosylation process was time consuming and also resulted in sample lost. Hence, the glycopeptides do not serve as suitable marker for hCG.

In terms of peptides, only the peptides that are differ in at least one amino acid residue were targeted in the analysis. This is possible as a different in one amino acid residue will be resulted in a different product ions masses upon tandem mass spectrometry analysis. An example is for a peptide sequence of V L Q G V L P *A* L P Q V V C N Y R of hCG. The amino acid residues in red are the different amino acid residues than those of LH. the expected product ions series of these closely resemblance peptides of hCG and LH is listed in Table 2.

It is clear that out of 32 product ions expected from the peptides, there were only three identical product ions between hCG and LH, namely b_1^+, b_2^+, b_3^+ product ions. The extensive difference in the product ions series between the seemingly similar peptides of hCG and LH was great, which allow distinctive identification of these two closely resemblance hormones. Due to the high similarity between the two hormones, false identification of LH as hCG in ELISA was reported as a result of cross reaction of hCG antibodies with LH (Bottger et al, 1993).

Peptide S *K* E P L R P *R* was formed from an incomplete digestion of β-subunit at the 2nd to 3rd amino acid residue. It is reasonable to suggest that the formation of this peptide was not due to insufficient enzyme used but rather it was due to the presence of a carboxyl side chain group (glutamic acid, E) on the digestion site, which remarkably reduced the rate of hydrolysis. The use of this peptide as hCG marker may not be suitable as its formation may vary due to incomplete trypsin digestion. In general, taking into the consideration of all the peptides and glycopeptides derived from hCG, only 4 peptides fulfilled the first two criteria (A and B) of peptide marker, these peptides were (V L Q G V L P *A* L P Q V V C N Y R), (G V N P V V S *Y A* V A L S C *Q* C *A L* C R) (S T *T* D C G G P K), (D H P L T C D *D* P *R*).

y-fragment ions	M/z LH	M/z hCG	b-fragment ions	M/z LH	M/z hCG
y_1^+	175.1	182.1	b_1^+	100.1	100.1
y_2^+	338.1	337.4	b_2^+	213.2	213.3
y_3^+	439.2	451.5	b_3^+	341.2	341.4
y_4^+	560.2	613.7	b_4^+	412.3	398.5
y_5^+	699.3	712.9	b_5^+	511.3	497.6
y_6^+	780.3	812.0	b_6^+	624.4	610.8
y_7^+	926.4	940.1	b_7^+	721.4	707.9
y_8^+	1023.5	1037.2	b_8^+	818.5	779.0
y_9^+	1136.5	1150.4	b_9^+	931.6	892.1
y_{10}^+	1233.6	1221.5	b_{10}^+	1028.6	989.3
y_{11}^+	1330.6	1318.6	b_{11}^+	1156.7	1117.4
y_{12}^+	1443.7	1431.7	b_{12}^+	1255.8	1216.5
y_{13}^+	1542.8	1530.9	b_{13}^+	1300.8	1315.6
y_{14}^+	1613.8	1587.9	b_{14}^+	1515.8	1476.8
y_{15}^+	1683.9	1716.1	b_{15}^+	1616.9	1590.9
y_{16}^+	1855.0	1829.1	b_{16}^+	1779.9	1754.1

Table 2. The predicted product ions of V L Q G V L P A L P Q V V C N Y R and V L Q A V L P P L P Q V V C T Y R of hCG and LH, respectively.

Figure 4 shows the total ion chromatogram (TIC) of the tryptic digested peptides of hCG, a total of 3 out of 4 potential peptide markers were detected, where V L Q G V L P *A* L P Q V V C *N* Y R were presented as the most abundant ion followed by G V *N* P V V S *Y A* V A L S C *Q* C *A L* C R as the second most abundant ion in the analysis. Nevertheless, the detection of G V *N* P V V S *Y A* V A L S C *Q* C *A L* C R peptide by mass spectrometry is anomalous. In some analyses, the peptide cannot be detected. The rationale for this irregularity in the detection of this peptide is not understood, therefore the use of this peptide as hCG marker is not indicated. Hence, in terms of the ease of peptide ionization (criteria C), V L Q G V L P *A* L P Q V V C *N* Y R was found to be the most suitable peptide marker for hCG.

The molecular ion of V L Q G V L P *A* L P Q V V C *N* Y R peptide was $[964.7]^{2+}$ while the corresponding peptide of LH has an amino acid sequence of V L Q A V L P P L P Q V V C T Y R and the doubly charged ion was $[977.6]^{2+}$. This LH peptide differs from the hCG peptide by 3 amino acid residues (red). With these minor differences, a completely different set of parent ions and product ion spectra were expected (Table 2). These differences can be used to distinguish hCG from LH. Therefore, this peptide is unique to hCG for tandem mass spectrometric qualitative and quantitative analysis. Furthermore, the long amino acid sequence (17 amino acid residues) of the peptide increases the method specificity by preventing false identification of the peptide. In addition, the high m/z ratio of its parent ion avoids interference from chemical noise, which normally found at low m/z ratios. All these features of the peptide fulfilled the requirement of peptide marker. Hence, this peptide was selected as the peptide marker for identification and quantification of hCG.

Nicking of hCG occurs at position between 47-48 amino acid residue of the β-subunit. After tryptic digestion, the nicked hCG peptide which comprises amino acid residue 48 to 60 was formed. This amino acid sequence is within the amino acid sequence of V L Q G V L P *A* L P

Q V V C *N* Y R peptide (from amino acid residue 44 to 60). For this reason, the nicked hCG peptide produces the y^+ product ions (y_1^+ to y_{12}^+) which are identical to some of the y^+ product ions of V L Q *G* V L P *A* L P Q V V C *N* Y R peptide. The similarity in the y^+ ions of the two peptides did not interfere with the quantitative result. This is because the specificity of the data dependent mass spectrometric method is able to eliminate the interference of nicked hCG peptide as the two peptides have distinctly different parent ion masses ($[964.7]^{2+}$ and $[765.9]^{2+}$).
Figure 2 (lower panel) shows the MS/MS spectrum of V L Q *G* V L P *A* L P Q V V C *N* Y R peptide, where the fragmentation of the peptide produced a comprehensive product ions spectrum, which provides a convincing identification of the peptide, and subsequently hCG. Once the targeted peptide markers were identified, the subsequently analysis of protein using tandem mass spectrometry can be carried out. Since this method only emphasizes tandem mass spectrometry analysis and its data interpretation, any peptides chromatography separation device can be applied. In our laboratory, we used either HPLC or 2D-LC dependent on the quantity of the protein being analyzed, in 2D-LC, we included an enrichment column prior to chromatographic separation by reversed-phase column suitable for separation of peptides. This device helps to concentrate the peptides and therefore increase the sensitivity of the analysis method. The parameter for HPLC separation method for hCG analysis was 20 µl/min through-column flow rate at a gradient of 5-95% B for 20 minutes and held at 95% B constant for 5 minutes. Mobile phase A was 0.05% TFA in H_2O and B was 0.05% TFA in ACN. The column used was C-18 Vydac column (300 Å, 5 µm, 1 mm X 50 mm).
As for the parameters for tandem mass spectrometry analysis, it is advice to apply the optimum parameters for respective mass spectrometers. In our laboratory, an ESI - ion trap mass spectrometer (LCQ, ThermoQuest) is used, where data dependent experiment method consisting of MS and MS/MS scan was created. The peptide markers ions were programmed and MS data were acquired at heating capillary temperature 200°C, sheath gas flow rate was 60 arb, spray voltage at 4 kV, tube lens offset was -60 V and capillary voltage was at 38 V. MS/MS scan was conducted at default collision energy of 25, charge state of 2, minimum signal acquired was $1x10^5$ counts, isolation width was 2 m/z.

4.2 Quantitative and qualitative analysis of hCG using selected peptide marker

The qualitative and quantitative experimental method involved the introduction of the marker to the tandem mass spectrometric analysis. This is to obtain product ions spectrum that reveals the amino acid sequence of the peptide marker, which will then confirm the presence of hCG. The experiment was carried out in the data dependent acquisition mode, where the parent ion mass of the marker was programmed in the parent ion list. The programmed software will run in such a way that as the parent ion is detected in the full scan MS, they will be selectively excited for the MS/MS scan. The resulting product ions produced by CID fragmentation of the parent ion will be shown as the MS/MS spectrum that provides characteristic of the amino acid sequence of the peptide.
The determination of amino acid sequence of the peptide marker was included in hCG confirmatory analysis to avoid the possibility of producing false positive results, which is more likely to happen in selected ion monitoring (SIM) data acquisition method. SIM depends solely on the parent ion mass and the retention time of the compound of interest, which results in the loss of qualitative information of the compound (Gam et al 2003).

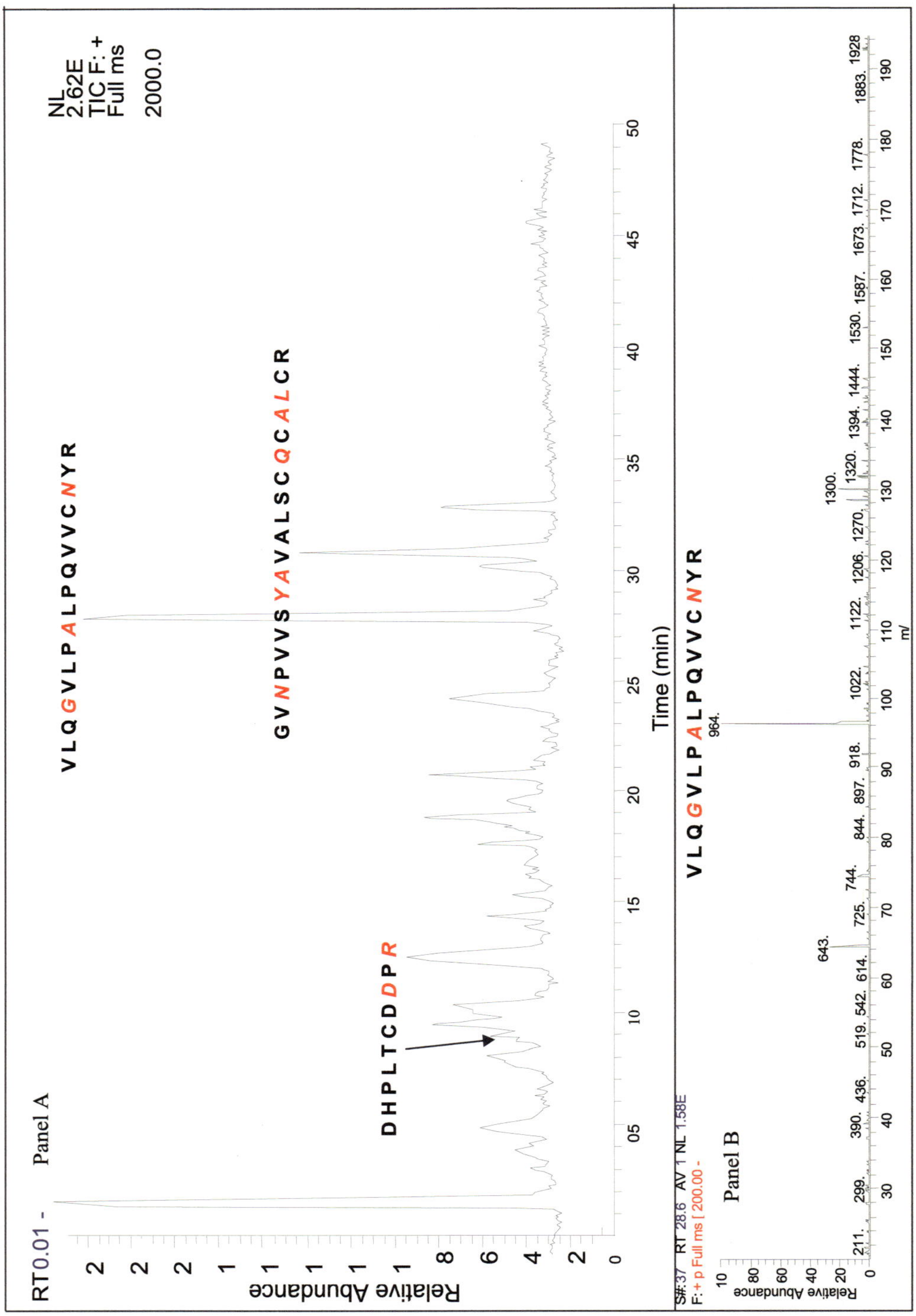

Fig. 2. Panel A; Based Peak Chromatogram for tryptic digested peptides. The y-axis is relative abundance and x-axis is retention time. Panel B; spectrum for peptide VLQGVLPAQVVCNYR. The y-axis is relative abundance and x-axis is mass to charge ratio.

Full scan MS/MS not only gives enough information for the qualitative data of hCG but at the same time, the intensity of the product ions of the peptide marker were used to quantify concentrations of hCG. The quantitative method using product ions spectrum allows relatively low quantification limit as compared to SIM method. This is because the MS/MS experiment is a technique that will minimize or eliminate all chemical and background noises.

For quantification purpose, three most abundant product ions of $[964.7]^{2+}$, namely b_6^+, b_9^+ and y_{11}^+ with the m/z ratios of $[610.3]^+$, $[891.5]^+$ and $[1317.8]^+$ respectively were selected as the quantitative markers. These ions can be evaluated individually by peak area display in the extracted product ion chromatogram or by the summation areas of the three product ions (Figure 3). The summation of three product ions increases the total peak area and therefore greatly reduced the quantification limit as the signal to noise ratio is tremendously increased.

At 5 mIU/mL which approximately equal to 1 pg/mL of hCG in urine matrix, the signal to noise ration for each product ions chromatogram of the peptide marker were exceeded one hundred (Figure 3, upper panel). The high signal to noise ratio reveals that this method would be able to detect a much lower concentrations of hCG. This detection limit of hCG (5 mIU/mL) using our current method is superior than SIM method where a detection limit of 25 mIU/mL was reported (Liu & Bowers 1997).

At 5 mIU/ml hCG, the parent ions of the peptide marker was indistinguishable from the background noise in full scan MS. Nevertheless, as long as the parent ion intensity surpasses the threshold set, the ion will be isolated and excited to data dependent scan to generate MS/MS data. The MS/MS data obtained not only gives identification to the peptide; furthermore, the product ions were used to quantify the protein concentration. At 5 mIU/ml hCG concentration, our currently developed method yields minimal, if there is any background interference. Quantitative analysis of hCG using SIM method at this concentration is not possible due to the same reason discussed above, where the parent ion was indistinguishable from the background noise, in this situation, the signal to noise ratio valid for quantification analysis could not be established.

Using this approach, we are able to conduct simultaneous qualitative and quantitative analysis on protein. The qualitative data (MS/MS scan data) confirmed the identity of the protein via its unique peptide marker while the product ions (quantifying ions) of the peptide marker were subjected to product ion extracted chromatogram to generate quantitative data of the protein. This approach avoids false quantification of ions, which is possible in SIM.

4.3 Method validation

A standard curve was constructed using hCG at 5 mIU/mL (1pg/mL), 8 mIU/mL (1.6 pg/mL), 10 mIU/mL (2 pg/mL), 15 mIU/ml (3 pg/mL), 20 mIU/mL (4 pg/mL) and 30 mIU/mL (6 pg/mL) concentrations. The protein was subjected to tryptic digestion and analyzed by HPLC/MS/MS according to the method described. In which, the peptide marker was eluted from the column at 12 minutes retention time.

Each of the standard points was performed in triplicate. Tables 3 to 6 show the peak areas (triplicate) of the extracted product ions chromatograms for the chosen quantifying ions. The reliability of each product ions as a quantitative marker was measured by their coefficient of variance (r^2) values. The coefficient of variance for $[610.0]^+$, $[891.8]^+$, $[1317.8]^+$ and the summation of $[610.0]^+ + [891.8]^+ + [1317.8]^+$ were 0.998 (Figure 4), 0.993 (Figure 5),

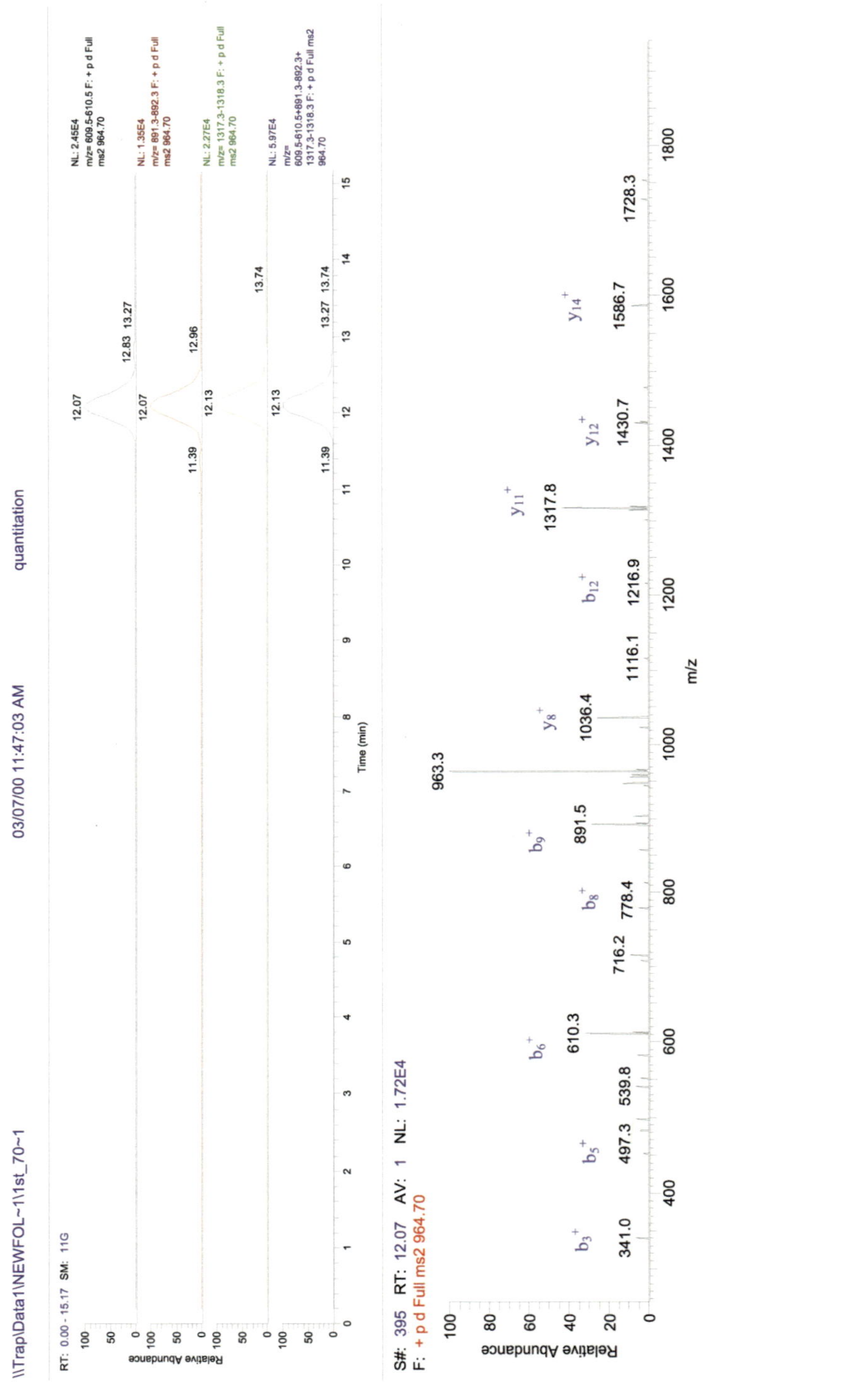

Fig. 3. Data for quantitative and qualitative analysis of hCG at 5 mIU/mL. Upper panel: extracted product ions chromatogram; Lower panel: MS/MS spectrum of the selected peptide marker (Gam *et al*, 2003).

0.997 (Figure 6) and 0.995 (Figure 7), respectively. Thus, it is obvious that the intensity of the product ions formed correlates well with the concentrations of hCG. The reliability of using the selected product ions as the quantitative markers is remarkable as the precision (C.V) of each standard point in triplicate is always < 10% (Tables 3 to 6). The linearity of the standard curve was only obtained in a narrow range of hCG concentration between 5 mIU/ml to 30 mIU/ml. Above this range, the curve deviates from linearity.

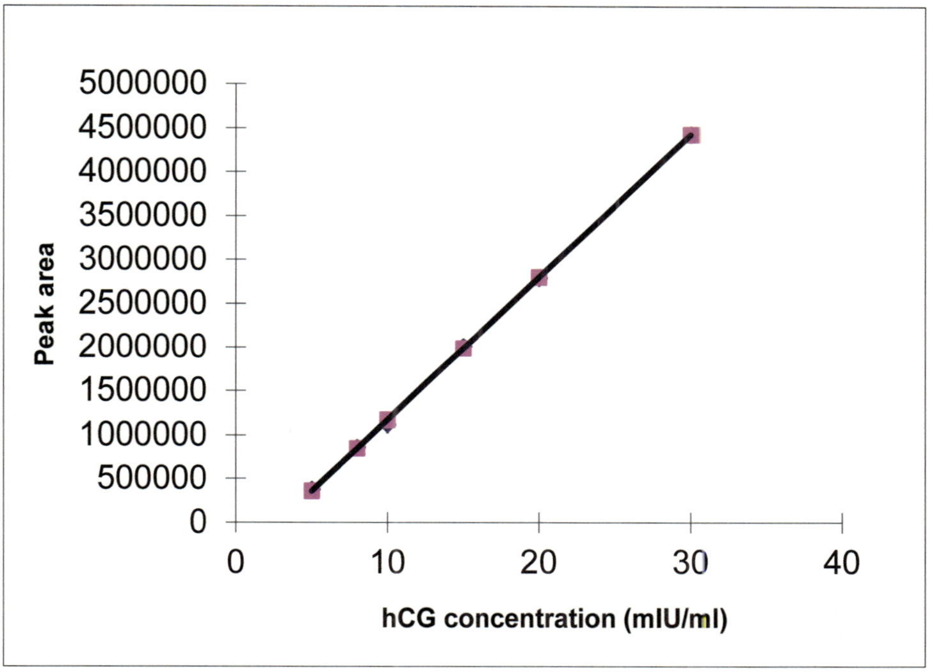

Fig. 4. The standard curve constructed using the b_6^+ product ion. Intercept = -457221, X variable = 1628623, $r^2 = 0.999647$.

hCG mIU/ml	Peak area			means	SD	CV
5	364576	395481	367517	375858	17057.52	5 %
8	856239	894572	826756	859189	34004.11	4%
10	1173659	1127543	1055078	1118760	59776.41	5%
15	1995623	1973176	2060943	2009914	45595.35	2%
20	2748931	2865973	2762191	2792365	64090.25	2%
30	4468624	4487696	4341096	4432472	79706.43	2%

Table 3. Standard curve data quantify using the b_6^+, $[610.0]^+$ product ion.

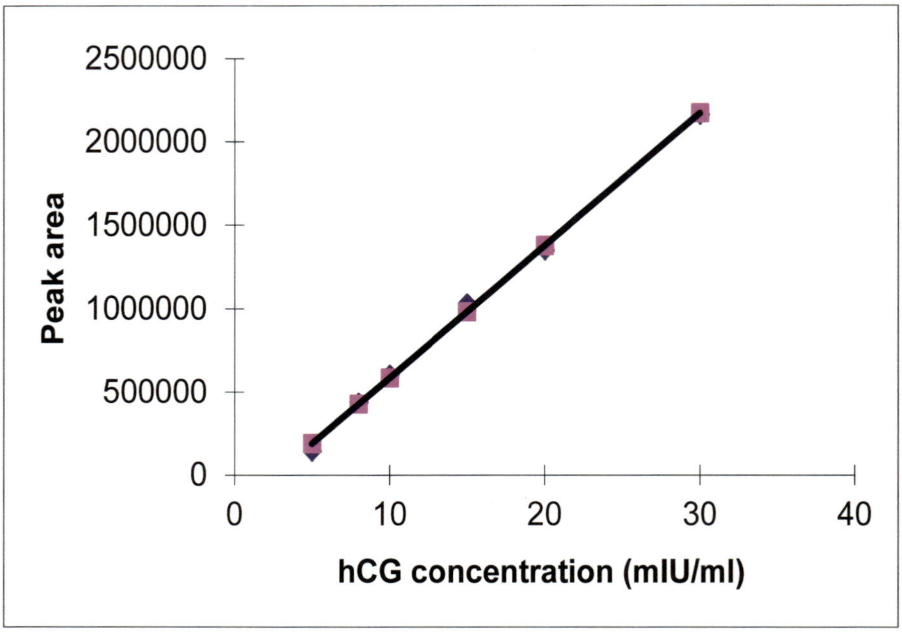

Fig. 5. The standard curve constructed using the b_9^+ product ion. Intercept = -207664, X variable = 79251.53, r^2 = 0.99765.

hCG mIU/ml	Peak area			means	SD	CV
5	140221	152539	134155	142305	9367.51	7%
8	419835	449395	442145	437125	15406.13	4%
10	623484	601678	594782	606648	4876.21	1%
15	1017695	1013689	1065890	1032425	36911.68	4%
20	1280853	1400054	1365535	1348814	24408.62	2%
30	2198756	2231539	2052210	2160835	12680.80	6%

Table 4. Standard curve data quantify using the b_9^+, $[891.8]^+$ product ion.

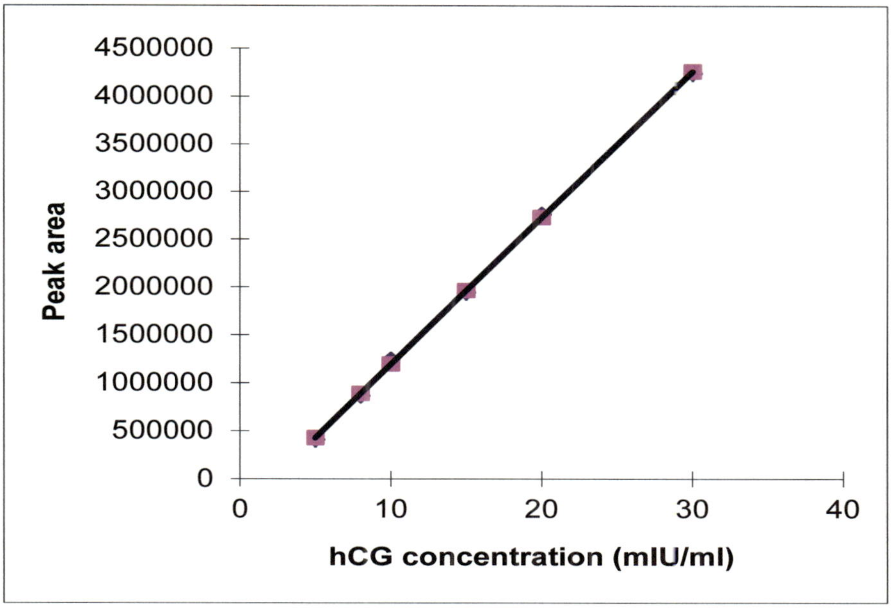

Fig. 6. The standard curve constructed using the y_{11}^{+} product ion. Intercept = -338677, X variable = 153377.7, r^2 = 0.999483.

hCG mIU/ml	Peak area			means	SD	CV
5	396514	419764	411420	409233	5900.10	1%
8	859176	883765	850385	864442	23603.22	3%
10	1327647	1262137	1141016	1243600	85645.48	7%
15	2032478	1863586	1929342	1941805	46496.51	2%
20	2759796	2789313	2737455	2762188	36669.14	1%
30	4245795	4353626	4132309	4243910	156494.8	4%

Table 5. Standard curve data quantify using the y_{11}^{+}, $[1317.8]^{+}$ product ion.

5. Future work

I believed the method demonstrated here will be of help to protein chemists whom struggle in protein quantitative analysis, especially for analysis of trace amount of protein in complex biological sample. This method may be useful not only in doping analysis for hCG, it can be applied to other doped proteins such as erythropoietin, growth hormone and ext. This is because an accurate quantitative data (definite amount of doped substances) is needed in doping analysis to differentiate between endogenous and exogenous protein, a fine line between doped and non-dope level. At this time, my works are mainly focused on the identification of biomarkers in diseases, where the biomarkers can be used as diagnostic markers or therapeutic markers for the diseases. The quantitative data of the expression of the biomarkers are important in determining the usefulness of individual biomarker, therefore, this developed method has created a good platform for conducting quantitative analysis on the identified biomarkers.

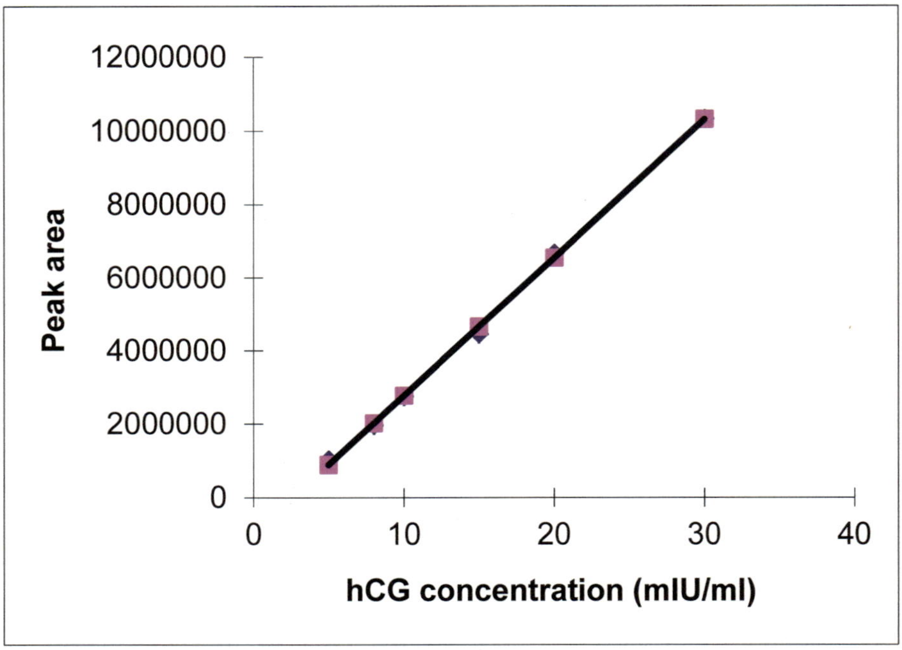

Fig. 7. The standard curve constructed using the summation of three product ions. Intercept = -1000044, X variable = 376994.7, r^2 = 0.998635.

hCG mIU/ml	Peak area			means	SD	CV
5	985703	1083863	998570	1022712	60311.26	6%
8	1903745	2014766	1954676	1957729	42490.05	2%
10	2869837	2729474	2664510	2754607	45963.48	2%
15	4375897	4400165	4563987	4446683	115839.60	3%
20	6542834	6742038	6713948	6666273	19862.62	0%
30	10428756	10514285	10038730	10327257	336238.20	3%

Table 6. Standard curve data quantify using the summation of three product ions: b_6^+, b_9^+ and y_{11}^+.

6. Conclusion

Using hCG as an example, the approach for simultaneous qualitative and quantitative analysis of protein by using tandem mass spectrometry has been demostrated. Since the qualitative data for identification of protein was carried out through product ions profiling which revealed the amino acid sequence of the protein, this analysis method give a high confident level of protein identity. Subsequently, the high abundant product ions are selected as quantifying ion for quantitative analysis of the protein. This quantification approach eliminates all background noises. This is because the quantifying ions were chosen from the fragmentation of the selected peptide marker, which was isolated from the remaining ions in MS scan and excited to collision induced dissociation. Using this approach, it is not possible to quantify a false ion, where only the correct parent ion will produce the expected product ions profile and subsequently the product ions were chosen as quantifying ions. Thus, our current method satisfies both the qualitative and quantitative requirements for protein analysis, which normally can be achieved for one aspect but not the other.

7. Acknowledgement

This work was carried out to accomplish of my Ph.D degree for a period of three years (1997-2000). I would like to express my gratitude to Universiti Sains Malaysia for providing me scholarship to pursue my study. Furthermore, I thank Prof Aishah Latiff, my Ph.D supervisor and also Doping Control Centre for provided me with support in terms of usage of chemicals and infrastructure for carrying out this research.

8. References

Bellar, T.A. & Budde, W.L. (1988). Interlaboratory comparison of methane electron capture negative ion mass spectra. *Anal. Chem.* 60:2076-2081.

Bottger, V., Micheel, B., Scharte, G., Wolf, G. & Schmechta, H. (1993). Monoclonal antibodies to hCG and their use in two-site binding enzyme immunoassays. *Hybridoma* 12 (1):81-91.

Canfield, R.E., Birken, S., Morse, J. H. & Morgan, F.J. (1976). Human Chorionic Gonadotroin. In: *Peptide Hormones*. Parsons J.A. eds. The Macmillan press LTP. Pp 299-385.

Canfield, R.E., Morgan, F.J., Kammerman, S., Bell, J. J. & Agosto, F.M. (1971). Studies of human chorionic gonadotropin. *Rec. Progr. Horm. Res.* 27:121-164.

Eichelberger, J.W., Kerns, E.H., Olynyk, P. & Budde, W.L. (1983). Precision and accuracy in the determination of organics in water by fused silica capillary column gas chromatography/mass spectrometry and packed column gas chromatography/ mass spectrometry. *Anal. Chem.* 55:1471-1478.

Endo, T. Nishimura, R. Saito S., Kanazawa, K., Namura, K., Katsuro, M., Shii, K., Mukhopaddyan, S., Baba, S. & Kobato, A. (1992). Carbohydrate structure of β-core fragment of hCG isolated from pregnant individual. *Endocrinol.* 130:2052-2058.

Gam LH and Latiff A. (2006) Tandem mass spectrometic analysis of glycopeptides derived from the tryptic digestion of human chorionic gonadotropin (hCG). Malaysian Journal of Science. 25(2): 87-95.

Gam LH, Tham SK and Aishah L. (2003), Immunoaffinity extraction and tandem mass spectrometric analysis of human chorionic gonadotropin in doping analysis. Journal of Chromatrography B, 79: 187-196.

Liu, C.L. & Bowers, L.D. (1997). Mass spectrometric characterization of the β-subunit of human chorionic gonadotropin. *J. Mass Spectrom.* 32: 33-42.

McCormack, AL, Jones, JL and Wysocki, VH. (1992). Surface-induced dissociation of multiply charged protonated peptides, J Am. Soc. Mass Spectrom. 3:859-862

Shoemaker, J. D. & Elliott, W.H. (1991). Automated screening of urine samples for carbohydrate, organic and amino acids after treatment with urease. *J. Chromatogr.* 562: 125-129.

Sweeley, C.C., Elliott, W.M., Fries, I. & Ryhage, R. (1966). Mass spectrometric determination of unresolved components in gas chromatography effluents. *Anal. Chem.* 38:1549-1953.

Vaitukaitis, J.L., Ross, G.T., Braunstein, G.D. & Rayford, P.L. (1976). Gonadotropins and their subunits: basic and clinical studies. *Recent Prog. Horm. Res.* 32:289-298.

9

Comparative Proteomics of Tandem Mass Spectrometry Analyses for Bacterial Strains Identification and Differentiation

Rabih E. Jabbour[1], Mary M. Wade[1], Samir V. Deshpande[2], Michael F. Stanford[1], Alan W. Zulich[1] and A. Peter Snyder[1]
[1]*U.S. Army Edgewood Chemical Biological Center, Aberdeen Proving Ground,*
[2]*Science and Technology Corporation, Edgewood, MD, USA*

1. Introduction

Bacterial proteome represents the collection of functional and structural proteins that are present in the cell. The bacterial proteome consists of diverse classes of proteins with different cellular functions. Overall, the protein content of the cell represents the majority of the cell dry weight, which makes it an ideal cellular component to be utilized for bacterial characterization (Loferer-Krobacher et al., 1998). The diversity of the bacterial proteome requires the determination, identification, and characterization of its protein content in order to understand their cellular functions (Costas et al., 1990). Moreover, studying the bacterial proteome is essential to identify pathological proteins for vaccine development, diagnose and provide counter measures to infectious diseases, and to the understanding of biological systems. The availability of microbial genomic sequencing information has led to an expansive area of researching bacterial proteomics. Proteomics studies allow addressing the functional proteins produced by the changes of genetic expressions. Using comparative proteomic studies allows the examination of bacterial strain differences, both phenotypic and genetic, bacterial growth under various nutrient and environmental conditions, i.e. nutrient type, growth phase, temperature, chemical compounds, such as antibiotics. Comparative Proteomics also provides the researcher with a tool to begin characterizing the functions of the vast proportion of "hypothetical" or "unknown" proteins elucidated from genome sequencing and database comparisons.
Comparative proteomics has been widely applied to microbial identification and characterization studies through the utilization of several mass spectrometry techniques, with tandem mass spectrometry techniques proving to be effective and reliable approach [Aebersold,2003; Anhalt & Fenselau, 1975; Dworzanski, 2006; Hillkamp,2000; Jabbour, 2005, Krishnamurthy, 2000). This chapter will address the utilization of comparative proteomics and the application of tandem mass spectrometry in the identification and differentiation of bacterial strains.

2. Overview of the utilization of tandem mass spectrometry in bacterial identification and differentiation

Mass Spectrometry techniques have been extensively used for rapid identification and differentiation of microbes in general and bacteria in particular. The most predominant mass spectrometry techniques that have been utilized for bacterial identification and differentiation include electrospray ionization tandem mass spectrometry/mass spectrometry (ESI-MS/MS); matrix-assisted laser desorption/ionization time-of-flight mass spectrometry (MALDI-TOF-MS); surface-enhanced laser desorption/ionization (SELDI) mass spectrometry; one- or two-dimensional sodium dodecyl sulfate-polyacrylamide gel electrophoresis (1- or 2-D SDS-PAGE); and hybrid techniques such as combination of mass spectrometry, gel electrophoresis, and bioinformatics. Those mass spectrometry methods provide either fingerprints of the bacterial proteins, i.e. MALDI-TOF-MS technique, or amino acid sequences, from tandem MS/MS analysis, of proteins from collision-induced dissociation (CID), Electron transfer dissociation, or post-source decay (PSD) of ionized tryptic peptides derived from bacterial proteins, i.e. ESI-MS/MS technique. This chapter will address the utilization of tandem mass spectrometry techniques in the differentiation of bacterial strains.

Tandem mass spectrometry techniques have witnessed significant utilization and success in the interrogation of the protein component of a biological species, virus proteins, protein toxins, and bacteria for identification and characterization purposes (Demirev & Fenselau,2008a, 2008b; Dworzanski & Snyder, 2005; Ho, 2002; Ecker, 2005; Fox, 2002, 2006; Hofstadler, 2005; Lambert,2005; Nagele, 2003; Pennigton, 1997; Sampath, 2007; Wilkins, 2006; Williams, 2002). Investigations of the protein component in biological systems constitute the realm of proteomics (Nagele, 2003; Pennigton, 1997). The LC- tandem MS technique is well-suited and equipped to handle the complex and very comprehensive suites of proteins, in a reproducible fashion (William 2002), present in biological threat microorganisms. The vast amount of protein and peptide data generated from a typical LC-tandem MS analysis needs to be addressed in an efficient and timely manner. Data reduction techniques have spawned a number of successful bioinformatics software analysis tools to efficiently address this task (Fox, 2002, 2006; Yates, 1998; Kuwana, 2002). Furthermore, new genomes are constantly being realized and resolved so as to increase the database of bacterial genomes to interrogate a biological sample (Dworzanski & Snyder, 2005). A major portion of the Centers for Disease Control (CDC) Category A, B, and C biological threats have their genomes fully sequenced and available for bioinformatics coupled to MS-based proteomics (NCBI website, 2010; Integrated genomic, 2010; Rotz, 2002).

The US Government has initiated extensive efforts in the detection and identification of biological threat species in their Defense Advanced Research Projects Agency (DARPA) programs that explore the "detect to protect" and "detect to treat" paradigms (National research Council [NRC], 2005; Demirev, 2005). Those initiatives cover areas of general health risk, bio-terrorism utility, Homeland Security, agricultural monitoring, quality of foodstuffs, environmental monitoring, and biological warfare agents in battlefield situations (Demirev & Fenselau,2008a). Some of the concerns include incidents such as a ricin attack (Bevilacqua, 2010) and the *Bacillus anthracis* spore attack on the US postal system in the fall of 2001 (Demirev & Fenselau,2008b; Dworzanski & Snyder, 2005; Friess, 2010; Ho, 2002; Wilkins, 2006).

Proteomic analyses by LC-MS have been used in the characterization of bacteria (Castanha,2006; Dworzanski, 2004, 2006;Lambert, 2005;). Given the degree of success for tandem MS-based proteomics in bacterial characterization, a comparative proteomic study was reported about the potential of the outer membrane protein (OMP) and whole cell protein extracts, independently, can distinguish between strains of the same species (Jabbour et al., 2010). Typically, whole cell protein extracts are usually investigated or select portions of the bacterium, such as the outer membrane, are isolated and the proteins extracted there from. In the membrane, the OMPs act as active mediators between the cell and its environment and are often associated with virulence in Gram-negative pathogens. In pathogenic *Escherichia coli*, there are multiple OMPs present which are required for intestinal colonization as well as those that play a role in the type III secretion system responsible for delivering effector proteins to host cells (Garmendia, 2005; Ide,2001; McDaniel, 1997; Wachter,1999).

3. Outer membrane proteins for bacterial strains differentiation

Outer membrane proteins (OMPs) of gram-negative bacteria act as active mediators between the cell and its environment and are often associated with virulence in gram-negative pathogens (Jerse et al., 1990; Kaper et al., 2004; Koebnik et al., 2000;). Avriulent strains often lack one or more of the plasmids or genes encoding proteins needed for virulence. These differences in OMP expression between virulent and avirulent strains of gram negative bacteria could potentially be exploited to distinguish among strains. Therefore, OMPs could prove to be potential biomarkers for Bacterial strain differentiation.
The off-line 2-D chromatofocussing and reverse phase LC with electrospray-time of flight (ESI-TOF)-MS and matrix-assisted laser desorption ionization (MALDI) TOF-MS detection instrumentation have been used to analyze whole cell protein extracts of non-pathogenic and pathogenic (O157:H7) *E. coli* strains (Zheng, 2005). Those analyses provided various proteins where, in addition to commonly shared proteins, seven unique proteins were found in a non-pathogenic *E. coli* strain, and five unique proteins were found to be expressed in the pathogenic O157:H7 strain. These intracellular, non-OMP proteins were the basis for distinguishing the *E. coli* strains; however, this information was not applied to bioinformatics cross-referencing with a proteome database.
A series of Enterobacteria were investigated and cross-referenced with on-line protein databases (Pribil, 2005). OMPs were investigated by MALDI-TOF- tandem MS where microgram amounts of cells were briefly subjected to trypsin digestion on a stainless steel target plate. Four Enterobacteria were investigated and protein mass spectra were analyzed. Peptide analyses provided protein identification, and multiple assignments allowed database searches for matching to the Enterobacteria species: *E. coli*, *E. herbicola*, *E. cloacae*, and *Salmonella typhimurium*. Some of the distinguishing proteins originated in the cellular milieu and unique OMPs were identified in all four species.
Top-down proteomics and matrix-assisted laser desorption/ionization time-of-flight (MALDI-TOF/TOF) tandem mass spectrometries were used to differentiate protein extracts of *E. coli* strains. Six ions found in a collection of mass spectra originated from proteins that could distinguish between pathogenic and non-pathogenic *E. coli* strains by tandem TOF mass spectrometry. A unique protein biomarker ion at m/z 7705.6 was found (putative uncharacterized YahO) in pathogenic O157:H7 and pathogenic nearest neighbor O55:H7 (infantile diarrhea) strains. Another ion at m/z 9737.5 indicative of the acid stress

chaperone-like protein: HdeA was found in the O157:H7 strain. An ion (m/z 9063.4) in the mass spectrum of non-pathogenic *E. coli* RM3061 was absent in the O157:H7 mass spectrum. Tandem TOF mass spectrometry analysis identified the peak as the HdeB acid stress chaperone-like protein which was useful in discrimination for this non-pathogenic *E. coli* strain.

In another study, the membranes of the *S. typhimurium* and *Klebsiella pneumoniae* Enterobacteria were isolated, and the proteins were extracted with subsequent 2-D electrophoresis (Fagerquist, 2010). The excised protein spots were digested with trypsin and analyzed by MALDI-TOF-MS and peptide mass fingerprinting. The masses predominately originated from OMP peptides and were searched against microorganism databases for identification purposes. Twenty-five and fourteen unique proteins were found in *S. typhimurium* and *K. pneumoniae,* respectively, in a reproducible fashion (Lamontagne, 2007). Pathogenic *E. coli,* such as the O157:H7 strain is a public health pathogen responsible for most common food borne and waterborne illnesses. This bacterium contains a full complement of OMP proteins.

Yersinia pestis is classified as a Category A pathogen and is an important potential biowarfare agent. Virulent *Y. pestis* contains three plasmids encoding multiple OMPs that are required for virulence (Ben-Gurion & Shafferman, 1981; Ferber, 1981; Filippov, 1990). For example, the pCD1 plasmid encodes several *Yersinia* OMPs and a type III secretion system, which are needed for survival and entry into host eukaryotic cells (Cornelis, 2002; Ramamurthi, 2002). Additionally, the pPCP1 plasmid encodes an OMP plasminogen activator that interferes with clotting and complements (Titball, 2003). Avirulent strains often lack one or more of the plasmids or genes encoding proteins needed for virulence, and it is these differences in OMP expression between virulent and avirulent strains of Gram-negative Enterobacteria that could potentially be exploited in order to distinguish among strains.

Alternatively, high-throughput tandem mass spectrometry-based proteomics was applied as a means for characterizing cellular proteins and producing amino acid sequence information for peptides derived from these proteins for *E. coli* and *Y. pestis*. Whole cell protein and cell membrane OMP extracts were compared and contrasted with the in-house BACid bioinformatics modeling tools for species and strain level discrimination (Jabbour, 2010).

4. Bioinformatics tools for bacterial strains differentiation using tandem mass spectrometry

Utilization of MS techniques for bacterial differentiation relies on the comparison of the proteomic information generated from either intact protein profiles (top-down) or the product ion mass spectra of digested peptide sequences (bottom-up) analyses (Warscheid, 2003; Washburn, 2001). For top-down analysis, bacterial differentiation is accomplished through the comparison of the MS data of intact proteins with an experimental mass spectral database containing the mass spectral fingerprints of the studied microorganisms (Craig, 2004). Conversely, bacterial differentiation using the product ion mass spectral data of digested peptide sequences is accomplished through the utilization of search engines against publically available sequence databases to infer identification (Eng, 1994; Warscheid 2004). Several peptide searching algorithms (i.e. SEQUEST and MASCOT) have been developed to address peptide identification using proteomics databases that were generated from either fully or partially genome sequenced organisms (Craig, 2004; Xiang, 2000).

Recent developments in the microbial differentiation field have focused on improving the selectivity of the MS data processing. The product ion mass spectrum-SEQUEST approach was reported for the identification of specific bacteria using a custom-made, limited database of sequences (Keller, 2002; VerBerkmoes, 2005). Another approach used open reading frame (ORF) translator programs to predict possible protein sequences from all probable ORFs and correlate them with the genomic sequences to establish an identification of microorganisms (Chen, 2001). This approach did not show advantages over the product ion mass spectrum method with regard to strain level discrimination (Wolters et al., 2001). However, a recent advancement in proteomics approaches to bacterial differentiation reported a hybrid approach combining protein profiling and sequence database searching using accurate mass tag (Lipton et al., 2002; Norbeck et al., 2006). This approach was used to probe defined mixtures of bacteria to evaluate its capabilities.
Alternatively, an emerging bioinformatics approach that is based on a cross correlation between the product ion spectra of the tryptic peptides and their corresponding bacterial proteins derived from an in-house comprehensive proteome database from genome sequenced microorganisms has been validated (Jabbour, 2010). The exploitation of this proteome database approach allowed for a faster search of the product ion spectra than that using genomic database searching. Also, it eliminates inconsistencies observed in publicly available protein databases due to the utilization of non-standardized gene finding programs during the process of constructing the proteome database. The proposed approach uses an ensemble of bioinformatics tools for the classification and potential identification of bacteria based on the peptide sequence information. This information is generated from the liquid chromatography tandem mass spectrometry (LC-MS-MS) analysis of tryptic digests of bacterial protein extracts and subsequent profiling of the sequenced peptides to create a matrix of sequence-to-microbe (STM) assignments. This proteomics approach is an unsupervised approach to reveal the relatedness between the analyzed samples and the database of microorganisms using a binary matrix approach. The binary matrix is analyzed using diverse visualization and multivariate statistical techniques for bacterial classification and identification.

5. Experimental methods

5.1 Bacterial strains growth and culture conditions

Pathogenic strains employed in the present study were *E. coli* O157:H7 and *Y. pestis* Colorado 92 (CO92). Non-pathogenic strains employed were *E. coli* K-12 and *Y. pestis* A1122. Working cultures were prepared by streaking cells from cryopreserved stocks onto tryptic soy agar (TSA) followed by incubation for approximately 18 hours at 37° C for *E. coli* and 30° C for *Y. pestis* strains. After incubation, all working culture plates were stored at 4° C. Cells from working cultures were used to inoculate broth cultures for each strain, which consisted of 100 mL of trypticase soy broth (TSB) for *E. coli* strains and 100 mL of brain heart infusion (BHI) for *Y. pestis* strains. Cultures were incubated for approximately 18 hours at 37° C for *E. coli* strains and 30° C for *Y. pestis* strains with rotary aeration at 180 rpm. After incubation, broth cultures were pelleted by centrifugation (2,300 RCF at 4° C for 10 min), washed, and resuspended in 10 mL HEPES buffer followed by heating at 95 °C for 1 hour to lyse the cells. After heating, a portion of each sample was plated onto TSA and incubated for five days at the appropriate temperature to ensure no growth prior to removing samples from the BSL-2 or BSL-3 laboratory. Total cellular protein samples (whole cell protein extracts) were heated for one hour to ensure that a no growth situation was confirmed on agar plates for safety concerns.

5.2 Isolation of the Outer Membrane Proteins (OMPs)

After lysis of the whole cells by heating at 95° C for one hour, the cell debris was pelleted by centrifugation at 2,300 RCF at 4° C for 10 min. The supernatant was then centrifuged at 100,000 x g for one hour to pellet the proteins. The pellet was resuspended in 1 mL of HEPES buffer, 1 mL of a 2% Sarkosyl solution (N-lauroylsarcosine sodium salt solution) was added, and the sample was incubated at room temperature for 30 min. Samples were centrifuged at 100,000 x g for one hour, and the pellet containing OMPs was resuspended in 1 mL of HEPES buffer.

5.3 Processing of whole cell lysates and OMPs samples

All protein samples were ultrasonicated (20 seconds pulse on, 5 seconds pulse off, and 25% amplitude for 5 min duration) and a small portion of the lysates was reserved for 1-D gel analysis. The lysates were centrifuged at 14,100 x g for 30 min to remove any debris. The supernatant was then added to a Microcon YM-3 filter unit (Millipore, Catalogue # 42404) and centrifuged at 14,100 x g for 30 min. The effluent was discarded. The filter membrane was washed with 100 mM ABC and centrifuged for 20 min at 14,100 x g. Proteins were denatured by adding 8 M urea and 3 µg/µL DTT to the filter and incubating overnight at 37° C on an orbital shaker at 60 rpm. Twenty microliters of 100% ACN was added to the tubes and allowed to incubate at room temperature for 5 min. The tubes were then centrifuged at 14,100 x g for 40 min and washed three times using 150 µL of 100 mM ABC solution. On the last wash, ABC was allowed to sit on the membrane for 20 min while shaking, followed by centrifugation at 14,100 x g for 40 min. The micron filter unit was then transferred to a new receptor tube and the proteins were digested with 5 µL trypsin in 240 µL of ABC solution + 5 µL ACN. Proteins were digested overnight at 37° C on an orbital shaker set to 55 rpm. Sixty microliters of 5% ACN/0.5% formic acid (FA) was added to each filter to quench the trypsin digestion followed by two minutes of vortexing for sample mixing. The tubes were centrifuged for 30 min at 14,100 x g. An additional 60 µl 5% ACN/0.5% FA mixture was added to the filter and centrifuged. The effluent was then analyzed using LC-ESI- tandem MS.

5.4 LC-tandem MS analysis of peptides

The tryptic peptides were separated using a capillary Hypersil C18 column (300 Å, 5 µm, 0.1 mm i.d. × 100 mm) by using the Surveyor LC from ThermoFisher (San Jose, CA 95101). The elution was performed using a linear gradient from 98% A (0.1% FA in water) and 2% B (0.1% FA in ACN) to 60% B over 60 min at a flow rate of 200 µL/min, followed by 20 minutes of isocratic elution. The resolved peptides were electrosprayed into a linear ion trap mass spectrometer (LTQ, Thermo Scientific, San Jose, CA 95101) at a flow rate of 0.8 µL/min. Product ion mass spectra were obtained in the data dependent acquisition mode that consisted of a survey scan over the m/z range of 400-2000 followed by seven scans on the most intense precursor ions activated for 30 ms by an excitation energy level of 35%. A dynamic exclusion was activated for 3 min after the first MS-MS spectrum acquisition for a given ion. Uninterpreted product ion mass spectra were searched against a microbial database with TurboSEQUEST (Bioworks 3.1, Thermo Scientific, San Jose, CA 95101) followed by application of an in-house proteomic algorithm for bacterial identification of the replicate analyses.

5.5 Protein database and database search engine

A protein database was constructed in a FASTA format using the annotated bacterial proteome sequences derived from fully sequenced chromosomes of 1433 bacteria, including their sequenced plasmids (as of May 2011). A PERL program (http://www.activestate.com/Products/ActivePerl) was written to automatically download these sequences from the National Institutes of Health National Center for Biotechnology (NCBI) site (http://www.ncbi.nlm.nih.gov). Each database protein sequence was supplemented with information about the source organism and genomic position of the respective open reading frame (ORF) embedded into a header line. The database of bacterial proteomes was constructed by translating putative protein-coding genes and consists of tens of millions of amino acid sequences of potential tryptic peptides obtained by the *in silico* digestion of all proteins (assuming up to two missed cleavages).

The experimental product ion mass spectra of bacterial peptides were searched using the SEQUEST (Warscheid, 2003) algorithm against a constructed proteome database of microorganisms. The SEQUEST thresholds for searching the product ion mass spectra of peptides were Xcorr, deltaCn, Sp, RSp, and deltaMpep. The search results were filtered by using Xcorr = 1.90, 2.20, and 3.75 thresholds for peptide ions of +1, +2, and +3 charges, respectively (Ma, 2009; Wu, 2003). These parameters provided a uniform matching score for all candidate peptides. The generated outfiles of these candidate peptides were then validated using the Peptide Prophet algorithm (Keller et al., 2002). Peptide sequences with a probability score of 95% and higher were retained in the dataset and used to generate a binary matrix of sequence-to-bacterium (STB) assignments. The binary matrix assignment was populated by matching the peptides with corresponding proteins in the database and assigning a score of one. A score of zero was assigned for a non-match. The column in the binary matrix represents the proteome of a given bacterium, and each row represents a tryptic peptide sequence from the LC product ion mass spectral analyses. A sample microorganism was matched with a database bacterium by the number of unique peptides that remained after filtering of degenerate peptides from the binary matrix. Verification of the classification and identification of candidate microorganisms was performed through hierarchical clustering analysis and taxonomic classification (Jabbour et al., 2010).

The SEQUEST-processed product ion mass spectra of the peptide ions were compared to an NCBI protein database with the in-house BACid developed software (Dworzanski et al., 2006). BACid provided a taxonomically meaningful and easy to interpret output. It calculated the probabilities that a peptide sequence assignment to a product ion mass spectrum was correct and used accepted spectrum–to-sequence matches to generate an STB binary matrix of assignments. Validated peptide sequences, either present or absent in various strains (STB matrices), were visualized as assignment bitmaps and analyzed by the BACid module that used phylogenetic relationships among bacterial species as part of a decision tree process. The bacterial classification and identification algorithm used assignments of organisms to taxonomic groups (phylogenetic classification) based on an organized scheme that begins at the phylum level and follows through the class, order, family, genus, and species to the strain level. BACid was developed in-house using PERL, MATLAB and Microsoft Visual Basic.

6. Results and discussion

6.1 Comparative proteomic differentiation between the whole cell and the OMP extracts for the *E. coli* O157:H7 strain

The whole cell protein extracts of *E. coli* strain O157:H7 were prepared and analyzed by LC-Tandem ESI-MS/MS. The bioinformatics analyses involved the nearest-neighbor analysis, using the Euclidean single linkage approach to arrive at a set of proteins for species and strain matching to the database.

Figure 1 shows the identification and classification of the experimental sample, whole cell extract, as *E. coli* O157:H7 strain. However, this identification is equally shared with *E. coli* UTI89, which is the causative agent of human urinary tract infections. Although *E. coli* UTI89 is related to *E. coli* O157:H7, it is missing certain proteins such as the OMP HU2 outer membrane and flagella related proteins that are distinctly expressed in *E. coli* O157:H7 (*vide infra*). A comparative proteome list of the strain-unique proteins and the total number of identified proteins for the mentioned *E. coli* O157:H7 extracts is shown in table 1. There are five and eight unique proteins resulted from the bioinformatics analysis of the peptide product ion mass spectra from the *E. coli* O157:H7 whole cell and OMPs extracts, respectively. Figure 2 shows the nearest neighbor similarity linkage results for the OMP extract of *E. coli* O157:H7. This dendogram shows an unambiguous strain level differentiation for the *E.coli* O157:H7 as compared together *E. coli* strains. It is worth mentioned that the next nearest neighbor, which is *E. coli* UT189, is relatively distant at approximately 2.2 linkage units unlike that from the whole cell protein extract (Figure 1). This result indicates that OMPs extract can potentially serves as strain-unique biomarkers for bacterial strain differentiation.

Whole Cell Extract		OMP Extract	
Accession Number	**Unique Protein Name**	**Accession Number**	**Unique Protein Name**
BAA35715	OMP HU2 protein	NP_310124	Acid sensitivity protein
NP_290616	50S ribosomal protein L10	NP_310689	Flagellin
NP_290256	Secreted protein EspA	NP_311482	Heat shock protein
NP_310689	Structural flagella protein	NP_308975	Hypothetical protein ECs0948
NP_312864	Two-component sensor protein	NP_309690	Outer membrane protease precursor
		NP_309226	Putative antirepressor protein
		NP_309783	Putative OMP
		NP_312404	Slp
Total Proteins	162		89

Table 1. Identified unique Proteins lists detected in the Whole Cell Protein and OMP Extracts of E. coli O157:H7.

Moreover, a closer look at the resulted bioinformatics data showed the total number of proteins identified between the two extraction techniques was such that the whole cell preparation had a significantly higher number of proteins of 162 as compared to the that of the number of OMP extract proteins of 89. However, the number of unique proteins that were identified from the OMP extract (eight proteins) was greater than that in the whole cell protein extract (five proteins) (Table 1). These numbers of unique proteins are very similar to that of the whole cell protein extracts for *E. coli* strains investigated (Zheng et al. 2005). That work found five unique proteins from the E. coli O157:H7 strain. However, this does

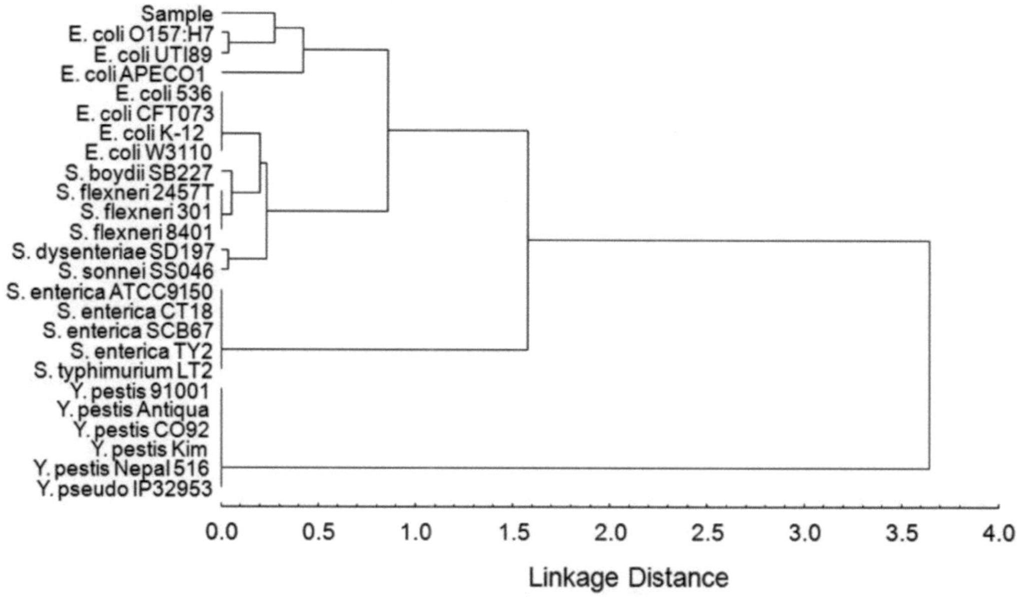

Fig. 1. Euclidean linkage similarity dendogram of the Nearest-neighbor classification of whole cell extract of E. coli O157 H:7.

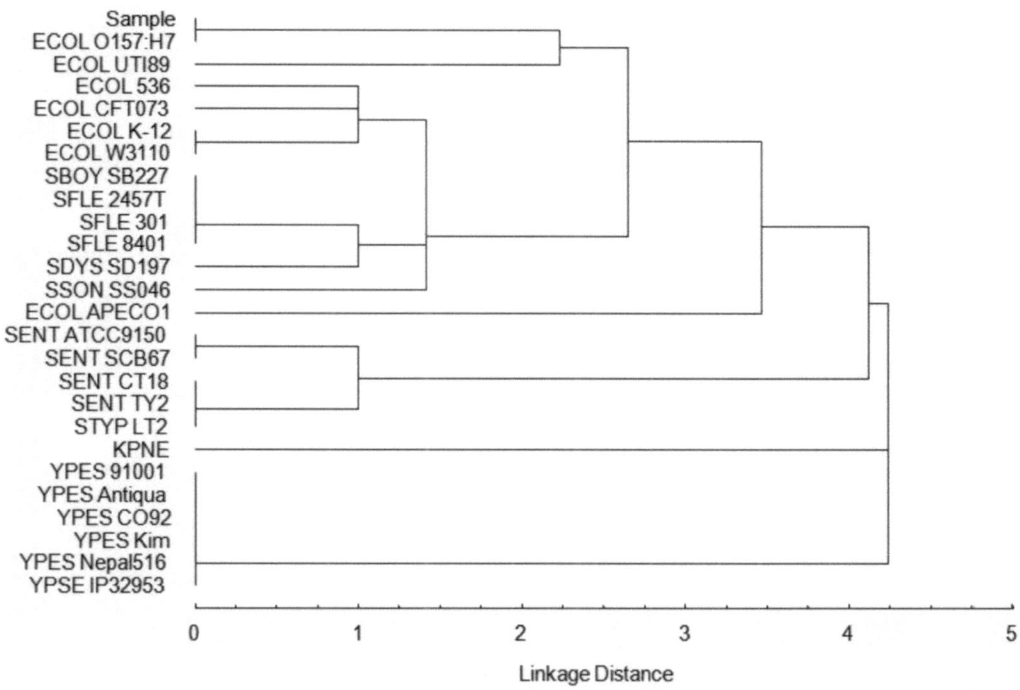

Fig. 2. Euclidean linkage similarity dendogram of the Nearest-neighbor classification of OMPs extract of *E. coli* O157 H:7.

not imply an absence of the additional OMPs in the whole cell extract. Rather it may be that a higher abundance of non-OMPs, or remaining protein in the cell, potentially suppressed the detection of the OMPs in the whole cell protein extracts by tandem MS. Mass spectral analysis can suffer from ionization suppression due to the presence of large numbers of ionizable species. Generally, a whole cell extract has a significantly larger number of ionizable peptides with a greater abundance of non-outer membrane tryptic peptides compared to that of an OMP extract. Therefore, whole cell protein extract analysis likely experiences a degree of ionization suppression during mass spectral analysis.

6.2 Comparative proteomic differentiation between the whole cell and the OMP extracts for the *E. coli* K-12 strain

The results of the bacterial strain level differentiation of the whole cell and OMPs extracts for *E. coli* K-12 are shown in Figures 3 and 4, respectively. The results indicate that those extracts provided sufficient number of identified proteins to correctly identify the *E. coli* K-12 strain. Figure 3 shows that the whole cell protein extract produced an equal similarity with the sample and the *E. coli* K-12 and W3110 strains. This is in agreement with the literature, which reported that *E. coli* W3110 is actually a substrain of K-12 (Baglioni et al., 2003; Yamada et al., 1993). It worth mentioning that the whole cell extract (Figure 3) is approximately 0.03 linkage units distant between the sample/K-12/W3110 *E. coli* group of strains and the next nearest-neighbor group that includes the *E. coli* 536/UT189/CFT73/O157:H7 strains. Hence, the whole cell protein extract was able to delineate the sample containing *E. coli* K-12 from that of the of the *E. coli* O157:H7 strain.

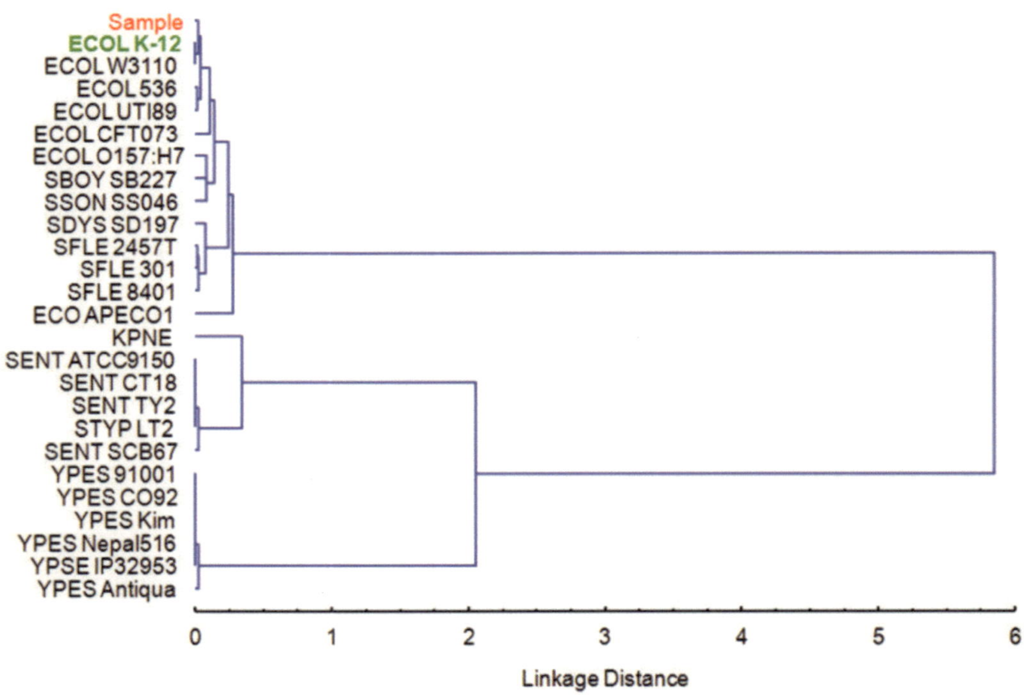

Fig. 3. Euclidean linkage similarity dendogram of the Nearest-neighbor classification of whole cell extract of E. coli K-12 strain.

Figure 4 shows the nearest neighbor Euclidean similarity linkage analysis for the OMP extracts of the *E. coli* K-12 sample. This dendogram shows that the OMP extract provided an enhancement of the strain differentiation as compared with that of whole cell extract. Although, a sample was matching with the non-pathogenic W3110 strain, however, the labels signify the same organism (*vide supra*). No ambiguity was observed in the strain differentiation. Moreover, there is a relatively larger linkage distance (0.10) between the sample/K-12/W3110 and the 536/UT189/CFT073/O157:H7 groups of *E. coli* strains from the OMP as compared to that from the whole cell extract, figure 3.

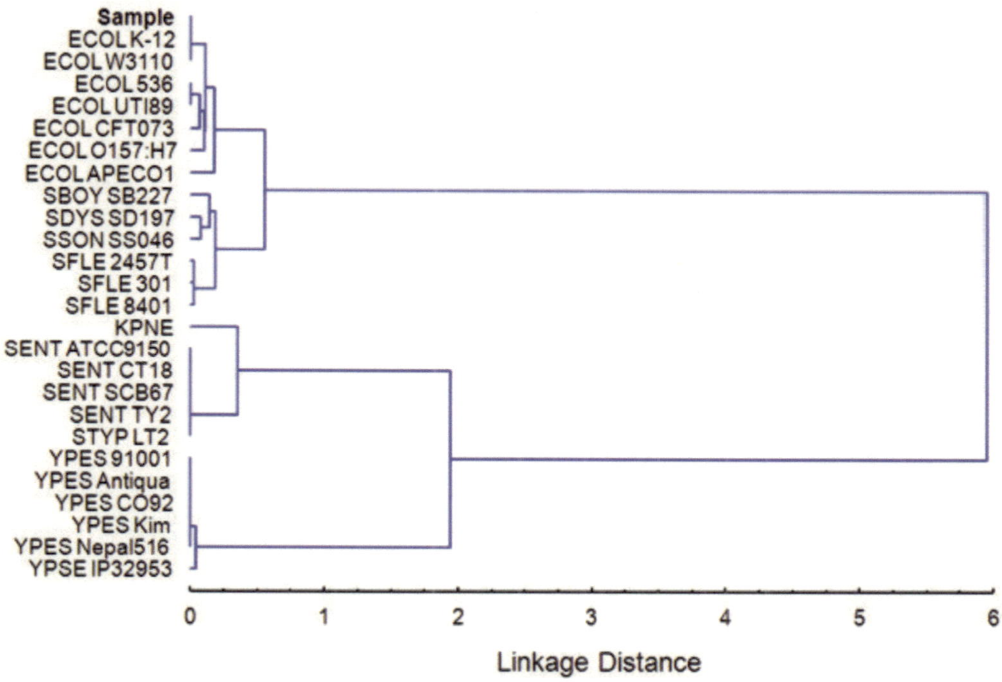

Fig. 4. Euclidean linkage similarity dendogram of the Nearest-neighbor classification of OMPs extract of E. coli K-12 strain.

Table 2 presents a list of the unique proteins. The total number of identified proteins found in the proteomics analysis for the K-12 strain was 194 and 112 for the whole cell protein and OMP extracts, respectively. The number of strain-unique proteins that were identified by the bioinformatics algorithm was greater in the OMP extracts (ten proteins) compared to that in the whole cell extracts (eight proteins). These numbers of unique proteins from the K-12 extracts are very similar to that of the whole cell protein extracts for *E. coli* strains investigated by Zheng et al. (Zheng et al., 2005). That work found seven unique proteins from the non-pathogenic *E. coli* 88-0447 (O136STa).

Overall, the comparative proteomic analyses of the E.coli whole cell extracts showed that there 162 proteins produced for *E. coli* O157:H7 strain vs. 194 for that of *E. coli* K-12 one, see tables 1-2. Upon removing the highly conserved, house-keeping, denigrate and energy transfer proteins from both strains, the number of strain-unique proteins was eight for *E. coli* K-12 and five for *E. coli* O157:H7. From analyses of the OMP protein extracts, a comparison of the total experimentally-determined number of proteins showed a difference

Whole Cell Extract		**OMPs Extract**	
Accession Number	**Protein Name**	**Accession Number**	**Protein Name**
YP_669714	Aspartyl-tRNA Synthetase	NP_415097	DLP12 prophage; outer membrane protease VII
NP_417795	Bacterioferrin	NP_415269	Peptidoglycan-associated outer membrane protein
NP_668903	Chorismate synthase	NP_417083	Protein disaggregation chaperone
NP_755058	GnsA/GnsB family	NP_415423	Pyruvate formate lyase I
NP_671573	Putative cytoplasmic protein	NP_415759	Oligopeptide transporter subunit
NP-415386	Lipoprotein	NP_416009	Predicted glutamate: gamma-aminobutyric acid antiporter
YP_670276	Hypothetical protein	NP_414968	Predicted lipoprotein
NP_415386	Predicted lipoprotein	NP_417320	5-keto-4-deoxyuronate isomerase
		NP_415772	OMP W
		NP_417963	Outer membrane lipoprotein
Total identified Proteins	**194**		**112**

Table 2. Identified unique proteins lists detected in the whole cell and OMP extracts of E. coli K-12 strain.

between the two *E. coli* strains. The O157:H7 strain had 89 total identified proteins compared to 112 for the K-12 strain. Upon removing the highly conserved, house-keeping, and energy transfer proteins from both strains, the number of strain-unique proteins for *E. coli* O157:H7 is eight and that for *E. coli* K-12 is ten in the OMPs extract of the studied *E. coli* strains as shown in table 2.

6.3 Comparative proteomic differentiation between the whole cell and the OMP extracts for the *Yersinia pestis* CO92 strain

A comparison of the LC-Tandem MS and bioinformatics results of the proteins present in the whole cell and OMP extracts of *Y. pestis* CO92 was performed. Figure 5 shows the identification results of the whole cell protein extract for *Y. pestis* CO92. The dendrogram indicates an ambiguous strain level differentiation between the experimental sample and the database *Y. pestis* CO92 entry.The bioinformatics analysis of the whole cell extracts of *Y. pestis* CO92 matched with five strains entries of *Yersinia* strains in the database. The CO92 experimental strain was matched to the only avirulent *Y. pestis* strain (91001) in the database as well as to the virulent Antiqua, CO92, Nepal 516, and IP32953 *Y. pestis* strains. However the Y. pestis KIM strain resided two linkage units distant from the sample and remaining five *Y. pestis* strains in the nearest neighbor similarity linkage analysis. The set of unique proteins for whole cell protein extracts of *Y. pestis* CO92 shows only four biomarkers associated with its reported virulence factors (Table 3).

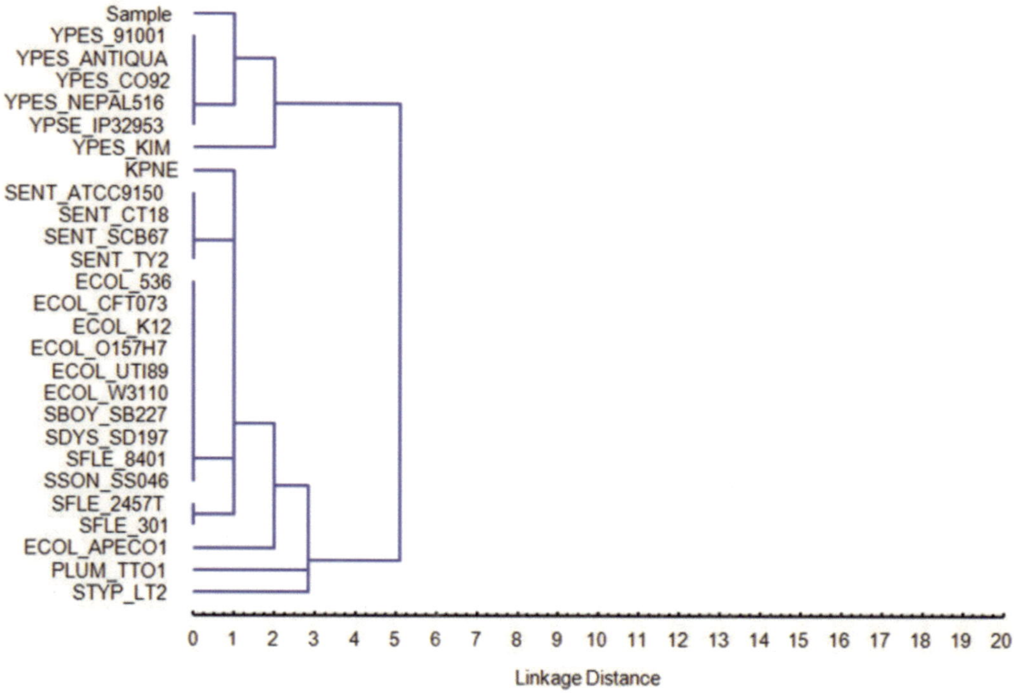

Fig. 5. Euclidean linkage similarity dendogram of the Nearest-neighbor classification of whole cell extract of *Yersinia pestis* CO92 strain.

Figure 6 shows the identification results for the OMP extracts of the *Y. pestis* CO92 sample. The dendrogram indicates an unambiguous, and correct, strain level identification with the *Y. pestis* CO92 strain in the proteome database. The experimental sample and Y. pestis entry of the *Y. pestis* CO92 strains are one linkage distance unit from the next nearest neighbor group consisting of the 91001/Antiqua/Nepal 516 strains. The set of unique proteins for virulent *Y. pestis* CO92 provides the presence of known biomarkers associated with virulence factors (Table 3). For example, virulence plasmids in *Y. pestis* such as pPCP1 that encodes for plasminogen activator protease precursor, pCD1 that encodes for low-calcium response protein, pMT1 that encodes for toxin protein and the structural gene for fraction 1 protein capsule (chaperonin protein) were found in the mass spectral analyses and are listed in Table 3. The chaperonin protein was present in higher abundance than that of the other protein biomarkers. The unique set of proteins had the closest match with *Y. pestis* strains compared to other similar bacteria in the database as seen in both dendrograms in Figures 5-6.

From analyses of both protein extracts, a comparison of the number of total, experimentally-determined number of proteins showed a difference between the two protein methods as applied to the *Y. pestis* sample. The whole cell protein and OMP approaches had 182 and 136, respectively, total identified proteins (Table 3). Upon removing the highly conserved, house-keeping, and energy transfer proteins from both strains, the number of strain-unique proteins (Table 3) for the whole cell protein and OMP approaches was four and thirteen, respectively. Even with a significant amount of unique proteins, the OMP differentiation capability did not provide a significant benefit (1.4 linkage units) with respect to the four proteins from the whole cell approach (1 linkage unit) as detailed in the dendograms in Figures 5-6.

Whole Cell Extract		OMP Extract	
Accession Number	**Protein Name**	**Accession Number**	**Protein Name**
NP_993129	Hypothetical protein YP_1779	CAL19718	Cationic 19 kDa OMP
NP_995559	Murine toxin	NP_991899	Fraction 1 protein capsule (chaperonin GroEL)
NP_994104	Periplasmic chaperone	YP_070861	Membrane bound lytic murein transglycosylase C precursor
NP_991935	30S ribosomal protein S6	NP_993916	Aminotransferase
		NP_395168	Low-calcium response protein
		CAL18706	Secreted thiol: disulfide interchange protein DsbA
		CAL18984	Tellurium resistance protein
		CAL19717	Putative surface antigen
		CAL21872	Putative sigma 54 modulation protein
		NP_395233	Plasminogen activator protease precursor
		CAL19882	OMP porin C
		NP_395420	Murine toxin
		YP_02420	Probable formyl transferase
Total Identified Proteins	**182**		**136**

Table 3. Identified unique proteins lists detected in the whole cell and OMP extracts of *Y. Pestis* CO92 strain.

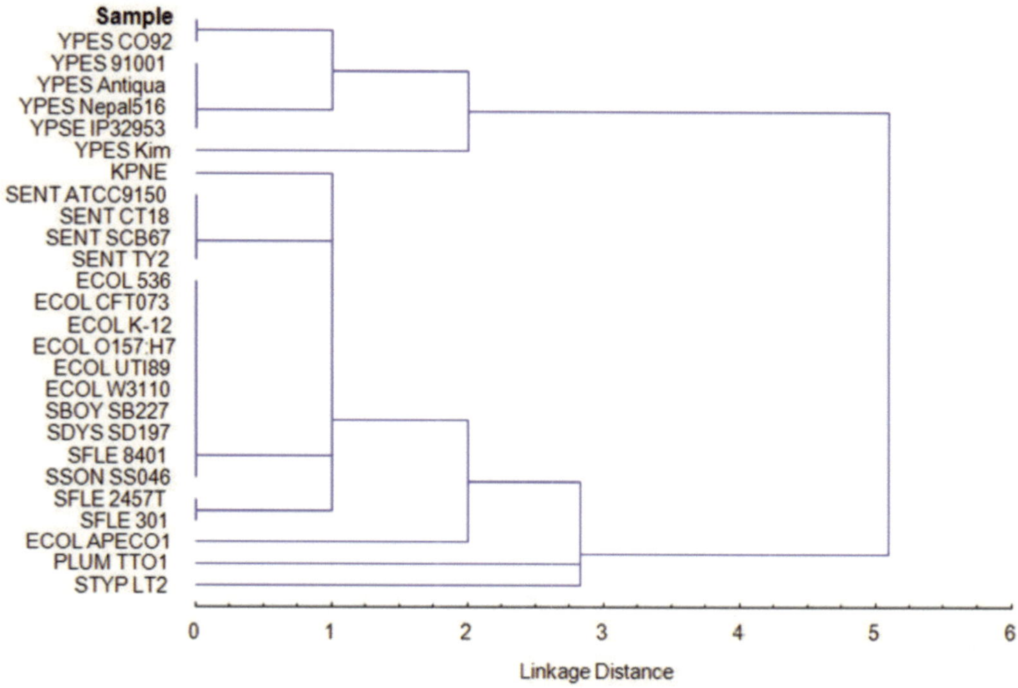

Fig. 6. Euclidean linkage similarity dendogram of the Nearest-neighbor classification of OMPs extract of *Yersinia pestis* CO92 strain.

6.4 Comparative proteomic differentiation between the whole cell and the OMP extracts for the *Y. pestis* A1122 strain

A comparison of the LC-Tandem MS and bioinformatics results of the proteins present in the whole cell and OMP extracts of the avirulent *Y. pestis* A1122 was performed. Figure 7 shows the nearest-neighbor similarity linkage analysis of the whole cell extract of the avirulent *Y. pestis* A1122 strain. A unique set of proteins for each extraction method had the closest match with *Y. pestis* strains compared to other similar Gram-negative bacteria in the database entries. In figure 7, the dendogram shows the similarity linkage for the whole cell protein extract from the *Y. pestis* A1122 in which the sample was identified to the pathogenic KIM, CO92 and Nepal 516 strains. Equidistant next nearest neighbors to this group are the 91001 and Antiqua strains. The linkage distance is minimal between these two groups of *Y. pestis* strains. On the basis of these results, the unique set of proteins (Table 4) from the experimental *Y. pestis* A1122 sample produced a closest similarity index to the CO92 and Nepal 516 virulent strains from whole cell protein extract preparations. A similar situation also was observed using whole cell protein extracts between the sample CO92 strain and the 91001/Antiqua/CO92/Nepal 516/IP32953 strains (Figure 5). As shown in table 4, there are three strain-unique proteins that were identified out of a total of 164 proteins from an analysis of the A1122 strain.

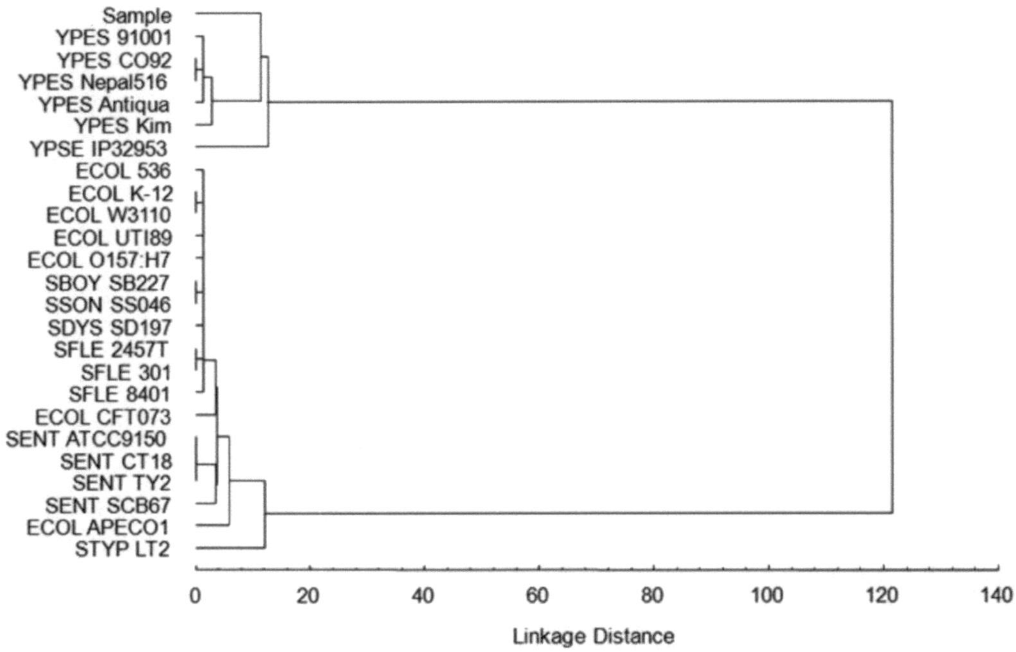

Fig. 7. Euclidean linkage similarity dendogram of the Nearest-neighbor classification of whole cell extract of *Yersinia pestis* A1122 strain.

On the other hand, the OMP analysis in Figure 8 shows that the sample was identified at the strain level as *Y. pestis* 91001. This finding is encouraging knowing that *Y. pestis* 91001 is the only avirulent strain in the proteome database which also includes several pathogenic *Y. pestis* strains. Because the avirulent *Y. pestis* A1122 strain has not been sequenced or is not publicly available, its absence from the database provided an indirect test of the robustness of the proteomics approach in the classification of a non-database bacterium to the database entries. It is worth mentioning that the constructed proteome data base consists of more than 1400 fully sequenced bacteria that had been translated into their complimentary protein expressions. All the samples studied were compared to all the proteomes in the constructed database and the top 20 closest near-neighbors were selected for further comparative proteomics analyses. This also provides confidence for identification at the species level (Figure 8). However, an equal similarity index is also shared with the Nepal 516 strain. The Antiqua strain is a very close nearest neighbor to the 91001 and Nepal 516 cluster of strains. The CO92 strain is observed to be relatively more removed from the 91001/Nepal 516 and Antiqua strains. On the basis of these results, the unique set of proteins for the experimental *Y. pestis* A1122 sample produced the same similarity index for the database *Y. pestis* 91001 and the Nepal 516 strains from the OMP extract preparation (Table 4). Figure 8 shows that there is a very small linkage distance between the groups of *Y. pestis* strains. Thus, the OMP analysis produces very similar classification results (very small linkage distances) for the six *Y. pestis* strains in the genome database. Table 4 lists the six unique proteins from a total of 94 proteins for the *Y. pestis* 91001 strain found in the OMP extract of the experimental A1122 strain. From analyses of the whole cell protein extracts, a comparison of the total number of proteins produced 182 (Table 3) and 164 (Table 4) for *Y. pestis* CO92 and *Y. pestis* A1122, respectively. Upon removing the highly conserved, housekeeping, and energy transfer proteins from both strains, the number of strain-unique proteins was four for *Y. pestis* CO92 and three for *Y. pestis* A1122. From analyses of the OMP protein extracts, a comparison of the number of total, experimentally determined number of proteins showed a difference between *Y. pestis* CO92 and *Y. pestis* A1122. The CO92 strain had 136 total identified proteins compared to 94 for the A1122 strain. Upon removing the highly conserved, housekeeping, and energy transfer proteins from both strains, the number of strain-unique proteins for *Y. pestis* CO92 was 13 and that for *Y. pestis* A1122 was 6.

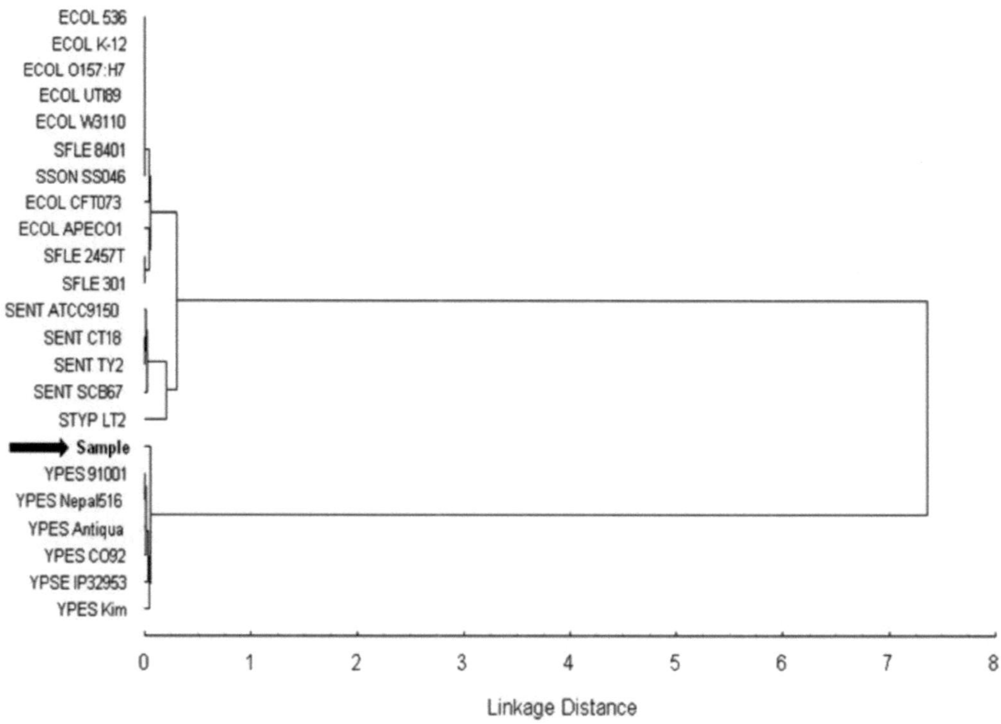

Fig. 8. Euclidean linkage similarity dendogram of the Nearest-neighbor classification of OMPs extract of *Yersinia pestis* A1122 strain.

Whole Cell Extract		OMP Extract	
Accession #	Protein Name	Accession #	Protein Name
NP_991849	Tellurium resistance protein	NP_991979	Transcription elongation factor NusA
NP_993230	arYPES_91001nyl-tRNA synthetase	NP_992082	Na(+)-translocating NADH-quinone reductase
NP_992224	Putative thioredoxin	NP_992120	Proline permease transport protein
		NP_993064	OMP porin
		NP_991484	Exported sulfate-binding protein
		NP_993650	OMP X
Total identified Proteins	**164**	**94**	

Table 4. Identified unique proteins lists detected in the whole cell and OMP extracts of *Y. Pestis* A1122 strain.

7. Conclusion

Comparative proteomics of tandem mass spectrometry data showed that the OMPs extract provided equal or better discrimination compared with the whole cell one with respect to the distance or similarity linkage with the next nearest neighbor(s). Also, the OMPs extracts of all studied strains showed correct database bacterial match with linkage similarity improved over the whole cell extract. The improved strain level differentiation using OMPs extract could be due to the possible ionization suppression experienced by whole cell that could shield the detection of important peptides that could be classified as unique biomarkers. However, whole cell lysates can be an appropriate option for the differentiation of Gram positive bacterial strains and the reported results herein support their potential application in bacterial species and potential strain differentiation. Also, Inclusion of more relevant bacteria such as *Francisella tularensis, Burkholderia,* and other Gram negative genera and species may provide a more comprehensive outlook on the importance of OMPs in comparison to the whole cell extract. These additions may also provide decision information as to the relative merit of applying OMP vs. whole cell protein extraction techniques in the analysis of an experimental bacterial sample for classification and diagnostic purposes.

Overall, Tandem MS-based proteomics and bioinformatics were shown to have utility in the comparative proteomics study for the differentiation of Gram-negative bacterial strains. Different numbers of distinguishing, unique proteins were obtained by the bioinformatics procedure between the whole cell and OMPs extracts. This resulted in different degrees of separation between the correctly determined database organism and the next nearest neighbor organism(s). Moreover, this approach relies on taxonomic correlation within the constructed proteome database and thus inferring an ID on sample organism not present in the genome database is possible. This capabilities is supported the fact that prokaryotic organism as they are arranged in hierarchal order their common proteins increase as we move from strain to phyla and vice versa. Such properties will allow the utilization of this approach to infer taxonomic class based on the depth of available genomic sequencing information for such strains, i.e. species vs. genus vs. family vs. order, etc.

8. References

Aebersold, R. (2003).A mass spectrometric journey into protein and proteome research, *J. Am. Soc. Mass Spectrom.* 14, 685-695.

Anhalt, J.P. & Fenselau, C. (1975). Identification of Bacteria using mass spectrometry, *Anal. Chem.* 47,219-225.

Baglioni, P.; Bini, L.; Liberatori, S.; Pallini, V.& Marri, L. (2003). Analysis of *E. coli* W3110 expressing heterologous sigma factor reported. *Proteomics, 3,* 1060-1065.

Ben-Gurion, R.; Shafferman, A. (1981). Essential virulence determinants of different *Yersinia* species are carried on a common plasmid. *Plasmid* , 5, 183-187.

Bevilacqua, V. L. H.; Nilles, J. M.; Rice, J. S.; Connell, T. R. ; Schenning, A. M.; Reilly, L. M.& Durst, H. D. (2010). Ricin Activity Assay by Direct Analysis in Real Time Mass Spectrometry Detection of Adenine Release. *Anal. Chem.*, 82(3): 798–800.

Castanha, E. R.; Fox, A.& Fox, K. F. (2006). Rapid discrimination of *Bacillus anthracis* from other members of the *B. cereus* group by mass and sequence of "intact" small acid soluble proteins (SASPs) using mass spectrometry. *J. Microbiol. Methods, 67,* 230-240

Chen, W.; Laidig, K. E.; Park, Y.; Park, K.; Yates III, J. R.; Lamont, R. J.& Hackett, M. (2001). Searching the *Porphyromonas gingivalis* genome with peptide fragmentation mass spectra. *Analyst,* 126, 52-57.

Cornelis, G. R. (2002). *Yersinia* type III secretion: send in the effectors. *J. Cell Biol., 158,* 401-408.

Costas; M. (1990). Numerical-Analysis of sodium dodecyl sulfate-polyacrylamide gel-electrophoretic protein-patterns for the classification, identification and typing of medically important bacteria. *Electrophoresis* 11: 382–391.

Craig, R.; & Beavis, R. (2004). TANDEM: matching proteins with tandem mass spectra. *Bioinformatics*, 20:1466-1467.

Demirev, P. A. & Fenselau, C. (2008). Mass Spectrometry for Rapid Characterization of Microorganisms. *Annu. Rev. Anal. Chem., 1,* 71-93.

Demirev, P. A. & Fenselau, C. (2008). Mass spectrometry in biodefense. *J. Mass Spectrom., 43,* 1441-1457

Demirev, P. A.; Feldman, A. B.& Lin, J. S. (2005). Chemical and biological weapons: current concepts for future defenses. *Johns Hopkins APL Tech. Digest, 26,* 321-333.

Dworzanski, J.P.; Deshpande, S.V.; Chen, R.; Jabbour, R.; Snyder, A.P.; Wick, C.H. & L. Li. (2006). Mass Spectrometry-Based Proteomics Combined with Bioinformatic Tools for Bacterial Classification, *Journal of Proteome Research,* 5, 76-87.

Dworzanski, J. P.& Snyder, A. P. (2005). Classification and identification of bacteria using mass spectrometry-based proteomics. *Expert Review Proteomics, 2,* 863-878.

Dworzanski, J. P.; Snyder, A. P.; Chen, R.; Zhang, H.; Wishart, D. & Li, L. (2004). Identification of Bacteria Using Tandem Mass Spectrometry Combined with a Proteome Database and Statistical Scoring, *Anal. Chem., 76,* 2355-2366.

Ecker, D. F.; Sampath, R.; Blyn, L. B.; Eshoo, M. W.; Ivy, C.; Ecker, J. A.; Libby, B.; Samant, V.; Sannes-Lowery, K. A.; Melton, R. E.; Russell, K. Freed, N.; Barrozo, C.; Wu, J.; Rudnick, K.; Desai, A.; Moradi, E.; Knize, D. J.; Robbins, D. W.; Hannis, J. C.; Harrell, P. M.; Massire, C.; Hall, T. A.; Jiang, Y.; Ranken, R.: Drader, J. J.; White, N.; McNeil, J. A.; Crooke, S. T. & Hofstadler, S. A. (2005). Rapid identification and strain-typing of respiratory pathogens for epidemic surveillance. *Proc. Natl. Acad. Sci., 102,* 8012-8017.

Eng, J. K.; McCormack, A. L. & Yates, J. R. (1994). An approach to correlate tandem mass-spectral data of peptides with amino-acid-sequences in a protein database. *J. Am. Soc. Mass. Spectrom.* 5:976-989.

Fagerquist, C. K.; Garbus, B. R. ; Miller, W. G. ; Williams, K. E.; Yee, E.; Bates, A. H.; Boyle, S.; Harden, L. A.; Cooley, M. B.& Mandrell, R. E. (2010). Rapid Identification of Protein Biomarkers of *Escherichia coli* O157:H7 by Matrix-Assisted Laser Desorption Ionization-Time-of-Flight−Time-of-Flight Mass Spectrometry and Top-Down Proteomics. *Anal. Chem., 82,* 2717-2725.

Ferber, D. M.& Brubaker, R. R. (1981). Plasmids in *Yersinia pestis. Infect. Immun., 31,* 839-841.

Filippov, A. A.; Solodovnikov, N. S.; Kookleva, L. M.; Protsenko, O. A. (1990). Plasmid content in Yersinia pestis of different origin. *FEMS Microbiol. Lett., 67,* 45-48.

Fox, A. (2006). Mass Spectrometry for Species or Strain Identification after Culture or without Culture: Past, Present, and Future. *J. Clin. Microbiol.,* 44, 2677-2680.

Fox, A.; Anderson, M.; Dunn, J.; Guenther, B.; Parks, L.; Pinnick, R.; Reed, C.; Rowe, J.; Luftig, R.; Long, G. W.; Lontz, R. Marchin, G. L.; McManus, A. T.; Setlow, P.; Siedow, J.; Sievers, A. J.; Tchikindas, M. L.& Tove, S. (2002) Report of the "Bioterrorism Workshop"—Duke University Thomas Center on 2–4 April 2002 organized by US Army Research Office. *J. Microbiol. Methods, 51,* 247-254.

Friess, R. (2008). The New York Times; August 5, 2008; http://www.nytimes.com/2008/08/05/us/05ricin.html?scp=1&sq=ricin&st=nyt. Accessed February 12, 2010

Garmendia, J.; Frankel, G. & Crepin, V. F. (2005). Enteropathogenic and Enterohemorrhagic *Escherichia coli* Infections: Translocation, Translocation, Translocation. *Infect. Immun., 73,* 2573-2585.

Hillenkamp, F.& Karas, M. (2000).Matrix-assisted laser desorption/ionization, an experience, *Int. J. Mass Spectrom.,* 200, 71-77.

Ho, Y.-P. & Hsu, P.-H. (2002).Investigating the effects of protein patterns on microorganism identification by high-performance liquid chromatography-mass spectrometry and protein database searches. *J. Chromatogr. A, 976,* 103-111

Hofstadler, S. A.; Sampath, R.; Blyn, L. B.; Eshoo, M. W.; Hall, T. A.; Jiang, Y.; Drader, J. J.; Hannis, J. C.; Sannes-Lowery, K. A., Cummins, L. L.; Libby, B.; Walcott, D. J.; Schink, A.; Massire, C.; Ranken, R.; Gutierrez, J.; Manalili, S.; Ivy, C.; Melton, R.; Levene, H.; Barrett-Wilt, G.; Li. F.; Zapp, V.; White, N.; Samant, V.; McNeil, J. A.; Knize, D.; Robbins, D.; Rudnick, K.; Desai, A.; Moradi, E. & Ecker, D. J. (2005). TIGER: the universal biosensor, *Intl. J. Mass Spectrom.,* 242, 23-41.

http://www.ncbi.nlm.gov/PMGifs/genomics/micr/html

http://wit.integratedgenomics.com/GOLD

Ide, T., Laarmann, S.; Greune, L.; Schillers, H.; Oberleithner, H.; Schmidt, M. A. (2001). Characterization of translocation pores inserted into plasma membranes by type III-secreted Esp proteins of enteropathogenic *Escherichia coli. Cell. Microbiol. 3,* 669-679

Jabbour, R.; Dworzanski, J.P.; Deshpande, S.V.; Snyder, A.P. & Wick, C.H. (2005).Effect of gas phase fractionation of peptide ions on bacterial identification using mass spectrometry-based proteomics approach. *Proceedings of the 53rd ASMS Conference on Mass Spectrometry and Allied Topics,* TX, USA, TP31

Jabbour, R.E.; Deshpande, S.V.; Wade, M.M; Stanford, M.F.; Wick, C.H.; Zulich, A.W.; Skowronski, E. W. & Snyder, A.P. (2010). Double Blind Characterization with non-Genome Sequenced Bacteria by Mass Spectrometry-based Proteomics. *Appl. Environ. Microbiol.* 2010, : 76(11):3637-3644

Jerse, A.E.; Yu, J.; Tall, B.D.; Kaper, J.B.; (1990). A genetic locus of enteropathogenic *Escherichia coli* necessary for the production of attaching and effacing lesions on tissue culture cells. *Proc Natl Acad Sci USA.*;87:7839–43.

Kaper, J.B.; Nataro, J.P.; Mobley, H.L.T. (2004). Pathogenic *Escherichia coli. Nat Rev Microbiol;* 2,123–138.

Keller, A.; Nesvizhskii, A. I.; Kolker, I. & Aebersold, R. (2002). Empirical statistical model to estimate the accuracy of peptide identifications made by MS-MS and database search. *Anal. Chem.* 74:5383-5392

Koebnik, R.; Locher, K.P.; Van Gelder, P. (2000). Structure and function of bacterial outer membrane proteins: barrels in a nutshell. *Mol Microbiol.*,37:239–253

Krishnamurthy, T.; Rajamani, U.; Ross, P. L.; Jabbour, R.; Nair, H.; Eng, J.; Yates, J.; Douglas, M. T. D.; Stahl, C. & Lee, T. D. (2000). Mass spectral investigation of microorganisms, *Toxin Reviews*,19:1, 95-117

Kuwana, R.; Kasahara, Y.; Fujibayashi, M.; Takamatsu, H.; Ogasawara, N.& Watabe, K. (2002). Proteomics characterization of novel spore proteins of *Bacillus subtilis. Microbiology, 148,* 3971-3982.

Lambert, J.-P.; Ethier, M.; Smith, J. C.& Figys, D. (2005) Multiplexed Proteomic Reactor for the Processing of Proteomic Samples. *Anal. Chem., 77,* 3771-3788.

Lamontagne, J.; Butler, H.; Chaves-Olarte, E.; Hunter, J.; Schirm, M.; Paquet, C.; Tian, M.; Kearney, P.; Hamaidi, L.; Chelsky, D.; Moriyon, M.; Moreno, E.& Paramithiotis, E. (2007). Extensive cell envelope modulation is associated with virulence in Brucella abortus. *J. Prot. Res., 6,* 1519-1529.

Loferer-Krobacher, M.; Klima, J. & Psenner R. (1998). Determination of Bacterial Cell Dry Mass by Transmission Electron Microscopy and Densitometric Image Analysis, *Applied Environ. Microbiol.*, 64, 688–694.

Lipton, M.S., Pasa-Tolic, L.; Anderson, G. A.; Anderson, D. J.; Auberry, D. L. ; Battista, J. R. ; Daly, M. J.; Fredrickson, J.; Hixson, K. K.; Ksotandarithes, H.; Masselon, C.; Markillie, L. M.; Moore, R. J.; Romine, M. F.; Shen, Y.; Stritmatter, E.; Tolic, N.; Udseth, H. R.; Venkaeswaran, A.; Wong, K. K.; Zhao, R.& Smith, R. D. (2002). Global analysis of the *Deinococcus radiodurans* proteome by using accurate mass tags. *Proc. Natl. Acad. Sci. U.S.A.* 99;11049-11054.

Ma, J.; Liu, J.; Sun, L.; Gao, L.; Liang, Z.; Zhang, L.& Zhang, Y. (2009). Online Integration of Multiple Sample Pretreatment Steps Involving Denaturation, Reduction, and Digestion with Microflow Reversed-Phase Liquid Chromatography-Electrospray Ionization Tandem Mass Spectrometry for High-Throughput Proteome Profiling *Anal. Chem., 81,* 6534-6540.

McDaniel, T. K.& Kaper, J. B. (1997). A cloned pathogenicity island from enteropathogenic *Escherichia coli* confers the attaching and effacing phenotype on E. coli K-12. *Mol. Microbiol., 23,* 399-407.

McDaniel, T. K.; Jarvis, K. G.; Donnenberg, M. S.; Kaper, J. B. (1995). A Genetic Locus of Enterocyte Effacement Conserved among Diverse Enterobacterial Pathogens. *Proc. Natl. Acad. Sci. USA, 92,* 1664-1668.

Nagele, E.; Vollmer, M.; Horth, P. (2003). Two-dimensional nano-liquid chromatography–mass spectrometry system for applications in proteomics. *J. Chromatogr. A, 1009,* 197-205

National Research Council. (2005). Sensor Systems for Biological Agent Attacks. Washington, DC: Natl. Acad. Press, ISBN-10: 0-309-09576-X.

Norbeck, A.D.; Callister, S. J. ; Monroe, M. E.; Jaitly, N.; Elias, D. A.; Lipton, M. S.& Smith, R. D. (2006). Proteomic approaches to bacterial differentiation. *J. Microbiol. Methods* 67,473-486.

Pennington, S. R.; Wilkins, M. R.; Hochstrasser, D. F. & Dunn, M. J. (1997) Proteome analysis: from protein characterization to biological function. *Trends Cell Biol.*, *7*, 168-173.

Pribil, P.; Fenselau, C. (2005). Characterization of Enterobacteria Using MALDI-TOF Mass Spectrometry. *Anal. Chem.*, *77*, 6092-6095.

Ramamurthi, K. S.& Schneewind, O. (2002). Type III protein secretion in *Yersinia* species. *Annu. Rev. Cell Dev. Biol.*, *18*, 107-133.

Rotz, L. D.; Khan, A. S.; Lillibridge, S. R.; Ostroff, S. M.& Hughes, J. M. (2002). Bioterrorism Preparedness for the Public Health and Medical Communities. *Emerging Infect. Dis.*, *8*, 225-230.

Sampath, R.; Hall, T. A.; Massire, C.; Li, F.; Blyn, L. B.; Eshoo, M. W.; Hofstadler, S. A. & Ecker, D. J. (2007). Rapid Identification of Emerging Infectious Agents Using PCR and Electrospray Ionization Mass Spectrometry. *Annals New York Acad. Sci.*, 1102, 109-120.

Titball, R. W.; Hill, J.; Lawton, D. G. & Brown, K. A. (2003). *Yersinia pestis* and plague *Biochem. Soc. Trans.*, *31*, 104-107.

VerBerkmoes, N.C.; Hervey, W. J.; Shah, M.; Land, M.; Hauser, L.; Larimer, F. W.; van Berkel, G. J. & Goeringer, D. E. (2005). Evaluation of "shotgun" proteomics for identification of biological threat agents in complex environmental matrices: experimental simulations. *Anal. Chem.*, 77:923-932.

Wachter, C.; Beinke, C.; Mattes, M. & Schmidt, M. A. (1999). Insertion of EspD into epithelial target cell membranes by infecting enteropathogenic Escherichia coli. *Mol. Microbiol.*, *31*, 1695-1702.

Warscheid, B, and C. Fenselau. (2003). Characterization of *Bacillus* spore species and their mixtures using post source decay with a curved-field reflectron. *Anal. Chem.* 75:5618-5627.

Warscheid, B.; and Fenselau, C. (2004). A targeted proteomics approach to the rapid identification of bacterial cell mixtures by matrix-assisted laser desorption/ionization mass spectrometry. *Proteomics* 4:2877-2892.

Washburn, M. P.; Wolters, D. & J. R. Yates III. (2001). Large-scale analysis of yeast proteome by multidimensional protein identification technology. *Nat. Biotechnol.* 19:242-247.

Wilkins, C. & Lay, J. (Eds.). (2006). Identification of Microorganisms by Mass Spectrometry. John Wiley & Sons: Hoboken; ISBN: 978-0-471-65442-1.

Williams, T. L.; Leopold, P.& Musser, S. (2002). Automated post processing of electrospray LC/MS data for profiling protein expression in bacteria *Anal. Chem.*, *74*, 5807-5813.

Wolters, D. A., Washburn, M. P. & J. R. Yates III. (2001). An automated multidimensional protein identification technology for shotgun proteomics. Anal. Chem. 73:5683-5690.

Wu, C. C.; MacCoss, M. J.; Howell, K. E.&Yates III, J. R. (2003). A method for the comprehensive proteomic analysis of membrane proteins. *Nat. Biotechnol.*, *21*, 532-538.

Xiang, F.; Anderson, G.; Veenstra, T.; Lipton, M.& Smith, R. (2000). Characterization of microorganisms and biomarker development from global ESI-MS-MS analyses of cell lysates. *Anal. Chem.*, 72:2475-2481.

Yamada, M.; Asaoka, S.; Saier, Jr., M. H. & Yamada, Y. (1993). Characterization of the gcd gene from Escherichia coli K-12 W3110 and regulation of its expression. *J. Bacteriol.*, 175, 568-571.

Yates, J. R. (1998). Mass Spectrometry and the Age of the Proteome. *J. Mass Spectrom.*, *33*, 1-19.

Zheng, S.; Schneider, K. A.; Barder, T. J.& Lubman, D. M. (2003). Two-dimensional liquid chromatography protein expression mapping for differential proteomic analysis of normal and O157:H7 Escherichia coli. *BioTechniques*, *35*, 1202-1212.

10

The Use of Mass Spectrometry for Characterization of Fungal Secretomes

Eduardo Callegari[1] and Mario Navarrete[2]
[1]Division of Basic Biomedical Sciences, Sanford School of Medicine, University of South Dakota
[2]Facultad de Ciencias Biológicas, Universidad Andres Bello, Santiago
[1]USA
[2]Chile

1. Introduction

Filamentous fungi are microorganisms with a great capacity of produce and export enzymes to the extracellular media. These enzymes have been studied from different point of views and many properties have been attributed, some of them with biotechnological applications. A traditional way of studying these enzymes has been through purification, characterization and sequencing of individual enzymes. This methodology has shown, so far, remarkable properties that help to understand biological phenomena. For example, the enzyme acetyl xylan esterase II is produced by the lignocellulolytic fungus *Penicillium purpurogenm* especially when the carbon source is acetylated xylan. Under optimal conditions this enzyme was purified and sequenced (Gutierrez et al. 1998) and then its structure was elucidated from its crystal with high resolution (Ghosh et al. 2001). In a later work the expression of this enzyme was evaluated in culture supernatants coming from the fungus grown on different carbon sources where the glucose was a condition of repression for this enzyme (Chavez et al. 2004). New questions can be made for this enzyme that participates in a complex process of degradation of xylan, one of the components of hemicelluloses, which degradation is a key step for later production of chemicals and bio fuels. Since this enzyme is an isoenzyme, is it possible that more isoenzymes can be expressed, under what conditions could this be? Is it possible that more isoforms of this isoenzyme are there? Is this the only enzyme varying when the carbon source changes? What other proteins are varying? And is it possible that this enzyme is part of multienzyme complexes? Theses new questions arise when one realizes the complexities of the proteome. In the late nineties it was propose, arguing that it is unlikely that an enzyme is alone and that it is very possible that a protein could be interacting with ten other proteins approximately(Alberts 1998). Given the emergence of new technologies that allowed the improvement of protein separation, genome sequencing and mass spectrometry, it has been possible to address these many questions from the point of view of proteomics. Extensive has been the use of proteomics in comparative studies in biomedicine and the future of this technologies in the solution of several diseases is promising (Plymoth and Hainaut 2011). These technologies are available to be applied in several processes and lately have been used in fungi. The

many alternatives that proteomics offer makes that there is not an established protocol and that there are different answer depending of the proteomic approach.
Filamentous fungi are able to grow in different carbon sources; the ability to sense the environment and then change protein expression inside the cell and in the extracellular media, makes this system very attractive for the study and understanding of protein expression (Aro, Pakula, and Penttila 2005). In this chapter we will focus in the use of these technologies in the study of the extracellular portion of fungi called the secretome.

2. The fungal secretome

The secretome is a portion of the cellular proteome which include all secreted proteins, either anchored to cell surface or in the extracellular milieu, and the proteins involved in secretory pathway (Tjalsma et al. 2000). The protein composition of that subgroup of proteins is susceptible to stimuli from the environment that produces drastic changes in protein profile. Among the principal factors that influences protein pattern in fungal secretomes are: carbon source (Paper et al. 2007) and liquid versus solid medium (Oda et al. 2006). The characterization of the secretome is very complex, the composition and size of it will depend on the conditions of growth and the tools employed. Carbon source is able to modify the size of a secretome, for example glucose exerts a strong repression over the hemicellulases of *Aspergilli, Thrichoderma* and *Penicillia* (Chavez, Bull, and Eyzaguirre 2006).

3. Displaying the secretome

Although the study of protein biochemistry is not a new area of research, it is proteomics; this integrates techniques of protein biochemistry introducing new techniques of protein separation and sequencing. Protein biochemistry focus on the study of structure and function of proteins involving the study of aminoacid sequence, structure determination and modeling in order to determine how structure rules the function. Biophysics and enzymologist make analysis of this kind to study single proteins, one at a time. Proteomics, on the other hand worries about the study of multiprotein systems focusing on the interaction of multiple and different proteins as part of a connected network. Analysis are directed to complex mixtures and the identification is not through the complete sequence but through partial sequences held by growing databases (Liebler 2002). The analysis of proteomes requires tools different from the employed on genetic analysis (Romiti 2006). At the present time there are several alternatives to separate the proteins from a secretome, being two-dimensional electrophoresis coupled to mass spectrometry the most widely used. An alternative to this technique is the gel free analysis using mass spectrometry. Here we will describe several cases from bibliography where the different approaches were applied and different results were obtained. A proteomic study should provide any of the following information regarding the identification of proteins, differential expression of enzymes, post-translational modifications, relative abundance and the possible association of enzymes producing high molecular weight multienzyme complexes.(Griffin, Goodlett, and Aebersold 2001).

4. The case of *Penicillium purpurogenum*

The fungus *Penicillium purpurogenum* has a great ability of adaptation to different environments. It is thought that this is due to changes in the protein profile allowing the

fungus to degrade more efficiently the hemicelluloses. An analysis of the secretome was performed using proteomics tools. The first approach was to grow the fungus in the following carbon sources: acetylated xylan, sugar beet pulp, pectin and fructose. Acetylated xylan is polyacetylated by chemical methods starting from oat spelt xylan (sigma)(Johnson et al. 1988) and it has been observed that is a powerful inducer of acetyl xylan esterase (Egana et al. 1996). Sugar beet pulp is an important agricultural residue in Chile and is composed by 50% pectin and 20% cellulose (Saulnier. and Thibault. 1999). Pectin was obtained from citrics (sigma). After four days of growth these supernatants coming from carbon sources composed of complex polysaccharides were submitted to two-dimensional electrophoresis and compared with a supernatant coming from a culture containing fructose as carbon source. The first two-dimensional electrophoresis were performed with pH 3-10 strips for the first dimension, where all the spots were located to the acidic zone of the gel, a feature observed in fungal secretomes (Bouws, Wattenberg, and Zorn 2008). After that it was decided to use pH 4-7 strips resulting in better resolution gels (Figure 1) with spots of different molecular weights ranging from 20 kDa to 200 kDa. At first sight it can be noted in the gels stained with Sypro ruby, that complex carbon sources like acetylated xylan, sugar beet pulp and pectin induce the fungus to produce and secrete a great variety of enzymes unlike fructose that produces a pattern with a few number of spots. The analysis of the spots was made by tandem mass spectrometry using a Q-TOF instrument. Experimental data produced by the instrument was submitted to MASCOT, where a search was performed in databases containing fungal genomes. Among the proteins identified there are several enzymes involved in the degradation of the substrate (Navarrete et al, submitted for publication) and other enzymes that are common for the complex polysaccharides. The effect of fructose is completely different, its behavior is very similar to the effect of catabolite repression that glucose exerts on filamentous fungi (Chavez, Bull, and Eyzaguirre 2006). It is evident that the carbon sources acetylated xylan and sugar beet pulp are very different from the point of view of the chemical composition; therefore a differential expression of hemicellulases is expected. There are some software available with the ability to compare gels and to establish differences in protein pattern like is the case of PDQuest (Bio-Rad). In two-dimensional electrophoresis is possible to observe that some spots are located very close to each other sharing molecular weight but differing pI. These spots generally correspond to the same protein with different post translational modifications (PTMs) and is usually observed in fungi as is the case for *Penicillium chrysogenum* (Jami et al. 2010). This particular phenomenon observed is due to nature of this kind of electrophoresis and to the resolution that is possible to achieve when the preparation of samples is adequate. This technique requires of good quality sample and denaturant conditions. Concentration of urea and thiourea are usually high in buffers previous to the first dimension. After isoelectric focusing, strips are submitted to two equilibration steps with buffer containing strong reducing agents like dithiotreitol (DTT) or tributylphosphine(TBP), and a second buffer containing alkylating agent usually iodoacetamide (IAA) (Simpson 2003). The combination of these chemicals produces denaturation of proteins to improve resolution. The obvious consequence is that under these conditions is not possible to measure enzyme activity and there is no possibility to know about protein-protein interactions.
Although the differences in protein expression are very clear in *Penicillium purpurogenum* grown on different carbon sources, in some cases the difference could be more difficult to appreciate and therefore other technologies must be applied. In order to make two-dimensional electrophoresis more reproducible it was developed the differential gel

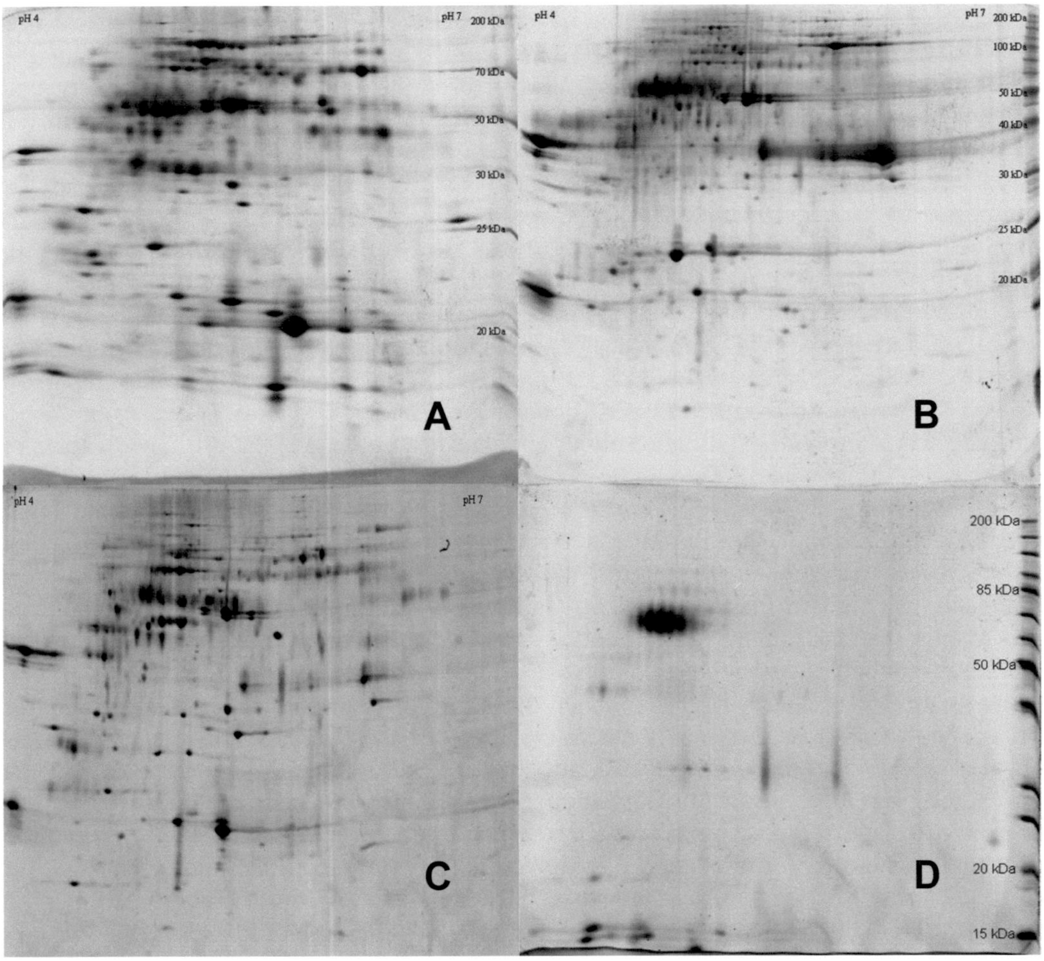

Fig. 1. Differential expression of hemicellulases secreted by the fungus *Penicillium purpurogenum*. The proteins of the different supernatants were separated by two-dimensional electrophoresis. The gels were stained with Sypro ruby. Proteins were extracted from supernatants coming from cultures using acetylated xylan (A), sugar beet pulp (B), pectin (C) and fructose (D). In a first dimension proteins were submitted to an isoelectric focusing using 17 cm pH 4-7 range strips. The second dimension corresponds to SDS-PAGE of 20 x 20 cm^2. The spots from gels were cut and analyzed by tandem mass spectrometry.

electrophoresis, which introduces the use of the fluorescent dyes Cy2, Cy3 and Cy5. In this technique two protein samples are pre labeled, then mixed and loaded in same gel (Unlu, Morgan, and Minden 1997). This technique was applied in *Aspergillus fumigatus* to evaluate metabolic changes under hypoxia (Vodisch et al. 2011). This was also applied to *P. purpurogenum* to analyze two conditions: acetylated xylan and sugar beet pulp. It is helpful to examine one gel where the two different groups of proteins are displayed in different colors (figure 2) where green is assigned to sugar beet pulp, red to acetylated xylan and in orange to yellow are the spots present in both conditions. The analysis by mass spectrometry of the spots and DIGE allowed us to clearly establish for example that β-xylosidase is present exclusively in acetylated xylan, and that pectate lyase is specific for

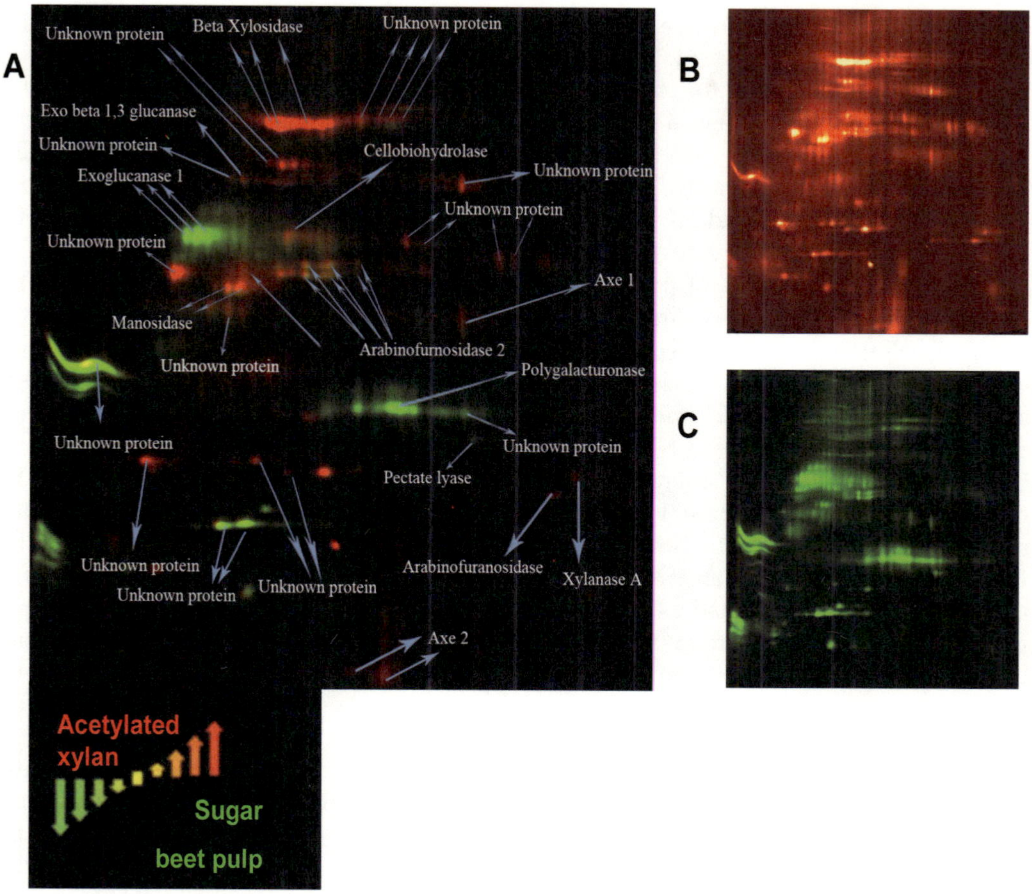

Fig. 2. Differential gel electrophoresis (DIGE) of the secretome of *Peniccilium purpurogenum* grown on different carbon source. Supernatant coming from acetylated xylan is labeled with Cy3 and the supernatant coming from sugar beet pulp is labeled with Cy 5, both supernatant are mixed together and loaded to the same 17 cm pH4-7 strip for isoelectric focusing. The second dimension is an SDS-PAGE of 20x20 cm^2. Images are acquired by a FX pro phosphor imager. Image B represents the protein pattern under acetylated xylan labeled with the chromophore Cy3, and image C is the protein pattern for sugar beet pulp secretome labeled with Cy5. The fusion of B and C result in image A. The most representative spots were excised from the gel and identified by tandem mass spectrometry (A). The most representative identities are marked in 1A.

sugar beet pulp. Mannosidases and arabinofuranosidase 2 is found in both conditions as indicated by the color. It is also possible to state that exoglucanase I present in sugar beet pulp and β-xylosidase present in acetylated xylan are submitted to post translational modifications producing the phenomenon of spot train since all of the spot in the train have the same identity.

So far we now know that the chemical composition of the carbon source produces differential expression of enzymes, and that the carbon source induces the expression of enzyme specific for the carbon source degradation. We can also assume that some post translational modifications could be occurring in this fungus like phosphorylations that modify pI, but more studies are needed. But a proteome is even more complex and other phenomenon could be present. The modulation of enzyme activity could be affected by protein-protein interaction for enzymes involved in xylan degradation, such is the case of *Fibrobacter succinogenes* where a xylanase is able to interact with an acetyl xylan esterase. This interaction generates synergy between these two activities (Kam et al. 2005). For another acetyl xylan esterase in *Termitomyces clypeatus*, it has been shown that the interaction with other enzymes provide a higher resistance to chemical denaturing agents that the resistance that offer the enzyme alone(Mukhopadhyay et al. 2003).

The most studied example of protein interaction involved in cellulose degradation is the cellulosome. These are structures formed by different proteins and have the ability to degrade cellulose and also hemicellulose due to the presence of all of the necessary enzymes. Their main characteristic is the presence of scaffolding proteins, which have been sequenced and very well studied, whose function is to keep the protein structure united(Ding et al. 1999).

So, this demonstrates that interactions are in direct benefit of the producer fungus. This experimental evidence along with the fact that hemicelluloses are chemically complex needing several enzyme activities acting on the same substrates makes suitable the idea that in fungi could be more than two proteins interacting. An analysis of the secretome of *P. purpurogenum* was performed using blue native (Wittig, Braun, and Schagger 2006). In this technique proteins are submitted to a gel in native conditions without interrupting protein interactions. In this case the samples used for this kind of analysis were culture supernatants coming from acetylated xylan, sugar beet pulp, corn cob and glucose. The sds is replaced by coomassie blue g-250 that is able to add negative charges to the proteins without denature the protein sample. Under these conditions is possible to do a zymogram over the gel, in this case we used methyl umbelliferyl acetate as substrate to evaluate acetyl esterase activities (Figure 3,I). This evidence shows that high molecular weight proteins are responsible for the activities, and according to two-dimensional electrophoresis in figure 1, the highest protein is about 200kDa. In figure 3 protein bands with enzyme activities are over 250 kDa and are present in the three complex polysaccharides. In this case is evident the catabolite repression that glucose exerts.

The pattern in figure 3,I shows that active high molecular weight complexes are present in the different conditions in changing molecular weights. This could be indicating that the complexes could be integrated by different subunits with different activities. This was evaluated by the use of antibodies that were used to co-immune precipitate the different complexes and by western identify subunits (Gonzalez-Vogel et al. 2011). A purification of one complex was performed by the used of different columns until a pure complex was obtained (Figure 3.II). The complex was submitted to SDS-PAGE and the subunits were identified by mass spectrometry. This approach has been used to analyze protein complexes in bacteria like *Escherichia coli* (Lasserre et al. 2006) and *Helicobacter pylori* (Pyndiah et al. 2007).

The discovery of protein complexes in *P. purpurogenum* harboring all of the necessary enzyme to degrade hemicellulose brings another point of view for the understanding of the complex process of degradation of lignocellulose by filamentous fungus. The interactions between subunits are strong and specific since they were not affected by treatment with high concentration of ammonium sulfate (Gonzalez-Vogel et al. 2011) still, new questions arise. The interactions could be mediated by direct protein-protein interactions, by the glycosylations, might be mediated by extracellular polysaccharides (Iwashita 2002) by

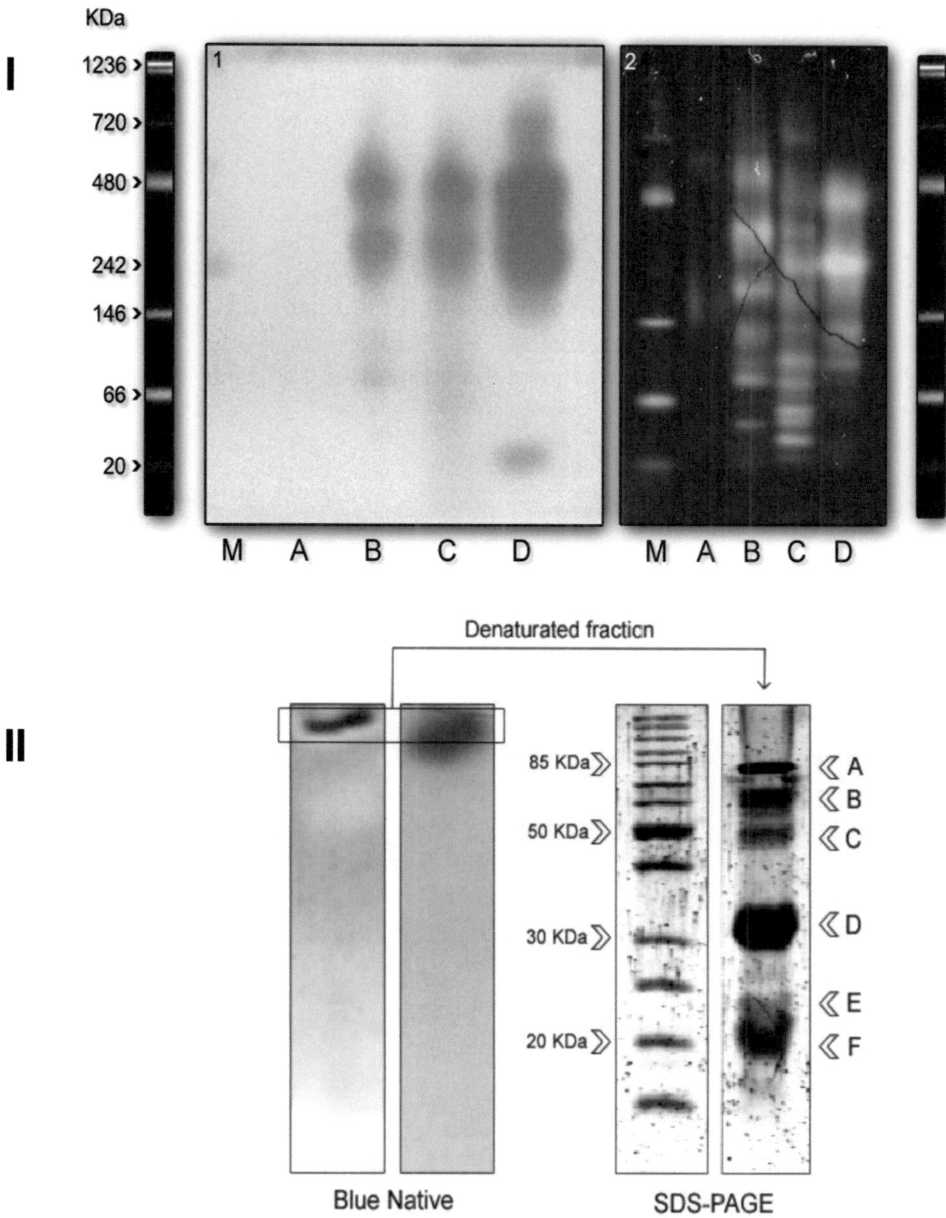

Fig. 3. Analysis of the secretome of *Penicillium purpurogenum* using Blue Native. Supernatants coming from cultures on different carbon sources were separated in native gels where acetyl esterase activity was measured (I,1) and the total protein pattern was stained with Sypro ruby (I,2). Lanes in I are: Molecular weight standard (M), supernatants coming from glucose (A), sugar beet pulp (B), corn cob (C) and acetylated xylan (D). Blue native of a pure complex from culture coming on sugar beet pulp. Sypro ruby staining indicates the purity of the fraction in the left lane, then in the right lane the zymogram indicates that is catalytically active. By SDS-PAGE the subunits A, B, C, D, E and F were separated and then identified by tandem mass spectrometry.

scaffolding proteins like dockerines in cellulosomes (Levasseur et al. 2004) or by core proteins (Bayer et al. 1998). All of these possibilities can be investigated by the point of view of proteomics that help to understand the fine mechanism of regulation in eukaryotes.

5. Analysis of glycosylations

Figures 1 and 2 show that is evident that post translational modifications are occurring in *P. purpurogenum* and also is evident that happens in other fungi (Bouws, Wattenberg, and Zorn 2008). This matter is very difficult to confront since the complexity of multienzyme systems is already difficult. Nevertheless, post translational modifications have been detected in cellulases of *Trichoderma reesei* (Fryksdale et al. 2002). In this study the enzymes were deglycosylated previous to two-dimensional electrophoresis; this made changes in protein pattern and also improved protein identification by mass spectrometry, because large glycosylation block trypsin.

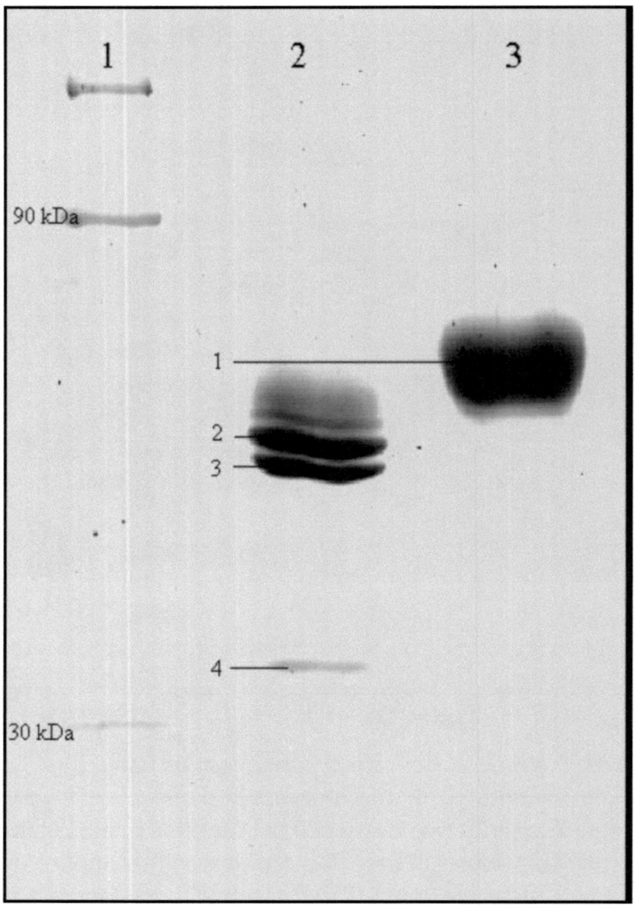

Fig. 4. Analysis of glycosylations in the secretome of *P. purpurogenum* under catabolite repression. The supernatant coming from glucose was loaded in lane 3. In lane 2 the same supernatant was treated with PNGase F. The gel was stained with Sypro ruby and the bands were identified by mass spectrometry. The identities are in table 1.

Band nº	Protein	Microorganism	GI number	Mowse score
1	FAD dependant oxygenase	*Aspergillus flavus*	238491442	120
2	FAD dependant oxygenase	*Aspergillus flavus*	238491442	129
3	FAD dependant oxygenase	*Aspergillus flavus*	238491442	152
4	PNGase F	*Flavobacterium Meningosepticum*	157833480	643

Table 1. List of proteins identified coming from the bands in figure 4.

In the case of *P. purpurogenum* we have started to deal with this problem. We have taken the sample of protein coming from a culture grown on glucose. Glucose exerts a strong catabolite repression producing a particular protein pattern in SDS-PAGE (Figure 4, lane 3). Only one major band is detected, this makes it suitable for the study of PTMs because of the simplicity of this secretome. The secretome was treated with PNGase F and the result is observed in figure 4, lane 2. The major band is not present but a smearing. All bands were identified by mass spectrometry; bands 1, 2, and 3 are the same protein, demonstrating that the difference in molecular weight is due to glycosylation. In this case, no glycosylation was sequenced by mass spectrometry, but by the use of mass spectrometry coupled to SDS-PAGE we were able to identify the same protein suffering molecular weight changes due to treatment with PNGase. This evidence is the first step for glycosylation occurring in this fungus, and is probably occurring in other carbon sources as well.

6. Results in other fungi

Studies on ascomycetes' secretomes identifying enzymes related to cellulose and hemicellulose degradation have been reported. Medina et al. (2004) analyzed the secretome of *Aspergillus flavus*. Utilizing 2D electrophoresis and MALDI-TOF mass spectrometry the authors attempted to identify the enzymes secreted by the fungus by comparing a rich medium (potato dextrose) with rutin (a glycoside) as sole carbon sources. They were able to identify a rather low number of spots (22 out of more than 100) from the gel (Medina, Kiernan, and Francisco 2004). Limiting factors in the identification were the lack of the genome sequence of this fungus, the presence of glycosylation (a common occurrence in extracellular fungal enzymes; (Peberdy 1994))which affects detection by MALDI-TOF, or the lack of similar sequences from other organisms in the databases. In a later publication (Medina et al. 2005), by incorporating the nanoLC-MS/MS technique, which does not have the limitations found in MALDI-TOF, they were able to identify 51 proteins. The extracellular proteome of *Aspergillus oryzae* was analyzed by comparing fungal cultures in liquid and solid media utilizing wheat bran as carbon source. By means of 1D and 2D electrophoresis and MALDI-TOF, plus the availability of expressed sequence tags from the sequenced genome, they were able to identify 29 proteins analyzing 85 spots from the solid culture and 110 from the liquid culture. An important finding is the difference in expression of proteins when comparing both media: in solid cultures the fungus secretes more protein and a larger variety of them(Oda et al. 2006).

Differences have also been found in the secretome when comparing solid and liquid cultures of *Aspergillus oryzae* grown on ground wheat (Te Biesebeke et al. 2006).
The secretome of some basidiomycetes active in lignocellulose biodegradation has also been reported. The secretome of the white-rot fungus *Phanerochaete chrysosporium* was grown on cellulose and have been compared with the open reading frames obtained from the sequenced genome. By means of liquid chromatography coupled with in-tandem mass spectrometry they have identified 32 glycosyl hydrolases out of 166 predicted, suggesting that probably many of these enzymes are expressed under different culture conditions (Wymelenberg et al. 2005). More recently, the same group (Vanden Wymelenberg et al. 2009) has extended this work reporting a transcriptome and secretome analyses of *P. chrysosporium* grown under defined ligninolytic and cellulolytic conditions so as to have a more complete view of gene expression. A total of 545 genes could be identified; however, 190 up regulated genes were predicted to encode proteins of unknown function, requiring further biochemical work for their proper identification. Additional work on the secretome of this fungus has been performed (Abbas et al. 2005) and (Ravalason et al. 2008). The first group used red oak wood chips as carbon source and could identify 16 proteins from 2D gels by LC/MS/MS sequencing, most of them acting on cellulose and hemicelluloses. Ravalason et al. (2008) grew the fungus on black pine (softwood) wood chips and identified seven glycosyl hydrolases.
The genome of the brown-rot fungus *Postia placenta* was analyzed and compared to that of the white-rot *P. chrysosporium*. Considerable differences were registered, indicating a much lower number of genes of cellulases in the brown-rot. White rot fungi degrade all components of the plant cell walls, while brown-rot can modify lignin but have a limited capacity to hydrolyze the polysaccharides of lignocellulose and have developed novel mechanisms for their degradation. An analysis of the secretome of *P. plancenta* grown on aspen wood, cellulose or cotton shows, among other enzymes, the presence of various hemicellulases (Martinez et al. 2009).
The state of the art on fungal secretome studies from both ascomycetes and basidiomycetes was reviewed (Bouws, Wattenberg, and Zorn 2008). They stress the fact that many more sequenced genomes and secretome analyses are necessary for a full understanding of the strategies used by fungi in extracellular biocatalysis and for a more efficient utilization of biomass in industrial applications. Although the study of fungal secretomes is increasing, many of the reports are still of a rather preliminary nature. They tend to be descriptive with still a limited discussion of the biological significance of the findings. This is an emerging area of research, which has to overcome certain limitations. This is due in part to the still low availability of sequenced and annotated fungal genomes and to a lack of knowledge of the function of many of the proteins deduced from the genome's open reading frames, since biochemical studies of these proteins lags far behind. However, despite of these limitations, a study of the proteome (and secretome) is essential for the understanding of numerous biological functions at the molecular level.

7. Conclusion

Filamentous fungi posses a very complex and fine mechanism of regulation. The application of proteomic methods can expose the consequences of these regulation. Through the optic of the fungus *Penicillium purpurogenum*, we have seen all of the possible mechanism by separating and identifying enzymes. In present time the number of proteomic studies

involving filamentous fungi are low but increasing. Comprehensive proteomic studies should consider PTMs and protein-protein interactions since these modifications affect the behavior of multienzyme systems. Two-dimensional electrophoresis is a powerful tool that allows to display the secretome where is possible to see the evidence of post translational modifications, but, by the other hand gel free analysis allows a more wide identification of proteins including those that do not enter in gels and the ones in low quantities.
The evidence presented here show that a combination of both a approaches should be use in order to get a wider picture of the secretome.

8. References

Abbas, A., H. Koc, F. Liu, and M. Tien. 2005. "Fungal degradation of wood: initial proteomic analysis of extracellular proteins of Phanerochaete chrysosporium grown on oak substrate." *Curr Genet* no. 47 (1):49-56. doi: 10.1007/s00294-004-0550-4.

Alberts, B. 1998. "The cell as a collection of protein machines: preparing the next generation of molecular biologists." *Cell* no. 92 (3):291-4. doi: S0092-8674(00)80922-8 [pii].

Aro, N., T. Pakula, and M. Penttila. 2005. "Transcriptional regulation of plant cell wall degradation by filamentous fungi." *FEMS Microbiol Rev* no. 29 (4):719-39. doi: S0168-6445(04)00092-0 [pii] 10.1016/j.femsre.2004.11.006.

Bayer, E. A., H. Chanzy, R. Lamed, and Y. Shoham. 1998. 'Cellulose, cellulases and cellulosomes." *Curr Opin Struct Biol* no. 8 (5):548-57. doi: S0959-440X(98)80143-7 [pii].

Bouws, H., A. Wattenberg, and H. Zorn. 2008. "Fungal secretomes--nature's toolbox for white biotechnology." *Appl Microbiol Biotechnol* no. 80 (3):381-8. doi: 10.1007/s00253-008-1572-5.

Chavez, R., P. Bull, and J. Eyzaguirre. 2006. "The xylanolytic enzyme system from the genus Penicillium." *J Biotechnol* no. 123 (4):413-33. doi: S0168-1656(06)00082-4 [pii] 10.1016/j.jbiotec.2005.12.036.

Chavez, R., K. Schachter, C. Navarro, A. Peirano, P. Bull, and J. Eyzaguirre. 2004. "The acetyl xylan esterase II gene from Penicillium purpurogenum is differentially expressed in several carbon sources, and tightly regulated by pH." *Biol Res* no. 37 (1):107-13.

Ding, S. Y., E. A. Bayer, D. Steiner, Y. Shoham, and R. Lamed. 1999. "A novel cellulosomal scaffoldin from Acetivibrio cellulolyticus that contains a family 9 glycosyl hydrolase." *J Bacteriol* no. 181 (21):6720-9.

Egana, L., R. Gutierrez, V. Caputo, A. Peirano, J. Steiner, and J. Eyzaguirre. 1996. "Purification and characterization of two acetyl xylan esterases from Penicillium purpurogenum." *Biotechnol Appl Biochem* no. 24 (Pt 1):33-9.

Fryksdale, B. G., P. T. Jedrzejewski, D. L. Wong, A. L. Gaertner, and B. S. Miller. 2002. "Impact of deglycosylation methods on two-dimensional gel electrophoresis and matrix assisted laser desorption/ionization-time of flight-mass spectrometry for proteomic analysis." *Electrophoresis* no. 23 (14):2184-93. doi: 10.1002/1522-2683(200207)23:14<2184::AID-ELPS2184>3.0.CO;2-1.

Ghosh, D., M. Sawicki, P. Lala, M. Erman, W. Pangborn, J. Eyzaguirre, R. Gutierrez, H. Jornvall, and D. J. Thiel. 2001. "Multiple conformations of catalytic serine and histidine in acetylxylan esterase at 0.90 A." *J Biol Chem* no. 276 (14):11159-66. doi: 10.1074/jbc.M008831200 [pii].

Gonzalez-Vogel, A., J. Eyzaguirre, G. Oleas, E. Callegari, and M. Navarrete. 2011. "Proteomic analysis in non-denaturing condition of the secretome reveals the presence of multienzyme complexes in Penicillium purpurogenum." *Appl Microbiol Biotechnol* no. 89 (1):145-55. doi: 10.1007/s00253-010-2953-0.

Griffin, T. J., D. R. Goodlett, and R. Aebersold. 2001. "Advances in proteome analysis by mass spectrometry." *Curr Opin Biotechnol* no. 12 (6):607-12. doi: S0958-1669(01)00268-3 [pii].

Gutierrez, R., E. Cederlund, L. Hjelmqvist, A. Peirano, F. Herrera, D. Ghosh, W. Duax, H. Jornvall, and J. Eyzaguirre. 1998. "Acetyl xylan esterase II from Penicillium purpurogenum is similar to an esterase from Trichoderma reesei but lacks a cellulose binding domain." *FEBS Lett* no. 423 (1):35-8. doi: S0014-5793(98)00055-6 [pii].

Iwashita, K. 2002. "Recent studies of protein secretion by filamentous fungi." *J Biosci Bioeng* no. 94 (6):530-5. doi: S1389-1723(02)80191-8 [pii].

Jami, M. S., C. Garcia-Estrada, C. Barreiro, A. A. Cuadrado, Z. Salehi-Najafabadi, and J. F. Martin. 2010. "The Penicillium chrysogenum extracellular proteome. Conversion from a food-rotting strain to a versatile cell factory for white biotechnology." *Mol Cell Proteomics* no. 9 (12):2729-44. doi: 10.1074/mcp.M110.001412.

Johnson, K.G., B.A. Harrisona, H. Schneidera, C.R. MacKenziea, and J.D. Fontanaa. 1988. "Xylan-hydrolysing enzymes from Streptomyces spp." *Enzyme and Microbial Technology* no. 10 (7):403-409 doi: 10.1016/0141-0229(88)90034-8.

Kam, D. K., H. S. Jun, J. K. Ha, G. D. Inglis, and C. W. Forsberg. 2005. "Characteristics of adjacent family 6 acetylxylan esterases from Fibrobacter succinogenes and the interaction with the Xyn10E xylanase in hydrolysis of acetylated xylan." *Can J Microbiol* no. 51 (10):821-32. doi: 10.1139/w05-074.

Lasserre, J. P., E. Beyne, S. Pyndiah, D. Lapaillerie, S. Claverol, and M. Bonneu. 2006. "A complexomic study of Escherichia coli using two-dimensional blue native/SDS polyacrylamide gel electrophoresis." *Electrophoresis* no. 27 (16):3306-21. doi: 10.1002/elps.200500912.

Levasseur, A., S. Pages, H. P. Fierobe, D. Navarro, P. Punt, J. P. Belaich, M. Asther, and E. Record. 2004. "Design and production in Aspergillus niger of a chimeric protein associating a fungal feruloyl esterase and a clostridial dockerin domain." *Appl Environ Microbiol* no. 70 (12):6984-91. doi: 70/12/6984 [pii] 10.1128/AEM.70.12.6984-6991.2004.

Liebler, Daniel C. 2002. *Introduction to Proteomics, tools for the new biology*. Totowa, New Jersey.: Humana Press.

Martinez, D., J. Challacombe, I. Morgenstern, D. Hibbett, M. Schmoll, C. P. Kubicek, P. Ferreira, F. J. Ruiz-Duenas, A. T. Martinez, P. Kersten, K. E. Hammel, A. Vanden Wymelenberg, J. Gaskell, E. Lindquist, G. Sabat, S. S. Bondurant, L. F. Larrondo, P. Canessa, R. Vicuna, J. Yadav, H. Doddapaneni, V. Subramanian, A. G. Pisabarro, J. L. Lavin, J. A. Oguiza, E. Master, B. Henrissat, P. M. Coutinho, P. Harris, J. K. Magnuson, S. E. Baker, K. Bruno, W. Kenealy, P. J. Hoegger, U. Kues, P. Ramaiya, S. Lucas, A. Salamov, H. Shapiro, H. Tu, C. L. Chee, M. Misra, G. Xie, S. Teter, D. Yaver, T. James, M. Mokrejs, M. Pospisek, I. V. Grigoriev, T. Brettin, D. Rokhsar, R. Berka, and D. Cullen. 2009. "Genome, transcriptome, and secretome analysis of wood decay fungus Postia placenta supports unique mechanisms of lignocellulose conversion." *Proc Natl Acad Sci U S A* no. 106 (6):1954-9. doi: 0809575106 [pii] 10.1073/pnas.0809575106.

Medina, M. L., P. A. Haynes, L. Breci, and W. A. Francisco. 2005. "Analysis of secreted proteins from Aspergillus flavus." *Proteomics* no. 5 (12):3153-61. doi: 10.1002/pmic.200401136.

Medina, M. L., U. A. Kiernan, and W. A. Francisco. 2004. "Proteomic analysis of rutin-induced secreted proteins from Aspergillus flavus." *Fungal Genet Biol* no. 41 (3):327-35. doi: 10.1016/j.fgb.2003.11.014 S108718450300210X [pii].

Mukhopadhyay, A., P. P. Hazra, T. Sengupta, R. Saha, R. Nandi, and S. Sengupta. 2003. "Protein-protein interaction conferring stability to an extracellular acetyl (xylan) esterase produced by Termitomyces clypeatus." *Biotechnol Prog* no. 19 (3):720-6. doi: 10.1021/bp0201307.

Oda, K., D. Kakizono, O. Yamada, H. Iefuji, O. Akita, and K. Iwashita. 2006. "Proteomic analysis of extracellular proteins from Aspergillus oryzae grown under submerged and solid-state culture conditions." *Appl Environ Microbiol* no. 72 (5):3448-57. doi: 72/5/3448 [pii] 10.1128/AEM.72.5.3448-3457.2006

Paper, J. M., J. S. Scott-Craig, N. D. Adhikari, C. A. Cuomo, and J. D. Walton. 2007. "Comparative proteomics of extracellular proteins in vitro and in planta from the pathogenic fungus Fusarium graminearum." *Proteomics* no. 7 (17):3171-83. doi: 10.1002/pmic.200700184.

Peberdy, J. F. 1994. "Protein secretion in filamentous fungi--trying to understand a highly productive black box." *Trends Biotechnol* no. 12 (2):50-7.

Plymoth, A., and P. Hainaut. 2011. "Proteomics beyond proteomics: toward clinical applications." *Curr Opin Oncol* no. 23 (1):77-82. doi: 10.1097/CCO.0b013e32834179c1.

Pyndiah, S., J. P. Lasserre, A. Menard, S. Claverol, V. Prouzet-Mauleon, F. Megraud, F. Zerbib, and M. Bonneu. 2007. "Two-dimensional blue native/SDS gel electrophoresis of multiprotein complexes from Helicobacter pylori." *Mol Cell Proteomics* no. 6 (2):193-206. doi: M600363-MCP200 [pii] 10.1074/mcp.M600363-MCP200.

Ravalason, H., G. Jan, D. Molle, M. Pasco, P. M. Coutinho, C. Lapierre, B. Pollet, F. Bertaud, M. Petit-Conil, S. Grisel, J. C. Sigoillot, M. Asther, and I. Herpoel-Gimbert. 2008. "Secretome analysis of Phanerochaete chrysosporium strain CIRM-BRFM41 grown on softwood." *Appl Microbiol Biotechnol* no. 80 (4):719-33. doi: 10.1007/s00253-008-1596-x.

Romiti, Monica. 2006. "Entrez Help." In, ed NCBI.

Saulnier., Luc, and Jean-Françoise Thibault. 1999. "Ferulic acid and diferulic acids as components of sugar-beet pectins and maize brand heteroxylans." *Journal of the science of food and agriculture* no. 79:396-402.

Simpson, Richard J. 2003. *Proteins and Proteomics: a laboratory manual*. New York: Cold Spring Harbor Laboratory Press.

Te Biesebeke, R., A. Boussier, N. van Biezen, C. A. van den Hondel, and P. J. Punt. 2006. "Identification of secreted proteins of Aspergillus oryzae associated with growth on solid cereal substrates." *J Biotechnol* no. 121 (4):482-5. doi: S0168-1656(05)00550-X [pii] 10.1016/j.jbiotec.2005.08.028.

Tjalsma, H., A. Bolhuis, J. D. Jongbloed, S. Bron, and J. M. van Dijl. 2000. "Signal peptide-dependent protein transport in Bacillus subtilis: a genome-based survey of the secretome." *Microbiol Mol Biol Rev* no. 64 (3):515-47.

Unlu, M., M. E. Morgan, and J. S. Minden. 1997. "Difference gel electrophoresis: a single gel method for detecting changes in protein extracts." *Electrophoresis* no. 18 (11):2071-7. doi: 10.1002/elps.1150181133.

Vanden Wymelenberg, A., J. Gaskell, M. Mozuch, P. Kersten, G. Sabat, D. Martinez, and D. Cullen. 2009. "Transcriptome and secretome analyses of Phanerochaete chrysosporium reveal complex patterns of gene expression." *Appl Environ Microbiol* no. 75 (12):4058-68. doi: AEM.00314-09 [pii] 10.1128/AEM.00314-09.

Vodisch, M., K. Scherlach, R. Winkler, C. Hertweck, H. P. Braun, M. Roth, H. Haas, E. R. Werner, A. A. Brakhage, and O. Kniemeyer. 2011. "Analysis of the Aspergillus fumigatus proteome reveals metabolic changes and the activation of the pseurotin A biosynthesis gene cluster in response to hypoxia." *J Proteome Res* no. 10 (5):2508-24. doi: 10.1021/pr1012812.

Wittig, I., H. P. Braun, and H. Schagger. 2006. "Blue native PAGE." *Nat Protoc* no. 1 (1):418-28. doi: nprot.2006.62 [pii] 10.1038/nprot.2006.62.

Wymelenberg, A. V., G. Sabat, D. Martinez, A. S. Rajangam, T. T. Teeri, J. Gaskell, P. J. Kersten, and D. Cullen. 2005. "The Phanerochaete chrysosporium secretome: database predictions and initial mass spectrometry peptide identifications in cellulose-grown medium." *J Biotechnol* no. 118 (1):17-34. doi: S0168-1656(05)00152-5 [pii] 10.1016/j.jbiotec.2005.03.010.

11

Strategies and Challenges in Measuring Protein Abundance Using Stable Isotope Labeling and Tandem Mass Spectrometry

Kolbrun Kristjansdottir, Satoe Takahashi,
Samuel L. Volchenboum and Stephen J. Kron
The University of Chicago, Chicago, IL
USA

1. Introduction

Mass spectrometry (MS) is a powerful method for identifying proteins, and modern mass spectrometers are capable of remarkable speed, resolution and sensitivity. A single tandem mass spectrometry experiment can now lead to the identification and quantitation of thousands of proteins down to sub-femtomolar concentrations. Tandem mass spectrometry experiments generally involve extraction of proteins from cells, biofluid, or tissue followed by digestion of proteins to peptides, separation of peptides on an HPLC, and direct injection into a mass spectrometer (LC-MS/MS). The mass spectrometer measures the mass of each peptide ion (MS1) and selected ions are fragmented (MS/MS or MS2). Mass and fragmentation spectra of each peptide are compared against predicted peptide fragmentation spectra from the known proteome by database search engines (reviewed in Aebersold and Mann 2003). LC-MS/MS instruments also record peptide ion intensities, offering the potential for direct measurement of peptide concentration and thereby protein abundance. However, the extent of ionization of peptides by electrospray ionization is dependent on peptide sequence and modification, elution conditions, complexity of the sample, and other factors. As a result, the absolute intensities of ions derived from non-identical peptides cannot provide accurate or direct quantitation. Approaches such as peptide ion chromatogram extraction and spectral counting have been developed to obtain relative quantitation of protein abundance (Ono et al. 2006; Fischer et al. 2006; Tang et al. 2006; Paoletti et al. 2006; Listgarten and Emili 2005; Wiener et al. 2004; Wang, Wu, Zeng, et al. 2006). Collectively termed "label-free" quantitation, these approaches require extensive analysis of reference samples and/or significant data redundancy, often requiring many hours of mass spectrometry time per sample. Although highly promising, label-free approaches remain impractical for users lacking access to dedicated mass spectrometry instrumentation and advanced informatic approaches.
Stable isotope labeling provides an attractive alternative to label-free approaches. Stable isotopes are sufficiently stable to be non-radioactive. They have equal numbers of protons as their parental element but they differ in mass by the difference in the number of neutrons. Carbon, hydrogen, oxygen, nitrogen and sulfur have two or more isotopes with measurable abundance in Nature. For example, carbon is found as the predominant "light" isotope ^{12}C

(98.89%), a stable "heavy" isotope of ^{13}C (1.11%) and a radioactive "heavy" isotope of ^{14}C (trace amounts) in Nature. Other stable isotopes relevant to protein mass spectrometry include Hydrogen ^{2}H (0.02%), also called deuterium, Nitrogen ^{15}N (0.37%), Oxygen ^{17}O (0.04%) and ^{18}O (0.02%), and Sulfur ^{33}S (0.76%), ^{34}S (4.29%), and ^{36}S (0.02%). Carbon and nitrogen are the most common atoms in peptides resulting in ^{13}C and ^{15}N being the predominant isotopes present in all naturally occurring proteins, and to a lesser extent, oxygen and sulfur isotopes. As a result, instead of each tryptic peptide injected into the mass spectrometer having a single mass, each peptide is represented as a collection of different masses in proportions that reflect the natural abundance of the stable isotopes. This collection of masses present in the mass spectrum is termed the "isotopic envelope" of a peptide (Fig. 3).

A stable isotope labeled peptide and its unlabeled counterpart have the same chemical formula and structure and thus nearly identical chemical properties, such that they are expected to elute together from reverse phase chromatography. Despite their similar chemical properties, the presence of the stable isotope facilitates independent assessment because of the mass difference. Combining the light (unlabeled) and heavy isotope labeled peptides in one sample allows for direct comparison of ion intensities. In principle, this offers highly accurate relative quantitation and avoids the need for significant data redundancy. With these and other advantages, stable isotope labeling would appear to satisfy the criteria for an ideal quantitative mass spectrometry strategy. However, challenges remain before stable isotope quantitation becomes a straightforward, robust, and reliable approach accessible both to non-experts and users of service laboratories. Here, we provide an up-to-date and critical review of stable isotope labeling methodologies, available software for data analysis, and emerging new applications of these powerful approaches.

2. Absolute versus relative quantitation

Stable isotope labeling can provide either absolute or relative quantitation. Absolute quantitation is obtained by comparing a known amount of added stable isotope labeled peptide and comparing directly with the unlabeled peptide counterpart, an expensive and laborious methodology for large proteomic studies. However, relative quantitation of proteins and peptides from complex samples can be performed by labeling the sample with stable isotope(s) and comparing to an unlabeled control. The levels are directly compared, providing a fold-change. Because the sample is compared to a control, changes due to the perturbation can be identified. Therefore, complex samples can be analyzed without any prior knowledge of the identity of peptides and proteins.

2.1 Using stable isotopes to achieve absolute quantitation

Stable isotopes can be incorporated into synthetic standards to obtain absolute quantitation. Isotope dilution and related approaches have been used in the small molecule field for decades (Baillie 1981). A known amount of stable isotope labeled analog of the compound of interest (internal standard) is spiked into a sample containing the unlabeled compound (Fig. 1). The intensity of the unlabeled molecule is compared directly to the intensity of the stable isotope labeled molecule, and the peak ratio is calculated. For optimal performance, a standard curve is generated from a range of concentrations of the internal standard. Some of the earliest peptide- and protein-based applications of mass spectrometry for tracking and

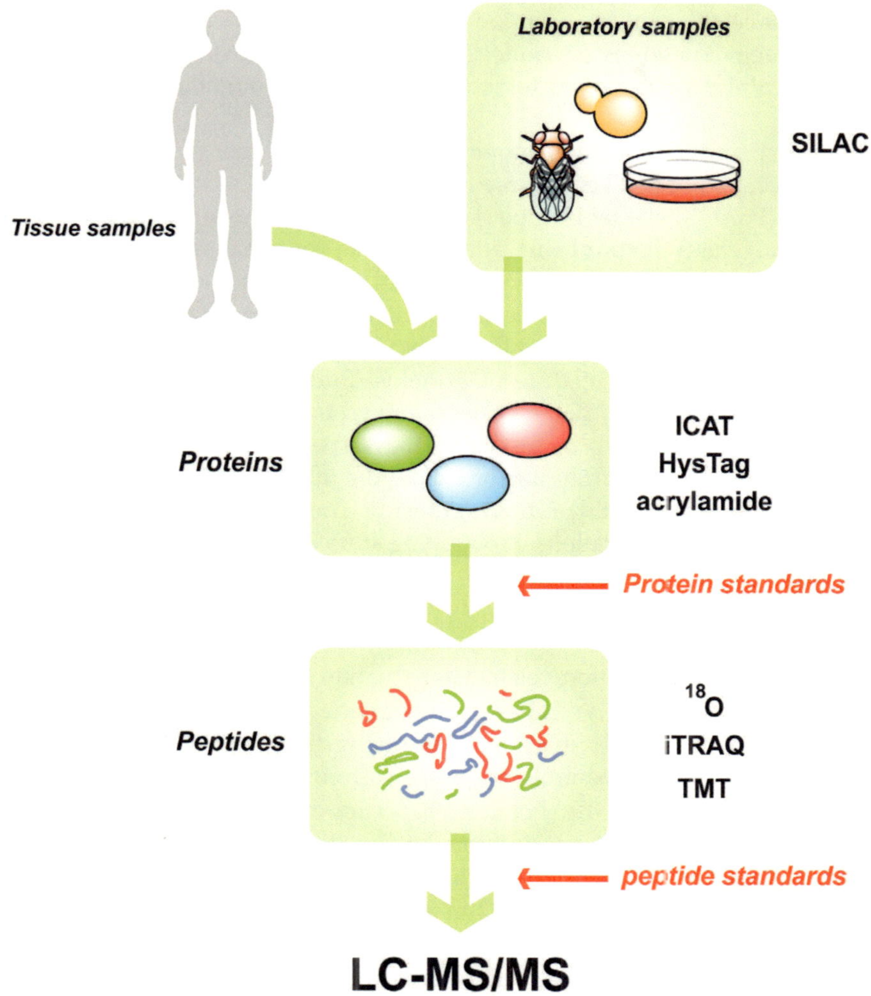

Fig. 1. Methods for stable isotope labeling. Stable isotope labeling for relative quantitation can occur through metabolic, chemical and enzymatic methods. Metabolic labeling is performed during cell culture growth or by feeding or growing organisms such as yeast and flies in stable isotope labeled food or media. An example of metabolic labeling is SILAC. Chemical labeling is possible at the protein level with methods such as ICAT, HysTag and acrylamide labeling and at the peptide level with iTRAQ and TMT. Finally, enzymatic labeling is performed at the peptide level by trypsin catalyzed ^{18}O labeling. Absolute quantitation involves spiking synthetic stable isotope labeled proteins or peptides (red arrows) into unlabeled samples.

quantitation were the use of enzymatically labeled peptides generated via trypsin ^{18}O-exchange (Desiderio and Kai 1983), peptides synthesized using ^{13}C, ^{2}H-labeled amino acids (Barr et al. 1996), and ^{15}N labeled peptide hormones (Kippen et al. 1997). To study pigments in the eye, a deuterium-containing peptide from rhodopsin was used as an internal peptide standard for determining the absolute amount present in rod outer segments (Barnidge et al. 2003). Taken to its logical extreme, it would be feasible to spike a sample with one or more heavy isotope labeled synthetic peptide reporters for every protein in the predicted proteome, a strategy known as Absolute Quantification (AQUA) (Gerber et al. 2003). This methodology can also be exploited to provide absolute quantitation of post-translational modifications.

Synthesizing or expressing stable isotope labeled proteins can generate several peptide standards that can be used even in fractionated samples. In Protein Standard Absolute Quantification (PSAQ), stable isotope labeled proteins are synthesized *in vitro* and purified to homogeneity before being added to the proteomic sample (Brun et al. 2007; Dupuis et al. 2008). Internal protein standards can also be obtained by metabolic labeling of organisms, such as *E. coli* (Hanke et al. 2008). Additionally, a single synthesized concatemer protein comprised of peptides from 20 proteins of interest (QconCAT) has been generated to quantify a mixture of proteins (Pratt et al. 2006; Johnson et al. 2008; Rivers et al. 2007; Beynon et al. 2005). Taken together, these studies show that the absolute quantitation of peptides and proteins using mass spectrometry is feasible (Brun et al. 2009). However, in order to synthesize or isolate the internal standard, the sequence and identity of the peptide or protein of interest must be known. The complexity of the sample is limited by practical considerations of obtaining sufficient numbers of internal standards for proteome coverage.

2.2 Relative quantitation

In contrast to absolute quantitation, relative quantitation requires no prior knowledge of the peptides and proteins. In a typical experiment, one sample is labeled with a stable isotope(s), while the other is left unlabeled. After perturbing or treating one of the samples, it is mixed with the untreated control and mass spectrometry analysis performed. Since the stable isotope does not change the chemical properties of the peptide, the intensities of the unlabeled and labeled ionized species can be directly compared and provide relative quantitation values between the samples. Although relative quantitation requires comparison to a control sample, it is not limited by sample complexity nor does it require prior knowledge of peptide identity. Because relative quantitation simply compares two (or more) samples, a wide range of stable isotope labeling methodologies can be used.

3. Stable isotope labeling methodologies for relative quantitation

Relative quantitation involves comparing unlabeled and labeled peptides or fragment ions. Quantitation can be performed at the MS1 or the MS2 level, depending on the nature of the stable isotope label. When labeling and quantitation occur at the MS1 level, the labeled sample is compared to an unlabeled (control) sample. Peptides are detected in the mass spectrometer as pairs, the heavy peptide shifted by the mass of the isotope(s). Comparison of light and heavy peptide peaks gives fold-difference or relative quantitation. For quantitation at the MS2 level, every sample, including the control, is labeled with an isobaric tag. The peptides co-elute and are undistinguishable at the MS1 level. However, fragmentation (MS2) of the peptide releases reporter ions that differ for each fraction and can be directly compared for relative quantitation. The method of labeling for MS1 level quantitation can occur via a metabolic, chemical or enzymatic process, but MS2 level quantitation is only feasible using chemical labeling. The method and timing of labeling is summarized in Fig. 1. Fig. 3 illustrates quantitation at MS1 and MS2 level.

3.1 Quantitation at the MS1 level

3.1.1 Metabolic labeling

Metabolic labeling for protein quantitation was demonstrated in yeast grown on commercial rich media derived from ^{15}N-enriched algal hydrolysate. The relative abundances of phosphopeptides in the light and heavy samples were then determined by MALDI mass spectrometry (Oda et al. 1999). Analogous approaches have been applied to a number of organisms including worms and flies, culminating with the metabolic labeling of a rat, accomplished through feeding with ^{15}N-enriched algae to produce tissue-specific internal standards for global quantitative proteomic analysis (Wu et al. 2004). While metabolic labeling with ^{15}N is inexpensive and simple to perform, the distribution of isotopic forms for each peptide depends on the amino acid composition, complicating quantitative analysis and manual validation.

Currently, the most widely used metabolic labeling approach for protein quantitation is stable isotope labeling with amino acids in cell culture (SILAC) (Ong, Kratchmarova, and Mann 2003; Ong et al. 2002; de Godoy et al. 2006; Amanchy et al. 2005). When cells are grown for several doublings in tissue culture with a stable isotope labeled form of an essential amino acid (e.g. lysine) as the sole source and at a small excess, this amino acid is incorporated into newly synthesized proteins until all are homogeneously labeled. Although any of the 20 naturally occurring amino acids could be used as a precursor for labeling, lysine, arginine and leucine are commonly used, with serine, glycine, histidine, methionine, valine, and tyrosine to a lesser extent (reviewed in (Beynon and Pratt 2005). The most common isotopes in SILAC are ^{13}C and ^{15}N, since they demonstrate less kinetic isotope effect than ^{2}H and do not change the elution profiles of labeled peptides in reverse phase HPLC chromatography (Zhang and Regnier 2002; Zhang et al. 2001; Zhang et al. 2002).

Trypsin is the most frequently used enzyme protease in proteomics, cleaving on the carboxyl-terminal side of lysine and arginine residues. Therefore, each non-C-terminal tryptic peptide is predicted to contain either a single carboxyl-terminal lysine or arginine. With a mass difference of 4 to 10 Da, due to labeling of the single terminal lysine or arginine, most pairs of peptides can be easily recognized by their offset envelopes of isotopic species (Fig. 3).

The advantages of SILAC using lysine and arginine as the labeled amino acids include the ease of complete labeling and that every peptide is labeled, with the exception of the C-terminal peptide. Although trypsin fails to cleave at some post-translationally modified lysine and arginine residues, this does not prevent quantitation. Stable isotope labeled amino acids (Cambridge Isotopes Laboratories) and several types of SILAC tissue culture media including DMEM, RPMI and IMEM (Thermo Scientific Pierce, Invitrogen) are commercially available.

SILAC is limited to organisms that can be grown on defined media. While this is straightforward for cell lines, bacteria, and yeast cells, it precludes most whole animal and patient studies. Super-SILAC, a method for quantitative proteomics of human tissues was recently described (Geiger et al. 2010). Here, unlabeled tissue samples are mixed with SILAC labeled human cell lines (Super-SILAC mix) and relative quantitation is performed. Examining several tissue samples mixed with the same internal control, the Super-SILAC mix, allows for relative comparison between different tumor samples. Similarly, stable isotope labeled mouse tissue (Mouse Express) is available from Cambridge Isotope Laboratories and can be used in combination with unlabeled mouse tissue for relative quantitation.

Finally, SILAC experiments usually consist of two samples, a control (heavy) and treatment (light) or *vice versa*. However, increased availability of labeled amino acids, now allows for comparisons of three (Blagoev et al. 2004; Andersen et al. 2005) to five samples (Molina et al. 2009).

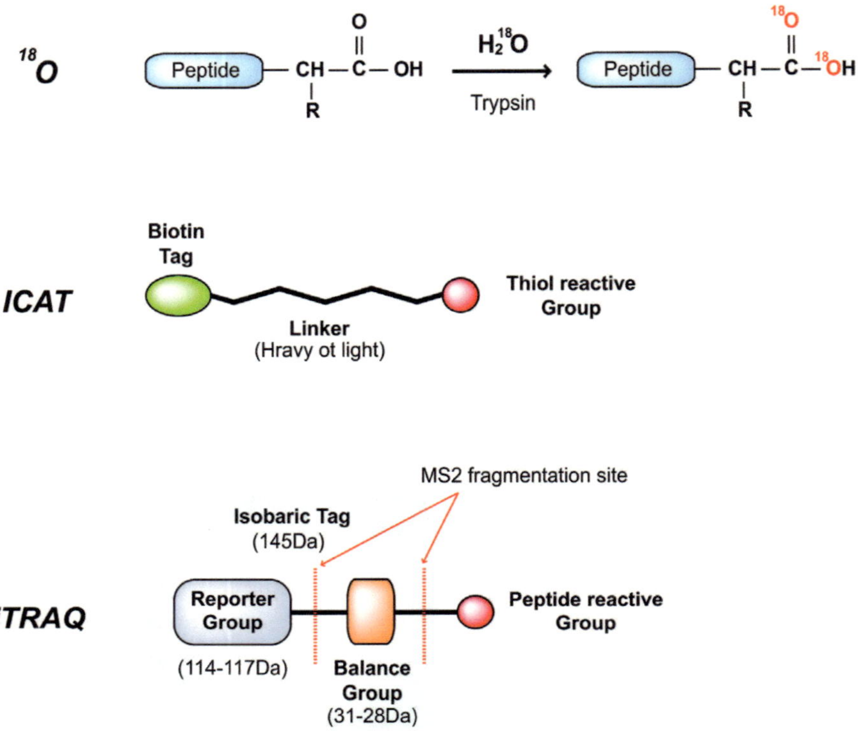

Fig. 2. Structure of ^{18}O, ICAT and iTRAQ labeling reagents. The figure shows the structure of several stable isotope labeling reagents. In ^{18}O labeling, tryptic preptides are incubated in $H_2{}^{18}O$ water and trypsin catalyzes an oxygen exchange reaction at the C-terminus of the peptide (two ^{18}O molecules incorporated). The sample is then mixed with an unlabeled sample and quantitation performed. In ICAT, a thiol reactive group reacts with the cysteine amino acids attaching the ICAT reagent, which includes a linker and biotin tag. Two samples are labeled, one with an ICAT reagent with a light linker, the other with a heavy linker. Biotinylated peptides are purified via the biotin tag and quantitation is performed. Finally, in iTRAQ, four (or eight) samples are labeled with isobaric reagents, all weighting 145 Da total but having different reporter and balance group. For example, a 114 Reporter and a 31 Da balance group or a 117 Da Reporter and 28 Da balance group all add up to 145 Da. The four labeled samples are mixed and fragmented together. The Reporter group is released and quantitation is performed by comparing the 114, 115, 116 and 117 peaks.

3.1.2 Chemical labeling

3.1.2.1 Isotope-coded affinity tag (ICAT)

One of the first commercialized stable isotope tagging reagents was isotope-coded affinity tag (ICAT) (Gygi et al. 1999). Since its introduction in 1999, the ICAT approach has been widely used (Gygi et al. 1999; Griffin et al. 2002; Zhou et al. 2002), and ICAT reagents are commercially available from Applied Biosystems. In ICAT, a pair of light and heavy reagents targets cysteine residues, adding a linker and a biotin tag for affinity purification (Fig. 2). While the linker region of the heavy reagent contains stable isotopes, the light reagent does not. Proteins are denatured, reduced and then labeled with heavy or light reagents, mixed and digested. The biotinylated peptides are purified using avidin affinity reagents, allowing for stringent washing that minimizes background binding. The main advantage of this method is that it leads to the enrichment of peptides containing cysteines, which are relatively rare, thereby significantly reducing the complexity of the peptide mixture and increasing the dynamic range of mass spectrometry analysis. But because only peptides and proteins containing cysteines are identified, the overall proteome coverage is low, resulting in less accurate quantitation. Finally, ICAT is limited to comparing two samples.

Several global quantitation experiments have been performed using the ICAT approach, such as a comparison of protein expression in yeast using ethanol or galactose as a carbon source (Gygi et al. 1999). Other ICAT studies include identification of proteins regulated by the Myc oncoprotein (Shiio et al. 2002) by comparing the protein expression patterns between myc-null and myc-expressing cells and identification of proteins regulated by interferon treatment in human liver cells (Yan et al. 2004).

3.1.2.2 Other cysteine-targeted methods

Several other methods have been developed for chemical labeling of cysteines, including HysTag (Olsen et al. 2004) and acrylamide labeling (Faca et al. 2006b). HysTag is a 10-mer derivatized peptide consisting of an affinity ligand (His6-tag), a tryptic cleavage site, an Ala-9 residue that contains either four (D_4) or no (D_0) deuterium atoms, and a thiol-reactive group that targets cysteines. The HysTag peptide is preserved in Lys-C digestion of proteins and allows subsequent charge-based selection of cysteine-containing peptides. To remove the HysTag, subsequent tryptic digestion reduces the labeling group to a dipeptide, which does not hinder effective MS2 fragmentation (Olsen et al. 2004). HysTag has many of the same advantages and disadvantages as ICAT.

Another method involves the alkylation of cysteines of intact proteins with acrylamide (Faca et al. 2006b). While cysteine alkylation with acrylamide *via* Michael addition is an undesired reaction that frequently occurs during polyacrylamide gel electrophoresis (Patterson 1994), several features make it a useful tagging approach for quantitative analysis with stable isotopes. First, because of its small size and hydrophilic nature, the acrylamide moiety does not introduce significant mass shift or charge changes in the protein and does not negatively affect protein solubility. Second, cysteine labeling is facile, allowing for complete labeling. Finally, the reagents are relatively inexpensive, making it practical to perform experiments starting with large amounts of protein as needed for extensive fractionation and in-depth analysis (Faca et al. 2006b). The acrylamide method does not include a cysteine peptide enrichment step. Since any peptide can be identified, protein coverage is increased, dynamic range is decreased. However, only cystein containing peptides are stable isotope labeled and can be quantified. protein coverage is increased, dynamic range decreased and quantitation is limited to cysteine containing peptides. A recent study used a combination of acrylamide label cysteines and succinic anhydride labeled lysines to increase the quantitative coverage (Wang et al. 2009). However, the mass shift is small (3 Da), resulting in frequent overlap between the isotopic envelopes of light and heavy peptides complicating analysis. Acrylamide labeling has been used extensively to study the proteome of serum in cancerous and non-cancerous samples (Faca et al. 2006a; Pitteri et al. 2008).

3.1.3 Enzymatic labeling - Trypsin-catalyzed ^{18}O labeling

An enzymatic method for stable isotope labeling, trypsin-catalyzed ^{18}O labeling, involves the transfer of ^{18}O from heavy water to the carboxyl terminal of peptides by an oxygen exchange reaction (Fig. 2) (Reynolds, Yao, and Fenselau 2002; Stewart, Thomson, and Figeys 2001; Heller et al. 2003; Yao et al. 2001; Mirgorodskaya et al. 2000). Trypsin is the most frequently used enzyme, though Lys-C and Arg-C are also capable of catalyzing this reaction. As trypsin digestion is the most common method of sample preparation before mass spectrometry, incubation of peptides with trypsin in ^{18}O-enriched water is a straightforward addition to the workflow. The samples are then mixed, and the ^{16}O and ^{18}O forms of each peptide elute together from the HPLC as pairs of ions, which are identical except for their carboxyl ends. Similar to SILAC and ICAT, the relative abundance of peptides can be inferred based on the relative intensity between the "light" ^{16}O and "heavy" ^{18}O ions in the MS1 spectra.

The reaction can proceed in the opposite direction, termed "back-exchange", resulting in decreased labeling yields. Although ^{18}O labeling is possible during digestion, a separate labeling exchange reaction after proteolysis is preferable, because it reduces the volume of $H_2{}^{18}O$ to be used, and the use of immobilized trypsin can minimize back-exchange (Yao et al. 2001). The overall advantages of protease-mediated ^{18}O exchange are that essentially any sample can be labeled, no chemical changes are introduced to the peptides, and the work-flow is simple and inexpensive. The disadvantages include that only 2 samples can be labeled and that samples must be kept separate throughout the lysis, enrichment and proteolysis steps, potentially introducing errors due to differences in sample handling. Another disadvantage is that labeling is not as reproducible as some chemical methods, as the exchange reaction is highly sequence specific, and relies heavily on the purity of the $H_2{}^{18}O$, the labeling time, buffer and temperature and the amount and activity of trypsin used. Trypsin-catalyzed ^{18}O labeling (Desiderio and Kai 1983; Heller et al. 2003) is a slow reaction and complete labeling is difficult to obtain.

3.2 Quantitation at the MS2 level – Chemical labeling with isobaric tags

The most common method for quantitation at the MS2 label is isobaric tags for relative and absolute quantitation (iTRAQ), developed by Pappin and colleagues (Ross et al. 2004). iTRAQ involves chemical labeling of amine groups on peptides. The iTRAQ reagent consists of a reporter group, a balance group and a reactive group that reacts with lysine side chains and N-terminal groups of peptides (Fig. 2). In the original 4-component version, the reporter group masses are 114, 115, 116 or 117 Da and the balance group masses are 31, 30, 29 or 28 Da resulting in a combined mass of 145 Da for all four reagents. Briefly, a control and three treated samples are labeled individually with one of the four iTRAQ reagents and then combined. Each isobaric tag has the same minor effect on the elution properties of the peptide resulting in co-elution of the four versions of the peptide. The peptides are indistinguishable at the MS1 and are selected to fragment within a single MS2 scan. During collision-induced fragmentation (CID) or other similar fragmentation methods, in addition to the conventional fragmentation at peptide bonds needed for peptide identification, the reporter group ions (114,115, 116 and 117 Da) also break away from the backbone peptides (Fig. 3). Relative quantitation for each of the treatment conditions being studied is obtained by comparing the intensities of the reporter group fragments. 4- and 8-component iTRAQ kits are available from Applied Biosystems.

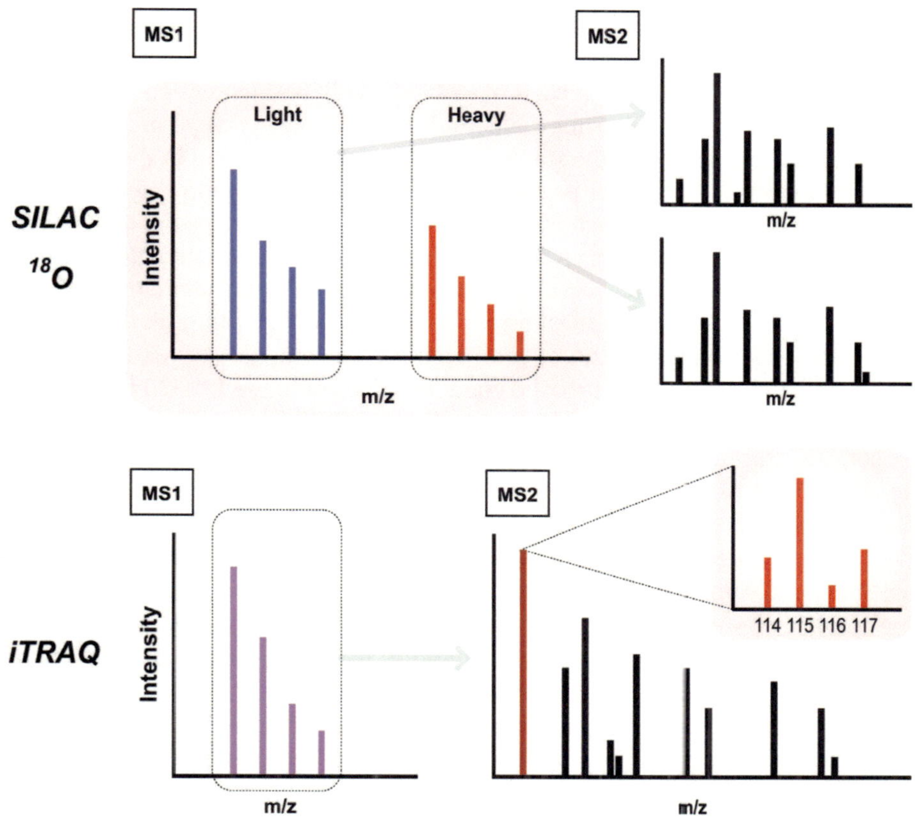

Fig. 3. Quantitation at MS1 and MS2 level. Upper panel illustrates quantitation at MS1 level as a result of labeling by SILAC or trypsin catalyzed ^{18}O labeling. One sample is labeled with heavy isotope while the other is not labeled. The samples are mixed and in the MS1 spectra each peptide is represented as a peptide pair with an unlabeled or "light" peptide (blue) and a labeled or "heavy" peptide (red). Each peptide can be subjected to fragmentation resulting in MS2 spectra with identical b-ions but y-ions are shifted by the weight of the heavy isotope (not shown). Lower panel illustrates labeling with iTRAQ reagents, which requires quantitation at MS2 level. In iTRAQ all samples are labeled, typically four or eight total. The samples are mixed and because the iTRAQ reagent is isobaric, the peptides all co-elute and overlap, resulting in a single peptide envelope (purple). The peptides are fragmented together and the MS2 spectra contains the conventional fragment ions that overlap for all peptides and also, unique reporter ions for each sample, which are used for quantitation (114, 115, 116, 117).

Tandem Isobaric Mass Tag (TMT) kits with two or six components that work by a similar principle are available from Thermo Scientific (Thompson et al 2003). Recently, Cystein-Reactive TMT reagents (cys-TMT) became available from Thermo Scientific. These cys-TMT reagents, like ICAT, target cysteines on proteins and allow for enrichment of cysteine containing peptides. As in iTRAQ and the conventional TMT, the reagents are isobaric and quantitation is performed at the MS2 level.

Another type of isobaric method is Isobaric Peptide Termini Labeling (IPTL) (Koehler et al. 2011; Koehler et al. 2009). In IPTL two non-isobaric reagents are used. In one sample, the C-terminal lysine residues are labeled with 2-methoxy-4,5-dihydro-1H-imidazole (MDHI) followed by N-terminal derivatization with tetradeuterated succinic anhydride (SA-d_4). In the second sample, the C-terminal lysines are labeled with tetradeuterated MDHI-d_4 and the N-terminal derivatized with SA. Thus the peptides are isobaric, each peptide is 4 Da heavier than an unlabeled peptide, with the stable isotopes on the C-terminal in the first group and N-terminal in the second group of peptides. Quantitation is then performed directly on the fragment ions at the MS2 level. Each fragment ion will have a 4 Da heavier counterpart, with deuterated C-terminal or y-ions from the first set of peptides deuterated N-terminal ions or b-ions from the second set. In IPTL each fragment ions provides a quantitative value.

Isobaric methods can be used to label any type of sample, including biofluids and tissue, and up to 8 samples/conditions can be compared concurrently. However, iTRAQ and TMT is limited to instruments that can provide good MS2 spectra in the 100-120 Da range, such as the QSTAR Quadrupole Time-of-Flight instrument (ABI). Pulsed Q dissociation (PQD) and higher energy C-trap dissociation (HCD) recently made it possible to detect the low mass isobaric tag reagent fragments on linear ion trap instruments including the LTQ-Orbitrap (Thermo Scientific) (Meany et al. 2007; Armenta, Hoeschele, and Lazar 2009; Kocher et al. 2009). The disadvantages of this type of chemical labeling are the presence of potential side reactions, the extra steps required to remove excess reagents and derivatization byproducts resulting in difficulty in achieving complete labeling. The iTRAQ approach has been used for several large scale proteomic quantitative studies including time resolved monitoring of kinase reactions (Zhang et al. 2005), comparison of organelle proteomes (Yan, Hwang, and Aebersold 2008) and monitoring of protein expression changes as cancer cells acquire increasing metastatic potential (Ho et al. 2009). Combining quantitation with phosphoproteomics, Aebersold and colleagues (Pflieger et al. 2008) recently described an iTRAQ method to simultaneously identify components and phosphorylation sites of protein complexes.

3.3 Considerations for designing stable isotope experiments

Factors to consider when choosing which stable isotope to work with include price and increased complexity by chromatographic seperation of ^{2}H (Ong, Kratchmarova, and Mann 2003). The most commonly used isotopes are ^{15}N, ^{13}C, and, to a lesser extent, ^{2}H and ^{18}O. A critical component to stable isotope labeling using chemical, enzymatic, or metabolic methods is achieving complete labeling. In metabolic labeling approaches, such as SILAC, stable isotopes are incorporated into proteins as they are expressed making complete labeling easily attainable. However, use of the method is limited to cultured cells or organisms that can be grown in the presence of heavy isotope. It is worth the effort to spend time optimizing and testing a labeled sample before starting an experiment, especially for non-metabolic labeling methods to achieve high levels of labeling. Although calculations can be done to normalize samples to extent of labeling, downstream analysis will be greatly simplified if labeling is complete. Unfortunately, even with optimization to achieve stoichiometric labeling of the majority of peptides, each of the methods is subject to one or more artifacts, resulting in a subset of peptides that display partial or unexpected labeling, thereby confounding analysis.

The timing of labeling is important in any quantitative experiment. The earlier the label is introduced and the sample can be mixed for downstream analysis, the better. Metabolic labeling methods allow for mixing of samples immediately following cell growth and before any protein and peptide enrichment methods. Protein labeling reagents such as ICAT and acrylamide allow for intermediate timing of labeling. In peptide labeling methods, such as iTRAQ, TMT and trypsin catalyzed ^{18}O labeling, the sample is not mixed until several steps including cell lysis, protein separation (if any) and digestion have been performed. This can lead to introduction of sample handling errors. However, these reagents also allow for labeling of samples that metabolic labeling reagents cannot, including a wide range of clinical samples such as urine and tissue. Thus selection of labeling method should take into account type of sample and enrichment methods. If the label is introduced at a late stage in sample preparation care should be taken to minimize any difference in sample handling.

All of the abovementioned methods of labeling, except for isobaric tags, result in generation of peptide pairs at the MS1 level, where the light and heavy peptides are separated by a predictable number of mass units. This allows background peaks to be readily distinguished from "real" peptides insofar as the "real" peptides are represented by both light and heavy forms with a characteristic mass offset. If the mass difference is small, the natural isotope distribution of the light form will overlap with the monoisotopic peak of heavy form, frustrating quantitiation. Trypsin-mediated ^{18}O exchange yields a 4 Da mass difference that leads to challenging quantitation of higher charged peptides and peptides over 20 residues, particularly if the labeling is incomplete. Indeed, incorporation of a single ^{18}O is common, leading to a mass difference of only 2 Da. In turn, even though acrylamide labeling is typically complete, it offers as little as a 3 Da mass shift. Although it is possible to deconvolute such overlapping distributions and quantify the heavy and light peaks, this is a complex and iterative process, requires high quality data, and is tedious. Thus, most commercial labeling reagents including SILAC and ICAT are generated to have ≥ 4 Da mass difference and avoid this complication. In addition, because increased number of peptide species results in more complexity at the MS1 level and mass spectrometers are limited in MS2 fragment scan speed and number, this can result in fewer protein identifications, especially when more than two samples are analyzed together.

For optimal quantitation, a standard calibration curve should be generated for each quantitative measurement, plotting the measured intensity against the amount of analyte. The limit of detection (LOD) or instrument detection limit (IDL) is the lowest amount of a substance that is distinguishable from background noise. Since there are often many steps prior to mass spectrometry analysis, additional error is imposed on each measurement and the method detection limit (MDL) accounts for these steps. The limit of quantitation (LOQ) is the point at which the mass spectrometer can distinguish between two different amounts of analyte. Ideally, measurements would be acquired along the linear dynamic range, the portion of the curve in which the intensity increases linearly with analyte concentration (Fig. 4, in green). The boundaries of the linear range are defined as the lower limit of quantitation (LLOQ) and the upper limit of quantitation (ULOQ). When performing absolute quantitation of a single or few peptides or proteins, generating standard curves for each analyte is straightforward. For complex and/or relative quantitation experiments individual standard curves are often not feasible. When validating quantitation from complex samples, one should consider that peptide measurements outside of the linear range may underestimate the change in abundance.

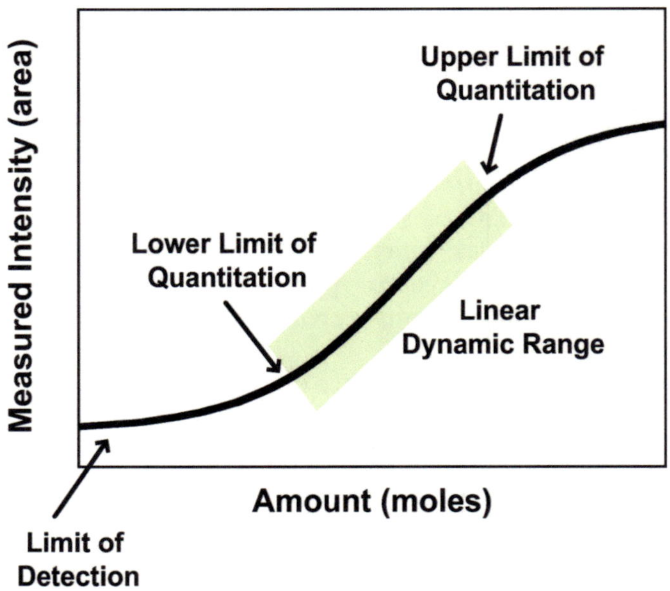

Fig. 4. An example of a standard curve showing the limits of detection and quantitation. The lowest amount of a peptide that can be detected in a mass spectrometer is defined as the limit of detection (LOD). Although the intensity of a peptide is dependent on the amount (number of moles present), this relationship is only linear over a certain range. This is the optimal range for quantitative measurements (in green) and its boundaries are defined as the lower limit of quantitation (LLOQ) and the upper limit of quantitation (ULOQ).

One of the most critical considerations for mass spectrometry analysis is the inclusion of replicate analyses and follow-up biological validation. Replicates allow the researcher to evaluate reproducibility in identification and quantitation and statistical significance of any identified proteome changes. Although technical replicates are helpful to identify variations due to mass spectrometry analysis, biological replicates should be performed for any proteomic experiment. The variations in sample handling, cell growth, labeling procedure can be high. This becomes vital in the case of clinical samples where biological variation can be very high. The exact number of replicates and statistical considerations vary between sample types, origin and type of experiment. An optimal experiment would involve consultation with a statistician before the experiment is started such that sufficient sample numbers and replicates can be included in the analysis.

4. Data analysis tools and suggestions to improve data quality

4.1 Quantitation software for MS1 level data

Selection of quantitation software depends on several factors including the stable isotope labeling method, the level of quantitation (MS1 versus MS2), the type of instrument used to obtain the data (ion traps versus Q-TOFs) and the availability of software (commercial versus open source) (Table 1). Additional problems arise when the stable isotope label results in a mass shift small enough to allow overlap of the resulting isotopic envelopes, but specialized software is available to compensate for this overlap.

Mass spectrometry manufacturers often provide proprietary software solutions for quantitation. Examples include Bioworks (Thermo Scientific), Peakpicker (Applied Biosystems) and WARP-LC™ 1.1 (Bruker). A large number of open-source software tools are available including AYUMS (Saito et al. 2007), ProRata (Pan et al. 2006), Mascot File Parsing and Quantification (MFPaQ) (Bouyssie et al. 2007), QUIL (Wang, Wu, Pisitkun, et al. 2006), MSQuant (Mortensen et al. 2010) and Uniquant (Huang et al. 2011). Compilations of software are available including Trans Proteomic Pipeline (TPP) developed at the Institute for Systems Biology (ISB) in Seattle (reviewed in (Deutsch et al. 2010). Modules for quantitation include XPRESS (Han et al. 2001) and ASAPratio (Li et al. 2003). The ISB tools have been incorporated into Computational Proteomics Analysis System (CPAS), a suite of database and analysis tools for managing proteomics based experimental workflows and integrating database search algorithms (Rauch et al. 2006) CPAS was originally developed in the Fred Hutchinson Cancer Research Center but is now distributed as part of Labkey Server, an open-source project managed by the Labkey Software Foundation. Another open-source integrated suite of algorithms, termed MaxQuant, was developed by Matthias Mann's group and was specifically developed for quantitation of high-resolution MS data (Cox and Mann 2008).

Separate software packages have been developed to account for the difficulties in interpreting spectra from isotopically labeled samples. An algorithm called regression analysis applied to mass spectrometry (RAAMS) corrects for partial ^{18}O labeling as well as incorporation of naturally occurring isotopes (Mason et al. 2007).

4.2 Quantitation software for MS2 level (isobaric) data

Quantitation software for isobaric tags include commercially available solutions such as ProteinPilot and ProQuant from Applied Biosystems, Spectrum Mill from Agilent, Proteome Discoverer from Thermo Scientific and Scaffold Q+ from Proteome Software (Table 1). Open-source software includes Libra, a software module used within the Trans Proteomic Pipeline (TPP), MassMatrix, a search engine that performs quantitation of TMT and iTRAQ (Warren et al. 2010), IsobariQ which was designed for IPTL, iTRAQ and TMT (Arntzen et al. 2011) and jTraqX, a platform independent tool for isobaric tag quantitation (Muth et al. 2010). COMPASS is an integrated suite of pre- and post-search proteomics software tools specific to the OMSSA database search engine (Wenger et al. 2011).

4.3 Common concerns and how to achieve high quality quantitation

Despite the available software packages for peptide and protein identification and quantitation, manual validation is often required for confirmation. Inaccurate or ambiguous results are common when too few peptides can be quantified from a protein or where the standard deviation or *p*-value between multiple quantified peptides from a protein is not statistically significant. High-abundance proteins that yield ratios close to 1:1 have the highest confidence levels but provide little or no biological insight. As with any mass spectrometry experiment, low-abundance proteins are difficult to study because of the limited dynamic range. In addition, if the sample is too complex (too many peptides are in the sample), overlapping peptide spectra can occur and bring about errors in peptide

	Type	Software	Reference
MS1			
	Commercial	Bioworks	Thermo Scientific
	Commercial	Peakpicker	Applied Biosystems
	Commercial	Proteinscape 2	Bruker
	Open source	AYUMS	Saito et al. 2007
	Open source	ProRata	Pan et al. 2006
	Open source	TPP -XPRESS and ASAP Ratio	Han et al. 2001; Li et al. 2003
	Open source	MaxQuant	Cox and Mann 2008
	Open source	MSQuant	Mortensen et al. 2010
	Open source	Quil	Wang, Wu, Pisitkun, et al. 2006
	Open source	RAAMS	Mason et al. 2007
	Open source	Uniquant	Huang et al. 2011
	Open source	MFPAQ	Bouyssie et al. 2007
MS2			
	Commercial	Proteome Discoverer	Thermo Scientific
	Commercial	Scaffold Q+	Proteome software
	Commercial	ProteinPilot and ProQuant	Applied Biosystems
	Commercial	Spectrum Mill	Agilent
	Commercial	Proteinscape 2	Bruker
	Open source	TPP- Libra	
	Open source	IsobariQ	Arntzen et al. 2011
	Open source	MassMatrix	Warren et al. 2010
	Open source	jTraqX	Muth et al. 2010
	Open source	COMPASS	Wenger et al. 2011

Table 1. A list of currently available software tools for quantitation with stable isotopes. Quantitation software is organized by the level of quantitation (MS1 versus MS2), type of software (Commercial or Open source), name of software and reference(s) listed.

quantitation both in MS1 and MS2. MS2 quantitation is currently much more sensitive to overlapping spectra than MS1 level quantitation methods. This is due to the large (>1 Da) selection window for peptide fragmentation at the MS2 level to obtain sufficient peptide signal. In iTRAQ and TMT, multiple co-fragmenting peptides can significantly alter the peptide ratio and because the reporter ions are cleaved they cannot be assigned to their originating peptides. Since fragment ions are used for quantitation in the IPTL method, this is likely to be less of an issue. Performing peptide and/or protein separations using immunoenrichment methods, chromatography, electrophoresis or by isolating cellular compartments can reduce sample complexity, improve quantitation at both the MS1 and MS2 level and increase detection of low abundance peptides. However, any fractionation method results in some sample loss and thus often requires more starting material. It is important to consider the smallest proteome subset appropriate for analysis. Since the mass spectrometer and reverse phase columns have limited loading capacity, reducing the sample complexity may improve both proteome coverage and the confidence of peptide identification and quantitation.

Accurate quantitation requires consideration of naturally occurring isotopes present in peptides, especially when small differences are measured using ^{13}C stable isotopes. For MS1 labeling, calculating the peptide ratio using the whole isotopic envelope is more accurate than performing quantitation on only the monoisotopic peak. For a given peptide sequence, the isotopic envelope can be calculated and correlated to the measured distribution, and the isotope enrichment can be measured. An overall enrichment correlation factor can ensure consistent metabolic labeling. This calculation needs to be performed for each quantitative measurement when using trypsin catalyzed ^{18}O labeling, which can vary with peptide sequence. This algorithm has been incorporated into quantitative software. RAAMS uses a method to estimate the isotopic envelope based on "averagine", the isotopic distribution in an average amino acid ($C_{4.9384}H_{7.7583}N_{1.3577}O_{1.4773}S_{0.0417}$) (Mason et al. 2007). In MS2 level quantitation, including iTRAQ, TMT and IPTL, quantitative measurements cannot be determined over the elution time of a peptide, only for single fragmentation spectra. Instead, replicate measurement should be performed to increase confidence in quantitation. Finally, for both MS1 and MS2 labeling, the measurements are performed at the peptide level and not protein level. Proteins are inferred from peptides. The peptide ratios from what appears to be a single protein can differ for several reasons including that peptides can be identical in several proteins and that peptides could be post-translationally modified and not visible in the mass spectrometry data. Thus the protein value is a weighted average of all protein forms and not a single protein ratio.

5. Harnessing the information obtained from stable isotope labeling

With the exception of isobaric methods, the MS1 spectra will contain isotopic peptide pairs consisting of an unlabeled and a labeled peptide. Optimally, the mass spectrometer would recognize these pairs and preferentially select only "light" monoisotopic ions of pairs for fragmentation, thereby avoiding background and/or contaminating ions and offsetting the added complexity in the sample. This is particularly important for the analysis of complex stable isotope labeled samples where the number of peptide pairs far exceeds the number of possible fragmentation scans. In principle, the existing user-defined, data-dependent scanning software provided on commercial mass spectrometers can be adapted to direct the mass spectrometer to flag ions that are separated by a pre-defined mass and subject only these to fragmentation. For example, such a setting is called "mass tag" in Xcalibur software for Orbitraps and FT-ICR mass spectrometers (Thermo Scientific). However, as of the writing of this review, "mass tag" remains to be fully implemented.

5.1 Other uses for stable isotope labeling

Stable isotope labeling has been used to distinguish contaminants from *bona fide* interactors in immunopurifications (I-DIRT) (Tackett et al. 2005). Yeast cells containing an affinity-tagged protein were grown in light SILAC media and control yeast cells in heavy media. After isolation of the affinity tagged protein complex, specific protein interactions were identified by mass spectrometry as a single unlabeled peptide (light), but background contaminant proteins present in both the control (heavy) and affinity-tag protein expressing cells (light) were identified as peptide pairs. Another clever use of stable isotope quantitation is to examine dynamic protein-protein complexes and protein-DNA complexes (Pflieger et al. 2008; Mittler, Butter, and Mann 2009) by combining affinity purification approaches with stable isotope tagging. Quantification of component stoichiometry of multiprotein complexes has also been performed using a peptide-concatenated standard (PCS) strategy (Kito et al. 2007). In this strategy, tryptic peptides suitable for quantification are selected from each component of the multiprotein complex and concatenated into a single synthetic protein, resulting in equimolar amounts of each "heavy" reference peptide. Other uses for stable isotope labeling include measuring the rate of protein turnover (Pratt et al. 2002) and identifying phosphorylation sites (Pflieger et al. 2008).

5.2 Use of stable isotopes to obtain faster and more accurate protein identification

A complementary advantage of stable isotope labeling is that when both heavy and light forms are subjected to fragmentation, mass shifts are observed in the MS2 spectra that facilitate deconvolution and peptide sequence analysis. When only the carboxyl terminus is labeled as in lysine/arginine SILAC or ^{18}O labeling, comparing the two fragmentation patterns reveals ions that derive from the carboxyl terminus (*y*-type ions) as those display characteristic mass shifts (e.g. 8 or 4 Da). Accordingly, comparison of spectra of labeled and unlabeled peptide fragments allows for assignment of peaks as shifting or non-shifting, facilitating *de novo* peptide sequence analysis (Hunt et al. 1986; Schnolzer, Jedrzejewski, and Lehmann 1996; Takao et al. 1991; Gray, Wojcik, and Futrell 1970; Rose et al. 1983). Peak assignment for validation of peptide identifications obtained by database search has been automated in the Validator software suite (Volchenboum et al. 2009), which recognizes isotopic peptide pairs from searched MS data and compares their identifications and fragmentation patterns. Because database search algorithms do not utilize the embedded information from comparison of labeled and unlabeled peptides, Validator software provides a direct and independent means to validate peptide identifications from database search algorithms.

6. Conclusions

Stable isotopes have become a versatile and useful tool in quantitative mass spectrometry. This review has described chemical, enzymatic and metabolic stable isotope labeling techniques while highlighting the advantages and disadvantages of each method. A wide variety of sample types can be labeled and analyzed from single proteins and complexes to bacteria, yeast, mammalian cells, biofluids and tissues. For optimal absolute quantitation, each peptide to be measured requires a labeled counterpart, making it a costly and laborious methodology for large proteomic studies. Complex samples are better suited to relative quantitation where fold-change is calculated by comparing peptide levels to a fully labeled control sample. Stable isotope labeling is also useful for identifying components and measuring the stoichiometry of protein-protein and protein-DNA complexes. Stable isotopes can also facilitate identification of posttranslational modifications and background contamination and to aid in peptide identification and validation.

The advent of modern mass spectrometers has allowed for precise quantitation of sub-femtomolar samples with their remarkable sensitivity, resolution, reproducibility and speed. However, many challenges remain, affecting the quality of results and resulting in pitfalls for experienced and naïve users alike. No isotopic method is free of the wide range of artifacts that arise due to biological variation, human error, design and implementation of instrumentation control and poorly written and implemented data analysis software. Confounding the situation, proteomics experiments provide spurious answers side-by-side with highly reliable results, often with no clear distinction among them.

Nonetheless, some common principles apply that will enhance the quality of every experiment. Achieving the most complete and consistent labeling feasible greatly simplifies downstream data analysis. Decreasing sample complexity to improve peptide statistics for each protein facilitates high confidence in identifications and the ready discovery of quantitation artifacts. Despite advantages in software design, manual validation through visual inspection of mass spectrometry spectra remains a critical step. Therefore, stable isotope labeling for protein quantitation by mass spectrometry is still an emerging technology and care must be taken to use appropriate controls, including biological and/or technical replicates, in order to identify potential problems with labeling, sample handling and/or data analysis.

7. References

Aebersold, R., and M. Mann. 2003. "Mass spectrometry-based proteomics." *Nature* no. 422 (6928):198-207. doi: 10.1038/nature01511.

Amanchy, R., D. E. Kalume, A. Iwahori, J. Zhong, and A. Pandey. 2005. "Phosphoproteome analysis of HeLa cells using stable isotope labeling with amino acids in cell culture (SILAC)." *J Proteome Res* no. 4 (5):1661-71. doi: 10.1021/pr050134h.

Andersen, J. S., Y. W. Lam, A. K. Leung, S. E. Ong, C. E. Lyon, A. I. Lamond, and M. Mann. 2005. "Nucleolar proteome dynamics." *Nature* no. 433 (7021):77-83.

Armenta, J. M., I. Hoeschele, and I. M. Lazar. 2009. "Differential protein expression analysis using stable isotope labeling and PQD linear ion trap MS technology." *J Am Soc Mass Spectrom* no. 20 (7):1287-302. doi: S1044-0305(09)00140-8 [pii] 10.1016/j.jasms.2009.02.029.

Arntzen, M. O., C. J. Koehler, H. Barsnes, F. S. Berven, A. Treumann, and B. Thiede. 2011. "IsobariQ: software for isobaric quantitative proteomics using IPTL, iTRAQ, and TMT." *J Proteome Res* no. 10 (2):913-20. doi: 10.1021/pr1009977.

Baillie, T. A. 1981. "The use of stable isotopes in pharmacological research." *Pharmacol Rev* no. 33 (2):81-132.

Barnidge, D. R., E. A. Dratz, T. Martin, L. E. Bonilla, L. B. Moran, and A. Lindall. 2003. "Absolute quantification of the G protein-coupled receptor rhodopsin by LC/MS/MS using proteolysis product peptides and synthetic peptide standards." *Anal Chem* no. 75 (3):445-51.

Barr, J. R., V. L. Maggio, D. G. Patterson, G. R. Cooper, L. O. Henderson, W. E. Turner, S. J. Smith, W. H. Hannon, L. L. Needham, and E. J. Sampson. 1996. "Isotope dilution--mass spectrometric quantification of specific proteins model application with apolipoprotein A-I." *Clin Chem* no. 42 (10):1676-82.

Beynon, R. J., M. K. Doherty, J. M. Pratt, and S. J. Gaskell. 2005. "Multiplexed absolute quantification in proteomics using artificial QCAT proteins of concatenated signature peptides." *Nat Methods* no. 2 (8):587-9.

Beynon, R. J., and J. M. Pratt. 2005. "Metabolic labeling of proteins for proteomics." *Mol Cell Proteomics* no. 4 (7):857-72. doi: 10.1074/mcp.R400010-MCP200.

Blagoev, B., S. E. Ong, I. Kratchmarova, and M. Mann. 2004. "Temporal analysis of phosphotyrosine-dependent signaling networks by quantitative proteomics." *Nat Biotechnol* no. 22 (9):1139-45. doi: 10.1038/nbt1005 nbt1005 [pii].

Bouyssie, D., A. Gonzalez de Peredo, E. Mouton, R. Albigot, L. Roussel, N. Ortega, C. Cayrol, O. Burlet-Schiltz, J. P. Girard, and B. Monsarrat. 2007. "Mascot file parsing and quantification (MFPaQ), a new software to parse, validate, and quantify proteomics data generated by ICAT and SILAC mass spectrometric analyses: application to the proteomics study of membrane proteins from primary human endothelial cells." *Mol Cell Proteomics* no. 6 (9):1621-37.

Brun, V., A. Dupuis, A. Adrait, M. Marcellin, D. Thomas, M. Court, F. Vandenesch, and J. Garin. 2007. "Isotope-labeled protein standards: toward absolute quantitative proteomics." *Mol Cell Proteomics* no. 6 (12):2139-49. doi: M700163-MCP200 [pii] 10.1074/mcp.M700163-MCP200.

Brun, V., C. Masselon, J. Garin, and A. Dupuis. 2009. "Isotope dilution strategies for absolute quantitative proteomics." *J Proteomics* no. 72 (5):740-9. doi: S1874-3919(09)00107-9 [pii] 10.1016/j.jprot.2009.03.007.

Cox, J., and M. Mann. 2008. "MaxQuant enables high peptide identification rates, individualized p.p.b.-range mass accuracies and proteome-wide protein quantification." *Nat Biotechnol* no. 26 (12):1367-72. doi: nbt.1511 [pii] 10.1038/nbt.1511.

de Godoy, L. M., J. V. Olsen, G. A. de Souza, G. Li, P. Mortensen, and M. Mann. 2006. "Status of complete proteome analysis by mass spectrometry: SILAC labeled yeast as a model system." *Genome Biol* no. 7 (6):R50.

Desiderio, D. M., and M. Kai. 1983. "Preparation of stable isotope-incorporated peptide internal standards for field desorption mass spectrometry quantification of peptides in biologic tissue." *Biomed Mass Spectrom* no. 10 (8):471-9.

Deutsch, E. W., L. Mendoza, D. Shteynberg, T. Farrah, H. Lam, N. Tasman, Z. Sun, E. Nilsson, B. Pratt, B. Prazen, J. K. Eng, D. B. Martin, A. I. Nesvizhskii, and R. Aebersold. 2010. "A guided tour of the Trans-Proteomic Pipeline." *Proteomics* no. 10 (6):1150-9. doi: 10.1002/pmic.200900375.

Dupuis, A., J. A. Hennekinne, J. Garin, and V. Brun. 2008. "Protein Standard Absolute Quantification (PSAQ) for improved investigation of staphylococcal food poisoning outbreaks." *Proteomics* no. 8 (22):4633-6. doi: 10.1002/pmic.200800326.

Faca, V., M. Coram, D. Phanstiel, V. Glukhova, Q. Zhang, M. Fitzgibbon, M. McIntosh, and S. Hanash. 2006a. "Quantitative analysis of acrylamide labeled serum proteins by LC-MS/MS." *J Proteome Res* no. 5 (8):2009-18. doi: 10.1021/pr060102+.

Faca, V., M. Coram, D. Phanstiel, V. Glukhova, Q. Zhang, M. Fitzgibbon, M. McIntosh, and S. Hanash. 2006b. "Quantitative analysis of acrylamide labeled serum proteins by LC-MS/MS." *Journal of proteome research* no. 5 (8):2009-18. doi: 10.1021/pr060102+.

Fischer, B., J. Grossmann, V. Roth, W. Gruissem, S. Baginsky, and J. M. Buhmann. 2006. "Semi-supervised LC/MS alignment for differential proteomics." *Bioinformatics* no. 22 (14):e132-40.

Geiger, T., J. Cox, P. Ostasiewicz, J. R. Wisniewski, and M. Mann. 2010. "Super-SILAC mix for quantitative proteomics of human tumor tissue." *Nat Methods* no. 7 (5):383-5. doi: nmeth.1446 [pii] 10.1038/nmeth.1446.

Gerber, S. A., J. Rush, O. Stemman, M. W. Kirschner, and S. P. Gygi. 2003. "Absolute quantification of proteins and phosphoproteins from cell lysates by tandem MS." *Proc Natl Acad Sci U S A* no. 100 (12):6940-5.

Gray, W. R., L. H. Wojcik, and J. H. Futrell. 1970. "Application of mass spectrometry to protein chemistry. II. Chemical ionization studies on acetylated permethylated peptides." *Biochem Biophys Res Commun* no. 41 (5):1111-9. doi: 0006-291X(70)90200-7 [pii].

Griffin, T. J., S. P. Gygi, T. Ideker, B. Rist, J. K. Eng, L. Hood, and R. Aebersold. 2002. "Complementary profiling of gene expression at the transcriptome and proteome levels in Saccharomyces cerevisiae." *Mol Cell Proteomics* no. 1 (4):323-33.

Gygi, S. P., B. Rist, S. A. Gerber, F. Turecek, M. H. Gelb, and R. Aebersold. 1999. "Quantitative analysis of complex protein mixtures using isotope-coded affinity tags." *Nat Biotechnol* no. 17 (10):994-9.

Han, D. K., J. K. Eng, H. Zhou, and R. Aebersold. 2001. "Quantitative profiling of differentiation-induced microsomal proteins using isotope-coded affinity tags and mass spectrometry." *Nat Biotechnol* no. 19 (10):946-51.

Hanke, S., H. Besir, D. Oesterhelt, and M. Mann. 2008. "Absolute SILAC for accurate quantitation of proteins in complex mixtures down to the attomole level." *J Proteome Res* no. 7 (3):1118-30. doi: 10.1021/pr7007175.

Heller, M., H. Mattou, C. Menzel, and X. Yao. 2003. "Trypsin catalyzed 16O-to-18O exchange for comparative proteomics: tandem mass spectrometry comparison using MALDI-TOF, ESI-QTOF, and ESI-ion trap mass spectrometers." *J Am Soc Mass Spectrom* no. 14 (7):704-18. doi: S1044030503002071 [pii].

Ho, J., J. W. Kong, L. Y. Choong, M. C. Loh, W. Toy, P. K. Chong, C. H. Wong, C. Y. Wong, N. Shah, and Y. P. Lim. 2009. "Novel Breast Cancer Metastasis-Associated Proteins." *J Proteome Res* no. 8 (2):583-594. doi: 10.1021/pr8007368 10.1021/pr8007368 [pii].

Huang, X., A. V. Tolmachev, Y. Shen, M. Liu, L. Huang, Z. Zhang, G. A. Anderson, R. D. Smith, W. C. Chan, S. H. Hinrichs, K. Fu, and S. J. Ding. 2011. "UNiquant, a program for quantitative proteomics analysis using stable isotope labeling." *J Proteome Res* no. 10 (3):1228-37. doi: 10.1021/pr1010058.

Hunt, D. F., J. R. Yates, 3rd, J. Shabanowitz, S. Winston, and C. R. Hauer. 1986. "Protein sequencing by tandem mass spectrometry." *Proc Natl Acad Sci U S A* no. 83 (17):6233-7.

Johnson, H., S. C. Wong, D. M. Simpson, R. J. Beynon, and S. J. Gaskell. 2008. "Protein quantification by selective isolation and fragmentation of isotopic pairs using FT-ICR MS." *Journal of the American Society for Mass Spectrometry* no. 19 (7):973-7. doi: 10.1016/j.jasms.2008.03.012.

Kippen, A. D., F. Cerini, L. Vadas, R. Stocklin, L. Vu, R. E. Offord, and K. Rose. 1997. "Development of an isotope dilution assay for precise determination of insulin, C-peptide, and proinsulin levels in non-diabetic and type II diabetic individuals with comparison to immunoassay." *J Biol Chem* no. 272 (19):12513-22.

Kito, K., K. Ota, T. Fujita, and T. Ito. 2007. "A synthetic protein approach toward accurate mass spectrometric quantification of component stoichiometry of multiprotein complexes." *J Proteome Res* no. 6 (2):792-800. doi: 10.1021/pr060447s.

Kocher, T., P. Pichler, M. Schutzbier, C. Stingl, A. Kaul, N. Teucher, G. Hasenfuss, J. M. Penninger, and K. Mechtler. 2009. "High precision quantitative proteomics using iTRAQ on an LTQ Orbitrap: a new mass spectrometric method combining the benefits of all." *J Proteome Res* no. 8 (10):4743-52. doi: 10.1021/pr900451u.

Koehler, C. J., M. O. Arntzen, M. Strozynski, A. Treumann, and B. Thiede. 2011. "Isobaric peptide termini labeling utilizing site-specific N-terminal succinylation." *Anal Chem* no. 83 (12):4775-81. doi: 10.1021/ac200229w.

Koehler, C. J., M. Strozynski, F. Kozielski, A. Treumann, and B. Thiede. 2009. "Isobaric peptide termini labeling for MS/MS-based quantitative proteomics." *J Proteome Res* no. 8 (9):4333-41. doi: 10.1021/pr900425n.

Li, X. J., H. Zhang, J. A. Ranish, and R. Aebersold. 2003. "Automated statistical analysis of protein abundance ratios from data generated by stable-isotope dilution and tandem mass spectrometry." *Anal Chem* no. 75 (23):6648-57. doi: 10.1021/ac034633i.

Listgarten, J., and A. Emili. 2005. "Statistical and computational methods for comparative proteomic profiling using liquid chromatography-tandem mass spectrometry." *Mol Cell Proteomics* no. 4 (4):419-34.

Mason, C. J., T. M. Therneau, J. E. Eckel-Passow, K. L. Johnson, A. L. Oberg, J. E. Olson, K. S. Nair, D. C. Muddiman, and H. R. Bergen, 3rd. 2007. "A method for automatically interpreting mass spectra of 18O-labeled isotopic clusters." *Mol Cell Proteomics* no. 6 (2):305-18. doi: M600148-MCP200 [pii] 10.1074/mcp.M600148-MCP200.

Meany, D. L., H. Xie, L. V. Thompson, E. A. Arriaga, and T. J. Griffin. 2007. "Identification of carbonylated proteins from enriched rat skeletal muscle mitochondria using affinity chromatography-stable isotope labeling and tandem mass spectrometry." *Proteomics* no. 7 (7):1150-63. doi: 10.1002/pmic.200600450.

Mirgorodskaya, O. A., Y. P. Kozmin, M. I. Titov, R. Korner, C. P. Sonksen, and P. Roepstorff. 2000. "Quantitation of peptides and proteins by matrix-assisted laser desorption/ionization mass spectrometry using (18)O-labeled internal standards." *Rapid Commun Mass Spectrom* no. 14 (14):1226-32. doi: 10.1002/1097-0231(20000730)14:14<1226::AID-RCM14>3.0.CO;2-V [pii] 10.1002/1097-0231(20000730)14:14<1226::AID-RCM14>3.0.CO;2-V.

Mittler, G., F. Butter, and M. Mann. 2009. "A SILAC-based DNA protein interaction screen that identifies candidate binding proteins to functional DNA elements." *Genome Res* no. 19 (2):284-93. doi: gr.081711.108 [pii] 10.1101/gr.081711.108.

Molina, H., Y. Yang, T. Ruch, J. W. Kim, P. Mortensen, T. Otto, A. Nalli, Q. Q. Tang, M. D. Lane, R. Chaerkady, and A. Pandey. 2009. "Temporal profiling of the adipocyte proteome during differentiation using a five-plex SILAC based strategy." *J Proteome Res* no. 8 (1):48-58. doi: 10.1021/pr800650r.

Mortensen, P., J. W. Gouw, J. V. Olsen, S. E. Ong, K. T. Rigbolt, J. Bunkenborg, J. Cox, L. J. Foster, A. J. Heck, B. Blagoev, J. S. Andersen, and M. Mann. 2010. "MSQuant, an open source platform for mass spectrometry-based quantitative proteomics." *J Proteome Res* no. 9 (1):393-403. doi: 10.1021/pr900721e.

Muth, T., D. Keller, S. M. Puetz, L. Martens, A. Sickmann, and A. M. Boehm. 2010. "jTraqX: a free, platform independent tool for isobaric tag quantitation at the protein level." *Proteomics* no. 10 (6):1223-5. doi: 10.1002/pmic.200900374.

Oda, Y., K. Huang, F. R. Cross, D. Cowburn, and B. T. Chait. 1999. "Accurate quantitation of protein expression and site-specific phosphorylation." *Proc Natl Acad Sci U S A* no. 96 (12):6591-6.

Olsen, J. V., J. S. Andersen, P. A. Nielsen, M. L. Nielsen, D. Figeys, M. Mann, and J. R. Wisniewski. 2004. "HysTag--a novel proteomic quantification tool applied to differential display analysis of membrane proteins from distinct areas of mouse brain." *Mol Cell Proteomics* no. 3 (1):82-92.

Ong, S. E., B. Blagoev, I. Kratchmarova, D. B. Kristensen, H. Steen, A. Pandey, and M. Mann. 2002. "Stable isotope labeling by amino acids in cell culture, SILAC, as a simple and accurate approach to expression proteomics." *Mol Cell Proteomics* no. 1 (5):376-86.

Ong, S. E., I. Kratchmarova, and M. Mann. 2003. "Properties of 13C-substituted arginine in stable isotope labeling by amino acids in cell culture (SILAC)." *J Proteome Res* no. 2 (2):173-81.

Ono, M., M. Shitashige, K. Honda, T. Isobe, H. Kuwabara, H. Matsuzuki, S. Hirohashi, and T. Yamada. 2006. "Label-free quantitative proteomics using large peptide data sets generated by nanoflow liquid chromatography and mass spectrometry." *Mol Cell Proteomics* no. 5 (7):1338-47.

Pan, C., G. Kora, W. H. McDonald, D. Tabb, N. C. VerBerkmoes, G. B. Hurst, D. A. Pelletier, N. F. Samatova, and R. L. Hettich. 2006. "ProRata: A quantitative proteomics program for accurate protein abundance ratio estimation with confidence interval evaluation." *Anal Chem* no. 78 (20):7121-31. doi: 10.1021/ac060654b.

Paoletti, A. C., T. J. Parmely, C. Tomomori-Sato, S. Sato, D. Zhu, R. C. Conaway, J. W. Conaway, L. Florens, and M. P. Washburn. 2006. "Quantitative proteomic analysis of distinct mammalian Mediator complexes using normalized spectral abundance factors." *Proc Natl Acad Sci U S A* no. 103 (50):18928-33.

Patterson, S. D. 1994. "From electrophoretically separated protein to identification: strategies for sequence and mass analysis." *Anal Biochem* no. 221 (1):1-15. doi: S0003-2697(84)71371-6 [pii] 10.1006/abio.1994.1371.

Pflieger, D., M. A. Junger, M. Muller, O. Rinner, H. Lee, P. M. Gehrig, M. Gstaiger, and R. Aebersold. 2008. "Quantitative proteomic analysis of protein complexes: concurrent identification of interactors and their state of phosphorylation." *Mol Cell Proteomics* no. 7 (2):326-46. doi: M700282-MCP200 [pii] 10.1074/mcp.M700282-MCP200.

Pitteri, S. J., V. M. Faca, K. S. Kelly-Spratt, A. E. Kasarda, H. Wang, Q. Zhang, L. Newcomb, A. Krasnoselsky, S. Paczesny, G. Choi, M. Fitzgibbon, M W. McIntosh, C. J. Kemp, and S. M. Hanash. 2008. "Plasma proteome profiling of a mouse model of breast cancer identifies a set of up-regulated proteins in common with human breast cancer cells." *Journal of proteome research* no. 7 (4):1481-9. doi: 10.1021/pr7007994.

Pratt, J. M., J. Petty, I. Riba-Garcia, D. H. Robertson, S. J. Gaskell, S. G. Oliver, and R. J. Beynon. 2002. "Dynamics of protein turnover, a missing dimension in proteomics." *Mol Cell Proteomics* no. 1 (8):579-91.

Pratt, J. M., D. M. Simpson, M. K. Doherty, J. Rivers, S. J. Gaskell, and R. J. Beynon. 2006. "Multiplexed absolute quantification for proteomics using concatenated signature peptides encoded by QconCAT genes." *Nat Protoc* no. 1 (2):1029-43.

Rauch, A., M. Bellew, J. Eng, M. Fitzgibbon, T. Holzman, P. Hussey, M. Igra, B. Maclean, C. W. Lin, A. Detter, R. Fang, V. Faca, P. Gafken, H. Zhang, J. Whiteaker, J. Whitaker, D. States, S. Hanash, A. Paulovich, and M. W. McIntosh. 2006. "Computational Proteomics Analysis System (CPAS): an extensible, open-source analytic system for evaluating and publishing proteomic data and high throughput biological experiments." *Journal of proteome research* no. 5 (1):112-21. doi: 10.1021/pr0503533.

Reynolds, K. J., X. Yao, and C. Fenselau. 2002. "Proteolytic 18O labeling for comparative proteomics: evaluation of endoprotease Glu-C as the catalytic agent." *J Proteome Res* no. 1 (1):27-33.

Rivers, J., D. M. Simpson, D. H. Robertson, S. J. Gaskell, and R. J. Beynon. 2007. "Absolute multiplexed quantitative analysis of protein expression during muscle development using QconCAT." *Mol Cell Proteomics* no. 6 (8):1416-27.

Rose, K., M. G. Simona, R. E. Offord, C. P. Prior, B. Otto, and D. R. Thatcher. 1983. "A new mass-spectrometric C-terminal sequencing technique finds a similarity between gamma-interferon and alpha 2-interferon and identifies a proteolytically clipped gamma-interferon that retains full antiviral activity." *Biochem J* no. 215 (2):273-7.

Ross, P. L., Y. N. Huang, J. N. Marchese, B. Williamson, K. C. Parker, S. Hattan, N. Khainovski, S. Pillai, S. Dey, S. Daniels, S. Purkayastha, P. Juhasz, S. Martin, M. Bartlet-Jones, F. He, A. Jacobson, and D. J. Pappin. 2004. "Multiplexed protein quantitation in Saccharomyces cerevisiae using amine-reactive isobaric tagging reagents." *Mol Cell Proteomics* no. 3 (12):1154-69.

Saito, A., M. Nagasaki, M. Oyama, H. Kozuka-Hata, K. Semba, S. Sugano, T. Yamamoto, and S. Miyano. 2007. "AYUMS: an algorithm for completely automatic quantitation based on LC-MS/MS proteome data and its application to the analysis of signal transduction." *BMC Bioinformatics* no. 8:15.

Schnolzer, M., P. Jedrzejewski, and W. D. Lehmann. 1996. "Protease-catalyzed incorporation of 18O into peptide fragments and its application for protein sequencing by electrospray and matrix-assisted laser desorption/ionization mass spectrometry." *Electrophoresis* no. 17 (5):945-53. doi: 10.1002/elps.1150170517.

Shiio, Y., S. Donohoe, E. C. Yi, D. R. Goodlett, R. Aebersold, and R. N. Eisenman. 2002. "Quantitative proteomic analysis of Myc oncoprotein function." *EMBO J* no. 21 (19):5088-96.

Stewart, I., T. Thomson, and D. Figeys. 2001. "18O labeling: a tool for proteomics." *Rapid Commun Mass Spectrom* no. 15 (24):2456-65.

Tackett, A. J., J. A. DeGrasse, M. D. Sekedat, M. Oeffinger, M. P. Rout, and B. T. Chait. 2005. "I-DIRT, a general method for distinguishing between specific and nonspecific protein interactions." *J Proteome Res* no. 4 (5):1752-6. doi: 10.1021/pr050225e.

Takao, T., H. Hori, K. Okamoto, A. Harada, M. Kamachi, and Y. Shimonishi. 1991. "Facile assignment of sequence ions of a peptide labelled with 18O at the carboxyl terminus." *Rapid Commun Mass Spectrom* no. 5 (7):312-5. doi: 10.1002/rcm.1290050703.

Tang, H., R. J. Arnold, P. Alves, Z. Xun, D. E. Clemmer, M. V. Novotny, J. P. Reilly, and P. Radivojac. 2006. "A computational approach toward label-free protein quantification using predicted peptide detectability." *Bioinformatics* no. 22 (14):e481-8.

Thompson, A., J. Schafer, K. Kuhn, S. Kienle, J. Schwarz, G. Schmidt, T. Neumann, R. Johnstone, A. K. Mohammed, and C. Hamon. 2003. "Tandem mass tags: a novel quantification strategy for comparative analysis of complex protein mixtures by MS/MS." *Anal Chem* no. 75 (8):1895-904.

Volchenboum, S. L., K. Kristjansdottir, D. Wolfgeher, and S. J. Kron. 2009. "Rapid validation of Mascot search results via stable isotope labeling, pair picking, and deconvolution of fragmentation patterns." *Mol Cell Proteomics* no. 8 (8):2011-22. doi: M800472-MCP200 [pii] 10.1074/mcp.M800472-MCP200.

Wang, G., W. W. Wu, T. Pisitkun, J. D. Hoffert, M. A. Knepper, and R. F. Shen. 2006. "Automated quantification tool for high-throughput proteomics using stable isotope labeling and LC-MSn." *Anal Chem* no. 78 (16):5752-61. doi: 10.1021/ac060611v.

Wang, G., W. W. Wu, W. Zeng, C. L. Chou, and R. F. Shen. 2006. "Label-free protein quantification using LC-coupled ion trap or FT mass spectrometry: Reproducibility, linearity, and application with complex proteomes." *J Proteome Res* no. 5 (5):1214-23.

Wang, H., C. H. Wong, A. Chin, J. Kennedy, Q. Zhang, and S. Hanash. 2009. "Quantitative serum proteomics using dual stable isotope coding and nano LC-MS/MSMS." *J Proteome Res* no. 8 (12):5412-22. doi: 10.1021/pr900158n.

Warren, C. M., D. L. Geenen, D. L. Helseth, Jr., H. Xu, and R. J. Solaro. 2010. "Sub-proteomic fractionation, iTRAQ, and OFFGEL-LC-MS/MS approaches to cardiac proteomics." *J Proteomics* no. 73 (8):1551-61. doi: S1874-3919(10)00109-0 [pii] 10.1016/j.jprot.2010.03.016.

Wenger, C. D., D. H. Phanstiel, M. V. Lee, D. J. Bailey, and J. J. Coon. 2011. "COMPASS: a suite of pre- and post-search proteomics software tools for OMSSA." *Proteomics* no. 11 (6):1064-74. doi: 10.1002/pmic.201000616.

Wiener, M. C., J. R. Sachs, E. G. Deyanova, and N. A. Yates. 2004. "Differential mass spectrometry: a label-free LC-MS method for finding significant differences in complex peptide and protein mixtures." *Anal Chem* no. 76 (20):6085-96.

Wu, C. C., M. J. MacCoss, K. E. Howell, D. E. Matthews, and J. R. Yates, 3rd. 2004. "Metabolic labeling of mammalian organisms with stable isotopes for quantitative proteomic analysis." *Anal Chem* no. 76 (17):4951-9. doi: 10.1021/ac049208j.

Yan, W., D. Hwang, and R. Aebersold. 2008. "Quantitative proteomic analysis to profile dynamic changes in the spatial distribution of cellular proteins." *Methods Mol Biol* no. 432:389-401. doi: 10.1007/978-1-59745-028-7_26.

Yan, W., H. Lee, E. C. Yi, D. Reiss, P. Shannon, B. K. Kwieciszewski, C. Coito, X. J. Li, A. Keller, J. K. Eng, T. Galitski, D. R. Goodlett, R. Aebersold, and M. G. Katze. 2004. "System-based proteomic analysis of the interferon response in human liver cells." *Genome Biol* no. 5 (8):R54.

Yao, X., A. Freas, J. Ramirez, P. A. Demirev, and C. Fenselau. 2001. "Proteolytic 18O labeling for comparative proteomics: model studies with two serotypes of adenovirus." *Anal Chem* no. 73 (13):2836-42.

Zhang, R., and F. E. Regnier. 2002. "Minimizing resolution of isotopically coded peptides in comparative proteomics." *J Proteome Res* no. 1 (2):139-47.

Zhang, R., C. S. Sioma, R. A. Thompson, L. Xiong, and F. E. Regnier. 2002. "Controlling deuterium isotope effects in comparative proteomics." *Anal Chem* no. 74 (15):3662-9.

Zhang, R., C. S. Sioma, S. Wang, and F. E. Regnier. 2001. "Fractionation of isotopically labeled peptides in quantitative proteomics." *Anal Chem* no. 73 (21):5142-9.

Zhang, Y., A. Wolf-Yadlin, P. L. Ross, D. J. Pappin, J. Rush, D. A. Lauffenburger, and F. M. White. 2005. "Time-resolved mass spectrometry of tyrosine phosphorylation sites in the epidermal growth factor receptor signaling network reveals dynamic modules." *Mol Cell Proteomics* no. 4 (9):1240-50. doi: M500089-MCP200 [pii] 10.1074/mcp.M500089-MCP200.

Zhou, H., J. A. Ranish, J. D. Watts, and R. Aebersold. 2002. "Quantitative proteome analysis by solid-phase isotope tagging and mass spectrometry." *Nat Biotechnol* no. 20 (5):512-5. doi: 10.1038/nbt0502-512 [pii].

12

The Use of Mass Spectrometry in Characterization of Bone Morphogenetic Proteins from Biological Samples

Genadij Razdorov and Slobodan Vukicevic
Laboratory of Mineralized Tissues, Center for Translational and Clinical Research, School of Medicine, University of Zagreb, Zagreb Croatia

1. Introduction

1.1 Bone morphogenetic proteins

Bone morphogenetic proteins (BMPs) are family of growth factors. (Chen et al., 2004; Vukicevic & Sampath, 2008) Discovered in context of bone biology, today they are recognized as important signaling molecules in wide range of biological processes, such as vertebrate embryonic development (Hogan, 1996), mesenchymal stem cell differentiation (Vukicevic & Grgurevic, 2009), kidney fibrosis, and more. For the last years BMP-2 and BMP-7 are used as therapeutics in orthopedics, harnessing their regenerative potential as growth factors.

From the onset of medicine scholars have been aware of the bone regenerative potential. In 1965. Urist was first to show that demineralized bone matrix (DBM) can induce bone growth if implanted into extraskeletal site. Active component from DBM was named *bone morphogenetic protein* by Urist & Strates (1971). Purification, cloning and sequencing of BMP was done almost 20 years later by Wozney et al. (1988). They showed that BMP is not a single protein but a family of growth factors.

From introduction of the BMP term through cloning and sequencing of individual BMPs in late 1980s, scientific output in the field has constantly grown and has exceeded 1500 papers in 2010. (Figure 1)

BMPs are part of transforming growth factor β (TGF-β) superfamily of proteins. In humans TGF-β superfamily constitutes of 37 proteins. (Figure 2) Beside BMPs, TGF-β superfamily includes TGF-β proteins, inhibins (INH), growth/differentiation factors (GDF) and few others: artemin (ARTN), glial cell line-derived neurotrophic factor (GDNF), left-right determination factor 1 (LFTY1), LFTY2, muellerian-inhibiting factor (MIS), nodal homolog (NODAL), neurturin (NRTN) and persephin (PSPN).

BMPs are functionally and structurally very conserved throughout animal kingdom. Their biological importance is reflected through functional and structural redundancy of different BMPs in single species. BMPs are translated as pre-propeptides. Signal peptide targets them for secretion out of cell. Prodomain is two thirds to four fifths of total peptide length and

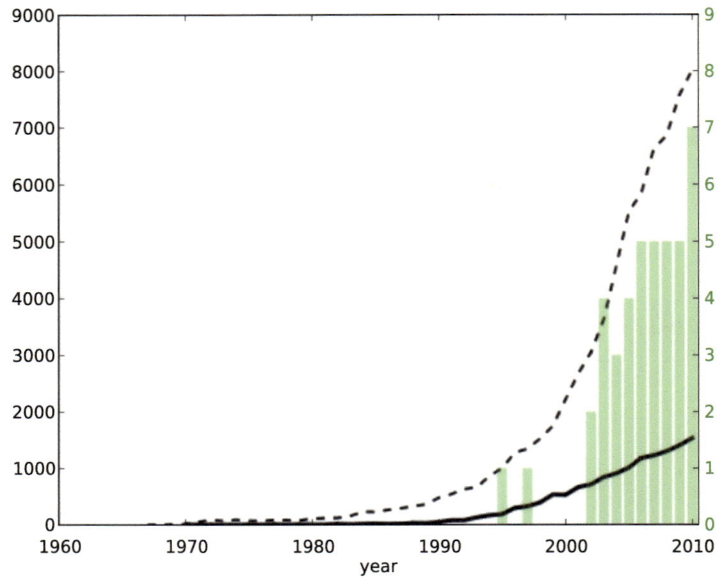

Fig. 1. Scientific output on *bone morphogenetic protein* (solid black line), *(mass spectrometry) protein OR proteomics* (dashed black line) and *(mass spectrometry) AND bone morphogenetic protein* (green bars). Data gathered from Pubmed on June 2011.

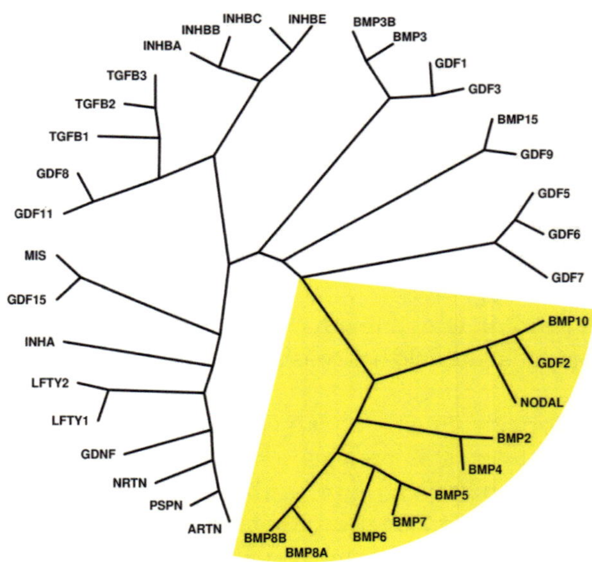

Fig. 2. Transforming growth factor β protein superfamily. Amino acid sequences were aligned by ClustalW and plotted by Cytoscape with PhyloTree plugin.

is cleaved off to produce mature peptide chain. Mature domain is highly conserved and has number of shared features. (Figure 3) There are seven conserved cysteines starting from position 446 (by unified position numbering displayed above aligned sequences), and then on positions 475, 479, 511, 512, 545 and 547. Counting from first conserved cysteine there are 30 completely conserved positions which is almost one third length of mature domain.

From purification and sequencing studies done in late 1980s, it was clear that the active form of a BMP is a homodimer made up of two mature domains connected by disulphide bridge. The first BMP crystal structure was resolved by Griffith et al. (1996). Until present there are more then 30 structures related to BMPs deposited into Protein Data Bank.

BMP mature domain is folded in the hand like structure. (Figure 4) Two fingers, made up of two beta sheets, are protruding from the cystine knot, which is made up of three disulphide bridges. On the opposite side of the cystine knot is alpha helix forming the heel of the hand. Active homodimer is made up of two hand structures pointing in the opposing directions palms facing toward each other.

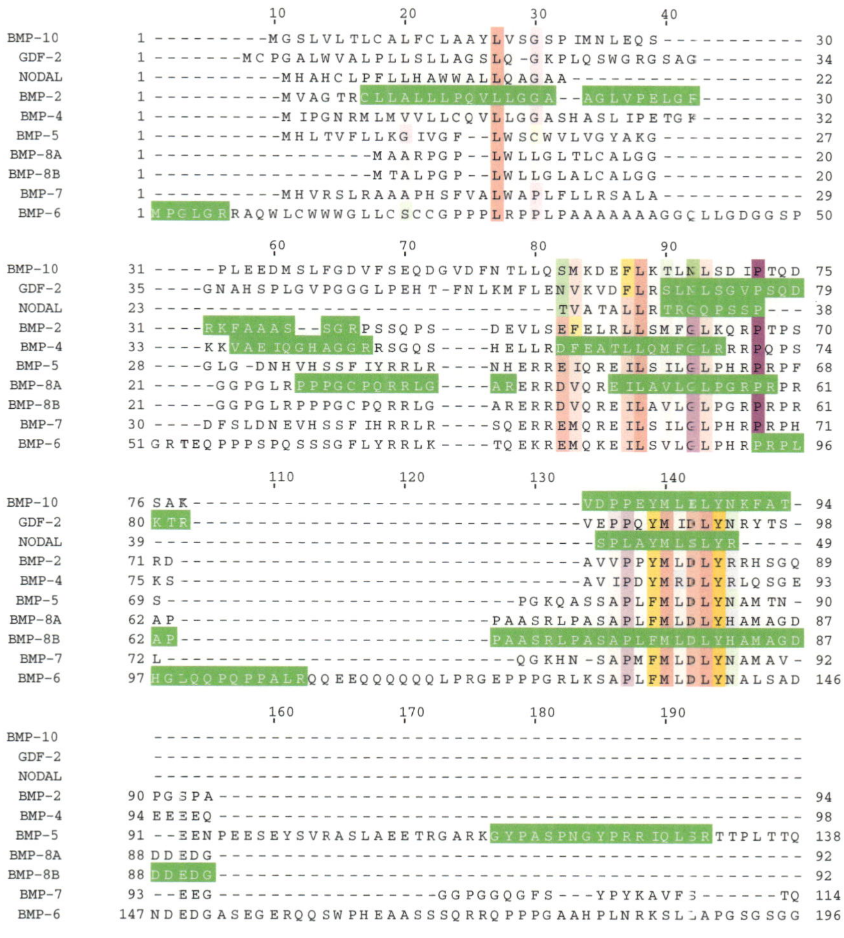

Fig. 3. BMPs sequence alignment. Ten most conserved BMPs aligned by ClustalW algorithm with most conserved residues color coded by their physicochemical properties (Zappo color scheme). Green sequence features are MS identified peptides deposited in PRIDE proteomics database.

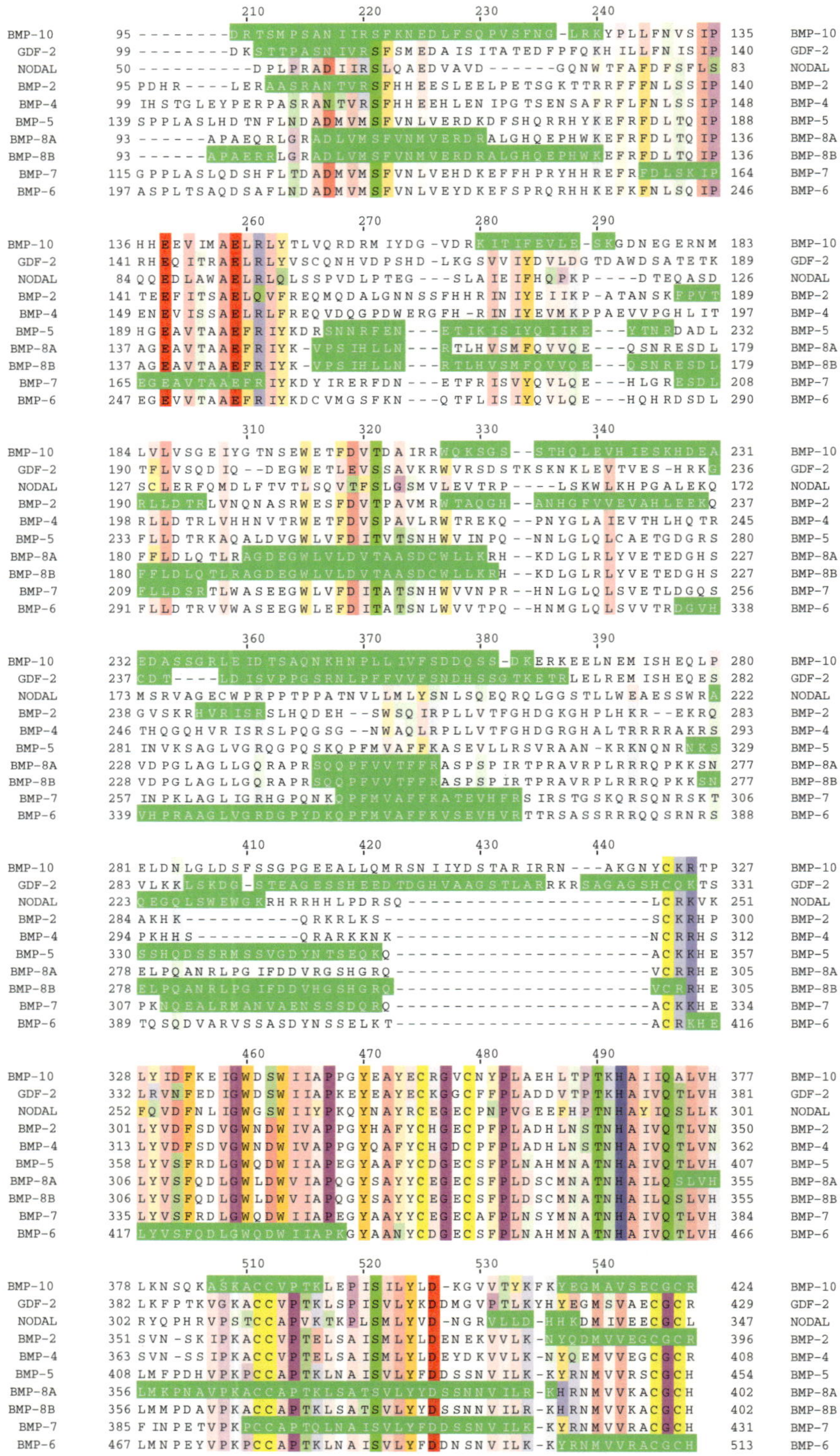

Fig. 3. ...continued.

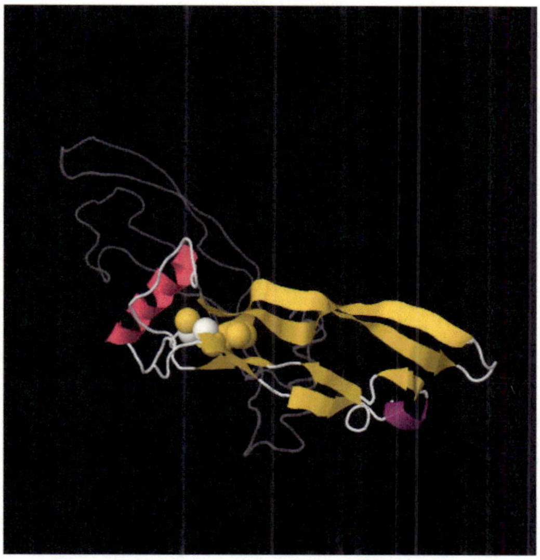

Fig. 4. Crystal structure of the mature homodimer BMP6. (Saremba et al., 2008) One monomer is cartoon rendered with sulfur atoms of cystine knot spacefill rendered. The structure with 2R52 accession number was downloaded from Protein Data Bank, and rendered by RasMol.

1.2 Mass spectrometry of proteins

Mass spectrometry (MS) of proteins exploded with the discovery of soft ionization techniques in late 1980s. Fenn et al. (1989) produced one of the first mass spectra of different proteins using electrospray ionization (ESI). They also introduced an algorithm for deconvolution of multiple charged envelope in the mass spectra of proteins. Hillenkamp & Karas (1990) did a similar thing with the matrix assisted laser dissociation ionization (MALDI).

Identification of proteins by peptide mass fingerprint (PMF) was introduced by Henzel et al. (1993). Proteins, separated by two dimensional gel electrophoresis (2DE), are in gel protease digested (usually by trypsin). Peptide masses are then measured by mass spectrometer. The measured peptide mass fingerprint is used for searching the protein sequence database by automated computer algorithm. The first large scale analysis using PMF was done by Shevchenko et al. (1996). They identified 128 different yeast genes from 150 2D gel spots. This scientific approach became *proteomics* after term *proteome* coined by Kahn (1995).

Protein identification is now done preferentially using tandem mass spectrometry. To avoid errors and to speed up process, algorithms for the protein database searching using tandem mass spectra of peptides had to be invented. (Eng et al., 1994; Perkins et al., 1999)

Except for the identification of proteins, mass spectrometry based quantification became reality with development of isotope-coded affinity tag (ICAT) strategy. (Gygi et al., 1999) After introduction, there have been devised different stable isotope labeling quantification strategies. Ong et al. (2002) developed stable isotope labeling by amino acids in cell culture (SILAC).

At last, proteomics community realized the importance of high mass resolving power, high mass accuracy, scanning speed and dynamic range as crucial parameters on which number and quality of the protein identifications depends. Mann & Kelleher (2008)

2. Bone morphogenetic proteins by mass spectrometry

Before 2000 there were only two papers published reporting on separation of the glycoforms of rhBMP-2 by capillary electrophoresis coupled to MS. (Yeung et al., 1997; Yim et al., 1995) From 2000 onward, there is a continuum of scientific research on the BMPs using a mass spectrometry. There are two main approaches to the research in the field: (a) isolation and identification of a BMP or related molecule from different biological sources; and (b) proteomic study of a system perturbed using a BMP or related molecule. We have reviewed papers based on their biological focus.

2.1 Bone and cartilage

BMPs have been discovered and studied first as bone and cartilage growth factors. After development and growth, bones are constantly remodeled to meet their body function. Resorption of the bone is done by osteoclasts and formation by osteoblasts. Therapeutic regeneration and skeletal tissue engineering depends on three elements: extracellular matrix, cells and growth factors. (Reddi, 1998)

Role of BMP-1-3 in bone fracture repair has recently been studied in our group. (Grgurevic et al., 2011) We have shown *in vivo* enhancement of bone healing by systemic and local administration of BMP-1-3. We used rat fractured femur and critical size defect of rabbit ulnae models and demonstrated a synergistic effect of BMP-1-3 and BMP-7 on bone regeneration. BMP-1-3 stimulates osteoblast differentiation *in vitro*. By combining isolation of growth factors with proteomic global approach we have identified potential biomarkers of acute bone fracture from plasma of 25 patients. (Grgurevic et al., 2007) To overcome inherent limitation of proteomics to cope with plasma protein dynamic range of more than 10 orders of magnitude, we have devised a growth factors enrichment strategy. Members of TGF-β protein superfamily have characteristic heparin binding domain as part of their structure, and can be affinity enriched using heparin chromatography. We enriched growth factors from 80 mL of pulled plasma, and than separeted them on minigel electrophoresis. To overcome loading capacity of analytical gels, proteins were loaded into consecutive wells, and afterwards combined in 12 fractions by electrophoretic mobility. After in-gel digestion, peptides were separated and measured by on-line nano LC coupled to Orbitrap working in data dependent scanning mode. Top five MS signals were isolated and fragmented by CID in linear IT. Global proteomic approach allowed us to identify 213 proteins, of which we have singled out 12 bone and cartilage related proteins. Four of them have never before been found in the circulation: Transforming growth factor β-induced protein ig-h3 (β ig-h3), cartilage acidic protein 1 (CRTAC1), procollagen C proteinase enhancer protein (PCPE-1) and TGF-β receptor type 3 (TGF-β-R-3). Nakasaki et al. (2008) have identified insulin-like growth factor I(IGF-I) as potent chemotactic factor for osteoblasts and showed IGF-I involvement in fracture healing. They used five step sequential chromatography to purify IGF-I from 18 L of MC3T3-E1 (mouse osteoblast-like cell line) serum free conditioned media (CM). After precipitation of CM with 80% saturated ammonium sulphate, they used gelatine-, Blue-, heparin-affinity chromatography, hydroxyapatite chromatography and gel filtration to purify IGF-I more then 200 fold. IGF-I was separated by silver stained SDS-PAGE, in gel digested, and measured by micro LC coupled to 3D IT working in data dependant mode.

Kim et al. (2009a) have studied transdifferentiation of murine premyoblast C2C12 cell line induced by BMP-2. They invented the Two-Stage Double-Technique Hybrid (TSDTH) proteomic strategy, for the analysis of early phosphoproteome and late proteome changes.

Authors have enriched more then 150 μg of phosphoproteins from 2 mg of the total protein cell lysate using commercial kit. Using a phosphoprotein enrichment, as opposed to a phosphopeptide enrichemnt, one should get a better sequence coverage of the individual phosphorylated proteins. Phosphoproteins were separated by SDS-PAGE, in gel digested, and analysed by micro LC coupled to linear IT. They measured MS3 fragmentation spectra of phosphopeptides showing dominant neutral loss of phospho moiety (minus 98 Th/49 Th/32.7 Th for singly, doubly or triply charged species respectively) using data dependant mode. For the proteome quantification, authors metabolically labelled all cell proteins using SILAC approach. After combining samples, they separated proteins using SDS-PAGE, in gel digested them, and analysed them by micro LC coupled to LTQ-FT working in data dependent mode. Top five signals from high resolution survey scan were measured individually using single ion monitoring (SIM) scan in FT-ICR analyser, and then CID fragmented in linear IT. From 1321 phosphoproteins identified and 433 proteins quantified authors selected 374 BMP-2 specific phosphoproteins and 54 differentially expressed proteins. By pathway enrichment analysis authors found IGF and calcium signaling pathways as well as TGF-β/BMP signaling proteins to be potentially involved in the early and long-term actions of BMP-2.

At present, BMP-2 and BMP-7 are the only BMPs commercially used for regenerative purposes. We have identified lysine 60 of BMP-6 as amino acid crucial for BMP-6 lower susceptibility to noggin inhibition in comparison to BMP-7. (Song et al., 2010) This is important because of the present therapeutic use of BMP-7 in large doses. In BMP-6, lysine 60 forms intramolecular hydrogen bonds with asparagine 65 connecting finger 1 and 2 in that way that it increases overall rigidity of the molecule.

Osteoporosis is one of the most prevalent bone disease. (Raisz, 2005) We have studied effects of systemically administered BMP-6 to osteoporotic rats. (Simic et al., 2006) We could not demonstrate the presence of BMPs in bone extract of ovariectomized rats. We next demonstrated that systemically administered recombinant human mature BMP-6 accumulates in the skeleton. BMP-6, systemically applied, restores the bone inductive capacity, microarchitecture, and quality of the skeleton in osteoporotic rats. Next, using gene expression profiling, we found that BMP-6 exerts its osteoinductive effect at least in part through the IGF-I and epidermal growth factor (EGF) pathways. (Grasser et al., 2007) Hong et al. (2010) have studied impairment of osteoblasts differentiation in context of the glucocorticoid-induced osteoporosis. They treated MC3T3-E1 cells with high dose of dexamethasone, which inhibited cell differentiation and proliferation and induced apoptosis. Authors used SILAC labelling with heavy/light lysine, SDS-PAGE protein separation and nano LC coupled to LTQ-Orbitrap proteomic approach. They found upregulation of tubulins (TUBA-1A, TUBB-2B, and TUBB-5), Ras GTPase-activating-like protein 1 (IQGAP1), S100 proteins (S100-A11, S100-A6, S100-A4, and S100-A10), myosins (MYH-9 and MYH-11), and apoptosis and stress proteins (BAX), and downregulation of ATP synthases (ATP5O, ATP5H, ATP5A1, and ATP5F1), G3BP-1, and Ras-related proteins (RAB-1A, RAB-2A, and RAB-7).

Hong et al. (2010a) have studied early osteoblast differentiation. They induced differentiation of mouse osteoprogenitor MC3T3-E1 cells by rhBMP-2 and have studied cells in an early differentiation stage critical for transformation of premature osteoblasts to mature osteoblasts. At that stage osteoblasts express ALP but do not form mineralized nodules yet. By label-free quantitative proteomic approach they identified numerous differentially expressed proteins important for actin skeleton regulation and/or focal adhesion. They used SDS-PAGE proteins separation, and nano LC coupled to LTQ-Orbitrap peptide analysis. By western blot they have confirmed upregulation of IQGAP1, isoform 1 of gelsolin (GSN), moesin, radixin, and cofilin 1 (CFL-1), and upregulation of focal adhesion proteins FLN-A, LAMA1, LAMA5, COL1A1, COL3A1, COL4A6, and COL5A2, as well as the downregulation of COL4A1, COL4A2, and COL4A4.

Park et al. (2009) have studied heparin-binding growth factor 2 (HBGF-2) effect on osteoblasts. Authors used 2DE/MALDI-TOF proteomic approach. They identified asparaginyl-tRNA synthetase (NARS), eukaryotic peptide chain release factor subunit 1 (ETF1), GDP-forming succinyl-CoA synthetase (SUCLG2), heat shock protein 84 (HSP 84), sorting nexin 9 (SNX9) and neutral α-glucosidase AB (GANAB) to be upregulated, and tropomyosin 2 to be downregulated. NARS, member of aminoacyl-tRNA synthetases (AARS), was upregulated more then 900 fold. NARS upregulation increased the cell proliferation. NARS downregulation suppressed cell proliferation and induced cell apoptosis. Also downregulation of NARS increased serum deprivation induced apoptosis. Reduction of NARS also reduced p-Akt activity and increased caspase-3. Authors shown that HBGF-2 induced NARS promotes osteoblast survival through PI3K/Akt pathway.

Kodaira et al. (2006) have purified BMP-like factor from fetal bovine serum (FBS), which they found to both inhibits myogenesis and stimulates differentiation of osteoblasts. They started with 5 L of FBS, which they separated by heparin, strong anion exchange (SAX), heparin again, strong cation exchange (SCX) and reverse phase chromatography. Using LC-MS/MS authors identified this factor as BMP4, and found it to be in over 100 kDa complex in circulation. Behnam et al. (2006) identified dermatopontin (DPT) as the most prominent BMP-2 co-purified demineralized bone matrix protein, using alkaline urea extraction of BM, separation of insoluble fraction by hydroxyapatite chromatography, SDS-PAGE of BMP active fraction and MALDI-TOF analysis. Kubota et al. (2002) have studied signal transduction from osteoclasts to osteoblasts. They have isolated and identified osteoblastogenesis inhibitory factor (OBIF) from RANKL treated RAW264.7 cell line conditioned medium. This OBIF inhibited osteoblastogenesis of MC3T3-E1 cell line, induced by BMP-4. After three step chromatography (heparin affinity, anion exchange and reverse phase chromatography) from 1.8 L of conditioned medium, they have managed to identify OBIF as platelet-derived growth factor BB (PDGF BB) homo-dimer using nano LC coupled to LCQ ion trap, working in data dependant mode..

2.1.1 Chondrogenesis

Ji et al. (2010) have studied mesenchymal stem cells chondrogenesis. They induced chondrogenic differentiation of C3H10T1/2 cells, murine embryonic mesenchymal cell line, by treatment with BMP-2. They validated chondrogenesis by following glycosaminoglycan and collagen type II. By isobaric tags for relative and absolute quantitation (iTRAQ) labeling and 2D nano LC-MS/MS (QSTAR XL) analysis they identified 100 differentially expressed proteins out of 1753 identified and quantified proteins. 83 proteins were downregulated, and 17 upregulated. This is in concordance with "stem state" concept, which assumes wider set of proteins expressed in cells which are less differentiated. Collagen types II and XI are upregulated, as are PAPSS 2 and LOX, enzymes involved in posttraslational modification of chondrocytes extracellular matrix components. Most enriched category of downregulated proteins is energy metabolism. Chondrocytes produce ATPs by anaerobic glycolysis, because cartilage is avascular and hypoxic. IGF-I was upregulated and was shown to exhibit additive effect on chondrogenesis with BMP-2. Fibulin-5, an ECM protein, was upregulated. BTF3L4, general transcription factor, was upregulated. Kim et al. (2010) have studied chondrogenesis of adipose tissue-derived stem cells (ASCs). They identified 756 proteins using nano LC coupled to LCQ MS/MS approach. From 33 chondrogenic factors or proteins identified authors especially mentioned type 2 collagen, biglycan, IGF-binding protein and TGF-β1.

2.2 Stem cells

One of the approaches in the regenerative medicine is the use of stem cells as cell therapies. (Keller, 2005; Murry & Keller, 2008; Wobus & Boheler, 2005) Their self renewal capability, the unlimited potential for differentiation, and signals that control differentiation fate are under broad scientific scrutiny. We have reviewed the role of BMP-6 in mesenchymal stem cell differentiation. (Vukicevic & Grgurevic, 2009) BMP-6 roles have been reviewed in bone remodeling, bone to pancreas coupling and kidney development.

Lee et al. (2010) have studied BMP-2 induced differentiation of bone marrow stem cells (BMSC). They identified and quantified 449 proteins using SILAC LC-MS/MS approach. Proteins were separated by SDS-PAGE and in gel digested. Peptides were separated by LC and analysed by LTQ-FT, working in the data dependant mode. Top five signals from the survey scan were measured in the FT analyser using SIM, and CID fragmented in the linear IT. 12 proteins were upregulated, 7 proteins were downregulated, and 19 proteins were only detected in BMP-2 induced cells. Authors singled out β-catenin (identified only in BMP-2 induced differentiated cells with 4 peptides). Wnt/β-catenin pathway is known to be important in osteoblast differentiation. They ruled out Smad and ERK BMP pathways on β-catenin upregulation. They identified PI3K pathway as crucial for BMP-2 induced BMSC differentiation. Willert et al. (2003) have for the first time isolated Wnt3a in an active form. They have expressed Wnt3a in mouse L cells, purified it from 2 L of conditioned medium by three step chromatography (Blue sepharose, gel filtration and heparin cation exchange) in the presence of CHAPS, due to hydrophobic nature of the Wnt protein. They have pinpointed hydrophobicity to lipid modification, and have identified cysteine 77 as place of palmitation, by MudPIT.

Hoof et al. (2009) studied phosphorylation dynamics during BMP4 induced differentiation of hESCs. They used SILAC labeling and phosphopeptide enrichment by SCX-TiO2 chromatography. High resolution Orbitrap, working in the data dependant top two mode, was used for the measurement of peptides in four time points (0, 30, 60, 240 min). Authors identified 5222 proteins, of which 1399 were phosphorylated on 3067 sites (2431 serines, 582 threonines, and 54 tyrosines). Half of the quantified phosphopeptides are regulated within the first hour of BMP-4 initiated differentiation. 586 identified proteins are regulated by the core transcriptional network. By a GO analysis prominent protein groups are associated with the epigenetic modification, transcription and translation. Authors found increased phosphorylation of SMADs, PI3K/AKT pathway activation and transient activation of JNKs. Regulators of pluripotency are also altered, but SOX2 did not changed its phosphorylation status. By analysis of the predicted kinases responsible for measured phosphorylation, authors found that CDK1/2 has central role.

Yocum et al. (2008) have identified and verified several potential markers of noggin-induced neural and BMP-4-induced epidermal ectoderm differentiation of hESCs. They developed hESCs cell culture system (on gelatin-coated dishes) free of mouse embryonic fibroblasts (MEF) feeder layer or conditioned media. In this way targeted proteome is not contaminated by proteins from MEFs and noggin/BMP4 effect on hESCs is not influenced by possible MEFs response. They used 2D LC/MALDI-TOF/TOF (ABI 4800 TOF/TOF) for global analysis and MRM on 4000 Q Trap for verification and targeted analysis. Beside verification of tubulin β-III and cytokeratin-8, previously known markers for neuronal and epithelial differentiation respectively, they proposed nuclear autoantigenic sperm protein (NASP) as marker for pluripotency and dihydropyrimidinase-related protein (DRP) 2 and 4 as markers for the early neuronal differentiation.

Nunomura et al. (2005) have identified 200 membrane proteins from mouse embryonic stem cells using biotin based cell surface labeling, biotin-avidin affinity peptide purification and 2D LC-MS/MS analysis on Q-Tof 2. Among 82 proteins involved in different cell signaling pathways, they identified leukemia inhibitory factor receptor (LIF-R), interleukin-6 receptor subunit β (IL-6Rβ), ciliary neurotrophic factor receptor (CNTFR-α), BMPR-1a and integrin α6β1.

Prowse et al. (2005) identified 102 proteins secreted by human neonatal fibroblasts, which are used for maintenance of undifferentiated growth of human embryonic stem cells, using 2D LC-MS and 2DE/MALDI-TOF. Among others they have identified proteins involved in cell adhesion, cell proliferation and inhibition of cell proliferation, Wnt signaling and inhibition of BMPs.

Kurisaki et al. (2005) have studied effects of the removal of leukemia inhibitory factor (LIF) from mouse embryonic stem cells. LIF maintains pluripotency of mouse ESCs, by activating JAK/STAT3 pathway. Only in cooperation with BMPs, can LIF inhibit differentiation of mouse ESCs. BMPs activate expression of Id genes and suppress neuronal differentiation. Authors identified more then 100 proteins by DIGE and MALDI-TOF analysis on AXIMA-CFR Plus and/or 4700 Proteomics Analyzer.

2.3 Cancer

BMPs, as part of the TGF-β protein superfamily, are involved in the cancer patophysiology. (Massague, 2008) Klose et al. (2011) have shown BMP-7 induced cell cycle arrest at G_1/S checkpoint of human glioma-derived Gli36ΔEGFR-LITG cells. Kim et al. (2009) have studied PRL-3-mediated metastasis using B16 and B16F10 mouse melanoma cell lines. Using 2D DIGE and MS they identified the heat shock protein 70, fascin-1, septin-6, ATP synthase beta subunit, and bone morphogenetic protein receptor type IB, as proteins that differ in cancer cells with low and high metastatic potential. We showed the inhibitory role of BMP-7 to bone metastases of breast cancer (Buijs et al., 2007) and prostate cancer (Buijs et al., 2007a). The cancer metastasis can be triggered by TGF-β stimulating epithelial-to-mesenchymal transition. Recombinant BMP-7 is inhibiting cancer bone metastases growth, so we proposed it as novel therapy for bone metastases. Rivera et al. (2007) have identified 43 differentially expressed proteins in melanoma cell line with breast cancer metastasis-suppressor 1 (BRMS1) gene silenced or over-expressed compared to wild type. They used 2D-DIGE for relative quantification and MALDI-TOF/TOF (ABI 4800 TOF/TOF Analyzer) for protein identification from preparative 2D gel. Among others they identified and confirmed BMPR-II. van Gils et al. (2005) reviewed serum and urine markers in European prostate cancer P-Mark project. BMP-6 was considered as bone metastasis prognostic serum marker, but later dismissed due to technical challenges in BMP-6 serum detection. Cheung et al. (2004) have identified 24 kDa SELDI-TOF signal specific to early prostate carcinogenesis as dimeric form of mature GDF-15. By laser capture micro-dissection

(LCM) they have isolated normal, early cancer (Gleason grade 2 to 4) and high-grade prostatic intraepithelial neoplasia (hPIN) epithelial cells from 22 patients. For SELDI-TOF analysis, authors used a copper IMAC chemistry, and proteins measured by Ciphergen Protein Biology System II spectrometer (Ciphergen Biosystems, Inc.). For the protein identification, authors first separated proteins by SDS-PAGE. 24 kDa band was analysed by oMALDI Q-TOF (ABI Q-Star).

2.4 Kidney

We have previously reviewed the role of BMPs on the development and homeostasis of kidneys. (Simic & Vukicevic, 2005) BMPs have been connected to kidney biology already in 1990s. The BMPs role in kidney development has been discovered by early localization studies using immunohistochemistry (Vukicevic et al., 1994a), autoradiography (Vukicevic et al., 1990) and *in situ* hybridization (Helder et al., 1995; Vukicevic et al., 1994). We have shown that BMP-7 is crucial for metanephric mesenchyme differentiation during kidney development. (Vukicevic et al., 1996)

BMP-7 is effective in treatment of acute renal failure, as we have shown in a ishemia/reperfusion rat model. (Vukicevic et al., 1998) Wang & Hirschberg (2011) have identified Y-box protein-1 (YB-1) as BMP-7 transcriptional activator in context of chronic kidney disease (CKD) using LC-MS/MS. They used promotor region of BMP7 gene to fish out factors from MDCK nuclear fraction. Proteins were separated by SDS-PAGE, and analysed by nano LC coupled to LTQ-Orbitrap XL. We have identified BMP-6, GDF-15 and BMP-1-3 from plasma of healthy volunteers and patients with CKD. (Grgurevic et al., 2011a) We showed an increased renal fibrosis in rats with CKD after systemic administration of BMP-1-3. Administration of anti BMP-1-3 antibody reduced the fibrosis.

2.5 Iron

Iron content is tightly controlled by individual cell and also systemically by hole organism. (Hentze et al., 2004; 2010) We found that BMP-6 is an endogenous regulator of iron metabolism *in vivo*. (Andriopoulos et al., 2009) *Bmp6* knockout mouse has reduced hepcidin expression and tissue iron overload, which resembles hemochromatosis. Hepcidin has a central role in maintaining iron body levels. Recombinant BMP6, when administered to mouse with hemochromatosis reduced the serum iron in a dose-dependent manner. Next, we found that BMP-6 treatment reduces hemochromatosis in *Hfe* knockout mouse. (Corradini et al., 2010) We also showed BMP-6 effect on TMPRSS6 expression, which is a negative regulator of hepcidin. (Meynard et al., 2011)

Kartikasari et al. (2008) used SELDI-TOF MS to detect and quantify hepcidin-25, 25 residues long hepcidin isoform which regulates the iron homeostasis. Authors used cation exchange chip for protein binding, and Ciphergen Protein Biology System IIc TOF mass spectrometer for the measurements. They found synergistic upregulation of hepcidin-25 by BMP-2/9 and IL-6, suggesting a crosstalk between iron and inflammatory pathways.

2.6 Other

BMP-15 is oocyte secreted growth factor important for female fertility. (Elvin et al., 2000) Saito et al. (2008) have characterized rhBMP-15 expressed in human embryonic kidney 293 cells by MS. Authors used MALDI-TOF (Reflex 3, Bruker) for the protein molecular weight measurement and micro LC coupled to LTQ-Orbitrap for the CID fragmentation and the neutral loss measurements. They found N-terminal amino acid to be pyroglutamic acid and C-terminal end to be truncated. 16 kDa BMP-15 form is phosphorylated on serine 6 and 17 kDa BMP-15 form is O-glycosilated on threonine 10. Tibaldi et al. (2010) have identified serine 6 from mature BMP-15 and BMP-9 to be phosphorylated by Golgi apparatus casein kinase (G-CK). Phosphorylation sites were identified using phosphopeptide enrichment and nano LC coupled to LCQ XL, or MALDI-TOF/TOF (ABI 4800 Plus) analysis.

Li et al. (2007) have studied ubiquitination of Smad1 mediated by carboxyl terminus of Hsc70-interacting protein (CHIP). They have confirmed N-termus mono-ubiquitination of Smad1 by MALDI-TOF analysis of *in vitro* ubiquinated Smad1.

Pulmonary arterial hypertension (PAH) was reviewed by Rubin (1997) and Rabinovitch (2008). Abdul-Salam et al. (2010) found 25 differentially expressed out of more than 300 identified proteins by SDS-PAGE and LC-MS/MS from lung tissue of 8 patients and 8 control subjects. They found for the first time increased expression of chloride intracellular channel 4 (CLIC4), receptor for advanced glycation end products, and periostin. Meyrick et al. (2008) found 16 differentially expressed proteins in familial pulmonary arterial hypertension (FPAH) patients compared to obligate carrier from the same family with known BMPR-2 mutation. They used 2D-DIGE/MALDI-TOF(/TOF) of EBV-transformed B lymphocytes. They connected adapter protein growth factor receptor bound protein (GRB2) to signal transduction of BMPR-2 receptor.

Pappano et al. (2003) have identified *in vivo* substrates of BMP-1 and mammalian Tolloid-like metalloproteinases (mTLL) using proteomics. They compared *in vivo* procollagen C-proteinase (pCP) activity of wild type *vs Bmp1 Tll* doubly homozygous null mice. Using nano ESI-QTof sequencing, they have identified four 2DE spots (absent in $Bmp1^{-/-}$ $Tll^{-/-}$ mouse embryo fibroblasts, MEF, conditioned media) as C-propeptides of the proα1 chain of type I procollagen, of the proα1 chain of type III procollagen, of the proα2 chain of type III procollagen and for the first time proline- and arginine-rich protein (PARP) subdomain of the N-terminal globular sequences of the proα1 chain of type XI collagen.

The low resolution structure of procollagen C-proteinase enhancer 1 (PCPE-1), an extracellular matrix glycoprotein that can stimulate the pCP action of tolloid metalloproteinases, was determined by Bernocco et al. (2003). Using MALDI-TOF they have determined MW of PCPE-1 to be 48628 Da. Wajih et al. (2004) have studied intracellular processing and transport of the matrix γ-carboxyglutamic acid protein (MGP) in human vascular smooth muscle cells (VSMCs) infected with adenovirus carrying the MGP construct. MGP is an inhibitor of arterial wall and cartilage calcification through the binding of BMP-2. They have identified bovine fetuin instead of MGP using SDS-PAGE followed by nano LC coupled to LCQ Deca XL ion trap MS. Using glutathion-S-transferase (GST) pull-down followed by 2DE/MALDI-TOF analysis, Hassel et al. (2004) identified 33 proteins interacting with BMPRII. Wermter et al. (2007) studied the substrate selectivity of BMP-1 using the full-length vs. isolated proteolytic domain of BMP-1. They managed to over-express BMP-1 catalytic domain in E. coli and to refold it properly. By MALDI-TOF analysis of reduced vs. unreduced tryptic peptides they confirmed the proper formation of three disulphide bridges (C65-C85, C43-C199 and C63-C66). Number of ligands that are binding to activin type II receptor (ActRII) and ActRIIB from human and mouse sera have been identified using affinity purification, SDS-PAGE and nano LC coupled to linear IT (LTQ). (Souza et al., 2008) They confirmed that in addition to myostatin, BMP-11 and activins-A, -B and -AB could regulate the muscle growth by inhibiting myblast-to-myotube differentiation. Meleady et al. (2008) have compared Chinese hamster ovary (CHO) cells expressing rhBMP-2 to cell co-expressing soluble exogenous paired basic amino acid cleaving enzyme (PACEsol), which improves post-translational processing of the mature rhBMP-2. They used 2D-DIGE/MALDI-TOF and found 60 differentially expressed proteins.

3. Conclusion

Growth factors MS based characterization from biofluids is almost exclusively done by CID on plethora of different tandem instruments. Very low levels of growth factors in circulation is limiting their MS based identification due to huge dynamic range of plasma/serum proteins. Different biochemical purification strategies are used to enrich individual growth factors enough to be able to analyse it by MS. Most of them use heparin affinity chromatography as one purification step. We have successfully identified number of growth factors from biofluids using heparin based enrichment, followed by the SDS-PAGE protein separation and peptide analysis on nano LC coupled to LTQ-Orbitrap.

4. Future prospects

Disparity between liters of starting material used in the isolation and identification studies of individual growth factors from biological fluids, and an unmet need to use small amounts of biological samples for high throughput clinically relevant studies, is ending by development of growth factors enrichment strategies and of course by development of new and better mass spectrometers, and/or new proteomic strategies.

In the future we can expect more and more proteomic studies with BMPs and other growth factors to be identified, quantified and/or characterized from different clinically relevant samples. This will help to better understand particular diseases, discover new biomarkers and to develop new therapeutic strategies.

5. Abrreviations

2DE	Two dimensional gel electrophoresis
β ig-h3	Transforming growth factor-β-induced protein ig-h3 (β ig-h3) (Kerato-epithelin) (RGD-containing collagen-associated protein) (RGD-CAP)
AARS	Aminoacyl-tRNA synthetases
ARTN	Artemin (Enovin) (Neublastin)
ASC	Adipose tissue-derived stem cells
ATP5F1	ATP synthase B chain, mitochondrial precursor
ATP5H	ATP synthase D chain, mitochondrial
ATP5O	ATP synthase O subunit, mitochondrial precursor
ATPA1	ATP synthase subunit alpha, mitochondrial precursor
BAX	Apoptosis regulator BAX
bFGF	Heparin-binding growth factor 2 (HBGF-2) (Basic fibroblast growth factor) (bFGF)
BMP	Bone morphogenetic protein

BMP-1	Bone morphogenetic protein 1 (BMP-1) (EC 3.4.24.19) (Mammalian tolloid protein) (mTld) (Procollagen C-proteinase) (PCP)
BMP-10	Bone morphogenetic protein 10 (BMP-10)
BMP-15	Bone morphogenetic protein 15 (BMP-15) (Growth/differentiation factor 9B) (GDF-9B)
BMP-2	Bone morphogenetic protein 2 (BMP-2) (Bone morphogenetic protein 2A) (BMP-2A)
BMP-3	Bone morphogenetic protein 3 (BMP-3) (Bone morphogenetic protein 3A) (BMP-3A) (Osteogenin)
BMP-3B	Bone morphogenetic protein 3B (BMP-3B) (Bone-inducing protein) (BIP) (Growth/differentiation factor 10) (GDF-10)
BMP-4	Bone morphogenetic protein 4 (BMP-4) (Bone morphogenetic protein 2B) (BMP-2B)
BMP-5	Bone morphogenetic protein 5 (BMP-5)
BMP-6	Bone morphogenetic protein 6 (BMP-6) (VG-1-related protein) (VG-1-R) (VGR-1)
BMP-7	Bone morphogenetic protein 7 (BMP-7) (Osteogenic protein 1) (OP-1) (Eptotermin α)
BMP-8A	Bone morphogenetic protein 8A (BMP-8A)
BMP-8B	Bone morphogenetic protein 8B (BMP-8) (BMP-8B) (Osteogenic protein 2) (OP-2)
BMPR-II	Bone morphogenetic protein receptor type-2 (BMP type-2 receptor) (BMPR-2) (EC 2.7.11.30) (Bone morphogenetic protein receptor type II) (BMP type II receptor) (BMPR-II)
BMSC	Bone marrow stem cell
BRMS1	Breast cancer metastasis-suppressor 1
BTF3L4	Transcription factor BTF3 homolog 4 (Basic transcription factor 3-like 4)
C3H10T1/2	Murine embryonic mesenchymal cell line
CDK1	Cyclin-dependent kinase 1 (CDK1) (EC 2.7.11.22) (EC 2.7.11.23) (Cell division control protein 2 homolog) (Cell division protein kinase 1) (p34 protein kinase)
CFL-1	Cofilin-1 (18 kDa phosphoprotein) (p18) (Cofilin, non-muscle isoform)
CHIP	E3 ubiquitin-protein ligase CHIP (EC 6.3.2.-) (Antigen NY-CO-7) (CLL-associated antigen KW-8) (Carboxy terminus of Hsp70-interacting protein) (STIP1 homology and U box-containing protein 1)
CID	Collision induced dissociation
CKD	Chronic kidney disease
CLIC4	Chloride intracellular channel protein 4 (Intracellular chloride ion channel protein p64H1)
CNTFR-α	Ciliary neurotrophic factor receptor subunit α (CNTF receptor subunit α) (CNTFR-α)
COL1A1	Collagen α-1(I) chain (α-1 type I collagen)

CRMP-2	Dihydropyrimidinase-related protein 2 (DRP-2) (Collapsin response mediator protein 2) (CRMP-2) (N2A3) (Unc-33-like phosphoprotein 2) (ULIP-2)
CRTAC1	Cartilage acidic protein 1 (68 kDa chondrocyte-expressed protein) (CEP-68) (ASPIC)
Da	Dalton
DBM	Demineralized bone matrix
DIGE	Difference gel electrophoresis
DPT	Dermatopontin (Tyrosine-rich acidic matrix protein) (TRAMP)
DRP-2	Dihydropyrimidinase-related protein 2 (DRP-2) (Collapsin response mediator protein 2) (CRMP-2) (N2A3) (Unc-33-like phosphoprotein 2) (ULIP-2)
EBV	Epstein-Barr virus
EGF	Epidermal growth factor
ESI	Electrospray ionisation
ETF1	Eukaryotic peptide chain release factor subunit 1 (Eukaryotic release factor 1) (eRF1) (Protein Cl1) (TB3-1)
FLN-A	Filamin-A (FLN-A) (Actin-binding protein 280) (ABP-280) (α-filamin) (Endothelial actin-binding protein) (Filamin-1) (Non-muscle filamin)
FT	Fourier transform mass spectrometry
G3BP-1	Ras GTPase-activating protein-binding protein 1
GANAB	Neutral α-glucosidase AB (EC 3.2.1.84) (α-glucosidase 2) (Glucosidase II subunit α) (α glucosidase 2α neutral subunit)
GDF	Growth/differentiation factor
GDF-1	Embryonic growth/differentiation factor 1 (GDF-1)
GDF-11	Growth/differentiation factor 11 (GDF-11) (Bone morphogenetic protein 11) (BMP-11)
GDF-15	Growth/differentiation factor 15 (GDF-15) (Macrophage inhibitory cytokine 1) (MIC-1) (NSAID-activated gene 1 protein) (NAG-1) (NSAID-regulated gene 1 protein) (NRG-1) (Placental TGF-β) (Placental bone morphogenetic protein) (Prostate derived factor)
GDF-2	Growth/differentiation factor 2 (GDF-2) (Bone morphogenetic protein 9) (BMP-9)
GDF-3	Growth/differentiation factor 3 (GDF-3)
GDF-5	Growth/differentiation factor 5 (GDF-5) (Cartilage-derived morphogenetic protein 1) (CDMP-1) (Radotermin)
GDF-6	Growth/differentiation factor 6 (GDF-6) (Growth/differentiation factor 16)
GDF-7	Growth/differentiation factor 7 (GDF-7)
GDF-8	Growth/differentiation factor 8 (GDF-8) (Myostatin)
GDF-9	Growth/differentiation factor 9 (GDF-9)
GDNF	Glial cell line-derived neurotrophic factor (hGDNF) (Astrocyte-derived trophic factor) (ATF)

GRB2	Growth factor receptor-bound protein 2 (Adapter protein GRB2) (Protein Ash) (SH2/SH3 adapter GRB2)
GSN	Gelsolin
GST	Glutathion-S-transferase
HBGF-2	Heparin-binding growth factor 2 (HBGF-2) (Basic fibroblast growth factor) (bFGF)
hPIN	High-grade prostatic intraepithelial neoplasia
HSP 84	Heat shock protein HSP 90-β (HSP 90) (Heat shock 84 kDa) (HSP 84) (HSP84)
ICAT	Isotope-coded affinity tag
ICR	Ion cyclotron resonance
IGF-I	Insulin-like growth factor I (IGF-I) (Mechano growth factor) (MGF) (Somatomedin-C)
IL-6Rβ	Interleukin-6 receptor subunit β (IL-6 receptor subunit β) (IL-6R subunit β) (IL-6R-β) (IL-6Rβ) (Interleukin-6 signal transducer) (Membrane glycoprotein 130) (gp130) (Oncostatin-M receptor subunit α) (CD antigen CD130)
IMAC	Immobilized metal ion affinity chromatography
INH	Inhibin
INH-α	Inhibin α chain
INH-β-A	Inhibin β A chain (Activin β-A chain) (Erythroid differentiation protein) (EDF)
INH-β-B	Inhibin β B chain (Activin β-B chain)
INH-β-C	Inhibin β C chain (Activin β-C chain)
INH-β-E	Inhibin beta E chain (Activin β-E chain)
IQGAP1	Ras GTPase-activating-like protein 1
IT	Ion trap
iTRAQ	Isobaric tags for relative and absolute quantitation
LAMA1	Laminin subunit α-1 (Laminin A chain) (Laminin-1 subunit α) (Laminin-3 subunit α) (S-laminin subunit α) (S-LAM α)
LC	Liquid chromatography
LC-MS/MS	Liquid chromatography coupled to tandem mass spectrometry
LCM	Laser capture microdissection
LCQ	3D ion trap from Thermo
LFTY1	Left-right determination factor 1 (Left-right determination factor B) (Protein lefty-1) (Protein lefty-B)
LFTY2	Left-right determination factor 2 (Endometrial bleeding-associated factor) (Left-right determination factor A) (Protein lefty-2) (Protein lefty-A) (Transforming growth factor β-4) (TGF-β-4)
LIF	Leukemia inhibitory factor
LIF-R	Leukemia inhibitory factor receptor (LIF receptor) (LIF-R) (D-factor/LIF receptor) (CD antigen CD118)
LOX	Protein-lysine 6-oxidase (EC 1.4.3.13) (Lysyl oxidase)

LTQ	Linear trap quadrupole
MALDI	Matrix assisted laser desorption/ionisation
MC3T3-E1	Mouse osteoblast precursor-like cell line
MEF	Mouse embryo fibroblasts
MGP	Matrix γ-carboxyglutamic acid protein
MIS	Muellerian-inhibiting factor
MRM	Multiple reaction monitoring
MS	Mass spectrometry
MS/MS	Tandem mass spectrometry
MS3	Triple stage mass spectrometry
MudPIT	Multidimensional protein identification technology
MYH	Myosin
NARS	Asparaginyl-tRNA synthetase, cytoplasmic (EC 6.1.1.22) (Asparagine–tRNA ligase) (AsnRS)
NASP	Nuclear autoantigenic sperm protein
NODAL	Nodal homolog
NRTN	Neurturin
OBIF	Osteoblastogenensis inhibitory factor
oMALDI	Orthogonal MALDI
PAPSS 2	Bifunctional 3′-phosphoadenosine 5′-phosphosulfate synthase 2 (PAPS synthase 2) (PAPSS 2) (Sulfurylase kinase 2) (SK 2) (SK2) [Includes: Sulfate adenylyltransferase (EC 2.7.7.4) (ATP-sulfurylase) (Sulfate adenylate transferase) (SAT); Adenylyl-sulfate kinase (EC 2.7 1.25) (3′-phosphoadenosine-5′-phosphosulfate synthase) (APS kinase) (Adenosine-5′-phosphosulfate 3′-phosphotransferase) (Adenylylsulfate 3′-phosphotransferase)]
PCPE-1	Procollagen C-endopeptidase enhancer 1 (Procollagen COOH-terminal proteinase enhancer 1) (PCPE-1) (Procollagen C-proteinase enhancer 1) (Type 1 procollagen C-proteinase enhancer protein) (Type I procollagen COOH-terminal proteinase enhancer)
PDGF	Platelet-derived growth factor
PMF	Peptide mass fingerprint
PRL-3	Protein tyrosine phosphatase type IVA 3 (EC 3.1.3.48) (PRL-R) (Protein-tyrosine phosphatase 4a3) (Protein-tyrosine phosphatase of regenerating liver 3) (PRL-3)
PSPN	Persephin (PSP)
Q-Tof	Quadrupole time-of-flight mass spectrometer
QSTAR	Q-Tof from Applied Biosystems
Q Trap	Hybrid triple quadrupole/linear ion trap mass spectrometer
RAB-1A	Ras-related protein Rab-1A (YPT1-related protein)

RANKL	Tumor necrosis factor ligand superfamily member 11 (Osteoclast differentiation factor) (ODF) (Osteoprotegerin ligand) (OPGL) (Receptor activator of nuclear factor kappa-B ligand) (RANKL) (TNF-related activation-induced cytokine) (TRANCE) (CD antigen CD254) [Cleaved into: Tumor necrosis factor ligand superfamily member 11, membrane form; Tumor necrosis factor ligand superfamily member 11, soluble form]
RAW264.7	Mouse osteoclast-like myeloma cell line
SAX	Strong anion exchange
SCX	Strong cation exchange
SDS-PAGE	Sodium dodecyl sulfate polyacrylamide gel electrophoresis
SELDI	Surface-enhanced laser desorption/ionization
SILAC	Stable isotope labeling by amino acids in cell culture
SIM	Single ion monitoring
SNX9	Sorting nexin-9 (SH3 and PX domain-containing protein 1) (Protein SDP1) (SH3 and PX domain-containing protein 3A)
SUCLG2	Succinyl-CoA ligase [GDP-forming] subunit β, mitochondrial (EC 6.2.1.4) (GTP-specific succinyl-CoA synthetase subunit β) (Succinyl-CoA synthetase β-G chain) (SCS-βG)
TGF-β	Transforming growth factor β
TGF-β-1	Transforming growth factor β-1 (TGF-β-1) [Cleaved into: Latency-associated peptide (LAP)]
TGF-β-2	Transforming growth factor beta-2 (TGF-β-2) (BSC-1 cell growth inhibitor) (Cetermin) (Glioblastoma-derived T-cell suppressor factor) (G-TSF) (Polyergin)
TGF-β-R-3	Transforming growth factor β receptor type 3 (TGF-β receptor type 3) (TGF-R-3) (βglycan) (Transforming growth factor β receptor III) (TGF-β receptor type III)
Th	Thomson
TMPRSS6	Transmembrane protease serine 6 (EC 3.4.21.-) (Matriptase-2)
TOF/TOF	Tandem time-of-flight mass spectrometry
VSMC	Vascular smooth muscle cells
YB-1	Nuclease-sensitive element-binding protein 1 (CCAAT-binding transcription factor I subunit A) (CBF-A) (DNA-binding protein B) (DBPB) (Enhancer factor I subunit A) (EFI-A) (Y-box transcription factor) (Y-box-binding protein 1) (YB-1)

6. References

Abdul-Salam, V. B., Wharton, J., Cupitt, J., Berryman, M., Edwards, R. J. & Wilkins, M. R. (2010). Proteomic analysis of lung tissues from patients with pulmonary arterial hypertension., *Circulation* 122(20): 2058–2067.
URL: *http://dx.doi.org/10.1161/CIRCULATIONAHA.110.972745*

Andriopoulos, B., Corradini, E., Xia, Y., Faasse, S. A., Chen, S., Grgurevic, L., Knutson, M. D., Pietrangelo, A., Vukicevic, S., Lin, H. Y. & Babitt, J. L. (2009). BMP6 is a key endogenous regulator of hepcidin expression and iron metabolism., *Nat Genet* 41(4): 482–487.
URL: *http://dx.doi.org/10.1038/ng.335*

Behnam, K., Murray, S. S. & Brochmann, E. J. (2006). BMP stimulation of alkaline phosphatase activity in pluripotent mouse C2C12 cells is inhibited by dermatopontin, one of the most abundant low molecular weight proteins in demineralized bone matrix., *Connect Tissue Res* 47(5): 271–277.
URL: *http://dx.doi.org/10.1080/03008200600995908*

Bernocco, S., Steiglitz, B. M., Svergun, D. I., Petoukhov, M. V., Ruggiero, F., Ricard-Blum, S., Ebel, C., Geourjon, C., Deleage, G., Font, B., Eichenberger, D., Greenspan, D. S. & Hulmes, D. J. S. (2003). Low resolution structure determination shows procollagen C-proteinase enhancer to be an elongated multidomain glycoprotein., *J Biol Chem* 278(9): 7199–7205.
URL: *http://dx.doi.org/10.1074/jbc.M210857200*

Buijs, J. T., Henriquez, N. V., van Overveld, P. G. M., van der Horst, G., Que, I., Schwaninger, R., Rentsch, C., Dijke, P. T., Cleton-Jansen, A.-M., Driouch, K., Lidereau, R., Bachelier, R., Vukicevic, S., Clezardin, P., Papapoulos, S. E., Cecchini, M. G., Lowik, C. W. G. M. & van der Pluijm, G. (2007). Bone morphogenetic protein 7 in the development and treatment of bone metastases from breast cancer., *Cancer Res* 67(18): 8742–8751.
URL: *http://dx.doi.org/10.1158/0008-5472.CAN-06-2490*

Buijs, J. T., Rentsch, C. A., van der Horst, G., van Overveld, P. G. M., Wetterwald, A., Schwaninger, R., Henriquez, N. V., Dijke, P. T., Borovecki, F., Markwalder, R., Thalmann, G. N., Papapoulos, S. E., Pelger, R. C. M., Vukicevic, S., Cecchini, M. G., Lowik, C. W. G. M. & van der Pluijm, G. (2007). BMP7, a putative regulator of epithelial homeostasis in the human prostate, is a potent inhibitor of prostate cancer bone metastasis in vivo., *Am J Pathol* 171(3): 1047–1057.
URL: *http://dx.doi.org/10.2353/ajpath.2007.070168*

Chen, D., Zhao, M. & Mundy, G. R. (2004). Bone morphogenetic proteins., *Growth Factors* 22(4): 233–241.
URL: *http://dx.doi.org/10.1080/08977190412331279890*

Cheung, P. K., Woolcock, B., Adomat, H., Sutcliffe, M., Bainbridge, T. C., Jones, E. C., Webber, D., Kinahan, T., Sadar, M., Gleave, M. E. & Vielkind, J. (2004). Protein profiling of microdissected prostate tissue links growth differentiation factor 15 to prostate carcinogenesis., *Cancer Res* 64(17): 5929–5933.
URL: *http://dx.doi.org/10.1158/0008-5472.CAN-04-1216*

Corradini, E., Schmidt, P. J., Meynard, D., Garuti, C., Montosi, G., Chen, S., Vukicevic, S., Pietrangelo, A., Lin, H. Y. & Babitt, J. L. (2010). BMP6 treatment compensates for the molecular defect and ameliorates hemochromatosis in Hfe knockout mice., *Gastroenterology* 139(5): 1721–1729.
URL: *http://dx.doi.org/10.1053/j.gastro.2010.07.044*

Elvin, J. A., Yan, C. & Matzuk, M. M. (2000). Oocyte-expressed TGF-beta superfamily members in female fertility., *Mol Cell Endocrinol* 159(1-2): 1–5.

Eng, J. K., McCormack, A. L. & Yates III, J. R. (1994). An approach to correlate tandem mass spectral data of peptides with amino acid sequences in a protein database, *Journal of the American Society for Mass Spectrometry* 5(11): 976 – 989.
URL: *http://www.sciencedirect.com/science/article/pii/1044030594800162*

Fenn, J. B., Mann, M., Meng, C. K., Wong, S. F. & Whitehouse, C. M. (1989). Electrospray ionization for mass spectrometry of large biomolecules., *Science* 246(4926): 64–71.

Grasser, W. A., Orlic, I., Borovecki, F., Riccardi, K. A., Simic, P., Vukicevic, S. & Paralkar, V. M. (2007). BMP-6 exerts its osteoinductive effect through activation of IGF-I and EGF pathways., *Int Orthop* 31(6): 759–765.
URL: *http://dx.doi.org/10.1007/s00264-007-0407-9*

Grgurevic, L., Macek, B., Durdevic, D. & Vukicevic, S. (2007). Detection of bone and cartilage-related proteins in plasma of patients with a bone fracture using liquid chromatography-mass spectrometry., *Int Orthop* 31(6): 743–751.
URL: *http://dx.doi.org/10.1007/s00264-007-0404-z*

Grgurevic, L., Macek, B., Healy, D. R., Brault, A. L., Erjavec, I., Cipcic, A., Grgurevic, I., Rogic, D., Galesic, K., Brkljacic, J., Stern-Padovan, R., Paralkar, V. M. & Vukicevic, S. (2011a). Circulating bone morphogenetic protein 1-3 isoform increases renal fibrosis., *J Am Soc Nephrol* 22(4): 681–692.
URL: *http://dx.doi.org/10.1681/ASN.2010070722*

Grgurevic, L., Macek, B., Mercep, M., Jelic, M., Smoljanovic, T., Erjavec, I., Dumic-Cule, I., Prgomet, S., Durdevic, D., Vnuk, D., Lipar, M., Stejskal, M., Kufner, V., Brkljacic, J., Maticic, D. & Vukicevic, S. (2011). Bone morphogenetic protein (BMP)1-3 enhances bone repair., *Biochem Biophys Res Commun* 408(1): 25–31.
URL: *http://dx.doi.org/10.1016/j.bbrc.2011.03.109*

Griffith, D. L., Keck, P. C., Sampath, T. K., Rueger, D. C. & Carlson, W. D. (1996). Three-dimensional structure of recombinant human osteogenic protein 1: structural paradigm for the transforming growth factor beta superfamily., *Proc Natl Acad Sci U S A* 93(2): 878–883.

Gygi, S., Rist, B., Gerber, S., Turecek, F., Gelb, M. & Aebersold, R. (1999). Quantitative analysis of complex protein mixtures using isotope-coded affinity tags, *Nat Biotechnol* 17(10): 994–999.

Hassel, S., Eichner, A., Yakymovych, M., Hellman, U., Knaus, P. & Souchelnytskyi, S. (2004). Proteins associated with type II bone morphogenetic protein receptor (BMPR-II) and identified by two-dimensional gel electrophoresis and mass spectrometry., *Proteomics* 4(5): 1346–1358.
URL: *http://dx.doi.org/10.1002/pmic.200300770*

Helder, M. N., Ozkaynak, E., Sampath, K. T., Luyten, F. P., Latin, V., Oppermann, H. & Vukicevic, S. (1995). Expression pattern of osteogenic protein-1 (bone morphogenetic protein-7) in human and mouse development., *J Histochem Cytochem* 43(10): 1035–1044.

Hentze, M. W., Muckenthaler, M. U. & Andrews, N. C. (2004). Balancing acts: molecular control of mammalian iron metabolism., *Cell* 117(3): 285–297.

Hentze, M. W., Muckenthaler, M. U., Galy, B. & Camaschella, C. (2010). Two to tango: regulation of mammalian iron metabolism., *Cell* 142(1): 24–38.
URL: *http://dx.doi.org/10.1016/j.cell.2010.06.028*

Henzel, W. J., Billeci, T. M., Stults, J. T., Wong, S. C., Grimley, C. & Watanabe, C. (1993). Identifying proteins from two-dimensional gels by molecular mass searching of peptide fragments in protein sequence databases., *Proc Natl Acad Sci U S A* 90(11): 5011–5015.

Hillenkamp, F. & Karas, M. (1990). Mass spectrometry of peptides and proteins by matrix-assisted ultraviolet laser desorption/ionization., *Methods Enzymol* 193: 280–295.

Hogan, B. L. (1996). Bone morphogenetic proteins: multifunctional regulators of vertebrate development., *Genes Dev* 10(13): 1580–1594.

Hong, D., Chen, H.-X., Yu, H.-Q., Liang, Y., Wang, C., Lian, Q.-Q., Deng, H.-T. & Ge, R.-S. (2010). Morphological and proteomic analysis of early stage of osteoblast differentiation in osteoblastic progenitor cells., *Exp Cell Res* 316(14): 2291–2300.
URL: *http://dx.doi.org/10.1016/j.yexcr.2010.05.011*

Hong, D., Chen, H.-X., Yu, H.-Q., Wang, C., Deng, H.-T., Lian, Q.-Q. & Ge, R.-S. (2010). Quantitative proteomic analysis of dexamethasone-induced effects on osteoblast differentiation, proliferation, and apoptosis in MC3T3-E1 cells using SILAC., *Osteoporos Int* .
URL: *http://dx.doi.org/10.1007/s00198-010-1434-8*

Hoof, D. V., Munoz, J., Braam, S. R., Pinkse, M. W. H., Linding, R., Heck, A. J. R., Mummery, C. L. & Krijgsveld, J. (2009). Phosphorylation dynamics during early differentiation of human embryonic stem cells., *Cell Stem Cell* 5(2): 214–226.
URL: *http://dx.doi.org/10.1016/j.stem.2009.05.021*

Ji, Y.-hua, Ji, J.-ling, Sun, F.-yong, Zeng, Y.-ying, He, X.-hui, Zhao, J.-xian, Yu, Y., Yu, S.-he & Wu, W. (2010). Quantitative proteomics analysis of chondrogenic differentiation of C3H10T1/2 mesenchymal stem cells by iTRAQ labeling coupled with on-line two-dimensional LC/MS/MS., *Mol Cell Proteomics* 9(3): 550–564.
URL: *http://dx.doi.org/10.1074/mcp.M900243-MCP200*

Kahn, P. (1995). From genome to proteome: looking at a cell's proteins., *Science* 270(5235): 369–370.

Kartikasari, A. E. R., Roelofs, R., Schaeps, R. M. J., Kemna, E. H. J. M., Peters, W. H. M., Swinkels, D. W. & Tjalsma, H. (2008). Secretion of bioactive hepcidin-25 by liver cells correlates with its gene transcription and points towards synergism between iron and inflammation signaling pathways., *Biochim Biophys Acta* 1784(12): 2029–2037.
URL: *http://dx.doi.org/10.1016/j.bbapap.2008.08.004*

Keller, G. (2005). Embryonic stem cell differentiation: emergence of a new era in biology and medicine., *Genes Dev* 19(10): 1129–1155.
URL: *http://dx.doi.org/10.1101/gad.1303605*

Kim, B.-G., Lee, J.-H., Ahn, J.-M., Park, S. K., Cho, J.-H., Hwang, D., Yoo, J.-S., Yates, J. R., Ryoo, H.-M. & Cho, J.-Y. (2009). 'Two-stage double-technique hybrid (TSDTH)' identification strategy for the analysis of BMP2-induced transdifferentiation of premyoblast C2C12 cells to osteoblast., *J Proteome Res* 8(10): 4441–4454.
URL: *http://dx.doi.org/10.1021/pr900231a*

Kim, B.-S., Kang, K.-S. & Kang, S.-K. (2010). Soluble factors from ASCs effectively direct control of chondrogenic fate., *Cell Prolif* 43(3): 249–261.
URL: *http://dx.doi.org/10.1111/j.1365-2184.2010.00680.x*

Kim, S. H., Kim, Y., Kim, M., Kim, D. S., Lee, S. C., Chi, S.-W., Lee, D. H., Park, S. G., Park, B. C., Bae, K.-H. & Kang, S. (2009). Comparative proteomic analysis of mouse melanoma cell line B16, a metastatic descendant B16F10, and B16 overexpressing the metastasis-associated tyrosine phosphatase PRK-3., *Oncol Res* 17(11-12): 601–612.

Klose, A., Waerzeggers, Y., Monfared, P., Vukicevic, S., Kaijzel, E. L., Winkeler, A., Wickenhauser, C., Lowik, C. W. G. M. & Jacobs, A. H. (2011). Imaging bone morphogenetic protein 7 induced cell cycle arrest in experimental gliomas., *Neoplasia* 13(3): 276–285.

Kodaira, K., Imada, M., Goto, M., Tomoyasu, A., Fukuda, T., Kamijo, R., Suda, T., Higashio, K. & Katagiri, T. (2006). Purification and identification of a BMP-like factor from bovine serum., *Biochem Biophys Res Commun* 345(3): 1224–1231.
URL: *http://dx.doi.org/10.1016/j.bbrc.2006.05.045*

Kubota, K., Sakikawa, C., Katsumata, M., Nakamura, T. & Wakabayashi, K. (2002). Platelet-derived growth factor BB secreted from osteoclasts acts as an osteoblastogenesis inhibitory factor., *J Bone Miner Res* 17(2): 257–265.
URL: *http://dx.doi.org/10.1359/jbmr.2002.17.2.257*

Kurisaki, A., Hamazaki, T. S., Okabayashi, K., Iida, T., Nishine, T., Chonan, R., Kido, H., Tsunasawa, S., Nishimura, O., Asashima, M. & Sugino, H. (2005). Chromatin-related proteins in pluripotent mouse embryonic stem cells are downregulated after removal of leukemia inhibitory factor., *Biochem Biophys Res Commun* 335(3): 667–675.
URL: *http://dx.doi.org/10.1016/j.bbrc.2005.07.128*

Lee, J.-H., Kim, B.-G., Ahn, J.-M., Park, H.-J., Park, S.-K., Yoo, J.-S., Yates, J. R. & Cho, J.-Y. (2010). Role of PI3K on the regulation of BMP2-induced beta-Catenin activation in human bone marrow stem cells., *Bone* 46(6): 1522–1532.
URL: *http://dx.doi.org/10.1016/j.bone.2010.02.013*

Li, R.-F., Shang, Y., Liu, D., Ren, Z.-S., Chang, Z. & Sui, S.-F. (2007). Differential ubiquitination of Smad1 mediated by CHIP: implications in the regulation of the bone morphogenetic protein signaling pathway., *J Mol Biol* 374(3): 777–790.
URL: *http://dx.doi.org/10.1016/j.jmb.2007.09.082*

Mann, M. & Kelleher, N. L. (2008). Precision proteomics: The case for high resolution and high mass accuracy, *Proc Natl Acad Sci U S A* 105(47): 18132–18138.

Massague, J. (2008). TGFbeta in Cancer., *Cell* 134(2): 215–230.
URL: *http://dx.doi.org/10.1016/j.cell.2008.07.001*

Meleady, P., Henry, M., Gammell, P., Doolan, P., Sinacore, M., Melville, M., Francullo, L., Leonard, M., Charlebois, T. & Clynes, M. (2008). Proteomic profiling of CHO cells with enhanced rhBMP-2 productivity following co-expression of PACEsol., *Proteomics* 8(13): 2611–2624.
URL: *http://dx.doi.org/10.1002/pmic.200700854*

Meynard, D., Vaja, V., Sun, C. C., Corradini, E., Chen, S., Lopez-Otin, C., Grgurevic, L., Hong, C. C., Stirnberg, M., Gutschow, M., Vukicevic, S., Babitt, J. L. & Lin, H. Y. (2011). Regulation of TMPRSS6 by BMP6 and iron in human cells and mice., *Blood* .
URL: *http://dx.doi.org/10.1182/blood-2011-04-348698*

Meyrick, B. O., Friedman, D. B., Billheimer, D. D., Cogan, J. D., Prince, M. A., Phillips, J. A. & Loyd, J. E. (2008). Proteomics of transformed lymphocytes from a family with

familial pulmonary arterial hypertension., *Am J Respir Crit Care Med* 177(1): 99–107.
URL: *http://dx.doi.org/10.1164/rccm.200703-499OC*

Murry, C. E. & Keller, G. (2008). Differentiation of embryonic stem cells to clinically relevant populations: lessons from embryonic development., *Cell* 132(4): 661–680.
URL: *http://dx.doi.org/10.1016/j.cell.2008.02.008*

Nakasaki, M., Yoshioka, K., Miyamoto, Y., Sasaki, T., Yoshikawa, H. & Itoh, K. (2008). IGF-I secreted by osteoblasts acts as a potent chemotactic factor for osteoblasts., *Bone* 43(5): 869–879.
URL: *http://dx.doi.org/10.1016/j.bone.2008.07.241*

Nunomura, K., Nagano, K., Itagaki, C., Taoka, M., Okamura, N., Yamauchi, Y., Sugano, S., Takahashi, N., Izumi, T. & Isobe, T. (2005). Cell surface labeling and mass spectrometry reveal diversity of cell surface markers and signaling molecules expressed in undifferentiated mouse embryonic stem cells., *Mol Cell Proteomics* 4(12): 1968–1976.
URL: *http://dx.doi.org/10.1074/mcp.M500216-MCP200*

Ong, S., Blagoev, B., Kratchmarova, I., Kristensen, D., Steen, H., Pandey, A. & Mann, M. (2002). Stable isotope labeling by amino acids in cell culture, SILAC, as a simple and accurate approach to expression proteomics, *Mol Cell Proteomics* 1(5): 376–386.

Pappano, W. N., Steiglitz, B. M., Scott, I. C., Keene, D. R. & Greenspan, D. S. (2003). Use of Mmp1/Tll1 doubly homozygous null mice and proteomics to identify and validate in vivo substrates of bone morphogenetic protein 1/tolloid-like metalloproteinases., *Mol Cell Biol* 23(13): 4428–4438.

Park, S. J., Kim, S. H., Choi, H. S., Rhee, Y. & Lim, S.-K. (2009). Fibroblast growth factor 2-induced cytoplasmic asparaginyl-tRNA synthetase promotes survival of osteoblasts by regulating anti-apoptotic PI3K/Akt signaling., *Bone* 45(5): 994–1003.
URL: *http://dx.doi.org/10.1016/j.bone.2009.07.018*

Perkins, D. N., Pappin, D. J., Creasy, D. M. & Cottrell, J. S. (1999). Probability-based protein identification by searching sequence databases using mass spectrometry data., *Electrophoresis* 20(18): 3551–3567.
URL: *http://dx.doi.org/3.0.CO;2-2*

Prowse, A. B. J., McQuade, L. R., Bryant, K. J., Dyk, D. D. V., Tuch, B. E. & Gray, P. P. (2005). A proteome analysis of conditioned media from human neonatal fibroblasts used in the maintenance of human embryonic stem cells., *Proteomics* 5(4): 978–989.
URL: *http://dx.doi.org/10.1002/pmic.200401087*

Rabinovitch, M. (2008). Molecular pathogenesis of pulmonary arterial hypertension., *J Clin Invest* 118(7): 2372–2379.
URL: *http://dx.doi.org/10.1172/JCI33452*

Raisz, L. G. (2005). Pathogenesis of osteoporosis: concepts, conflicts, and prospects., *J Clin Invest* 115(12): 3318–3325.
URL: *http://dx.doi.org/10.1172/JCI27071*

Reddi, A. H. (1998). Role of morphogenetic proteins in skeletal tissue engineering and regeneration., *Nat Biotechnol* 16(3): 247–252.
URL: *http://dx.doi.org/10.1038/nbt0398-247*

Rivera, J., Megias, D. & Bravo, J. (2007). Proteomics-based strategy to delineate the molecular mechanisms of the metastasis suppressor gene BRMS1., *J Proteome Res*

6(10): 4006–4018.
URL: *http://dx.doi.org/10.1021/pr0703167*

Rubin, L. J. (1997). Primary pulmonary hypertension., *N Engl J Med* 336(2): 111–117.
URL: *http://dx.doi.org/10.1056/NEJM199701093360207*

Saito, S., Yano, K., Sharma, S., McMahon, H. E. & Shimasaki, S. (2008). Characterization of the post-translational modification of recombinant human BMP-15 mature protein., *Protein Sci* 17(2): 362–370.
URL: *http://dx.doi.org/10.1110/ps.073232608*

Saremba, S., Nickel, J., Seher, A., Kotzsch, A., Sebald, W. & Mueller, T. D. (2008). Type I receptor binding of bone morphogenetic protein 6 is dependent on N-glycosylation of the ligand., *FEBS J* 275(1): 172–183.
URL: *http://dx.doi.org/10.1111/j.1742-4658.2007.06187.x*

Shevchenko, A., Jensen, O., Podtelejnikov, A., Sagliocco, F., Wilm, M., Vorm, O., Mortensen, P., Shevchenko, A., Boucherie, H. & Mann, M. (1996). Linking genome and proteome by mass spectrometry: Large-scale identification of yeast proteins from two dimensional gels, *Proc Natl Acad Sci U S A* 93(25): 14440–14445.

Simic, P., Culej, J. B., Orlic, I., Grgurevic, L., Draca, N., Spaventi, R. & Vukicevic, S. (2006). Systemically administered bone morphogenetic protein-6 restores bone in aged ovariectomized rats by increasing bone formation and suppressing bone resorption., *J Biol Chem* 281(35): 25509–25521.
URL: *http://dx.doi.org/10.1074/jbc.M513276200*

Simic, P. & Vukicevic, S. (2005). Bone morphogenetic proteins in development and homeostasis of kidney., *Cytokine Growth Factor Rev* 16(3): 299–308.
URL: *http://dx.doi.org/10.1016/j.cytogfr.2005.02.010*

Song, K., Krause, C., Shi, S., Patterson, M., Suto, R., Grgurevic, L., Vukicevic, S., van Dinther, M., Falb, D., Dijke, P. T. & Alaoui-Ismaili, M. H. (2010). Identification of a key residue mediating bone morphogenetic protein (BMP)-6 resistance to noggin inhibition allows for engineered BMPs with superior agonist activity., *J Biol Chem* 285(16): 12169–12180.
URL: *http://dx.doi.org/10.1074/jbc.M109.087197*

Souza, T. A., Chen, X., Guo, Y., Sava, P., Zhang, J., Hill, J. J., Yaworsky, P. J. & Qiu, Y. (2008). Proteomic identification and functional validation of activins and bone morphogenetic protein 11 as candidate novel muscle mass regulators., *Mol Endocrinol* 22(12): 2689–2702.
URL: *http://dx.doi.org/10.1210/me.2008-0290*

Tibaldi, E., Arrigoni, G., Martinez, H. M., Inagaki, K., Shimasaki, S. & Pinna, L. A. (2010). Golgi apparatus casein kinase phosphorylates bioactive Ser-6 of bone morphogenetic protein 15 and growth and differentiation factor 9., *FEBS Lett* 584(4): 801–805.
URL: *http://dx.doi.org/10.1016/j.febslet.2009.12.052*

Urist, M. R. (1965). Bone: formation by autoinduction., *Science* 150(698): 893–899.

Urist, M. R. & Strates, B. S. (1971). Bone morphogenetic protein., *J Dent Res* 50(6): 1392–1406.

van Gils, M. P. M. Q., Stenman, U. H., Schalken, J. A., Schroder, F. H., Luider, T. M., Lilja, H., Bjartell, A., Hamdy, F. C., Pettersson, K. S. I., Bischoff, R., Takalo, H., Nilsson, O., Mulders, P. F. A. & Bangma, C. H. (2005). Innovations in serum and urine markers in prostate cancer current European research in the P-Mark project., *Eur Urol*

48(6): 1031–1041.
URL: *http://dx.doi.org/10.1016/j.eururo.2005.06.009*

Vukicevic, S., Basic, V., Rogic, D., Basic, N., Shih, M. S., Shepard, A., Jin, D., Dattatreyamurty, B., Jones, W., Dorai, H., Ryan, S., Griffiths, D., Maliakal, J., Jelic, M., Pastorcic, M., Stavljenic, A. & Sampath, T. K. (1998). Osteogenic protein-1 (bone morphogenetic protein-7) reduces severity of injury after ischemic acute renal failure in rat., *J Clin Invest* 102(1): 202–214.
URL: *http://dx.doi.org/10.1172/JCI2237*

Vukicevic, S. & Grgurevic, L. (2009). BMP-6 and mesenchymal stem cell differentiation., *Cytokine Growth Factor Rev* 20(5-6): 441–448.
URL: *http://dx.doi.org/10.1016/j.cytogfr.2009.10.020*

Vukicevic, S., Helder, M. N. & Luyten, F. P. (1994). Developing human lung and kidney are major sites for synthesis of bone morphogenetic protein-3 (osteogenin)., *J Histochem Cytochem* 42(7): 869–875.

Vukicevic, S., Kopp, J. B., Luyten, F. P. & Sampath, T. K. (1996). Induction of nephrogenic mesenchyme by osteogenic protein 1 (bone morphogenetic protein 7)., *Proc Natl Acad Sci U S A* 93(17): 9021–9026.

Vukicevic, S., Latin, V., Chen, P., Batorsky, R., Reddi, A. H. & Sampath, T. K. (1994). Localization of osteogenic protein-1 (bone morphogenetic protein-7) during human embryonic development: high affinity binding to basement membranes., *Biochem Biophys Res Commun* 198(2): 693–700.
URL: *http://dx.doi.org/10.1006/bbrc.1994.1100*

Vukicevic, S., Paralkar, V. M., Cunningham, N. S., Gutkind, J. S. & Reddi, A. H. (1990). Autoradiographic localization of osteogenin binding sites in cartilage and bone during rat embryonic development., *Dev Biol* 140(1): 209–214.

Vukicevic, S. & Sampath, K. T. (eds) (2008). *Bone morphogenetic proteins: From Local to Systemic Therapeutics*, Springer.
URL: *http://www.springer.com/978-3-7643-8551-4*

Wajih, N., Borras, T., Xue, W., Hutson, S. M. & Wallin, R. (2004). Processing and transport of matrix gamma-carboxyglutamic acid protein and bone morphogenetic protein-2 in cultured human vascular smooth muscle cells: evidence for an uptake mechanism for serum fetuin., *J Biol Chem* 279(41): 43052–43060.
URL: *http://dx.doi.org/10.1074/jbc.M407180200*

Wang, S. & Hirschberg, R. (2011). Y-box protein-1 is a transcriptional regulator of BMP7., *J Cell Biochem* 112(4): 1130–1137.
URL: *http://dx.doi.org/10.1002/jcb.23027*

Wermter, C., Howel, M., Hintze, V., Bombosch, B., Aufenvenne, K., Yiallouros, I. & Stocker, W. (2007). The protease domain of procollagen C-proteinase (BMP1) lacks substrate selectivity, which is conferred by non-proteolytic domains., *Biol Chem* 388(5): 513–521.
URL: *http://dx.doi.org/10.1515/BC.2007.054*

Willert, K., Brown, J. D., Danenberg, E., Duncan, A. W., Weissman, I. L., Reya, T., Yates, J. R. & Nusse, R. (2003). Wnt proteins are lipid-modified and can act as stem cell growth factors., *Nature* 423(6938): 448–452.
URL: *http://dx.doi.org/10.1038/nature01611*

Wobus, A. M. & Boheler, K. R. (2005). Embryonic stem cells: prospects for developmental biology and cell therapy., *Physiol Rev* 85(2): 635–678.
URL: *http://dx.doi.org/10.1152/physrev.00054.2003*

Wozney, J. M., Rosen, V., Celeste, A. J., Mitsock, L. M., Whitters, M. J., Kriz, R. W., Hewick, R. M. & Wang, E. A. (1988). Novel regulators of bone formation: molecular clones and activities., *Science* 242(4885): 1528–1534.

Yeung, B., Porter, T. J. & Vath, J. E. (1997). Direct isoform analysis of high-mannose-containing glycoproteins by on-line capillary electrophoresis electrospray mass spectrometry., *Anal Chem* 69(13): 2510–2516.

Yim, K., Abrams, J. & Hsu, A. (1995). Capillary zone electrophoretic resolution of recombinant human bone morphogenetic protein 2 glycoforms. An investigation into the separation mechanisms for an exquisite separation., *J Chromatogr A* 716(1-2): 401–412.

Yocum, A. K., Gratsch, T. E., Leff, N., Strahler, J. R., Hunter, C. L., Walker, A. K., Michailidis, G., Omenn, G. S., O'Shea, K. S. & Andrews, P. C. (2008). Coupled global and targeted proteomics of human embryonic stem cells during induced differentiation., *Mol Cell Proteomics* 7(4): 750–767.
URL: *http://dx.doi.org/10.1074/mcp.M700399-MCP200*

13

Tandem Mass Spectrometry of Tagged and Permethylated Polysaccharides

Wen-Bin Yang
Genomics Research Center, Academia Sinica
Taiwan R.O.C.

1. Introduction

Polysaccharides are the most abundant materials occurring in organism. In addition, the increasing biological importance of saccharides is difficult to detect because their lack an ionization natural and low sensitivity in mass spectrometry than other biomass (Zaia 2010; Chang et al., 2011a; Harvey, 2011; Mischnick 2011). The development of methods for mass spectrometry of isolated glycan from glycoconjugates and polysaccharides are commonly used chemical operation to form the tagged or/and permethylated glycan for mass determination and structural analysis (Fig. 1). Because of most the hydrophobic aldo-derivatives gives a higher signals than the native glycan in instrumental determination so in recent years, matrix-assisted laser desorption/ionization (MALDI) and electrospray ionization (ESI) MS with tandem mass instrument has become a powerful tool for the structural determination of the non-derivatived/derivatived glycans.

MS is capable of providing structural information for oligosaccharides, although complete structural determination typically required several analytical technologies including tandem mass, GC/MS linkage analysis, *exo-/endo*-glycosidase digestions, and NMR (Yoo & Yoon 2005; Kim et al., 2006; Kukushkin et al., 2011; Yang et al., 2011). MS methods are used to have molecular weights and fragment ions information producing with tiny sample loading, on the other hand, MS provides a higher sensitivity than other glycan analytic methods. Combinations of chemical labeling, tandem mass spectrometry may be used to build signals to define the structures of glycans present from biological sources. Here we introduce four commonly used labeling methods for glycan. One is reductive amination labeling (Harvey 2011), a number of regents are commercially available for reductive amination reaction such as 2AB, 2AA and AP tags. Second is a new glycan tagging method with NAIM tag (Lin et al., 2008; Lin et al., 2010a; Chang et al., 2011; Lin et al., 2011). Aromatic *otho*-diamines are used to label reducing end of glycan presence with catalytic amount of iodine to form a serial of glycan-NAIM derivatives. These glycan-NAIMs provide superiority in enhancement ionization of saccharides base on their size, molecular weight and linkage since saccharide-NAIM derivatives exhibit a stronger mass intensity than native sugar in MALDI time-of-flight (TOF) MS analysis. The third is 1-phenyl-3-methyl-5-pyrazolone (PMP) derivation (Taga et al., 2001; Kodama et al., 2006), which condensd PMP with some monosaccharides to form aldo-(PMP)2 derivatives, and these derivatives were resolved by micellar electrokinetic chromatography (MEKC) using (*S*)- or (*R*)-dodecoxycarbonylvaline as the chiral selector for enantiomeric analysis. The last is permethylation of glycan

(Hakamori 1964) results in the conversion of all hydrogen atoms that are bound to oxygen and nitrogen atoms to methyl groups and serves to render glycans hydrophobic. Permethylated carbohydrates are considerably more stable than native glycans and produce more information on tandem mass spectra. Tandem MS with MALDI and ESI applied permethylated polysaccharides in glycan mixtures is powerful tool for saccharides' linkage analysis. In this context we described mass based approaches for chemical derivatized glycans such as tagged and permethylated oligo-/poly-saccharides. And also these linkages information of permethylated glycans can be elucidated by tandem mass for structural determination.

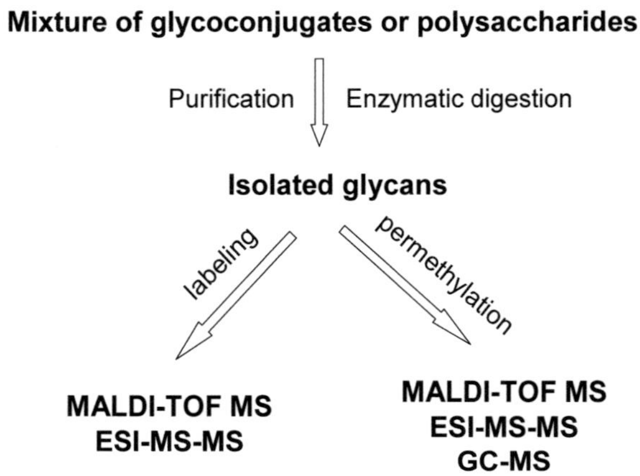

Fig. 1. The strategy for the mass determination of the glycans.

2. Mass ion production and ionization methods

Mass spectrometric ionization of carbohydrates has been reviewed recently (Zaia, 2008; Havery, 2011). Therefore, this section will provide a brief description of two main ionization methods, MALDI and ESI, and highlight some specific requirements for glycan sample preparation that improve overall efficiency of the MS ionization process. The analysis of complex polysaccharides by mass spectrometry is limited by the phenomenon of ion suppression. MALDI entails mixing analyte molecules with an organic matrix, such as 2,5-dihydroxybenzoic acid (2,5-DHB) and 2′,4′,6′-trihydroxyacetophenone (THAP).

Singly charged ions predominate in MALDI-TOF determination is common. It is typically used for analysis of neutral or charged saccharides. MALDI MS has the advantage of sample preparation and a relatively high tolerance to salts and other contaminants. However, acidic carbohydrates (such as phosphorylated, sulfated, or sialylated glycans) are quite different during MALDI ionization and may be in various ionization models and mechanism.

ESI entails spraying a solution containing the analyte appropriately charged droplets move toward the mass spectrometer orifice while undergoing solvent evaporation. This process results in the formation of multiple charged gas phase ions that are analyzed in the mass spectrometer. ESI is used with many types of mass spectrometers and is well suited to the analysis of methylated saccharides. ESI produces inherently better resolved peaks for glycans due to the absence of matrix adduct peaks and provides better sugar-ring fragmentation ability than MALDI (Harvey 2000). In addition, permethylation protects all the hydroxyl groups on glycans to provide superior ionization than native glucans due to the low volatility of hydroxyl groups for ionization during the MALDI and ESI process.

2.1 MALDI mass spectrometry of polysaccharides

The MALDI-TOF ionization efficiency for neutral carbohydrates oligomers has been reported (Harvey et al., 1996; Zaia, 2004), where the ionization efficiency decreases with an increasing molecular weight. Therefore, chemical or enzymatic degradation and purification are essential prior to MALDI-TOF MS analysis for subsequent identification and structure elucidation. Analysis of *N*- and *O*-linked glycans from glycoproteins using MALDI-TOF mass spectrometry is common manipulation. The clean-up profiles of native and permethylated oligosaccharides for an efficient MALDI-TOF MS analysis have been described (Morelle et al., 2007; Zaia, 2010).

There are less of native polysaccharides giving MALDI-TOF MS analysis without degradation and derivation. For examples, Sturiale1 et al. (2005) resolved and identified of Gram-negative bacteria through MALDI-MS of native R-type LPSs. The simples can be successfully and systematically adopted for the analysis of these complex biomolecules without prior chemical degradation. This is quite important since a bacterial R-type LPS is actually a mixture of similar molecules and a MALDI mass spectrum provides the relative intensities among the different species. Structure determination of β-glucans from *Ganoderma lucidum* with MALDI mass spectrometry also reported (Hung et al., 2008). The mass range of detectable polysaccharides is 2000 Da in average.

Pullulans are polysaccharides produced from different strains of the fungus *Aureobasidium pullulans*. Pullulans play an important role in analytical chemistry since they are commonly used as calibration standards in aqueous SEC. Masses below 5 kDa are detectable by mass spectrometry as an alternative method providing direct data on the molar masses. NanoESI-MS analysis of pullulans was successfully carried out with a sample of weights average molar mass of approximately 5900 Da (Bahr et al., 1997). It was possible to obtain a more or less uniform charge state by addition of three sodium ions per molecule. For samples with higher masses it is increasingly difficult to get such simple spectra. However, the MALDI process is known to produce predominately singly charged ions. Different matrices have been used for the analysis of pullulans, like 2,5-dihydroxybenzoic acid (Stahl et al., 1991; Garozzo et al., 2000), 2,4,6-trihydroxyacetophenone (Hsu et al., 2007) and nor-harmane (Fukuyama et al., 2005). The use of the ionic liquid matrix like 2,5-dihydroxybenzoic acid butylamine (DHBB) turned out to be well suited for the analysis of pullulan samples in terms of signal intensities of very high mass polymers (Chang et al., 2011b).

In recent years, MALDI MS has become a powerful tool for the determination of the characteristic molecular weights of polymers. Other techniques like electrospray ionization are known to generate multiple charged ions and provide complex spectra due to a superposition of mass and charge distribution. Due to the huge mass range of polymers with broad distributions the commonly used combination of MALDI ion source with TOF that is well suited for the determination of molar mass distributions. Moreover, TOF instruments have a nearly unlimited mass range if the polysaccharides can be departed with matrix by laser beams. There are some further effects that influence mass like the voltage parameters of the MALDI-TOF instrument, the laser power, the choice of the matrix and ionizing agent and the nature of the analytes are also involved.

Some polysaccharides containing charges are easier to determine than neutral glycans. For examples, the characterization of polysialic acids by a high diversity mass based method allows a rapid, highly sensitive, and unambiguous identification of native polysialic acid as well as fluorescently labeled sialic acid polymers without the need of standard substances due to exact mass determination (Galuska et al., 2007). In addition, some tandem mass spectrometry for structural determination of permethylated sialic acid oligosaccharides are also reported (Wheeler & Harvey, 2000; Yoo & Yoon, 2005; Pabst & Altmann, 2008) in recent years.

2.2 Effect of matrix in MALDI MS for polysaccharides determination

The first investigation of non-derivatized oligosaccharides by MALDI with 3-amino-4-hydroxybenzoic acid as matrix was reported by Mock et al. (1991). Stahl et al. (1991) subsequently discovered that 2,5-DHB yielded better reproducibility and higher signal-to-noise ratio than 3-amino-4-hydroxybenzoic acid. Since then, DHB has become the primary choice of matrix for oligosaccharides. Improvements in sensitivity with a concomitant improvement in resolution were achieved with the addition of 10% 2-hydroxy-5-methoxybenzoic acids to DHB (Wang et al., 1993) and these co-matrices were referred as 'super-DHB'. 1-Hydroxyisoquinoline (Mohr et al., 1995) was also found to be an effective additive to DHB to produce more homogeneous samples for MALDI detection. DHB with or without additives has been broadly used as matrix on MALDI for neutral saccharides. Several other matrices have also been reported (Nonami et al., 1998; Harvey, 1999; Mirza et al., 2004). Recently, Hsu et al. (2007) reported THAP as matrix for MALDI of neutral polysaccharides with molecular weight up to approximately 5,000 Da. For example, a linear neutral polysaccharide with m/z higher than 47,000 was readily detected by MALDI using THAP. Use of THAP as matrix always yielded high quality spectra with good reproducibility in their study.

In addition, positive-/negative-ion MALDI MS of saccharides such as dextran 8,000 Da with 2,5-DHB as matrix (Hao, et al., 1998). The matrix-to-analyte mole ratio was about 10,000. The matrix plays a more important role in the ionization process for oligosaccharides, while in the desorption process for polysaccharides (Chang et al., 2007). There have been only a few matrices reported for detection of polysaccharides with molecular weight higher than 3,000 daltons by MALDI mass spectrometry (Hao, et al., 1998; Hsu et al., 2007). Large polysaccharides, dextrans, glycoproteins and polysialic acids are still under challenge to detect by MALDI MS with various matrices.

2.3 Ionic liquid-assisted electrospray ionization of polysaccharides

Ionic liquids are organic or semiorganic salts with a low vapor pressure. Due to their ability to dissolve a wide range of analytes they have been used in a number of analytical techniques (Schnöll-Bitai, et al., 2008). In 2001, ionic liquids were introduced as matrices in MALDI MS for the analysis of biomolecules and synthetic polymers (Tholey & Heinzle, 2006). The first ionic liquid matrices (ILMs) were synthesized of the commonly used matrix substances sinapinic acid (SA) and α-cyano-4-hydroxycinnamic acid (CHCA) combined with a variety of cations based on amine structures (Armstrong et al., 2001). An example of spectroscopic application is the determination of pullulans. The combination of 2,5-DHB with butylamine (DHBB) turned out to be well suited for the analysis of oligosaccharides and glycolipids (Mank et al., 2004; Laremore et al., 2006, 2007). The same ILM was used for the determination of the molecular mass distribution of polysaccharides (pullulans) which are used as calibration standards in aqueous size-exclusion chromatography (Schnöll-Bitai et al., 2008). In most cases the ionization process in ILMs led to protonated species of the analytes, very similar to solid matrices. However, the analysis of large glycans by ILMs seems to be impeded by the fact that these molecules tend to form complexes with the cations or anions of the matrix.

Recently, ionic liquid-assisted ESI (ILA-ESI) mass spectrometry has been improved for detection of large neutral polysaccharides (Chang et al., 2011). Detection sensitivity of polysaccharides by adding various ionic liquid compounds into samples was improved by ESI or MALDI-TOF mass spectra. Mass spectra obtained were greatly simplified and appeared to be similar to spectra from MALDI due to the narrow charge number distribution.

2.4 ESI mass spectrometry of saccharides

Electrospray ionization techniques for saccharides determination are used in the large amount of published work from 1980 era (Whitehouse et al., 1985; Meng et al., 1988; Zaia, 2007). Conventional ESI-MS (Meng et al., 1988; Fenn et al.. 1990) involves the pumping of a solution into the ion source, and has been observed to produce relatively weak ion signals for native oligosaccharides compared to those for peptides and proteins (Burlingame et al., 1994; Reinhold et al., 1994; 1995). ESI produces ion signals that are comparable between the peptide and carbohydrate compound classes. Therefore, it appears that the hydrophilicity of oligosaccharides limits the surface activity in ESI droplets and their sensitivity is significantly enhanced. The sensitivity increase observed when oligosaccharides are derivatized cause by reducing their hydrophilicity that increased their volatility in the surface (Karas et al., 2000). In fact the ESI of carbohydrates appears to be less effective upon the nano grams than MALDI-TOF for glycans. Interfaces for on-line ESI LC/MS typically produce droplet sizes to those produced by the use of MALDI for neutral oligosaccharide analysis, particularly for applications that involve the profiling of mixtures released from glycoproteins. Although fragmentation allows the analysis of carbohydrate ions, which is caused by the higher internal energies imparted to the ions for structural analysis using ESI method, however, ESI is low efficiency in its ionization process in native oligo-/polysaccharides.

2.5 LC-MS/MS spectrometry of glycoproteins

Mass spectrometry evolved as a key technique in the analysis of proteins and their post translational modifications. *N*-linked oligosaccharide provides a relative chromatographic quantification via HPLC and subsequent identification via MS. The procedure demonstrates that the glycan hydrolysis, derivitization, and chromatographic separation. Subsequent analysis of the chromatographic peaks via LC/MS or LC/MS/MS will yield additional data to confirm the identity of the glycan, and allow deconstruction of the glycan for additional information on its branch and sub-unit structure. Some of peptide mapping methods will give glycosylation sites, identification of the glycan and its structures. The information may provide insights into the heterogeneity of the glycosylation Sequence-based peptide analysis by LC-ESI-MS/MS is most often applied for identification or quantification of proteins in typical "proteomics" projects. Careful evaluation of the peptide-mass fingerprinting data allows determining the glycan composition at individual glycosylation sites. A precious side product is the possibility of confirming the protein termini and of eventually revealing other post-translational modifications. Both MALDI-TOF MS and LC-ESI-MS can generate such data but LC-ESI generally provides higher sequence coverage. The multiple charged ions formed by ESI also facilitate MS/MS experiments, which substantiate any conclusions on the nature of presumed glycopeptides.

Recently, Wang et al. (2011) describled consists of 2D HPLC fractionation of intact proteins and liquid chromatography multistage tandem mass spectrometry (LC-MS/MS^n) analysis of digested protein fractions. A digital ion trap mass spectrometer with a wide mass range is then used for LC-MS/MS^n analysis of intact glycopeptides from the 2D HPLC fractions. The standard approach for peptide-based glycoprotein analysis starts with bands of Coomassie-stained polyacrylamide gel. After S-alkylation and digestion with trypsin the resulting mixture of peptides and glycopeptides is separated by capillary reversed-phase HPLC and analyzed by ESI-MS and/or ESI-MS/MS. The free glycan analysis from isolated cells or from whole tissues are preferred by MALDI-TOF MS especially in the case of neutral glycans, e.g. from plant, fungi or bacteria cell wall polysaccharides. LC-ESI-MS is chose for sialylated oligomers or mixtures of sialylated and neutral glycans especially when structural information is desirable.

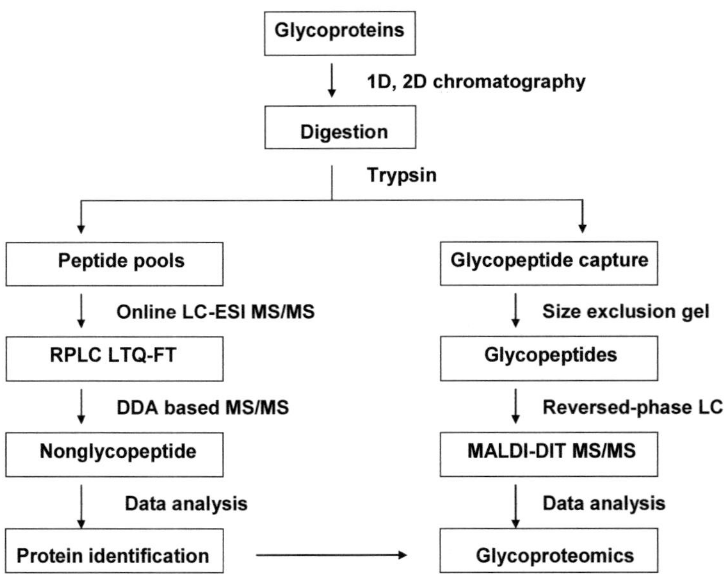

Fig. 2. A general overview of the procedure for profiling glycoproteins. Sample is subjected to an immuno-depletion chromatography followed by 2D HPLC fractionation. The digested 2D HPLC fractions are then analyzed by LC-MS/MS (Wang, et al., 2011). Abbreviation: DDA represents data dependent acquisition, DIT represents digital ion trap, LTQ-FT represents linear trap quadrupole Fourier transformation, and RPLC represents reversed phase liquid chromatography.

3. Strategy for the mass determination of glycans

Tandem MS of native and tagged glycans may be acquired using either positive or negative ions. Commonly, positive ion model of tandem MS is abundant glycosidic bond product ions that occur adjacent to HexNAc, sialic acid, and Fuc particularly abundant (Zaia, 2010; Harvey 2011). In summary, tandem MS produces the greatest structural detail on permethylated glycans. Negative ion tandem MS is effective for producing useful structural information on native and tagged glycans, such as sialic acid containing, sulfated and phosphorylated glycan and those classes are commonly not compatible with permethylated glycans (Wheeler & Harvey, 2000; Larsen et al., 2006; Miller et al., 2006; Mechref et al., 2006; Lei et al., 2009; Yu et al., 2009; Barnes et al., 2011). For tagged glycans the mass shift varies according to the glycan's derivatives. The differentially labeled samples are combined prior to MALDI MS analysis to minimize sample-to-sample variation in peak abundances and maximize the ability to perform the comparison of two samples. Thus, it is possible to analyze the mixtures directly in the MS mode, or to select the nominal masses for subsequent tandem mass spectrometric analysis of glycoform mixtures.
Stable labels are used commonly in saccharide analysis to improve determination of glycans in the sample. Chemical derivatized saccharides increase volatility and stability for MS analysis and helpful to purify them in chromatography when they were labeled. Depending on the sample and extent of the information needed, several different labels can be used as described in subsections below.

3.1 Tagged polysaccharide with reductive amination labeling

Reductive amination derivatization is very useful tool for mass spectral analysis of glycans. A number of reductive amination reagents are commercially available (Yuan et al., 2005; Hitchcock et al., 2006; Zaia, 2008). During the derivatization, an aromatic amine forms a Schiff base at the acyclic reducing sugar residue. The resulting Schiff bass is then chemically reduced by sodium cyanoborohydride ($NaBH_3CN$) to form a stable labeled glycan. Both steps of the derivatization can be performed in a single reaction (Klapoetke et al., 2010). Nevertheless, glycan tagging has higher ionization efficiency than native glucans. A recent review for derivatization of carbohydrates for analysis by chromatography and mass spectrometry was published by Harvey (2011). The most approach is labeling the reducing end of the glycans with reductive amination to generate fluorescence for glycan profile by LC-fluorescence and by MS for glycan identification. In this approach, glycan profiles were readily obtained due to high fluorescent sensitivity imparted by the labeling agent.

3.2 Tagged polysaccharide with aldo-NAIM labeling

An alternative method for the conversion of native aldose to aldo-naphthimidazole (aldo-NAIM) has been developed (Lin et al., 2008; Lin et al., 2010a; Lin et al., 2010b; Lin et al., 2011). Using iodine as a catalyst in acetic acid solution, a series of mono-, oligo-, and polysaccharides, including those containing carboxyl and acetamido groups, progresses an oxidative condensation reaction with aromatic vicinal diamines at room temperature to give the corresponding aldo-NAIM products in high yields (Fig. 3). In addition, a series of aldo-NAIMs was determined by MALDI–TOF MS to analyze molecular weight and ion intensity. For instance, 2,3-naphthalene diamine-labeled Ling-zhi polysaccharides showed enhanced signals in MALDI–TOF MS (Fig. 4; Lin et al., 2010b).

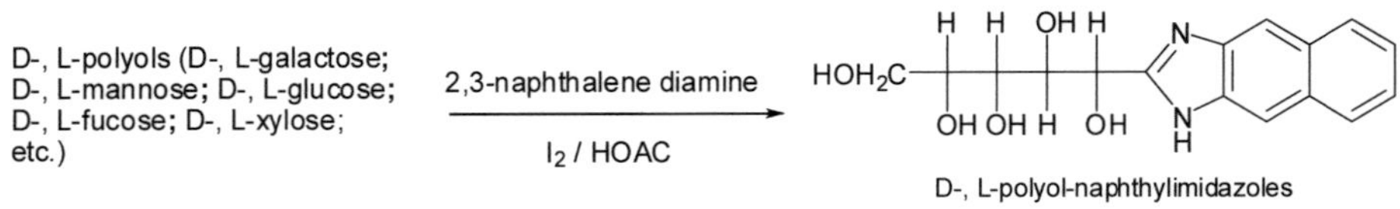

Fig. 3. The preparative method for the conversion of native aldose to aldo-NAIM.

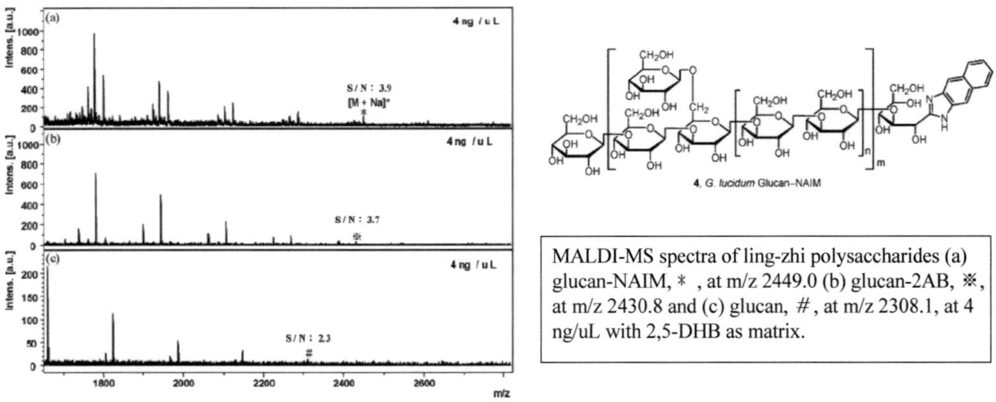

Fig. 4. The corresponding naphthimidazole derivative of Ling-zhi polysaccharides showed the enhanced signals property in the MALDI-MS spectrum.

These aldo-tagged derivatives give a higher sensitivity than the native glycan in common. Not only mono-/disaccharides but also oligo-/polysaccharides were labelled in straightway method. For example, pullulan (molecular mass distribution 2,500~6,000) was tagged with 2,3-naphthalene diamine to obtain pullulan-NAIM derivatives for MALDI-TOF mass spectrometry analysis (Fig. 5). Because this pullulan sample is a polydispersable natural polysaccharide with 15–40 DP (degree of polymerization), pullulan-NAIM displayed their mass signals with a difference of 162.1 Da between neighboring peaks. For instance, the characteristic signals [pullulan-NAIM (DP = 18 + Na]$^+$ at *m/z* 3096 and [pullulan (DP = 36) + Na] $^+$ at *m/z* 6012 are shown in Fig. 5, respectively. Even with as little amount of analyte the signal is still measurable.

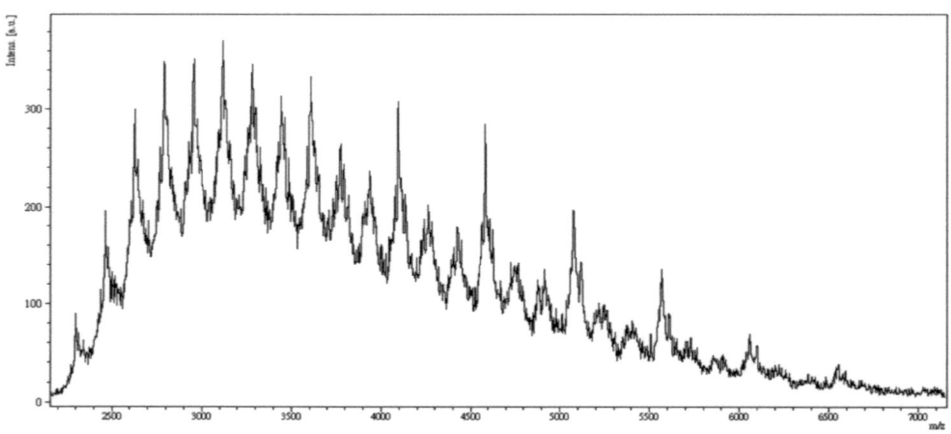

Fig. 5. The corresponding NAIM derivative of pullulan polysaccharides in the MALDI-MS.

3.3 Tagged saccharide with PMP and indole

Glycan labeling strategies are useful in identification and quantification of saccharides. (Ruhaak et al., 2010). Honda have reported the condensation derivatives of some monosaccharides with 1-phenyl-3-methyl-5-pyrazolone (PMP) in Fig. 6, which were resolved by micellar electrokinetic chromatography (MEKC) using (*S*)- or (*R*)-dodecoxycarbonylvaline as the chiral selector (Honda et al., 1997; Taga et al., 2006). The racemic monosaccharides such as PMP-D-/L-Man, PMP-D-/L-Gal, and PMP-D-/L-Fuc were enantioseparated by ligand-exchange capillary electrophoresis (Kodama et al., 2001). Kuo et al. (2011) reported a series of aldo-bis-indole (aldo-BIN) derivatives (Fig. 6) was prepared by aromatic C-alkylation reaction to condense aldose with two molecular indoles in water/acetic acid solution for enantioseparation of racemic monosaccharides. Common monosaccharides were derivatized smoothly to form the UV absorbable aldo-BINs. However, both tagging reagents failed in polysaccharide labeling due to the low reactivity.

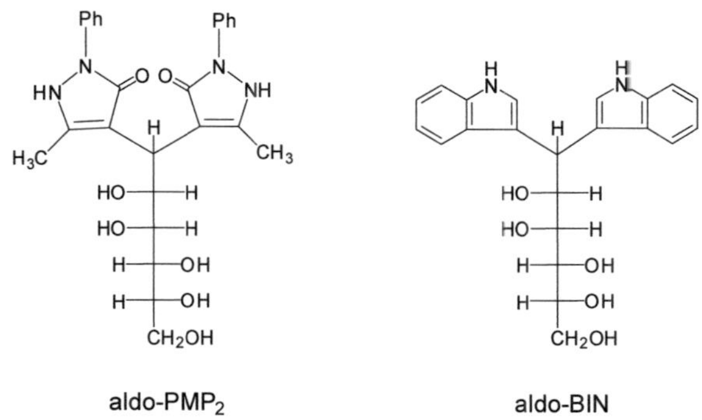

Fig. 6. The chemical structures of aldo-$(PMP)_2$ and aldo-BIN.

3.4 Permethylation labeling of polysaccharides

Permethylation and reductive amination derivatization is most common tool for mass spectral analysis of glycans. Permethylation approach required methylation of all hydroxyl groups on saccharides. Permethylation improves MS sensitivity and stabilizes saccharides as well as glycuronic acids and sialic acids by converting the carboxylic groups into methyl esters. However, permethylation involves complicated sample preparation and clean up with liquid-liquid extraction. In brief, the desiccated samples were dissolved in dimethyl sulfoxide (DMSO) suspension, which was prepared by vortexing DMSO and powdered sodium hydroxide (NaOH) or sodium hydride (NaH) at room temperature, excess of methyl iodide (MeI) was added, and the solution was kept for hours at room temperature with occasional vortexing (Ciucanu & Kerek, 1984). After finished reaction, the sample was partitioned by adding chloroform, the suspension was extracted times with diluted acetic acid aqueous solution, and the chloroform layer was concentrated. The sample was stored at -20 °C prior to analysis. An example of this labeling approach is demonstrated using a comparative glycoform mapping method (C-GlycoMAP), developed based on differential stable isotope labeling (Kang et al., 2007). The differentially isotope labeled samples are combined prior to MALDI-TOF MS analysis to minimize sample-to-sample variation in peak abundances and maximize the ability to perform the comparison of samples. However, to purify permethylated glycans (no chromophore) is quite difficult, so for smaller amounts of samples, they were analyzed directly after extraction and washed with aqueous dilute bicarbonate solutions.

Permethylation is also reported to be particularly useful for in-depth analysis of glycans as it provides information on linkage and branching. So, methylation is a traditional method that provides GC/MS and tandem mass for structural determination of glycans. Strategies for acquisition and interpretation of multistage MS have been most fully developed for permethylated glycans (Ashline et al., 2005). For example, Hung et al. (2008) measured permethylated *G. lucidum* glucans using 2,5-DHB as a matrix. The *G. lucidum* glucans were observed as sodium attached ions and molecular masses were calculated as 219.13 (a terminal sugar) + (204.13)n + 22.99 Da and 219.13 + (204.13)n + 31.02 (mass of reducing end residue) + 22.99 Da, respectively (Fig. 7).

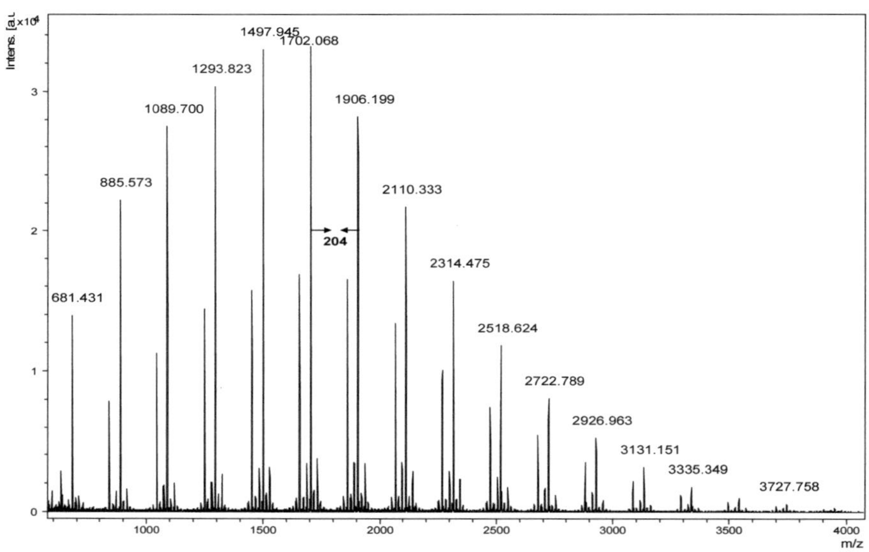

Fig. 7. MALDI-TOF mass spectrum of permethylated *G. lucidum* glucans with 2,5-DHB as matrix. A peak-to-peak mass difference of 204.1 Da is observed. The molecular masses were measured as sodium attached ions at 681.4; 855.6; 1089.7; 1293.8; 1497.9; 1702.1; 1906.2; 2110.3; 2314.5; 2518.6; 2722.8; 2927.0; 3131.2; 3335.4; 3539.6 and 3727.8 (DP = 3~18), respectively.

In addition, permethylated xylans were observed as sodium attached ions with peak-to-peak mass difference of 160.1 Da (Fig. 8). One of polysaccharides from alginic acid, which is a kind of polysaccharide mixture of hexose and aldouronic acid was derivatived by NaH/MeI in DMSO to get permethylated alginic acids and following determined by MALDI-TOF MS. The peak-to-peak mass difference was observed in two series 204.1 Da (Hex) and 218.2 Da (HexA), respectively (Fig. 9).

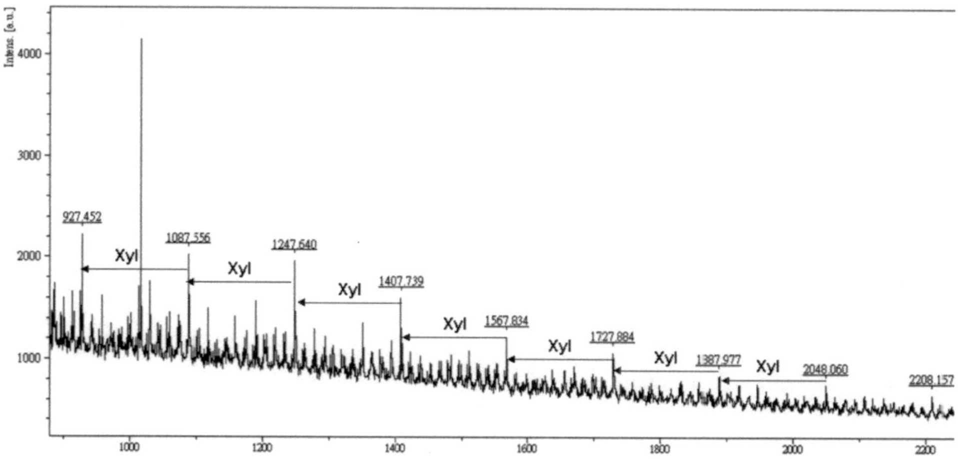

Fig. 8. MALDI-TOF mass spectrum of permethylated xylans with 2,5-DHB as matrix. A peak-to-peak mass difference of 160.1 Da is observed.

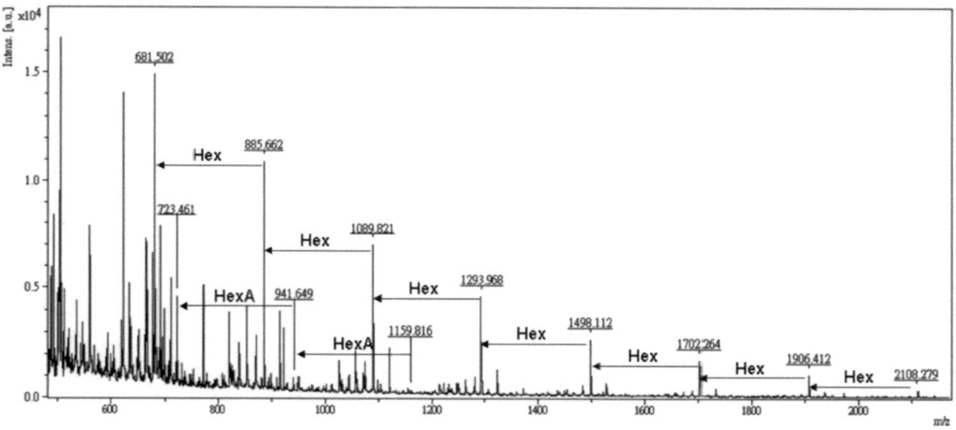

Fig. 9. MALDI-TOF mass spectrum of permethylated alginic acids with 2,5-DHB as matrix. The peak-to-peak mass difference of two series is 204.1 Da and 218.2 Da, respectively.

The advantage to this approach is that tandem mass spectrometric dissociation of a glycosidic bond leaves a site that lacks a methyl group that is clearly indicated by mass. Thus, the linkage position is indicated by the mass of crossring cleavage ions (A- or X-types). It is possible to differentiate some types of positional isomers based on the formation of specific product ion types. Tandem MS of permethylated glycans produces more structural detail than does that of native and reductively aminated glycans.

4. Tandem mass of methylated polysaccharides for structural determination

The challenges to polysaccharides determination are that the glycan-moiety has different chemical properties than proteins or nucleic acids (Forsberg et al., 2000; Faber et al., 2001; Lattová et al., 2005; Mischnick et al., 2005; Nielsen et al., 2010) and most of them are insoluble that makes problems in chemical manipulation such as permethylation and reducing end labeling. Since tandem MS for polysaccharide hasn't been reviewed in detailed, only a few of methodology and reports have been described. However the key aspects of fragmentation procedures are same with those glycoproteomics studies on oligosaccharides and may applied for further study on other source of polysaccharides. Here we introduce and review some results on tagged and permethylated polysaccharides by tandem mass spectrometry.

Tandem MS with MALDI and ESI used permethylated polysaccharides is powerful tool for saccharides' linkage analysis. The use of tandem MS is driven by the need to produce structural information on glycans (Harvey, 1999; Zaia, 2004). The tandem MS experiment is to analyze a mixture of positional isomers directly. Sequential stages of tandem MS are performed in series and the stages of MS are abbreviated MS^n. The masses of the product ions of glycan substructures may be selected for dissociation in further stages. In addition, tandem mass analysis of permethylated glucan can be refered to GC/MS on their methylated acetyled alditols and in comparison with the spectra of synthetic standards for structural analysis of polysaccharides. Most glycan tandem mass spectra are produced by collisional induced dissociation (CAD), a technique in which selected precursor ions are dissociated by collision with gas atoms in a collision cell. Typically, the weakest bonds rupture to produce the most abundant product ions. It is possible to dissociate permethylated glycans using high energy CAD that uses a MALDI TOF/TOF or ESI MS instrument, under which conditions bond rupture is kinetically controlled and cross-ring cleavage ions are more abundant for structural analysis.

4.1 Nomenclature for tandem mass spectrometric ions of glycans

The nomenclature for glycan and glycoconjugate product ions (Domon & Costello, 1988) is shown in Fig. 10 and will be used throughout this section. Product ions containing a non-reducing terminus are labeled as A, B, C, and those containing the reducing end are labeled X, Y, Z. Cleavages across residue rings (A-and X-type ions) are particularly useful for determining linkages.

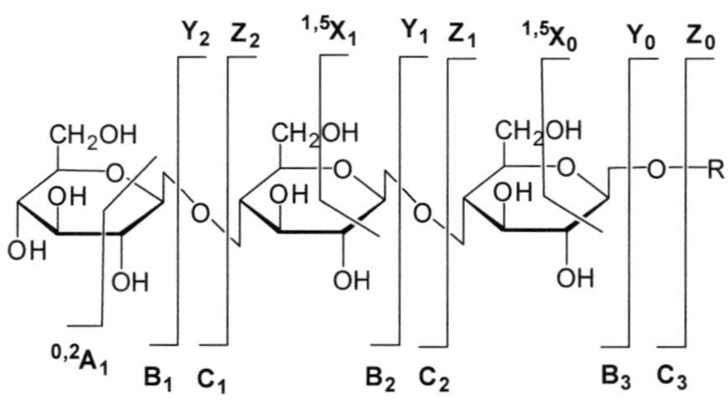

Fig. 10. Nomenclature of tandem mass spectromic ions of glycans and glycoconjugates (Domon & Costello, 1988).

4.2 Characterization of *N*-, *O*-linked glycans from glycoproteins

Recently, Zaia (2010) has reported tandem MS of isomeric glycan mixtures. For glycoconjugates tandem mass spectrometric ions are cleavaged glycan occurs by rupture of glycosidic bonds (B, C, X, Y types) or across rings (A and X types) based on defined by

Domon & Costello (1988). Product ions containing the original nonreducing oligosaccharide end are A, B, and C types, and those containing the original reducing end or aglycon are X, Y, and Z types. Permethylation of glycans has particular advantages for tandem mass spectrometric structural analysis because all glycan OH and NH groups are derivatized. As a result, glycan bond scission occurring during tandem MS creates unmodified sites, which indicate which bond has been cleaved. The linkage position for a given monosaccharide is therefore indicated by the masses of the crossring cleavage ions (A or X types). The crossring cleavage ion masses are useful for determining the linkages and masses of substituents. These principles have been developed into a strategy for determination of glycan linkage and branching structure (Ashline et al., 2005). Tandem mass spectra of native and reductively aminated glycans produce less definitive structural information because the glycosidic bonds cleaved during dissociation do not leave a mass of ring fragments. Thus, it is only possible to determine residue linkage sites when a crossring cleavage is observed to that residue in which the substituents remain intact.

For example, in positive ion mode, native and reductively aminated glycans form abundant product ions from cleavages adjacent to labile NeuAc, HexNAc, and Fuc residues. Crossring cleavages to branching residues are typically low in abundances. Such linkages may also be differentiated in the negative mode using MS^2 of deprotonated ions in which an ion corresponding to A ions correlates with an α-2,6 isomer (Wheeler & Harvey, 2000). In addition, an MS^n series may be used to differentiate glycan linkages by dissociation of C- or D-ions in sialic acid linkage isomers (Deguchi et al., 2007; Ito et al., 2007). For permethylated glycans, the masses of specific A-type ions occurring to the saccharide residues. These fragment ions have been used to differentiate sialic acid linkages using modern MALDI-TOF instruments (Mechref et al., 2006). The determination of the glycan branching structures using tandem MS entails the observation of crossring cleavage ions occurring about the branching residues. Native or reductively aminated glycans observed as sodiated precursor ions dissociate to form A-type ions to the core branching mannose residue of *N*-glycans that may be used to determine the compositions of the three and six branches (Harvey, 2000). Such ions are often observed in increasing abundances for deprotonated precursor ions in the negative mode, as has been observed for branched milk oligosaccharides (Chai et al., 2002) and *N*-glycans (Harvey, 2005).

The formation of the D-ion is particularly useful since it occurs only for three-linked residues. Electron detachment dissociation (EDD) of native glycans has been shown to produce tandem mass spectrometric patterns that are particularly useful for deriving structural information (Adamson & Hakansson, 2007). On the other hand, a single stage of dissociation gives rise to B- and Y- type ions. Single-stage MS of permethylated glycans is a well-established approach for determining molecular weights. A-type crossring cleavages are often abundant, and serve to define the compositions of antenna when they occur to branching residues. Tandem MS of permethylated glycans carries the advantage that ions formed from cleavage of glycosidic bonds have a unique mass value that distinguishes an internal fragment from a single bond cleavage. Multistage dissociation of permethylated glycans has been used to determine detailed linkage information for saccharide units formed by gas phase disassembly of the glycans (Zhang et al., 2005; Prien et al., 2008, 2009; Jiao et al., 2011). The key to multistage dissociation of glycans is the selection of a series of precursor ions that isolate structural branches for subsequent stages. Subsequent dissociation of B-type ions yields crossring cleavages that are useful for determining linkage.

4.3 Characterization of plant glycans

Naturally occurring plant saccharides are huge and comprise mixtures of various conjugated polymers. More recently ESI and MALDI combined with tandem mass spectrometry has been shown to provide valuable structural information for plant polysaccharides. For instance, three D-xylan type per-*O*-methylated saccharides with various types of linkages between the D-xylopyranose units were examined by mass spectrometry (Bagag et al., 2008). In addition, polysaccharides are the principal components of the plant cell wall from its main structural framework. In high plants the three main polysaccharides of the cell wall are cellulose, pectin and hemicellulose. Fernández et al. (2003) has reported the structural characterization of arabinoxylans from wheat by MALDI-TOF and ESI-Q-TOF MS. An arabinoxylan sample digested with endoxylanase was analyzed, the resulting in identification of molecular ions for saccharide residues with up to 22 DP. The permethylated arabinoxylan was also performed to obtain structural information regarding arabinose branching and xylose backbone.

4.4 Characterization of fungi glycans from *Ganoderma lucidum*

Ganoderma lucidum (a medicinal fungi in Asia) has been used for a long time to prevent and treat various human diseases (Lin et al., 2006; Cheng et al., 2007; Hua et al., 2007; Ji et al., 2007; Zhu et al., 2007; Chuang et al., 2009; Chen et al., 2010; Lai et al., 2010). The mainly components of *G. lucidum* polysaccharides are (1→3) and (1→6)-β-D-glucan. Beta-D-glucan is a carbohydrate polymer with chains of glucose molecules linked together by β-glycosidic linkages (Sone et al., 1985; Usui et al., 1983; Wang et al., 2002). Beta-glucan isolated from *G. lucidum* having quite huge of molecular weights and is challenge for studies using the MALDI-MS method. Hung et al. (2008) have analyzed non-derivatized and through permethylated derivatized *G. lucidum* polysaccharides. The permethylated *G. lucidum* glucan was measured, which is derived from acidic degradation (Fig. 11 up). And its tandem mass MS^2 at m/z = 1293.7 of this permethylated *G. lucidum* hexasaccharide was dominated by peaks resulting from cleavage at glycosidic bonds, giving the C/Y ion series and a less intense series of B/Z ions (Fig. 11 middle), which are same as the observation in the curdlan (one of fungi polysaccharide with β-1→3-D-glucan). The MS^2 spectrum of permethylated *G. lucidum* glucan B5 ion at m/z 1058.8 is shown in Fig. 11 bottom. The major fragment ions are the Y and C ions (1277.1, 1075.5, 871.3, 667.2, 463.1 m/z), respectively. However, the B5 ions differ between the permethylated *G. lucidum* hexasaccharide and other linkaged permethylated hexasaccharide such as malto- or dextro-hexaose. The fragmented ions from m/z at 1058.8 ~ 871.3 are 940.7 ($^{0,3}A_5$), 928.7 ($^{1,3}A_5$/$^{2,4}A_5$), 912.3 ($^{0,4}A_5$), 898.7 ($^{1,5}X_5$) and 883.1 ($^{2,3}A_5$), respectively. For the ions with m/z at 912.3 ($^{0,4}A_5$) and 898.7 ($^{4,5}A_5$), it indicates that *G. lucidum* glucan has 1→6 linkage and 883.1 ($^{2,3}A_5$) indicates out the presence of 1→3 linkage. Based on A and X ions in Fig. 11 bottom, we confirm that this methylated glucan has 1→3 and 1→6 linkage between glycosidic bonds.

The aforementioned example of the utility of multistage fragmentation of B-ions generated from permethylated *G. lucidum* glycan. A retro-Diels-Alder reaction in a 1,3-linked B5-type pentaose shows formation of fragment ions. The generic cross ring cleavages that may be formed from B-type ions of various linkages.

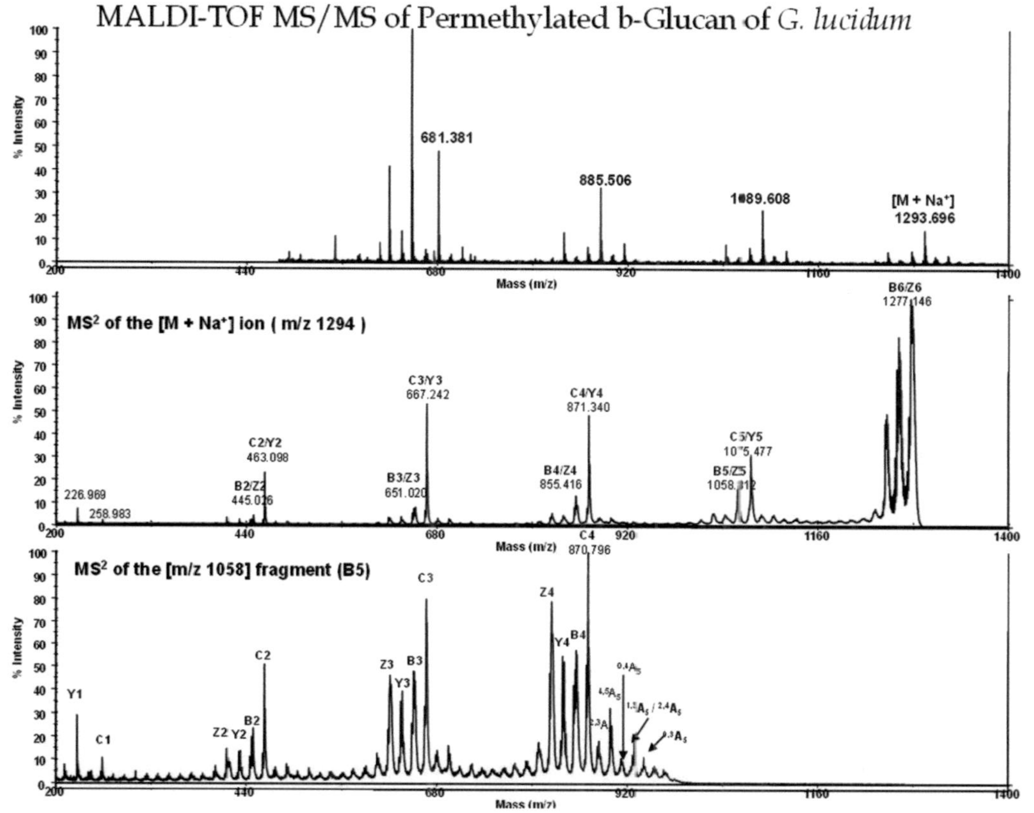

Fig. 11. MALDI-TOF MS/MS of permethylated *G. lucidum* glucans. The ion of permethylated *G. lucidum* hexasaccharide (m/z = 1293.7, $[M + Na]^+$) shown in up panel. MS^2 at the hexasaccharide ion ($[M + Na]^+$, m/z = 1293.8) shown in middle panel and MS^2 at the B5 sodiated fragment ion (m/z = 1058.8) shown in bottom panel. The major cross-ring fragment ions at B5~C4 region are 940.7 ($^{0,3}A_5$), 928.7 ($^{1,3}A_5$/$^{2,4}A_5$), 912.3 ($^{0,4}A_5$), 898.7 ($^{1,5}X_5$/$^{4,5}A_5$) and 883.1 ($^{2,3}A_5$), respectively.

5. Conclusion

This review introduced mass spectrometry of tagged glycans and the uses tandem mass spectrometry for permethylated glycans. These chemical derivatives are useful for the structural analysis of glycans and have been used to study the glycosylation of isolated complex glycoconjugates or polysaccharides in medicinal herbs or fungi. The structural information obtained from tandem mass studies is complicate but useful for glycan linkage information. The glycan in tandem mass also compatible with the derivatization conditions, permethylation remains the alternative choice. Using permethylation, ionization responses are increased over those of underivatized glycans, and the chemical stability improved. Multistage tandem mass spectrometric dissociation of permethylated glycans produces the greatest level of detail possible when using mass spectral techniques. Glycan classes modified with sulfate or other fragile substituents are not compatible with permethylation, but may use reductive amination method to label glycan in tandem mass determination. A number of LC/MS and CE/MS approaches have been incorporated into comprehensive analysis for tagged and permethylated glycans. Biological important biomass glycans and glycoconjugates may be analyzed using tandem mass a combination of various electron dissociation methods.

6. Acknowledgment

The author appreciated a Grant-in-Aid for scientific research from the Genomics Research Center, Academia Sinica, Taiwan R.O.C.

7. References

Adamson, J.T. & Hakansson, K. (2007). Electron detachment dissociation of neutral and sialylated oligosaccharides. *J. Am. Soc. Mass Spectrom.*, 18, 2162–2172.

Armstrong, D.W.; Zhang, L.-K.; He, L. & Gross, M.L. (2001). Ionic liquids as matrixes for matrixassisted laser desorption/ionization mass spectrometry. *Anal. Chem.*, 73, 3679–3686.

Ashline, D.; Singh, S.; Hanneman, A. & Reinhold, V. (2005). Congruent strategies for carbohydrate sequencing: mining structural details by MS^n. *Anal. Chem.*, 77. 6250–6262.

Bagag, A.; Laprévote, O.; Hirsch, J. & Kováčik, V. (2008). Atmospheric pressure photoionization mass spectrometry of per-*O*-methylated oligosaccharides related to D-xylans. *Carbohydr. Res.*, 343, 2813–2818.

Bahr, U.; Pfenninger, A.; Karas, M. & Stahl, B. (1997). High-sensitivity analysis of neutral underivatized oligosaccharides by nanoelectrospray mass spectrometry. *Anal. Chem.*, 69, 4530–4535.

Barnes, J.; Lim, J.-M.; Godard, A.; Blanchard, F.; Wells, L. & Steet, R. (2011). Extensive mannose phosphorylation on leukemia inhibitory factor (LIF) controls its extracellular levels by multiple mechanisms. *J. Biol. Chem.*, 286, 24855–24864.

Burlingame, A.L.; Boyd, R.K. & Gaskell, S.J. (1994). Mass spectrometry. *Anal. Chem.*, 66, 634R–683R.

Chai, W.; Piskarev, V. & Lawson, A.M. (2002). Branching pattern and sequence analysis of underivatized oligosaccharides by combined MS/MS of singly and doubly charged molecular ions in negative-ion electrospray mass spectrometry. *J. Am. Soc. Mass Spectrom.*, 13, 670–679.

Chang, W.C.; Huang, L.C.L.; Wang, Y.-S.; Peng, W.-P.; Chang, H.C.; Hsu, N.Y.; Yang, W.B. & Chen, C.H. (2007). Matrix-assisted laser desorption/ionization (MALDI) mechanism revisited. *Anal. Chim. Acta.*, 582, 1–9.

Chang, Y.-L.; Liao, S.K.-S.; Chen, Y.-C.; Hung, W.-T.; Yu, H.-M.; Yang, W.-B.; Fang, J.-M.; Chen, C.-H. & Lee, Y.C. (2011a). Tagging saccharides for signal enhancement in mass spectrometric analysis. *J. Mass. Spectrom.*, 46, 247–255.

Chang, Y.-L.; Lee, Y.-C.; Yang, W.-B. & Chen, C.-H. (2011b). Ionic liquid-assisted electrospray ionization of polysaccharides. *J. Mass Spectrom.*, 46, 367–375.

Chen, W.Y.; Yang, W.B.; Wong, C.H. & Shih, D.T.B. (2010). Effect of Reishi polysaccharides on human stem/progenitor cells. *Bioorg. Med. Chem.*, 18, 8583–8591.

Cheng, K.C.; Huang, H.C.; Chen, J.H.; Hsu, J.W.; Cheng, H.C.; Ou, C.H.; Yang, W.B.; Wong, C.H. & Juan, H.F. (2007). *Ganoderma lucidum* polysaccharides in human monocytic leukemia cells: from gene expression to network construction. *BMC Genomics*, 8:411, 1–17.

Chuang, M.H.; Chiou, S.H.; Huang, C.H.; Yang, W.B. & Wong, C.H. (2009). The lifespan-promoting effect of acetic acid and Reishi polysaccharide. *Bioorg. Med. Chem.*, 17, 7831–7840.

Ciucanu, I. & Kerek, F. (1984). A simple and rapid method for the permethylation of carbohydrates. *Carbohydr. Res.*, 131, 209–217.

Deguchi, K.; Ito, H.; Baba, T.; Hirabayashi, A.; Nakagawa, H.; Fumoto, M.; Hinou, H. & Nishimura, S.-I. (2007). Structural analysis of *O*-glycopeptides employing negative- and positive-ion multi-stage mass spectra obtained by collision-induced and electron-capture dissociations in linear ion trap time-of-flight mass spectrometry. *Rapid Commun. Mass Spectrom.*, 21, 691–698.

Domon, B. & Costello, C.E. (1988). A systematic nomenclature for carbohydrate fragmentations in FAB-MS/MS spectra of glycoconjugates. *Glycoconj. J.*, 5, 397–409.

Faber, E.J.; van den Haak, M.J.; Kamerling, J.P. & Vliegenthart, J.F.G. (2001). Structure of the exopolysaccharide produced by *Streptococcus thermophilus* S3. *Carbohydr. Res.*, 331, 173–182.

Fenn, J.B.; Mann, M.; Meng, C.K. & Wong, S.F. (1990). Electrospray ionization—Principles and practice. *Mass Spectrom. Rev.*, 9, 37–70.

Fernández, L.E.M.; Obel, N.; Scheller, H.V. & Roepstorff, P. (2003). Characterization of plant oligosaccharides by matrix-assisted laser desorption/ionization and electrospray mass spectrometry. *J. Mass Spectrom.*, 38, 427–437.

Forsberg, L.S.; Bhat, U.R. & Carlson, R.W. (2000). Structural characterization of the *O*-antigenic polysaccharide of the lipopolysaccharide from *Rhizobium etli* strain CE3. *J. Biol. Chem.*, 275, 18851–18863.

Fukuyama, Y.; Kolender, A.A.; Nishioka, M.; Nonami, H.; Matulewicz, M.C.; Erra-Balsells, R. & Cerezo, A.S. (2005). Matrix-assisted ultraviolet laser desorption/ionization time-of-flight mass spectrometry of β-(1→3), β-(1 → 4)-xylans from *Nothogenia fastigiata* using nor-harmane as matrix. *Rapid Commun. Mass Spectrom.*, 19, 349–358.

Galuska, S.P.; Geyer, R.; Hlenhoff, M.M. & Geyer, H. (2007). Characterization of oligo- and polysialic acids by MALDI-TOF MS. *Anal. Chem.*, 79, 7161–7169.

Garozzo, D.; Spina, E.; Cozzolino, R.; Cescutti, P. & Fett, W.F. (2000). Studies on the primary structure of short polysaccharides using SEC MALDI mass spectroscopy. *Carbohydr. Res.*, 323, 139–146.

Hakamori, S.I. (1964). A rapid permethylation of glycolipid and polysaccharide catalyze by methylsulfinyl carbanion in dimethyl sulfoxide. *J. Biochem.*, 55, 250–208.

Hao, C.; Ma, X.; Fang, S.; Liu, Z.; Liu, S.; Song F. & Liu, J. (1998). Positive and negative-ion matrix-assisted laser desorption/ionization mass spectrometry of saccharides. *Rapid Commun. Mass Spectrom.*, 12, 345–348.

Harvey, D.J. (1999). Matrix-assisted laser desorption/ionization mass spectrometry of carbohydrates. *Mass Spectrom. Rev.*, 18, 349-450.

Harvey, D.J. (2000). Electrospray mass spectrometry and fragmentation of *N*-linked carbohydrates derivatized at the reducing terminus. *J. Am. Soc. Mass Spectrom.*, 11, 900–915.

Harvey, D.J. (2005). Fragmentation of negative ions from carbohydrates: part 1~3. Use of nitrate and other anionic adducts for the production of negative ion electrospray spectra from *N*-linked carbohydrates. *J. Am. Soc. Mass Spectrom.*, 16, 622–659.

Harvey, D.J. (2011). Derivatization of carbohydrates for analysis by chromatography; electrophoresis and mass spectrometry. *J. Chromatogr. B*, 879, 1196–1225.

Harvey, D.J.; Naven, T.J. & Küster, B. (1996). Identification of oligosaccharides by matrix-assisted laser desorption ionization and electrospray MS. *Biochem. Soc. Trans.*, 24, 905–912.

Hitchcock, A.M.; Costello, C.E. & Zaia, J. (2006). Glycoform quantification of chondroitin/dermatan sulfate using an LC/MS/MS platform. *Biochemistry*, 45, 2350–2361.

Honda, S.; Taga, A.; Kotani, M. & Grover, E.S. (1997). Separation of aldose enantiomers by capillary electrophoresis in the presence of optically active *N*-dodecoxycarbonylvalines. *J. Chromatogr. A*, 792, 385–391.

Hsu, N.-Y.; Yang, W.-B.; Wong, C.-H.; Lee, Y.-C.; Lee, R.T.; Wang, Y.-S. & Chen, C.-H. (2007). Matrix-assisted laser desorption/ionization mass spectrometry of polysaccharides with 2′,4′,6′-trihydroxyacetophen one as matrix. *Rapid Commun. Mass Spectrom.*, 21, 2137–2146.

Hua, K.F.; Hsu, H.Y.; Chao, L.K.; Chen, S.T.; Yang, W.B.; Hsu, J. & Wong, C.H. (2007). *Ganoderma lucidum* polysaccharides enhance CD14 endocytosis of LPS and promote TLR4 signal transduction of cytokine expression. *J. Cell. Physiol.*, 212, 537–550.

Hung, W.-T.; Wang, S.-H.; Chen, C.-H. & Yang, W.-B. (2008). Structure determination of β-glucans from *Ganoderma lucidum* with matrix-assisted laser desorption/ionization (MALDI) mass spectrometry. *Molecules*, 13, 1538–1550.

Ito, H.; Yamada, K.; Deguchi, K.; Nakagawa, H. & Nishimura, S.-I. (2007). Structural assignment of disialylated biantennary *N*-glycan isomers derivatized with 2-aminopyridine using negative-ion multistage tandem mass spectral matching. *Rapid Commun. Mass Spectrom.*, 21, 212–218.

Ji, Z.; Tang, Q.; Zhang, J.; Yang, Y.; Jia, W. & Pan, Y. (2007). Immunomodulation of RAW264.7 macrophages by GLIS, a proteopolysaccharide from *Ganoderma lucidum*. *J. Ethnopharmacol.* 112, 445–450.

Jiao, J.; Zhang, H. & Reinhold, V.N. (2011). High performance IT-MS^n sequencing of glycans: spatial resolution of ovalbumin isomers. *Int. J. Mass Spectrom.*, 109–117.

Kang, P.; Mechref, Y.; Kyselova, Z.; Goetz, J.A. & Novotny, M.V. (2007). Comparative glycomic mapping through quantitative permethylation and stable-isotope labeling. *Anal. Chem.*, 79, 6064–6073.

Karas, M.; Bahr, U. & Dulcks, T. (2000). Nano-electrospray ionization mass spectrometry: addressing analytical problems beyond routine. *Fresenius J. Anal. Chem.*, 366, 669–676.

Kim, J.S.; Laskowich, E.R.; Michon, F.; Kaiser, R.E. & Arumugham, R.G. (2006). Monitoring activation sites on polysaccharides by GC-MS. *Anal. Biochem.*, 358, 136–142.

Klapoetke, S.; Zhang, J.; Becht, S.; Gu, X. & Ding, X. (2010). The evaluation of a novel approach for the profiling and identification of *N*-linked glycan with a procainamide tag by HPLC with fluorescent and mass spectrometric detection. *J. Pharm. Biomed. Anal.*, 53, 315–324.

Kodama, S.; Aizawa, S.; Taga, A.; Yamashita, T. & Yamamoto, A. (2006). Chiral resolution of monosaccharides as 1-phenyl-3-methyl-5-pyrazolone derivatives by ligand-exchange CE using borate anion as a central ion of the chiral selector. *Electrophoresis*, 27, 4730–4734.

Kukushkin, N.V.; Alonzi, D.S.; Dwek, R.A. & Butters, T.D. (2011). Demonstration that endoplasmic reticulum-associated degradation of glycoproteins can occur downstream of processing by endomannosidase. *Biochem. J.*, 438, 133–142.

Kuo, C.-Y.; Liao, K.-S.; Liu, Y.-C. & Yang, W.-B. (2011). Bis-indole derivatives for polysaccharide compositional analysis and chiral resolution of D-, L-monosaccharides by ligand exchange capillary electrophoresis using borate-cyclodextrin as a chiral selector. *Molecules*, 16, 1682–1694.

Lai, C.Y.; Hung, J.T.; Lin, H.H.; Yu, A.L.; Chen, S.H.; Tsai, Y.C.; Shao, L.E.; Yang, W.B. & Yu, J. (2010). Immunomodulatory and adjuvant activities of a polysaccharide extract of *Ganoderma lucidum* in vivo and in vitro. *Vaccine*, 28, 4945–4954.

Laremore, T.N.; Murugesan, S.; Park, T.-J.; Avci, F.Y.; Zagorevski, D.V. & Linhardt, R.J. (2006). Matrix-assisted laser desorption/ionization mass spectrometric analysis of uncomplexed highly sulfated oligosaccharides using ionic liquid matrices. *Anal. Chem.*, 78, 1774–1779.

Laremore, T.N.; Zhang, F. & Linhardt, R.J. (2007). Ionic liquid matrix for direct UV-MALDI-TOF-MS analysis of dermatan sulfate and chondroitin sulfate oligosaccharides. *Anal. Chem.*, 79, 1604–1610.

Larsen, M.R.; Trelle, M.B.; Thingholm, T.E. & Jensen, C.N. (2006). Analysis of posttranslational modifications of proteins by tandem mass spectrometry. *BioTechniques*, 40, 790–798.

Lattová, E.; Snovida, S.; Perreault, H. & Krokhin, O. (2005). Influence of the labeling group on ionization and fragmentation of carbohydrates in mass spectrometry. *J. Am. Soc. Mass Spectrom.*, 16, 683–696.

Lei, M.; Mechref, Y. & Novotny, M.V. (2009). Structural analysis of sulfated glycans by sequential double-permethylation using methyl iodide and deuteromethyl iodide. *J. Am. Soc. Mass Spectrom.*, 20, 1660–1671.

Lin, K.I.; Kao, Y.Y.; Kuo, H.K.; Yang, W.B.; Chou,A.; Lin, H.H.; Yu, A.L. & Wong, C.H. (2006). Reishi polysaccharides induce immunoglobulin production through the TLR4/TLR2-mediated induction of transcription factor Blimp-1. *J. Biol. Chem.*, 281, 24111-24123.

Lin, C.; Hung, W.-T.; Chen, C.-H.; Fang, J.-M. & Yang, W.-B. (2010a). A new naphthimidazole derivative for saccharide labeling with enhanced sensitivity in mass spectrometry detection. *Rapid Commun. Mass Spectrom.*, 24, 85–94.

Lin, C.; Hung, W.-T.; Kuo, C.-Y.; Liao, K.-S.; Liu, Y.-C. & Yang, W.-B. (2010b). I_2-catalyzed oxidative condensation of aldoses with diamines: synthesis of aldo-naphthimidazoles for carbohydrate analysis. *Molecules*, 15, 1340–1353.

Lin, C.; Kuo, C.-Y.; Liao, K.-S. & Yang, W.-B. (2011). Monosaccharide-NAIM derivatives for D-, L-configurational analysis. *Molecules*, 16, 652–664.

Lin, C.; Lai, P.-T.; Liao, S.K.-S.; Hung, W.-T.; Yang, W.-B. & Fang, J.-M. (2008). Using molecular iodine in direct oxidative condensation of aldoses with diamines: an improved synthesis of aldo-benzimidazoles and aldo-naphthimidazoles for carbohydrate analysis. *J. Org. Chem.*, 73, 3848–3853.

Mank, M.; Stahl, B. & Boehm, G. (2004). 2,5-Dihydroxybenzoic acid butylamine and other ionic liquid matrixes for enhanced MALDI-MS analysis of biomolecules. *Anal.Chem.*, 76, 2938–2950.

Mechref, Y.; Kang, P. & Novotny, M.V. (2006). Differentiating structural isomers of sialylated glycans by matrix-assisted laser desorption/ionization time-of-flight/time-of-flight tandem mass spectrometry. *Rapid Commun Mass Spectrom.*, 20, 1381-1389.

Meng, C.K.; Mann, M. & Fenn, J.B. (1988). Of protons or proteins. *Atoms Mol. Clust.*, 10, 361–368.

Nielsen, T.C.; Rozek, T.; Hopwood, J.J. & Fuller, M. (2010). Determination of urinary oligosaccharides by high-performance liquid chromatography/electrospray ionization–tandem mass spectrometry: application to Hunter syndrome. *Anal. Biochem.*, 402, 113–120.

Miller, M.J.C.; Costello, C.E.; Malmström, A. & Zaia, J. (2006). A tandem mass spectrometric approach to determination of chondroitin/dermatan sulfate oligosaccharide glycoforms. *Glycobiology*, 16, 502–513.

Mirza, S.P.; Raju, N.P.; Madhavendra, S.S. & Vairamani, M. (2004). 5-Amino-2-mercapto-1,3,4-thiadiazole: a new matrix for the efficient matrix-assisted laser desorption/ionization of neutral carbohydrates. *Rapid Commun. Mass Spectrom.*, 18, 1666–1674.

Mischnick, P. (2011). Mass spectrometric characterization of oligo- and polysaccharides and their derivatives. *Adv. Polym. Sci.*, DOI: 10.1007/12-2011-134, pp.1–70.

Mischnick, P.; Niedner, W. & Adden, R. (2005). Possibilities of mass spectrometry and tandem-mass spectrometry in the analysis of cellulose ethers. *Macromol. Symp.*, 223, 67–77.

Mock, K.K.; Davey, M. & Cottrell, J.S. (1991). The analysis of underivatised oligosaccharides by matrix-assisted laser desorption mass spectrometry. *Biochem. Biophys. Res. Commun.*, 177, 644–651.

Mohr, M.D.; Börnsen, K.O. & Widmer, H.M. (1995). Matrix-assisted laser desorption /ionization mass spectrometry: improved matrix for oligosaccharides. *Rapid Commun. Mass Spectrom.*, 9, 809–814.

Nonami, H.; Tanaka, K.; Fukuyama, Y. & Erra-Balsells, R. (1998). β-Carboline alkaloids as matrices for UV-matrix assisted laser desorption/ionization time-of-flight mass spectrometry in positive and negative ion modes. Analysis of proteins of high molecular mass, and of cyclic and acyclic oligosaccharides. *Rapid Commun. Mass Spectrom.*, 12, 285–296.

Morelle, W. & Michalski, J.-C. (2007). Analysis of protein glycosylation by mass spectrometry. *Nat. Protoc.*, 2, 1585–1602.

Pabst, M. & Altmann, F. (2008). Influence of electrosorption, solvent, temperature, and ion polarity on the performance of LC-ESI-MS using graphitic carbon for acidic oligosaccharides. *Anal. Chem.*, 80, 7534–7542.

Reinhold, B.B.; Chan, S.Y.; Reuber, L.; Marra, A.; Walker, G.C. & Reinhold, V.N. (1994). Detailed structural characterization of succinoglycan, the major exopolysaccharide of *Rhizobium meliloti* RmlO21. *J. Bacteriol.*, 176, 1997–2002.

Reinhold, V.N.; Reinhold, B.B. & Costello, C.E. (1995). Carbohydrate molecular weight profiling, sequence, linkage, and branching data: ES-MS and CID. *Anal. Chem.*, 67, 1772–1784.

Prien, J.M.; Ashline, D.J.; Lapadula, A.J.; Zhang, H. & Reinhold, V.N. (2009). The high mannose glycans from bovine ribonuclease B isomer characterization by ion trap MS. *J. Am. Soc. Mass Spectrom.*, 20, 539–556.

Prien, J.M.; Huysentruyt, L.C.; Ashline, D.J.; Lapadula, A.J.; Seyfried, T.N. & Reinhold, V.N. (2008). Differentiating *N*-linked glycan structural isomers in metastatic and nonmetastatic tumor cells using sequential mass spectrometry. *Glycobiology*, 18, 353–366.

Ruhaak, L.R.; Zauner, G.; Huhn, C.; Bruggink, C.; Deelder, A.M. & Wuhrer, M. (2010). Glycan labeling strategies and their use in identification and quantification. *Anal. Bioanal. Chem.*, 397, 3457–3481.

Schnöll-Bitai, I.; Ullmer, R.; Hrebicek, T.; Rizzi, A. & Lacik, I. (2008). Characterization of the molecular mass distribution of pullulans by matrix-assisted laser desorption/ionization time-of-flight mass spectrometry using 2,5-dihydroxybenzoic acid butylamine (DHBB) as liquid matrix. *Rapid Commun. Mass Spectrom.*, 22, 2961–2970.

Sone, Y.; Okuda, R.; Wada, N.; Kishida, E. & Misaki, A. (1985). Structures and antitumor activities of the polysaccharides isolated from fruiting body and the growing culture of mycelium of *Ganoderma lucidum*. *Agric. Biol. Chem.*, 49, 2641–2653.

Stahl, B.; Steup, M.; Karas, M. & Hillenkamp, F. (1991). Analysis of neutral oligosaccharides by matrix-assisted laser desorption ionization mass spectrometry. *Anal. Chem.*, 63, 1463–1466.

Sturiale1, L.; Garozzo1, D.; Silipo, A.; Lanzetta, R.; Parrilli, M. & Molinaro, A. (2005). New conditions for matrix-assisted laser desorption/ionization mass spectrometry of native bacterial R-type lipopolysaccharides. *Rapid Commun. Mass Spectrom.*, 19, 1829–1834.

Taga, A.; Suzuki, S. & Honda, S. (2001). Capillary electrophoretic analysis of carbohydrates derivatized by in-capillary condensation with 1-phenyl-3-methyl-5-pyrazolone. *J. Chromatogr. A*, 911, 259–267.

Tholey, A. & Heinzle, E. (2006). Ionic (liquid) matrices for matrix-assisted laser desorption/ionization mass spectrometry—applications and perspectives. *Anal. Bioanal. Chem.*, 386, 24–37.

Usui, T.; Iwasaki, Y. & Mizuno, T. (1983). Isolation and characterization of antitumor active β-D-glucans from the fruit bodies of *Ganoderma applanatum*. *Carbohydr. Res.*, 115, 273–280.

Wang, B.H.; Dreisewerd, K.; Bahr, U.; Karas, M. & Hillenkamp, F. (1993). Gas-phase cationization and protonation of neutrals generated by matrix-assisted laser desorption. *J. Am. Soc. Mass Spectrom.*, 4, 393–398.

Wang, Y.Y.; Khoo, K.H.; Chen, S.T.; Lin, C.C.; Wong, C.H. & Lin, C.H. (2002). Studies on the immuno-modulating and antitumor activities of *Ganoderma lucidum* (Reishi) polysaccharides: functional and proteomic analyses of a fucose-containing glycoprotein fraction responsible for the activities. *Bioorg. Med. Chem.*, 10, 1057–1062.

Wang, H.; Wong, C.H.; Chin, A.; Taguchi, A.; Taylor, A.; Hanash, S.; Sekiya, S.; Takahashi, H.; Murase, M.; Kajihara, S.; Iwamoto, S. & Tanaka, K. (2011). Integrated mass spectrometry-based analysis of plasma glycoproteins and their glycan modifications. *Nat. Protoc.*, 6, 253–269.

Wheeler, S.F. & Harvey, D.J. (2000). Negative ion mass spectrometry of sialylated carbohydrates: discrimination of *N*-acetylneuraminic acid linkages by MALDI-TOF and ESI-TOF mass spectrometry. *Anal. Chem.*, 72, 5027–5039.

Whitehouse, C.M.; Dreyer, R.N.; Yamashita, M. & Fenn, J.B. (1985). Electrospray interface for liquid chromatographs and mass spectrometers. *Anal Chem.*, 57, 675–679.

Yang, F.-L.; Yang, Y.-L.; Liao, P.-C.; Chou, J.-C.; Tsai, K.-C.; Yang, A.-S.; Sheu, F.; Lin, T.-L.; Hsieh, P.-F.; Wang, J.-T.; Hua, K.-F. & Wu, S.-H. (2011). Structure and immunological characterization of the capsular polysaccharide of a pyrogenic liver abscess caused by *Klebsiella pneumoniae. J. Biol. Chem.*, 286, 21041–21051.

Yoo, E. & Yoon, L. (2005). Applications of tandem mass spectrometry in the structure determination of permethylated sialic acid-containing oligosaccharides. *Bull. Korean Chem. Soc.*, 26, 1347–1353.

Yu, S.-Y.; Wu, S.-W.; Hsiao, H.-H. & Khoo, K.-H. (2009). Enabling techniques and strategic workflow for sulfoglycomics based on mass spectrometry mapping and sequencing of permethylated sulfated glycans. *Glycobiology*, 19, 1136–1149.

Yuan, J.; Hashii, N.; Kawasaki, N.; Itoh, S.; Kawanishi, T. & hayakawa, T. (2005). Isotope tag method for quantitative analysis of carbohydrates by liquid chromatography-mass spectrometry. *J. Chromatogr. A*, 1067, 154–152.

Zaia, J., 2010. Mass spectrometry and glycomics. OMICS Journal of Integrative Biology, vol. 14, no. 4, pp. 401–418, Department of Biochemistry, Boston University, Boston, Massachusetts.

Zaia, J. (2004). Mass Spectrometry of oligosaccharides. *Mass Spectrom. Rev.*, 2004, 23, 161–227.

Zaia, J. (2008). Mass spectrometry and the emerging field of glycomics. *Chem. Biol. Rev.*, 15, 881–892.

Zaia, J., 2007. Mass spectrometric ionization of carbohydrates. In Encyclopedia of Mass Spectrometry, vol. 6, Gross, M.G. & Caprioli, R.M. Eds. (Elsevier, New York).

Zhang, H.; Singh, S. & Reinhold, V.N. (2005). Congruent strategies for carbohydrate sequencing. A MS^n spectral library. *Anal. Chem.*, 77, 6263–6270.

Zhu, X. L.; Chen, A. F. & Lin, Z. B. *Ganoderma lucidum* polysaccharides inhance the function of immunological effector cells in immunosuppressed mice. *J. Ethnopharmacol.*, 111, 219–226.

Part 3

Metabolite Identification and Quantification

14

Metabolomics Research with Tandem Mass Spectrometry

Guo-Fang Zhang[1], Qingling Li[1], Ling Li[2] and Takhar Kasumov[2]
[1]*Department of Nutrition, Case Western Reserve University, Cleveland, Ohio;*
[2]*Department of Gastroenterology and Hepatology, Cleveland Clinic, Cleveland, Ohio, USA*

1. Introduction

Metabolomics, the large-scale analysis of metabolites generated in metabolic pathways, has recently joined with "omics" integration studies (Tan *et al.*, 2008;Martinez-Pinna *et al.*, 2010;Gehlenborg *et al.*, 2010;Kint *et al.*, 2010;Vlaanderen *et al.*, 2010). In addition to metabolite profiling characteristic to metabolomics, metabonomics describes the wide spectrum of metabolites changing under certain conditions, such as disease, interventions and functional gene changes. Since these two terms (metabolomics and metabonomics) used interchangeably in literature (Mishur and Rea, 2011), in future discussion we will use only metabolomics for describing both quantitative metabolic profiling and changes in response to stimulus. Biochemically, metabolomics is the ultimate endpoint measurement of biological events and capturing the influences of nutrition, environmental influences, responses to pharmaceuticals, and many more. The history of metabolomics starts with the pioneering work of Horning who in the early 1960s applied gas chromatography (GC) for the metabolites profiling in urine (SWEELEY and Horning, 1960;Horning and VANDENHEUVEL, 1963). A few years later Pauling et al. expanded application of GC for metabolites analysis in both urine and breath (Pauling *et al.*, 1971). Current advancements in bioanalytical instrumentations in combination with sample preparation techniques enables metabolomics studies virtually in all kinds of biological tissues and fluids. Ultimately clinical application of metabolic profiling became an important clinical tool in diagnosis and screening for inherited metabolic disorders (IMD), a group of over 200 single gene disorders.

The vital progress of metabolomics usually occurs when new instrumentation or technique appears. Various state of the art techniques have been applied to metabolomics studies, including: nuclear magnetic resonance (NMR) spectroscopy (Reo, 2002;Ward *et al.*, 2007;Jordan and Cheng, 2007;Serkova *et al.*, 2007;Duarte *et al.*, 2009;Maher *et al.*, 2009;Sofia *et al.*, 2011), magnetic resonance imaging (MRI) (Elias *et al.*, 2008;De Leon-Rodriguez *et al.*, 2009), infrared spectroscopy (IR) (Singh and Sinclair, 2007;Shaw *et al.*, 2009;Corte *et al.*, 2010), gas chromatography mass spectrometry (GC-MS) (Styczynski *et al.*, 2007;Xu *et al.*, 2009;Lin *et al.*, 2010;Dehaven *et al.*, 2010;Carroll *et al.*, 2010;Tsugawa *et al.*, 2011), capillary electrophoresis-MS (Lee *et al.*, 2007;Soga, 2007;Lapainis *et al.*, 2009;Barbas *et al.*, 2011;Britz-McKibbin, 2011) and HPLC-MS (Chen *et al.*, 2007;Lu *et al.*, 2008;Llorach *et al.*, 2009;Gika and Theodoridis, 2011;Bajad and Shulaev, 2011) . Current metabolomics and metabolic profiling studies rely almost exclusively on ^{1}H NMR, GC-MS and LC-MS due to the technological maturity of these instrumentations.

Although the concept of the application of NMR to metabolomics studies dates back to early nineteen eighties (Nicholson *et al.*, 1983a;Nicholson *et al.*, 1983b;Bales *et al.*, 1984), Nicholson and colleagues laid the foundation of the contemporary NMR based metabolomics in 1999 (Nicholson *et al.*, 1999). In this elegant study, a wide range of biochemical, clinical and toxicological problems were addressed using a high resolution NMR based matabolomics approach. Nowadays NMR (mainly ^{1}H NMR) is widely used for metabolomics studies. NMR has the following advantages in the metabolic studies: (i) sample preparation for NMR is much simplified; (ii) NMR is non-invasive and non-destructive so that the sample could be used in subsequent analysis by an alternative technique; (iii) the data is easily quantified and highly reproducible (Zhang *et al.*, 2010); and (iv) various NMR spectral libraries are available for metabolites identifications, such as Human Metabolome Database (HMDB), the Madison Metabolomics Consortium Database (MMCD), the Biological Magnetic Resonance Data Bank (BMRB), the Magnetic Resonance Metabolomics Database or the Chenomx Database and either open source or publicly unavailable software tools (e.g. Analysis of MIXtures (AMIX) program, MetaboMiner, and rNMR which also permits to quantify the metabolites) (Malet-Martino *et al.*, 2011). However, the low sensitivity of NMR analysis limits its application in metabolomics research. Recent effort has been put on improving the sensitivity of NMR by improving Larmor frequency, decreasing probe diameter and reducing noise (Malet-Martino *et al.*, 2011).

GC-MS technique has been seen as a traditional and standard approach for metabolomics studies because of its high separation ability, high sensitivity and easy analyte identification. Typical GC column has a length of 10 to 100 meters with the theoretical plate number of 1,000 to 1000,000. This enables the GC-MS to analyze very complex biological samples. The high sensitivity of the GC-MS allows metabolomics profiling of a complex mixture with a small volume of sample injection (1 µl) into GC-MS. Most compounds can be detected at 1 pmol or lower levels by GC-MS. Many public spectral libraries are available for electron-impact (EI) GC-MS. The available libraries can save huge amount of work and time to identify the metabolites. The national institute of standardization and technology mass spectral library (NIST 11) contains 243,893 carefully evaluated spectra of 212,961 unique compounds, with identifications, nearly all with chemical structures. Other available GC-MS libraries are Fiehn libraries for GC-quadrupole mass spectrometers from Agilent, and for GC-TOF mass spectrometers from Leco. These libraries are collected using EI mass spectrometers that combined with GC. Since only volatile compounds can be analyzed by GC-MS, therefore GC-MS data only covers part of metabolites from biological samples although some nonvolatile metabolites could become more volatile after derivatization.

Notwithstanding the specific strengths of those different methodologies, the unique role of tandem mass spectrometry in metabolomics research is indisputable. Tandem mass spectrometry for metabolomics study has the following merits: (i) the high sensitivity of tandem mass spectrometry provides more information on trace amounts of metabolites; (ii) specific mass scan features of tandem mass spectrometry, such as precursor ion scan, neutral loss scan and multiple reaction monitoring scan, can simplify ion chromatograms. The application of liquid chromatography-coupled tandem mass spectrometry has revolutionized not only metabolic profiling of inherited metabolic disorders (IMD) but also the entire metabolic research, particularly the allied field of metabolomics.

In turn, recent advances in metabolomics have aided major discoveries made in several areas, including the identification of new metabolites and biomarkers of different diseases in both animal models and humans. In addition, metabolomics has been utilized in the investigation of metabolic pathways, biomarker identifications and molecular interactions and regulations. More recently, in combination with stable isotope technology, metabolomics has been successfully used for the identification of new metabolic pathways and the quantification of metabolic fluxes. Because of the objective to profile thousands or more metabolites in complex biological samples, untargeted metabolomics research requires a very elegant methodology or multiple combined methodologies. In this chapter we will focus on our recent metabolomics studies by LC-MS/MS as well as some seminal literature as an introduction to the current applications of tandem mass spectrometry in metabolomics research.

2. Targeted and untargeted metabolite profiling with tandem mass spectrometry

The tandem mass spectrometry is a vital technique in identifying and quantifying different metabolites. We will use our recent work on lipid and folate metabolism as examples to highlight the unique role of tandem mass spectrometry in both targeted and untargeted metabolite profiling. Briefly, the targeted metabolomics experiment with tandem mass spectrometry measures defined ion transitions from known metabolites. The untargeted metabolomics experiment records all ions within a certain mass range, including the ions belonging to structurally novel metabolites (Vinayavekhin and Saghatelian, 2010).

As a part of our research on lipid metabolism we applied tandem mass spectrometry to the analysis of different acyl-CoA derivatives. Acyl-CoAs are intermediates in both fatty acid oxidation and synthesis. Acyl-CoAs are also formed during metabolism of some drugs containing carboxyl groups and therefore they can be used to track the metabolic pathways of several drugs. Structurally acyl-CoAs belong to a class of a large number of diverse compounds that have coenzyme A part and different acyl moieties. Despite the difference in acyl groups, all acyl-CoAs are catabolized through two distinct oxidation pathways in mitochondria and peroxisomes. Mitochondrial beta oxidation is the primary pathway of fatty acid oxidation that produces energy. Each cycle of mitochondrial beta oxidation generates enoyl-CoA, 3-hydroxyacyl-CoA and 3-ketoacyl-CoA intermediates before losing one acetyl-CoA. In contrast to mitochondrial oxidation, peroxisomal fatty acid oxidation can operate relatively independent of the cellular energy demand. This characterizes peroxisomal fatty acid oxidation as the pathway that eliminates poorly metabolized compounds, including the spillover of fatty acids and drug metabolites. Peroxisomal oxidation is primarily responsible for chain shortenings of very-long-chain fatty acids, methyl-branched fatty acids and dicarboxylic acids. Except the first step, all other steps in peroxisomal oxidation are very similar. The first and rate limiting step in peroxisomal beta oxidation is catalyzed by acyl-CoA oxidase utilizing molecular oxygen with the production of hydrogen peroxide. In this step, instead of ATP formation, the energy produced is dissipated as heat. The acetyl-CoA generated in peroxisomes is transferred into mitochondria for complete oxidation to CO_2. The concentrations of acyl-CoAs formed in these metabolic pathways are quite different, some acyl-CoAs can reach 50-100 nmol/g, some of them are only 0.1-1 nmol/g (Gu *et al.*, 2010;Harris *et al.*, 2011). Therefore, acyl-CoA analysis is a challenge to the method development. Tandem mass spectrometry offers several unique capacities for the acyl-CoA analysis. First of all, multiple reaction monitoring (MRM) has high sensitivity to acyl-CoA, because acyl-CoA containing multiple nitrogen atoms has very good ionization yield in positive electrospray ionization (ESI) mass spectrometry. Second, the collision induced dissociation (CID) of all acyl-CoA derivatives in tandem mass spectrometry generate the the common fragments at m/z 428 and 261 derived from coenzyme A moiety. Third, all acyl-CoA molecules in tandem mass spectrometry has the same neutral loss of 507. Most importantly, the neutral loss of 507 is the most abundant daughter ion for most of acyl-CoAs.

2.1 Untargeted metabolites profile by tandem mass spectrometry

The tandem mass spectrometer is designed for sensitive quantification of analytes. To quantify an analyte by tandem mass spectrometer, some compound information and parameters including precursor ion, daughter ion, declustering potential, and collision energy etc., should be known and optimized first. All these seem to determine that tandem mass spectrometer is only used for targeted metabolomics analysis. However, tandem mass spectrometry provides additional features that can be applied to untargeted metabolomics studies. Two types of tandem mass spectrometric untargeted metabolic profiling approaches are introduced here. The first one is the precursor ion scan that profiles all the metabolites having the same fragment. The second approach is MRM scan that is basically similar to neutral loss scan.

2.1.1 Precursor ion scan

Tandem mass spectrometric characteristics of the acyl-CoA molecules can be used for the comprehensive analysis of fatty acid oxidation intermediates. The typical fragmentation of acyl-CoA is shown in Fig 1. As a survey analysis of all CoA derivatives, we developed acyl-CoA precursor ion scan method using fragment at either 428 or 261 (Dalluge *et al.*, 2002). Application of tandem mass spectrometry for discovery of new metabolites, including phospho-hydroxyacyl-CoA derivatives will be discussed in detail in section 3. The precursor ion scan is extremely helpful for this type of untargeted metabolite profiling because it profoundly filters out interferences by focusing on the metabolites that gives specific daughter ion(s).

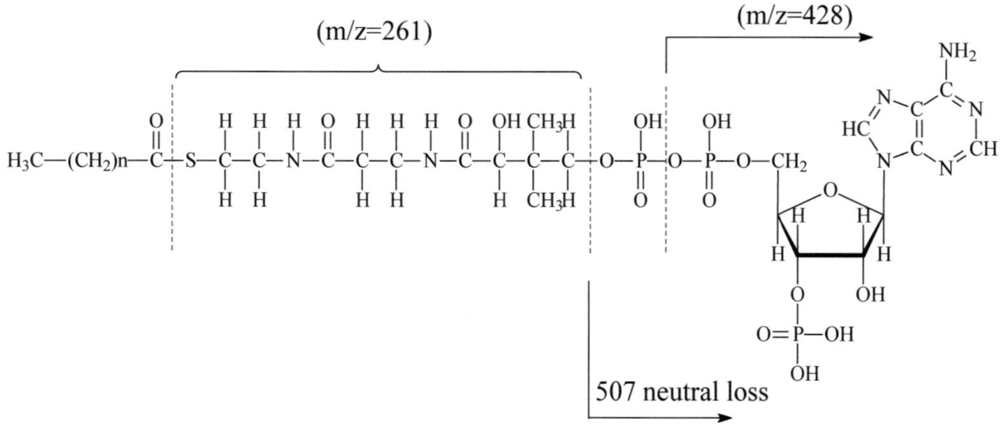

Fig. 1. The acyl-CoA structure and typical fragmentation pattern in positive ESI.

The applications of precursor ion scan to profile metabolites have been reported by others (Millington *et al.*, 1989;Wen *et al.*, 2008;Millington and Stevens, 2011). Acylcarnitines that are from their corresponding acyl-CoAs are used for diagnostic test for inherited disorders of fatty acid and branched-chain amino acid catabolism. Millington et al. used precursor ion scan mode to profile acylcarnitines in plasma and whole blood using acylcarnitine methyl esters typical fragment ion at m/z 99 (CH_2-CH=CH-$COOCH_3$, positive ion) (Millington *et al.*, 1989;Millington *et al.*, 2011). In addition to qualitative profiling, precursor ion scan was also used for acylcarnitines quantiation by adding deuterium labeled acyl-carnitines in their work. Another fragment (m/z 85, CH_2-CH=CH-COOH, positive ion) of acylcarnitine butyl esters was chosen for precursor ion scan to profile acylcarnitines in blood. Ten common acylcarnitines and 40 rare acylcarnitines were detected in defined mutant mice (Rolinski *et al.*, 2000). Wen et al. screened unknown glutathione conjugate by negative precursor ion scan using fragment m/z 272 that corresponds to deprotonated γ-glutamyl-dehydroalanyl-glycine originating from the glutathione moiety. The new metabolite, 2-hydroxy-3-(glutathione-S-yl)-deschloro-meclofenamic acid, from meclofenamic acid was found by this method (Wen *et al.*, 2008).

2.1.2 MRM scan

The advantage of MRM scan is its high sensitivity suitable for targeted quantification purposes. To establish a method in MRM mode one needs to obtain both the precursor and daughter ions information. This requires direct infusion of standard (precursor) compound into mass spectrometer and optimization of ionization and fragmentation parameters. Here we present our strategy to perform non-targeted analysis of acyl-CoA metabolites by MRM. As it is mentioned earlier, all acyl-CoAs have a neutral loss of 507 in tandem mass spectrometry (Fig 1). In addition, the mass of most endogenous acyl-CoA molecules ranges from 767 (non-esterified free CoA) to 1117 (C24 acyl-CoA, the largest physiological CoA derivative). For the comprehensive analysis of all acyl-CoA one can set the precursor ion scan from 768 ($[M+H]^+$ for free CoA) to 1118 ($[M+H]^+$ for C24 acyl-CoA), and their corresponding daughter ions' m/z values are their precursor ions' m/z data minus 507 at positive ESI. Therefore, MRM scan is programmed based on the neutral loss. However, MRM mode is more sensitive than neutral loss scan mode.

After validation this MRM method with known acyl-CoAs, we performed a non-targeted acyl-CoA profiling from the isolated rat liver perfused with levulinic acid (4-ketopentanoic acid) (Harris *et al.*, 2011). "Head-to-head" comparison of acyl-CoA profiles with the sham liver tissue enabled identification of the following acyl-CoA metabolites in the rat livers perfused with levulinic acid: levulinyl-CoA, 4-hydroxypentanoyl-CoA; 4-phosphopentanoyl-CoA, 3-hydroxypentanaoyl-CoA, 3-ketopentanoyl-CoA, 4-hydroxy-2-pentenoyl-CoA, 4-keto-2-pentenoyl-CoA, 3,4-dihydroxypentanoyl-CoA, 4-keto-3-hydroxy-pentanoyl-CoA, 3-keto-4-hydroxypentanoyl-CoA and 3,4-diketopentanoyl-CoA. With such detailed acyl-CoA metabolites profile information, we can deduce three parallel catabolic pathways of levulinic acid in rat livers (Harris *et al.*, 2011). Pathway A is the phosphorylation of 4-hydroxypentanoyl-CoA to form 4-phosphopentanoyl-CoA followed by isomerization to form 3-hyroxypentanoyl-CoA. The latter compound is a regular beta oxidation intermediate that can be further oxidized completely to form acetyl-CoA and propionyl-CoA. In pathway B, levulinic acid is reduced to 4-hydroxypentanoic acid and activated to 4-hydroxypentanoyl-CoA. The latter intermediate is cleaved to form lactyl-CoA and acetyl-CoA via one cycle of beta oxidation. In pathway C, levulinic acid can be directly activated to form levulinyl-CoA that is further metabolized to 3,4-diketopentanoyl-CoA. Partial enzymatic reduction of this metabolite generates 3-keto-4-hydroxypentanoyl-CoA. Further degradation of 3-Keto-4-hydroxypentanoyl-CoA forms lactyl-CoA and acetyl-CoA (Harris *et al.*, 2011). Some of these intermediates are in small amount. For example, 4-Keto-2-petanoyl-CoA is approximately 500 fold less than acetyl-CoA. Therefore, only a sensitive method utilizing the state-of-the-art technique like tandem mass spectrometry can identify such small quantities of these intermediates.

2.2 Targeted metabolic profiling of ceramide species by tandem mass spectrometry

Recently we utilized LC-MS/MS for the simultaneous profiling and quantification of different ceramide species in biological samples. Ceramides represent a class of compounds characterized with acyl group attached to amino group of sphingosine backbone. The carbon number before the name, such as C14-ceramide for N-myrystyl-sphyngosine, specifies fatty acyl chain length. The fatty acids with chain length of C14-C26 are most common ceramide species existing in nature, although ceramides with shorter (C10, C12) and longer (C30) chains are also found.

Ceramides play an important role in insulin resistance, cell signaling, cell differentiation, proliferation and apoptosis and serve as a precursor for many other sphingolipids (Hannun and Obeid, 2008;Kewalramani *et al.*, 2010). Interestingly, ceramide and distinct ceramide metabolites have different and sometimes opposing functions in cell survival and apoptosis, glucose uptake and insulin resistance (Fig.2). Particularly, it is known that ceramide causes apoptosis and inhibits insulin signaling while sphingosine-1-phosphate increases intracellular glucose uptake and inhibits apoptic pathways through the activation of Akt. Because of these opposing effects ceramide/sphigosine-1-phosphate rheostat notion was formulated (Hannun *et al.*, 2008).

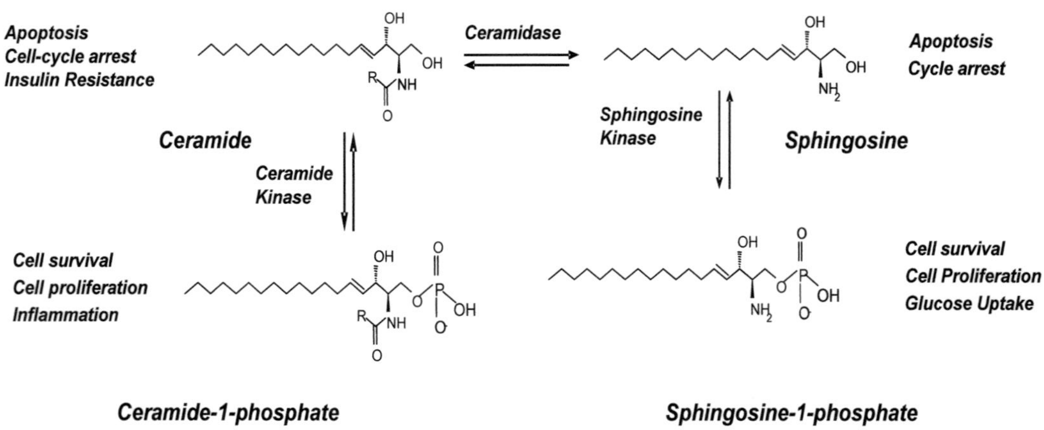

Fig. 2. Role of ceramide and its metabolites in cell signaling.

Although it is not known how the structure of individual ceramide species defines their physiological functions, it has been shown that ceramide containing specific fatty acids are generated in response to certain stimuli, underscoring the structure/function relationship for different ceramide species (Pewzner-Jung *et al.*, 2006). For example, it is known that C16 and C24 ceramide species are involved in cell death (Osawa *et al.*, 2005), while C18 ceramide inhibits cell growth (Koybasi *et al.*, 2004). Investigation of the physiological function of distinct ceramides and their metabolites requires an accurate and sensitive method for their quantification in biological samples. Ceramides were analyzed by an enzymatic diacylglycerol kinase assay (Preiss *et al.*, 1987), by high performance liquid chromatography (HPLC) (Yano *et al.*, 1998), and GC-MS analysis after derivatization (Tserng and Griffin, 2003). However, these methods are labor intensive and time consuming. Although, some of these approaches may be used to analyze total ceramide or fatty acid moiety of ceramide, they do not provide information on individual ceramide species. In addition, these methods generate discrepant results on ceramide levels in different biological samples (Turinsky *et al.*, 1990;Yamaguchi *et al.*, 2004). The tandem mass spectrometry offers several advantages for the analysis of ceramides and their metabolites. The MS/MS mode enables high specificity of identification of the structurally related isomeric and isobaric complex compounds through the precursor/product relationship of ionized molecular and fragment ions. Monitoring of selected precursor and daughter ions in MRM mode improves the sensitivity of the analysis. A wide dynamic range of tandem mass spectrometry allows analysis of both low and high abundant ceramides and their metabolites in a single run. However, structural similarities, wide range biological levels of related sphingolipids and different physico-chemical properties (solubility, ionization) of ceramides and metabolites requires very careful selection of extraction methods, internal standards, instrumentation and mass spectrometric parameters.

Sphingolipids, including ceramides, are easily ionized and they produce several product ions characteristic for the backbone and fatty acids attached to the backbone. Most of the basic and complex sphingolipids form $(M-H)^+$ and $(M+H\text{-}H_2O)^+$ positive ions, while phosphorylated sphinglipids also form negative $(M\text{-}H)^-$ ions in negative ESI mode. Recently lipidomic studies have used ESI-MS/MS technologies to analyze ceramides and related sphingolipids in biological samples (Schmelzer *et al.*, 2007;Sullards *et al.*, 2007).

We have optimized and validated a reverse-phase liquid chromatography coupled ESI-MS/MS technique for the simultaneous measurement of multiple ceramide species in different biological matrices (Kasumov *et al.*, 2010). The method of analysis of tissue samples is based on Bligh and Dyer extraction (BLIGH and DYER, 1959), reverse-phase HPLC separation and MRM of ceramides. Preparation of plasma samples also requires isolation of sphingolipids by silica gel column chromatography prior to LC-ESI-MS/MS analysis. The limits of detection and quantification are in a range of 5-50 pg/ml for distinct ceramides. The method is reliable for inter-assay and intra-assay precision, accuracy and linearity. The separation and quantification of several endogenous long-chain and very-long-chain ceramides using two non-physiological odd chain ceramide (C17 and C25) as internal standards is achieved on a C8 reverse-phase column in less than 5 min during a single 21 min chromatographic run. This method took advantage of the formation of m/z 264.3 daughter ion representing the sphingosine backbone of ceramide (Fig. 3).

To perform the survey analysis of different subspecies a scan of precursor ions was performed over a wide range of collision energies (10-65 eV) in a triple quadrupole tandem mass spectrometry. The MRM method was composed based on existing species in biological samples. The following transitions were selected for sensitive and selective analysis of biological ceramide species: 482.3/264.3 (C12:0), 510.3/264.3 (C14:0), 538.3/264.3 (C16:0), 552.3/264.3 (C17:0), 564.3/264.3 (C18:1), 566.3/264.3 (C18:0), 592.6/264.3 (C20:1), 594.6/264.3 (C20:0), 620.6/264.3 (C22:1), 622.6/264.3 (C22:0), 648.6/264.3 (C24:1), 650.6/264.3 (C24:0), 676.6/264.3 (C26:1) and 678.6/264.3 (C26:0). In addition, the following transitions were used for non-physiological internal standards: 552.3/264.3 (C17:0) and 664.6/264.3 (C25:0).

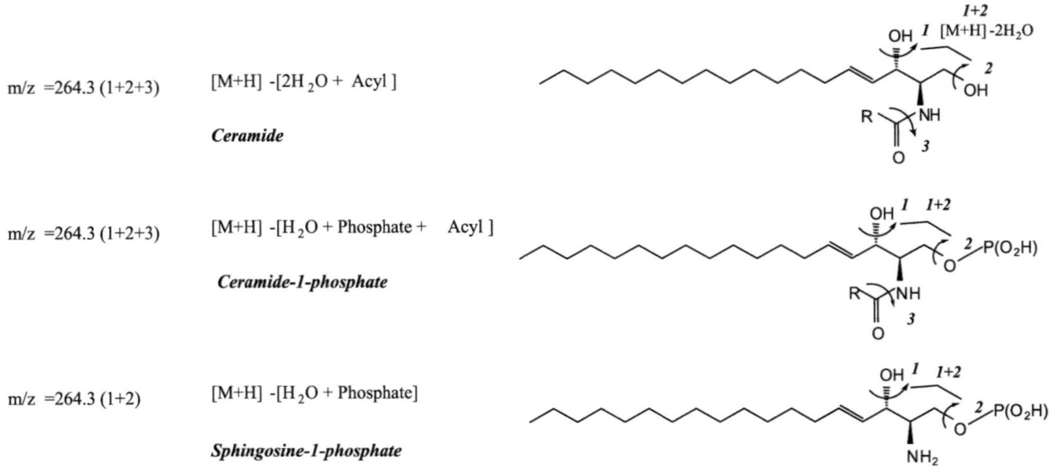

Fig. 3. Structure and fragmentation of ceramide, ceramide-1-phosphate and sphingosine-1-phosphate with m/z 264.3 as the most abundant product ion during collision induced ionization.

The technique was applied to quantify distinct ceramide species and metabolites in different rat tissues and in human plasma. Using this analytical technique we demonstrated that plasma and muscle ceramides are increased in obese subjects with type 2 diabetes and are associated with reduced insulin sensitivity (Haus *et al.*, 2009). This technique was extended for the quantification of ceramide metabolites ceramide-1-phosphate and sphingosine-1-phosphate which utilized 380.3/264.3, 618.7/264.3 and 366.3/250.3 transitions for the sphingosine-1-phosphate and ceramide-1-phosphate and C17-sphingosine-1-phosphate (internal standard), respectively (Fig. 3). This method also could be applied to the analysis of other sphingolipids without significant modification.

2.3 Targeted folate derivatives quantitation by tandem mass spectrometry

Another good example of the application of tandem mass spectrometry for metabolite profiling is folate analysis. Folate is a water-soluble vitamin (B9), which plays a key role in the methylation cycle and in DNA biosynthesis. Folate deficiency has been implicated in hyperhomocysteinemia, which results in an increased risk of cardiovascular disease and dementia, and in neural tube defects. The folate analysis is challenging because of the complexity of folate derivatives in biological samples (see Fig 4) (Zhang *et al.*, 2003;Zhang *et al.*, 2005). Different substitutes (R, Fig 4)) and various glutamate units account for the most of the structural diversity of folate derivatives. It is especially challenging for the most methods to differentiate such large number of folate derivatives. The classical gold method for folate analysis is a microbiological method that analyzes bacterial growth dependent on the available amount of folate (Koontz *et al.*, 2005). Although microbiological method for folate analysis is very sensitive, it suffers from being not reproducible. In addition, only the total amount of folate (but not individual species) is measured by microbiological method.

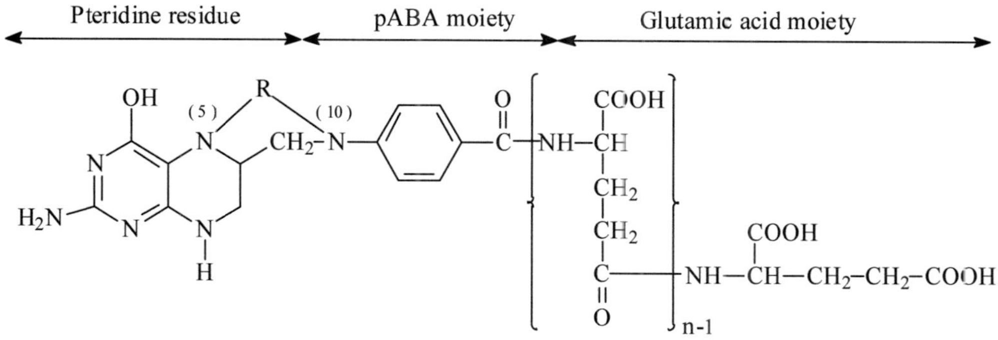

Substituent (R)	Position
—CH_3 (methyl)	5
—CHO (formyl)	5 or 10
—CH_2—(methylene)	5 and 10
=CH—(methenyl)	5 and 10

Tetrahydropteroylpolyglutamates

Fig. 4. The structure of folate derivatives with different substituent and different number of glutamate.

Tandem mass spectrometry has the unique capacity to differentiate the compounds as long as they have different masses of precursor ions and/or product ions. Mono-glutamate folates have different precursor ion and product ion masses, poly-glutamate folates have the same product ion as their corresponding mono-glutamate folate product ion with different precursor ion masses. For example: precursor/product transitions for 5-methyltetrahydrofolate mono-glutamate, 5-methyltetrahydrofolate di-glutamate, 5-methyltetrahydrofolate tri-glutamate, 5-methyltetrahydrofolate tetra-glutamate, 5-methyltetrahydrofolate penta-glutamate and 5-methyltetrahydrofolate hexa-glutamate are 460/313, 589/313, 718/313, 847/313, 976/313 and 1105/313, respectively. Folate molecules with different number of glutamate have the following tandem mass spectrometric characteristics: (i) each additional glutamate has an additional 129 mass unit increase at the m/z of the precursor ion, (ii) they all have the same product ion.

Antifolates are among the first anti-microbial agents invented. Clinical resistance to anti-folates has been mainly attributed to mutations that alter the structure or the expression of enzymes involved in *de novo* folate synthesis. Our recent study showed that the deficiency of 5,10-methenyltetrahydrofolate synthase was found to be hyper susceptible to anti-folate agents. We applied the tandem mass spectrometry technique to measure folates in the *Mycobacterium smegmatis* with the mutation of 5,10-methenyltetrahydrofolate synthase. The folate tandem mass spectrometry method is briefly summarized here. A Thermo Scientific Hypersil GOLD C18 column (150 × 2.1 mm), protected by a guard column (Hypersil GOLD C18 5 μm, 10 × 2.1 mm) was used to separate folates by applying gradient elution of two mobile phases. Mobile phase A is 98% H_2O with 2% acetonitrile containing 0.1% formic acid and mobile phase B is 98% acetonitrile with 2% H_2O containing 0.1% formic acid. The starting eluent was 98% A / 2% B. Mobile phase B was linearly increased to 30% in 15 minutes, then further increased to 90% in 2 minutes. The mobile B was kept at 90% for 4 minutes and then adjusted to initial condition. The column was re-equilibrated at initial condition for 10 minutes before the next injection. The column oven and auto sampler were kept at 35 and 4°C, respectively. The 4000 Qtrap mass spectrometer (AB Sciex, Foster City, CA) source parameters are set as follows: Turbo ion-spray source at 600 °C, gas 1: 60 psi, gas 2: 70 psi, curtain gas at 30 psi. The mono-glutamate folate mass transitions and compound parameters are shown in Table 1. Poly-glutamate folate parent m/z is 129 more for one additional glutamate but with the same product ion m/z as the one corresponding to mono-glutamate folate. Profiling of the folate derivatives showed that mono-glutamate folates had no change compared to the wild type *Mycobacterium smegmatis*, but dramatically decreased poly-glutamate 5-formyltetrahydrofolate, which indicates the important physiological role of poly-glutamate 5-formyltetrahydrofolate in the folate metabolism (Ogwang *et al.*, 2011).

Folates	Parent *m/z*	Product *m/z*	CE (V)	DP (V)	EP (V)	CXP (V)
Folate	442.1	295.0	27	75	10	10
Dihydrofolate	444.1	178.0	27	70	10	10
Tetrahydrofolate	446.1	299.1	20	76	10	10
5,10-methenyltetrahydroflate	456.1	412.1	31	60	10	10
5,10-methylenetetrahydrofolate	458.1	311.1	23	70	10	10
5-methyltetrahydrofolate	460.1	313.1	25	60	10	10
5/10-formyltetrahydrofolate	474.1	327.1	20	71	10	10

Table 1. Mono-glutamate folate compound parameters and mass transitions in tandem mass spectrometry.

3. New metabolite discovery by tandem mass spectrometry

Identification of new metabolites is a challenge due to (i) the complexity of biological samples, and (ii) the fact that novel metabolites usually have no reference compound or existing library data. In addition, it is difficult to characterize compound identities based on molecular mass alone. In this section we will show how tandem mass spectrometry can provide more fragmentation information to help elucidating the structure of molecules. In addition we will examine how high resolution tandem mass spectrometry confirms the hypothesized molecule and fragment formula.

Tandem mass spectrometry provides information on the fragmentation, which is critical in the identification of new unknown metabolites. For example, the identity of 4-phosphopentanoyl-CoA, a metabolite of 4-hydroxypentanoic acid, is deduced by careful analyzing its fragmentation pattern in positive ionization mode of tandem mass spectrometry (Zhang *et al.*, 2009;Sadhukhan *et al.*, 2010). 4-Phosphopentanoyl-CoA (m/z 948) is 80 mass unit heavier than 4-hydroxypentanoyl-CoA (m/z=868). This suggests a possible phosphorylation. In addition, 4-phosphopentanoyl-CoA has fragments with m/z of

441, 428, 343, 312, 261, 214 and 160. The m/z at 428 and 261 confirms that this unknown compound is an acyl-CoA. The neutral loss of 507 from precursor ion (m/z=948) yields a fragment ion at m/z=441. As we discussed earlier, neutral loss of 507 fragmentation pattern is one of the acyl-CoA characteristics in positive ionization tandem mass spectrometry. Two steps of further fragmentation with a loss of 98 (one phosphoric acid) from fragments of 441 and 312 to form daughter ions 343 and 214. This fragmentation pattern confirms the phosphorylation at 4-hydroxy group of 4-hydroxypentanoyl-CoA. The detailed scheme of fragmentation and chemical structures of fragment ions are shown in Fig 5.

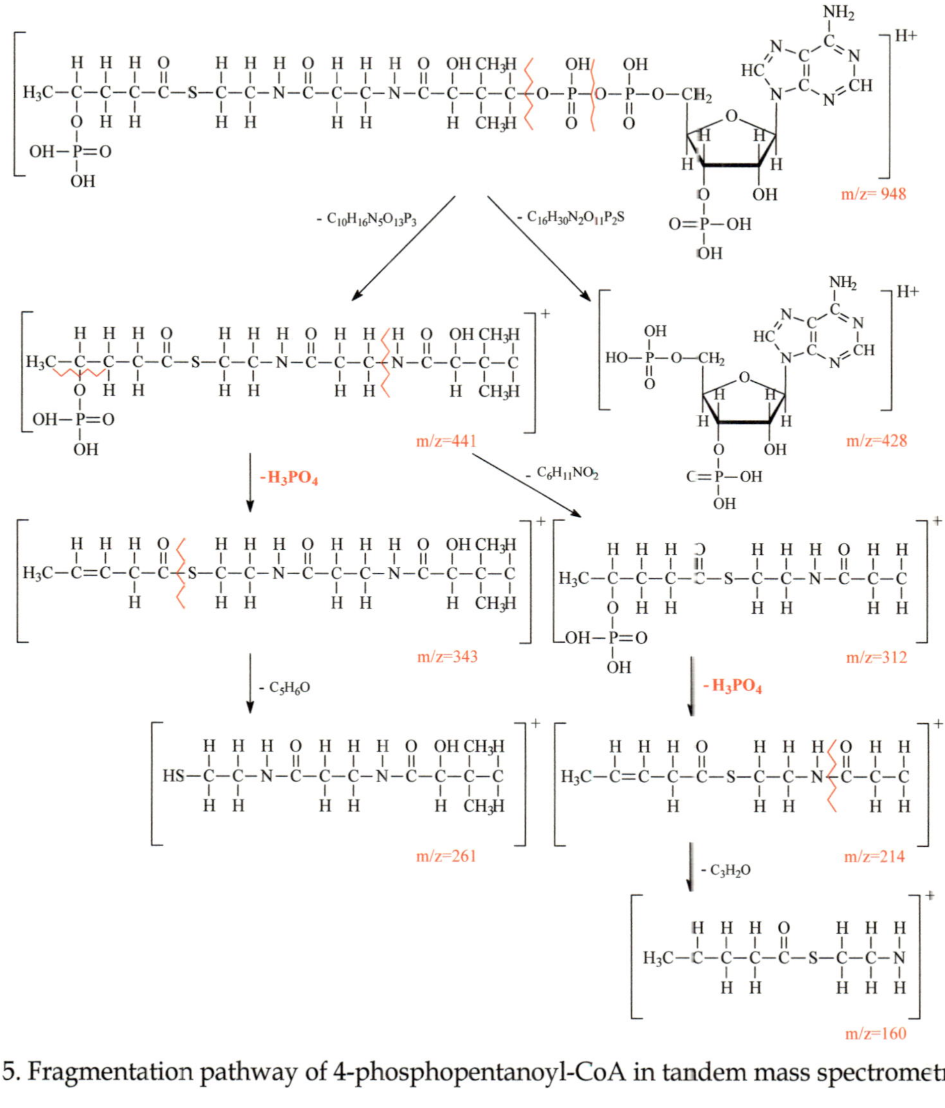

Fig. 5. Fragmentation pathway of 4-phosphopentanoyl-CoA in tandem mass spectrometry.

This example demonstrates the utility of product ion scan for identification of new metabolites. However, the success rate of determining metabolite structures based on MRM fragmentation data will ultimately depend on the experience of the individual investigator, the availability of in house or public fragmentation libraries of the same or similar class of compounds, and the chemical nature of the metabolite itself, in terms of fragmentation information produced. Compared to EI NIST library data, LC-tandem mass spectrometry library is still at initial stage.

Single or triple quadrupole mass spectrometers usually have relatively low mass resolution (unit mass resolution). However, instruments such as time of flight (TOF), Orbitrap™, and Fourier transform ion cyclotron resonance mass spectrometry (FTICR) have high mass resolution of up to 100,000 (at m/z 400). The high mass resolution and mass accuracy can ensure mass error less than 3 ppm (Metz *et al.*, 2007). From such accurate mass measurements, candidate empirical formula of detected species may be determined. One or more empirical formula may be generated for each metabolite. The high resolution tandem mass spectrometry provides additional information on the identity of unknown metabolites. The excellent example of using accurate mass spectrometer is the identification of the unknown metabolite (4-phosphopentanoyl-CoA) from 4-hydroxypentanoic acid catabolism (see section 2.1). The exact mass of 4-phosphopentanoyl-CoA and its three fragments (A, B and C) were measured by high resolution mass spectrometry to obtain their exact masses. Table 2 represents both measured exact mass and the theoretical mass, calculated from chemical formula of hypothesized metabolite (4-phosphopentanoyl-CoA), and its three fragments (m/z values of fragments are 441, 428 and 343). The differences between measured exact mass and theoretical values are all lower than 0.5 ppm. Thus, the exact mass measurements by high resolution tandem mass spectrometry confirmed the hypothesized chemical formula.

	Hypothesized formula	Theoretical mass	Measured mass	Difference (ppm)
[4-Phospho-pentanoyl-CoA-H]$^+$	$C_{26}H_{46}N_7O_{21}P_4S$	948.141803	948.14141	0.41449
Fragment A	$C_{16}H_{30}N_2O_8PS$	441.146051	441.14612	0.15641
Fragment B	$C_{10}H_{16}N_5O_{10}P_2$	428.037244	428.03732	0.17755
Fragment C	$C_{16}H_{27}N_2O_4S$	343.169154	343.16905	0.30306

Table 2. The exact mass of 4-phosphopentanoyl-CoA and its three fragments.

4. Identification of biomarkers for specific diseases

A biomarker is a substance that is objectively measured for indication of the presence of an abnormal condition in a patient and allows disease progression and/or therapeutic response to be monitored. In this section we will review selected examples of the applications of tandem mass spectrometry to the discoveries of metabolic biomarkers characteristic of several pathological conditions.

Recently Sreekumar et al. found that sarcosine was significantly increased in the invasive prostate cancer cell line (Sreekumar *et al.*, 2009). If this finding is true, the sarcosine analysis in urine sample has significance in prostate cancer diagnosis. However, the relation between sarcosine and prostate cancer are still subject to dispute (Pavlou and Diamandis, 2009;Jentzmik *et al.*, 2010;Struys *et al.*, 2010). Soga et al. identified ophthalmic acid as a new oxidative stress biomarker in mice treated with acetaminophen (Soga *et al.*, 2006). Ophthalmic acid, L-γ-glutamyl-L-α-aminobutyrylglycine, is a tripeptide analogue of glutathione in which the cysteine group is replaced by L-2-aminobutyrate. Glutathione is a major cellular anti-oxidant and is involved in the second phase detoxification of several drugs, including acetaminophen. As expected, acetaminophen treated mice had a sudden drop in hepatic glutathione levels which was paralleled with the appearance of an unknown compound with m/z of 290 in positive ionization mass spectrometry. The product ion spectrum of this unknown metabolite was comparable with that of the product ion spectra of glutathione but with 17.957 differences in mass. Based on these results authors deduced that the SH group of the cysteine residue of glutathione was replaced by a methyl (CH_3) group. This non polar side chain corresponds to 2-aminobutyrate, the side product of *de novo* cysteine synthesis in transsulfuration pathway. The replacement of cysteine residue by 2-aminobutyrate in glutathione molecule forms ophthalmic acid. The reduced glutathione and cysteine are decreased in conditions associated with oxidative stress. Sequestration of glutathione and cysteine induces both cysteine and glutathione synthesis. Stimulated cysteine production generates 2-aminobutyrate while increased demand for glutathione induces glycine cysteine synthase that utilizes 2-aminobutyric acid when cysteine levels are below the Km of this enzyme. The mechanism of ophthalmate formation in conditions associated with oxidative stress is outlined in Fig 6. Abbas et al. found increased ophthalmate formation in the rabbit liver with implanted tumor which was paralleled with the reduced glutathione synthesis (Abbas *et al.*, 2011).

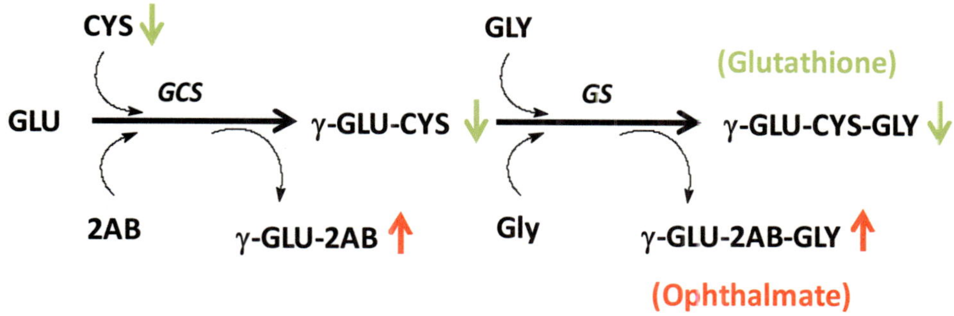

Fig. 6. The relationship between ophthalmate and glutathione concentration changes in the oxidative stress.

Recently the tandem mass spectrometry assisted biomarker discovery was applied to predict the development of diabetes (Wang *et al.*, 2011). Among 2422 normoglycemic individuals followed by 12 years, 201 developed diabetes. In these cases, 5 branched-chain and aromatic amino acids, i.e. isoleucine, leucine, valine, tyrosine and phenylalanine, were found to have significant associations with future development of diabetes. The risk of future diabetes was elevated at least 4-fold in subjects with high plasma amino acid concentrations.

5. Tandem mass spectrometry combined with the mass isotopomer analysis in metabolomics research

Because of the natural existence of low proportions of heavy ^{13}C (1.10%), ^{2}H (0.015%), ^{15}N (0.37%), and ^{18}O (0.20%) isotopes, biological molecules are identified by the number and position(s) of these isotopes. Mass isotopomers differ by the number of heavy atoms in their molecules, resulting in different molecular weights. In mass spectrometric jargon, they are referred to as M, M1, M2, …, Mn (Brunengraber *et al.*, 1997). The application of stable isotope labeled compounds in metabolomics research is not only useful for the identification and the characterization of metabolic pathways but also provides a powerful strategy to measure metabolite fluxes, i.e. the synthesis and degradation rates of those metabolites. In this section, we will briefly discuss the application of ^{13}C and ^{2}H isotope containing compounds to the quantification of metabolic fluxes through fatty acid oxidation and protein synthesis.

First we consider $^{2}H_2O$-metabolic labeling based approach utilizing tandem mass spectrometry for protein turnover studies (Kasumov *et al.*, 2011). Proteins exist in a dynamic state of equilibrium with their surrounding environment. The net static protein expression is determined by changes in a protein turnover, i.e. both synthesis and degradation. It is difficult to detect small changes in protein abundance, whereas the changes in the turnover rates are easily measurable. Advances in isotopic tracer methods and improvements in subcellular isolation methods have enabled studies of various protein fractions, i.e. total membrane, mitochondrial, sarcoplasmic or cytosolic proteins. However, understanding the pathologies related to the regulations of protein metabolism requires methods for studying the synthesis of an individual protein. Classical protein turnover studies relied on precursor-product relationships and involved the administration of a labeled amino acid. These methods require a long-term oral consumption or intravenous infusion of labeled amino acids, isolation mixed or individual proteins at different time points and degradation of proteins to individual amino acids. After column purification amino acids are derivatized and their labeling measured by GC-MS. This is a labor-intensive protocol that requires extensive purification of proteins and amino acids and often associated with contaminations. Recently we used a high resolution ion-trap tandem mass spectrometry, i.e. FTICR MS, to study the protein synthesis. Predictable fragmentation pattern of peptides in an ion-trap as a result of CID allows *de novo* sequencing of a protein. Particularly, the cleavage at the C-N amide (along with Cα-C and N-Cα bonds) in an ion-trap MS results in the mass difference between consecutive ions within a series corresponding to the specific amino acid and allows deduction of the peptide sequence. In addition, the high resolution of the ICR detector allows an accurate mass isotopomer analysis of a selected peptide. We took advantage of these capacities of the high resolution ion-trap tandem mass spectrometry for the measurement of ^{2}H-enrichment of protein-bound amino acids in our $^{2}H_2O$-based protein turnover studies. Oral administration of $^{2}H_2O$ (a safe, non-radioactive isotope) in drinking water results in a rapid steady state labeling of body water in free living organisms. $^{2}H_2O$ rapidly labels proteogenic amino acids, indicating that amino acids transfer to protein chain is a rate liming step in a protein biosynthesis (Rachdaoui *et al.*, 2009). Incorporation of multiple copies of ^{2}H from $^{2}H_2O$ to non essential amino acids results in the amplification of the isotopic enrichment in a product and enhances the measurements of their labeling. A time-domain stepwise fragmentation of a peptide in a selective reaction monitoring mode with the zoom scans of the precursor peptide and two of its consecutive fragments allows measurement of the ^{2}H-labeling of a protein bound amino acids. Thus, instead of purification of an individual protein and following degradation to amino acids for labeling measurement, this technique allows the measurement of an individual protein-bound amino acid even in a complex mixture of proteins. This approach was applied to measure the turnover rate constant of albumin in rats with the total body water enrichment of ~2.65% (Kasumov *et al.*, 2011). Fig 7 demonstrates the time dependent labeling observed in daughter ions derived from tryptic albumin peptide LVQEVTDFAK. The difference between the labeling of two consecutive fragment ions, i.e. QEVTDFAK *vs* EVTDFAK represents the labeling of peptide-bound glutamine. As expected, glutamine with several exchangeable hydrogen atoms incorporates a substantial quantity of deuterium. Thus, measurement of a peptide-bound amino acid labeling allows a reliable estimation of albumin kinetic parameters, i.e. the fractional synthesis rate (k= 0.37±0.05 day^{-1}) and the half-life ($t_{1/2}$ = 1.71 day) when compared with data obtained using the precursor ion (k= 0.36±0.06 day^{-1} and $t_{1/2}$ = 1.76 day). Assessment of the labeling of individual protein-bound amino acids via tandem mass spectrometry is especially important when calculating the kinetics of a protein with shorter half-life. The major assumption for a protein turnover study is that amino acids precursor labeling reaches the steady state much faster than that of their incorporation into proteins. Therefore, simultaneous measurements of labeling of proteolytic peptides and a peptide-bound amino acid allow both the validation of this assumption and the calculation the kinetic parameters.

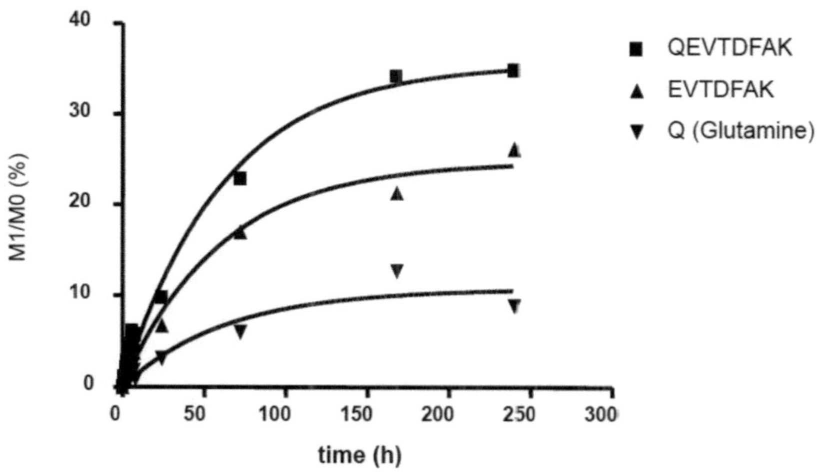

Fig. 7. Time course labeling of measured tryptic peptides QEVTDFAK and EVTDFAK and calculated labeling of peptide-bound glutamine (Q).

The second example highlights the utilization of the mass isotopomer analysis coupled metabolomics in the metabolic pathway discovery. In this case metabolomics in combination with stable isotopes is used to dissect 4-hydroxy-2-trans-nonenal (HNE) catabolism. HNE is a reactive unsaturated aldehyde derived from lipid peroxidation. HNE modified proteins, lipids and DNAs are the pathogenic factors. The detoxification of HNE is mainly via glutathione conjugation or reduction to 1,4-dihydroxynonene. HNE oxidation to 4-hydroxynonanoic acid is another major detoxification pathway. We identified two parallel catabolic pathways of 4-hydroxynonanoic acid in rat liver (Zhang *et al.*, 2009;Sadhukhan *et al.*, 2010). Two pathways are shown in Fig 8. The first round beta oxidation is the same for both pathways. The difference between pathway A and B starts from carbon 3 and 4 of 4-hydroxynonanoic acid. With a specifically isotope positional labeled 4-hydroxynonanoic acid, i.e. [3,4-$^{13}C_2$]-4-hydroxynonanoic acid, one can track the catabolic pathways of 4-hydroxynonante. In addition, quantification of the isotopically labeled products also allows calculating the relative rate of two pathways. [3,4-$^{13}C_2$]-4-Hydroxynonanoic acid is activated to M2 4-hydroxynonanoyl-CoA in the rat liver. In pathway A, M2 4-hydroxynonanoyl-CoA is isomerized to M2 3-hydroxynonanoyl-CoA that undergoes regular beta oxidations. Continuous beta oxidations of 3-hydroxynonanoyl-CoA complete HNE degradation in the pathway A and generate one M2 acetyl-CoA. In the pathway B, 4-hydroxynonanoyl-CoA splits one acetyl-CoA through first round beta oxidation to form M2 2-hydroxyheptanoyl-CoA. M2 2-hydroxyheptanoyl-CoA is alpha oxidized to form M1 formic acid and M1 hexanoyl-CoA. M1 hexanoyl-CoA is further beta oxidized to form one M1 acetyl-CoA. Therefore, M1 and M2 acetyl-CoA are generated through pathway B and A, respectively. One can analyze mass isotopomer enrichment of acetyl-CoA and determine the rate of two pathways based on M2 acetyl-CoA/M1 acetyl-CoA ratio if M1 and M2 acetyl-CoA are exclusively from pathway B and A.

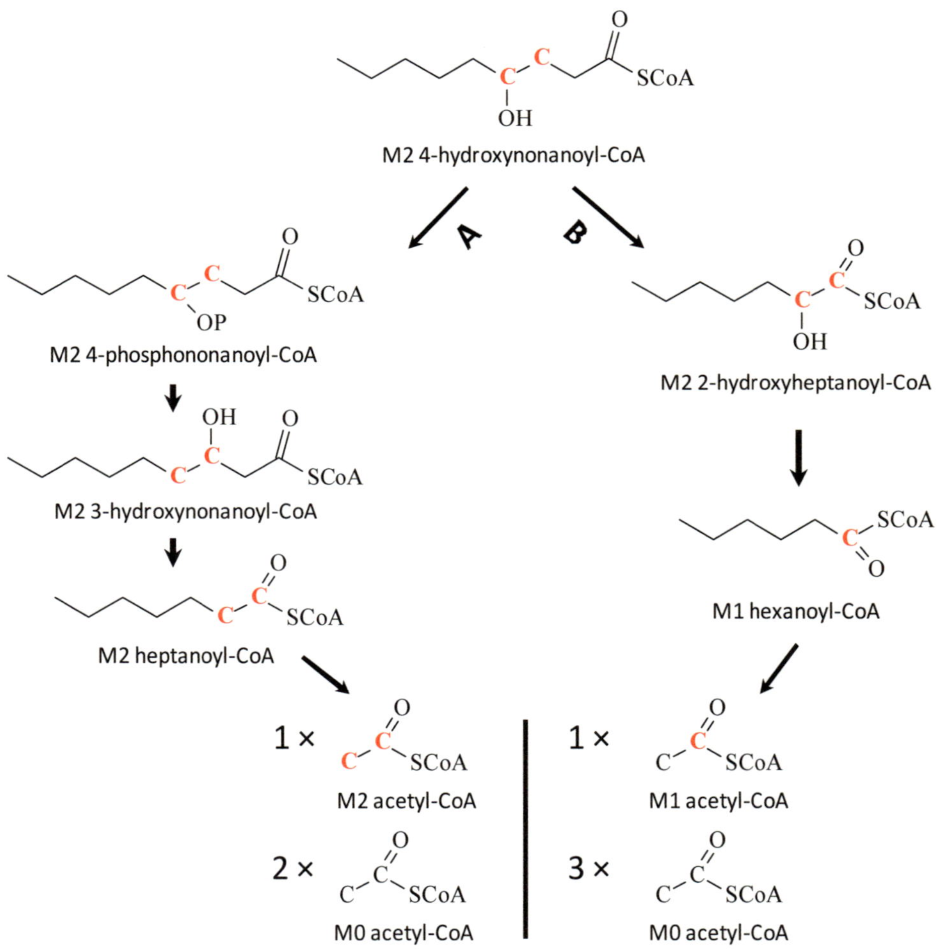

Fig. 8. The formation of M2 and M1 acetyl-CoA from two parallel catabolic pathways (A and B) of [3,4-$^{13}C_2$]-4-hydroxynonanoyl-CoA (M2 4-hydroxynonanoyl-CoA).

In both examples discussed above tandem mass spectrometry was used to measure the labeling of the fragment ions of the analytes. The mass isotopomer analysis of fragment ions, rather than whole molecule has following advantages: (i) the lower natural mass isotopomer distribution background of fragment improves the accuracy of mass isotopomer measurement, and (ii) the fragment mass isotopomer analysis provides positional labeling information. The positional mass isotope labeling measured by tandem mass spectrometry provides additional information on metabolic pathways and enables flux calculation that enriches metabolomics studies (Jeffrey *et al.*, 2002;Antoniewicz *et al.*, 2007;Kiefer *et al.*, 2007;Choi and Antoniewicz, 2011).

6. Conclusions

Tandem mass spectrometry is a sensitive and accurate technique that has been applied for targeted metabolites quantification. Traditional metabolomics investigation was limited by the sensitivity of analytical methods. Therefore the application of tandem mass spectrometry in metaboliomics study has emerged recently. This chapter mainly discussed the major applications of tandem mass spectrometry in recent metabolic studies: targeted metabolites quantification, untargeted analysis of a class of metabolites, biomarker assay, unknown metabolites identification and mass isotopomer analysis coupled tandem mass spectrometry. More insightful metabolic findings have been achieved with the help of sensitive tandem mass spectrometric technique. Current tandem mass spectrometry is limited to the targeted or semi un-targeted metabolic profiling. The single cell metabolism study is a challenging task because of the limitations related to the sensitivity of all existing methods. Achieving these goals in metabolomics studies requires the following improvements: (i) a more general un-targeted metabolomics methodology using tandem mass spectrometry is needed, (ii) a standardized tandem mass spectrometric condition is needed so that a public tandem mass spectrometric library could be generated, and (iii) further improvement of sensitivity. The future advancement of tandem mass spectrometry and bioinformatics tools should address these and other related issues in the metabolomics field.

7. Acknowledgement

This work was supported, in whole or in part, by National Institutes of Health Grant 5R21RR025346-02 (TK, Cleveland Clinic), and Cleveland Mt. Sinai Health Care Foundation (GZ, Case Western Reserve University).

8. References

Abbas R, Kombu RS, Ibarra RA, Goyal KK, Brunengraber H, and Sanabria JR (2011) The dynamics of glutathione species and ophthalmate concentrations in plasma from the VX2 rabbit model of secondary liver tumors. *HPB Surg,* 2011, 709052.

Antoniewicz MR, Kelleher JK, and Stephanopoulos G (2007) Accurate assessment of amino acid mass isotopomer distributions for metabolic flux analysis. *Anal Chem,* 79, 7554-7559.

Bajad S and Shulaev V (2011) LC-MS-based metabolomics. *Methods Mol Biol,* 708, 213-228.

Bales JR, Higham DP, Howe I, Nicholson JK, and Sadler PJ (1984) Use of high-resolution proton nuclear magnetic resonance spectroscopy for rapid multi-component analysis of urine. *Clin Chem,* 30, 426-432.

Barbas C, Moraes EP, and Villasenor A (2011) Capillary electrophoresis as a metabolomics tool for non-targeted fingerprinting of biological samples. *J Pharm Biomed Anal,* 55, 823-831.

BLIGH EG and DYER WJ (1959) A rapid method of total lipid extraction and purification. *Can J Biochem Physiol,* 37, 911-917.

Britz-McKibbin P (2011) Capillary electrophoresis-electrospray ionization-mass spectrometry (CE-ESI-MS)-based metabolomics. *Methods Mol Biol,* 708, 229-246.

Brunengraber H, Kelleher JK, and Des RC (1997) Applications of mass isotopomer analysis to nutrition research. *Annu Rev Nutr,* 17, 559-596.

Carroll AJ, Badger MR, and Harvey MA (2010) The MetabolomeExpress Project: enabling web-based processing, analysis and transparent dissemination of GC/MS metabolomics datasets. *BMC Bioinformatics,* 11, 376.

Chen C, Gonzalez FJ, and Idle JR (2007) LC-MS-based metabolomics in drug metabolism. *Drug Metab Rev,* 39, 581-597.

Choi J and Antoniewicz MR (2011) Tandem mass spectrometry: a novel approach for metabolic flux analysis. *Metab Eng,* 13, 225-233.

Corte L, Rellini P, Roscini L, Fatichenti F, and Cardinali G (2010) Development of a novel, FTIR (Fourier transform infrared spectroscopy) based, yeast bioassay for toxicity testing and stress response study. *Anal Chim Acta,* 659, 258-265.

Dalluge JJ, Gort S, Hobson R, Selifonova O, Amore F, and Gokarn R (2002) Separation and identification of organic acid-coenzyme A thioesters using liquid chromatography/electrospray ionization-mass spectrometry. *Anal Bioanal Chem,* 374, 835-840.

De Leon-Rodriguez LM, Lubag AJ, Malloy CR, Martinez GV, Gillies RJ, and Sherry AD (2009) Responsive MRI agents for sensing metabolism in vivo. *Acc Chem Res,* 42, 948-957.

Dehaven CD, Evans AM, Dai H, and Lawton KA (2010) Organization of GC/MS and LC/MS metabolomics data into chemical libraries. *J Cheminform,* 2, 9.

Duarte IF, Lamego I, Rocha C, and Gil AM (2009) NMR metabonomics for mammalian cell metabolism studies. *Bioanalysis,* 1, 1597-1614.

Elias DR, Thorek DL, Chen AK, Czupryna J, and Tsourkas A (2008) In vivo imaging of cancer biomarkers using activatable molecular probes. *Cancer Biomark,* 4, 287-305.

Gehlenborg N, O'Donoghue SI, Baliga NS, Goesmann A, Hibbs MA, Kitano H, Kohlbacher O, Neuweger H, Schneider R, Tenenbaum D, and Gavin AC (2010) Visualization of omics data for systems biology. *Nat Methods,* 7, S56-S68.

Gika H and Theodoridis G (2011) Sample preparation prior to the LC-MS-based metabolomics/metabonomics of blood-derived samples. *Bioanalysis,* 3, 1647-1661.

Gu L, Zhang GF, Kombu RS, Allen F, Kutz G, Brewer WU, Roe CR, and Brunengraber H (2010) Parenteral and enteral metabolism of anaplerotic triheptanoin in normal rats. II. Effects on lipolysis, glucose production, and liver acyl-CoA profile. *Am J Physiol Endocrinol Metab,* 298, E362-E371.

Hannun YA and Obeid LM (2008) Principles of bioactive lipid signalling: lessons from sphingolipids. *Nat Rev Mol Cell Biol,* 9, 139-150.

Harris SR, Zhang GF, Sadhukhan S, Murphy AM, Tomcik KA, Vazquez EJ, Anderson VE, Tochtrop GP, and Brunengraber H (2011) Metabolism of levulinate in perfused rat livers and live rats: conversion to the drug of abuse 4-hydroxypentanoate. *J Biol Chem,* 286, 5895-5904.

Haus JM, Kashyap SR, Kasumov T, Zhang R, Kelly KR, Defronzo RA, and Kirwan JP (2009) Plasma ceramides are elevated in obese subjects with type 2 diabetes and correlate with the severity of insulin resistance. *Diabetes,* 58, 337-343.

Horning EC and VANDENHEUVEL JA (1963) GAS CHROMATOGRAPHY. *Annu Rev Biochem,* 32, 709-754.

Jeffrey FM, Roach JS, Storey CJ, Sherry AD, and Malloy CR (2002) 13C isotopomer analysis of glutamate by tandem mass spectrometry. *Anal Biochem,* 300, 192-205.

Jentzmik F, Stephan C, Miller K, Schrader M, Erbersdobler A, Kristiansen G, Lein M, and Jung K (2010) Sarcosine in urine after digital rectal examination fails as a marker in prostate cancer detection and identification of aggressive tumours. *Eur Urol,* 58, 12-18.

Jordan KW and Cheng LL (2007) NMR-based metabolomics approach to target biomarkers for human prostate cancer. *Expert Rev Proteomics,* 4, 389-400.

Kasumov T, Huang H, Chung YM, Zhang R, McCullough AJ, and Kirwan JP (2010) Quantification of ceramide species in biological samples by liquid chromatography electrospray ionization tandem mass spectrometry. *Anal Biochem,* 401, 154-161.

Kasumov T, Ilchenko S, Li L, Rachdaoui N, Sadygov RG, Willard B, McCullough AJ, and Previs S (2011) Measuring protein synthesis using metabolic (2)H labeling, high-resolution mass spectrometry, and an algorithm. *Anal Biochem,* 412, 47-55.

Kewalramani G, Bilan PJ, and Klip A (2010) Muscle insulin resistance: assault by lipids, cytokines and local macrophages. *Curr Opin Clin Nutr Metab Care,* 13, 382-390.

Kiefer P, Nicolas C, Letisse F, and Portais JC (2007) Determination of carbon labeling distribution of intracellular metabolites from single fragment ions by ion chromatography tandem mass spectrometry. *Anal Biochem,* 360, 182-188.

Kint G, Fierro C, Marchal K, Vanderleyden J, and De Keersmaecker SC (2010) Integration of 'omics' data: does it lead to new insights into host-microbe interactions? *Future Microbiol,* 5, 313-328.

Koontz JL, Phillips KM, Wunderlich KM, Exler J, Holden JM, Gebhardt SE, and Haytowitz DB (2005) Comparison of total folate concentrations in foods determined by microbiological assay at several experienced U.S. commercial laboratories. *J AOAC Int,* 88, 805-813.

Koybasi S, Senkal CE, Sundararaj K, Spassieva S, Bielawski J, Osta W, Day TA, Jiang JC, Jazwinski SM, Hannun YA, Obeid LM, and Ogretmen B (2004) Defects in cell growth regulation by C18:0-ceramide and longevity assurance gene 1 in human head and neck squamous cell carcinomas. *J Biol Chem,* 279, 44311-44319.

Lapainis T, Rubakhin SS, and Sweedler JV (2009) Capillary electrophoresis with electrospray ionization mass spectrometric detection for single-cell metabolomics. *Anal Chem,* 81, 5858-5864.

Lee R, Ptolemy AS, Niewczas L, and Britz-McKibbin P (2007) Integrative metabolomics for characterizing unknown low-abundance metabolites by capillary electrophoresis-mass spectrometry with computer simulations. *Anal Chem,* 79, 403-415.

Lin S, Liu N, Yang Z, Song W, Wang P, Chen H, Lucio M, Schmitt-Kopplin P, Chen G, and Cai Z (2010) GC/MS-based metabolomics reveals fatty acid biosynthesis and cholesterol metabolism in cell lines infected with influenza A virus. *Talanta,* 83, 262-268.

Llorach R, Urpi-Sarda M, Jauregui O, Monagas M, and ndres-Lacueva C (2009) An LC-MS-based metabolomics approach for exploring urinary metabolome modifications after cocoa consumption. *J Proteome Res,* 8, 5060-5068.

Lu W, Bennett BD, and Rabinowitz JD (2008) Analytical strategies for LC-MS-based targeted metabolomics. *J Chromatogr B Analyt Technol Biomed Life Sci,* 871, 236-242.

Maher AD, Lindon JC, and Nicholson JK (2009) (1)H NMR-based metabonomics for investigating diabetes. *Future Med Chem,* 1, 737-747.

Malet-Martino M and Holzgrabe U (2011) NMR techniques in biomedical and pharmaceutical analysis. *J Pharm Biomed Anal,* 55, 1-15.

Martinez-Pinna R, Barbas C, Blanco-Colio LM, Tunon J, Ramos-Mozo P, Lopez JA, Meilhac O, Michel JB, Egido J, and Martin-Ventura JL (2010) Proteomic and metabolomic profiles in atherothrombotic vascular disease. *Curr Atheroscler Rep,* 12, 202-208.

Metz TO, Zhang Q, Page JS, Shen Y, Callister SJ, Jacobs JM, and Smith RD (2007) The future of liquid chromatography-mass spectrometry (LC-MS) in metabolic profiling and metabolomic studies for biomarker discovery. *Biomark Med,* 1, 159-185.

Millington DS, Norwood DL, Kodo N, Roe CR, and Inoue F (1989) Application of fast atom bombardment with tandem mass spectrometry and liquid chromatography/mass spectrometry to the analysis of acylcarnitines in human urine, blood, and tissue. *Anal Biochem,* 180, 331-339.

Millington DS and Stevens RD (2011) Acylcarnitines: analysis in plasma and whole blood using tandem mass spectrometry. *Methods Mol Biol,* 708, 55-72.

Mishur RJ and Rea SL (2011) Applications of mass spectrometry to metabolomics and metabonomics: Detection of biomarkers of aging and of age-related diseases. *Mass Spectrom Rev.*

Nicholson JK, Buckingham MJ, and Sadler PJ (1983a) High resolution 1H n.m.r. studies of vertebrate blood and plasma. *Biochem J,* 211, 605-615.

Nicholson JK, Lindon JC, and Holmes E (1999) 'Metabonomics': understanding the metabolic responses of living systems to pathophysiological stimuli via multivariate statistical analysis of biological NMR spectroscopic data. *Xenobiotica,* 29, 1181-1189.

Nicholson JK, Sadler PJ, Cain K, Holt DE, Webb M, and Hawkes GE (1983b) 88MHz 113Cd-n.m.r. studies of native rat liver metallothioneins. *Biochem J,* 211, 251-255.

Ogwang S, Nguyen HT, Sherman M, Bajaksouzian S, Jacobs MR, Boom WH, Zhang GF, and Nguyen L (2011) Bacterial conversion of folinic acid is required for antifolate resistance. *J Biol Chem,* 286, 15377-15390.

Osawa Y, Uchinami H, Bielawski J, Schwabe RF, Hannun YA, and Brenner DA (2005) Roles for C16-ceramide and sphingosine 1-phosphate in regulating hepatocyte apoptosis in response to tumor necrosis factor-alpha. *J Biol Chem,* 280, 27879-27887.

Pauling L, Robinson AB, Teranishi R, and Cary P (1971) Quantitative analysis of urine vapor and breath by gas-liquid partition chromatography. *Proc Natl Acad Sci U S A,* 68, 2374-2376.

Pavlou M and Diamandis EP (2009) The search for new prostate cancer biomarkers continues. *Clin Chem,* 55, 1277-1279.

Pewzner-Jung Y, Ben-Dor S, and Futerman AH (2006) When do Lasses (longevity assurance genes) become CerS (ceramide synthases)?: Insights into the regulation of ceramide synthesis. *J Biol Chem,* 281, 25001-25005.

Preiss JE, Loomis CR, Bell RM, and Niedel JE (1987) Quantitative measurement of sn-1,2-diacylglycerols. *Methods Enzymol,* 141, 294-300.

Rachdaoui N, Austin L, Kramer E, Previs MJ, Anderson VE, Kasumov T, and Previs SF (2009) Measuring proteome dynamics in vivo: as easy as adding water? *Mol Cell Proteomics,* 8, 2653-2663.

Reo NV (2002) NMR-based metabolomics. *Drug Chem Toxicol,* 25, 375-382.

Rolinski B, Arnecke R, Dame T, Kreischer J, Olgemoller B, Wolf E, Balling R, Hrabe de AM, and Roscher AA (2000) The biochemical metabolite screen in the Munich ENU Mouse Mutagenesis Project: determination of amino acids and acylcarnitines by tandem mass spectrometry. *Mamm Genome,* 11, 547-551.

Sadhukhan S, Han Y, Zhang GF, Brunengraber H, and Tochtrop GP (2010) Using isotopic tools to dissect and quantitate parallel metabolic pathways. *J Am Chem Soc,* 132, 6309-6311.

Schmelzer K, Fahy E, Subramaniam S, and Dennis EA (2007) The lipid maps initiative in lipidomics. *Methods Enzymol,* 432, 171-183.

Serkova NJ, Spratlin JL, and Eckhardt SG (2007) NMR-based metabolomics: translational application and treatment of cancer. *Curr Opin Mol Ther,* 9, 572-585.

Shaw RA, Rigatto C, Reslerova M, Ying SL, Man A, Schattka B, Battrell CF, Matthewson J, and Mansfield C (2009) Toward point-of-care diagnostic metabolic fingerprinting: quantification of plasma creatinine by infrared spectroscopy of microfluidic-preprocessed samples. *Analyst,* 134, 1224-1231.

Singh R and Sinclair KD (2007) Metabolomics: approaches to assessing oocyte and embryo quality. *Theriogenology,* 68 Suppl 1, S56-S62.

Sofia M, Maniscalco M, de LG, Paris D, Melck D, and Motta A (2011) Exploring airway diseases by NMR-based metabonomics: a review of application to exhaled breath condensate. *J Biomed Biotechnol,* 2011, 403260.

Soga T (2007) Capillary electrophoresis-mass spectrometry for metabolomics. *Methods Mol Biol,* 358, 129-137.

Soga T, Baran R, Suematsu M, Ueno Y, Ikeda S, Sakurakawa T, Kakazu Y, Ishikawa T, Robert M, Nishioka T, and Tomita M (2006) Differential metabolomics reveals ophthalmic acid as an oxidative stress biomarker indicating hepatic glutathione consumption. *J Biol Chem,* 281, 16768-16776.

Sreekumar A, Poisson LM, Rajendiran TM, Khan AP, Cao Q, Yu J, Laxman B, Mehra R, Lonigro RJ, Li Y, Nyati MK, Ahsan A, Kalyana-Sundaram S, Han B, Cao X, Byun J, Omenn GS, Ghosh D, Pennathur S, Alexander DC, Berger A, Shuster JR, Wei JT, Varambally S, Beecher C, and Chinnaiyan AM (2009) Metabolomic profiles delineate potential role for sarcosine in prostate cancer progression. *Nature,* 457, 910-914.

Struys EA, Heijboer AC, van MJ, Jakobs C, and Blankenstein MA (2010) Serum sarcosine is not a marker for prostate cancer. *Ann Clin Biochem,* 47, 282.

Styczynski MP, Moxley JF, Tong LV, Walther JL, Jensen KL, and Stephanopoulos GN (2007) Systematic identification of conserved metabolites in GC/MS data for metabolomics and biomarker discovery. *Anal Chem,* 79, 966-973.

Sullards MC, Allegood JC, Kelly S, Wang E, Haynes CA, Park H, Chen Y, and Merrill AH, Jr. (2007) Structure-specific, quantitative methods for analysis of sphingolipids by liquid chromatography-tandem mass spectrometry: "inside-out" sphingolipidomics. *Methods Enzymol,* 432, 83-115.

SWEELEY CC and Horning EC (1960) Microanalytical separation of steroids by gas chromatography. *Nature,* 187, 144-145.

Tan K, Tegner J, and Ravasi T (2008) Integrated approaches to uncovering transcription regulatory networks in mammalian cells. *Genomics,* 91, 219-231.

Tserng KY and Griffin R (2003) Quantitation and molecular species determination of diacylglycerols, phosphatidylcholines, ceramides, and sphingomyelins with gas chromatography. *Anal Biochem,* 323, 84-93.

Tsugawa H, Tsujimoto Y, Arita M, Bamba T, and Fukusaki E (2011) GC/MS based metabolomics: development of a data mining system for metabolite identification by using soft independent modeling of class analogy (SIMCA). *BMC Bioinformatics,* 12, 131.

Turinsky J, O'Sullivan DM, and Bayly BP (1990) 1,2-Diacylglycerol and ceramide levels in insulin-resistant tissues of the rat in vivo. *J Biol Chem,* 265, 16880-16885.

Vinayavekhin N and Saghatelian A (2010) Untargeted metabolomics. *Curr Protoc Mol Biol,* Chapter 30, Unit-24.

Vlaanderen J, Moore LE, Smith MT, Lan Q, Zhang L, Skibola CF, Rothman N, and Vermeulen R (2010) Application of OMICS technologies in occupational and environmental health research; current status and projections. *Occup Environ Med,* 67, 136-143.

Wang TJ, Larson MG, Vasan RS, Cheng S, Rhee EP, McCabe E, Lewis GD, Fox CS, Jacques PF, Fernandez C, O'Donnell CJ, Carr SA, Mootha VK, Florez JC, Souza A, Melander O, Clish CB, and Gerszten RE (2011) Metabolite profiles and the risk of developing diabetes. *Nat Med,* 17, 448-453.

Ward JL, Baker JM, and Beale MH (2007) Recent applications of NMR spectroscopy in plant metabolomics. *FEBS J,* 274, 1126-1131.

Wen B, Ma L, Nelson SD, and Zhu M (2008) High-throughput screening and characterization of reactive metabolites using polarity switching of hybrid triple quadrupole linear ion trap mass spectrometry. *Anal Chem,* 80, 1788-1799.

Xu F, Zou L, and Ong CN (2009) Multiorigination of chromatographic peaks in derivatized GC/MS metabolomics: a confounder that influences metabolic pathway interpretation. *J Proteome Res,* 8, 5657-5665.

Yamaguchi M, Miyashita Y, Kumagai Y, and Kojo S (2004) Change in liver and plasma ceramides during D-galactosamine-induced acute hepatic injury by LC-MS/MS. *Bioorg Med Chem Lett,* 14, 4061-4064.

Yano M, Kishida E, Muneyuki Y, and Masuzawa Y (1998) Quantitative analysis of ceramide molecular species by high performance liquid chromatography. *J Lipid Res,* 39, 2091-2098.

Zhang GF, Kombu RS, Kasumov T, Han Y, Sadhukhan S, Zhang J, Sayre LM, Ray D, Gibson KM, Anderson VA, Tochtrop GP, and Brunengraber H (2009) Catabolism of 4-hydroxyacids and 4-hydroxynonenal via 4-hydroxy-4-phosphoacyl-CoAs. *J Biol Chem,* 284, 33521-33534.

Zhang GF, Maudens KE, Storozhenko S, Mortier KA, Van Der SD, and Lambert WE (2003) Determination of total folate in plant material by chemical conversion into para-aminobenzoic acid followed by high performance liquid chromatography combined with on-line postcolumn derivatization and fluorescence detection. *J Agric Food Chem,* 51, 7872-7878.

Zhang GF, Storozhenko S, Van Der SD, and Lambert WE (2005) Investigation of the extraction behavior of the main monoglutamate folates from spinach by liquid chromatography-electrospray ionization tandem mass spectrometry. *J Chromatogr A,* 1078, 59-66.

Zhang S, Nagana Gowda GA, Ye T, and Raftery D (2010) Advances in NMR-based biofluid analysis and metabolite profiling. *Analyst,* 135, 1490-1498.

15

Determination of Ractopamine Residues in Pigs by Ultra Performance Liquid Chromatography Tandem Mass Spectrometry

Jelka Pleadin[1], Ana Vulić[1], Nina Perši[1] and Wolfgang Radeck[2]
[1]Croatian Veterinary Institute,
[2]Federal Office of Consumer Protection and Food Safety,
[1]Croatia
[2]Germany

1. Introduction

Ractopamine hydrochloride is a xenobiotic that belongs to a large group of $ß_2$-adrenergic agonist compounds. $ß_2$-Agonists are used in human and veterinary medicine for treatment of lung diseases as bronchodilators, tocolytics and heart tonics (Courtheyn et al., 2002; Malucelli et al., 1994; Meyer & Rinke, 1991). Besides their legal use, these drugs are often misused as growth promoters, to improve carcass composition by decreasing fat to the benefit of muscle mass, gaining higher economic benefit to producers (Anderson et al., 2009; Moody et al., 2000). Ractopamine hydrochloride increases the amount of lean meat and decreases the amount of carcass fat when fed to swine during the last 50 kg of gain, also increasing the rate of weight gain and feed conversion (Anderson et al., 1989; Merkel et al., 1987; Watkins et al., 1990; Williams et al., 1994). The biochemical basis of ractopamine effects is increasing the nitrogen retention, protein synthesis, enhancing lipolysis and suppressing lipogenesis (Apple et al., 2007; Armstrong et al., 2004; Carr et al., 2005; Mills, 2002; Mitchell et al., 1990; Mitchell, 2009).

Illegal use of $ß_2$-agonists in 5- to 10-fold therapeutic doses leads to accumulation of these compounds in animal tissues such as liver, kidney and muscle (Smith, 1998; Smith & Shelver, 2002). High amounts of $ß_2$-agonist residues in meat and meat products led to a number of cases of food poisoning in humans in the last 20 years (Brambilla et al. 1997; Garay et al., 1997; Martinez-Navarro, 1990; Pulce at al., 1991; Ramos et al., 2003), although the Council Directive 96/22/EC banned the use of these substances in the European Union. Consequently, detection of $ß_2$-agonists in biological material from farm animals is a high priority because of the public health concern; relatively large numbers of samples have to be analyzed and more stringent criteria used in view of the serious public health implications of positive results. In order to provide quality assurance for the consumer and to satisfy legal testing obligations, the ability to detect drug residues at low concentrations has become a very important issue.

Although all incidents of poisoning were caused by clenbuterol toxicity, the European Union has placed ban upon the use of all β-agonists, thus requiring strict monitoring for the illegal use of this and other β-agonists. Ractopamine was approved by the U.S. Food and

Drug Administration for use in finishing swine in 1999 (Anderson et al., 2009), whereas in European Union the use of ractopamine like other β-agonists is completely banned.

As β-agonists are classified into group A substances having an anabolic effect and unauthorised substances, confirmation of the target analyte is necessary according to analytical performance criteria of the Commission Decision 2002/657/EC. As a result of the requirement of analytical criteria on confirmatory methods, mass spectrometry techniques are widely used to identify trace levels of organic residues and contaminants (Shao et al., 2009). A number of methods, in most cases gas chromatography (GC) or liquid chromatography (LC) coupled with mass spectrometry (MS), are available for the analysis of ractopamine in a variety of matrices. On this, a minimum of four characteristic ions are required to satisfy the four identification points for the GC-MS or LC-MS techniques and one precursor ion and two daughter ions can provide four identification points for tandem mass spectrometry techniques including triple quadropole mass spectrometry and ion trapping techniques (e.g., LC-MS/MS and GC-MS/MS). Fragmentation pathway of ractopamine is shown in Figure 1. However, GC/MS methods require derivation because of their high polarity and low volatility, which is time-consuming, tedious, laborious and expensive (Shao et al., 2009). Because of its high sensitivity and selectivity, LC-MS/MS is often the method of choice in the analysis of trace levels of polar contaminants (Nielen et al., 2008; Pleadin et al., 2011) and recently many authors demonstrated its superior performance in ractopamine analyses (Blanca et al., 2005; Churchwell et al., 2002). So far, ultra performance liquid chromatography tandem mass spectrometry (UPLC-MS/MS) is one of the most efficient methods (Nielen et al., 2008; Shao et al., 2009) because of the high resolution, rapid separation of UPLC, and the selectivity and sensitivity characteristic of MS-MS detection (Dong et al., 2011; Zheng et al., 2010).

Analysis of ractopamine residues is mostly performed in urine and plasma samples taken from living animals as well as in tissue samples after slaughtering (muscle, liver, kidney and fat) (Antignac et al., 2002; Gratacós-Cubarsí et al., 2006; Shao et al., 2009; Thompson et al., 2008). Besides tissues and body fluids, hair has been shown to be an excellent site for accumulation of different drugs including ß-agonists and therefore also an appropriate matrix to monitor the presence of drug residues. Another advantage of hair as a matrix of choice to monitor the presence of ß-agonists is the ability to collect hair samples from living animals. On the other hand, there are little data on ractopamine accumulation in the hair of farm animals. Literature data show that in swine, ractopamine hydrochloride is very rapidly eliminated resulting in very low tissue residues, while major elimination route is *via* conjugation with glucuronic acid and urinary excretion (Dalidowicz, 1992).

Bearing in mind rapid urinary excretion of ractopamine hydrochloride and hair affinity for $ß_2$-agonist accumulation, the aim of this study was to determine ractopamine residues using UPLC-MS/MS and to compare residue levels of ractopamine hydrochloride in swine urine and hair samples after sub-chronic treatment with low ractopamine doses.

2. Materials and methods

2.1 Chemicals and apparatus

Ractopamine hydrochloride (Sigma-Aldrich-Chemie, Steinheim, Germany) was used for animal treatment and method validation. Ractopamine-D5-hydrochloride (RIKILT, Wageningen, The Netherlands) was applied as internal standard. Protease Type XIV, from *Streptomyces grisues* (Sigma-Aldrich-Chemie, Steinheim, Germany) was used for hair sample

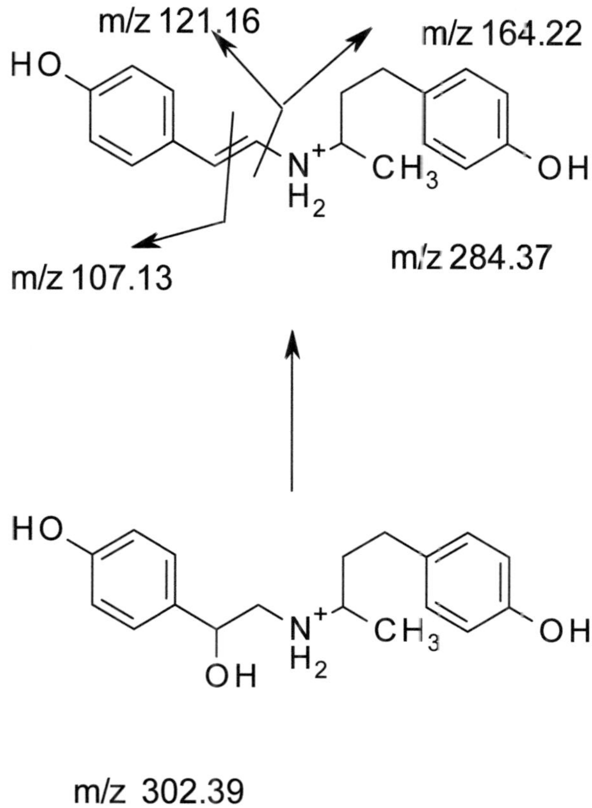

Fig. 1. Fragmentation pathway of ractopamine.

analysis. ß-Glucuronidase/aryl sulfatase (Merck Chemicals, Darmstadt, Germany) was used for urine sample analysis. All solvents used were of HPLC grade. Screen Dau solid phase extraction columns (500 mg, 6 mL) used for clean up were from Amchro (Hattersheim, Germany). The analyses were performed on UPLC/MS/MS Xevo® TQ-S (Waters, En Yvelines Cedex, France). Hair samples were homogenized by means of a MM400 mixer mill (Retsch, Haan, Germany). Filtration and centrifugation of hair samples were performed on Amicon Ultra Centrifugal Filters Ultracell 10 K (Millipore, Carrigtwohill, Ireland). Hair samples were dried in a vacuum drying cabinet VT 6060 M (Heraeus, Hanau, Germany).

2.2 Animals and sampling procedure

The study included 12 male pigs (9 treated and 3 controls), Zegers hybrid type (white hair), 55 kg body weight, farm-bred and kept under same conditions. Animals (n=9) were randomly divided into 3 groups and treated with ractopamine hydrochloride in a dose of 1 mg daily (absolute) *per os* during 28 days (0.51 mg/kg b.w.). Treated animals were orally administered ractopamine hydrochloride in the form of a capsule filled with pure chemical admixed to feed. Three animals served as a control group and were left untreated. On days 1, 3 and 8 after treatment withdrawal, treated animals were sacrificed in groups of 3. Control group animals were sacrificed on day 8 after experimental animal treatment withdrawal. Hair samples were obtained by shaving pigs with a razor blade and stored at room temperature. Urine samples after collection on slaughtering were stored at -20 °C until analysis. All experiments were performed according to the Croatian Animal Protection Act (Official Gazette of the Republic of Croatia 135/06).

2.3 Hair sample preparation and clean up

Hair samples were washed with 2x20 mL of water and dried overnight. Dry hair was homogenized in mixer mill for 2x2 minutes and then mixed "head over head" for 60 minutes. Internal standard of ractopamine-D5-hydrochloride (0.1 ng/µL), 5 mL of tris-buffer (pH=8) and 100 µL of protease solution (50 mg/mL in water) were added to the portion of 500 mg of hair to obtain spiking level of 5 ng/g. The mixtures were incubated overnight at 55 °C in shaker water bath. After incubation, 2 mL of phosphate buffer (pH=6) was added and pH was adjusted to 6. Samples were then shaken in ultrasonic bath at room temperature and centrifuged at 4 °C and 4000 rpm using Amicon filter units. The supernatants were transferred to another tube with addition of 200 µL of methanol, followed by centrifugation at 4 °C and 4000 rpm. The centrifuged extracts were loaded to SPE cartridges conditioned with 2 mL of methanol, 2 mL of water and 2 mL of phosphate buffer pH=6. Cartridges were washed with 1 mL of 1 M acetic acid and evaporated to dryness followed by washing with 2 mL of methanol and evaporating to dryness. The elution was performed with 6 mL of mixture consisting of ethyl acetate and 25% ammonia at a 97:3 ratio. The samples were evaporated to dryness under stream of nitrogen at 35 °C. Residues were then dissolved in 200 µL of HPLC mobile phase consisting of 0.1% formic acid in water (A)/0.1% formic acid in acetonitrile at a 95:5 ratio (B).

2.4 Urine sample preparation and clean up

A portion of 10 mL of urine was spiked with internal standard of ractopamine-D5-hydrochloride (0.1 ng/µL) with addition of 5 mL of sodium acetate buffer (pH=5) and 50 µL of glucuronidase/aryl sulfatase to obtain spiking level of 5 ng/mL. The same steps were performed also without hydrolysis. The samples were shaken and incubated overnight at 37 °C. After cooling at room temperature, 5 mL of phosphate buffer (pH=6) was added. The hydrolyzed solution was centrifuged followed by addition of 200 µL of methanol to obtained supernatants. The supernatants were loaded to SPE cartridges conditioned with 2 mL of methanol, 2 mL of water and 2 mL of phosphate buffer (pH=6). Cartridges were then washed with 1 mL of 1 M acetic acid and evaporated to dryness followed by washing with 2 mL of methanol and evaporating to dryness. The elution was performed with 6 mL of a mixture consisting of ethyl acetate and 25% ammonia at a 97:3 ratio. The samples were evaporated to dryness under stream of nitrogen at 35 °C. Residues were then dissolved in 200 µL of HPLC mobile phase consisting of 0.1% formic acid in water (A)/0.1% formic acid(B) in acetonitrile at a 95:5 ratio.

2.5 Liquid chromatography tandem mass spectrometry conditions

The UPLC separation was performed on Acquity HSS C18 columns (150x2, 1.8 µm particle size) at a flow rate of 0.45 mL/min and temperature 40 °C. The mobile phase consisted of constituent A (0.1% formic acid in water) and constituent B (0.1% formic acid in acetonitrile). A gradient elution program was employed as follows: 0-5 min 95% A, 15 min 50% A, 17 min 50% A, 18 min 10% A, 19 min 10% A, 20 min 95% A and 25 min 95% A. The injection volume was 10 µL. The mass spectrometry conditions were as follows: electrospray ionization, positive polarity, capillary voltage 0.65 kV, source temperature 150 °C, desolvation temperature 550 °C, cone gas 20 L/h, desolvation gas 1200 L/h, and collision gas 0.1 L/h. The mass spectrometer was operated in multiple reaction monitoring mode, the protonated molecular ion of ractopamine at m/z = 302.2 being the precursor ion. Four product ions at m/z = 284.2, m/z = 164.2, m/z = 121.2 and m/z = 107.1 were monitored. Quantitation was performed with most intensive transition (m/z 302.2 → 164.2) *versus* internal standard monitored (ractopamine-D5, m/z 307.1>167.1) and extrapolation using a six point calibration curves.

2.6 Validation process

Validation was carried out according to Commission Decision 2002/657/EC by an alternative approach of matrix comprehensive in-house validation by means of a factorial design software used for factorial design and calculation was InterVal Plus (quo data, Gesellschaft für Qualitätsmanagement und Statistik GmbH, Dresden, Germany). In validation process, decision limit (CCα), detection capability (CCβ), precision, recovery, repeatability, in-house reproducibility, matrix effects, specificity and ruggedness were studied. Validation process started with factorial design for both matrices. Factors and their levels for hair and urine are presented in Table 1 and Table 3, respectively.

Factor	Level
Species	Calf / Cow
Matrix condition	White / Black
Operator	Analyst 1 / Analyst 2
Storage of extracts (injection solution)	2 days, 4 °C before injection/ without
Instrument	Xevo /Q-TOF
Amount of matrix	200 mg / 500mg

Table 1. Factors of interest and their levels used for validation of ractopamine in hair.

For validation of ractopamine in hair, 16 runs, each for 5 concentration levels, were conducted within 16 days and with different factor combinations. In total, 80 measurements were performed. Table 2 shows experimental design for validation of the method of hair analysis as an example.

Within each run, blank samples were fortified at five concentration levels: 2, 5, 8, 11 and 15 ng/g. In addition, a matrix blank sample and reagent blank sample were included in each run.

Validation for ractopamine in urine was preformed with 32 runs, 8 runs *per* species. Within each run, blank samples were fortified at six concentration levels: 0.125, 0.25, 0.375, 0.50, 0.625 and 0.75 ng/mL. In addition, a matrix blank sample and reagent blank sample were included in each run.

			Factor level combination					
Run	Sample Number	Run name	Species	Matrix cond.	Operator	Storage of extracts	Instrument	Sample amount
Run 15	P100260	AS100502	cow	white	Analyst 1	without	Q-TOF	500 mg
Run 02	P100262	AS100475	cow	white	Analyst 1	2 days 4 °C	Xevo	200 mg
Run 09	P100248	AS100496	calf	black	Analyst 1	without	Xevo	200 mg
Run 04	P100257	AS100477	cow	black	Analyst 1	without	Xevo	500 mg
Run 08	P100246	AS100481	calf	black	Analyst 1	2 days 4 °C	Q-TOF	500 mg
Run 10	P100247	AS100497	calf	black	Analyst 2	without	Q-TOF	500 mg
Run 11	P100250	AS100498	calf	white	Analyst 1	2 days 4 °C	Xevo	500 mg
Run 06	P100253	AS100479	calf	white	Analyst 1	without	Q-TOF	200 mg
Run 05	P100252	AS100478	calf	white	Analyst 2	without	Xevo	500 mg
Run 13	P100259	AS100500	cow	black	Analyst 1	2 days 4 °C	Q-TOF	200 mg
Run 14	P100256	AS100501	cow	black	Analyst 2	2 days 4 °C	Xevo	500 mg
Run 07	P100249	AS100480	calf	black	Analyst 2	2 days 4 °C	Xevo	200 mg
Run 03	P100258	AS100476	cow	black	Analyst 2	without	Q-TOF	200 mg
Run 01	P100261	AS100474	cow	white	Analyst 2	2 days 4 °C	Q-TOF	500 mg
Run 12	P100251	AS100499	calf	white	Analyst 2	2 days 4 °C	Q-TOF	200 mg
Run 16	P100263	AS100503	cow	white	Analyst 2	without	Xevo	200 mg

Table 2. Experimental plan for hair.

Factor	Level
Species	Bull / Calf / Cattle / Pig
Ultrafiltration of injection solution	Without / With
Operator	Analyst 1 / Analyst 2
Storage of matrix before analysis	1 week at 4 °C / 1 week frozen
SPE cartridges	Small / Large
Injection of final solutions	Immediately / 2 days after finishing sample preparation, if stored at 4 °C
Enzyme	Sigma / Merck
Cartridges	Starta X/ CSD

Table 3. Factors of interest and their levels used for validation of ractopamine in urine.

3. Results and discussion

3.1 Applicability of study results

There are literature reports on ractopamine determination using different techniques (Shelver & Smith, 2003; Smith et al., 1993; Thompson et al., 2008; Turberg et al., 1995). Studies suggest the use of LC/MS/MS systems as probably the best methods to improve sensitivity in determination of β-adrenergic agonists, while retaining excellent selectivity (Smith & Shelver, 2002), pointing to UPLC-MS/MS as one of the most efficient methods.

As the analysis of ractopamine in samples from all stages of production is important for monitoring illegal use in European Union, development of sensitive and selective methodologies in different matrices is required. Control and monitoring programs mandated by government have also necessitated implementation of assays for determination of ractopamine accumulation and excretion from tissues and body fluids in farm animals.

Studies of ractopamine residue detection after different treatment schedules in different animal species have been reported by several authors (Elliot et al., 1998; Qiang et al., 2007; Smith & Shelver, 2002; Thompson et al., 2008). Published studies report on tissue residues of ractopamine and its urinary excretion (Antignac et al., 2002; Blanca et al., 2005; Dickson et al., 2005; Moragues & Igualada, 2009; Nielen et al., 2008; Thompson et al., 2008; Van Hoof et al., 2005) and ractopamine residues in hair (Nielen et al., 2008). However, there are little data on the accumulation of ractopamine in pig hair (white or black) as a novel matrix for the control of ractopamine illegal use.

The aim of our study was to determine residue levels in swine urine (without and with sample hydrolysis) and hair samples after sub-chronic treatment of animals with ractopamine hydrochloride, using UPLC/MS/MS as a sensitive and reliable analytical method for determination of low ractopamine concentrations. Our study provided additional data on the ractopamine residue excretion and accumulation in swine.

3.2 Method validation

No interference on ractopamine identification was found owing to the highly specific MRM acquisition method and the use of an appropriate deuterated internal standard. The validation results for both analytical matrices are shown in Table 4 and Table 5. It is concluded that the methods showed relevant decision limit (CCα) and detection capability (CCβ), with values 0.37 and 0.51 ng/g for urine and 2.53 and 2.98 ng/g for hair, respectively. The mean recoveries ranged from 93.9% to 94.6% for urine and from 103.0% to 103.6% for hair. Also, the methods showed good repeatability, in-house reproducibility and linearity, and met the validation criteria set for quantitative residue analysis methods according to Decision. Successful validation of the method according to the European Union requirements and its application to real samples demonstrated its efficiency for veterinary control of ractopamine as a β-agonist in hair and urine.

Matrix	Urine	Hair
CCα [ng/g]	0.37	2.53
CCβ [ng/g]	0.51	2.98
Specificity	passed	passed
Ruggedness	passed	passed

Table 4. Validation results for ractopamine in urine and hair.

As part of the whole validation process the short time stability of ractopamine in urine and moreover the long-term stability was investigated. For that purpose a isochronic approach were applied (Lamberty 1998). The results are summarized in Figure 2.

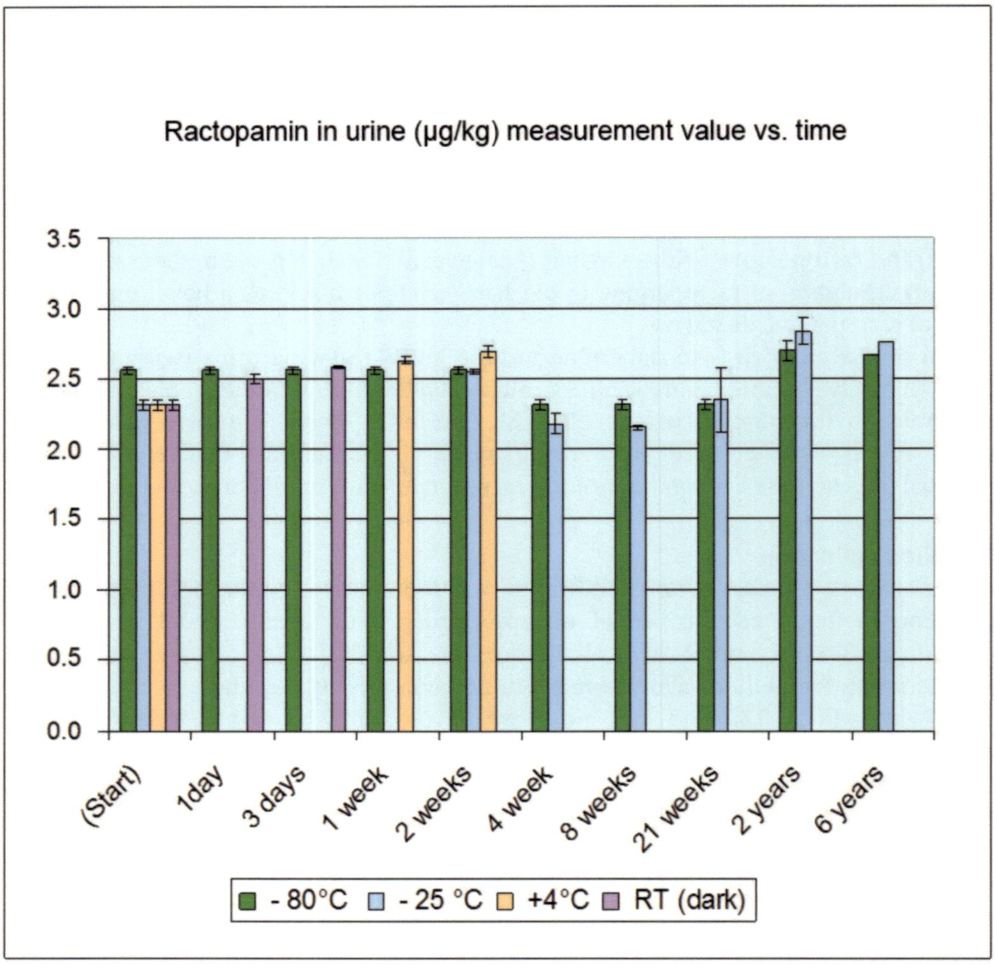

Fig. 2. Short time and long-term stability of ractopamine in urine.

Matrix	Spiked concentration (ng/mL)	s_r (ng/mL)	RSD (%)	s_{WR} (ng/mL)	RSD (%)	Recovery (%)
Urine	0.125	0.008	6.1	0.045	35.8	93.9
	0.250	0.008	3.1	0.045	17.9	94.3
	0.375	0.011	3.0	0.062	16.6	94.5
	0.500	0.015	3.0	0.081	16.2	94.5
	0.625	0.019	3.0	0.100	16.0	94.6
	0.750	0.023	3.0	0.119	15.9	94.6
Hair	2.000	0.076	3.8	0.184	9.2	103.0
	5.000	0.190	3.8	0.374	7.5	103.4
	8.000	0.304	3.8	0.600	7.5	103.5
	11.00	0.418	3.8	0.837	7.6	103.5
	15.00	0.570	3.8	1.155	7.7	103.6

Table 5. Repeatability, in-house reproducibility and recovery.

3.3 Concentration of ractopamine residues

3.3.1 Urine residues

Ractopamine concentrations were determined in urine on days 1, 3 and 8 after 28 days of continuous treatment of pigs. The mean (±SD) ractopamine concentrations in urine samples without and with enzyme hydrolysis on days after treatment discontinuation in the experimental group of animals are shown in Table 6.

Withdrawal time (days)	Urine ractopamine concentrations (mean±SD, ng/mL)	
	Without hydrolisis	With hydrolisis
1	1.52±0.17	7.63±0.19
	0.29±0.03	4.47±0.04
	0.76±0.04	8.15±0.10
3	0.15±0.01	2.40±0.08
	0.47±0.08	6.32±0.15
	a	a
8	0.38±0.01	6.53±0.12
	0.22±0.00	4.94±0.01
	a	a

[a] Insufficient urine obtained for analysis

Table 6. Concentrations of ractopamine detected in urine by UPLC-MS/MS on days after withdrawal.

As $ß_2$-adrenergic agonist compounds are extensively metabolized to β-glucuronide and/or sulfate conjugates in humans and animals, in the present study deconjugation was used as a second step on urine sample preparation. The ractopamine concentrations determined in swine urine samples were much greater after β-glucuronidase hydrolysis than those determined without this analytical step (Table 6). Figure 3 shows the UPLC-MRM chromatograms of confirmatory analysis of ractopamine in pig urine on days after withdrawal determined with hydrolysis. The concentration of ractopamine in urine samples processed with enzyme hydrolysis was almost 10-fold that recorded in urine samples analyzed without enzyme hydrolysis. Deconjugation step confirmed ractopamine to be excreted mainly in the form of glucuronide metabolites, as reported previously (Qiang et al., 2007; Shelver & Smith, 2003). Therefore, further analyses (hair) were performed exclusively with sample hydrolysis and values obtained with sample enzyme hydrolysis were used on interpretation of the urine ractopamine concentrations.

Earlier investigations performed in pigs with the use of a higher ractopamine anabolic dose (18 mg ractopamine hydrochloride/kg of feed) showed significantly higher concentrations of ractopamine in urine, with mostly twofold concentrations determined with sample enzyme hydrolysis (Qiang et al., 2007).

Studies carried out in cattle and sheep report on detectable ractopamine residues in urine 5 to 7 days after the last exposure to dietary ractopamine, pointing that hydrolysis of ractopamine metabolites may extend the period in which it is detected in cattle (Smith & Shelver, 2002). In their study, Thompson et al. (2008) fed pigs a feed containing 18 mg/kg ractopamine once daily for 10 days (180 mg of ractopamine in total). Ractopamine residues in pig urine were detectable by both screening and confirmatory methods until day 21 of treatment withdrawal. Urine sample analyses showed high concentrations of ractopamine ranging from 473.6 ng/mL to 1131.6 ng/mL on day 1 of withdrawal period. After seven days, the concentration of ractopamine was considerably lower, ranging from 3.4 ng/mL to 6.2 ng/mL. In our study, which was also performed in pigs but with a dose approximately 6 times lower (28 mg of ractopamine in total), the mean ractopamine residues in urine ranged from 6.7±1.8 ng/mL on day 1 of withdrawal to 5.7±0.9 ng/mL on the last day of withdrawal. Our study indicated that in spite of rapid urinary excretion, ractopamine residues can be detected in urine samples eight days after treatment cessation. Elliott et al. (1998) report ractopamine depletion in calves; their study showed the high concentration of ractopamine detected on day 1 of withdrawal (280 ng/mL) to be followed by a decline to the level of 18 ng/mL (day 3 of withdrawal), and no ractopamine residue detectable on day 14. In calf, the ractopamine residue concentrations found after drug withdrawal were also substantially lower than during the medication period, and were only detectable in one animal 2 weeks after removal of medication from the diet (Elliot et al., 1998). In their study on cattle treated with 20 mg/kg dietary ractopamine for seven days, Smith and Shelver (2002) obtained similar results concerning rapid urinary excretion of ractopamine. Ractopamine residues were 2523±1367 ng/mL on day 1 of withdrawal and on day 6 residue levels were below the limit of quantification. This study was simultaneously, in the same conditions and treatment schedule, conducted in sheep. Results obtained by analyzing sheep urine samples were rather different. The concentration of ractopamine in urine samples after 7 days of withdrawal was still detectable, 178±78 ng/mL (Smith & Shelver, 2002). These results indicate that urinary excretion of ractopamine is species dependent, meaning that even with similar doses, different animal species excrete ractopamine in different time frames. Literature data indicate that swine eliminate nearly 85% of the ractopamine administered during the first day, resulting in relatively low tissue residues (Dalidowicz et al, 1992). In comparison with some literature data (Qiang et al., 2007; Smith & Shelver, 2002), low concentrations of ractopamine urinary residues determined in our study could be explained by low exposure of animals to ractopamine.

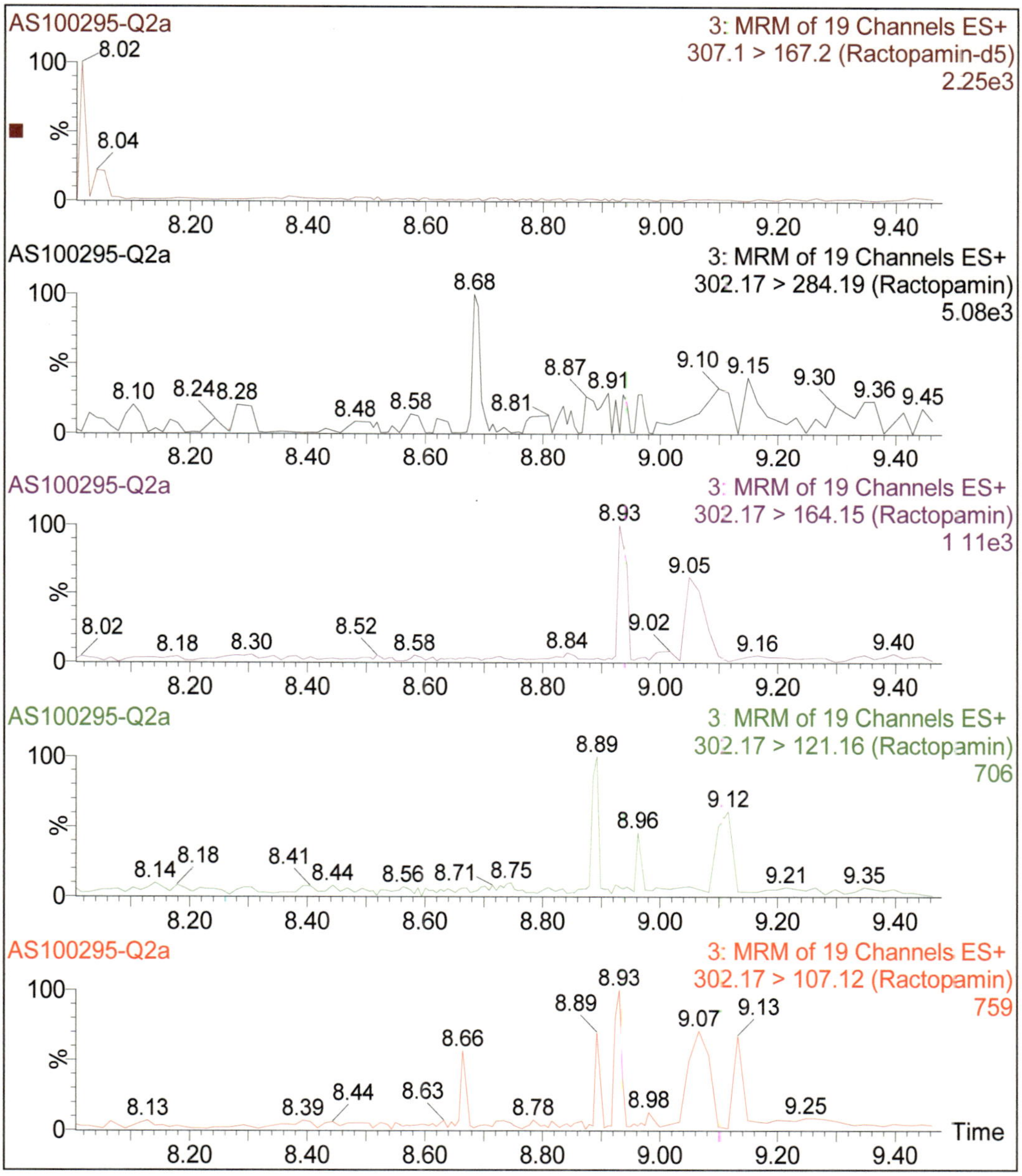
AS100295-Q2a
3: MRM of 19 Channels ES+
307.1 > 167.2 (Ractopamin-d5)
2.25e3
8.02
8.04
AS100295-Q2a
3: MRM of 19 Channels ES+
302.17 > 284.19 (Ractopamin)
5.08e3
8.68
AS100295-Q2a
3: MRM of 19 Channels ES+
302.17 > 164.15 (Ractopamin)
1 11e3
8.93
9.05
AS100295-Q2a
3: MRM of 19 Channels ES+
302.17 > 121.16 (Ractopamin)
706
8.89
8.96
9.12
AS100295-Q2a
3: MRM of 19 Channels ES+
302.17 > 107.12 (Ractopamin)
759
8.93
8.89
9.07
9.13
8.66
Time

A

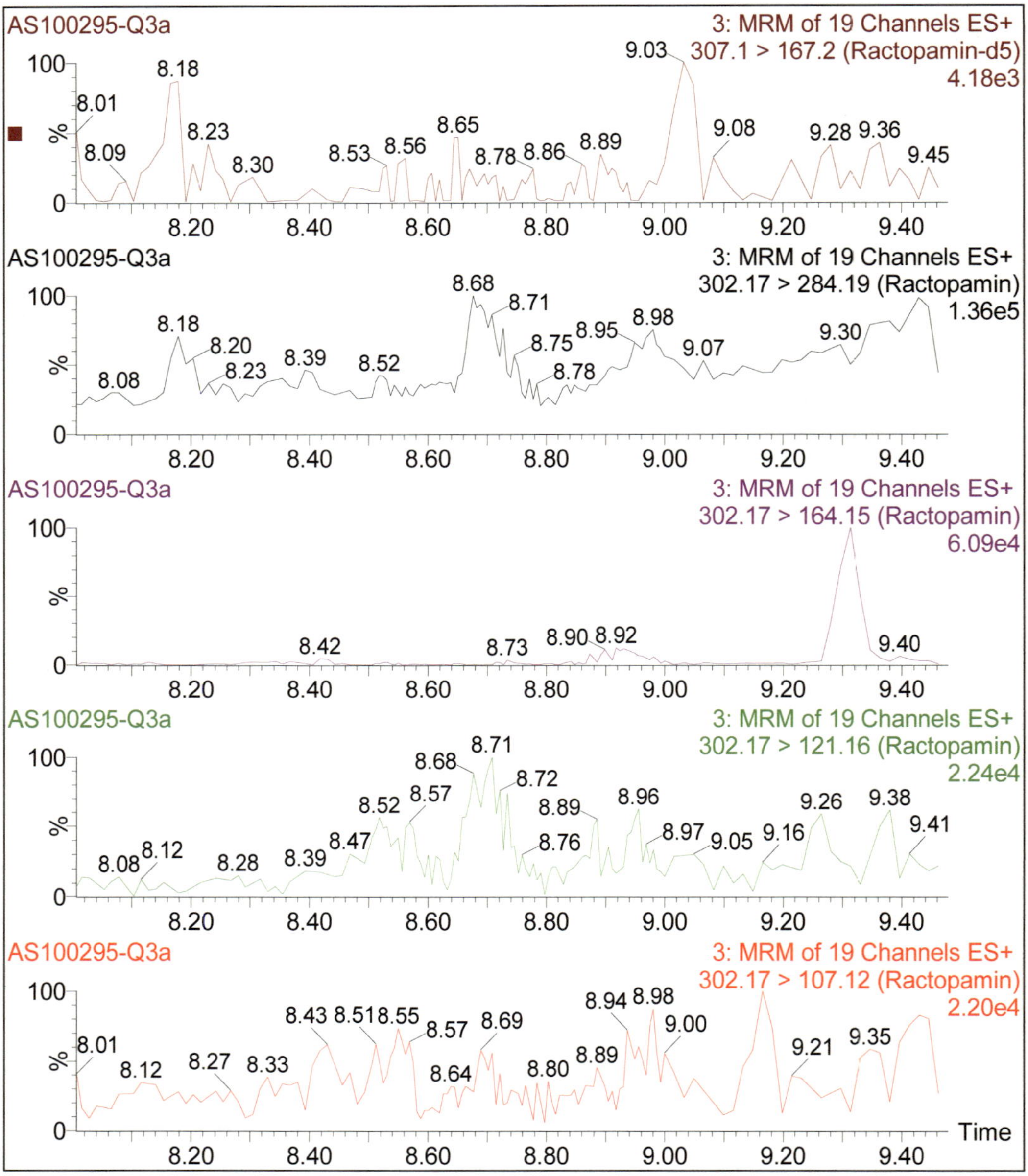
AS100295-Q3a
3: MRM of 19 Channels ES+
307.1 > 167.2 (Ractopamin-d5)
4.18e3
AS100295-Q3a
3: MRM of 19 Channels ES+
302.17 > 284.19 (Ractopamin)
1.36e5
AS100295-Q3a
3: MRM of 19 Channels ES+
302.17 > 164.15 (Ractopamin)
6.09e4
AS100295-Q3a
3: MRM of 19 Channels ES+
302.17 > 121.16 (Ractopamin)
2.24e4
AS100295-Q3a
3: MRM of 19 Channels ES+
302.17 > 107.12 (Ractopamin)
2.20e4
%
Time

B

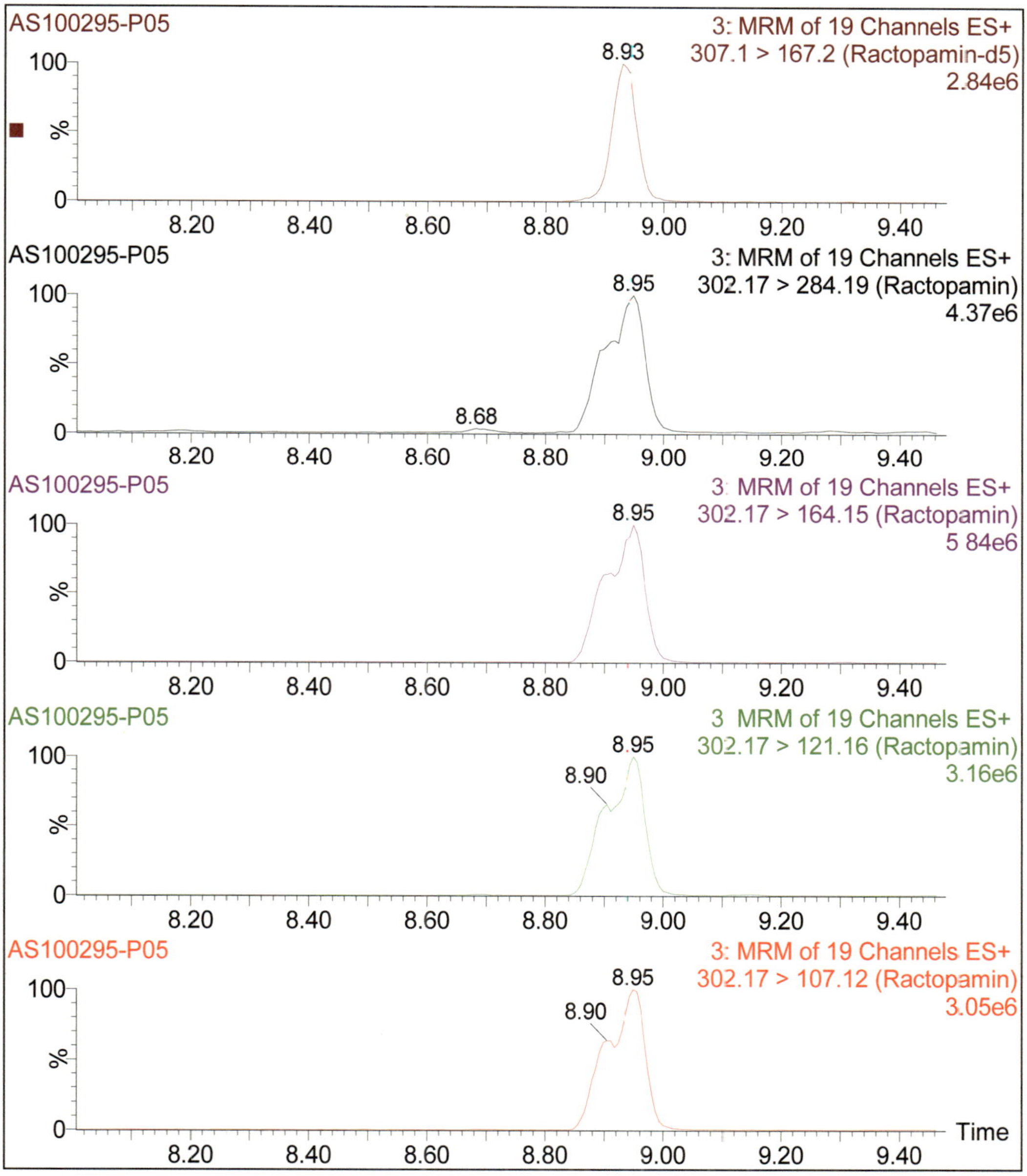
AS100295-P05
3: MRM of 19 Channels ES+
307.1 > 167.2 (Ractopamin-d5)
2.84e6
8.93
AS100295-P05
3: MRM of 19 Channels ES+
302.17 > 284.19 (Ractopamin)
4.37e6
8.95
8.68
AS100295-P05
3: MRM of 19 Channels ES+
302.17 > 164.15 (Ractopamin)
5.84e6
8.95
AS100295-P05
3: MRM of 19 Channels ES+
302.17 > 121.16 (Ractopamin)
3.16e6
8.95
8.90
AS100295-P05
3: MRM of 19 Channels ES+
302.17 > 107.12 (Ractopamin)
3.05e6
8.95
8.90
100
%
0
8.20
8.40
8.60
8.80
9.00
9.20
9.40
Time

C

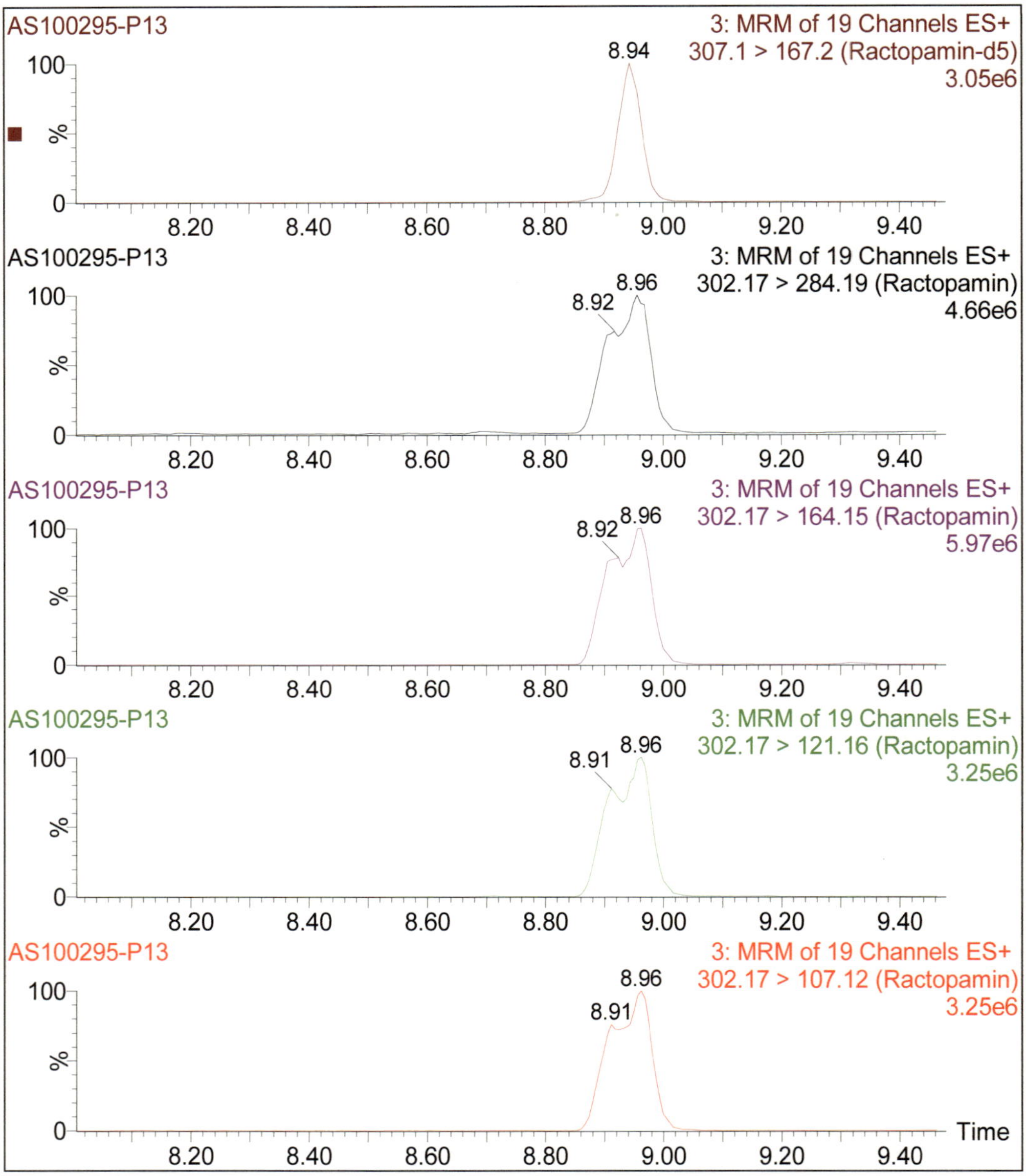
AS100295-P13
3: MRM of 19 Channels ES+
307.1 > 167.2 (Ractopamin-d5)
3.05e6
8.94
AS100295-P13
3: MRM of 19 Channels ES+
302.17 > 284.19 (Ractopamin)
4.66e6
8.92
8.96
AS100295-P13
3: MRM of 19 Channels ES+
302.17 > 164.15 (Ractopamin)
5.97e6
8.92
8.96
AS100295-P13
3: MRM of 19 Channels ES+
302.17 > 121.16 (Ractopamin)
3.25e6
8.91
8.96
AS100295-P13
3: MRM of 19 Channels ES+
302.17 > 107.12 (Ractopamin)
3.25e6
8.91
8.96
100
%
0
8.20
8.40
8.60
8.80
9.00
9.20
9.40
Time

D

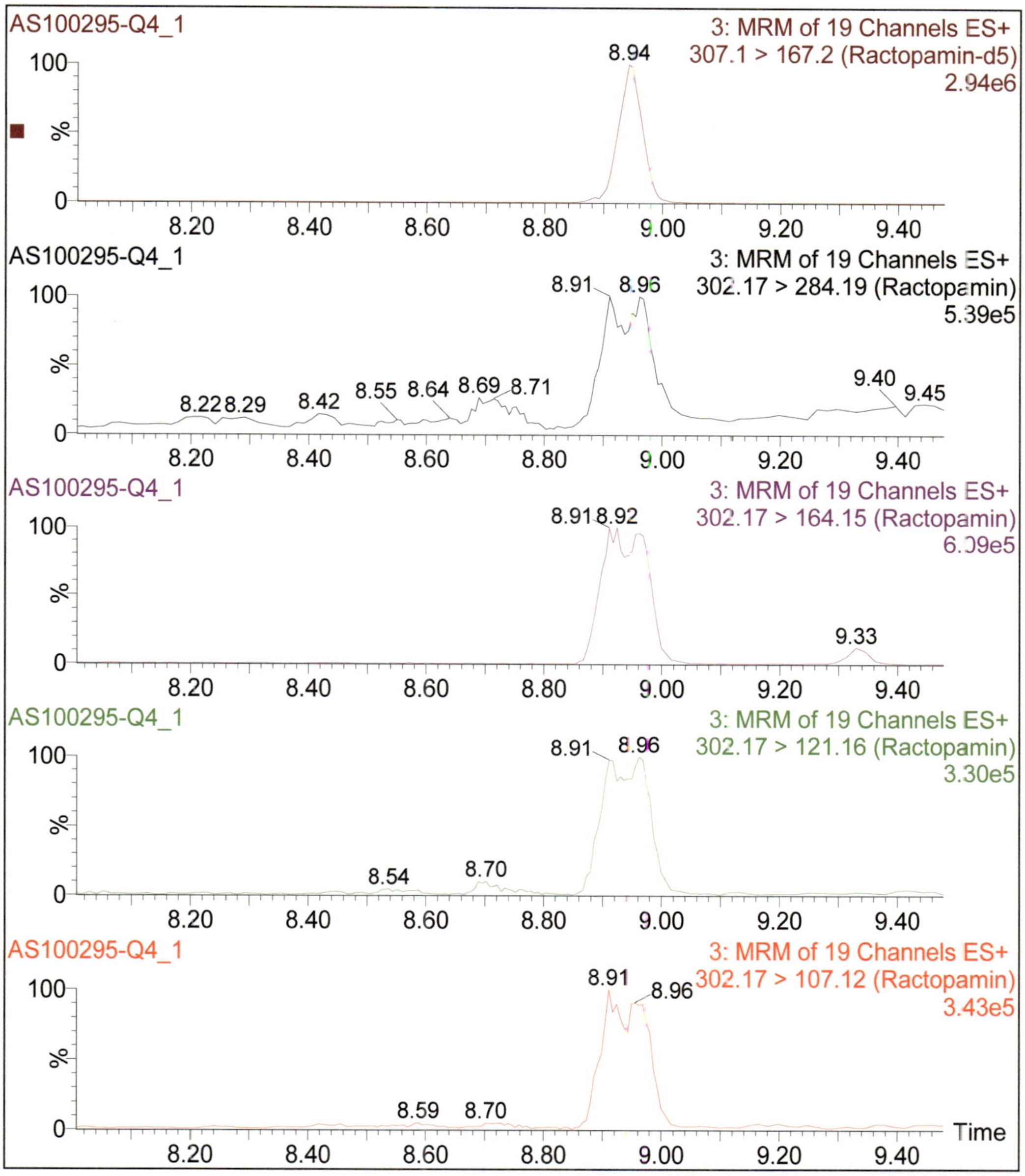

E

Fig. 3. UPLC-MRM chromatograms of confirmatory analysis of ractopamine in pig urine: (A) urine reagent blank; (B) urine matrix blank; (C) urine sample of treated animal on day 1 after withdrawal; (D) urine sample of treated animal on day 8 after withdrawal; (E) blank urine sample spiked with 0.1 μg/kg ractopamine and 0.5 μg/kg ractopamine-d5.

3.3.2 Hair residues

On days 1, 3 and 8 after 28 days of continuous treatment of pigs, ractopamine concentrations were determined in hair (white) samples. The mean (±SD) ractopamine concentrations in hair on days after treatment discontinuation in the experimental group of animals are shown in Table 7. In spite of the low ractopamine dose administered to pigs in our study, residues were determined in hair with the UPLC-MS/MS method, showing its selectivity and sensitivity, i.e. applicability in the control of ractopamine illegal use using hair as a matrix. Chromatograms of the UPLC-MS/MS method of confirmatory analysis of ractopamine in pig hair on days after withdrawal are shown in Figure 4.

Withdrawal time (days)	Hair ractopamine concentrations (mean±SD, ng/g)
	9.34±1.58
1	13.31±0.16
	13.72±0.68
	11.06±0.21
3	13.7±0.45
	9.8±0.41
	9.95±0.05
8	8.65±0.95
	7.70±0.10

Table 7. Concentrations of ractopamine detected in hair by UPLC-MS/MS on days after withdrawal.

In our study, the mean concentrations of ractopamine determined in hair samples were 12.12±2.42 ng/g on day 1 after withdrawal, 11.52±1.99 ng/g on day 3 and 8.77±1.13 ng/g on the last day after withdrawal. Analyses of hair samples showed the concentrations of ractopamine on the same days of withdrawal to be significantly higher in hair samples than in urine. The hair/urine concentration ratios on days 1, 3 and 8 of withdrawal were 1.8, 2.6 and 1.5, respectively. Radeck and Gowik (2010) conducted a study on non-lactating cows treated with 6 different $ß_2$-agonists, including ractopamine in a dose of 1500 mg overall. That study revealed the concentration of ractopamine on days after withdrawal to be higher in urine samples than in hair samples, both white and black. Results of our study performed on pigs indicated the concentration of ractopamine on days after withdrawal to be higher in white hair samples than in urine samples. This discordance of results can be interpreted with regard to different experimental animal species, the dose of ractopamine used and different ractopamine application schemes employed. Bearing in mind the scarce data on ractopamine residues in swine hair, both black and white, there is a need to perform another study that would include swine with black hair in order to yield data on ractopamine residues in both hair types. Also, a long-term depletion study of ractopamine on swine would be useful to determine maximum withdrawal time with detectable residues in hair samples.

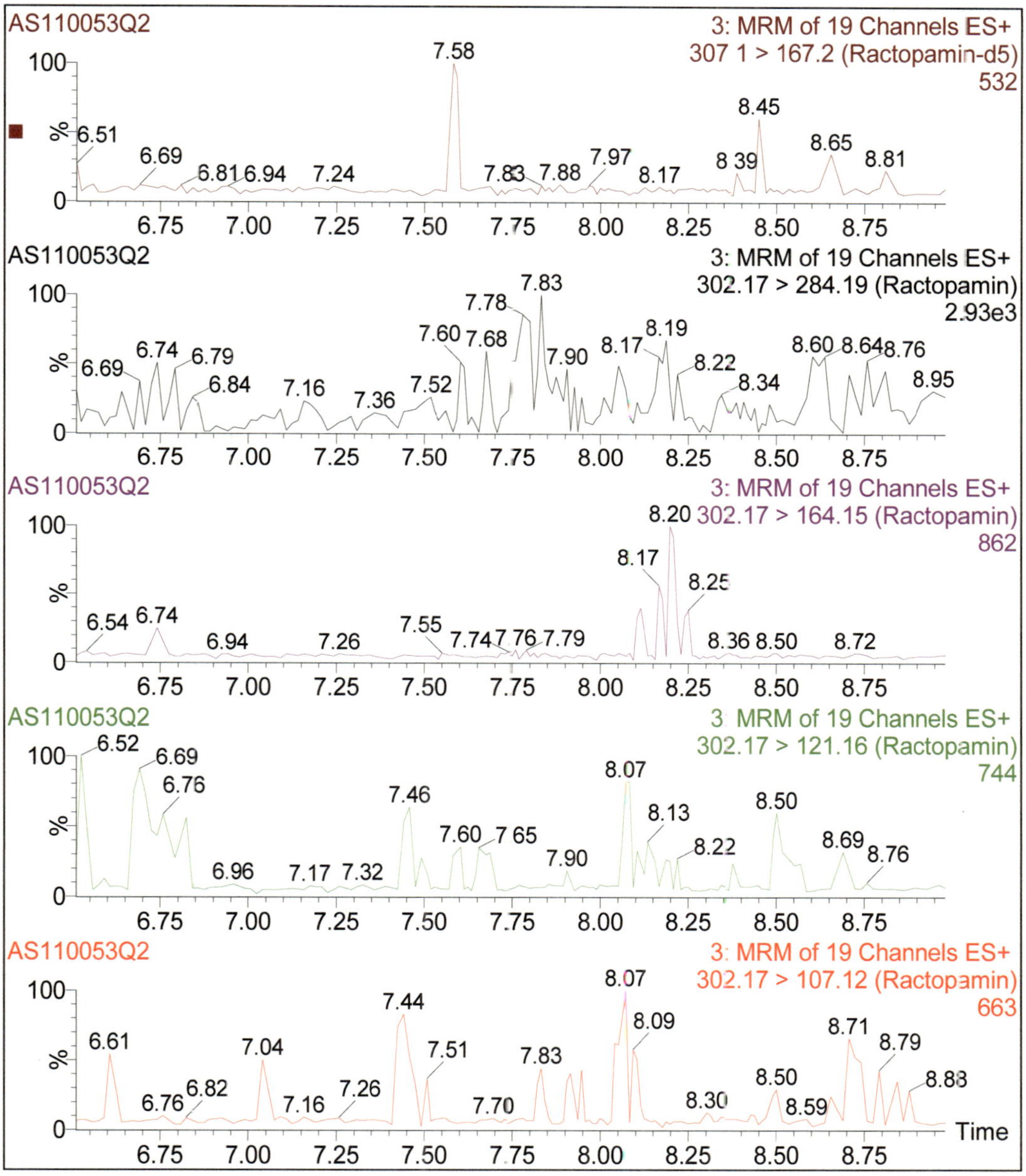
AS110053Q2
3: MRM of 19 Channels ES+
307.1 > 167.2 (Ractopamin-d5)
532
AS110053Q2
3: MRM of 19 Channels ES+
302.17 > 284.19 (Ractopamin)
2.93e3
AS110053Q2
3: MRM of 19 Channels ES+
302.17 > 164.15 (Ractopamin)
862
AS110053Q2
3: MRM of 19 Channels ES+
302.17 > 121.16 (Ractopamin)
744
AS110053Q2
3: MRM of 19 Channels ES+
302.17 > 107.12 (Ractopamin)
663
%
Time

A

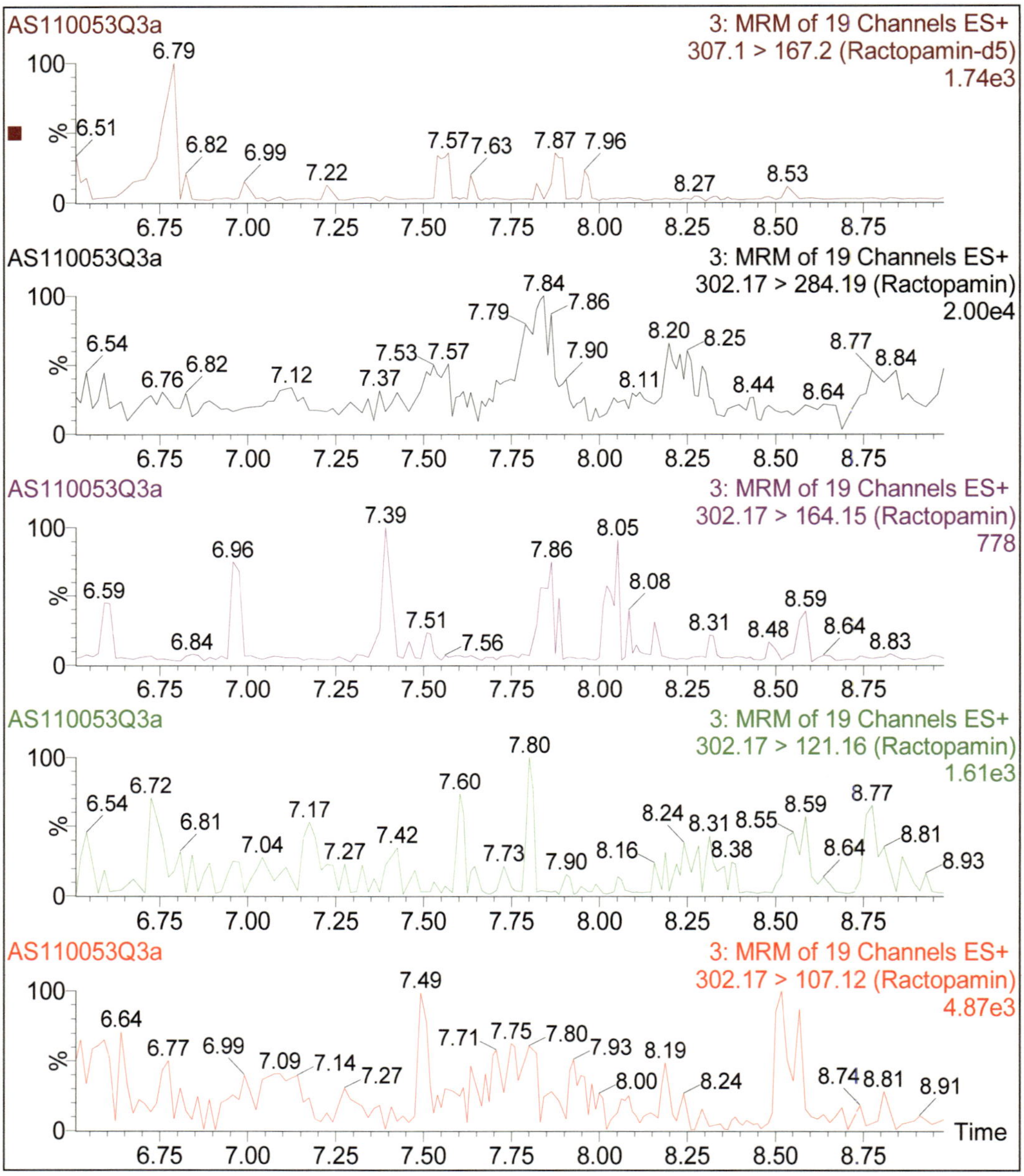
AS110053Q3a
3: MRM of 19 Channels ES+
307.1 > 167.2 (Ractopamin-d5)
1.74e3
AS110053Q3a
3: MRM of 19 Channels ES+
302.17 > 284.19 (Ractopamin)
2.00e4
AS110053Q3a
3: MRM of 19 Channels ES+
302.17 > 164.15 (Ractopamin)
778
AS110053Q3a
3: MRM of 19 Channels ES+
302.17 > 121.16 (Ractopamin)
1.61e3
AS110053Q3a
3: MRM of 19 Channels ES+
302.17 > 107.12 (Ractopamin)
4.87e3
%
Time

B

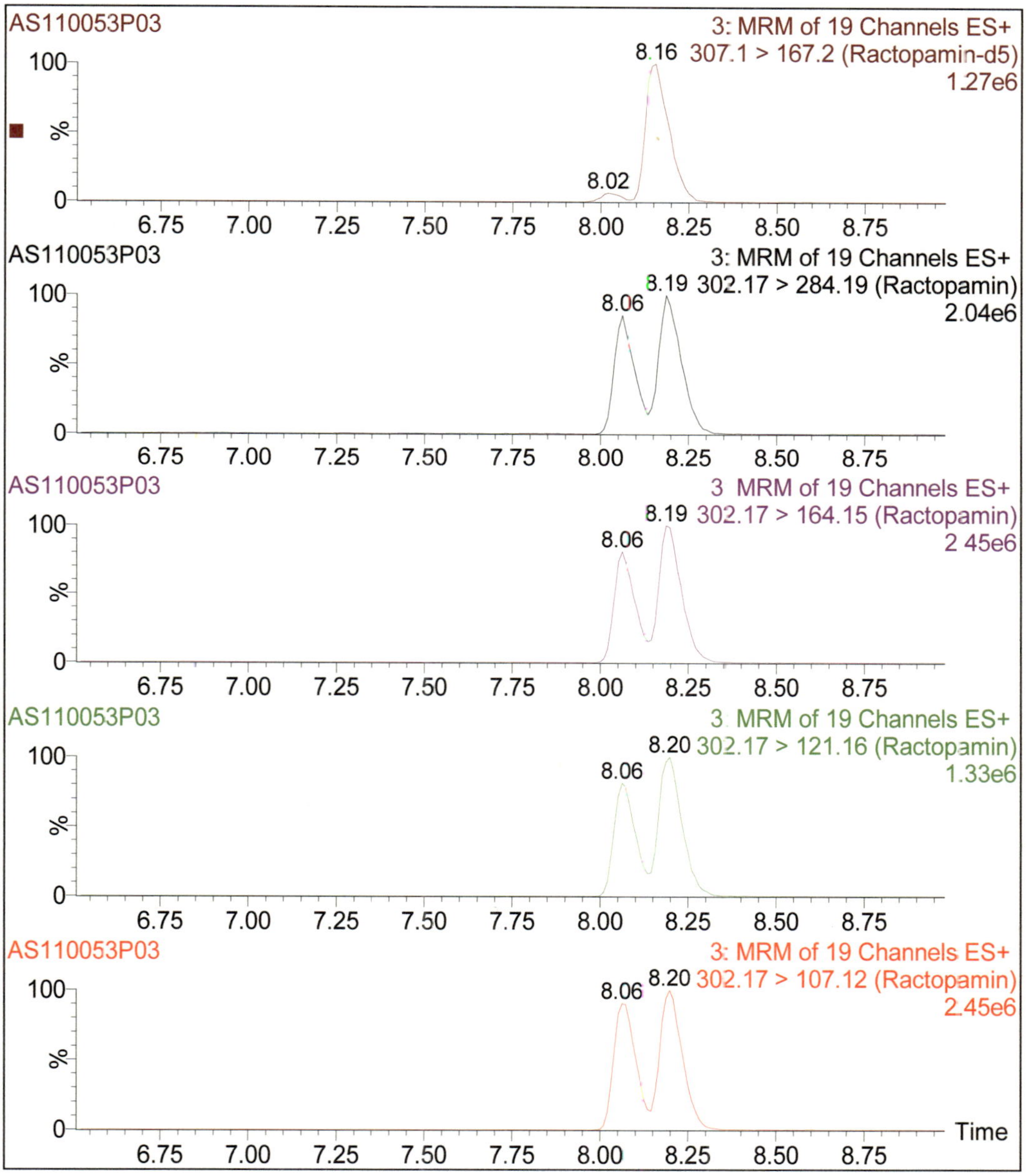
AS110053P03
3: MRM of 19 Channels ES+
307.1 > 167.2 (Ractopamin-d5)
1.27e6
8.16
8.02
AS110053P03
3: MRM of 19 Channels ES+
302.17 > 284.19 (Ractopamin)
2.04e6
8.19
8.06
AS110053P03
3 MRM of 19 Channels ES+
302.17 > 164.15 (Ractopamin)
2 45e6
8.19
8.06
AS110053P03
3 MRM of 19 Channels ES+
302.17 > 121.16 (Ractopamin)
1.33e6
8.20
8.06
AS110053P03
3: MRM of 19 Channels ES+
302.17 > 107.12 (Ractopamin)
2.45e6
8.20
8.06
100
%
0
6.75 7.00 7.25 7.50 7.75 8.00 8.25 8.50 8.75
Time

C

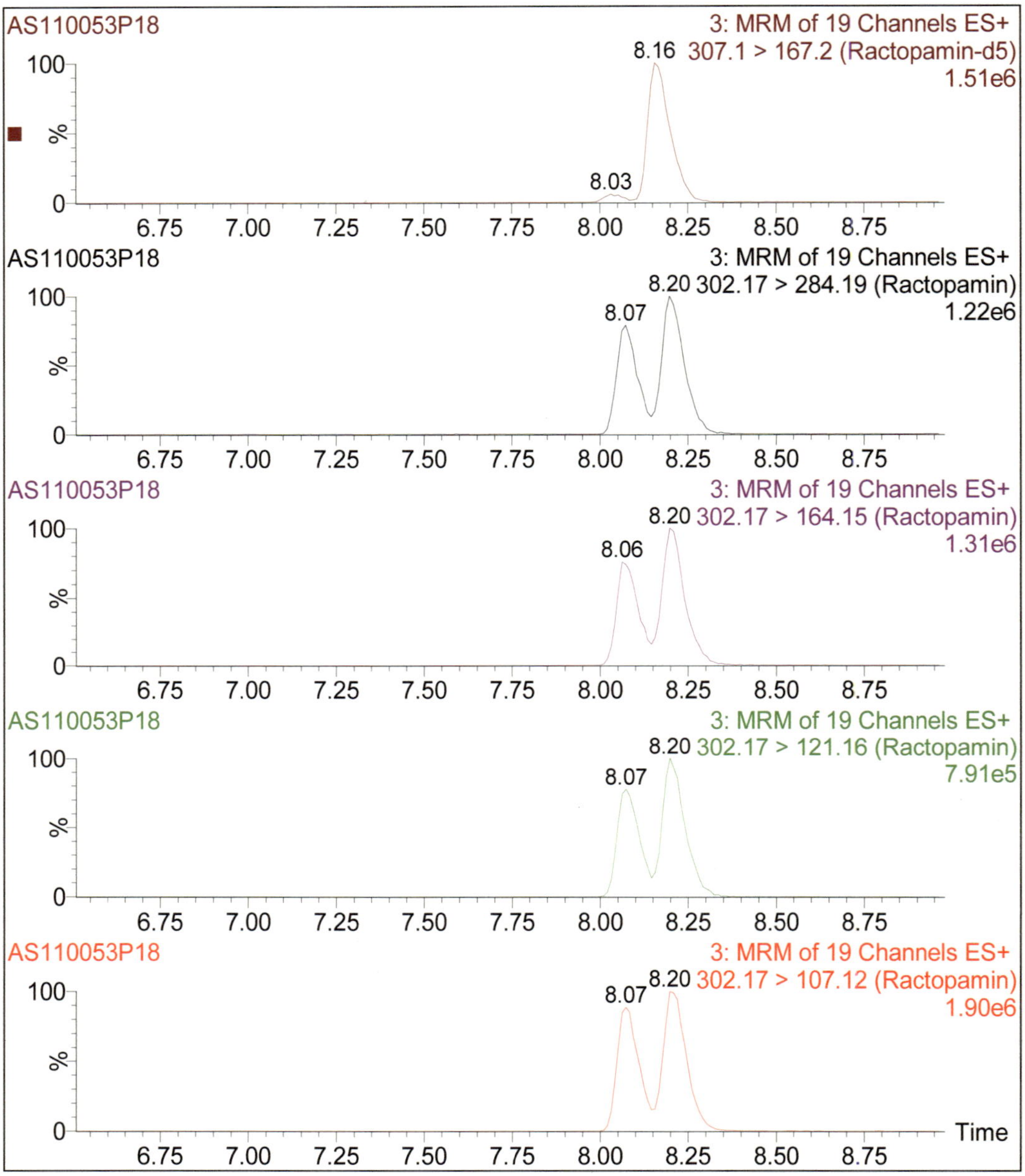

D

Fig. 4. UPLC-MRM chromatograms of confirmatory analysis of ractopamine in pig hair: (A) hair reagent blank; (B) hair matrix blank; (C) hair sample of treated animal on day 1 after withdrawal; (D) hair sample of treated animal on day 8 after withdrawal.

4. Conclusion

A validated UPLC-MS/MS method was employed for determination of ractopamine in pig urine and hair at trace levels. The method features were found to be fit-for-purpose, with successful method validation according to the European Union requirements and its suitability for determination of low ractopamine residues in real samples. Study results indicated that the excretion of ractopamine in pig urine and accumulation in hair could clearly point to its abuse in pigs as food producing animals, in particular when using sample hydrolysis with β-glucuronidase on ractopamine determination, which extended the period in which ractopamine could be detected. Results of our study indicated the concentration of ractopamine on days after withdrawal to be higher in white hair samples than in urine samples.

5. Acknowledgments

We gratefully acknowledge great support to this study by Prof. PhD. Željko Cvetnić, Director of the Croatian Veterinary Institute, and financial support from the Croatian Veterinary Institute. We acknowledge the Federal Office of Consumer Protection and Food Safety, EU and National Reference Laboratory, Berlin, for useful education and performance of analyses.

6. References

Act on Animal Welfare No.135 (2006). Official gazette of the Republic of Croatia.

Anderson, D.B.; Veenhuizen, E.L.; Wagner, J.F.; Wray, M.I. & Mowrey, D.H. (1989). The effect of ractopamine hydrochloride on nitrogen retention, growth performance, and carcass composition of beef cattle. *Journal of Animal Science*, 67, (Supp. 1), pp. (222)

Anderson, D.B.; Moody, D.E. & Hancock, D.L. (2009). Beta adrenergic agonists, In: *Encyclopedia of Animal Science*, Pond, W.G. & Bell, A.W., pp. (104-107). New York, Marcel Dekker, Inc.

Antignac, J.-P.; Marchand, P.; Le Bizec, B. & Andre, F. (2002). Identification of ractopamine residues in tissue and urine samples at ultra-trace level using liquid chromatography-positive electrospray tandem mass spectrometry. *Journal of Chromatography B*, 774, pp. (59-66)

Apple, J.K.; Rincker, P.J.; McKieth, F.K.; Carr, S.N.; Armstrong, T.A.; PAS & Matzat, P. D. (2007). Meta-analysis of the ractopamine response in finishing swine. *The Professional Animal Scientist*, 23, pp. (179-196)

Armstrong, T.A.; Ivers, D.J.; Wagner, J.R.; Anderson, D.B.; Weldon, W.C. & Berg, E.P. (2004). The effect of dietary ractopamine concentration and duration of feeding on growth performance, carcass characteristics, and meat quality of finishing pigs. *Journal of Animal Science*, 82, pp. (3245-3253)

Blanca, J. ; Muñoz, P. ; Morgado, M. ; Méndez, N. ; Aranda, A. ; Reuvers, T. & Hooghuis, H. (2005). Determination of clenbuterol, ractopamine and zilpaterol in liver and urine by liquid chromatography tandem mass spectrometry. *Analytica Chimica Acta*, 529, pp. (199-205)

Brambilla, G.; Loizzo, A.; Fontana, L.; Strozzi, M.; Guarino, A. & Soprano, V. (1997). Food poisoning following consumption of clenbuterol-treated veal in Italy. *The Journal of the American Medical Association*, 278, pp. (635-640)

Carr, S.N.; Ivers, D.J.; Anderson, D.B.; Jones, D.J.; Mowrey, D.H.; England, M.B.; Killefer, J.; Rincker, P.J. & McKeith, F.K. (2005). The effects of ractopamine hydrochloride on lean carcass yields and pork quality characteristics. *Journal of Animal Science*, 83, pp. (2886-2893)

Churchwell, M.I.; Holder, C.L.; Little, D.; Preece, S.; Smith, D.J. & Doerge, D.R. (2002). Liquid chromatography/electrospray tandem mass spectrometric analysis of incurred ractopamine residues in livestock tissues. *Rapid Communications in Mass Spectrometry*, 16, pp. (1261-1265)

Commission of the European Communities (1996). Council Directive 96/22/EC on the prohibition of the use of certain substances having a hormonal and thyreostatic action and β-agonists in animal husbandry. Official Journal of the European Communities, L 125.

Commission of the European Communities (2002). Commission Decision 2002/657/EC of 12th August 2002 implementing Council Directive 96/23/EC concerning the performance of analytical methods and the interpretation of results. Official Journal of the European Communities, L 221.

Courtheyn, D. ; Le Bizec, B. ; Brambilla, G. ; De Brabander, H.F. ; Cobbaert, E. ; Van de Wiele, M. ; Vercammen, J. & De Wasch, K. (2002). Recent developments in the use and abuse of growth promoters. *Analytica Chimica Acta*, 473, pp. (71-82)

Dalidowicz, J.E.; Thomson, T.D. & Babbitt, G.E. (1992). Ractopamine hydrochloride, a phenethanolamine repartitioning agent: Metabolism and tissue residues, In: *Xenobiotics and food-producing animals,* Hutson, D.H.; Hawkins, D.R.; Paulson, G.D. & Struble, C.B., pp. (234-243), Washington, ACS Symposium Series 503, American Chemical Society

Dickson, L.C.; McNeil, J.D.; Lee, S. & Fesser, A.C.E. (2005). Determination of β-agonist residues in bovine urine using liquid chromatography-tandem mass spectrometry. *Journal of AOAC International,* 88, pp. (46-56)

Dong, Y.; Xia, X.; Wang, X.; Ding, S.; Li, X.; Zhang, S.; Jiang, H.; Liu, J.; Li, J.; Feng, Z.; Ye, N.; Zhou, M. & Shen, J. (2011). Validation of an ultra-performance liquid chromatography-tandem mass spectrometry method for determination of ractopamine: Application to residue depletion study in swine. *Food Chemistry*, 127, pp. (327-332)

Elliot, C.T.; Thompson, C.S.; Arts, C.J.M.; Crooks, S.R.H.; van Baak, M.J.; Verheij, E.R. & Baxter, G.A. (1998). Screening and confirmatory determination of ractopamine residues in calves treated with growth promoting doses of the β-agonist. *Analyst,* 123, pp. (1103-1107)

Garay, J.B.; Jimenez, J.E.H.; Jimenez, M.L.; Sebastian, M.V.; Matesanz, J.P.; Moreno, P.M. & Galiana, J.R. (1997). Clenbuterol poisoning - clinical manifestations and analytical findings in an epidemic outbreak in Mostoles, Madrid. *Revista Clinica Española,* 197, pp. (92-95)

Gratacós-Cubarsí, M.; Castellari, M.; Valero, A. & García-Regueiro, J.A. (2006). Hair analysis for veterinary drug monitoring in livestock production. *Journal of Chromatography B*, 834, pp. (14-25)

Lamberty, A.; Schimmel, H. & Pauwels, J. (1998). The study of the stability of reference materials by isochronous measurements. *Fresenius' Journal of Analytical Chemistry*, 360, pp. (359)

Malucelli, A.; Ellendorff, F. & Meyer, H.H.D. (1994). Tissue distribution and residues of clenbuterol, salbutamol, and terbutaline in tissues of treated broiler chickens. *Journal of Animal Science*, 72, pp. (1555-1560)

Martinez-Navarro, J.F. (1990). Food poisoning related to consumption of illicit beta-agonist in liver. *Lancet*, 336, pp. (1311)

Merkel, R.A.; Dickerson, P.S.; Johnson, S.E.; Burnett, A.L.; Schroeder, A.L.; Bergen, W.G. & Anderson, D.B. (1987). The effect of ractopamine on lipid metabolism in pigs. *Federation Proceedings*, 46, pp. (1177)

Meyer, H.H.D. & Rinke, L.M. (1991). The pharmacokinetics and residues of clenbuterol in veal calves. *Journal of Animal Science*, 69, pp. (4538-4544)

Mills, S.E. (2002). Biological basis of the ractopamine response. *Journal of Animal Science*, 80, E. Suppl. 2, pp. (E28-E32)

Mitchell, A.D.; Solomon, M.B. & Steele, N.C. (1990). Response of low and high protein select lines of pigs to the feeding of the beta-adrenergic agonist ractopamine (phenethanolamine). *Journal of Animal Science*, 68, pp. (3226-3232)

Mitchell, A.D. (2009). Effect of ractopamine on growth and body composition of pigs during compensatory growth. *Animal*, 3, pp. (173-180)

Moody, D.E.; Hancock, D.L. & Anderson, D.B. (2000). Phenethanolamine repartitioning agents. In: *Farm animal metabolism and nutrition*, D'Mello, J.P.F, pp. (65-96), WallingfordOxon, CAB International

Moragues, F. & Igualada, C. (2009). How to decrese ion suppression in a multiresidue determination of β-agonists in animal liver and urine by liquid chromatography-mass spectrometry with ion-trap detector. *Analytica Chemica Acta*, 637, pp. (193-195)

Nielen, M.W.F.; Lasaroms, J.J.P.; Essers, M.L.; Oosterink, J.E.; Meijer, T.; Sanders, M.B.; Zuidema, T. & Stolker, A.A.M. (2008). Multiresidue analysis of beta-agonists in bovine and porcine urine, feed, and hair using liquid chromatography electrospray ionization tandem mass spectrometry. *Analytical and Bioanalytical Chemistry*, 391, pp. (199-210)

Pleadin, J.; Bratoš, I.; Vulić, A.; Perši, N. & Đugum, J. (2011). Analysis of clenbuterol residues in pig liver using liquid chromatography electrospray ionization tandem mass spectrometry. *Reviews in Analytical Chemistry*, 30, pp. (5-9)

Pulce, C.; Lamaison, D.; Keck, G.; Bostvironnois, C.; Nicolas, J. & Descotes, J. (1991). Collective and human food poisonings by clenbuterol residues in veal liver. *Veterinary and Human Toxicology*, 33, pp. (480-481)

Qiang, Z.; Shentu, F.; Wang, B.; Wang, J.; Chang, J. & Shen, J. (2007). Residue depletion of ractopamine and its metabolites in swine tissues, urine, and serum. *Journal of Agricultural and Food Chemistry*, 55, pp. (4319-4326)

Radeck, W. & Gowik, P. (2010). A long-term depletion study of β-agonists in bovine hair and urine. *Proceedings of 6th International Symposium on Hormone and Veterinary Drug Residue Analysis*, Ghent, June 2010

Ramos, F.; Cristino, A.; Carrola, P.; Eloy, T.; Manuel Silva, J.; Castilho, A.C. & Silveira, M.I.N. (2003). Clenbuterol food poisoning diagnosis by gas chromatography-mass spectrometric serum analysis. *Analytica Chimica Acta*, 483, pp. (207-213)

Shao, B.; Jia, X.; Zhang, J.; Meng, J.; Wu, Y.; Duan, H. & Tu, X. (2009). Multi-residual analysis of 16 β-agonists in pig liver, kidney and muscle by ultra performance liquid chromatography tandem mass spectrometry. *Food Chemistry*, 114, pp. (1115-1121)

Shelver, W.L. & Smith, D.J. (2003). Determination of ractopamine in cattle and sheep urine samples using an optical biosensor analysis: comparative study with HPLC and ELISA. *Journal of Agricultural and Food Chemistry*, 51, pp. (3715-3721)

Smith, D.J.; Feil, V.J.; Huwe, J.K. & Paulson, G.D. (1993). Metabolism and disposition of ractopamine hydrochloride by turkey poults. *Drug Metabolism and Disposition*, 21, pp. (624-633)

Smith, D.J. (1998). The pharmacokinetics, metabolism, and tissue residues of β-adrenergic agonists in livestock. *Journal of Animal Science*, 76, pp. (173-194)

Smith, D.J. & Shelver, W.L. (2002). Tissue residues of ractopamine and urinary excretion of ractopamine and metabolites in animals treated for 7 days with dietary ractopamine. *Journal of Animal Science*, 80, pp. (1240-1249)

Thompson, C.S.; Haughey, S.A.; Traynor, I.M.; Fodey, T.L.; Elliott, C.T.; Antignac, J.-P.; Le Bizec, B. & Crooks, S.R.H. (2008). Effective monitoring for ractopamine residues in samples of animal origin by SPR biosensor and mass spectrometry. *Analytica Chimica Acta*, 608, pp. 217-225.

Turberg, M.P.; Macy, T.D.; Lewis, J.J. & Coleman, M.R. (1995). Determination of ractopamine hydrochloride in swine and turkey tissues by liquid chromatography with coulometric detection. *Journal of AOAC International*, 78, pp. (1394-1402)

Zheng, H.; Deng, L.-G.; Lu, X.; Zhao, S.-C.; Guo, C.-Y.; Mao, J.-S.; Wang, Y.-T.; Yang, G.-S. & Aboul-Enein, H.Y. (2010). UPLC-ESI-MS-MS determination of three β_2-agonists in pork. *Chromatographia*, 72, pp. (79-84)

Van Hoof, N.; Courtheyn, D.; Antignac, J.-P.; Van der Wiele, M.; Poelmans, S.; Noppe, H. & De Brabander, H. (2005). Multi-residue liquid chromatography/tandem mass spectrometric analysis of beta-agonists in urine using molecular imprinted polymers. *Rapid Communications in Mass Spectrometry*, 19, pp. (2801-2808)

Watkins, L.E.; Jones, D.J.; Mowrey, H.; Anderson, D.B. & Veenhuizen, E.L. (1990). The effect of various levels of ractopamine hydrochloride on the performance and carcass characteristics of finishing swine. *Journal of Animal Science*, 68, pp. (3588-3595)

Williams, N.H.; Cline, T.R.; Schinckel, A.P. & Jones, D.J. (1994). The impact of ractopamine, energy intake, and dietary fat on finisher pig growth performance and carcass merit. *Journal of Animal Science*, 72, pp. (3152-3162)

16

Simultaneous LC-MS/MS Determination of Racemic Warfarin and Etravirine in Rat Plasma and Its Application in Pharmacokinetic Studies

Jyothy John, Keila Robinson, Mathew John,
Jason Caballero, Jing Ma, Dong Liang and Cyril Abobo
Texas Southern University
USA

1. Introduction

Warfarin, one of the most commonly used oral anticoagulant in the US and across the globe, is a drug of choice for millions. It's a unique drug of its kind due to the multiple pharmacological and pharmacokinetic properties. The posology of warfarin cannot be generalized. Having the right dose for the right patient makes this drug follow a pattern that is unique among the lot. Warfarin has a high interpatient variability (Min-Jung Kwon et al., 2009) and narrow therapeutic index which requires continuous monitoring of its plasma concentration, the prothrombin time and the international normalized ratio followed by a dosage adjustment. Warfarin in pure form exists as a racemic mixture consisting of equal amounts of R and S enantiomers (Porter et al., 1986). The S enantiomer is more potent than the R form (Breckenridge et al., 1974). Warfarin is highly metabolized in the body in a stereo specific pathway catalyzed by cytochrome P450. R-warfarin is metabolized primarily by CYP1A2 to 6- and 8-hydroxy warfarin, and by CYP3A4 to 10-hydroxy warfarin, while S-warfarin is metabolized primarily by CYP2C9 to 7-hydroxy warfarin (Kaminsky and Zhang., 1997). The properties of warfarin, such as narrow therapeutic index, high protein binding, CYP dependent metabolism and a very high elimination half life render to be prone to many drug interactions (Chan et al., 2009). Elimination half life of warfarin is relatively long (10-16 hours in animals and 40-46 hours in humans), causing a dramatic increase in the anticoagulant effect upon concomitant administration of warfarin with other drugs causing drug-drug interactions (Alexander and Areg et al., 2009).

Etravirine is the first drug in the second generation of non-nucleoside reverse transcriptase inhibitors (NNRTIs) for HIV/AIDS. The drug was recently marketed for the treatment of HIV infection. It is of great advantage in combination with other antiretrovirals in the treatment of patients who are on this regimen for a considerable period of time. (Martha Boffito et al., 2009). Etravirine is highly bound to plasma proteins and is primarily metabolized by CYP450; 3A4, 2C9 and 2C19 iso-enzymes. Potential drug interactions of warfarin and etravirine are expected due to their high protein binding and similar hepatic metabolic characteristics. For most antiretrovirals it's critical that drug concentrations are maintained above the suggested minimum effective concentration throughout the dosing interval. Suboptimal antiretroviral exposure may permit viral replication and predispose to the selection of drug-resistant virus. Resistant strains may hence lead to disease progression and treatment failure (Martha Boffito et al., 2009). On the other hand, long term use of NNRTIs can potentially cause HIV related thromboembolic events, which requires the use of the anticoagulant warfarin. A clinically significant etravirine-warfarin interaction could occur after their concomitant administration (Welzen et al., 2011).

Several LC-MS/MS methods have been developed for the analysis of etravirine in human plasma samples. For example, (Heine et al., 2009) developed an LC-MS/MS method quantifying etravirine in human plasma, dry blood spot, and peripheral blood mononuclear cell lysate. LC-tandem MS methods were also available for simultaneous analysis of etravirine with other NNRTIs (Fayet et al., 2009; Else et al., 2010) in human plasma and with protease inhibitors (Quaranta et al., 2009). All the above referenced methods used a direct plasma protein precipitation using organic co-solvent and had etravirine detection limit of $\geq$ 10 ng/mL, except for a study by (Rezk et al., 2009), where a 2 ng/mL detection limit was reported. Most of these LC-MS/MS methods were aimed for and been used for clinical drug monitoring. Few have been used for pharmacokinetic characterization of potential drug-drug interactions between etravirine and other concomitantly administered drugs. For example, Scholler-Gyure et al., 2008 validated and applied an LC/MS/MS method of etravirine for a pharmacokinetic drug-drug interaction study between etravirine and acid-suppressing agents in healthy human volunteers. However, very limited analytical methods are currently available for the determination of etravirine in rat plasma. Furthermore, there are no analytical methods till date for the simultaneous determination of racemic warfarin and etravirine in the rat plasma. Developing such analytical method is necessary for further characterization and evaluation of factors that affect the absorption and disposition of etravirine and warfarin. Hence it is a highly significant to develop an analytical method for the simultaneous determination and validation of racemic warfarin and etravirine to extrapolate its applicability in pharmacokinetic studies of these compounds. In the current study, we were successful in coming up with an LC-MS/MS determination of racemic warfarin and etravirine in rat plasma which can be of high relevance in the pharmacokinetic profiling of both etravirine and warfarin. The limit of detection was as low as 1ng/mL and the injection volume used was 10ul without any interference with the sensitivity of the assay. The total run time of 11 minutes permitted the assay to be carried out in a relatively short time period with minimal consumption of the solvents and reagents.

Therefore, the objective of this study is to develop a simultaneous LC-MS/MS method suitable for determination and quantification of R-warfarin, S-warfarin, and etravirine in rat plasma. The method has been validated and applied to a pharmacokinetic evaluation of the drug-drug interaction between warfarin and etravirine using rat as an animal model.

2. Materials and methods

2.1 Chemicals

Racemic warfarin standard powder was purchased from Sigma Aldrich Co. (Missiouri, St. Louis, USA). Etravirine as the standard powder was purchased from Toronto research chemicals Inc. (Canada). Acetaminophen and acetic acid was purchased from Sigma Aldrich Co. (Missiouri, St. Louis, USA). HPLC graded water was purchased from Mallinckrodt Baker, Inc. (New Jersey, Phillipsburg, USA). Acetonitrile HPLC grade was purchased from VWR international Chemicals (Pennsylvania, West Chester, USA).The chemical structures and fragmentation patterns for racemic warfarin, etravirine, and the internal standard acetaminophen are presented in Figure 1.

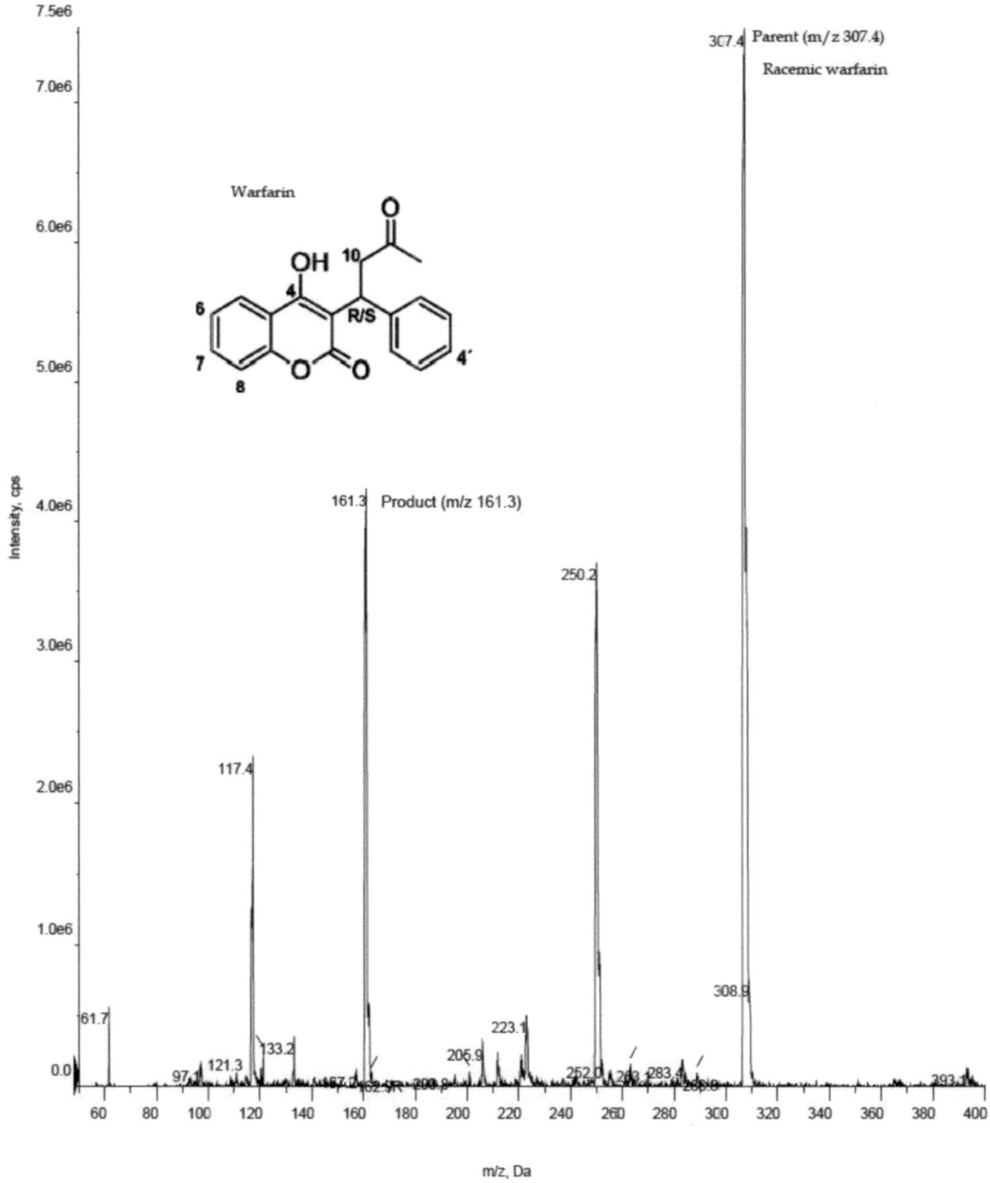

Fig. 1. Part I.

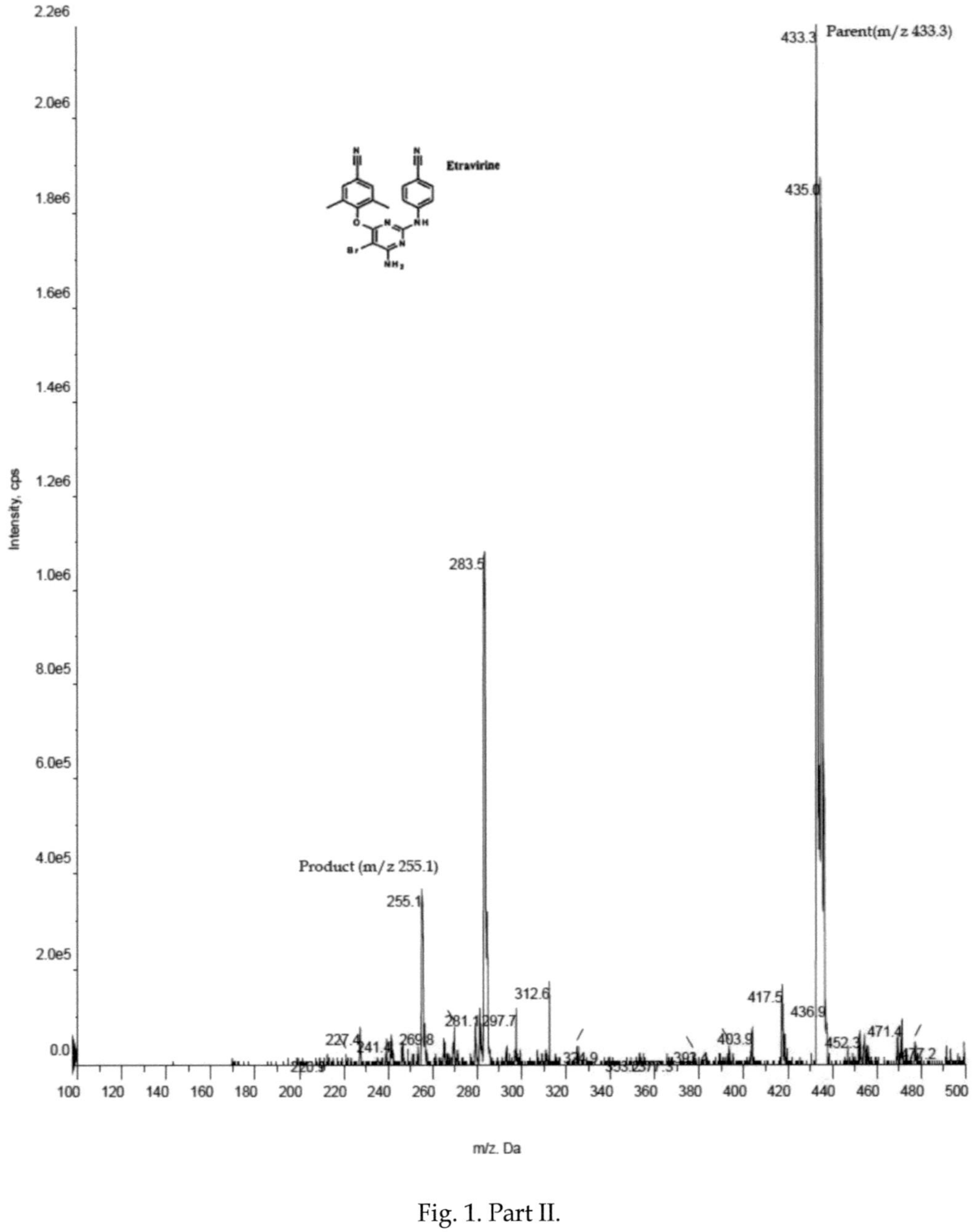

Parent(m/z 433.3)
433.3
435.0
Etravirine
283.5
Product (m/z 255.1)
255.1
312.6
417.5
436.9
281.1
297.7
227.4
241.4
269.8
403.9
452.3
471.4
Intensity, cps
m/z, Da

Fig. 1. Part II.

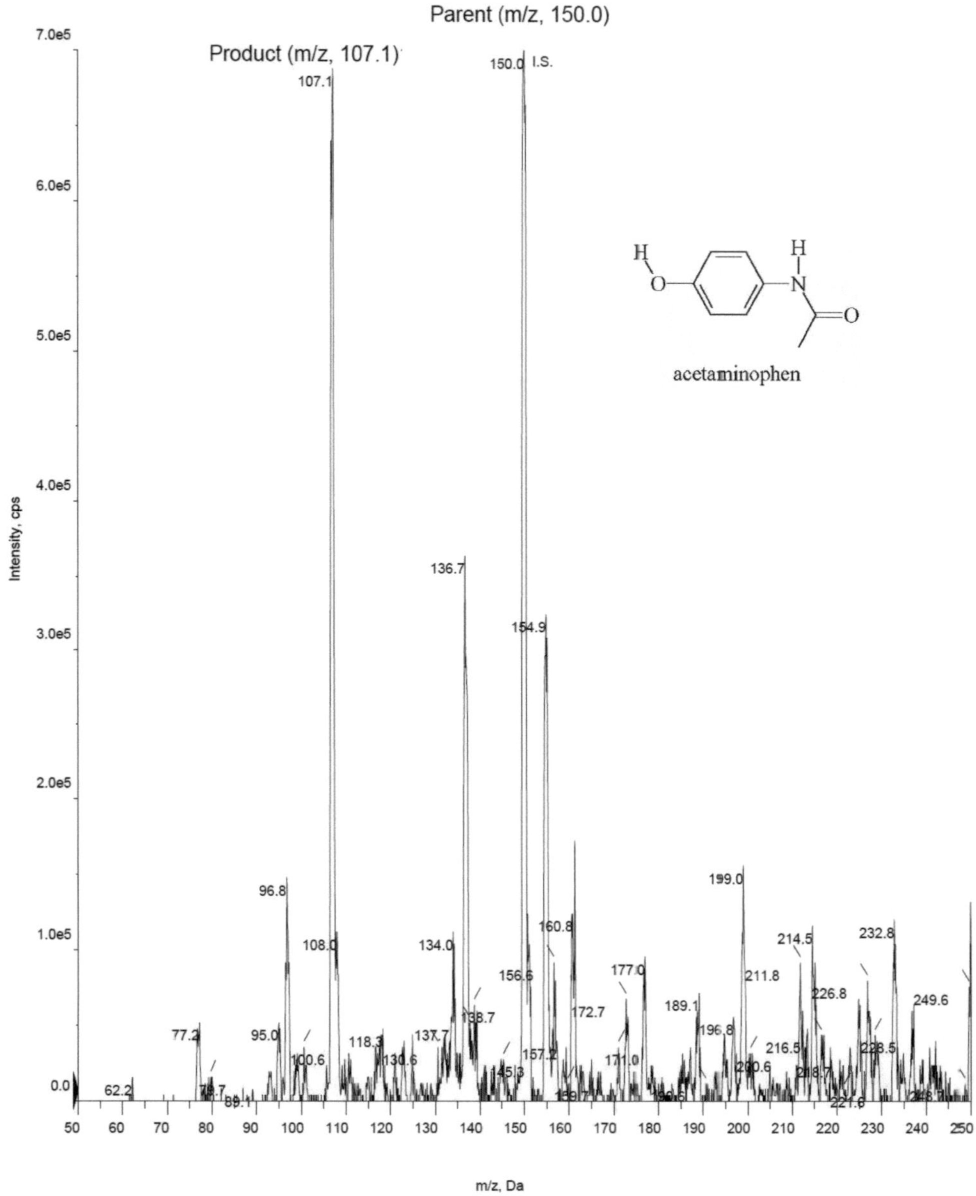

Part III.

Fig. 1. Chemical structures and fragmentation patterns for racemic warfarin (A), etravirine (B) and internal standard, acetaminophen (C).

2.2 LC/MS/MS conditions

Warfarin, etravirine and the IS were analyzed by using an Agilent 1200 series HPLC system (Foster city, CA) using a commercially available chiral column, Lux cellulose -1 with a dimension of 250X4.6mm i.d. packed with 5μm particles (Phenomenex, Torrance, CA, USA) in conjunction with a Lux Cellulose-1 guard column with a dimension of 4 x 3.0mm (Phenomenex, Torrance, CA, USA). A gradient elution was used, consisting of 0 .1% acetic acid in water (mobile phase A) and 100% acetonitrile (mobile phase B). The gradient started with mobile phase A and B being mixed at a ratio of 40:60 %v/v respectively, at a flow rate of 1.5mL/min till 4.5 minutes. At time 4.5-6 minutes, A and B were mixed at a ratio of 40:60 % v/v respectively, at a flow rate of 1.3 mL/min. From 6-9.1 minutes, the mobile phase was solely comprised of B, at a flow rate of 1.5mL/min. At 9.1-11 minutes, A and B were mixed at a ratio of 40:60 %v/v, at a flow rate of 1.5 mL/min. Two minutes was allowed at the end of each run for equilibration. The injection volume was 10ul and the total run time was 11 minutes. The column effluent was monitored using a 3200 QTRAP ® LC-MS/MS (AB Sciex, Foster city, CA) which is a hybrid triple quadrupole linear ion trap equipped with a TurboIonspray ion source. A Parker Balston Source 5000Tri Gas generator was used to generate pure nitrogen. The nebulizer gas and the heater gas were maintained at 80psi and the ion spray heater was set at 650^{0}C. Ion spray needle voltage was 4500 V, the curtain gas was 30psi and the collision gas was set to medium. The mass spectrophotometer was set at the negative mode. The transition ions were detected using multiple reaction monitoring from a specific parent ion to product ion for etravirine (m/z 433.3→142.1), racemic warfarin 307.4→161.3 and acetaminophen 150→107.1. Peak areas and other compound parameters were determined by Analyst ^{R} software, ver.1.5.

2.3 Preparation of standards and calibration curves

Individual stock solutions of racemic warfarin, acetaminophen and etravirine (1mg/mL) were prepared by dissolving 25mg of each substance in 25mL of respective solvents (acetonitrile for racemic warfarin and acetaminophen, DMSO for etravirine) and stored in refrigerator at 4°C. All the working solutions were freshly prepared daily. For quantitative analysis, a series of working standard mixtures were prepared by mixing and diluting the stock solutions of racemic warfarin and etravirine with mobile phase to yield concentrations of 0.025, 0.05, 0.1, 0.25, 0.5, 1, 2.5 and 5 μg/mL. A working solution of IS was prepared by diluting the stock solution of acetaminophen with acetonitrile to yield a final concentration of 500ng/mL. In vivo standard curves were prepared by spiking blank plasma with warfarin and etravirine standard solutions to yield calibration standards in plasma at concentrations of 2.5, 5, 10, 25, 50, 100, 250 and 500 ng/mL.

2.4 In vivo sample preparation

The rat plasma was collected from male Sprague Dawley rats in our laboratory and stored at -80^{0}C until its use. Each warfarin and etravirine containing rat plasma sample (100μl) was extracted and deproteinized by mixing it with 100μl of IS working solution containing 500ng of acetaminophen per mL of acetonitrile. The mixture was briefly vortexed for 1 minute and centrifuged at 13,000 rpm for 10 minutes. A 10 μl of the supernatant was directly injected in to the HPLC column. Calibration curves were plotted by peak area ratio of each analyte and internal standard vs. concentrations in rat plasma.

2.5 Assay validation

The "Guidance for Industry - Bioanalytical Method Validation" document from FDA was used as a guide for the assay validation described as follows (FDA. 2001).

2.5.1 Linearity, accuracy, precision, and recovery

Linear calibration curves in rat plasma were generated by plotting peak area ratios of racemic warfarin and etravirine to the IS versus seven known plasma racemic warfarin and etravirine concentrations over the range of 2.5-500 ng/mL. Slope, intercept, and coefficient of determination values were estimated using least square regression analysis. Quality control plasma samples containing low (10ng/mL), medium (100ng/mL), and high (400ng/mL) racemic warfarin and etravirine concentrations were used to evaluate the precision and accuracy of the assay method. The intra-day assay precision and accuracy were obtained by analyzing six replicates of the quality control samples using calibration curves constructed on the same day. The inter-day assay precision and accuracy were obtained by analyzing six quality control samples using calibration curves constructed on 3 different days. The assay precision was reflected by the relative standard deviation and the assay accuracy was reflected by the relative percentage error from the theoretical drug concentrations. The limit of detection (LOD) was defined as the plasma concentration that yielded a peak height equal to three times that of baseline noise. The lowest limit of quantification (LLOQ) was selected as the lowest racemic warfarin and etravirine plasma level on the calibration curve. The extraction recoveries of racemic warfarin and etravirine from rat plasma (expressed as a percentage) were calculated as the ratio of the slope of a calibration curve for racemic warfarin and etravirine in spiked plasma to that of spiked mobile phase.

2.5.2 Stability

Three aliquots of the low, medium, high concentration QC samples were used to conduct each of the following sets of stability tests: three freeze-thaw cycles, storage under refrigeration conditions (4°C), and storage at room temperature in autosampler.

2.5.3 Matrix effects

Matrix effects from endogenous substances present in extracted rat plasma may cause ion suppression or enhancement of the signal. Matrix effects were assessed by comparing the peak areas of racemic warfarin and etravirine after addition of low, medium, and high (n=3 each) concentrations of racemic warfarin and etravirine to (A) mobile phase and (B) and supernatant of extracted blank plasma. The peak area ratio of B/A (as a percentage) or the percentage matrix factor was used as a quantitative measure of matrix effect. The absolute matrix effect was quantified by using the "post extraction spike method" which compares the average peak area ratios of a group of standard samples diluted in mobile phase (group A) and another group spiked into plasma after plasma extraction (group B). Concentrations of the standard solutions were established low, medium and high values (n=3 each). The formula utilized to the compare the peak area ratios was "B" PAR/ "A" PAR = %Matrix effect.

2.5.4 Pharmacokinetic study

The animal experiment and protocol were reviewed and approved by the Institutional Animal Care and Use Committee at Texas Southern University. The jugular veins of six male adult Sprague Dawley rats (Harlan Sprague Dawley, Inc., Indianapolis, IN) weighing 250-300g were cannulated under anesthesia the day before the study. Each rat was given a 1mg/kg of racemic warfarin intravenous followed by 25mg/kg etravirine intravenous.

Serial heparinized blood samples (approximately 250µl) were collected from the jugular vein cannula for up to 144h. After centrifugation, the plasma samples were collected, immediately stored in -80^{0}C and analyzed within a week.

3. Results and discussion

3.1 Chromatographic conditions

The chromatographic run times for the protein precipitated plasma samples were 11 minutes. Retention times 2, 6.02, 7.76, and 9.5 minutes were observed for internal standard, R-warfarin, S-warfarin and etravirine, respectively. Figure 2 illustrates chromatograms obtained from blank rat plasma, spiked drug concentrations in plasma and after administration of the drug at 8 hours. These chromatograms demonstrate peak separation at the baseline excluding in vivo matrix interference.

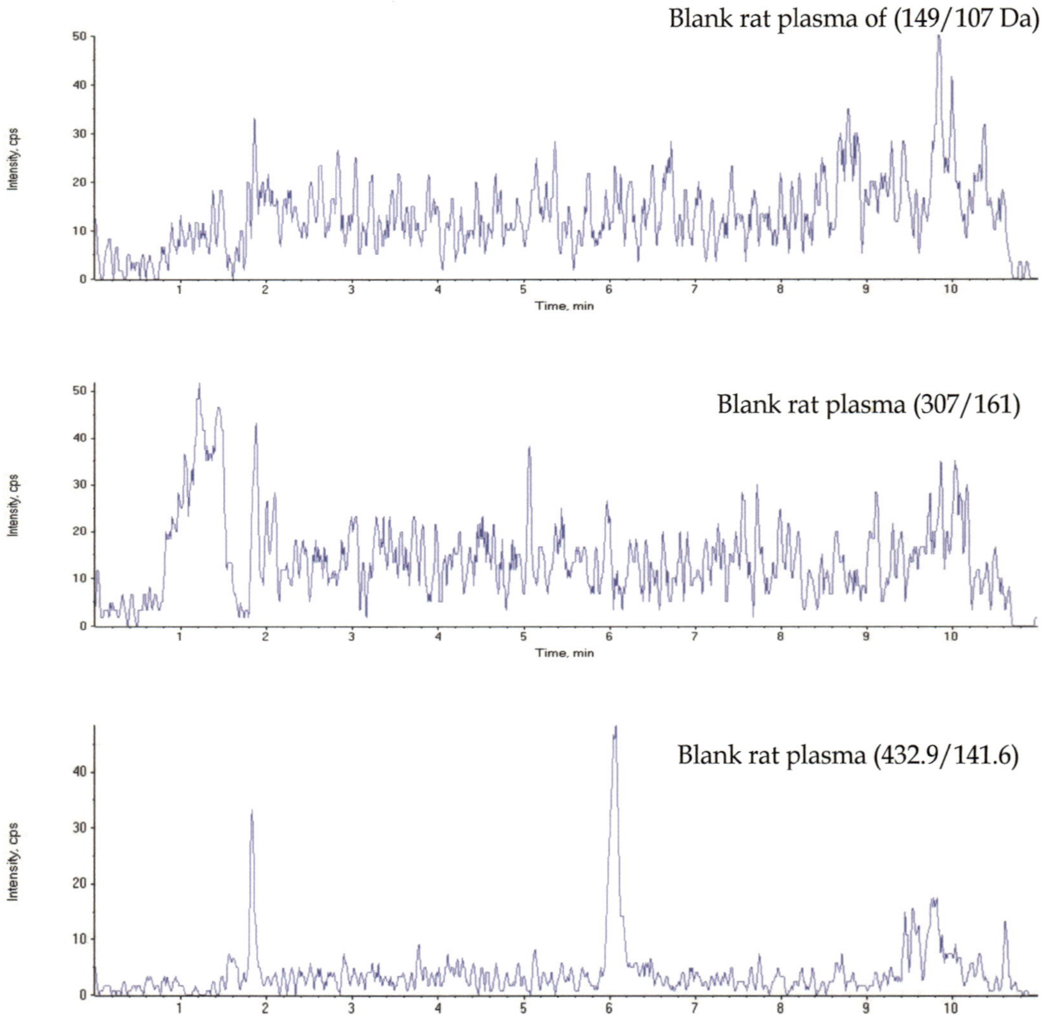

Fig. 2. (A)

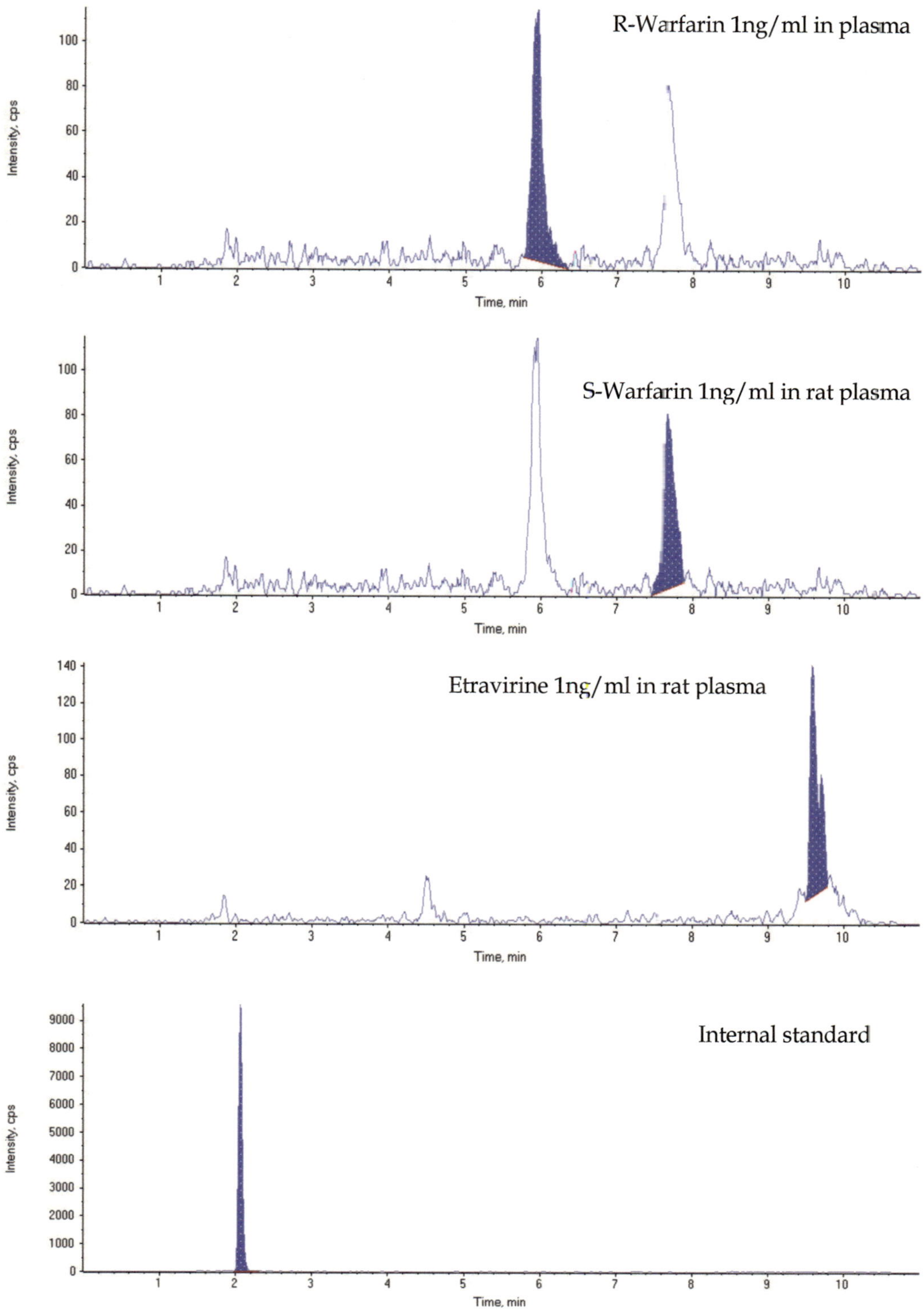
R-Warfarin 1ng/ml in plasma
S-Warfarin 1ng/ml in rat plasma
Etravirine 1ng/ml in rat plasma
Internal standard
Intensity, cps
Time, min

Fig. 2. (B)

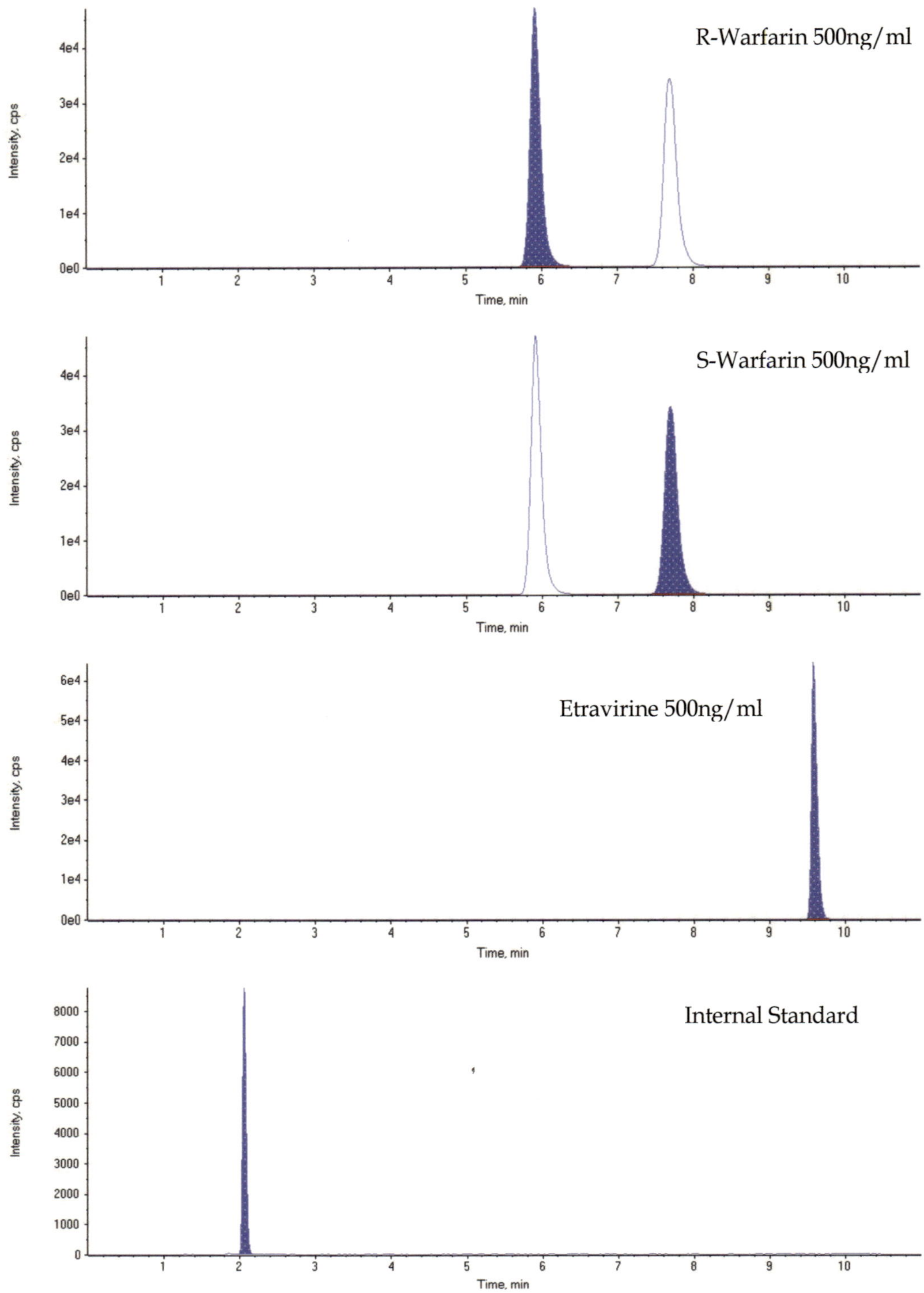
R-Warfarin 500ng/ml
Intensity, cps
Time, min
S-Warfarin 500ng/ml
Intensity, cps
Time, min
Etravirine 500ng/ml
Intensity, cps
Time, min
Internal Standard
Intensity, cps
Time, min

Fig. 2. (C)

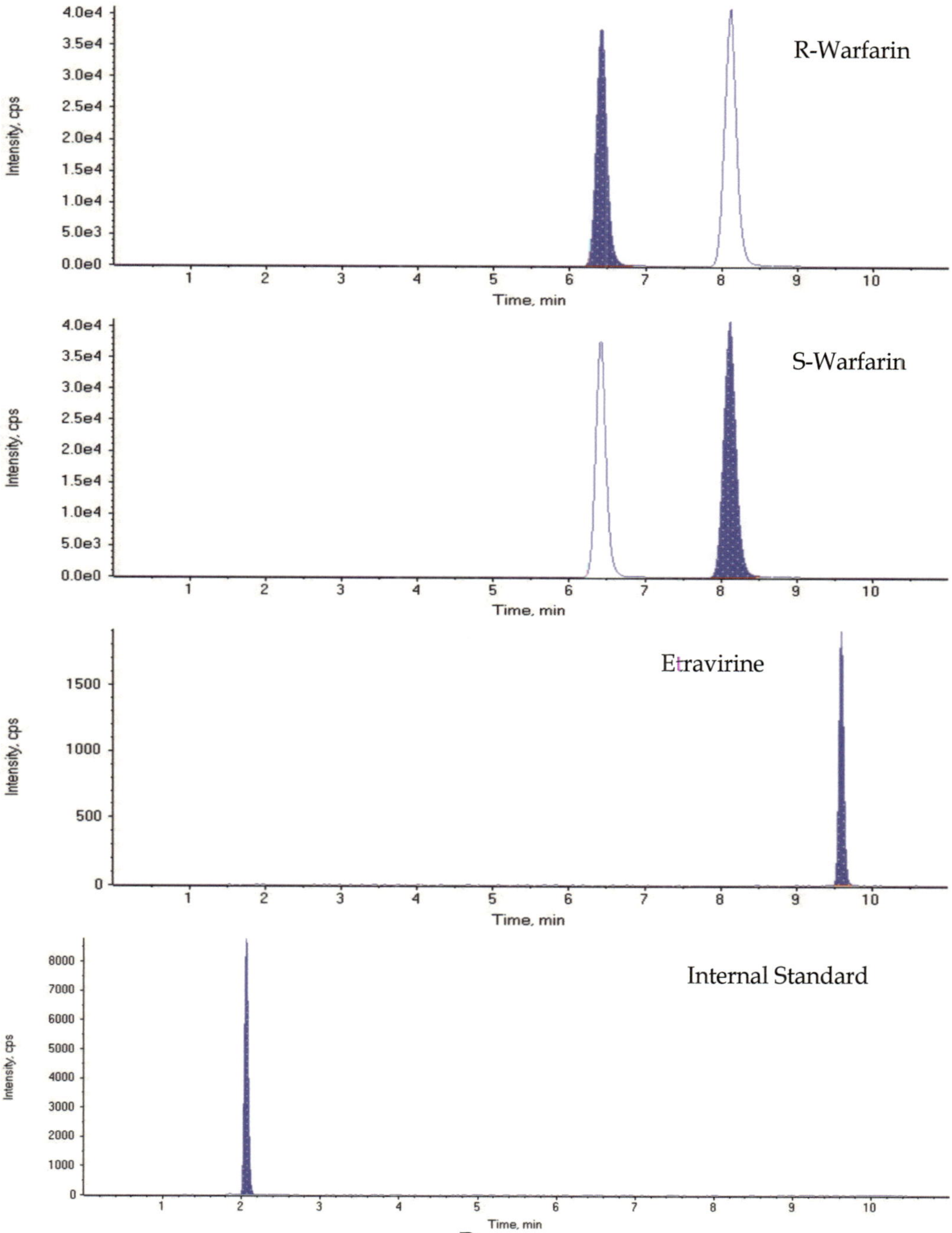

D

Fig. 2. Chromatograms of: (A) blank rat plasma; (B) Blank rat plasma spiked with 1ng/mL racemic warfarin, etravirine and the internal standard; (C) Blank rat plasma spiked with 500ng/mL racemic warfarin, etravirine and the internal standard; (D) Plasma sample from a rat at 8 hours after concomitant intravenous administration of 1mg/kg dose of racemic warfarin and intravenous 25mg/kg dose of etravirine.

3.2 Linearity, sensitivity and detection limit of the assay

Linear relationships were determined by the ratios of the peak areas of analytes to that of the internal standard over the range of 2.5-500ng/mL. Table 1 shows the intra- and inter-day assay changeability of the calibration curves. The lowest limits of detection for all analytes were 1 ng/mL. The lowest limit of quantitation for R-warfarin, S-warfarin, and etravirine were 5, 5 and 2.5 ng/mL, respectively.

	Y-intercept	Slope	R^2
R-Warfarin			
Intra-day (n=3)			
Mean	0.0360	0.0640	0.9990
SD	0.0090	0.0020	0.0010
Precision (CV%)	25.73	5.6	0.1
Inter-day (n=5)			
Mean	0.0256	0.0707	0.998
SD	0.0031	0.0029	0.002
Precision (CV%)	12.06	4.0	0.2
S-Warfarin			
Intra-day (n=3)			
Mean	0.1681	0.0580	0.9980
SD	0.0503	0.0019	0.0010
Precision (CV%)	30	6.4	0.1
Inter-day (n=5)			
Mean	0.0255	0.0634	0.9970
SD	0.0216	0.0046	0.0040
Precision (CV%)	84.4	7.3	0.4
Etravirine			
Intra-day (n=3)			
Mean	0.0453	0.0062	0.9970
SD	0.0012	0.0029	0.0030
Precision (CV%)	2.6	23.4	0.3
Inter-day (n=3)			
Mean	0.0160	0.0059	0.998
SD	0.0055	0.0005	0.003
Precision (CV%)	34.58	9.3	0.3

CV% = (standard deviation/mean)×100.

Table 1. Linearity of calibration curves for warfarin-R in rat plasma.

	Theoretical concentration		
	Low (10 ng/mL)	Medium (100ng/mL)	High (400 ng/mL)
R-Warfarin			
Intra-day (n=6)			
Mean	9.84	94.19	435.68
SD	0.41	6.60	33.70
Precision (CV%)	4.1%	7%	7.7%
Accuracy (% bias)	1.6%	5.8%	8.9%
Inter-day (n=18)			
Mean	8.90	97.34	406.44
SD	0.89	10.0	5.65
Precision (CV%)	10.0%	10.27%	1.40%
Accuracy (% bias)	11%	2.7%	1.6%
S-Warfarin			
Intra-day (n=6)			
Mean	9.93	97.7	409.18
SD	0.59	6.15	34.5
Precision (CV%)	6%	6.3%	8.4%
Accuracy (% bias)	0.7%	2.3%	2.3%
Inter-day (n=18)			
Mean	9.20	101.55	403.11
SD	0.74	3.84	5.26
Precision (CV%)	8.08%	3.7%	1.31%
Accuracy (% bias)	8%	1.2%	0.78%
Etravirine			
Intra-day (n=6)			
Mean	12.7	109.18	438.3
SD	1.42	11.5	63.73
Precision (CV%)	11.07%	10.5%	14.5%
Accuracy (% bias)	27%	9.1%	9.5%
Inter-day (n=18)			
Mean	10.88	98.87	416.00
SD	0.47	8.99	23.15
Precision (CV%)	4.32%	9%	5.6%
Accuracy (% bias)	8.8%	1.13%	4%

CV% = (standard deviation/mean)×100.
% bias = [(measured concentration-mean theoretical concentration)/measured concentration]×100.

Table 2. Precision and accuracy of the simultaneous assay method for R- warfarin, S-warfarin and etravirine in rat plasma.

3.3 Extraction recovery

The plasma concentration range of 2.5-500 ng/mL displayed a mean recovery rate (n=3) of R-warfarin, S-warfarin, etravirine 121%± 0.06, 125.5% ± 0.08, 108.6%± 0.40, respectively. The mean recovery was calculated from the ratio of the slopes of the analytes calibration curves in plasma to that in mobile phase.

3.4 Accuracy and precision of the assay

To investigate the intra-day and inter-day precision and accuracy in rat plasma; low, medium, and high quality control spiked plasma samples were used. Table 2 illustrates the intra-day (n=6) and inter-day (n=3) quality control samples for racemic warfarin and etravirine.

3.5 Stability

Stability results in plasma were summarized in Table 3. The racemic warfarin and etravirine quality control samples were not affected by the three freeze-thaw cycles. Racemic warfarin and etravirine were stable in plasma under refrigeration conditions and room temperature for up to 12 hours.

Storage condition	Nominal concentration	R-Warfarin Mean (± SD) percent of nominal concentration remaining (n=3)	S-Warfarin Mean(± SD) percent of nominal concentration remaining (n=3)	Etravirine Mean(± SD) percent of nominal concentration remaining (n=3)
4°C temperature for 12 hours in rat plasma	10 ng/mL 100 ng/mL 400ng/mL	97.0 ± 0.46 96.0 ± 2.47 99.5 ± 11.79	103.0 ± 0.38 99.0 ± 3.27 105.5 ± 18.16	108.5 ± 0.89 95.2 ± 0.17 98.8 ± 11.69
Room temperature in the autosampler for 12 hours in rat plasma	10 ng/mL 100 ng/mL 400ng/mL	96.3 ± 0.37 97.2 ± 12.29 103 ± 63.4	102.2 ± 0.16 96.6 ± 10.17 103.5 ± 66.6	102 ± 1.24 114 ± 22.1 83 ± 110.8
Freeze-thaw Cycle 1	10 ng/mL 100ng/mL 400 ng/mL	74.29 ± 0.37 69.30 ± 0.63 92.58± 8.38	81.88 ± 0.73 77.71 ± 1.52 82.64 ± 3.59	107.43 ± 0.31 92.93 ± 1.41 91.90 ± 25.63
Cycle 2	10 ng/mL 100ng/mL 400 ng/mL	78.8 ± 0.03 71.58 ± 0.43 100.31 ± 4.23	89.6 ± 0.31 81.5 ± 2.38 90.39 ± 24.33	121.88 ± 0.18 102.45 ± 5.22 107.63 ± 10.73
Cycle 3	10 ng/mL 100ng/mL 400 ng/mL	76.78 ± 0.21 82.28 ± 3.62 112.82 ± 68.23	85.63 ± 0.14 91.46 ± 4.67 111.17 ± 76.69	133.4± 2.7 150.97 ± 19.24 94.55 ± 19.47

Table 3. Stability of warfarin and etravirine in rat plasma.

3.6 Matrix effect

The matrix factor percentages of 118.2±10.9, 114±5.8, 143.6±17.57 for warfarin-R, warfarin-S, and etravirine were obtained. These results are in agreement with international guidelines (FDA, 2001).

3.7 Application of the assay method

The analytical method was applied to study the pharmacokinetics of concomitant administration of racemic warfarin and etravirine using the rat as an animal model (Figure 3 and 4).

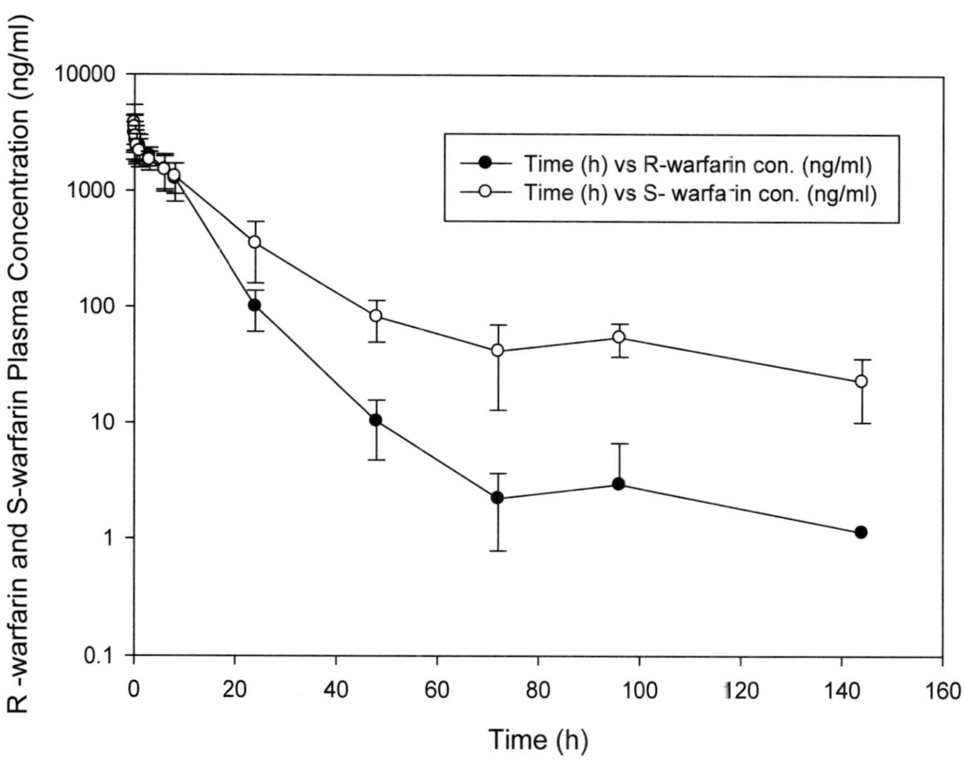

Fig. 3. Mean (± SD; n=6) plasma concentration – time profile of R-warfarin and S- Warfarin after an intravenous 1mg/kg dose of racemic warfarin to male Sprague Dawley rats.

The mean (± SD) areas under the plasma concentration curve for R-warfarin, S-warfarin, and etravirine were 30000 ±5584.4, 32009 ±10231, and 5203.5 ± 500 ng x hr/mL, respectively. The half-lives for these analytes were 7.5 ± 10.7, 16 ± 7, and 22 ± 11 hrs, respectively. Furthermore, these analytes exhibited clearances of 39.5 ±12.8, 31.6± 12 and 4.6 ± 0.0002 mL/hr, respectively. The volumes of distribution for R-warfarin, S-warfarin, and etravirine were calculated to be 414 ±162.4, 637.5 ± 554, and 144 ±72 mL, respectively. Based upon these findings it can be suggested that R-warfarin is metabolized and eliminated quickly. S-warfarin has a longer half-life and is not cleared as quickly as R-warfarin. Therefore, this demonstrates that S-warfarin is the more potent enantiomer, circulating longer throughout the body, and has the potential for eliciting more therapeutic efficacy. Etravirine demonstrated a decrease in clearance, suggesting the saturation of the metabolic pathways responsible for it's clearance upon concomitant administration with warfarin. Further pharmacokinetic studies are underway to examine the steady state drug interactions upon concomitant administration of racemic warfarin and etravirine.

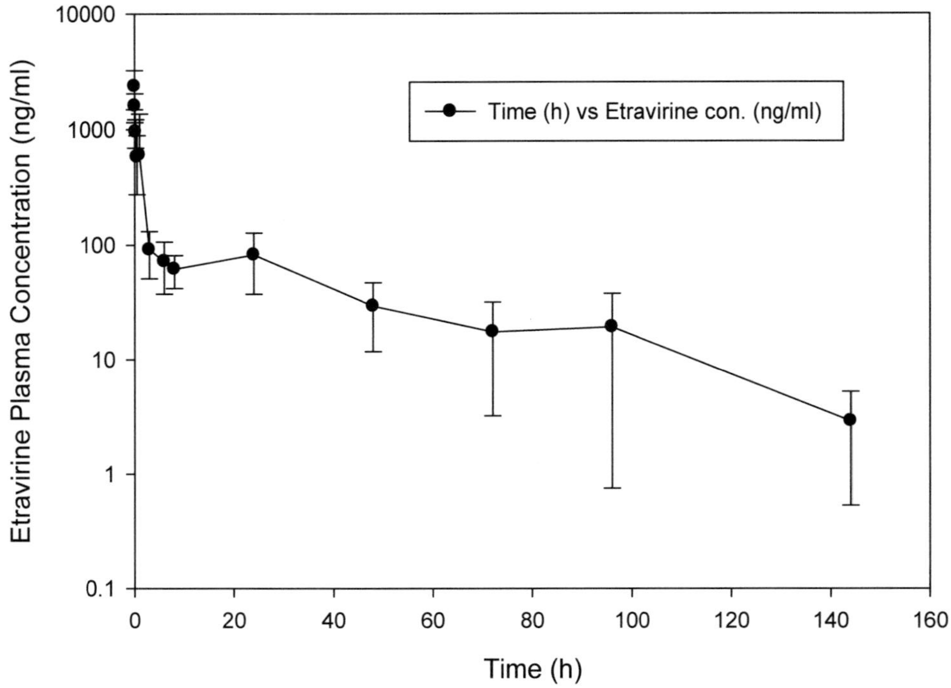

Fig. 4. Mean (± SD; n=6) plasma concentration – time profile of etravirine after an intravenous 25mg/kg dose of etravirine to male Sprague Dawley rats.

4. Conclusion

In conclusion, we have developed a rapid, simple and sensitive LC-MS/MS assay using a chiral column, for the simultaneous detection and quantification of R-warfarin, S- warfarin and etravirine in small volumes (100µl) of rat plasma. This method enabled us to do the simultaneous pharmacokinetic evaluation of intravenously administered racemic warfarin and etravirine using Sprague-Dawley rat as an animal model. Among the two enantiomers of warfarin, the S warfarin is more potent and stays in the body for a longer period of time as observed from its longer half life. This observation is in agreement with the previous studies on the pharmacokinetics of racemic warfarin reported by Eli Chan et al. 2009 and others. Furthermore, upon concomitant administration of etravirine and racemic warfarin, there is a statistically significant increase in the half life of etravirine which is a clear indication of a drug-drug interaction between these two commonly administered pharmaceuticals. Hence the concomitant administration of warfarin and etravirine could increase the efficacy of this NNRTI due to the above mentioned reasons. Due to the limited pharmacokinetic studies for etravirine in the literature, we need to proceed further with similar studies to fully understand the therapeutic impacts of this combination drug regimen. Hence this study can be used for preclinical analysis which can be applied for further pharmacokinetic characterization of racemic warfarin with etravirine.

5. Acknowledgements

This publication was made possible in part by NIH/NCRR/RCMI grant 5G12RR003045-21 and a Texas Southern University Research seed grant.

6. References

Abobo, C.V., & Lei, W., & John, J., & Joseph, M., & Bates, T., & Liang, D. (2010). LC-MS/MS determination of etravirine in rat plasma and its application in pharmacokinetic studies. *Journal of Chromatography B*, 878: 3181-3186.

Alexander P, Areg H., 2009. Pharmacokinetic and Pharmacodynamic Study of Interaction of Rhodiola Rosea SHR-5 Extract with warfarin and theophylline in Rats. *Phytotherapy research* 23: 351-357.

Barclay, V.K., & Tyrefors, N.L., & Johansson, I.M., & Pettersson, C.E. (2011). Trace analysis of fluoxetine and its metabolite norfluoxetine. Part I: Development of a chiral liquid chromatography-tandem mass spectrometry method for wastewater samples. *Journal of Chromatography A*, 1218: 5587-5596.

Breckenridge, A., & Orme, M., & Wesseling, H., & Lewis, R.J., & Gibbons, R. (1974). Pharmacokinetics and pharmacodynamics of the enantiomers of warfarin in man. *Clinical Pharmacology and Therapeutics*, 15: 424-430.

Boffito, M., & Jackson, A., & Lamorde, M., & Back, D., & Watson, V., & Taylor, J., & Waters, L., & Asboe, D., & Gazzard, B., & Pozniak, A. (2008). Pharmacokinetics and safety of etravirine administered once or twice daily after 2 weeks treatment with efavirenz in healthy volunteers. Journal of acquired immune deficiency syndromes (1999), 52(2): 222-227.

Chan, E., & Hegde, A., & Chen, X. (2009). Effect of rutin on warfarin anticoagulation and pharmacokinetics of warfarin enatiomers in rats. *The Journal of Pharmacy and Pharmacology*, 61:451-458.

Else, L. & Watson, V., & Tji, J., & Hughes, J., & Siccardi, M., & Khoo, S., & Back, D. (2010). Validation of a rapid and sensitive high-performance liquid chromatography-tandem mass spectrometry (HPLC-MS/MS) assay for the simultaneous determination of existing and new antiretroviral compounds. *Journal of chromatography B, analytical technologies in the biomedical and life sciences*, 878:1455-1465.

Fayet, A., & Béguin, A., & Zanolari, B., & Cruchon, S., & Guignard, N., & Telenti, A., Cavassini, M., & Günthard, H.F., & Buclin, T., & Biollaz, J., & Rochat, B., & Decosterd, L.A. (2009). A LC-tandem MS assay for the imultaneous measurement of new antiretroviral agents: Raltegravir, maraviroc, darunavir, and etravirine. *Journal of chromatography B, analytical technologies in the biomedical and life sciences*, 877:1057-1069

Kakuda,T.N., & Schöller-Gyüre, M., & Hotelmans, R.M. (2011) . Pharmacokinetic interactions between warfarin and non-antiretroviral drugs. *Clinical Pharmacokinetics*, 50: 25-39.

Kaminsky, L.S., & Zhang, Z.Y. (1997). Human p450 metabolism of warfarin. *Pharmacology & Therapeutics*, 73: 67-74.

Quaranta, S., & Woloch, C., & Paccou, A., & Giocanti, M., & Solas, C., & Lacarelle, B. (2009). *Therapeutic Drug Monitoring*, 31(6):695-702.

Rezk, N. L., & White, N.R., & Jennings, S.H., & Kashuba, A.D. (2009). A novel LC-ESI-MS method for the simultaneous determination of etravirine, darunavir and ritonavir in human blood plasma. *Talanta*, 79(5): 1372-1378.

Radwan, M.A., & AlQuadeib, B.T., & Aloudah, N.M., & Aboul Enein, H.Y. (2010). Pharmacokinetics of ketorolac loaded to polyethylcyanoacrylate nanoparticles using UPLC MS/MS for its determination in rats. *International Journal of Pharmaceutics*, 397(1-2):173-178.

Schöller-Gyüre, M., & Kankuda, T.N., & De Smedt, G., & Vanaken, H., & Bouche, M.P., & Peeters, M., & Woodfall, B., & Hoetelmans, R.M. (2008). A pharmacokinetic study of etravirine (TMC125) co-administered with ranitidine and omeprazole in HIV-negative volunteers. *British Journal of Clinical Pharmacology*, 66(4): 508.

Sharma, K., & Kandaswamy, M., & Mithra, C., & Meena, A.K., & Giri, S., & Rajagopal, S., & Mullangi, R. (2011). Highly sensitive LC-MS/MS-ESI method for simultaneous quantitation of albendazole and ricobendazole in rat plasma and its application to a rat pharmacokinetic study. *Biomedical Chromatography: BMC*, doi:10.1002/ bmc:1654.

Ter Heine, R., & Rosing, H., & van Gorp, E.C., & Mulder, J.W., & Beijnen, J.H., & Huitema, A.D. (2009). Quantification of etravirine (TMC125) in plasma, dried blood spots and peripheral blood mononuclear cell lysate by liquid chromatography tandem mass spectrometry. *Journal of pharmaceutical and biomedical analysis*, 49: 393-400.

U.S. Food and Drug Administration. (2001). Guidance for industry, bioanalytical method validation, Available from: < http://www.fda.gov/cder/guidance >

Wei, B., & Dong, L., & Bates, T.R. (2008). Development and validation of an HPLC method to determine griseofulvin in rat plasma: Application to pharmacokinetic studies. Analytical Chemistry Insights, 3: 103-109.

Welzen, M.E., & van den Berk, G.E., & Hamers, R.L., & Burger, D.M. (2011). Interaction between antiretroviral drugs and acenocoumarol. Antiviral Therapy, 16(2):249-52.

17

Application of Tandem Mass Spectrometry for Analyzing Melamine

Wei-Chih Cheng[1] and I-Jen Wang[2,3,4,5]
[1]*Food and Drug Administration, Department of Health, Executive Yuan, Taipei,*
[2]*Department of Pediatrics, Taipei Hospital Department of Health, Taipei,*
[3]*College of Public Health, China Medical University, Taichung,*
[4]*College of Medicine, Fu Jen Catholic University, Taipei,*
[5]*National Taiwan University College of Public Health*
Taiwan

1. Introduction

Melamine (MEL; 1,3,5-triazine-2,4,6-triamine, CAS No. 108-78-1) is a heterocyclic triazine compound containing 67% nitrogen by weight (Figure 1). It is a polar compound and slightly soluble in water and ethanol. MEL forms synthetic resins with formaldehyde which are used in the fabrication of laminates, glues, adhesives, and surface coating resins (WHO, 2008). The pesticide cyromazine (CYRO) can degrade to form MEL (Chou et al., 2003).
In terms of toxicity, MEL is not metabolized and is rapidly eliminated in the urine. Animal studies have shown that the oral LD50 in rats is 3,161 mg/kg body weight. High doses of MEL have an effect on the urinary bladder, in particular causing inflammation, the formation of bladder stones and crystals in urine. With regards to carcinogenicity, the International Agency for Research on Cancer (IARC) has concluded there is inadequate evidence for carcinogenicity in humans (WHO, 2008).

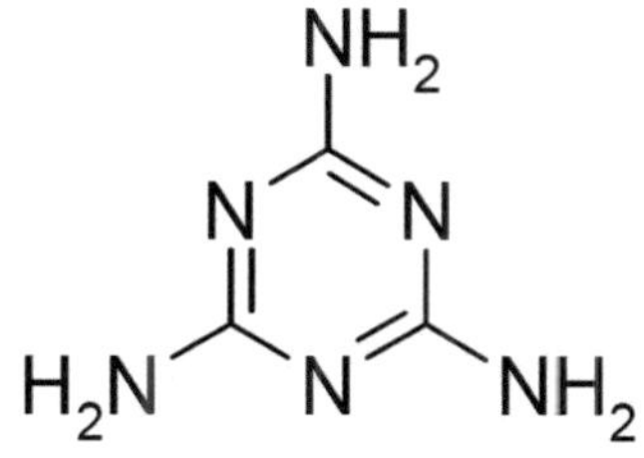

Fig. 1. Structure of melamine.

There are no approved uses for the direct addition of MEL to food or feed. However, in 2007, MEL was found in pet food causing the death of a large number of dogs and cats due to kidney failure. In September of 2008, MEL tainted milk resulted in nephrolithiasis and renal failure in infants in China. More than 50,000 infants and young children were hospitalized for urinary problems, possible renal tube blockages and possible kidney stones related to the consumption of melamine contaminated infant formula and related dairy

products. Six infant deaths were confirmed in mainland China (WHO, 2008). The adverse renal effects of melamine-tainted formula have also found in children in Taiwan, where cross-strait diary trade with China had bloomed in recent years (Wang et al., 2009). The dietary exposure based on the consumption of MEL tainted infant formula in China at the median levels of MEL reported in the most contaminated brand was estimated to be 40-120 times the tolerable daily intake (TDI) of 0.2 mg/kg-body weight, explaining the dramatic health outcomes in Chinese infants (WHO, 2009).
Currently, the amount of protein in foodstuffs is measured indirectly using the Kjeldahl method, which is based only on nitrogen content in samples regardless of whether or not it is incorporated into protein. For example, the factor is 6.38 for milk, of which the protein content is around 16%. However, since MEL is intentionally added to foodstuffs and pet food in order to increase the nitrogen (protein) content, other factors clearly need to be considered for the calibration of protein content.
Many countries have reported finding MEL in milk containing products. Samples testing positive for MEL include biscuits, cakes and confectionery (0.6–945.86 mg/kg); liquid milk and yoghurt products (0.5–648 mg/kg); frozen desserts (39–60.8 mg/kg); powdered milk and cereal products (0.38–1143 mg/kg); processed foodstuffs (0.6–41 mg/kg); food-processing ingredients (1.5–6694 mg/kg); and animal feed (116.2–410 mg/kg) (Hilts & Pelletier, 2009).

2. Analytical method

Many methods have been developed to analyze MEL since the contamination of pet food in 2007 and milk in 2008, including screening and selective quantitative methods (Table 1). Screening methods have the advantages of simplicity, cost-effectiveness, time saving, and labor-saving, where selective methods provide much more information on identification with reliable, reproducible results. Current methods focus on MEL, cyanuric acid (CYA), and related compounds, such as ammelide (AMD) and ammeline (AME) content in food, feed and biological tissues or body fluids.

2.1 Screening and qualitative methods

In screening methods, commercial enzyme-linked immunosorbent assay (ELISA) has been used to detect MEL in dog food (Garber, 2008), infant formula (Garber & Brewer, 2010), muscles, the liver, kidney, plasma, and urine (Wang et al., 2010). Choi and Lee (2010) developed a competitive chemiluminescent enzyme immunoassay (CLEIA) with detection of 1,1′-oxalydimidazole (ODI) derivatives to analyze MEL in milk. This method is capable of rapidly quantifying and screening MEL.
Other studies have introduced alternative qualitative methods, such as surface-enhanced Raman spectroscopy (SERS), matrix-assisted laser desorption ionization mass spectrometry (MALDI-MS), and direct analysis in real time (DART) coupled with time of flight mass (TOFMS). SERS has been used to screen MEL in a number of food matrices including gluten, chicken feed, and processed foods (Lin et al., 2008). Vail et al., (2007) used DART ion source coupled with TOFMS to detect MEL in pet food. Dane and Cody (2010) also employed DART to selectively ionize MEL directly from powdered milk. To eliminate the interference of 5-hydroxymethylfurfual (5-HMF) produced during heating of the milk, TOFMS with a higher resolving power has been used to distinguish MEL (*m/z* 127.0732) and 5-HMF (*m/z* 127.0395) which have very close molecular weights. Tang et al., (2009) employed MALDI-

MS for direct analysis of biochemical crystals containing melamine/cyanuric acid complexes in urine samples. MEL can be desorbed/ionized upon N_2 laser irradiation by simply mixing the commonly adopted MALDI matrixes with melamine cyanurate in urine residue. The high-throughput MEL analysis method using low-temperature plasma (LTP) coupled with an ion trap mass spectrometer of bench top type (Huang et al., 2009) or handheld mass spectrometer (Huang et al., 2010) have also been introduced recently. The mechanism involves thermally-assisted (from 20 °C to 170 °C) vaporization and plasma ionization of MEL. The characteristics are direct sampling, no solvents used, air serving as the plasma support gas, low power consumption, small size and rapid analysis.

Screening and quantitative methods provide an effective alternative for the analysis of samples suspected of containing MEL without extensive sample preparation or expensive instrumentation. The cost, selectivity and sensitivity are low, but the speed is fast. However, positive results should be confirmed using a confirmatory method in most cases.

2.2 Selective quantitative methods

For selective methods, high performance liquid chromatography (HPLC) coupled with ultraviolet absorption (UV) and diode array detection (DAD) have been used to analyze food and feed. MEL has UV absorbance at approximately 240 nm. In addition to retention time confirmation, the spectrometric profile of MEL recorded between 200 and 400 nm has been used to confirm the identity of a selected peak. However, the detection limit is around the parts per million (ppm) level (Kim et al., 2008; Muniz-Valencia et al., 2008; Chou et al., 2003; Lutter et al., 2011; Venkatasami & Sowa, 2010).

Gas chromatography (GC), and HPLC coupled with single stage mass spectrometry (MS) or tandem mass spectrometry (MS/MS) are frequently used for the analysis of MEL and related compounds. In March 2007, the Forensic Chemistry Center was asked to assist in the investigation of pet foods responsible for the adverse effects in cats and dog. GC-MS methods to detect the presence of MEL and related compounds were developed for rapid screening. The selected ion monitoring mode (SIM) was used for quantification and full scan mode for further identification (Litzau et at, 2008).

LC-MS/MS has many applications in MEL analysis in food, feed and urine (Andersen et al., 2008; Heller & Nochetto, 2008; Somker & Krynitsky, 2008; Turnipseed, 2008; Cheng et al., 2009a; Ibáñez et al., 2009; Zhang et al., 2010; Lutter et al., 2011) LC is suitable for the separation of polar MEL, and double stage mass (MS/MS) analysis provides high selectivity and sensitivity, with a reported detection limit range from 0.01 to 0.1 μg/g.

3. Sample preparation and tandem mass spectrometry analysis

Biological samples, especially food and feed samples containing proteins, fats, mineral salts, carbohydrates and others are complex. The food constituents may result in contamination of the separation column and interface between inlet and detector. Co-eluents with MEL can result in severe interference during analysis, and matrix effects can affect the MEL recovery from a sample during preparation. For these reasons, sample preparation is performed in most studies before instrumental analysis, although some high-throughput methods using directly handheld mass spectrometers have also been introduced. This section firstly introduces extraction and clean-up procedures in sample preparation. The solvent used for extraction steps from food, which is the most complicated matrix, is discussed, as well as the ion-exchange cartridge used for clean up. Moreover, the raising the sensitivity and

	Matrices	Analytes	LOD (μg/g)	S/C	Ref.
LC-DAD	Wet pet food, animal feed	MEL, CYA, AML, AMD	1.2-113	C	Kim et al., 2008; Muñiz-Valencia et al., 2008
LC-UV	Poultry, eggs, milk, milk-based infant formula,	MEL	0.06-0.36	C	Chou et al., 2003; Lutter et al., 2011; Venkatasami & Sowa Jr, 2010
LC-MS/MS	Catfish, trout, tilapia, salmon, shrimp, infant formula, ,milk-based products, beverage products, animal feed	MEL, CYA	0.01-0.1	C	Andersen et al., 2008; Heller & Nochetto, 2008; Smoker & Krynitsky, 2008; Turnipseed, 2008; Ibáñez et al., 2009; Lutter et al., 2011
GC-MS	milk, milk products, milk-based powdered infant formula	MEL, CYA, AML, AMD	0.002-0.009	C	Lutter et al., 2011; Miao et al, 2009.
GC-MS	Wheat gluten, rice protein, wet pet food, corn gluten, soybean meal, dry cat food, salmon favor	MEL, AML, AMD, CYA	10	S,C	Litzau et al., 2008
ELISA	Infant formula, wheat food products, dog food	MEL	1-2.5	S	Garber & Brewer, 2010; Garber, 2008
ODI CLEIA	Milk	MEL	1.12 ppb	S	Choi & Lee, 2010
DAPCI-MS	Milk products	MEL	$1.6X10^{-11}$ g/mm^2	S	Yang et al., 2009
SERS	Gluten, chicken feed, processed foods	MEL	0.05-0.1%	S	Lin et al., 2008
DART-TOFMS	Powdered milk	MEL	-	S	Dane & Cody, 2010
UPLC-MS/MS	Urine	MEL	0.01	C	Cheng et al., 2009a
LC-MS/MS	Urine	MEL, CYA	0.01	C	Zhang et al., 2010
LTP-MS/MS	Urine	MEL	0.01	S	Huang et al., 2009
MALDI-MS	Urine	MEL cyanurate	1.25	S	Tang et al., 2009
ELISA	Tissue, body fluid samples (urine)	MEL	0.05	S	Wang et al., 2010

LOD: Limits of detection; MEL: Melamine; CYA: Cyanuric acid; CYRO: Cyromazine; AMD: Ammelide; AME: Ammeline.
S: Screening method; C: Confirmation method.
DAPCI: Desorption atmospheric pressure chemical ionization; SERS: Surface enhanced Raman spectroscopy; DART: Direct analysis in real time; LPT: Low-temperature plasma; ELISA: Enzyme-linked immunosorbent assay; ODI: 1,1'-oxalydimidazole; CLEIA: Chemiluminescent enzyme immunoassay;

Table 1. Comparison of the methods used to analyze melamine in food, feed and urine.

strengthening the resolution power by high pressure liquid chromatography is also addressed. Secondly, the selective analytical method of tandem mass spectrometry is the most important analytical tool to trace contaminants in a biological matrix. In this section, the principle of the tandem mass spectrometric technique, including ionization, mass analysis, fragmentation, and quantification, are also illustrated, along with detailed information on the advantages and drawbacks of tandem mass spectrometry.

3.1 Sample preparation

To analyze MEL, two steps are performed in most studies, including extraction and clean up in LC techniques. Derivation of the extracted MEL is necessary in GC techniques to improve the sensitivity. Sample preparation has positive effects on the instrumental analysis, and especially limits of detection (LOD).

Because MEL is a slightly polar molecule, 50% acetonitrile (Cheng et al., 2009a; Fujita et al., 2009; Goscinny et al., 2011; Mondal et al., 2010; Smoker & Krynitsky, 2008), 1-5% trichloroacetic acid (Juan et al., 2009; Xia et al., 2009), and 2.5% formic acid (Turnipseed et al, 2008) are used for extraction. Acetonitrile is the most frequently used organic solvent in reversed-phase liquid chromatography. Acetonitrile has the characteristics of miscibility with water, low viscosity and results in protein denaturation which is useful in food or feed preparation to eliminate protein in the sample solution. Some studies have also reported using 30-50% methanol to reduce the amount of acetonitrile combined with sonication or two step centrifugation to extract the MEL in liquid formula, dry powder or pet foods (Tran et al., 2010; Vendatasami & Sowa Jr 2010). In addition, the non-polar solvents dichloromethane and hexane are used in lipid-lipid extraction to reduce lipid effects in high-fat bio-samples, such as animal-derived food or bakery goods. (Deng et al., 2010; Goscinny et al., 2011, Zhu et al., 2009). To enhance the protonation of MEL before it is applied to cation solid phase extraction (SPE), hydrochloric acid, formic acid, and acetic acid are added in most cases. Acid also can denature protein and reduce interference.

For urine samples, solvent extraction is generally omitted because the level of proteins in urine is relatively low. Nevertheless, a clean up procedure is still necessary for the isolation and concentration of MEL before analysis. Hydrochloric acid is added to make sure MEL is protonated before SPE clean up. To eliminate renal stones in urine, especially for nephrolithiasis cases, centrifugation is performed at around 3,000 g for 10 minutes at 5 °C (Cheng et al., 2009a).

Aside from the solvent extraction, cation exchange SPE such as Oasis® MCX SPE cartridge (Waters Corp., Malford, MA, USA), Bond Elut®-SCX (Agilent Technologies, Inc. Santa Clara, CA, USA) is used to clean up extracts (Fujita et al., 2009; Smoker & Krynitsky, 2008; Cheng et al., 2009a) from food, feed, or urine samples. The Oasis® MCX cartridge contains a strong sulfonic (HSO_3) group bonded onto a poly(divinylbenzene-coN-polyvinyl-pyrrolidone) copolymer to extract the basic compounds with cation-exchange groups. Bond Elut-SCX contains a benzene ring in the functional group. The non-polar characteristic becomes particularly important when conducting ion-exchange from aqueous systems, where selectivity towards compounds exhibiting cationic and non-polar character is seen.

To shorten the preparation time, several improvements have been developed. MALDI-MS has been applied to the direct analysis of melamine cyanurate and MEL in urine (Tang et al., 2009). A low temperature plasma ambient ionization source, coupled to a portable mass spectrometer (Mini 10.5), has been used to determine MEL contamination in whole milk and related materials, such as synthetic urine and fish meat (Huang et al., 2010). However, the urine matrix results in interfering ion peaks and suppresses the ion intensity of MEL, while a clean up process consisting of simply washing with water eliminates such interference and enhances the ion intensity (Tang et al., 2009).

3.2 Chromatographic and mass analysis (ionization, fragmentation and mass analysis)

Mass spectrometry is a sensitive analytical technique that has been used extensively in chemical, biological, and environmental applications. Since public concern about food safety is growing, mass spectrometry is nowadays regarded as the most important and novel technique to detect trace residues or contaminates in food and the environment, such as pesticides in vegetables, veterinary drugs in animal derived foods, adulteration (e.g. MEL in pet food and milk), and environmental or processed food contamination (e.g. dioxins in fatty foods, acrylamide in starchy foods, 3-MCPD in soy sauce etc.) (Cheng et al., 2004; 2009b).

The basic structure of a mass spectrometer consists of an ionization source, mass analyzer(s) and detector. Inlets separate the desired compounds to deliver the analyte to the ion source. The devices should be chromatographs, such as GC, LC, HPLC, UPLC or CE. A direct probe is also an alternatively method to provide the samples in certain circumstances.

HPLC is the most frequently used inlet device to separate MEL from other compounds. However, MEL is not easily retained in a reverse phase chromatographic column. Mostly, zwitterionic HILIC (hydrophilic interaction chromatography) columns are chosen because of the high polarity and hydrophilic properties of MEL (Cheng et al., 2009a; Heller & Nochetto, 2008; Goscinny et al., 2011; Smoker & Krynitsky, 2008; Turnipseed et al; 2008, Xia et al, 2010; Zhou et al., 2009). HILIC is a variation of normal-phase chromatography where the stationary phase is polar material, such as silica, cyano, amino, or diol. Retention times are prolonged for polar compounds and sensitivity is enhanced in mass spectrometry. Other methods use reversed-phase C18 or C8 columns (Lin et al., 2008; Venkatasami & Sowa Jr, 2010; Ibánez et al., 2009; Lutter et al., 2011). Recently, UPLC has been introduced in this field. The particle size in the column in UPLC is 1.7 μm, which can be used under high pressure and increases the sensitivity to MEL (Cheng et al., 2009a).

GC is used in the separation step; however, derivatization is needed to improve the sensitivity for involatile compounds such as MEL (Zhu et al., 2009; Miao et al., 2009). However, to reduce solvent waste, simplify sample preparation and avoid the need for hazardous reagents, a simple and reliable injection-port derivatization method coupled with GC-furan-CI-MS/MS has been developed to determine MEL and cyanuric acid in powdered milk samples (Tzing & Ding; 2010). Recently, direct determination (no derivation) of MEL in dairy products by GC-MS has been developed with coupled column separation. Thirty meters of DB-5ms ((5%-phenyl)-methylpolysioxane, 0.25 mm i.d., 0.25 μm df) coupled with 1.5 m of Innowax (polyethylene glycol, 0.32 mm i.d., 0.25 μm df) by a quartz capillary column connector has been introduced as separation column (Xu et al., 2009).

During ionization, samples migrate from the inlet system into the ionization chamber. A variety of ionization techniques are used for mass spectrometry, such as atmospheric pressure chemical ionization (APCI), chemical ionization (CI), electron ionization (EI), electrospray ionization (ESI), and MALDI. The ionization methods used for the majority of MEL analyses are EI and ESI.

EI is suitable for gas-phase ionization, although its use is limited to compounds that are sufficiently volatile and thermally stable. Samples are usually heated to increase the vapor pressure for analysis. The EI source consists of a heated filament giving off electrons. EI is gained by bombardment with electrons at 70 electron volts of energy, with the energy being transferred to the sample molecule leading to the ejection of an electron to form a molecular ion with a positive charge. The excess energy absorbed by the molecular ion will

continuously lead to extensive fragmentation to produce characteristic ions. If the molecular ion is unstable, nearly all of the molecular ions will decompose into fragment ions; if not, an intense molecular ion will be recorded. The characteristic ions of MEL derivatives (derived with *N, O*-bis(trimethylsily)trifluoroacetamide)) are *m/z* 327, 171, 99, and the molecular ion *m/z* 329. The most abundant ion is *m/z* 327, and it is considered to be the quantifying ion (Zhu et al., 2009). The characteristic ions for MEL alone are found at *m/z* 126, 85, 68, and 43. The molecular ion *m/z* 126 is most abundant and used as the quantitative ion. Because the full scan spectrum of EI is quite specific, single-stage mass analysis is normally enough to elucidate results. However, the selectivity is less than that for MS/MS methods. In MS/MS experiments, a first analyzer is used to isolate a precursor ion, which then undergoes spontaneous (or by some activation) fragmentation to yield product ions and neutral fragments.
For compounds that are thermally labile or are not volatile, ions of these compound must be extracted from the condensed to the gas phase. ESI and APCI are two examples of atmospheric pressure ionization sources. Such sources ionize the sample at atmospheric pressure and then transfer the ions into the mass spectrometer (De Hoffmann & Strocbant, 2001).
The molecular weight of MEL is low and not easily analyzed by GC-MS directly. Moreover, matrix interference will mask the signal unless there is a clear extract. However, fragmentation of MEL is sufficient for it to be detected using LC-MS/MS by collision-induced dissociation (CID). MEL has been analyzed using ESI in positive ion (ESI+) mode. The protonated molecular ion of MEL in ESI+ mode is *m/z* 127. The fragments of *m/z* 127 are *m/z* 127>85 ([M+H-CH_2N_2]+) and *m/z* 127>68 ([M+H-CH_5N_3]+), where *m/z* 85 is the dominant ion, as shown in Figure 2 in product ion scan mode.

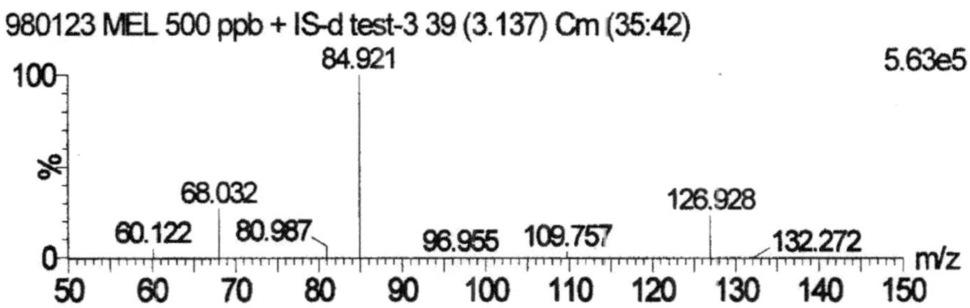

Fig. 2. Mass spectrum of melamine (Cheng et al., 2009a).

4. Validation of analytical methods

To establish an accurate and precise method, the process of validation is necessary before an analytical method is adapted in routine practice. Generally, there are four validation steps; qualitative analysis, quantitative analysis, recoveries test, and detection limit evaluation. After the method has been developed, inter-laboratory or international precision test results are also applied for verification.

4.1 Qualitative analysis

Retention time is the major index used to identify analytes in chromatography coupled with a UV detector. However, the selectivity is low because of interference of the same retention time, and even a specific wavelength that are selected for some compounds. Mass spectrometry is more selective, where not only the retention time, but also the mass spectrum can provide information on the chemical structure for identification.

The four main scan modes which are available using MS/MS include product ion scan, precursor ion scan, neutral loss scan, and selected reaction monitoring. For tandem spectrometry, the mass fragments of the analytes offer the most important information for qualitative evaluation. However, the criteria of the precursor ion and product ion should be carefully considered. For example, the multiple reaction monitoring (MRM) mode is used during LC-MS/MS for MEL, with the characteristic fragmentation transition of *m/z* 127>85 for quantitative analysis and *m/z* 127>68 for confirmatory analysis. The ion ratio of transition ions *m/z* 127>68 and 127>85 is calculated for confirmation. For mass spectrometry techniques, the confirmation criteria applied are according to Commission Decision 2002/657/EC. Retention time, two transition reactions, and the peak area ratio from the different transition reactions are evaluated to identify the target compounds. Table 2 shows the maximum permitted tolerances for relative ion intensities using a range of mass spectrometric techniques.

Relative intensity (% of base peak)	EI-GC-MS (relative)	CI-GC-MS, GC-MS^n LC-MS, LC-MS^n (relative)
> 50 %	± 10 %	± 20 %
> 20 % to 50 %	± 15 %	± 25 %
> 10 % to 20 %	± 20 %	± 30 %
≤ 10 %	± 50 %	± 50 %

Table 2. Maximum permitted tolerances for relative ion intensities using a range of mass spectrometric techniques (2002/657/EC).

4.2 Quantitative analysis

Because of the anlayts losing in sample preparation and the presence of matrix effects, external standard method, matrix-matched external standard method or internal standard methods are used to correct analytical differences during analysis. Matrix effects, especially are relatively significant in LC-MS/MS analysis where the ionization of analysts may be enhanced or suppressed. Matrix effects are expressed as the ratio of the mean peak area of analytes between the post-extraction fortification and solvent standards (Heller & Nochetto, 2008; Xia et al., 2010). Matrix effects are more variable among different feeds with the most pronounced suppression of CYA in wheat gluten; of MEL in pelletized hog feed and fish feed respectively (Heller & Nochetto, 2008). Xia et al., (2010) reported that the matrix effects are in general around 10%for MEL, AML, AME and CYA in several matrices including cat feed, dog feed, egg, milk, yoghurt, ice cream and milk powder. Nevertheless, matrix-matched external standard or internal standards can be used to improve the accuracy and precision of the data. The application of isotope-labeled internal standards is the most ideal choice for MS based methods because their physical and chemical properties are close to the analytes. Generally, the internal standards are added prior to sample preparation to account for the loss of analytes during the sample preparation process, such as extraction, clean up, or concentration and for the ionization step in LC-MS/MS. Stable isotope-labeled internal standards are currently used for MEL analysis including $^{13}C_3{}^{15}N$-MEL, $^{13}C_3$-MEL and $^{15}N_3$-MEL (Jaco et al., 2011; Deng et al., 2010; Ding et al., 2009; Smoker & Krynitsky, 2008; Cheng et al., 2009a; Goscinny et al., 2011). For example, the internal standard $^{13}C_3N_3(^{15}NH_2)_3$ gives a protonated molecular ion of *m/z* 133, where *m/z* 133>89 and *m/z* 133>71 are the predominant fragments by CID during MS/MS experiments. The *m/z* 89 product ion is monitored for routine analysis. However, the isotopic internal standards are not economic and only can be used in MS spectrometry. Some of matrix matched standards or internal standards are alternatively used, such as resorcine in HPLC-DAD methods (Ehling et al., 2007; Muñiz-Valencia et al., 2008).

4.3 Recoveries test

In the recoveries test, samples are fortified with standard (or target) compounds at different concentrations, and then undergo preparation, clean up procedures and finally instrumental analysis. The calculated recoveries show the performance of the developed method and whether it is feasible for routine use. As shown in Table 3, yellow fish and infant formula were fortified with 50 or 200 µg/kg of MEL, followed by the application of internal standards, homogenization, extraction, clean up procedure and finally injection into the instrument, in this example UPLC-MS/MS (Cheng et al., 2009c). The results showed that the recovery was quite good in the range of 101 to 107%, and the coefficient of variation (CV) was also lower than 5%.

Sample blank (MEL, µg/kg)	MEL added (µg/kg)	MEL measured (µg/kg wet weight)	Recovery (%)	CV (%)
Yellow fish (ND)	50	52,56,53,52	107 ± 4	4
	200	208, 200, 200, 201	101 ± 2	2
Infant formula (ND)	200	211, 209, 214, 201, 204, 199	103 ± 3	3

ND: None detected

Table 3. Recovery test of the UPLC-MS/MS method (Cheng et al., 2009c).

According to the Commission Decision 2002/657/EC, in cases of repeated analysis of a sample carried out under within-laboratory reproducible conditions, the intra-laboratory coefficient of variation of the mean should not exceed the values shown in Table 4. The CV results fell in the range of 15% showing excellent reproducibility in this study.

Mass fraction	CV (%)
≥ 10 µg/kg to 100 µg/kg	20
> 100 µg/kg to 1 000 µg/kg	15
≥ 1 000 µg/kg	10

Table 4. CVs for quantitative methods by element mass fractions (2002/657/EC).

4.4 Detection limit

The detection limit is the smallest sample quantity that yields a signal that can be distinguished from the background noise (generally a signal equal to ten times the background noise) (Hoffmann & Stroobant, 2001). In method detection limit (MDL) evaluation, the spiked (signal-to-noise, S/N ratio) and statistic method are usually used. The limit of quantification (LOQ) suggested by the Food and Drug Administration (FDA) in food is the level at which a 10:1 peak to peak S/N ratio is observed for the analyte quantification ion transition, and a 3:1 peak-to-peak S/N ratio observed for the analyte secondary ion transition. In the statistic method, matrices fortified with MEL are used to evaluate the MDL by using Students' values at the 99% confidence level for 6 degrees of freedom (MDL=t (6,0.99) * SD) as shown in Table 5 (Cheng et al., 2009a). In this example, the MDL was 0.006 μg/mL in the sample. The calculated MDL concentration of MEL can be further fortified into the matrix and analyzed again to guarantee that the statistical results are feasible.

Test number	MEL* measured (μg/mL)
1	0.051
2	0.056
3	0.050
4	0.050
5	0.051
6	0.053
7	0.053
Mean ± SD	0.052 ± 0.02
CV%	4
MDL (3 x SD)**	0.006

* MEL spiked (50 ng/mL) urine
** MDL = $t_{(n-1,\, 1-\alpha\, =\, 0.99)}$ x SD, t: Student's value, SD: Standard deviation

Table 5. The method detection limit of UPLC-MS/MS methods (n=7) (Cheng et al., 2009a).

4.5 Performance test

Precision evaluation is measured as repeatability (intra-day) and reproducibility (inter-day) of samples as demonstrated by injection (in triplicate) of the sample on the initial day and three consecutive days (Venkatasami & Sowa, 2010).

Moreover, participation in proficiency tests internationally or nationally can show the performance of the method. For example, the Food Analysis Performance Assessment Scheme (FAPAS) conducted by the Central Science Laboratory in the UK ran several tests involving MEL in pet food in 2008, chocolate in 2009/10, and they aim to test milk powder in 2012 (FAPAS, 2011). The European Commission's Joint Research Centre, and Institute for Reference Materials and Measure under a request by the European Commission's Directorate-General for Health and Consumer Protection also conducted melamine proficiency testing in 2009 (Breidbach et al., 2009). Laboratories from 31 countries including Taiwan and 21 of the 27 member states of the European Union participated. The purpose was to assess the capabilities of control laboratories to measure MEL in food samples, including skimmed milk powder and starch-containing foods. The method we developed during the MEL food safety scare in 2008 using UPLC-MS/MS was also applied to analyze the MEL content in these two samples (Cheng et al., 2009d). The method was found to have good precision and accuracy as shown in Table 6. The SD and CV were quite low, and the Z-score fell between +/- 2. The report compared the methods used in the tests and concluded that isotope dilution mass spectrometry with stable-isotope labeled MEL was clearly advantageous with regards to the accuracy of the results (Breidbach et al., 2009).

	Material A Milk powder	Material B Baking mix
Mean (mg/kg) ± uncertainty	9.2 ± 0.40	3.0 ± 0.13
SD	0.1	0.02
CV	1.3	0.7
Z-score	-0.7	-0.4
Theoretical value (mg/kg) ± uncertainty	10.6 ± 0.6	3.18 ± 0.17

Table 6. Results obtained in the EU Melamine proficiency test 2009 (Cheng et al., 2009d).

5. Other applications of tandem mass spectrometry

Tandem mass spectrometry is frequently used in industry and in academic research. It has applied to identify protein, peptides, oligonucleotides in biotechnology, polycyclic aromatic hydrocarbons (PAHs), polychlorinated biphenyls (PCBs), polybrominated diphenyl ethers (PBDEs), and endocrine disrupting chemicals in the environment and food. After 2008 MEL tainted milk event, the tandem mass spectrometry becomes more and more important in food safety. Tandem spectrometry coupled with gas or liquid chromatography has been applied extensively in the determination of residues or contaminants, and even unknown contaminants in food. Moreover, it can be performed to measure biologically active compounds and their metabolites in the body fluid for clinical applications. The development of tandem mass spectrometry newborn screening led to a large expansion of potentially detectable congenital metabolic diseases that affect blood levels of organic acids (Chace et al., 2003; Wang et al., 2003). We have checked cord blood cotinine by HPLC-MS/MS for objectively measuring the prenatal tobacco smoke exposure (Wang et al., 2008). We have also checked cord blood perfluorinated compounds by UPLC-MS/MS and correlated its concentration with atopic disorders (Wang et al., 2011). The future of the analytical chemistry and food safety is closely associated with the development of mass spectrometry.

6. Summary

The MEL tainted milk event in 2008 raised public concerns about food safety. Because food is increasingly transported worldwide, governments are now paying more and more attention to safeguarding food products, and many novel methods have been introduced for this purpose. A lot of researchers are working on developing screening methods and selective methods to detect contaminants. Both methods have applications in different respects. To achieve reliable results, most studies have used selective methods to analyze MEL. Because of the chemical properties of MEL, organic solvents such as acetonitrile or methanol are used to extract it from food, feed or biological matrices followed by cleaning up of SPE in the preparation steps. Gas or liquid chromatography are responsible for the separation of analytes from interference before the analytes enter the detection areas. To answer the demand for high selectivity, mass spectrometry has become the most popular method used to indentify MEL and related compounds. The ionization process can charge molecules by using EI, ESI, CI, APCI, or MALDI. Either single stage mass or tandem mass are used in spectrometry. The application of tandem mass can increase the selectivity and raise the *S/N*, resulting in low LOD and LOQ. HPLC coupled with tandem mass has become the major analytical method in the analysis of MEL and related compounds. Before applying the method as a routine practice, researchers should perform a series of validation evaluations, including qualification, quantification, recoveries, reproducibility, and detection tests. International or inter-laboratory precision tests will provide confidence in the ability of the analytical method. As the technique of tandem mass spectrometry progresses, we can apply it in various biological and clinical fields. It is a small step for the technique, but a giant leap for the science. We look ahead to these and other emerging applications as the benefits of this technology become incorporated into current and future human health.

7. Acknowledgements

This work was supported by the National Science Council of Taiwan (NSC 98-2314-B-192 -001 -MY3).

8. References

Andersen, WC.; Turnipseed, SB.; Karbiwnyk, CM.; Clark, SB.; Madson, MR.; Gieseker, CM.; Miller, RA.; Rummel, NG. & Reimschuessel, R. (2008). Determination of confirmation of melamine residues in catfish, trout, tilapia, salmon, and shrimp by liquid chromatography with tandem spectrometry. *Journal of Agricultural and Food Chemistry*, Vol.56, No.12, (June 2008), pp. 4340-4347

Breidbach, A.; Bouten, K.; Kröger, K.& Ulberth, F. (2009). Melamine Proficiency Test 2009-Assessing the capabilities of control laboratories to measure melamine in skimmed milk powder and starch-containing foods. *JRC Scientific and Technical Reports.* European Commission Joint Research Centre Institute for Reference Materials and Measurements. pp. 4. ISBN 1018-5593. European Communities.

Chace, DH.; Kalas, TA.& Naylor, EW. (2003). Use of tandem mass spectrometry for multianalyte screening of dried blood specimens from newborns. *Clinical Chemistry*, Vol.49, No.11, (November 2003), pp. 1797-1817.

Cheng, WC.; Chen, HC.; Lin, YP.; Lee, HF.; Chang, PC.& Chou, SS. (2004). Survey on 3-monochloro-1,2-propandiol (3-MCPD) contents of soy sauce products during fiscal year 2002 in Taiwan. *Journal of Food and Drug Analysis,* Vol.12, No.4, (December 2004), pp. 336-341

Cheng, WC.; Chen, SK.; Lin, TJ.; Wang, IJ.; Kao, YM. & Shih, DYC. (2009a). Determination of urine melamine by validated isotopic ultra-performance liquid chromatography/ tandem/mass spectrometry. *Rapid Communication in Mass Spectrometry,* Vol.23, No.12, (June 2009), pp. 1776-1782

Cheng, WC.; Kao, YM.; Shih, DY.; Chou, SS.& Yeh, AI. (2009b). Validation of an improved LC/MS/MS method for acrylamide analysis in foods. *Journal of Food and Drug Analysis,* Vol.17, No.3, (June 2009), pp. 190-197

Cheng WC., Chang MH, Liu HY., Yan WL., Kao YM., Shih YC. (2009c). Validation of UPLC/MS/MS method for analyzing melamine content in fishes and surveillance. Symposium on Food Sanitation and Inspection 2009, pp. 110. Taoyuan, Taiwan, October, 2009

Cheng WC., Chang MH., Hsu CL., Chiu YC., Liu HY., Kao YM., Shih YC. (2009d). UPLC/MS/MS method for analyzing melamine content in milk and baked powder and its' uncertainty evaluation. Symposium on Food Sanitation and Inspection 2009, pp. 111. Taoyuan, Taiwan, October, 2009

Chou, SS.; Hwang, DF.& Lee, HF. (2003). High performance liquid chromatographic determination of Cyromazine and its derivative melamine in poultry meats and eggs. *Journal of Food and Drug Analysis,* Vol.11, No.4, (December 2003), pp. 290-295

Choi, J.; Kim, YT.& Lee, JH. (2010). Rapid quantification of melamine in milk using competitive 1,1'-oxalyldiimidazole chemiluminescent enzyme immunoassay. *Analyst,* Vol.135, No. 9, (September 2010), pp. 2445-2450

Commission Decision 2002/657/EC. (August 12, 2002). Implementing Council Directive 96/23/EC concerning the performance of analytical methods and the interpretation of results. *Official Journal of the European Communities,* 2002(L221), pp. 8-36

Dane, AJ.& Cody, RB. (2010). Selective ionization of melamine in powdered milk by using argon direct analysis in real time (DART) mass spectrometry. *Analyst,* Vol.135, No.4, (April 2010), pp. 696-699

De, Hoffamann E.& Stroobant, V. (2001). In: *Mass spectrometry: principles and applications,* E. de Hoffamann, V. Stroobant (2nd Ed.), 183-205, John Wiley & Sons Ltd, , ISBN 0-471-48566-7, Vukovar, West Sussex, England.

Ehling S.; Tefera S.& Ho IP. (2007). High-performance liquid chromatographic method for the simultaneous detection of the adulteration of cereal flours with melamine and related triazine by-products ammeline, ammelide, and cyanuric acid. *Food Additives and Contaminants,*Vol.24, No.12, (December 2007), pp. 1319-1325

FAPAS. (August 2011). 2011-2012 FAPAS programme. Available from http://www.fapas.com/prog.cfm?currsch=fapas

Fujita, M.; Kakimoto, K.; Nagayoshi, H.; Konishi, Y.; Uchida, K.; Osakada, M.; Okihashi, M.& Obana, H. (2009). Determination of melamine in Chinese-made processed food]. *Shokuhin Eiseigaku Zasshi,* Vol.50, No.3, (June 2009), pp.131-134

Garber, EA. (2008). Detection of melamine using commercial enzyme-linked immunosorbent assay technology. *Journal of Food Protection.* Vol.71, No.3, (March 2008),pp. 590-594

Garber, EA. & Brewer, VA. (2010). Enzyme-linked immunosorbent assay detection of melamine in infant formula and wheat food products. *Journal of Food Protection.* Vol.73, No.4, (April 2010), pp. 701-707

Goscinny, S.; Hanot, V.; Halbardier, J.; Michelet, J.& Loco, J. (2011). Rapid analysis of melamine residue in milk, milk products, bakery goods and flour by ultra-performance liquid chromatography/tandem mass spectrometry: from food crisis to accreditation. *Food Control,* Vol.22, pp. 226-230

Heller, DN.& Nochetto, CB. (2008). Simultaneous determination and confirmation of melamine and cyanuric acid in animal feed by zwitterionic hydrophilic interaction chromatography and tandem mass spectrometry. *Rapid Communication in Mass Spectrometry*, Vol.22, No.22, (November 2008), pp. 3624-3632

Hilts, C.& Pelletier, L. (2009) Background paper on occurrence of melamine in foods and feed. Available from http://www.who.int/foodsafety/fs_management/Melamine_3.pdf

Huang, G.; Ouyan, Z.& Cooks, RG. (2009). High-throughout trace melamine analysis is complex mixtures. *Chemical Communications*, Vol.7, No.5, (February 2009), pp. 556-558

Huang, G.; Xu, W.; Visbal-Onufrak, MA.; Ouyang, Z.& Cooks, RG. (2010). Direct analysis of melamine in complex matrices using a handheld mass spectrometer. *The Analyst*, Vol.135, No.4, (April 2010), pp. 705-711

Ibáñez, M.; Sancho, JV.& Hernández, F. (2009). Determination of melamine in milk-based products and other food and beverage products by ion-pair liquid chromatography-tandem mass spectrometry. *Analytica Chimica Acta*, Vol.649, No.1, (September 2009), pp. 91-97

Jacob, CC.& Gamboa, da Costa G. (2011). Low-level quantification of melamine and cyanuric acid in limited samples of rat serum by UPLC-electrospray tandem mass spectrometry. *Journal of Chromatography B*, Vol.879, No.9-10, (March 2011), pp. 652-656

Kim, B.; Perkins, LB.; Bushway, RJ.; Nesbit, S.; Fan, T.; Sheridan, R.& Greene, V. (2008). Determination of melamine in pet food by enzyme immunoassay, high-performance liquid chromatography with diode array detection, and ultra-performance liquid chromatography with tandem mass spectrometry. *Journal of AOAC International*, Vol.91, No.2, (March 2008), pp. 408-413

Kim, H.; Roh, JS.& Oh, CH. (2010). Investigation of the cyanuric acid contamination caused by acetonitrile and alkali detergent for the LC-ESI MS/MS analysis. *Bulletin of Environmental Contamination and Toxicology*, Vol.85, No.2, (August 2010), pp. 165-169

Lin, M.; He, L.; Awika, J.; Yang, L.; Ledoux, DR.; Li, H.& Mustapha, A. (2008). Detection of melamine in gluten, chicken feed, and processed foods using surface enhanced Raman spectroscopy and HPLC. *Journal of Food Science*, Vol.73, No.8, (October 2008), pp. T129-T134

Litzau, JJ.; Mercer, GE.& Mulligan, KJ. (October 2009). GC-MS screen for the presence of melamine, ammeline, ammelide and cyanuric acid. Available from http://www.fda.gov/AboutFDA/CentersOffices/CVM/WhatWeDo/ucm134742.htm

Lutter, P.; Savoy-Perroud, M.; Campos-Gimenez, E.; Meyer, L.; Goldmann, T.; Bertholet, M.; Mottier, P.; Desmarchelier, A.; Monard, F.; Perrin, C.; Robert, F.& Delatour, T. (2011). Screening and confirmatory methods for the determination in cow's milk and milk-based powdered infant formula: Validation and proficiency-tests of ELISA, HPLC-UV, GC-MS and LC-MS/MS. *Food Control*, Vol.22, pp. 903-913

Miao, H.; Fan, S.; Wu, YN.; Zhang, L.; Zhou, PP.; Li, JG.; Chen, HJ.& Zhao, YF. (2009). Simultaneous determination of melamine, ammelide, ammeline, and cyanuric acid in milk and milk products by gas chromatography-tandem mass spectrometry. *Biomedical and environmental Sciences*, Vol.22, No2, (April 2009), pp. 87-97

Mondal, AMR.; Desmarchelier, A.; Konings, E.; Acheson-Shalom, R.& Delatour, T. (2010). Liquid chromatography-tandem mass spectrometry (LC-MS/MS) method extension to quantify simultaneously melamine and cyanuric acid in egg powder and soy protein in addition to milk products. *Journal of Agricultural and Food Chemistry*, Vol.58, No.22, (November 2010), pp. 11574-11579

Muñiz-Valencia, R.; Ceballos-Magaña, SG. ; Rosales-Martinez, D.; Gonzalo-Lumbreras, R.; Santos-Montes, A.; Cubedo-Fernandez-Trapiella, A.& Izquierdo-Hornillos, RC. (2008). Method development and validation for melamine and its derivatives in rice concentrates by liquid chromatography. Application to animal feed samples. *Analytical and Bioanalytical Chemistry*. Vo.392, No.3, (October 2008), pp.523-531.

Smoker, MS.& Krynitsky, AJ (2008). Interim method for determination of melamine and cyanuric acid residues In Foods using LC-MS/MS: Version 1.0, Available from http://www.fda.gov/Food/ScienceResearch/LaboratoryMethods/DrugChemicalResiduesMethodology/ucm071673.htm

Tang, HW.; Ng, KM.; Chui, SS.; Che, CM.; Lam, CW.; Yuen, KY.; Siu, TS.; Lan, LC.& Che, X. (2009). Analysis of melamine cyanurate in urine using matrix-assisted laser desorption/ionization mass spectrometry. *Analytical chemistry*, Vol.81, No.9, (May 2009), pp. 3676-3682

Tittlemier, S. (2010). Methods for the analysis of melamine and related compounds in foods: a review. *Food Additives and Contaminants*, Vol.27, No.2, (February 2010), pp. 129-145, ISSN 0265-203X

Tran, BN.; Okoniewski, RO.; Storm, R.; Jansing, R.& Aldous, KM. (2010). Use of methanol for the efficient extraction and analysis of melamine and cyanuric acid residues in dairy products and pet foods. *Journal of Agricultural and Food Chemistry*, Vol.58, No.1, (January 2010), pp. 101-107

Tzing, SH.& Ding, WH. (2010). Determination of melamine and cyanuric acid in powdered milk using injection-port derivatization an gas chromatography-tandem mass spectrometry with furan chemical ionization. *Journal of Chromatography A*, Vol.1217, pp. 6267-6273

Turnipseed, S.; Casey, C.; Nochetto, C.& Heller, DN. (2008). Determination of melamine and cyanuric acid residues in infant formula using LC-MS/MS. Available from http://www.fda.gov/Food/ScienceResearch/LaboratoryMethods/DrugChemicalResiduesMethodology/ucm071637.htm

Venkatasami, G.; & Sowa, Jr. (2010). A rapid, acetonitrile-free, HPLC method for determination of melamine in infant formula. *Analytica Chimica Acta*, Vol.665. No.2, (April 2010), pp 227-230

Vail, T.; Jones, PR.& Sparkman, OD. (2007). Rapid and unambiguous identification of melamine in contaminated pet food based on mass spectrometry with four degree of confirmation. Journal of Analytical Toxicology, Vol.31. No.6, pp. 304-312

Wang, IJ.; Hsieh, WS.; Chen, CY.; Fletcher, T.; Lien, GW.; Chiang, HL.; Chiang, CF.; Wu, TN.& Chen, PC. (2011). The effect of prenatal perfluorinated chemicals exposures on pediatric atopy. *Environmental Ressearch*. Vol.111, No.6, (august 2011), pp.785-791.

Wang, IJ.; Chen, PC.& Hwang, KC. (2009). Melamine and nephrolithiasis in children in Taiwan. *The New England Journal of Medicine.* Vol.360, No.11, (March 2009); pp.1157-1158.

Wang, IJ.; Hsieh, WS.; Wu,KY.; Guo, YL.; Hwang, YH.; Jee, SH.& Chen PC.(2008). Effect of gestational smoke exposure on atopic dermatitis in the offspring. *Pediatric Allergy Immunology.* Vol.19, No.7, (November 2008); pp.580-586.

Wang, IJ.; Chu, SY.; Wang, CY.; Wang, PJ.&Hwu, WL. (2003). Maple Syrup Urine Disease Presenting with Status Epilepticus, A Case Report. *Acta Paediatrica Taiwanica.* Vol.44, No. 4: (July 2003); pp.246-248.

Wang, Z.; Ma, X.; Zhang, L.; Yang, W.; Gong, L.; He, P.& Li, Z. (2010). Screening and determination of melamine residues in tissue and body fluid samples. *Analytica Chimica Acta.* Vol.662, No.1, (March 2010), pp. 69-75

WHO. (2008). Expert meeting to review toxicological aspects of melamine and cyanuric acid. Available from http://www.who.int/foodsafety/fs_management/Exec_Summary_melamine.pdf

WHO. (2009). Toxicological and Health Aspects of Melamine and Cyanuric Acid. Available from http://whqlibdoc.who.int/publications/2009/9789241597951_eng.pdf

Xia, J.; Zhou, N.; Zhou, C.; Chen, B.; Wu, Y.& Yao, S. (2010). Simultaneous determination of melamine and related compounds by hydrophilic interaction liquid chromatography-electrospray mass spectrometry. *Journal of Separation Science,* Vol.33, No. 17-18, (September 2010), pp. 2688-2697

Xia, X.; Ding, S.; Li, X.; Gong, X.; Zhang, S.; Jiang, H.; Li, J.& Shen, J. (2009). Validation of a confirmatory method for the determination of melamine in egg by gas chromatography-mass spectrometry and ultra-performance liquid chromatography-tandem mass spectrometry. *Analytica Chimic Acta,* Vol.65, No.12, (October 2009), pp. 196-200

Xu, XM.; Ren, YP.; Zhu, Y.; Cai, ZX.; Han, JL.; Huang, BF.& Zhu, Y. (2009). Direct determination of melamine in dairy products by gas chromatography/mass spectrometry with coupled column separation. *Analytica Chimica Acta,* Vol.650, No.1, (September 2009), pp. 39-43

Zhou, J.; Zhao, J.; Xue, X.; Zhang, J.; Chen, F.; Li, Y.; Wu, L.; Li, Ch.& Mi, J. (2009). Hydrophilic interaction chromatography/tandem mass spectrometry for the determination of melamine in royal jelly and royal jelly lyophilized powder. *Journal of Chromatography B,* Vol.877, No.32, (December 2009), pp. 4164-4170

Zhu, X.; Wang, S.; Liu, Q.; Xu, Q.; Xu, S.& Chen, H. (2009). Determination of residues of cyromazine and its metabolite, melamine, in animal-derived food by gas chromatography-mass spectrometry with derivatization. *Journal of Agricultural and Food Chemistry,* Vol.57, No.23, (December 2009), pp. 11075-11082

Zhang, M.; Li, S.; Yu, C.; Liu, G.; Jia, J.; Lu, C.; He, J.; Ma, Y.; Zhu, J.& Yu, C. (2010). Determination of melamine and cyanuric acid in human urine by a liquid chromatography tandem mass spectrometry. *Journal of Chromatography B,* Vol.878, No.9-10, (March 2010), pp. 758–762

18

Identifying and Overcoming Matrix Effects in Drug Discovery and Development

Terence G. Hall, Inese Smukste, Karen R. Bresciano,
Yunxia Wang, David McKearn and Ronald E. Savage
ArQule Inc.
USA

1. Introduction

Liquid chromatography/tandem mass spectrometry (LC-MS/MS) with electrospray ionization is a highly specific and sensitive analytical technique that has become the industry standard for quantifying drugs, metabolites, and endogenous compounds in biological matrices (e.g. plasma). The technique is widely used because of its ability to accurately quantitate analytes of interest with minimal sample clean-up and rapid LC separation. Despite these advantages, LC-MS/MS methodology occasionally encounters problems, some of which are caused by matrix effects.

The "matrix" refers to all components in the sample other than the analyte(s) of interest. Some common matrices typically encountered by bioanalytical scientists include blood, plasma, urine, bile, feces, and tissue samples. Although these complex matrices have a number of common components, not all are known and levels may vary among individuals. For example, plasma samples obtained from different patients enrolled in clinical trial studies may contain different levels of endogenous components based on their genetics and/or disease state as well as different drugs used to manage their disease. Consequently, each patient's plasma may have its own particular set of matrix components and therefore is viewed as being a unique sample. Matrix effects can arise from a number of matrix components including, but not limited to:

- Endogenous biological components such as phospholipids, carbohydrates, and endogenous metabolites (bilirubin)
- Residual formulation components from intraperitoneal (ip), intravenous (iv), or oral dosing (po) experiments; for example, polyethylene glycol (PEG), solutol, polysorbate (Tween 80), etc.
- An interaction between the analyte of interest and the matrix, such as covalent binding to plasma proteins or the enzymatic degradation of a prodrug
- Co-eluting drug metabolites
- Concomitant medications
- Mobile phase additives

A matrix effect is generally encountered when a component, for example a phospholipid (present in plasma) co-elutes with the analyte of interest and causes a decrease (suppression) or increase (enhancement) in ionization efficiency relative to the analyte eluting in the absence of the matrix component. This may result in an erroneous reporting of sample concentrations. In the case of clinical trails, where dosing regimens are being optimized based on pharmacokinetic profiles and are dependant on accurate determination of drug plasma concentrations such miscalculations can lead to errors in determining optimum dosing regimens and in extreme cases failure of a drug in clinical trials.

The degree of enhancement or suppression of ionization of an analyte by a given matrix component can be dependant on the physicochemical properties of the analyte. For example, highly polar compounds generally appear to be affected to a greater degree than less polar molecules which may be due in part to their co-elution with other polar components causing matrix effects.

In most cases, matrix components which might cause ionization enhancement or suppression are removed during the sample clean-up process. The simplest form of sample clean-up is protein precipitation (PPT), in which an organic solvent, commonly acetonitrile (ACN), is combined with the sample of interest enabling most of the protein to be removed from the sample. Two additional methods of sample clean-up that will be discussed in greater detail later in this Chapter are liquid-liquid extraction (LLE) and solid phase extraction (SPE). These methods are more labor intensive, but generally result in better removal of matrix components. Along with sample clean-up, sample dilution is another simple technique effective in minimizing matrix effects.

The exact mechanisms by which matrix components cause ionization suppression (or enhancement) are not known. However, King et al. have postulated that matrix components interfere with the processes involved in the transfer of the charged analyte (ion) into the gas phase thereby increasing or decreasing the ionization efficiency (King et al., 2000).

Preparation of standards in the same matrix as the sample and use of internal standards (ISTDs) which have similar or nearly identical chemical and chromatographic properties to the analyte help in further minimizing matrix effects. For this reason, a stable-isotope-labeled analyte is typically the best choice for an internal standard and in many cases corrects for almost all matrix effects (Xu et al., 2007).

In many cases, it may not be readily apparent that ionization suppression is occurring in a given LC-MS/MS method. Hence, methodologies have been developed to remedy this. A commonly employed technique for detecting ionization suppression involves comparing the LC-MS/MS chromatographic profiles of injections of blank extracted matrix to neat blank matrix (water) which are obtained while continuously infusing diluted analyte solutions post column (prior to entering the MS detector).

Our understanding of matrix effects and how to handle them has continually progressed over the last decade. This Chapter will provide a short review of the current industry perspective on matrix effects.

2. Detection of matrix effects

There are several methodologies available to analytical scientists for detection and quantification of matrix effects. Selection of the particular methodology employed typically depends on the stage of the program (discovery or development). For example, compounds in early discovery generally receive a limited qualitative matrix effect evaluation, due to strict timelines for data generation, while compounds in development warrant more detailed quantitative evaluation.

2.1 Qualitative determination of matrix effects

Post-column infusion of an analyte, as diagramed in Figure 1, is a fast and easy technique that can be used to qualitatively identify regions of ion suppression or enhancement in a particular matrix extract. In this technique, an extracted matrix sample is injected onto the HPLC column using the LC-MS/MS method for the analyte, while a steady flow of that analyte is infused into the effluent flow between the column and the mass spectrometer source. Additionally, a blank solution such as water, buffer, or the initial mobile phase mixture must also be injected to determine the baseline for the analysis. The regions of suppression or enhancement can be visualized in the resulting chromatograms by comparing the baseline obtained from the blank with each of the matrices tested (Figure 2). It is important to note that the degree of the effect will depend on the concentration of the analyte being infused. If the concentration of analyte being infused is too high, matrix effects could be masked. Any regions of enhancement or suppression must be compared with the retention time of the analyte. Investigation of matrix effects using this methodology with an internal standard is also highly recommended.

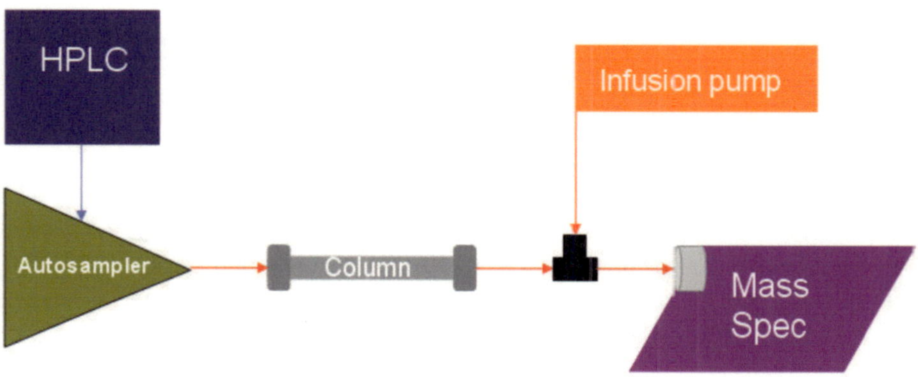

Fig. 1. Post-column infusion schematic.

2.2 Quantitative determination of matrix effects

Post-extraction analyte spiking provides a more quantitative measure of matrix effects. One technique involves extracting two sets of samples; one set contains the analyte added to an extracted matrix (post-extraction sample), and the other contains the analyte in mobile phase, solvent, or buffer (external solution). Both sets of samples are prepared with equivalent concentrations of the analyte and then are processed identically. One can quantitate the degree of enhancement or suppression caused by the matrix effect(s) by use of the equations shown below (Matuszewski et al., 2003):

- **Matrix effect (%)** = B / A * 100 (can also be calculated using (B-A/A) * 100)
- **Recovery (%)** = C / B * 100
- **Extraction Efficiency (%)** = C / A * 100

A = external solution peak area, **B** = post-extraction sample peak area, **C** = extracted matrix peak area

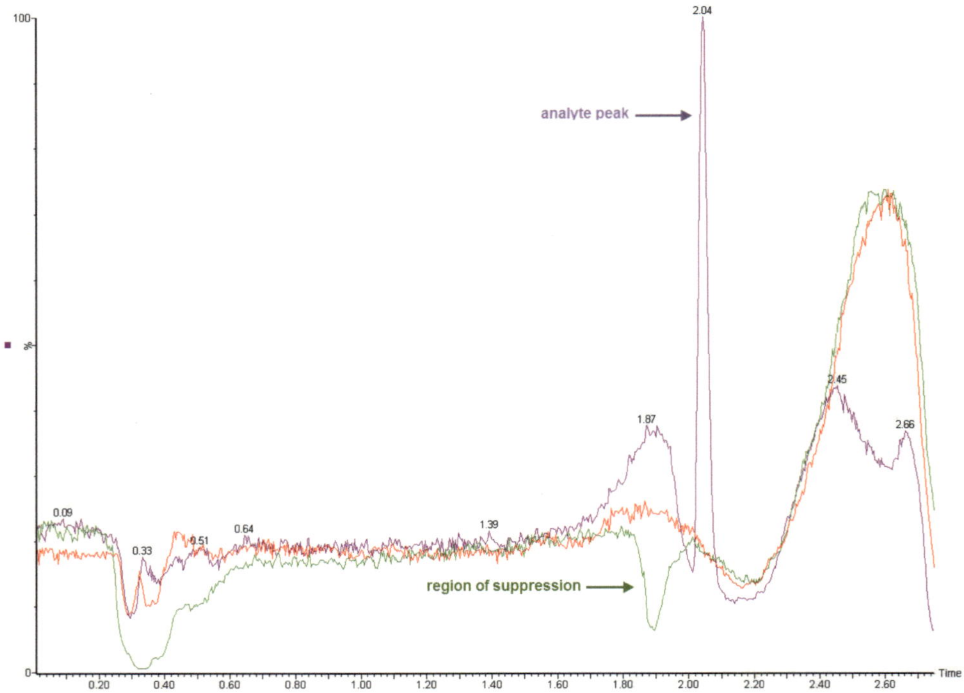

Fig. 2. Infusion test: the red trace is the baseline chromatogram for a solution of 50% acetonitrile in water (reagent blank), the green trace is the extracted matrix blank, and the purple trace is an analyte sample in reagent blank. In this example, the region of suppression does not coincide with the analyte peak. Therefore, this LC-MS/MS method could be used for the analysis of this single analyte, but if metabolite analysis was needed, this method could prove to be inadequate. These experiments would need to be repeated as metabolite standards became available.

	Analyte		Internal Standard	
Nominal Concentration (ng/mL)	**External solution peak area**	**Post-extraction peak area**	**External solution peak area**	**Post-extraction peak area**
250	5483.6	5674.5	21700.3	21312.5
	5491.5	5478.7	22144.9	21916.0
	5554.8	5278.1	21034.5	19840.3
		5619.4		21238.6
		5171.7		20322.3
		5061.1		20447.8
n	3	6	3	6
Mean	5510.0	5380.6	21626.6	20846.3
SD	39.0	248.7	558.9	769.5
RSD (%)				
Matrix Effect* (%)	-2.3		-3.6	

*Positive value indicates percent enhancement, negative value indicates percent suppression

Table 1. Example of matrix effect evaluation.

	Analyte		Internal Standard	
Nominal Concentration (ng/mL)	**Extracted matrix peak area**	**Post-extraction peak area**	**Extracted matrix peak area**	**Post-extraction peak area**
15	12155.0	14473.2	74739.9	88933.0
	12212.1	14139.8	74226.6	90050.2
	11467.6	14473.4	71490.1	92964.1
	9920.8		64027.5	
	12196.4		79932.9	
	13774.6		84727.1	
n	6	3	6	3
Mean	11954.4	14362.1	74857.4	90649.1
SD	1253.1	192.5	7102.2	2081.2
RSD (%)	10.5	1.3	9.5	2.3
Recovery (%)		83.2		82.6
90	76391.8	80913.1	70177.3	87189.2
	74081.7	84259.3	74564.5	89830.9
	77464.8	82794.6	77328.6	91620.0
	80999.7		78844.2	
	79771.5		79697.3	
	79202.9		81323.1	
n	6	3	6	3
Mean	77985.4	82655.7	76989.2	89546.7
SD	2522.2	1677.4	4048.6	2229.0
RSD (%)	3.2	2.0	5.3	2.5
Recovery (%)		94.3		86.0
750	649097.6	743563.2	75758.1	92358.5
	650131.4	755794.7	78693.2	92230.1
	632140.7	764104.4	76311.1	90620.6
	654594.0		81492.2	
	664467.5		77712.3	
	614953.9		72628.6	
n	6	3	6	3
Mean	644230.9	754487.4	77099.3	91736.4
SD	17769.2	10332.8	2987.3	968.4
RSD (%)	2.8	1.4	3.9	1.1
Recovery (%)		85.4		84.0
Overall Recovery (%)	**87.6**		**84.2**	

Table 2. Recovery for an analyte, at various concentrations, and internal standard (ISTD). The "extracted matrix" samples were spiked with analyte or ISTD prior to extraction, and the "post-extraction" samples were spiked with analyte or ISTD after extraction. The recovery of the analyte does not appear to be affected by concentration.

Matrix effect, as previously stated, is a measure of percent change in signal caused by matrix components. Table 1 shows results from an experiment designed to determine matrix effects. The calculated *matrix effect (%)* shows that the analyte and the internal standard responses in this experiment are not affected by matrix effects. Extraction efficiency is a measure of the percent of total response in an extracted sample when compared to the external solution. Sample recovery describes the percent of analyte that was recovered during the extraction process. Table 2 shows an evaluation of recovery for an analyte and internal standard for three concentrations.

The most rigorous technique for determining matrix effects is typically not performed until a compound advances into regulated animal studies where a Good Laboratory Practice (GLP) method validation is required. It involves analysis of three sets of samples for each matrix being evaluated. Each set is prepared as a series of calibration standards with internal standard; one set is prepared in mobile phase, solvent, or buffer (external solution), the second is prepared by addition of analyte to extracted matrix (post-extraction sample), and the third is prepared by addition of analyte directly into the study matrix (Matuszewski et al., 2003).

3. Mechanisms of matrix effects

3.1 Mechanisms of matrix effects

Matrix effects arise at the interface between the LC system and the MS system (King et al., 2000). This interface is referred to as the 'ion source', and it is here that analytes are desolvated and charged. The principals of MS detection dictate that only charged (positive or negative) gas phase ions are detectable. Anything that interferes with either the charging or the desolvation of the analyte will produce a matrix effect.

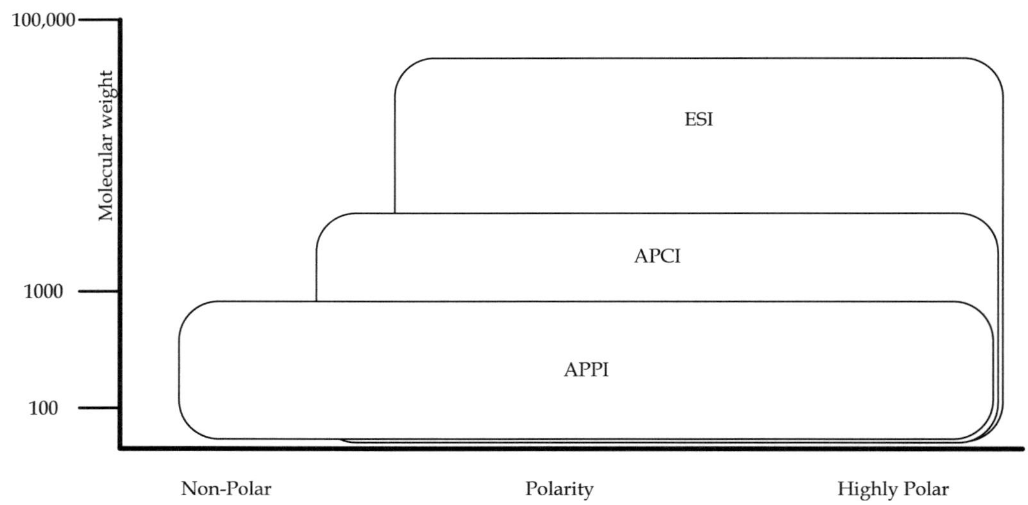

Fig. 3. Applicability of various ion sources depending on analyte polarity and molecular weight.

A complete overview of MS interface design is beyond the scope of the present work; however, some understanding of the different operating principles of the major ion sources is useful for examining the root causes of matrix effects. In modern LC-MS there are a variety ion sources available from which the analyst can choose. This choice often depends on a number of factors including source availability, type of analyte, and sensitivity of the instrument. Figure 3 shows the three most commonly used ion sources, and their applicability for use based on analyte polarity and molecular weight. Electrospray ionization (ESI) is the most widely used ion source. It has broad applicability to a wide spectrum of analytes of varying molecular weight and polarity, and it is capable of ionizing larger macromolecules such as peptides and proteins. Atmospheric pressure chemical ionization (APCI) is the next most widely used ionization method. Additionally, there are a number of alternative source designs such as atmospheric pressure photo ionization (APPI), nano-spray (low volume ESI), and inductively coupled plasma mass spectrometry (ICPMS). Regardless of the type of source, all share a single goal of transitioning uncharged analytes in solution to detectable gas phase ions.

3.2 Electrospray Ionization

In the ESI source, analytes must acquire a charge in solution and then successfully transition to gas phase while maintaining their charge. Figure 4 depicts a generic ESI source. The acquisition of charge in the solution phase and successful transitioning to the gas phase makes the ESI source the most vulnerable to matrix effects when compared to either APCI or APPI (Jessome and Volmer, 2006; King et al., 2000; Trufelli et al., 2011).

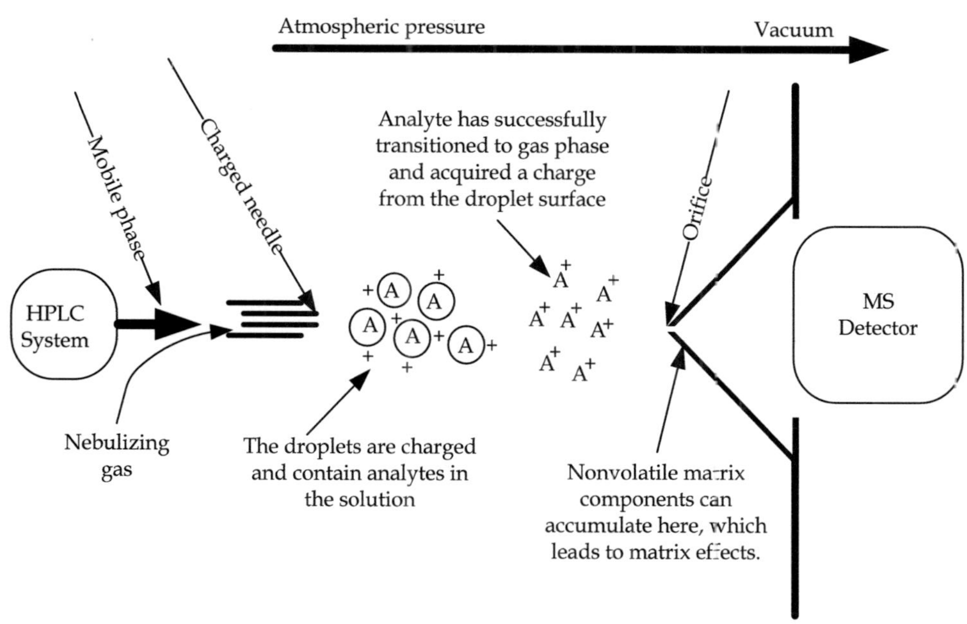

Fig. 4. ESI source.

The mobile phase containing the analyte along with appropriate mobile phase additives such as formic acid or ammonium hydroxide are passed through a charged needle, which transfers this charge to the exterior surface of the solution. At the tip of this charged needle, the solution takes the form of a narrow cone of charged liquid called a Taylor cone. Through the use of nebulising gasses and heat, the droplets are aerosolized. As these charged droplets transverse the ESI source they are progressively reduced in size through evaporation until they become completely desolvated gas-phase ions. These ions are then directed into the orifice of the mass spectrometer. Figure 5 depicts the desolvation of the charged droplets (containing the uncharged analyte) to the charged gas phase ions that are detected by the mass spectrometer.

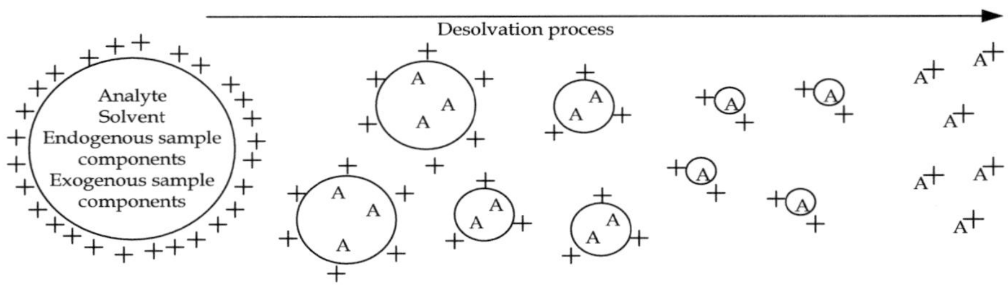

Fig. 5. ESI desolvation process.

Within the ESI source, there are a number of factors which can affect the ability of the analyte to be consistently ionized. Figure 6 depicts some of the causes of matrix effects that may occur while the analyte is still in the liquid phase in an ESI source. Endogenous and exogenous co-eluting sample components can cause suboptimal droplet formation. Oversized droplets, or droplets with non-volatile components, are difficult to completely desolvate and can result in solution phase analyte collecting around the orifice or inside the quadrupole chamber, which can cause signal suppression (Bonfiglio et al., 1999; King et al., 2000). Certain sample components such as the surfactants Tween 80 or PEG 400 have a high affinity for the air-liquid interface of a droplet. As the charge is also localized at the surface of the droplet, this affinity for the surface of the droplet can limit an analyte with less surface affinity from gaining access to the charge (Xu et al., 2005). In addition to surfactants, phospholipids can also interfere with an analyte's access to the surface of the droplet, and thus hinder access to the charges located on the droplet's surface (Bennett and Liang, 2004; Chambers et al., 2007). Ion-pairing reagents, typically used to improve chromatography, can interfere with an analyte's ability to accept a charge. In ESI, the charge is localized on the surface of the droplet; therefore any co-eluting sample component can interfere with an analyte's migration to the surface of the droplet. This interference occurs via competition with the analyte for the charge or repulsion of the analyte away from the surface thereby preventing ionization of the analyte. In charge competition scenarios, the charge will go to analyte or matrix component with the greatest ionization potential at the pH of the mobile phase.

Non-volatile materials, such as phosphate (a common HPLC mobile phase buffer that is generally avoided for LC-MS/MS work), can accumulate inside the MS source housing, on the orifice of the detector, or on the front face of the mass spectrometer's quadrupoles. Samples may also contain components (for example endogenous lipids) that can precipitate inside an ion source. The accumulation of non-volatile materials can increase electrical

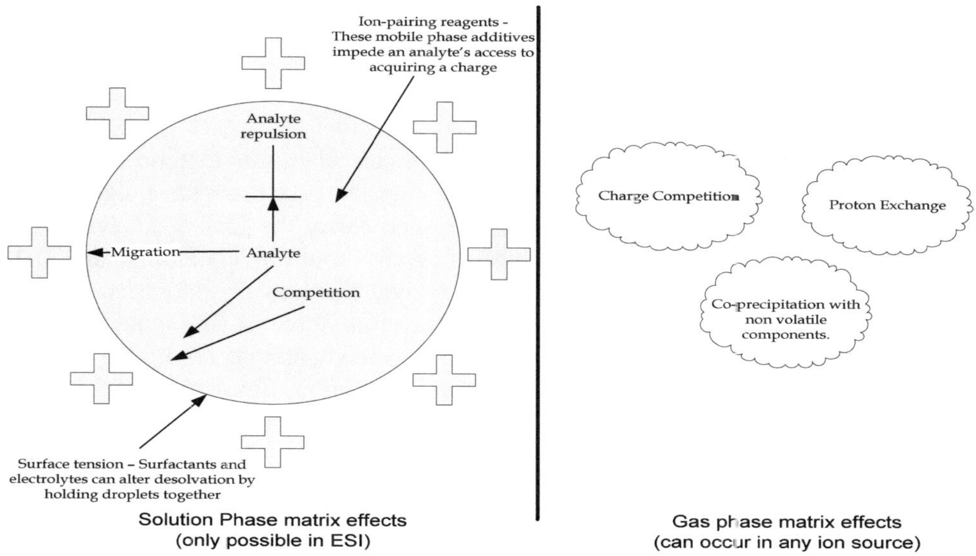

Fig. 6. Mechanisms of ion suppression in LC-MS.

resistance, thereby preventing the ions from following the electromagnetic gradient into the detector, resulting in signal loss (Mei, 2005). These deposits can also block the orifice, physically preventing entrance into the MS detector.

Fig. 7. Accumulation of non-volatile components in ESI sources: A: shows a recently cleaned source with no accumulations; B: shows a crystalline ring of deposits resulting from polar matrix components and endogenous electrolytes; C: shows a slightly greasy smear around the orifice that may be the result of insufficient sample clean-up of endogenous lipids.

In modern mass spectrometers, the LC inlet is orthogonal to the MS orifice. This configuration allows ions to follow the electrical charge gradient to the orifice and into the MS. Previous generations of LC-MS/MS instruments were designed with the LC inlet and the MS orifice aligned in a linear placement. This linear source design tended to have more issues with matrix effects and non-volatile accumulation in the source and the front of the MS quadrupoles, which often resulted in degradation of MS performance more rapidly. Figure 7 shows the orifices of several LC-MS/MS systems including clean and with build up of non-volatile components.

Nano-scale ESI, a variant of conventional ESI sources that relies on ultra low amounts of solvent, has been reported to reduce matrix effects (Chiu et al., 2010). The reduced volumes used in nano-scale ESI sources diminish the potential impact of matrix effects on the droplet formation and desolvation.

3.3 Atmospheric pressure chemical ionization

Atmospheric pressure chemical ionization (APCI) is an alternate method of analyte ionization to ESI. Although less susceptible to matrix effects, APCI can have significantly lower ionization efficiencies for some analytes. In such cases, the loss of analyte signal sensitivity must be weighed against the reduction of matrix effects (Trufelli et al., 2011). APCI, unlike ESI, does not rely on solution phase analyte charging, therefore many of the matrix effects due to droplet formation and phase transformation are eliminated. Instead, desolvation of the analyte occurs almost instantly in a heated ceramic vaporizing collar at the inlet from the LC. However, the high temperature required for desolvation makes APCI unsuitable for the quantitative analysis of thermally labile molecules. Figure 8 depicts a typical APCI ion source.

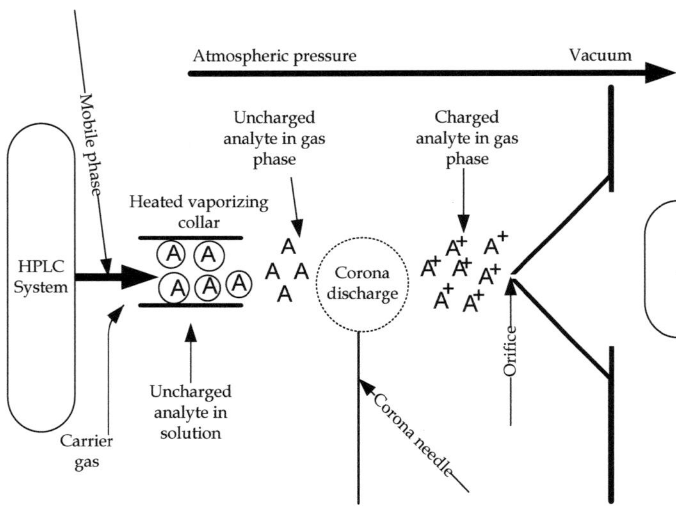

Fig. 8. APCI source.

APCI is not immune to matrix effect issues of its own (Gosetti et al., 2010). In APCI, ion creation occurs as the desolvated neutral analyte and solvent molecules pass through the gas phase solvent ions (plasma) produced by the corona needle. The corona needle discharges electrons that initially charge the gas phase solvent molecules, and this charge is then transferred to the analyte. The ionization process must occur with a limited amount of transferrable charge in a very short amount of time. Thus, co-eluting species passing simultaneously through the corona often compete with the analyte for available charge. In a positive ionization mode, any component with a higher pKa can affect analyte signal intensity (van Hout et al., 2003). For example, Tween 80, a common excipient in drug formulations, competes with the analyte for charge thereby lowering the ionization efficiency of the analyte (Xu et al., 2005). Co-precipitation of analyte with non-volatile matrix components can also play a role in APCI signal suppression (van Hout et al., 2003).

3.4 Atmospheric pressure photo ionization

Atmospheric pressure photo ionization (APPI) is a recently introduced ionization source that achieves ionization by channeling the uncharged gas phase sample molecules through a charged photon beam. APPI has been demonstrated to be less susceptible to matrix effects than ESI and APCI. APPI, like APCI, ionizes the analyte in the gas phase, eliminating potential issues that arise from solvent phase ionization. Additionally, APPI produces higher energy protons that can overcome potential charge competition between the analyte and solvent or extraneous materials, which sometimes occurs in APCI sources (Gosetti et al., 2010). Figure 9 shows an APPI source diagram.

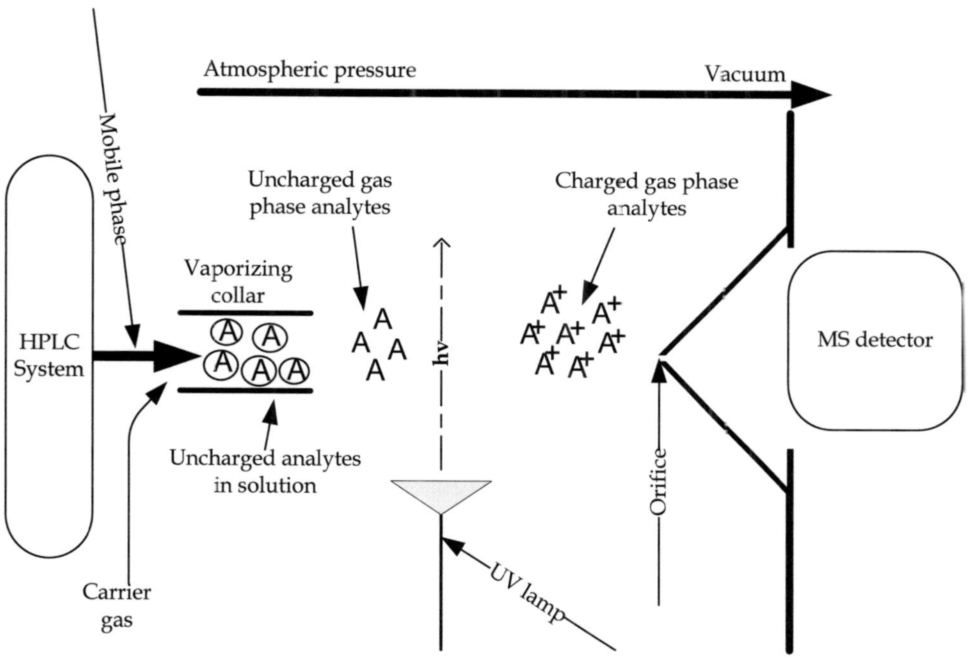

Fig. 9. APPI interface.

3.5 General Considerations

In experiments with multiple analytes, it is necessary to evaluate potential matrix effects for each analyte. Many factors, including pKa, solvation energy, solvent properties, etc., affect electrospray ion formation. Furthermore, each of these factors can impact various analytes within a mixture differently (King et al., 2000).

In general, the consensus is that APCI and APPI systems are less prone to suffer from matrix effects than ESI, however, each method has its advantages and disadvantages (King et al., 2000). In certain matrices and for certain analytes, no significant improvement over ESI sources was observed, emphasizing that matrix effects should be assessed on a compound-by-compound and matrix-by-matrix basis (Lien et al., 2009).

Although it may be possible to resolve an observed matrix effect by changing sources, this is not always an option. For example, analytes often exhibit different ionization efficiencies depending on the mode of ionization. Moreover, changing ion sources does not guarantee elimination of matrix effects. A better option is to understand the specific cause of the matrix effect and then address it directly.

4. Sources of matrix effects

Ionization suppression and/or enhancement due to sample matrices have become one of the most important causes for failures and errors in bioanalysis. Matrix effects are subdivided into two groups: 1) endogenous matrix effects caused by components naturally occurring in the biological or environmental sample and 2) exogenous matrix effects caused by components introduced prior to or during sample collection and analysis.

Commonly encountered sources of matrix effects are summarized below:

- Endogenous components of matrices such as lipids, phospholipids, proteins, and bile salts
- Exogenous components of matrices that are introduced during analysis such as formulation excipients, leachables from the labware, anticoagulants, analyte stabilizers used during sample collection, and reagents used in the preparation of bioanalytical samples
- Degradation products of the analyte created during sample preparation and analysis (e.g. degradation products of prodrugs and compounds sensitive to pH, temperature, or light)
- Impurities and salts contained in analytes and ISTDs
- Poor recovery of analytes due to binding to biological matrices or containers (e.g. non-specific binding of hydrophobic analytes to plastic)
- Solvents and additives used for LC
- Xenobiotics and their metabolites present in analytical samples (e.g. other drugs present in patient samples)

The sources of matrix effects are extremely diverse and are analyte-, LC-MS/MS method- and ion source-dependent. Therefore, great care should be taken during method development and validation to identify potential issues (Matuszewski, 2006; Taylor, 2005; Trufelli et al., 2011; Vogeser and Seger, 2010).

During the bioanalytical method development and validation for GLP studies, many of the exogenous matrix effects are addressed by careful evaluation of the reagents and supplies used for sample collection, preparation, and analysis. However, some matrix effects are difficult to identify and prevent during early stage drug discovery when a large number of diverse compounds are evaluated in a high-throughput manner and general LC-MS/MS methods are used. The problem is exacerbated by matrix effects resulting from impurities present in investigational compounds, such as inorganic and organic salts, and degradation products. Some exogenous and endogenous matrix effects are well known and have been described in literature, but many remain undetermined and are rarely explored in detail. A limited number of extensive and systematic studies have been performed to assess the broad nature of matrix effects providing much needed information regarding mechanisms of matrix effects in general and methods for their elimination (Chambers et al., 2007; Ismaiel et al., 2010; Little et al., 2006; Mallet et al., 2004; Marchi et al., 2010; Muller et al., 2002; Tong et al., 2002; Xu et al., 2005). In contrast, most of the current knowledge is based on studies focused on specific matrix effects. In every day practice, ion suppression is often overcome using empirical methods such as modification of LC-MS/MS methods, substitution of

HPLC columns, more rigorous sample clean-up, and optimization of labware and reagents without identifying or fully exploring the underlying causes.

4.1 Exogenous components leading to matrix effects

4.1.1 Eluents and additives

The composition of the mobile phase profoundly influences the ionization efficiency of all sample components due to the mechanisms discussed previously (Gao et al., 2005; Kostiainen and Kauppila, 2009). Multi-fold increases in signal of the analyte can be achieved by modifying composition of the mobile phase during the LC-MS/MS analysis. For example, an increase in the organic solvent portion of the mobile phase usually leads to increased ionization efficiency (Dams et al., 2002). Such modifications can also lead to changes in retention times of analytes and matrix components thereby altering elution patterns and potentially introducing matrix effects.

Acidic, basic, ion-pairing or buffer salt additives in the mobile phase may have signal suppression or enhancement effects. Mallet and colleagues performed a systematic analysis of formic acid (FA), acetic acid (AA), trifluoroacetic acid (TFA), ammonium hydroxide, ammonium formate, ammonium biphosphate, ammonium bicarbonate, and nonafluoropentadecanoic acid as mobile phase additives and studied their effect on the MS signal intensity of a diverse set of compounds (Mallet et al., 2004). Ionization enhancement or suppression was highly dependent upon each additive, analyte and ionization mode. Consistent with numerous other reports, TFA strongly suppressed ionization of all tested compounds in both positive and negative ionization modes. The signal suppressing effects of TFA and other fluorinated acids are due to the ion-pairing and surface tension effect (as described previously), which disrupts the ionization of the analyte (Gustavsson et al., 2001). The post-column addition of a propionic acid and 2-propanol mixture may counteract the deleterious effects of TFA by facilitating TFA evaporation during ionization (Apffel et al., 1995). Benijts and colleagues reported that addition of acids resulted in significant signal suppression in the analysis of environmental water samples for 35 endocrine disrupting chemicals in negative and positive ESI, while 1 mM ammonium formate reduced the matrix effect. Further improvements were achieved by sample clean-up and use of SIL-ISTDs (Benijts et al., 2004).

Additionally, reagents used during sample collection and preparation that are retained in the analytical sample may lead to matrix effects. Mei et al. reported that commonly used anticoagulant Li-heparin enhanced ionization efficiency for some of the analytes (Mei et al., 2003). Tris(hydroxymethyl)aminomethane (TRIS) buffer and nicotinamide adenine dinucleotide phosphate (NADPH) led to significant matrix effects during the analysis of 27 highly diversified pharmaceutical compounds in microsomal samples when PPT was utilized for sample clean-up (Zheng et al., 2002). SPE based sample clean-up was shown to significantly reduce these matrix effects. Phosphate buffer and non-volatile ion-pairing reagents such as sodium dodecyl sulfate (SDS) lead to severe ion suppression due to accumulation in the source as discussed in Section 3.2.

4.1.2 Leachables and impurities

Matrix effects caused by polymers and plasticizers leaching from common labware are often overlooked (Guo et al., 2006; Mei et al., 2003; van Hout et al., 2003). Importantly, ion source contamination with these polymers may lead to long-lasting interference with analysis.

Particular care should be exercised when organic solvents such as ethyl acetate, dichloromethane, or ethers are used with polymer-based containers, caps, or solid phase materials, as they may solubilize polymers and plasticizers. Minor solvent impurities and additives may also affect the accuracy of analyses (Annesley, 2007).

4.1.3 Formulation agents

Some formulation agents may also lead to matrix effects. In particular, excipients used in early drug discovery for solution or suspension formulations are known to interfere with analyses. These excipients are present in plasma at high concentrations (>1 mg/mL) in early PK sampling time points. In studies conducted by Xu and colleagues, 20% hydroxypropyl-β-cyclodextrin (HPβCD) or 0.4% methyl cellulose vehicles did not lead to matrix effects either during iv or po dosing. Conversely, when 0.1% Tween 80 was used as a vehicle, 50-80% ion suppression was observed for both iv and po administration routes (Xu et al., 2005). These results are supported by several other researchers (Larger et al., 2005; Shou and Naidong, 2003; Tong et al., 2002). Moreover, polysorbates including Tween 80 are complex mixtures of components containing polyoxyethylene (POE) sorbitan and POE sorbitan monoesters (Figure 10). POE sorbitan contains approximately twenty ethylene oxide subunits arranged in four chains of various lengths. The oleate monoester accounts for 58–85% of esters in Tween 80, while the remaining esters have alkyl chains from C14 to C18 and include stearate, linoleate, and linoleneate esters. Therefore, the composition of Tween 80 is quite variable and depends on the manufacturing process (Hewitt et al., 2011). Additionally, Tween 80 is rapidly hydrolyzed to oleic acid and polyethoxylated sorbitan by esterases in rodent plasma leading to further variability in analyses (Larger et al., 2005; van Tellingen et al., 1999). Dosing vehicles containing PEG 400 also lead to ionization suppression, in particular, for early eluting compounds (Shou and Naidong, 2003; Tong et al., 2002; Weaver and Riley, 2006; Xu et al., 2005). Co-infusion of PEG 400, Tween 80, or HPCD with an analyte in 70% ACN solution (pH 8) led to significant LC-MS signal suppression in positive and negative ionization modes, while co-infusion with propylene glycol (PG) resulted in the lowest interference among of the four (Tong et al., 2002).

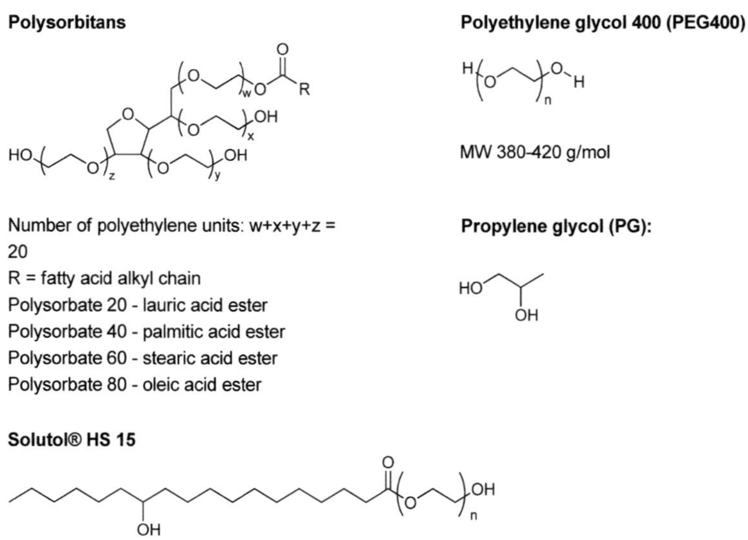

Fig. 10. Commonly used formulation agents.

4.1.4 Cross–talk

Cross-talk occurs with compounds that produce fragments with similar masses. Typically, MRM dwell times and inter scan times are very short. Hence, fragment ions from one transition can still be present in the collision cell when the next MRM transition is monitored, leading to false positive signal. Most commonly cross-talk is observed when structural analogs or SIL-ISTDs are used as internal standards, or when metabolites of the analyte are present in the sample. On rare occasions, structurally unrelated compounds may produce ion fragments with an identical mass leading to cross-talk. Separate analysis of analytes and ISTDs while monitoring all MRM ion transitions is the most reliable method for cross-talk detection. Similar evaluation should be done with known metabolites or analyte degradation products when feasible. The simplest way to eliminate cross talk is to enter a "dummy" or blank MRM transition between mass transitions where cross-talk is observed. The dummy or blank transitions allow the collision cell to completely clear the ions from the previous mass transition. Additionally, cross-talk can be eliminated by reducing the amount of ions entering the collision cell. Therefore, dilution of the analyzed sample, reduced concentrations of ISTDs, or chromatographic resolution of the analyte and the ISTD is typically used to eliminate cross-talk (Morin et al., 2011).

4.1.5 Stable-isotope-labeled internal standards

Stable-isotope-labeled internal standards (SIL-ISTDs) are routinely used in pharmacokinetic analyses and in clinical and forensic toxicology to mitigate matrix effects during LC-MS/MS analysis (Stokvis et al., 2005). SIL-ISTDs are compounds where several atoms of the analyte molecule have been replaced with their stable isotopes. Most commonly hydrogen (^{1}H) is exchanged for deuterium (^{2}H) and carbon (^{12}C) is exchanged for carbon (^{13}C), however, nitrogen (^{15}N) and oxygen (^{17}O) labeled SIL-ISTDs also can be used. At least three atoms should be exchanged during the labeling to avoid interference between analyte and SIL-ISTD signals due to cross-talk or signal contribution caused by isotope distribution. Furthermore, SIL-ISTDs should be of high isotopic purity and stable during the analysis. For example, hydrogen-deuterium exchange has been observed in aqueous solutions, therefore ^{13}C labeling is preferred (Chavez-Eng et al., 2002).

SIL-ISTDs possess nearly identical physicochemical properties compared to their non-labeled counterparts. Therefore, one may expect nearly identical retention times during chromatographic seperation and similar behavior in the ion source. Consequently, SIL-ISTD and an analyte should be exposed to the same ionization conditions and normalization relative to SIL-ISTD should minimize variability during the sample analysis. However, on occasion SIL-ISTDs themselves may lead to ion suppression or enhancement of the analyte (Liang et al., 2003; Remane et al., 2010b). To minimize SIL-ISTDs matrix effects, appropriate concentrations of SIL-ISTDs should be employed and linearity of the response should be tested. In addition, there have been reports that SIL-ISTDs are affected differently by matrix effects than the analyte leading to inaccurate quantification (Lindegardh et al., 2008; Jemal et al., 2003; Wang et al., 2007).

4.1.6 In-source fragmentations

Compounds with weak bonds, such as glucuronide- and sulfate-conjugated metabolites, may fragment in the source during the ionization process, thereby regenerating the parent molecule (Figure 11). When the analyte and its metabolite(s) are not resolved via chromatography, in-source fragmentations may lead to inaccurate quantification of the analyte (Vogeser and Seger, 2010; Vogeser et al., 2001; Yan et al., 2003).

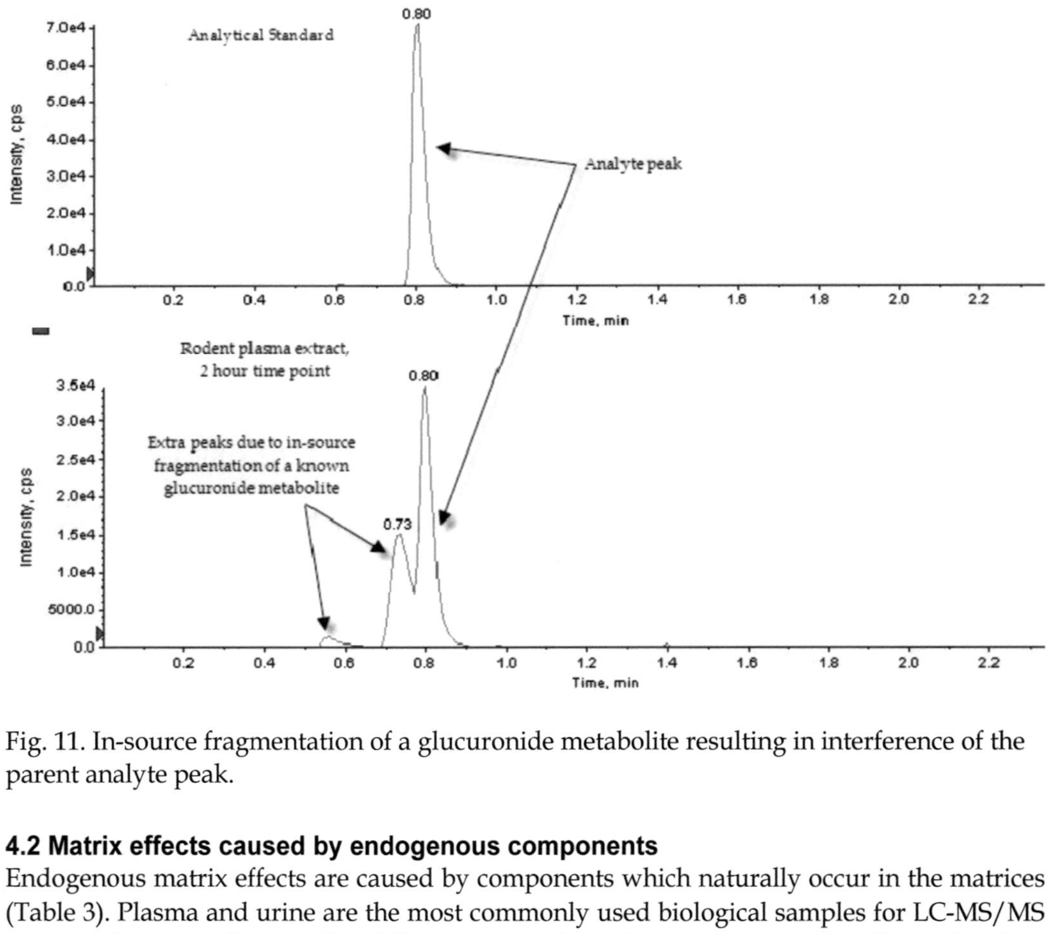

Fig. 11. In-source fragmentation of a glucuronide metabolite resulting in interference of the parent analyte peak.

4.2 Matrix effects caused by endogenous components

Endogenous matrix effects are caused by components which naturally occur in the matrices (Table 3). Plasma and urine are the most commonly used biological samples for LC-MS/MS analysis, however, feces, saliva, bile, and tissue homogenates are occasionally analyzed as well. Each of these biological matrices contains distinct endogenous components that may lead to different matrix effects and require specific protocols for sample preparation. Furthermore, substantial variations in matrix effects can be observed in samples from different individual subjects due to genetic variation, disease state and/or the presence of other xenobiotic compounds (Remane et al., 2010a). Detection of endogenous substances, compounds with limited stability in biological matrices, and trace analytes are particularly prone to matrix effects.

In general, polar compounds are affected more by endogenous matrix effects than non-polar compounds, because many components of biological matrices and environmental samples are polar, water-soluble compounds and are eluted early during a reverse phase (RP) chromatography (Bonfiglio et al., 1999; Muller et al., 2002).

Matrix	Components	Amount
Blood	Plasma	~55%
	Red blood cells	~45%
	White blood cells	>1%
Plasma	Water	~90%
	Protein	~8%
	Inorganic salts	0.9%
	Organic substances (lipids, hormones, vitamins etc.)	1.1%
Urine	Water	95%
	Urea	9.3 g/L
	Chloride	1.87 g/L
	Sodium	1.17 g/L
	Potassium	0.75 g/L
	Creatine	0.67 g/L
	Other ions and compounds	lesser amounts
Liver bile*	Water	90-95%
	Total solids	5-10%
	Bile salts (bile acids)	3-45 mM
	Bilirubin	1-2 mM
	Phospholipids	150-800 mg/dL
	Free Cholesterol	80-200 mg/dL
	Protein (total)	2-20 mg/dL
	Glutathione	0-5 mM
	Sodium	140-170 mM
	Potassium	2.7-6.7 mM
	Calcium	2.5-6.4 mM
	HCO_3^-	12-55 mM

*Data adopted from (Dancygier, 2010)

Table 3. General composition of biological matrices.

4.2.1 Plasma

Plasma contains dissolved proteins, amino acids, peptides, glucose, carbohydrates, vitamins, electrolytes, hormones and lipids, all of which may lead to matrix effects. Serum protein is a major component of plasma with concentrations ranging from 6.0 to 8.3 g/dL. Albumins, globulins, and fibrinogens constitute approximately 60%, 18% and 4% of the total serum plasma protein, respectively. Albumin serves as a transport protein for carrying large organic anions, such as fatty acids, bilirubin, drugs, and hormones, such as cortisol and thyroxine. Albumin concentrations may vary among individual subjects and are affected by dehydration, protein malnutrition, kidney and liver disease, etc. Non-specific binding of analytes to albumin and globulins or specific binding to particular plasma proteins may impact analyte recovery during the analytical sample preparation. High affinity, specific binding may be of particular concern during analysis of naturally occurring compounds, their analogues, and highly potent drugs. For example, analysis of vitamins D2, D3, and their metabolites requires disruption of protein binding and longer chromatography methods to avoid matrix effects (Casetta et al., 2010; Hollis, 2007; Vogeser and Seger, 2010).

In most cases, analyte binding to proteins can be disrupted by organic solvent, acid, inorganic salt, or metal ion treatment. In MS/MS analysis acids, salts, or metal ions may lead to undesired ion suppression on their own and therefore are rarely used. Organic solvent based PPT efficiently removes most of the protein and does not require post-precipitation processing of samples. In a study conducted by Polson et al., a volume ratio of 2.5:1 of precipitant to plasma was found to maximize protein removal while minimizing sample dilution. Under these conditions, ACN precipitation removed >97% of protein, while methanol removed ~94% of protein (Polson et al., 2003). LLE and SPE also rely on use of organic solvents to efficiently disrupt analyte-protein interactions and remove proteins from the samples; therefore, proteins rarely lead to matrix effects in LC-MS/MS analysis of small molecules. However, endogenous amino acids and peptides can interfere with protein, peptide and peptidomimetic analysis. Albumins and immunoglobulins along with other abundant plasma proteins lead to significant ion suppression in proteomics and interfere with the detection of less abundant proteins (Ahmed et al., 2003; Lo et al., 2009). Recently, significant efforts have been devoted to the development of LC-MS/MS-based biomarker assays for diagnostic purposes (Ahmed et al., 2003; Apweiler et al., 2009; Pusch et al., 2003). Affinity chromatography and sodium dodecyl sulfate-polyacrylamide gel electrophoresis (SDS-page) can be utilized to deplete plasma samples of highly abundant proteins and improve detection limits of biomarkers (Ahmed et al., 2003; Borg et al., 2011; Liu et al., 2009).

Other plasma matrix components such as salts, sugars, amino acids, lipids, and vitamins are not removed as efficiently by PPT, while LLE or SPE are better suited for this purpose as will be discussed in section 5.1. The highly polar components of plasma such as electrolytes, amino acids, glucose, and vitamins rarely interfere with analysis of non-polar organic compounds typically encountered in the drug development process, as they are easily separated by HPLC. However, they may lead to matrix effects when very rapid LC methods are used.

Lipids are considered the main culprit of matrix effects in blood and plasma samples. Plasma contains thousands of distinct lipids comprised of six main categories including fatty acyls, glycerolipids, glycerophospholipids, sphingolipids, sterols, and prenols (Figure 12) (Quehenberger et al., 2010). Phospholipids are present in plasma at extremely high concentrations, with glycerophosphocholines and lysophospholipids constituting up to 70% and 10% of the total plasma phospholipids, respectively. Phospholipids have a polar head group that contains an ionizable negatively charged phosphate group and a positively charged amine group that are responsible for strong ion suppression in both positive and negative ionization modes (Chambers et al., 2007). In addition, phospholipids contain one or two fatty acid esters that are responsible for the hydrophobicity of phospholipids. In general, ACN extracts contain significantly lower amounts of residual lipids than methanol or acetone extracts (Chambers et al., 2007; Ismaiel et al., 2010). LLE can provide cleaner samples, but lipid separation depends heavily on pH and organic solvents used for the extraction (Chambers et al., 2007; Ismaiel et al., 2010; Muller et al., 2002). Non-polar lipids such as triacylglycerols, cholesterol, and cholesterol esters have good solubility in hexane, chloroform, and ethers and will be extracted with these solvents from plasma. On the other hand, polar lipids, such as phospholipids, are more soluble in polar solvents, such as methanol and ethyl acetate. Therefore, it is important to determine which lipids lead to matrix effects and select solvents for LLE or SPE that will not extract these lipids. Since phospholipids are more likely to lead to pronounced matrix effects, the use of less polar solvents, such as methyl tert-butyl ether (MTBE) and MTBE-hexane mixtures for LLE can help to minimize matrix effects caused by phospholipids (Bennett and Liang, 2004; Ismaiel et al., 2010). Similar considerations apply to SPE sample clean-up, but derivatized solid-phase columns can provide efficient phospholipid removal (Aurand and Trinh, 2009; Chambers et al., 2007).

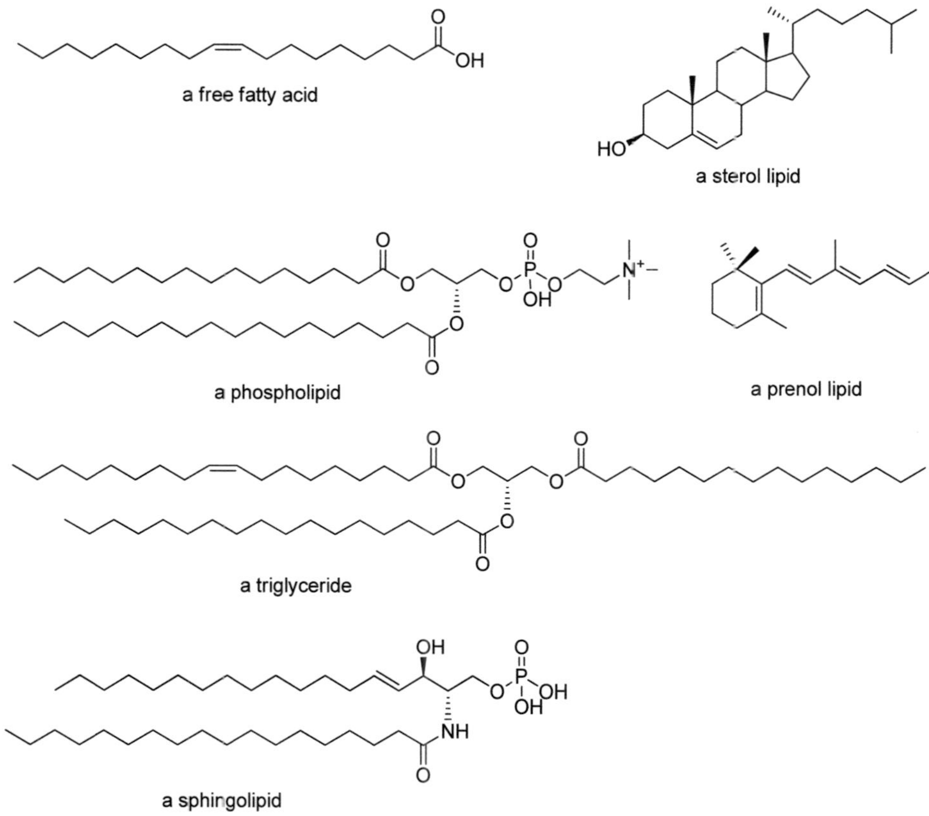

Fig. 12. Representative structures of endogenous lipid classes in human plasma (Fahy et al., 2005).

Chromatographic resolution of non-polar analytes from phospholipids is complicated by several aspects. Among them, elution of phospholipids from RP columns requires long run times and high organic solvent concentrations (60-90%) which may lead to co-elution with many non-polar drug-like compounds. Moreover, plasma contains numerous phospholipids that elute at various retention times collectively leading to broad ion suppression regions. Finally, lipid and phospholipid concentrations and composition in plasma can vary greatly between individuals leading to high variability in the observed matrix effects. When rapid gradients and short LC methods are used, phospholipids are retained on the column leading to accumulation and subsequent degradation of the column, as well as sporadic elution of lipids. Ultimately, a more thorough sample preparation/clean-up method may be required to remove these lipids.

4.2.2 Urine

In comparison to plasma, urine has much lower protein content and fewer matrix components. It primarily contains highly polar urea, inorganic salts, creatinine, and low quantities of carbohydrates, hormones, proteins, and metabolites of endogenous compounds (e.g. steroid metabolites, 2-oxoglutarate, succinate, citrate, dimethylglycine, trimethylaminoxide and taurine) and xenobiotics (e.g. xenobiotic glucuronides). Therefore, matrix effects are observed less frequently and simple sample dilution is sufficient to resolve observed matrix effects (Bell et al., 2011; Schreiber et al., 2007; Schreiber and Formal, 2007). However, it is important to note that, the composition and pH of urine is considerably more variable compared to plasma and is affected by age, gender, disease state, and dietary habits. Thus, pH adjustment of samples may be needed and calibration samples may not be representative of the matrix from individual samples.

5. Overcoming matrix effects

Strategies for overcoming issues associated with matrix effects will be reviewed in this section. Figure 13 and Table 4 summarize the various method modifications that can be applied to eliminate matrix effects.

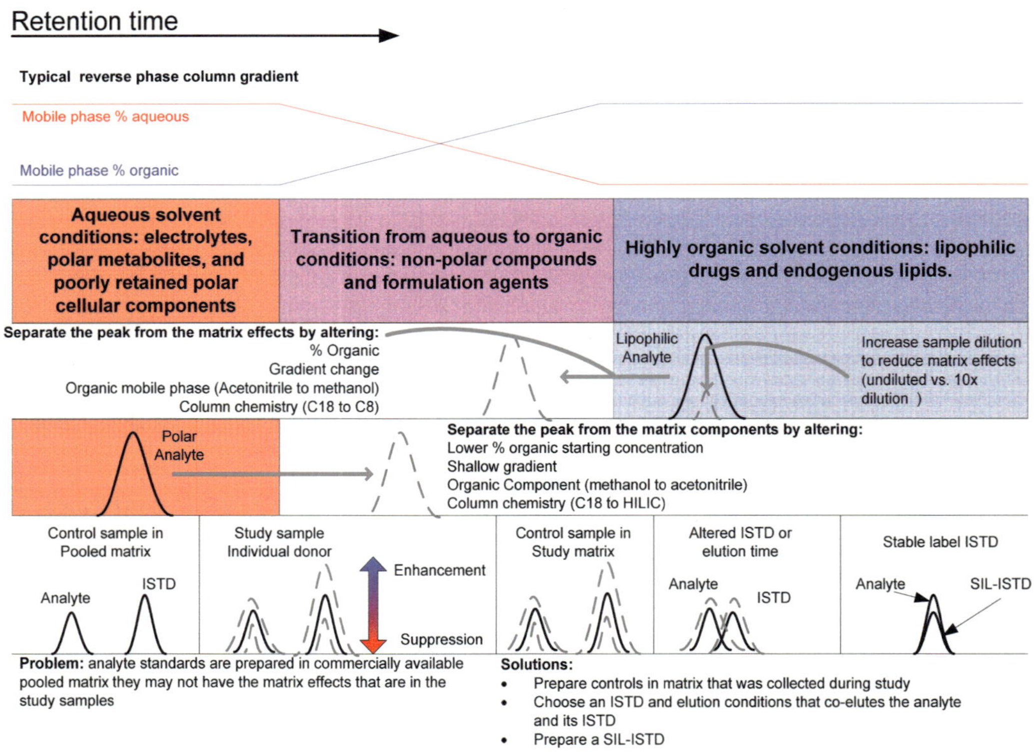

Fig. 13. Addressing matrix effects.

Source of Matrix Effect		Example	Impact	Potential solutions	References
Endogenous	Lipids	Lyso-Phosphotidylcholine 18:0	Suppression; late elution (if not fully eluted it can impact subsequent runs)	Adjust gradient	King et al., 2000
	Polar matrix components	K+, Ca++	Suppression, especially early in the run	HILIC column	Hsieh, 2008; Ji et al. 2008
Exogenous	Formulation components	Tween 80, PEG 400	Suppression, especially for early eluters	Increase mobile phase pH	Weaver & Riley, 2006
	Labware	Plasticizers	Suppression, later in run, if not fully eluted it can impact subsequent runs	Lengthen LC run; change labware supplier	Guo, 2006; Mei, 2003; van Hout, 2003
	In-source fragmentation	Glucuronide- and sulfate-conjugate metabolites	Enhancement if it co-elutes with parent	Modify source parameters (temperature or voltage); Additionally, changes can be made to the LC gradient or column temperature	Vogeser, 2001; Vogeser, 2010; Yan, 2003

Table 4. Examples of matrix effects, their impact, and suggested solutions.

5.1 Sample preparation

Protein precipitation (PPT) is a rapid, nonspecific method that can be utilized for sample clean-up in a high-throughput, automated manner. PPT-based purification relies on reduced solubility of proteins and highly polar matrix components in aqueous-organic solvent solutions. Acetonitrile, methanol or acetonitrile-methanol mixtures are most commonly used for PPT. More than 90% of plasma proteins can be removed from samples using PPT when the plasma to organic solvent ratio is at least 1 to 2.5 (Chambers et al., 2007; Polson et al., 2003). Unfortunately, many matrix components such as lipids, formulation agents, and other substances remain in the supernatant following centrifugation. These components often cause ion suppression, leading to increased variability between samples. Alternative sample preparation and extraction techniques include SPE and LLE. However, these extraction methods are much more labor intensive and therefore they are not typically used in early drug discovery. A comprehensive discussion of LLE and SPE is beyond the scope of this manuscript, and only a brief description of each will follow.

Liquid-liquid extraction (LLE) is based on the partitioning of an analyte into two separate liquids. The technique works by taking advantage of the differential solubility of an analyte in two immiscible liquids. One of the phases usually is water or a buffer solution, while the other is an organic solvent such as toluene, diethyl ether, hexane, dichloromethane, or MTBE. Selection of the proper organic solvent to obtain maximum recovery should be based on the analyte's solubility in the particular solvent. Chambers et al. compared LLE methods to several SPE and PPT methods in terms of each technique's overall cleanliness, matrix effects, and analyte recovery. They found that using MTBE and basified MTBE with a single extraction technique resulted in clean extracts that were similar to those obtained from cation exchange SPE and were better than PPT. However, analyte recovery using these two LLE methods significantly decreased compared to cation exchange SPE with basified methanol - average % recovery values for MTBE LLE, basified MTBE LLE, and cation exchange SPE with basified methanol were 43%, 38%, and 94%, respectively. Yet, when they used basified MTBE with a two step extraction procedure, both the cleanliness and analyte recovery (average of 87%) increased (Chambers et al., 2007). While LLE provides clean extracts and decent analyte recovery, it is much more labor intensive than PPT. Furthermore, as evidenced by the Chambers' investigation, multiple extractions may be necessary to obtain a sufficient quantity of analyte, decreasing overall efficiency that is essential for high-throughput sample clean-up and analysis in early drug discovery.

Solid phase extraction (SPE) methods rely on the affinity of an analyte for a stationary phase and are often used to isolate analyte(s) of interest from a wide range of matrices including urine, blood, tissue homogenates, etc. Depending on the properties of the analyte and the solid phase, either the analyte of interest is retained while the unwanted matrix components elute with the solvent wash. Or the unwanted matrix components are retained and the analyte elutes with the solvent wash. In the first case, the retained analyte is subsequently eluted with a different solvent. There are numerous SPE stationary phases available, including normal phase, reversed phase, and ion exchange (Chambers et al., 2007; Supelco, 1998). In addition, more specialized solid supports such as HILIC, mixed-mode resins, and zirconium coated particles for phospholipid removal are also commercially available. Table 5 shows typical analyte, matrix, and stationary phase/sorbent examples. The stationary phase and eluent can be adjusted to achieve the optimal sample clean-up and analyte recovery. For example, extraction of primary, secondary, and tertiary amines from biological fluids would be best accomplished by using strong cation exchange (SCX stationary phase); whereas extraction of large, hydrophobic molecules from biological matrices or water should be performed via reversed phase SPE with a C18-T (wide pore) stationary phase (Phenomenex, 2009). Optimal SPE conditions depend upon physicochemical properties of analytes and matrix components in the samples and require extensive method development. Therefore, SPE is less useful for a high-throughput analysis of a diverse set of compounds encountered in the early stages of the drug discovery but is widely used for clinical sample analysis. A more thorough discussion on LLE and SPE can be found elsewhere (Wilson et al., 2000).

SPE Phase	Analyte Properties	Eluting Solvent Properties	Stationary Phase / Sorbent
Normal Phase	Moderate to highly polar	Mid- to non-polar organic solvent (hexane, toluene, chloroform)	Polar (CN, NH_2, EPH, Silica, Alumina-N)
Reversed Phase	Low to moderately polar, hydrophobic	Non-polar or polar organic solvents with or without water, buffer, and/or strong acid or base	Non-polar (C18, C8, SDB-L, Phenyl, CN)
Ion Exchange	Ionized/charge compounds	Aqueous & low ionic strength buffers (biological fluids plus buffers)	Polar (cation exchange: WCX, SCX, Screen-C; anion exchange: WAX/NH_2, SAX, Screen-A)

Table 5. Analyte, matrix, and stationary phase properties associated with three commonly used SPE methods (Phenomenex, 2009; Supelco, 1998).

Relatively quick and simple approaches to overcome ion suppression are available and include reducing the volume of the sample injected and/or diluting the samples prior to injection. Sample dilution and reducing the injection volume simply decreases the amount of interfering matrix components introduced into the ion source (Choi et al., 2001; Taylor, 2005). Schuhmacher et al. has demonstrated that dilution of samples with mobile phase results in reduced ion suppression. In their study, dilution from 10- to 100-fold allowed investigators to identify that the matrix effects were caused by the vehicles (PEG 400 and Solutol) used for dosing (Schuhmacher et al., 2003).

Some commonly encountered matrix effects are caused by formulation agents. The easiest approach for reducing these matrix effects is to avoid formulation agents that are known to lead to significant ionization suppression (e.g. Tween 80, PEG 400, Solutol). Alternatively, modifying chromatographic conditions and/or implementing SPE can be used to overcome formulation agent-caused matrix effects. One rather unique approach for eliminating vehicle-related matrix effects was reported by Schuhmacher et al. in which drugs for iv studies were formulated in the respective species' plasma (Schuhmacher et al., 2003).

Column overload is caused by accumulation of matrix components which are not adequately removed during sample clean-up and is a common occurrence when a column has been used extensively and these components are not sufficiently eluted by the HPLC method. While reducing the injection volume may fix the issue for a short period of time, a better long term solution is to add a highly (95-100%) organic wash step after analyte elution. In extreme cases where washing does not resolve the problem, the LC column may need to be replaced. Figure 14 shows chromatograms before and after column replacement. The top two chromatograms show peak tailing caused by column overload. After the LC column was replaced, the bottom two chromatograms showed much sharper analyte and ISTD peaks. Additionally, column replacement increased the analyte signal for both the compound of interest and the ISTD.

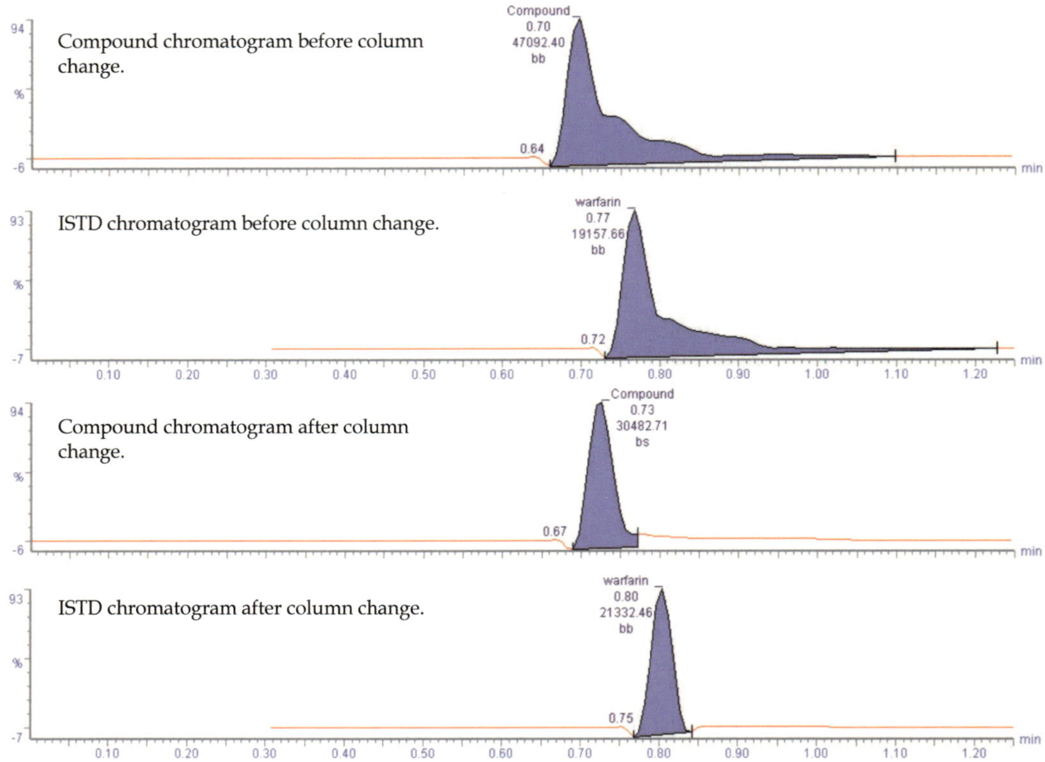

Fig. 14. Chromatograms illustrating an example of column matrix overload, and the subsequent column replacement to overcome matrix effects.

5.2 Chromatographic conditions

Another widely used strategy for the reduction of matrix effects is the optimization of chromatographic conditions. In most cases, ion-suppression is caused by the co-elution of the matrix components with the analyte of interest. Therefore, with increased chromatographic separation between the analyte(s) and the matrix components, fewer matrix effects are likely to be encountered.

Optimizing a gradient method involves modification of chromatographic parameters, such as initial and final eluent strength and gradient duration, all of which may prove advantageous for chromatographic resolution of matrix components away from the analyte (Schellinger and Carr, 2006). When Gallart-Ayala et al. developed the analytical method for determining bisphenols in soft drinks using LC-ESI-MS/MS, they observed that the responses of the analytes were 80-95% lower in the matrix than those obtained from a standard solution at the same concentration level. They then monitored the samples through both LC-UV and LC-MS/MS, and found that signal suppression was likely caused by the co-elution of the matrix components with the analytes. To reduce the ionization suppression, several strategies were successfully applied: 1) extension of the linear gradient duration from 1 min to 3 min; 2) lowering of the initial mobile phase organic content from 50% to 15% methanol; 3) adjustment of the final mobile phase composition from 100% to 80% methanol. The chromatographic peaks of the analytes were almost completely resolved from the matrix components and the matrix effects were greatly reduced (Gallart-Ayala et al., 2011). A shorter gradient duration normally leads to greater matrix effects due to the reduction of chromatographic resolution between analytes and matrix components (Chambers et al., 2007). When matrix effects are caused by co-elution of matrix components and the analyte of interest, changing to another column type or manufacturer, increasing the column length and/or extending the gradient duration will typically resolve the issue. Alternatively, fine tuning the mobile phase solvent strength can prove to be effective to achieve chromatographic separation. Ye et al. used a mixture of methanol and ACN as the organic mobile phase, to minimize phospholipid-related matrix effects and maintain high sample throughput (Ye et al., 2011).

Recently, an ultra-fast gradient approach also referred to as a "ballistic" gradient, allowed for analysis cycle times of 2 minutes or less (Romanyshyn and Tiller, 2001; Romanyshyn et al., 2000). Tiller and Romanyshyn studied matrix effects encountered using ultra-fast gradients and fast isocratic methods. They demonstrated that while both methods are subject to similar matrix effects, gradient elution had several advantages in reducing matrix effects. Gradient elution gave better separation of the analyte from the poorly retained polar components, and a more complete elution of highly retained contaminants from the column, thereby reducing the potential matrix effects during sample analysis. Moreover, the early and late chromatographic regions can be diverted to waste, thus reducing matrix components that can foul the MS source (Tiller and Romanyshyn, 2002).

Ultra Performance Liquid Chromatography (UPLC) provides advantages over traditional HPLC in speed, sensitivity, and resolution of analytes (Novakova et al., 2006; Wren and Tchelitcheff, 2006). When analyzing nine different drugs in three surface water samples, Van De Steene and Lambert observed severe matrix effects using HPLC. After implementing analogue ISTDs, significant matrix effects were still observed. However, through UPLC implementation, these matrix effects were significantly reduced, and accurate quantitation of all nine compounds using analogue ISTDs became feasible (Van De Steene and Lambert, 2008). With improved sensitivity and resolution, UPLC generally encounters fewer matrix effects and affords a more robust analytical method than HPLC.

Mobile phase pH can influence the retention times of ionizable analytes (basic or acidic) by changing the ionization equilibrium. Under HPLC conditions with an acidified mobile phase, basic compounds are present as charged species. As a result, they are poorly retained on the column and elute early with a highly aqueous mobile phase (Chambers et al., 2007). Under acidic conditions, basic compounds may encounter matrix effects from salts and highly polar, poorly retained matrix components. Conversely, at basic pH, basic compounds stay neutral, are better retained, elute with high organic content mobile phase, and generate stronger MS signals. In a study of matrix effects caused by PEG 400, the basic analyte co-eluted directly with PEG 400 under acidic mobile phase conditions. To overcome this, the mobile phase pH was increased to 10.5, and improved separation between the analyte and PEG 400 was achieved. This resulted in greatly reduced matrix effects and allowed for more accurate quantification of the analyte (Weaver and Riley, 2006).

Analytical columns of the same stationary phase (for example C18) from different manufacturers are generally not equivalent and will often result in slightly different separation of sample components. Hence, the degree of a matrix effect may differ when columns from different manufacturers are used. For example, in developing a ballistic gradient, De Nardi & Bonelli compared matrix effects after using an ACE C18-300 30x2.1mm (5μ) and a POLARITY dC18 30x2.1mm (5μ) analytical column. The analyte peaks had no overlap with the PEG 400 peaks when using the ACE column, while partial overlap occurred when using the POLARITY column (De Nardi and Bonelli, 2006).

A relatively new column, the hydrophilic interaction LC (HILIC) column, consists of bare silica or a polar phase (amino, diol, cyano, etc.) bonded to silica and provides more favorable conditions for polar compound retention and ionization (Hsieh, 2008) (Ji et al., 2008). With HILIC chromatography, polar analytes have prolonged retention times and elute at high organic mobile phase content; therefore, they are separated from ion suppressing-early eluters and have high ionization efficiencies. Ji et al. developed a quantitation method for doxazosin using HILIC-MS/MS with a mobile phase consisting of ACN/ammonium formate (100 mM, pH 4.5) (93:7 v/v). The method was free of matrix effects assessed by post extraction analyte spiking (Ji et al., 2008). However, it is important to note that under HILIC conditions, certain polar endogenous components could be strongly retained on a HILIC column, potentially causing matrix effects that are not observed in RP conditions (Jian et al., 2010). Although water is usually used as the strong solvent in HILIC, by switching from water to a weaker solvent (methanol, ethanol, propanol etc.), HILIC columns can be used to effectively separate less polar compounds (Jian et al., 2010; Xu et al., 2007).

As discussed earlier, cleaner samples are obtained from SPE or multiple LLE. However, with the ever-increasing pace in drug discovery study, minimal sample pretreatment and automation are desired. Consequently, on-line 2-D chromatography with valve switching technology has seen more and more use lately (Mullett, 2007; Pascoe et al., 2001; Pol and Hyotylainen, 2008). Two-dimensional chromatography involves the transfer of a fraction or fractions of sample from a primary column to a secondary column for further separation. In orthogonal 2-D chromatography two columns with different stationary phases are utilized to increase the separation between the analyte and the matrix components. Dodgen et al. developed a simple method for simultaneous quantitation of dextromethorphan, omeprazole, and their metabolites in plasma for the phenotypic profiling of cytochrome P450 2D6 and 2C19. In this study, the supernatant generated from PPT was loaded directly to a SPE capture column for purification. After the initial unretained matrix components eluted, a switching valve directed the trapped analytes to the analytical column. The primary SPE column removed almost all ion suppression/enhancement effects caused by the matrix components. The total run time was 6 min (Dodgen et al., 2011).

5.3 Calibration with internal standards

Perhaps the most common method used to compensate for matrix effects involves the appropriate choice of ISTDs. When performing PPT, the ISTD is typically included with the organic solvent that is used to precipitate the protein. It is added to the precipitating solvent so that there is a uniform concentration throughout the samples to be analyzed. For an ISTD to effectively compensate for matrix effects it should have a retention time similar to the analyte of interest. The analyte response can be normalized to the ISTD peak simply by dividing the analyte peak area by the ISTD peak area. Since the ISTD concentration is equal across all of the samples, comparing the analyte's peak area to the ISTD's peak area serves to normalize the data and compensate for signal response variability caused by matrix effects.

Ideally, a stable-isotope-labeled analyte analog can be used as an ISTD. Since this SIL-ISTD would have nearly identical chemical and structural properties as the analyte of interest, the two compounds should behave similarly during sample preparation and LC-MS analysis

(Jessome and Volmer, 2006). However, availability and cost concerns can severely limit the extent to which SIL-ISTDs are employed (Schuhmacher et al., 2003; Xu et al., 2007).

6. Regulatory perspectives

A complete discussion regarding the regulatory perspective on matrix effects is beyond the scope of this review and has been discussed in detail elsewhere (Viswanathan et al., 2007). In short, analysis of samples from GLP and clinical studies require a validated LC-MS/MS method. The Food and Drug Administration (FDA) has provided a guidance document for validating methods for the analysis of bioanalytical samples (FDA, 2001). This guidance provides details on the parameters required for a complete validation including method linearity, precision, accuracy, sample stability, evaluation of matrix effects, etc. The FDA recently required incurred sample analysis in which a subset of samples (~10%) from clinical and GLP studies are reanalyzed. The reanalysis results are compared to the original results, and that at least two-thirds of the reanalysis batch results must be within 20% of the original sample values. Failure to meet this criterion will initiate an investigation to determine the source of the discrepancy in the data.

Although the parameters described above almost always lead to a robust, validated method, there are certain situations where matrix effects may persist. In such cases, sample reanalysis generally leads to widely different results from what was originally reported. These cases are frequently associated with unknown matrix effects, which are often unique to a particular patient's sample(s). As mentioned previously, clinical trial samples are distinctive in their composition, thus a universal LC-MS/MS method will not always remove or compensate for the components that may interfere with analyte quantitation. Further, method validation across different patients' samples is impractical due to this inherent uniqueness. These situations, when they occur, are handled on a case by case basis. Typically, an investigation into the discrepancy is conducted and documented and the reporting of the data is specified by Standard Operating Procedures (SOPs) that are written to handle such situations.

7. Conclusion

Matrix effects can be a complicated and time consuming challenge for the analytical chemist charged with developing robust, reproducible, and accurate analytical methods. There are a diversity of sample and system conditions that lead to matrix effects, and an equally diverse set of potential options to remedy them.

The mechanisms by which matrix components cause ionization suppression (or enhancement) are still not well-understood. This serves as a testament to the challenge they provide to the analytical chemist. In the years since the first published articles describing matrix effects, analytical chemists have recognized the importance of understanding and mitigating matrix effects.

8. References

Annesley, T.M. (2007). Methanol-associated matrix effects in electrospray ionization tandem mass spectrometry. Clin Chem 53, 1827-1834.

Apweiler, R., Aslanidis, C., Deufel, T., Gerstner, A., Hansen, J., Hochstrasser, D., Kellner, R., Kubicek, M., Lottspeich, F., Maser, E., *et al.* (2009). Approaching clinical proteomics:

current state and future fields of application in fluid proteomics. Clin Chem Lab Med *47*, 724-744.

Aurand, C., and Trinh, A. (2009). Selective Depletion of Phosphoslipids in Bioanalysis Using HybridSPE-PPT Technology. Chromatography Today, 38-41.

Bell, C., George, C., Kicman, A.T., and Traynor, A. (2011). Development of a rapid LC-MS/MS method for direct urinalysis of designer drugs. Drug Test Anal *3*, 496-504.

Benijts, T., Dams, R., Lambert, W., and De Leenheer, A. (2004). Countering matrix effects in environmental liquid chromatography-electrospray ionization tandem mass spectrometry water analysis for endocrine disrupting chemicals. J Chromatogr A *1029*, 153-159.

Bennett, P., and Liang, H. (2004). Overcoming Matrix Effects Resulting from Biological Phosphoslipids Through Selective Extraction in Quantitative LC/MS/MS. In ASMS Conference (Nashville, TN, Tandem Labs).

Bonfiglio, R., King, R.C., Olah, T.V., and Merkle, K. (1999). The effects of sample preparation methods on the variability of the electrospray ionization response for model drug compounds. Rapid Commun Mass Spectrom *13*, 1175-1185.

Borg, J., Campos, A., Diema, C., Omenaca, N., de Oliveira, E., Guinovart, J., and Vilaseca, M. (2011). Spectral counting assessment of protein dynamic range in cerebrospinal fluid following depletion with plasma-designed immunoaffinity columns. Clin Proteomics *8*, 6.

Casetta, B., Jans, I., Billen, J., Vanderschueren, D., and Bouillon, R. (2010). Development of a method for the quantification of 1alpha,25(OH)2-vitamin D3 in serum by liquid chromatography tandem mass spectrometry without derivatization. Eur J Mass Spectrom (Chichester, Eng) *16*, 81-89.

Chambers, E., Wagrowski-Diehl, D.M., Lu, Z., and Mazzeo, J.R. (2007). Systematic and comprehensive strategy for reducing matrix effects in LC/MS/MS analyses. J Chromatogr B Analyt Technol Biomed Life Sci *852*, 22-34.

Chavez-Eng, C.M., Constanzer, M.L., and Matuszewski, B.K. (2002). High-performance liquid chromatographic-tandem mass spectrometric evaluation and determination of stable isotope labeled analogs of rofecoxib in human plasma samples from oral bioavailability studies. J Chromatogr B Analyt Technol Biomed Life Sci *767*, 117-129.

Chiu, M.L., Lawi, W., Snyder, S.T., Wong, P.K., Liao, J.C., and Gau, V. (2010). Matrix Effects - A Challenge Toward Automation of Molecular Analysis. Journal of the Association for Laboratory Automation *15*, 233-242.

Choi, B.K., Hercules, D.M., and Gusev, A.I. (2001). Effect of liquid chromatography separation of complex matrices on liquid chromatography-tandem mass spectrometry signal suppression. J Chromatogr A *907*, 337-342.

Dams, R., Benijts, T., Gunther, W., Lambert, W., and De Leenheer, A. (2002). Influence of the eluent composition on the ionization efficiency for morphine of pneumatically assisted electrospray, atmospheric-pressure chemical ionization and sonic spray. Rapid Commun Mass Spectrom *16*, 1072-1077.

Dancygier, H. (2010). Clinical Hepatology: Principles and Practice of Hepatobiliary Diseases, Vol 1 (New York, NY, Springer-Verlag).

De Nardi, C., and Bonelli, F. (2006). Moving from fast to ballistic gradient in liquid chromatography/tandem mass spectrometry pharmaceutical bioanalysis: Matrix effect and chromatographic evaluations. Rapid Commun Mass Spectrom *20*, 2709-2716.

Dodgen, T.M., Cromarty, A.D., and Pepper, M.S. (2011). Quantitative plasma analysis using automated online solid-phase extraction with column switching LC-MS/MS for characterising cytochrome P450 2D6 and 2C19 metabolism. J Sep Sci *34*, 1102-1110.

Fahy, E., Subramaniam, S., Brown, H.A., Glass, C.K., Merrill, A.H., Jr., Murphy, R.C., Raetz, C.R., Russell, D.W., Seyama, Y., Shaw, W., *et al.* (2005). A comprehensive classification system for lipids. J Lipid Res *46*, 839-861.

FDA (2001). Guidance for Industry: Bioanalytical Method Validation (Rockville, MD, Department of Health and Human Services, FDA, Center for Drug Evaluation and Research).

Gallart-Ayala, H., Moyano, E., and Galceran, M.T. (2011). Analysis of bisphenols in soft drinks by on-line solid phase extraction fast liquid chromatography-tandem mass spectrometry. Anal Chim Acta *683*, 227-233.

Gosetti, F., Mazzucco, E., Zampieri, D., and Gennaro, M.C. (2010). Signal suppression/enhancement in high-performance liquid chromatography tandem mass spectrometry. J Chromatogr A *1217*, 3929-3937.

Guo, X., Bruins, A.P., and Covey, T.R. (2006). Characterization of typical chemical background interferences in atmospheric pressure ionization liquid chromatography-mass spectrometry. Rapid Commun Mass Spectrom *20*, 3145-3150.

Gustavsson, S.A., Samskog, J., Markides, K.E., and Langstrom, B. (2001). Studies of signal suppression in liquid chromatography-electrospray ionization mass spectrometry using volatile ion-pairing reagents. J Chromatogr A *937*, 41-47.

Hewitt, D., Alvarez, M., Robinson, K., Ji, J., Wang, Y.J., Kao, Y.H., and Zhang, T. (2011). Mixed-mode and reversed-phase liquid chromatography-tandem mass spectrometry methodologies to study composition and base hydrolysis of polysorbate 20 and 80. J Chromatogr A *1218*, 2138-2145.

Hollis, B.W. (2007). Assessment of circulating 25(OH)D and 1,25(OH)2D: emergence as clinically important diagnostic tools. Nutr Rev *65*, S87-90.

Hsieh, Y. (2008). Potential of HILIC-MS in quantitative bioanalysis of drugs and drug metabolites. J Sep Sci *31*, 1481-1491.

Ismaiel, O.A., Zhang, T., Jenkins, R.G., and Karnes, H.T. (2010). Investigation of endogenous blood plasma phospholipids, cholesterol and glycerides that contribute to matrix effects in bioanalysis by liquid chromatography/mass spectrometry. J Chromatogr B Analyt Technol Biomed Life Sci *878*, 3303-3316.

Jemal, M., Schuster, A., and Whigan, D.B. (2003). Liquid chromatography/tandem mass spectrometry methods for quantitation of mevalonic acid in human plasma and urine: method validation, demonstration of using a surrogate analyte, and demonstration of unacceptable matrix effect in spite of use of a stable isotope analog internal standard. Rapid Commun Mass Spectrom *17*, 1723-1734.

Jessome, L.L., and Volmer, D.A. (2006). Ion Suppression: A Major Concern in Mass Spectrometry. LCGC North America *24*, 498-510.

Ji, H.Y., Park, E.J., Lee, K.C., and Lee, H.S. (2008). Quantification of doxazosin in human plasma using hydrophilic interaction liquid chromatography with tandem mass spectrometry. J Sep Sci *31*, 1628-1633.

Jian, W., Edom, R.W., Xu, Y., and Weng, N. (2010). Recent advances in application of hydrophilic interaction chromatography for quantitative bioanalysis. J Sep Sci *33*, 681-697.

King, R., Bonfiglio, R., Fernandez-Metzler, C., Miller-Stein, C., and Olah, T. (2000). Mechanistic investigation of ionization suppression in electrospray ionization. J Am Soc Mass Spectrom *11*, 942-950.

Kostiainen, R., and Kauppila, T.J. (2009). Effect of eluent on the ionization process in liquid chromatography-mass spectrometry. J Chromatogr A *1216*, 685-699.

Larger, P.J., Breda, M., Fraier, D., Hughes, H., and James, C.A. (2005). Ion-suppression effects in liquid chromatography-tandem mass spectrometry due to a formulation agent, a case study in drug discovery bioanalysis. J Pharm Biomed Anal *39*, 206-216.

Liang, H.R., Foltz, R.L., Meng, M., and Bennett, P. (2003). Ionization enhancement in atmospheric pressure chemical ionization and suppression in electrospray ionization between target drugs and stable-isotope-labeled internal standards in quantitative liquid chromatography/tandem mass spectrometry. Rapid Commun Mass Spectrom *17*, 2815-2821.

Lindegardh, N., Annerberg, A., White, N.J., and Day, N.P. (2008). Development and validation of a liquid chromatographic-tandem mass spectrometric method for determination of piperaquine in plasma stable isotope labeled internal standard does not always compensate for matrix effects. J Chromatogr B Analyt Technol Biomed Life Sci *862*, 227-236.

Little, J.L., Wempe, M.F., and Buchanan, C.M. (2006). Liquid chromatography-mass spectrometry/mass spectrometry method development for drug metabolism studies: Examining lipid matrix ionization effects in plasma. J Chromatogr B Analyt Technol Biomed Life Sci *833*, 219-230.

Liu, A., Tweed, J., and Wujcik, C.E. (2009). Investigation of an on-line two-dimensional chromatographic approach for peptide analysis in plasma by LC-MS-MS. J Chromatogr B Analyt Technol Biomed Life Sci *877*, 1873-1881.

Mallet, C.R., Lu, Z., and Mazzeo, J.R. (2004). A study of ion suppression effects in electrospray ionization from mobile phase additives and solid-phase extracts. Rapid Commun Mass Spectrom *18*, 49-58.

Marchi, I., Viette, V., Badoud, F., Fathi, M., Saugy, M., Rudaz, S., and Veuthey, J.L. (2010). Characterization and classification of matrix effects in biological samples analyses. J Chromatogr A *1217*, 4071-4078.

Matuszewski, B.K. (2006). Standard line slopes as a measure of a relative matrix effect in quantitative HPLC-MS bioanalysis. J Chromatogr B Analyt Technol Biomed Life Sci *830*, 293-300.

Matuszewski, B.K., Constanzer, M.L., and Chavez-Eng, C.M. (2003). Strategies for the assessment of matrix effect in quantitative bioanalytical methods based on HPLC-MS/MS. Anal Chem *75*, 3019-3030.

Mei, H. (2005). Matrix Effects: Causes and Solutions. In Using Mass Spectrometry For Drug Metabolism Studies, W.A. Korfmacher, ed. (CRC Press), pp. 103-149.

Mei, H., Hsieh, Y., Nardo, C., Xu, X., Wang, S., Ng, K., and Korfmacher, W.A. (2003). Investigation of matrix effects in bioanalytical high-performance liquid chromatography/tandem mass spectrometric assays: application to drug discovery. Rapid Commun Mass Spectrom *17*, 97-103.

Morin, L.P., Mess, J.N., Furtado, M., and Garofolo, F. (2011). Reliable procedures to evaluate and repair crosstalk for bioanalytical MS/MS assays. Bioanalysis *3*, 275-283.

Muller, C., Schafer, P., Stortzel, M., Vogt, S., and Weinmann, W. (2002). Ion suppression effects in liquid chromatography-electrospray-ionisation transport-region collision induced dissociation mass spectrometry with different serum extraction methods for systematic toxicological analysis with mass spectra libraries. J Chromatogr B Analyt Technol Biomed Life Sci *773*, 47-52.

Mullett, W.M. (2007). Determination of drugs in biological fluids by direct injection of samples for liquid-chromatographic analysis. J Biochem Biophys Methods *70*, 263-273.

Novakova, L., Solichova, D., and Solich, P. (2006). Advantages of ultra performance liquid chromatography over high-performance liquid chromatography: comparison of different analytical approaches during analysis of diclofenac gel. J Sep Sci *29*, 2433-2443.

Pascoe, R., Foley, J.P., and Gusev, A.I. (2001). Reduction in matrix-related signal suppression effects in electrospray ionization mass spectrometry using on-line two-dimensional liquid chromatography. Anal Chem *73*, 6014-6023.

Phenomenex, I. (2009). Chromatography Product Guide 09/10. In HPLC, GC, SPE (Torrance, CA, Phenomenex, Inc.).

Pol, J., and Hyotylainen, T. (2008). Comprehensive two-dimensional liquid chromatography coupled with mass spectrometry. Anal Bioanal Chem *391*, 21-31.

Polson, C., Sarkar, P., Incledon, B., Raguvaran, V., and Grant, R. (2003). Optimization of protein precipitation based upon effectiveness of protein removal and ionization effect in liquid chromatography-tandem mass spectrometry. J Chromatogr B Analyt Technol Biomed Life Sci *785*, 263-275.

Pusch, W., Flocco, M.T., Leung, S.M., Thiele, H., and Kostrzewa, M. (2003). Mass spectrometry-based clinical proteomics. Pharmacogenomics *4*, 463-476.

Quehenberger, O., Armando, A.M., Brown, A.H., Milne, S.B., Myers, D.S., Merrill, A.H., Bandyopadhyay, S., Jones, K.N., Kelly, S., Shaner, R.L., *et al.* (2010). Lipidomics reveals a remarkable diversity of lipids in human plasma. J Lipid Res *51*, 3299-3305.

Remane, D., Meyer, M.R., Wissenbach, D.K., and Maurer, H.H. (2010a). Ion suppression and enhancement effects of co-eluting analytes in multi-analyte approaches: systematic investigation using ultra-high-performance liquid chromatography/mass spectrometry with atmospheric-pressure chemical ionization or electrospray ionization. Rapid Commun Mass Spectrom *24*, 3103-3108.

Remane, D., Wissenbach, D.K., Meyer, M.R., and Maurer, H.H. (2010b). Systematic investigation of ion suppression and enhancement effects of fourteen stable-isotope-labeled internal standards by their native analogues using atmospheric-pressure chemical ionization and electrospray ionization and the relevance for multi-analyte liquid chromatographic/mass spectrometric procedures. Rapid Commun Mass Spectrom *24*, 859-867.

Romanyshyn, L., Tiller, P.R., and Hop, C.E. (2000). Bioanalytical applications of 'fast chromatography' to high-throughput liquid chromatography/tandem mass spectrometric quantitation. Rapid Commun Mass Spectrom *14*, 1662-1668.

Romanyshyn, L.A., and Tiller, P.R. (2001). Ultra-short columns and ballistic gradients: considerations for ultra-fast chromatographic liquid chromatographic-tandem mass spectrometric analysis. J Chromatogr A *928*, 41-51.

Schellinger, A.P., and Carr, P.W. (2006). Isocratic and gradient elution chromatography: a comparison in terms of speed, retention reproducibility and quantitation. J Chromatogr A *1109*, 253-266.

Schuhmacher, J., Zimmer, D., Tesche, F., and Pickard, V. (2003). Matrix effects during analysis of plasma samples by electrospray and atmospheric pressure chemical ionization mass spectrometry: practical approaches to their elimination. Rapid Commun Mass Spectrom *17*, 1950-1957.

Shou, W.Z., and Naidong, W. (2003). Post-column infusion study of the 'dosing vehicle effect' in the liquid chromatography/tandem mass spectrometric analysis of discovery pharmacokinetic samples. Rapid Commun Mass Spectrom *17*, 589-597.

Supelco (1998). Guide to Solid Phase Extraction. In Bulletin 910 (Bellefonte, PA, Supelco).

Taylor, P.J. (2005). Matrix effects: the Achilles heel of quantitative high-performance liquid chromatography-electrospray-tandem mass spectrometry. Clin Biochem *38*, 328-334.

Tiller, P.R., and Romanyshyn, L.A. (2002). Implications of matrix effects in ultra-fast gradient or fast isocratic liquid chromatography with mass spectrometry in drug discovery. Rapid Commun Mass Spectrom *16*, 92-98.

Trufelli, H., Palma, P., Famiglini, G., and Cappiello, A. (2011). An overview of matrix effects in liquid chromatography-mass spectrometry. Mass Spectrom Rev *30*, 491-509.

Van De Steene, J.C., and Lambert, W.E. (2008). Comparison of matrix effects in HPLC-MS/MS and UPLC-MS/MS analysis of nine basic pharmaceuticals in surface waters. J Am Soc Mass Spectrom *19*, 713-718.

van Hout, M.W., Niederlander, H.A., de Zeeuw, R.A., and de Jong, G.J. (2003). Ion suppression in the determination of clenbuterol in urine by solid-phase extraction atmospheric pressure chemical ionisation ion-trap mass spectrometry. Rapid Commun Mass Spectrom *17*, 245-250.

van Tellingen, O., Beijnen, J.H., Verweij, J., Scherrenburg, E.J., Nooijen, W.J., and Sparreboom, A. (1999). Rapid esterase-sensitive breakdown of polysorbate 80 and its impact on the plasma pharmacokinetics of docetaxel and metabolites in mice. Clin Cancer Res *5*, 2918-2924.

Viswanathan, C.T., Bansal, S., Booth, B., DeStefano, A.J., Rose, M.J., Sailstad, J., Shah, V.P., Skelly, J.P., Swann, P.G., and Weiner, R. (2007). Quantitative bioanalytical methods validation and implementation: best practices for chromatographic and ligand binding assays. Pharm Res *24*, 1962-1973.

Vogeser, M., and Seger, C. (2010). Pitfalls associated with the use of liquid chromatography-tandem mass spectrometry in the clinical laboratory. Clin Chem *56*, 1234-1244.

Vogeser, M., Zachoval, R., Spohrer, U., and Jacob, K. (2001). Potential lack of specificity using electrospray tandem-mass spectrometry for the analysis of mycophenolic acid in serum. Ther Drug Monit *23*, 722-724.

Wang, S., Cyronak, M., and Yang, E. (2007). Does a stable isotopically labeled internal standard always correct analyte response? A matrix effect study on a LC/MS/MS method for the determination of carvedilol enantiomers in human plasma. J Pharm Biomed Anal *43*, 701-707.

Weaver, R., and Riley, R.J. (2006). Identification and reduction of ion suppression effects on pharmacokinetic parameters by polyethylene glycol 400. Rapid Commun Mass Spectrom *20*, 2559-2564.

Wilson, I.D., Adlard, E.R., Cooke, M., and Poole, C.F. (2000). Encyclopedia of Separation Science, I.D. Wilson, Adlard, E.R., Cooke, M., and Poole, C.F., ed. (San Diego, Academic Press).

Wren, S.A., and Tchelitcheff, P. (2006). Use of ultra-performance liquid chromatography in pharmaceutical development. J Chromatogr A *1119*, 140-146.

Xu, R.N., Fan, L., Rieser, M.J., and El-Shourbagy, T.A. (2007). Recent advances in high-throughput quantitative bioanalysis by LC-MS/MS. J Pharm Biomed Anal *44*, 342-355.

Xu, X., Mei, H., Wang, S., Zhou, Q., Wang, G., Broske, L., Pena, A., and Korfmacher, W.A. (2005). A study of common discovery dosing formulation components and their potential for causing time-dependent matrix effects in high-performance liquid chromatography tandem mass spectrometry assays. Rapid Commun Mass Spectrom *19*, 2643-2650.

Yan, Z., Caldwell, G.W., Jones, W.J., and Masucci, J.A. (2003). Cone voltage induced in-source dissociation of glucuronides in electrospray and implications in biological analyses. Rapid Commun Mass Spectrom *17*, 1433-1442.

Ye, Z., Tsao, H., Gao, H., and Brummel, C.L. (2011). Minimizing matrix effects while preserving throughput in LC-MS/MS bioanalysis. Bioanalysis *3*, 1587-1601.

Zheng, J.J., Lynch, E.D., and Unger, S.E. (2002). Comparison of SPE and fast LC to eliminate mass spectrometric matrix effects from microsomal incubation products. J Pharm Biomed Anal *28*, 279-285.

19

Evaluating PK/PD Relationship of CNS Drug by Using Liquid Chromtography/ Tandem Mass Spectrometry Coupled to *In Vivo* Microdialysis

Ying Qu[1*], Loren Olson[2], Xiaohui Jiang[1], Leah Aluisio[1], Christopher King[1], Elliott B. Jones[2] and Timothy W. Lovenberg[1]

[1]*Johnson & Johnson Pharmaceutical Research & Development, L. L. C., San Diego, California*

[2]*AB SCIEX, Foster City, California*

U.S.A.

1. Introduction

Pharmacokinetic (PK) characterization and *in vivo* pharmacological properties of new chemical entities are important components during lead compound selection and optimization in the drug discovery process. Accordingly, reliable techniques are needed that can generate the requisite pharmacokinetic/pharmacodynamic (PK/PD) information for an increased number of compounds. When dealing with compounds targeting the central nervous system (CNS), biophase PK may differ significantly from plasma PK, because blood-brain barrier (BBB) transport and brain distribution often do not occur instantaneously and to a full extent. Therefore the brain distributional behaviors are important determinants of *in vivo* drug effects. *In vivo* microdialysis technique has been used increasingly over the past years for *in vivo* sampling of extracellular exogenous compounds in brains of freely moving rats (1-3). The extracellular brain concentration of free drug which measured by microdialysis may reflect the amount of drug available at the pharmacological target. As the microdialysis technique also allows the simultaneous determination of different endogenous substances, such as neurotransmitters, which sometimes represent the efficacy biomarker of a CNS drug, in the same local interstitial environment. Microdialysis is an attractive tool for PK/PD investigations of CNS active drugs (4). However, the extracellular concentration of neurotransmitters in the synaptic cleft is very low which can be in the range of fg/µl; the sampling recovery of microdialysis probe is less than 20%. The application of this technique was highly limited by lack the proper sensitive analytical methods to determine the endogenous substance and exogenous drug.

For analysis of endogenous neurotransmitters, such as, serotonin (5-HT), dopamine (DA), noradrenaline (NE), acetylcholine (ACh), histamine, γ-aminobutyric acid (GABA) and glutamate (Glu), microdialysis in combination with high performance liquid chromatography (HPLC) - electrochemical detection (ED) or fluorescence detection has been

---

* Corresponding Author

widely used in last two decades (5-6). 5-HT, DA and NE are electrochemically active and can be directly detected in electrochemical detector. However, histamine, Glu and GABA need a pre-column derivatization to produce electrochemical or fluorescence derivatives for their measurement (7-10). The sensitivity of this latter method is further limited by the instability of the final derivatives and the yield of derivatizations. ACh in microdialysate are measured by HPLC-ED following a post-column enzyme reactor and frequently below the detection limit (11). Most investigators have used acetylcholinesterase inhibitors in the perfusion medium to increase basal extracellular ACh concentrations to readily detectable levels (12-13). However, the artificially increased concentration exerts a significant influence on the cholinergic system, thereby making interpretation of drug effects problematic (11, 14-16). For analysis of exogenous drug, HPLC-ED was sensitive, but not applicable since most of drug candidates are electrochemically inactive.

Liquid chromatography/tandem mass spectrometry (LC-MS/MS) technique improvement provides a direct, structural-specific measurement of individual components with very high sensitivity. The mass spectrometer has minimal baseline drift and can be equilibrated very rapidly. However, only a few LC-MS/MS methods for analysis of drugs and individual neurotransmitter in microdialysis sample have been reported (15-19) and analysis all neurotransmitters have never been reported. There are two major problems for the application (20). First, the critical requirement for the use of the LC-MS/MS systems is that the mobile phase must be volatile, because non-volatile ions could significantly reduce the ionization efficiency by ion suppression. The high ionic strength of microdialysis samples generates high background noise and suppresses the ionization of neurotransmitters resulting in considerable reductions in sensitivity and changes in peak shape. Second, these neurotransmitters are polar compounds with low molecular weights; their retentions on standard reversed phase column are generally poor. In the present study, we have used two approaches to solve these problems and developed a series of LC-MS/MS methods which enable us to monitor drug and all neurotransmitters, 5-HT, DA, NE, ACh, Histamine, GABA, Glu, in single microdialysis sample. The methods were applied to study donepezil, a selective acetylcholinesterase inhibitor for treatment of Alzheimer's disease; and citalopram, an antidepressant of selective serotonin reuptake inhibitor. These applications demonstrated *in vivo* microdialysis coupled with LC-MS/MS could 1). Examine the drug efficacy by measuring the changes of extracellular concentration of all neurotransmitters after acute systematic administration of a drug; 2). Estimate the BBB penetration of a drug by measuring the concentration of free drug in the brain interstitial fluid; 3). The time course of free drug in the targeted brain region provided a biophase PK information which was used to evaluate PK/PD relationship by comparing the time course of free drug versus biomarker.

2. Methods

2.1 *In vivo* microdialysis experiments

Male Sprague Dawley rats (Charles River Laboratories, Wilmington, MA), weighing 300-400 g and were single housed under standard laboratory conditions. Rats were implanted with a probe guide cannula (Eicom, Japan) at the prefrontal cortex (PFC) (21): Incisor bar, -3.5mm, A (anterior to bregma) 3.2 mm, L (lateral from the mid-sagittal suture) 0.8 mm, V (ventral to bregma) 1 mm. Microdialysis experiments were performed at least two days after surgery. Rats were placed in their home cages and connected to a dual channel microdialysis liquid swivel (Instech Solomon, Plymouth Meeting, PA, USA) mounted on a spring loaded counter balanced arm which allowed animal freely moving. Microdialysis probes have a molecular weight cut-off of 50,000 Da. The exposed tubing of dialysis probe was 4 mm length with an outer diameter of 0.22 mm (Eicom, Japan). The input tube of the dialysis probe was connected to a CMA/102 Microdialysis Pump (CMA, Sweden), which delivered an artificial cerebrospinal fluid (aCSF) containing 147 mM NaCl, 4 mM KCl, 0.85 mM $MgCl_2$, 2.3 mM $CaCl_2$. The probe was perfused with aCSF at a rate of 0.6 µl/min overnight to obtain the stable basal level of neurotransmitters. The following morning the flow rate was increased to 1.0 µl/min and microdialysis samples were collected every 60 min to a 96 well plate, starting from 8:00 am for 20 hours. New probes were used every time without determining *in vitro* recovery. Fifteen µl antioxidant which contained 3 mM L-cysteine and 1 mM oxalic acid in 0.1 M acetic acid was added to each sample to prevent serotonin oxidation. Donapazil (2 mg/kg s.c.) or citalopram (10 mg/kg s.c.) were administrated to animal after four baseline microdialysis samples were obtained. The 60 µl microdialysis sample plus 15 µl antioxidant can be aliquot to each 5 µl, then diluted to 50 µl. Ten µl was used for each LC-MS/MS injection. Five to six rats were used for each treatment group. Vehicle animals received 5% N-methyl-2-pyrrolidone solution (v/v, 1 ml/kg, s.c.) injection. The drug concentration in the microdialysis samples were calculated by using *in vitro* recovery of microdialysis probe at a perfusion rate of 1.0 µl/min.

2.2 LC-MS/MS methods

The LC-MS/MS system consisted of two Shimadzu LC-10AD pumps (Shimadzu, Kyoto, Japan) coupled to an Applied Biosystems API-4000™ triple-quadrupole mass spectrometer equipped with a TURBO V™ Ion Source (Applied Biosystems/MDS SCIEX, Toronto, Canada). A CTC Analytics HTS PAL autosampler (Leap Technologies, Carrboro, NC, USA) fitted with a cooled sample tray and a six-port injection valve with a 20 µl injection loop. A ten-port diverter valve was equipped between column and mass spectrometer to divert salt to waste. The mass spectrometer was operated in electrospray positive ion mode and monitored with multiple reaction monitoring (MRM) using the parameters described in Table 1. Data was acquired using Analyst® software (version 1.4.1). The mass spectrometric (ion path) conditions for neurotransmitters and drugs were determined by using the quantitative optimization function of Analyst® software. Different HPLC Methods were developed for each neurotransmitter and drug.

2.2.1 ACh, 5-HT and DA method

100 mm x 2.1 mm i.d., 3 µm, Dicovery HS F5 column (Supelco, Bellefonte, PA, USA). Mobile Phase A and B were composed of formic acid (0.1%, v/v) in water and acetronitrile, respectively. An elution profile was composed of an isocratic step of 100% A for 1.5 min and then a linear gradient to 90% B over 3.5 min to separate DA, 5-HT and ACh, kept at 90% B for 1.5 min then equilibrated with 100% A for 1 min. The flow rate was 0.5 ml/min. Valco valve was diverted to MS from 2.5 to 5 min.

2.2.2 Histamine method

100 mm x 2.1 mm i.d., 3 µm, Dicovery HS F5 column (Supelco, Bellefonte, PA, USA). Mobile Phase A and B were composed of formic acid (0.05%, v/v) in water and acetronitrile, respectively. An elution profile was composed of an isocratic step of 95% B for 2.0 min and then a linear gradient to 5% B over 0.5 min to let salts elute first, held at 5% B for 4.0 min,

Analyate	DA	5-HT	NA	ACh	Histemine	GABA	Glu	Donepezil	Citalopram
Transition (m/z)	154→137	177→160	170→152	146→60	112→95	104→87	148→84	380→243	325→109
Declustering potential (DP, V)	41	36	31	41	41	21	36	81	76
Collision cell entrance potential (CE, V)	15	16	13	17	21	15	23	37	39
Collision cell exit potential (CXP, V)	10	11	12	4	8	6	6	22	8
Entrance potential (EP)	10	10	10	10	10	10	10	10	10
Collision gas ion energy (CAD)	6	6	5	6	5	5	5	5	5
Curtain gas (CUR)	14	14	15	14	15	15	15	15	15
Ion Source Gas 1 (GS1)	55	55	50	55	50	50	50	50	50
IonSpray Voltage (GS2)	50	50	50	50	50	50	50	50	50
IonSpray Voltage (IS)	3000	3000	3000	3000	3000	3000	3000	3000	3000
Temperature (TEM)	500	500	500	500	650	500	650	600	600

Table 1. The system parameters of mass spectrometry.

then equilibrated with 95% B for 1.5 min. The flow rate was 0.5 ml/min. Valco valve was diverted to MS from 3.8 to 5 min.

2.2.3 Glu method

150 mm x 4 mm i.d., 3 μm, Dicovery HS F5 column (Supelco, Bellefonte, PA, USA). Mobile Phase A and B were composed of formic acid (0.05%, v/v) in water and acetronitrile, respectively. An elution profile was composed of an isocratic step of 100% A for 1.5 min and then a linear gradient to 95% B over 0.5 min, held at 95% B for 1.0 min then equilibrated

with 100% A for 3.5 min. The flow rate was 1.2 ml/min. Valco valve was diverted to MS from 1.3 to 2 min.

2.2.4 GABA and NE method

50 mm x 2.1 mm i.d., 5 µm, 100 Å Primesep 200 column (SIELC Technologies, Prospect Heights, IL, USA). Mobile Phase A and B were composed of acetic acid (0.2%, v/v) in water and acetronitrile, respectively. An elution profile composed of an isocratic step of 90% A for 1.5 min and then a linear gradient to 90% B over 3.5 min for separation, hold at 90% B for 1.5 min then equilibrate with 100% A for 1 min. The flow rate was 0.5 ml/min. Valco valve was diverted to MS from 2.5 to 5 min.

2.2.5 Donepezil and citalopram method

30 mm x 2.0 mm i.d., Synergi 4µ Hydro-RP 80 Å column (Phenomenex. Torrance, CA, USA). Mobile Phase A and B were composed of formic acid (0.05%, v/v) in water and acetronitrile, respectively. An elution profile was composed of an isocratic step of 90% A for 1.5 min and then a linear gradient to 90% B over 3.5 min for separation, held at 90% B for 1.5 min then equilibrated with 100% A for 1 min. The flow rate was 0.5 ml/min. Valco valve was diverted to MS from 2.5 to 5 min.

2.3 Analysis of extracellular 5-HT and NE levels by HPLC-ED

Compared to LC-MS/MS method, 5-HT and NE are easier to measure by high-performance liquid chromatography (ESA Model 582) coupled to an electrochemical detector (CoulArray Coulometric, ESA) with dual channel coulometric microdialysis cell (ESA 5014B). Separation was performed on a C18 column (Hypersil, 150 x 3.2 mm I.D.) at room temperature. The mobile phase consisted of 75 mM NaH_2PO_4, 0.5 mM Disodium-EDTA, 350 mg/L 1-octanesulfonic acid, pH 3.1, 1.0% THF, 9.0% ACN. The flow rate was 0.4 ml/min. 22µl microdialysate was injected by an autosampler (ESA Model 540). The first electrode of the detector was set at -90 mV (reduction) and the second at +280 mV (oxidation).

2.4 Statistical analysis

All values for microdialysis studies were calculated as percentage change at each time point compared with the average of four baseline values. The extracellular neurotransmitter concentration at each time point after vehicle injection was compared with the average of four baseline values by one way ANOVA - Dunnett's test for their significant differences.
The overall effect of drug treatments on extracellular neurotransmitter levels was determined by a two way ANOVA with treatment as the independent variable and time as the repeated measurement. If significant, the ANOVA was followed by post-hoc Duncan's multiple range test (SigmaStat, SPSS Inc., www.spss.com).

3. Results

Figure 1 shows the mass spectra and molecular structures of precursor ions and associated daughter ions of neurotransmitters, donepezil and citalopram. The spectra show several prominent fragment ions. The ion that was selected for the detection of each neurotransmitter and drug was always the highest intensity ion except ACh. For ACh the lower intensity ion at m/z = 60 gave greater signal to noise than the higher intensity

product ion at m/z = 86. A summary of MRM parameters for all positive mode analytes was given in table 1.
Figure 2 shows the representative chromatogram of each neurotransmitter and drug in standard mixture solution (1 pg/µl standard in 1:10 dilution of aCSF) and in 1:10 dilution of basal level microdialysates from frontal cortex of freely moving rat.

3.1 Drug effect on the extracellular neurotransmitters levels in rat frontal cortex

The LC-MS/MS methods were applied to evaluate the effects of acute administration of either donepezil or citalopram on the extracellular concentration of ACh, Histamine, GABA, Glu, DA, NE and 5-HT in rat frontal cortex (as shown in Figure 3). Before dosing the animal, four 60 µl microdialysis samples were collected and the basal extracellular concentration of neurotransmitters in rat frontal cortex which were measured by LC-MS/MS methods are (mean ± S.E.): ACh, 1.93 ± 0.14 pg/µl (n=15); Histamine, 1.2 ± 0.15 pg/µl (n=15); GABA, 6.25 ± 0.35 pg/µl (n=15); Glu, 2.44 ± 0.25 ng/µl (n=15); DA, 0.15 ± 0.008 pg/µl (n=15); NE, 0.40 ± 0.015 pg/µl (n=8) and 5-HT, 0.20 ± 0.016 pg/µl (n=6). Since the LC-MS/MS methods presented here required a 10-fold dilution of the microdialysis sample, the 5-HT and NE detection limit was challenging and the HPLC-ED was a preferred method for these analytes. The basal extracellular concentrations of 5-HT and NE (Mean ± S.E.) in frontal cortex which measured by HPLC-ED were NE: 0.36 ± 0.15 pg/ml (n=20) and 5-HT, 0.10 ± 0.06 pg/µl (n=50).
Vehicle administration (n=5) did not significantly alter basal extracellular ACh, Histamine, GABA, Glu, DA, NE and 5-HT concentration in rat frontal cortex as shown in Figure 3. The acute administration of donepezil (2.0 mg/kg, s.c., n=5) evoked a significant increase of extracellular ACh from basal levels in the frontal cortex of rats compared with vehicle animal, to a maximum of 645 ± 69% at the first hours post dose and the average over the 3-6 hour treatment period was sustained at 178 ± 12% of basal values (as shown in Figure 3A). Donepezil did not significantly change the extracellular concentration of Histamine, GABA, DA, NE and 5-HT (Figure 3B, 3C, 3E, 3F and 3G), but significantly decreased the Glu level in frontal cortex of rats compared with vehicle animal. Glu concentration continually decreased to a maximum of 57 ± 17% at the sixth hour post dose (Figure 3D).
The acute administration of citalopram (10 mg/kg, s.c., n=5) immediately evoked a significant increase of extracellular 5-HT from basal levels in the frontal cortex of rats compared to vehicle animals at the first hour post dose and the average over the 6 hour treatment period was sustained at 412 ± 14% of basal values (as shown in Figure 3G). Citalopram did not significantly change the extracellular concentration of ACh, Histamine, Glu, DA and NE, (Figure 3A, 3B, 3D, 3E and 3F), but significantly decreased GABA level in frontal cortex of rats compared with vehicle animal. GABA level decreased to a maximum of 53 ± 9.4% at the forth hour post dose and the average over the 2-6 hour treatment period was sustained at 69% ± 4.6% of basal values (as shown in Figure 3C).

3.2 Biophase PK profile and the PK/PD relationship

The biophase PK profiles of donepezil and citalopram (as shown in Figure 4) were generated by measuring the extracellular concentration of free donepezil or citalopram in midialysate of rat frontal cortex using LC-MS/MS methods. The PK profile was compared with the time course of extracellular concentration of neurotransmitters in the same microdialysate of frontal cortex to evaluate their PK/PD relationship (as shown in Figure 4).

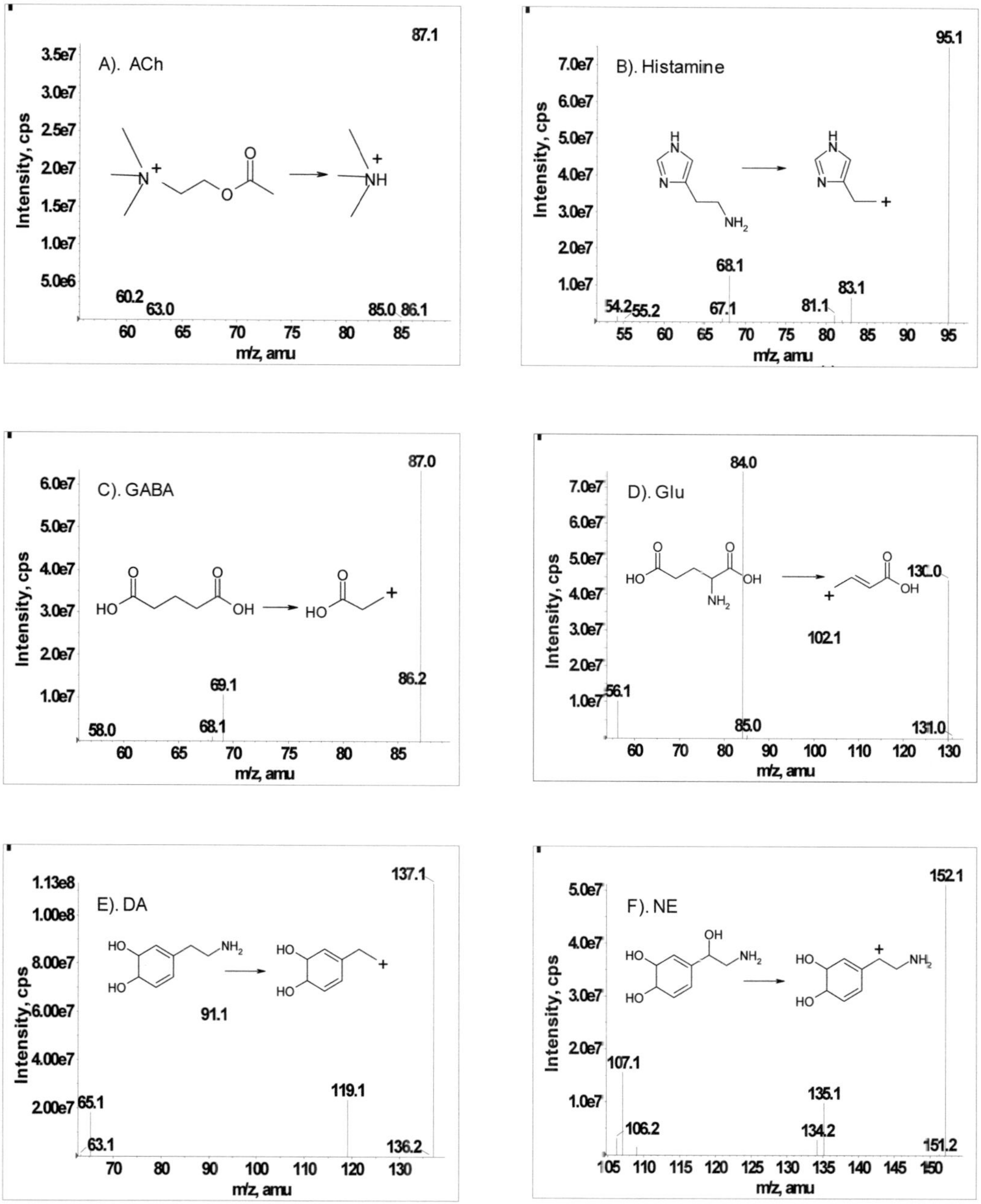
A). ACh
Intensity, cps
3.5e7
3.0e7
2.5e7
2.0e7
1.5e7
1.0e7
5.0e6
87.1
60.2
63.0
85.0
86.1
60
65
70
75
80
85
m/z, amu
B). Histamine
7.0e7
6.0e7
5.0e7
4.0e7
3.0e7
2.0e7
1.0e7
95.1
68.1
83.1
54.2
55.2
67.1
81.1
55
60
65
70
75
80
85
90
95
C). GABA
6.0e7
5.0e7
4.0e7
3.0e7
2.0e7
1.0e7
87.0
69.1
86.2
58.0
68.1
D). Glu
84.0
130.0
102.1
56.1
85.0
131.0
60
70
80
90
100
110
120
130
E). DA
1.13e8
1.00e8
8.00e7
6.00e7
4.00e7
2.00e7
137.1
91.1
119.1
65.1
63.1
136.2
F). NE
5.0e7
4.0e7
3.0e7
2.0e7
1.0e7
152.1
107.1
106.2
135.1
134.2
151.2
105
110
115
120
125
130
135
140
145
150

Fig. 1. Part I

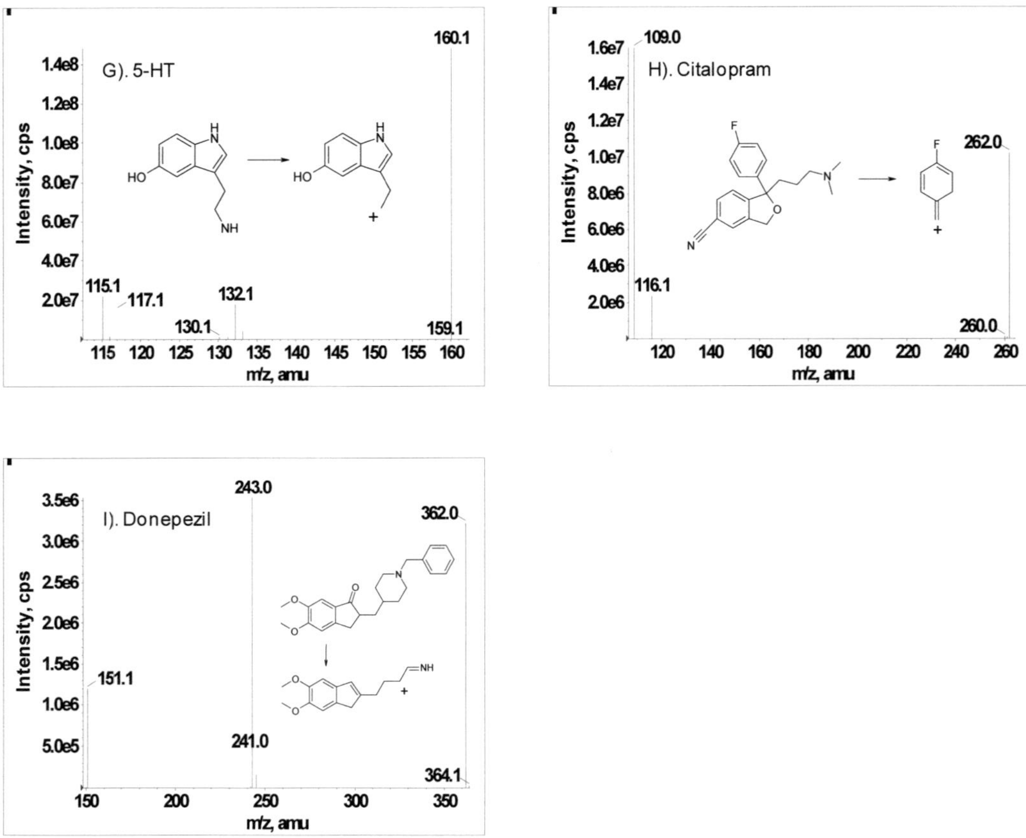

Part II

Fig. 1. Mass spectra and molecular structures of precursor ions and associated daughter ions for A). ACh; B). Histamine; C). GABA; D). Glu; E). DA; F). NE; G). 5-HT; H). citalopram; I). donepezil.

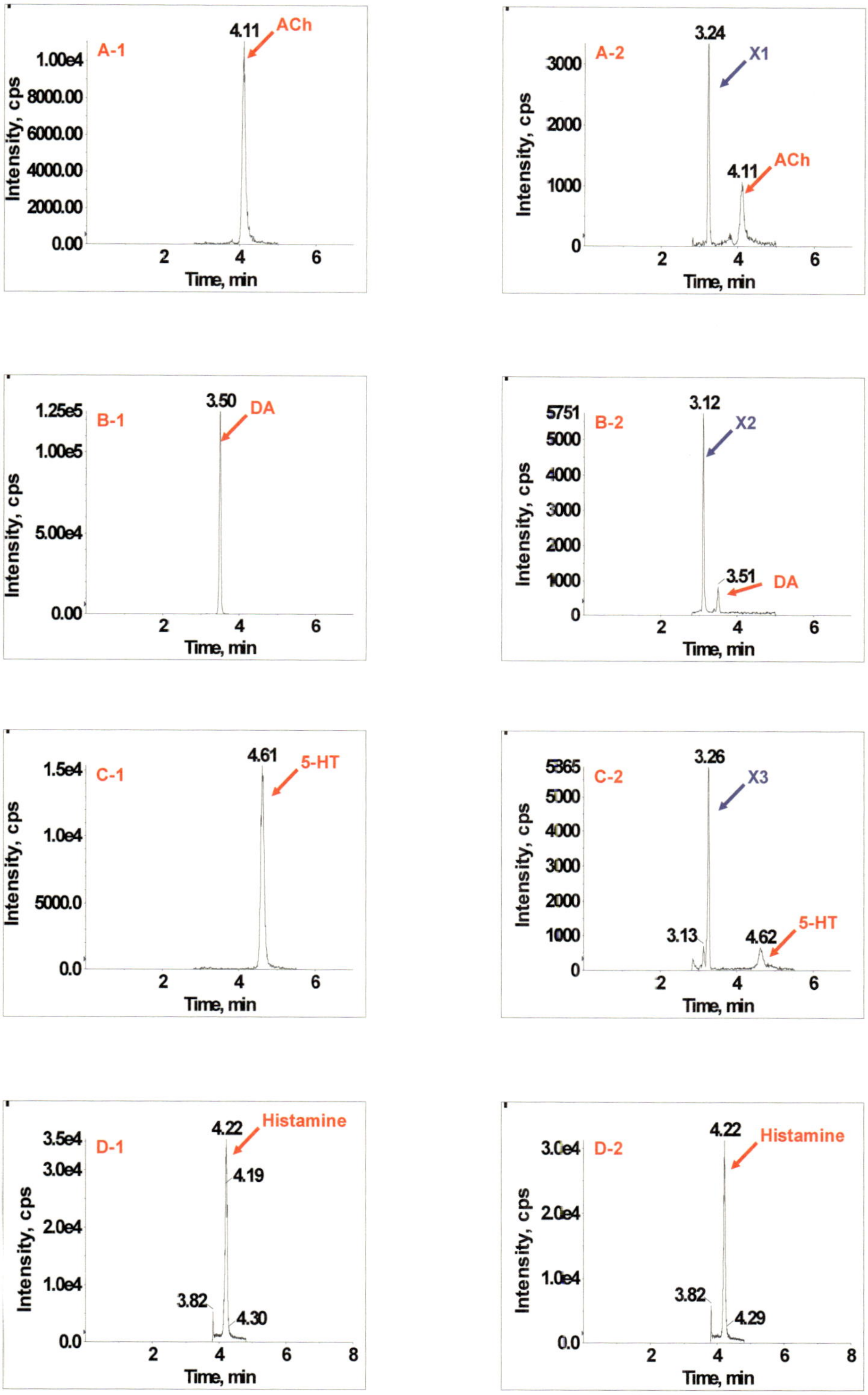
A-1
4.11
ACh
1.00e4
8000.00
6000.00
4000.00
2000.00
0.00
Intensity, cps
2
4
6
Time, min
A-2
3.24
X1
4.11
ACh
3000
2000
1000
0
Intensity, cps
Time, min
B-1
3.50
DA
1.25e5
1.00e5
5.00e4
0.00
Intensity, cps
Time, min
B-2
3.12
X2
3.51
DA
5751
5000
4000
3000
2000
1000
0
Intensity, cps
Time, min
C-1
4.61
5-HT
1.5e4
1.0e4
5000.0
0.0
Intensity, cps
Time, min
C-2
3.26
X3
3.13
4.62
5-HT
5365
5000
4000
3000
2000
1000
0
Intensity, cps
Time, min
D-1
4.22
Histamine
4.19
3.82
4.30
3.5e4
3.0e4
2.0e4
1.0e4
0.0
Intensity, cps
8
Time, min
D-2
4.22
Histamine
3.82
4.29
3.0e4
2.0e4
1.0e4
0.0
Intensity, cps
Time, min

Fig. 2. Part I

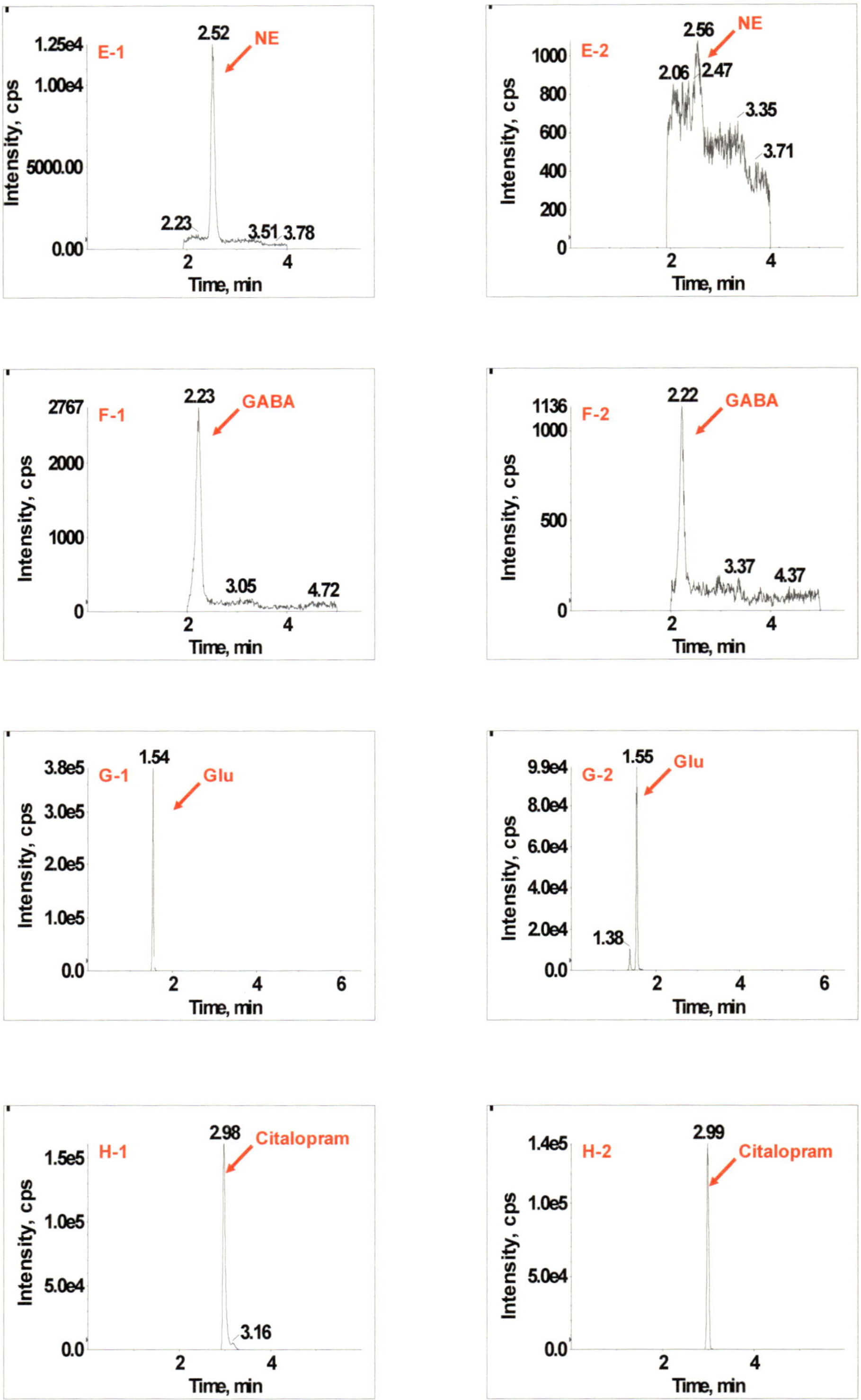
E-1
NE
2.52
2.23
3.51
3.78
1.25e4
1.00e4
5000.00
0.00
Intensity, cps
Time, min
E-2
NE
2.56
2.06
2.47
3.35
3.71
1000
800
600
400
200
0
Intensity, cps
Time, min
F-1
GABA
2.23
3.05
4.72
2767
2000
1000
0
Intensity, cps
Time, min
F-2
GABA
2.22
3.37
4.37
1136
1000
500
0
Intensity, cps
Time, min
G-1
Glu
1.54
3.8e5
3.0e5
2.0e5
1.0e5
0.0
Intensity, cps
Time, min
G-2
Glu
1.55
1.38
9.9e4
8.0e4
6.0e4
4.0e4
2.0e4
0.0
Intensity, cps
Time, min
H-1
Citalopram
2.98
3.16
1.5e5
1.0e5
5.0e4
0.0
Intensity, cps
Time, min
H-2
Citalopram
2.99
1.4e5
1.0e5
5.0e4
0.0
Intensity, cps
Time, min

Fig. 2. Part II

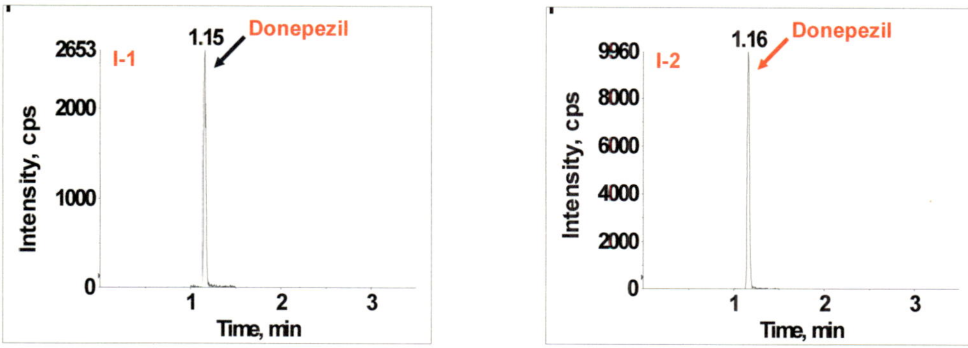

Part III

Fig. 2. The representative chromatograms of a 1 pg/µl standard mixture of neurotransmitters and drugs in 1:10 dilution of aCSF. A-1. ACh; B-1. DA; C-1. 5-HT; D-1. Histamine; E-1. NE; F-1. GABA; G-1. Glu; H-1. Citalopram; I-1. Donepezil. The representative chromatograms of 1:10 dilution of basal level microdialysates from frontal cortex of freely moving rat for each neurotransmitter and drug at the first hour after drug adminstration: A-2. ACh; B-2. DA; C-2. 5-HT; D-2. Hstamine; E-2. NE; F-2. GABA; G-2. Glu; H-2. Citalopram; I-2. Donepezil. X1: the unknown molecule share the same transition with ACh (m/z 146→87); X2: the unknown molecule share the same transition with DA (m/z 154→137); X3: the unknown molecule share the same transition with 5-HT (m/z 177→160).

Donepezil is a centrally acting reversible acetylcholinesterase inhibitor. Its main therapeutic use is in the treatment of Alzheimer's disease where it is used to increase cortical ACh release and modest benefits in cognition and/or behavior. Therefore the extracellular ACh level is the efficacy biomarker for donepezil. Figure 4A shows the comparison of time course of donepezil and its effect on ACh efflux in frontal cortex after donepezil (2 mg/kg, s.c.) administration. The immediate increase in donepezil concentrations was associated directly with a concomitant increase in extracellular ACh levels within the first hour, which closely followed the fast clearance of donepezil from extracellular fluid. The donepezil concentration in microdialysate increased to the peak concentration, 157 ± 23 ng/ml and gradually decreased to minimum concentration 5.5 ± 0.51 ng/ml at end of experiment (18 hours after administration).

Citalopram is an antidepressant used to treat major depression associated with mood disorders. Citalopram belongs to selective serotonin reuptake inhibitors, which exert their therapeutic effects by increasing extracellular 5-HT level in the synaptic cleft. Figure 4B shows the comparison of time course of citalopram concentration and its effect on 5-HT efflux in frontal cortex after citalopram (10 mg/kg, s.c.) administration. Interestingly, the immediate increase in citalopram concentrations was associated directly with a increase in extracellular 5-HT levels, however 5-HT level did not follow the fast clearance of citalopram from extracellular fluid and kept increasing to the peak release (545 ± 50% of basal level) at 9 hours after dosing, while remaining elevated for at least 18 hours. The citalopram concentration in microdialysate increased to the peak concentration, 281 ± 37 ng/ml within 1 hour of administration and was eliminated within 17 hours. The persistent increase in 5-HT concentrations may be related to the effect of citalopram metabolite on 5-HT release.

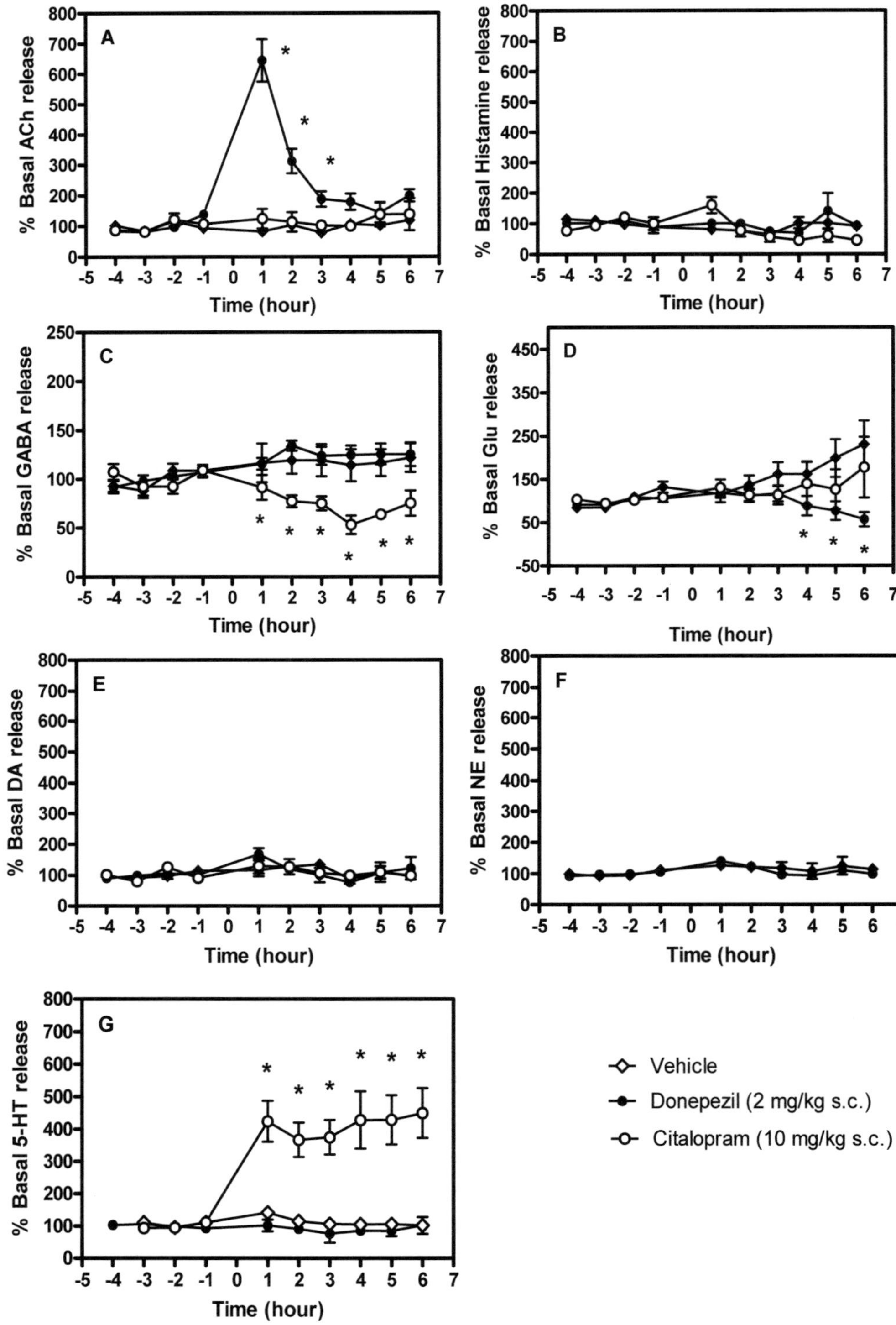

Fig. 3. Effect of donepezil (2 mg/kg, s.c., n=5) and citalopram (10 mg/kg, s.c., n=5) on extracellular concentration of neurotransmitters in the frontal cortex of rat: A. ACh; B.

Histamine; C. GABA; D. Glu; E. DA; F. NE and G. 5-HT. Values are mean ± S.E.M.. The neurotransmitter levels expressed as a percentage of the average of three baseline samples (defined as 100%). Two ways ANOVA - post-hoc Duncan's multiple range test were used for comparison. (Asterisks indicate significance of overall effect of drug treatment versus vehicle: * P<0.05, ** P<0.01, *** P<0.001.)

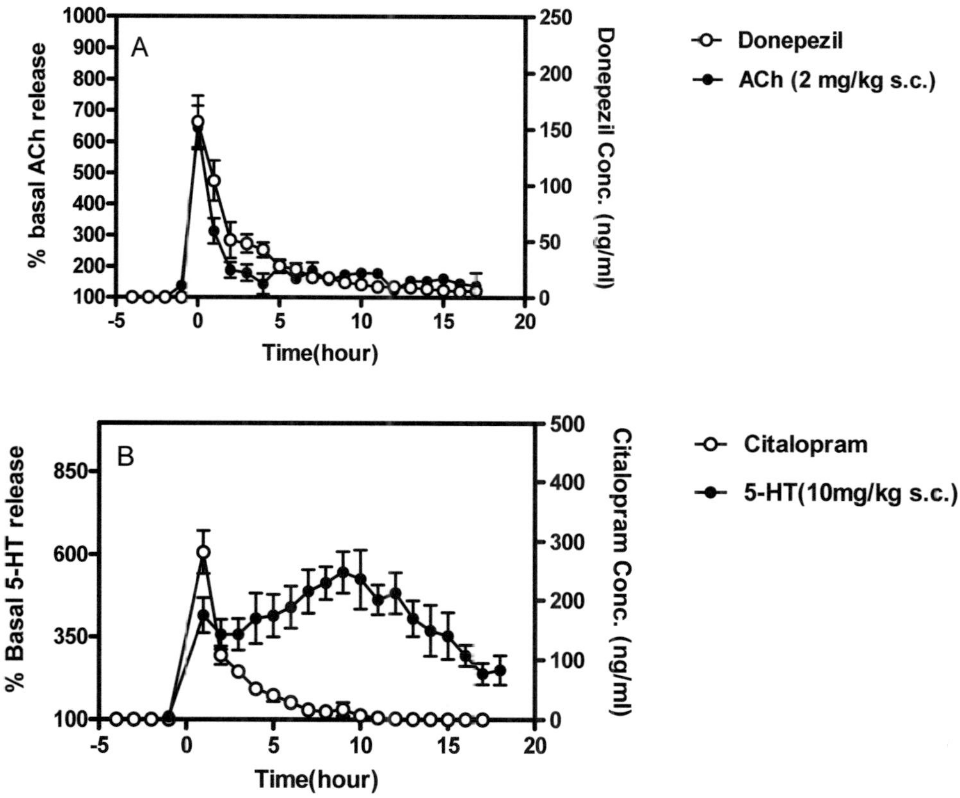

Fig. 4. PK/PD profiles of Donepezil and Citalopram: A. The extracellular concentration of donepezil and ACh in the frontal cortex of rats after Donepezil (2 mg/kg, s.c.) administration; B. The extracellular concentration of citalopram and 5-HT in the frontal cortex of rats after citalopram (10 mg/kg, s.c.) administration. Values are mean ± S.E.M. The neurotransmitter levels was expressed as a percentage of the average of three baseline samples (defined as 100%) and the drug concentration was expressed as ng/ml.

4. Discussion

In these studies, *in vivo* microdialysis in conjunction with automated sampling for 24 hours in freely moving rats provided a very useful technique for collecting the endogenous neurochemicals and exogenous drugs in the specific brain region without stress from sampling procedures or anesthesia. Analysis of these samples with LC-MS/MS methods, which provide high sensitivity and structural specificity, make it possible to monitor all neurotransmitter release in single sample collection and is a valuable tool for evaluating PK/PD relationship of new chemical entities in drug discovery.

LC-MS/MS applications in measuring neurotransmitters in microdialysis samples were limited by poor retention on reversed column and high salted aCSF matrix (20). To solve the first problem, two novel stationary phase columns were used in this study. One column is a Discovery F5 column, which uses pentafluorophenyl (PEP)(22) as a function group to attach on the silica via a propyl chain. The PEP type phase provides greater retention of polar solutes without ion-pairing or ion-suppressing agents in the mobile phase. Another column is Primesep 200 (23), a reverse-phase analytical column with embedded weak acidic ion-pairing groups. It improves retention of strong basic compounds by cation-exchange mechanism and retains neutral compounds by reverse-phase mechanism. All mobile phases for these two columns were LC-MS mobile phase (acetonitrile with formic acid or acetic acid). ACh, DA, 5-HT, histamine, NE, GABA and Glu were retained in these columns for 2-5 min, which is sufficient to separate them from high salt and endogenous interferences in microdialysis sample. The effluent that contained salts and the major ion-suppressing endogenous interferences were diverted to waste for the first 2-3 min as the online desalt step. In addition, the microdialysis sample (60 μl in each collection) was diluted to 1:10 by water to further reduce ion-suppressing influence from salt and endogenous interferences. 600 μl sample volume could make several LC-MS/MS injections for monitoring different biomarkers or drug in different methods, since the typical LC injection volume is 10 μl.

In contrast to the neurotransmitters, the average molecular weight of drug was 300-600 and drug molecule is very easy to retain on reverse phase column and separated with salt and ion-suppression agents. Therefore reverse phase column was used for drug LC-MS/MS method. We were unable to improve the 5-HT and NE sensitivity for LC-MS/MS method to monitor them at a 1 to 10 dilution of microdialysate. As discussed before, they have very good electrochemical activity and therefore the best way to monitor 5-HT and NE is still HPLC-ED.

The mass spectrometry conditions for each neurotransmitter and drug in this study are similar with other studies (15-19). In addition, compared with the chromatogram of neurotransmitter's standard solution (as shown in Figure 2), some unknown peaks in microdialysis were observed by LC-MS/MS. These unknown peaks always had the same transition (m/z) as the neurotransmitters, as shown in Figure 2. ACh (m/z 146→87) was eluted at 4.11 min and peak X1 (m/z 146→87) was eluted at 3.24 min (Figure 2A-2). DA (m/z 154→137) was eluted at 3.51 min and peak X2 (m/z 154→137) at 3.12 min (Figure 2B-2); 5-HT (m/z 177→160) was eluted at 4.62 min and peak X3 (m/z m/z 177→160) at 3.26 min (Figure 2C-2). A similar phenomenon has been observed in a study by Zhu et al. (17). In their full scan MS/MS experiments, the (3-carboxypropyl) trimethylammonium was identified, which was the unknown peak and had the same transition as ACh (m/z 146→87). (3-carboxypropyl) trimethylammonium has been reported to be a substrate in the production of γ-betaine hydroxylase, an enzyme in the biosynthesis of carnithine. A further study to identify these unknown peaks will be very interesting and may lead us to find new endogenous substance in the brain and expand our knowledge of neurochemistry.

In this study, for the first time we report the effect of citalopram (and donepzile) on seven neurotransmitter release. In agreement with literatures, 10 mg/kg citalopram in this study preferentially elevated levels of 5-HT in the frontal cortex of freely moving rats (24-25). DA and NE were not changed by citalopram at this dose; this agrees with Millan et al.'s results (26). It only evoked a slight rise in levels of NE and DA even at the highest dose (40 mg/kg) (26). Interestingly 10 mg/kg citalopram also markedly decreased extracellular GABA concentration in the frontal cortex in the same animals. To our knowledge, others have not reported this result, which is a favorable effect for depression patients. 10 mg/kg citalopram did not change the extracellular concentration of ACh, histamine and Glu in the present study.

Donepezil, a potent and selective acetylcholinesterase inhibitor for the treatment of Alzheimer's disease. Our results show that subcutaneous injection of donepezil (2 mg/kg) produced a marked (up to 7-fold) elevation of extracellular ACh in the frontal cortex. In previously studies, Giacobini et al. (27) examined the effect of subcutaneous injection of donepezil (0.5 and 2 mg/kg) on extracellular ACh concentration in the cerebral cortex of rats by using transcortical microdialysis. In their experimental, 2 mg/kg donepezil produced 20-fold elevation of extracellular ACh in the cortex. It had the similar time to the maximum plasma concentration (1 hour after dosing) and duration (3-6 hours) (28). Our study is the first to report changes in extracellular Glu cortical levels in parallel with ACh after donepezil administration. A similar result has been reported: ENA713 (29), a novel acetylcholinesterase inhibitor, significantly decreased extracellular glutamate level in rat hippocampus by using *in vivo* microdialysis. Decreased Glu was a favorable effect for treatment of Alzheimer's disease, since increased cerebrospinal fluid Glu concentration has been reported in Alzheimer's disease patients (30). These applications provided a very good example to screen CNS drug by comparing their drug effects on multiple neurotransmitters release with LC-MS/MS coupled to *in vivo* microdialysis. The results may help us to find the new therapeutic indication for an old drug.

In present study, the basal level 5-HT, NE and DA in the frontal cortex of rats which were measured by LC-MS/MS were 0.20 ± 0.016 pg/μl (n=6), 0.40 ± 0.015 pg/μl (n=8), 0.15 ± 0.008 pg/μl respectively, These results are correlated with the results in our lab that measured by HPLC-ED: 0.10 ± 0.06 pg/μl (n=50) for 5-HT; 0.36 ± 0.15 pg/μl (n=20) for NE and 0.06 ± 0.05 pg/μl (n=20) for DA. Due to a different route of administration and possible difference in the microdialysis techniques, the basal levels of other neurotransmitters in the frontal cortex of rats were slightly different in our study than in others, but were in similar ranges.

Since the method development of LC-MS/MS for drugs is relatively easier than for neurotransmitters and is amenable to a 1/10 dilution of microdialysis samples, this provides a unique opportunity to estimate drug CNS penetration and elucidate PK/PD profiles by monitoring concentrations of a drug simultaneous with drug induced neurotransmitter release in the same animal. In Figure 4A, administration of 2 mg/kg donepezil caused a rapid increase in ACh. The extracellular donepezil concentration in the frontal cortex of rats was associated with a concomitant increased in extracelluar ACh level and reached to the maximal concentration in the same time. The time-course of denepezil induced ACh efflux, which represents drug's efficacy or PD property, correlated well with the concentration curve of donepezil. This result suggested that donepezil had very fast blood-brain barrier penetration and inhibited the ChE, which directly elevated ACh release on the site of drug action. This result demonstrates good agreement with drug PK/PD properties.

In Figure 4B, administration of 10 mg/kg citalopram caused an immediate increase of 5-HT in rat frontal cortex. However, the time-course of citalopram induced 5-HT efflux did not parallel the biophase PK profile of citalopram. This finding indicates that the time-course for extracellular 5-HT in the frontal cortex reflects a more complex mechanism than direct action of citalopram on 5-HT reuptake inhibition at its dose. Citalopram has several metabolites, demethylcitalopram (DCIT), didemethylcitalopram (DDCIT), and citalopram N-oxide (CIT-NO) in human plasma and urine (31-32). Due to lack of commercial standards of citalopram metabolite, their concentration in microdialysate could not be measured in this study.

These examples illustrate the applicability of the LC-MS/MS coupled to *in vivo* microdialysis technique to rapidly estimate the PK/PD profiles of novel substances. This technique will be a very useful tool to study the drug penetration of BBB in lead optimization of drug discovery. Compared to the traditional method for the study of drug BBB penetration (33), this technique avoids having to euthanize the animal at each time point to measure drug concentration in whole brain tissue and provides continually monitoring extracellular drug concentration in specific brain regions at each hour for a 24-hour period by using a single animal as shown in table 2.

	Traditional Method	*In vivo* microdialysis-LC/MS/MS method
Specificity	• Using whole brain tissue • Measure total (intracellular plus extracellular) drug concentration	• Sample from individual brain region (such as frontal cortex, striatum, hippocampus, hypothalamus) • Measure extracellular drug concentration
Efficiency	• Using 8 animals to get 8 time points • More labor to sacrifice each animal, homogenate brain • Sample preparation for LC injection	• Using 1 animal to get 24 time points • Automatic sample collection for 24 hours without supervision. Only need to implant microdialysis probe guide by brain surgery. • Sample ready for LC injection
Information	• PK data (drug concentration in the brain)	• PK and PD data (both drug and biomarker concentration)

Table 2. Comparison of drug blood-brain barrier (BBB) penetration study by using traditional method versus *in vivo* microdialysis-LC/MS/MS method.

5. Conclusion

This study demonstrated that liquid Chromatography/tandem mass spectrometry (LC-MS/MS) coupled with *in vivo* microdialysis provides a powerful method for the measurement of endogenous and exogenous substances in the brain interstitial fluid (ISF) surrounding the probe and so it represents an important tool in CNS drug discovery. 1). It can be used to measure the pharmacodynamic response of neuroactive compounds, represents the **drug efficacy**, by measuring neurotransmitters and other biomarkers; 2). A core requirement for an effective CNS drug is an ability to cross the BBB and remain in the brain ISF for sufficient duration and concentration to evoke the desired therapeutic effect. LC-MS/MS coupled with *in vivo* microdialysis can be used to evaluate the **BBB penetration** of drug candidate by measuring the free concentration of CNS drug in brain ISF; 3). The profile of time-free concentration of CNS drug in the targeted brain region provides **biophase pharmacokinetic** information which can differ significantly from pharmacokinetics in plasma. Establishing **PK-PD relationship** of the drug, especially biophase PK, allow better understanding of exposure-response relationships and help the selection of drug; 4). In some case, not parent drug, but **drug metabolite** contributes to efficacy or toxicity. LC-MS/MS coupled with *in vivo* microdialysis also can be used to study drug metabolism during drug discovery.

6. Acknowledgments

This work was supported in part by Excellence in Science Grant of the Johnson & Johnson Corporate Office of Science and Technology. The authors gratefully acknowledge Dr. Michael Ye (Supelco/Sigma-Aldrich Inc.) for his suggestion of using pentafluorophenyl (Discovery HS F5) column; Dr.Jiejun Wu and Kenway Hoey for LC-MS/MS technique training and support; Dr. Pascal Bonaventure for helpful discussions during the experiments and writing the chapter draft.

7. References

Alavijeh, M., and Palmer, A. (2010) Measurement of the pharmacokinetics and pharmacodynamics of neuroactive compounds, *Neurobiol Dis. 37*, 38-47.

Barbier, A., Aluisio, L., Lord, B., Qu, Y., Wilson, S., Boggs, J., Bonaventure, P., Miller, K., Fraser, I., Dvorak, L., Pudiak, C., Dugovic, C., Shelton, J., Mazur, C., Letavic, M., Carruthers, N., and Lovenberg, T. (2007) Barbier, A.J. et al. Pharmacological characterization of JNJ-28583867, a histamine H(3) receptor antagonist and serotonin reuptake inhibitor *Eur J Pharmacol 576*, 43-54

Berzas-Nevado, J., Villasenor-Llerena, M., Guiberteau-Cabanillas, C., and Rodriguez-Robledo, V. (2006) Enantiomeric screening of racemic citalopram and metabolites in human urine by entangled polymer solution capillary electrophoresis: an innovatory robustness/ruggedness study *Electrophoresis 27*, 905-917.

Blakeley, J., and Portnow, J. (2010) Microdialysis for assessing intratumoral drug disposition in brain cancers: a tool for rational drug development, *Expert Opin Drug Metab Toxicol. 6*, 1477-1491.

Ceglia, I., Carlik, M., Baviera, M., Renoldi, G., Calcagno, E., and Invernizzi, R. (2004) The 5-HT receptor antagonist M100,907 prevents extracellular glutamate rising in response to NMDA receptor blockade in the mPFC. , *J Neurochem 91*, 189-199.

Chu, M., Huang, Z., Qu, W., Eguchi, N., Yao, M., and Urade, Y. (2004) Chu, M. et al. Extracellular histamine level in the frontal cortex is positively correlated with the amount of wakefulness in rats, *Neurosci Res 49*, 417-420.

Csernansky, J., Bardgett, M., Sheline, Y., Morris, J., and Olney, J. (1996) CSF excitatory amino acids and severity of illness in Alzheimer's disease *Neurology 46*, 1715-1720.

David, D., Bourin, M., Jego, G., Przybylski, C., Jolliet, P., and Gardier, A. (2003) David, D.J. et al. Effects of acute treatment with paroxetine, citalopram and venlafaxine in vivo on noradrenaline and serotonin outflow: a microdialysis study in Swiss mice, *Br J Pharmacol 140,* 1128-1136.

de Lange, E., Ravenstijn, P., Groenendaal, D., and van Steeg, T. (2005) Toward the prediction of CNS drug-effect profiles in physiological and pathological conditions using microdialysis and mechanism-based pharmacokinetic-pharmacodynamic modeling, *AAPS J 7,* E532-543.

Del Arco, A., and Mora, F. (2000) Endogenous dopamine potentiates the effects of glutamate on extracellular GABA in the prefrontal cortex of the freely moving rat, *Brain Res Bull 53,* 339-345.

Felton, T., Kang, T., Hjorth, S., and Auerbach, S. (2003) Effects of selective serotonin and serotonin/noradrenaline reuptake inhibitors on extracellular serotonin in rat diencephalon and frontal cortex, *Naunyn Schmiedebergs Arch Pharmacol 367,* 297-305.

Giacobini, E., Zhu, X., Williams, E., and Sherman, K. (1996) The effect of the selective reversible acetylcholinesterase inhibitor E2020 on extracellular acetylcholine and biogenic amine levels in rat cortex *Neuropharmacology 35,* 205-211.

Hows, M., Lacroix, L., Heidbreder, C., Organ, A., and Shah, A. (2004) High-performance liquid chromatography/tandem mass spectrometric assay for the simultaneous measurement of dopamine, norepinephrine, 5-hydroxytryptamine and cocaine in biological samples *J Neurosci Methods 138,* 123-132.

Hows, M., Organ, A., Murray, S., Dawson, L., Foxton, R., Heidbreder, C., Hughes, Z., Lacroix, L., and Shah, A. (2002) High-performance liquid chromatography/tandem mass spectrometry assay for the rapid high sensitivity measurement of basal acetylcholine from microdialysates *J Neurosci Methods 121,* 33-39.

Ichikawa, J., Dai, J., and Meltzer, H. (2000) Acetylcholinesterase Inhibitors are neither necessary nor desirable for microdialysis studies of brain acetylcholine *Current Separation 19,* 37-43.

Ichikawa, J., Dai, J., O'Laughlin, I., Fowler, W., and Meltzer, H. (2002) Atypical, but not typical, antipsychotic drugs increase cortical acetylcholine release without an effect in the nucleus accumbens or striatum *Neuropsychopharmacology 26,* 325-339.

Kawashima, K., Sato, A., Yoshizawa, M., Fujii, T., Fujimoto, K., and Suzuki, T. (1994) Effects of the centrally acting cholinesterase inhibitors tetrahydroaminoacridine and E2020 on the basal concentration of extracellular acetylcholine in the hippocampus of freely moving rats *Naunyn Schmiedebergs Arch Pharmacol 350,* 523-528.

Keski-Rahkonen, P., Lehtonen, M., Ihalainen, J., Sarajarvi, T., and Auriola, S. (2007) Quantitative determination of acetylcholine in microdialysis samples using liquid chromatography/atmospheric pressure spray ionization mass spectrometry *Rapid Commun Mass Spectrom 21,* 2933-2943.

Kissinger, P. (1989) Biomedical applications of liquid chromatography-electrochemistry, *J Chromatogr. 488,* 31-52.

Kosasa, T., Kuriya, Y., and Yamanishi, Y. (1999) Effect of donepezil hydrochloride (E2020) on extracellular acetylcholine concentration in the cerebral cortex of rats *Jpn J Pharmacol 81*, 216-222.

Lacroix, L., Hows, M., Shah, A., Hagan, J., and Heidbreder, C. (2003) Selective antagonism at dopamine D3 receptors enhances monoaminergic and cholinergic neurotransmission in the rat anterior cingulate cortex *Neuropsychopharmacology 28*, 839-849.

Li, J., Shao, S., Jaworsky, M., and Kurtulik, P. (2008) Simultaneous determination of cations, zwitterions and neutral compounds using mixed-mode reversed-phase and cation-exchange high-performance liquid chromatography *J Chromatogr A 1185*, 185-193.

Maura, G., and Raiteri, M. (1986) Cholinergic terminals in rat hippocampus possess 5-HT1B receptors mediating inhibition of acetylcholine release, *Eur J Pharmacol 129*, 333-337.

Millan, M., Dekeyne, A., Papp, M., La Rochelle, C., MacSweeny, C., Peglion, J., and Brocco, M. (2001) S33005, a novel ligand at both serotonin and norepinephrine transporters: II. Behavioral profile in comparison with venlafaxine, reboxetine, citalopram, and clomipramine, *J Pharmacol Exp Ther 298*, 581-591.

Needham, S., Brown, P., and Duff, K. (1999) Phenyl ring structures as stationary phases for the high performance liquid chromatography electrospray ionization mass spectrometric analysis of basic pharmaceuticals, *Rapid Commun Mass Spectrom 13*, 2231-2236.

Paxinos, G., and Watson, C. (1997) *The Rat Brain in Stereotaxic Coordinates,* Academic Press San Diego.

Plock, N., and Kloft, C. (2005) Microdialysis--theoretical background and recent implementation in applied life-sciences, *Eur J Pharm Sci 25*, 1-24.

Titier, K., Castaing, N., Scotto-Gomez, E., Pehourcq, F., Moore, N., and Molimard, M. (2003) Titier, K. et al. High-performance liquid chromatographic method with diode array detection for identification and quantification of the eight new antidepressants and five of their active metabolites in plasma after overdose *Ther Drug Monit 25*, 581-587.

Trabace, L., Cassano, T., Cagiano, R., Tattoli, M., Pietra, C., Steardo, L., Kendrick, K., and Cuomo, V. (2001) Trabace, L. et al. Effects of ENA713 and CHF2819, two anti-Alzheimer's disease drugs, on rat amino acid levels, *Brain Res. 910*, 182-186.

Ungerstedt, U. (1991) Microdialysis--principles and applications for studies in animals and man, *J Intern Med 230*, 365-373.

Yoshitake, T., Yamaguchi, M., Nohta, H., Ichinose, F., Yoshida, H., Yoshitake, S., Fuxe, K., and Kehr, J. (2003) Yoshitake, T. et al. Determination of histamine in microdialysis samples from rat brain by microbore column liquid chromatography following intramolecular excimer-forming derivatization with pyrene-labeling reagent *J Neurosci Methods 127*, 11-17.

Zhang, M. Y. B., C.E. (2006) Measurement of neurotransmitters from extracellular fluid in brain by in vivo microdialysis and chromatography-mass spectrometry *J Pharm Biomed Anal 40*, 492-499.

Zhu, Y., Wong, P., Cregor, M., Gitzen, J., Coury, L., and Kissinger, P. (2000) Zhu, Y. et al. In vivo microdialysis and reverse phase ion pair liquid chromatography/tandem mass spectrometry for the determination and identification of acetylcholine and related compounds in rat brain. Rapid Commun Mass Spectrom *Rapid Commun Mass Spectrom 14*, 1695-1700.

20

Principles and Applications of LC-MS/MS for the Quantitative Bioanalysis of Analytes in Various Biological Samples

Ju-Seop Kang
Department of Pharmacology & Clinical Pharmacology Laboratory, College of Medicine; Division of Molecular Therapeutics Development, Hanyang Biomedical Research Institute, Hanyang University, Seoul, South Korea

1. Introduction

Standard technology for analyte detection in clinical chemistry fields rely on indirect characteristics of an analyte, e.g. its absorption capability of light, chemical reactivity or physicochemical interaction with macro-molecules as like antigen-antibody interaction. Otherwise, analytes are detected directly from molecular characteristics as molecular mass and molecular disintegration patterns in mass spectrometric method. Thus, mass spectrometric technology are very attractive for the quantification of biomarkers or chemicals in the context of diagnostic procedures since those techniques can be provide higher quality of analysis of much higher specificity compared to standard technologies such as photometry or ligand-binding tests. Even if gas chromatography-mass spectrometry (GC-MS) is first introduced mass spectrometric methods to laboratory medicine about 40 years ago and provide highly specific and sensitive quantification of thermo-stable molecules below a molecular weight of about 500, application of GC-MS method remained restricted to few specialized institution in laboratory medicine such as mainly toxicology laboratories, medical centres, and reference laboratories. The handling and maintenance of GC-MS instruments is very demanding and time-consuming; sample preparation is very laborious and includes procedures of sample extraction and analyte derivatisations and long run time with a typical sample throughput of less than 50 samples per day. Over the past decade with introduction of atmospheric pressure ionization (API) techniques, liquid chromatography-mass spectrometry (LC-MS) has undergone tremendous technological improvement allowing for its applications to endogenous components such as proteins, peptides, carbohydrates, DNA, and drugs or metabolites. Furthermore, powerful new technologies of ion-analyses (tandem MS, time-of-flight MS, ion-trap MS) substantially increased the capabilities of MS analyzers with respect to specificity and to the extent of data read out. These developments suggest a more widespread use of MS techniques superior to other analytical methods in routine laboratory medicine.

The aims of this chapter is to comprehensively characterize the basic principles of mass spectrometric detection, recent development of MS, bottleneck of LC-MS method associate with matrix effect, application fields of LC with tandem MS in basic and clinical laboratory fields such as drug discovery and development, drug metabolism and toxicology studies, quantification of biogenic amines, doping control, TDM and pharmacokinetic (PK) studies.

2. Basic principles of mass spectrometric detection

Mass spectrometry (MS) has been described as the smallest scale in the world, not because of its size of what it weighs a molecule and a microanalytical technique that can be used selectively to detect and determine the amount of a given analyte (Watson & Sparkman, 2007; Chiu & Muddiman, 2008). MS is also used to determine the elemental composition and some aspect of the molecular structure of an analyte. Unique features of MS include its capacity for direct determination of the nominal mass of an analyte, and to produce and detect fragments of the molecule that correspond to discrete groups of atoms of different elements that reveal structure features (Watson & Sparkman, 2007). The tools of MS are mass spectrometers, and data are called mass spectra that can be displayed in many different ways, which allow the desired information about the analyte to be easily extracted (Watson & Sparkman, 2007). A MS is an apparatus which produces a beam of gaseous ions from a sample, sorts out the resulting mixture of ions according to their mass-to-charge ratios, and provides output signals which are measures of relative abundance of each ionic species present. MS are usually classified on the basis of how the mass separation is accomplished, but they all can be described as ion optical devices which separate ions according to their mass-to-charge (*m/z*) ratios by utilizing electric and/or magnetic force fields (Figure. 1). The concept of MS is to form ions from a sample, to separate the ions based on their *m/z* ratio (this can be considered to be the same as the mass because the ion has only a single charge in most cases), and to measure the abundance of the ions. In modern MS instrumentation used in environmental analyses, all of the functions (ionization separation of the ions, rate of data acquisition, detection of the ions, and storage of the data) are under computer control. Gaseous molecules are ionized in the ion source to form molecular ions which some of that will fragment. By various processes, ions of differing *m/z* values pass through the mass analyzer one at a time to reach the detector. When the ions strike the detector, they are converted into an electrical signal which, in turn, is converted into a digital response that can be stored by the computer (Sparkman, 2000).

A mass specterometer does not directly determine mass but, determines the mass of a molecule by measuring the *m/z* of its ion. The knowledge of the *m/z* of the ions enables one to determine what is present, while the measured ion intensities answer the question of how much is present. In addition, systematic interpretation of the mass spectra provides a detailed picture of the ionization process which, in turn, may be utilized in the elucidation of molecular structures. This definition of the term *m/z* is important to understanding of MS. It should be noted that the *m/z* value is a dimensionless number that is always used as an adjective, e.g. the ions with *m/z* 256, or the ion has an *m/z* value of 256. A recording of the number of ions (abundance) of a given *m/z* value as a function of the *m/z* value is a mass spectrum (Watson & Sparkman, 2007). The mass component that makes up the dimensionless *m/z* unit is based on an atomic scale rather than the physical scale normally considered as mass. Only ions are detected in mass specterometer and any nonionic particles that have no charge are removed from the mass spectrometer by the continuous pumping that maintains the vacuum. The MS first must produce a collection of ions in the gas phase. These ions are separated according to their *m/z* values in a vacuum where the

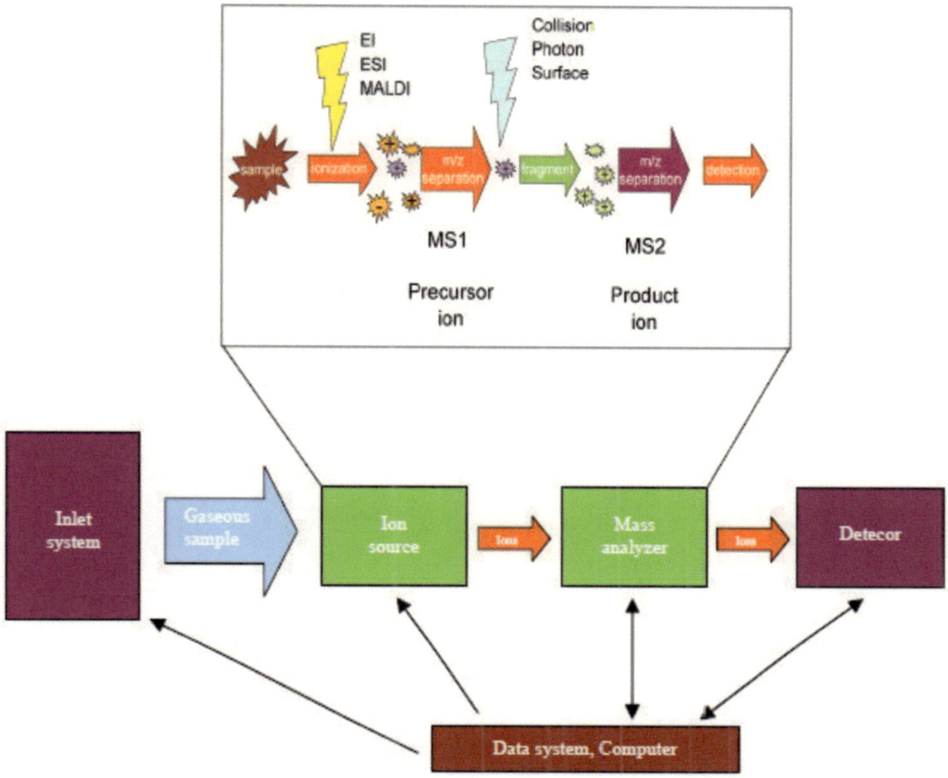

Fig. 1. Diagram of mass spectrometer system.

ions cannot collide with any other forms of matter during the separation process. Ions of individual *m/z* values are separated and detected in order to obtain the mass spectrum. Separation of ions in an evacuated environment is mandatory. If an ion collides with neutrals in an elastic collision during ion separation process, the ion's direction of travel could be altered and ion might not reach the detector. If an ion's collision with neutral is inelastic, sufficient energy transfer may cause it to decompose, meaning that the original ion will not be detected. Close encounters between ions of the same charge can be cause deflection in the path of each. Direct contact between ions of opposite charge sign will result in neutralization.

Ions are positively or negatively charged atoms, groups of atoms, or molecules. The process whereby an electrically neutral atom or molecule becomes electrically charged, due to losing or gaining one or more of its extra nuclear electrons, is called ionization (Chiu & Muddiman, 2008). Although both positive and negative ions can be analyzed by MS, the majority of instruments are used to investigate positive ions because in most ion sources they are produced in larger number than negative ions. (Chiu & Muddiman, 2008)

There is a minimum amount of energy, characterized by the "ionization potential," that must be provided in order for ion formation to occur. The first ionization potential of an atom or molecule is defined as the energy input required removing (to infinite distance) a valence electron from the highest occupied atomic or molecular orbital of the neutral particle to form the corresponding atomic or molecular ion, also in its ground state. When only one electron is removed the ion is called an atomic or molecular ion; often the term "parent ion" is used. The formation of parent ions may be considered as ionization without cleavage. The numerical magnitude of the ionization potential is influenced by such factors as the charge upon the nucleus, the atomic or molecular radius, the shielding effect of the inner electronic shells, and the extent to which the most loosely bound electrons penetrate the cloud of electric charge of the inner shells (Figure 2).

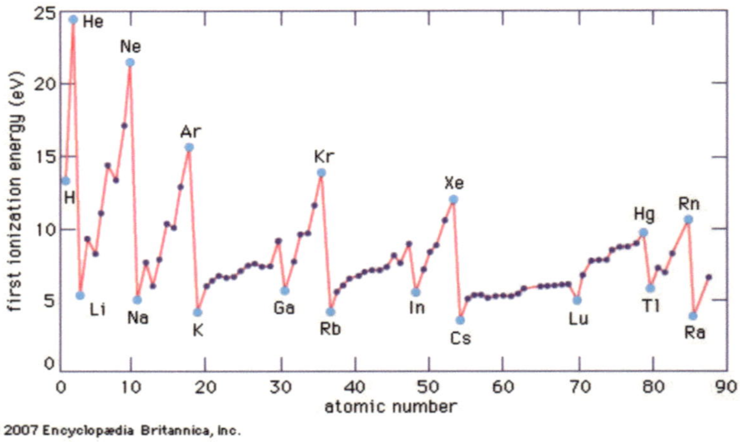

Fig. 2. Ionization potential (http://www.britannica.com/EBchecked/media/647/First-ionization-energies-of-the-elements).

Because only ions can be detected in MS, any particles that are not ionic (molecules or radicals) are removed from the MS by the continuous pumping that maintains the vacuum. When only individual ions are present, they can be grouped according to their unique properties (mass and number of charges) and moved freely from one point to another. In order to have individual ions free from any other forms of matter, it is necessary to analyze them in a vacuum, which means that the ions must be in the gas phase. It is a fundamental requirement of MS that ions be in the gas phase before they can be separated according to their individual *m/z* values and detected (Watson & Sparkman , 2007).

Due to ionization sources such as electrospray ionization (ESI) and matrix-assisted laser desorption/ionization (MALDI), MS has become an irreplaceable tool in the biological sciences. Over the past decade, MS has undergone tremendous technological improvements allowing for its application to proteins, peptides, carbohydrates, DNA, drugs, and many other biologically relevant molecules (Chiu & Muddiman, 2008).

2.1 Instrumentation of mass spectrometric detection

Mass spectrometry is a particularly powerful scientific technique because it can be successfully applied even if you have only a tiny quantity available for analysis—as little as 10^{-12} g, 10^{-15} moles for a compound of mass 1000 Daltons(Da). Compounds can be identified through mass spectrometry at very low concentrations (one part in 10^{12}) in chemically complex mixtures. The basic mass spectrometry processes of instrumentation are consisted of (1) introduction of sample; a sample which can be a solid, liquid, or vapor is loaded onto a mass spectrometry device and is vaporized, (2) ionization; sample components are ionized by one of several available methods to create ions, (3) analyzer sorting; the ions are sorted in an analyzer according to their *m/z* ratios through the use of electromagnetic fields, (4) detector; the ions then pass through a detector where the ion flux is converted into a propertional electrical current and (5) date conversion; the magnitude of the ion/electrical signals is converted into a mass spectrum (Watson & Sparkman , 2007).

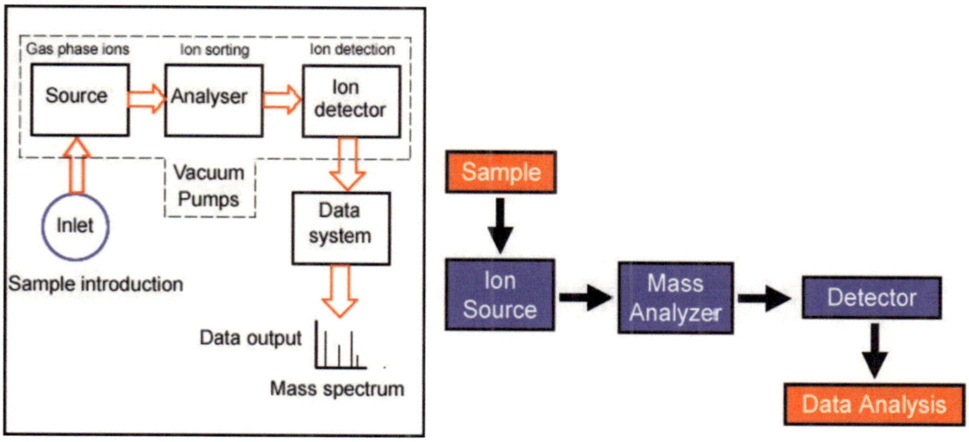

Fig. 3. Simplified diagram and function of instrumentation of typical mass spectromer (Image Source: The University of Hull, Mass Spectrometry Principles and Interpretation, Encyclopedia).

MS instruments consist of three modules: an ion source, which can convert gas phase sample molecules into ions (or, in the case of ESI, move ions that exist in solution into the gas phase); a mass analyzer, which sorts the ions by their masses by applying electromagnetic fields; and *a* detector, which measures the value of an indicator quantity and thus provides data for calculating the abundances of each ion present. The technique has both qualitative and quantitative uses. These include identifying unknown compounds, determining the isotopic composition of elements in a molecule, and determining the structure of a compound by observing its fragmentation. Other uses include quantifying the amount of a compound in a sample or studying the fundamentals of gas phase ion chemistry (the chemistry of ions and neutrals in a vacuum). MS is now in very common use in analytical laboratories that study physical, chemical, or biological properties of a great variety of compounds.

2.2 Basic principles of mass spectrometric detection

2.2.1 Sample Introduction Techniques

Sample introduction was an early challenge in MS. In order to perform mass analysis on a sample, which is initially at atmospheric pressure (760 torr), it must be introduced into the instrument in such a way that the vacuum inside the instrument remains relatively unchanged (~10^{-6} torr). The most common methods of sample introduction are direct insertion with a probe or plate commonly used with MALDI-MS, direct infusion or injection into the ionization source such as ESI-MS (Chiu & Muddiman, 2008).

1) Direct insertion: Using an insertion probe/plate is a very simple way to introduce a sample into an instrument. The sample is first placed onto a probe and then directly inserted into the ionization region of the mass spectrometer, typically through a vacuum interlock.

Vaccum interlock allows for the vacuum of the mass spectrometer to be maintained while the instrument is not in use. It also allows for the sample (at atomopheric pressure) to be introduced into high vacuum of the mass specterometer. The sample is then subjected to any number of desorption processes, such as laser desorption or direct heating, to facilitate vaporization and ionization.
2) Direct infusion: A simple capillary or a capillary column is used to introduce a sample as a gas or in solution. Direct infusion is also useful because it can efficiently introduce small quantities of sample into a mass spectrometer without compromising the vacuum. Capillary columns are routinely used to interface separation techniques with the ionization source of mass spectrometry. These techniques, including gas chromatography (GC) and liquid chromatography (LC), also serve to separate a solution's different components prior to mass analysis. In GC, separation of different components occurs within a glass capillary column. As the vaporized sample exits the GC, it is directly introduced into the MS.
In the 1980s the incapability of LC with MS was due largely to the ionization techniques being unable to handle the continuous flow of LC. However, ESI, APCI and atmospheric pressure photoionization (APPI) now allows LC/MS to be performed routinely. LC-MS ion chromatogram and the corresponding electrospray mass spectrum are typical technique. GC-MS produces results in much the same way as LC-MS, however, GC-MS uses an electron ionization source, which is limited by thermal vaporization (UV refers to ultraviolet and TIC is the total ion current) (Chiu & Muddiman, 2008).

2.2.2 Ionization method

Ionization method refers to the mechanism of ionization while the ion source is the mechanical device that allows ionization to occur. The different ionization methods work by either ionizing a neutral molecule through electron ejection, electron capture, protonation, cationization, or deprotonation, or by transferring a charged molecule from a condensed phase to the gas phase (Watson & Sparkman, 2007).
Protonation is a method of ionization by which a proton is added to a molecule, producing a net positive charge of 1^+ for every proton added. Positive charges tend to reside on the more basic residues of the molecule, such as amines, to form stable cations. Peptides are often ionized via protonation. Pronation can be achieved via MALDI, ESI, and APCI.
Deprotonation is an ionization method by which the net negative charge of 1^- is achieved through the removal of a proton from a molecule. This mechanism of ion-ization, commonly achieved via MALDI, ESI, and APCI is very useful for acidic species including phenols, carboxylic acids, and sulfonic acids.
Cationization is a method of ionization that produces a charged complex by non-covalently adding a positively charged ion to a neutral molecule. While protonation could fall under this same definition, cationization is distinct for its addition of a cation adduct other than a proton (e.g. alkali, ammonium). Moreover, it is known to be useful with molecules unstable to protonation. The binding of cations other than protons to a molecule is naturally less covalent, therefore, the charge remains localized on the cation. This minimizes delocalization of the charge and fragmentation of the molecule. Cationization is commonly achieved via MALDI, ESI, and APCI. Carbohydrates are excellent candidates for this ionization mechanism, with Na^+ a common cation adduct.
The transfer of compounds already charged(=precharged) in solution is normally achieved through the desorption or ejection of the positive or negative charged species from the

condensed phase into the gas phase. This transfer is commonly achieved via ESI, APCI, FAB and MALDI.

Electron ejection achieves ionization through the ejection of an electron to produce a 1^+ net positive charge, often forming radical cations. Observed most commonly with electron ionization (EI) sources, electron ejection is usually performed on relatively nonpolar compounds with low molecular weights and it is also known to generate significant fragment ions. It can provide observation of EI and can provide molecular mass as well as fragmentation information. However, it often generate too much fragmentation and can be unclear whether the highest mass ion is the molecular ion or a fragment.

With the electron capture ionization method, a net negative charge of -1 is achieved with the absorption or capture of an electron. It is a mechanism of ionization primarily observed for molecules with a high electron affinity, such as halogenated compounds. It can provide observation of electron ionization and can provide molecular mass as well as fragmentation information. However, it often generate too much fragmentation and can be unclear whether the highest mass ion is the molecular ion or a fragment (Chiu & Muddiman, 2008).

2.2.3 Ionization sources

The ion source is the part of the mass spectrometer that ionizes the material under analysis (the analyte). Ionization methods include the following: APCI, CI, EI, ESI, fast atom/ion bombardment (FAB), field desorption/field ionization (FD/FI), MALDI, thermospray ionization (TSP). With most ionisation methods there is the possibility of creating both positively and negatively charged sample ions, depending on the proton affinity of the sample. Before embarking on an analysis, the user must decide whether to detect the positively or negatively charged ions (Ashcroft, 2011). Prior to the 1980s, EI was the primary ionization source for mass analysis. However, EI limited chemists and biochemists to small molecules well below the mass range of common bio-organic compounds. These techniques have revolutionized biomolecular analyses, especially for large molecules. Among them, ESI and MALDI have clearly evolved to be the methods of choice when it comes to biomolecular analysis. ESI and MALDI are now the most common ionization sources for biomolecular mass spectrometry, offering excellent mass range and sensitivity.

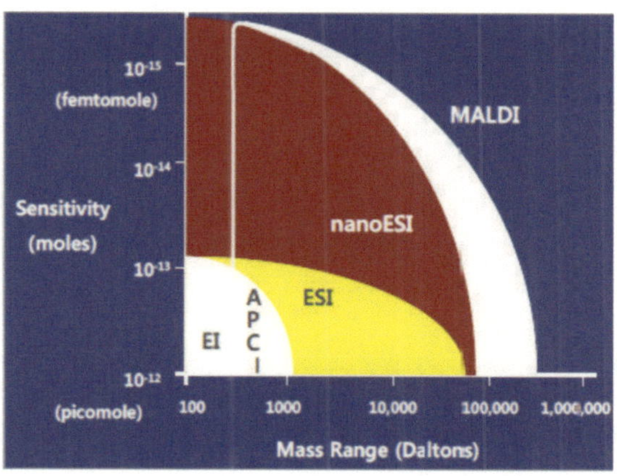

Fig. 4. A glance at the typical sensitivity and mass ranges allowed by different ionization techniques provides a clear answer to the question of which are most useful; EI, APCI and DIOS are somewhat limiting in terms of upper mass range, while ESI, nanoESI, and MALDI have a high practical mass range (Chiu & Muddiman, 2008).

Ion loses between major components of a mass spectrometer are inevitable and occur between the source and the analyzer, within the analyzer, and between the analyzer and the detector. Atmospheric pressure ion sources, such as encountered with ESI, APCI, APPI, etc., present a particular challenge with respect to efficient transfer of ions from the ion source into the *m/z* analyzer. Techniques for ionization have been key to determining what types of samples can be analyzed by MS. EI and chemical ionization(CI) are used for gases and vapors. Two techniques often used with liquid and solid biological samples include ESI and MALDI and these techniques have clearly evolved to be the methods of choice when it comes to biomolecular analysis. The idea of ESI, while not new, has been rejuvenated with its recent application to biomolecules.

Ionization Source	Event
Electrosray ionization(ESI)	Evaporation of charged droplets
Nanoelectrospray ionization(nanoESI)	Evaporation of charged droplets
Atmospheric pressure chemical ionization (APCI)	Corona discharge and proton transfer
matrix-assisted laser desorption ionization (MALDI)	Photon absorption/proton transfer
Desorption/ionization on silicon(DIOS)	Photon absorption/proton transfer
Fast atom/ion bormbardment(FAB)	Ion desorption/proton transfer
Electron ionization(EI)	Electron beam/electron transfer
Chemical Ionization(CI)	Proton transfer

Table X. Summary of event according to ionization sources (Chiu & Muddiman, 2008).

Inductively coupled plasma (ICP) sources are used primarily for cation analysis of a wide array of sample types. In the ion source technology, a 'flame' of plasma that is electrically neutral overall, but that has had a substantial fraction of its atoms ionized by high temperature, is used to atomize introduced sample molecules and to further strip the outer electrons from those atoms. The plasma is usually generated from argon gas, since the first ionization energy of argon atoms is higher than the first of any other elements except He, O, F and Ne, but lower than the second ionization energy of all except the most electropositive metals. The heating is achieved by a radio-frequency current passed through a coil surrounding the plasma. Others include glow discharge, field desorption (FD), FAB, thermospray, DIOS, Direct Analysis in Real Time (DART), APCI, secondary ion mass spectrometry (SIMS), spark ionization and thermal ionization (TIMS) (Bruins, 1991). Ion attachment ionization is a newer soft ionization technique that allows for fragmentation free analysis (Wikipedia, 2011).

2.2.4 Mass analyzers

The *m/z* analyzer (mass analyzer) is used to separate the ions according to their *m/z* ratio based on their characteristic behavior in electric and/or magnetic fields (Jennings & Dolnikowski, 1990; Farmer & McDowell, 1963). MS takes advantages of these different behaviors to separate the ions of different *m/z* values in space or time so that their abundances can be determined. Together with the particle's initial conditions, it completely determines the particle's motion in space and time in terms of *m/q*. When presenting data, it is common to use the dimensionless *m/z*, where *z* is the number of elementary charges (*e*) on the ion ($z=q/e$). This quantity, although it is informally called the *m/z* ratio, more accurately speaking represents the ratio of the mass number, *m* and the charge number, *z*. Many MS use two or more mass analyzers for tandem MS (MS/MS).

With the advent of ionization sources that can vaporize and ionize biomolecules, it has become necessary to improve mass analyzer performance with respect to speed, accuracy, and resolution. More specifically, quadrupoles, quadrupole ion traps, time-to-flight (TOF), time-to-flight reflection (TOFR), and ion cyclotron resonance (ICR) mass analyzers have undergone numerous modifications and improvements over the past decade in order to be interfaced with MALDI and ESI. In the most general terms, a mass analyzer measures gas phase ions with respect to their *m/z*, where the charge is produced by the addition or loss of a proton(s), cation(s), anion(s) or electron(s). The addition of charge allows the molecule to be affected by electric fields thus allowing its mass measurement. This is important aspect to remember about mass analyzers that they measure the *m/z* ratio, not the mass itself. Multiple charging is especially common with ESI, yielding numerous peaks that correspond to the same species yet are observed at different *m/z*.

The performance of a mass analyzers can be typically be defined by the following characteristics such as accuracy, resolution, mass range, tandem analysis capabilities, and scan speed. Accuracy is the ability with which the analyzer can be accurately provide *m/z* information and is largely a function of an instrument's stability and resolution. For example, an instrument with 0.01% accuracy can provide information on a 1000 Da peptide to±0.1 Da or a 10,000 Da protein±1.0 Da. (1) The accuracy varies dramatically from analyzer to analyzer depending on the analyzer type and resolution. (2) Resolution so called resolving power is the ability of a mass spectrometer to distinguish between ions of different *m/z* ratios. Therefore, greater resolution corresponds directly to the increased ability to differentiate ions. (3) The mass range is the *m/z* range of the mass analyzer. For instance, quadrupole analyzers typically scan up to *m/z* 3000. A magnetic sector analyzer typically scans up to *m/z* 10,000 and TOF analyzers have virtually unlimited *m/z* range. (4) Tandem MS analysis is the ability of the analyzer to separate different molecular ions, generate fragment ions from a selected ion, and then mass measure the fragmented ions. The fragmented ions are used to for structural determination of original molecular ions. Typically, MS/MS experiments are performed by collision of a selected ion with inert gas molecules such as argon or helium, and the resulting fragments are mass analyzed. MS/MS analysis is used to sequence peptides, and structurally characterize carbohydrates, small oligonucleotides, and lipids. Tandem mass analysis in space is performed by consecutive analyzers whereas tandem mass in time is performed with the same analyzer, which isolates the ion of interest, fragments it, and analyzes the fragment ions. (5) Scan speed refers to the rate at which the analyzer scans over a particular mass range. Most instruments require performing a full scan; however this can vary widely depending on the analyzer. TOF analyzers, for example, complete analyses within milliseconds or less (Watson & Sparkman, 2007; Chiu & Muddiman, 2008).

2.2.4.1 Sector

A **sector field mass analyzer** uses an electric and/or magnetic field to affect the path and/or velocity of the charged particles in some way. The sector instruments bend the trajectories of the ions as they pass through the mass analyzer, according to their *m/z* ratios, deflecting the more charged and faster-moving, lighter ions more. The analyzer can be used to select a narrow range of *m/z* or to scan through a range of *m/z* to catalog the ions present (Paul & Steinwedel, 1953).

2.2.4.2 Time-of-flight (TOF)

The operating principles of the TOF-MS involve measuring the time required for an ion to travel from ion source to a detector usually located 1 to 2 meter from the source (Graff, 1995; Guilaus, 1995). The TOF analyzer uses an electric field to accelerate the ions through the same potential, and then measures the time they take to reach the detector. If the particles all have the same charge, the kinetic energies will be identical, and their velocities will depend only on their masses. As the ions traverse the "field-free" region between the ion source and detector, they separate into groups or packets according to velocity, which is a function of their *m/z* values, i.e. lighter ions will reach the detector first (March, 2000). As will be seen, all other types of *m/z* analyzers have upper limits for *m/z* values that can be transmitted and separated based on the way the electric and/or magnetic fields are used to separate ions of different *m/z* ratios.

2.2.4.3 Quadrupole

Quadrupole mass analyzers use oscillating electrical fields to selectively stabilize or destabilize the paths of ions passing through a radio frequency (RF) quadrupole field. Only a single *m/z* ratio is passed through the system at any time, but changes to the potentials on magnetic lenses allows a wide range of *m/z* values to be swept rapidly, either continuously or in a succession of discrete hops. A quadrupole mass spectrometer acts as a mass-selective filter and is closely related to the quadrupole ion trap, particularly the linear quadrupole ion trap except that it is designed to pass the untrapped ions rather than collect the trapped ones, and is for that reason referred to as a transmission quadrupole. A common variation of the quadrupole is the triple quadrupole. Quadrupoles offer three main advantages such as tolerance to relatively high pressures, significant mass range with the capability of analyzing up to an *m/z* 4000, which is useful because ESI of proteins and other biomolecules commonly produce charge distributions from *m/z* 1000 to 3500 and relatively low cost instruments. In order to perform tandem MS analysis with a quadrupole instrument, it is necessary to place three quadrupoles in series. Each quadrupole has a separate function; the first (Q1) is used to scan across a preset *m/z* range and select an ion of interest. The second (Q2), also known as the collison cell, focuses and transmits the ions while introducing a collison gas (argon or helium) into the flight path of the selected ion. The third (Q3) serves to analyze the fragment ions generated in the collison cell (Q2) (Watson & Sparkman, 2007).

2.2.4.4 Quadrupole ion trap (QIT)

The name quadrupole is derived the fact that an electric field is created between four opposing electrical poles (Watson & Sparkman, 2007). The QIT works on the same physical principles as the quadrupole mass analyzer, but the ions are trapped and sequentially ejected. Ions are created and trapped in a mainly quadrupole RF potential and separated by *m/q*, non-destructively or destructively. There are many *m/z* separation and isolation methods but most commonly used is the mass instability mode in which the RF potential is ramped so that the orbit of ions with a mass $a > b$ are stable while ions with mass b become unstable and are ejected on the z-axis onto a detector. Ions may also be ejected by the resonance excitation method, whereby a supplemental oscillatory excitation voltage is applied to the endcap electrodes and the trapping voltage amplitude and/or excitation voltage frequency is varied to bring ions into a resonance condition in order of their *m/z* ratio (Schwartz et al., 2002; Lammert et al., 2006). Two method of using an ion trap for MS are involved to generate ions internally with EI, followed by mass analysis and, more popular method involving in generation of ions externally with ESI or MALDI and using ion optics for sample injection into the trapping volume. The QIT typically consists of a ring electrode and two hyperbolic endcap electrodes. The isolated ions can subsequently be fragmented by collisional activation and the fragments detected. The primary advantage of QIT is that multiple collision induced dissociation experiments can be performed quickly without having multiple analyzers, such that real time LC-MS/MS is now routine. Other important advantages of QIT include their compact size, and thier ability to trap and accumulate ions to provide a better ion signal. QIT have been utilized in a number of applications ranging from electrospray ionization MS of biomolecules to their recent interface with MALDI. LC-MS/MS experiments are performed on proteolytic digests which provide both MS and MS/MS information. This information allows for protein identification and post-translational modification characterization. The mass range (~ *m/z* 4000) of commercial LC-traps is well matched to *m/z* values generated from the ESI of peptides and the resolution allows for charge state identification of multiple-charged peptide ions. QIT-MS can analyze peptides from a tryptic digest present at the 20~100 fmol level. Another asset of the ion trap technique for peptide analysis is the ability to perform multiple stages of MS, which can significantly increase the amount of structural information. The cylindrical ion trap MS is a derivative of the QIT mass spectrometer (Watson & Sparkman, 2007).

2.2.4.5 Linear quadrupole ion trap (LIT)

The linear ion trap (LIT), sometimes referred to as the 2-dimensional quadrupole ion trap (2D QIT), is one of the more recent additions to the single *m/z* analyzer group (Douglas et al, 2005; Hager, 2002). The LIT makes use of the basic structure of a transmission quadrupole; i.e., an array of four electrical surfaces; however, instead of being used to filter ions of all *m/z* values except for those of a desired value from an unresolved ion beam, they are used for trapping, manipulation of ion trajectories, and *m/z*-selective ion ejection (Londry & Hager, 2003). A LIT is similar to a QIT, but it traps ions in a 2D quadrupole field, instead of a 3-dimensional (3D) quadrupole field as in a 3D QIT. The LIT differs from the 3D ion trap as it confines ions along the axis of a quadrupole mass analyzer using a 2D radio frequency (RF) field with potentials applied to end electrodes. The primary advantage to the linear trap over the 3D trap is the larger analyzer volume lends itself to a greater dynamic ranges and an improved range of quantitative analysis (Chiu & Muddiman, 2008). This toroidal shaped trap is a configuration that allows the increased miniaturization of an ion trap mass analyzer. The LIT has higher injection efficiencies and storage capacities than conventional 3D QIT (or Paul trap). 2D multipole fields are well known for their capacity to trap and manipulate ions (Gerlich, 1992). Given the power of the ion trap the major limitations of this device that keep it from being the ultimate tool for pharmacokinetics and proteomics include the following: (1) the ability to perform high sensitivity triple quadrupole-type precursor ion scanning and neutral loss scanning experiments is not possible with ion traps. (2) The upper limit on the ratio between precursor *m/z* and the lowest trapped fragment ion is ~0.3 (also known as the "one third rules"). (3) The dynamic range of ion traps is limited because when too many ions are in the trap, space charge effects diminish the performance of the ion trap analyzer. To get around this, automated scans can rapidly count ions before they go into the trap, therefore limiting the number of ions getting in. Yet this approach presents a problem when an ion of interest is accompanied by a large background ion population (Chiu & Muddiman, 2008).

2.2.4.6 Orbitrap

The orbitrap mass spectrometer is the latest development in trapping devices used as an *m/z* analyzer. The orbitrap ia a new mass analyzer (Hardman & Makarov, 2003; Hu et al., 2005; Makarov et al., 2006); however, it is useful to consider the orbitrap as a modified Knight-style Kingdon trap with specially shaped inner spindle-like (axial) electrode and a outer barrel-like (coaxial) electrodes that form an electrostatic field with quadro-logarithmic potential distribution (Makarov, 2000; Makarov et al., 2006a; Hu et al., 2005). The frequency of these harmonic oscillations is independent of the ion velocity and is inversely proportional to the square root of the mass-to-charge ratio (*m/z* or *m/q*). By sensing the ion oscillation in a manner similar to that used in Fourier transform ion cyclotron resonance (FTICR)-MS, the trap can be used as a mass analyzer. Orbitraps have a high mass accuracy (1–2 ppm), a high resolving power (up to 200,000) and a high dynamic range (around 5000) (Makarov et al, 2006a; Makarov et al., 2006b). Very similar nonmagnetic FTMS has been performed, where ions are electrostatically trapped in an orbit around a central, spindle shaped electrode. Mass spectra are obtained by Fourier transformation of the recorded image currents. Similar to Fourier transform ion cyclotron resonance mass spectrometers, Orbitraps have a high mass accuracy, and high sensitivity and a good dynamic range (Hu et al., 2005).

2.2.4.7 Fourier transform ion cyclotron resonance

Fourier transform mass spectrometry (FT-MS), or more precisely Fourier transform ion cyclotron resonance MS, measures mass by detecting the image current produced by ions cyclotroning in the presence of a magnetic field. Instead of measuring the deflection of ions with a detector such as an electron multiplier, the ions are injected into a Penning trap (a static electric/magnetic ion trap) where they effectively form part of a circuit. Since the frequency of an ion's cycling is determined by its *m/z* ratio, this can be deconvoluted by performing a Fourier transform on the signal. FTMS has the advantage of high sensitivity (since each ion is "counted" more than once) and much higher resolution and thus precision (Comisarow & Marshall, 1974; Marshall et al., 1998). Ion cyclotron resonance (ICR) is an older mass analysis technique similar to FT-MS except that ions are detected with a traditional detector. Ions trapped in a Penning trap are excited by an RF electric field until they impact the wall of the trap, where the detector is located. Ions of different mass are resolved according to impact time.

2.2.5 Detectors

The final element of the mass spectrometer is the detector that records either the charge induced or the current produced when an ion passes by or hits a surface. In a scanning instrument, the signal produced in the detector during the course of the scan versus where the instrument is in the scan (at what *m/q*) will produce a mass spectrum, a record of ions as a function of *m/q* (Wikipedia). Typically, some type of electron multiplier is used, though other detectors including Faraday cups and ion-to-photon detectors are also used. Because the number of ions leaving the mass analyzer at a particular instant is typically quite small, considerable amplification is often necessary to get a signal. Microchannel plate detectors are commonly used in modern commercial instruments (Dubois et al., 1999). In FT-MS and Orbitraps, the detector consists of a pair of metal surfaces within the mass analyzer/ion trap region which the ions only pass near as they oscillate. No DC current is produced, only a weak AC image current is produced in a circuit between the electrodes. Other inductive detectors have also been used (Park et al., 1994). Sensitivity, accuracy, and response time are importantparameters that distinguish different ion detection system (Koppenaal et al., 2005). High accuracy and fast-response times usually are mutually exclusive features, therefore, the fast-response electron multiplier required to follow rapid scans (e.g. *m/z* 50 to 500 in 0.1 sec) probably will not provide the accuracy desired (better than ±0.1 %) in isotope ratio measurements. Such isotope ratio measurements are usually made with the Faraday cup detector (Nygren et al., 2006; Harris et al., 1984; Mckinney et al., 1950). Ideally, a statement of sensitivity should explicity define the ion current received by detector for a specified rate of sample consumption. The definition should include (1) the name of the calibration compound, (2) the *m/z* value of the ion current that is measured, (3) the resolving power of the instrument, (4) the quantity of calibration compound consumed per second, (5) the intensity of the electron beam and pressure in the ion source for CI, and (6) the ion current arriving at the detector (e.g., that impinging the conversion dynode of an electron multiplier.

2.2.5.1 Electron Multiplier (EM)

The EM uses the principle of secondary-electron emission to effect amplification (Harris et al., 1984). Perhaps the most common means of detecting ions involves an EM, which is made up of a series (12 to 24) of aluminum oxide (Al_2O_3) dynodes maintained at ever increasing potentials. Ions strike the first dynode surface causing an emission of electrons. These electrons are then attracted to the next dynode held at a higher potential and therefore more secondary electrons are generated. Ultimately, as numerous dynodes are involved, a cascade of electrons is formed that results in an overall current gain on the order of one million or higher. The high energy dynode (HED) uses an accelerating electrostatic field to increase the velocity of the ions. In the discrete-dynode version of an EM, The ion beam from the *m/z* analyzer is focused onto the conversion dynode, which emits electrons in direct proportion to the number of bombarding ions. The secondary electrons from the conversion dynode (ions to electrons) are accelerated and focused onto a second dynode, which itself emits secondary electrons (electrons to electrons). This means, amplification is accomplished through a ''cascading effect" of secondary electrons from dynode to dynode, because the number of electrons impinging on it (Fies, 1988). The efficiency of EM is dependent upon the velocity of the impinging particle (Westmacott et al., 2000), a feature responsible for the disappointing performance of channel EM arrays in MALDI-TOF/MS for high-mass ions (Chen et al., 2003). The secondary electrons are attracted along the positive electrical gradient farther on into the cornucopia. Each time these electrons collide with the wall, additional secondary electrons are expelled, thereby providing amplification. This form of the continuous dynode is curved so as to prevent positive ions from causing spurious signals or feedback signals due to seconadry ionization of residue gas molecule (Watson & Sparkman , 2007).

2.2.5.2 Faraday cup

In this conventional electrical detector, positive ions impinging on the collector are neutralized by electrons dawn from groud after passage through a high-ohmic resistor. A Faraday cup involves an ion striking the dynode (BeO, GaP, or CsSb) surface which causes secondary electrons to be ejected. This temporary electron emission induces a positive charge on the detector and therefore a current of electrons flowing toward the detector. This detector is not particularly sensitive, offering limited amplification of signal, yet it is tolerant of relatively high pressure (Chiu & Muddiman, 2008). Faraday cup detectors have high accuracy, constant sensitivity, and low electrical noise. The principal disadvantage of the classical Faraday is tis relatively long time constant inherent in the amplification system associated with the use of a high-ohmic resistor in the circuitry (Watson & Sparkman, 2007).

2.2.5.3 Negative-ion detection

Modification of a magnetic mass spectrometer to detect negative ions instead of positive ions is relatively simple and the polarity of the accelerating voltage and the magnet current must be reversed. The negative ions have sufficient injection kinetic energy (typically 3~4 kV) to impinge on the EM even though it has a negative bias on its first dynode so that secondary electrons will experience an overall positive gradient toward the last stage(Watson & Sparkman, 2007). In mass spectrometer that uses quadrupole fields no change is required in the analyzer part of the instrument because the field acts on positive and negative ions equally. In the transmission quadrupole, the repeller and acceleration voltages have to be changed from positive to negative.

2.2.5.4 Post-acceleration detector

With the advent of the desorption/inonization technique MALDI, it has been possible to generate ions of extraordinarily high *m/z* values, however, it is often difficult to detect such ions efficiently because their velocity is relatively low if it depends only on the accelerating voltage of the ion source (Beuhler et al., 1991; Matthew et al., 1986). One technique for improving the detection efficiency of high *m/z* value ions is to provide a "post-acceleration" electric field to increase their velocity prior to impacting the conversion dynode of the detector (Hedin et al., 1987). Detection of macro-ions (those heavier than hundreds of kDa having a single or double charge) has been challenge in modern MS. Conventional MS has relied on ionization-based detectors, namely those producing and multiplying secondary electrons associated with particle bombardment of a metal surface; such detectors are restricted to ions having a mass less than 10^6 Da. An ion-to-proton conversion detector (IPD) shows promise for detecting macro-ions (Dubois et al., 1997; Dubois et al., 1999). Photosensitive detectors, relying on laser-indued fluorescence (LIF) or elastic light scattering (ELS), show promise in this area.

2.2.5.5 Channel Electron Multiplier Array (CEMA)

CEMA consists of a honeycomb arrangement of many channels up to several hundred per square inch, each having a diameter on the order of 10 μm made possible with fiber optic technology using metal-doped glass. A potential difference applied to opposite ends of the channels with top and bottom of the honeycomb creates an electrical gradient along the resistive but conducting surface of each channel. The ions enter the individual channels slightly off-axis so that they impinge the wall of the channel. This primary event (an ion colliding with the surface to expel electrons) has very poor efficiency; the efficiency is directly proportional to the velocity of the impacting ion, which explains the exceptionally poor response of the CEMA to massive ions. The secondary electrons resulting from ions impinging the wall continue to ricochet down the channel, producing more secondary electrons upon each impact, thereby amplifying the original ion beam (Watson & Sparkman, 2007). The CEMA is also employed in TOF-MS because it presents a simple flat surface that can be arranged perpendicular to the ion beam to preserve velocity resolution of the ion packets (Price & Milnes, 1984).

2.2.5.6 Photomultiplier conversion dynode

The photomultiplier conversion dynode detector is not as commonly used at the EM yet it is similar in design where the secondary electrons strike a phosphorus screen instead of a dynode. The phosphorus screen releases photons which are detected by the photomultiplier. One advantage of the conversion dynode is that the photomultiplier tube is sealed in a vacuum, unexposed to the environment of the mass spectrometer and thus the possibility of contamination is removed. This improves the lifetimes of these detectors over electron multipliers. A five-year or greater lifetime is typical, and they have a similar sensitivity to the EM (Chiu & Muddiman, 2008).

2.2.5.7 The daly detector

The positive ion beam passing the detector slit is attracted toward an aluminized cathode (the Daly knob) held at a very large negative potential (e.g. -15,000 V) (Daly et al, 1968a; Daly et al, 1968b). Positive ions impacting the Daly knob, which essentially serves ac a conversion dynode, produces up to eight secondary electrons, which are attracted to scintillator unit held at ground potential. The Daly detector offers two significant advantages over other similar detectors. First, most of components (all except the Daly knob) are located outside the vacuum chamber, and thus can be served without distrupting the pressure regime of the mass spectrometer. Second, the large potential difference used between the Daly knob and the slit is particularly advantageous for detecting ions of high mass as a post-accelerator detector. In addition, a deceleration lens can be installed in front of Daly knob to distinguish between stable ions and those that are ionic products of metastable decay (Watson & Sparkman, 2007)

2.2.5.8 Array detector

An array detector is a group of individual detectors aligned in an array format. The array detector, which spatially detects ions according to their different *m/z*, has been typically used on magnetic sector mass analyzers. Spatially differentiated ions can be detected simultaneously by an array detector. The primary advantage of this approach is that, over a small mass range, scanning is not necessary and therefore sensitivity is improved (Chiu & Muddiman, 2008).

2.2.6 Vacuum techniques and systems

All mass spectrometer need a vacuum to allow ions to reach the detector without colliding with other gaseous molecules or atoms. If some collisions between molecules did occur, the instrument would suffer from reduced resolution and sensitivity. Low pressure (i.e. infrequent ion/molecule or molecule/molecule collisions) is also essential for preservation of resolving power in *m/z* analyzers. Once the direction of an ion's path has been established, any interaction with other matter can cause that direction to change (Watson & Sparkman, 2007). One of the first obstacles faced by the originators of mass spectrometer was coupling the sample source to a mass spectrometer. The sample is initially at atmospheric pressure (760 torr) before being transferred into the mass spectrometer's vacuum (~10^{-6} torr), which represents approximately a billion-fold difference in pressure. One approach is to introduce the sample through a capillary column (GC) or through a small orifice directly into the instrument. Another approach is to evacuate the sample chamber through a vacuum lock (MALDI) and once a reasonable vacuum is achieved (< 10^{-2} torr) the sample can be presented to the primary vacuum chamber (< 10^{-5} torr). The mechanical pump serves as a general workhorse for most MS and allows for an initial vacuum of about 10^{-3} torr to be obtained. Once a 10^{-3} torr vacuum is achieved, the other pumping systems, such as diffusion, cryogenic and turbomolecular can be activated to obtain pressures as low as 10^{-11} torr (Figure 5) (Chiu & Muddiman, 2008).

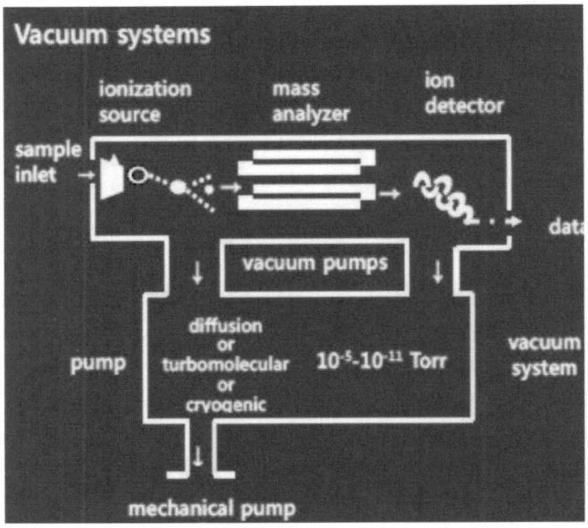

Fig. 5. A well-maintained vacuum is essential to the function of a mass spectrometer (Chiu & Muddiman, 2008).

2.2.7 Conclusion

The mass spectrometer as a whole can be separated into distinct sections that include the sample inlet, ion source, mass analyzer, and detector. A sample is introduced into the mass specterometer and is then ionized. The ion source produces ions either by electron ejection, electron capture, cationization, deprotonation or the transfer of a charged molecule from the condensed to the gas phase. MALDI and ESI have had a profound effect on mass spectrometry because they generate charged intact biomolecules into the gas phase. In comparison to other ionization sources such as APCI, EI, FAB, and CI, the techniques of MALDI and ESI have greatly extended the analysis capabilities of mass spexterometry to a wide range of compounds with detection capabilities ranging from the picomole to the zeptomole level. The mass analyzer is a critical component to the performance of any mass specterometry. Among the most commonly used are the quadrupole, quadrupole ion trap, TOF, TOFR, and FTMS. However, the list is growing as more specialized analyzers allow for more difficult questions to be addressed. For example, the development of the quad-TOF has demonstrated its superior capabilities in high accuracy tandem mass spectrometry experiments. Once the ions are separated by the mass analyzer they reach the ion detector, which is ultimately responsible for the signal we observe in the mass spectrum.

3. Recent development of in high-throughput quantitative bioanalysis by liquid chromatography (LC) coupled with mass spectrometry (MS or MS/MS)

Development of standard techniques of analytical detection in clinical chemistry relies on indirect characteristics of an analyte, e.g. its absorption of light, chemical reactivity or physical interaction with macro-molecules (Vogeser & Kirchhoff, 2011). During the past decade, LC-MS/MS technologies have substantially extended the methodologic armamentarium of clinical laboratories (Vogeser & Seger, 2008). It has become one of the essential basic technologies used in laboratory medicine that shown in the analytical technology mindmap (Figure 6).

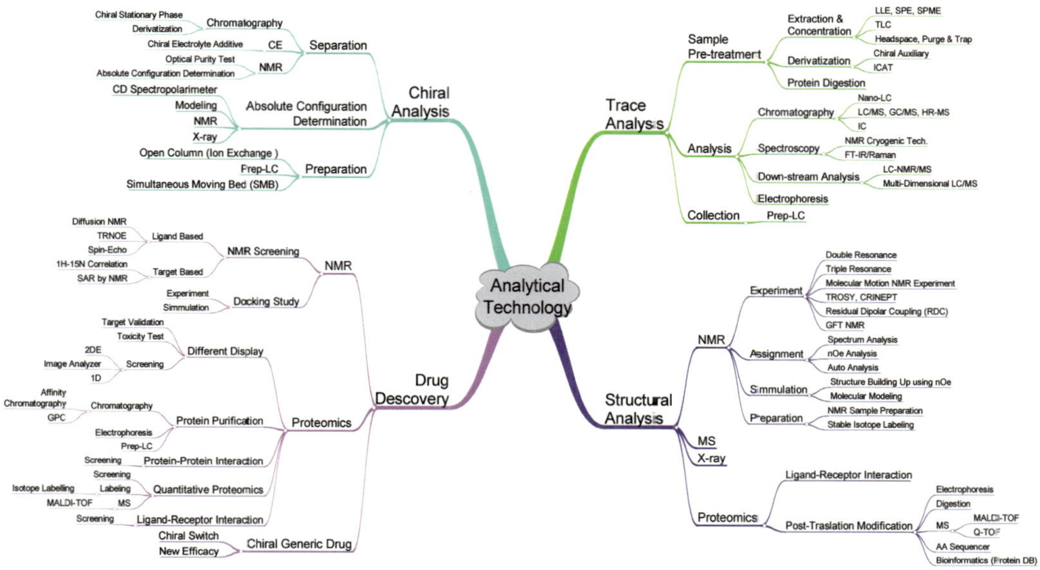

Fig. 6. Mindmap for description of analytical technology (www.bene-technology.com).

In the MS method, in contrast, analytes are detected directly from the molecular characteristics as molecular mass and molecular disintegration patterns. Thus, MS techniques are very attractive for quantification of biomarkers or xenobiotics in the context of diagnostic procedures, since those techniques can enable analyses of much higher specificity compared to standard technologies such as photometry or ligand binding tests. With GC-MS, first MS was introduced to laboratory medicine about 40 years ago. GC-MS allowed the highly specific and sensitive quantification of thermo-stable molecules below a molecular weight of about 500 and became a key method in the toxicology field. With the respect to standardisation and quality assurance of small molecule analytical routine methods the introduction of GC-MS as a reference method was an essential progress, in particular for endocrinology. However, for the several reasons the application of GC-MS remained restricts too few specialized institution in laboratory medicine including mainly toxicological laboratories, metabolism centre, and reference laboratories. The handling and maintenance of GC-MS instruments is very demanding and time-consuming; sample preparation is very laborious and includes sample extraction and analyte derivatization; the analytical run times are long with a typical sample throughput of less than 50 samples per day (Vogeser & Kirchhoff, 2011).

Introduction of API techniques about 20 years ago made practically all potential bio-medical analytes amenable for MS. Furthermore, powerful new technologies of ion-analyses (tandem MS, TOF-MS, ion-trap MS) substantially increased the capabilities of MS analyzers with respect to specificity and to the extent of data read-out in the 1990s. These developments suggested a widespread use of MS methods in routine laboratory medicine. In particular tandem MS instruments hyphenated to LC systems used for sample introduction and pre-fractionation have been implemented in a constantly growing number of clinical laboratories worldwide now (Vogeser & Seger, 2008). LC-MS/MS is attractive for laboratory medicine for three main reasons. (1) The development of new methods is in general straightforward and independent from the diagnostic industry, without the need e.g. to develop analytical antibodies. (2) The highly multiplexed analyses are feasible with very low current costs; the range of potential analytes is practically unlimited; individual "metabolomic analyses" addressing hundreds of analytes from different biochemical pathways and from different chemical classes are possible, as well as a comprehensive and individual description of xenobiotics ("xenobiom"). (3) When applying the principle of isotope dilution internal standardisation, analyses on a reference method-level of accuracy can be performed in a routine laboratory setting (Vogeser & Kirchhoff, 2011). LC-MS/MS today holds enormous potentials for improvements in the pharmaceutical fields and laboratory medicine mainly including TDM, endocrinology, toxicology and metabolomic analyses, therefore, hyphenated techniques are examples of new tools that adoped for developing fast and cost-effective analytical methods. One of the most prevalent hyphenated techniques, LC-MS/MS, has led to major breakthroughs in the field of quantitative bioanalysis since the 1990s due to its inherent specificity, sensitivity, and speed. It is now generally accepted as the preferred technology for quantitating small molecule drugs, metabolites, and other xenobiotic biomolecules in biological matrices as like plasma, blood, serum, urine, and tissue (Xu et al., 2007).

Because samples from biological matrices are usually not directly suitable with LC-MS/MS analysis, sample preparation is inevitable and has traditionally been done using protein precipitation (PPT), liquid-liquid extraction (LLE), or solid-phase extraction (SPE) processes. Manual operations associated with sample preparation are very laborious and time-consuming. Parallel sample processing in 96-well format using robotic liquid handlers and direct injection of plasma using an on-line extraction method have significantly shortened the time for analysis and generated a lot of interests in recent years. A major advantage of on-line SPE over off-line extraction techniques is that the sample preparation step is included into the chromatographic separation and thus eliminates most of the sample preparation time classically performed at the bench-side. Fast gradients and short columns were first utilized in early applications of high-throughput LC-MS/MS assays to reduce run times. Better understanding of how matrix effects can compromise the integrity of bioanalytical methods has reemphasized the need for adequate chromatographic separation of analytes from endogenous biological components in quantitative bioanalysis using LC-MS/MS method (Chiu & Muddiman, 2008).

New developments from chromatographic techniques such as ultra-performance LC with sub-2 mm particles and monolithic chromatography are showing promise in delivering higher speed, better resolution and sensitivity for high-throughput analysis while minimizing matrix effects. On the other hand, automation in LC-MS/MS is great advance in instrumentation field and has to address a number of different processes in laboratory medicine application of LC-MS/MS that includes management of primary samples, assay-specific work-up of samples prior to actual MS analysis, integrated control of the subunits of the LC-MS/MS systems, processing of primary read-outs, and further handling of result data.

3.1 Sample preparation

3.1.1 Basic principles of sample preparation

Adequate sample preparation is a key aspect of quantitative bioanalysis and can often be the cause of bottlenecks during high-throughput analysis. Sample preparation techniques in 96-well format have been well adopted in high-thoughput quantitative bioanalysis. Four main principles of sample extraction are applied in LC: PPT by addition of organic solvents, inorganic acids and/or chaotropic salts, protein filtration, LLE, and SPE. Typically not a very high degree of protein removal is achieved. The increasing demand for high-throughput causes a unique situation of balancing cost versus analysis speed as each sample preparation technique offers unique advantages. Sample preparation with PPT is widely used in bioanalysis of plasma samples. The method has been extended to quantitation of drug and metabolites from whole blood. A sensitive and specific LC–MS/MS method for the simultaneous determination of cyclosporine A and its three main metabolites (AM1, AM4N, and AM9) in human blood have been developed (Koseki et al., 2006). Overall, PPT offers a generic and fast sample preparation technique that can be easily automated. However, when analyzing supernatant from a plasma sample using PPT, salts and endogenous material are still present and cause ion suppression or enhancement that will lead to higher variation from sample to sample. Solvent extraction and SPE typically result in very clean extracts; ion-suppression effects due to residual matrix components are reduced with solvent extraction and SPE compared to mere PPT (Annesley, 2003; Taylor, 2005). While LLE is predominantly based on differential polarity, a variety of different extraction surfaces is available for SPE such as carbohydrates (C2, C8 and C18), ion exchange materials (De Jong et al., 2007; De Jong et al., 2010a), phenylic groups, amino groups, co-polymer mixed mode materials (Dunér et al., 2007), immobilized on particles packed in cartridges. SPE materials can have added particular functions, as removing of phospholipids. Both LLE and SPE allow up to ten-fold concentration of analytes; however, these methods are technically far more demanding compared to PPT. The optimum choice of one of the four main principles of sample preparation is specific for the respective analyte, but also for the individual MS/MS system.

3.1.2 Automated off-line sample preparation

The most widely used principle for automated sample preparation in LC-MS/MS is so far SPE. A wide range of SPE materials is available today and a highly efficient extraction protocol can be tailored for practically all analytes by optimizing the content of organic solvents, pH or ion strength. The use of SPE often gives superior results to those by a PPT method but may not be as cost-effective as PPT due to the labor and material costs associated with the process. A novel 96-well SPE plate that was designed to minimize the elution volume required for quantitative elution of analytes and the plate was packed with 2 mg of a high-capacity SPE sorbent that allows loading of up to 750 µL of plasma. The novel design permitted elution with as little as 25 µL solvent (Mallet et al., 2003). The evaporation and reconstitution step that is typically required in SPE is avoided due to the concentrating ability of the sorbent. A sensitive µElution-SPE-LC–MS/MS method were developed for the determination of M^{+4} stable isotope labeled cortisone and cortisol in human plasma (Yang et al., 2006). In the method, analytes were extracted from 0.3 mL of human plasma samples using a Waters Oasis HLB 96-well µElution SPE plate with 70 µL methanol as the elution solvent. The lower limit of quantitation was 0.1 ng/mL and the linear calibration range was from 0.1 to 100 ng/mL for both analytes. A direct coupling of a pre-analytics module directly with a LC-MS/MS instrument is feasible as well, either by a dedicated system (Alnouti et al., 2005; Koal et al., 2006) or by implementing a HPLC injection valve onto a generic liquid handling system (Daurel-Receveur et al., 2006). A 96-well LLE LC-MS/MS method for determination of a basic and polar drug candidate from plasma samples developed (Xu et al., 2004). A LLE procedure for sample preparation of dextromethaphan, an active ingredient in many OTC regimens, and dextrophan, an active metabolite of dextromethaphan in human plasma were reported (Bolden, 2002). Combination of sample preparation techniques have been developed to achieve desired sample extract purity with high throughput. A simplified PPT/SPE procedure was investiged (Xue et al., 2006). A mixture of acetonitrile and methanol along with formic acid was used to precipitate plasma proteins prior to selectively extracting the basic drug. After vortexing and centrifugation, supernatants were directly loaded onto an unconditioned Oasis MCX µElution 96-well extraction plate, where the protonated drug was retained on the negatively charged sorbant while interfering neutral lipids, steroids or other endogenous materials were eliminated. Additional wash steps were deemed unneccessary and not performed prior to sample elution (Xu et al., 2007). However, the system primarily aims to the analysis of research samples in micro-vials and is not yet useful for the direct application of standards sample containers as used in clinical laboratories.

3.1.3 On-line solid-phase extraction (SPE)

The on-line SPE technique offers speed, high sensitivity by the pre-concentration factor, and low extraction cost per sample, but typically requires the use of program controlled switch valves and column re-configuration. However, the on-line techniques can be fully automated and several generic approaches have recently been developed for on-line sample extraction coupled to LC-MS (Veuthey et al., 2004; Xu et al., 2005; Zang et al., 2005; Xu et al., 2006b). Various column demensions can be configured for the fast analysis of drug and their metabolites in biological matrix at the ng/mL level or lower. A generic method for the fast determination of a wide range of drugs in serum or plasma has been presented for the Spark Holland system (Schellen et al., 2003). The method comprises generic SPE with HySphere particles, on-line coupled to gradient HPLC-MS/MS detection. The optimized generic SPE-LC-MS/MS protocol was evaluated for 11 drugs with different physicochemical properties. An approach for on-line introduction of internal standard (IS) for quantitative analysis was developed on the Spark Holland system (Alnouti et al., 2006). This new technique was applied for direct analysis of model compounds in rat plasma using on-line SPE-LC-MS/MS quantification. On-line IS introduction allows for non-volumetric sample (plasma) collection and direct analysis without the need of measuring and aliquoting a fixed sample volume prior to the on-line SPE-LC-MS/MS that enables direct sample (plasma) analysis without any sample manipulation and preparation (Koal et al., 2006; Alnouti et al., 2005). On-line SPE with high flow rate (normally, 4~6 mL/min) has been achieved by using extraction columns packed with large diameter particles. Sample extraction occurs with very high solvent linear velocity without significant backpressure. Turbulent flow chromatography (TFC) columns marketed by Cohesive Technologies are widely used for this purpose. Minimum or no sample pre-treatment is required and significant sample preparation time is saved (Chassaing et al., 2005). Another commonly used on-line SPE sorbent material is restricted access material (RAM). With a small pore size, RAM works by eliminating the access of large molecules such as proteins to the inner surface of the particles, otherwise, small molecules can freely bind to the sorbent in the normal hydrophobic interaction mode.

Protein molecules quickly pass through the column and are washed out to waste. RAM columns have been used as the SPE and analytical column in the single column mode or coupled with another analytical column in column-switching mode (Vintiloiu et al., 2005; Kawano et al., 2005). A polar functionalized polymer(Strata-X, Phenomenex) has been explored as the extraction support in an on-line SPE LC-MS/MS assay that allows direct analysis of plasma samples containing multiple analytes (Zang et al., 2005). Beside rapid chromatographic seperation, new monolithic-phases have been investigated as extraction support for on-line SPE and an good examples of it was automated procedure using on-line extraction with monolithic sorbent for pharmaceutical component analysis in plasma by LC-MS/MS (Xu et al., 2006a). Endogenous materials from urine contains a great deal of amount of metabolic products that may present a significant challenge to assay developers and often require tedious sample preparation to remove the interfering small molecules. Method development for determining drug or metabolite concentrations from urine samples has been simplified with the implementation of on-line SPE. A sensitive method using on-line SPE and LC-MS/MS system for quantification of urinary cortisol metabolite and cortisol was developed and human urine sample were injected directly onto an on-line SPE apparatus followed by HPLC separation and LC-MS/MS detection (Barrett et al., 2005).

3.1.4 New techniques for sample preparations

While SPE in a 96 position array, PPT in an array, and on-line SPE is used in many clinical MS laboratories now, several further methods of sample preparation may become useful for clinical LC-MS/MS applications in the future. Some of them are already in environmental and food analyses. SPE materials can be packed into pipetting tips; the steps of SPE can be performed within these tips by automated pipetting of sample, washing solution and finally elution medium with direct injection of the eluate into an injection port (Erve et al., 2009). A clear separation between sample preparation and chromatographic fractionation can indeed be overcome. A quadrupole TOF-MS was interfaced with a NanoMate system for immunosuppressant quantification is described with automated clean-up of the hemolysed samples by chromatographic zip-tips and direct injection to an 400 nozzle nano flow-ESI chip source without further chromatographic separation (Almeida et al, 2008). SPE materials can be packed in permanently used syringes of autosamplers as well and also referred to as solid phase micro extraction (SPME) or microextraction by packed sorbents (MEPS) (Blomberg et al., 2009; Vuckovic et al., 2010). Reliable methods can be applied in modern high-end autosampler devices (e.g., CTC PAL) and can be used an alternative to on-line SPE. Ferromagnetic micro-particles with modified surfaces (e.g., C-18 material) might be interesting for the automation of sample preparation for LC methods as well (Vogeser et al, 2008). Such particles represent a solid phase, however, can be handled in a suspension as a liquid. The use of such particles in mass spectrometry might parallel the achievements in the automation of heterogonous immunoassays which also require a convenient handling of solid phases. Ferromagnetic particles may also be handled within a HPLC system (Vogeser M., 2009). Another methods of sample preparation for mass spectrometry also include miniaturized "lab-on-the-chip" solutions for microfluidic applications (Koster & Verpoorte, 2007).

3.2. Separation

3.2.1 Ultra-performance liquid chromatography(UPLC) with sub-2 μm particles

The use of smaller particles in packed-column LC is a well-known approach to shorten the diffusion path for a given analyte. Recent technology advances have made available reverse phase chromatography media with sub-2 µm particle size along with liquid handling systems that can operate such columns at much higher pressures. This technology termed UPLC, offers significant theoretical advantages in resolution, speed, and sensitivity for analytical applications, particularly when coupled with mass spectrometers capable of high-speed acquisitions (Xu et al, 2007; Mazzeo et al, 2005). Today, two driving forces continue to test the limits of HPLC. One is the need for faster separations, such as analyses of either simple samples or a few constituents in a complex sample (Romanyshyn, 2001). The second is the desire to achieve greater separation power to quantify or identify all the constituents of a complex sample or to compare the contents of complex samples with each other (van der Horst & Schoenmakers, 2003; Liu et al. 2002; Plumb et al. 2003). The same driving forces resulted in the overwhelming breakthrough in the past decade of LC-MS techniques, which continue to spawn new approaches for faster or more powerful separations (Niessen, 2003; Tiller et al, 2003). The trend in LC has been the continued reduction in particle size. Harnessing the chromatographic potential of sub-2-µm particles leads to significant improvements in terms of resolution, analysis speed, and detection sensitivity. The van Deemter plots for hexylbenzene demonstrate the performance improvements that 1.7-µm particles offer over the currently used 5.0- and 3.5-µm sizes. The 1.7-µm particles give 2~3 x lower plate-height values. The particles also achieve the lower plate height at higher linear velocities and over a wider range of linear velocities. The result is better resolution and sensitivity as well as reduced analysis time (Niessen, 2003; Tiller et al, 2003). In isocratic separations, the resolution is proportional to the square root of efficiency. Particle size and efficiency at the optimum linear velocity are inversely proportional. Therefore, resolution is inversely proportional to the square root of particle size. With this knowledge, we can calculate that 1.7-µm particles will offer 1.7 x and 1.4 x greater resolution than 5.0-µm and 3.5-µm particles, respectively, at equal column lengths. Because analysis time in isocratic separations is inversely proportional to flow rate, 1.7-µm particles offer 1.7 x higher resolution than 5.0-µm particles in a third of the time, or 5 x higher productivity (resolution per unit time). The benefit of 1.7-µm particles over 3.5-µm ones is 1.4 x higher resolution in half of the time, or 3 x higher productivity. More efficient peaks translate to narrower and taller peaks. Peak width is inversely proportional to the square root of efficiency; the peak height is inversely proportional to peak width. Therefore, when smaller particles are used to make the peaks narrower, the peak height is also increased. If the detector is assumed to be concentration-sensitive (as is the case for UV detectors) and the detector noise remains constant, then sensitivity, as defined by S/N, will also increase. Specifically, in comparison with 5.0- µm particles, 1.7-µm particles offer 1.7 x higher sensitivity. When compared with 3.5-µm particles, 1.7-µm particles provide 1.4 x higher sensitivity for the same column length. Several practical issues must be addressed to achieve optimum performance. At constant L and optimum flow rate, back pressure is inversely proportional to the third power of dp (Martin et al, 1974). Therefore, the 3-fold reduction in dp in UPLC translates to back pressures that are 27 x higher compared with HPLC separations when 5-µm particles are used. Therefore, UPLC is an exciting new area of LC. A natural extension of HPLC, this technique is easy to take full advantage of and requires minimal training. Although we have demonstrated the use of UPLC for reversed-phase separations, we expect that it will also be beneficial in the areas of normal-phase, hydrophilic interaction and ion-exchange chromatographies as well as chiral separation modes that applied for the determination of doxazosine in human plasma by UPLC–MS/MS (Al-Dirbashi et al, 2006) and investigated

UPLC as an alternative to HPLC for the analysis of pharmaceutical development compounds (Wren & Tchelitcheff, 2006). Data on three compounds were presented showing that significant reductions in separation time can be achieved without compromising the quality of separation. Using a poly-drug reference standard and whole blood extracts, the authors successfully separated and identified amphetamine, methamphetamine, ephedrine, pseudoephedrine, phentermine, MDA, MDMA, MDEA, and ketamine in less than 3 min by the Acquity UPLC-Micromass Quattro Micro API MS system (Waters Corporation, USA) (Apollonio et al, 2006). The approach of orthogonal extraction/chromatography and UPLC significantly improves assay performance while also increasing sample throughput for drug development studies (Shen et al, 2006). Other direct comparison experiments using UPLC–MS/MS and HPLC–MS/MS have shown that the UPLC–MS/MS improved cycle time by 50–100% with increased sensitivity In the study about the differences in LC–MS performance by conducting a side-by-side comparison of UPLC for several methods previously optimized for HPLC-based separation and quantification of multiple analytes with maximum throughput (Churchwell, 2005). Sensitivity increases with UPLC, which were found to be analyte-dependent, were as large as 10-fold and improvements in method speed were as large as 5-fold under conditions of comparable peak separations. Improvements in chromatographic resolution with UPLC were apparent from generally narrower peak widths and from a separation of diastereomers not possible using HPLC. A similar HPLC-MS/MS quantification protocol was developed for comparison purposes (Yu et al, 2006). Both UPLC-MS/MS and HPLC-MS/MS analyses were performed in both positive and negative ion modes during a single injection. Peak widths for most standards were 4.8 sec for the HPLC analysis and 2.4 sec for the UPLC analysis. Compared with the HPLC-MS/MS method, the UPLC-MS/MS method offered 3-fold decrease in retention time, up to 10-fold increase in detected peak height, with 2-fold decrease in peak width. Limits of quantification (LOQ) for both HPLC and UPLC methods were evaluated.

3.2.2 Monolithic chromatography

There is considerable interest to improve throughput by using monolithic columns because they exhibit higher seperation efficiency at high flow velocities when compared to conventional LC columns (Zhou et al, 2005; Ikegami & Tanaka, 2004; Cabrera, 2004; Li et al., 2005; Wang et al, 2006).The structural characteristics of the monoliths and those of the conventional beds of particular packing materials are very different that are their high external porosity resulting from the structure of the network of through macropores and the structure of the stationary phase skeleton that consists of a network of small, thin threads of porous silica. Two types of monolithic supports are currently available organic polymers such as polymethacrylates, polystyrenes, or polyacrylamide and inorganic polymers based on silica, carbon and zirconia (Xu et al, 2007). A high-throughput LC-MS/MS method using a Chromolith RP-18 monolithic column was developed for the determination of bupropion, an antidepressant drug, and its metabolites, hydrobupropion and threo-hydropupropion in human, mouse, and rat plasma (Borges et al, 2004). The monolithic column performance as a function of a column backpressure, peak asymmetry, and retention time reproducibility are adquately maintained over 864 extracted plasma injections like as a high-througput LC-MS/MS method for the determination of methylphenidate (MPH), a central nervous system stimulant, and its de-esterified metabolite, ritalinic acid (RA) in rat plasma samples (Barbarin et al, 2003). A monolithic column was directly compared to a conventional C18 column as the analytical column in method validation of a drug and its epimer metabilite (Huang et al, 2006). Because the chosen drug and its epimer metabolite have same selected reaction monitoring (SRM) transitions, chromatographic baseline seperation of these two compounds was required. Sample preparation, mobile phases and MS conditions were kept the same in the column comparison experiment. The methods on the two systems were found to be equivalent in validation parameters and chromatographic seperation, but the monolithic column method increased the sample throughput by a factor of two (Xu et al, 2007). The significantly improved seperation speed by monolithic columns demanded higher throughput on sample extraction. An attractive approach using monolithic separation is to combine it with high-flow on-line extraction, which allowed for the fast extraction and separation of samples (Zeng et al, 2003; Zhou et al, 2005). A normal phase method on silica was adapted for the high-throughput analysis with separations and run times less than 1 min. Another high-throughput approach using a monolithic column showed tha analysis of over 1100 plasma samples preparaed by PPT in approximately 9 h (Mawwhinney & Henion, 2002). Some disadvanages for using monolithic columns for high-speed methods have been reported(Chen et al, 2002) that a lack of separation for very polar analytes (nicotinic acid and five metabolites) using a monolithic RP-18e column.

3.2.3 Hydrophilic interaction chromatography (HILIC)

HILIC-MS has been gaining recognition as a valuable technique for analyzing polar molecules in biological matrix in recent years (Eerkes et al, 2003; Song et al, 2006; Deng et al, 2005). Polar compounds typically have very limited retention on reverse-phase(RP)-columns. Reverse phased(RP)-HPLC mobile phase with a very low organic content must be used to seperate the anlayte from matrix interference. When using ESI-MS, the very high aqueous mobile phase can cause low ionization efficiency. The highly volatile organic mobile phases used in HILIC provide increased ESI-MS sensitivity. A bioanalytical method using automated sample transferring, automated LLE and HILIC-MS/MS for the determination of fluconazole in himan plasma are developed (Eerkes et al, 2003). After LLE, the extracts was evaporated to dryness, reconstituted, and injected onto a silica column using an acquous-organic mobile phase. The chromatographic run time was 2.0 min per injection. The use of HILIC could eliminate the evaporation and recostitution steps that hamper improvement of throughput and automation. In a validated a single-pot LLE with HILIC-MS/MS method for the determination of Muraglitazar, a hydrophobic diabetic drug, in human plasma, organic layer was then directly injected into an LC-MS/MS after extraction with acetonitrile and toluene. Incomparison with a reversed-phase LC-MS/MS, this single-pot LLE, HILIC-MS/MS imporved the detection sensitivity by greater than 4-fold based on the LLOQ signal to noise ratio.

3.2.4 Turbulent flow chromatography

Turbulent flow chromatography(TFC) is a high flow chromatographic technique, which allows high-throughput bioanalysis by requiring little sample preparation and potentially fast cycle times. The TFC system takes advantage of unique flow dynamics that allow us to analyze compounds in biological matrices with very little sample preparation (Ayton et al, 1998). High-throughput LC-MS/MS approaches using TFC employ small internal diameter columns (typically 1 mm or 0.18 µm) with large particles (20~60 µm), which allow high flow rates to be utilized without impractical pressure increases. The large particles serve to trap the analytes and thus allow unretained compounds to be washed out. The resulting trapped analyte is typically eluted onto an analytical column (Ayrton et al, 1997). First reports of high-throughput bioanalysis using TFC coupled to MS/MS did not use an analytical column but a 50 mm x 1 mm, 50 µm column to perform cleanup and analysis that describes the direct analysis of plasma samples using a commercially available TFC system; the total run time was 2.5 min, and the method was validated from 5 to 1000 ng/mL(Ayrton et al, 1997) and later extended the use of "ultra-high flow" to a capillary LC column packed with large particles of 30 µm (Ayrton et al, 1999). Although there are advantages in using ultra-high flow on capillary (0.18 mm i.d.) versus narrow bore (1.0 mm i.d.) columns, trade offs must be in sample handling, column capacity, and robustness. Comparison of the methods showed the LLOQ of the TFC and SPE methods was 1 ng/mL; the TFC method used a much smaller plasma aliquot. It should be noted that drug B analyzed using a PPT method, which achieved an LLOQ of 0.5 ng/mL (Zimmer et al, 1999). In several reports about comparison of methods, these reports show that TFC is an attractive approach for high-throughput work involving metabolites (Ramos et al, 2000; Lim et al, 2001; Hopfgartner et al, 2002; Herman, 2002). Commercial systems are available which acommodate 1 to 4 seperate TFC systems and one mass spectrometer (Berna et al, 2004).

4. Bottlenecks of liquid chromatography with mass spectrometric method associate with matrix effect

4.1 What is the matrix effect?

HPLC-MS/MS detection has been demonstrated to be a powerful technique of choice for the quantitative determination of drugs and metabolites in biological fluids (Matuszewski et al, 2003; Taylor, 2005a; Kebarle & Tang, 1993). However, the common perception that utilization of HPLC-MS/MS guarantees selectivity has been challenged by a number of reported examples of lack of selectivity due to ion suppression or enhancement caused by the sample matrix (Clarke et al, 1996; Buhrman et al, 1996; Matuszewski et al, 2003) and interferences from metabolites (Constanzer et al, 1997; Jemal & Xia, 1999). The central issue is what experiments, in addition to the validation data usually provided for bioanalytical methods, need to be conducted to demonstrate the absence of a relative ("lot-to-lot") matrix effect and to confirm HPLC-MS/MS assays selectivity. Matrix effects occur when molecules coeluting with the compounds of interest alter the ionization efficiency of the electrospray interface. The matrix effects is probably originates from the competition between an analyte and the coeluting, undetected matrix components. Matrix effects are the result of competition between nonvolatile matrix components and analyte ions for access to the droplet surface for transfer to the gas phase (King et al, 2000). Better understanding of how matrix effects can compromise the integrity of bioanalytical methods has re-emphasized the need for adequate chromatographic separation of analytes from endogenous biological components in quantitative bioanalysis using LC-MS/MS (Xu et al, 2007). Inherent specificity of LC-MS/MS methods results in chromatograms that do not present any apparent interference, although relatively high concentrations of matrix components are sometimes present (Hernández et al, 2005). Matrix effects are also compound-dependent and the chemical nature of a compound has a significant effect on the degree of matrix effects (Bonfiglio et al, 1999). In a study of 4 compounds of different polarities under the same mass spectrometric conditions, the most polar was found to have the largest ion suppression and the least polar was affected less by ion suppression. These findings of differential matrix effects have important ramifications particularly when selecting a suitable IS for quantification purposes. For example, if a drug and a glucuronide metabolite were quantified by IS against a close analogue of the parent drug and matrix effects were slightly different between samples, then the change in ionization of the more polar glucuronide metabolite would probably not be compensated by the IS. Thus if there are multiple analytes to be quantified, with varying degrees of polarity, there may be requirements for multiple IS (Lagerwerf et al, 2000). The importance of matrix effects on the reliability of HPLC-ESI-MS/MS has been shown in terms of accuracy and precision (Matuszewski et al, 1998), and when ion suppression occurs, the sensitivity and lower limit of quantification of a method may be adversely affected (Buhrman et al, 1996). Thus to develop a reliable HPLC-ESI-MS/MS method, experiments should be performed to understand these matrix effects by careful consideration to evaluate and eliminate matrix effects.

4.2 Evaluation of matrix effects

The two main techniques used to determine the degree of matrix effects on an HPLC-ESI-MS/MS method are postextraction addition and postcolumn infusion. The postextraction addition technique requires sample extracts with the analyte of interest added postextraction compared with pure solutions prepared in mobile phase containing equivalent amounts of the analyte of interest (Annesley, 2003; Matuszewski et al, 2003; Matuszewski, 1998). The difference in response between the postextraction sample and the pure solution divided by the pure solution response determines the degree of matrix effect occurring to the analyte in question under chromatographic conditions. The post-column infusion method provides a qualitative assessment of matrix effects, identifying chromatographic regions most likely to experience matrix effects (Bonfiglio et al, 1999). Any endogenous compound that elutes from the column and causes a variation in ESI response of the infused analyte is seen as a suppression or enhancement in the response of the infused analyte (Bakhtiar & Majumdar, 2007; Bonfiglio et al, 1999). This approach, however, does not provide a quantitative understanding of the level of matrix effect observed for specific analytes. In addition, if several compounds are determined in one method, all compounds should be infused separately to investigate possible matrix effects for every analyte. Therefore, matrix effects are not investigated for lower than LLOQ. In contrast, the post-extraction spike method quantitatively assesses matrix effects by comparing the response of an analyte in neat solution to the response of the analyte spiked into a blank matrix sample that has been carried through the sample preparation process (Matuszewski et al, 2003; Chambers et al, 2007). The absolute matrix effect was defined as the comparison of the signal response of a standard present in a sample extract from one single lot to the response of a standard in neat solution. However, even more important is the evaluation of the relative matrix effect, which is the comparison of matrix effect values between different lots of biofluids and therefore proposed that matrix effects should be investigated in biofluid samples from at least five different sources (Matuszewski et al, 2003). In a more recent paper, they have suggested to use the precision of the calibration line slopes in five different lots of a biofluid as an indicator of relative matrix effects. The relative standard deviation should not exceed 3–4% for the method to be considered practically free from relative matrix effects (Matuszewski, 2006). Recently, Heller presented a new concept, namely matrix effect maps, for visualizing the impact of various parameters on matrix effects associated with a given method. In this approach matrix effects are studied as a function of the amount of co-injected matrix extract (Heller, 2007). In Heller's approach, two sets of mixtures were prepared in different formats and, to provide different matrix effect conditions for testing these mixtures, chromatographic conditions were altered as well. Furthermore, operational variables like desolvation gas flow and temperature could also be included in these matrix effect maps. However, these issues remain to be fully tested. In the future, this approach could help to determine the ruggedness of a developed method.

4.3 Methodology to overcome matrix effects

Different actions can be taken to overcome matrix effects. Matrix effects may be reduced by simply injecting smaller volumes or diluting the samples, which is useful as long as instrumental sensitivity remains adequate (Heller, 2007). Other possibilities to reduce or eliminate matrix effects are the optimization of sample preparation and/or chromatographic parameters (Xu et al, 2007; Hernández et al, 2005; Niessen et al, 2006). Another approach is the use of an internal standard (IS) to compensate for the alteration in signal. If sensitivity is not an issue, an alternative ionization source, less sensitive to matrix effects, can be used, e.g. APCI (Chambers et al, 2007) or electron ionization. In many cases, several approaches are combined to achieve adequate quantitative results (Niessen et al, 2006). Because the presence of coeluting compounds cause matrix effect, there is need to remove or minimize their presence to obtain a robust LC-MS/MS.

4.3.1 Sample preparation techniques

In general, matrix effects may be reduced by simply injecting smaller volumes or diluting the sample. However, these solutions will clearly influence the sensitivity of the method and are therefore in many cases not appropriate (Antignac et al, 2005). Proper sample clean-up is therefore of primordial importance. Even if PPT is the simplest and fastest method for preparing samples, however, it does not result in a very clean extract and is most likely to cause ion suppression in ESI. Co-elution of these compounds with the compound of interest affects the ESI droplet desolvation process (Dams et al, 2003; Bakhtiar & Majumdar, 2007; Chambers et al, 2007). In comparison with PPT, the extracts obtained from SPE are relatively cleaner (Bakhtiar & Majumdar, 2007; Chambers et al, 2007). LLE often yields rather clean extracts, but the procedures are usually cumbersome and have many pitfalls. Multiple extraction steps are commonly needed to increase analyte recovery and to obtain cleaner extracts (Jessome & Volmer, 2006). Supported LLE can be used to decrease sample preparation time and improve analyte recovery. Rapid extraction of analyte occurs during this intimate contact between the two immiscible phases. The solvent moves through the packing by gravity flow or use of a gentle vacuum(Majors, 2006).

4.3.2 Chromatographic sepreration

Improved chromatography is a straightforward way to separate interfering compounds from analytes. Gradient elution can help to wash the column after injection and prevent late-eluting compounds from the previous injection to interfere. If the analytes are ionizable, the pH of the mobile phase can have a significant impact on the retention, selectivity and sensitivity of the separation (Chamber et al, 2007). It was also shown that fast gradient LC promotes matrix effects by reducing chromatographic separation between analytes and endogenous compounds. If high throughput is required, effective sample pretreatment becomes critical, since the chromatographic conditions are not able to reduce matrix effects. Another possibility is the use of a stationary phase with a different selectivity, for example hydrophilic-interaction liquid chromatography (HILIC). HILIC which combines the use of bare silica or polar bonded stationary phases and mobile phases with a high content of organic solvents, has been proven to be a valuable tool for the analysis of polar compounds in biological samples (Ji et al, 2008). UPLC retains the practicality and principles of classical LC, but increases the speed, resolution and sensitivity of the method by using columns with small diameter (1.7-µm) particles and high pressures. The improved resolution might provide a benefit with respect to matrix effects, through improved separation from endogenous components and obtained a statistically significant reduction in matrix effects under a variety of chromatographic conditions and with multiple basic analytes, using the UPLC technology (Chambers et al, 2007).

4.3.3 Mass spectrometric analysis

APCI-MS is less susceptible to matrix effects than ESI-MS and, however, the occurrence of matrix effects has also been shown with APCI (Dams et al, 2003; Sangster et al, 2004; Niessen et al, 2006). In addition, the APCI interface was more susceptible to matrix effects than the ESI interface for the same instrument. Even with this source, matrix effects were observed for most analytes and internal standards at all concentrations tested, with values ranging from 85.2 to 149.4% (Mei et al, 2003). Miniaturized ESI methods are proven to be more tolerant towards contaminations in the analyte solution. Dialysate matrix effects were estimated at different concentration levels of oxcarbazepine and its major metabolite, using a column switching microbore, capillary and nanoLC–MS/MS system (Lanckmans et al, 2006). Only at the lowest level of the microbore system, a significant matrix effect was observed. Since a lower flow rate reduces the size of the charged droplets, fewer droplet fission events and less solvent evaporation are required for ion release in the gas phase. This leads to a reduction in contaminant concentration(Lanckmans et al, 2006; Schmidt et al, 2003). The ionization suppression caused by matrix effects after plasma or urine injection have been observed in a conventional restricted-access media-LC system (Georgi & Boos, 2006), in contrast, assaying similar compounds with a capillary chromatographic setup, such matrix effects were not present and this attribute to the better characteristics of ESI-MS/MS under low flow rate (Santos-Neto et al, 2008). The extent of matrix effects also depends on the source design of the LC–MS system used. In some cases, problems observed with matrix effects can be solved by using a MS instrument from another manufacturer (Niessen et al, 2006; Mei et al, 2003). Regarding the ionization polarity, the negative mode is usually considered as more specific and consequently less subjected to ion suppression, but in practice, this is of course not possible for all analytes. (Antignac et al, 2005; Niessen et al, 2006).

4.3.4 Role of internal standard

As IS either a structural analogue or a stable isotope labeled-IS can be applied. However, the ionization of the analogue IS and the analyte may be differently affected by the matrix.

This can be solved by using a stable isotope labeled-IS which co-elutes with the drug, since matrix effects should not affect the relative efficiency of ionization of the drug and its stable isotope labeled-IS. The stable isotope labeled-IS are compounds in which several atoms in the analyte are replaced by their stable isotopes, such as ^{2}H (D), ^{13}C, ^{15}N or ^{17}O with ^{2}H being the most frequently used isotope. It is important that the mass difference between the analyte and the stable isotope labeled-IS is at least 3 mass units (Stokvis et al, 2005), in order to avoid signal contribution of the abundance of the natural isotopes to the signal of the IS. If the compound and IS are not separated adequately by mass, this will result in quadratic standard curves (http://www.ionsource.com/tutorial/msquan). However, issues like isotopic purity of compounds, cross-contamination and cross-talk between MS/MS channels, isotopic integrity of the label in biological fluid and during sample processing, etc. should be carefully addressed (Matuszewski et al, 2003). The ^{13}C, ^{15}N or ^{17}O-labeled IS may be more ideal than the ^{2}H- labeled ones, since deutrium and hydrogen have greater differences in their physical properties than for example ^{12}C and ^{13}C (Wang et al, 2007). It is of primordial importance to choose an appropriate IS during method development and to closely monitor the method performance in routine use, since only limited lots of biological matrix are tested during method validation (Wang et al, 2007). If ion suppression significantly reduces the signal of analyte and/or of the IS, the signal to noise may decrease to a poit where accuracy and precision may be negatively affected (Annesley, 2003). Even if this is not always practically feasible, a number of labeled IS identical to the number of compounds to be analyzed would be required (Hernandez et al, 2005).

4.3.5 Matrix effect on the biological method validation

Studies about matrix effects in quantitative bioanalysis revealed that the ion suppression or ion enhancement is frequently accompanied by significantly deterioration of the precision and accuracy of the method. The precision, expressed as relative standard deviation, is plotted as a function of the analyte concentration. While for a single plasma lot the precision is acceptable, this is not the case when five different lots of plasma are taken into account (Matuszewski et al, 2003; Niessen et al, 2006). An absolute matrix effects on the other hand will primarily affect the accuracy of the method (Matuszewski et al, 2003). Even if matrix effects can be compensated by an appropriate IS, efforts should be made to eliminate these co-eluting compounds, since their presence will be reduce mathod sensitivity. When analyzing low concentrated samples, this can lead to false negative results. Biofluids such as plasma and especially urine represent highly complex matrices and its composition can be various significantly between individuals and species, but also within an individauls (Georgi & Boos, 2006). Most method validations are performed using calibration standards and QC samples prepared from the same pool of blank matrix. Using these homogenous samples for validation does not take into account the inter-and intrapatient matrix variability (Taylor, 2005a). Checking the quality of an assay using QC samples, which are prepared in the same matrix as calibration standards, will not reveal matrix effects observed in the incurred samples (Dewe et al, 2007). Repetitive analysis incurred samples is one of the best strategies to evaluate any hidden analytical effect in the method. While it is not practical to prepare calibration standards and QC standards for each individual matrix source, some assessment of patients variability must be undertaken (Taylor, 2005b).

5. Application of LC-MS/MS in basic and clinical laboratory fields

5.1 LC-MS/MS in drug metabolism & toxicology studies

Studies of the metabolic fate of drugs and other xenobiotics in living systems may be divided into three broad areas such as qualitative studies for elucidation of metabolic pathways through identification of circulatory and excretory metabolites, quantitative studies for determination of pharmacokinetics of the parent drug and/or its primary metabolites, and mechanistic studies for identification of chemically-reactive metabolites, which play a key role as mediators of drug-induced toxicities. The mass spectrometry has been regarded as one of the most important analytical tools in studies of drug metabolism, pharmacokinetics and biochemical toxicology. With the commercial introduction of new ionization methods such as API techniques and the combination of LC-MS or LC-MS/MS, it has now become a truly indispensable technique in pharmaceutical research. Triple stage quadrupole and ion trap mass spectrometers are presently used for this purpose, because of their sensitivity and selectivity. API-TOF mass spectrometry has also been very attractive due to its enhanced full-scan sensitivity, scan speed, improved resolution and ability to measure the accurate masses for protonated molecules and fragment ions (Kamel & Prakash, 2006). The study of the metabolic fate of drugs is an essential and important part of the drug development process. The analysis of metabolites is a challenging task and several different analytical methods have been used in these studies. However, after the introduction of the API technique, ESI and APCI, LC-MS has become an important and widely used method in the analysis of metabolites owing to its superior specificity, sensitivity and efficiency. Among the pharmacokinetic properties, metabolic characterization is a key issue and nowadays it is integrated into the early discovery phase. Biotransformation can lead also to some unwanted consequences, such as rapid clearance of the drug from the body, formation of active metabolites, drug–drug interactions due to enzyme induction or competition and formation of reactive or other toxic metabolites (Gibson & Skett, 1994). Metabolic pathways are divided into phase I and phase II reactions, and both classes of reaction often occur in parallel for particular compounds. The information required to determine the metabolic fate of an NCE includes detection of metabolites, structure characterization and quantitative analysis. Since the introduction of the ESI and APCI techniques, LC/MS has become an ideal and widely used method in the analysis of metabolites owing to its superior specificity, sensitivity and efficiency (Sinz & Podoll, 2002). API techniques are compatible with reversed-phase eluent systems, taking into account the use of volatile solvents and additives in chromatographic separation, thus preserving all the advantages of LC. Together, ESI and APCI provide efficient ionization for very different type of molecules including polar, labile, and high molecular mass drugs and metabolites (Kostiainen et al, 2003). Most work in metabolite analysis is carried out by using triple-quadrupole mass spectrometers (QQQ), as their MS/MS scan types (Schwartz et al, 1990) are highly helpful in the identification of metabolites and provide the required specificity and sensitivity (Clarke et al, 2001). Product ion scans are used for identification and MRM provides the high sensitivity required in quantitative analysis. The unique feature of QQQ is its capability to identify families ofmetabolites by using neutralloss and precursor ion scans.33 However, the sensitivity of identification of metabolites by the full-scan mode may not always suffice and instead of QQQ the use of IT and TOF-MS has increased. Both of these techniques provide high full-scan sensitivity. In addition, modern IT-MS technology with its MS*n* (Schwartz et al, 1990) capability is highly efficient in the structural analysis of metabolites. The API-TOF mass spectrometry provides high-resolution analysis with a mass accuracy better than 10 ppm, and hence the possibility of the determination of the elemental compositions of metabolites and high specificity in their detection. (Zhang et al, 2000a; Zhang et al, 2000b). Furthermore, the quadrupole-TOF mass spectrometer (Q-TOF) provides high sensitivity for the determination of full-scan product ion spectra of metabolites (Zhang et al, 2000a). The first step in metabolite profiling for dertection of metabolitea and deferentiation of the site of biotransformation is to identify all the possible metabolites. The second step is their structural characterization and finally quantitation. The classical method in metabolic analysis is radioactive labeling (^{14}C, ^{3}H) of a parent drug and detection of metabolites by LC with radioactivity detection (Veltkamp et al, 1987; Egnash & Ramanathan, 2002). Themethod is especially powerful when combinedwith on-line MS detection. Radioactivity detection provides localization of the metabolites in a chromatogram and MS ensures structure specific identification of the metabolites. However, this approach has several disadvantages. For example, synthesis and purification of radioactive compounds are expensive and time-consuming, radiation is a potential health risk for humans and the requirements for handling radioactive material and wastes make the use of radiolabeled compounds very costly. For these reasons and the fact that radioactively labeled compounds are only very rarely available in early discovery phase, different, simpler, MS techniques are increasingly used in metabolite identification (Kostiainen et al, 2003). MS/MS offers more specific detection of metabolites in complex matrices than unit resolution full-scan MS. MS/MS scanning methods include product ion, precursor ion and neutral loss scans (Schwartz et al, 1990). Metabolites are derivatives of the parent drug and as such it can be assumed that many of the metabolites show the same fragment ions or neutral losses as the parent drug. Therefore, the precursor ion and neutral loss scan modes with QQQ are especially useful in group-specific detection of metabolites (Clarke et al, 2001). To overcome the problems in metabolic analysis, new technologies are continuously being developed. The recently introduced LC-MS/NMR technology provides unambiguous structure characterization of metabolites. Although the sensitivity of NMR does not suffice for the analysis of metabolites in trace quantities, the sensitivity of the technology is continuously being improved. The use of standards or radiolabeled compounds in quantitative analysis and thus the time-consuming synthesis step of reference compounds can be avoided, when on-line coupling of LC-MS to detection techniques that provide equimolar responses, such as ICP-MS or chemiluminescent nitrogen detection, are used. Microfluidic systems offer possibilities to integrate all the experimental steps of metabolite analysis on one microchip, providing complete analysis steps (e.g. sample pretreatment, chemical reactions, analytical separation, and detection and data processing steps) on a single device with a high level of automation. Progress in microfluidics gives reason to assume that the metabolite analysis will be carried out by miniaturized lab-on-a-chip techniques integrated with miniaturized mass spectrometers in the near future (Kostiainen et al, 2003).

5.2 LC-MS/MS as quantification method for biogenic amines

The term biogenic amines refer to amine containing biogenic substrates such as catecholamines, serotonin and histamine. Biogenic amines function throughout the body, both in the central and peripheral nervous system. Disorders affecting their metabolism or action can have devastating effects on homeostasis of the human body (Hyland, 1999). In clinical chemistry quantification of biogenic amines is mainly used for diagnosis of neuroendocrine tumors such as pheochromcytoma and carcinoids. HPLC coupled tandem MS is becoming an indispensible technique in the special chemistry laboratories in clinical chemistry as it greatly increases sensitivity and specificity of test results. Applying this technique will result in improved biochemical diagnosis of endocrine disorders, and opens new roads to gain insight in pathophysiological processes. For the measurement of biogenic amines in biological matrices, LC-MS/MS needs to compete with conventional HPLC with UV, fluorescence or ECD and to a lesser extend GC methods and immunoassay (Yi & Brown, 1991). In pharmaceutical industries and toxicology laboratories, LC-MS/MS is the method of choice for the development and the measurement of drugs (Maurer, 2007). For clinical chemical analyses LC-MS/MS is rapidly emerging and is applicable to a broad selection of compounds, especially for the diagnosis of aberrations within the endocrine system, such as biogenic amines (Chan et al, 2000; Kushnir et al, 2002; Lionetto et al, 2008; de Jong et al, 2008; de Jong et al, 2009a), but also steroids (Soldin and Soldin, 2009), thyroid hormones (Yue et al, 2008), and vitamin D (Higashi et al, 2008). Furthermore LC-MS/MS enable the use of sophisticated sample pretreatment techniques and automation of the whole process by on-line coupling of the separate techniques. To correct for losses during sample pretreatment, analyte separation and detection, stable isotopes of analytes are used as internal standard with mass spectrometry. LC-MS/MS combines the physical separation capabilities of HPLC with the high analytical sensitivity, specificity and accuracy of mass spectrometric detection. In recent years, LC-MS/MS equipment has been improved in performance. Due to its superior specificity, shorter runtimes and less laborious sample preparation, LC-MS/MS methods replace more and more of the conventional HPLC, GC-MS and immunoassay techniques (Taylor, 2005b; Vogeser & Seger, 2008b; Maurer, 2007). LC-MS/MS has, just as HPLC and GC-MS, the advantage that several compounds can be measures simultaneously (Holst et al, 2007).

Introduction of on-line solid-phase extraction coupled to LC-MS/MS shortened chromatographic run times and allowed automation of sample preparation (de Jong et al, 2007). Urinary deconjugated metanephrines have been analyzed with LC-MS/MS for many years. Since these markers occur in higher concentration ranges and require less sensitive assays. Catecholamines assist in the diagnosis of neuroendocrine catecholamine-producing tumors, such as pheochromocytoma and neuroblastoma, in addition to metanephrine and HVA (Lenders et al, 2002; Sawka et al, 2003). Recently, an on-line SPE LC-MS/MS method has been described for catecholamines in urine, especially improving specificity, sensitivity and run time compared to the conventional HPLC-ECD method used on the same laboratory (de Jong et al, 2010a).

Serotonin in blood is mainly stored in platelets. Free serotonin occurs in low concentrations in plasma because of active reuptake and fast metabolism which complicates detection with most conventional techniques (de Jong et al, 2011). Platelet serotonin is specifically measured for the detection of carcinoid tumors that secrete little serotonin. With LC-MS/MS, it is possible to measure accurate and reproducible serotonin both in platelet-rich and plasma-poor plasma (de Jong et al, 2009b; Monaghan et al, 2009; de Jong et al, 2010b). For this purpose, protein precipitation, using acetonitrile, combined with chromatography based on strong cation exchange and reversed-phase interaction, was used with a total run time of 6 min (Monaghan et al, 2009) and a detection limit of 5 nmol/L. Solid phase extraction based on weak cation exchange and HILIC, comparable to the method described for plasma metanephrines, resulted in the same run time with detection limits even below 1 nmol/L (de Jong et al, 2010). Therefore, LC-MS/MS is becoming an indispensible tool for low-molecular weight biomarker quantification also in the field of special clinical chemistry. It overcomes drawbacks of conventional techniques, such as long analysis times and chance of interferences, has a broad analyte compatibility and high analytical performance. It enables more sensitive and specific measurement of biogenic amines and metabolites, and the routine quantification of biomarkers in low concentration ranges.

5.3 Usefulness of LC-MS/MS in doping control

LC-MS(/MS) has become an integral part of modern sports drug testing as it offers unique capabilities complementing immunological and GC-MS(/MS)-based detection methods for prohibited compounds. The improved options of fast and sensitive targeted analysis as well as untargeted screening procedures utilizing high resolution/high accuracy MS have considerably expanded the tools available to anti-doping laboratories for initial testing and confirmation methods. One approach is to focus on preselected target analytes that are measured with utmost specificity and sensitivity using diagnostic precursor–product ion pairs in low resolution tandem mass spectrometers. The other scenario is to measure and plot extracted ion chromatograms of protonated or deprotonated molecules as well as product ions as recorded in the full scan mode with high resolution/high accuracy MS (Thevis et al, 2011).

For modern doping control laboratories, the use of LC-MS(/MS) has become obligatory to meet the needs of fast, robust, sensitive, and specific detection methods in sports drug testing. GC with or without low or high resolution GC-MS held a superior position in routine doping controls until almost the end of the last century, and various analytical challenges such as those presented by heavy volatile or polar target analytes were successfully overcome, e.g., by means of sophisticated derivatization strategies; a few aspects, however, remained unsolved by mass spectrometry, particularly concerning high molecular weight analytes. In addition, the demands with regard to time and manpower as well as the necessity to use hazardous derivatizing reagents in some applications have further strengthened the position of LC-MS(/MS) in the sports drug testing arena (Hemmersbach, 2008; Thevis & Schänzer, 2007). Traditionally, most of test methods in doping control laboratories were established for urine samples as dictated by the class of analytes, e.g., one assay for stimulants, one for narcotics, one for anabolic-androgenic steroids, etc., because most of these compounds within one category share common physicochemical properties. Because derivatization and structure-specific pre-concentration and purification are not mandatory anymore for a considerable number of analytes that are detectable by LC-MS(/MS) at relevant concentration levels, screening procedures have become (at least partly) independent of such groupings and allow comprehensive analyses within significantly shortened analytical times. Nevertheless, besides its enormous value for doping control purposes, the complementary strategy of LC-MS(/MS) bears its own challenges such as the needs to account for positive and negative ionization, the generation of multiply charged ions (particularly with molecules of higher molecular mass) and, consequently, adequate identification criteria, as well as the higher susceptibility of LC-MS(/ MS) to ion suppression/matrix effects compared to GC-MS(/MS) methods.

With the constantly improving resolving power of modern LC systems and enhanced robustness and sensitivity of tandem mass spectrometers interfaced with atmospheric pressure ionization, a considerable trend towards multianalyte testing procedures with or even without prior sample preparation and purification steps has been recognized. This trend is evidenced by a variety of earlier studies, the common goal of which was the increase of sample throughput, thus meeting the constantly growing demands concerning the implementation of new drugs, short reporting times, and cost-effectiveness (Thevis et al, 2011). Following an established extraction protocol based on alkaline (pH=9.5), LLE of urine with diethyl ether, ultrahigh-performance liquid chromatography (UHPLC) and MS/MS with scan-to-scan polarity switching was employed for the combined analysis of 34 target compounds, namely 25 diuretics, 5 stimulants, and 4 other substances (Ventura et al, 2008). Using a comparable instrumental setup composed of a 2.1×50 mm BEH C-18 column (1.7-µm particle size), gradient elution with 10 mM ammonium acetate (solvent A) and methanol (solvent B), and scan-to-scan polarity switching, a total of 133 target analytes directly from diluted urine were measured (Thörngren et al, 2008). The target analytes were 37 diuretics and masking agents, 24 narcotics, and 72 stimulants, all of which were analyzed in a total runtime of 7.5 min with LODs between 1 and 50 ng/mL (i.e., well below the corresponding MRPLs) while a minimum of 10 data points/peak was maintained. Considering the rapid (and automated) sample preparation and very short runtimes, a high-throughput option for initial doping control tests is provided. As an alternative to targeted MRM analyses, the use of high resolution/high accuracy mass spectrometry as a screening tool for sports drug testing purposes has been thoroughly investigated and initial studies and approaches were successfully pursued. One of the major advantages of these procedures over purely targeted measurements with dedicated and fixed ion transitions is the complete collection of raw data, which provides the retrospective option to mine the analytical data for formerly unknown compounds as well as new drugs when they become relevant for doping controls. Because the International Standard for Laboratories (ISL) allows forre-processing and re-analysis of doping control samples (that were initially reported negative) in the case of new relevant information, the evaluation of electronic data is a fast and cost-effective means to filter the enormous amount of samples for those being suspicious in light of new intelligence (World Anti-Doping Agency, 2011). The combined use of considerably improved instrumental options has initiated the development of a variety of multi-analyte test methods in sports drug testing laboratories and these assays cover a great variety of compounds with highly diverse physicochemical properties.

Conventional HPLC was still in use in the cases of several low resolution mass spectrometry-based approaches, whereas procedures relying on high resolution/high accuracy mass spectrometry used either monolithic columns or UHPLC capable stationary phases to fully exploit the analytical power provided by these technologies. The strategy to investigate the metabolic fate of therapeutics as well as non-approved designer drugs in a doping control context deviates slightly from clinical settings.

Analytical goals of sports drug testing usually involve the long-term detection of drug administration, most often targeting inactive metabolites. However, temporal indications of drug abuse are supportive in order to distinguish between long-term and acute use, e.g., when only the in-competition use of the drug is prohibited. LC-MS(/MS) has been an integral part of sports drug testing efforts for more than a decade, and the continuously improving instrumentation with a gradually expanding number of supportive features has considerably enhanced the quality and speed of doping control analysis for low and high molecular weight compounds (Thevis et al, 2011).

5.4 LC-MS/MS in therapeutic drug monitoring for immunosuppressants

The outcome of post-transplantation patient care has been improved dramatically over the last decades, primarily due to the availability of appropriate immunosuppressive regimens (Yang et al., 2005). The immunosuppressants present toxicity and have narrow therapeutic ranges. For example, CsA showed numerous side effects including immunological, renal, hepatic and neurological complications, requiring dose adjustments or discontinuations in a significant percentage of patients (Rezzani, 2004). In addition, blood levels of the active drugs vary significantly in different individuals and ethnicities as well as different combinations of immunosuppressants (Holt, 2002; Yang et al, 2005). Therefore, the success of the post-transplant patient care largely depends on optimization of immunosuppressive therapy based on routine therapeutic drug monitoring (TDM). The available analytical methods for monitoring immunosuppressant levels in patient specimens can be divided into two categories: immunoassays, such as microparticle enzyme immunoassay (MEIA), enzyme multiplied immunoassay technique (EMIT), fluorescent polarization immunoassay (FPIA), cloned enzyme donor immunoassay (CEDIA), and liquid chromatography-based methods (Yang et al, 2005). Immunoassays are widely employed to measure mycophenolic acid (MPA, active metabolite of MMF), CsA, tacrolimus, and sirolimus in human blood or serum/plasma. Chromatography-based methods include high performance liquid chromatography (HPLC) with ultraviolet detection, HPLC-MS, and HPLC-MS/MS. HPLC-MS/MS has gained increasing popularity in clinical laboratories due to the advantages of the technology over other methods while the capital cost for instruments has been decreased. HPLC-MS/MS provides high specificity and sensitivity for the above mentioned immunosuppressants. In addition, HPLC-MS/MS is able to simultaneously measure several drugs and/or their major metabolites in one single analytical run. The combination of HPLC and MS has revolutionized the analytical society during the past decades (Wilson & Brinkman, 2003). MS measures the abundance of charged particles based on mass over charge ratio (m/z). MS/MS technique utilizes multiple (two or more) MS with collision cell in between resulting in improved specificity by providing characteristic molecular fragments (fingerprints) generated by collision-induced dissociation. Because of high specificity of MS/MS detection, the baseline separation of target compounds from their potential interferents by HPLC, which is a must for many HPLC-UV or HPLC-MS assays, becomes unnecessary leading to significant savings in analytical time, sample purification effort and chemical reagents. Usually a typical HPLC-MS/MS run can be accomplished within a few minutes in immunosuppressant monitoring (Yang et al., 2005; Ceglarek et al., 2006). In addition MS/MS can measure multiple ions or transitions to determine multiple drugs in a single analytical run. Like any other technique, however, HPLC-MS/MS has its own limitations. HPLC-MS/MS instruments require high initial capital investment and need highly trained analysts to appropriately operate and maintain the systems. Thus HPLC-MS/MS instruments are only available in a limited number of clinical laboratories. The most widely used ESI source for MS/MS is vulnerable to ion suppression (Annesley, 2003), which could lead to significant sensitivity loss and erroneous results. Ion suppression occurs when co-eluting compounds suppress the ionization of the target compounds in the ionization sources. One effective way to eliminate ion suppression is to remove extraneous matrix components through sample cleanup procedures. However, it adds extra time and effort that many clinical laboratories cannot afford. In addition, HPLC-MS/MS is not completely immune to interferences caused by isomers or isobaric molecules (Vogeser & Spohrer, 2005).

There have been a few new HPLC-MS/MS methods developed in recent years to measure single immunosuppressive drug in patient specimens, including sample types other than blood, plasma or serum. There are also recent development on application of new technologies in immunosuppressant TDM including sample preparation and ionization (Yang & Wang, 2008). Dried blood spot sampling is an alternative to venous blood sampling for immunosuppressant TDM due to the convenient sample collection from fingerprick and small blood volume collected. A preliminary investigation of an HPLC-MS/MS assay of tacrolimus in dried blood spot formed on sample paper by fingerprick blood sampling was reported (Hoogtanders et al., 2007a). There was no significant bias observed for dried blood sampling compared to venous blood sampling using 34 samples in 26 stable renal transplant outpatients measured by HPLC-MS/MS (Hoogtanders et al., 2007b). Because of various targets of action from different immunosuppressants, combinations of immunosuppressants are often used together to achieve synergistic therapeutic effects with reduced toxicity (Yang et al., 2005). Therefore, it is ideal to simultaneously measure these drugs in one analytical run. A rapid HPLC-MS/MS method was developed for determination of MPA and MPAG in plasma using the same sample preparation and HPLC-MS/MS conditions as used in simultaneous measurement of CsA, tacrolimus, sirolimus, and everolimus in whole blood (Ceglarek et al., 2006). For the practical issues with application of HPLC-MS/MS in immunosuppressant monitoring, Internal standards should be carefully selected when using HPLC-MS/MS to measure immunosuppressants, because they are critical to compensating for sample preparation variations, ionization efficiency differences, and matrix effects with MS/MS detection. The ideal internal standards are isotope-labeled target compounds, which have the exact same physical and chemical properties as the target compounds. In many cases, however, the isotope-labeled compounds are not always available and the alternative structural analogues should be considered and evaluated carefully before use including the consideration of inter-subject differences(Taylor, 2007). Nine commercial methanol used in HPLC eluting solvents were evaluated for signal suppression of sirolimus, tacrolimus, and MPA in MS/MS detection (Annesley, 2007). Product ion intensity was found to vary by 10 folds among the methanol tested. Though appropriate internal standards could compensate for the signal loss, performance of the assay (e.g., LLOQ) could be compromised. Therefore, HPLC-MS/MS continues being considered as the method of choice in TDM of immunosuppressants due to its high sensitivity and specificity while random access immunoassays, though may give significant different results compared to HPLC-MS/MS, also play an important role in this area. However, one should carefully select the internal standards and organic solvents used in HPLC-MS/MS method and evaluate their effects on the assay performance prior to implementation of an HPLC-MS/MS method.

6. Conclusion

In this chapter, we reviewed basic principles and most recent advances of LC-MS/MS methdology including sample preparation, seperation and MS/MS detection and appications in the several areas such as quantification of biogenic amines, pharmacokinetic and TDM for immunosuppresants and doping control. Until now, together with advancement including automation in the LC-MS/MS instrumentations along with pararell sample processing, column switching, and usage of more efficient supports for SPE, they drive the trend towards less sample clean-up times and total run times–high-throughput methodology-in today's quantitative bioanalysis area. Newly introduced techniques such as ultra-performance liquid chromatography with small particles (sub-2μm) and monolithicchromatography offer improvements in speed, resolution and sensitivity compared to conventional chromatographic techniques. Hydrophilic interaction chromatography (HILIC) on silica columns with low aqueous/high organic mobile phase is emerging as a valuable supplement to the reversed-phase LC-MS/MS. Sample preparation formatted to 96-well plates has allowed for semi-automation of off-line sample preparation techniques, significantly impacting throughput. On-line SPE utilizing column-switching techniques is rapidly gaining acceptance in bioanalytical applications to reduce both time and labor required producing bioanalytical results. Extraction sorbents for on-line SPE extend to an array of media including large particles for turbulent flow chromatography, restricted access materials (RAM), monolithic materials, and disposable cartridges utilizing traditional packings such as those used in Spark Holland systems. Also this chapter also discusses recent studies of evaluation and overcome of matrix effect in LC-MS/MS analysis and how to reduce/eliminate matrix effect in method development and validation and clinical applications in the several areas as like biogenic amines analysis, therapeutic drug monitoring of immunosuppressant and doping control.

7. Acknowledgment

I thank Dr. Y.S. Park, Ms. S.Y. Kim and Ms. H.J. Kim for preparation of chapter.

8. References

Al-Dirbashi, O.Y., Aboul-Enein, H.Y., Jacob, M., Al-Qahtani, K. & Rashed, M.S. (2006) UPLC-MS/MS determination of doxazosine in human plasma. *Analytical and Bioanalytical Chemistry*, Vol.385, No.8, (August 2006), pp. (1439-1443), ISSN 1618-2642.

Almeida, R., Mosoarca, C., Chirita, M., Udrescu, V., Dinca, N., Vukelić, Z., Allen, M. & Zamfir, A.D.(2008). Coupling of fully automated chip-based electrospray ionization to high-capacity ion trap mass spectrometer for ganglioside analysis. *Analytical Biochemistry*, Vol. 378, No.1, (July 2008), pp.(43-52), ISSN 0003-2697.

Alnouti, Y., Li, M., Kavetskaia, O., Bi, H., Hop, C.E. & Gusev, A.I. (2006). Method for internal standard introduction for quantitative analysis using on-line solid-phase extraction LC-MS/MS. *Analytical Chemistry*, Vol.78, No.4, (Feburary 2006), pp. (1331-1336), ISSN 0003-2700.

Alnouti, Y., Srinivasan, K., Waddell, D., Bi, H., Kavetskaia, O. & Gusev, A.I.(2005). Development and application of a new on-line SPE system combined with LC-MS/MS detection for high throughput direct analysis of pharmaceutical compouns in plasma. *Journal of Chromatography. A*, Vol.1080, No.2, (July 2005), pp. (99-106), ISSN 0021-9673.

Annesley, T.M. (2003). Ion suppression in mass spectrometry. *Clinical Chemistry*, Vol.49, No.7, (July 2003), pp.(1041–1044), ISSN 0009-9147.

Annesley, T.M. (2007). Methanol-associated matrix effects in electrospray ionization tandem mass spectrometry. *Clinical Chemistry*, Vol.53, No.10, (October 2007), pp. (1827-1834), ISSN 0009-9147.

Antignac, J., de Wasch, K., Monteau, F., De Brabander, H., Andre F. and Le Bizec. B. (2005). The ion suppression phenomenon in liquid chromatography–mass spectrometry and its consequences in the field of residue analysis. *Analytica Chimica Acta*, Vol. 529, No.1-2, pp. (129-136), ISSN 0003-2670.

Apollonio, L.G., Pianca, D.J., Whittall, I.R., Maher, W.A. & Kyd, J.M. (2006). A demonstration of the use of ultra-performance liquid chromatography–mass spectrometry [UPLC/MS] in the determinationof amphetamine-type substances and ketamine for forensic and toxicological analysis. *Journal of Chromatography B: Analytical Technologies in the Biomedical and Life Sciences,* Vol. 836, No.1-2, (May 2006), pp. (111-115), ISSN 1570-0232.

Ashcroft, A.E. (2011). An Introduction to Mass Spectrometry, (http://www.astbury.leeds.ac.uk/facil/MStut/msturorial.htm.)

Ayrton, J., Dear, G.J., Leavens, W.J., Mallet, D.N. & Plumb, R.S. (1997). The use of turbulent flow chromatography/mass spectrometry for the rapid, direct analysis of a novel pharmaceutical compound in plasma. *Rapid Communication in the Mass Spectrometry*, Vol. 11, No.18, (December 1997), pp. (1953-1958), ISSN 0951-4198.

Ayrton, J., Dear, G.J., Leavens, W.J., Mallet, D.N. & Plumb, R.S. (1998). Optimisation and routine use of generic ultra-high flow-rate liquid chromatography with mass spectrometric detection for the direct on-line analysis of pharmaceuticals in plasma. *Journal of Chromatography A*, Vol. 828, No.1-2, (December 1998), pp.(199-207), ISSN 0021-9673.

Ayrton, J., Clare R.A., Dear, G.J., Mallet, D.N. & Plumb, R.S. (1999). Ultra-high flow rate capillary liquid chromatography with mass spectrometric detection for the direct analysis of pharmaceuticals in plasma at sub-nanogram per millilitre concentrations. *Rapid Communications in Mass Spectrometry*, Vol.13, No.16, (August 1999), pp. (1657-1662), ISSN 0951-4198.

Bakhtiar, R. and Majumdar, T. (2007). Tracking problems and possible solutions in the quantitative determination of small molecule drugs and metabolites in biological fluids using liquid chromatography-mass spectrometry. *Journal of Pharmacological and Toxicological Methods*, Vol. 55 , No.3, pp. (262-278), ISSN 1056-8719.

Barbarin, N., Mawhinney, D.B., Black, R., & Henion, J.(2003). High-throughput selected reaction monitoring liquid chromatography–mass spectrometry determination of methylphenidate and its major metabolite, ritalinic acid, in rat plasma employing monolithic columns. *Journal of Chromatography B: Analytical Technologies in the Biomedical and Life Sciences.*Vol.783, No.1, (Januarl 2003), pp. (73-83), ISSN 1570-0232.

Barrett, Y.C., Akinsanya, B., Chang, S.Y. & Vesterqvist, O. (2005). Automated on-line SPE LC-MS/MS method to quantitate 6beta-hydroxycortisol and cortisol in human urine: use of the 6beta-hydroxycortisol to cortisol ratio as an indicator of CYP3A4 activity.*Journal of Chromatography. B: Analytical Technologies in the Biomedical and Life Sciences*, Vol.821, No.2, (July 2005), pp. (159-165), ISSN 1570-0232.

Berna, M.J., Ackermann, B.L. & Murphy, A.T. (2004). High-throughput chromatographic approaches to liquid chromatographic/tandem mass spectrometric analysis to support drug discovery and development. *Analytica Chimica Acta*, Vol.509, No.1,(April 2004), pp.(1-9), ISSN 0003-2670.

Beuhler, R.J., Chu, Y.Y., Friedlander, G., Friedman, L., Alessi, J.G., LoDestro, V. V. & Thomas, J.P.(1991). Cluster-impact fusion: time-to-flight experiments. *Physical Review Letters,* Vol.67, No.4, (July 1991) pp. (473-476), ISSN 0031-9007.

Blomberg, L.G. (2009). Two new techniques for sample preparation in bioanalysis: microextraction in packed sorbent (MEPS) and use of a bonded monolith as sorbent for sample preparation in polypropylene tips for 96-well plates. *Analytical and Bioanalytical Chemistry,* Vol.393, No.3, (February 2009), pp. (797-807), ISSN 1618-2642.

Bolden, R.D., Hoke 2nd, S.H., Eichhold, T.H., McCauley-Myers, D.L. & Wehmeyer, K.R. (2002). Semi-automated liquid–liquid back-extraction in a 96-well format to decrease sample preparation time for the determination of dextromethorphan and dextrorphan in human plasma. *Journal of Chromatography B: Analytical Technologies in the Biomedical and Life Sciences.* Vol.772, No.1, (May 2002), pp. (1-10), ISSN 1570-0232.

Bonfiglio, R., King, R.C., Olah T.V. & Merkle, K. (1999).The effects of sample preparation methods on variability of the electrospray ionization response for model drug compounds. *Rapid Communications in Mass Spectrometry,* Vol.13, No.12, (June 1999), pp. (1175–1185), ISSN 0951-4198.

Borges, V., Yang, E., Dunn, J. & Henion, J. (2004). High-throughput liquid chromatography-tandem mass spectrometry determination of bupropion and its metabolites in human, mouse and rat plasma using a monolithic column. *Journal of Chromatography B: Analytical Technologies in the Biomedical and Life Sciences,* Vol.804, No.4, (May 2004), pp. (277-287), ISSN 1570-0232.

Bruins, A.P. (1991). Mass spectrometry with ion sources operating at atmospheric pressure. *Mass Spectrometry Reviews,* Vol.10, No.1, (January 1991), pp. (53–77), ISSN 0277-7037.

Buhrman, D., Price, P.I. & Rudewicz, P. J. (1996). Quantitation of SR 27417 in human plasma using electrospray liquid chromatography–tandem mass spectrometry: a study of ion suppression. *Journal of the American Society for Mass Spectrometry,*Vol.7, No.11, (November 1996),pp. (1099–1105), ISSN 1044-0305.

Cabrera, K. (2004). Applications of silica-based monolithic HPLC columns. *Journal of Separation Science,* Vol.27, No.10-11, (July 2004),pp. (843-852), ISSN 1615-9306.

Ceglarek, U., Casetta, B., Lembeke, J., Baumann, S., Fiedler, G.M., Thiery, J. (2006). Inclusion of MPA and in a rapid multi-drug LC-tandem mass spectrometric method for simultaneous determination of immunosuppressants. *Clinica Chimica Acta,* Vol.373, No.1-2, (November 2006), pp. (168-171), ISSN 0009-8981.

Chambers, E., Wagrowski-Diehl, D., Lu Z. and Mazzeo, J. (2007). Systematic and comprehensive strategy for reducing matrix effects in LC/MS/MS analyses. *Journal of Chromatography B: Analytical Technologies in the Biomedical and Life Sciences,* Vol. 852, No.1-2, pp. (22-34), ISSN 1570-0232.

Chan, E.C. & Ho, P.C. (2000). High-performance liquid chromatograpy/atmospheric pressure chemical ionization mass spectrometric method for the analysis of catecholamines and metanephrines in human urine. *Rapid Commmunications in Mass Spectrometry,* Vol.14, No.21, pp.(1959-1964), ISSN 0951-4198.

Chassaing, C., Stafford, H., Luckwell, J., Wright, A. & Edgington, A. (2005). A Parallel Micro Turbulent Flow Chromatography-Tandem Mass Spectrometry Method for the

Analysis of a Pharmaceutical Compound in Plasma. *Chromatographia,* Vol. 62, No.1-2, (July 2005), pp. (17-24), ISSN 0009-5893.

Chen, X., Westphall, M.S. & Smith, L.M. (2003). MS of DNA mixtures: Instrumental effects responsible for decreased sensitivity with increasing mass. *Analytical Chemistry,* Vol.75, No.21, (November 2003), pp. (5944-5952), ISSN 0003-2700.

Chen, Y., Junga, H., Jiang, X., Weng, N. (2002). In: Presented at the 50th ASMS Conference, Orlendo, FL.

Chiu, C.M. & Muddiman, D.C. (2008). What is Mass Spectrometry?, American Society for Mass Spectrometry, http://www.asms.org/whatisms.

Churchwell, M.I., Twaddle, N.C., Meeker L.R. & Doerge. D.R. (2005). Improving LC–MS sensitivity through increases in chromatographic performance: Comparisons of UPLC–ES/MS/MS to HPLC–ES/MS/MS. *Journal of Chromatography. B, Analytical Technologies in the Biomedical and Life Sciences,* Vol.825, No.2, (October 2005), pp. (134-143), ISSN 1570-0232.

Clarke, S.D., Hill, H.M., Noctor, T.A.G. & Thomas, D. (1996). Matrix-related Modification of Ionization in Bioanalytical Liquid Chromatography-Atmospheric Pressure Ionization Tandem Mass Spectrometry. *Pharmaceutical Sciences,* Vol.2, No.4, (April 1996), pp. (203-207), ISSN 2042-7158.

Clarke, N.J., Rindgen, D., Korfmacher,W.A., Cox, K.A. (2001). Systematic LC/MS metabolite dentification in drug discovery. *Analytical Chemistry, Vol.* 73, No.15, pp. (430A-439A), ISSN 0003-27000.

Comisarow, M.B. & Marshall, A.G. (1974). Fourier transform ion cyclotron resonance spectroscopy. *Chemical Physics Letters,* Vol.25, No.2, (March 1974), pp. 282–283, ISSN 0009-2614.

Constanzer, M.L., Chavez, C.M., Matuszewski, B.K., Carlin J. & Graham. D. (1997). Low level determination of a novel 4-azasteroid and its carboxylic acid metabolite in human plasma and semen using high-performance liquid chromatography with atmospheric pressure chemical ionization tandem mass spectrometry. *Journal of Chromatography B: Biomedical Sciences and Applications,* Vol. 693, No.1, (May 1997), pp. (117-129), ISSN 1387-2273.

Costello, C.E. (1997). Time, life…and mass spectrometry. New techniques to address biological questions. *Biophysical Chemistry,* Vol.68, No.1-3, (October 1997), pp. (173-188), ISSN 0301-4622.

Daly, N.R., McCormick, A. & Powell, R.E. (1968a) Detector for the metastable ions observed in the mass spectra of organic compounds. *Review of Scientific Instruments* Vol.39, No.8, (Feburary 1968), pp. (1163-1167), ISSN 0034-6748

Daly, N.R., McCormick, A. & Powell, R.E. (1968b) Metastable spectra of cis- and trans-butanes. *Organic Mass Spectrometry,* Vol. 1, No.1, pp. (167-168), ISSN 0030-493X.

Dams, R., Huestis, M.A., Lambert, W.E. & Murphy, C.M.(2003). Matrix effect in bio-analysis of illicit drugs with LC-MS/MS: influence of ionization type, sample preparation, and biofluid. *Journal of American Society for Mass Spectrometry,* Vol.14, No.11, (November 2003), pp. (1290-1294), ISSN 1044-0305.

Daurel-Receveur, M., Titier, K., Picard, S.P., Ducint, D., Moore. N, & Molimard, M. (2006). Fully automated analytical method for mycophenolic acid quantification in human plasma using on-line solid phase extraction and high performance liquid chromatography with diode array detection. *Therapeutic Drug Monitoring,* Vol.28, No.4, (August 2006), pp. (505-511), ISSN 0163-4356.

de Jong, W.H., Graham, K.S., van der Molen, J.C., Links, T.P., Morris, M.R., Ross, H.A., De Vries, E.G. & Kema, I.P. (2007). Plasma free metanephrine measurement using automated online solid-phase extraction HPLC tandem mass spectrometry. *Clinical Chemistry,* Vol.53, No.9, (September 2007), pp. (1684-1693), ISSN 0009-9147.

de Jong, W.H., Graham, K.S., de Vries, E.G., Kema, I.P. (2008). Urinary 5-HIAA measurement using automated on-line solid extraction-high-performance liquid chromatography-tandem mass spectrometry. *Journal of Chromatography B: Analytical Technologies in the Biomedical and Life Sciences,* Vol.868, Vol.1-2, (June 2008) pp. (28-33), ISSN 1570-0232.

de Jong, W.H., de Vries E.G., Kema, I.P. (2009a). Serotonin rising, *The New England Journal of Medicine,* Vol.360, No.24, (June 2009), pp. (2580-2581), ISSN 0028-4793.

de Jong, W.H., Smit, R., Bakker, S.J., de Vries, E.G., Kema, I.P. (2009b). Plasma tryptophan, kynurenine and 3-hydroxykynurenine measurement using automated on-line solid-phase extraction HPLC-tandem mass spectrometry. *Journal of Chromatography B: Analytical Technologies in the Biomedical and Life Sciences,* Vol.877, Vol.7, (March 2008), pp. (603-609), ISSN 1570-0232.

de Jong, W.H., de Vries, E.G., Wolffenbuttel, B.H., Kema, I.P. (2010a). Automated mass spectrometric analysis of urinary free catecholamines using on-line solid phase extraction. *Jourmal of Chromatography B: Analytical Technologies in the Biomedical and Life Sciences.* Vol.878, No.19, (June 20010), pp. (1506-1512), ISSN 1570-0232.

de Jong, W.H., Wilkens, M.H., De Vries, E.G. & Kema, I.P. (2010b). Automated mass spectrometric analysis of urinary and plasma serotonin. *Analytical amd Bioanalytical Chemistry,* Vol.396, No.7, (April 2010), pp. (2609–2616), ISSN 1618-2642.

de Jong, W.H., de Vries, E.G., Kema,I.P. (2011). Current status and future developments of LC-MS/MS in clinical chemistry for quantification of biogenic amines, *Clinical Biochemistry,* Vol.44, No.1, (January 2010), pp. (95-103), ISSN 0009-9120.

Deng, Y., Zhang, H., Wu, J.T. & Olah, T.V. (2005). Tandem mass spectrometry with online high-flow reversed-phase extraction and normal-phase chromatography on silica columns with aqueous-organic mobile phase for quantitation of polar compounds in biological fluids. *Rapid Communications in Mass Spectrometry,* Vol.19, No.20, (October 2005), pp. (2929-2934), ISSN 1097-0231.

Dewe, W., De Smet, M., Evrard, N., Culot, B., Lastelle, M., Saucez, M., Ingels, D., Taillieu, X. (2007). Partial within-animal calibration: a new calibration approach in lead optimization high-throughput bioanalysis. *Journal of Chromatography B: Analytical Technologies in the Biomedical and Life Sciences,* Vol.854, No.1-2, (July 2007), pp.(183-191), ISSN 1570-0232.

Douglas, D., Frank, A. & Mao, D. (2005). Linear ion traps in mass spectrometry. Mass Spectrometry Reviews, Vol. 24, No.1, (January-Feburary 2005), pp.(1-29), ISSN 0277-7037.

Dubois, F., Knochenmuss, R. & Zenobi, R. (1997). An ion-to-photon conversion detector for mass spectrometry. *International Journal of Mass Spectrometry and Ion Processes,* Vol.169-170, (December 1997), pp. (89-98), ISSN : 1387-3806.

Dubois, F., Knochenmuss, R., Zenobi, R., Brunelle, A., Deprun, C. & Beyec, Y.L. (1999). A comparison between ion-to-photon and microchannel plate detectors. *Rapid Communications in Mass Spectrometry,* Vol.13, No.9, pp. 786–791, ISSN 1097-0231.

Dunér, K., Bäckström, J., Magnell, N., Svennberg, H., Ahnoff, M. & Logren, U. (2007). Determination of ximelagatran, melagatran and two intermediary metabolites in plasma by mixed-mode solid phase extraction and LC-MS/MS. *Journal of Chromatography B: Analytical Technologies in the Biomedical and Life Sciences*, Vol.852, No.1-2, (June 2007), pp. (317–324), ISSN 1570-0232.

Eerkes, A., Shou, W.Z. & Naidong, W. (2003). Liquid/liquid extraction using 96-well plate format in conjunction with hydrophilic interaction liquid chromatography–tandem mass spectrometry method for the analysis of fluconazole in human plasma. *Journal of Pharmaceutical and Biomedical Analysis,* Vol.31, No.5, (April 2003), pp. (917-928), ISSN 0731-7085.

Egnash. L.A. & Ramanathan, R. (2002). Comparison of heterogeneous and homogeneous radioactivity flow detectors for simultaneous profiling and LC–MS/MS characterization of metabolites. *Journal of Pharmaceutical and Biomededical Analysis.* Vol.27, No.1-2, (January 2002), pp. (271-284), ISSN 0731-7085.

Erve, J.C. , Beyer, C.E., Manzino L. & Talaat, R.E. (2009). Metabolite identification in rat brain microdialysates by direct infusion nanoelectrospray ionization after desalting on a ZipTip and LTQ/Orbitrap mass spectrometry. *Rapid Communications in Mass Spectrometry,* Vol.23, No.24, (December 2009), pp. (4003-4012), ISSN 1097-0231.

Farmer, J. B. In Mass Spectrometry, McDowell, C. A., Ed.; McGraw-Hill: New York, 1963; pp 10-11

Fies, W. J. (1988). A Method for measuring the gain of an electron multiplier. *International Journal of Mass Spectrometry and Ion Processes,* Vol.82, No.1-2, (January 1988), pp. (111-129), ISSN 0168-1176.

Gerlich, D. (1992). *Inhomogenous RF fields: A versatile tool for the study of processes with slow ions, In Advances in chemical physices, Vol.82,* (eds C.-Y. Ng, M. Baer, I. Prigogine and S. A. Rice), pp. 1-176, John Wiley & Sons, Inc, ISBN 0-471-53258-3, Hoboken, NJ, USA.

Georgi, K. & Boos, K. (2006). Multidimensional on-line SPE for undisturbed LC-MS-MS analysis of basic drugs in biofluids. *Chromatographia,* Vol.63, No.11-12, (June 2006), pp. (523-531), ISSN 0009-5893.

Gibson, G.G. & Skett, P. Pathways of Drug Metabolism. In *Introduction to Drug Metabolism,* Gibson GG, Skett P (eds). Blackie Academic and Professional: London, 1994; 1–34.

Graff, D.K. (1995). Fourier and Hadamard: Transforms in Spectroscopy. *Journal of Chemical Education,* Vol.72, No.4, (April 1995) pp. 304-309, ISSN 0021-9584.

Guilaus, M. (1995). *Special feature: Tutorial.* Principles and instrumentation in time-of-flight mass spectrometry. Physical and instrumental concepts. *Journal of Mass Spectrometry,* Vol.30, No.11, (November 1995), pp. (1519-1532), ISSN 1076-5147.

Hager, J.W. (2002). A new linear ion trap mass spectrometer. *Rapid Communications in Mass Spectrometry,* Vol.16, No.6, (March 2002), pp. (512-526), ISSN 1097-0231.

Hardman, M. & Makarov, A.A. (2003). Interfacing the orbitrap mass analyzer to an electrospray ion source. *Analytical Chemistry,* Vol.75, No.7, (April 2003), pp. (1699-1705), ISSN 0003-2700.

Harris, F.M., Trott, G.W., Morgan, T.G., Brenton, A.G., Kingston, E.E. & Beynon, J.H. (1984). Signal-to-noise ratios in the measurement of low ion currents using electron multipliers. *Mass Spectrometry Reviews,* Vol.3, No.2, (Summer 1984), pp. (209-229), ISSN 0277-7037.

Hedin, A., Haakansson, P. & Sundqvist, B.U.R. (1987). On the detection of the large organic ions by secondary electron production. *International Journal of Mass Spectrometry and Ion Processes,* Vol.75, No.3, (April 1987), pp. (275-289), ISSN 0168-1176.

Heller, D.N. (2007). Ruggedness testing of quantitative atmospheric pressure ionization mass spectrometry methods: the effect of co-injected matrix on matrix effects. *Rapid Communications in Mass Spectrometry,* Vol.21, No.5, (March 2007), pp. (644-652), ISSN:1097-0231.

Hemmersbach, P. (2008). History of mass spectrometry at the Olympic Games. *Journal of Mass Spectrometry,* Vol.43, No.7, (July 2008), pp. (839–853), ISSN 1076-5174.

Herman, J.L. (2002). Generic method for on-line extraction of drug substances in the presence of biological matrices using turbulent flow chromatography. *Rapid Communications in Mass Spectrometry,* Vol.16, No.5, (March 2002), pp. (421-426), ISSN 1097-0231.

Hernández, F., Sancho, J.V. & Pozo, O.J. (2005). Critical review of the application of liquid chromatography/mass spectrometry to the determination of pesticide residues in biological samples. *Analytical amd Bioanalytical Chemistry,* Vol.382, No.4, (June 2005), pp. (934-946), ISSN 1618-2642.

Higashi, T., Shibayama, Y., Fuji, M., Shimada, K. (2008). Liquid chromatography-tandem mass spectrometric method for the determination of salivary 25-hydroxyvitamin D3: a noninvasive tool for the assessment of vitamin D status. *Analytical and Bioanalytical Chemistry,* Vol.391, No.1, (May 2009), pp. (229-238), ISSN 1618-2642.

Hoogtanders, K., van der Heijden, J. , Christiaans, M., Edelbroek, P., van Hooff J.P., Stolk, L.M.L. (2007a). Therapeutic drug monitoring of tacrolimus with the dried blood spot method. *Journal of Pharmaceutical and Biomedical Analysis,* Vol.44, No.3, (July 2007), pp. (658-664), ISSN 0731-7085.

Hoogtanders, K., van der Heijden, J., Christiaans, M., van de Plas, A., van Hooff J., Stolk, L. 2007b). Dried blood spot measurement of tacrolimus is promising for patient monitoring. *Transplantation,* Vol. 83, No.2, (January 2007), pp. (237-238), ISSN 0041-1337.

Holst, J.P., Soldin, S.J., Tractenberg, R.E., Guo, T., Kundra, P., Verbalis, J.G., Jonklaas, J. (2007). Use of steroid profiles in determining the cause of adrenal insufficiency. *Steroids,* Vol.72, No.1, (January 2007), pp. (71-84), ISSN 0039-128X.

Holt, D.W. (2002). Therapeutic drug monitoring of immunosuppressive drugs in kidney transplantation. *Current Opinion in Nephrology and Hypertension,* Vol.11, No.6, (November 2002), pp. (657-663), ISSN 1062-4821.

Hopfgartner, G., Husser, C. & Zell, M. (2002). High-Throughput Quantification of Drugs and Their Metabolites in Biosamples by LC-MS/MS and CE-MS/MS: Possibilities and Limitations. *Therapeutic Drug Monitoring,* Vol. 24, No.1, (February 2002), pp.(134-143), ISSN 0163-4356.

Hu, Q., Noll, R.J., Li, H., Makarov, A., Hardman, M. & Cooks, R.G. (2005). The Orbitrap: a new mass spectrometer. *Journal of Mass Spectrometry,* Vol.40, No.4, (April 2005), pp. (430-443), ISSN 1076-5147.

Huang, M.Q., Mao, Y., Jemal, M. & Arnold, M. (2006). Increased productivity in quantitative bioanalysis using a monolithic column coupled with high-flow direct-injection liquid chromatography/tandem mass spectrometry. *Rapid Communications in Mass Spectrometry,* Vol.20, No.11, (June 2006), pp. (1709-1714), ISSN:1097-0231.

Hyland, K. (1999). Neurochemistry and defects of biogenic amine neurotransmitter metabolism. Journal of Inherited Metabolic Disease, Vol.22, No.4, (June 1999), pp. (353-363), ISSN 0141-8955.

Ikegami, T. & Tanaka, N. (2004).Monolithic columns for high-efficiency HPLC separations. *Current Opinion in Chemical Biology*, Vol.8, No.5, (October 2004), pp. (527-533), ISSN 1367-5931.

Jemal, M., Ouyang, Z. & Powell, M.L.(2002). A strategy for a post-method-validation use of incurred biological samples for establishing the acceptability of a liquid chromatography/tandem mass-spectrometric method for quantitation of drugs in biological samples.*Rapid Communications in Mass Spectrometry,* Vol.16, No.16, (August 2002), pp. (1538-1547), ISSN 1097-0231.

Jennings, K.R. & Dolnikowski, G.G.(1990). Mass analyzers. *Methods in Enzymology,* Vol.193, pp. (37-61), ISSN:0076-6879.

Jessome, L.L. and Volmer, D.A. (2006). Ion suppression: A major concern in Mass Spectrometry. *LCGC North America,* Vol. 24, No.5, (May 2006), pp. (498-506), ISSN 1527-5949.

Ji, H, Park, E., Lee, K. & Lee. H. (2008). Quantification of doxazosin in human plasma using hydrophilic interaction liquid chromatography with tandem mass spectrometry. *Journal of Seperation Science.* Vol. 31, No.9 (May 2008), pp. (1628-1633), ISSN 1615-9314.

Kamel, A. & Prakash, C. (2006). High performance liquid chromatography/atmospheric pressure ionization/tandem mass spectrometry (HPLC/API/MS/MS) in drug metabolism and toxicology. Current Drug Metabolism, Vol.7, No.8, (Dec 2006), pp. (837-852), ISSN 1389-2002.

Karas, M., Bachmann, D., Bahr, U. & Hillenkamp, F. (1987). Matrix-assisted ultraviolet laser desorption of non-volatile compounds. *International Journal of Mass Spectrometry and Ion Processes,* Vol.78, (September 1987), pp. (53-68), ISSN 0168-1176.

Kawano, S., Takahashi, M., Hine, T., Yamamoto, E. & Asakawa, N. (2005). On-line pretreatment using methylcellulose-immobilized cation-exchange restricted access media for direct liquid chromatography/mass spectrometric determination of basic drugs in plasma.*Rapid Communications in Mass Spectrometry,* Vol.19, No.19, (October 2005), pp. (2827-2832), ISSN 1097-0231.

Kebarle, P. & Tang, L. (1993). From ions in solution to ions in the gas phase - the mechanism of electrospray mass spectrometry. *Analytical Chemistry,* Vol.65, No.22, (November 1993), pp. (972A-986A), ISSN 0003-2700.

King,R., Bonfiglio, R., Fernandez-Metzler, C., Miller-Stein, C. & Olah, T. (2000), Mechanistic investigation of ionization suppression in electrospray ionization. *Journal of the American Society for Mass Spectrometry,* Vol.11, No.11, (November 2000), pp. (942–950), ISSN 1044-0305.

Koal, T., Sibum, M., Koster, E., Resch, K. & Kaever, V. (2006). Direct and fast determination of antiretroviral drugs by automated online solid-phase extraction-liquid chromatography-tandem mass spectrometry in human plasma. *Clinical Chemical Laboratory Medicine,*Vol.44, No.3,(July 2006), pp.(299-305), ISSN 1434-6621.

Koppenaal, D.W., Barinaga, C.J., Denton, M.B., Sperline, R.P., Hieftje, G.M., Schilling, G.D., Andrade, F.J. & Barnes, J.H.IV. (2005). MS detectors. *Analytical Chemistry,* Vol.77, No.21, (November 2005), pp. (418A-427A), ISSN 0003-2700.

Koseki, N., Nakashima, A., Nagae, Y. & Masuda, N. (2006). Simultaneous quantitative determination of cyclosporine A and its three main metabolites (AM1, AM4N and AM9) in human blood by liquid chromatography/mass spectrometry using a rapid sample processing method. *Rapid Communications in Mass Spectrometry*, Vol.20, No.5, (March 2006), pp. (733-740), ISSN 1097-0231.

Koster, S. & Verpoorte, E. (2007). A decade of microfluidic analysis coupled with electrospray mass spectrometry: an overview. *Lab on a Chip*, Vol.7, No.11, (September 2007), pp. (1394-1412), ISSN: 1473-0197.

Kostiainen, R., Kotiaho, T., Kuuranne, T., Auriola, S. (2003). Liquid chromatography/atmospheric pressure ionization–mass spectrometry in drug metabolism studies. *Journal of Mass Spectrometry*, Vol.38, No.4, (March 2003), pp. (357-372), ISSN 1076-5174.

Kushnir, M.M., Urry, F.M., Frank, E.L., Roberts, W.L., Shushan, B. (2002). Analysis of catecholamines in urine by positive-ion electro-spray tandem mass spectrometry. *Clinical Chemistry*, Vol.48, No.2, (Feburary 2002), pp. (323-331), ISSN 0009-9147.

Lagerwerf, F.M., van Dongen, W.D., Steenvoorden, R.J.J.M., Honing, M. & Jonkman, J.H.G. (2000). Exploring the boundaries of bioanalytical quantitative LC–MS–MS. *TrAC Trends in Analytical Chemistry*, Vol.19, No.7, (July 2000), pp. (418-427), ISSN: 0165-9936.

Lanckmans, K., Van Eeckhaut, A., Sarre, S., Smolders, I. & Michotte. Y. (2006). Capillary and nano-liquid chromatography-tandem mass spectrometry for the quantification of small molecules in microdialysis samples: Comparison with microbore dimensions. *Journal of Chromatography A*, Vol. 1131, No.1-2, (October 2006), pp. (166-175), ISSN 0021-9673.

Lammert, S.A., Rockwood, A.A., Wang, M., Lee. M,L,, Lee. E,D,, Tolley. S,E., Oliphant, J.R., Jones, J.L. & Waite, R.W.(2006). Miniature Toroidal Radio Frequency Ion Trap Mass Analyzer. *Journal of the American Society for Mass Spectrometry*, Vol.17, No.7, (July 2006), pp. (916–922), ISSN 1044-0305.

Lenders, J.W., Pacak, K, Walther, M.M., Linehan, W.M., Mannelli, M., Friberg, P., Keiser, H.R., Goldstein, D.S., Eisenhofer, G. (2002). Biochemical diagnosis of pheochromocytoma: Which test is best? *JAMA*, Vol.287, No.11, pp. (1427-1434), ISSN 0098-7484.

Li, R., Dong, L. & Huang, J. (2005). Ultra performance liquid chromatography–tandem mass spectrometry for the determination of epirubicin in human plasma. *Analytica Chimica Acta*, Vol.546, No.2,(August 2005), pp. (167-173), ISSN:0003-2670.

Lim, H.K., Chan, K.W., Sisenwine, S. & Scatina, J.A., (2001). Simultaneous Screen for Microsomal Stability and Metabolite Profile by Direct Injection Turbulent-Laminar Flow LC−LC and Automated Tandem Mass Spectrometry. *Analytical Chemistry*, Vol.73, No.9, (May 2001), pp. (2140-2146), ISSN:0003-2700.

Lionetto, L., Lostia, A.M., Stigliano, A., Cardelli, P., Simmaco, M. (2008). HPLC-mass spectrometry method for quantitative detection of neuroendocrine tumor markers: vanillyl-mandelic acid, homovanillic acid and 5-hydroxyindoleacetic acid. *Clinica Chimica Acta*, Vol.398, No.1-2, pp. (53-56), ISSN 0009-8981.

Liu, H., Berger, S.J., Chakraborty, A.B., Plumb, R.S. & Cohen, S.A.(2002). Multidimensional chromatography coupled to electrospray ionization time-of-flight mass spectrometry as an alternative to two-dimensional gels for the identification and analysis of complex mixtures of intact proteins. *Journal of Chromatography. B*,

Analytical Technologies in the Biomedical and Life Sciences, Vol.782, No.1-2, (December 2002), pp. (267-289), ISSN 1570-0232.

Londry, F.A. & Hager, J.W. (2003). Mass selective axial ion ejection from a linear quadrupole ion trap.*Journal of the American Society for Mass Spectrometry,* Vol.14, No.10,(Octorber 2003), pp. (1130-1147), ISSN 1044-0305.

Majors, R.E. (2006). Miniaturized approaches to conventional liquid-liquid extraction. *LC-GC North America,* Vol.24, No.2,(February 2006), pp.(118-130), ISSN 1527-5949.

Makarov, A. (2000). Electrostatic axially harmonic orbital trapping: A high-performance technique of mass analysis. *Analytical Chemistry,* Vol.72, No.6, (March 2000), pp. (1156-1162), ISSN 0003-2700.

Makarov, A., Denisov, E., Lange, O. & Horning, S. (2006a). Dynamic range of mass accuracy in LTQ Orbitrap hybrid mass spectrometer. *Journal of the American Society for Mass Spectrometry,* Vol.17, No.7,(July 2006), pp. (977-982), ISSN 1044-0305.

Makarov, A., Denisov, E., Kholomeev, A., Balschun, W., Lange, O., Strupat, K. & Horning, S. (2006b). Performance evaluation of a hybrid linear ion trap/orbitrap mass spectrometer. *Analytical Chemistry,* Vol.78, No.7, (April 2006), pp. (2113-2120), ISSN:0003-2700.

Mallet, C.R., Lu, Z., Fisk, R., Mazzeo, J.R. & Neue, U.D. (2003). Performance of an ultra-low elution-volume 96-well plate: drug discovery and development applications. *Rapid Communications in Mass Spectrometry,* Vol.17, No.2, (January 2003), pp. (163-170), ISSN 1097-0231.

March, R.E. (2000). Quadrupole ion trap mass spectrometry: a view at the turn of the century. *International Journal of Mass Spectrometry,* Vol.200, No.1–3, (December 2000), pp. (285–312), ISSN 1387-3806.

Marshall, A.G., Hendrickson, C.L. & Jackson, G.S. (1998). Fourier transform ion cyclotron resonance mass spectrometry: a primer. *Mass Spectrometry Reviews,* Vol.17, No.1, (January-February 1998) pp. (1–34), ISSN:0277-7037.

Martin, M., Eon, C. & Guiochon, G. (1974). Study of the pertinency of pressure in liquid chromatography: I. theoretical analysis. *Journal of Chromatography. A,* Vol. 99, pp. (357-376), ISSN 0021-9673.

Matthew, M.W., Beuhler, R.J., Ledbetter, M. & Friedman, L. (1986). Large, energetic impacts on surface. *Nuclear Instruments & Methods in Physics Research, Section B: Beam interactions with Materials and Atoms B,* Vol.14, No.4-6, (April 1986), pp. (448-460), ISSN 0168-583X

Matuszewski, B.K. (2006). Standard line slopes as a measure of a relative matrix effect in quantitative HPLC–MS bioanalysis. *Journal of Chromatography B: Analytical Technologies in the Biomedical and Life Sciences,* Vol.830, No.2, (January 2006), pp. (293-300), ISSN 1570-0232.

Matuszewski, B.K., Constanzer M.L. & Chavez-Eng, C.M. (1998). Matrix effect in quantitative LC/MS/MS analyses of biological fluids: a method for determination of finasteride in human plasma at picogram per millilitre concentrations. *Analytical Chemistry,* Vol.70, No.5, (March 1998), pp. (882-889), ISSN 0003-2700.

Matuszewski, B.K., Constanzer, M.L. & Chavez-Eng, C.M. (2003). Strategies for the assessment of matrix effect in quantitative bioanalytical methods based on HPLC–MS/MS. *Analytical Chemistry,* Vol.75, No.13, (July 2003), pp. (3019-3030), ISSN 0003-2700.

Maurer, H.H. (2007). Current role of liquid chromatography-mass spectrometry in clinical and forsenic toxicology. Analytical and Bioanalytical Chemistry Vol.388, No.7, (August 2007), pp. (1315-1325), ISSN 1618-2642.

Mawwhinney, D. & Henion, J. (2002), in: Presented at the 50th ASMS conference, Orlando, FL,

Mazzeo, J.R., Neue, U.D., Kele, M. & Plumb R.S. (2005).Advancing LC Performance with Smaller Particles and Higher Pressure. *Analytical Chemistry,* Vol.77, No.23, (December 2005), pp. (460A-467A), ISSN:0003-2700.

Mckinney, C.R., McCrea, J.M., Epstein, S., Allen, H.A. & Urey, H.C. (1950). Improvements in mass spectrometers for the measurement of small differences in isotope abundance ratios. *The Review of Scientific Instruments,* Vol.21, No.8, (August 1950), pp. (724-730), ISSN: 0034-6748.

Mei, H., Hsieh, Y., Nardo, C., Xu, X,. Wang, S., Ng K. & Korfmacher. W. (2003), Investigation of matrix effects in bioanalytical high-performance liquid chromatography/tandem mass spectrometric assays: Application to drug discovery. *Rapid Communication in Mass Spectrometry,* Vol. 17, No.1, pp. (97-103), ISSN 0951-4198.

Monaghan, P.J., Brown, H.A., Houghton, L.A., Keevil, B.G. (2009). Measurement of serotonin in platelet depleted plasma by liquid chromatography tandem mass spectrometry. *Journal of Chromatography B: Analytical Technologies in the Biomedical and Life Sciences,* Vol.877, No.22, pp. (2163-2167), ISSN 1570-0232.

Niessen, W. M. A. J. (2003). Progress in liquid chromatography–mass spectrometry instrumentation and its impact on high-throughput screening. *Journal of chromatography. A,* Vol.1000, No.1-2, (June 2003), pp. (413-436), ISSN:0021-9673.

Niessen, W., Manini P. & Andreoli. R. (2006), Matrix effects in quantitative pesticide analysis using liquid chromatography-mass spectrometry. *Mass Spectrometry Reviews,* Vol.25, No.6, (November-December 2006), pp. (881-899), ISSN 0277-7037.

Niessen, W. M. A. J. (2003). Progress in liquid chromatography–mass spectrometry instrumentation and its impact on high-throughput screening. *Journal of chromatography. A,* Vol.1000, No.1-2, (June 2003), pp. (413-436), ISSN 0021-9673.

Nygren, U., Ramebaeck, H., Berglund, M. & Baxter, D.C. (2006). The importance of a correct dead time setting in isotope ratio mass spectrometry: Implementation of an electronically determined dead time to reduce measurement uncertainty.*Journal of Mass Spectrometry,*Vol.257, No.1-3, (November 2006), pp. (12-15), ISSN:1387-3806.

Park, M.A., Callahan, J.H. & Vertes, A. (1994). An inductive detector for time-of-flight mass spectrometry. *Rapid Communications in Mass Spectrometry,* Vol.8, No.4, (April 1994), pp. (317-322), ISSN 1097-0231.

Paul, W. & Steinwedel, H. (1953). Ein neues Massenspektrometer ohne Magnetfeld. *Zeitschrift für Naturforschung A,* Vol.8, No.7, (June 1953). pp. 448–450, ISSN : 0932-0784.

Plumb, R.S., Stumpf, C.L., Granger, J.H, Castro-Perez, J., Haselden, J.N. & Dear, G.J. (2003). Use of liquid chromatography/time-of-flight mass spectrometry and multivariate statistical analysis shows promise for the detection of drug metabolites in biological fluids. *Rapid Communications in Mass Spectrometry,* Vol.17, No.23, (December 2003), pp. (2632-2638), ISSN:1097-0231.

Price, D. & Milnes, G.J. (1984). Recent developments in techniques utilizing TOF MS. *International Journal of Mass Spectrometry and Ion Processes,* Vol.60, No.7, (September 1984), pp. (61-81), ISSN :1387-3806.

Ramos, L., Brignol, N., Bakhtiar, R., Ray, T., McMahon, L.M. & Tse, F.L.S. (2000). High-throughput approaches to the quantitative analysis of ketoconazole, a potent inhibitor of cytochrome P450 3A4, in human plasma. *Rapid Communications in Mass Spectrometry,* Vol.14, No.23, (December 2000), pp. (2282-2293), ISSN 1097-0231.

Rezzani, R. (2004). Cyclosporine A and adverse effects on organs: histochemical studies. *Progress in Histochemistry and Cytochemistry,* Vo.39, No.2, (August 2004), pp. (85-128), ISSN 0079-6336.

Romanyshyn, L., Tiller, P.R., Alvaro, R., Pereira, A., Hop, C.E. (2001). Ultra-fast gradient vs. fast isocratic chromatography in bioanalytical quantification by liquid chromatography/tandem mass spectrometry. *Rapid Communications in Mass Spectrometry,* Vol.15, No.5, pp. (313–319), ISSN 0951-4198.

Sangster, T., Spence, M., Sinclair, P., Payne, R. & Smith. C. (2004). Unexpected observation of ion suppression in a liquid chromatography/atmospheric pressure chemical ionization mass spectrometric bioanalytical method. *Rapid Communications in Mass Spectrometery,* Vol. 18, No.12, pp. (1361-1364), ISSN 0951-4198.

Santos-Neto, A., Bergquist, J., Lanças F. & Sjöberg, P. (2008). Simultaneous analysis of five antidepressant drugs using direct injection of biofluids in a capillary restricted-access media-liquid chromatography-tandem mass spectrometry system. *Journal of Chromatography A,* Vol.1189, No.1-2, (May 2008), pp. (514-522), ISSN 0021-9673.

Sawka, A.M., Jaeschke, R., Singh, R.J., Young Jr, W.F. (2003). A comparison of biochemical teats for pheochromocytoma: measurement of fractionated plasma metanephrines compared with the combination of 24-hour urinary metanephrines and catecholamines. *The Journal of Clinical Endocrinology and Metabolism.* Vol.88, No.2, (Febrary 2003), pp. (553-558), ISSN 0021-972X.

Schellen, A., Ooms, B., Van de Lagemaart, D., Vreeken, R., van Dongen, W.D. (2003). Generic solid phase extraction-liquid chromatography-tandem mass spectrometry method for fast determination of drugs in biological fluids. *Journal of Chromatography B: Analytical Technologies in the Biomedical and Life Sciences,* Vol.788, No.2, (May 2003), pp. (251-259), ISSN 1570-0232.

Schmidt, A., Karas, M. & Dülcks. T. (2003). Effect of different solution flow rates on analyte ion signals in nano-ESI MS, or: When does ESI turn into nano-ESI? *Journal of the American Society for Mass Spectrometry,* Vol. 14, No.5 (May 2003), pp. (492-500), ISSN 1044-0305.

Schwartz, J.C., Wade, A.P., Enke, C.G., Cooks, R.G. (1990). Systematic delineation of scan modes in multidimensional mass spectrometry. *Analytical Chemistry, Vol.* 62, No.17, (September 1990), pp. (1809-1818), ISSN 0003-2700.

Schwartz, J.C., Senko, M.W., Syka, J.E. (2002). A two-dimensional quadrupole ion trap mass spectrometer. *Journal of the American Society for Mass Spectrometry,* Vol.13, No.6, (June 2002), pp. (659–669), ISSN 1044-0305.

Shen J.X., Wang H., Tadros S., Hayes R.N. (2006). Orthogonal extraction/chromatography and UPLC, two powerful new techniques for bioanalytical quantification of desloratadine and 3-hydroxydesloratadine at 25 pg/mL, *Journal of Pharmaceutical and Biomedical Analysis.* Vol.40, No.3, (Feburary 2006), pp. (689-706), ISSN 0731-7085.

Sinz, M.W., Podoll, T. The mass spectrometer in drug metabolism. In *Mass Spectrometry in Drug Discovery,* RossiDT, SinzMW (eds). Marcel Dekker: New York, 2002; 271–335.

Soldin, S.J. & Soldin, O.P. (2009). Steroid hormone analysis by tandem mass spectrometry. *Clinical Chemistry,* Vol.55, No.6, (July 2009), pp. (1061-1066), ISSN 0009-9147.

Song, Q., Naidong, W. (2006). Analysis of omeprazole and 5-OH omeprazole in human plasma using hydrophilic interaction chromatography with tandem mass spectrometry (HILIC-MS/MS)--eliminating evaporation and reconstitution steps in 96-well liquid/liquid extraction. *Journal of Chromatography B: Analytical Technologies in the Biomedical and Life Sciences,* Vol.830, No.1, (January 2006), pp. (135-142), ISSN 1570-0232.

Sparkman, O.D. (2000). Mass Spectrometry: Overview and History, Mass Spectrometry Section, In: *Encyclopedia of Analytical Chemistry,* R.A. Meyers, (Ed.), pp. 11501-11557, Wiley, ISBN 3-540-40739-1, Chichester, England.

Taylor, P.J. (2005a). High-performance liquid chromatography-mass spectrometry in the clinical laboratory, *Therpeutic Drug Monitoring,* Vol.27, No.6, (December 2005), pp. (689-693), ISSN 0163-4356.

Taylor, P.J. (2005b). Matrix effects: the Achilles heel of quantitative high-performance liquid chromatography-electrospray-tandem mass spectrometry. *Clinical Biochemistry,* Vol.38, No.4, (April 2005), pp. (328–334), ISSN 0009-9120.

Taylor, P.J. (2007). Internal standard selection for immunosuppressant drugs measured by high-performance liquid chromatography tandem mass spectrometry. *Therapeutic Drug Monitoring,* Vol.29, No.1, (Faburary 2007), pp. (131-132), ISSN 0163-4356.

Thevis, M., Thomas, A., Schanzer, W. (2011). Current role of LC-MS(/MS) in doping control. *Anal Bioanl Chem* Vol.301, No.2, (August 2011), pp. (405-420), ISSN 1618-2642

Thevis, M., Schänzer, W. (2007). Current role of LC-MS (/MS) in doping control. *Analytical Bioanalytical Chememistry,* Vol.388, No.7. (August 2007), pp. (1351–1358), ISSN 1618-2642.

Thörngren, J.O., Östervall, F., Garle, M. (2008). A high-throughput multicomponent screening method for diuretics, masking agents, central nervous system (CNS) stimulants and opiates in human urine by UPLC-MS/MS. *Journal of Mass Spectrometry,* Vol.43, No.7, pp. (980–992), ISSN 1076-5174.

Tiller, P. R., Romanyshyn, L. A., Neue, U. D. (2003). Fast LC/MS in the analysis of small molecules. *Analytical and Bioanalytical Chemistry,* Vol.377, No.5, (November 2003), pp. (788-802), ISSN 1618-2642.

van der Horst, A. and Schoenmakers, P. J. (2003). Comprehensive two-dimensional liquid chromatography of polymers. Journal of Chromatography A, Vol.1000, No.1-2, (June 2003), pp. (693–709), ISSN 0021-9673.

Veltkamp, A.C., Das, H.A., Frei, R.W., Brinkman, U.A. (1987). On-line low-level radiometric detection of [^{14}C]remoxipride in liquid chromatographic effluents. Application to urine samples. *Journal of Chromatography. Vol.* 384, (January 1987), pp. (357-369), ISSN 0021-9673.

Ventura, R., Roig, M., Montfort, N., Saez, P., Berges, R., Segura, J. (2008). High-throughput and sensitive screening by ultraperformance liquid chromatography tandem mass spectrometry of diuretics and other doping agents. *European Journal of Mass Spectrometry (Chichester, Eng)* Vol.14, No.3, pp. (191–200), ISSN 1469-0667.

Veuthey, J.L., Souverain, S. & Rudaz, S. (2004). Column-switching procedures for the fast analysis of drugs in biologic samples. *Ther Drug Monit,* Vol.26, No.2, (April 2004), pp. (161-166), ISSN 0163-4356.

Vintiloiu, A., Mullett, W.M., Papp, R., Lubda, D., Kwong, E. (2005). Combining restricted access material (RAM) and turbulent flow for the rapid on-line extraction of the cyclooxygenase-2 inhibitor rofecoxib in plasma samples. *Journal of Chromatography A*, Vol.1082, No.2, (August 2005), pp. (2660-2668), ISSN 0021-9673.

Vogeser, M. & Kirchhoff, F. (2011). Progression in automation of LC-MS in laboratory medicine. *Clinical Biochemistry*, Vol.44, No.1, (January 2011), pp. (4-13), ISSN 0009-9120.

Vogeser, M. & Seger, C. (2008). A decade of HPLC-MS/MS in the routine clinical laboratory: goals for further developments. *Clinical Biochemistry,* Vol.41, No.9, (June 2008), pp. (649-662), ISSN 0009-9120.

Vogeser, M., Geiger, A., Herrmann R. And Kobold, U. (2008). Sample preparation for liquid chromatography-tandem mass spectrometry using functionalized ferromagnetic micro-particles. *Clinical Biochemistry,* Vol. 41, No.16-17, (November 2008), pp. (1417–1419), ISSN 0009-9120.

Vogeser, M. & Spohrer, U. (2005). Pitfall in the high-throughput quantification of whole blood cyclosporin A using liquid chromatography-tandem mass spectrometry. *Clinical Chemistry and Laboratory Medicine*, Vol.43, No.4, (Feburay 2005), pp. (400-402), ISSN 1434-6621.

Vuckovic, D., Zhang, X., Cudjoe, E., Pawliszyn J. (2010). . Solid-phase microextraction in bioanalysis: New devices and directions. *Journal of Chromatography A*. Vol.1217, No.25, (June 2010), pp. (4041–4060), ISSN 0021-9673.

Wang, G., Hsieh, Y., Cui, X., Cheng, K.C., Korfmacher, W.A., (2006). Ultra-performance liquid chromatography/tandem mass spectrometric determination of testosterone and its metabolites in in vitro samples. *Rapid Communicatios in Mass Spectrometry,* Vol.20, No.14, pp. (2215-2221), ISSN 0951-4198.

Wang, S., Cyronak, M, Yang, E. (2007). Does a stable isotopically labeled internal standard always correct analyte response? A matrix effect study on a LC/MS/MS method for the determination of carvedilol enantiomers in human plasma. *Journal of Pharmaceutical and Biomedical Analysis,* Vol.43, No.2, (January 2007), pp. (701-707), ISSN 0731-7085.

Watson, J.T. & Sparkman, O.D. (2007). Introduction of Mass Spectrometry: Instrumentation, Applications and Strategies for Data Interpretation. (4th Eds), John Wiley & Sons Ltd, ISBN 978-0-470-51634-8, West Sussex, England

Westmacott, G., Frank, M., Labov, S.E. & Benner, W.H. (2000). Using a superconducting tunnel junction detector to measure the secondary electron emission efficiency for a microchannel plate detector bombarded by large molecular ions. *Rapid Communicatios in Mass Spectrometry*, Vol.14, No.19, pp. (1854-1861), ISSN 0951-4198.

Wilson, I.D. & Brinkman, U.A.T. (2003). Hyphenation and hypernation — the practice and prospects of multiple hyphenation. *Journal of Chromatography A*, Vol.1000, No.1-2, (June 2003), pp. (325-356), ISSN 0012-9673.

Wikipedia, http://en.wikipedia.org/wiki/Mass_spectrometry.

World Anti-Doping Agency (2009). The world anti-doping code -international standard for laboratories. http://www.wada-ama.org/rtecontent/document/International_Standard_for_ laboratories_v6_0_January_2009.pdf. Accessed 24 Nov 2008.

Wren, S.A. & Tchelitcheff, P. (2006) Use of ultra-performance liquid chromatography in pharmaceutical development. *Journal of Chromatography A*, Vol.1119, No.1-2, (June 2006), pp. (140–146), ISSN 0021-9673.

Xu, R.N., Fan, L., Rieser, M.J. & El-Shourbagy, T.A. (2007). Recent advances in high-throughput quantitative bioanalysis by LC-MS/MS. *Journal of Pharmaceutical and Biomedical Analysis,* Vol.44, No.2, (June 2007), pp. (342-355), ISSN 0731-7085.

Xu, R.N., Kim, G.E., Gregg, H., Wagdy, A., Swaine, B.A. & Chang, M.S., El-Shourbagy, T.A. (2004). Automated 96-well liquid-liquid back extraction liquid chromatography-tandem mass spectrometry method for the determination of ABT-202 in human plasma. *Journal of Pharmaceutical and Biomedical Analysis,* Vol.36, No.1, (September 2004), pp. (189-195), ISSN 0731-7085.

Xu, R.N., Fan, L., Rieser, M.J., El-Shourbagy, T.A. (2007). Recent advances in high-throughput quantitative bioanalysis by LC-MS/MS. *Journal of Pharmaceutical and Biomedical Analysis,* Vol.44, No.2, (June 2007), pp. (342-355), ISSN 0731-7085.

Xu, X.S., Yan, K.X., Song, H. & Lo, M.W. (2005). Quantitative determination of a novel dual PPAR alpha/gamma agonist using on-line turbulent flow extraction with liquid chromatography-tandem mass spectrometry. *Journal of Chromatography B: Analytical Technologies in the Biomedical and Life Sciences,* Vol.814, No.1, (January 2005), pp. (29-36), ISSN 1570-0232.

Xue, Y.J., Akinsanya, J.B., Liu, J. & Unger, S.E. (2006). A simplified protein precipitation/mixed-mode cation-exchange solid-phase extraction, followed by high-speed liquid chromatography/mass spectrometry, for the determination of a basic drug in human plasma, *Rapid Communications in Mass Spectrometry,* Vol.20, No.18, pp. (2660-2668), ISSN 0951-4198.

Yang, A.Y., Sun, L., Musson, D.G. & Zhao, J.J. (2006). Determination of M+4 stable isotope labeled cortisone and cortisol in human plasma by microElution solid-phase extraction and liquid chromatography/tandem mass spectrometry *Rapid Communications in Mass Spectrometry,* Vol.20, No.2, pp. 233–240, ISSN 0951-4198.

Yang, Z., Peng, Y., Wang, S. (2005). Immunosuppressants: pharmacokinetics, methods of monitoring and role of high performance liquid chromatography/mass spectrometry. *Clinical and Applied Immunology Reviews,* Vol.5, No.6, (December 2005), pp. (405-430), ISSN 1529-1049.

Yang, Z. & Wang, S. (2008). Recent development in application of high performance liquid chromatography-tandem mass spectrometry in therapeutic drug monitoring of immunosuppressants, *Journal of Immunological Methods,* Vol.331, No.2, (July 2008), pp. (98-103), ISSN 0022-1759.

Yi, Z. & Brown, P.R. (1991). Chromatographic methods for the analysis of basic neurotransmitters and their acidic metabolites. *Biomedical Chromatography,* Vol.5, No.3, (May 1991), pp.(101-107), ISSN 0269-3879.

Yu, K., Little, D., Plumb, R. & Smith, B. (2006) High-throughput quantification for a drug mixture in rat plasma - a comparison of Ultra Performance™ liquid chromatography/tandem mass spectrometry with high-performance liquid chromatography/tandem mass spectrometry. *Rapid Communications in Mass Spectrometry,* Vol.20, No.4, pp. (544–552), ISSN 0951-4198.

Yue, B., Rockwood, A.L., Sandrock, T., La'ulu, S.L., Kushnir, M.M., Meikle, A.W. (2008). Free thyroid hormones in serum by direct equilibrium dialysis and online solid-phase extraction-liquid chromatography/tandem mass spectrometry. *Clinical Chemistry,* Vol.54, No.4, (April 2008), pp. (642-651), ISSN 0009-9147.

Zang, X., Luo, R., Song, N., Chen, T.K. & Bozigian, H. (2005). A novel on-line solid-phase extraction approach integrated with a monolithic column and tandem mass spectrometry for direct plasma analysis of multiple drugs and metabolites. *Rapid Communications in Mass Spectrometry,* Vol.19, No.22, pp. (3259-3268), ISSN 0951-4198.

Zeng, H., Deng, Y., Wu, J.T. (2003). Fast analysis using monolithic columns coupled with high-flow on-line extraction and electrospray mass spectrometric detection for the direct and simultaneous quantitation of multiple components in plasma. *Journal of Chromatography B: Analytical Technologies in the Biomedical and Life Sciences,* Vol.788, No.2, (May 2003), pp. (331-337), ISSN 1570-0232.

Zhang, N., Fountain, S.T., Bi, H., Rossi, D.T. (2000a). Quantification and rapid metabolite identification in drug discovery using API time-of-flight LC/MS. *Analytical Chemistry,* Vol. 72, Vol.4, (Feburary 2000), pp. (800-806), ISSN 0003-2700.

Zhang, H., Henion, J., Yang, Y., Spooner, N. (2000b). Application of atmospheric pressure ionisation time-of-flight mass spectrometry coupled with liquid chromatography for the characterization of in vitro drug metabolites. *Analytical Chemistry.* Vol.72, No.14, (July 2000), pp. (3342-3348), ISSN 0003-2700.

Zhou, S., Zhou, H., Larson, M., Miller, D.L., Mao, D., Jiang, X., Naidong, W. (2005). High-throughput biological sample analysis using on-line turbulent flow extraction combined with monolithic column liquid chromatography/tandem mass spectrometry. *Rapid Communication in Mass Spectrometry,* Vol.19, No.15, (August 2005), pp. (2144-2150), ISSN 0951-4198.

Zimmer, D., Pickard, V., Czembor, W., Muller, C., (1999). Comparison of turbulent-flow chromatography with automated solid-phase extraction in 96-well plates and liquid-liquid extraction used as plasma sample preparation techniques for liquid chromatography-tandem mass spectrometry. *Journal of Chromatography A,* Vol.854, No.1-2, (August 2007) pp. (23-35), ISSN 0021-9673.

21

HPLC-MS/MS of Highly Polar Compounds

Luigi Silvestro[1], Isabela Tarcomnicu[2] and Simona Rizea Savu[1]
[1]3S-Pharmacological Consultation & Research GmbH, Harpstedt,
[2]Pharma Serv Int'l SRL, Bucharest,
[1]Germany
[2]Romania

1. Introduction

Complicate high-performance liquid chromatographic (HPLC) interfaces like the moving belt, direct liquid interface, thermospray (Arpino, 1985) fast atom bombardment (FAB) (Caprioli et al., 1986; Garcia & Barceló, 1993; Dell et al., 1988; Larry et al., 1989) characterized the early days of HPLC used in tandem with mass-spectrometry (MS); all of them, apart FAB, were more effective with moderately polar compounds (Strege et al., 1999). After years of difficulties and struggles the introduction of atmospheric pressure ionization by electrospray (ESI) or chemical ionization marked a clear breakthrough: an effective, robust and user friendly HPLC interface was finally available.
As a matter of fact, considering the physicochemical principles of the source, the polar compounds were the first compounds successfully ionized (Griffits, 2008).
Despite the huge potential of ESI for polar and highly polar compounds the applications for the latter proved to be critical due to the fact that effective HPLC separations often require non volatile buffers and/or high water content that are not suitable in MS detection.
Along the years, the development of HPLC columns, specific derivatization methods, new chromatographic techniques and sample preparation approaches have increased the number of applications with highly polar compounds. Three classes of not so widely studied molecules have been selected in this study in order to illustrate the application of HPLC-MS/MS to highly polar compounds; updates as well a review of past data will be presented.
The groups of compounds taken in consideration are: glycosaminoglycans, bisphosphonates and amynoglycoside antibiotics; the first-ones are both endogenous and exogenous while the last 2 classes are of typical pharmaceutical interest.
Obviously other more well-known molecules like peptides and polynucleotides are also included in this class of highly polar compounds but HPLC-MS/MS methods in these particular fields are widely described and probably represent the most common areas of application. It is also important to keep in mind that this chapter doesn't intend to be an exhaustive presentation of the published analytical methods for heparins, bisphosphonates or aminoglycosides, but these compounds are used as models to emphasize the potential of HPLC-MS/MS in the analysis of highly polar molecules.

2. Glycosaminoglycans

2.1 Chemical structure and main features of glycosaminoglycans

Glycosaminoglycans (GAGs) are a family of highly sulphated, complex, polydispersed linear polysaccharides that display a variety of important biological roles. They contain amino-sugars with alternating disaccharide units, which comprise an acidic monosaccharide such as glucuronic or iduronic acid and a basic monosaccharide such as N-acetylglucosamine or N-acetyl-galactosamine; both monosaccharides can be sulphated. In natural biological systems GAGs are mainly linked to proteins to form proteoglycans, with great interest for medical science being structural proteins or receptor as well adhesion molecules (just to mention a few examples) and can be categorized, based on the GAGs structure, into three main structural groups: heparin/heparan sulphate group, dermatan/chondroitin sulphate group and hyaluronan (Mao et al., 2002; Tiayu Peng, 2002).
Heparin and heparan sulphate occur as proteoglycans in which the glycosaminoglycan chains have different lengths and are composed of disaccharide units consisting of L-iduronic or D-glucuronic acids α(1-4) linked to glucosamine units that can be N-acetylated or N-sulphated (Figure 1).
Heparin and heparan sulphate have closely related structures and consist of similar disaccharide composition except that heparin has a greater content of iduronic acid and is more highly sulphated per polysaccharide chain than is heparan sulphate. The differences between the chemical structure of the sequences of heparin and heparan sulphate explain the biological function related to them. As an example, the specific interaction of heparin with the protein antithrombin III, determining its anticoagulant activity, is mediated by a characteristic pentasaccharide that is not normally present in heparan sulphate and it accounts for just a few percent of the heparin mass (Saad & Leary, 2003).

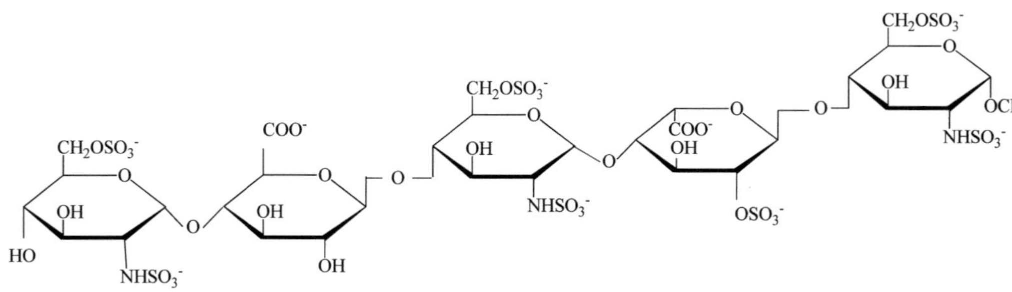

Fig. 1. Main sequence structure of heparin/heparan sulphate.

The dermatan/chondroitin sulphate are the main components of connective tissues and the chemical structure comprises alternating uronic acid and N-acetyl-galactosamine residues; depending on their composition 3 main types of chondroitin sulphates have been identified (Gunay & Linhardt, 1999). The structures of these types of chondroitin sulphate are presented in Figure 2.

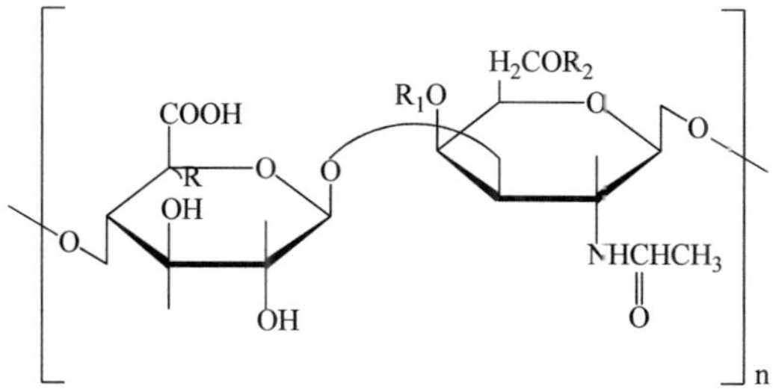

Chondroitin Sulphate Type A: R=H, R_1=SO_3H, R_2=H, R_3=COOH
Chondroitin Sulphate Type B: R=COOH, R_1= SO_3H, R_2=H, R_3=H
Chondroitin Sulphate Type C: R=H, R_1=H, R_2=SO_3H, R_3=COOH

Fig. 2. Structure of chondroitin sulphate.

Chondroitin sulphate A and chondroitin sulphate C contain D-glucuronic acid, N-acetylgalactosamine and sulphate residues in equimolar quantities but differ in the position of the sulphate ester group (Saito et al., 1968). Dermatan sulphate formerly called chondroitin sulphate B is similar to chondroitin sulphate A but instead of D-glucuronic acid it consists mainly of L-iduronic acid.

Hyaluronic acid has alternating repeating units of the structure, (1→3) β-N-acetyl-D-glucosamine (1→4) β-D-glucuronic acid, but tipically it is not sulphated and the molecules are larger than the other glycosaminoglycans, often several million Da, with special rheologic characteristic (Mao et al., 2002).

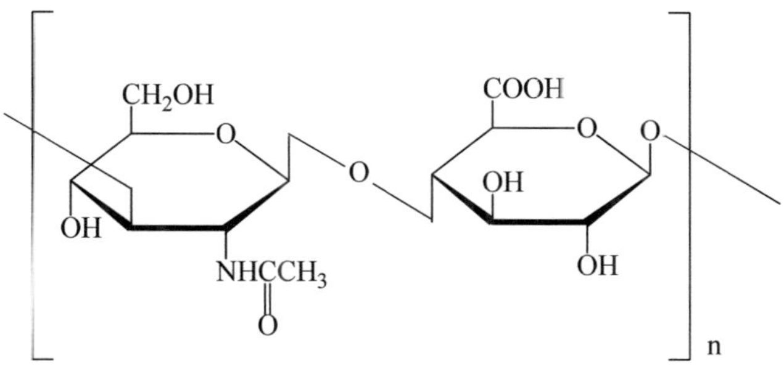

Fig. 3. Structure of the repeating units of Hyaluronic acid.

2.2 Special problems of these molecules and potential approach to solve them

GAGs are extremely difficult to analyze because of their polydispersity and microheterogeneity of chemical structure combined with high polarity. The first two aspects (polydispersity and microheterogeneity) push to get an optimal separation of all species presented in a sample in order to may accurately characterize each component. It is in fact important to consider that these molecules are closely related polymers with common disaccharides/oligosaccharides and if not well separated it is also impossible to get detailed structure information.

In the meantime the high polarity of GAGs make them difficult compounds for optimal chromatographic separation and an adequate separation of all components present in a GAGs sample is virtually impossible.

In such conditions two approaches are feasible:

1. Application of analytical techniques able to gather structure data also of unfractionated or partially fractionated complex mixtures
2. Structure specific degradation of the GAGs in order to reduce such complex samples with almost infinite variant of similar polymers to more discrete mixtures of oligomers corresponding to their "building blocks" that can be properly analyzed and quantified obtaining quite detailed information of the polymers themselves intended as a mean structure or a fingerprint.

Regarding the first approach, only NMR permits to get in-depth chemical structure information of almost unfractionated materials but the instrumental sensitivity is a major drawback when studying limited biological samples and in any case the information obtained are practically a mean of the individual structures of the components present in the sample. Despite the fact that hyphenation of NMR and HPLC (as separation method) is feasible, due to sensitivity problems its application on such complex samples is more theoretical then practical so far.

Before to close this parenthesis on NMR, outside the scope of this chapter, it is evident that NMR spectroscopy is also a very effective technique for the characterization of the oligosaccharides (Yung, 2011). It gives valuable information on monosaccharide composition, glycosidic linkage, uronic acid type and sulfation patterns (Bo et al., 2011), but it requires always relatively large amount of material and is not suitable for the analysis of very small samples, like most biological samples. By coupling capillary isotachophoresis with on-line microcoil NMR detection (CITP-NMR) the sensitivity for the characterization of heparin - derived oligosaccharides has been improved (Korir & Larive, 2007).

Considering the limitations related to this first analytical approach, based on intact molecules, the focus in the next paragraphs will be on presenting methods to analyze GAGs following depolimerization; clearly that in rare cases of small oligosaccharide (i.e. from synthesis) or highly purified fractions such step can be avoided.

2.3 Overview on methods of GAGs depolimerization

As previously pointed out most often the structure characterization of GAGs involves degradation steps to obtain disaccharides and/or a range of oligosaccharide fragments allowing an efficient disaccharide/oligosaccharide mapping (Bo et al., 2011). These structural analyses of GAGs may involve complete or partial chemical degradation, as well as enzymatic degradation permitting a quite selective cut of the polysaccharide chains (Ruiz-Calero et al., 2003; Gatti et al., 2010). Highly specific enzymes have been isolated and they are now commercially available making their application viable (Johnson, 1982; Jandik et al., 1996; Cohen & Linhardt, 1990). In figure 4 the main used enzymes are shown.

An important aspect of these enzymes is the high specificity for the substrate, fact that if it is welcomed in biochemical studies can be critical sometimes when analyzing semisynthetic derivatives.

Chemical degradation is a very interesting alternative to enzymatic degradation giving also quite specific structure information. The degradation with nitrous acid and the periodic oxidation followed by Smith degradation are two classical tools for GAGs analysis.

In several cases depolimerization methods are followed by derivatization procedures generally to enhance the analytical detection when using conventional HPLC or CE detectors.

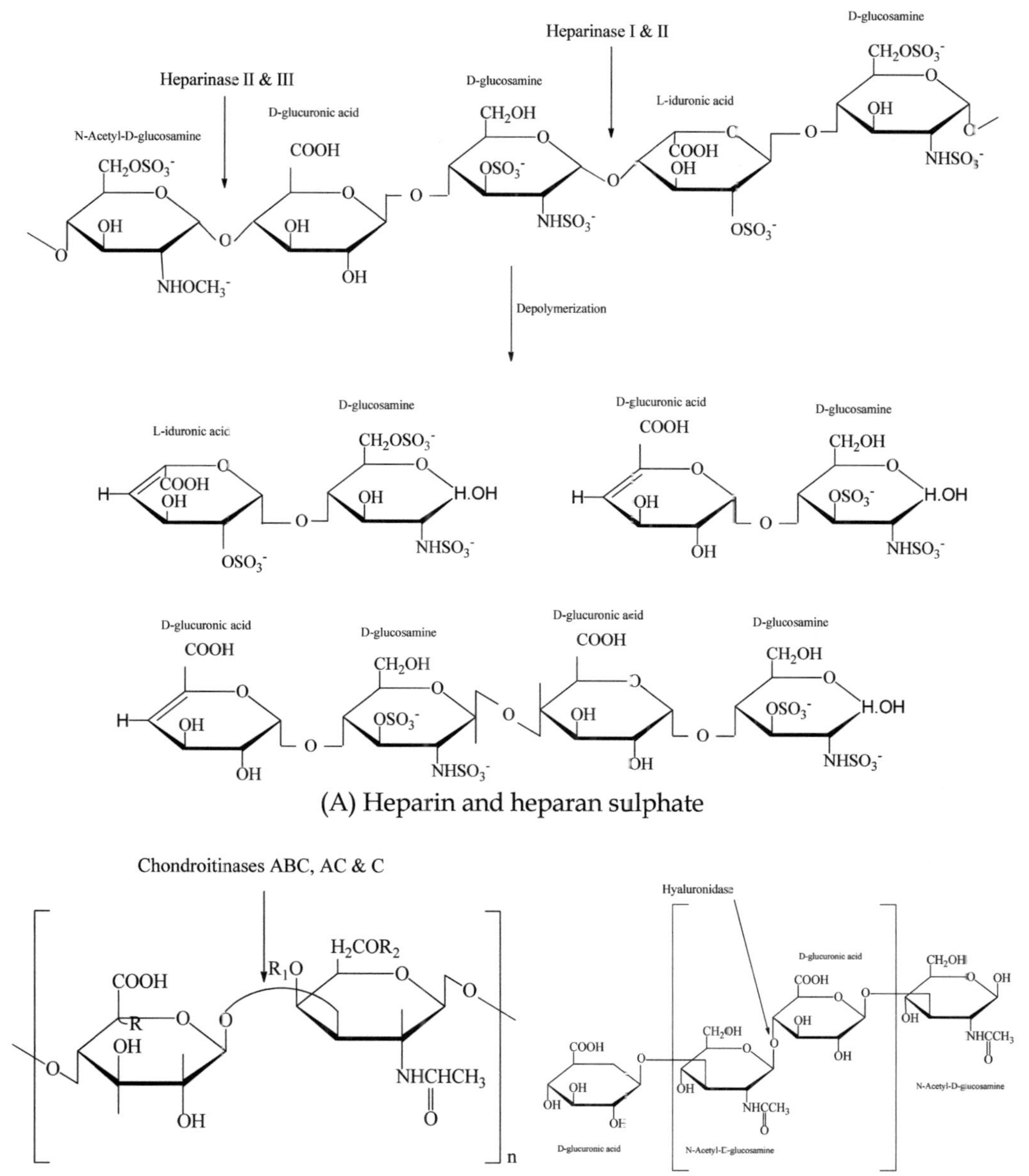

Fig. 4. The enzymatic cleavage of GAGs; A) heparin, heparan sulphate and the specific enzymes: Heparinase I, Heparinase II and Heparinase III; B) Chondroitin sulphate and the specific enzymes: Chondroitinase ABC for Chondroitin sulphate A, B and C, Chondroitinase AC for Chondroitin sulphate A and C and Chondroitinase C for Chondroitin sulphate A and C; C) Hyaluronic acid and the specific enzyme: Hyaluronidase

2.4 Overview on HPLC analysis of GAGs

For the analysis of intact GAGs, a variety of chromatographic techniques have been developed mainly to separate different group of GAGS and/or to characterize the molecular weight distribution. Concerning this last aspect gel permeation chromatography (both as HPLC or low pressure chromatography) is the preferred approach.

Ion exchange chromatography and reversed phase ion-pair chromatography are the most employed techniques to separate the main classes of intact GAGs, which present differences in polarity, both as low pressure chromatography (Patel et al., 2009; Jin et al., 2009) or high performance liquid chromatography (Shao et al., 2004; Trehy et al., 2009). More recently also capillary electrophoresis (CE) (Malsch & Harenberg, 1996; Malsch et al., 1996) have been used successfully. The main interest is, however, on methods for the separation of depolymerisation fragments of GAG, obtained by different digestion methods (chemical or enzymatic), by HPLC (Du & Eddington, 2002; Plaas et al., 1996; Toyoda et al., 2000) or CE (Malavaki et al., 2008; Ruiz-Calero, et al., 2003) followed by detection with conventional spectral system (UV or fluorescence in case of adequate derivatization), as in the previous papers, but in the last years often by mass spectrometry (Oguma et al., 2007; Yang et al., 2011; Barosso et al., 2005; Silvestro et al., 1992; Da Col et al., 1993; Hemstrom & Irgum, 2006; Silvestro et al., 1996).

The peculiar aspects of separation in connection with MS detection will be detailed in the following paragraph.

2.5 Overview on mass-spectrometric analysis of GAGs

Methodologies using fast atom bombardment (FAB) were first employed; however a direct coupling to HPLC separation was not achieved, as in the case of peptides, due to limitations in mobile phase composition (Dell et al., 1988; Mallis et al., 1989). The introduction of soft ionization methods (ESI), permitting an easy interfacing to HPLC has first provided an on-line separation and structure elucidation of complex oligosaccharide mixtures (Henriksen et al., 2006; Barosso et al., 2005; Thanawiroon et al., 2004). As a complement, matrix-assisted laser desorption ionization (MALDI) has proved to be a valuable tool for the analysis of protein/peptide-heparin/HS oligosaccharide complexes (Ori et al., 2009; Venkataram et al., 1999).

The first important point to consider is the kind of oligosaccharides and disaccharides to be analyzed in particular if they are sulphated or not sulphated.

2.5.1 Sulphated oligosaccharides

As discussed previously *(Chapter 2.4),* sulphated oligosaccharides and disaccharides are separated, when using conventional detectors (UV, fluorescence, electrochemical), employing strong ion exchange columns with gradients of molarity of salts from strong acids (sulphates, phosphates, chloride etc.). It is evident that such mobile phases, being non-volatile, cannot be used in case of MS interfacing and alternative solutions have been developed.

An option is the application of weak anion exchange columns permitting the elution of sulphated derivatives also with gradients of ammonium acetate, compatible with an ESI source; nonetheless it cannot be excluded that highly sulphated large oligomers are not eluted from this kind of columns.

Another possibility, probably the most widely used, is the separation on a reversed phase column in presence of an ion-pairing reagent. In an early paper by our group (Da Col et al., 1993) tetrapropylammonium (TPA) was selected as ion pair reagent for this kind of separation; smaller quaternary ammonium derivatives (tetramethylammonium and tetraethylammonium) didn't improve at all the chromatographic separation while larger derivative (tetrabutylammonium and higher) showed critical problems of volatility in the ionization source without offering other advantages. It is important to note that the choice of the stationary phase can be quite critical: we observed in fact that in case of reversed phase columns with longer aliphatic chain (C_8 or longer) the sulphated oligosaccharides couldn't be eluted being probably retained by an HILIC-like retention mechanism. Best conditions were obtained on C_4 - C_6 columns with gradients of acetonitrile water (from 100% to 50% water) in presence of 3.3 mM tetrapropylammonium at pH 4.5 with formic acid. To prepare the mobile phase most common tetrapropylammonium salts like the sulphate (common in conventional HPLC) were avoided due to volatility problems and preparations were made starting from tetrapropylammonium hydroxide. An example of TIC chromatograms recorded on a digest of heparin is presented in Figure 5.

More recently, several groups have successfully introduced the use of tertiary aliphatic amines (tributylammonium acetate in acetonitrile as ion pair reagents) claiming that a better volatility can be obtained (Thanawiroon et al., 2004). It must be observed that short chain amines like triethylamine, widely used in HPLC, are not able to form ion-pair with these sulphated oligosaccharides.

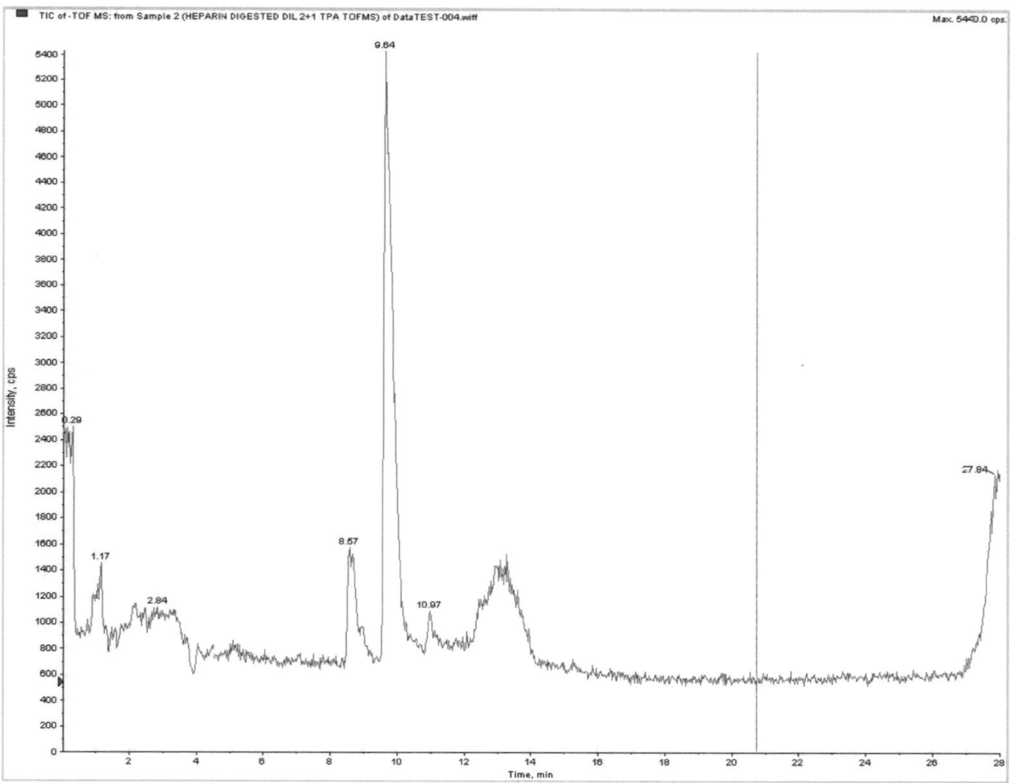

Fig. 5. Total ion chromatogram obtained in TOF full scan after injecting a sample of heparin digested with heparinase II. Column: Kromasil C4 (100x2.1mm, 3μm); mobile phase: (A) TPA 3.3 mM in water, pH 4 and (B) TPA 3.3 mM in acetonitrile/water (90/10, v/v), pH 4. Mass spectrometer: QTOF Qstar XL, operated in negative ESI mode.

The addition of an ion pair reagent to sulphated oligosaccharides brings to the formation of complexes stable enough to be ionized as such. Complex pseudomolecular ions with one or more molecules of the ion-pair reagent, sometimes multiple charged, are predominant in spectra of chromatographic separations of such di- and oligo-saccharides; 2 examples of similar spectra are presented in Figure 6 and Figure 7.

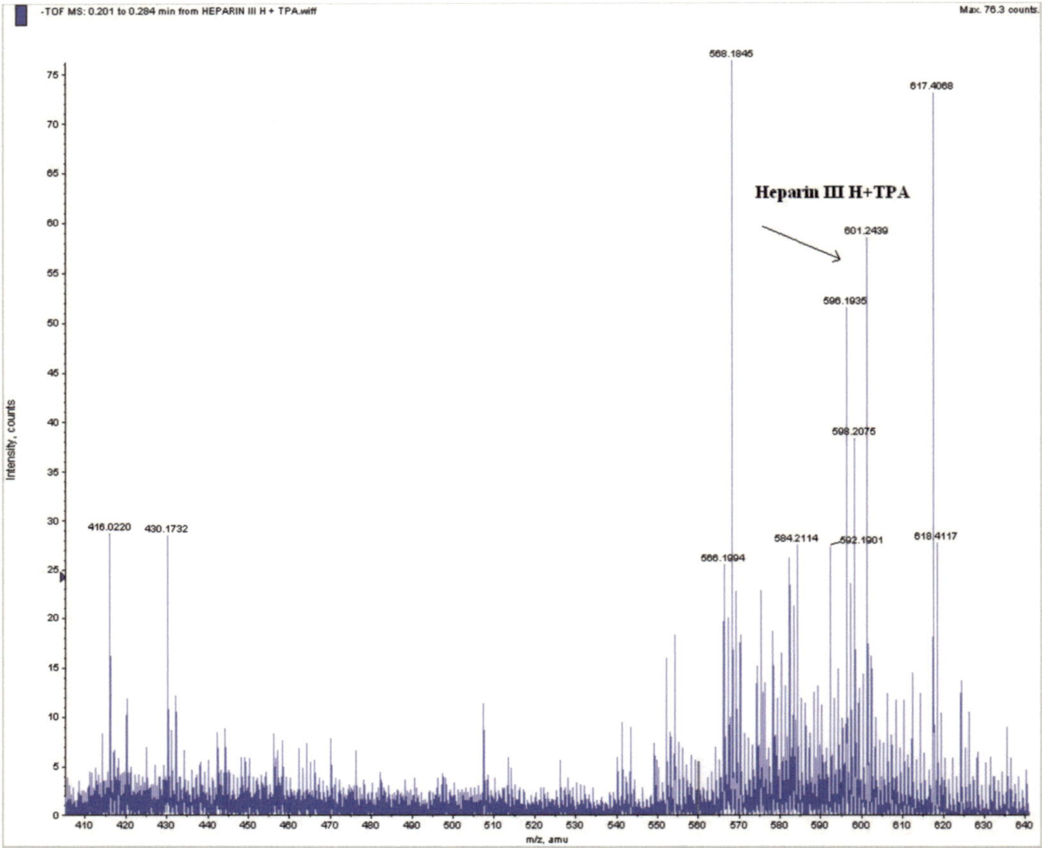

Fig. 6. The molecular ion of the adduct formed by the disaccharide Heparin III H and TPA $[M+TPA-H]^-$ = 601.2439. Spectrum recorded by direct infusion with the syringe of a solution diluted at 1 µg/mL. Mass spectrometer: QTOF Qstar XL, operated in negative ESI mode.

It is evident that the use of high resolution instruments is a valuable tool to clarify the charge state of these ions while to clarify the stoichiometry of ion-pair reagent/oligosaccharide ratio the utilization of a labelled ion-pairing reagent is very interesting (Silvestro et al., 1996).

In an effort to obtain more easily interpretable spectra an attractive approach is also the use model MMPC of the ion suppressor (i.e. Dionex…) to remove the tetrapropylammonium from the chromatographic mobile phase. The experiments carried out have really showed an efficient removal of tetrapropylammonium (an example is shown in Figure 8). Anyway, the oligosaccharides as free acidic compounds are much more prone to non-sequence related in source fragmentation during ionization (mainly desulphation), making sometimes very difficult to understand the oligosaccharide real structure.

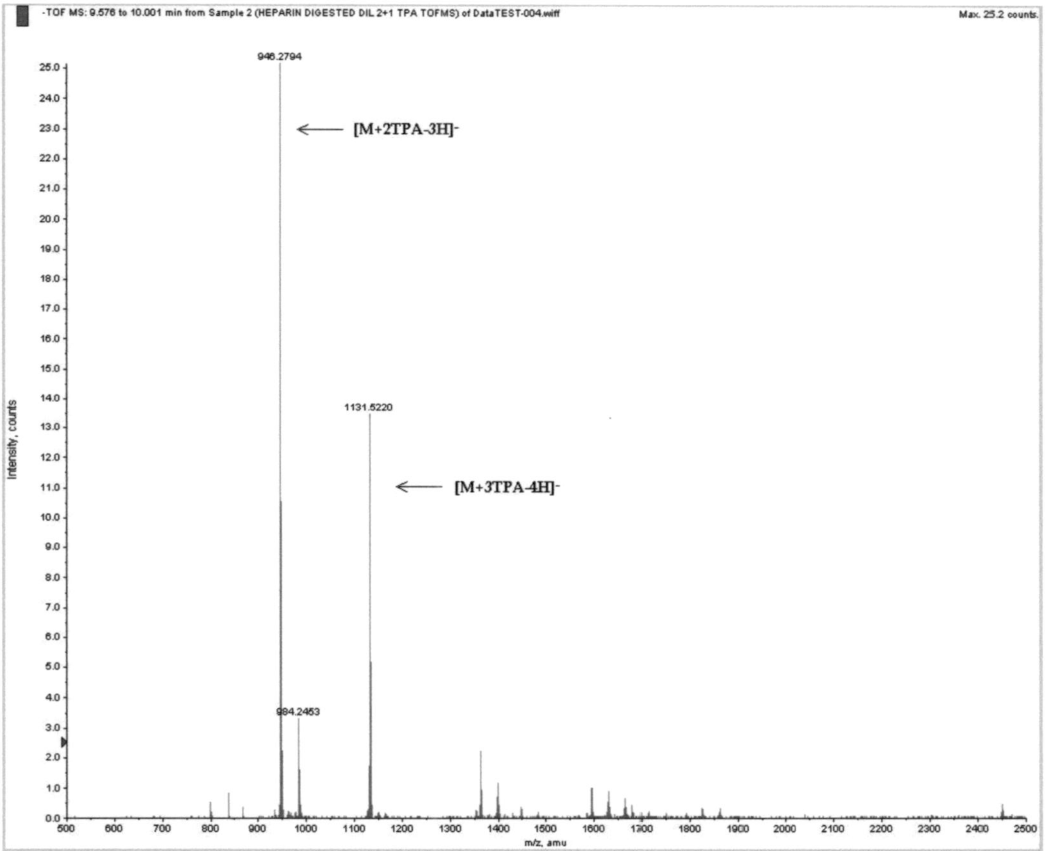

Fig. 7. Extracted spectrum obtained after injecting a sample of heparin digested with heparinase II. Column: Kromasil C4 (100x2.1mm, 3µm); mobile phase: (A) TPA 3.3 mM in water, pH 4 and (B) TPA 3.3 mM in acetonitrile/water (90/10, v/v), pH 4. Mass spectrometer: QTOF Qstar XL, operated in negative ESI mode. M: trisulphated disaccharide obtained from heparin by enzymatic digestion.

The MS/MS properties of these di- and oligo-saccharides are also noteworthy: in case of parent ions (single or multiple charged) the desulphation fragments are generally the most abundant being easily formed and generally very few weak signals coming from fragmentation of the sugar back bone can be observed. The situation is similar in case of TPA complexes with the tendency to an even less favourable fragmentation.

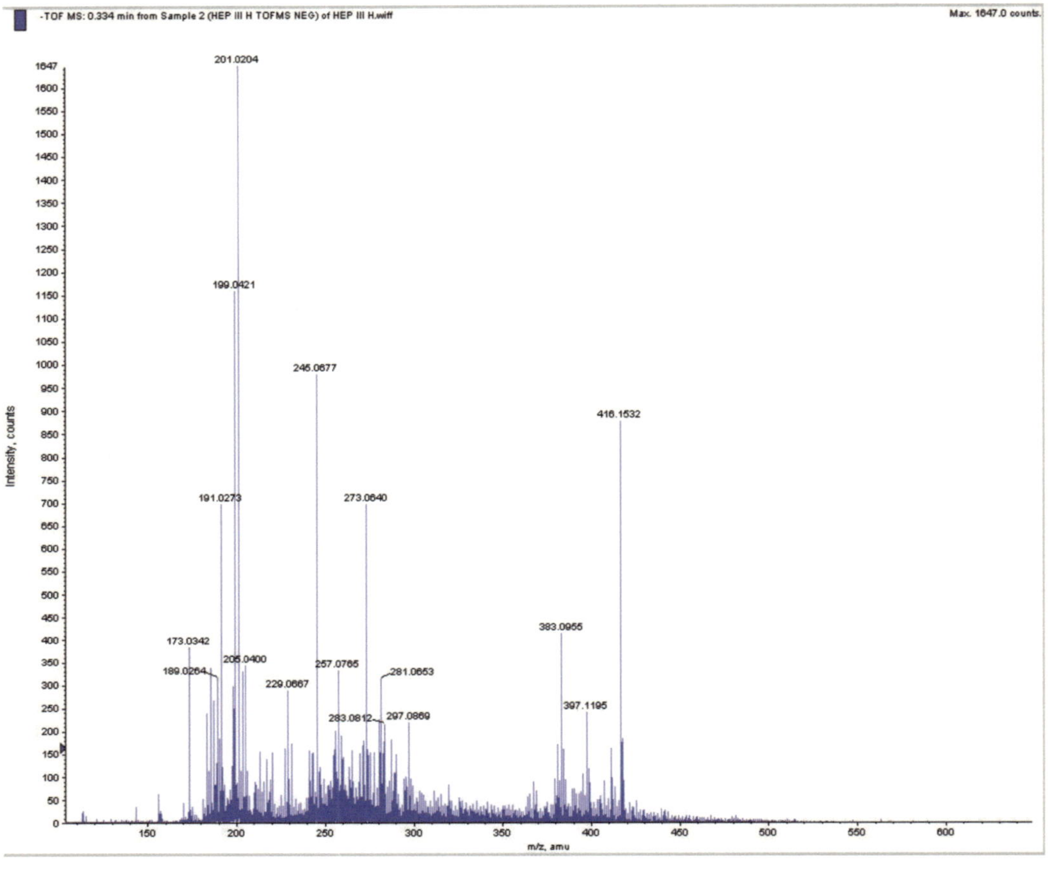

Fig. 8. The molecular ion of the oligomer Heparin III H after ion suppression; $[M-H]^-$ = 416.1532 can be easily seen. Spectrum recorded by direct infusion with the syringe of a solution diluted at 1 µg/mL. Mass spectrometer: QTOF Qstar XL, operated in negative ESI mode.

2.5.2 Non sulphated oligosaccharides

The analytical separations of these molecules are less challenging than those of the more charged sulphated products. In this case the use of ion exchange columns eluted with gradients of volatile buffers (i.e. ammonium acetate) at high ionic strength is effective and permits a good separation of oligomers with different chain length. The application of ion-pair chromatography is practically not needed and in general the formation of ion-pair complexes having only carboxylic groups is not effective.

Deprotonated molecular ions, sometimes multiple charged, are typically observed, as shown in Figure 9.

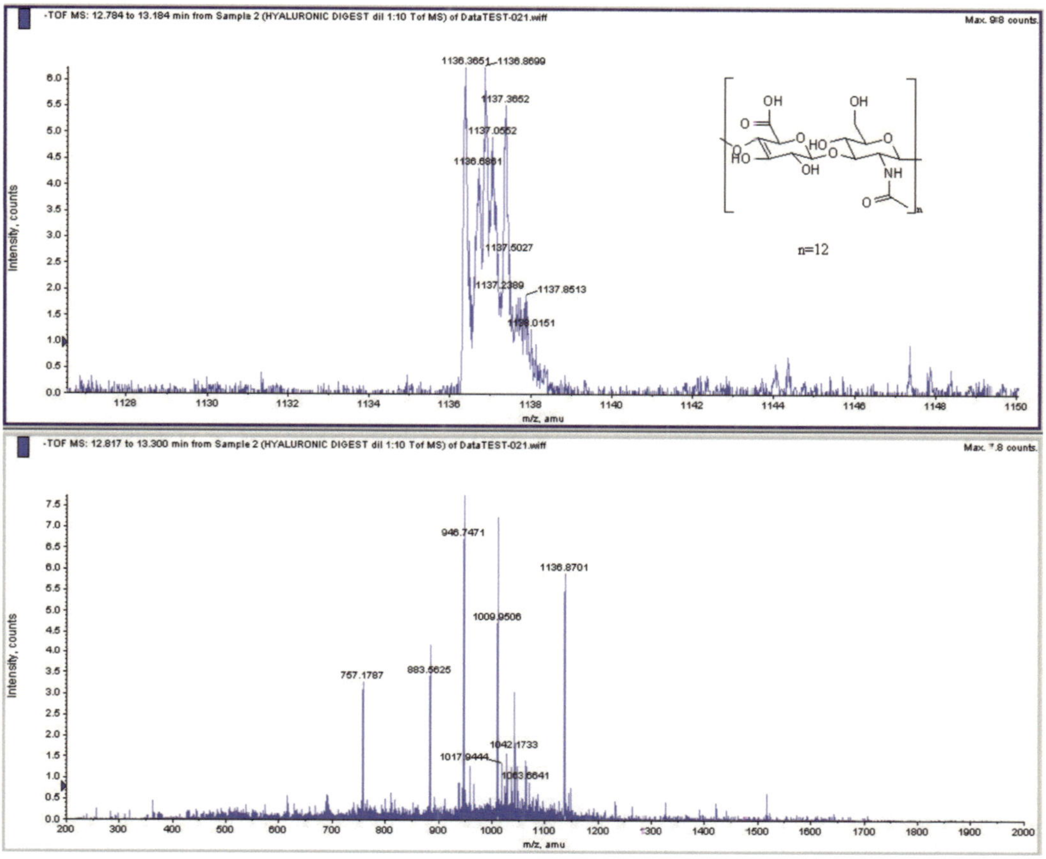

Fig. 9. Mass spectrum obtained in TOF full scan mode after injecting a sample of a hyaluronic acid digested with hyaluronidase. Column: Supelcosil LC-SAX 1 (250x4.6mm, 5µm); mobile phase: (A) water/methanol (1/1, v/v) (B) ammonium acetate 0.5 M in water/methanol (1/1, v/v). Mass spectrometer: QTOF Qstar XL, operated in negative ESI mode. The main ions in the spectrum are multiple charged; they all correspond to dissacahride oligomers resulted from hyaluronic acid degradation. A detail can be seen in the spectra above for $[M-2H]^{2-}$=1136.8, a doubly charged ion corresponding to a dodecasaccharide.

Daughter spectra can be obtained as it is shown in Figure 10. However, it can be seen, these kinds of molecules are hard to break and only scarce structural information can be gathered.

3. Bisphosphonates analysis

3.1 Chemical structure and main features of bisphosphonates

Bisphosphonates, polar structures with two phosphonate groups covalently bonded to carbon (called P-C-P bridge, Fig. 11) are a class of drugs used in the treatment of osteoporosis, osteolytic metastasis, Paget's disease (osteitis deformans), and other disorders involving bone fragility. They have also non-therapeutic usage, based on their chelating properties (as water softeners, in agriculture, in paper, detergent and cosmetics industries).

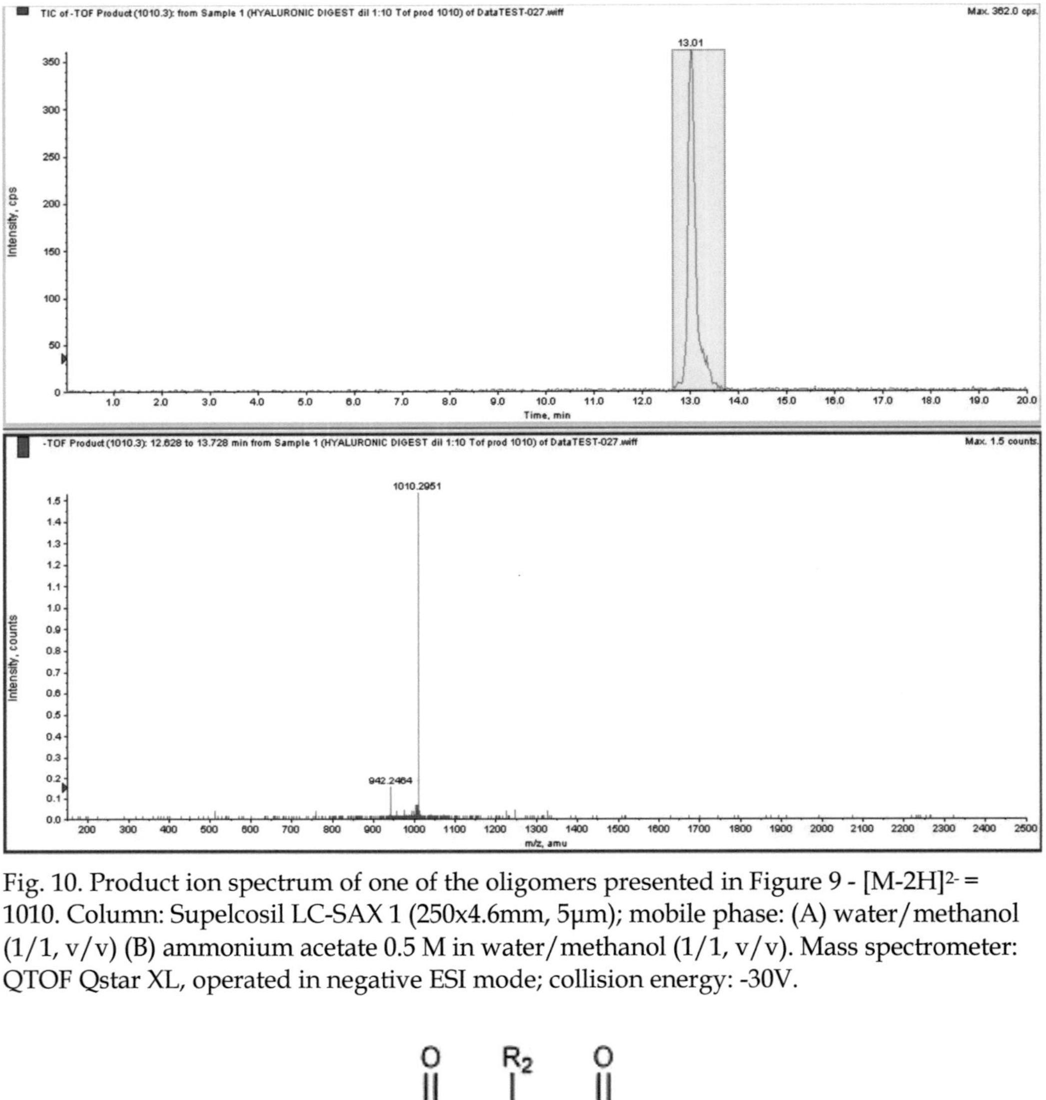

Fig. 10. Product ion spectrum of one of the oligomers presented in Figure 9 - $[M-2H]^{2-}$ = 1010. Column: Supelcosil LC-SAX 1 (250x4.6mm, 5µm); mobile phase: (A) water/methanol (1/1, v/v) (B) ammonium acetate 0.5 M in water/methanol (1/1, v/v). Mass spectrometer: QTOF Qstar XL, operated in negative ESI mode; collision energy: -30V.

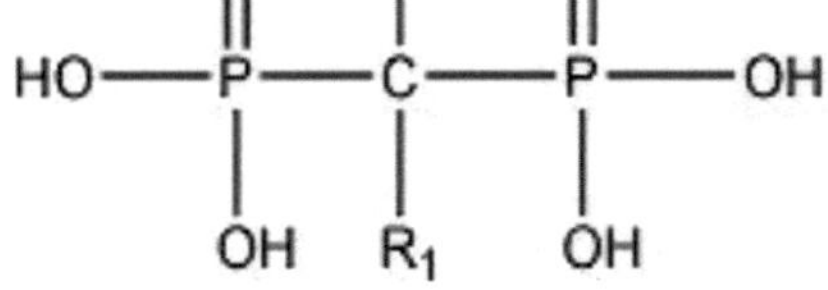

Fig. 11. General structure of bisphosphonates.

The long-chain substituent (R_2 in figure) determines the chemical properties and mechanism of action, while the short-chain substituent (R1) influences mainly the pharmacokinetics of the bisphophonate drug (Fleish 2004). Obviously, bisphosphonates are extremely hydrophilic and their structures similar to endogenous phosphorylated compounds. They can be grouped in two classes: nitrogen (N) - containing (pamidronate, neridronate, olpadronate, alendronate, ibandronate, risedronate, zoledronate) and non-N-containing (etidronate, clodronate, tiludronate) bisphosphonates.

3.2 Special problems of these molecules and potential approach to solve them: derivatization or not

Analysis of bishphosphonates from biological matrices is very challenging, due to the aforementioned features of these molecules: high polarity and similarity with endogenous phosphates. Finding appropriate conditions for their extraction from plasma or urine is very difficult; chromatographic separation raise typical problems of hydrophilic compounds and last but not least, bisphosphonate molecules often lack chromophores or fluorescent groups needed for sensitive UV or fluorescence detection in liquid chromatography. As an approach to facilitate either the isolation from biological samples, or the chromatographic separation, or the detection, or all of them, derivatization is widely used in the analysis of bisphosphonates. Mass spectrometric detection was used more recently to develop sensitive quantification methods, but also most of the times in combination with derivatization procedures. In gas-chromatography (GC), derivatization is needed to increase the volatility and thermal stability of the molecules. The N-containing bisphosphonate undergo reaction of the amino- group, and for all compounds the derivatization of the hydroxyls from the phosphonate group can give good results.

3.3 HPLC analysis of bisphosphonates

Various analytical procedures for the determination of bisphosphonates from biological samples or pharmaceutical formulations have been presented so far; review papers were previously published by Sparidans and den Hartigh (1999). Zacharis and Tzanavaras (2007). One of the first methods, developed for dichloromethylene bisphposphonates, was described by Chester et al (1981). The separation is based on ion-exchange chromatography, without derivatization, using an AG1 X8 resin column and a mobile phase containing hydrochloric acid and a flame-photometric detector. The coprecipitation with calcium phosphate was used to extract the target compounds from urine. This method has a limit of detection of 2μmol/l.

Separation procedures without derivatization are performed on ion-exchange columns like Waters IC-Pak HR (Tsai et al, 1992) or Dionex OmniPac PAX-100; the mobile phases contain mineral acids (mostly diluted nitric acid without or with small percentage of organic modifier) and the detection is realized through flame-photometry (Chester et al, 1981), conductivity (Tsai et al, 1992) or inductive coupled plasma mass spectrometry (ICP-MS) (Kovacevic et al, 2004).

Mineral acids are replaced with organic acids (especially formic acid) in researches that use ion-chromatography in tandem with mass-spectrometry (Ip et al, 1994). An alternative to ion-exchange chromatography is ion-pairing reversed-phase chromatography. Xie et al (2006) used ion-pair chromatography with evaporative light-scattering detection to determine ibandronate in pharmaceutical formulations. The C8 Inertsil column was eluted in isocratic conditions with a mobile phase consisted of ammonium acetate containing amylamine/acetonitrile/methanol. Tetrabutyl ammonium hydrogen sulphate (Rao et al, 2005), tetrabutyl-ammonium phosphate (Aluoch et al, 2005), 1-octyltrimethylammonium phosphate (Vallano et al, 2003) are among the ion-pair reagents employed in separations without derivatization.

As mentioned above, a pre- or post-column derivatization step is often needed to introduce in the molecule a chromophore group, in order to increase the sensitivity of the detection by UV or fluorescence. The pre-column derivatization, if possible directly in biological samples, is also very important for the isolation of bisphosphonates, introducing in the hydrophilic molecule a large non-polar group that facilitates their subsequent extraction in organic solvents.

Most of the published methods involve therefore a derivatization approach, either of the amino, or of the hydroxyl groups, or of both of them in some cases.

Fluorescamine was one of the most used derivatization reagents of the amino group (Flesh et al, 1991, Wong et al, 2004). Aminoalkylphosphonic acids were isolated from urine or plasma by coprecipitation with calcium phosphate; then the precipitate was dissolved in Na_2EDTA pH10 and reacted with fluorescamine. The reaction products were chromatographed on an octadecyl stationary phase (Nucleosil C18) and detected by fluorescence.

9-Fluorenylmethyl-chloroformate was also widely applied in the determination of alendronate from different matrices like pharmaceutical preparations, plasma or urine. The derivative was chromatographed on C18 columns, with mobile phases consisting of acetonitrile, methanol and pyrophosphate buffer, and detection in UV or fluorescence (de Marco et al, 1989, Ptacek et al, 2002, Yun et al, 2006, Apostolou et al, 2007). As an example, Ptacek et al (2002) obtained with this derivatization a limit of quantification as low as 5 ng/ml from urine, with reproducible results that could be used for bioavailability studies. The reaction was conducted at alkaline pH after bisphosphonates isolation from biological sample by coprecipitation with calcium phosphate and the calcium ions removed on ion-exchange resins; the chromatography was carried out on a C18 stationary phase, in gradient conditions, with fluorescence detection.

The derivatization of the amino group with isobutyl chloroformate was proposed by Sakyiama et al (1996) prior to GC separation. A fused-silica capillary column HP-1 (5m x 0.53 mm i.d., 2.65μm film thickness) was used and flame photometric detection. The same reaction was later used by our group directly on biological samples (urine); this reduced the polarity of the molecule and allowed the extraction with organic solvent (Tarcomnicu et al, 2007).

An interesting reaction described by Kline et al (1992) uses 2,3 naphtalene-dicarboxaldehyde, in presence of cyanide ions as nucleophiles. The substituted cyanobenzo[f]indole obtained was measured by fluorescence. Alendronate was determined from urine samples using this technique. The same authors studied also the use of acetyl-penicillamine as nucleophile for improved sensitivity; electrochemical detection was preferred over fluorescence.

Introducing a chromophore in order to get sensitive UV or fluorescence measurements is also possible post-column, after the separation by ion-exchange chromatography. Daley-Yates et al (1989) described such a derivatization with ammonium persulphate and molybdenum ascorbate; a phosphomolibdate detected in visible (820 nm) was obtained.

Nevertheless, in the LC methods above presented, although the separation is performed on octadecyl stationary phase, the mobile phases consisted of phosphate, pyrophosphate, citrate buffers in combination with various percentages of organic modifiers like methanol or acetonitrile. These inorganic buffers, as well as mineral acids, are not suitable for LC-MS.

3.4 Overview on mass spectrometric analysis of biphosphonates

When coupling gas chromatography with mass spectrometry, derivatization is needed in order to obtain volatile compounds. In this case the hydroxyls of the phosphonate group are silylated or alkylated. Sakiyama et al (2005) described a method for alendronate analysis in urine that employs double derivatization: first the amino group was reacted directly in the biological matrix with isobutyl chloroformate and extracted with organic solvent, then diazomethane was used to methylate the phosphonate groups. A fused silica capillary column containing cross-linked OV-1 stationary phase (Quadrex 12mx25mm id, 0.25µm film thikness) eluted with helium at 1 ml/min was employed for chromatographic separation. The limit of detection was 20 ng/ml.

The studies conducted for (aminoalkan)phosphonic acids, widely used either as pesticides or as pharmaceuticals, are good starting points also for bisphoshonate analysis. A large variety of derivatization approaches is described in the literature.

Silylation with N-methyl N-(tert-butyldimethylsilyl)trifluoroacetamide (MTBSTFA) was proposed by Moye and Deyrup (1984) for GC/MS analysis of the herbicide glyphosate and its metabolite, aminomethylphosphonic acid. In order to obtain high yield of the reaction, coating of the glassware by exposing it to a diluted solution of phosphoric acid in ethanol was necessary; however authors reported problems at low ppb levels.

Also in GC/MS pentafluorobenzyl bromide (Palit et al, 2004) and trimethyl orthoacetate (Kudzin et al, 2003) were used. The two techniques were applied in the determination of alkyl and aryl bisphosphonic drugs and in trace analysis of organophosphorous pesticides or nerve agents and their metabolites from biological or environmental samples. Palit et al (2004) studied the derivatization of methyl phosphonic and pinacolyl methyl phosphonic acid, degradation products of organophosphorous compounds, with pentafluorobenzyl bromide (PFBBr) at alkaline pH, in different conditions. They further compared the stability of the pentafluorobenzyl phosphonic esters with that of the sylylated derivatives, the later more widely used in GC/MS, but more sensitive to humidity.

The same reaction was tested by our group on two drugs from the bisphosphophonate class, and the obtained compounds were separated by LC-MS/MS (unpublished data). A phenyl stationary phase (Supelcosil LC-DP 15cmx3mm, 5µm) was used, eluted in gradient with a mobile phase containing 50mM ammonium acetate and acetonitrile. High resolution mass spectrometric detection on a quadrupole-time of flight (QTOF) instrument model Qstar Pulsar i (AB Sciex), equipped with an electrospray ionisation source (ESI) operated in negative ions mode was employed. Di- and triesters resulted mainly from the derivatization of clodronate, as shown in figure 12; further optimisation is needed to increase the yield of one of the derivatives in order to use this approach for quantitative purposes.

Zhu et al (2006) reported a methodology for the quantification of risedronate and alendronate by LC-MS/MS in biological samples after derivatization of the phosphonate groups with diazomethane. As a general approach, the urine or serum samples were loaded onto Bond-Elut SAX solid-phase extraction (SPE) columns that were washed with water prior to on-column derivatization with diazomethane. The methylated compounds were eluted with methanol and separated on an ion-exchange Zorbax 300-SCX column, with ammonium formate pH 2.5 and acetonitrile (75/25, v/v) as mobile phase. Triple quadrupole mass-spectrometers model API 4000 or 5000 were used, operated in ESI positive mode. Low limits of quantification (LLOQs) as of 0.2 ng/ml for risedronate and 1 ng/ml for alendronate in urine were achieved using this technique.

A general drawback of the methods using diazomethane is the toxic and explosive nature of the reagent; special care needs to be taken to ensure safety of all operations.

Alternative alkylations of aminoalkyl phosphonic acids could be performed with fluorinated alcohols/perfluorinated anhydrides mixtures (Deyrup et al, 1985) or trimethyl orthoacetate (Royer et al, 2000, Kudzin et al (2003).

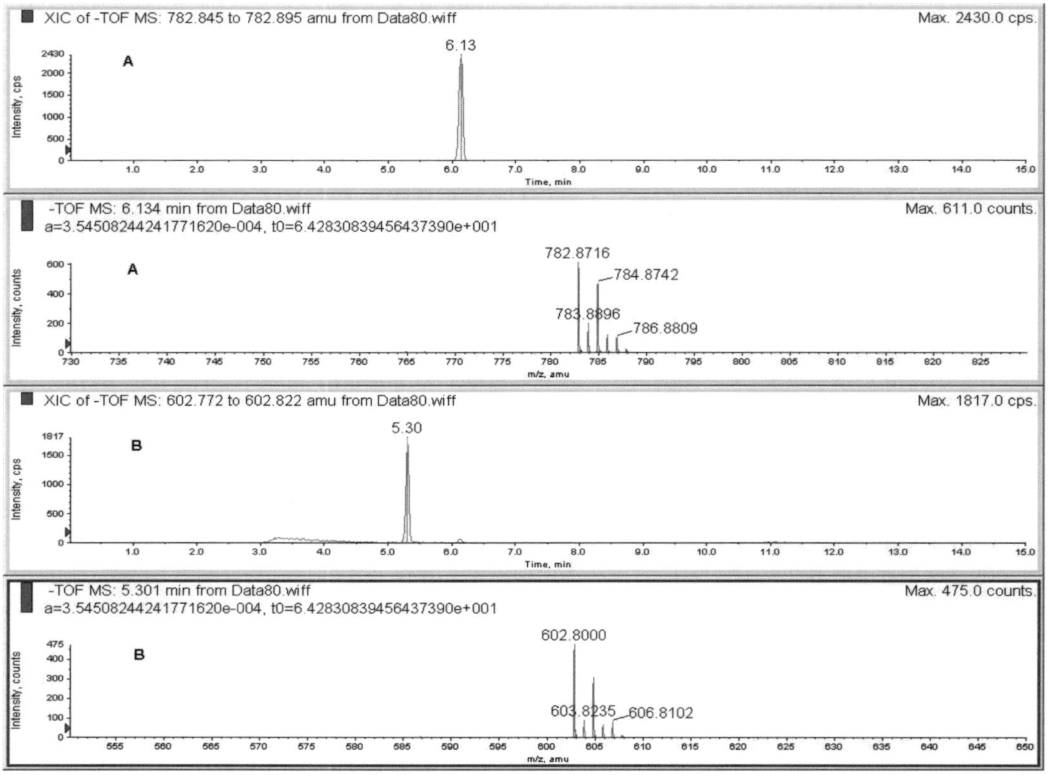

Fig. 12. Extracted chromatogram of the negative molecular ions corresponding to the triester - $[MH]^-$ =782.8716 (A) and diester - $[MH]^-$ =602.800 (B) obtained in the reaction of clodronate and PFBBr. Column: Supelcosil LC-DP 15cmx3mm, 5µm, mobile phase: ammonium acetate and acetonitrile, in gradient. Mass spectrometer: QTOF Qstar Pulsar i, operated in negative ESI mode.

Deyrup et al (1985) analysed glyphosate (N-phosphonomethylglycine) and its major metabolite, aminomethylphosphonic acid, from water samples by GC with flamephotometric, electron capture or mass spectrometric detection, after derivatization with trifluoroethanol (TFE) and trifluoroacetic anhydride (TFAA). With this approach the amino and hydroxyl groups are derivatized in the same step. No special glass coating is needed, as in the case of MTBSTFA, and no particular safety issues are raised, as in the case of diazomethane derivatization. An LLOQ of 0.5 µg/L in deionized water was obtained with electron capture detection.

Trimethylorthoacetate (TMOA) was preferred by Royer et al (2000) for the derivatization of another herbicide, gluphosinate (DL homoalanine-4-yl (methyl)phosphonic acid), and two of its metabolites, 3-methylphosphonico-propionic and 2-methylphosphonico-acetic acids, prior to analysis by GC-MS/MS with chemical ionization. This method was applied for the quantification of gluphosinate and its metabolites in water samples, over a range of 10-150 µg/L.

Kudzin et al (2003) combined anhydride and TMOA approaches for the derivatization of a series of biologically active aminoalcanphosphonic acids. Briefly, the neat standards or dried water extracts were dissolved in 0.1 ml of trifluoroacetic acid/trifluoroacetic anhydride mixture (1/1, v/v), incubated for 90 min at 40 °C, then 0.4 ml TMOA were added, followed by another incubation step at 100 °C for 120 min. Both amino and hydroxyl groups were derivatized and the reaction mixtures were analized after preconcentration under vacuum, by ^{31}P NMR or GC-MS. The derivatization products were stable for several weeks at ambient temperature and the quantifiable levels by GC-FID (flame ionization detection) were in the low µmol/L range.

More recently our group studied the application of TMOA derivatization in the analysis of bisphosphonate drugs. The hydroxyl groups reacted with TMOA in acidic conditions independently from the presence of amino groups and all tested drugs (alendronate, pamidronate, etidronate, clodronate, risedronate, ibandronate) were successfully derivatized. Triethylorthoacetate (TEOA), trimethylorthoformiate (TMOF), trimethylorthobutirate (TMOB) were also tested as alkylation reagents and found effective (unpublished data). All reagents do not require special safety measures which is an important advantage.

Studying these reactions, the expected alkylation products were found, but also other compounds with higher molecular mass were obtained in the case of bisphosphonates containing a hydroxyl group at the carbon in the P-C-P bridge. This hydroxyl is also involved in the reaction probably resulting in cyclic structures. In the case of clodronate, the carbon in the bridge is substituted with two chlorine atoms and only the tetraalkylated structure is formed, as it can be seen in Figure 13. When the reagent was TMOF, an increase of 28 mass units was observed; for TMOA and TEOA the shift was 42, and for TMOB 70 Da.

Based on these results, we have further developed a double derivatization method for the quantification of alendronate in urine by LC-MS/MS (Tarcomnicu et al, 2007). First, the amino group was reacted with isobutyl chlorcformate (IBCF), to facilitate the extraction of the polar drug from biological matrix with organic solvent. The dried extract was redissolved in acetic acid, and incubated with TMOA for 1 h at 100 °C. The methylation of the phosphonate groups significantly reduced the polarity of the molecule permitting the separation on a reversed-phase column (Supelco Discovery HSC18) with a gradient of mobile phase containing formic acid 0.1% in water/formic acid 0.1% in acetonitrile, which was suitable for LC/MS and enhanced the ionization in positive mode. The clyclic compounds, as described above, were obtained with highest yield for alendronate and pamidronate; a product ion spectrum of the alendronate derivative is presented in Figure 14.

The method vas fully validated according to FDA guidelines over a range of concentrations of 6.667 - 4860 ng/mL alendronate in urine, and it was applied to a bioequivalence study. With the help of a switching valve 2 columns were run in parallel, resulting in an analysis time of 5 min/sample, that was very short compared with HPLC-fluorescence or with GC methods.

Trimethylsilyl diazomethane (TMSD) was another derivatization reagent successfully used in LC-MS/MS, in order to analyze ibandronate from plasma samples for pharmacokinetic applications (Tarcomnicu et al, 2009). A high-throughput method with an LLOQ of 200 pg/mL was optimized and fully validated according to FDA guidelines. Owing to its structure (N-substituted aminobisphosphonate), ibandronate was isolated from plasma extracts with the same technique previously proposed for alendronate (Tarcomnicu et al, 2007). To reduce the polarity of the molecule and make it suitable for reversed-phase LC, in the next step we decided to methylate the phosphonic groups. First, TMOA was tried as derivatization reagent, but in the case of plasma extracts this was found not as effective as for urine extracts. TMSD proved to be very reactive also with plasma extracts, as already

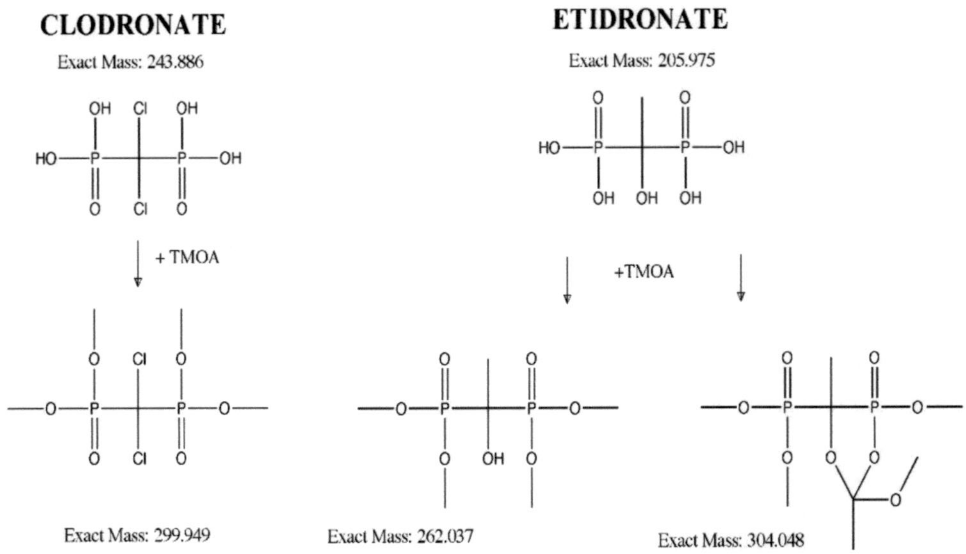

Fig. 13. Clodronate and etidronate derivatization with TMOA; for the latter both hypotheses (methylation and methylation plus cyclization) are shown.

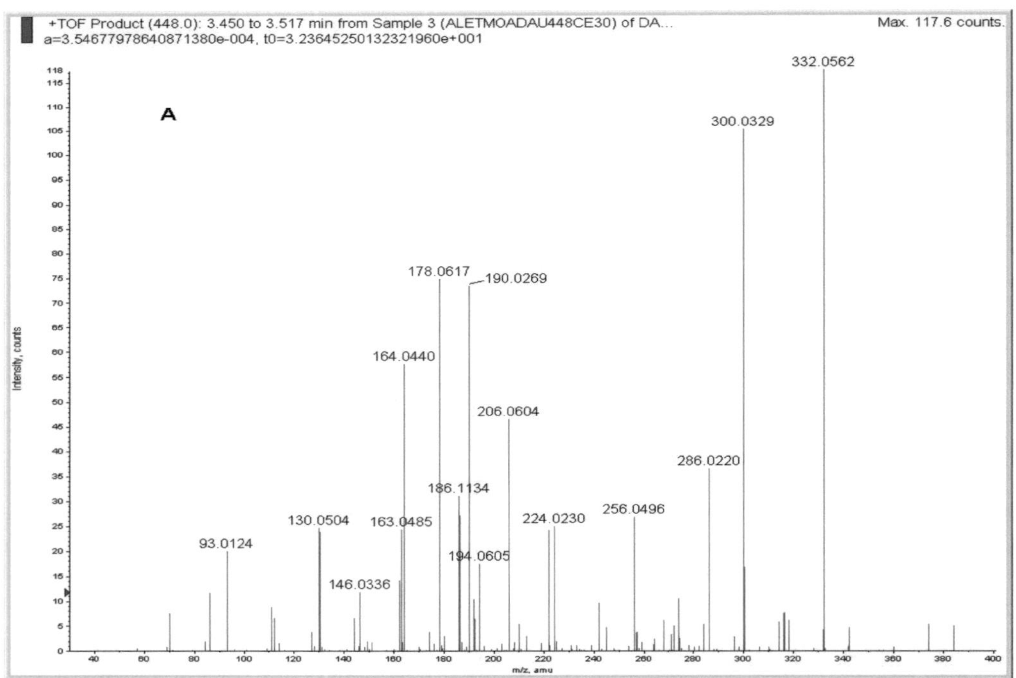

Fig. 14. Product ion spectrum of the alendronate derivative with IBCF and TMOA ($[M+H]^+$= 448), during chromatographic separation. Column: Supelco Discovery HSC18, 10cmx2.1mm, 5µm, mobile phase: (A) formic acid 0.1% in water and (B) formic acid 0.1% in acetonitrile; gradient conditions. Mass spectrometer: QTOF Qstar Pulsar i, operated in positive ESI mode.

studied by another group (Ranz et al 2008) and compared to diazomethane it is a stable and safe reagent. The alkylation reaction was carried out for 30 min at 70 °C and the fully methylated compound was obtained. The derivatization scheme is presented in Figure 15, the product ion spectrum of the ibandronate derivative in Figure 16, and Figure 17 shows selected chromatograms obtained on a blank, calibrator 1 (LLOQ) and calibrator 8 extracted from plasma. The separation was performed on a Discovery HSC18 column (10cm×2.1mm i.d., 5µm particle size) and a mobile phase consisting of formic acid 0.1% in water and formic acid 0.1% in acetonitrile, with a composition gradient. An alternative satisfactory approach using alkaline mobile phase (ammonium hydroxide 0.05% in water/ammonium hydroxide 0.05% in acetonitrile) on a Purospher Star RP-18e (3cm×2.1mm, 2.7µm) was also tested; good peak intensities, similar to the ones with acidic mobile phase, were achieved. Elution at acidic pH was preferred for this study due to better background, but the basic mobile phase can be also considered.

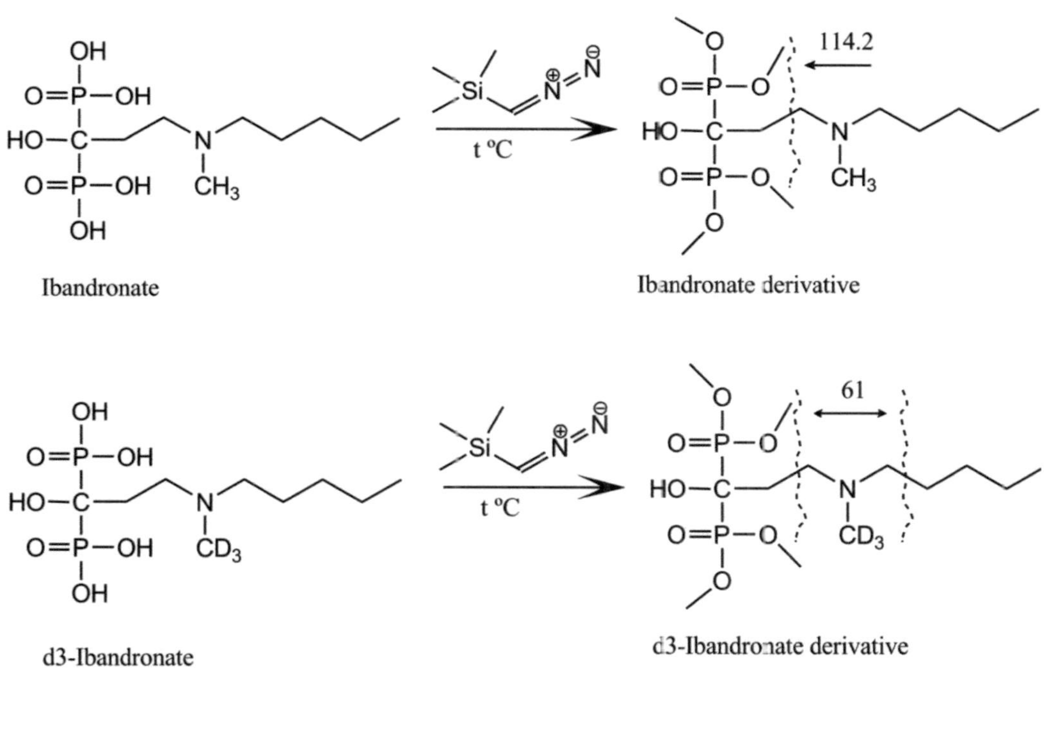

Fig. 15. Derivatization scheme of ibandronate and its internal standard, d3-ibandronate.

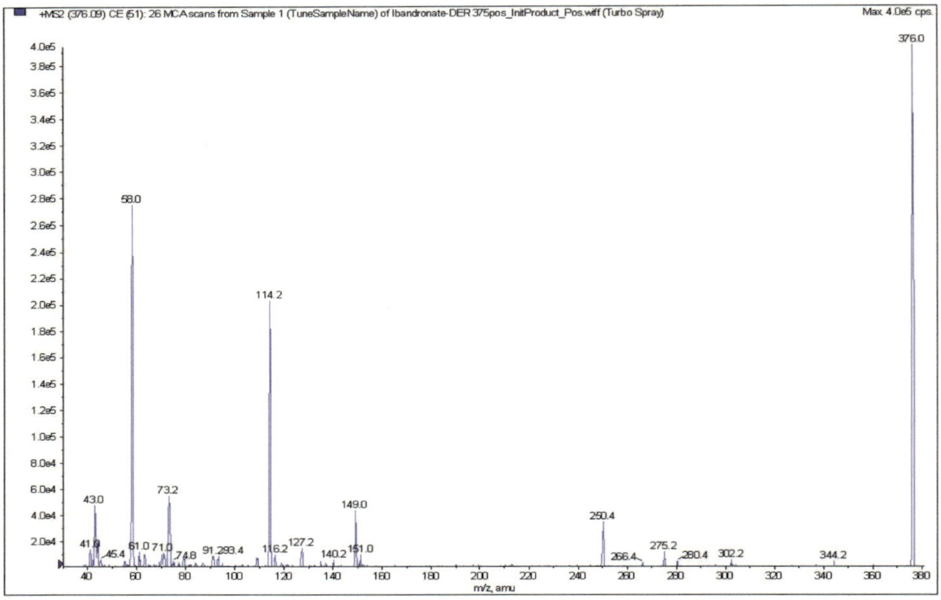

Fig. 16. Product ion spectrum of the ibandronate derivative with TMSD - $[M+H]^+ = 376$ obtained with direct infusion via syringe of a 1μg/mL solution water/methanol (1/1, v/v). Mass spectrometer: quadrupole-linear ion trap API4000QTrap, operated in positive ESI mode; collision energy: 25V.

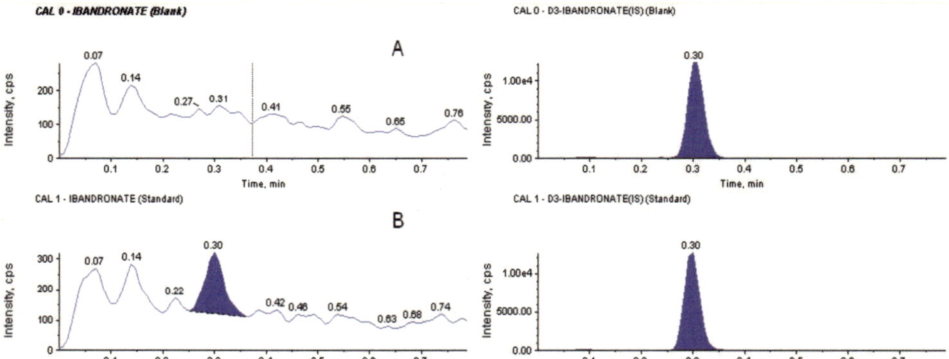

Fig. 17. HPLC traces of the MRM transitions selected for ibandronate (left) and d3-ibandronate (right) recorded on a blank plasma sample (A), plasma spiked with ibandronate for CAL1 – 0.2 ng/ml (B) and CAL8 – 175.0 ng/ml (C) concentrations. Note: The retention time shown in the figure is relative, as the analyses were performed in overlapping mode; 1.5 min must be added to get the total retention time.

3.5 Remarks

Considering existing data for the analysis of bisphosphonates, especially with MS detection, the derivatization approach has given so far the best results. Nevertheless, the new developments in column manufacturing could open possibilities for the analysis of underivatized bisphosphonates. A promising direction is represented by HILIC (hydrophilic interaction liquid chromatography) which is gaining interest in the analysis of polar compounds, with applications in various fields (A.J. Alpert, 2011, Van Nuijs et al, 2011). HILIC is a version of normal phase liquid chromatography; the typical mobile phase consists of acetonitrile with a small percentage of water, while the stationary phases are polar surfaces (silica, diol, amino, amide, zwitterionic bonded phases). The retention of polar compounds is not based on liquid-solid partition like in reversed phase separations, but on a liquid-liquid partition mechanism between the water-deficient bulk eluent and the water-enriched layer immobilized on the HILIC stationary phases (Alpert et al, 1990). Therefore polar analytes early eluting in reversed-phase are strongly retained in HILIC conditions. HILIC is particularly useful in the analysis of aminoglycoside antibiotics (more details will be provided in the following chapter), other various pharmaceuticals (e.g. metformin, salicylic and acetyl salicylic acid, tetracyclines), nucleosides, organic phosphates, peptides and proteins, sugars, warfare agents or drugs of abuse (cocaine, benzoylecgonine) etc. The use of mobile phases with high percentage of organic solvent in HILIC separation is beneficial for MS detection, because of enhanced ionization which results in an increased sensitivity. Methanol, ethanol, 2-propanol, tetrahydrofuran can be used for some applications. The most common buffers used to control pH and ionic strength are ammonium acetate or ammonium formate, at low molarities, also favouring the ionization in MS; only rarely high molarities or non-volatile buffers are required. Another advantage is the simpler sample preparation, being no need for derivatization. The functional groups on the phase surface influence the selectivity of the separation, thus the capacity to differentiate similar compounds. Other interactions which could be involved in HILIC are electrostatic interaction, hydrogen-bonding, dipole-dipole interaction, molecular shape selectivity, and even hydrophobic interaction.

The retention mechanisms have been studied by several groups in order to characterize and classify HILIC stationary phases (Dinh et al, 2011, Kawachi et al, 2011).

4. Analysis of aminoglycoside antibiotics

4.1 Chemical structure and main features

Aminoglycosides are antimicrobial agents used in the treatment of both animals and humans against aerobic gram-negative bacteria. These molecules are natural products or semisynthetic derivatives of compounds produced by different varieties of actinomycetes isolated from soil (Higgins and Kastner, 1967); the class include very important antibiotics as gentamicin, tobramycin, amikacin, netilmicin, kanamycin, streptomycin and neomycin.

The first drug of the class, Streptomycin, was isolated from a strain of *Streptomyces griseus* by Waksman and coworkers in 1944 and it is an important agent for the treatment of tuberculosis. Gentamicin and netilmicin are derived from species of the actinomycete *Micromonospora*. Tobramycin is produced by S.tenebrarius (Chambers, 2006). Amikacin, a derivative of kanamycin, and netilmicin, a derivative of sisomicin, are semisynthetic products.

The antimicrobial activity of aminoglycosides is based on their ability to selectively inhibit protein synthesis, in bacteria; the most important pathogens treated with aminoglycosides are pseudomonas, enterococci, coliforms, and salmonellae.

The molecule of aminoglycosides presents two or more amino sugars linked by a glycosidic bridge to a hexose nucleus (aminocyclitol) that is either streptidine (found in streptomycin) or 2-deoxystreptamine (in all other available aminoglycosides).

The aminoglycosides are divided into the following subclasses (Chambers, 2006):

a. *The neomycin family,* which includes neomycin B and paromomycin, presents three amino sugars attached to the central 2-deoxystreptamine.

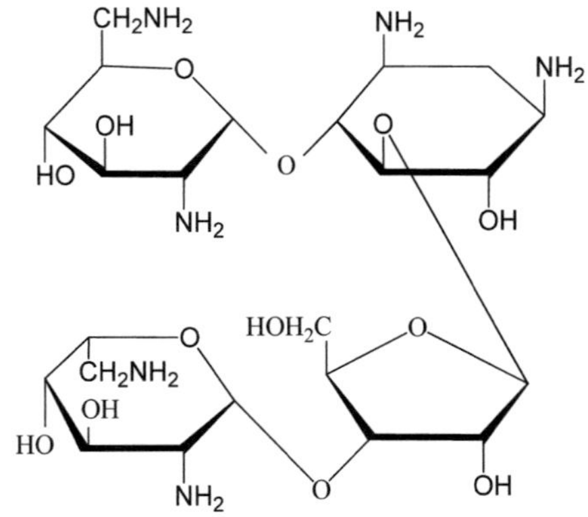

Neomycin B

b. *The kanamycin family* (kanamycin A - B, amikacin and tobramycin) presents two amino sugars attached to the central 2-deoxystreptamine moiety; one of these being a 3-aminohexose.

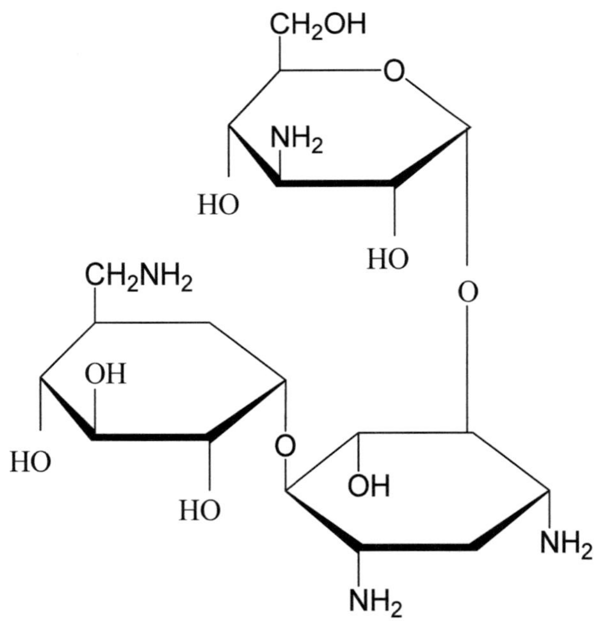

Kanamycin A

c. *The gentamicin family,* including gentamicin, sisomicin and netilmicin, contains two amino sugars attached to the central 2-deoxystreptamine, one of these being garosamine.

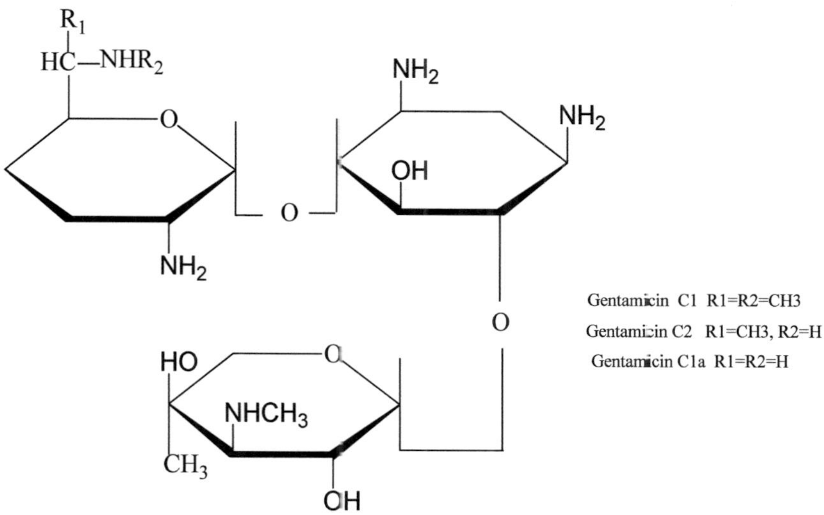

Gentamicin C

d. *Streptomycin* contains streptidine that is not in the central position like in the other aminoglycosides.

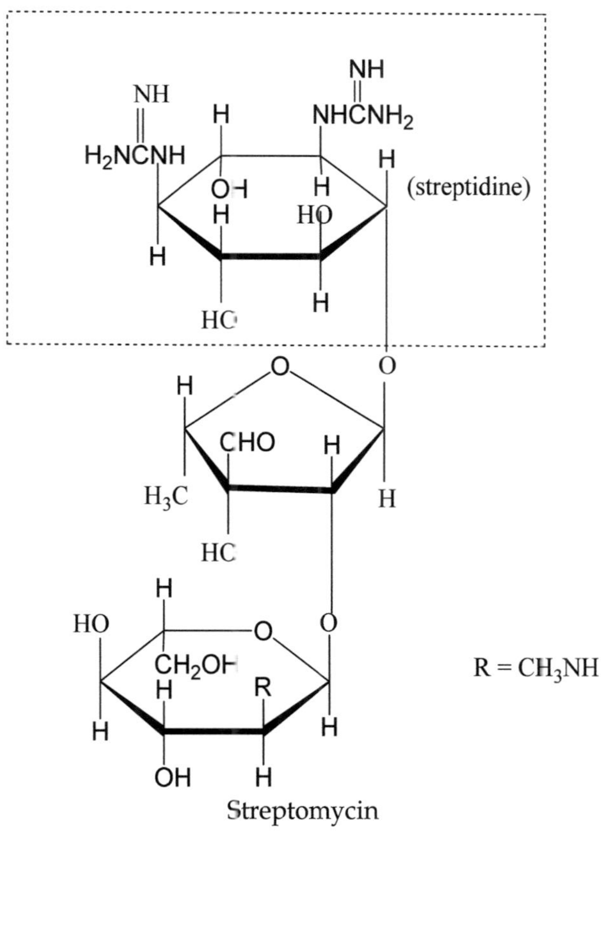

Streptomycin

4.2 Chromatography of aminoglycosides

Aminoglycoside antibiotics are very polar compounds, highly water soluble, practically insoluble in hydrophobic organic solvents, lacking chromophores for UV absorption or fluorescent groups. They have been determined by bioassays, GC–MS and HPLC, in this case often following derivatization (to improve detection and/or separation), adding ion-pair agents in the mobile phase or by ion-exchange techniques. More recently, HILIC was also proposed as worthwhile approach and there is an increasing number of applications; in literature brief reviews about the application of these methods for the separation of aminoglycoside antibiotics can be found (Tawa et al, 1998, Isoherranen and Sobak, 1999, Stead, 2000).

Because of the nonvolatile and very polar nature of aminoglycosides the gas chromatographic analysis invariably involves derivatization of the amino and hydroxyl groups of these molecules. trimethylsilylimidazole (TMSI) and heptafluoro-butyrylimidazole (HFBI) have been used as derivatizing agents for the analysis of gentamicin, tobramycin, netilmicin, amikacin and paromomycin in serum by Mayhew et al (1978). Trimethylsilyldiethylamine (TMSDEA) was also tested as derivatization reagent for both amino and hydroxyl groups but with poor results as repeatability and linearity (Margosis and Tsuji, 1978).

Owing to their hydrophilic properties, aminoglycosides are not adequately separated also by reversed-phase liquid chromatography, and therefore derivatization with non polar agents or mobile phase containing counter-ions reagents, forming ion pairs, have been widely employed in HPLC.

As derivatization reagents o-phthalaldehyde (OPA), 1-fluoro-2,4-dinitrobenzene (FDNB) and dansyl chloride are the most commonly used for the analysis of aminoglycosides (Isoherranen and Sobak, 1999). OPA reacts in the presence of mercaptan or other strong reducing agents with primary amines to form fluorescent derivatives. The reagent is stable in different buffer solutions and the reaction has a fast rate at room temperature allowing the derivatization to be performed either pre- or post- column. FDNB easily reacts with both primary and secondary amines, in basic conditions, producing UV absorbing derivatives with maximum absorption wavelength at 365 nm.

The separation columns used for the derivatized aminoglycoside generally contain C8 or C18 silica as stationary phase; the mobile phase consist of acidic buffers with methanol and/or acetonitrile.

Gentamicins C1, C1a and C2 were determined as their 2,4-dinitrophenyl derivatives in plasma and urine using a C18 RP column and as mobile phase 680ml/l acetonitrile-320ml/l Tris buffer (8.3mmol/l titrated to pH 7.0) (Isoherranen and Sobak, 2000). A HPLC method for the analysis of tobramycin in urine samples used pre-column derivatization of tobramycin with fluorescein isothyociante followed by fluorescence detection; the chromatographic separation was carried out on a Phenomenex Luna C18 column and the limit of quantitation of the method in urine was 250 ng/mL (Mashat et al, 2008).

Another derivatization agent used for the RP-HPLC analysis of aminoglycoside antibiotics was phenylisocyanate in the presence of triethylamine. Phenylisocianate groups easily react with amino groups of aminoglycosides like kanamicin, neomycin and gentamicin (Kim et al, 2001 and 2003); the complex structures were confirmed by ESI-MS.

Clarot et al (2005) described a method for the determination of neomycin sulphate, framycetin sulphate and other related compounds by evaporative light scattering detection (ELSD) thus avoiding the need of derivatization for detection. The chromatographic separation was performed on a Polaris C18 column using a mobile phase with 170 mM trifluoroacetic acid; in this case trifluoroacetate ions act like counter-ions forming ion pairs with aminoglycosides. The ELSD elution of the complexes with antibiotics was confirmed by MS.

Ion-pair chromatography is probably the most popular approach for aminoglycosides analysis; when using standard detectors mainly alkyl sulphonates (pentane, heptane and hexane sulphonates) in acetate or phosphate buffers are selected as counter-ions. For HPLC-MS applications the volatile fluorinated carboxylic acids are preferred. Pre-column and post column derivatization with o-phthalaldehyde are often used for the detection of these molecules also when separated by ion-pair HPLC. As an example, Aubin et al separated nine aminoglycoside antibiotics (streptomycin, dihydrostreptomycin, neomycin, sisomicin gentamicin C1, C1a, C2, C2a, netilmicin) on an Atlantis C18 column eluted with 0.02M potassium dihydrogen phosphate (adjusted to pH 3.0 with phosphoric acid) containing 35 g/l sodium sulfate, 500mg/l sodium octanesulfonate and 15ml/l tetrahydrofuran; the detection was achieved with a pulsed electrochemical detector. Tobramycin and colistin sulphate were simultaneous determined from pharmaceutical formulations using a Zorbax SBC18 column eluted with acetonitrile / water / trifluoroacetic acid and evaporative light scattering detection (Clarot et al, 2009).

Pulsed electrochemical detection was applied also for the analysis of underivatized tobramycin (Shruti et al, 2010) and kanamycin by ion-pairing with octansulfonate in phosphate buffer (Manyanga et al, 2010). A system for the determination of gentamicin by ion chromatography using pulsed amperometric detector after isocratic elution on a polystyrene column was developed by Metrohm in 2005.

4.3 Analysis of aminoglycosides by mass spectrometry

The concentrations usually monitored in plasma or serum of patients treated with aminoglycoside antibiotics are in the low ng/mL range. Due to their non negligible toxicity in biological systems and to side-effects like inducing resistance in bacteria or alteration of the normal microbial flora (thus influencing various biological systems), antibiotics are also carefully monitored in environmental or food and feed samples. Maximum residue limits (MRLs) have been set in most countries and they are in the low ppb ranges, too. Therefore, mass spectrometry, being a very sensitive and selective analytical technique is extremely important in the analysis of these antibiotics. With respect to aminoglycosides, the GC/MS methods are based on complex and lengthy derivatizations as above mentioned; Preu et al, 1998 adapted the method using trimehtylsilylimidazole and heptafluorobutyrylimidazole for capillary GC/MS.

LC-MS/MS methods have been preferred recently owing to their high sensitivity and ease of use compared to GC/MS. Two non-derivatization approaches have been developed in parallel, reversed phase ion-pair chromatography and HILIC.

Aminoglycoside antibiotics are among the best candidates to be analyzed with HILIC and such methods were applied for difficult matrices like plasma, kidney or meat. Oertel et al (2004) described an automated method for the simultaneous quantification of amikacin, gentamicin, kanamycin, neomycin, paromomycin, and tobramycin in human serum by HILIC-MS/MS. Separation was carried out on a zwitterionic SeQuant ZIC-HILIC column, 100mm × 2.1mm with a Phenomenex SecurityGuard C18, 4mm × 2mm i.d. The mobile phase was composed of acetonitrile, 2mM ammonium acetate and formic acid, in gradient conditions, and data acquisition was performed with a Quattro Micro triple quadrupole mass spectrometer equipped with ESI source, operated in positive ions mode. Low limits of quantification (LLOQ) of 100ng/mL were obtained using 500µL serum extracted by solid phase extraction (SPE).

Ishii et al (2008) reported a quantification method for seven aminoglycosides from swine and bovine meat and kidney that also employed a SeQuant ZIC–HILIC column (100 mm × 2.1 mm, 5 µm) eluted with a mobile phase composed of (A) ammonium acetate 150 mM + 1% formic acid in water and (B) AcN, at a flow rate of 0.3 mL/min. MS detection was performed in positive mode. LOQs in swine bovine kidney were 25 ng/g for gentamicin, 50 ng/g for spectinomycin, dihydrostreptomycin, kanamycin and apramycin, and 100 ng/g for streptomycin and neomycin, well below the existing FAO MRLs.

Beside HILIC the most common separation technique for aminoglycoside antibiotics is by ion-pair chromatography, performed mostly on octyl (C8) or octadecyl (C18) columns, eluted with mobile phases containing trifluoroacetic (TFA), pentafluoropropionic (PFPA), heptafluorobutiric (HFBA) or nonafluoropentanoic (NFPA) acids as ion-pair reagents. Good peak shape and high sensitivities are obtained, thus compensating the negative aspect that the ion-pair reagents are serious contaminants for the MS interface.

Shen et al (2008) studied the separation of amikacin, streptomycin, spectinomicyn and gentamicin by ion-pair and HILIC. The ion-pair chromatography was performed on an Agilent Zorbax SB-C8 (30 X 2.0 mm, 5 µm column), eluted with a mobile phase consisting of (A) 10 mM NFPA with 10 mM ammonium hydroxide in water and (B) 5 mM NFPA acid with 5 mM ammonium hydroxide in a 10:90 mixture of water:acetonitrile, with a gradient from 20% to 90% (B). HILIC was carried out on a SeQuant Zic-HILIC (50 X 2.0 mm, 5 µm) column, with a mobile phase composed of (A) 5 or 25 mM ammonium formate in water with formic acid (pH =~2.5) and (B) acetonitrile with 1% (v:v) formic acid; the gradient started from 90% B and was decreased to 10%B. A simple protein precipitation with cold acetonitrile was used for clean-up of the different samples (mouse, rat or Guinea pig plasma) and triple quadrupole (API 4000 or 5000) mass spectrometers were used for detection. An LLOQ of 20ng/mL for all four antibiotics was achieved with the ion pair method, while the HILIC method was highly sensitive only for spectinomycin, needing further optimization for the other analytes.

Amikacin, neomycin and gentamicin in plasma or tissue samples were also analyzed by Zimmer et al (2008) using a similar approach. They tested HILIC on a Varian Inertsil 5 Si (50X2.0mm column) with a mobile phase composed of acetonitrile /ammonium acetate 100 mM in water, pH 3, and ion pair separation on a Discovery HS-C18 (2.1x50mm, 3µm) column eluted with 0.1% HFBA in water / 0.1% HFBA in acetonitrile (all methods were carried out using a composition gradient). LLOQs of 5 ng/mL were obtained using ion-pair LC, which also in this case has proven to be more sensitive than HILIC.

Granja et al (2009) described a method for the determination of streptomycin residues in honey that employs ion-pairing with HFBA in water/acetonitrile (85:15, v/v) as a mobile phase (isocratic conditions), on a C18 column (Gemini 5 micron C18, 50x2mm). Also Hammel et al (2009) presented a multi-screening approach for 42 antibiotic residues in honey and 3 aminoglycosides where among them. HPLC analyses were run on a Zorbax SB-C18 reverse phase column (2.1×50 mm, 1.8µm). The mobile phases were constituted with solvent A:water containing 1mM NFPA mixed with 0.5% formic acid (v/v) and solvent B: acetonitrile/methanol (50/50, v/v) containing 0.5% formic acid (v/v).

The following paragraphs present the practical experiments of our group on aminoglycoside antibiotics by LC-MS/MS (unpublished data). The first method was developed for residues of streptomycin in honey, by ion-exchange chromatography, on a Grom-Sil 300WCX

(100x4.6mm, 7µm) eluted in gradient with (A) methanol/water (60/40, v/v) and (B) ammonium acetate 500mM in methanol/water (60/40, v/v); the initial mobile phase was 100% A. The honey sample was simply diluted 1/1 with water and injected; detection was achieved with an API 4000 triple quadrupole mass-spectrometer operated in positive ESI ionization. The calibration curve in honey was built over a range of 10 - 250 ng/mL and gentamicin was used as internal standard. In order to improve the sensitivity and robustness of the determination, later we switched to ion-pair chromatography on a pentafluorophenylpropyl column (Discovery HSF5 100x2.1 mm, 5µm), using a gradient of PFPA 0.2% in water/PFPA 0.2% in acetonitrile as mobile phase. An LLOQ of 3.3 ng/mL was obtained with the same instrument set-up and same sample preparation; gentamicin was replaced as internal standard by cimetidine. Figure 18 shows two examples of chromatograms, recorded on honey samples.

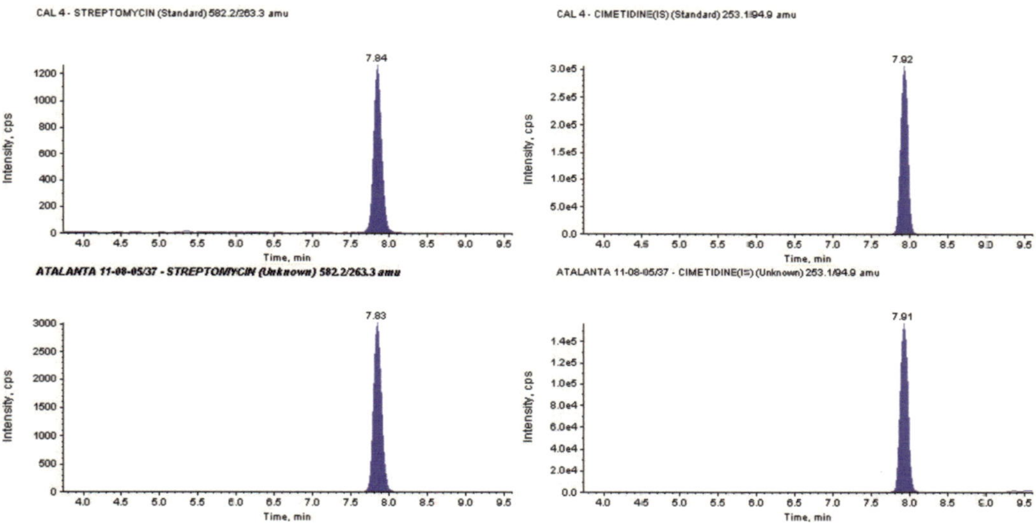

Fig. 18. HPLC traces of the MRM transitions selected for streptomycin and its internal standard cimetidine, recorded on a spiked honey sample - concentration 90 ng/mL (top), and on an unknown sample (bottom chromatograms) that was obviously contaminated with streptomycin. Column: Discovery HSF5 (10cmx2.1mm, 5µm), mobile phase: A) 0.2% PFPA in water and B) 0.2% PFPA in acetonitrile; flow: 0.3 ml/min; injection volume: 50µL. Mass spectrometer: Triple quadrupole API4000, operated in ESI positive mode.

The same method was later adapted for the analysis of neomycin in ophthalmic solutions, and streptomycin was used as internal standard. The samples were just diluted 1:1000 with the initial mobile phase (water/acetonitrile (95/5, v/v) + 0.2 % pentafluoropropionic acid) and injected in the LC-MS system; being the concentrations quite high (1-5 µg/mL) no further optimization was needed. An example of chromatogram is presented in Figure 19.
Another approach in aminoglycoside analysis by HPLC remains pre-column derivatization, aiming to reduce polarity and make the compounds suitable for reversed-

phase separations. Based on a previous research of Lauser and Bergner-Lang (1995), we tested in our laboratory the derivatization of neomycin with FMOC. The reaction needs to be optimized, because partial and fully substituted derivatives (all six amino groups reacted with FMOC) are formed and in order to obtain a better sensitivity the equilibrium should be ideally directed towards one product. We performed the chromatographic separation on a C8 column eluted with water, acetonitrile and formic acid and preliminary tests on plasma extracts have given promising results. The chromatographic traces corresponding to the fully substituted neomycin and paromomycin derivatives (the selected internal standard) on plasma samples are shown in Figure 20, while the product ion spectrum of neomycin derivative in Figure 21.

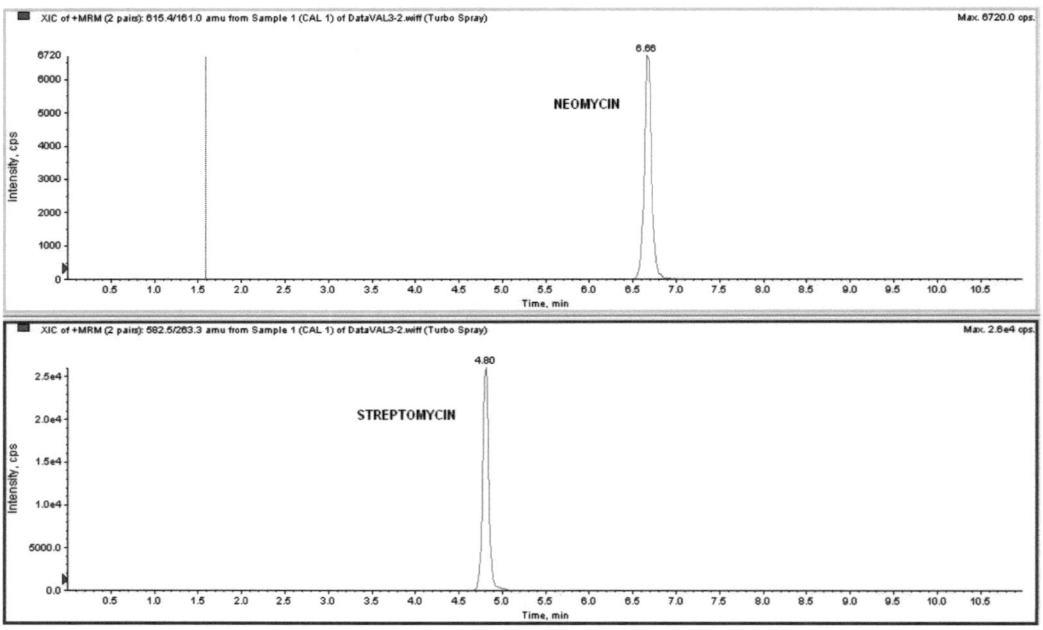

Fig. 19. Chromatographic traces of neomycin and its internal standard streptomycin, recorded on spiked sample (concentration 1 μg/ml). Column: Discovery HSF5 (10cmx2.1mm, 5μm), mobile phase: A) 0.2% PFPA in water and B) 0.2% PFPA in acetonitrile; flow: 0.3 ml/min; injection volume: 10μL. Mass spectrometer: quadrupole-linear ion trap API4000 QTrap, operated in ESI positive mode.

Last by not least, we have studied HILIC for the determination of vancomycin in plasma, using neomycin as internal standard. The column employed was an Ascentis Express HILIC (100x2.1mm, 2.7μm), the separation being carried out in isocratic conditions with acetonitrile/water (90/10, v/v) containing 0.1% formic acid. A good sensitivity in the range of few ng/mL, adequate for therapeutic drug monitoring, was obtained.

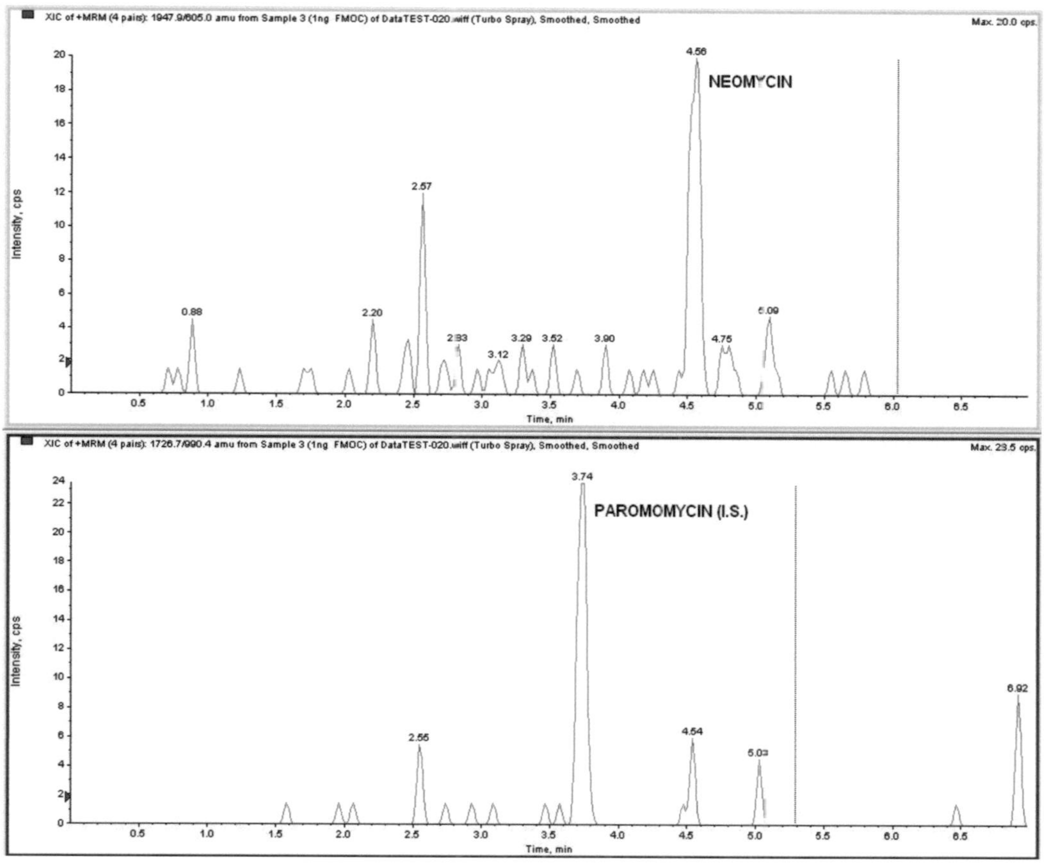

Fig. 20. The MRM transitions of neomycin and paromomycin (internal standard) derivatives with FMOC, obtained after injection of a derivatized plasma extract (spiked concentration 1ng/mL). Column: Ascentis Express C8 (100x2.1mm, 2.7 µm); mobile phase: A) water/acetonitrile (90/10, v/v) and B) acetonitrile, with composition gradient; injection volume: 50µL. Mass spectrometer: triple quadrupole API 4000 operated in ESI positive mode.

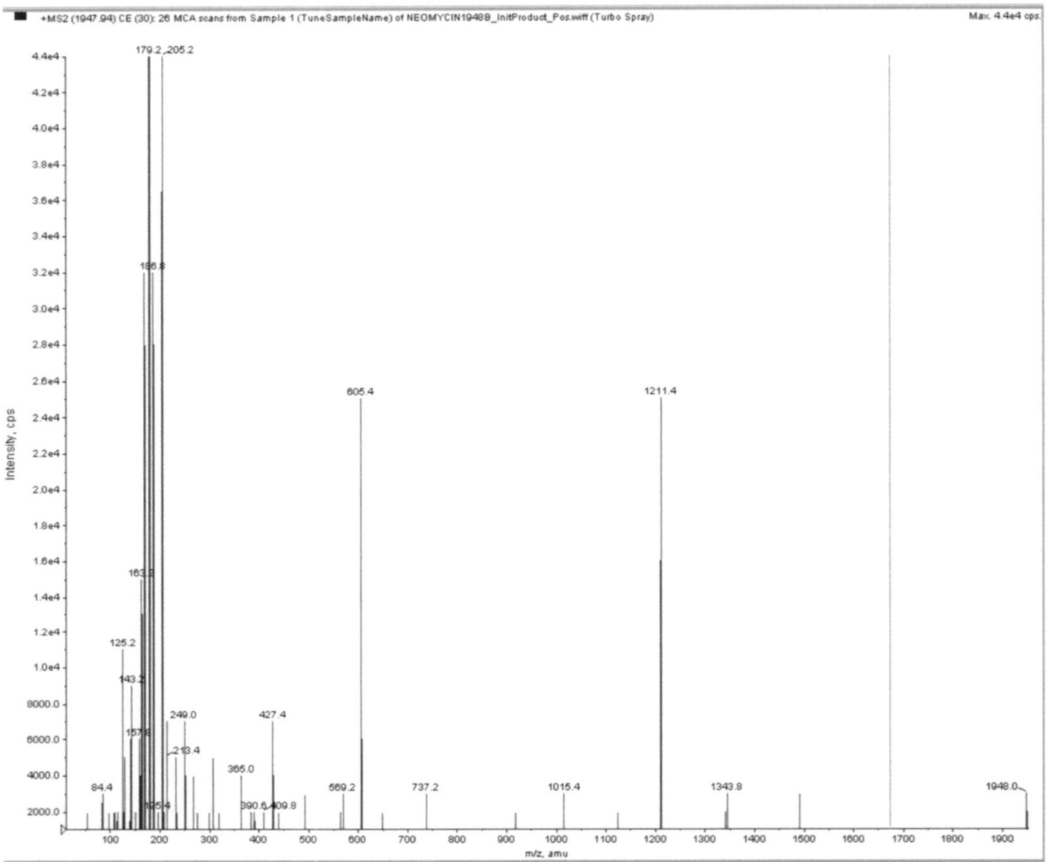

Fig. 21. Product ion spectrum of the neomycin derivative with FMOC - $[M+H]^+ = 1948$ obtained with direct infusion via syringe of a 1µg/mL solution in water/methanol (1/1, v/v). Mass spectrometer: triple quadrupole API4000, operated in positive ESI mode; collision energy: 30V.

5. Conclusions

The data presented above clearly show that, despite the fact that highly polar compounds present optimal characteristics to be analyzed by MS, they are difficult molecules due to critical separation problems generated by the need to employ mobile phases compatible with the ionization source and process. It is therefore essential, when developing a new method for such compounds, to first consider the available options to get an optimal separation compatible with the MS interfacing.

Between the various options HILIC is gaining a lot of interest as a separation for highly polar compounds; the development of several columns dedicated to this technique are widening a lot its application and the growing mass of publications in the field is a prove. HILIC employs high percentage of organic solvent in the mobile phase usually leading to enhanced ionization and lesser contamination of the ion source.

Ion exchange remains an interesting alternative especially when considering weak exchanger stationary phases; in comparison to HILIC however the technical development is moving much slower. Mixed mode columns (ion exchange + reversed phase) are an interesting opportunity to get improved separation methods.

The use of ion-pair reagents in connection with reversed-phase columns is another opportunity but it presents a few pitfalls: restrictions on ionization polarity (in general ion-pair suppress ionization in the same polarity, i.e. quaternary ammonium salt and positive ionization), often complex spectra, reduced analytical sensitivity and finally a tendency to contaminate the ionization source and quadrupoles. It is however interesting, see the GAGs examples, that ion-pair reagents can, in some cases, positively help the ionization reducing source fragmentation.

Globally the previously indicated strategies are preferable whenever possible. As a final option derivatization has to be considered, without meaning that it is the last desirable approach. It has to be considered especially in case of amphoteric compounds bearing both positive and negative charges; such compounds are generally difficult to extract and the ionization is not optimal; aminobisphosphonates are a good example in this context. The development of derivatization methods can be time consuming but often very solid methods and a quite good analytical sensitivity can be obtained. It is interesting to see a continuous development for obtaining optimal derivatization reagents for LC-MS detection. GC derivatization agents are indeed aiming to get non-polar and volatile compounds, aspects not always desirable in LC-MS, while fluorescence and UV detection reagents can be chemically very complex without favoring the ionization process.

The analytical approach described can be usefully applied as well to other classes of highly polar compounds (for examples polynucleotides, phosphorylated carbohydrates, several pharmaceuticals).

6. References

Alpert, A.J. (1990). Hydrophilic-interaction chromatography for the separation of peptides, nucleic acids and other polar compounds. *Journal of Chromatography* 499: 177–196

Alpert, A.J. (2011). HILIC at 21: Reflections and perspective. *Journal of Chromatography A 1218, 5879*

Aluoch, A., Tatini, R., Parsons, D.M. & Sadik, O. (2005). Stability Indicating Ion-Pair HPLC Methods for the Determination of Risedronate in A Commercial Formulation, *Journal of Liquid Chromatography Related Technologies, 27, 2799–2813*

Apostolou, C., Dotsikas, Y., Kousoulos, C., Tsatsou, G., Colocouri, F., Soumelas, G.S. & Loukas, Y. (2007). Application of a Semi-Automated 96-well Format Solid-Phase Extraction, Column-Switching, Fluorescence Detection Protocol for the Determination of Alendronate in Human Urine Samples Obtained from a Bioequivalence Study, *Journal of Pharmaceutical and Biomedical Analysis*, 43 (3), 1151

Arpino, P. J. (1985). Ten years of liquid chromatography-mass spectrometry, *Journal of Chromatography A 323 (1), 3-11*

Aubin, A., Analysis of aminoglycoside antibiotics with Waters 2465 electrochemical detector, *Waters Application Note*

Barosso, B., Didraga, M. & Bischoff, R. (2005). Analysis of proteoglycans derived sulphated disaccharides by liquid chromatography/mass spectrometry, *Journal of Chromatography A, 1080(1), 43-48*

Caprioli, R.M., Fan, T. & Cottrell,J.S. (1986). A continuous flow sample probe for fast atom bombardment mass spectrometry, *Analytical Chemistry, 58, 2949-2954*

Chambers, H. F., (2006). Aminoglycosides, in *Goodman & Gilman's The Pharmacological Basis of Therapeutics, 11e, pag.1155, Brunton, L.L., Lazo, J.S. & Parker., K.L. (Ed) McGraw Hill Companies Inc.*

Chester, T.L., Lewis, E.C., Benedict, J.J., Sunberg, R.J. & Tettenhorst, W.C. (1981). Determination of (dichloromethylene) Diphosphonate in Physiological Fluids by Ion-exchange Chromatography with Phosphorous-selective Detection, *Journal of Chromatography, 225, 17*

Chopra, S., Vanderheyden, G., Hoogmartens, J., Schepdael, A. & Adams E. (2010). Comparative study on the performance of different detectors for the LC analysis of tobramycin, Journal of Pharmaceutical and Biomedical Analysis, 53(2), 151-157

Clarot, I., Storme-Paris, I., Chaminade, P., Estevenon, O., Nicolas, A. & Rieutord, A. (2009). Simultaneous quantitation of tobramycin and colistin sulphate by HPLC with evaporative light scattering detection, *Journal of Pharmaceutical and Biomedical Analysis, 50(1), 2009, 64-67*

Clarot, I., Regazzeti, A., Auzeil, N., Laadani, F., Citton, M., Netter, P. & Nicolas, A. (2005). Analysis of neomycin sulphate and framycetin sulphate by HPLC using evaporative light scattering detection , *Journal of Chromatography A, 1087(1-2), 236-244.*

Cohen, D.M. & Linhardt, R.J. (1990). Randomness in the heparin polymer: Computer simulations of alternative action patterns of heparin lyase, *Biopolymers 30(7-8), 733-741*

Da Col, R., Silvestro, L., Naggi, A., Torri, G., Baiocchi, C., Moltrasio, D., Cedro, A. & Viano, I. (1993). Characterization of the chemical structure of sulphated glycosaminoglycans after enzymatic digestion: Application of liquid chromatography—mass spectrometry with an atmospheric pressure interface, *Journal of Chromatography, 647, 289-300*

Daley-Yates, P.T., Gifford, L.A. & Hoggarth, C.R. (1989). Assay of 1-hydroxy-3-aminopropylidene-1,1-bisphosphonate and related bisphosphonates in human urine and plasma by high-performance ion chromatography, *Journal of Chromatog*raphy, 490 (2), 329

Dell, A., Rogers, M.E. & Thomas-Oates, J.E. (1988). Fast-atom-bombardment mass-spectrometric strategies for sequencing sulphated oligosaccharides, *Carbohydrate Researches, 179, 7*

Deyrup, C.L., Chang, S.M., Weintraub, R. & Moye, H.A. (1985). Simultaneous esterification and acylation of pesticides for analysis by gas chromatography. 1. Derivatization of glyphosate and (aminomethyl)phosphonic acid with fluorinated alcohols-perfluorinate anhydrides, *Journal of Agricultural and Food Chemistry, 33, 944-947*

Dinh, N.P, Jonsson, T, Irgum, K. (2011). Probing the interaction mode in hydrophilic interaction chromatography. *Journal of Chromatography A 1218, 5880-5891*

Du,J. & Eddington, N. (2002). Determination of the Chondroitin Sulphate Disaccharides in Dog and Horse Plasma by HPLC Using Chondroitinase Digestion, Precolumn Derivatization, and Fluorescence Detection, *Analytical Biochemistry, 306, 252-258*

Flesch, G., Tominaga, N. & Degen ,P.(1991). Improved Determination of the Bisphosphonate Pamidronate Disodium in Plasma and Urine by Pre-Column Derivatization with Fluorescamine, High-Performance Liquid Chromatography and Fluorescence Detection, *Journal of Chromatography*, 568, 261

Fleisch, H. (2004). Development of biphosphonates, *Breast Cancer Research, 4 (1), 30*

Food and Drug Administration, *Guidance for Industry / Bioanalytical method validation*, May 2001

Garcia, J. F. & Barceló, D. (1993). An overview of LC–MS interfacing systems with selected applications, *Journal of High, Resolution Chromatography, 16(11), 633–641*

Gatti, R., Andreatta, P, Gioia, M.G. & Boscchetti, S. (2010). A Simple and validated LC method for the Simultaneous Analysis of Glucosamine and Chondroitin Sulphate Equivalent in Dietary products, *Journal of Liquid Chromatography and Related Technologies, 33(19), 1760-1775*

Granja, RH., Nino, AM, Zucchetti RA, Nino RE., Patel, R., Salerno, AG. (2009). Determination of streptomycin residues in honey by liquid chromatography-tandem mass spectrometry. *Analitica Chimica Acta 637, (1-2), 64-47.*

Gunay, N.S. & Linhardt, R.J. (1999). Heparinoids: Structure, Biological Activity and Therapeutic Applications, *Planta Medica, 65, 301-306*

Hammel, Y-A., Mohamed, R., Gremaud, E., LeBreton, M-H., Guy, P. (2008). Multi-screening approach to monitor and quantify 42 antibiotic residues in honey by liquid chromatography–tandem mass spectrometry. *Journal of Chromatography A, 1177, 58-76*

Hemstrom, P. & Irgum, K., (2006). Hydrophilic interaction chromatography, *Journal of Separation Sciences, 29(12),1784-1821*

Henriksen, J., Roepstorff , P. & Ringborg, L.H. (2006). Ion-pairing reversed-phased chromatography/mass spectrometry of heparin, *Carbohydrate Researches, 341(3), 382-387*

Higgins, C.E. & Kastner, R.E. (1967). Nebramycin, a new broad-spectrum antibiotic complex. II. Description of Streptomyces tenebrarius, *Antimicrobial Agents Chemotherapy, 7, 324-31.*

Ip, D.P., Qin, X.Z., Sakuma, T. & Tsai, E.W. (1994). Pharmaceutical Application of Liquid Chromatography-Mass spectrometry .2. Ion Chromatography Ion-spray Mass-Spectrometric Characterization of Alendronate, *Journal of Chromatography A, 686, 205*

Ishii, R., Horie, M., Chan, W., MacNeil, J. (2008). Multi-residue quantitation of aminoglycoside antibiotics in kidney and meat by liquid chromatography with tandem mass spectrometry. *Food Addit. Contam. A 25 (2008) 1509.*

Isoherranen, N. & Soback, S. (1999). Chromatographic methods for analysis of aminoglycoside antibiotics, *Journal of Association of Analytical Chemists International, 82 (5), 1017-1045*

Isoherranen, N. & Soback, S. (2000). Determination of Gentamicins C1, C1a and C2 in plasma and urine by HPLC, *Clinical Chemistry 46(6) 837-842*

Jandik, K.A., Kruep, D., Cartier, M. & Linhardt, R.J. (1995). Accelerated stability studies of heparin, *Journal of Pharmaceutical Sciences, 85(1), 45-51*

Jin, P., Ma,J., Wu,X., Zou,D., Sun,C. & Hu,X. (2009). Simultaneous determination of chondroitin sulphate sodium, allantoin and pyridoxine hydrochloride in pharmaceutical eye drops by an ion-pair high-performance liquid chromatography, *Journal Pharmaceutical and Biomedical Analysis 50(3), 293-297*

Johnson, E.A. (1982). Characterisation and separation of sulphated glycosaminoglycuronans, *Pharmacological Research Communications, 14(4), 289-320*

Kawachi, Y., Tohru Ikegami, T., Hirotaka Takubo, H, Ikegami, Y., Miyamoto, M., Tanaka, N. (2011). Chromatographic characterization of hydrophilic interaction liquid chromatography stationary phases: Hydrophilicity, charge effects, structural selectivity, and separation efficiency. *Journal of Chromatography A 1218, 5903-5919*

Kim, B.H., Lee, S.C., Lee, H.J. & Ok, J.H. (2003). Reversed-phase liquid chromatographic method for the analysis of aminoglycoside antibiotics using pre-column derivatisation with phenylisocyanate, *Biomedical Chromatography 17(6), 396-403*

Kim, B.H., Lee, S.C., Lee, H.J. & Ok, J.H. (2001). Development of liquid chromatographic method for the analysis of kanamycin residues in varicella vaccine using phenylisocyanate as a derivatization agent, *Biomedical Chromatography, 752(1), 173-7*

Kline, W.F. & Matuszewski, B.K. (1992). Improved determination of the bisphosphonate alendronate in human plasma and urine by automated precolumn derivatization and high-performance liquid chromatography with fluorescence and electrochemical detection, *Journal of Chromatography,* 583, 183

Korir, A.K. & Larive, C.K. (2007). On-line NMR detection of microgram quantities of heparin-derived oligosaccharides and their structure elucidation by microcoil NMR, *Analytical and Bioanalytical Chemistry, 388(8), 1707-1716*

Kovacevic, M., Gartner, A. & Novic, M.(2004). Determination of Bisphosphonates by Ion Chromatography–Inductively Coupled Plasma Mass Spectrometry, *Journal of chromatography A,* 1039, 77

Kudzin, Z.H., Gralak, D.K., Andrijewski, G., Drabowicz J. & Luczak, J.(2003). Simultaneous analysis of biologically active aminoalkanephosphonic acids, *Journal of Chromatography A,* 998, 183

Lauser, G., Bergner-Lang, B. (1995). Flurimetrische Bestimmung von Aminoglykosid Antibiotika nach Vörsaulenderivatisierung im Autosampler mit 9-Fluorenylmethyl Chloroformiat und anschliessender HPLC-Trennung. *Deutsche Lebensmittel Rundschau 91 (12), 390-396*

Malavaki, C. J., Asimakopoulou, A.P, Lamari, F.N., Teocharis, A.D.,.Tzanakakis, G.N. & Karamanos, N.K. (2008). Capillary electrophoresis for the quality control of chondroitin sulphates in raw materials and formulations, *Analytical Biochemistry, 374(1), 213-220*

Mallis, L.M., Wang, H.M., Loganathan, D. & Linhardt, R.J. (1989). Sequence analysis of highly sulphated, heparin-derived oligosaccharides using fast atom bombardment mass spectrometry, *Analytical Chemistry, 61, 1453*

Malsch, R. & Harenberg, J. (1996) Purity of glycosaminoglycan-related compounds using capillary electrophoresis, *Electrophoresis, 17, 401-405*

Malsch, R., Harenberg, J., Piazolo, L., Huhle, G. & Heene, D.L. (1996). Chromatographic and electrophoretic applications for the analysis of heparin and dermatan sulphate, *Journal of Chromatography B, 685, 223-231*

Manyanga, V., Dhulipalla, R. L., Hoogmartens, J. & Adams, E. (2010). Improved LC method with pulsed electrochemical detection for the analysis of kanamycin, *Journal of Chromatography A, 1217(24), 3748-3753.*

Mao, W., Thanawiroon, C. & Linhardt, R.J. (2002). Capillary electrophoresis for the Analysis of Glycosaminoglycan-Derived Oligosaccharides, *Biomedical Chromatography,* 16, 77-94

Mayhew, J.W. & Gorbach, S. L. (1978). Gas-liquid chromatographic method for the assay of aminoglycoside antibiotics in serum, *Journal of Chromatography, 151, 133-146*

de Marco, J.D., Biffar, S.E., Reed, D.G. & Brooks, M.A. (1989). The Determination of 4-amino-1-hydroxybutane-1,1-diphosphonic Acid Monosodium Salt Trihydrate in Pharmaceutical Dosage Forms by High-Performance Liquid Chromatography, *Journal of Pharmaceutical and Biomedical Analusis, 7(12),1719–1727.*

Margosis, M. &Tsuji, K. (1973), Journal of Pharmaceutical Sciences, 62, 1873

Mashat, M., Chrystyn, H., Clark, B.J. & Assi, K.H. (2008). Development and validation of HPLC method for the determination of tobramycin in urine samples post-inhalation using pre-column derivatization with fluorescein isothiocyanate, *Journal of Chromatography B, 869(1-2), 59-66*

Metrohm. Gentamicin determination by ion chromatography. *Laboratorytalk editorial team,* (2005). http://www.laboratorytalk.com/news/mea/mea456.html

Moye, HA, Deyrup, CL, (1984). Simultaneous esterification and acylation of pesticides for analysis by gas chromatography. 1. Derivatization of glyphosate and (aminomethyl)phosphonic acid with fluorinated alcohols-perfluorinated anhydrides, *Journal of Agricultural and Food Chemistry, 32, 192*

Oertel, R,. Neumeister, V., Kirch, W. (2004). Hydrophilic interaction chromatography combined with mass spectrometry to determine six aminoglycosides in serum. *Journal of Chromatography A, 1058, 197-201*

Oguma, T., Tomatsu, S., Montano, A. M. & Okazaki, O. (2007). Analytical method for the determination of disaccharides derived from keratan, heparan, and dermatan sulphates in human serum and plasma by high-performance liquid chromatography/turbo ionspray ionization tandem mass spectrometry, *Analytical Biochemistry 368(1), 79-86*

Ori, A., Free, P., Courty, J., Wilkinson & M.C., Fernig, D.G. (2009). Identification of Heparin-binding Sites in Proteins by Selective Labeling, *Molecular & Cellular Proteomics 8(10), 2256-2265*

Palit, M., Gupta, A., Jain, R. & Raza, S.K. (2004). Determination of pentafluorobenzyl derivatives of phosphonic and phosphonothioic acids by gas-chromatography-mass spectrometry, *Journal of Chromatography A*, 1043, 275

Patel, R.P., Narkowicz, C. & Jacobson, G.A. (2009) Effective reversed-phase ion pair high-performance liquid chromatography method for the separation and characterization of intact low-molecular-weight heparins, *Analytical Biochemistry 367(1), 113-121*

Plaas, A.H., Hascall, V.C. & Midura, R.J. (1996). Ion exchange HPLC microanalysis of chondroitin sulphate: quantitative derivatization of chondroitin lyase digestion products with 2-aminopyridine, *Glycobiology, 6(8), 823-9*

Ptacek, P., Klima, J. & Macek, J. (2002). Determination of Alendronate in Human Urine as 9-fluorenylmethyl Derivative by High Performance Liquid Chromatography, *Journal of Chromatography B*, 767, 111

Rao, B.M., Srinivasu, M.K., Rani, Ch.P., Kumar, S.S., Kumar, P.R., Chandrasekhar, K.B. & Veerender, M., (2005). A Validated Stability Indicating Ion-Pair RP-LC Method for Zoledronic Acid, *Journal of Pharmaceutical and Biomedical Analysis, 39(3-4), 781–790*

Ranz, A., Korpecka, J. & Lankmayr, E. (2008) Optimized derivatization of acidic herbicides with trimethylsilyldiazomethane for GC analysis, *Journal of Separation Science 31 (4), 746*

Royer, A., Beguin, S., Sochor, H., Communal, P., (2000). Determination of gluphosinate ammonium ant its metabolite (AE F064619 and AE F061517) residues in water by gas-chormatography with tandem mass spectrometry after ion exchange clean-up and derivatization, *Journal of Agricultural and Food Chemistry, 48, 5184-5189*

Ruiz-Calero, V., Puignou, L. & Galceran, M.T., Determination of glycosaminoglycan monosaccharides by capillary electrophoresis using laser-induced fluorescence detection, *Journal of Chromatography B, 791(1-2), 193-202*

Yang, B., Solakyidirim, K., Chang, Y. & Linhardt, R.J. (2011). Hyphenated techniques for the analysis of heparin and heparan sulphate, *Analytical and Bioanalytical Chemistry*, 399, 541-557

Yun, M.H. & Kwon, K.L. (2006). High performance Liquid Chromatography for Determining Alendronate Sodium in Human Plasma by Detecting Fluorescence: Application to a Pharmacokinetic Study, *Journal of Pharmaceutical and Biomedical Analysis, 40, 168–172*Saad, O.M & Leary, J. A. (2003). Composition Analysis and Quantification of Heparin and Heparan Sulphate by Electrospray Ionization Ion Trap Mass Spectrometry, *Analytical Chemistry, 75(13), 2985-2995*

Saito, H., Yamagata, T. & Suzuki, S. (1968). Enzymatic Methods for the Determination of Small Quantities of Isomeric Chondroitin Sulphates, *Jouranl of Biological Chemistry, 243, 1536-1542*

Shao,Y., Alluri,R., Mummert,M., Koetter,U. & Lech, S. (2004). A stability-indicating HPLC method for the determination of glucosamine in pharmaceutical formulations, *Journal of Pharmaceutical and Biomedical Analysis, 35(3), 625-631*

Silvestro, L., Rizea Savu, S., van Veelen, P.A. & Jacobs, P.L. (1996). Application of Mass Spectrometry to the Analysis of Natural and Synthetic sulphated Oligosaccharides, in Nonanticoagulant Actions of Glycosaminoglycans, *Harenberg, J. & Casu, B. (Ed), Plenum Press, New York*

Silvestro, L., Viano, I., Naggi, A., Torri, G., Col, R.Da. & Baiocchi, C. (1992). High-performance liquid chromatographic-mass spectrometric analysis of oligosaccharides from enzymatic digestion of glycosaminoglycans: Application to human samples, Journal of Chromatography, 591, 225-232

Shen, A. Q., Morgan, L., Barroso, M.L., Zhang X., & Nguyen, T. (June 2008). Method development of LC-MS/MS analysis of aminoglycoside drugs: challenges and solutions, Presented at the 2008 ASMS Conference, Denver, CO, June 2008, Available from: http://tandemlabs.com/documents/ASMS08-Angela-Web.pdf

Sakiyama, N., Kataoka, H. & Makita, M. (1996). Gas chromatographic analysis of 3-amino-1-hydroxypropilidende-1,1-bisphosphonate and related bisphosphonate as their N-isobutoxycarbonyl methyl ester derivatives, *Journal of Chromatography A*, 724, 279

Sakiyama, N., Kataoka, H.& Makita, M. (1995). Selective and sensitive determination of pamidronate in human plasma and urine by gas-chromatography with flame photometric detection, *Biomedical Chromatography* 9 (5), 243

Sparidans, R.W. & den Hartigh, J. (1999). Chromatographic Analysis of Biphosphonates, *Pharmaceutical World & Science, 21(1), 1–10*

Stead, D.A. *(2000).* Current methodologies for the analysis of aminoglycosides, *Journal of Chromatography B, 747(1-2), 69-93*

Strege, M. A., (1999). High-performance liquid chromatographic–electrospray ionization mass spectrometric analyses for the integration of natural products with modern high-throughput screening, *Journal of Chromatography B, 725 (1-2), 67-78*

Tarcomnicu, I., Silvestro, L., Rizea Savu, S., Gherase, A. & Dulea, C. (2007). Development and application of an HPLC-MS/MS method to determine alendronate in human urine, *Journal of Chromatography A*, 1160, 21-33

Tarcomnicu, I., Gheorghe, MC., Silvestro, L., Rizea Savu, S., Boaru, I., Tudoroniu, A. (2009). High-throughput HPLC-MS/MS method to determine ibandronate in humanplasma for pharmacokinetic applications. *Journal of Chromatography B, 877, 3159-3168*

Tawa, R., Matsunaga, H. & Fujimoto, T. (1998). HPLC analysis of aminoglycoside antibiotics, *Journal of Chromatography A, 812(1-2), 141-150*

Thanawiroon, C., Rice, K. G., Toida, T. & Linhardt, R. J. (2004). Liquid Chromatography/Mass Spectrometry Sequencing Approach for Highly Sulphated Heparin-derived Oligosaccharides, *Journal of Biological Chemistry, 279(4), 2608-2615*

The Merck Index, (2006) Fourtheenth Edition, Published by Merck Research Laboratories, Merck and Co. Inc., Whitehouse Station NJ, USA

Tiayu Peng, (2002). Analysis of Saccharides by Capillary Electrodriven Separation Methods, *Ph.D Thesis, Department of organic Chemistry, Faculty of Science, University of Ghent, Belgium*

Toyoda, H., Kinoshita-Toyoda, A., Fox, B., Selleck, S.B. (2000). Structural Analysis of Glycosaminoglycans in*Drosophila* and *Caenorhabditis elegans* and Demonstration That *tout-velu*, a *Drosophila* Gene Related to EXT Tumor Suppressors, Affects Heparan Sulphate *in Vivo, Journal Biological Chemistry , 275(4), 2269-2275*

Trehy, M.L.,.Reepmeyer, J.C., Kolinski, R.E., Westenberger, B.J. & Buhse, L.F. (2009). Analysis of heparin sodium by SAX/HPLC for contaminants and impurities, *Journal of Pharmaceutical and Biomedical Analysis 49(3), 670-673*

Tsai, E. W., Ip, D.P. & Brooks, M.A. (1992). Determination of Alendronate in Pharmaceutical Dosage Formulations by Ion Chromatography with Conductivity Detection, *Journal of Chromatography A, 556, 217*

Vallano, P.T., Shugarts, S.B., Kline, W.F., Woolf E.J. & Matuszewski, B.K. (2003). Determination of Risedronate in Human Urine by Column-Switching Ion-Pair High-Performance Liquid Chromatography with Ultraviolet Detection *Journal of Chromatography B. 794, 23–33*

Van Nuijs A., Tarcomnicu I., Covaci A. (2011) Application of hydrophilic interaction chromatography for the analysis of polar contaminants in food and environmental samples. *Journal of Chromatography A*, 1218, 5964-5974

Venkataram, G., Shriver, Z., Raman, R., & Sasisekharan, R. (1999) Sequencing Complex Polysaccharides, *Science, 286, 537-542*

Xie, Z., Jiang, Y. & D. Zhang (2006). Simple Analysis of Four Bisphosphonates Simultaneously by Reverse Phase Liquid Chromatography Using *n*-amylamine as Volatile Ion-Pairing Agent, *Journal of Chromatography A*, 1004(1-2), 173

Zhu, L.S. , Lapko, V.N., Lee, J.W., Basir, Y.J., Kafonek, C., Olsen, R. & Briscoe, C. (2006). A general approach for the quantitative analysis of bisphosphonates in human serum and urine by high-performance liquid chromatography/tandem mass spectrometry, *Rapid Communications. Mass Spectrometry 20, 2426–3421*

Zimmer, J. S.D., Needham, S. R., McKinnell, J., Cass, B. & Karr, D. (June 2008). Development of a Bioanalytical LC/MS/MS Assay for the Quantitative Analysis of Amikacin, Neomycin and Gentamicin in Plasma and Tissue Samples. Presented at the 2008 ASMS Conference, Denver, CO, June 2008. Available from: http://www.alturasanalytics.com/home/pdf/ASMS2008AlturasAnalyticsPosterReprint.pdf

Wong, J.A., Renton, K.W., Crocker, J.F., O'Regan, R.A. & Acott, P.D. (2004). Determination of Pamidronate in Human Whole Blood and Urine by Reversed-Phase HPLC with Fluorescence Detection, *Biomedical Chromatogr*aphy, 18, 98

22

Quantification of Glucuronide Metabolites in Biological Matrices by LC-MS/MS

Jurij Trontelj
Faculty of Pharmacy, University of Ljubljana
Slovenia

1. Introduction

For many drugs and xenobiotics glucuronidation is the major metabolic pathway of detoxification. There, a highly hydrophilic glucuronic acid is coupled to aliphatic / aromatic alcohols, thiols or carboxylic acids or even amines. Although the glucuronidation is usually considered a deactivation reaction, electrophilic acyl-glucuronides can be formed from a number of acidic drugs, such as diclofenac, zomepirac, diflunisal, valproate, ibuprofen, gemfibrozil and clofibrate, among others. Acyl-glucuronides have been linked with idiosyncratic adverse drug reactions due to the chemical reactivity towards plasma and tissue proteins and nucleic acids (Shipkova, Armstrong et al. 2003). Even the most common ether O-glucuronides, which are generally pharmacologically inactive, can be cleaved back to the active parent compounds inside the human body or in the in the environment. Similarly, some N-glucuronides have been linked to bladder cancer due to their fast deconjugation which releases the toxic aglycone inside the bladder. Therefore, on many occasions in drug metabolism studies or in toxicology, it is necessary or prudent to detect and to quantify glucuronide metabolites in biological samples.

2. Overview of glucuronide quantification approaches

In the past, the majority of the quantification methods for glucuronides involved an enzymatic (β-glucuronidase) or an acid/base hydrolysis of the glucuronide ether bond prior to HPLC analysis, mainly because the hydrophilic glucuronides were difficult to separate from the interfering matrix components (Kadi and Hefnawy 2009). Nowadays, the highly selective and sensitive triple quadrupole LC-MS/MS instruments enable us to directly measure the glucuronides without the need to de-conjugate the glucuronide moiety before the analysis. This offers some significant benefits: quicker sample preparation and better accuracy and precision, as the glucuronide hydrolysis may be incomplete or the aglycon is unstable in hydrolytic conditions or binds non-specifically to the β-glucuronidase or matrix components (Kadi and Hefnawy 2009). Furthermore, the glucuronide moiety can be bound to different nucleophilic functional groups on the parent aglycone, giving rise to different glucuronide structural isomers, which can then be selectively quantified. Differentiation among glucuronide isomers might be important as they can have profoundly different pharmacological activity, for example the morphine-6-glucuronide has a many times greater affinity towards the μ-opioid receptor than the morphine-3-glucuronide or the morphine-6,3-diglucuronide (Loser, Meyer et al. 1996).

On the other hand, a direct measurement of the glucuronides in biological samples requires the use of their authentic analytical standards for the preparation of the calibration and the quality control samples in an appropriate matrix. Quite often, these analytical standards are not easily obtainable and even if they are commercially available, a few milligrams may cost up to several thousand of US Dollars or even more. Their isotope labeled analogs (for use as internal standards) can cost even more. To overcome this difficulty, several possibilities exist (Soars, Mattiuz et al. 2002; Sanceau, Larouche et al. 2007). The first is a chemical synthesis of the glucuronide standards, which can be performed either by a glucopyranoside synthesis or by the use of glucuronide donors or glucuronide precursors and often involves a several step reaction, which is time-consuming and the glucuronide yield is often very low (Yu, Zhu et al. 2000; Burkon and Somoza 2008). The second approach is an enzymatic synthesis of the glucuronides, where the aglycone is incubated with uridine diphosphoglucuronosyl transferases (UGTs) in the presence of a co-factor uridine-diphospho-glucuronic acid (UDPGA). Recombinant human UGTs or human microsomes can be used in these incubations in order to obtain the same glucuronide isomers as present in human body fluids (Trontelj, Bogataj et al. 2007). Some alternative methods to microsome incubations have also been described where the UGT enzymes are adsorbed or covalently linked onto silica or to a monolithic carrier surface. There, the enzymes can be re-used many times or can continuously catalyze the glucuronidation reaction by a substrate flow-through technique called immobilized enzyme reactors (IMER) (Kim and Wainer 2005).

The third approach is a bioproduction process, where specially modified strains of bacteria (for example *Streptomyces sp.*) can be used to produce large amounts of certain glucuronides (Trdan, Roskar et al. 2011). However, the latter approach needs an extensive clean-up and purification procedures in order to obtain high purity analytical standards (Briggs, Baker et al. 1999). The fourth approach is the semi-quantitative method where one aliquot of the sample containing the glucuronides is subjected to a quantitative beta-glucuronidase cleavage and the other aliquot is not. After the analysis of first and second aliquot, the glucuronide *response factor* is calculated as the ratio between the glucuronide signal from the second aliquot and the rise of the signal of aglycon in the first aliquot (originating from the cleaved glucuronide). Then, in subsequent analyses, the glucuronide *response factor* is used to quantify the glucuronides without the β-glucuronidase treatment and by using the calibration curve of the easily obtainable aglycone (Sanceau, Larouche et al. 2007; Farrell, Poquet et al. 2011; Trdan, Roskar et al. 2011). In such scenarios, a careful study of the possible relative matrix effects on glucuronides is warranted. The semi-quantitative approach can also be used if a radiolabeled parent is available by correlating the radioactive responses from the radiolabeled metabolite and parent to the corresponding LC-MS/MS responses (Yu, Chen et al. 2007). If a radiolabeled parent is not available, the labeled glucuronide can be synthesized by microsomes using a radiolabeled UDP-glucuronic acid (UDP-[U-^{14}C]-GlcA) (Maurel 2010). The semi-quantitative approach is based on the assumption that the concentration-response relationship for glucuronide is linear. However, on many occasions in LC-MS analyses, this is not the case. Therefore, it is advisable to check the response factors across the expected glucuronide concentration range (Trdan, Roskar et al. 2011) .

3. Sample preparation techniques

This step is often the most crucial in the whole analysis. The choice of the sample preparation method is based on the glucuronide concentration, the available sample volume, the matrix type and the analyte stability. In this section, the most commonly used methods for sample preparation are presented, including: deconjugation with β-glucuronidase, solid and liquid extractions, protein precipitation, on-line SPE and others. In order to ensure adequate stability of the glucuronides prone to hydrolysis or acyl-group migration, a proper sample collection / storage / buffering is required. Therefore, the stability considerations are discussed as well.

3.1 Deconjugation with β-glucuronidase

This is not a complete sample preparation procedure; it is just one of the optional steps within the sample handling workflow which is used to liberate the aglycone from glucuronide in order to quantify it more easily. Historically, the β-glucuronidase treatment was the most commonly used approach to quantify glucuronides of xenobiotics in biological samples (Kadi and Hefnawy 2009). However, this procedure has some drawbacks and considerations that have to be taken into account.

First, not all glucuronides can be cleaved by the commercially available enzymes, such as the β-glucuronidase from *E. coli* or *Helix pomatia*, bovine liver glucuronidase etc. Some N-glucuronides may remain quite stable in such incubation mixtures and require a chemical cleavage with, for example, hydrazine hydrate (Frandsen 2007) or a hydrolysis with a dilute acid. On the other hand, the acyl (ester) glucuronides can be easily cleaved by a dilute base (0.1-1.0 M NaOH) with mild heating (Wen, Stern et al. 2006; Zhang, Zhu et al. 2008). β-glucuronidases isolated from different species may have different enzyme activities toward various substrates and even for structural isomers of glucuronides (Choi, Ha et al. 2010). In addition, some acyl-glucuronides or their isomers formed after acyl-migration can be β-glucuronidase stable and require a basic hydrolysis (with 1M NaOH) (Wen, Stern et al. 2006). Furthermore, β-glucuronidases isolated from mollusks (like *Helix pomatia* or *Patella vulgata*) may posses a significant sulfatase activity as well. Note that molluskan β-glucuronidases may be less effective in hydrolyzing ether glucuronides than E. coli β-glucuronidases (Smith and Athanaselis 2005).

Secondly, in most cases, there is at least some amount of aglycone present in the starting material. Therefore, the measured aglycone concentration after deconjugation has to be subtracted by the initial aglycone concentration in order to quantify the amount of present glucuronide. Of course, in such calculations, it is necessary to use the molar concentrations and not mass/volume concentrations because the molecular mass of the freed aglycone is lower by 176 amu compared to the mono glucuronide in positive ESI, which corresponds to the loss of monodehydrated glucuronic acid.

Thirdly, the binding of glucuronides to the matrix components can significantly reduce the β-glucuronidase cleavage efficiency. For example, a complete cleavage of raloxifene glucuronides dissolved in aqueous solution of acetate buffer takes only a few minutes if not seconds, but if plasma proteins are present, the glucuronides can be >95% bound, and consequently, it can take up to even 24-48 hours to completely deconjugate the same amount of glucuronides (*author's own experience, data not published*). Similarly, when performing a glucuronide quantification in tissue samples, the β-glucuronidase may not have a complete access to all subcellular compartments, such as microsomes, resulting in an underestimation of glucuronides. In such instances, better results may be obtained after an acid hydrolysis (Gu, Laly et al. 2005). Additionally, some β-glucuronidases may posses other catalytic properties as well, like for example steroid conversion with preparations from *H. pomatia* (Gomes, Meredith et al. 2009).

Alternatively, instead of deconjugation with beta glucuronidases, also a methanolysis of glucuronides can be used, which frees both glucuronide and sulfate conjugates. The downsides of this approach are an increase in baseline noise in mass spectrometer and a poor efficiency of deconjugation (Cooper, Currie et al. 2001; Yu, Ho et al. 2005).

Therefore, if the β-glucuronidase treatment turns out to be effective on the studied glucuronide within its sample matrix, then this should be the method of choice for its deconjugation. The most common procedure for a β-glucuronidase treatment is as follows.

3.1.1 Sample pre-treatment

Biological samples, such as urine, plasma or feces are simply diluted in 1:1 (volume/volume) ratio with purified water or buffer. Note that the buffer can be added later on as well. A feces sample has to be homogenized prior to the dilution step. If an internal standard is to be used, this is the correct moment for its addition. However, in some instances, for example, if a deuterated glucuronide internal standard is not available and another glucuronide or a structural isomer or a compound with similar retention factor is used instead, then it may be beneficial to add the internal standard solution *after* the extraction procedure in order to minimize the IS content variability. In such cases, the aim of internal standard addition is not to correct for the sample preparation inconsistencies, but rather to correct the possible matrix effects. Furthermore, if an internal standard is not a deuterated or a C-14 labeled analogue, then it may as well be extracted differently than the analyte, thereby increasing the method variation (RSD). Rarely, the samples are extracted or treated with an organic solvent prior to the β-glucuronidase treatment. This shortens the cleavage incubation time as the protein binding is usually eliminated by the extraction with an organic solvent. However, the organic solvent must be evaporated before the enzyme is added to the sample.

3.1.2 Sample buffering

The pH of incubation buffer may be important for the reaction velocity and the selectivity towards the glucuronides or the sulfates. Quite high buffer concentrations can be used (0.1-1.0 M). The optimal pH range depends on the enzyme source. If molluskan β-glucuronidase is used, then the pH should be 4.0 - 4.5 for maximal glucuronidase activity. However, if maximal sulfatase activity is desired, then the pH should be 6.5 (Tephly, Green et al. 1997). If bovine liver glucuronidase is used, then the maximal glucuronidase and sulfatase activity is achieved at pH 4.4 (Himeno, Hashiguchi et al. 1974). If *E. coli* glucuronidase is used, then there should be no sulfatase activity and the maximal velocity is achieved at pH range of 6-7 (DongHyun, YoungHo et al. 1995). The selection of buffer salts depends on the desired pH range. For acidic buffers, acetate buffer is most commonly used, for more neutral pH values, phosphate or carbonate buffers can be used.

3.1.3 Sample incubation

The sample incubations are usually performed in at least two parallels. Sometimes, a separate incubation of a commercially obtainable substrate (for example 4-methylumbelliferone glucuronide) in the same matrix is used to monitor the β-glucuronidase activity (Blount, Milgram et al. 2000). Each incubation should contain 200 -

25 000 units of β-glucuronidase per 1 mL of incubation medium (Gu, Laly et al. 2005; Belanger, Caron et al. 2009). Note that vendors define the glucuronidase activity in Roy or in Fishman units, where 1 Fishman unit releases 1 μg phenolphthalein from phenolphthalein-β-glucuronide in 1 h at 38°C. Typically, the incubations are left in a thermostated shaker for 1-5 hours at 37-60°C. For slow deconjugation reactions, 24 hours and longer incubations are needed in order to quantitatively deconjugate all the present glucuronides. It is noteworthy, that an excess of enzyme may lead to incomplete deconjugation, therefore it is important to determine the appropriate incubation conditions for the specified conjugate in a given sample matrix (Gomes, Meredith et al. 2009).

3.1.4 Reaction termination and final sample preparation

Prior to the analysis, the proteins have to be denatured and removed. This is usually accomplished by a subsequent liquid-liquid extraction with an organic solvent like ethyl acetate, cyclohexane, diethyl ether, methyl tert-butyl ether. Note that if the extraction solvent is highly non-polar, like cyclohexane alone, then it may fail to extract the possibly remaining hydrophilic glucuronide from the sample, resulting in an unnoticed glucuronide left-over and consequently its under prediction. Therefore, a combination of organic solvents can be used, like for example, ethyl acetate/diethyl ether (1:1, vol/vol), in order to extract both, aglycone and glucuronide. Alternatively, a solid phase extraction can be used (Weinmann, Vogt et al. 2000; Stolker, Niesing et al. 2004). After the protein removal, a pH re-adjustment may be necessary in order to gain an acceptable recovery, especially for basic aglycones, which tend to be protonated at acidic pHs of β-glucuronidase incubations. After evaporating the organic solvent, the samples are reconstituted in a solvent similar to mobile phase and injected into LC-MS/MS. Alternatively, instead of a liquid-liquid extraction, it is also possible to denature the proteins with an equal volume of ice-cold acetonitrile, or by adding approximately 1/10 of incubation volume of trichloroacetic acid solution (TCA) (10% vol/vol) (Lu, Palmer et al. 2003), followed by a brief vortex mixing, centrifugation (15 min at 3 000 – 15 000 × *g*) and injection of the supernatant into LC-MS/MS. Such sample preparations procedures, without an extraction, allow a confident detection of the possibly still remaining glucuronides; however, the analyte responses may be lower due to the sample dilution and the peak broadening, in contrast to the liquid-liquid and solid phase extractions, where the analytes can be concentrated many times and reconstituted in a weaker elution solvent.

3.2 Liquid-liquid extraction

Sometimes, a liquid-liquid extraction (LLE) may be preferable over a solid phase extraction (see paragraph 3.3) because of its greater simplicity and applicability to almost all laboratories and low cost (Jemal, Ouyang et al. 2010). However, the extraction of glucuronides into a lipophilic organic solvent can be difficult because of the glucuronide's highly polar and hydrophilic nature (both the glucuronic acid and hydroxyl groups can be ionized). In order to de-ionize the glucuronic acid moiety, acidic conditions should be ensured during the extraction. For example, if 1 mL urine sample is to be extracted, then some 200 μL of 1 M HCl should be used to sufficiently protonate the glucuronic acid (pKa 3.1-3.2) (Paradkar 2008). Note that if extraction of both glucuronide and aglycone is required, then acidic extraction conditions may diminish the extraction efficiency of the basic aglycones. The extraction efficiency for both types of analytes may be improved if a combination of organic solvents is used (see subheading 3.1.4 for details and references). The choice of the extraction solvents may have a strong impact on the matrix effects later in LC-MS analysis. Methyl-tert-butyl ether (MTBE) and n-butyl chloride seem to extract only negligible amounts of lyso-phosphatidyl choline, which is the matrix component that elutes early from a reverse phase (RP) column and tends to interfere with ionization of drugs and their metabolites (Jemal, Ouyang et al. 2010). However, both solvents are highly non-polar and the recovery of glucuronides is expected to be low. Nevertheless, if aglycone should be extracted after a β-glucuronidase treatment, then these two solvents can be used successfully for that purpose.

3.3 Solid Phase Extraction (SPE)

SPE has become a very popular sample preparation technique in bioanalysis due to low sample volume requirements, low organic solvents consumption, high recovery, good reproducibility, easy automation, high speed and high throughput. There are also some drawbacks, like a possible low recovery of hydrophilic analytes (some glucuronides included), column blockage due to sample viscosity or precipitation, analyte instability in elution solvent, and lot-lot variability in column packaging (Telepchak, August et al. 2010).
The principle mechanism of separation in a SPE is similar to analytical chromatography, i.e. the analytes in sample solution bind with a greater affinity to the solid surface than the matrix components which are washed away. Afterwards, a stronger elution solvent is used to elute the analytes from the solid surface and collect them for further analysis (HPLC, LC-MS/MS).
The main goal while developing a SPE method is to achieve the highest possible reproducibility, good matrix clean-up and a high recovery, if possible. The retention of glucuronides on a solid phase sorbent can be facilitated by a number of mechanisms, from hydrophobic to ion-pairing. The chemistry and hydro-lipophilic properties of the aglycone play a major role. There are a number of possible combinations of solid phases / eluents and the analyst should choose the right combination that gives the best performance for the samples to be analyzed. A good starting point is the sorbent selection kit or the method development kit, which is offered by the majority of leading SPE cartridge producers (like Phenomenex, Waters, Agilent, Varian, Supelco, Macherey-Nagel, Thermo, etc.). Usually, in such a package, there are at least four different cartridge chemistries: strong cation exchange, a weak cation exchange, a strong anion exchange and a weak anion exchange. There are also classical reversed phases available, like C-18, C-8, C-4 and polymeric phases for example divinylbenzene and N-vinylpyrrolidone. A mixed-mode SPE of glucuronides is also possible, where an ionized analyte binds to an ion-exchange sorbent due to its charged state whereas neutrals can be retained at the same time due to their lipophilic interactions with the polymeric or reversed phase part of the sorbent. The manufacturers also propose the generic condition/wash/elution solvent protocols. After subjecting the pre-spiked, post-spiked, and solvent-only samples to various sorbents (in replicates), the subsequent LC-MS/MS analysis shows which sorbent chemistry gives the highest and the most reproducible recovery (RE) and can also show the matrix effects (ME). See the paragraph 5.5.
In glucuronide extraction with SPE, special attention should be paid to the washing step if the analytes are retained by a reverse phase mechanism. Too high percentage of organic solvent can easily wash the hydrophilic glucuronides into waste (author's own experience, data not published). If silica based or polymeric reversed phase cartridges are used,

decreasing the methanol content in the elution solvent can reduce the elution of phospholipids which can cause unwanted matrix effects later in LC-MS. On the other hand, 100% acetonitrile as an elution solvent drastically minimized the phospholipid elution on silica based cartridges (Lahaie, Mess et al. 2010). If endogenous glucuronides are present in "blank" samples , like for instance steroid glucuronides, then it is necessary to spike the blanks with labeled glucuronides for matrix effect and recovery calculation (Pozo, Van Eenoo et al. 2008). The glucuronides can be retained to various extents on a reversed phase (Trontelj, Bogataj et al. 2007), on a strong anion-exchange (Pu, McKinney et al. 2004), on a mixed-mode (Kakimoto, Toriba et al. 2008) and even on strong and weak cation exchange sorbents (Okura, Komiyama et al. 2007), depending on the ionization and lipophilic properties of the aglycone. Even highly polar glucuronides and sulfates, like paracetamol conjugates can have a high recovery on mixed-mode weak anion exchange sorbents like Waters Oasis WAX (Sunkara and Wells 2010). Some lipophilic conjugates, like steroid glucuronides and sulfates may not be eluted fully with only methanol and may require addition of ion-pairing reagents like 5 mM triethylamine (Isobe, Serizawa et al. 2006). It is noteworthy that the ion-exchange or mixed mode SPE is preferred over the reversed phase mechanism of retention since the former two methods can include a rigorous washing of the cartridge with a strong solvent, for example 100% methanol or ethyl acetate, without any detrimental effect on the analyte recovery while producing the lowest possible matrix effects (Chambers, Wagrowski-Diehl et al. 2007). With the mixed mode anion-exchange SPE sorbents it is even possible to separate not only the conjugates from the parent compounds, but also to differentiate between the various types of conjugates, like glucuronides from sulfates (Fontanals, Marcé et al. 2010).

3.4 Protein precipitation

Protein presence in the samples can cause significant problems in glucuronide analysis. First, the glucuronides can be more than 50% protein bound (Hochner-Celnikier 1999; Burkon and Somoza 2008) which may negatively impact the absolute method recovery. In high throughput analysis, there is a tendency to minimize the sample preparation steps. Protein precipitation (PP) is one of such fast and easy-to-automate procedures. Therefore, PP is quite a popular sample preparation technique, especially for plasma samples and *in vitro* incubations, where this step terminates the enzyme reactions as well (Rosenfeld 2004).
However, samples obtained with a PP procedure may contain higher amounts of problematic matrix components than samples after a SPE or LLE. Furthermore, chromatography problems may arise, since the percentage of organic solvent in the sample is higher, worse peak shape can be expected, shorter column life-time, worse separation and lower responses due to the sample dilution with a precipitating agent. In addition, more frequent ESI spray chamber and capillary cleaning may be required (Bakhtiar and Majumdar ; Elbarbry and Shoker 2007), (author's experience). Most importantly, worse matrix effects can be expected with PP than with either LLE or SPE. The PP is most commonly accomplished with an addition of an organic solvent in 3:1 volume ratio to the sample. The most commonly used organic solvents are acetonitrile or methanol, which should precipitate at least 99% of the proteins. Acidification of the samples is another effective alternative to the organic solvents; a 6:10 ratio of 10% TCA should precipitate more than 99% of the proteins, however the resulting drop in pH can cause analyte instability. A 1:1 (vol/vol) addition of 2 M perchloric acid to a plasma sample precipitates the proteins but

may not release all of the protein bound glucuronides and may not ensure the stabilization of acyl-glucuronides despite the acidic pH. Better results have been obtained with an addition of 1:2 (vol/vol) 15% metaphosphoric acid to a plasma sample (de Loor, Naesens et al. 2008). Addition of metal ions (like zinc sulfate) or hydrophilic salts (like ammonium sulfate) is used less frequently to precipitate the proteins in LC-MS analyses (Rosenfeld 2004). After the precipitation, a centrifugation is used to separate the liquid supernatant from the proteins. Centrifugal field of 15.000 × *g* for 10 minutes is usually enough for complete sedimentation. Supernatant can be directly injected or evaporated and then reconstituted.

3.5 Hybrid SPE

A novel sample clean-up procedure has been recently introduced, the Hybrid SPE, which should combine the advantages of both the protein precipitation and SPE. It is said to be quick, simple and should provide extracts almost free of phospholipids, one of the main culprits for matrix effects. The proteins in samples are precipitated with a triple volume of acetonitrile, then passed through the hybrid SPE cartridges, and directly injected into the LC-MS. The hybrid SPE cartridges should firmly bind phospholipids and eliminate them from the sample. Unfortunately, the glucuronides can bind to the same sorbent very efficiently; therefore the glucuronide recovery may be close to 0% (author's own experience with haloperidol glucuronide, (Silvestro, Gheorghe et al. 2011)). Hence, this method may be used in applications, where a complete elimination of glucuronide is needed.

3.6 Direct injection of biological samples

The recent advances in increasing the sensitivity of triple quad mass spectrometers for 10 times and more made it possible to directly inject some liquid biological samples (like for example urine or saliva) into LC-MS/MS systems almost without any sample pre-treatment. Actually, at least the particulates have to be removed from the samples somehow prior to their injection to prevent injector fluid path blockage, column deterioration and pressure build-up. Narrow bore UHPLC capillaries and columns are especially sensitive to particulate impurities because of their finer pore diameters and dense inlet filters. Therefore, at least some sort of sample pre-treatment is necessary, like diluting the samples 1:10 with water in combination with filtering through 0.2 µm pore filters or centrifugation (McMaster 2007; Kaklamanos, Theodoridis et al. 2011). Such an approach without any form of extraction greatly reduces the time needed for an analysis; however the method and system robustness parameters should be carefully monitored. Furthermore, the majority of biological samples may contain at least some amount of proteins, which may pass through the aforementioned filters and precipitate later on the column or in the tubing when the percentage of organic modifier increases. This difficulty can be overcome by two approaches. First, an on-line SPE can be used to wash the endogenous impurities from the sample into waste before they can reach and contaminate the analytical column and mass spectrometer (see subheading 3.6 (On line-SPE) for details). Second, the samples can be incubated with proteases in order to degrade the proteins into peptides which are no longer a threat to a LC-MS system (Yu, Ho et al. 2005).

3.7 On-line SPE

On-line SPE is a system which allows injecting crude samples or diluted samples onto an extraction column, which is first washed with a weak solvent to waste to remove the salts and proteins and then eluted with a stronger solvent onto the analytical HPLC or UHPLC column. The key to this separation procedure is the turboflow chromatography. Sometimes, it is also designated as a size-exclusion chromatography or a restricted access material (RAM) or a high turbulence liquid chromatography (HTLC). This process takes place in a short extraction column (for example 30x1, 25x4 mm) with RP silica or polymeric based stationary phase with large particles and pores (25-50 µm) for greater robustness (Bentayeb, Batlle et al. 2008). After the injection, the flow rate is quite high (1-5 mL/min), so the proteins and other macromolecules cannot enter the pores of the stationary phase and are quickly washed away into waste. Smaller molecules however can enter the pores and are therefore retained until a stronger solvent elutes them (usually in reverse flow direction) onto the analytical column. A dilution step in a high pressure mixing tee may be employed during the transfer of the analytes from the extraction onto the analytical column in order to concentrate the analytes in a narrow band on analytical column for narrower peak shape, resulting in better sensitivity (Ye, Kuklenyik et al. 2005). The hardware setup for such an on-line SPE is usually a modified HPLC with a larger sample loop (injection volumes are usually 10-900 µL), an additional binary or quaternary pump, and an additional six- or ten-port valve. If a 10-port valve is used, two extraction columns can be used at the same time to increase throughput; this allows a combination of sample loading, washing, equilibration and elution steps simultaneously (Ferreirós Bouzas, Dresen et al. 2009). Typically, on-line SPE columns can withstand a few hundred injections of diluted plasma or urine samples, it depends on the sample matrix and injection volume (Rosenfeld 2004). Care must be taken to disrupt the protein binding prior to the injection; otherwise the recovery may be diminished. Therefore, some authors use a protein precipitation step, followed by evaporation and reconstitution before the samples are injected into an on-line SPE (Ferreirós Bouzas, Dresen et al. 2009)

3.8 Dried blood spots

Dried blood spots (DBS) and dried plasma spots (DPS) are novel approaches for quantitative determination of analytes in systemic circulation. Very low sample volumes are needed (as low as 15 µL up to 100 µL) for a successful quantitative HPLC-MS/MS method (Spooner, Lad et al. 2009). Plasma or blood is applied onto a paper sorbent; after drying, a punch from DBS (or DPS) sample is extracted with a solvent, which is injected in HPLC-MS/MS. This method offers some significant benefits over the classical sample preparation techniques: very low volume of plasma or blood is needed, which minimizes the patient burden (finger prick or heel prick sampling instead of venous cannula) and is very appropriate for animal and pediatric studies; the stability/transport/storage of samples are a lot simpler because no refrigeration is needed and in case of DBS, also no centrifugation at the site of the clinical trial is required. Furthermore, when the stability of paracetamol glucuronide and sulfate was studied in DBS samples, no noticeable degradation and paracetamol formation was observed (Spooner, Lad et al. 2009). This technique can be automated for high throughput analysis in pharmaceutical drug discovery and clinical trials (Barfield and Wheller 2010). Apart from classical HPLC-ESI-MS configuration, some direct DPS analysis tools have been introduced, such as the thin layer chromatography MS interface (CAMAG) which does not require punching of DPS samples (Abu-Rabie and Spooner 2009), direct elution or on-line desorption (Deglon, Thomas et al. 2009), desorption electrospray ionization (DESI) and direct analysis in real time (DART) (Takats, Wiseman et al. 2004). It should be emphasized however, that all the presented direct DPS analysis techniques without any chromatographic separation are principally unsuitable for glucuronide analysis due to the possible in-source deconjugation of glucuronides or sulfates or even N-oxides back to their parent molecules and ion channel cross-talk (Wong, Pham et al. 2010).

3.9 Glucuronide derivatization

Some of the more hydrophilic glucuronide conjugates (like paracetamol-glucuronide) exhibit a poor ionization either due to their hard-to-ionize aglycones or due to the slow desolvation caused by the low percentage of organic solvent in mobile phase composition required for their chromatographic retention. In such instances, two approaches may be explored to overcome the low glucuronide signals. First, the hydrophilic interaction chromatography (HILIC) or a Hypercarb™ column may be a good option to increase the percentage of organic solvent for elution of glucuronides (see subheading 4.4). Secondly, after the extraction, the glucuronides can be derivatized at their carboxylic moiety using an ethereal diazomethane in methanol. Methyl esters are formed at room temperature. Afterwards, the samples are dried and reconstituted with methanol and injected into LC-APCI-MS, where a positive or negative ionization mode can be used to detect either $[M+H]^+$ or $[M+O_2]^-$ ions, respectively (Ohta, Kawakami et al. 2003).

3.10 Glucuronide stability issues

The stability of any type of glucuronide (O-, N-, S-, acyl-) should be thoroughly investigated at all stages of the analysis: sample acquisition and storage, during the sample preparation and after the sample preparation (autosampler time included). It has to be emphasized that even if the glucuronides are quantified indirectly, with β-glucuronidase treatment, the stability of glucuronides is crucial prior to this step, otherwise the glucuronides may be underestimated and the aglycone overestimated. The stability study is especially important for acyl- and N-glucuronides, which can be highly labile compounds in slightly alkaline and even neutral solutions (Mullangi, Bhamidipati et al. 2005). Even acidic additives like perchloric acid may not provide a good-enough stabilization of the acyl-glucuronides. With acyl-glucuronides, either a hydrolysis or an acyl migration can occur. Acyl migration is a process, where the aglycone is being transferred from the C-1 on glucuronide ring to C-2, C-3, C-4 carbon atoms. Such isomeric glucuronides become more stable towards β-glucuronidases and require alkaline conditions for their hydrolysis. Usually, the samples are acidified if acyl-glucuronides are present. Low temperature, such as 4°C may play a major role for acyl-glucuronide stability (Tan, Jin et al. 2009). Plasma and tissue samples may contain endogenous β-glucuronidases which may compromise even O-glucuronide stability.

4. Chromatographic methods

It is imperative to achieve a good chromatographic separation of conjugates (glucuronides) not just from the solvent peak but also from their parent molecules, otherwise they can de-conjugate in MS ion source (ESI, APCI) or in the collision cell (ion channel cross-talk), thereby increasing the signal of the parent, which may cause its false over-estimation (Naidong, Lee et al. 1999). This is why Jemal et al. suggest to use an incurred plasma sample (plasma after dose which still contains metabolites) during the method development in order to ascertain that the developed method for a drug is not compromised by their presence either by chromatographic co-elution and channel cross-talk or by deconjugation during the sample handling (Jemal, Ouyang et al. 2010). The glucuronides often prove to be difficult analytes for HPLC separations. First, in most instances, they are rather hydrophilic, and therefore they tend to elute near the peak of the solvent or the hydrophilic matrix components. At the same time, the strongest ion suppression is often observed at the beginning of the chromatographic run. Secondly, if more than one glucuronide structural isomers are present in a sample, i.e. the glucuronide acid moiety can be bound to different nucleophilic centers on the aglycon, then the chromatographic separation may be difficult and requires a careful gradient elution (Trontelj, Bogataj et al. 2007; Trdan, Roskar et al. 2011). Furthermore, the chromatographic separation of glucuronide structural isomers is crucial, as they share the same parent mass to charge ratio and the fragmentation pattern; therefore a tandem mass analyzer cannot distinguish among the different co-eluting isomers. Ultra high pressure liquid chromatography (UHPLC) seems to offer some significant advantages over the classical HPLC in terms of higher selectivity towards the glucuronide isomers, a greater sensitivity and a higher speed of analysis (Korfmacher 2009).

4.1 Reversed phase chromatography

Reversed phase HPLC and UHPLC on C-18 columns has been used quite successfully for glucuronide separation (Kemp, Fan et al. 2002; Trontelj, Bogataj et al. 2007; Trdan, Roskar et al. 2011). The mobile phase is usually neutral to slightly acidic (pH 2.5-6) and the buffers used are 0.01 – 0.1% formic or acetic acid. 2-5 mM ammonium acetate with or without acetic acid addition may also be used. The organic phase is usually acetonitrile because in comparison to methanol, acetonitrile may provide a better peak shape of basic compounds and glucuronides, a better resolution among the structural glucuronide isomers and a higher MS response (author's own experience). Usually, a gradient is started at a very low percentage of organic modifier (2-10%) in order to assure an adequate retention of the glucuronides and a good separation from the hydrophilic matrix components (Kemp, Fan et al. 2002; Fayet, Béguin et al. 2009). Ballistic gradients are not optimal for glucuronide HPLC methods if the samples may also contain the parent molecule or glucuronide isomers. Ion channel cross-talk and in-source deconjugation, can both cause inaccuracy in quantification of the parent if it is not adequately separated from conjugates (Mei and Morrison 2009). Furthermore, greater matrix effects may also be observed in fast or ballistic gradients due to the poorer separation of analytes from the matrix components (Chambers, Wagrowski-Diehl et al. 2007). A more efficient chromatographic separation of analytes from the endogenous or other interfering components, such as drug vehicles, PPG, PEG or even over-the-counter medicines like analgesics, may be achieved with UHPLC columns resulting in lesser matrix effects (Marín, Gracia-Lor et al. 2009).

4.2 Ion-pairing reagents

Although ion-pairing reagents have proven to be beneficial for glucuronide retention (Yau, Vathsala et al. 2004), such techniques have rarely been used in LC-MS analyses due to the limited compatibility of ion-pairing reagents with ionization and desolvatation in LC-MS interface. A successful and robust application of hexylamine as an ion pairing reagent together with a careful control of mobile phase pH has been described for acidic metabolites retention and detection with LC-negative-ESI-MS. In this case, no interference was observed from hexylamine as it is a volatile reagent and ionizes only in the positive mode (Coulier, Bas et al. 2006). With a similar ion pairing reagent, dimethylhexylamine, a switch from the negative to the positive ESI is possible, which increases the sensitivity and selectivity of the method (Pruvost, Théodoro et al. 2008). Interestingly, even the addition of ion-pairing reagents just to the sample solution or to a reconstitution solvent (and not to the mobile phase) may significantly affect the retention of polar analytes. For example, a retention time of methadone has been doubled when the heptafluorobutanoic acid (HFBA) had been added to the reconstitution solvent after the extraction. Moreover, the addition of this ion-pairing reagent significantly reduced the carry-over, probably because it reduced the attachment of the analyte to LC surfaces (Gao, Bhoopathy et al. 2006).

4.3 Other reversed stationary phases

Apart from the classic RP stationary phases, like the C-18 or C-8, phenyl group containing sorbents can be used in order to enhance the retention of mainly aromatic analytes through the formation of PI-PI bonds in addition to the regular mechanisms of RP chromatography (adsorption, partition, lipo- and solvophilic interactions) (Rafferty, Zhang et al. 2007). Additional selectivity enhancement may be observed when the percentage of methanol is increased in the organic modifier phase (in combination with acetonitrile), since the methanol can increase the formation of the aforementioned PI-PI interactions. Therefore, phenyl phases, such as the phenyl-hexyl, phenyl-butyl, biphenyl, pentafluorophenyl can be used to effectively increase the retention of hydrophilic aromatic metabolites, like for example paracetamol-glucuronide and sulfate (Xiong, Jin et al. 2010), or cyanidine-glucuside-glucuronides (Marczylo, Cooke et al. 2009). Phenyl columns have been successfully used for more lipophilic metabolites as well, like for example cannabinoid glucuronides (Mazur, Lichti et al. 2009).

4.4 Zwitter ionic and hydrophilic interaction chromatography (HILIC)

HILIC is a mode of separation similar to the normal phase chromatography, but it can be performed on an ordinary RP HPLC instrument using a HILIC column and a mobile phase similar to the RP chromatography. In HILIC, the stationary phase is polar and the mobile phase is an aqueous/polar organic solvent, where water (or buffer) plays the role of a stronger eluting solvent. The order of peaks eluting from a HILIC column is reverse compared to RP chromatography, and glucuronides elute after their aglycones. Therefore, HILIC can be used primarily for the increased retention and separation of highly polar analytes. For example, a successful application of HILIC mechanism for morphine glucuronides separation and ionization has been described more than a decade ago (Naidong, Lee et al. 1999). The retention mechanism in underivatized silica is not only hydrophilic, but also hydrophobic and ion-exchange, which may cause a mixed-mode retention, and is sometimes difficult to predict (Nguyen and Schug 2008). In derivatized silica with a zwitterionic bonded phase (commercially known as ZIC-HILIC), there is no ion-exchange mechanism of retention (the zwitterions in stationary phase cancel each other out and there is no net charge of ion-exchange (Ahuja and Dong 2005)). For some phase I metabolites and parent compounds that are highly hydrophobic or less polar than glucuronides, the HILIC column may not provide enough retention and these compounds may be eluted close to the solvent peak. To overcome this issue, a dual column orthogonal approach has been described by (Qin, Zhao et al. 2008), where the estrogens have been derivatized with dansyl chloride (for better ionization) and separated on a RP column, while the polar estrogen conjugates were separated on a ZIC-HILIC column. In step one, the sample is divided into two fractions: polar and non-polar: first, both columns are in series, and a sample is injected onto a RP column; the minimally retained polar conjugates are eluted from a RP column onto a HILIC column for subsequent separation in the next step, while the non-polar derivatized estrogens remain parked on the RP column. In step two, the mobile phase flows only through the HILIC column, where the polar analytes are separated and eluted to MS. In step three, the flow is directed only through the RP column, where the non-polars are separated and eluted to MS. This 2D orthogonal approach combines the benefits of both RP and HILIC methods and provides an excellent sensitivity (Thomas, Déglon et al. 2010).

5. Mass spectrometry

Glucuronides can be analyzed by LC-MS/MS in a number of different MS fragmentation experiments. The analyst will choose an appropriate MS experiment based on the purpose of the study.

5.1 Ionization of glucuronides

Before an analyte can enter a mass spectrometer and produce a signal, it must be efficiently ionized first. The majority of documented methods for glucuronide LC-MS detection employ the electrospray ionization (ESI) as it is the softest and the most suitable method for ionization without the in-source fragmentation of the relatively weak glucuronide ether or ester bonds (Kadi and Hefnawy 2009). On the other hand, sometimes it is desired to achieve an in-source fragmentation in order to record just one Selected Reaction Monitoring (SRM) trace for the glucuronide and the parent to achieve a better sensitivity (Felli, Martello et al. 2011), (Schwartz, Desai et al. 2006). Of course, in such cases it is imperative to achieve a good chromatographic separation from the parent. The conjugate metabolites (glucuronides included) are usually ionized in positive ionization with electrospray (+ESI) (Keski-Hynnilä, Kurkela et al. 2002; Levsen, Schiebel et al. 2005). Generally, the glucuronides should not be problematic also in the negative ionization mode due to the presence of the acidic carboxylic group on glucuronic moiety with the reported pKa of around 3 (Farrell, Poquet et al. 2011). However, the negative ionization is often less sensitive than the positive one. Furthermore, fragmentation mechanisms in negative ion mode are frequently hard to predict and understand. Moreover, in negative ionization the specificity may be worse than in the positive mode (Pruvost, Théodoro et al. 2008). On many occasions, it is therefore advisable to try to develop a method for glucuronide quantification using the positive ionization mode, especially if the aglycone is a proton-acceptor (a weak base). In positive-ion mode, an abundant fragment $[M + H - 176]+$ is detected in most cases after a glucuronide fragmentation; the neutral loss of 176 Da corresponds to a monodehydrated glucuronic acid. Acyl- or benzyl glucuronides, however, may undergo a loss of 194 Da (glucuronic acid) from the pseudo- or quasi-molecular ions, either in addition to the loss of 176 Da or even exclusively (Levsen, Schiebel et al. 2005).

Sometimes however, the positive ionization mode cannot be used if the aglycone is not a base or does not accept a proton or any other positive adduct-forming ions, like for example Na^+, K^+, NH_4^+ etc. In such cases, it is necessary to use the negative ionization or switch to another type of ionization. If analyte derivatization has been employed to increase the ionization efficiency like for example with steroids or with paracetamol glucuronide (Ohta,

Kawakami et al. 2003), then the negative ionization mode may be even superior in sensitivity to the positive mode because the noise in negative mode may be lower, thereby increasing the signal/noise ratio (Higashi and Shimada 2004).

Atmospheric pressure chemical ionization (APCI) interface may sometimes be beneficial over the ESI if the principally lower sensitivity of APCI is still adequate for the intended study. The APCI may be less prone to matrix effects due to plasma components and salts (Schwartz, Desai et al. 2006), especially if isotope labeled internal standard for glucuronide is not available (Matuszewski, Constanzer et al. 1998; Matuszewski, Constanzer et al. 2003). Furthermore, APCI seems to have a wider dynamic response range than ESI (Ramanathan, Comezoglu et al. 2009).

Novel LC-MS interfaces include APPI (atmospheric photo ionization, which is similar to APCI and may be important for less polar analytes; further, nanospray ionization (NSI) coupled with nano flow LC may provide a superior limit of quantification and reduced matrix effects (Ramanathan, Comezoglu et al. 2009)

5.2 Mass spectrometry operation modes for glucuronide detection

Tandem MS is the key principle for metabolite structure elucidation and quantification (Clarke, Rindgen et al. 2001). There are many layouts and types of mass filters or analyzers for selecting the parent and daughter ions and the collision cells. For example just to name a few major types of mass analyzers: quadrupole/linear ion trap, triple stage quadrupole (QQQ), time of flight (TOF), hybrid instruments (QTOF, QTRAP) and others. For detection and quantification of glucuronides in complex biological matrices, the triple quadrupole mass spectrometer is usually the instrument of choice due to the high selectivity and superior sensitivity (Mullangi, Bhamidipati et al. 2005). Therefore, our brief review will focus on mass experiments performed typically in a QQQ type of instrument.

Precursor ion scan mode is usually used at the start of conjugate identification; aglycone is selected as the product by Q3 mass filter and Q1 is scanning for the possible conjugates that may give rise to the selected parent fragment after a collision induced dissociation (CID). Such an experiment may reveal for example glucuronides, glucosides, sulfates and glutathione conjugates with mass shifts of m/z 176, 80, 162, 305 Da, respectively (Levsen, Schiebel et al. 2005; Ramanathan, Comezoglu et al. 2009).

Constant neutral-loss experiment is the most suitable mode experiment for searching the selected or expected conjugates in a sample as it detects the loss of a neutral fragment. Here, both Q1 and Q3 are scanning at the same time, with a pre-selected mass shift between them. Therefore, the analyst does not need to know the exact mass of the conjugate or its parent (it may undergo an unknown phase I metabolic conversion prior to the conjugation). The result of such an experiment is the *m/z* of the parent molecule before the CID and the resulting mass shift of 176 Da (for glucuronic acid), similar to the precursor ion scan experiment, described in the upper paragraph. In neutral loss experiment and in positive ionization mode, the glucuronides will produce a nominal shift of *m/z* 176 (176.0321 on accurate mass analyzers) (Ramanathan, Comezoglu et al. 2009). In negative ionization, the glucuronides produce ions at *m/z* 175 (anhydrous glucuronic acid) and 113 (a fragment of glucuronic acid) (Fay 2010).

Product ion scan experiment is performed on a single pre-selected ion in Q1, which is fragmented in the collision cell and daughter ions are scanned by Q3. The result of such an experiment is a full scan product ion spectrum, which may also contain unfragmented parent ion (figure 1).

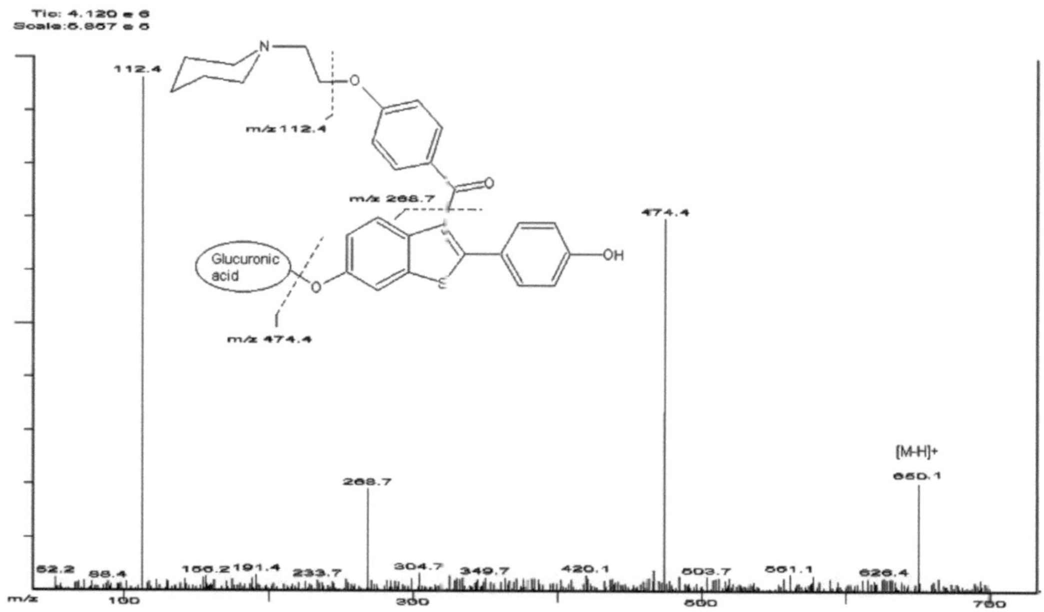

Fig. 1. A product ion scan from raloxifene-6-glucuronide (*m/z* 650.1), aglycone fragment (*m/z* 474.4), phenyl benzotiophene ring (*m/z* 268.7) and N-ethyl-piperidine fragment (*m/z* 112) (Trontelj, Bogataj et al. 2007). With permission.

Usually, the full scan mode does not provide sufficient sensitivity in a triple quadrupole MS for a proper glucuronide detection or quantification in a complex matrix (Fay 2010).
The Selected Reaction Monitoring (SRM) and the Multiple Reaction Monitoring (MRM) modes are the techniques of choice for selective and sensitive glucuronide quantification. Here the Q1 and Q3 are set at the selected *m/z* values for conjugate (glucuronide) and its parent molecule, respectively. Additional specificity for isobaric ions in MS detector can be provided by the high-field asymmetric waveform ion mobility spectrometry (FAIMS) module, which is inserted between the ion source and the high vacuum MS inlet. The principle of this additional separation technique is that the ion mobility at high electric fields can differ according to the shape of the ions. Therefore, FAIMS can separate isomeric or isobaric analytes that the LC and ordinary MS cannot (Hatsis and Kapron 2008). On a model acyl-glucuronide of ifetroban Xia et al. have demonstrated the use of FAIMS for locating the place where the acyl glucuronide is being fragmented to its parent inside the mass spectrometer (Xia and Jemal 2009). FAIMS method is also one of the confirmatory methods for detection of testosterone glucuronide in urine after a hydrolysis and immuno-affinity separation coupled to LC-MS/MS (Fong-Ha and et al. 2010). Guddat et al. reported a better interference removal with FAIMS in an anti-doping testing of human urine, which should effectively extend the post-dose detection time of androgen steroids and their glucuronides (Guddat, Thevis et al. 2009).

5.3 Other mass spectrometry approaches used in glucuronide characterization

The described mass experiments in previous subparagraph can be performed also on some other tandem instruments as well, like for example on ion traps, where the CID happens in time rather than in space, as in triple quads. The sensitivity of ion traps is usually lower compared to QQQ instruments, however, they tend to be cheaper. Furthermore, with ion traps, MS^n experiments can be performed for possible structure elucidation.

QTRAP, which is a hybrid between a triple quad and an ion trap, can perform all the described tandem mass experiments and more - there is the possibility to perform MS^n experiments (multiple stages of fragmentation) which may be beneficial for glucuronide structure elucidation, but lacks the mass accuracy of a Q-TOF. Information dependent acquisition (IDA) can trigger the product ion scans (MS^2 and MS^3) when a certain ion intensity threshold is reached in full scan or neutral loss scan experiment. Therefore, IDA should shorten the cycle time and provide some structural information in a minimal number of analytical runs.

The mass spectrometers that enable accurate mass measurements in LC/MS are time-of-flight instruments and their hybrids (TOF), Q-TOF, TOF-TOF and Orbitrap or Fourier transform ion cyclotron resonance mass spectrometers (FTICRMS) (Tolonen, Turpeinen et al. 2009). It is noteworthy that an accurate mass instrument like Orbitrap can be used for quantification and glucuronide detection comparable to a middle range QQQ because of the Orbitrap's extreme selectivity due to its high resolution of 15000 and more (mass/FWHM) (Zhang, Yu et al. 2009).

Q-TOF instruments are also capable of tandem mass experiments. These hybrid systems offer an increased mass accuracy and mass resolution of TOF and provide a good fragmentation capability. Furthermore, full scan mode in TOF mass analyzers provides an increased sensitivity compared to a QQQ. Therefore, the product ion scan experiment on Q-TOF offers significant benefits in terms of accuracy and sensitivity, which may help in element and structural elucidation for isomeric glucuronides, like for example multiple estriol glucuronides (Lampinen-Salomonsson, Bondesson et al. 2006). On the other hand, precursor ion scan and neutral loss experiments cannot be performed on a Q-TOF (Hoffmann and Stroobant 2007).

A good example of complimentary approach by different mass spectrometers (QQQ, ion trap, Q-TOF) is presented on a complicated mixture of estriol glucuronide isomers and their structure elucidation (Lampinen-Salomonsson, Bondesson et al. 2006).

5.4 Data acquisition and ion channel cross-talk

For glucuronide quantification, the ion source settings are important for efficient ionization. However, the parameters of quadrupoles and detector in MRM mode are also important for LC-MS selectivity and especially, sensitivity. If analyte concentration level and sensitivity of the instrument permit it, the mass resolution of both Q1 and Q3 should be set to at least one unit (0,7-1.0) Da full width at half maximum (FWHM)). If the concentration level of glucuronides is very low, then increasing the mass window of both quadrupoles to up to 3 units may increase the signal intensity up to 10 times and more (author's own experience) without the equal increase in noise at the high *m/z* values as can be expected for glucuronide conjugates. However, the selectivity of the method may suffer. Nowadays, triple quadrupoles with enhanced resolution exist which offer a resolution of ≤ 0.1 FWHM and may provide a significant increase in selectivity from the interfering substances (like PPG, PEG) and a minimal loss in sensitivity (3 times lower signals, but lower noise as well) (Yang, Amad et al. 2002).

Another important parameter is the dwell time (the time at which quadrupole stays at one *m/z* value). The greater the dwell time, the better is the sensitivity and the slower rate of data

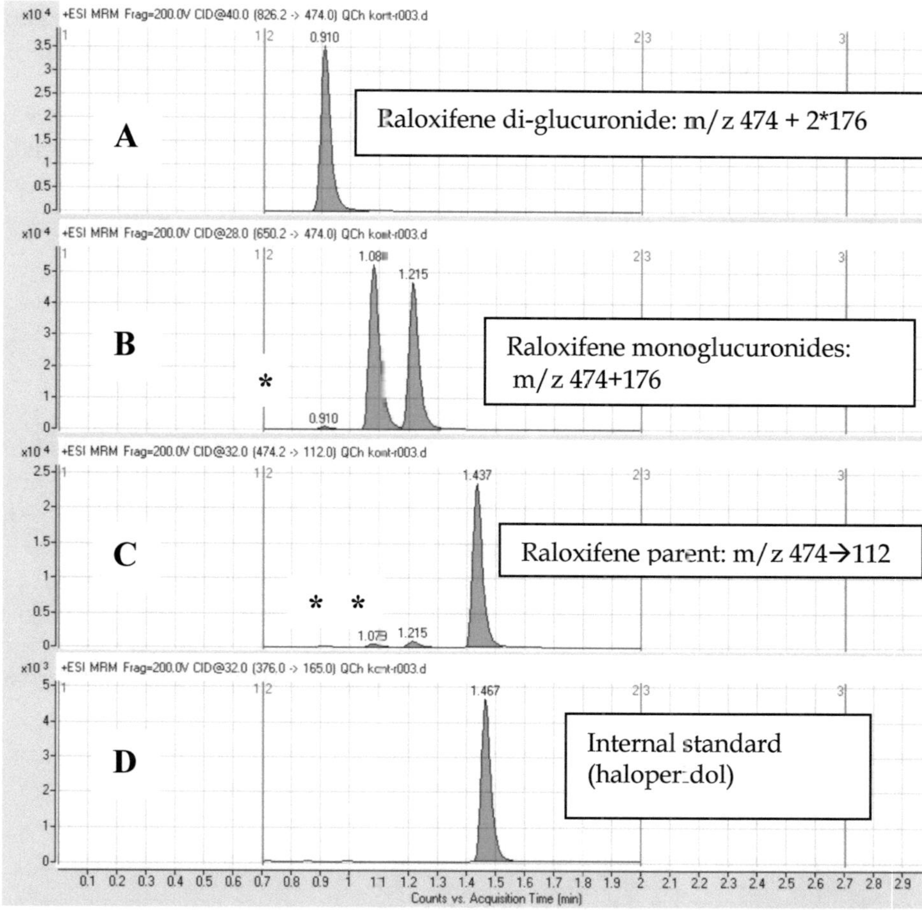

Fig. 2. MRM chromatogram of an extracted urine sample containing raloxifene di-glucuronide (trace A), two isobaric raloxifene monoglucuronides (trace B), parent raloxifene (trace C) and haloperidol (D) as an internal standard. Ion channel cross-talk is evident in traces B in C, where the little peaks in front of analytes (marked as stars *) present cross talk and originate from different analytes (producing the same daughter ion fragments).

acquisition. With shorter dwell times, the probability of ion-channel cross talk also increases. Ion-channel cross talk is a false positive signal in a SRM channel that originates from another analyte. A typical example is the signal of aglycone in the SRM trace of its conjugate (Figure 2, trace C with two peaks originating from cross-talk). Typical values of dwell times are 1 to 250 ms (Lee, Zhu et al. 2011) However, with modern UHPLC columns which can produce very high and narrow peaks (often less than 5 s wide at half height) (Jemal, Ouyang et al. 2010), the data acquisition rate should be fast enough to provide at least 10-15 points per peak. Not only the dwell time, also the cycle time (the summation of dwell times of all MRMs and pause times between them) is extremely important for data acquisition rate. Therefore, for the fastest cycle times with long-enough dwell times it is advisable to use the scheduled MRM mode if ultimate sensitivity is required. Again, an effective and robust chromatographic separation of analytes is vital. In glucuronide analysis, ion channel cross-talk is often observed even with high-end QQQ instruments where the commercial

brochures claim that there is no cross talk even at 1 ms dwell time (author's own experience, see figure 2). This however is of no real concern if the analytes are baseline separated chromatographically.
The detector voltage (or delta EMV as it is sometimes called) has also a big influence on the signal intensity. However, with increasing the detector voltage, the rise in background noise is almost certain as well. Furthermore, the dynamic range of the detector may be narrower with higher voltage and the lifetime of detector may be shortened by using high voltages over longer periods of time.

5.5 Matrix effects

The most problematic matrix components for LC-MS/MS analyses are supposed to be the phospholipids, which are rather lipophilic and therefore, late-eluting (Chambers, Wagrowski-Diehl et al. 2007). The glucuronides on the other hand, are usually more hydrophilic and fast eluting. Yet, they may still be very susceptible to interferences from the co-eluting matrix components (author's own in-house experiences, data not shown). Not only the phospholipids, but also earlier eluting lysophospholipids can cause substantial matrix effects (Xia and Jemal 2009), which may interfere with the quantification of drugs and their metabolites (Jemal, Ouyang et al. 2010). Especially difficult matrices are plasma and urine, where even after a LLE or SPE clean-up, major matrix effects can be observed.
Recovery (RE) and matrix effect (ME) can be calculated from extraction experiments of blank matrix which is pre- and post- spiked by using equations 1 and 2 (Matuszewski, Constanzer et al. 2003), see also paragraph 3.2.

$$RE = \frac{\text{pre-spiked}}{\text{post-spiked}} * 100\% \qquad \text{Eq. 1}$$

$$ME = \left(1 - \frac{\text{post-spiked}}{\text{solvent-only}}\right) * 100\% \qquad \text{Eq. 2}$$

A similar method for calculation of the matrix effect and recovery can also be used with other types of extraction and protein precipitation. Note that the presented method implies the calculation of absolute matrix effect. However, much more important for bioanalytical quantification is the *relative* matrix effect, caused by interindividual variability in the sample matrix (Matuszewski, Constanzer et al. 2003). Therefore, at least 5 matrices from individual donors have to be used to asses the relative matrix effect either by a post spike method presented above or by comparing the slopes of calibration lines in those matrices. The coefficient of variation of slopes should not exceed 4-5% (Matuszewski, Constanzer et al. 2003). The use of an isotope labeled aglycone as an internal standard does not necessarily correct the matrix effect for glucuronide as well, because the retention times between both analytes are supposed to be different and the nature and/or concentration of the co-eluting matrix components are very likely different as well (Taylor 2005). Therefore, to *thoroughly* compensate the matrix effects for glucuronides, their labeled analogues should be used. Even with labeled internal standards, the analyst should be cautious, since the purity of a labeled standard may not be 100% and studies should be performed to confirm the isotope integrity of the label in a sample matrix and during the sample handling. Furthermore, the possible presence of an ion channel cross-talk between the analyte and the labeled standard

SRMs must be investigated as same fragments can be formed from the analyte and the internal standard (Chavez-Eng, Constanzer et al. 2002; Matuszewski, Constanzer et al. 2003). Alternatively, the β-glucuronidase approach (paragraph 3.1) can be used to transform the glucuronide into its parent molecule, for which an isotope labeled analogue may be available. If more than one glucuronide isomers are present in a sample then the β-glucuronidase approach cannot be used directly to quantify each of the present glucuronides. In such cases, the response factors for each glucuronide has to be determined based on isolated β-glucuronidase incubations for each isolated glucuronide. The matrix effects in such cases cannot be overcome with a simple use of the labeled analogue of the parent aglycone, because the quantification of glucuronides has to be performed directly with their MRMs and not *via* their degradation to parent aglycone.

6. Some selected examples

In this section, a brief overview of the current literature is given, regarding the use of a LC-tandem-MS instrument for the quantification of glucuronides in various samples from both *in vitro* and *in vivo* studies, such as microsome and tissue incubations and pharmacokinetic studies or forensic applications (Tables 1, 2).

Analyte	Matrix	Sample preparation	separation technique	Ionization/ type of MS	Range	Reference
Nicotine-N glucuronide[D]	microsome incubation	PP with perchloric acid		(+) ESI / QQQ	10-1000 nM	(Guo, Zhou et al. 2011)
ethyl-glucuronide	human hair	water extraction followed by SPE on activated carbon	UHPLC	(-) ESI QQQ	2pg/g - 300 pg/g hair	(Kronstrand, Brinkhagen et al. 2011)
raloxifene O-glucuronides	human urine	RP polymeric SPE	UHPLC C18	(+)ESI / QQQ	1 -4800 nM*	(Trdan, Roskar et al. 2011)
morphine-O-glucuronides	human urine	RP (C18) SPE	HPLC-HILIC	(+)ESI / TOF	50-5000 μg/L	(Kolmonen, Leinonen et al. 2010)
sorafenib-O-glucuronide[D]	mouse plasma and liver homogenate	PP	HPLC C18	(+)ESI / QQQ	1-500 μg/L	(Sparidans, Vlaming et al. 2009)
rampiril-acyl-glucuronide	plasma	RP (C18) SPE	HPLC C18	(+)ESI / QQQ	1-500 μg/L	(Tan, Jin et al. 2009)
cediranib-N⁺ -glucuronide[D]	micorosomes/ feces	PP / SPE	HPLC C18	(+)ESI / ion trap	?	(Lenz, Spear et al. 2010)
THC-acyl-glucuronide	blood, liver, muscle, urine, bile	ACN homogenization /LLE with ethylacetate/hexane	HPLC C18	(+)ESI	100-1000 μg/L	(Gronewold and Skopp 2011)

[D] signifies the use of a labeled internal standard for glucuronide
* The range was divided into two calibration curves

Table 1. Some selected recent examples of a direct glucuronide measurement approach. For descriptions of abbreviations, please see the text.

Analyte	Matrix	Sample preparation	separation technique	Ionization/ type of MS	Range	Reference
SN-38-O-glucuronide	plasma	PP / β-glucuronid.	HPLC	(+) ESI QTRAP	0.5-200 μg/L	(Zhang, Dutschman et al. 2009)
ezetimibe-O-glucuronide	feces	LLE / tert butyl ether	HPLC C18	(-) APCI /QQQ	0.1-20 mg/L for total ezetimibe	(Oswald, Scheuch et al. 2006)
quercetine-O-glucuronide	rat brain tissue	β-glucuronid. / LLE (ethyl acetate)	HPLC C18	(-) ESI / QQQ	?	(Ishisaka, Ichikawa et al. 2011)
bisphenol A- O-glucuronide	human urine	β-glucuronid. / PP / On-line SPE	HPLC C18	(-) APCI / QTRAP	1-500 pg/μL	(Völkel, Kiranoglu et al. 2008)

Table 2. Some selected recent examples of an indirect glucuronide measurements (by deconjugation). For descriptions of abbreviations, please see the text.

7. Summary

The glucuronide metabolites are important both for toxicology and for pharmacokinetics of many drugs and xenobiotics. Even though glucuronides often lack the pharmacological activity, after de-conjugation, the free aglycone can regain that activity and may present an environmental burden. Furthermore, glucuronides may be used as markers of the past substance exposure, like for example the ethyl-glucuronide in body hair for alcohol abuse or the cannabinoid-glucuronides for cannabis abuse. Therefore, quantification of glucuronide metabolites in biological matrices may be very important. However, due to their highly hydrophilic nature, the separation of glucuronides from the matrix (body fluids, hair, and environmental samples) may be difficult. For that purpose, many sample preparation techniques can be used, including the liquid extraction, solid phase extraction, dried blood spots, protein precipitation, direct injection, on-line solid phase extraction and others. Furthermore, some glucuronides, such as acyl- and N-glucuronides can be highly unstable. Glucuronides can be measured either by a direct or an indirect approach. With the latter, the glucuronides are cleaved back to their parent aglycone by enzymes known as the β-glucuronidases, afterwards the freed aglycone is quantified according to its calibration curve. With direct measurement, unchanged glucuronides are quantified by their MRMs, where some significant benefits can be expected, namely, quicker sample preparation, better accuracy and precision, and selectivity towards glucuronide isomers. Tandem mass spectrometry is the corner stone for direct glucuronide quantification because it offers both exceptional selectivity and sensitivity. However, this approach requires the use of authentic glucuronide standards. This necessity can be avoided with the semi-quantitative approach, where the response factor for glucuronide is used to quantify it against the calibration curve for the parent. The recent advancement of mass spectrometers' speed and sensitivity coupled with higher mass accuracy promises an even easier measurement of glucuronides in complex biological matrices, allowing a faster method development, robust response, and quicker sample preparation. Notwithstanding the described improvements, the glucuronide identification and quantification in difficult matrices will still remain a challenge for the analyst.

8. References

Abu-Rabie, P. and N. Spooner (2009). "Direct quantitative bioanalysis of drugs in dried blood spot samples using a thin-layer chromatography mass spectrometer interface." *Anal Chem.* 81(24): 10275-10284.

Ahuja, S. and M. W. Dong (2005). *Handbook of pharmaceutical analysis by HPLC*. Amsterdam; Boston, Elsevier Academic Press.

Bakhtiar, R. and T. K. Majumdar "Tracking problems and possible solutions in the quantitative determination of small molecule drugs and metabolites in biological fluids using liquid chromatography-mass spectrometry." *Journal of Pharmacological and Toxicological Methods* 55(3): 227-243.

Barfield, M. and R. Wheller (2010). "Use of Dried Plasma Spots in the Determination of Pharmacokinetics in Clinical Studies: Validation of a Quantitative Bioanalytical Method." *Analytical Chemistry* 83(1): 118-124.

Belanger, A. S., P. Caron, et al. (2009). "Glucuronidation of the antiretroviral drug efavirenz by UGT2B7 and an in vitro investigation of drug-drug interaction with zidovudine." *Drug Metab Dispos.* 37(9): 1793-1796. Epub 2009 Jun 1791.

Bentayeb, K., R. Batlle, et al. (2008). "Determination of bile acids in human serum by on-line restricted access material-ultra high-performance liquid chromatography-mass spectrometry." *Journal of Chromatography B* 869(1-2): 1-8.

Blount, B. C., K. E. Milgram, et al. (2000). "Quantitative Detection of Eight Phthalate Metabolites in Human Urine Using HPLC−APCI-MS/MS." *Analytical Chemistry* 72(17): 4127-4134.

Briggs, B. S., P. J. Baker, et al. (1999). "Microbial process for preparation of glucuronides of raloxifene." *Journal of Industrial Microbiology & Biotechnology* 23(3): 194-197.

Burkon, A. and V. Somoza (2008). "Quantification of free and protein-bound trans-resveratrol metabolites and identification of trans-resveratrol-C/O-conjugated diglucuronides - Two novel resveratrol metabolites in human plasma." *Molecular Nutrition & Food Research* 52(5): 549-557.

Chambers, E., D. M. Wagrowski-Diehl, et al. (2007). "Systematic and comprehensive strategy for reducing matrix effects in LC/MS/MS analyses." *Journal of Chromatography B* 852(1-2): 22-34.

Chambers, E., D. M. Wagrowski-Diehl, et al. (2007). "Systematic and comprehensive strategy for reducing matrix effects in LC/MS/MS analyses." *J Chromatogr B Analyt Technol Biomed Life Sci* 852(1-2): 22-34.

Chavez-Eng, C. M., M. L. Constanzer, et al. (2002). "High-performance liquid chromatographic-tandem mass spectrometric evaluation and determination of stable isotope labeled analogs of rofecoxib in human plasma samples from oral bioavailability studies." *Journal of Chromatography B: Analytical Technologies in the Biomedical and Life Sciences* 767(1): 117-129.

Choi, R. J., I. J. Ha, et al. (2010). "Biotransformation of flavonoid-7-O-glucuronides by β-glucuronidases." *Natural Product Sciences* 16(1): 1-5.

Clarke, N. J., D. Rindgen, et al. (2001). "Peer Reviewed: Systematic LC/MS Metabolite Identification in Drug Discovery." *Analytical Chemistry* 73(15): 430 A-439 A.

Cooper, J., W. Currie, et al. (2001). "Comparison of the efficiences of enzymatic and chemical hydrolysis of (nortestosterone and diethylstilboestrol) glucuronides in bovine urine." *Journal of Chromatography B: Biomedical Sciences and Applications* 757(2): 221-227.

Coulier, L., R. Bas, et al. (2006). "Simultaneous Quantitative Analysis of Metabolites Using Ion-Pair Liquid Chromatography−Electrospray Ionization Mass Spectrometry." *Analytical Chemistry* 78(18): 6573-6582.

de Loor, H., M. Naesens, et al. (2008). "Stability of mycophenolic acid and glucuronide metabolites in human plasma and the impact of deproteinization methodology." *Clinica Chimica Acta* 389(1-2): 87-92.

Deglon, J., A. Thomas, et al. (2009). "On-line desorption of dried blood spot: A novel approach for the direct LC/MS analysis of micro-whole blood samples." *J Pharm Biomed Anal.* 49(4): 1034-1039. Epub 2009 Feb 1013.

DongHyun, K. I. M., J. I. N. YoungHo, et al. (1995). "Purification and Characterization of β-Glucuronidase from Escherichia coli HGU-3,a Human Intestinal Bacterium." *Biological & pharmaceutical bulletin* 18(9): 1184-1188 %U http://ci.nii.ac.jp/naid/110003641017/en/.

Elbarbry, F. A. and A. S. Shoker (2007). "Therapeutic drug measurement of mycophenolic acid derivatives in transplant patients." *Clinical Biochemistry* 40(11): 752-764.

Farrell, T., L. Poquet, et al. (2011). "Characterization of hydroxycinnamic acid glucuronide and sulfate conjugates by HPLC-DAD-MS2: Enhancing chromatographic quantification and application in Caco-2 cell metabolism." *Journal of Pharmaceutical and Biomedical Analysis* 55(5): 1245-1254.

Fay, L.-B. (2010). *Mass spectometry and nutrition.* [S.l.], Royal Soc Of Chemistry.

Fayet, A., A. Béguin, et al. (2009). "A LC-tandem MS assay for the simultaneous measurement of new antiretroviral agents: Raltegravir, maraviroc, darunavir, and etravirine." *Journal of Chromatography B* 877(11-12): 1057-1069.

Felli, M., S. Martello, et al. (2011). "LC-MS-MS method for simultaneous determination of THCCOOH and THCCOOH-glucuronide in urine: Application to workplace confirmation tests." *Forensic Science International* 204(1-3): 67-73.

Ferreirós Bouzas, N., S. Dresen, et al. (2009). "Determination of basic drugs of abuse in human serum by online extraction and LC–MS/MS." *Analytical and Bioanalytical Chemistry* 395(8): 2499-2507.

Fong-Ha, L. and et al. (2010). "Final report on CCQM-K69 key comparison: Testosterone glucuronide in human urine." *Metrologia* 47(1A): 08018.

Fontanals, N., R. M. Marcé, et al. (2010). "Mixed-mode ion-exchange polymeric sorbents: dual-phase materials that improve selectivity and capacity." *TrAC Trends in Analytical Chemistry* 29(7): 765-779.

Frandsen, H. (2007). "Deconjugation of N-glucuronide conjugated metabolites with hydrazine hydrate - Biomarkers for exposure to the food-borne carcinogen 2-amino-1-methyl-6-phenylimidazo[4,5-b]pyridine (PhIP)." *Food and Chemical Toxicology* 45(5): 863-870.

Gao, S., S. Bhoopathy, et al. (2006). "Evaluation of volatile ion-pair reagents for the liquid chromatography-mass spectrometry analysis of polar compounds and its application to the determination of methadone in human plasma." *Journal of Pharmaceutical and Biomedical Analysis* 40(3): 679-688.

Gomes, R. L., W. Meredith, et al. (2009). "Analysis of conjugated steroid androgens: Deconjugation, derivatisation and associated issues." *Journal of Pharmaceutical and Biomedical Analysis* 49(5): 1133-1140.

Gronewold, A. and G. Skopp (2011). "A preliminary investigation on the distribution of cannabinoids in man." *Forensic Science International* 210(1-3): e7-e11.

Gu, L., M. Laly, et al. (2005). "Isoflavone Conjugates Are Underestimated in Tissues Using Enzymatic Hydrolysis." *Journal of Agricultural and Food Chemistry* 53(17): 6858-6863.

Guddat, S., M. Thevis, et al. (2009). "Application of FAIMS to anabolic androgenic steroids in sport drug testing." *Drug Testing and Analysis* 1(11-12): 545-553.

Guo, J., D. Zhou, et al. (2011). "Liquid chromatography-tandem mass spectrometry method for measurement of nicotine N-glucuronide: A marker for human UGT2B10 inhibition." *Journal of Pharmaceutical and Biomedical Analysis* 55(5): 964-971.

Hatsis, P. and J. T. Kapron (2008). "A review on the application of high-field asymmetric waveform ion mobility spectrometry (FAIMS) in drug discovery." *Rapid Communications in Mass Spectrometry* 22(5): 735-738.

Higashi, T. and K. Shimada (2004). "Derivatization of neutral steroids to enhance their detection characteristics in liquid chromatography–mass spectrometry." *Analytical and Bioanalytical Chemistry* 378(4): 875-882.

Himeno, M., Y. Hashiguchi, et al. (1974). "β-Glucuronidase of Bovine Liver." *Journal of Biochemistry* 76(6): 1243-1252.

Hochner-Celnikier, D. (1999). "Pharmacokinetics of raloxifene and its clinical application." *Eur J Obstet Gynecol Reprod Biol.* 85(1): 23-29.

Hoffmann, E. d. and V. Stroobant (2007). *Mass spectrometry : principles and applications.* Chichester, J. Wiley & Sons.

Ishisaka, A., S. Ichikawa, et al. (2011). "Accumulation of orally administered quercetin in brain tissue and its antioxidative effects in rats." *Free Radical Biology and Medicine* In Press, Corrected Proof.

Isobe, T., S. Serizawa, et al. (2006). "Horizontal distribution of steroid estrogens in surface sediments in Tokyo Bay." *Environmental Pollution* 144(2): 632-638.

Jemal, M., Z. Ouyang, et al. (2010). "Systematic LC-MS/MS bioanalytical method development that incorporates plasma phospholipids risk avoidance, usage of incurred sample and well thought-out chromatography." *Biomedical Chromatography* 24(1): 2-19.

Kadi, A. A. and M. M. Hefnawy (2009). "Biological Fluids: Glucuronides from LC/MS." *Encyclopedia of Chromatography, Third Edition*: 203 - 209.

Kakimoto, K., A. Toriba, et al. (2008). "Direct measurement of the glucuronide conjugate of 1-hydroxypyrene in human urine by using liquid chromatography with tandem mass spectrometry." *Journal of Chromatography B* 867(2): 259-263.

Kaklamanos, G., G. A. Theodoridis, et al. (2011). "Determination of anabolic steroids in bovine serum by liquid chromatography-tandem mass spectrometry." *Journal of Chromatography B* 879(2): 225-229.

Kemp, D. C., P. W. Fan, et al. (2002). "Characterization of raloxifene glucuronidation in vitro: contribution of intestinal metabolism to presystemic clearance." *Drug Metab Dispos.* 30(6): 694-700.

Keski-Hynnilä, H., M. Kurkela, et al. (2002). "Comparison of Electrospray, Atmospheric Pressure Chemical Ionization, and Atmospheric Pressure Photoionization in the

Identification of Apomorphine, Dobutamine, and Entacapone Phase II Metabolites in Biological Samples." *Analytical Chemistry* 74(14): 3449-3457.

Kim, H. S. and I. W. Wainer (2005). "The covalent immobilization of microsomal uridine diphospho-glucuronosyltransferase (UDPGT): Initial synthesis and characterization of an UDPGT immobilized enzyme reactor for the on-line study of glucuronidation." *Journal of Chromatography B* 823(2): 158-166.

Kolmonen, M., A. Leinonen, et al. (2010). "Hydrophilic interaction liquid chromatography and accurate mass measurement for quantification and confirmation of morphine, codeine and their glucuronide conjugates in human urine." *Journal of Chromatography B* 878(29): 2959-2966.

Korfmacher, W. A. (2009). "Using Mass Spectrometry for Drug Metabolism Studies." *Taylor and Francis.*

Kronstrand, R., L. Brinkhagen, et al. (2011). "Ethyl glucuronide in human hair after daily consumption of 16 or 32 g of ethanol for 3 months." *Forensic Science International* In Press, Corrected Proof.

Lahaie, M., J.-N. Mess, et al. (2010). "Elimination of LC–MS/MS matrix effect due to phospholipids using specific solid-phase extraction elution conditions." *Bioanalysis* 2(6): 1011-1021.

Lampinen-Salomonsson, M., U. Bondesson, et al. (2006). "Differentiation of estriol glucuronide isomers by chemical derivatization and electrospray tandem mass spectrometry." *Rapid Communications in Mass Spectrometry* 20(9): 1429-1440.

Lee, M. S., M. Zhu, et al. (2011). "Mass spectrometry in drug metabolism and disposition basic principles and applications." from http://site.ebrary.com/id/10456208.

Lenz, E. M., M. Spear, et al. (2010). "Characterisation and identification of the human N+-glucuronide metabolite of cediranib." *Journal of Pharmaceutical and Biomedical Analysis* 53(3): 526-536.

Levsen, K., H.-M. Schiebel, et al. (2005). "Structure elucidation of phase II metabolites by tandem mass spectrometry: an overview." *Journal of Chromatography A* 1067(1-2): 55-72.

Loser, S. V., J. Meyer, et al. (1996). "Morphine-6-O-beta-D-glucuronide but not morphine-3-O-beta-D-glucuronide binds to mu-, delta- and kappa- specific opioid binding sites in cerebral membranes." *Naunyn Schmiedebergs Arch Pharmacol* 354(2): 192-197.

Lu, J., B. D. Palmer, et al. (2003). "Thalidomide Metabolites in Mice and Patients with Multiple Myeloma." *Clinical Cancer Research* 9(5): 1680-1688.

Marczylo, T., D. Cooke, et al. (2009). "Pharmacokinetics and metabolism of the putative cancer chemopreventive agent cyanidin-3-glucoside in mice." *Cancer Chemotherapy and Pharmacology* 64(6): 1261-1268.

Marín, J. M., E. Gracia-Lor, et al. (2009). "Application of ultra-high-pressure liquid chromatography-tandem mass spectrometry to the determination of multi-class pesticides in environmental and wastewater samples: Study of matrix effects." *Journal of Chromatography A* 1216(9): 1410-1420.

Matuszewski, B. K., M. L. Constanzer, et al. (1998). "Matrix Effect in Quantitative LC/MS/MS Analyses of Biological Fluids: A Method for Determination of Finasteride in Human Plasma at Picogram Per Milliliter Concentrations." *Analytical Chemistry* 70(5): 882-889.

Matuszewski, B. K., M. L. Constanzer, et al. (2003). "Strategies for the assessment of matrix effect in quantitative bioanalytical methods based on HPLC-MS/MS." *Anal Chem.* 75(13): 3019-3030.

Maurel, P. E. S. F.-G., Michael W.H. Coughtrie, Mohamed Ouzzine and Jacques Magdalou (2010). Hepatocytes methods and protocols. New York, Humana: 309-326.

Mazur, A., C. F. Lichti, et al. (2009). "Characterization of human hepatic and extrahepatic UDP-glucuronosyltransferase enzymes involved in the metabolism of classic cannabinoids." *Drug Metab Dispos.* 37(7): 1496-1504. Epub 2009 Apr 1491.

McMaster, M. C. (2007). *HPLC : a practical user's guide*. Hoboken [NJ, USA], Wiley-Interscience.

Mei, H. and R. A. Morrison (2009). PK Principles and PK/PD Applications. *Using Mass Spectrometry for Drug Metabolism Studies, Second Edition*, CRC Press: 59-98.

Mullangi, R., R. K. Bhamidipati, et al. (2005). "Bioanalytical Aspects in Characterization and Quantification of Glucuronide Conjugates in Various Biological Matrices." *Current Pharmaceutical Analysis* 1(3): 251-264.

Naidong, W., J. W. Lee, et al. (1999). "Simultaneous assay of morphine, morphine-3-glucuronide and morphine-6-glucuronide in human plasma using normal-phase liquid chromatography-tandem mass spectrometry with a silica column and an aqueous organic mobile phase." *Journal of Chromatography B: Biomedical Sciences and Applications* 735(2): 255-269.

Nguyen, H. P. and K. A. Schug (2008). "The advantages of ESI-MS detection in conjunction with HILIC mode separations: Fundamentals and applications." *Journal of Separation Science* 31(9): 1465-1480.

Ohta, M., N. Kawakami, et al. (2003). "Analysis of acetaminophen glucuronide conjugate accompanied by adduct ion production by liquid chromatography-atmospheric pressure chemical ionization-mass spectrometry." *Journal of Pharmaceutical and Biomedical Analysis* 30(6): 1759-1764.

Okura, T., N. Komiyama, et al. (2007). "Comparative measurement of spinal CSF microdialysate concentrations and concomitant antinociception of morphine and morphine-6[beta]-glucuronide in rats." *Life Sciences* 80(14): 1319-1326.

Oswald, S., E. Scheuch, et al. (2006). "A LC-MS/MS method to quantify the novel cholesterol lowering drug ezetimibe in human serum, urine and feces in healthy subjects genotyped for SLCO1B1." *Journal of Chromatography B* 830(1): 143-150.

Paradkar, A. R. (2008). "Biopharmaceutics and Pharmacokinetics."

Pozo, O. J., P. Van Eenoo, et al. (2008). "Direct quantification of steroid glucuronides in human urine by liquid chromatography-electrospray tandem mass spectrometry." *Journal of Chromatography A* 1183(1-2): 108-118.

Pruvost, A., F. Théodoro, et al. (2008). "Specificity enhancement with LC-positive ESI-MS/MS for the measurement of nucleotides: application to the quantitative determination of carbovir triphosphate, lamivudine triphosphate and tenofovir diphosphate in human peripheral blood mononuclear cells." *Journal of Mass Spectrometry* 43(2): 224-233.

Pu, F., A. R. McKinney, et al. (2004). "Direct detection of boldenone sulfate and glucuronide conjugates in horse urine by ion trap liquid chromatography-mass spectrometry." *Journal of Chromatography B* 813(1-2): 241-246.

Qin, F., Y. Y. Zhao, et al. (2008). "Column-switching reversed phase-hydrophilic interaction liquid chromatography/tandem mass spectrometry method for determination of free estrogens and their conjugates in river water." *Analytica Chimica Acta* 627(1): 91-98.

Rafferty, J. L., L. Zhang, et al. (2007). "Retention Mechanism in Reversed-Phase Liquid Chromatography: A Molecular Perspective." *Analytical Chemistry* 79(17): 6551-6558.

Ramanathan, R., S. N. Comezoglu, et al. (2009). Metabolite Identification Strategies and Procedures. *Using Mass Spectrometry for Drug Metabolism Studies, Second Edition*, CRC Press: 127-203.

Rosenfeld, J. M. (2004). *Sample preparation for hyphenated analytical techniques*. Oxford; Boca Raton, Blackwell ; CRC Press.

Sanceau, J. Y., D. Larouche, et al. (2007). "Synthesis and deuterium labelling of the pure selective estrogen receptor modulator (SERM) acolbifene glucuronides." *Journal of Labelled Compounds & Radiopharmaceuticals* 50(3-4): 197-206.

Schwartz, M. S., R. B. Desai, et al. (2006). "Determination of a prostaglandin D2 antagonist and its acyl glucuronide metabolite in human plasma by high performance liquid chromatography with tandem mass spectrometric detection--A lack of MS/MS selectivity between a glucuronide conjugate and a phase I metabolite." *Journal of Chromatography B* 837(1-2): 116-124.

Shipkova, M., V. W. Armstrong, et al. (2003). "Acyl Glucuronide Drug Metabolites: Toxicological and Analytical Implications." *Therapeutic Drug Monitoring* 25(1): 1-16.

Silvestro, L., M. Gheorghe, et al. (2011). "Development and validation of an HPLC–MS/MS method to quantify clopidogrel acyl glucuronide, clopidogrel acid metabolite, and clopidogrel in plasma samples avoiding analyte back-conversion." *Analytical and Bioanalytical Chemistry* 401(3): 1023-1034.

Smith, F. P. and S. A. Athanaselis (2005). *Handbook of forensic drug analysis*. Amsterdam; Boston, Elsevier Academic Press.

Soars, M. G., E. L. Mattiuz, et al. (2002). "Biosynthesis of drug glucuronides for use as authentic standards." *J Pharmacol Toxicol Methods* 47(3): 161-168.

Sparidans, R. W., M. L. H. Vlaming, et al. (2009). "Liquid chromatography-tandem mass spectrometric assay for sorafenib and sorafenib-glucuronide in mouse plasma and liver homogenate and identification of the glucuronide metabolite." *Journal of Chromatography B* 877(3): 269-276.

Spooner, N., R. Lad, et al. (2009). "Dried Blood Spots as a Sample Collection Technique for the Determination of Pharmacokinetics in Clinical Studies: Considerations for the Validation of a Quantitative Bioanalytical Method." *Analytical Chemistry* 81(4): 1557-1563.

Stolker, A. M., W. Niesing, et al. (2004). "Liquid chromatography with triple-quadrupole and quadrupole-time-of-flight mass spectrometry for the determination of micro-constituents - a comparison." *Analytical and Bioanalytical Chemistry* 378(7): 1754-1761.

Sunkara, M. and M. J. M. Wells (2010). "Phase II pharmaceutical metabolites acetaminophen glucuronide and acetaminophen sulfate in wastewater." *Environmental Chemistry* 7(1): 111-122.

Takats, Z., J. M. Wiseman, et al. (2004). "Mass spectrometry sampling under ambient conditions with desorption electrospray ionization." *Science.* 306(5695): 471-473.

Tan, A., W. Jin, et al. (2009). "Bioanalytical method development and validation using incurred samples--Simultaneous quantitation of ramipril and ramiprilat in human EDTA plasma by LC-MS/MS." *Journal of Chromatography B* 877(29): 3673-3680.

Taylor, P. J. (2005). "Matrix effects: the Achilles heel of quantitative high-performance liquid chromatography-electrospray-tandem mass spectrometry." *Clinical Biochemistry* 38(4): 328-334.

Telepchak, M. J., T. F. August, et al., Eds. (2010). *Forensic and Clinical Applications of Solid Phase Extraction*. Totowa, New Jersey, Humana Press Inc.

Tephly, T. R., M. D. Green, et al. (1997). Metabolism of Endobiotics and Xenobiotics by UDP-Glucuronosyltransferase. *Advances in Pharmacology*. G. E. David S. Goldstein and M. Richard, Academic Press. Volume 42: 343-346.

Thomas, A., J. Déglon, et al. (2010). 'On-line desorption of dried blood spots coupled to hydrophilic interaction/reversed-phase LC/MS/MS system for the simultaneous analysis of drugs and their polar metabolites." *Journal of Separation Science* 33(6-7): 873-879.

Tolonen, A., M. Turpeinen, et al. (2009). "Liquid chromatography-mass spectrometry in in vitro drug metabolite screening." *Drug Discovery Today* 14(3-4): 120-133.

Trdan, T., R. Roskar, et al. (2011). "Determination of raloxifene and its glucuronides in human urine by liquid chromatography-tandem mass spectrometry assay." *Journal of Chromatography B* 879(23): 2323-2331.

Trontelj, J., M. Bogataj, et al. (2007). "Development and validation of a liquid chromatography-tandem mass spectrometry assay for determination of raloxifene and its metabolites in human plasma." *J. Chromatogr., B· Anal. Technol. Biomed. Life Sci.* 855(2): 220-227.

Völkel, W., M. Kiranoglu, et al. (2008). "Determination of free and total bisphenol A in human urine to assess daily uptake as a basis for a valid risk assessment." *Toxicology Letters* 179(3): 155-162.

Weinmann, W., S. Vogt, et al. (2000). "Simultaneous determination of THC-COOH and THC-COOH-glucuronide in urine samples by LC/MS/MS." *Forensic Science International* 113(1-3): 381-387

Wen, Z., S. T. Stern, et al. (2006). "Structural Characterization of Anti-HIV Drug Candidate PA-457 [3-O-(3',3'-Dimethylsuccinyl)-betulinic acid] and Its Acyl Glucuronides in Rat Bile and Evaluation of in Vitro Stability in Human and Animal Liver Microsomes and Plasma." *Drug Metabolism and Disposition* 34(9): 1436-1442.

Wong, P., R. Pham, et al. (2010). "Increasing efficiency for dried blood spot analysis: prospects for automation and simplified sample analysis." *Bioanalysis* 2(11): 1787-1789.

Xia, Y.-Q. and M. Jemal (2009). "High-Field Asymmetric Waveform Ion Mobility Spectrometry for Determining the Location of In-Source Collision-Induced Dissociation in Electrospray Ionization Mass Spectrometry." *Analytical Chemistry* 81(18): 7839-7843.

Xia, Y.-Q. and M. Jemal (2009). "Phospholipids in liquid chromatography/mass spectrometry bioanalysis: comparison of three tandem mass spectrometric techniques for monitoring plasma phospholipids, the effect of mobile phase composition on phospholipids elution and the association of phospholipids with matrix effects." *Rapid Communications in Mass Spectrometry* 23(14): 2125-2138.

Xiong, W. J., H. Jin, et al. (2010). "Evaluating human liver reserve function by measuring serum concentrations of phenacetin and its metabolites." *Journal of Digestive Diseases* 11(6): 358-363.

Yang, L., M. a. Amad, et al. (2002). "Investigation of an enhanced resolution triple quadrupole mass spectrometer for high-throughput liquid chromatography/ tandem mass spectrometry assays." *Rapid Communications in Mass Spectrometry* 16(21): 2060-2066.

Yau, W.-P., A. Vathsala, et al. (2004). "Simple reversed-phase ion-pair liquid chromatography assay for the simultaneous determination of mycophenolic acid and its glucuronide metabolite in human plasma and urine." *Journal of Chromatography B* 805(1): 101-112.

Ye, X., Z. Kuklenyik, et al. (2005). "Automated On-Line Column-Switching HPLC-MS/MS Method with Peak Focusing for the Determination of Nine Environmental Phenols in Urine." *Analytical Chemistry* 77(16): 5407-5413.

Yu, B., X. Zhu, et al. (2000). "A Novel and Effective Procedure for the Preparation of Glucuronides." *Organic Letters* 2(16): 2539-2541.

Yu, C., C. L. Chen, et al. (2007). "A rapid method for quantitatively estimating metabolites in human plasma in the absence of synthetic standards using a combination of liquid chromatography/mass spectrometry and radiometric detection." *Rapid Communications in Mass Spectrometry* 21(4): 497-502.

Yu, N. H., E. N. M. Ho, et al. (2005). "Screening of anabolic steroids in horse urine by liquid chromatography-tandem mass spectrometry." *Journal of Pharmaceutical and Biomedical Analysis* 37(5): 1031-1038.

Zhang, D., M. Zhu, et al. (2008). Drug metabolism in drug design and development : basic concepts and practice. Hoboken, Wiley-Interscience.

Zhang, N. R., S. Yu, et al. (2009). "Quantitation of small molecules using high-resolution accurate mass spectrometers - a different approach for analysis of biological samples." *Rapid Communications in Mass Spectrometry* 23(7): 1085-1094.

Zhang, W., G. E. Dutschman, et al. (2009). "Quantitation of Irinotecan and its two major metabolites using a liquid chromatography-electrospray ionization tandem mass spectrometric." *Journal of Chromatography B* 877(27): 3038-3044.

Tandem Mass Spectrometry of Alkanolamines in Environmental Samples

John Headley[1], Kerry Peru[1], Adeola Adenugba[2] and Dena McMartin[2]
[1]National Water Research Institute, Environment Canada,
[2]Faculty of Engineering and Applied Science, University of Regina, Canada

1. Introduction

Alkanolamines are anthropogenic compounds used extensively not only in the processing of sour natural gas but in other operations such as the petrochemical and fossil-fuel fired industries to reduce and prevent the emission of green house gases into the environment. Alkanolamines also find uses as corrosion inhibitors, emulsifying agents and in pharmaceutical formulations (Dobberpuhl and Johnson, 1995). Over many years of operation, there have been inadvertent leaks of these chemicals from the surrounding gas processing facilities to groundwater, wetlands soil, and associated uptake by vegetation (Luther et al., 1998; Headley et al., 1999; Hamborg et al., 2011). Since alkanolamines are highly water miscible (Davis and Carpenter, 1997), their analysis has presented some challenges (Jagota et al., 1996) particularly the requirements for suitable extraction from biological matrices and soil, along with sensitive detection using commonly available instrumentation. Early methods of analyzes were focused on direct determination of the analytes with little or no derivatization steps (Witzaney and Fedorak, 1996). Alkanolamines were determined primarily using gas chromatography (GC) with flame ionization (FI) or mass spectrometric (MS) detection (Shahi et al., 1994; Dawodu and Meisen, 1991, 1994, Niitsue et al., 1993 as cited in Peru et al., 2004). Positive-ion electrospray ionization tandem mass spectrometry (ESI-MS-MS) is well suited for the confirmation of uptake of alkanolamines and transformation products in wetland vegetation (Headley et al., 1999). The objective of this chapter is to enumerate the use of tandem mass spectrometry for the determination of alkanolamines from environmental samples exemplified by wetland vegetation.

2. Occurrence and uses of alkanolamines

In the last 25 years, alkanolamines have been studied extensively because of their industrial importance in natural gas processing plants, synthetic ammonia plants, fossil-fuel-fired power plants, chemical, and petrochemical industries (Mundhwa and Henni, 2007). Alkanolamines are used as chemical intermediates and as surface-active agents in cosmetics, pharmaceuticals, agricultural products, and in various industrial products and applications. Aqueous solutions of alkanolamines are commonly used on a large scale in the natural gas industry to strip and remove acid gases such as carbon dioxide, sulfur dioxide and hydrogen sulfide) from natural gases and hydrocarbon liquids (Carroll and Marther, 1997; Kohl and Nielsen, 1997 as cited in Hansen et al., 2010). Other uses and applications of alkanolamines are summarized in the Table 1.

Application	Function	Type of alkanolamines
Coating	Used in both water and solvent based coatings. Increase the solubility of other components and enhance solution stability	DEEA and DMEA
Emulsifying and Dispersing agents	Emulsifying additives in textile, lubricants, polishes, detergents, pesticides and personal care products - hand lotions, shaving creams and shampoos.	Fatty acid soaps of DEEA and DMEA
Gas Treating agents	Stripping of undesirable gases such as SO_2, CO_2, H_2S during natural and refinery processes	MEA, DEA, TEA, DIPA*
Pharmaceuticals	Used as intermediates for the production of active pharmaceutical ingredients	DMEA - synthesis of procanine and procaine penicillin G, Tamoxifen DMEA and MEA - synthesis of antihistamines.
Catalysts	Promotes foam rise and gel strength properties in the production of urethane foam used in refrigerator and other insulation applications	DMEA
Water Treatment	Used widely in the production of water treatment products such as water-soluble polymeric flocculants and ion exchange resins and corrosion inhibitors	DMEA

Table 1. Uses and application of alkanolamines [DOW Application note (2003); *Shih et al., 2002]. The acronyms: MEA, DEA, TEA, DIPA, DEEA, DMEA are defined below in Figure 1.

3. Structure and classification of alkanolamines

Alkanolamines can be classified according to the number of alkyl groups bonded to the nitrogen atom of the amino group as: (1) Primary e.g. diglycolamine (DGA); monoethanolamine (MEA); monoisopropanolamine (MIPA); (2) Secondary e.g. diethanolamine(DEA); diisopropanolamine (DIPA); (3) Tertiary e.g. diethylethanolamine (DEEA), dimethylethanolamine (DMEA), methyldiethanolamine (MDEA), triethanolamine (TEA); and (4) sterically hindered amine, 2-amino-2-methyl-1-propanol (AMP) (Padurean et al., 2011). In accordance with the convention used in the alkanolamine literature (Jamal et al., 2006 as cited in Padurean et al., 2011), MEA is represented as R_1NH_2,, where R_1 signifies CH_2CH_2OH; DEA is represented as R_1R_2NH, where R_1-R_2 denotes CH_2CH_2OH; MDEA is represented as R_1R_2 R_3NH, where R_1-R_2 implies CH_2CH_2OH and R_3 is CH_3; AMP is represented as R_4NH_2 where R_4 is $-C(CH_3)_2CH_2OH$ (Metz et al., 2005; Gibbins and Chalmers, 2008 as cited in Padurean et al., 2011). Chemical structures of some common alkanolamines are given in Figure 1. Of all the alkanolamines four are commonly used for capture of CO_2 from industrial processes of fossil fuel. They are MEA, DEA, MDEA, AMP, (Libralato et al., 2010; Padurean et al., 2011). Padurean et al., (2011), expounded on the criteria on which, the consideration was based. A less common alkanolamines that is also used in the acid gas removal from natural gas industry is 2-piperidinethanol (2-PE) (Shih et al., 2002). Table 2 presents some of the physical properties of common alkanolamines.

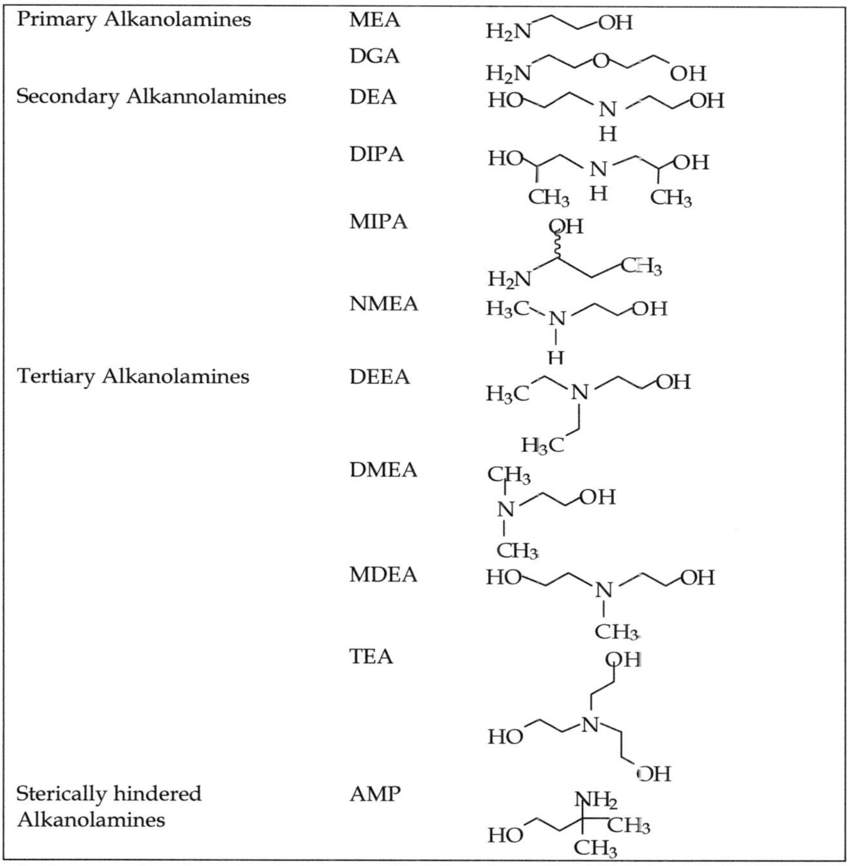

Fig. 1. Chemical structures of some common alkanolamines.

4. Production and environmental distribution

In 1993, approximately 20 million kg of alkanolamines were used for natural gas purification in the United States (Hawthone et al., 2005 cited in Lu et al., 2009); the figure is expected to have increased in recent times. Such a high demand in the use of alkanolamines has led to a surge in the prices of production. (BASF, 2006). Figure 2, illustrates the potential environmental fate of alkanolamines following various industrial applications.

Compound	CAS Number	Melting Point (°C)	Boiling Point (°C)	Water Solubility (mg/L)	Log P (octanol-water)	Vapor Pressure (mm Hg)	pKa Dissociation Constant	Henry's Law Constant (atm-m³/mole)
MEA	141-43-5	10.5	171	1×10^6	-1.31	0.404	9.5	3.25×10^{-8}
DEA	111-42-2	28	268.8	1×10^6	-1.43	2.8×10^{-4}	8.96	3.87×10^{-11}
TEA	102-71-6	149.2	335.4	1×10^6	-1.00	3.59×10^{-6}	7.76	7.05×10^{-13}
MDEA	105-59-9	-21	247	1×10^6	-1.50	2.00×10^{-4}	8.52	3.14×10^{-11}
DIPA	110-97-4	44.5	250	8.7×10^5	-0.82	1.25×10^{-4}	9.1	6.91×10^{-11}

Table 2. Physical and chemical properties of some selected alkanolamines (Modified: Headley et al., 2002).

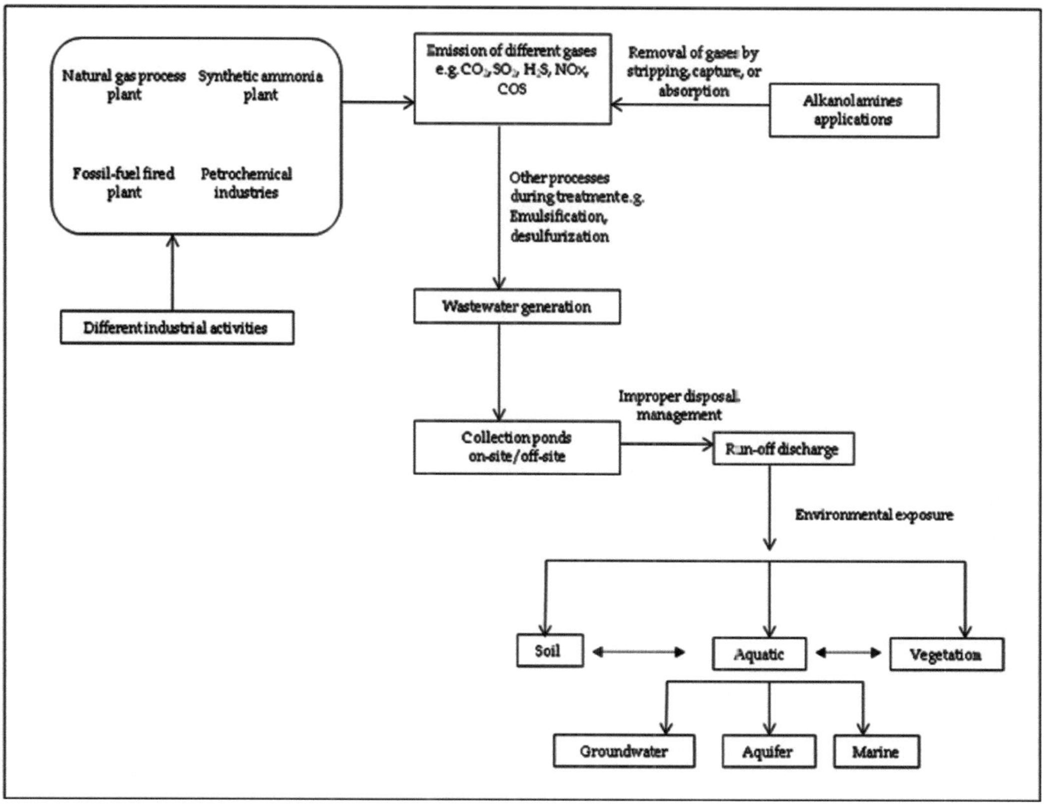

Fig. 2. Potential environmental distribution of alkanolamines.

Improper desulfurization processes results in a corrosive plant installations creating technological malfunction in transporting generated wastewater from the treatment plants. The wastewater not only composed of lyes, rinse water and column refluxes, but also process amines and their degradation products, including ammonium ion (Chackma and Meisen, 1997; Rooney and Bacon, 1998; Rooney and Bacon, 1996; 1997, as cited in Kaminski et al., 2002). One of the most commonly used alkanolamines, diethanolamine (DEA), is miscible in water due to its small molecular weight (mw 106) and high polarity (Headley et al., 2002). Such physicochemical properties mean DEA and related compounds have the potential for rapid and widespread transport in the environment and exposure to organisms. However, the concentrations of alkanolamines in natural waters are expected to be low because of dispersion processes (Hansen et al., 2010). For instance, Greene et al., (1999) (as cited in Hansen et al., 2010), reported that concentrations of the alkanolamine diisopropanolamine (DIPA) were 350 mg/L in groundwater near the source of contamination at a sour gas plant. The concentration decreased to less than 50µg/L down-gradient from the plume.

5. Methods for ionization of alkanolamines in environmental samples

5.1 Electrospray ionization (ESI)

ESI sources for use with mass spectrometers were introduced by Fenn and co-workers in 1988 (Yamashita and Fenn, 1984; Fenn et al., 1989 as cited in Rodrigues et al., 2008). Electrospray allows desorption of large, non-volatile intact analytes directly from solution. Generally, an electrospray is produced by spraying a sample solution through a capillary into a strong electric field in the presence of a flow of nitrogen, transferring the ions into the gas phase As a result of the applied electric field, the surface of the liquid emerging from the capillary is highly charged and assumes a conical shape known as the Taylor cone. When the Rayleigh limit is reached (namely the point at which the surface tension of the liquid is exceeded by the repulsive forces between the charges) the surface of the cone breaks into droplets that further fragment into smaller and smaller drops, eventually producing ions in the gas phase (Kebarle and Tang, 1993 as cited Rodrigues et al., 2008). The ESI source is easy to use, has a wide polarity range and can be applied to thermally labile compounds (Mallet et al., 2004 as cited in Raffi et al., 2009).

5.2 Atmospheric pressure chemical ionization (APCI)

APCI was originally developed as an analytical technique for analysis of trace components in the gas phase. APCI is only possible when the analyte exists in the gaseous state as ions. Since the instrument of separation for example, liquid chromatography (LC), high pressure liquid chromatography (HPLC) or GC is coupled to the MS, it is necessary to volatilize the molecules from the eluent. In APCI, the analyte is sprayed by a heated pneumatic nebulizer probe at atmospheric pressure for nebulization and a high voltage needle produces a corona discharge for ionization of the evaporated solvent. This dissociates the analyte molecule and generates ion (Raffi et al., 2009). Figure 3 gives an overview of tandem mass spectrometry operation.

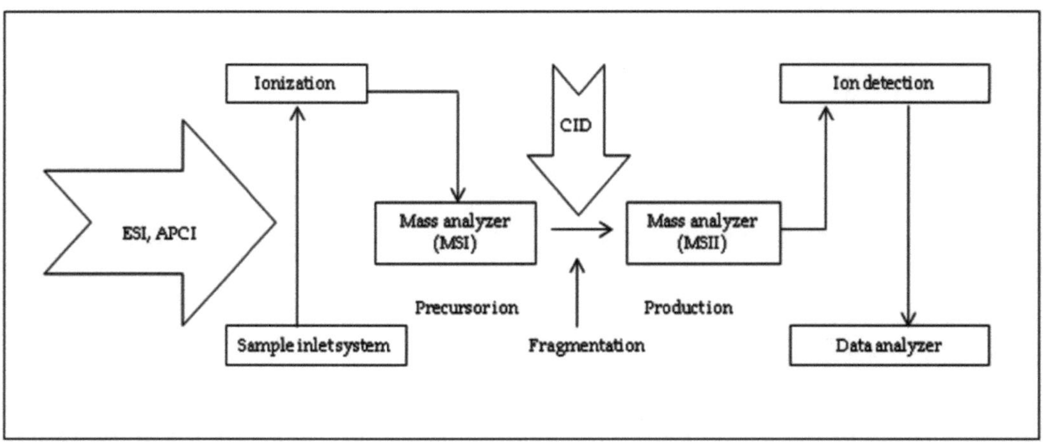

Fig. 3. Diagramatic representation of tandem mass spectrometry [Modified: Rodrigues et al., (2008)].

6. Mass analyzers for tandem MS analyzes of alkanolamines

The instrumentation used for tandem mass spectrometry of alkanolamines can be divided into three main types namely ion-trapping (quadrupole ion trap (QIT); triple quadrupoles and sector or time-of-flight instruments respectively (Payne and Glish, 2005,). The notable characteristic that differentiate ion-trapping instrumentation from other types of mass spectrometers (MSs) is that tandem mass spectrometry (MS-MS) is performed through a tandem-in-time method, instead of tandem in space. This implies that each stage of mass spectrometry is conducted in succession in the same analyzer in time. A triple quadrupole (QqQ) by contrast has each step of MS-MS operated in different analyzer separated in succession in space, i.e. an analyzer for each level of MS-MS operation (Payne and Glish, 2005). In sector instruments, each subsequent level requires the addition of another reaction region and mass analyzer. An immediately apparent advantage of trapping instruments is that multiple stages of MS-MS (MSn) can be performed without instrumental modifications. The ion intensity is the only limitation to the number of MS-MS levels possible in a trapping instrument (Louris et al., 1990 as cited in Payne and Glish, 2005). In addition, 80-90% of MS-MS product ions can be trapped unlike in linear quadrupole (LQq) where efficiencies are generally an order of magnitude lower (Johnson et al., 1990 as cited in Payne and Glish, 2005).

As much of the tandem MS applications of alkanolamines were conducted by use of triple quadrupoles, a brief discussion of these analyzers will now be given. Quadrupole mass analyzers are of two types (1) linear quadrupole and Triple quadrupole mass analyzer. By design triple-quadrupole mass spectrometers consist of two resolving quadrupoles (Q1 and Q3), and a collision cell (Q2) positioned in-between (Rodrigues et al., 2008). Triple-quadrupole mass spectrometers can be operated in four modes of acquisition: multiple-reaction monitoring (MRM), product-ion scan, precursor-ion scan, and neutral loss. MRM is the most common type of acquisition utilized for quantitative analysis. Collision induced dissociation (CID) on different brands of triple-quadrupole mass spectrometers typically results in similar fragment patterns, while significant differences may be observed in the relative abundances of the fragment ions, mainly due to the differences in instrumental design. The reason for the high sensitivity of the MRM acquisition is that the mass analyzer is monitoring the same mass to charge ratios during the entire time of the data acquisition, and the specificity is based on monitoring molecular weight and compound-specific fragmentation (Rodrigues et al., 2008; Payne and Glish, 2005).

7. Analysis of alkanolamines from environmental samples

A summary of work reported on analysis of alkanolamines in the last two decades is presented in Table 3. This table was extracted from Science Direct to identify existing peer reviewed literature published (Peters, 2011), from 1991-2011. Interestingly, only 6% of the total hits (206) of the original research had a connection with alkanolamines in relation to ecotoxicity, toxicity, biodegradation assays. Most work involved the fortification (Peru et al., 2004) or spiking, dosing of the samples with alkanolamines before the testing procedure. For each of the study, different instrumental operations were carried out to quantify (the fortified concentrations) of the test compounds (alkanolamines). It is worth mentioning that some guidelines have been established for some alkanolamines in water, soil and plants (Appendix 1). However, vis-à-vis tandem mass spectrometry, more attention will be given on the work by Peru et al., (2004) where alkanolamines were determined in cattails (*Typha latifolia*) utilizing ESI with selected reaction monitoring and ion-exchange chromatography (IEC).

Title	Subject	Environmental samples	Alkanolamines	Instrumentation	References
Seawater ecotoxicity of monoethanolamine, diethanolamine and triethanolamine	Ecotoxicity	Aquatic species: alga, oyster, mussel, crustacea	MEA, DEA, TEA	Compact Ion Chromatography (CCI)	Libralato et al., 2010
Molecular effects of diethanolamine exposure on *Calanus finmarchicus* (Crustacea: Copepoda)	Toxicity	Aquatic species: *Calamus finamarchicus*	MEA, DEA	High resolution magic angle spinning, nuclear magnetic resonance (HR-MAS- NMR)	Hansen et al., 2010
Gas chromatographic-mass spectrometric determination of sulfonane in wetland vegetation exposed to sour gas-contaminated groundwater	Plant Uptake	Wetland vegetation: roots, shoots, berries, leaves, seeds, grasses,	Sulfolane containing DIPA	Gas chromatography mass spectrometry (GC-MS)	Headley et al., 1999
Absorption, distribution, metabolism and excretion of intravenously and dermally administered triethanolamine in mice	Toxicity	Urine, feces, blood, kidney	MEA, DEA, TEA	GC-MS, LC-MS	Stott et al., 2000
Evaporation and air-stripping to assess and reduce ethanolamines toxicity in oily wasterwater	Toxicity	Wastewater	MEA, DEA, TEA	Ion chromatography system (ICS)	Libralato et al., 2008
Environmental impact of amines	Bio-degradation	Marine using alga Skeletonema costatum	8 amines including MEA, DEA, TEA e.t.c.	Organization for economic cooperation and development (OECD) guidelines 306	Eide-Haugmon et al., 2009
Investigation of the formation of N-nitrosodiethanolamine in B6C3F1 mice following topical administration of triethanolamine	Toxicity	Blood and urine	DEA, TEA	GC-MS	Saghir et al., 2005
Degradation of MMEA at absorber and stripper conditions	De-gradation	Different experimental conditions	MMEA	LC-MS; GC-MS	epaumier et al., 2011
The pharmacokinetics of diethanolamine in Sprague-Dawley rats following intravenous administration	Toxicity	Blood, liver, kidneys, heart, brain, carcass	DEA	GC-MS	Mendrala et al., 2001
In vitro human skin penetration of diethanolamine	Toxicity	Human skin	[^{14}C]-DEA*	Liquid scintillation spectroscopy (LSC)	Kraeling et al., 2004
Evaluation of the genotoxic potential of alkylalkanolamines	Genotoxicity	Blood of mice	DMEA, MDEA, t-BDEA	Biological assays	Leung and Ballantyne, 1997
The inhalation of di- and triethanolamine upon repeated exposure	Toxicity Viscera of mice	Liver, kidneys, brain, lungs of mice	DEA, TEA	Gravimetric determination	Gamer et al., 2008

Table 3. Instrumentation used in analyzes of alkanolamines from environmental matrices.

For tandem mass spectrometry of alkanolamines from environmental samples, certain procedures must be adhered to for maximum level of detection and high quality data generation. The key steps include: extraction of the compound of interest from the matrix of containment, cleanup of the extract to reduce interference by other co-exiting compounds, pre-concentration to reduce the solvent of extraction hence increase the sensitivity and therefore the detectability of the analyte by instrumental analysis, and finally the actual determination to quantify the precise amount or concentration of the compounds in the original sample. A thorough outline of the sampling site, how much sample, and the representativeness of the sample for statistical analysis must be thought through before the sampling operations.

The application by Peru et al., (2004) will now be discussed as an example of tandem MS for the detection of alkanolamines in environmental samples. The method employed ion chromatography separation (Peru et al., 2004). Their work documented improvements in the selectivity and detection limits of a selected reaction monitoring (SRM) method for the quantification of DIPA in vegetation tissue. They extended the sphere of the method to the determination of other related alkanolamines including MEA, DEA, MDEA, MIPA, and TEA) in the upper (shoots) and lower (roots) tissues of *Typha latifolia* (cattails) grown hydroponically with various levels of exposure to DIPA under controlled laboratory conditions (Peru et al., 2004).

8. Tandem MS for determination of alkanolamines in cattails

Doucette et al., (2002) (as cited in Peru et al., 2004), has elaborated on the sampling, fortification and laboratory growing conditions of cattails analyzed for alkanolamines. In summary, the plants were obtained as bare-root plants from the Aquatic and Wetland Company (Fort Lupton, CO, USA). Following their reception, they were transplanted into 12 separate glass reactors and grown hydroponically under aerobic conditions for several weeks prior to analysis (Peru et al., 2004).

8.1 Extraction of plants

Samples were prepared as described previously by Headley et al., (1999). Briefly, tissues were prepared and extracted using different methods where: (a) about 1g of subsample of soil soil-free plant was homogenized after thawing with organic-free Milli-Q water and (b) the ample portion of tissue sample was ground under liquid nitrogen to a free-flowing powder and then centrifuge. The sample was allowed to warm to room temperature before adding Milli-Q water to avoid ice formation in the following extraction step. In both cases, further centrifugation was carried out for 45 minutes at 2500rpm to remove large debris. The supernatant was filtered using a 0.45µm surfactant-free cellulose acetate membrane filter. Sub-samples of 500µL of the extract were taken, and 10µL of 20% formic acid were added to each to ascertain complete ionization of DIPA. TEA was used as the internal standard of final extract concentration of 0.5µg/mL. TEA was chosen as the internal standard (IS) because of its similar chemical properties to those of DIPA (Headley et al., 2004). For other matrices such as sediment or soil, the extraction of sample might require more rigorous methods as mentioned above and may require some clean up steps before direct injection or derivatization (to be discussed later) of the extract.

8.2 Calibration of standards

Standard solutions for calibration or spiking are very important in the entire analytical procedure using MS-MS technique. They must be prepared from original stock of high quality with high purity. Similarly, a suitable solvent (s) must be prepared used and water of an analytical grade and purity. The standard alkanolamines must be prepared in stock solution and the necessary dilution carried out for fortification and calibration. Acetonitrile (ACN), acetone (ACE) or a mixture of the two is commonly used as versatile solvent for chromatographic analysis of most compounds (Bu et al., 2000, Nödler et al., 2010). Again, the choice of solvent depends on the matrix from the alkanolamines are to be extracted. In the study under review, individual and mixed stock solutions of alkanolamines (1000µg/mL) were prepared and appropriate dilutions made with water: methanol (1:3 v/v) with the water containing 0.1% formic acid. All standard solutions are usually stored at 4°C. It is vital that all glassware is cleaned and rinsed with methanol, dried in the oven at 100°C to minimize contamination (Peru et al., 2004).

8.3 Instrumentation

There have been few developments in the use of LC-MS for the analysis of alkanolamines. RP-HPLC with thermospray (TSP) interface has been used to detect primary, secondary, and tertiary amines in water (Imago et al., 1993 as cited in Headley et al., 2002). The use of the soft ionization techniques such as direct insertion probe-positive ion chemical ionization (DIP-PICI) and ESI in combination with MS and tandem MS-MS for the detection and confirmation of alkanolamines in water has been reported (Dickson et al., 1996). Under soft ionization conditions, alkanolamines give intense protonated molecular-ions with little or no fragmentation (Dickson et al., 1996), which is well suited for MS-MS analyzes. Increased specificity was obtained using MS-MS as compared to MS. The benefits of MS-MS mixture analyzes include increased confidence in identification of unknowns and analyte confirmations, improved detection limits through reduction in chemical noise, and the availability of a variety of operating modes which can be used for screening, target compound analyzes, and confirmation (Harrison, 1983; Yost and Enke, 1979, as cited in Headley et al., 1999, 2002).

Among the methods available for analyzes of alkanolamines, techniques employing GC or LC with direct aqueous injections have received the most attention (Headley et al., 2002). To this end, Peru et al., (2004) conducted IEC using a Waters 2695 separations module with a Dionex IonPac CS14, 2mm x 25cm cation-exchange column to determine alkanolamines in cattail tissue. The eluent consisted of 75:25 methanol/water containing 0.1% formic acid (isocratic) with a flow rate of 200µL/min. An injection volume of 10µL was used for all samples and standards.

In Peru et al., (2004) study, a Quattro Ultima triple-quadrupole mass spectrometer (Micromass, UK) using positive ESI was used for all analyzes. For comparison, both selected ion monitoring (SIM) and selected reaction monitoring (SRM) modes were employed. For ESI-MS, instrumental parameters were as follows: source temperature 90C, cone voltage setting 26V, capillary voltage setting 1.81 kV, hexapole 1 4.0V, hexapole 2 and aperture 0V, desolvation temperature 220°C, desolvation gas N_2 flow 488L/h, cone gas N_2 flow 145L/h, and nebulizer gas N_2 flow was at maximum. Peak width was set to achieve unit mass resolution, and the multiplier was set to 650V. All analyzes were performed using ESI in the positive-ion mode. Initial quantitative analyzes used selected ion recording, monitoring $[M+H]^+$ ions at *m/z* 134 for DIPA and *m/z* 150 for TEA (IS), with an inter-channel delay of 0.10s and a dwell time of

0.50s. For ESI-MS-MS, used to provide additional selectivity, SRM was utilized as the final method of quantification. Instrumental conditions were as above for ESI-MS with the following exceptions/additions: argon was used as the collision gas at a pressure sufficient to increase the Pirani gauge reading to 3.53x10^{-4} Torr. Collision energy was set at 17V (laboratory frame of reference). Reaction monitoring was used for quantification, monitoring losses of either one or two molecules of water. Although quantification of only DIPA (*m/z* 134→98) along with TEA (*m/z* 150→132) as the internal standard) was required for this study, other relevant alkanolamines were examined to determine the applicability of the method for future investigations. Product ion scans (Figures 4a-f) were acquired to determine suitable transitions for reaction monitoring, and are listed in Table 4.

Compound	MW	Precursor ion *m/z* $[M+H]^+$	SRM transition* monitored for quantification
MEA	61	62	62>44
MIPA	75	76	76>58
DEA	105	106	106>88
MDEA	119	120	120>102
DIPA	133	134	134>98
TEA	149	150	150>132

Table 4. Transitions monitored during SRM analysis (Peru et al., 2004) * All SRM analysis completed using 18eV collision energy (Q2), argon as collision gas (cell pressure 3.53 x 10^{-4} Torr).

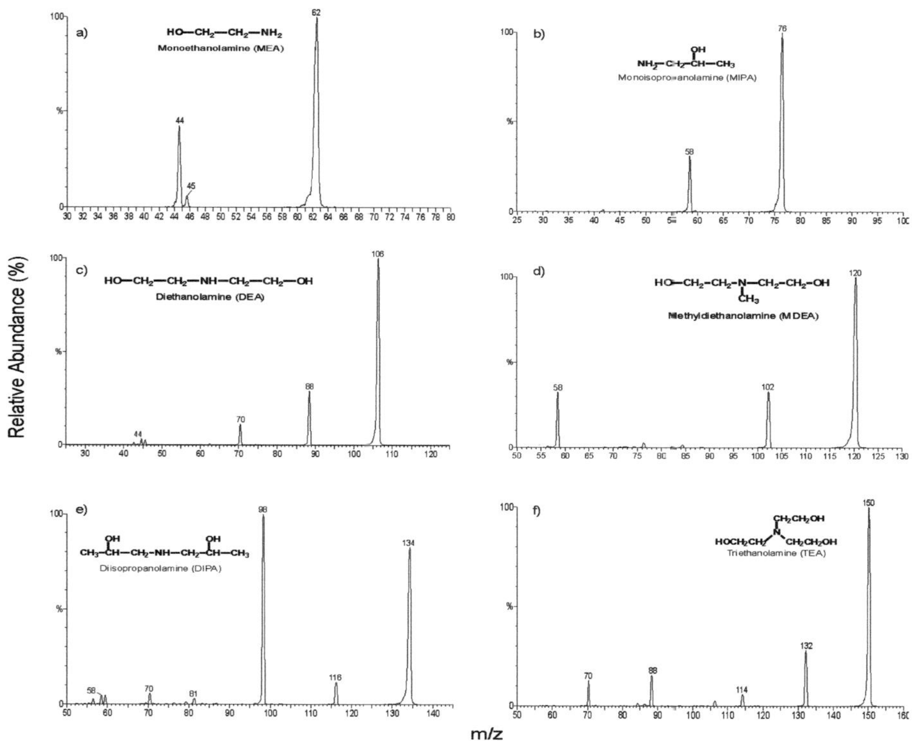

Fig. 4. Product ion spectra of $[M+H]^+$ ions of six alkanolamines (MEA, MIPA, DEA, MDEA, DIPA, TEA) (Peru et al., 2004).

As demonstrated in Figures 4 (a-f), the main product ions were formed via water loss. Typically, the loss of one or two water molecules is not considered a highly specific transition to monitor for SRM analysis. Nonetheless, in this case, there were no other options. Moreover, SRM using the loss of 2(H_2O) for DIPA and H_2O for TEA (IS) provided adequate specificity to exclude SIM interferences experienced during the study (Peru et al., 2004).

The improved sensitivity and resolution of SRM mode in the study of Peru et al., (2004) is demonstrated in Figure 5 for DIPA. In Figure 6A, approximately 10% of the samples had prohibitive mass interference that could not be resolved either chromatographically or by mass spectrometer. Product ion scans were acquired to determine whether suitable product ions were formed that could be used for MRM to improve selectivity.

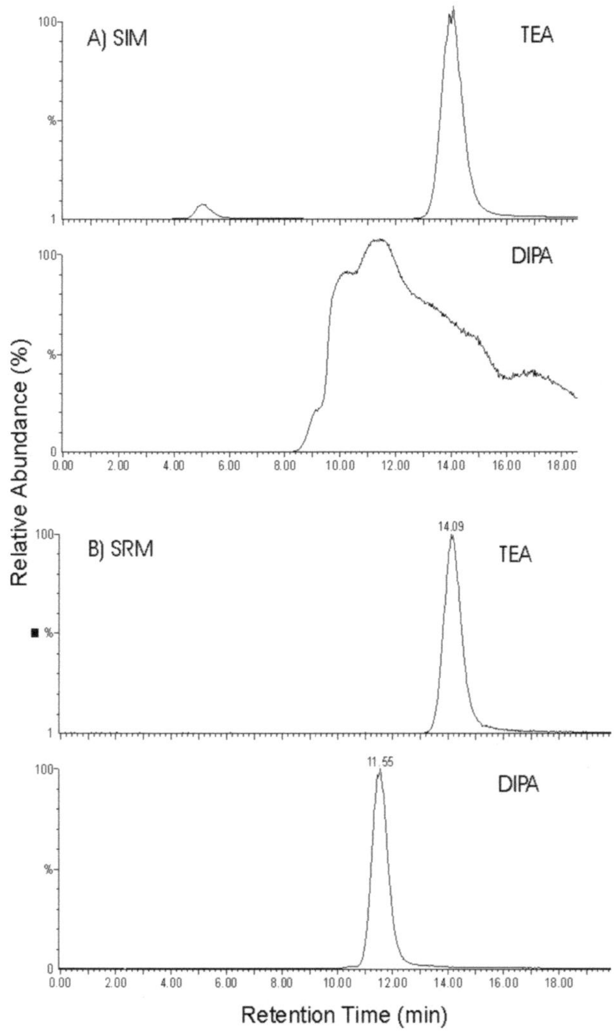

Fig. 5. (A) Selected ion monitoring chromatogram illustrating prohibitive interference during the analysis of DIPA. (B) Chromatogram of the same extract in used (A) illustrating the selectivity of MS-MS using SRM (Peru et al., 2004).

9. Overcoming matrix effect- enhancement and suppression of ions

Though LC-MS-MS is a useful analytical characterization tool, there is one significant setback of electrospray mass spectrometry. This is its high susceptibility to matrix signal suppression or enhancement which is a major challenge for quantitative LC-MS application of environmental samples (Choi et al., 2001). Matrix effects occur when molecules coeluting with the compound/s of interest alter the ionization efficiency of the electrospray interface. The exact mechanism of matrix effects is unknown, but it probably originates from the competition between an analyte and the coeluting, undetected matrix components (Taylor, 2005). King et al., (2000) (as cited in Taylor, 2005) have shown through a series of experiments that matrix effects are the result of competition between nonvolatile matrix components and analyte ions for access to the droplet surface for transfer to the gas phase. They conclude that the exact mechanism of the alteration of analyte release into the gas phase by these nonvolatile components is unclear. They postulate a likely list of effects relating to the attractive force holding the drop together and keeping smaller droplets from forming should account for a large proportion of the ionization suppression observed with electrospray ionization. Depending on the environment in which the ionization and ion evaporation processes take place, this competition may effectively decrease (commonly known as ion suppression) or increase (ion enhancement) the efficiency of formation of the desired analyte ions present at the same concentrations in the interface. Thus the efficiency of analyte ions to form is very much dependent on the matrix entering the electrospray ion source (Taylor, 2005). However, King et al. (2000) (as cited in Taylor et al., 2005) have demonstrated that ion suppression is much more severe with ESI than APCI. The increased ion suppression is a result of high concentrations of nonvolatile materials present in the spray with the analyte. Another reason for the higher ion suppression in the ESI mode could be due to the fact that in this mode, unlike in the APCI mode, the analyte is ionized in the liquid phase inside the electrically charged droplets. As the solvent evaporates, the analyte precipitates from solution either as solid compound or as a co-precipitate with other non-volatile sample components (Raffi et al., 2009).The two main techniques used to determine the degree of matrix effects on an ESI-MS-MS method are: (1) post-extraction addition which is considered to be a static technique providing information about matrix effect at the point of elution of the analyte, and (2) post-column infusion, a more dynamic technique involving an infusion pump that delivers a constant flow of analyte into the LC effluent at a point after the chromatographic column and before the mass spectrometer ionization source (Taylor, 2005).

Different plan of action can be employed to overcome matrix effects during tandem mass spectrometry. These include but not limited to: (a) modification of the sample extraction procedure and (b) improved chromatographic separation (Avery, 2003 as cited in Taylor, 2005). In addition, the application of internal standards has been especially useful in addressing quantitative signal reproducibility issues between standard and matrix analytes (Boyd, 1993; Temesi, 1999 as cited in Choi et al., 2001). In order to determine if there were matrix effects during the analysis of alkanolamines in their study, Peru et al., (2004) utilized the post-column technique as illustrated in Figure 6.

9.1 Post infusion technique for alkanolamines determination in cattails

A mixture of the alkanolamines at a concentration of 1μg/mL for each component was infused post column at a rate of 25μL/min. This gave rise to an elevated baseline. Blank extracted cattail root matrix was then injected via the LC system and the multiple reaction monitoring transitions (Table 4) were monitored (Figure 7). If ionization enhancement or suppression exists, a positive or negative deflection in the elevated baseline would be observed. As shown in Figure 8, some suppression was observed (symbolized with an *) prior to the retention times of the analytes of interest. These results indicated that suppression/enhancement matrix effects were not significant for the analytes investigated.

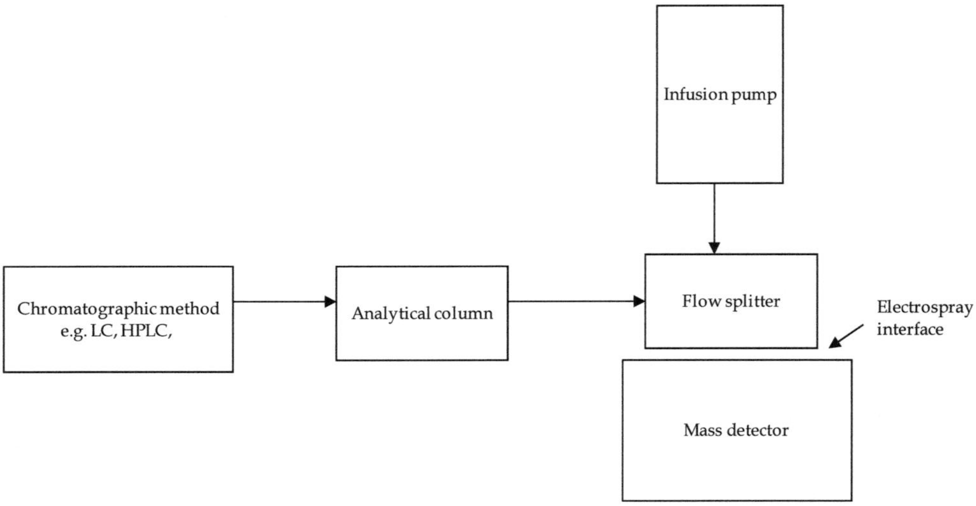

Fig. 6. Schematic of post-column infusion system [Modified from Taylor, (2005)].

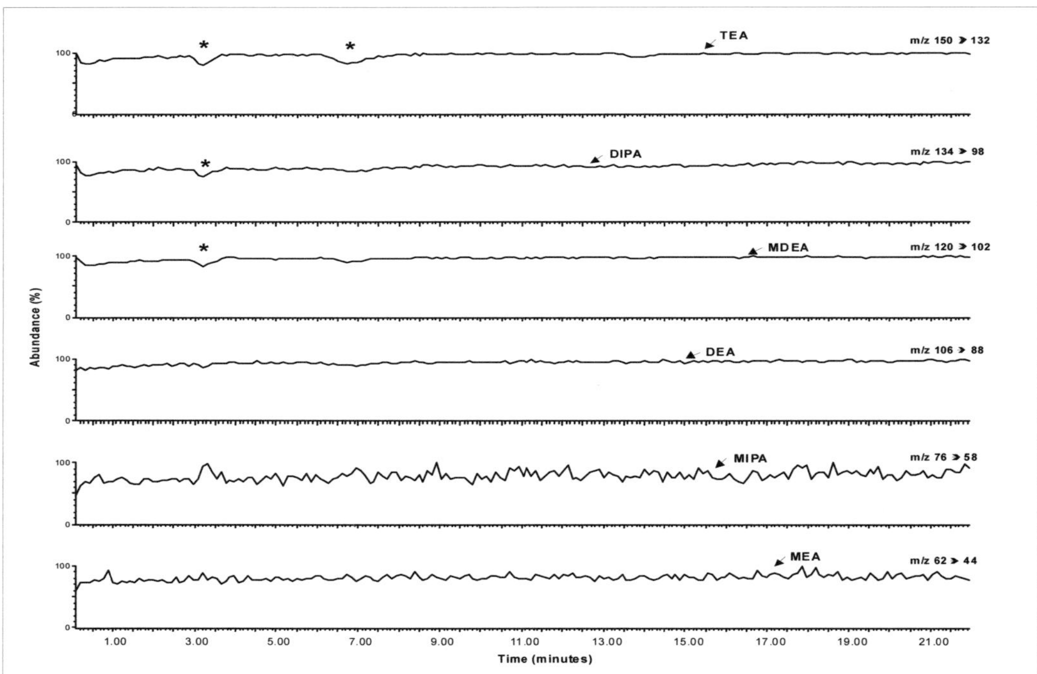

Fig. 7. Multiple reaction monitoring chromatogram (transitions monitored as per Table 4) of a mixture of six alkanolamines post column infusion to ascertain ionization enhancement/suppression (Peru et al., 2004).

Improved chromatographic separation was achieved using a Dionex CS14 cation-exchange column for reliable quantification of the selected alkanolamines (Figure 8). Chromatography and instrumental sensitivity did not decline significantly during the course of analyzes (over 80 vegetation extractions analyzed during a 3-week duration), while evaluation of between-day precision based on replicate samples gave a RSD values of <10% indicating an overall method robustness (Peru et al., 2004).

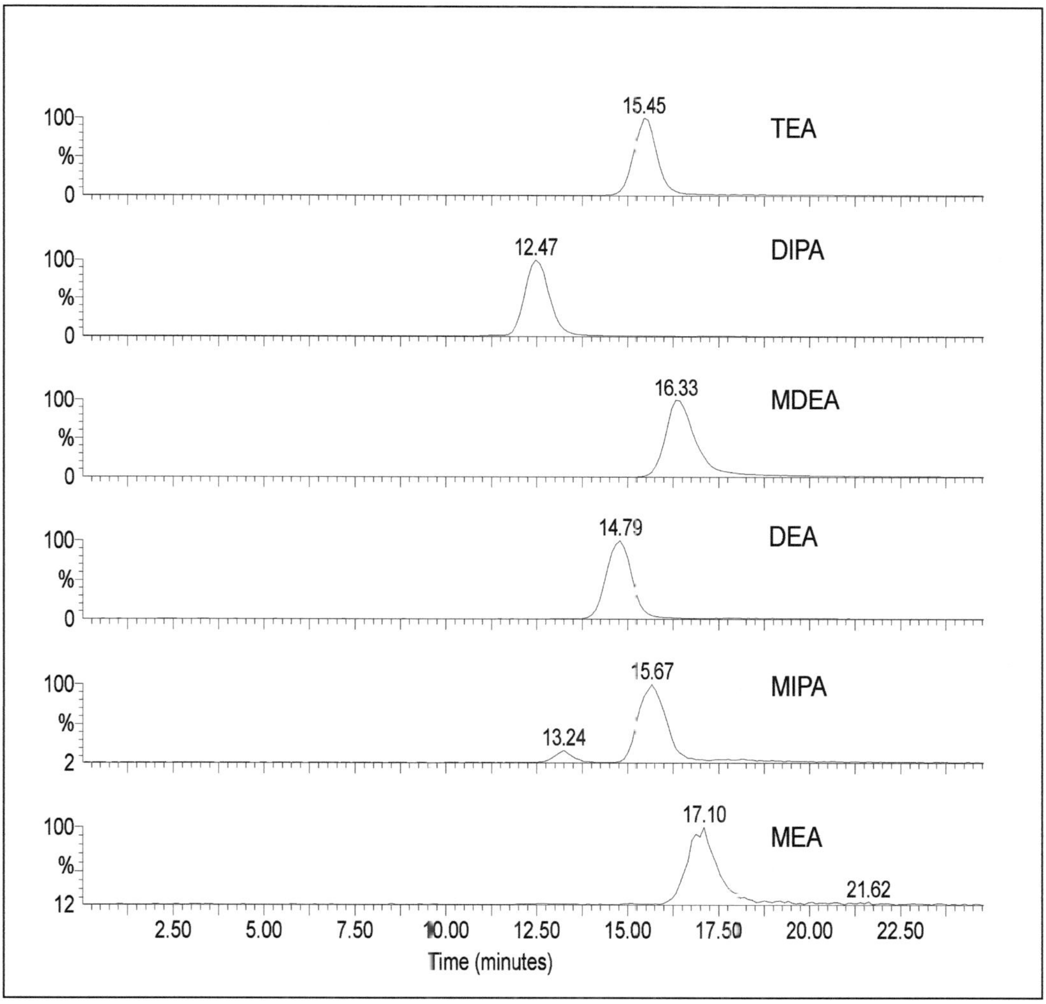

Fig. 8. Improved chromatographic separation of six alkanolamines using ion-exchange chromatography and MRM (transitions monitored as per Table 4) (Peru et al., 2004).

9.3 Quantification calculation of alkanolamines in cattails

Signal to noise ratio (S/N) is one of the most important performance characteristics considered in assessment to measure low analyte levels in environmental samples. In routine practice, the S/N value is used for prediction of LOQ (Kovalczuk et al., 2006).

Instrumental detection limits (DL), based on three times the signal-to-noise (S/N) ratio obtained from spiked blank matrix extract, improved using SRM for three of the five alkanolamines previously reported using SIM; DIPA 6 pg vs. 20 pg, TEA 4 pg vs. 30 pg and MDEA 14 pg vs. 32 pg injected. MEA and DEA provided poorer detection limits as compared with those previously reported; MEA 300 pg vs. 40 pg and DEA 47 pg vs. 40 pg injected. MIPA was not included in the SIM method therefore its DL of 104pg injected could not be compared. Likewise, limit of quantification (LOQs) for DIPA (2.6ng/g), TEA (5.2ng/g) and MDEA (18.6ng/g) represented were an improvement on the reported value of 20ng/g for each alkanolamine obtained using the SIM method. For both MEA (200 ng/g) and DEA (6 ng/g), sensitivity was compromised for selectivity. MIPA gave a LOQ of 69 ng/g. LOQs were calculated using an S/N ratio of 6:1 from data obtained for extracts of spiked root matrix. At this S/N ratio, variability in the reported LOQs was <10%relative standard deviation (RSD). The linearity for calibration standards, ranging from 0.01-1.00mg/mL, had r^2 values ranging from 0.9966-0.9999. For all the alkanolamines investigated, the response began to plateau (deviate from linearity) at concentrations above 3.00 mg/mL. Blank root tissues were spiked at two levels based on upper and lower levels expected in samples (2.5 and 10μg/g). Recoveries ranged from 76-116% for the 2.5mg/g level and 68-98% for the 10mg/g level with RSD <10%, n=6 (Table 5) (Peru et al., 2004).

Alkanolamines	Recovery (%) in root matrix ±SD (2.5μg/g)	RSD (%) (n=6)	Recovery (%) in root matrix ±SD (10.0μg/g)	RSD (%) (n=6)
TEA	92±1.1	4.4	83±3.2	3.9
DIPA	76±1.3	5.2	77±6.1	7.9
MDEA	106±2.4	6.0	98±3.5	3.6
DEA	77±1.2	3.6	68±3.6	5.3
MIPA	84±2.6	8.1	84±6.7	7.9
MEA	116±2.3	7.6	94±7.4	7.8

Table 5. Recovery data of six alkanolamines from fortified *Typha latifolia* root (fortification concentration in parenthesis) (Peru et al., 2004).

9.4 Derivatization of alkanolamines

To date, there are a few methods and techniques that use derivatization steps to improve the detectability of alkanolamines from environmental samples. These protocols however, do not use tandem mass spectrometry for quantification. They involve GC-MS, LC, and /or IEC as separation tool to mobilize the analytes from the matrices. The detection, confirmation and quantification of alkanolamines either use fluorescence or conductivity techniques (Serbin and Birkholz, 1995 as cited in Headley et al., 2002).

10. Conclusions

According to Headley, et al., (2002), the choice of which procedure is adopted in a given laboratory, for the determination of alkanolamines will be based primarily on the expertise and instrumentation available in such laboratory. Where appropriate, methods employing direct aqueous injections of samples are best suited for analyzes of alkanolamines in environmental samples. These methods eliminate the time and expense of organic solvent extraction procedures. While capillary GC columns have also been used for the analysis of the alkanolamines, best results have been achieved using LC methods with or without derivatization steps. Because procedures employing advances in positive-ion electrospray ionization techniques with IC- MS-MS detection have proven useful for the confirmation and recovery of alkanolamines in wetland vegetation such as cattails, it is of essence that future research should focus on the determination of the native compounds from different samples from the environment. Further developments will likely centre on mass spectrometric analyzes for identification of metabolites and transformation products in aquatic environments. Such work would assist in policy making to control the level of discharge of these compounds into the environment.

11. Appendix 1

11.1 Guidelines for some common alkanolamines in water, soil and plants

Though the preceding section enumerated studies carried out by different academic research groups on alkanolamines, it should be mentioned that some governmental projects have been carried out in Canada by different agencies such as the Canadian Council of Ministers of the Environment (CCME), the Canadian Association of Petroleum Producers (CAPP) and Alberta Environment (AENV) to set guidelines for the levels of alkanolamines that are allowed into environmental compartments (water, soil and plant/vegetation) from discharges generated by chemical industries where green house gases are emitted and the application of alkanolamines are required. These guidelines originated basically by collating different research works that have been carried out on alkanolamines and limits were set based on a wide range of data collected from toxicological, ecotoxological and genotoxicological studies.

Reports on the presence of anthropogenic DIPA in the environment are limited to data collected at sour gas processing facilities in western Canada (CAPP 1997; Wrubleski and Drury 1997 as cited in CCME, 2005). The maximum measured DIPA concentration in groundwater was 590mg/L in a shallow till aquifer (Greene et al. 1999 as cited in CCME, 2005). No studies were found that had detected DIPA as a naturally occurring compound in the environment (CCME, 2005).

Uptake of DIPA by wetland vegetation was studied as part of a research program to evaluate natural attenuation processes in contaminated wetlands (CAPP 1998, 1999, 2000 as cited in CCME 2005). Roots, stems, leaves, flower heads, seed heads, and berries of cattail, dogwood, sedge, marsh reed grass, cow parsnip, and smooth brome growing in a DIPA impacted wetland were included in the study (CAPP 1999, 2000; Headley et al. 1999a,b as cited in CCME, 2005). Analytical results indicated highly variable DIPA concentrations for different parts of the same species (e.g., roots versus leaves), between different plant species (e.g., cattail leaves versus sedge leaves), and even between different samples of the same part of the same species. Although the maximum measured DIPA concentration in water in the wetland was only 13mg/L, DIPA concentration as high as 208mg/kg were measured in the plants (CCME, 2005).

Environmental sample	Alkanolamines	Guide line values (mg/L)
Aquatic		
Fresh water	DIPA*	1.6[a]
Marine		NRG
Human drinking water	DEA	0.06
	MEA	0.6
Fresh water aquatic life	DEA	0.45
	MEA	0.075
Agricultural Soil		Guideline values (mg/kg)
Fine soil	DEA	2.0
	MEA	20
Coarse soil	DEA	3.5
	MEA	10

Table 6. Guidelines values for some common alkanolamines (Collated from CCME, 2005 and Government of Alberta, 2010). *Data sourced from CCME, 2005. [a]Interim guideline; [b]No recommended guideline.

12. Acknowledgement

We would like to thank the Program of Energy and Research Development (PERD) for financial support of this project.

13. References

BASF the Chemical Company (2006). BASF increases prices for alkanolamines, alkylamines, In: *Focus on Surfactant,* Press release from: BASF AG, Ludwigshafen, Germany, 03.06.2011, Available from http://www.basf.com

Bu, H-Z., Poglod, M., Micetich, R., & Khan, J. (2000). Novel sample preparation method facilitating identification of urinary drug metabolites by liquid chromatography–

tandem mass spectrometry. *Journal of Chromatography B: Biomedical Sciences and Applications*, Vol.738, No. 2, (November 1999), pp. 259-265, ISSN 0378-4347

Canadian Council of Ministers of the Environment (CCME) (2005). Canadian water quality guidelines for the protection of aquatic life: Diisopropanolamine, 08.07.2011, pp. 1-3, ISBN 189-6997-34-1, Available from http://www.ec.gc.ca/ceqg-rcqe

Choi, B., Hercules, D., & Gusev, A. (2001). Effect of liquid chromatography separation of complex matrices on liquid chromatography-tandem mass spectrometry signal suppression. *Journal of Chromatography A*, Vol.907, No.1-2, (October 2000), pp. 337-342, ISSN 0021-9673

Davis, J., Carpenter, C. (1997). Environmental assessment of alkanolamines. *Reviews of Environmental Contamination and Toxicology*, Vol.149, (January, 1997), pp. 87-132, ISSN 1498-7137

Dickson, L.C., Headley, J., Peru, K., Spiegel, K.., & Gandrass, J. (1996). HPLC and MS/MS study of polar contaminants in a wetland adjoining a sour-gas plant. *SETAC 17th Annual Meeting*, Washington DC, November 1996

Dobberpuhl, D., & Johnson, D. (1995). Pulsed electrochemical detection of alkanolamines separated by multimodal high-performance liquid chromatography. Journal of Chromatography A, Vol.694, (November 1994), pp. 391–398, ISSN 0021-9673

DOW the Chemical Company (2003). Alkyl Alkanolamines, application note. No. 111-01376-303AMS, (01.06.2011), Available from http://msdssearch.dow.com/PublishedLiteratureDOWCOM

Eide-Haugmo, I., Brakstad, O., Hoff, K., Sørheim, K., da Silva, E., & Svendsen, H. (2009). Environmental impact of amines. *Energy Procedia*, Vol.1, No.1, pp. 1297-1304

Gamer, A., Rossbacher, R., Kaufmann, W., & van Ravenzwaay, B. (2008). The inhalation of di- and triethanolamine upon repeated exposure. *Food and Chemical Toxicology*, Vol. 46, No.6, (February 2008), pp. 2173-2183, ISSN 0278-6915

Government of Alberta (2010). Soil and groundwater remediation guidelines for monoethanolamine and diethanolamine, 08.07.2011, Available from http://environment.gov.ab.ca/info/ Online ISBN 978-0778-5900-6

Hamborg, E., Derks, P., van Elk, E., & Versteeg, G. (2011). Carbon dioxide removal by alkanolamines in aqueous organic solvents: a method for enhancing the desorption process. *Energy Procedia*, Vol.4, pp.197-194

Hansen, B., Altin, D., Andy Booth, A , Vang, S-H., Frenzel, M., Sørheim, K., Brakstad, O., Størseth, T. (2010). Molecular effects of diethanolamine exposure on *Calanus finmarchicus* (Crustacea: Copepoda. *Aquatic Toxicology*, Vol.99, No.2, (April 2010), pp. 212-222, ISSN 0166-445X

Headley, J., Fedorak, P., Dickson, L. (2002). A review of analytical methods for the determination of sulfolane and alkanolamines in environmental studies. *Journal of AOAC International*, Vol.85, No.1, pp. 154–162

Headley, J., Peru K., & Dickson L. (1999). Gas chromatographic-mass spectrometric determination of sulfolane in wetland vegetation exposed to sour gas-contaminated groundwater. *Journal of Chromatography A,* Vol.859, (July 1999), pp. 69-75, ISSN 0021-9673

Jagota, N., Chetram, A., & Nair, J. (1996). Ion chromatography of amylamine and *tert*-butylamine in pharmaceuticals. Journal of Chromatography A, Vol.739, No.1-2, (July 1996), pp.343-349, ISSN 0021-9673

Kamiński, M., Jastrzebski, D., Przyjazny, A., & Kartanowicz, R. (2002). Determination of the amount of wash amines and ammonium ion in desulfurization products of process gases and results of related studies. *Journal of Chromatography A,* Vol.947, No.2, (December 2001), pp. 217-225, ISSN 0021-9673

Kovalczuk, T., Jech, M., Poustka, J., & Hajšlová, J. (2006). Ultra-performance liquid chromatography–tandem mass spectrometry: a novel challenge in multiresidue pesticide analysis in food. *Analytica Chimica Acta,* Vol. 577, No.1, (June 2006), pp. 8-17, ISSN 0003-2670

Kraeling, M., Yourick, J., & Bronaugh, R. (2004). *In vitro* human skin penetration of diethanolamine. *Food and Chemical Toxicology,* Vol.42, No.10, (April 2004), pp. 1553-1561, ISSN 0278-6915

Lepaumier, H., Grimstvedt, A., Vernstad, K., Zahisen, K., & Svendsen, H. (2011). Degradation of MMEA at absorber and stripper conditions. *Chemical Engineering Science,* Vol.66, No.15, (April 2011), pp. 3491-3498, ISSN 0009-2509

Leung, H-W., Ballantyne, B. (1997). Evaluation of the genotoxic potential of alkylalkanolamines. *Mutation Research/Genetic Toxicology and Environmental Mutagenesis,* Vol.393, No.1-2, (September 1997), pp. 7-15, ISSN 1383-5718

Libralato, G., Ghirardini, A., & Avezzù, F. (2008). Evaporation and air-stripping to assess and reduce ethanolamines toxicity in oily wastewater. *Journal of Hazardous Materials,* Vol.153, No.3, (September 2007), pp.928-936, ISSN 0304-3894

Libralato, G., Ghirardini, A., & Avezzù, F. (2010). Seawater ecotoxicity of monoethanolamine, diethanolamine and triethanolamine. *Journal of Hazardous Materials,* Vol.176, No.1-3, pp. 535-539, ISSN 0304-3894

Lu, C-S., Chen, C-C., Mai, F-D., Li, H-H. (2009). Identification of the degradation pathways of alkanolamines with TiO_2 photocatalysis. *Journal of Hazardous Materials,* Vol. 165, No.1-3, (September 2009), pp. 306-316, ISSN 0304-3894

Luther, S., Dudas, M., & Fedorak, P. (1998). Sorption of sulfolane and diisopropanolamine by soils, clays and aquifer materials. *Journal of Contaminant Hydrology,* Vol.32, No.1-2, (December 1997), pp. 159-176, ISSN 0169-7722

Mendrala, A., Waechter Jr., J., Bormett, G., Bartels, M., & Stott, W. (2001). The pharmacokinetics of diethanolamine in Sprague–Dawley rats following intravenous administration. *Food and Chemical Toxicology,* Vol.39, No.9, (March 2001), pp. 931-939, ISSN 0278-6915

Mundhwa, M., & Henni, A. (2007). Molar excess enthalpy (H^E_m) for various {alkanolamine (1) + water (2)} systems at T= (298.15, 313.15, and 323.15) K. Journal of Chemical Thermodynamics, Vol.39, (March 2007), pp. 1439-1451, ISSN 0021-9614

Nödler, K., Licha, T., Bester, K., & Sauter, M. (2010). Development of a multi-residue analytical method, based on liquid chromatography–tandem mass spectrometry, for the simultaneous determination of 46 micro-contaminants in aqueous samples. *Journal of Chromatography A*, Vol.1217, No.42, (August 2010), pp. 6511-6521, ISSN 0021-9673

Padurean, A., Cormos, C-C., Cormos, A-M., & Paul-Serban Agachi, P-S. (2011), Multicriterial analysis of post-combustion carbon dioxide capture using alkanolamine. *International Journal of Greenhouse Gas Control, In Press, Corrected Proof, Available online 2 March 2011,* (February 2011), ISSN 1750-5836

Payne, A., & Glish, G. (2005).Tandem mass spectrometry in quadrupole ion trap and ion cyclotron resonance mass spectrometers. *Methods in Enzymology,* Vol.402, pp. 109-148

Peru, K., Headley, J., & Doucette, W. (2004). Determination of alkanolamines in cattails (Typha latifolia) utilizing electrospray ionization with selected reaction monitoring and ion-exchange chromatography. *Rapid Communications in Mass spectrometry,* Vol.18, (May 2004), pp. 1629-1634

Peters, F. (2011). Recent advances of liquid chromatography-(tandem) mass spectrometry in clinical and forensic toxicology. Clinical Biochemistry, Vol.44, No.1, (August 2010), pp. 54-65, ISSN 0090-9120

Raffi, M., Elango, R., House, J., Courtney-Martin, G., Darling, P., Fisher, L., & Pencharz, P. (2009). Measurement of homocysteine and related metabolites in human plasma and urine by liquid chromatography electrospray tandem mass spectrometry. *Journal of Chromatography B,* Vol.877, No.28, (May 2009), pp. 3282-3291, ISSN 1570-0232

Rodrigues, J., Taylor, A., Sumptom, D., Reynolds, J., Pickford, R., & Thomas-Oates, J. (2008). Mass spectrometry of carbohydrates: new aspects. *Advances in Carbohydrates Chemistry and Biochemistry,* Vol.61, pp. 59-141, ISBN 978-0-12-373920-9

Saghir, S., Brzak, K., Markham, D., Bartels, M., & Stott, W. (2005). Investigation of the formation of *N*-nitrosodiethanolamine in B6C3F1 mice following topical administration of triethanolamine. *Regulatory Toxicology and Pharmacology,* Vol.43, No.1, (January 2005), pp. 10-18, ISSN 0273-2300

Shih, T-W., Chen, Y-J., & Li, M-H. (2002). Heat capacity of aqueous mixtures of monoethanolamine with 2-piperidineethanol. *Thermochimica Acta,* Vol.389, (January 2002), pp. 33-41, ISSN 0040-6031

Stott, W., Waechter Jr, J., Rick, D., & Mendrala, A. (2000). Absorption, distribution, metabolism and excretion of intravenously and dermally administered triethanolamine in mice. *Food and Chemical Toxicology,* Vol.38, No.11, (June 2000), pp. 1043-1051, ISSN 0278-6915

Taylor, A., Linforth, R., Harvey, B., & Blake, A. (2000). Atmospheric pressure chemical ionisation mass spectrometry for in vivo analysis of volatile flavour release. *Food Chemistry,* Vol.71, No.3, (April 2000), pp. 327-338, ISSN 0308-8146

Vaidya, P., & Kenig, E. (2009). Kinetics of carbonyl sulfide reaction with alkanolamines: a review. Chemical Engineering Journal, Vol.148, (August 2008), pp. 207-211, ISSN 1385-8947

Witzaney, A.M. and Fedorak, P.M., (1996). A Review of the Characteristics, Analyses and Biodegradability of Sulfolane and the Alkanolamines used in Sour Gas Processing, Consulting report submitted to Shell Canada Limited, February 1996

24

Application of Tandem Mass Spectrometry in Chemical Kinetics

Jinping Qiao
College of Chemistry, Beijing Normal University
China

1. Introduction

Chemical kinetics, also known as reaction kinetics, is the study of rates of chemical processes. Chemical kinetics includes investigations of how different experimental conditions can influence the speed of a chemical reaction and yield information about the reaction's mechanism and transition states, as well as the construction of mathematical models that can describe the characteristics of a chemical reaction.

The mathematical models that describe chemical reaction kinetics provide chemists and chemical engineers with tools to better understand and describe chemical processes such as food decomposition, microorganism growth, stratospheric ozone decomposition, and the complex chemistry of biological systems. These models can also be used in the design or modification of chemical reactors to optimize product yield, more efficiently separate products, and eliminate environmentally harmful by-products. When performing catalytic cracking of heavy hydrocarbons into gasoline and light gas, for example, kinetic models can be used to find the temperature and pressure at which the highest yield of heavy hydrocarbons into gasoline will occur.

Chemical kinetics can be studied by experimental determination. Kinetic measurements represent quite a challenge to the experimentalist. Firstly, reactions proceed on a vast range of different timescales-varying from the almost geological to sub nano-second. We need all sorts of different strategies for making measurements over this range. Secondly, many reactions involve complex mixtures, perhaps with the species in vastly different concentrations; we want to be able to measure the concentrations of all these species individually. Thirdly, we want to be able to do all this without interfering with the reaction mixture- this point to the use of physical methods of measuring concentration, which are non-invasive. Finally, it would be nice to be able to automate taking concentration readings.

The basic measurements we can make are concentration as a function of time. We then use various methods to determine the rate law from this raw data. Rate laws are essentially differential equations, and so need to be integrated (solved) in order to see if the data fits the law. If the fit is acceptable to within the errors of the experimental data we say that the proposed rate law is consistent with the data. If the fit is not good enough, another law will have to be proposed and tested against the data. A few simple rate laws can be solved "by hand", but most can only be solved numerically using a computer program. There are many computer algorithms available for tackling this problem.

Many methods have been used to measure concentration, for example, UV absorptions, IR measurements, conductivity measurements, classical methods of chemical analysis, such as titrations. UV absorption and IR measurement tend to be rather broad, so it is possible that more than one species will absorb at a given wavelength. NMR is not very sensitive so as an analysis method it is not very fast. The relationship between conductance and concentration can be rather involved, so it is really only convenient to use conductivity measurement for relative measurements of concentration i.e. appropriate for first order kinetics. Classical methods are rather slow, and so it is usually necessary to extract some of the reaction mixture and then stop the reaction so that no further reaction takes place during the analysis (Teng et al., 2001; Chancellor et al., 2008; Katsuda& Toshiro, 2009, Pierre et al., 2009; Stepensky et al., 2004).

A more sophisticated method is mass spectrometry, gas chromatogram (GC) in conjunction with a mass spectrometer and high-performance Liquid Chromatography combined with Mass Spectrometry (LC-MS or LC-MS/MS) have been used to detect concentrations. A GC or LC separates the components of a mixture, and different components travel at different speeds down the column and so are detected at different times after injecting the sample; their concentration can thus be measured separately. The mass spectrometer can not only detect the presence of molecules but also identify them. LC-MS is a powerful technique used for many applications which has very high sensitivity and selectivity. A tandem mass spectrometer can also be used on its own to measure and identify species directly-that is without the use of a GC or LC (Gadkari, T.; 2010; Yuan et al., 2009; Larsen et al., 2009).

In this chapter, the application of mass spectrometry in chemical kinetics will be introduced.

2. Chemical kinetics

In 1864, Peter Waage and Cato Guldberg pioneered the development of chemical kinetics by formulating the law of mass action, which states that the speed of a chemical reaction is proportional to the quantity of the reacting substances (Denisov et al., 2003; Cox, 2003; Grimes, 2001).

Chemical reaction rates are the rates of change in concentrations or amounts of either reactants or products. For changes in amounts, the units can be one of mol/s, g/s, lb/s, kg/day etc. For changes in concentrations, the units can be one of mol/(L s), g/(L s), %/s etc.

With respect to reaction rates, we may deal with average rates, instantaneous rates, or initial rates depending on the experimental conditions.

Thermodynamics and kinetics are two factors that affect reaction rates. The study of energy gained or released in chemical reactions is called thermodynamics, and such energy data are called thermodynamic data. However, thermodynamic data have no direct correlation with reaction rates, for which the kinetic factor is perhaps more important. For example, at room temperature (a wide range of temperatures), thermodynamic data indicates that diamond shall convert to graphite, but in reality, the conversion rate is so slow that most people think that diamond is forever (Denisov et al., 2003; Cox, 2003; Grimes et al., 2001).

2.1 Factors affecting reaction rate

Many factors influence rates of chemical reactions, and these are summarized below.

2.1.1 Nature of Reactants

Depending upon what substances are reacting, the reaction rate varies. Acid/base reactions, the formation of salts, and ion exchange are fast reactions. When covalent bond formation takes place between the molecules and when large molecules are formed, the reactions tend to be very slow. Nature and strength of bonds in reactant molecules greatly influences the rate of its transformation into products. The reactions which involve lesser bond rearrangement proceed faster than the reactions which involve larger bond rearrangement (Laidler, 1989; Tur'yan et al., 1998; John& Schlegel, 1998).

2.1.2 Temperature

Temperature usually has a major effect on the rate of a chemical reaction. Molecules at a higher temperature have more thermal energy. Although collision frequency is greater at higher temperatures, this alone contributes only a very small proportion to the increase in rate of reaction. Much more important is the fact that the proportion of reactant molecules with sufficient energy to react (energy greater than activation energy: $E > E_a$) is significantly higher and is explained in detail by the Maxwell-Boltzmann distribution of molecular energies.

2.1.3 concentration effect

Concentration plays a very important role in reactions, because according to the collision theory of chemical reactions, molecules must collide in order to react together. As the concentration of the reactants increases, the frequency of the molecules colliding increases, striking each other more frequently by being in closer contact at any given point in time. Think of two reactants being in a closed container. All the molecules contained within are colliding constantly. By increasing the amount of one or more of the reactants it causes these collisions to happen more often, increasing the reaction rate.

2.1.4 Physical state

The physical state (solid, liquid, or gas) of a reactant is also an important factor of the rate of change. When reactants are in the same phase, as in aqueous solution, thermal motion brings them into contact. However, when they are in different phases, the reaction is limited to the interface between the reactants. Reaction can only occur at their area of contact, in the case of a liquid and a gas, at the surface of the liquid. Vigorous shaking and stirring may be needed to bring the reaction to completion. This means that the more finely divided a solid or liquid reactant, the greater its surface area per unit volume, and the more contact it makes with the other reactant, thus the faster the reaction. To make an analogy, for example, when one starts a fire, one uses wood chips and small branches—one doesn't start with large logs right away. In organic chemistry, on water reactions are the exception to the rule that homogeneous reactions take place faster than heterogeneous reactions.

2.1.5 Catalysts

By the nature of the term, catalysts play important roles in chemical reactions. Generic potential energy diagram shows the effect of a catalyst in a hypothetical endothermic chemical reaction. The presence of the catalyst opens a different reaction pathway with lower activation energy. The final result and the overall thermodynamics are the same.

A catalyst is a substance that accelerates the rate of a chemical reaction but remains chemically unchanged afterwards. The catalyst increases rate reaction by providing a different reaction mechanism to occur with lower activation energy. In autocatalysis a reaction product is itself a catalyst for that reaction leading to positive feedback. Proteins that act as catalysts in biochemical reactions are called enzymes. Michaelis-Menten kinetics describes the rate of enzyme mediated reactions. A catalyst does not affect the position of the equilibrium, as the catalyst speeds up the backward and forward reactions equally.

2.1.6 Pressure

Increasing the pressure in a gaseous reaction will increase the number of collisions between reactants, increasing the rate of reaction. This is because the activity of a gas is directly proportional to the partial pressure of the gas. This is similar to the effect of increasing the concentration of a solution.

2.1.7 Equilibrium

While chemical kinetics is concerned with the rate of a chemical reaction, thermodynamics determines the extent to which reactions occur. In a reversible reaction, chemical equilibrium is reached when the rates of the forward and reverse reactions are equal and the concentrations of the reactants and products no longer change. This is demonstrated by, for example, the Haber–Bosch process for combining nitrogen and hydrogen to produce ammonia. Chemical clock reactions such as the Belousov–Zhabotinsky reaction demonstrate that component concentrations can oscillate for a long time before finally attaining the equilibrium.

2.1.8 Free energy

In general terms, the free energy change (ΔE) of a reaction determines whether a chemical change will take place, but kinetics describes how fast the reaction is. A reaction can be very exothermic and have a very positive entropy change but will not happen in practice if the reaction is too slow. If a reactant can produce two different products, the thermodynamically most stable one will generally form except in special circumstances when the reaction is said to be under kinetic reaction control. The Curtin–Hammett principle applies when determining the product ratio for two reactants interconverting rapidly, each going to a different product. It is possible to make predictions about reaction rate constants for a reaction from free-energy relationships (Saunders et al., 1999; Laidler, 1989; Tur'yan et al., 1998; John& Schlegel, 1998).

2.2 Rate laws and rate constants

Chemical kinetics deals with the experimental determination of reaction rates from which rate laws and rate constants are derived. Relatively simple rate laws exist for zero-order reactions (for which reaction rates are independent of concentration), first-order reactions, and second-order reactions, and can be derived for others. In consecutive reactions the rate-determining step often determines the kinetics. In consecutive first-order reactions, a steady state approximation can simplify the rate law. The activation energy for a reaction is experimentally determined through the Arrhenius equation and the Eyring equation (Hagrman et al., 2004).

2.2.1 Differential method

If the law is of the form

$$r = k\,[C]^n \tag{1}$$

then an appealing method of finding the value of the order, n, is to plot the log of the rate against the log of [C]

$$\ln r = \ln k + n\ln[C] \tag{2}$$

such a graph will have slope n. this method, called the differential method.
The drawback is that rather than plotting a function of concentration we have to plot rates, and rates are much harder to measure than concentrations. The rate is the slope of a graph of concentration against time, and as such a graph is usually curved taking an accurate slope is not at all easy.

2.2.2 Zero-order reaction

For a zero-order reaction, the rate of reaction is a constant. When the limiting reactant is completely consumed, the reaction abrupt stops.
Differential Rate Law:

$$r = k \tag{3}$$

The rate constant, k, has units of mole L^{-1} sec^{-1}.

2.2.3 First-order reaction

For a first-order reaction, the rate of reaction is directly proportional to the concentration of one of the reactants.
Differential Rate Law:

$$r = k\,[C] \tag{4}$$

The rate constant, k, has units of sec^{-1}.
The reaction rate can be described by the following equation (Su et al., 2009):

$$-\frac{d[C]}{dt} = k[C] \tag{5}$$

The equation can be integrated as follows:

$$\ln\frac{C_t}{C_0} = -kt \tag{6}$$

Where C_t is the concentration at time t, and C_0 is the initial concentration, k is the rate constant.

2.2.4 Second-order reaction

For a second-order reaction, the rate of reaction is directly proportional to the square of the concentration of one of the reactants.
Differential Rate Law:

$$r = k\,[C]^2 \tag{7}$$

The rate constant, k, has units of L mole^{-1} sec^{-1}.
The reaction rate can be described by the following equation:

$$-\frac{dc_A}{dt} = k_2 c_A^2 \tag{8}$$

2.2.5 Half lives

The half life of a reaction is defined as the time it takes for the concentration of a specified reagent to fall to half of its initial value.

Order of reaction	Reaction Equation	Initial Concentration	Rate Formula	Half-life
Zero-Order	A → Production	$C_{A,0} = a$	$-\frac{dc_A}{dt} = k_0$	$\frac{a}{2k_0}$
First-Order	A → Production	$C_{A,0} = a$	$-\frac{dc_A}{dt} = k_1 c_A$	$\frac{\ln 2}{k_1}$
Second-Order	A+B → Production	$C_{A,0} = C_{B,0} = a$	$-\frac{dc_A}{dt} = k_2 c_A^2$	$\frac{1}{k_2 a}$

Table 1. Summary of rate laws and rate constants of chemical kinetics.

3. Arrhenius equation

It is well-known that raising the temperature increases the reaction rate. Quantitatively this relationship between the rate a reaction procession and its temperature is determined by the Arrhenius Equation:
The influence of temperature on the rate constant (k) could be described through the Arrhenius equation (Boys& Konermann, 2007).

$$k = Ae^{(-E_a/RT)} \tag{9}$$

The Arrhenius equation is often written in the logarithmic form:

$$\ln k = \ln A - \frac{E_a}{RT} \tag{10}$$

where k is the reaction rate constant, A is the frequency factor, E_a is the activation energy (Jmol^{-1}), R is the universal gas constant (8.314 Jmol^{-1}K^{-1}), and T is the absolute temperature (K).
Thus the frequency factor is a constant, specific for each reaction.
The Arrhenius equation is based on the collision theory which supposes that particles must collide with both the correct orientation and with sufficient kinetic energy if the reactants are to be converted into products.

4. Application of tandem mass spectrometry in chemical kinetics

Many papers have been reported to perform chemical kinetics using mass spectrometry. Kumar et al (2011) used LC-MS/MS to study the kinetic modeling of phenol oxidation in a medium suitable for bioremediation of organic pollutants. The reaction mechanism used for kinetic modeling is based on the intermediate oxidation products identified in this study using LC-MS/MS and ion chromatography. Progress of the chemical oxidation by Fenton's reagent was monitored by determining the residual phenol concentration and concentrations of evolved intermediate compounds (catechol and hydroquinone) at regular time intervals. The rate of phenol oxidation and ultimate conversion of phenol were found to increase with increase in hydrogen peroxide concentration. The increase in temperatures has a positive effect on phenol oxidation and the rate of phenol oxidation was found to increase with temperature in the range of 5-35°C. Kinetic parameters, namely rate constants and activation energies for reactions involved, were determined by best-fitting the experimental data to the proposed reaction model.

Additionally, Clopyralid (3,6-dichloropyridine-2-carboxylic acid) is a herbicide, LC-MS/MS (ESI+) was used to study its pathways of the photocatalytic degradation. It is found the investigated concentration range (0.5-3.0mM) the photocatalytic degradation kinetics of clopyralid in the first stage of the reaction follows approximately a pseudo-first kinetic order. The effect of the presence of hydrogen peroxide, potassium bromate, and ammonium persulfate, acting as electron acceptors along with molecular oxygen, were also studied. The reaction intermediates (3,6-dichloropyridin-2-ol, 3,6-dichloro hydroxypyridine-2-carboxylic acid, and 3,3',6,6'-tetrachloro-2,4'-bipyridine-2'-carboxylic acid) were identified and the kinetics of their appearance/disappearance was obtained (Sojic et al., 2009).

The photochemical behavior of the antifouling agent zinc pyrithione (ZnPT) was studied in aqueous media of different composition under simulated solar irradiation using a xenon light source. The influence of important constituents of natural water (dissolved organic matter and nitrate) was also examined using a multivariate kinetic model. It was found that photodegradation proceeds via a pseudo first-order reaction. Kinetic experiments were monitored by LC-MS (Sakkas et al., 2007).

Diaminodithiol (N_2S_2)-type compounds readily oxidize to produce disulfide. The ethyl cysteinate dimer (ECD) was an important N_2S_2 ligand, ^{99m}Tc complexes of ECD have extensive utility in medicine as brain perfusion agents and in renal function studies. An ultra-performance liquid chromatography/tandem mass spectrometry (UPLC-MS/MS) method in multiple reaction monitoring mode was developed, and ECD and its oxidized product were quantitated in solution. The dynamic oxidation process of ECD in solution was studied in detail. The full time course of the decrease in ECD and the increase in its oxide was observed, the oxidation procedure followed the first-order kinetics, and the half-life time of ECD was 51min (Qiao et al., 2009).

Fig 1 shows the structures of ECD and its oxidized product. In the full-scan spectrum (Fig 2), peaks of m/z 325 and m/z 323 were observed for the $[M+H]^+$ peaks of ECD and its oxide, respectively. The peaks of m/z 645 and m/z 967 were from the $[2M+H]^+$ and $[3M+H]^+$ peaks of the oxidized product, respectively. Then the product scans were performed. Fig 3 illustrates the product ion mass spectra of m/z 325 and 323. The product ion spectrum of ECD (m/z 325) is shown in Figure 3(a). The most prominent product ion was at m/z 176, which resulted from the loss of the $C_5H_{11}NO_2S$ group from the precursor ion. For the MRM analysis, the m/z 325→176 transition was used to monitor ECD. Figure 3(b) shows the MS/MS spectrum of m/z323, which is the $[M+H]^+$ peak of the oxidized product. The fragment ion at m/z 174 corresponds to the loss of the $C_5H_{11}NO_2S$ group. The MRM mode was applied by monitoring the transition between m/z 323 and m/z 174 for the oxide.

UPLC was performed with a Waters Acquity UPLC™ system (Waters, Milford, MA, USA), equipped with a binary solvent delivery manager and a sample manager. LC-MS/MS analysis was performed with a Waters Micromass Quattro Micro™ tandem quadrupole mass spectrometer (Micromass Manchester, UK). The system was controlled with a Masslynx™ 4.1 with a QuanLynx™ Application Manager.

The UPLC separation was performed on a Waters Acquity ethylene-bridged (BEH™) C_{18} column (2.1 × 50 mm, 1.7 μm particle size) at ambient temperature at a flow rate of 0.35 mL/min. The mobile phase consisted of (A) water containing 0.1% (v/v) formic acid and (B) acetonitrile. Both eluents (A) and (B) were filtered through a 0.22 μm membrane filter and degassed for 5 min in an ultrasonic bath. An isocratic elution was performed using 80% B and 20% A. The injection volume was 2 μL.

The quantification analysis was performed in multiple reaction monitoring (MRM) mode using a Quattro Micro mass spectrometer. The sample was ionized by electrospray in positive mode (ESI+), and nitrogen was used as the nebulizing gas, desolvation gas, and cone gas. Argon at a pressure of $2.5e^{-3}$ mbar was used as the collision gas. The capillary voltage was 3.2 kV, the source temperature was 110 °C, the desolvation temperature was 350 °C, the desolvation gas flow was 400 L/h, the cone gas flow was 30 L/h, and the collision energy was 30 eV. The monitoring ions for the MRM analysis were set to m/z 325→176 and 323→174, and the dwell time was set to 0.3 s per transition.

Because MRM detection was highly selective for the compounds of interest, and it can monitor several analytes simultaneously and requires limited sample separation, there was no intention to establish the chromatographic separation of ECD and its oxidized product in this experiment. The retention time was 3.36 min and the representative MRM chromatograms are shown in Fig. 4. Fig 4(a) shows the MRM chromatogram m/z 325→176 for ECD; Fig 4(b) shows the MRM chromatogram m/z 325→174 for the oxidized product; Fig 4(c) shows the total MRM chromatogram.

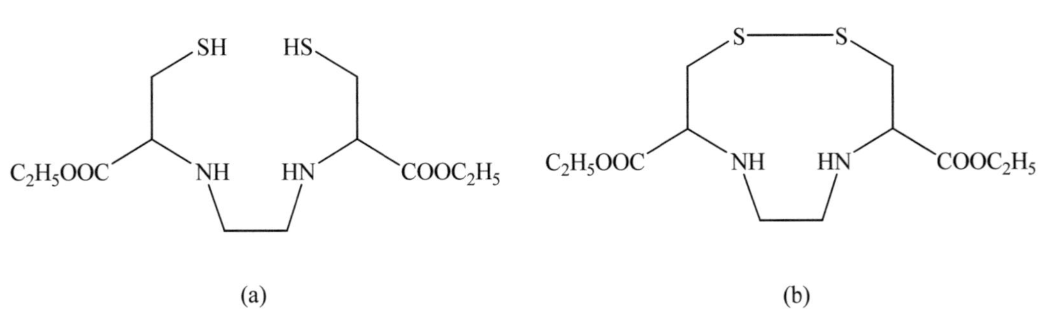

Fig. 1. Chemical structures of ECD and its oxidized product. (a) ECD, (b) the oxidized product.

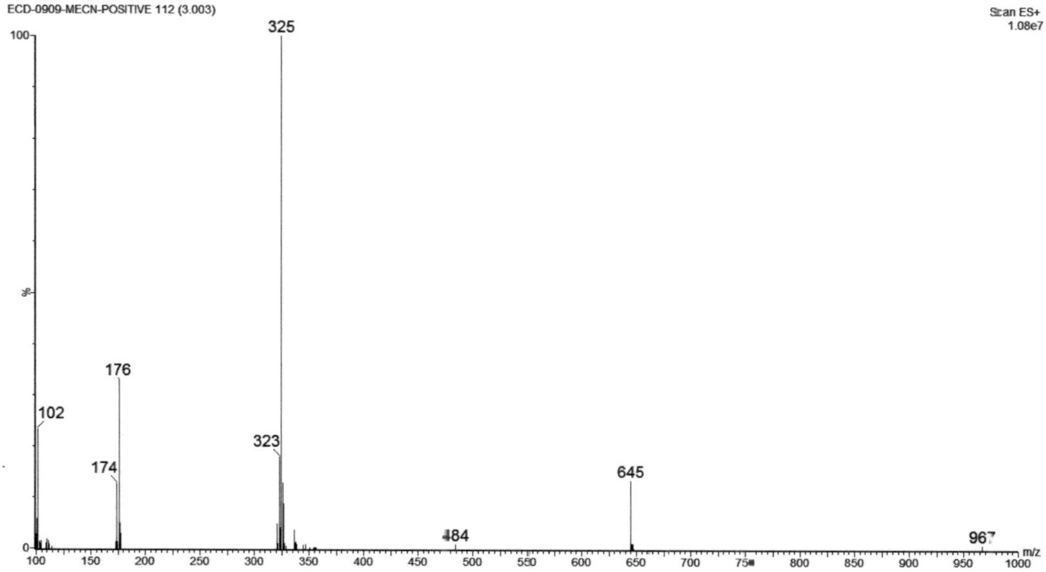

Fig. 2. Full-scan spectrum of the sample obtained using the Quattro Micro triple-quadrupole mass spectrometer.

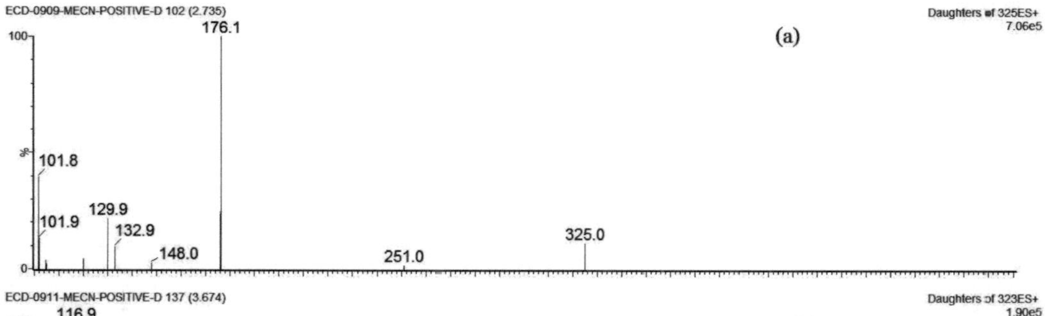

Fig. 3. Product ion mass spectra of m/z 325 and 323. (a) Product ion spectrum of ECD (m/z 325). (b) Product ion spectrum of m/z 323, which is the $[M+H]^+$ peak of the oxidized product.

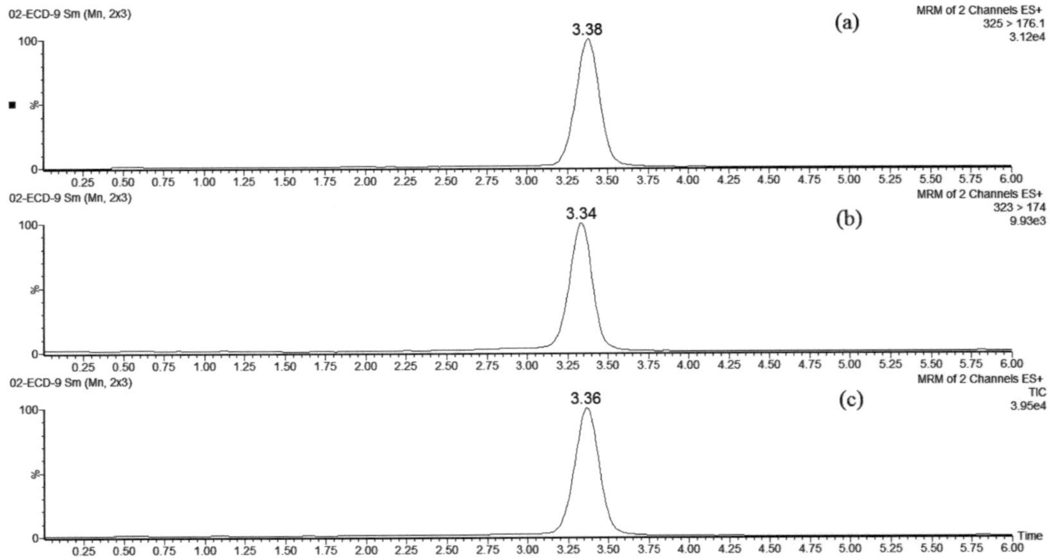

Fig. 4. Representative MRM chromatograms. (a) MRM chromatogram m/z 325→176 for ECD; (b) MRM chromatogram m/z 323→174 for the oxidized product; (c) total MRM chromatogram.

Another example, Technetium-99m-L,L-ethylenedicysteine (^{99m}Tc-L,L-EC) is a renal imaging agent, a UPLC-MS/MS method was developed to perform a stress testing and study oxidative stability to estimate the potential shelf-life of the ligand L,L-EC under normal storage temperature condition (20-25°C). L,L-EC was detected as a function of time at four different temperatures. The degradation of L,L-EC followed the first order kinetics, and the temperature-dependent kinetics was well described by the linear Arrhenius equation. The activation energy (Ea) was calculated, and the shelf-life at 25 and 4°C was predicted. The results are useful for the proper storage and quality evaluation of L,L-EC (Sun et al., 2010).

5. Conclusion and future direction

Tandem mass spectrometry, especially LC-MS/MS, is a powerful analytical instrument, chemical kinetics research is very important in many fields. It is promising to study chemical kinetics using tandem mass spectrometry, especially LC-MS/MS.

6. References

Boys, BL. & Konermann, L. (2007). A temperature-jump stopped-flow system for monitoring chemical kinetics triggered by rapid cooling. *Talanta*, Vol. 71, No. 3, (February 2007), pp. (1276-1281), ISSN 0039-9140

Chancellor, TF.; Russell, RJ.; Dravid, V. & Lele, TP. (2008). Toward in Situ Biochemistry: Combining Chemical Kinetics Approaches with Biomolecular Imaging in Living Cells. *Biotechnology Progress*, Vol. 24, No. 1, (January/February 2008), pp. (89-95), ISSN 1520-6033

Cox, RA (2003). Chemical Kinetics and Atmospheric Chemistry: Role of data Evaluation. *Chemical Reviews*, Vol. 103, No. 12, (November 2003), pp (4533-4548), ISSN 0009-2665

Denisov, ET.; Sarkisov, OM. & Likhtenshtein, GI. (2003). *Chemical Kinetics Fundamentals and New Development* (First Edition). Elsevier Science B.V., ISBN 0444509380, Amsterdam

Gadkari, T.; Chandrachood, P.; Ruikar, A.; Tele, S.; Deshpande, N.; Salvekar, J. & Sonawane, S. (2010). Validated stability indicating LC-PDA-MS method to investigate pH rate profile and degradation kinetics of efavirenz and identification of hydrolysis product by LCMS. *International Journal of Pharmacy and Pharmaceutical Sciences*, Vol. 2, No. 1, (January-March 2010), pp. (169-176), ISSN 0974-1491

Grimes, SM.; Mehta, LK. & Ngwang, HC. (2001). Chemical kinetics of the photocatalytic degradation of Trans-cinnamic, dihydrocinnamic, Trans-caffeic, and dihydrocaffeic acids. *Journal of Environmental Science and Health, Part A: Toxic/Hazardous Substances and Environmental Engineering*, Vol. 36, No. 10, (2001), pp. (1891-1904), ISSN 1093-4529

Hagrman, D.; Goodisman, J. & Souid, AK. (2004). Kinetic Study on the Reactions of Platinum Drugs with Glutathione. *J Pharmacol Exp Ther*, Vol. 308, No. 2, (February 2004) pp. (658-666), ISSN 1521-0103

John, H. & Schlegel, W. (1998). Thermodynamic and structural characterization of cis-trans isomerization of 12-(S)-hydroxy-(5Z,8E,10E)-heptadecatrienoic acid by high-performance liquid chromatography and gaschromatography-mass spectrometry. *Chemistry and Physics of Lipids*, Vol. 95, No. 2, (October 1998), pp. (181-188), ISSN 0009-3084

Katsuda, T. & Toshiro, M. (2009). Chemical-kinetics model for the growth of a multicellular tumor spheroid. *Journal of Chemical Engineering of Japan*, Vol. 42 , No. 3 , (2009) pp. (198-203), ISSN 0021-9592

Kumar, P.; Nikaktari, H.; Nemati, M. & Hill, GA. (2011). Kinetic Modelling of Phenol Oxidation in a Bioremediation Medium Using Fenton's Reagent. *Internal Journal of chemical reactor engineering*, Vol. 9, A38, ISSN 1542-6580

Laidler, KJ. (1989). The Contribution of Electrochemistry to the Development of Chemical Kinetics, in: *Electrochemistry, Past and Present*, American Chemical Society, Vol. 390, Chapter 5, pp 63-74, ISBN13 9780841215726

Larsen, J.; Cornett, C.; Jaroszewski, JW. & Hansen, SH. (2009). Reaction between drug substances and pharmaceutical excipients: Formation of citric acid esters and amides of carvedilol in the solid state. *Journal of pharmaceutical and biomedical analysis*, Vol 49, No. 1, (January 2009), pp. (11-17), ISSN 0731-7085

Pierre, ZN.; Field, CR. & Scheeline A. (2009). Sample Handling and Chemical Kinetics in an Acoustically Levitated Drop Microreactor. *Analytical Chemistry*, Vol. 81, No. 20, (September 2009), pp (8496-8502), ISSN 1520-6882

Qiao, JP.; Sun, XT. & Zhu L. (2009). Study of the oxidation dynamics of ethyl cysteinate dimer in solution by ultra-performance liquid chromatography with tandem mass spectrometry. *Rapid Communications in Mass Spectrometry*, Vol. 23, No. 9, (March 2009), pp. (1264-1268), ISSN 1097-0231

Sakkas, VA.; Shibata, K.; Yamaguchi, Y.; Sugasawa, S. & Albanis T. (2007). Aqueous phototransformation of zinc pyrithione-Degradation kinetics and byproduct identification by liquid chromatography-atmospheric pressure chemical ionisation mass spectrometry. *Journal of Chromatography A,* Vol. 1144, No. 2, (March 2007). pp. (175-182), ISSN 0021-9673

Saunders, SC. & Martin JW. (1999). The Relationship Between Chemical Kinetics and Stochastic Processes in Photolytic Degradation, in: *Service Life Prediction of Organic Coatings, American Chemical Society,* Chapter 15, pp. (233-245), ACS Symposium Series, Vol 722, ISBN13 9780841235977

Sojic, DV.; Anderluh, VB.; Orcic DZ. & Abramovic, BF. (2009). Photodegradation of clopyralid in TiO_2 suspensions: Identification of intermediates and reaction pathways. *Journal of hazardous materials,* Vol. 168, No. 1, (August 2009), pp. (94-101), ISSN 0304-3894

Stepensky, D.; Chorny, M.; Dabour, Z. & Schumacher, I. (2004). Long-term stability study of L-Adrenaline injections: Kinetics of sulfonation and racemization pathways of drug degradation. *Journal of Pharmaceutical Sciences,* Vol. 93, No. 4, (January 2004), pp. (969-980), ISSN 1520-6017

Su, YS.; Morrison, DT. & Ogle, RA. (2009). Chemical kinetics of calcium hypochlorite decomposition in aqueous solutions. *Journal of Chemical Health and Safety,* Vol. 16, No. 3, (May-June 2009), pp. (21-25), ISSN 1871-5532

Sun, XT.; Qiao, JP.; Zhu, L.; Qiao, HW. & Zhong, JG. (2010). Kinetic study for a stress testing of l,l-ethylenedicysteine by ultra-performance liquid chromatography/tandem mass spectrometry analysis. *Applied Radiation and Isotopes,* Vol. 68, No. 12, (December 2010), pp. (2153-2156), ISSN 0969-8043

Teng, XW.; Foe, K.; Brown, KF.; Cutler, DJ. & Davies, NM. (2001). High-performance liquid chromatographic analysis of mometasone furoate and its degradation products: Application to in vitro degradation studies. *Journal of Pharmaceutical and Biomedical Analysis,* Vol. 26, No. 2, (September), pp. (313-319), ISSN 0731-7085

Tur'yan, YI. (1998). Kinetics and equilibrium of the dehydration-hydration and recombination-dissociation reactions of glyoxylic acid investigated by electrochemical methods. *Croatica Chemica Acta,* Vol. 71, No. 3, (June 1998). pp. (727-743), ISSN 0011-1643

Yuan, W.; Heinecke, JW. & Fu, XY. (2009). Hypochlorous Acid Converts the gamma-Glutamyl Group of Glutathione Disulfide to 5-Hydroxybutyrolactam, a Potential Marker for Neutrophil Activation. *Journal of Biological Chemistry,* Vol. 284, No. 39, (July 2009), pp. (26908-26917), ISSN 0021-9258

Part 4

Natural Products Analysis

Electrospray Ionization Tandem Mass Spectrometry as a Tool for the Structural Elucidation and Dereplication of Natural Products: An Overview

Herbert Júnior Dias, Nathalya Isabel de Melo and Antônio Eduardo Miller Crotti
Núcleo de Pesquisas em Ciências Exatas e Tecnológicas, Universidade de Franca, Franca, SP Brazil

1. Introduction

Natural products (NPs) obtained from plants and microorganisms often exhibit biological activities of interest for the discovery of new drugs as well as pharmaceutical and agrochemical products (Fredenhagen et al., 2005; Konishi et al., 2007; Lang et al., 2008). For instance, over 60% of the anticancer drugs have been discovered directly from NPs or are semi-synthetic derivatives of these compounds (Butler, 2008; Costa-Lotufo et al., 2010; Harvey, 2008; Newman & Cragg, 2007). However, the research for new bioactive compounds is usually laborious and slow, once the biological evaluation processes are preceded by isolation, purification, and structural elucidation steps, which are usually expensive and time-consuming (Konishi et al., 2007). This fact, in combination with the large number of known compounds, has led the scientific community to develop new techniques for the direct identification of NPs from extracts and natural broths, thus avoiding reisolation of already known compounds (Bindseil et al., 2001; Konishi et al., 2007; Lee, 2004; Newman et al., 2000; Newman et al., 2003; Shu, 1998).
Dereplication is the process that allows for the rapid identification of bioactive metabolites in crude extracts by distinguishing previously identified compounds from novel ones (Crotti et al., 2006). This technique avoids repetitive work of isolation of already known NPs (Wolf & Siems, 2007), promoting chemical screening or metabolite profiling (Crotti et al., 2006). The dereplication process involves separation of single metabolites by chromatographic methods, identification of these compounds by spectroscopic methods, bioassays for evaluation of the biological activity, and searches in databases for verification of the novelty of these compounds (Konishi et al., 2007; Wolf & Siems, 2007).
Hyphenated techniques have played a key role in the identification of NPs and other organic compounds (Crotti et al., 2006; Prasain et al., 2003). These techniques combine a separation method (i.e., gas chromatography, GC; liquid chromatography, LC) with a structural identification technique (i.e., mass spectrometry, MS; ultraviolet-visible spectroscopy, UV-vis; nuclear magnetic resonance, NMR). Although a number of

approaches dealing with the dereplication of NPs have been based on GC-MS (gas chromatography mass spectrometry), LC-UV (liquid chromatography ultraviolet spectroscopy), and LC-NMR (liquid chromatography nuclear magnetic resonance), LC-MS (liquid chromatography mass spectrometry) has been the most widely employed technique for this purpose (Lang et al., 2008; Lee, 2002; Oliveira & Watson, 2000; Wolfender & Hostettmann, 1996; Wolfender et al., 1994).

In this chapter we present an overview on the use of LC-MS and electrospray ionization tandem mass spectrometry (ESI-MS/MS) techniques for the dereplication of NPs belonging to different classes, such as sesquiterpene lactones, lignans, caffeoyl quinic acid derivatives, flavonoids, and alkaloids.

2. Why use LC-MS?

LC-MS combines the versatility of LC (usually high-pressure liquid chromatography, HPLC), which enables analysis of a wide range of compounds (Degani et al., 1998), with the sensitivity of MS (Niessen, 2006). About 60% to 80% of all the existing compounds are amenable to HPLC analysis, whereas about 15% are analyzable by GC. The use of GC is often limited to thermally stable and volatile compounds, as well as those that can be analyzed after derivatization reactions. On the other hand, HPLC is employed to separate macromolecules, ionic species, polar and high-molecular weight compounds, on condition that they are soluble in the mobile phase. These factors have made HPLC the technique of choice for dereplication and metabolite profiling studies of NPs in complex mixtures (Dong, 2006).

Figure 1 shows a schematic representation of LC-hyphenated techniques for the structural identification of organic compounds. In the HPLC system, the sample is injected into the mobile-phase stream and transported through the column with the stationary-phase where the separation takes place. This separation occurs by selective interaction between the sample and the mobile and stationary phases, and it is monitored with the aid of a flow-through detector, usually UV, NMR, or MS. Although NMR is the most powerful technique for the structural elucidation of organic compounds, some drawbacks of coupling NMR to LC emerges from the high costs of using deuterated solvents as mobile phase in LC and the low sensitivity as compared to MS, thereby limiting its use for the identification of NPs at trace levels. UV-vis does not provide conclusive data for structural elucidation alone, but development of the diode array detector (DAD) has increased its detection power and effectiveness. UV-DAD has been more and more often applied as a first detection method in dereplication studies using LC because it is rapid and inexpensive, as compared to the other techniques (Crotti et al., 2006).

Liquid chromatography mass spectrometry (LC-MS) is a highly sensitive and selective method for identification of NPs in complex mixtures (Fredenhagen et al., 2005; Furtado et al., 2007). Compounds are separated in the column of the LC system and directed to the mass spectrometer by a flow separator, where they are ionized and further separated in the mass analyzer according to their mass-to-charge (m/z) ratio (Niessen, 2006). However, three major difficulties have limited the coupling of LC to MS for many years, namely I) apparent incompatibility of the flow-rate as expressed by the need for introducing a large volume of a liquid effluent from a conventional LC column into the high vacuum of the mass spectrometer; II) incompatibility of the effluent composition as a result of the frequent use of non-volatile mobile phase additives during the LC separation process, and (III) ionization of

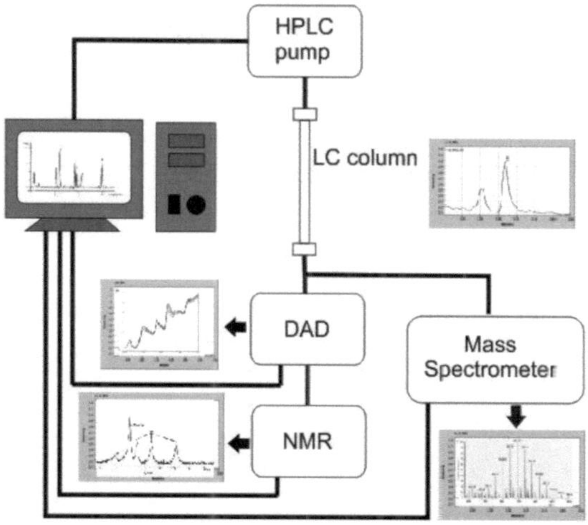

Fig. 1. LC-hyphenated techniques for structural identification.

non-volatile and/or thermally labile analytes (Niessen & Tinke, 1995). These difficulties were overcome with the advent of the atmospheric pressure ionization (API) techniques, more specifically electrospray ionization (ESI) and atmospheric pressure chemical ionization (APCI) (Tomer, 2001).

In ESI, the sample is dissolved in a polar, volatile solvent and pumped through a narrow stainless steel capillary (75-150 μm i. d.). A sufficiently high voltage, commonly 3-5 kV, is applied at the capillary tip, which is situated within the ionization source of the spectrometer. This high voltage causes charge migration of the species in solution to the metal/solution interface, resulting in an electric double layer. In the capillary tip, electrostatic forces acting on the ion preformed in the double layer counterbalance the surface tension, culminating in a conical surface known as Taylor cone (Vessecchi et al., 2007). The small charged-droplets of the fine spray that emerge from the capillary tip are diminished in size due to a co-axially introduced nebulizing gas (usually nitrogen) flowing around the outside of the capillary, thus leading to solvent evaporation (Herbert & Johnstone, 2003). When the Rayleigh limit is reached, the Coulombic repulsion between charges in the droplet surface overcomes the surface tension, thereby giving rise to the droplet "Coulombic explosion" (Fig. 2). Finally, ions are produced by the droplets and flow into the mass analyzer (Ardrey, 2003; Niessen, 2006; Smeraglia et al., 2002; Vékey, 2001).

As a result of ESI, protonated ($[M+H]^+$)/deprotonated ($[M-H]^-$) molecules, and cationized (usually $[M+Na]^+$ or $[M+K]^+$)/anionized molecules (i.e, $[M+Cl]^-$) can be produced, depending on the molecular structure of the analyte (Herbert & Johnstone, 2003; Todd, 1995). Natural products exhibiting acidic groups (i.e., flavonoids, and caffeoyl quinic acid derivatives) are easily deprotonated, which enables their analysis in the negative ion mode, whereas for NPs with basic groups (i.e., alkaloids), which can be easily protonated, the positive ion mode is more adequate (Ardrey, 2003). The major advantage of interfacing ESI with LC is related to the ion formation, which occurs at atmospheric pressure in the condensed phase outside the high vacuum of the mass spectrometer, thereby eliminating one of the greatest difficulties inherent to the analysis: the negative influence of the vacuum system (Ackermann et al., 1996). Furthermore, some other characteristics make the ESI an important ionization process: elimination of solvent ejected from droplets by the electrical field and difference in pressure; generation of multiply charged ions, which is important for molecules of high weight because the effective mass range of the mass spectrometer is increased; detection of molecular weight; and low solvent flow, which is required for operation at optimum sensitivity, thereby reducing sample consumption (Crotti et al., 2006).

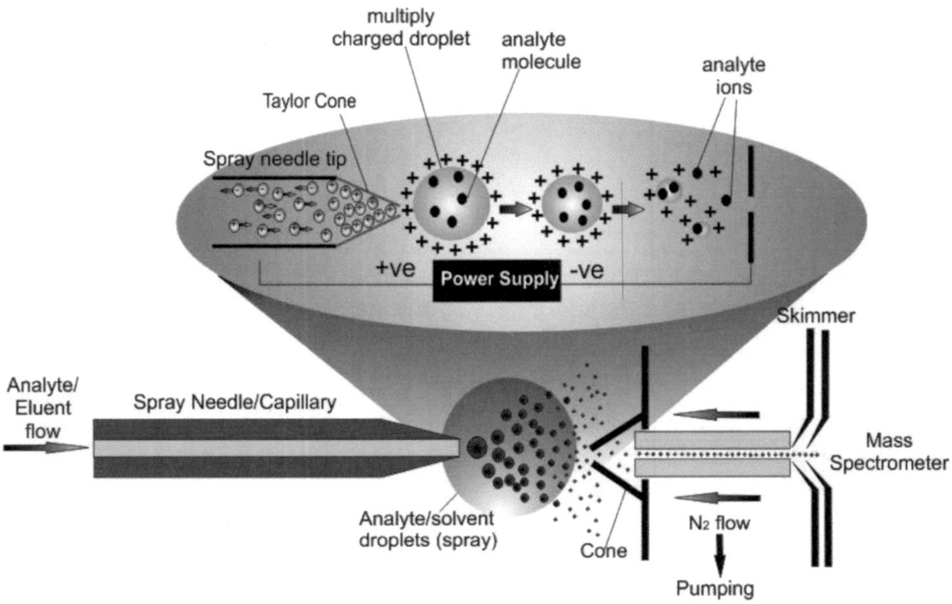

Fig. 2. Representation of an electrospray ion source and the electrospray process (Crotti et al., 2006).

3. How to use LC-MS and ESI-MS/MS for the dereplication of NPs?

ESI is considered to be a "soft ionization" process, since the internal energy of the formed ions is low, and little or no fragmentation takes place (Ardrey, 2003; Núñez et al., 2005; Vékey, 2001). This feature is very interesting for identification of the molecular weight or molecular formula. However, it represents an inconvenience for structural elucidation studies, which are based on fragment ions (Herbert & Johnstone, 2003; Wolfender et al., 2000). In tandem mass spectrometry (MS/MS), an ion (called precursor ion) from the first stage of MS is selected and activated, to produce fragment ions, which are then analyzed in the second stage of MS. The most widely employed ion activation method is the collision-induced dissociation (CID), which consists in promoting the energy-controlled collision of a chemically inert gas, (e.g. Ar, He, N_2, or CO_2) with the precursor ion (Fig. 3). The collision energy may be chosen, in order to optimize the MS/MS spectrum. Low collision energy values promote soft fragmentation and produce few fragments, whereas high collision energy values prompt extensive fragmentation, so the produced fragment ions can be used to obtain structural information. On the other hand, important structural information can be obtained and eventual comparison with spectral libraries may be made when moderate collision energy values are utilized (Gates et al., 2006). MS spectrometers that are used in MS/MS experiments can be of two main types: i) instruments that are able to store the ions, thus allowing for selection of the target ions by injection with authentic patterns, followed by fragmentation, hence generating the mass spectra (e.g. ion cyclotron resonance, ICR; and quadrupole ion trap, IT); and ii) instruments that use a sequence of mass spectrometers in space consisting of quadrupole mass analyzers in sequence (e.g, triple quadrupole, QqQ), or a hybrid conformation with quadrupole in sequence (e.g., quadrupole-time-of-flight, Q-TOF) (De Hoffmann & Stroobant, 2007).

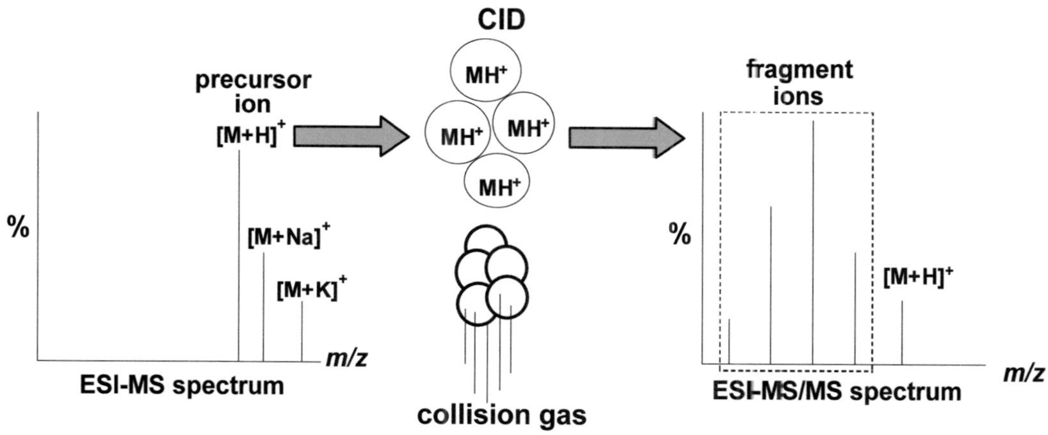

Fig. 3. Representation of the CID process in tandem mass spectrometry (MS/MS).

Dereplication of NPs by means of LC-MS and ESI-MS/MS has often been accomplished on the basis of retention times, UV-DAD spectra, and MS/MS data obtained for each peak of the chromatogram of the crude extract as compared to those of previously analyzed authentic standards. MS/MS data of the authentic standards are obtained using the same collision energy value, which is also employed in LC-MS/MS experiments involving the crude extract (Gobbo-Neto & Lopes, 2008). In the case of standards belonging to the same class of NPs and with similar structures, the acquisition of the MS/MS spectra using the same collision energy value ensures that differences in the fragmentation are the sole result of differences in their structures, so that structure-fragmentation relationships can be established (Crotti et al., 2005). The presence of structure diagnostic fragment ions (DFI) is very important for distinction between structural features (Aguiar et al., 2010) or maybe for characterization of a class of NPs. This methodology enables direct online identification of already known compounds from crude extracts.

Besides the importance of ESI-MS/MS for the online identification of NPs using LC-MS, this technique can also be used for offline identification. In this case, authentic standards are also previously investigated by MS/MS, and diagnostic fragment ions are identified. Accurate mass data, which allows for identification of the molecular formulas of the compounds, are especially important for this purpose. After that, the crude extracts are analyzed directly by MS without undergoing previous chromatographic separation in the LC system. The MS spectrum is analyzed, and those ions with the same *m/z* as the authentic standards are selected as precursor ions for the MS/MS experiments. Finally, the MS/MS spectra of the standards and those obtained from the selected precursor ions are compared. The disadvantage of this offline identification as compared to online identification by LC-MS is that data from the LC system, such as retention time and UV-DAD spectra, are lost. A good example of offline dereplication of NPs is the identification of plumeran alkaloids in the crude methanol extract of stem bark from *Aspidosperma spruceanum* (Aguiar et al., 2010). In this study, Aguiar and co-workers employed ESI-MS/MS and accurate-mass data of six authentic standards (**1-6**) to identify important diagnostic fragment ions, which were then used to distinguish aspidoscarpine (**1**) from aspidolimidine (**2**) in the crude methanol extract (Fig. 4).

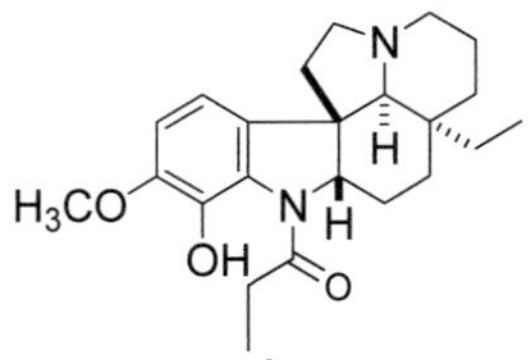

1
$C_{23}H_{32}N_2O_3$
Exact Mass: 384,2413

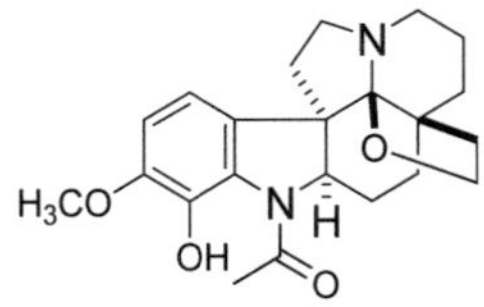

2
$C_{22}H_{28}N_2O_4$
Exact Mass: 384,2049

Fig. 4. Structures of aspidoscarpine (**1**) and aspidolimidine (**2**).

4. Applications of MS/MS in the dereplication of NPs

4.1 Sesquiterpene lactones (STL)

Sesquiterpene lactones (STL) are an important class of natural products that occur in several plant families, mainly in Asteraceae (Crotti et al., 2005). These compounds are interesting due to their several biological activities (Picman, 1986) and relevance as chemotaxonomic markers (Crotti et al., 2005). The biosynthesis of STLs is initiated by condensation of three isoprene molecules followed by oxidation, and their chemical structures are divided into groups according to their carbocyclic skeleton (e.g., germacranolides, eudesmanolides, guaianolides, glaucolides, cadinanolides, hirsutinolides, and furanoheliangolides) (Crotti et al., 2005; Dewick, 2004).

The fragmentation pattern of the protonated STL belonging to the goyazensolide type of furanoheliangolides has been investigated by Crotti and co-workers using ESI-MS/MS (Crotti et al., 2005). Firstly, the authors selected ten STL standards exhibiting the same structural core, but differing in terms of the presence/absence of a hydroxyl group at C-15, a single/double bond between C-4 and C-5, and the presence of an acyloxy group or hydroxyl group at C-8 (Fig. 5). All the MS/MS spectra were obtained at 10 eV, as optimized by varying the collision energies between 5 and 50 eV. The authors reported that the fragment ion $[M+H-R_2CO_2H]^+$ is diagnostic for compounds that exhibit an acyloxy group at C-8 (**3-7**, **9**, **11** and **12**), whereas the fragment ion $[M+H-CO_2]^+$ indicates the presence of a hydroxyl group at C-8. The acylium ion R_2^+, which is formed for compounds **3-7**, **9**, **11**, and **12**, was useful for identification of the ester bound at C-8. Moreover, the relative configuration of C-8 of centratherin (**7**) and budlein A (**9**) could also be identified on the basis of the relative intensity of the fragment ion $[M+H-R_2CO_2H]^+$. These data were further used by Gobbo-Neto & Lopes, in combination with retention times and UV-DAD spectra, for the online identification of 36 compounds in the methanol extract of leaves from *Lychnophora ericoides* (Asteraceae), including STLs **6**, **7**, and **12** (Gobbo-Neto & Lopes, 2008).

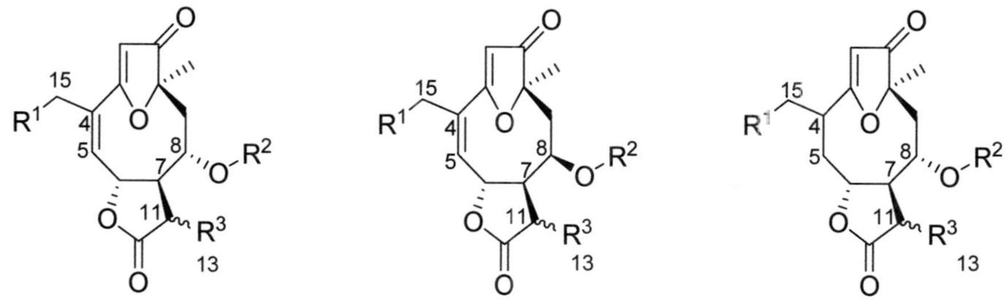

3-8 **9-10** **11-12**

	3	4	5	6	7	8	9	10	11	12
R^1	H	H	OH	H	OH	H	OH	H	H	H
R^2	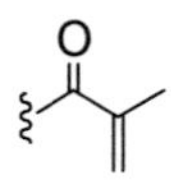	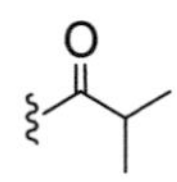	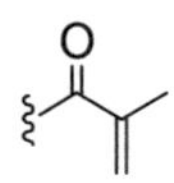	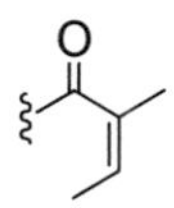	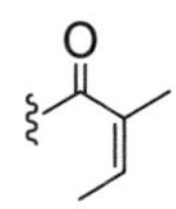	H	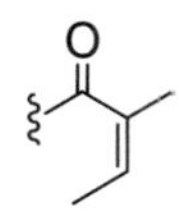	H	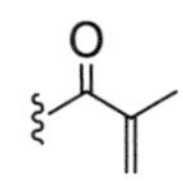	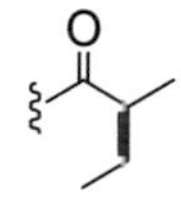
R^3	CH_2	CH_2	CH_2	CH_2	CH_2	α-CH_3	CH_2	CH_2	CH_2	CH_2
MW	344	346	360	358	374	278	374	276	346	360

Fig. 5. Sesquiterpene lactones of the goyazensolide type (Crotti et al., 2005).

4.2 Phenylpropanoids

Phenylpropanoids is one of the largest and most important classes of NPs. It includes a vast range of phenolic compounds, from small and simple forms to complex molecular structures. Moreover, from a biosynthetic viewpoint phenylpropanoids are the precursors of important secondary metabolites, such as aromatic acids, benzoic acids, cinnamic acids, lignins, lignans, phenylpropenes, coumarins, styrylpyrones, flavonoids, stilbenes, flavolignans, isoflavonoids, terpenoid quinones, and tannins, among others.

4.2.1 Aromatic acids

Gómes-Romero and co-workers have identified several phenol derivatives in foods used in diets (e.g., propolis, lemon, borage, cabbage-broccoli, garlic, onion, etc) by using a MS/MS library previously established from commercially available standard phenolic compounds (Gómez-Romero et al., 2011). The authors reported that the negative ion mode electrospray ionization was more adequate than the positive ion mode for this purpose, although the latter was also utilized when necessary. They described that loss of CO_2 (44 Da) or H_2O (18 Da) from the $[M-H]^-$ ion, which was used as precursor ion in MS/MS experiments, are the major fragmentation routes for the selected compounds. However, they did not result in diagnostic fragment ions (DFI). Thus, in order to adjust the intensity of some specific and diagnostic fragment ions, the collision energy values were varied.

Samples of these foods were freeze-dried, powdered, and extracted with methanol 80% in ultrasonic bath, which was followed by centrifugation, filtration, and dilution with water/acetonitrile 1:1 (v/v). After that, samples were analyzed by LC coupled with a diode array detector (DAD) set at 254 nm, and a quadrupole orthogonal acceleration time-of-flight mass spectrometer (micrOTOF-Q™) equipped with an electrospray (ESI) ion source. The authors compared the MS/MS data of each peak of the chromatogram with those of the previously established library in both the positive and negative ion modes, and they identified nine hydroxybenzoic acid derivatives, nine cinnamic acid derivative, and six simple phenolic compounds, apart from eleven flavonoids. Vanillic (**13**) and syringic (**14**) acids were identified on the basis of their DI *m/z* 152 and 182, respectively, as well as on the DI *m/z* 92, which is common for both compounds in the negative ion mode (Fig. 6). Moreover, syringic acid (**14**) produced the ions *m/z* 166 and 123 as DI, via direct loss of two $CH_3^{\bullet}$ from the deprotonated molecule, and loss of $CH_3^{\bullet}$ elimination followed by CO_2 elimination, respectively. *Trans*-cinnamic (**15**), caffeic (**16**), and three coumaric acid isomers (**17-19**) were also shown to eliminate CO_2 from the deprotonated molecule (Gómez-Romero et al., 2011).

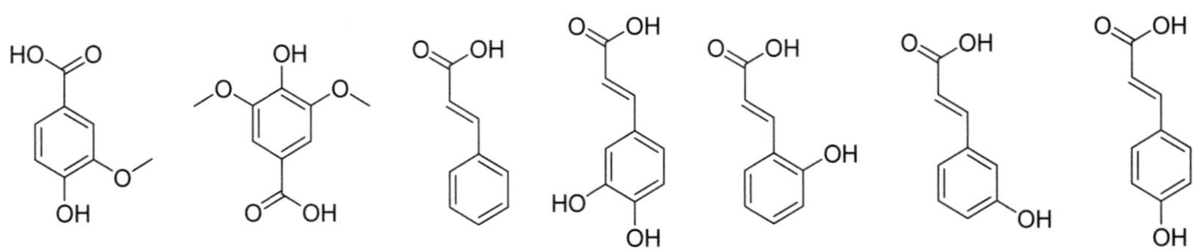

13 **14** **15** **16** **17** **18** **19**

Fig. 6. Structures of some hydroxybenzoic acids (**13** and **14**) and cinnamic acid derivatives (**15-19**).

4.2.2 Caffeoylquinic acid (CQA) and feruloyl quinic (FQA) derivatives

Miketova and co-workers have published a study on the fragmentation of protonated and deprotonated 3,5- and 4,5-dicaffeoylquinic acid (di-CQA) derivatives using electrospray ionization tandem mass spectrometry (Miketova et al., 1999). They demonstrated that both the positive and negative ion modes of analysis allow for identification of the ester groups bound at the quinic acid moiety, although the negative ion mode is the most informative method of analysis for the free compounds. The presence of a phenolic hydroxyl at the *ortho* or *para* position was proposed to be important for the formation of some diagnostic ions in the negative ion mode, as shown in Figure 7.

Gobbo-Neto & Lopes have reported on a sensitive analytical method for the dereplication of various classes of secondary metabolites found in the *L. ericoides* leaf extracts, including caffeoylquinic acid (CQA) and feruloyl quinic (FQA) derivatives (Gobbo-Neto & Lopes, 2008). The authors employed a methodology based on HPLC coupled with a diode array detector (HPLC-DAD) and HPLC coupled with electrospray ionization tandem mass spectrometry (ESI-MS/MS) using collision-energy values ranging between 10 and 25 eV, in combination with co-injection of authentic standards and accurate-mass measurements. The chemical structures of the identified compounds are depicted in Fig. 8.

For chlorogenic acids (**20**, **22** and **23**), the authors utilized the ion with *m/z* 353 as precursor ion and reported that the fragment ion at *m/z* 173, which is a result of water elimination from the quinic acid moiety, is diagnostic for CQA derivatives esterified at position 4. The CQA derivatives substituted at position 5 and di-CQA isomers (3,5- and 4,5-substituted) were confirmed by co-elution with authentic standards. HPLC-DAD analysis was used in combination accurate mass measurements for identification of feruloylquinic acids (FQA) and feruloyl-caffeoylquinic acids (FCQA). The ion at *m/z* 367 was used as the precursor ion, yielding the fragment ion with *m/z* 173, which is diagnostic for 4-FQA derivatives. 3-FQA and 5-FQA were identified on the basis of the ions *m/z* 193 and 191, respectively, as well as by comparison with previously reported studies (Clifford et al., 2003; Clifford et al., 2005). On the other hand, the FCQA isomers could not be identified, even when DAD, ESI-MS/MS in the positive and negative ion modes, and accurate mass data were employed.

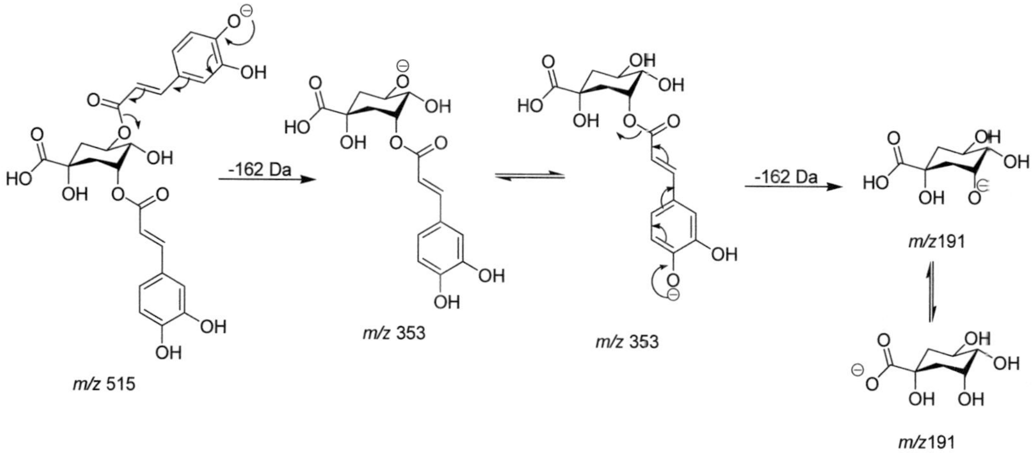

Fig. 7. Fragmentation of 3,5-dicaffeoylquinic acid (Miketova et al., 1999).

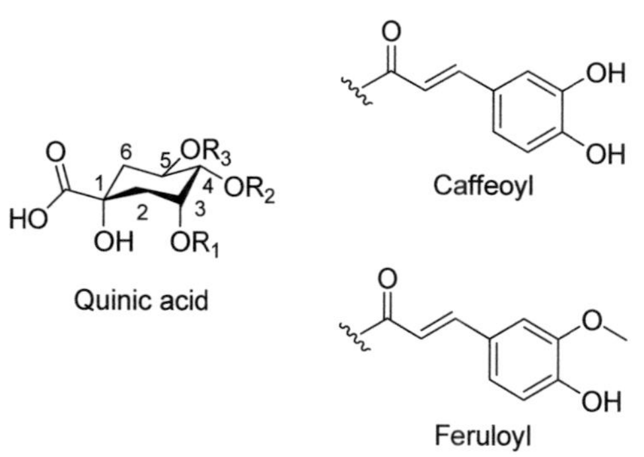

	R_1	R_2	R_3
20	caffeoyl	H	H
21	feruloyl	H	H
22	H	caffeoyl	H
23	H	H	caffeoyl
24	H	feruloyl	H
25	H	H	feruloyl
26	caffeoyl	caffeoyl	H
27	caffeoyl	H	caffeoyl
28	H	caffeoyl	caffeoyl
29	feruloyl	feruloyl	H
30	caffeoyl	caffeoyl	caffeoyl

Fig. 8. Structures of the caffeoylquinic and feruloylquinic acid derivatives identified in the *L. ericoides* leaf extract.

4.2.3 Lignans

Lignans are a group of secondary metabolites consisting of compounds containing two phenylpropanoid units linked by oxidative coupling between C-8 and C-8′ in their structures (Willfor et al., 2006). Several biological activities, such as anti-tumoral, fungicide, and anti-viral actions, have been attributed to the presence of these compounds, (Ayres & Loike, 1990; Botta et al., 2001; Pan et al., 2009).

Shengmai San (SMS), a prescription comprised of *Panax ginseng*, *Schisandra chinensis*, and *Ophiopogon japonicus*, has long been used in traditional Chinese medicine. Wang and co-workers have employed HPLC-DAD-MS/MS for identification of the multiple SMS constituents (Wang et al., 2011). These authors identified 53 compounds, including 21 lignans, 6 steroidal glycosides, and 12 homoisoflavonoids. Twenty-seven compounds were identified by comparison of their retention times, MS data, and UV spectra with those of authentic compounds, and the other twenty-six were tentatively identified by comparing their UV spectra, molecular weights, and structural information from MS/MS spectra with those previously published in the literature. Studies on the dereplication of lignans in the ethanol extract from the fruits of *Schisandra chinensis* have also been conducted by He and co-workers using high-performance liquid chromatography coupled with a photodiode-array detector and an electrospray ionization ion source. The authors identified seventeen dibenzylcyclooctadiene lignans on the basis of the relative intensities of protonated and cationized molecules, their UV spectra, and the respective retention times (He et al., 1997).

Zheng and co-workers have utilized HPLC coupled with an ESI ion source and a hybrid ion IT-TOF mass analyzer for dereplication of lignans in *Panax ginseng, Radix ophiopogonis,* and *Schisandra chinensis baill* extracts, which are used in traditional Chinese medicine for treatment of tumoral diseases, coronary atherosclerosis, and some other cardiopathies (You et al., 2006; Yu et al., 2007). They reported that elimination of $CH_3^{\bullet}$ or $^{\bullet}OCH_3$ from the protonated molecule is diagnostic of a methoxyl group at rings A or B of lignans, and that there is not an oxygen atom on ring C, as in the case of schizandrin A (**31,** Fig. 9). Elimination of C_5H_{10} by cleavage of the eight-membered ring to produce a five-membered ring was also reported. On the other hand, elimination of a water molecule (loss of 18 Da) produces the diagnostic product ion (DPI) for lignans that have a hydroxyl group at C-6 or C-7 of ring C, such as shizandrol A (**32**).

Lignans that have a hydroxyl group at C-7 of ring C and an ester function at rings A or B, such as angeloyl gomisin H (**33**) , have been reported to fragment by elimination of H_2O and an olefinic ketene (C_5H_6O or C_7H_4O), the latter being associated with the presence of the ester at rings A or B. Elimination of the phenol ester as an olefinic ketene from lignans using electron ionization (EI) has been previously reported by Zhai & Cong (Zhai & Cong, 1990). In addition, the authors described that the formation of the sodiated molecule ($[M+Na]^+$) is a more favored ionization process, as compared to protonation, for lignans that have a hydroxyl group at C-7 and an ester bound at C-6 of ring C, such as schisantherin A (**34**). In the case of compound **34**, elimination of the corresponding carboxylic acid (e.g., benzoic acid, $C_7H_6O_2$) is the major fragmentation process (Zheng et al., 2009).

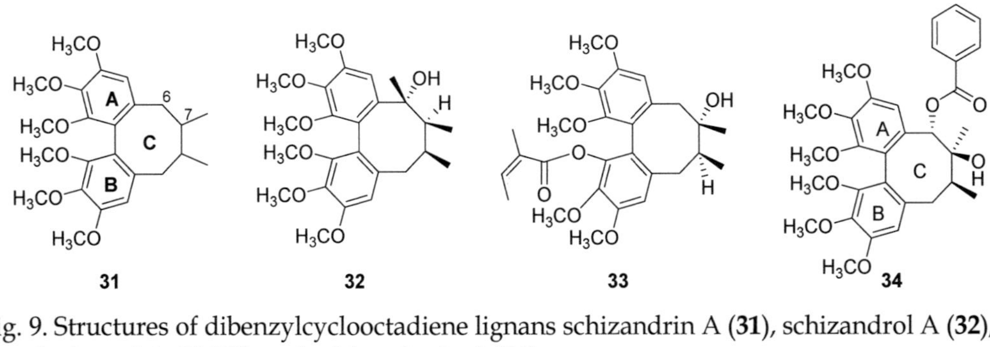

Fig. 9. Structures of dibenzylcyclooctadiene lignans schizandrin A (**31**), schizandrol A (**32**), angeloyl gomisin H (**33**), and schisantherin A (**34**).

4.2.4 Flavonoids

Flavonoids are an important class of secondary metabolites that are biosynthesized by plants. They are related to protection against predators and UV-Vis light, attraction of

pollinators, and antioxidant and hormonal control, among other functions (Dewick, 2004). They also have economic importance because they can be used as pigments, tanning substances, nutritional complements, and food flavors. Moreover, they display pharmacological properties, such as anticarcinogenic, anti-inflammatory, allergen, antiviral, and anti-ulcerogenic actions, among others (Simões et al., 2004).
The fragmentation of aglicone and glycoside flavonoids by MS/MS has been extensively investigated. The major fragment ions result from different retrocyclization cleavages (e.g., retro-Diels-Alder reactions, RDA), as shown in Fig. 10. The nomenclature adopted for the RDA cleavages was firstly proposed by Ma and co-workers (Ma et al., 1997; Ma et al., 1999). The superscripts on the left of rings A or B indicate the bonds that have been broken.

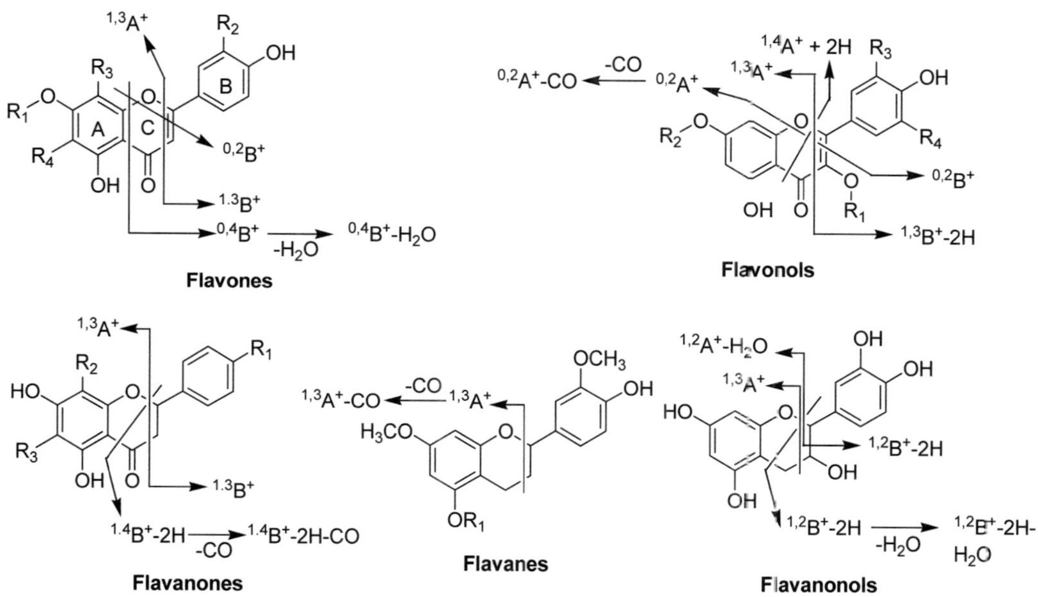

Fig. 10. Main retrocyclization cleavages of some classes of flavonoids.

Wolfender and co-workers have proposed an LC-MS/MS methodology for the online characterization and dereplication of selected commercial and isolated flavonoids (Wolfender et al., 2000). ESI-MS/MS experiments were carried out in hybrid quadrupole time-of-flight (Q-TOF) and ion trap (IT) mass spectrometers. In the IT mass analyzer, the precursor ions are trapped in a short space for varied periods of time, which allows for an increased number of collisions to take place between the collision gas and the precursor ions, thus raising the internal energy of these ions. On the other hand, in the q-TOF mass analyzer the interactions between the precursor ion and the collision gas occur during short periods of time, so that a spatial separation between the ionization and the collision induced dissociation (CID) process comes into effect, thereby diminishing the collision energy. Due to these differences, the MS/MS spectra obtained using IT and q-TOF mass analyzers are different from each other. In order to improve the efficiency of the transmission energy, the precursor ions are accelerated in a linear beam instrument, promoting formation of different fragment ions in both cases. The authors reported that the higher energy in Q-TOF causes hard fragmentation of flavonoids, so that some fragment ions may not be observed using IT, which compromises the feasibility of using both techniques for dereplication. Considering the flavone apigenin (**35,** Fig. 11) described in that work, the authors firstly approached the generation of ring cleavages in the positive ion mode. In the positive IT-MS mode at 35% energy, only the ion with *m/z* 153 was detected, while the same fragments produced during Q-TOF were achieved when the energy was amplified up to 50%(Wolfender et al., 2000).

Rak and co-workers have put forward a strategy for the dereplication of flavonoids in a sample of commercial black currant juice without any preliminary study about its components. This strategy was based on the fact that most of the flavonoid derivatives have an aglycone part. Firstly, the authors developed a procedure for detection of aglycone flavonoids in the juice chemical constituents using multiple reaction monitoring (MRM) in the negative ion mode. In this type of scan, both the precursor and the product ion are specified for the detection of only one pair at the detector (Sleno & Volmer, 2004). Considering the great number of possible aglycone derivatives, the authors chose apigenin (**35**), luteolin (**36**), quercetin (**37**), myricetin (**38**), and naringenin (**39**), which are amongst the commonest aglycone flavonoids (Fig. 11). In addition, the authors performed the experiment using a high declustering potential (DP) with some standard solutions of aglycone analogues, aiming to minimize formation of cluster ions from solvents. The authors showed that the major compounds in the black currant juice can be identified by selecting the characteristic *m/z* values, in combination with their retention times. They also conducted full scan MS experiments, which evidenced that odd-electron fragment ions resulting from homolytic cleavages were more abundant than even-electron ion fragment ions when the highest negative DP values were employed. The tendency toward radical fragmentation reactions was confirmed by MRM experiments, which were accomplished in parallel with the chromatographic run (Rak et al., 2010). A third chromatographic run was carried out on a quadrupole linear ion trap apparatus, in order to confirm the structure of the precursor ions on the basis of diagnostic fragment ions. Twelve flavonoid derivatives were identified in the sample of black currant juice, demonstrating the great versatility of these MS techniques for the dereplication of flavonoids (Rak et al., 2010). However, the authors reported some difficulties in distinguishing between isomer flavonoids using this methodology.

	CID	Precursor ion	Fragment ions					
			$^{1,3}A^+$	$^{1,3}B^+$	$^{0,2}B^+$	$^{0,4}B^+$	$^{0,4}B^+-H_2O$	Main ion
q-TOF MS-MS (POS)	25	271	153 (25)	119 (5)	121 (2)	163 (2)	145 (2)	271 (100)
	30		153 (85)	119 (20)	121 (10)	163 (10)	145 (10)	271 (100)
	35		153 (100)	119 (35)	121 (20)	163 (10)	145 (15)	153 (100)
IT POS MS	35	271	153 (4)	-----	-----	-----	-----	271 (100)
	40		153 (35)	119 (5)	121 (3)	-----	145 (5)	217 (100)
	50		153 (100)	119 (15)	121 (10)	163 (5)	145 (15)	153 (100)
	60		153 (100)	119 (20)	121 (10)	163 (3)	-----	153 (100)
IT NEG MS	50	269	151 (20)	117 (10)	-----	-----	-----	225 (100)

Table 1. MS/MS data of protonated and deprotonated apigenin (**35**).

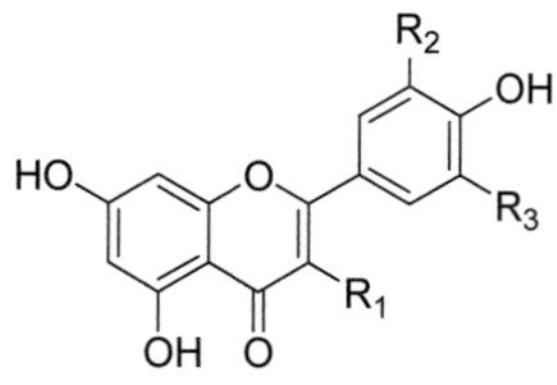

35-38

	R_1	R_2	R_3
35	H	H	H
36	H	OH	H
37	OH	OH	H
38	OH	OH	OH

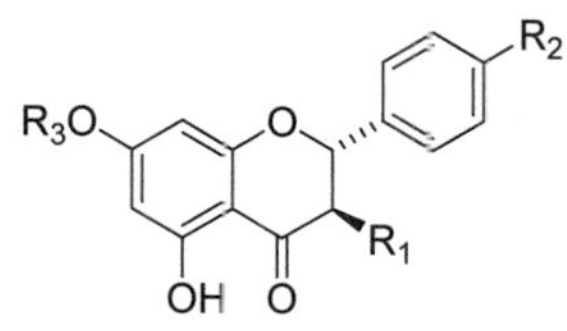

39-43

	R_1	R_2	R_3
39	H	H	H
40	H	H	CH_3
41	H	OH	H
42	OH	H	CH_3
43	OAc	H	CH_3

Fig. 11. Structure of some aglycone flavonoids.

For the online identification of aglycon flavonoids present in the hydroalcoholic extract of *L. ericoides* leaves, Gobbo & Lopes have compared the retention time, the UV spectra and the accurate mass of each peak of the chromatogram with authentic standards, besides comparing the MS data with those previously published in the literature (Gobbo-Neto & Lopes, 2008). A peak exhibiting absorbance maximum above 340 nm, which is characteristic of flavanones previously identified in *L. ericoides* (e.g, pinocembrin, **40;** and pinostrombin, **41**) was identified in the MS/MS spectrum of the deprotonated molecule, as well as by comparison with literature results (Cuyckens & Claeys, 2004; Fabre et al., 2001; Ma et al., 1997; Zhang & Brodbelt, 2003). MS/MS spectrum of the corresponding protonated molecule displayed the base peak at *m/z* 153, which indicates a dihydroxyl substituent at ring A of pinocembrin (**40**). On the other hand, the presence of a fragment ion at *m/z* 167, together with the absence of diagnostic ion at *m/z* 153, indicates a hydroxyl and a methoxyl substituent at ring A, indicative of the presence of the flavanone pinostrombin (**41**) (Gobbo-Neto & Lopes, 2008). Pinobanksin (**42**) was identified on the basis of its UV spectra, comparison with retention time of previously isolated standards, and accurate mass data. 3-*O*-acetylpinobaskin (**43**) was elucidated by comparison with pinobanksin ($[M+H]^+$ at *m/z* 313), being the mass difference due to the methyl group atom bound at the phenol oxygen. Clearly, the substance was assigned as 3-*O*-acetylpinobaskin not only because of the difference of 14 mass units at PI, but also because of the appearance of the same product ions in 3-*O*-acetylpinobaskin, which differs by 14 mass units from the pinobaskin product ions (*e.g.*, the diagnostic product ion of pinobaskin, *m/z* 153, presented a difference of 14 mass units from 3-*O*-acetylpinobaskin, *m/z* 167, thus confirming the presence of an acetyl group instead of a hydroxyl group in ring A).

Waridel and co-workers have proposed an LC-MS/MS methodology for the differentiation between C-6 and C-8 glycoside flavonoids (Waridel et al., 2001). Firstly, the authors analyzed standards of selected flavonoids (Fig. 12) in a reverse-phase C_{18} column using isocratic elution with acetonitrile/water 4:1 containing 0.5% acetic acid. Low energy CID experiments were performed on ion-trap (IT) and hybrid quadrupole time-of-flight (Q-TOF) instruments. ESI and APCI were used in both the positive and negative ion modes of analysis. The optimal CID collision energy was chosen so that the same MS/MS spectral profile would be generated and the relative intensities and *m/z* values of the fragment ions could be compared as a result of the structural differences. The authors postulated the distinction between C-8 and C-6 glycoside flavonoids on the basis of the relative intensity of some peaks of the MS/MS spectra. Taking into account the stabilities of the fragment ions, which were considered to be associated with their respective relative intensities, the authors reported that loss of H_2O and formation of *m/z* 379 are processes that are more favored for C-6 isomers (e.g., isovitexin, **44**; and isoorientin, **45**) than for C-8 isomers (e.g., vitexin, **46**; and orientin, **47**). Another proposition to distinguish between C-6 and C-8 glicoside flavonoid isomers was based on the intensity of the product ion [M+H–120]$^+$, which had low intensity when the skimmer voltage (to produce *in source* dissociation) was not employed, but was the base peak when the skimmer voltage was used. The authors selected the ion [M+H–120]$^+$ as the precursor ion in Q-TOF and used the multiple-stage mass spectrometry (MS^n) in IT. The results revealed a large difference between the isomers for the collision energy at 30 eV or 50%, which enabled distinction by hard fragmentation and some elimination reactions, specific for each isomer (Fig. 12). In the case of Q-TOF, the product ion *m/z* 283, which results from CH_2O loss from [M+H-120]$^+$, was the base peak in the MS/MS spectra of both C-6 and C-8 isomers. On the other hand, when IT was employed, the product ion at m/z 283 was the base peak for C-8 isomers, whereas the product ion at *m/z* 295 [M+H-120-H_2O]$^+$ was the base peak in the MS/MS spectrum of C-6 isomers. The authors also reported that differentiation between C-6 and C-8 glycoside flavonoid isomers in the negative ion mode was not possible when the MS/MS experiments were performed in the ion trap equipment, once CO elimination from [M-H-120]$^-$ is the only fragmentation process for both isomers. The fragmentation map proposed by the authors is illustrated in Fig. 13 (Waridel et al., 2001). Nevertheless, the main difficulty reported by the authors was method standardization, because screening and online identification of these compounds requires previous optimization of the parameters of both IT and q-TOF apparatus.

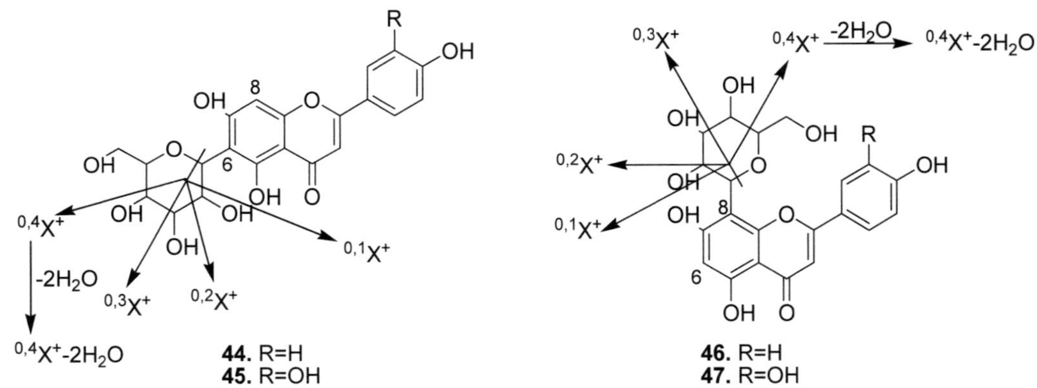

Fig. 12. Structure of C-6 (**44** and **45**) and C-8 (**46** and **47**) glicoside flavonoid isomers.

One of the most important steps in dereplication studies of natural products by LC-MS is the chromatographic separation process. Resolution (e.g, the distance between two adjacent peaks in the chromatogram) is essential for the correct interpretation of the subsequent UV-DAD, MS and MS/MS data. The addition of formic or acetic acid to the LC mobile phase (usually methanol/water or acetonitrile/water) has been the most used analytical strategy to improve the chromatographic resolution in HPLC. Shi and co-workers have identified glycoside flavonoid isomers in crude extracts of *Fructus aurantii* and *Fructus aurantii immaturus* using HPLC-UV-MS^n (Shi et al., 2007). The chromatographic separation between the compounds **48-50** on reverse phase C_{18} column was achieved by using a mixture of acetonitrile and water/formic acid (0.1%) and applying a linear gradient from 10% and 95% of water/formic acid (0.1%). The differentiation between the compounds was made on the basis of the comparison of retention times, maximum UV wavelengths, and MS and MS/MS data of each individual peak with those of authentic standards. MS/MS spectrum of deprotonated **49** (*m/z* 593) displayed the fragment ion [M-H-308]$^-$, suggesting elimination of

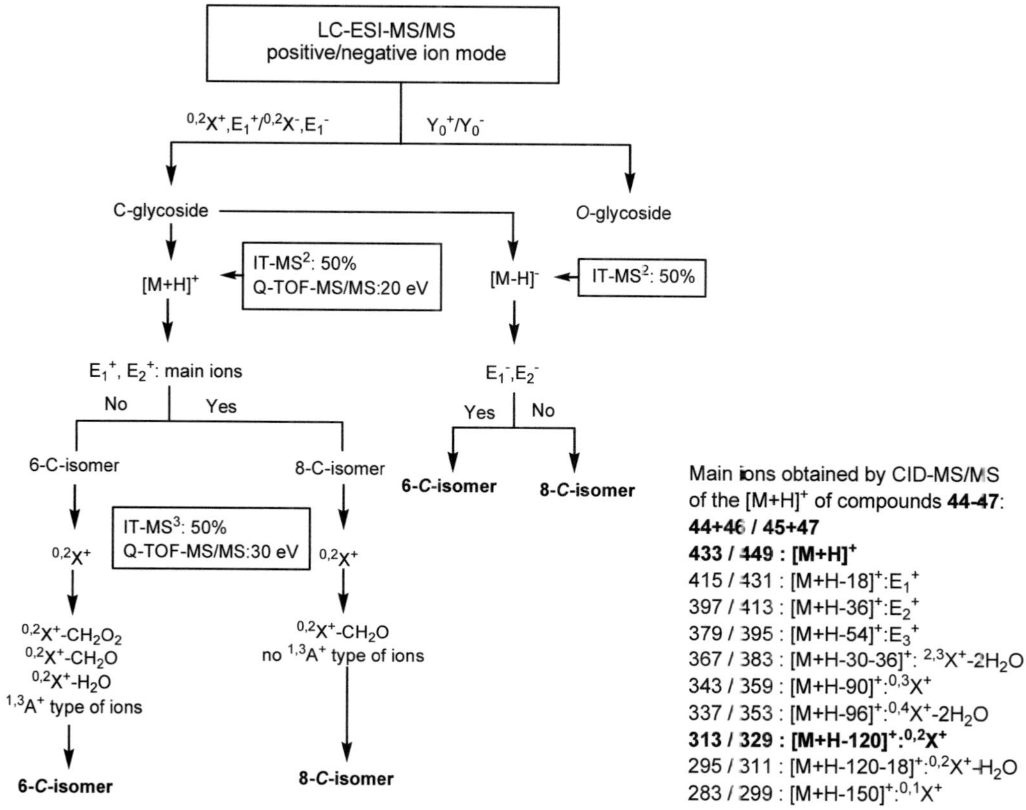

Fig. 13. Fragmentation map proposed for the differentiation between C-6 and C-8 glicoside flavonoid isomers (Waridel et al., 2001).

a rutinose molecule, whereas its MS³ spectrum showed the same profile as that of the aglycone of poncirin (**50**) (Figure 14). Moreover, the retention time of didymin (**49**, t_R=46,1 min) in the total ion chromatogram was higher than hespiridin (**48**, t_R=32,9 min), thus indicating a structure with fewer oxygen atoms as compared to **48**.

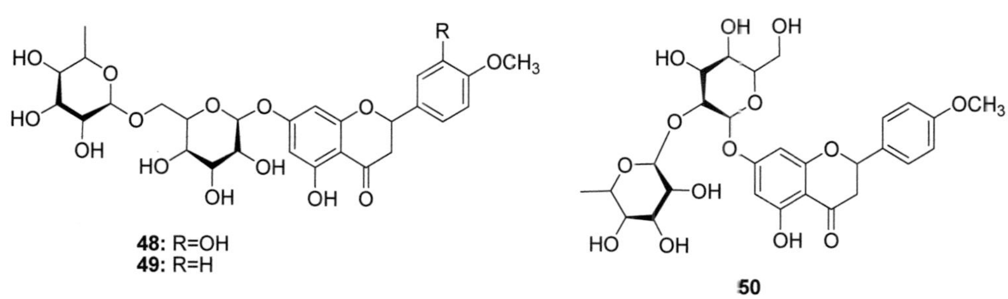

Fig. 14. Chemical structures of hesperidin (**48**), didymin (**49**) and poncirin (**50**).

4.3 Alkaloids

Alkaloids comprise a vast group of secondary metabolites found mainly in higher plants, such as angiosperms (Cordell et al., 2001). Although there is no general definition that encompasses the great structural diversity of this class of NPs, it is known that these compounds exhibit alkaline properties due to the nitrogen atoms found in their cyclic skeleton. Another chemical property of alkaloids is their solubility in aqueous solutions when salt complexes are formed in the presence of mineral acids (Cordell et al., 2001; Dewick, 2004).

Zhou and co-workers have proposed a methodology involving ultra-performance liquid chromatography tandem mass spectrometry (UPLC-MS/MS) for the screening of some pyrrolizidine alkaloids exhibiting toxic effects against human beings, such as fetal problems and hepatotoxic and tumorigenic effects (Zhou et al., 2010). The pyrrolizidine alkaloids were extracted directly from the powdered plant using diluted hydrochloric acid, followed by mixed-phase cation exchange (MCX) and solid-phase extraction (SPE) and elution with methanol/ammonia 3:1 (v/v). After solvent removal, the residue was analyzed by UPLC coupled with DAD and triple quadrupole tandem mass spectrometry, equipped with an electrospray ion source operating in the positive ion mode. The authors demonstrated that the fragment ions with *m/z* 150 and *m/z* 168 are diagnostic for pyrrolizidine alkaloids belonging to the Otonecine type group (e. g., **51** and **52**), whereas the fragment ions *m/z* 120 and *m/z* 138 are diagnostic for compounds of the Retronecine type (e.g., **53-57**), as represented in Fig. 15. Some pyrrolizidine alkaloids analogs (e.g., compounds senkirkine, **51;** and clivorine, **52**) that have fragment ions in common could not be distinguished from each other on the basis of their precursor ion spectra only. In this case, dereplication was achieved by combining data from the spectra of the precursor ion with those of multiple reaction monitoring (MRM) experiments.

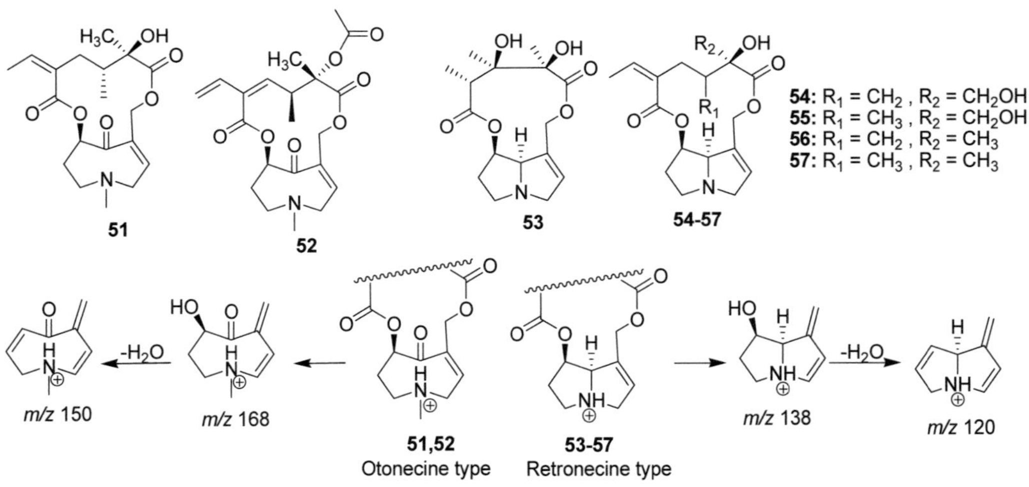

Fig. 15. Structure of pyrrolizidine alkaloids of the Otonecine and Retronecine type, and formation of their diagnostic fragment ions.

Pivatto and co-workers have utilized accurate mass electrospray ionization tandem mass spectrometry for the offline dereplication of selected piperidine alkaloids from flowers and fruit extracts of *Senna spectabilis*, which is commonly used in traditional medicine as anti-inflammatory, analgesic, laxative, antimicrobial, and antiulceral agent in some countries (Samy & Ignacimuthu, 2000; Viegas Jr. et al., 2004). The authors obtained the ethanol extract from powdered dried flowers and green fruits, followed by redissolution with aqueous

H_2SO_4 and washing with *n*-hexane. The acid residue was basified with NH_4OH (pH 9) and then extracted with dichloromethane. The analysis showed that some protonated compounds have common mass losses (e.g., loss of H_2O) and known fragment patterns (e.g., the charge-induced fragmentation that forms the acylium ion in *p*-coumaroyl derivatives), as shown in Fig. 16. Also, the authors reported that the extracts obtained from flowers contained more oxygen atoms than those obtained from fruits. On the other hand, the formation of some ions as evidenced in the spectrum of the precursor ion could not be explained on the basis of their ESI-MS/MS data, but they can also be of great importance when this technique is used in combination with NMR (Pivatto et al., 2005).

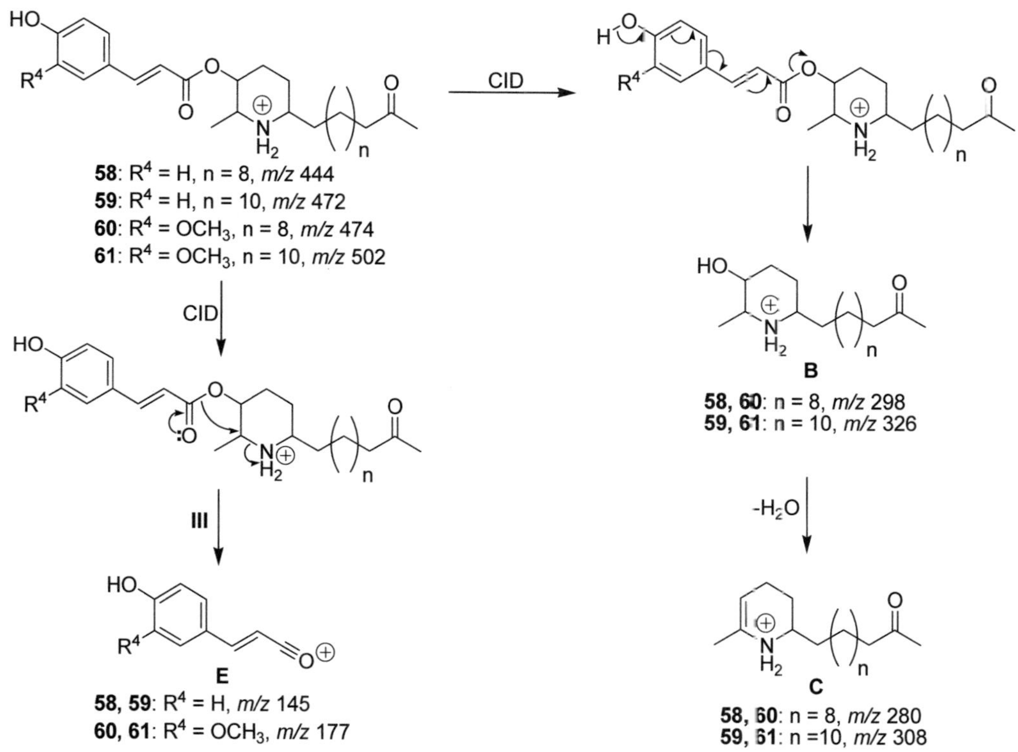

Fig. 16. Fragmentation of piperidine alkaloids identified in *Senna spectabilis* by ESI-MS/MS.

The dereplication of alkaloids has also been achieved by means of UV-Vis spectroscopy in combination with tandem mass spectrometry (MS/MS). Fabre and co-workers have studied a direct and fast characterization of isoquinoline alkaloids from the aerial parts of *Eschscholtzia californica*, which is used in European folk medicine as analgesic, anodyne, diaphoretic, diuretic, soporific, and spasmolytic agent (Fabre et al., 2000). Firstly, the authors analyzed authentic standards of the isoquinoline alkaloids berberine (**62**) and papaverine (**63**) by HPLC and ESI-MS/MS using direct infusion and collision-induced dissociation, aiming to identify possible diagnostic ions for each compound. After that, the powdered aerial parts of *E. californica* were extracted with MeOH for 15 min at 60°C. The residue obtained after solvent elimination was dissolved in aqueous HCl (pH 1), then dodecylsulfate sodium salt was added and further extracted with chloroform (three times). The organic phase was concentrated and dried, then suspended in MeOH and analyzed by high-performance liquid chromatography coupled with a diode array detector (HPLC-DAD) and HPLC-ESI-MS/MS. The UV and the MS/MS spectra of each peak of the chromatogram were compared with those of the authentic standards. They found that the fragment ions with *m/z* 177, 205, 235, 283, and 293 are diagnostic of compounds **64** and **65**; whereas the fragment ions *m/z* 235 and 263 are diagnostic for compounds **66** and **67** (Fabre et al., 2000). The major compounds identified in the crude extract were **64**, **65**, and **71**. The structures of the identified alkaloids are summarized in Fig. 17.

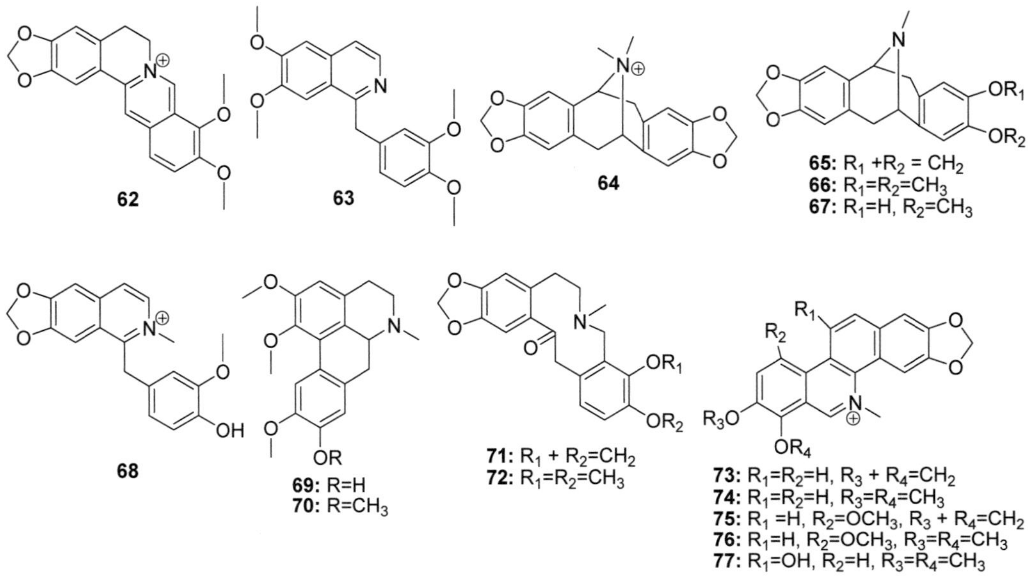

Fig. 17. Structures of the isoquinoline alkaloids identified in the crude methanol extract of aerial parts of *Eschscholtzia californica* (Fabre et al., 2000).

5. Concluding remarks

The use of electrospray ionization tandem mass spectrometry (ESI-MS/MS) for the online and offline dereplication of natural products (NPs) has been discussed herein. Online identification using ESI-MS/MS coupled with liquid chromatography with UV-DAD as first detection has shown to be a more powerful technique as compared to the sole use of ESI-MS/MS. Structural determination is usually achieved on the basis of a combination between retention times, UV and MS/MS spectra and comparison with data of a previously established library of authentic standards. However, this strategy is still limited because previously isolated (or commercially acquired) standards are necessary, not to mention the difficulty in establishing reliable libraries using MS/MS data acquired on different equipments. These are the main challenges to be overcome in coming years in this research field.

6. Acknowledgment

The authors thank the Brazilian Foundations FAPESP (Proc. FAPESP 2007/54241-8), for financial support, and CAPES and CNPq, for fellowships.

7. References

Ackermann, B. L.; Regg, B. T.; Colombo, L.; Stella, S. & Coutant, J. E. (1996). Rapid Analysis of Antibiotic-Containing Mixtures from Fermentation Broths by Using Liquid Chromatography-Electrospray Ionization-Mass Spectrometry and Matrix-Assisted Laser Desorption Ionization-Time-of-Flight-Mass Spectrometry. *Journal of the American Society for Mass Spectrometry*, Vol.7, No.12, (December 1996), pp. 1227-1237, ISSN 1044-0305.

Aguiar, G. P.; Wakabayashi, K. A. L.; Luz, G. F.; Oliveira, V. B.; Mathias, L.; Vieira, I. J. C.; Braz-Filho, R. & Crotti, A. E. M. (2010). Fragmentation of Plumeran Indole Alkaloids from *Aspidosperma spruceanum* by Electrospray Ionization Tandem Mass Spectrometry. *Rapid Communications in Mass Spectrometry*, Vol.24, No.3, (February 2010), pp. 295-308, ISSN 0951-4198.

Ardrey, R. E. (2003). *Liquid Chromatography-Mass Spectrometry: An Introduction.* , Wiley, ISBN 978-0-471-49801-8, England.

Ayres, D. C. & Loike, J. D. (1990). *Lignans: Chemical, Biological, and Clinical Properties*, Cambridge University Press, ISBN 9780521304214, Cambridge.

Bindseil, K. U.; Jakupovic, J.; Wolf, D.; Lavayre, J.; Leboul, J. & van der Pyl, D. (2001). Pure Compound Libraries: A New Perspective for Natural Product Based Drug Discovery. *Drug Discovery Today*, Vol.6, No.16, (August 2001), pp. 840-847, ISSN 1359-6446.

Botta, B.; Monache, G. D.; Misiti, D.; Vitali, A. & Zappia, G. (2001). Aryltetralin Lignans: Chemistry, Pharmacology and Biotransformations. *Current Medicinal Chemistry*, Vol.8, No.11, (September 2001), pp. 1363-1381, ISSN 0929-8673.

Butler, M. S. (2008). Natural Products to Drugs: Natural Product-Derived Compounds in Clinical Trials. *Natural Product Reports*, Vol.25, No.3, (May 2008), pp. 475-516, ISSN 1460-4752.

Clifford, M. N.; Johnston, K. L.; Knight, S. & Kuhnert, N. (2003). Hierarchical Scheme for LC-MS^n Identification of Chlorogenic Acids. *Journal of Agricultural and Food Chemistry*, Vol.51, No.10, (May 2003), pp. 2900-2011, ISSN 0021-8561.

Clifford, M. N.; Knight, S. & Kuhnert, N. (2005). Discriminating Between the Six Isomers of Dicaffeoylquinic Acid by LC-MS^n. *Journal of Agricultural and Food Chemistry*, Vol.53, (April 2005), pp. 3821-3832, ISSN 0021-8561.

Cordell, G. A.; Quinn-Beattie, M. L. & Farnsworth, N. R. (2001). The Potential of Alkaloids in Drug Discovery. *Phytotherapy Research*, Vol.15, (May 2001), pp. 183-205, ISSN 1099-1573.

Costa-Lotufo, L. V.; Montenegro, R. C.; Alves, A. P. N. N.; Madeira, S. V. F.; Pessoa, C.; Moraes, M. E. A. & Moraes, M. O. (2010). The Contribution of Natural Products as Source of New Anticancer Drugs: Studies Carried Out at the National Experimental Oncology Laboratory from the Federal University of Ceará *Revista Virtual de Química*, Vol.2, No.1, (August 2010), pp. 47-58, ISSN 1984-6835.

Crotti, A. E. M.; Carollo, C. A.; Gobbo-Neto, L.; Santos, M. D.; Gates, P. J. & Lopes, N. P. (2006). LC-Hyphenated Techniques: Uses in the Structural Elucidation of Low and High Molecular Weight Compounds, In: *Modern Biotechnology in Medicinal Chemistry and Industry*, C. A. Taft, pp. 1-43, Signpost Research, ISBN 81-308-0132-9, Kerala, India.

Crotti, A. E. M.; Cunha, W. R.; Lopes, N. P. & Lopes, J. L. C. (2005). Sesquiterpene lactones from *Minasia alpestris*. *Journal of the Brazilian Chemical Society*, Vol.16, No.3B, (May/June 2005), pp. 677-680, ISSN 0103-5053.

Crotti, A. E. M.; Lopes, J. L. C. & Lopes, N. P. (2005). Triple Quadrupole Tandem Mass Spectrometry of Sesquiterpene Lactones: A Study of Goyazensolide and its Congeners. *Journal of Mass Spectrometry*, Vol.40, No.8, (June 2005), pp. 1030-1034, ISSN 1096-9888.

Cuyckens, F. & Claeys, M. (2004). Mass Spectrometry in the Structural Analysis of Flavonoids. *Journal of Mass Spectrometry*, Vol.39, No.1, (January 2004), pp. 1-15, ISSN 1096-9888.

De Hoffmann, E. & Stroobant, V. (2007). *Mass Spectrometry: Principles and Applications*, John Wiley & Sons, ISBN 978-0-470-03310-4, West Sussex, UK.

Degani, A. L. G.; Cass, Q. B. & Vieira, P. C. (1998). Cromatografia: Um Breve Ensaio. *Quimica Nova na Escola*, Vol.7, (May 1998), pp. 21-25, ISSN 0104-8899.

Dewick, P. M. (2004). *Medicinal Natural Produtcs: A Biosynthetic Approach*, John Wiley & Sons ISBN 9780470741672, Chichester.

Dong, M. W. (2006). *Modern HPLC for Practicing Scientists*, Wiley Interscience, ISBN 978-0-471-72789-7, New Jersey

Fabre, N.; Claparols, C.; Richelme, S.; Angelin, M. L.; Fouraste, I. & Moulis, C. (2000). Direct Characterization of Isoquinoline Alkaloids in a Crude Plant Extract by Ion-Pair Liquid Chromatography–Electrospray Ionization Tandem Mass Spectrometry: Example of *Eschscholtzia californica*. *Journal of Chromatography A*, Vol.904, No.1, (December 2000), pp. 35-46, ISSN 0021-9673.

Fabre, N.; Rustan, I.; Hoffmann, E. & Quetin-Leclerq, J. (2001). Determination of Flavone, Flavonol, and Flavonone Aglycones by Negative Ion Liquid Chromatography Electrospray Ion Trap Mass Spectrometry. *Journal of the American Society for Mass Spectrometry*, Vol.12, No.6, (May 2001), pp. 707-715, ISSN 1387-3806.

Fredenhagen, A.; Derrien, C. & Gassmann, E. (2005). An MS/MS Library on an Ion-Trap Instrument for Efficient Dereplication of Natural Products. Different Fragmentation Patterns for $[M + H]^+$ and $[M + Na]^+$ Ion. *Journal of Natural Products*, Vol.68, No.3, (March 2005), pp. 385-391, ISSN 0163-3864.

Furtado, N. A. J. C.; Vessecchi, R.; Tomaz, J. C.; Galembeck, S. E.; Bastos, J. K.; Lopes, N. P. & Crotti, A. E. M. (2007). Fragmentation of Diketopiperazines from Aspergillus Fumigatus by Electrospray Ionization Tandem Mass Spectrometry (ESI-MS/MS). *Journal of Mass Spectrometry*, Vol.42, No.10, (October 2007), pp. 1279-1286, ISSN 1096-9888.

Gates, P. J.; Santos, M. D.; Gobbo-Neto, L.; Carollo, C. A.; Crotti, A. E. M.; Vessecchi, R. & Lopes, N. P. (2006). Mass Spectrometry Ionisation Techniques and Applications for the Analysis of Organic Compounds, In: *Current Methods in Medicinal Chemistry and Biological Physics*, C. A. Taft and C. H. T. P. Silva, pp. 187-206, Research Signpost, ISBN 81-308-0141-8, Kerala, India.

Gobbo-Neto, L. & Lopes, N. P. (2008). Online Identification of Chlorogenic Acids, Sesquiterpene Lactones, and Flavonoids in the Brazilian Arnica *Lychnophora ericoides* Mart.(Asteraceae) Leaves by HPLC-DAD-MS and HPLC-DAD-MS/MS and a Validated HPLC-DAD Method for their Simultaneous Analysis. *Journal of*

Agricultural and Food Chemistry, Vol.56, No.4, (February 2008), pp. 1193-1204, ISSN 0021-8561.

Gómez-Romero, M.; Zurek, G.; Schneider, B.; Baessmann, C.; Segura-Carretero, A. & Fernandez-Gutierrez, A. (2011). Automated Identification of Phenolics in Plant-Derived Foods by Using a Library Search Approach. *Food Chemistry*, Vol.124, No.1, (January 2011), pp. 379-386, ISSN 0308-8146.

Harvey, A. L. (2008). Natural Products in Drug Discovery. *Drug Discovery Today*, Vol.13, No.19-20, (October 2008), pp. 894-901, ISSN 1359-6446.

He, X. G.; Lian, L. Z. & Lin, L. Z. (1997). Analysis of Lignan Constituents from *Schisandra chinensis* by Liquid Chromatography-Electrospray Mass Spectrometry. *Journal of Chromatography A*, Vol.757, No.1-2, (January 1997), pp. 81-87, ISSN 0021-9673.

Herbert, C. G. & Johnstone, R. A. W. (2003). *Mass Spectrometry Basics*, CRC Press, ISBN 978-0849313457, Boca Raton.

Konishi, Y.; Kiyota, T.; Draghici, C.; Gao, J. M.; Yeboah, F.; Acoca, S.; Jarussophon, S. & Purisima, E. (2007). Molecular Formula Analysis by an MS/MS/MS Technique To Expedite Dereplication of Natural Products. *Analytical Chemistry*, Vol.79, No.3, (December 2006), pp. 1187-1197, ISSN 0003-2700.

Lang, G.; Mayhudin, N. A.; Mitova, M. I.; Sun, L.; Van der Sar, S.; Blunt, J. W.; Cole, A. L. J.; Ellis, G.; Laatsch, H. & Munro, M. H. G. (2008). Evolving Trends in the Dereplication of Natural Product Extracts: New Methodology for Rapid, Small-Scale Investigation of Natural Product Extracts. *Journal of Natural Products*, Vol.71, No.9, (September 2008), pp. 1595-1599, ISSN 0163-3864.

Lee, K. L. (2004). Current Developments in the Discovery and Design of New Drug Candidates from Plant Natural Product Leads. *Journal of Natural Products*, Vol.67, No.2, (December 2003), pp.273-283, ISSN 0163-3864.

Lee, M. S. (2002). *LC/MS Applications in Drug Development*, Wiley Intersciences, ISBN 978-0471405207, New York.

Ma, Y. L.; Li, Q. M.; Van der Heuvel, H. & Claeys, M. (1997). Characterization of Flavone and Flavonol Aglicones by Collision Induced Dissociation Tandem Mass Spectrometry. *Rapid Communications in Mass Spectrometry*, Vol.11, No.12, (August 1997), pp. 1357-1364, ISSN 1097-0231.

Ma, Y. L.; Van den Heuvel, H. & Clayes, M. (1999). Characterization of 3-Methoxyflavones Using Fast-Atom Bombardment and Collision-Induced Dissociation Tandem Mass Spectrometry. *Rapid Communications in Mass Spectrometry*, Vol.13, No.19, (September 1999), pp. 1932-1942, ISSN 19991015.

Miketova, P.; Schram, K. H.; Whitney, J.; Kearns, E. H. & Timmermann, B. N. (1999). Mass Spectrometry of 3,5- and 4,5-Dicaffeoylquinic Acids and Selected Derivatives. *Journal of Mass Spectrometry*, Vol.34, No.12, (December 1999), pp. 1240-1252, ISSN 1096-9888.

Newman, D. J. & Cragg, G. M. (2007). Natural Products as Sources of New Drugs Over the Last 25 Years. *Journal of Natural Products*, Vol.70, No.3, (February 2007), pp. 461-477, ISSN 0163-3864.

Newman, D. J.; Cragg, G. M. & Snader, K. M. (2000). The Influence of Natural Products Upon Drug Discovery (Antiquity to Late 1999). *Natural Product Reports*, Vol.17, No.3, (June 2000), pp. 215-234, ISSN 1460-4752.

Newman, D. J.; Cragg, G. M. & Snader, K. M. (2003). Natural Products as Sources of New Drugs Over the Period 1981-2002. *Journal of Natural Products*, Vol.66, No.7, (July 2003), pp. 1022-1037, ISSN 0163-3864.

Niessen, W. M. A. (2006). *Liquid Chromatography-Mass Spectrometry*, CRC Press, ISBN 978-0824719364, Boca Raton.

Niessen, W. M. W. & Tinke, A. P. (1995). Liquid Chromatography-Mass Spectrometry: General Principles and Instrumentation. *Journal of Chromatography A*, Vol.703, No.1-2, (June 1995), pp. 37-57, ISSN 0021-9673.

Núñez, O.; Moyano, E. & Galceran, M. T. (2005). LC–MS/MS Analysis of Organic Toxics in Food. *Trends in Analytical Chemistry*, Vol.24, No.7, (July 2005), pp. 683-703, ISSN 0165-9936.

Oliveira, E. J. & Watson, D. G. (2000). Liquid Chromatography-Mass Spectrometry in the Study of the Metabolism of Drugs and Other Xenobiotics. *Biomedical Chromatography*, Vol.14, No.6, (October 2000), pp. 351-372, ISSN 1099-0801.

Pan, J. Y.; Chen, S. L.; Yang, M. H.; Wu, J.; Sinkkonen, J. & Zou, K. (2009). An Update on Lignans: Natural Products and Synthesis. *Natural Product Reports*, Vol.26, No.10, (October 2009), pp. 1251-1292, ISSN 1460-4752.

Picman, A. K. (1986). Biological Activities of Sesquiterpene Lactones. *Biochemical Systematics and Ecology*, Vol.14, No.3, (May 1986), pp. 255-281, ISSN 0305-1978.

Pivatto, M.; Crotti, A. E. M.; Lopes, N. P.; Gamboal, I. C.; Rezende, A.; Viegas Jr., C.; Young, M. C. M.; Furlan, M. & Bolzani, V. S. (2005). Electrospray Ionization Mass Spectrometry Screening of Piperidine Alkaloids from *Senna spectabilis* (Fabaceae) Extracts: Fast Identification of New Constituents and Co-Metabolites. *Journal of the Brazilian Chemical Society*, Vol.16, No.6b, (December 2005), pp. 1431-1438, ISSN 0103-5053

Prasain, J. K.; Jones, K.; Kirk, M.; Wilson, L.; Smith-Johnson, M.; Weaver, C. & Barnes, S. (2003). Profiling and Quantification of Isoflavonoids in Kudzu Dietary Supplements by High-Performance Liquid Chromatography and Electrospray Ionization Tandem Mass Spectrometry. *Journal of Agricultural and Food Chemistry*, Vol.51, No.15, (July 2003), pp. 4213-4218, ISSN 0021-8561.

Rak, G.; Fodor, P. & Abrankó, L. (2010). Three-Step HPLC–ESI-MS/MS Procedure for Screening and Identifying Non-Target Flavonoid Derivatives. *International Journal of Mass Spectrometry*, Vol.290, No.1, (November 2011), pp. 32-38, ISSN 1387-3806.

Samy, R. P. & Ignacimuthu, S. (2000). Antibacterial Activity of Some Folklore Medicinal Plants Used by Tribals in Western Ghats of India. *Journal of Ethnopharmacology*, Vol.69, No.1, (January 2000), pp. 63-71, ISSN 0378-8741.

Shi, P.; He, Q.; Song, Y.; Qu, H. & Cheng, Y. (2007). Characterization and Identification of Isomeric Flavonoid *O*-diglycosides from Genus *Citrus* in Negative Electrospray Ionization by Ion Trap Mass Spectrometry and Time-of-Flight Mass Spectrometry. *Analytica Chimica Acta*, Vol.598, No.1, (August 2007), pp. 110-118, 0003-2670.

Shu, Y. Z. (1998). Recent Natural Products Based Drug Development: a Pharmaceutical Industry Perspective. *Journal of Natural Products*, Vol.61, No.8, (August 1998), pp. 1053-1071, ISSN 0163-3864.

Simões, C. M. O.; Schenkel, E. P.; Gosmann, G.; Mello, J. C. P.; Mentz, A. L. & Petrovick, P. R. (2004). *Farmacognosia: da Planta ao Medicamento*, Editora da UFSC, ISBN 978-85-7025-927-1, Porto Alegre / Florianópolis.

Sleno, L. & Volmer, D. A. (2004). Ion Activation Methods for Tandem Mass Spectrometry. *Journal of Mass Spectrometry*, Vol.39, No.10, (October 2004), pp. 1091-1112,

Smeraglia, J.; Baldrey, S. F. & Watson, D. (2002). Matrix Effects and Selectivity Issues in LC-MS-MS. *Chromatographia*, Vol.55, (July 2001), pp. 95-99, ISSN 0009-5893.

Todd, J. F. J. (1995). Recommendations for Nomenclature and Symbolism for Mass Spectroscopy. *International Journal of Mass Spectrometry and Ion Processes*, Vol.142, No.3, (April 1995), pp. 209-240, ISSN 0168-1176.

Tomer, K. B. (2001). Separations Combined with Mass Spectrometry. *Chemical Reviews*, Vol.101, No.2, (January 2001), pp. 297-328, ISSN 0009-2665.

Vékey, K. (2001). Mass Spectrometry and Mass-Selective Detection in Chromatography. *Journal of Chromatography A*, Vol.921, No.2, (July 2001), pp. 227-236, ISSN 0021-9673.

Vessecchi, R.; Crotti, A. E. M.; Guaratini, T.; Colepicolo, P.; Galembeck, S. E. & Lopes, N. P. (2007). Radical Ion Generation Processes of Organic Compounds in Electrospray Ionization Mass Spectrometry. *Mini-Reviews in Organic Chemistry*, Vol.4, No.1, (February 2007), pp. 75-87, ISSN 1570-193X.

Viegas Jr., C.; Bolzani, V. S.; Furlan, M.; Barreiro, E. J.; Young, M. C. M.; Tomazela, D. & Eberlin, M. N. (2004). Further Bioactive Piperidine Alkaloids from the Flowers and Green Fruits of *Cassia spectabilis*. *Journal of Natural Products*, Vol.67, No.5, (April 2004), pp. 908-910, ISSN 0163-3864.

Wang, Y. H.; Qiu, C.; Wang, D. W.; Hu, Z. F.; Yu, B. Y. & Zhu, D. N. (2011). Identification of Multiple Constituents in the Traditional Chinese Medicine Formula Sheng-Mai San and Rat Plasma after Oral Administration by HPLC–DAD–MS/MS. *Journal of Pharmaceutical and Biomedical Analysis*, Vol.54, No.5, (April 2011), pp. 1110–1127, ISSN 0731-7085.

Waridel, P.; Wolfender, J. L.; Ndjoko, K.; Hobby, K. R.; Major, H. J. & Hostettmann, K. (2001). Evaluation of Quadrupole Time-of-Flight Tandem Mass Spectrometry and Ion-Trap Multiple-Stage Mass Spectrometry for the Differentiation of C-Glycosidic Flavonoid Isomers. *Journal of Chromatography A*, Vol.926, No.1, (August 2001), pp. 29-41, ISSN 0021-9673.

Willfor, S. M.; Smeds, A. I. & Holmbom, B. R. (2006). Chromatographic Analysis of Lignans. *Journal of Chromatography A*, Vol.1112, No.1-2, (April 2006), pp. 64–77, ISSN 0021-9673.

Wolf, D. & Siems, K. (2007). Burning the Hay to Find the Needle - Data Mining Strategies in Natural Product Dereplication. *Natural Products in Drug Discovery*, Vol.61, No.6, (June 2007), pp. 339–345, ISSN 1359-6446.

Wolfender, J.-L. & Hostettmann, K. (1996). Importance of LC/MS in Plant Analysis. *Spectroscopy Europe*, Vol.8, No.4, (April 1996), pp. 7-12, ISSN 0966-0941.

Wolfender, J.-L.; Maillard, M. & Hostettmann, K. (1994) Thermospray Liquid Chromatography-Mass Spectrometry in Phytochemical Analysis. *Phytochemical Analysis*, Vol.5, No.4, (August 1994), pp. 153-182, ISSN 0958-0344.

Wolfender, J. L.; Waridel, P.; Ndjoko, K.; Hobby, K. R.; Major, H. J. & Hostettmann, K. (2000). Evaluation of Q-TOF-MS/MS and Multiple Stage IT-MSn for the Dereplication of Flavonoids and Related Compounds in Crude Plant Extracts. *Analusis*, Vol.28, No.10, (December 2000), pp. 895-906, ISSN 1286-482X.

You, J. S.; Huany.g, H. F.; Chang, Y. L. & Lee, Y. S. (2006). Sheng-Mai-San Reduces Adriamycin-Induced Cardiomyopathy in Rats. *The American Journal of Chinese Medicine*, Vol.34, No.2, (February 2006), pp. 295-305, ISSN 1793-6853.

Yu, K.; Ma, Y. L.; Shao, Q.; Qu, H. & Cheng, Y. (2007). Simultaneously Determination of Five Ginsenosides in Rabbit Plasma Using Solidphase Extraction and HPLC/MS Technique After Intravenous Administration of 'SHENMAI' Injection. *Journal of Pharmaceutical and Biomedical Analysis*, Vol.44, No.2, (June 2007), pp. 532-539, ISSN 0731-7085.

Zhai, H. B. & Cong, P. Z. (1990). Mass Spectrometric Studies on Dibenocyclooctadiene Lignan Compounds from Schisandraceae Plants. *Acta Pharmacologica Sinica*, Vol.25, No.2, (February 1990), pp. 110-122, ISSN 1745-7254.

Zhang, J. & Brodbelt, J. S. (2003). Structural Characterization and Isomer Differentiation of Chalcones by Electrospray Ionization Tanden Mass Spectrometry. *Journal of Mass Spectrometry*, Vol.38, No.5, (May 2003), pp. 555-572, ISSN 1096-9888.

Zheng, C.; Hao, H.; Wang, X.; Wu, X.; Wang, G.; Sang, G.; Liang, Y.; Xie, L.; Xia, C. & Yao, X. (2009). Diagnostic Fragment-Ion-Based Extension Strategy for Rapid Screening and Identification of Serial Components of Homologous Families Contained in Traditional Chinese Medicine Prescription Using High-Resolution LC-ESI IT-TOF/MS: Shengmai Injection as an Example. *Journal of Mass Spectrometry*, Vol.44, No.2, (February 2009), pp. 230-244, ISSN 1096-9888.

Zhou, Y.; Li, N.; Choi, F. F. K.; Qiao, C. F.; Song, J. Z.; Li, S. L.; Liu, X.; Cai, Z. W.; Fu, P. P.; Lin, G. & Xu, H. X. (2010). A New Approach for Simultaneous Screening and Quantification of Toxic Pyrrolizidine Alkaloids in Some Potential Pyrrolizidine Alkaloid-Containing Plants by Using Ultra Performance Liquid Chromatography–Tandem Quadrupole Mass Spectrometry. *Analytica Chimica Acta*, Vol.681, No.1-2, (November 2010), pp. 33-40, ISSN 0003-2670.

From the Collisionally Induced Dissociation to the Enzyme-Mediated Reactions: The Electron Flux Within the Lignan Furanic Ring

Andreina Ricci and Simona Piccolella
Second University of Naples
Italy

1. Introduction

The coupling of high-performance liquid chromatography (HPLC) with tandem mass spectrometry (MS/MS) is nowadays routinely used for the quali- and quantitative analysis of complex organic mixtures such as plant extracts, biological fluids, foods, etc. In this context the *MS/MS detector*, through the MRM or SMR techniques, ensures an excellent selectivity and sensitivity of basic importance in quantitative trace analysis. However, a reliable quantitative analysis assumes the knowledge of the analyte mass fragmentation pattern and of its energy resolved CAD mass spectra, allowing the accurate and effective choice of both the *m/z* ratio fragments and collision energies for the monitored transitions. Literature on mass spectrometric data concerning the fragmentation pattern of many plant secondary metabolites is almost limited to few classes of compounds. This issue represents the main object of our experimental research (Ricci et al, 2008; Ricci et al, 2010; Ricci et al, 2010).

In this chapter first we will focus on the structural investigation studies by tandem mass spectrometry of plant secondary metabolites belonging to the two classes of furofuranic lignans (FUR) and tetrahydrofuranic lignans (THF).

Further topics discussed will include gas phase ion molecules evidences emerging from MS/MS studies and their contribution to the comprehension of enzyme mediated reactions. In fact, gas phase ion chemistry studies, performed by investigating the mechanisms of predominant Collisionally Activated Dissociations by Mass Spectrometry (CAD/MS), can offer the possibility to investigate the "intrinsic" electronic properties of organic molecules. The isomerization/fragmentation processes induced in the MS/MS experiments can be influenced by a number of environmental factors, and their effect on the stability/reactivity of analytes can be extrapolated to different solvent-free environments. It is known that the effectiveness of enzyme mediated reactions depends dramatically on the high specificity of steric and electronic interactions between the substrate and the biological molecule. In this context, the "intrinsic reactivity" of the substrate plays a fundamental role and its study represents an essential tool for understanding the mechanism of enzymatic processes and for addressing theoretical effort toward the comprehension of the most important interactions leading to the enzymatic reactions.

2. Furofuranic and tetrahydrofuranic lignans and their glycosides

2.1 The importance of lignans

Lignans are a widespread group of important plant secondary metabolites, sharing a structure in which two cinnamic acid residues or their biogenetic equivalents are joined. The most abundant type of lignans found in nature consists of two phenylpropanoid units, linked by a carbon–carbon bond between the carbon chains (positions 8 and 8′). Lignans are found in roots, stems, leaves, seeds and fruits of vascular plants and are related to lignin. Despite their not complex chemical structure, the several hundreds of lignans up to now identified belong to different classes of chemical compounds, such as furofurans, tetrahydrofurans, dibenzylbutanes, dibenzylbutyrolactones, aryltetrahydronaphthalenes, arylnaphthalenes and dibenzocyclooctadienes. It depends on the way the oxygen atom is incorporated into the molecule and on the kind of cyclization. They occur either in free form or glycosidically linked to different carbohydrates.

In particular the skeleton of furofuranic lignans, depicted in figure 1, shows two tetrahydrofuranic rings connected to two aromatic rings which can have one or more substituents at the 3, 4, 3′, 4′ positions. Tetrahydrofuranic lignans have only one tetrahydrofuranic ring which is bound to a phenyl and benzyl ring at the 7 and 8′ positions, respectively (figure 1).

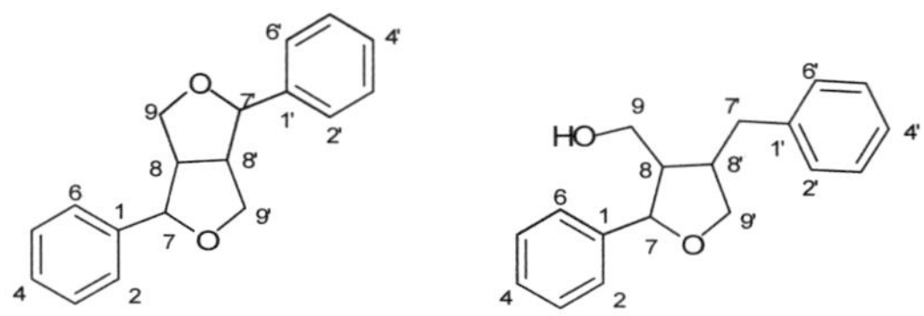

Fig. 1. Chemical skeleton of investigated FUR and THF lignans.

Lignans possess significant pharmacological activities, such as, e.g. for tetrahydrofuranic lignans, anti-parasite, antibacterial, antifungal, anti-tumoral, anti-inflammatory and platelet-activating factor (PAF) inhibition.

Some plant lignans can be converted by intestinal bacteria into the so-called enterolignans, enterodiol and enterolactone. Secolariciresinol and matairesinol are known to undergo this metabolism, but recently the number of lignans probed to be the enterolignans precursors has been enlarged, including also pinoresinol, lariciresinol and sesamin.

2.2 Tandem mass spectrometry in the structural characterization of furanic lignans

Due to their possible role in human health, the isolation and identification of new bioactive lignans from complex matrices is an important analytical issue. The coupling of high-performance liquid chromatography (HPLC) with mass spectrometry (MS) is routinely used for the characterization of lignans and their glycosides in organic mixtures. In spite of the importance of lignans, systematic ESI tandem mass spectrometric investigations on these compounds are yet limited to specific lignan sub-classes and rather incomplete (Eklud et al, 2008). However this technique can make efficient, straightforward and unambiguous their identification in plant extracts. Systematic tandem mass spectrometric studies on different types of lignans and their glycosides have been only recently reported. Most of the mass spectrometric data concerns the fragmentation patterns of their $[M-H]^-$ ions.

In figure 2 the chemical structures of FUR and THF lignans investigated by tandem mass spectrometry in our laboratory are presented.

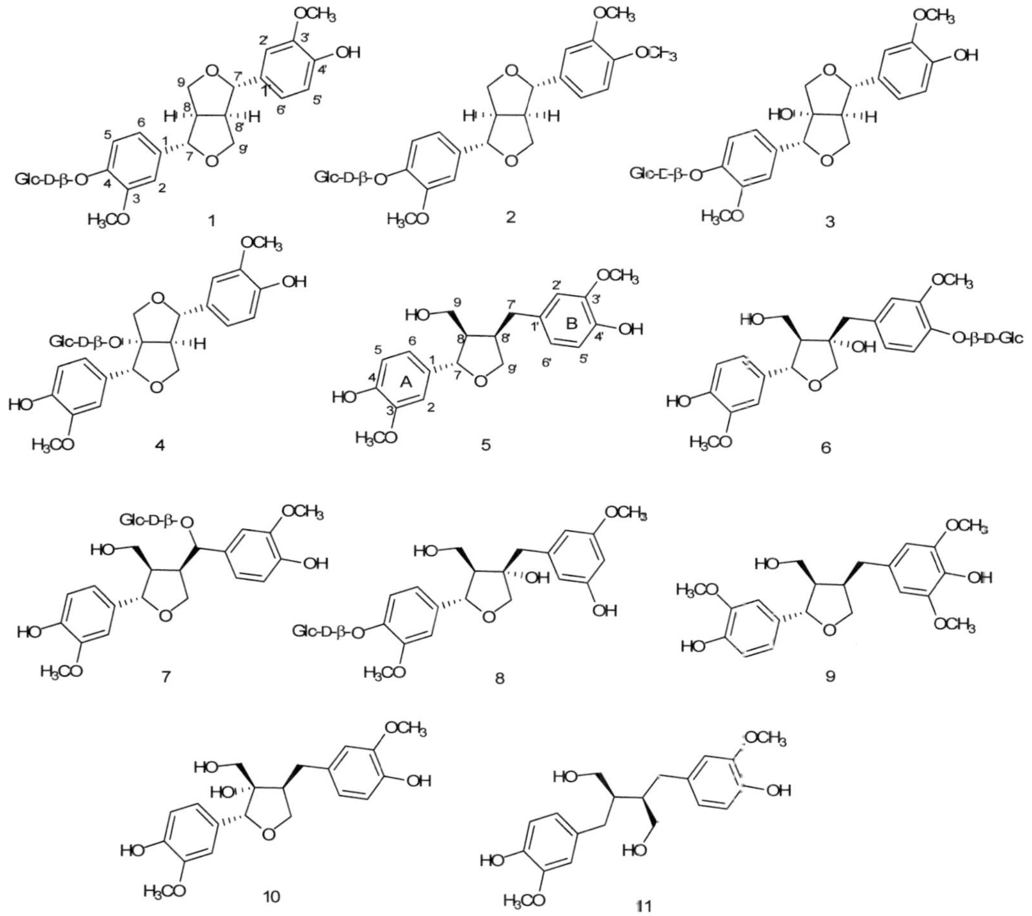

Fig. 2. Chemical structures of investigated FUR and THF lignans.

2.2.1 Positive ion mode

Useful information were obtained in the positive ion mode and CAD/MS spectra of $[M+H]^+$ ions and of alkali metal cation adducts were proposed for their rapid and unambiguous identification in plant extracts (Ricci et al, 2008; Ricci et al, 2010).

The CAD mass spectra of glycosylated lignans coordinated to different alkali metal cations (Alk^+, e.g. Li^+, Na^+, K^+, Rb^+, Cs^+) showed that when $Alk^+=Cs^+$, Rb^+, K^+ the adducts dissociated into the Alk^+ cation and the lignan glycoside, thus no providing structural information. However, the $[M+Li]^+$ and $[M+Na]^+$ CAD mass spectra revealed the presence of structurally diagnostic product ions.

All the $[M+Na]^+$ CAD mass spectra of the investigated glycosylated lignans showed ions arising from the loss of the hexose residue as -162 Da when the sugar is bound to a phenolic oxygen (figure 3a), and as -180 Da when glycosylation occurs at an alcoholic OH group (figure 3b).

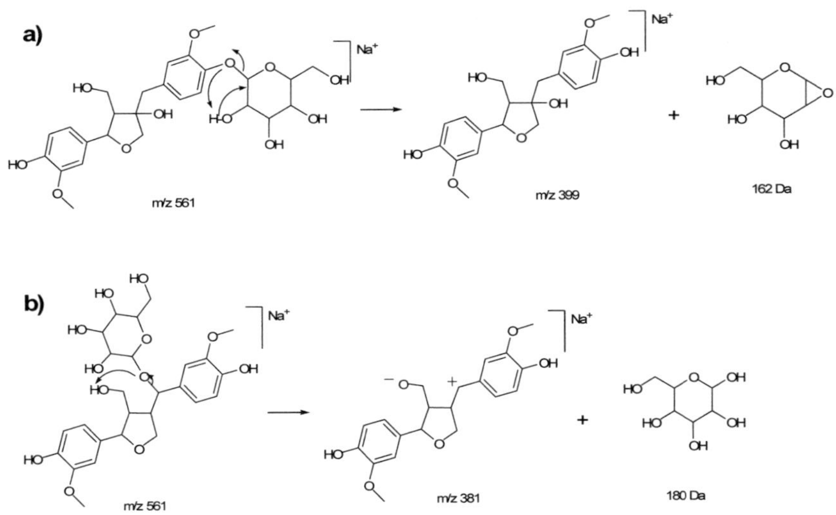

Fig. 3. Hypothesized mechanism for the loss of the hexose as (a) -162 Da, and (b) -180 Da.

If the Na^+ cation is retained by the sugar moiety the ion at *m/z* 185 (162+23), corresponding to the sodiated hexose moiety, is present in the spectrum (figure 4).

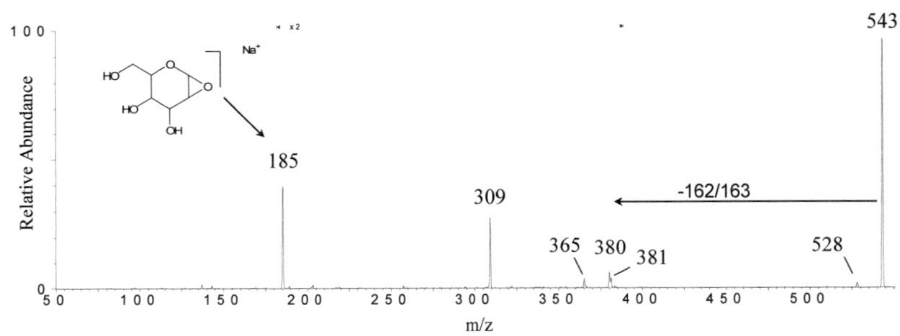

Fig. 4. $[M+Na]^+$ CAD mass spectrum of pinoresinol (compound **1**).

In addition, for the sodiated adduct of pinoresinol (compound **1**), homolytic cleavage of the glycosidic bond, leading to the elimination of a radical glucosyl moiety (163 Da) and generating the sodiated radical aglycone ion, was also observed (figure 4). The nature and the intensities of the fragments observed in the $[M+Alk]^+$ CAD mass spectra of FUR and THF lignan glucosides reasonably reflects the gas-phase strength of the multisite coordinative bond between the alkali metal cations and the cation-coordinating groups of lignan glycosides. For example if the intensities of the sodiated aglycone ion, arising from a neutral loss of 162 Da, and of the sodiated sugar fragment at *m/z* 185 are comparable, the strength of the interaction between the alkali metal cation with the coordinating group and the sugar residue could be comparable.

To obtain information on the nature of product ions generated during TQ/CAD processes, thus confirming our fragmentation mechanism hypotheses, the H/D solution exchange proved to be a very effective tool. In these experiments lignan glucosides were dissolved in a D_2O/CD_3OD solution and all the acidic H atoms, namely those bonded to the oxygen atoms, undergo selectively D exchange. For this reason the *m/z* ratio of precursor and fragment ions

increases according to the number of incorporated D atoms. Therefore, when the glucose is present in the molecule, four D atoms belong to the sugar moiety, and if it is bound to a phenolic OH group the loss of 165 Da (instead of 162) occurs. The fourth deuterium is transferred to the aglycone in accordance with the mechanism described in figure 3a.
H/D exchange experiments played a key role in explaining a different dissociation mechanism observed for lignan **6**. In the CAD mass spectrum of the sodiated adduct of this compound (at *m/z* 561) a fragment ion (at *m/z* 381), derived from the neutral loss of 180 Da, was detected, but in this case it was found to not correspond to the sugar moiety owing to the lower number of deuterium atoms incorporated in the fragment. The hypothesized mechanism is proposed in figure 5.

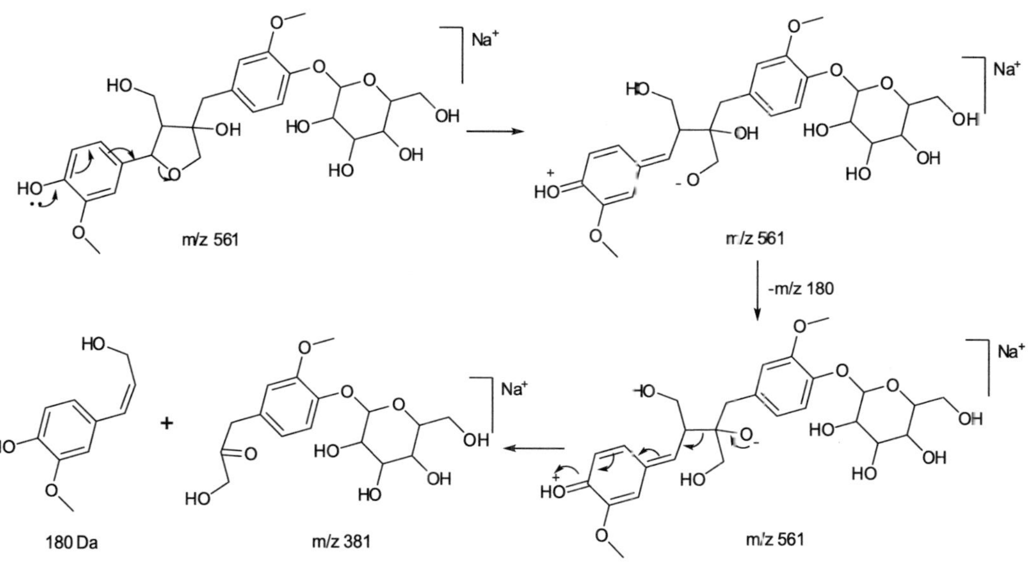

Fig. 5. Proposed mechanism for the loss of a 180 Da moiety not containing the hexose residue.

Moreover, in the positive ion mode, the mass spectra of sodiated adducts of compounds **1-3** and **8**, that is of FUR and THF lignan glucosides bonding the sugar at the oxygen in the C-4 position of the guaiacyl moiety, showed a structurally diagnostic ion at *m/z* 309, never described before and arising from the loss of the [CH_3O-C_6H_4-O-glucosyl] moiety, namely the glucosilated guaiacyl group, coordinated to the Na^+ ion. The nature of this fragment was hypothesized on the basis of the following evidences:

1. this ion was not detected in the $[M+Na]^+$ CAD mass spectra of lignans which bound the glucose moiety in a position different from C-4;
2. the ion shifted from *m/z* 309 to 293 when Li^+ instead of Na^+ was used as chelating ion;
3. after H/D solution exchange in the $[M+Na]^+$ CAD spectra of deuterated lignans **1-3** and **8**, the ion at *m/z* 309 shifted to 313 according to the presence of four deuterium atoms.

In this case, our mechanistic hypothesis involves that the Alk^+ cation interacts mainly with the glycosidic bridging oxygen, as depicted in figure 6. This hampers the proton transfer from the sugar to the aglycone because the Na^+ chelation decreases the basicity of the bridging glycosidic oxygen atom, preventing the glycosidic bond cleavage.

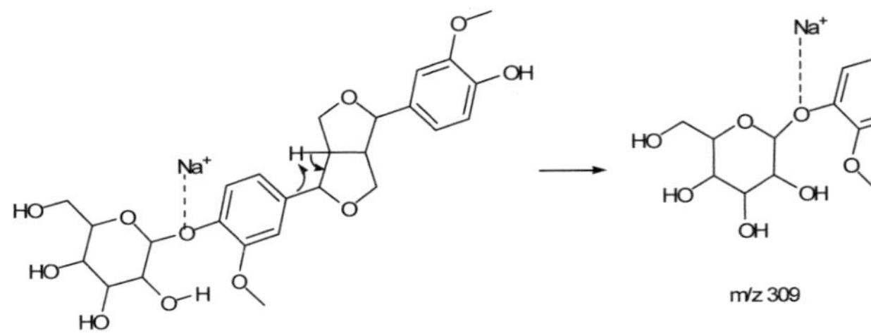

Fig. 6. Proposed mechanism for the formation of the fragment ion at *m/z* 309.

2.2.2 Negative ion mode

In the negative ion mode CAD mass spectra of deprotonated glycosylated lignans showed the typical neutral loss of the glucose moiety as 162 or 180 Da through the same mechanisms described above in the positive ion mode. Furthermore, if deprotonation occurs at the sugar residue, the ion at *m/z* 161 is present in the spectrum (figure 7).

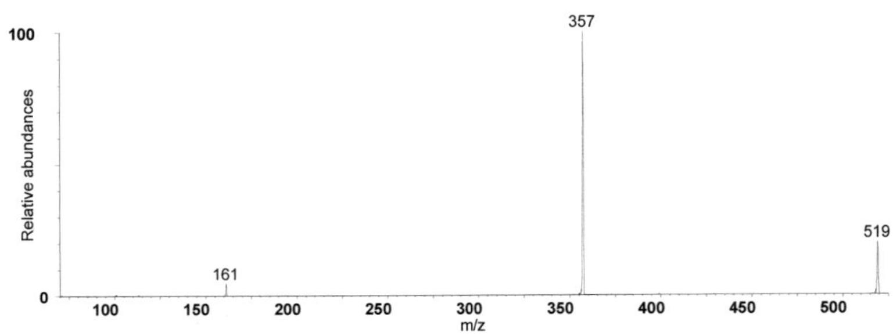

Fig. 7. CAD mass spectrum of deprotonated pinoresinol (lignan **1**) at *m/z* 519.

In the negative ion mode our attention has been mainly devoted to the fragmentation of the lignan aglycone ions arising from the in-source CAD of glucosylated species in order to study their most significant fragmentation patterns. For this purpose we recorded energy resolved TQ/CAD mass spectra of deprotonated aglycone lignans (**1-4** and **6-8**) that were used in a qualitative way to infer information on the integrated energetic picture of main CAD fragmentations and to investigate the mechanism of the predominant dissociation/isomerisation processes. On the basis of the hypothesized fragmentation mechanisms, gas-phase features of the furanic ring were derived. As an example in figure 8 the energy resolved CAD mass spectra of deprotonated pinoresinol glucoside and of its aglycone, generated by in-source CAD, are reported.

In the case of pinoresinol the symmetry of the molecule makes the nature of fragmentation products independent from the negative charge localization. According to the mechanism depicted in figure 8, the formation of the ion at *m/z* 151, the main fragmentation pathway of pinoresinol, arose from the cleavage of the C7′-C8′ (or C7-C8) and O-C9 (or O-C9′) bonds of the tetrahydrofuran ring. As in the positive, also in the negative ion mode H/D exchange experiments allowed to confirm the nature of fragmentation products. In fact the ion at *m/z* 151 does not contain deuterium atoms, confirming the mechanism that involves charge delocalization from the aromatic ring not containing the deuterium atom. In agreement with this hypothesis the CAD mass spectra of deprotonated medioresinol, syringaresinol and 3,4-demethyleudesmin 4-O-glucoside, reported in literature (Eklud et al, 2008; Guo et al, 2007),

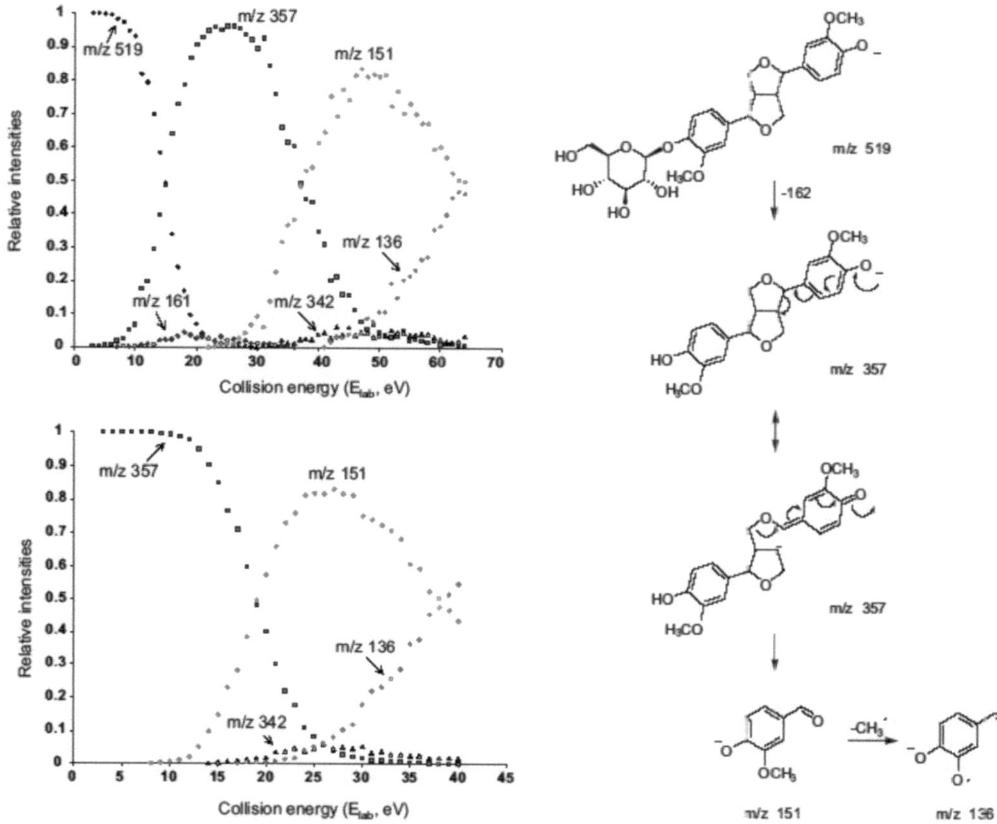

Fig. 8. Energy-resolved TQ/CAD mass spectrum of deprotonated pinoresinol 4-O-β-D-glucopyranoside (at *m/z* 519) and of deprotonated pinoresinol (at *m/z* 357) generated by in-source CAD, and its hypothesized fragmentation mechanism.

exhibit syringyl (*m/z* 181) and guaiacyl (*m/z* 151) ions, or ions whose formation can be explained by this mechanism.

The presence in the C-8 position of an OH substituent deeply modifies the fragmentation pathway. In fact likely it causes the C–O bond cleavages in the furanic ring with the loss of CH_2O units. It can be supposed that charge delocalization from the phenolic oxygen through the aromatic ring increases the electronic density charge at the C-7 carbon and the fragmentation pathways of furanic lignans reflect the effect of the R substituent in the C-8 position. So, if R=H, cleavage of the C7′–C8′ bond takes place, but when R=OH the lower charge density in the C-8′ position, due to the electron-withdrawing effect of this substituent, is assisted by charge delocalization from the ring through a hyperconjugative effect and the C7′–C8′ bond gets stronger. Subsequently, C7′–O bond cleavage occurs and leads to the loss of the CH_2O unit (figure 9).

In order to get information on the collision-activated fragmentation mechanisms of THF lignans we first investigated the $[M-H]^-$ CAD fragmentation pattern of secoisolariciresinol (lignan **11**), a lignan showing, instead of the tetrahydrofuran ring, a butanediol structure. Figure 10 shows the energy resolved CAD mass spectrum of deprotonated secoisolariciresinol at *m/z* 361.

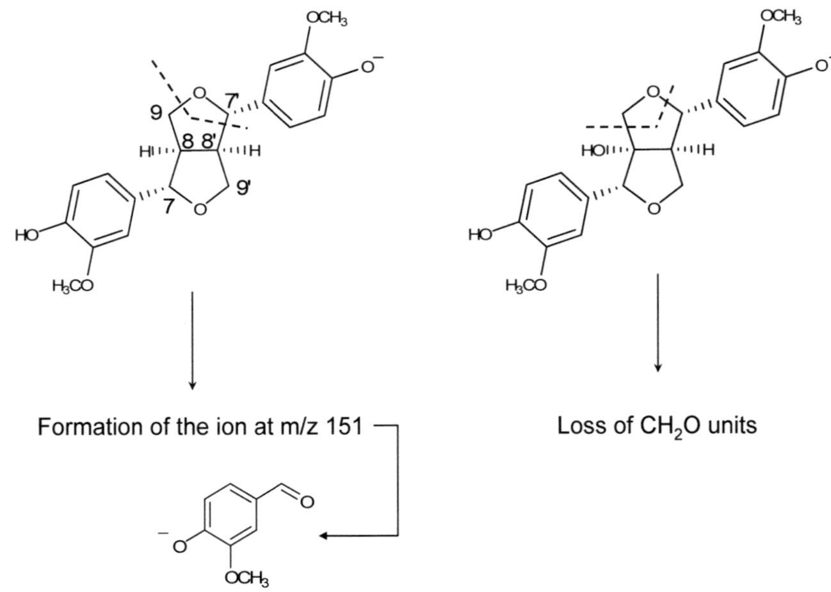

Fig. 9. The effect of the R substituent in the C-8 position on the fragmentation mechanism.

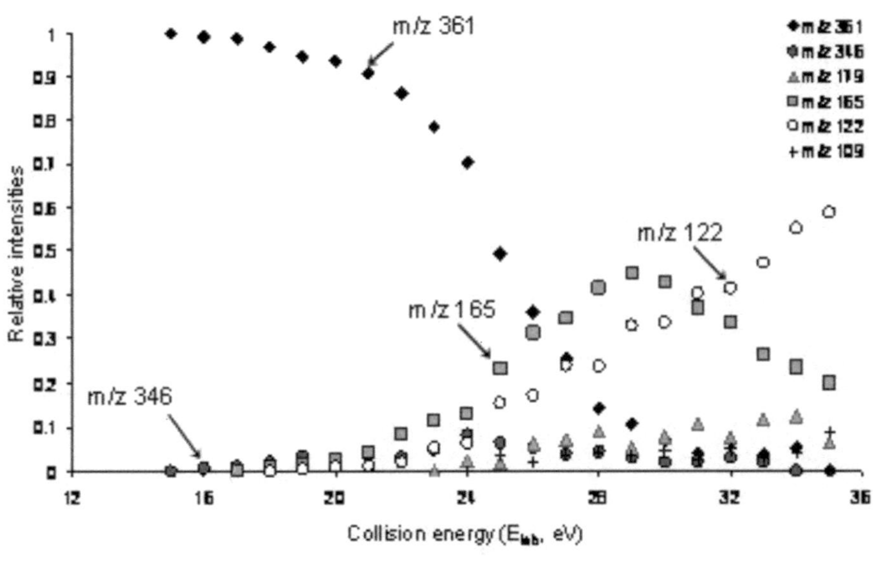

Fig. 10. [M-H]- energy resolved CAD mass spectrum of deprotonated secoisolariciresinol at *m/z* 361.

As previously reported (Eklund et al, 2008; Struijs et al, 2008) the loss of a methyl radical, occurring at relatively high collision energies (<15 eV), leads to the formation of the ion at *m/z* 346 which, in turn, after β-β cleavage (C8-C8′ bond) yields the ion at *m/z* 165 and at *m/z* 179, the main fragmentation channels together with the ion at *m/z* 122 at higher collision energies. The ion at *m/z* 346, generated by in-source CAD, is found to be the precursor ion of all these fragments. After H/D exchange experiments the *m/z* ratio of the precursor ion shifts to *m/z* 364, according to the presence of three acidic hydrogen atoms and the ions at *m/z* 346, 179 and 165 shift to *m/z* 349, 180 and 166. On the basis of these evidences the possible fragmentation mechanism is reported in figure 11.

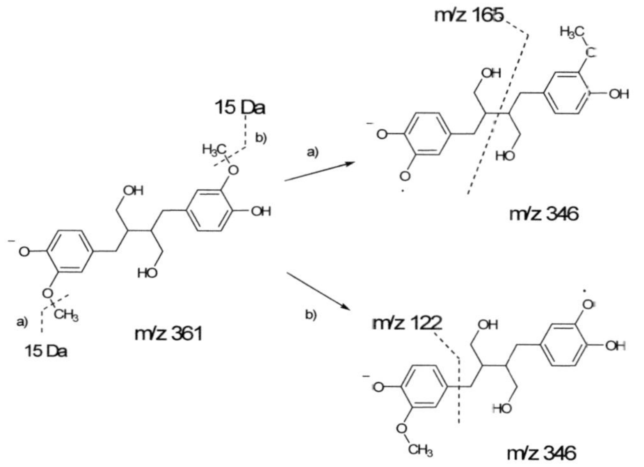

Fig. 11. The hypothesized fragmentation mechanism proposed for secoisolariciresinol.

According to figure 11, the CAD fragmentation pathway of secoisolariciresinol starts from the loss of a CH_3 radical followed by the breaking of the β-β (C8-C8') bond. It should be noted that only little intensities of the ion at *m/z* 331, deriving from the loss of 30 Da (CH_2O), are observed (see later). To confirm our mechanistic hypothesis and to get further information, we considered the CAD fragmentation pattern, previously reported by Eklund and co-workers (Eklund et al, 2008), of deprotonated 7-hydroxysecoisolariciresinol (at *m/z* 377) (figure 12).

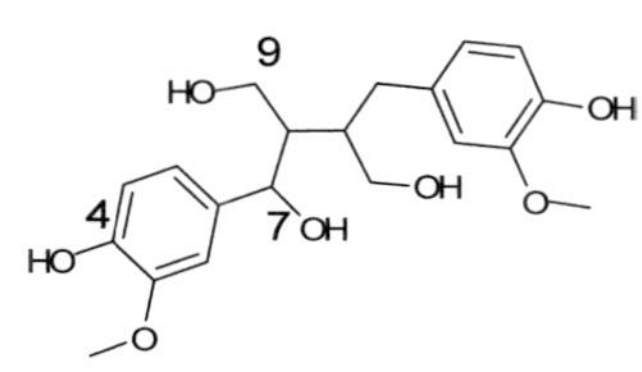

Fig. 12. The chemical structure of 7-hydroxysecoisolariciresinol.

For this lignan no loss of $CH_3\cdot$ was observed, but a loss of 48 Da takes place, which corresponds to the elimination of a water molecule (18 Da) and of a CH_2O moiety (30 Da). We supposed that the loss of CH_2O was not due to the initial deprotonation of the $C(9)H_2OH$ group because it does not occur in secoisolariciresinol, but it involves a preliminary deprotonation at the C-4 phenolic group and a charge delocalization through the aromatic ring. The occurrence of the C(7)OH deprotonation can be considered, at least in principle, rather unlike by taking into account the high gas phase acidity difference between the phenolic and the secondary alcoholic positions.

Thus, it can be concluded that in secoisolariciresinol the lack of oxidized benzylic positions compels an alternative fragmentation channel, namely the loss of a CH_3 radical, which promotes the homolytic cleavage of the C8-C8' bond allowing the formation, during CAD processes, of stabilized radicals. Confirming this conclusion, the CAD fragmentation pattern of all the lignans having a butanediol structure reported by Eklund (Eklund et al, 2008) exhibit the same fragmentation pathways.

To investigate the fragmentation mechanisms of THF lignans, the $[M-H]^-$ CAD mass spectra of lariciresinol (lignan **5**) and its OH derivatives (lignans **6-10**) were compared in order to derive the influence of the OH group on the fragmentation mechanism, as explained for furofuranic lignans. The energy resolved CAD mass spectrum of deprotonated lariciresinol at *m/z* 359 (reported in figure 13) shows as the main fragmentation channel the loss of 30 Da, reasonably CH_2O, giving the ion at *m/z* 329 appearing at a collision energy threshold value (~ 4 eV) lower than that observed for the loss of the $CH_3\cdot$ in secoisolariciresinol.

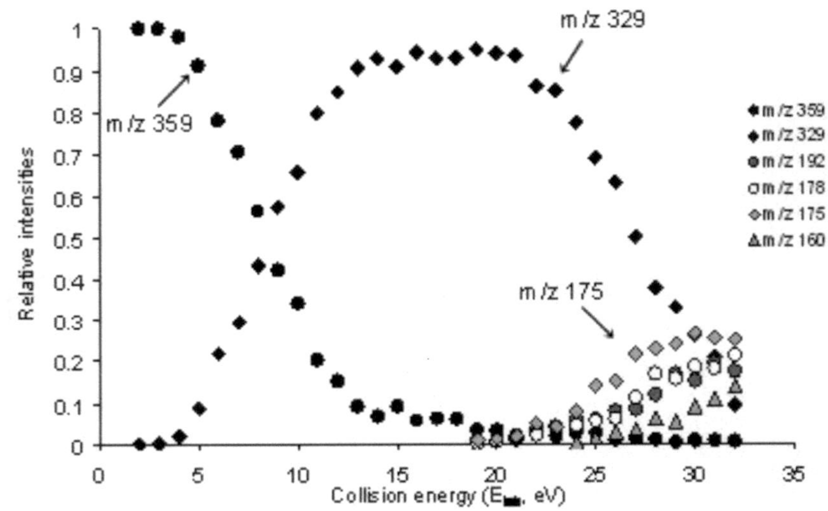

Fig. 13. Energy resolved CAD mass spectrum of deprotonated lariciresinol at *m/z* 359.

For each investigated THF lignan the loss of 30 Da occurs at low collision energies, except for compound **6** and **8**, where it represents a minor fragmentation channel, and for secoisolariciresinol, where it is completely absent.

Therefore it is likely to hypothesize a generalized mechanism, namely the cleavage of the furanic ring and not of the $C(9)H_2O$ chain. In confirmation, lignan **9**, having a 3',5'-dimethoxy,4'-hydroxyphenyl instead of the guaiacyl moiety, showed the same fragmentation pattern of deprotonated lariciresinol. In fact, from deprotonated lignan **9** (*m/z* 389) after the loss of 30 Da the ion at *m/z* 359 is generated.

Finally, according to our mechanistic hypothesis on the basis of these experimental evidences and by considering the CAD mass spectra of deprotonated secoisolariciresinol we concluded that in lariciresinol and its derivatives the presence of the tetrahydrofuran ring instead of the butanediolic structure allows the breaking of the C7-O bond, due to charge delocalization from the oxygen bonded to the C-4 atom of the *A* ring to the C-7 atom, leading to the loss of a formaldehyde molecule (figure 14).

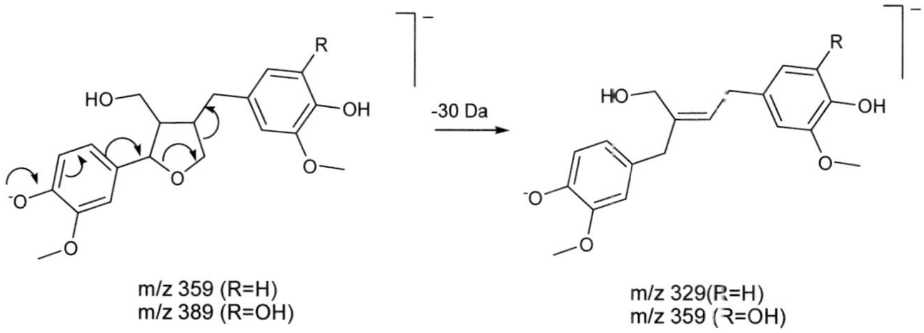

Fig. 14. Hypothesized mechanism for the loss of formaldehyde.

Lignan **10** is a lariciresinol derivative which differs for the presence of a hydroxyl substituent at the C-8 position. Despite the presence of the OH group, also in this case a loss of 30 Da (CH_2O) is observed at low collision energy values, comparable to those observed in lariciresinol for the same fragmentation. In figure 15 we report the energy resolved CAD mass spectrum of deprotonated lignan **10** at *m/z* 375.

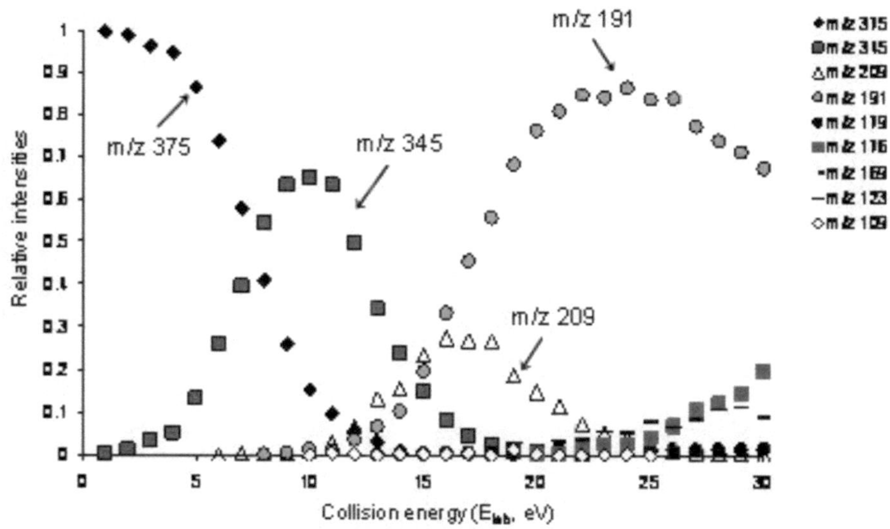

Fig. 15. Energy resolved CAD mass spectrum of deprotonated lignan **10** at *m/z* 375.

It seems to suggest that the presence of the OH substituent does not influence the mechanism for the loss of 30 Da. Therefore also in this case we hypothesized that the C7-O bond breaking takes place as a result of charge delocalization from the C-4 deprotonated position to the C-7 position and the cleavage of the C8′-C9′ bond leads to the loss of formaldehyde (figure 16).

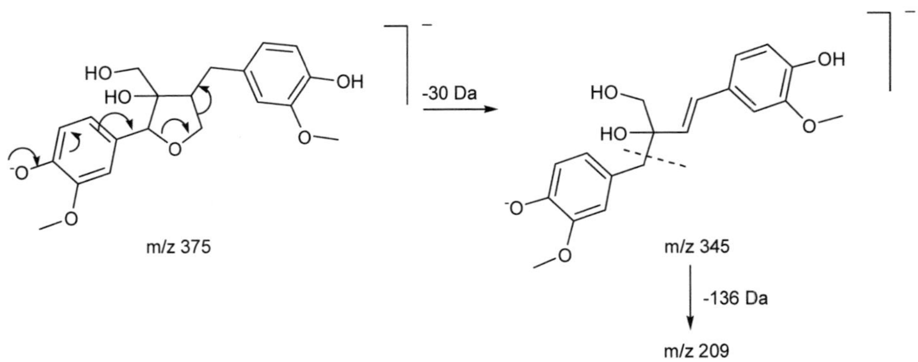

Fig. 16. Hypothesized fragmentation mechanism of deprotonated lignan **10.**

The 7′-hydroxylariciresinol, arising from lignan **7** after the loss of 162 Da, has an OH group bound at the C-7′ position. The energy resolved CAD mass spectrum of deprotonated aglycone **7** at *m/z* 375 reported in figure 17 showed that both the fragment ions at *m/z* 357 (-18 Da) and 345 (-30 Da), which structures are depicted in figure 18, are formed at comparable collision energies. In turn, these ions can lose CH_2O and H_2O, respectively, giving the ion at *m/z* 327. The latter ion can still lose CH_2O yielding the fragment at *m/z* 297.

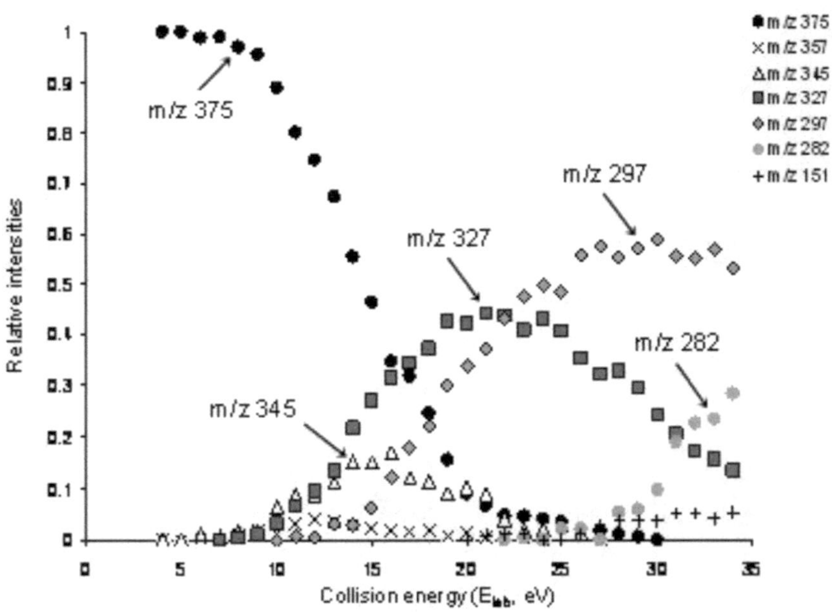

Fig. 17. Energy resolved CAD mass spectrum of deprotonated 7′-hydroxylariciresinol at *m/z* 375.

In spite of the other THF lignans a charge delocalization from C-4′ position (besides the delocalization from C-4 position) can be considered, and it is able likewise to justify the formation of the fragment ion at *m/z* 327 (figure 18).

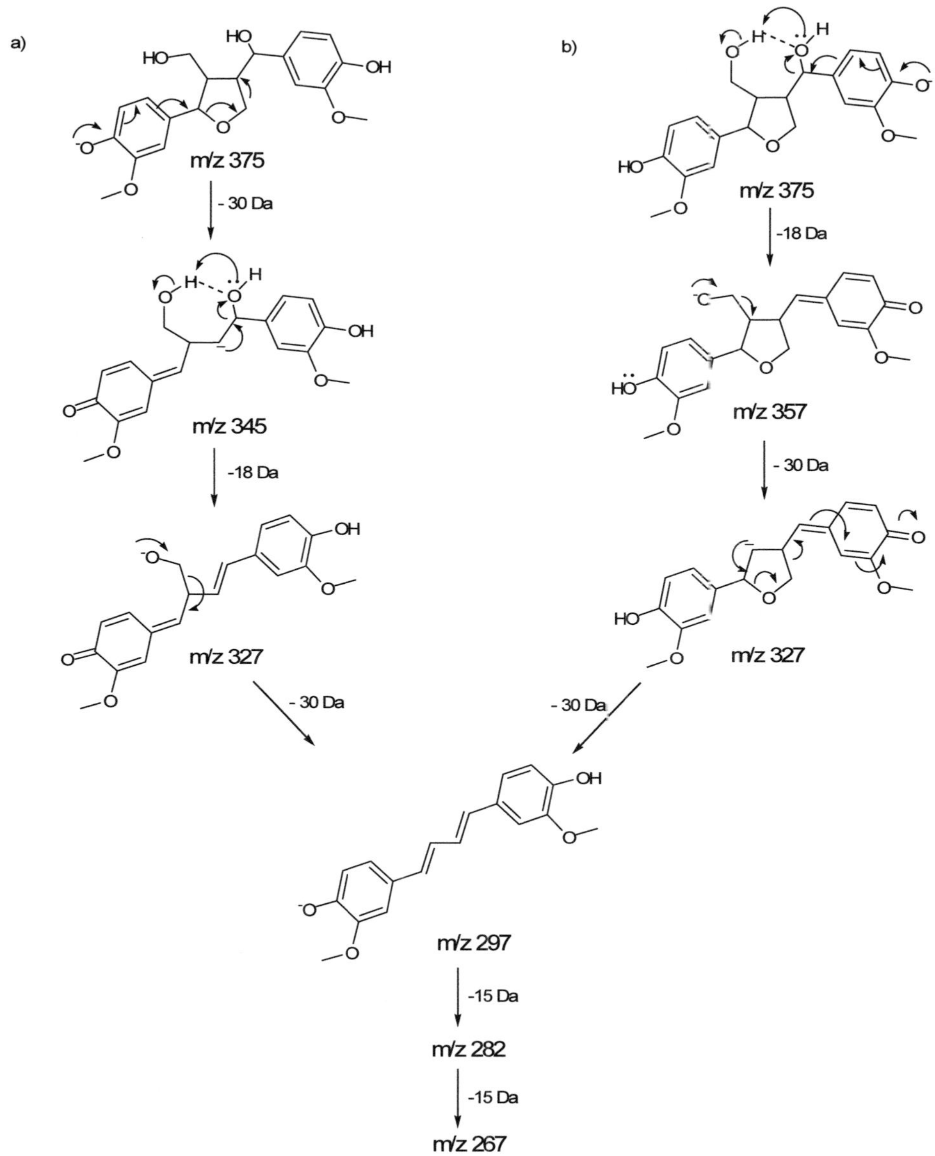

Fig. 18. Hypothesized fragmentation mechanisms for deprotonated 7′-hydroxylariciresinol.

Deprotonated aglycone ions from lignans **6** and **8** (figure 19) showed very similar fragmentation patterns. The energy resolved CAD mass spectrum reported in figure 19 refers to lignan aglycone **6**, but that of lignan aglycone **8** can be easily overlapped.

In both cases the spectrum seems to be deeply influenced by the presence of the OH group at the C-8' position. In fact it should be noted that only low intensities of the ion at *m/z* 345, arising from the loss of CH_2O, are present. We hypothesized that charge delocalization from the C-4 position causes the breaking of the C7-O bond but the loss of CH_2O is prevented by a fast proton transfer from the tertiary OH alcoholic group at the C-8′ position to the C(9′)H_2O chain (figure 20).

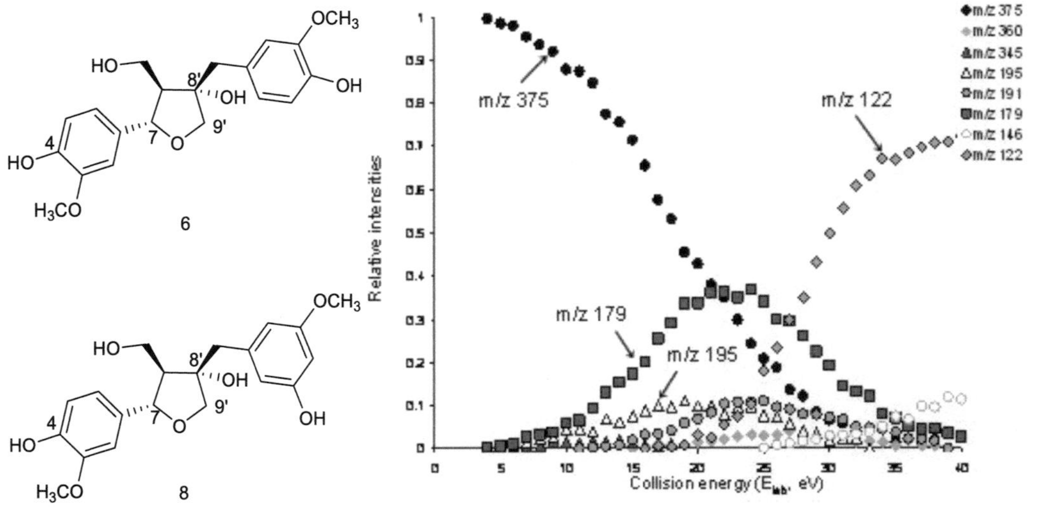

Fig. 19. Energy resolved CAD mass spectrum of deprotonated lignan **6** at *m/z* 375.

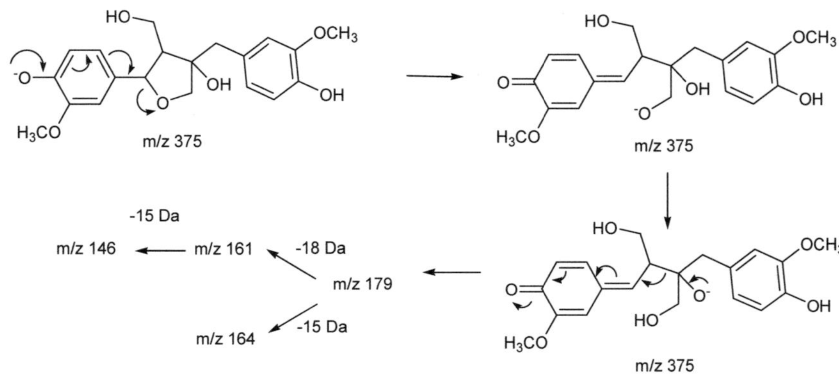

Fig. 20. Hypothesized fragmentation mechanism of deprotonated lignan **6**.

3. Tandem mass spectrometry and enzyme-mediated reactions: the Pinoresinol-Lariciresinol reductase

Despite their importance for human health as well as for plant biology, systematic studies on the enzymes involved in the biosynthesis of lignans and isoflavonoids have only recently reported likely because only fairly recently have been identified several genes, proteins and enzymes involved in their biosynthetic pathways. Indeed , no NMR, and X- ray crystal structures up to now were available for structural analysis. The study of the biosynthetic pathways of monolignol derived dimers found as 8-5′, or 8-8′, or 8-O-4′ linked moieties that, depending upon the specific plant species may be either optically active or racemic, leads to the isolation and characterization of various NADPH-dependent aromatic alcohol reductases, namely: i) pinoresinol-lariciresinol reductase (PLR) which catalyzes reduction of the 8-8′ linked lignan, pinoresinol generating the secoisolariciresinol diglucoside in flaxseed (*Linum usitatissimum*) an important cancer preventive natural compound, as depicted in figure 21; ii) phenylcoumaran benzylic ether reductase (PCBER), and iii) isoflavone reductase (IFR), which are involved in central biosinthetic steps of various important bioactive lignans and isoflavonoids in a wide variety of plant species (Min et al, 2003).

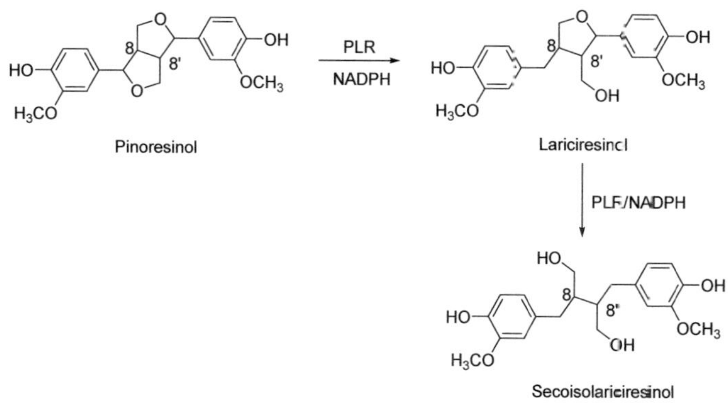

Fig. 21. Enzyme-mediated reduction from pinoresinol to secoisolariciresinol.

Interestingly, the enzyme reduction is obtained through the breaking of the C7-O bond, the same process occurring in the CAD/MS experiments and leading as final step to the loss of the neutral fragment of 30 Da, CH_2O.

It can be envisaged that both the enzymatic and mass spectrometric processes could be favourite by the same factors and occur by a similar mechanism. In this context, gas phase ion chemistry evidences by tandem mass spectrometry of $[M\text{-}H]^-$ furanic lignan ions seems to attribute a key role to the negative charge delocalization from the oxygen atom at the C-4′ position of guaiacyl moiety to the C-7′ position of tetrahydrofuranic ring, namely to the electron flux within the furanic ring.

Although very little is known about aromatic alcohol dehydrogenases, substrate modelling studies of the three related aromatic dehydrogenases PLR, PCBER, and IFR, carried out to predict both substrate binding and enzyme catalytic mechanisms suggest a [4R] -hydride transfer from the C-4 atom of the nicotinamide ring to the target bond of the corresponding substrate to be reduced, the nicotinamide ring resulting in a stacking mode to respect the aromatic ring which in turn is partially stabilized by stacking interactions and by additional interactions with residues within the active site. Here in particular there are conserved lysine residues which can be predicted to serve as the general base for catalysis. From a biochemical perspective, enzymatic reductions of PLR, IFR, and PCBER appear to be analogous and occur through formation and reduction of hypothetical enzyme-bound enone intermediates, in agreement with the hypothesized mechanism of CAD/MS fragmentation pathways.

4. Conclusion

The collisionally activated dissociation (CAD) mass spectra of $[M\text{-}H]^-$ tetrahydrofuran and furofuran lignans, M=THF and M=FUR, and of their alkali metal adducts $[Alk+FUR]^+$, $[Alk+THF]^+$ generated by electrospray ionization were investigated by Triple Quadrupole (TQ) mass spectrometry. The energy resolved TQ/CAD mass spectra were used in a qualitative way to infer information on the integrated energetic picture of CAD fragmentations and by means of a comparative approach, to get highlight into the mechanisms of the predominant dissociation/isomerization pathways of THF and FUR lignans. H/D solution exchange experiments, performed by introducing the investigated compounds in D_2O/CD_3OD solutions, allowed the selective H/D exchange of all the more acidic hydrogen atoms and were employed to obtain information on the nature of fragments generated during TQ/CAD processes and on the mechanisms of the most important fragmentation pathways. The whole picture emerging from tandem mass spectrometric investigation on furofuran and tetrahydrofuran lignans allowed to infer generalized conclusion on the gas phase properties of their deprotonated ions to be extended to their enzymatic biosynthesis pathways.

5. Acknowledgment

The authors are grateful to the organic chemistry group of the Second University of Naples, to Prof. P. Monaco and A. Fiorentino, for the isolation and structural characterization by NMR spectroscopy of all the THF and FUR lignans, whose MS/MS spectra were reported in this manuscript. The authors are also grateful to Prof. F. Pepi of the University of Rome "Sapienza" for his QIT mass spectra.

6. References

Eklund, PC, Backman, MJ, Kronberg, LÅ, Smeds, AI, & Sjöholm, RE. (2008). Identification of lignans by liquid chromatography-electrospray ionization ion-trap mass spectrometry. *Journal of Mass Spectrometry,* Vol.43, No.1, pp. 97-107

Guo, H, Liu, AH, Ye, M, Yang, M, & Guo, D. (2007). Characterization of phenolic compounds in the fruits of Forsythia suspensa by high-performance liquid chromatography coupled with electrospray ionization tandem mass spectrometry. *Rapid communications in Mass Spectrometry,* Vol.21, No.5, pp. 715-729

Min, T, Kasahara, H, Bedgar, DL, Youn, B, Lawrence, PK, Gang, DR, Halls, SC, Park, H, Hilsenbeck, JL, Davin, LB, Lewis, NG, & Kang, C. (2003). Crystal Structures of Pinoresinol-Lariciresinol and Phenylcoumaran Benzylic Ether Reductases and Their Relationship to Isoflavone Reductases. *The journal of biological chemistry,* Vol.278, No.12, pp. 50714–50723

Ricci, A, Fiorentino, A, Piccolella, S, Golino, A, Pepi, F, D'Abrosca, B, Letizia, M, & Monaco, P. (2008). Furofuranic glycosylated lignans: a gas-phase ion chemistry investigation by tandem mass spectrometry. *Rapid communications in Mass Spectrometry,* Vol.22, No.21, pp. 3382-3392

Ricci, A, Fiorentino, A, Piccolella, S, D'Abrosca, B, Pacifico, S, & Monaco, P. (2010). Structural discrimination of isomeric tetrahydrofuran lignan glucosides by tandem mass spectrometry. *Rapid communications in Mass Spectrometry,* Vol.24, No.7, pp. 979-985

Ricci, A, Piccolella, S, Fiorentino, A, Pepi, F, D'Abrosca, B, & Monaco, P. (2010). A tandem mass spectrometric investigation of the low-energy collision-activated fragmentation of *neo*-clerodane diterpenes. *Rapid communications in Mass Spectrometry,* Vol.24, No.11, pp. 1543-1556

Struijs, K, Vincken, JP, & Gruppen, H. (2008). Comparison of atmospheric pressure chemical ionization and electrospray ionization mass spectrometry for the detection of lignans from sesame seeds. *Rapid communications in Mass Spectrometry,* Vol.22, No.22, pp. 3615-3623

Part 5

Lipidomics and Clinical Applications

27

Characterization of Phospholipid Molecular Species by Means of HPLC-Tandem Mass Spectrometry

Natale G. Frega, Deborah Pacetti and Emanuele Boselli
Department of Agricultural, Food and Environmental Sciences,
Marche Polytechnic University
Ancona,
Italy

1. Introduction

Phospholipids are the main constituents of the permeability barrier of cells and subcellular organelles. The phospholipid bilayer is the environment in which isolated proteins or dynamic nanoassemblies of sterols, sphingolipids, and proteins, called lipid rafts, act their vital functions, such as energy transduction, signal transduction, solute transport, DNA replication, protein targeting and trafficking, cell-cell recognition, secretion and many others (Lingwood et al., 2010). Christie et al. (2011) reported a detailed description of the general structure of phospholipids. According to Sud et al., 2007, phoshopholipids represent the category of lipids with the highest variety of structures (7775), followed in descending order, by polyketides (6713), fatty acyls (3942), sphingolipids (3936), glycerolipids (3044), sterol lipids (2196) and others.

The positive effects of dietary phospholipids (PL) on hepatic lipid metabolism, atherosclerosis, obesity-related disorders and cardiovascular disease is a consistent experimental evidence (Shirouchi et al., 2007). The daily intake of PL can vary from 2 to 8 g per day and represents 1-10% of total daily fat intake (Cohn et al., 2008). The main natural phospholipid is phosphatidylcholine (lecithin) which is completely absorbed in humans. Natural or synthetic PLs are extracted from eggs and soybean and used as drug delivery systems since decades (Papahadjopoulos, 1978).

During brain development, one of the most efficient forms of supplying ω3-fatty acids such as DHA (docosahexaenoic acid) are phospholipids (Bourre & Dumont, 2002). In animal studies, it was reported that krill oil (rich in ω3-containing PLs) had stronger effects with respect to fish oil (rich in ω3-containing triacylglycerols) in increasing the DHA level in rat brain (Di Marzo et al., 2010). Although ω3 are predominantly linked in the position Sn-1,3 of triacylglycerols (TG) in seal blubber oil, they are esterified in the Sn-2 position of the TGs and PLs of eggs obtained after feeding laying hens with enriched diets. Moreover, more ω3 fatty acids are incorporated in phosphatidylethanolamine (PE) and phosphatidylcholine (PC) than in TGs (Pacetti et al., 2005). Apoptotic events and age-related diseases are strictly related to oxidized phospholipids. Much research remains to be done for the PL characterization by Liquid chromatography (LC) – tandem mass spectrometry (MS) aimed to understand their biological and physiopathological activity (Domingues et al., 2008). In the food industry, phospholipids are considered among the best emulsifying agents. Commercial preparations derived from soybean and corn oils (commercial *lecithins*) are used extensively in manufactured foods such as bakery items, frostings, non-dairy creamers, confectionery products and ice creams. Staling and off-flavors are often related to the deterioration of the functional lipids in foods (Weihrauch et al., 1983). PLs extracts from eggs and/or fish products can play a pivotal role as an innovative food ingredient (Dumay et al., 2009; Commission Decision, 2000). Recently, a lipid extract rich of ω3-containing PLs from krill (*Euphausia superba*) has been authorised by the EC as novel food/food ingredient (Commission Decision, 2009). Species, geographical origin, and production method of fish (i.e. wild/farmed) are the information which must be labelled in fishery and aquaculture products according to EU labelling regulations (Commission Regulation (EC) No 2065/2001). TGs are considered as markers for wild and farmed fish since their profile reflects the diet lipids (Standal et al., 2010), whereas the phospholipid profile is less affected by the diet and can be related to other variables, such as the species and stock (Joensen et al., 2000).

2. Classes of the main phospholipids and their abundance

The typical range of composition of a mammalian nucleated cell was reported by Vance & Steenbergen (2005). According to these authors, phosphatidylcholine and phosphatidylethanolamine, the quantitatively major phospholipids, range 45–55 % and 15–25 % of total lipids, respectively. Other quantitatively minor phospholipids are phosphatidylinositol (10–15%), phosphatidylserine (2–10%), phosphatidic acid (1–2%), sphingomyelin (5–10%), cardiolipin (2–5%) and glycosphingolipids (2–5).

Cardiolipin (CL) can be found predominantly in the mitochondrial inner membrane (Xiao et al., 2011). CL shows different physiological roles, including the maintainance of the structure of membrane-embedded proteins and proton trapping during oxidative phosphorylation (Houtkooper & Vaz, 2008). Abnormal cardiolipin metabolism was related to a life-threatening inherited disease, the Barth syndrome, as well as other important diseases, such as ischemia/reperfusion injury, heart failure, neurodegeneration, and diabetes (Chicco & Sparagna, 2007). Anticardiolipin (aCL) antibody of IgG and/or IgM isotype in serum or plasma, present in medium or high titer, is one of the classification criteria for the antiphospholipid syndrome (APS). APS is associated with thrombosis, heart valve lesions, renal small-artery vasculopathy, chronic renal ischemia and thrombocytopenia (Miyakis et al., 2006).

Phosphatidic acid (PA) has recently been reported as a lipid messenger involved in various cellular functions in plants, animals, and microorganisms. PA is linked to different regulatory processes, such as signaling pathways in cell growth, proliferation and reproduction, as well as in responses to hormones and biotic and abiotic stresses (Wang et al., 2006).

Phosphatidylcholine (PC) and phosphatidylethanolamine (PE) are the most abundant phospholipids in eukaryotic cells and thus have major roles in the formation and maintenance of vesicular membranes (Henneberry et al., 2001). In mammalian cell membranes, PC constitutes 40–50% of total phospholipids and can be synthesized by either of two pathways, the methylation pathway or the CDP-choline pathway. PE is the second most abundant mammalian membrane phospholipid and constitutes about 45% of total phospholipids in the brain, but only about 20%of total phospholipids in the liver (Morita et al., 2010). Both PC and PE are considered a source of arachidonic acid (together with phosphatidylinositol), which is liberated by phospholipase A2. Arachidonic acid is converted via the cyclo-oxygenase and lipoxygenase pathways to eicosanoids, including prostaglandins, thromboxanes, prostacyclins, leukotrienes and lipoxins (Billah & Anthes, 1990). Although many prokaryotes lack it, PC can be found in significant amounts in membranes of rather diverse bacteria. It is assumed that more than 10% of all bacteria possess PC (Sohlenkamp et al., 2003). In prokaryotic cells, in which PE is the most abundant membrane phospholipid, all of the PE is derived from PS decarboxylation (Vance & Steenbergen, 2005).

Lysophospholipids such as **Lysophosphatidylcholine (LPC)** are released by phospholipase A2 together with the fatty acid. According to Qiao et al., 2006, LPC is proinflammatory and atherogenic. In the vascular endothelium, LPC increased permeability and expression of proinflammatory mediators such as adhesion molecules and cytokines. However, saturated LysoPC in high concentrations reduced melanoma cell adhesion *in vitro* (Jantscheff et al., 2011). Certain lysophospholipids such as ether-linked lysophosphatidylcholine are acetylated by a specific acetyltransferase to form platelet activating factors (Hanahan, 1986).

Ether-linked PE includes plasmanylethanolamine and plasmenylethanolamine which are characterized by an ether bond (not an ester) at the *sn*-1 position of the glycerol moiety. Plasmenylethanolamine is also called **phosphatidylethanolamine plasmalogen (pPE)** and displays a *cis* double bond on the alkyl chain, adjacent to the ether bond, forming a vinyl-ether linkage. Plasmalogens of PE and PC range approx. 18% of phospholipids in humans, but the plasmalogen content of individual tissues or cell types varies widely. It was demonstrated that liver cells export plasmenylethanolamine with lipoproteins. In erythrocytes, kidney, lung, testes and skeletal muscle, the plasmenyl forms account for 20-40% of the total amount of ethanolamine phosholipids. These values were found to be higher in brain, heart, lymphocytes, spleen, macrophages and polymorphonuclear leukocytes.

Phosphatidylinositol (PI) is the precursor of phosphatidylinositol phosphates upon the activation of PI-kinases, which are implicated in thrombin-mediated signaling processes in platelets (Holinstat et al., 2009). Phosphatidylinositol transfer proteins (PITPs) bind and facilitate the transport of phosphatidylinositol (PI) and phosphatidylcholine between membrane compartments; they are able to sequester PI in their hydrophobic pocket. Their knockdown affects signal transduction, membrane trafficking, stem cell viability, neurite outgrowth and cytokinesis (Cockroft et al., 2011).

Phosphatidylserine (PS) is a quantitatively minor membrane phospholipid. It is produced both by prokaryotic and eukaryotic cells. PS is not simmetrically exposed across the cell membrane. It is generally exposed on the outside surface of cell membranes and is widely believed to play a key role in the removal of apoptotic cells and in initiation of the blood clotting cascade. Upon decarboxylation, PS is also converted into phosphatidylethanolamine in bacteria, yeast and mammalian cells. PS is present in brain in higher amounts with respect to other mammalian tissues such as the liver, particularly in molecular species containing docosahexaenoic acid, which accounts for 30-40% of the total fatty acid content of the human grey matter. Moreover, PS is a cofactor of important enzymes, such as some protein kinases (Vance et al., 2005).

Sphingomyelin (Sph) is one of the main phospholipid component of the lipid rafts (Simons et al., 1997) which affect the membrane fluidity and membrane protein trafficking. Sphingolipid-cholesterol nanoassemblies are essential to the activity of some membrane proteins because they regulate the subcompartmentalization propensity of the phospholipid bilayer at little energetic cost (Lingwood and Simons, 2010).

3. Methods of analysis of phospholipids

The characterization of the phospholipid profile is a strategic tool in order to evaluate the nutritional properties of several food products, food ingredients, nutraceuticals, functional foods, cosmetics and pharmaceutical products containing polar lipids (Cohn et al., 2008) or liposomes (Henna Lu et al., 2010). Moreover, the increased commercial use of phospholipids as ingredients for functional food, baby-food and pet food, has lead to the development of specific analytical methods in order to separate and identify phospholipids (Le Grandois et al., 2009; Standal et al., 2010; Winther et al., 2010).

For these reasons, the terms 'phospholipidomics' and 'phospholipidomic analysis' have been recently introduced to define the study of the phospholipid classes in their biological and natural environment in order to detect characteristic fingerprints of metabolic processes (Viola et al., 2007). FTIR was used for the determination of the total content of phospholipids in *biceps femoris* of pigs (Villé et al., 1995). However, the fingerprint of the phospholipid classes and possibly of the molecular species of PL is often the necessary premise for biological, medical and food research studies.

3.1 Extraction

The isolation of phospholipids at an analytical scale is usually accomplished upon solvent/solvent extraction of the total lipids of a sample and further purification of the PL fraction. In food products, the main fraction of total lipids is usually formed by neutral lipids, i.e., triacylglycerols. Phospholipids are the minority of the total lipids, ranging from less than 0.1% in some plant oils, to about 5-25% in some dairy products, 20-30% of the total lipids in egg derived products and more in lean meats. In soy, lecithin is bound to proteins. A detailed survey of the phospholipid content of several kinds of food was reviewed by Weihrauch & Son (1983). **Table 1** reports the total phospholipid content of egg, dairy and meat products as reported by more recent literature.

For analytical purposes, the extraction of the total lipid fraction according to the procedures firstly described by Folch et al. (1957) and Bligh & Dyer (1959) and their successive modifications more than 50 years ago are still extensively used, because the extraction yield is practically quantitative. Thus, these methods are preferred to the extraction with petroleum ether or Soxhlet extraction. Novel methods of extraction of total lipids are based on pressurized liquid extraction (PLE) (Boselli et al., 2001) and even neat supercritical CO_2 (Boselli et al., 2000).

3.2 Isolation and purification

The clean-up of the PL fraction is needed prior to HPLC analysis in order to increase the sensitivity of the determination when the PL content is relatively low compared to the neutral lipids. Usually, membrane lipids (lean meat or fish or egg products and cell membranes) are supposed to contain a PL content (Table 1) which is suitable for direct HPLC analysis. This is also the case of commercial lecithins (AOCS, 2007; ISO, 2009). However, if a better sensitivity is needed, the main ancillary techniques used for PL cleanup prior to HPLC analysis are column chromatography, solid phase extraction (SPE) and thin layer chromatography (TLC) with a normal phase approach.

	Food product	mg PLs/100 mg total fat	mg PLs/1000 mg food product	Reference
Egg products	Egg yolk powder	31		Aro et al., 2009
Dairy products	Milk (in vat)	0.4	0.09	Avalli et al., 2005
	Raw milk	0.71	0.29	Rombaut et al., 2005
	Butter	0.2	1.6	Avalli et al., 2005
	Butter	0.22	1.81	Rombaut et al., 2005
	Cream (natural)	0.9	1.98	Avalli et al., 2005
	Cream	0.35	1.39	Rombaut et al., 2005
	Cream (centrifuged)	0.5	0.9	Avalli et al., 2005
	Buttermilk	4.5	0.9	Avalli et al., 2005
	Buttermilk	21.8	0.91	Rombaut et al., 2005
	Whey (Cheddar)	5.3	0.18	Rombaut et al., 2005
	Quarg	24.7	0.32	Rombaut et al., 2005
	Cheese (Cheddar)	0.47	1.5	Rombaut et al., 2005
	Cheese (Grana Padano)	0.16-0.59	n.r.	Caboni et al., 2000
Meat	Goat meat	n.r.	0.062-0.082	Almeida et al., 1997
	Rabbit meat	9-19	n.r.	Cambero et al., 1991
	Pork muscle (glycolitic)	n.r.	4.8	Leseigneur-Meynier et al., 1991
	Pork muscle (oxidative)	n.r.	8.6	Leseigneur-Meynier et al., 1991
	Pig (commercial meat)	56.1	8.8	Villé et al., 1995
	Pig (biceps femoris)	74.6	10.5	Villé et al., 1995
	Beef (steer)	4.4-22.2	4.6-5.6	Duckett et al., 1993
	Lambs (*longissimus toracis*)	24.7-41.8	5.8-6.6	Aurousseau et al., 2004

Table 1. Phospholipid content of some food products of animal origin (n.r., not reported).

Hoischen et al. (1997) described a procedure aimed to separate the main lipid classes by using column chromatography. Total lipids dissolved in chloroform were fractionated on a silica gel stationary phase into neutral lipids (eluted with chloroform), glycolipids (acetone-chloroform, 95:5 by vol.), and phospholipids (methanol- chloroform 90:10 by vol.).
An example of a SPE clean-up procedure used for both the neutral lipids and polar lipids of vegetable origin is reported by Pacetti et al. (2007). The Bligh& Dyer lipid extract (10 mg) was dissolved in 200 μl of chloroform/methanol (2:1, v/v) and subjected to SPE by using a silica gel stationary phase packed in a cartridge of 6-ml volume (1 g of adsorbent). The purification was performed with sequential elution of 5ml hexane/diethyl-ether (4:1, v/v), 5ml hexane/diethyl ether (1:1, v/v), 5ml of methanol and 5ml of chloroform/ methanol/water (3:5:2, v/v/v). The fractions of hexane and diethyl ether, containing neutral lipids (NL), were combined, dried and used for fatty acid (FAME) analysis after transmethylation. The fractions of methanol and chloroform/methanol/water, containing polar lipids (PoL), such as glycolipids and phospholipids, were combined, dried and used for HPLC/MS and FAME analysis of the polar fraction. Caboni et al., 1996 applied SPE columns packed with several different stationary phases (silica, aminopropyl, C18 and C8) to the extraction of PL from fatty foods.
A simple TLC procedure described by Ferioli & Caboni (2010) for the purification of PL prior to GC analysis of FAME can be potentially applied to purify the PL fraction prior to HPLC injection. These authors dried 20 mg of lipids under nitrogen. Successively, the lipids were dissolved in 100–150 μL of chloroform and were loaded on a Silica gel TLC plate. The eluent was a mixture of n-hexane/diethyl ether 3/2 (v/v). The total PL band was located at the origin and was visualised under UV light (254 nm) by spraying the TLC plate with a 0.02% (w/v)

ethanolic solution of 2,7-dichlorofluorescein (sodium salt) and then was scraped off and collected. Successively, PL were extracted three times with chloroform (1 mL each).

4. HPLC Separation, detection and quantitation

4.1 HPLC separation

The separation of PL into different classes can be accomplished either with normal phase (NP) or reversed phase (RP) high performance liquid chromatography.

NP-HPLC is more extensively applied than RP-HPLC to characterize PL from biological tissue and food products. With NP-HPLC, phospholipids are separated essentially on the basis of the polarity of the head group. They elute by order of increasing polarity: phosphatidylglycerol (PG), phosphatidylinositol (PI), phosphatidylethanolamine (PE), phosphatidylserine (PS), phosphatidylcholine (PC), sphingomyelin (Sph). The lyso forms elute after their corresponding parent phospholipid. Almost all the published methods use a porous silica stationary phase, usually 100 or 250 mm in length, with an internal diameter ranging from 2.0 to 4.6 mm. Recently, monolithic silica gel columns have been used increasingly more often for the separation of lipids, and several applications have been published (Jakab & Forgács, 2002; Merlin et al, 2006). Graeve and Janseen (2009) improved the HPLC–evaporative light scattering (ELSD) method, which permits the separation and quantification of lipid classes ranging from squalene (with a very low polarity) to highly polar lysophosphatidylcholine, using a monolithic silica stationary phase (100mm×4.6mm I.D. column). Other methods use silica gels that are chemically modified with diol (Silversand & Haux, 1997; Sas et al, 1999), cyanopropyl or amminopropyl moieties to alter the polarity. However, the acidic phospholipids will not be recovered using an amminopropyl column with the same efficiency as can be achieved by using an unmodified silica column. Yoon and Kim (2002) investigated the HPLC separation of phosphatidylcholine originated from egg yolk. Column temperature, mobile phase composition (isopropanol/hexane/methanol; methanol/water; methanol) and kinds of stationary phase (silica, bonded phases like C1, NH_2, CN and diol) were varied to understand the effectiveness of PC separation. Pure methanol as mobile phase, silica as a stationary phase and temperature of 45°C represented the best HPLC operating condition. When the NH_2-bonded and diol column are used, the peaks of PC and Sph overlapped. Anyway, Neron at al. (2004) successfully applied the diol column in order to separate and quantitate the wheat flour phospholipids during dough mixing in the presence of phospholipase. They showed that the diol column allowed an efficient separation in less than 16 min of all PL classes of wheat flour (N-acyllysophosphatidylethanolamine, N-acylphosphatidylethanolamine, PE, PC, PG, Lyso-PC and PI). Pang et al (2009) achieved the optimal separation and quantification of seven major phospholipid classes in human blood (PS, PE, PG, PI, PC, Lyso-PC and Sph) by using elution with mobile phase hexane (A) and 2-propanol with water, formic acid and ammonia as modifiers (B) and using a HPLC diol column (250mm×3.0mm, i.d., 5.0 μm, particle size). An isocratic elution method (A, 30%; B, 70%) was used for better repeatability and no balance time.

The choice of mobile phase is strictly dependent on the method of detection. The mobile phases containing phosphate or sulfate buffer salts are incompatible with electrospray ionization. The selectivity of the solvents used in the mobile phase can exert a marked effect on the separation of individual phospholipids, and in particular it can change the order of elution of specific components. Specific examples of NP-HPLC separation of PL in biological samples were provided by Hayakawa et al (2004) using acetonitrile/methanol/1M ammonium formiate. Kakela et al. (2003) used hexane/isopropanol/formic acid/triethylamine. Malavolta et al. (2004) used chloroform/methanol/water/ammmonium hydroxide. Commonly, NP HPLC requires an eluent gradient or one or more changes of the eluent mixture (Becart et al 1990; Fagan & Wijesundera, 2004; Hemström & Irgum, 2006; Bang et al, 2007). Good separation of all PL classes is possible by using gradient elution starting with a solvent of low polarity and ending with a solvent mixture of higher polarity. Anyway, the isocratic condition is often used in order to reduce time-consuming analysis and to avoid column reconditioning between subsequent runs. Kivini et al (2004) studied qualitative and quantitative HPLC chromatographic methods to determine the effects of feed supplements on hen egg yolk phospholipids. They used silica as stationary phase and they tested the isocratic elution using methanol/acetonitrile/concentrated sulphuric acid (30:100:0.05, v/v) and the ternary gradient elution using n-hexane, isopropanol:chloroform (4:1) and isopropanol:water (1:1). The isocratic condition allowed for a faster elution of Sph, PE and PC classes with respect to gradient elution (6 min vs 26 min, respectively) but the peaks obtained were too wide for reliable area calculation. So, the isocratic method was suitable for qualitative separation of PL while the gradient elution was found to be useful in the quantitative determination. Pacetti et al. (2005) employed NP-HPLC coupled with a tandem mass spectrometric detector in order to characterize the molecular species of PL of egg from hens fed diets enriched in seal blubber oil. A silica column was used as stationary phase. The mobile phase was a gradient of solvent A ($CHCl_3/MeOH/NH_3$ 80:19.5:0.5, v/v) and solvent B ($CHCl_3/MeOH/H_2O/NH_3$, 60:34:5.5:0.5, v/v). The total lipid fraction was injected without prior clean-up. This procedure reduced analysis time, manipulations, solvent amounts and consumable parts (e.g., TLC plates and derivatization reagents) to a minimum. The separation of five main classes from the egg lipid extracts was obtained as shown in **Fig. 1**. The retention time for the different classes increased in the order PE, PI, PC, Sph and Lyso-PC. All classes eluted within 18 min. The chromatographic system mainly separated the different phospholipid classes and also allowed some minor separation among different species within a given class. Thus, it is possible to highlight that a binary eluent system consisting of chloroform, methanol and water allowed a separation of five PL classes whereas the ternary eluent system (Kivini et al, 2004) allowed the separation of three PL classes, although the separation was faster.

Some care must be exercised in the use of normal phase HPLC because the presence of water in the solvent system can alter the affinity of PL for silica in replicated runs. Although the order of elution typically remains the same, the exact retention time can be difficult to reproduce (Pulfer & Murphy, 2003). In addition, in NP separations often rather extreme pH conditions (basic or acidic) are required to obtain baseline separation whereby potential hydrolysis can occur during the separation process. In order to avoid this disadvantageous condition, recent studies have developed a method for separation of PL classes based on a silica hydrophilic interaction liquid chromatography (HILIC). This LC technique is typically employed for polar compounds. It uses a polar stationary phase that can retain polar compounds on the column without the disadvantages of using solvents that are immiscible with water, as used in traditional normal phase HPLC. The highly polar organic mobile phase (often acetonitrile) used in HILIC is especially compatible with ESI-MS, resulting in high sensitivity. The HILIC separation is based on hydrophilicity of compounds: amphiphilic molecules like phospholipids elute in order of increasing hydrophylicity. Schwalbe-Herrmann et al. (2010) developed an isocratic HILIC method to separate the

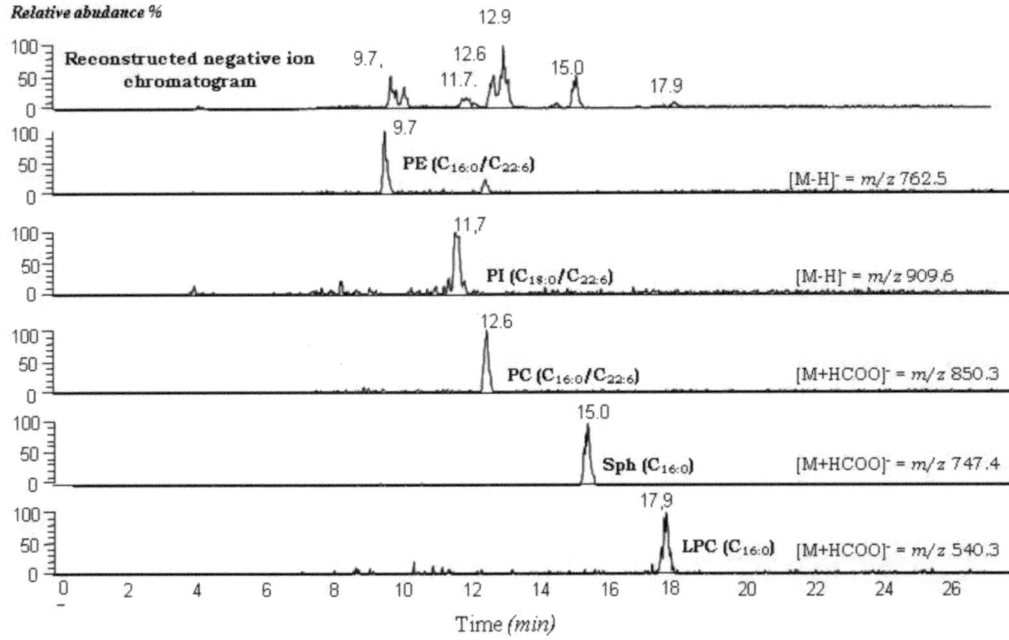

Fig. 1. Negative ion HPLC-ESI-MS analysis of PL of eggs from hens fed diets enriched in seal blubber oil with the MS operating in scan mode.

phospholipid classes in human blood plasma and in tissue extract (swine brain). The detected classes in both cases were PE, PC, Sph and LPC and their chromatographic baseline separation was achieved with a mobile flow of acetonitrile, methanol and ammonium acetate 10mM (55:35:10) at a flow of 0.6 mL/min instead of 1 mL/min which is the typical flow used with silica columns. Zhao et al. (2011) reported a good separation and quantification of 11 lipid classes including all major choline-containing compounds in egg yolk, such as acetylcholine, betaine, choline, glycerophosphocholine, lysoPC, lysoPE, PC, PE, PI, phosphocholine and Sph. They varied the mobile phase using a binary gradient elution with acetonitrile (phase A) and 10 mM ammonium formate in water at pH 3.0 (phase B) as eluents.

RP-HPLC systems are used for PL separation as well (Nakanishi et al. 2010; Ogiso et al. 2008; Lin et al 2004). The separation is based on the lipophilicity of the combined chain length and number of double bonds present in the fatty acid side chains (Peterson et al., 2006). An advantage of RP chromatography is the high resolving power. Two lipid species that only differ in the intramolecular position of a single double bond can be separated (Brouwers et al, 1998). Because of the complexity of closely related structure with ionic charges, mixture of PL molecular species have not been readily resolved by conventional RP with a simple solvent. In the absence of mobile phase electrolyte additives, negatively charged PL, such as fosfatidic acid, elute through an alkylsilyl RP column with little retention (Abidi, 1991), while neutrally charged PL (i.e PC, PE) are partially adsorbed on the alkylsilica phase (e.g. octadecylsilica – ODS) resulting in unusually long retention times and peak broadening. (Abidi, 1992). This problem can be avoided by means of the incorporation of electrolytes into mobile phases, such as quaternary ammonium phosphates or by using reversed phase ion-pair and ion-interaction (Abidi, 1996; Ma et al, 1995; Dodbiba et al, 2011).

However using ESI with ion pair reagents in the mobile phase is not desirable because of their strong suppression effect on the analyte response and possible contamination of the ion source. Recently, Barroso et al. (2005), enhanced the RP-HPLC performance in the separation of PL in human bronchoalveolar lavage fluid. They used a C8 pellicular packing material at elevated temperature with an increasing gradient of acetonitrile containing 0.1% formic acid. A good separation of phospholipids (PC, PG, PI) and their lysophospholipids was achieved without peak tailing. Vernooij et al. (2002) obtained a good chromatographic resolution for egg yolk PC molecular species by means of RP-HPLC system consisting of a Licrospher end-capped RP-18 and acetonitrile/methanol/triethylamine (550/1000/25) as mobile phase. The end-capping of the stationary phase in combination with the use of triethylamine in the eluent prevents the secondary interaction of PL with underivatized silanol groups. As a consequence, even molecular species with the same numbers of carbon atoms and double bonds showed baseline separation.

4.2 Detection and quantitation

The detection of PLs can be accomplished by using various apparatus such as refractive index (RI) (Itoh et al 1985), ultraviolet (UV), mass spectrometry (MS) and evaporative light-scattering (ELS) detectors. RI shows a low sensitivity and high degree of variability depending on the solvent. Thus, it is only applied when the isocratic elution is performed. The UV detection can be carried out at 200-220 nm since PLs have no chromophoric groups absorbing light in higher UV or visibile range. UV detector presents high sensitivity but it does not allow the use of mobile phase that absorb in UV range, such as chloroform widely used in NP system. Generally the solvents system used are limited to two systems: acetonitrile–methanol–water (Chen & Kou, 1982; Wang et al, 2003) and n-hexane-isopropanol–water (Yandrasitz et al, 1981; Dugan et al, 1986; Guan et al, 2001; Seppanen-Laakso et al, 2001; Neron et al 2004) on silica gel-based columns. Nevertheless, the quantification is difficult using UV detector due to the response of PL dependent on the unsaturated double bonds present in the acyl groups. Saturated PL yielded lower responses than unsaturated PL. In order to enhance UV absorbance, PL can be derivatized with benzoates, dinitrobenzoates, pentafluorobenzoate and nicotinic acid. Dobson et al (2001) identified the molecular species of PC from soybean, egg yolk and bovine liver after conversion to UV-absorbing diacylglycerol nicotinate derivatives. Anyway, they found the overlapping of several components.

In the last decade, the mass spectrometric detection has become the most popular method of PL molecular species analysis because it can overcome many of the problems reported above and it has strong identification ability, high sensitivity and specificity. The several types of spectrometers and their ionisation sources available for PL analysis were recently overviewed elsewhere (Peterson & Cummings, 2006). Xia & Jemal (2009) evaluated optimal mass spectrometric techniques to monitor the fate of PL during development of bionalytical LC MS/MS method since plasma PL may cause matrix effects. The electrospray ionization (ESI) (Murphy et al 2001; Smith, 2006; Pulfer & Murphy 2003) and matrix-assisted laser desorption ionization (MALDI) (Schiller et al., 2004; Fuchs et al. 2010) are the ionization methods more frequently applied. These are soft ionization methods which allow the ionization of non-volatile and thermolabile compounds. They preserve the information based on the relative position of acyl radicals on the glycerol backbone since a minimal fragmentation of molecules is produced. Compared to other ionization methods ESI is very useful for analyzing polar lipids, because polar lipids such as phospholipids are sensitive to soft ionization without a derivatization procedure. Furthermore, ESI can be coupled with nano-flow HPLC easily (Bang et al, 2006; Isaac et al, 2003; Ito et al, 2008).Two different approaches can be adopted for analysis of PLs by means of ESI. The direct infusion of crude lipids extracts to the MS instrument was proposed by Benfenati & Reginato (1985), whereas the other approach employs liquid chromatography coupled on line with mass detector (LC-ESI-MS). The latest is more frequently applied compared to direct infusion, since the combination of HPLC and ESI MS reduces the ionisation suppression effect of low abundance PL species generated by the presence of polar impurities (Annesley, 2003; Issac et al 2003; Hermansson et al 2005; Houjou et al, 2005). Positive or negative ES ionization will occur at either the phosphate group or on the head group of PL. Normally, positively charged phospholipids such as PC and Sph appear as ions in the positive mode, whereas PE, PI, PG, PS and PA form ions in negative mode. It is possible to detect PG in the positive ion if ammonium adducts are formed. The quantitative analyses performed by using ESI MS are not straightforward because it is not possible to make the assumption that the intensity of any observed ion is proportional to the concentration of PL in the mixture. MS signals of PL are influenced by different effects related to acyl chain length, the degree of acyl chain unsaturation and concentration. Several authors (Berdeaux et al; Ahn et al 2007) have shown that the ESI-MS response increases with increasing degree of insaturation of the fatty acid side chains whereas it decreases with an increase in chain length when the phospholipids concentration range in the samples is higher than picomolar. Only the picomolar range concentration results, in fact, in a linear response. This non-linearity is more severe in negative ionization rather than positive mode (Zacarias et al, 2002). To overcome the dependence of ESI signal to the PL structures, in order to quantify the species, an internal standard chemically similar to the sample of interest can be used. A deuterated standard of each PL molecular species should ideally be used (Koc et al, 2002; Enjalbal et al., 2004). However, deuterated standards are very expensive and not always available for each PL molecular species (several hundreds of different phospholipid molecular species are present in biological samples considering all the subclasses and individual molecular species within a class). Moreover, even if an isotopic internal standard is applied, the scaling response implies a nonlinear relationship between intensity and concentration. The larger the discrepancy in intensity between the internal standard and the analyte of interest is correlated with a greater error in estimating the concentration when a linear approximation is used. Thus, the other possible way is to use a synthetic PL standard that has a low abundance in the sample, since PL species of the same class show similar ESI MS peak intensities. The standard should have a molecular weight outside of the molecular weight envelope that encompasses the naturally occurring molecular species. Pacetti et al. (2004) achieved the quantification of seven PC molecular species from the serum of cystic fibrosis subjects using ESI MS in the negative scan mode with 1,2-diundecanoyl-sn-glycero-phosphocholine as the internal standard. They showed that the use of mass spectrometry as detector allowed a low limit of quantification (12 ppm of PC). Han and Gross (2005) used 15:0-15:0 diacyl PS and 14:1-14:1 PC. When is sufficient to observe differences in phospholipid composition among different samples, a semi-quantitative approach can be also performed. The relative abundance of individual molecular species within a phospholipid class can be calculated from the single ion current responses. This strategy is widely applied to address a specific biochemical issue or to characterize the phospholipid composition in foods.

Otherwise to the ESI MS detection, for quantification purposes the use of an Evaporative Light Scattering Detector (ELSD) resulted more fast and easy. It allows the non-selective determination of all the non volatile components in the polar lipid extract and a precise quantification of the PL classes, when a calibration curve is calculated for each phospholipid class. It is very sensitive and it can quantitate nanomolar amounts of PL. ELSD usually gives a stable flat baseline. Its response is independent of the number of double bonds in the molecules. The ELSD baseline signal is scarcely affected by solvent changes (Stith et al, 2000; Becart et al, 1990) and thus allows the use of gradient elution needed to manage complex separations. Differently, the ELSD response can be affected by parameters of the nebulization step and particularly the mobile phase composition that affects the size of the droplets (Godoy Ramos et al 2008). **Fig. 2** shows how the evaporation temperature affects the ELSD response: the temperature increase is correlated to a change of the baseline profile.

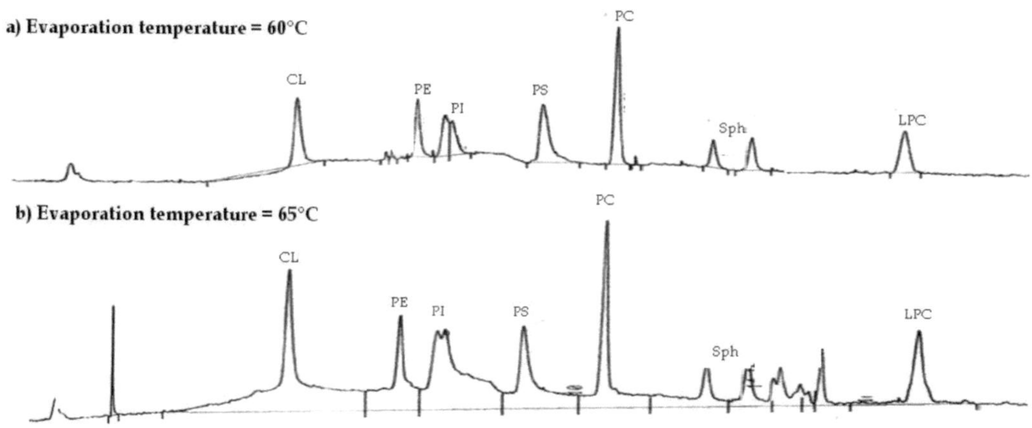

Fig. 2. NP-HPLC- ELSD trace of a standard PL mixture (CL,PE,PI,PS,PC, Sph, LPC). The gas flow (from air compressor) of the ELSD was 7.5 L/min and the vaporization temperature was set: a) at 60 °C b) 65 °C.

The effect of the different mobile phase composition on variation of baseline is reported in figure 3.

In the literature, several papers reported the quantification of PL with ELSD (Rodríguez-Alcalá & Fontecha, 2010; Seri et al, 2010; Wang et al, 2009; Rombaut et al 2007). Yan et al 2010 have successfully quantified PL extracted from human, porcine, and bovine erythrocyte ghost membranes using isocratic elution on a silica column coupled to an ELS detector. The detection limits for PS, PE, PC and Sph were 50, 50, 80 and 150 ng, respectively. Caboni et al. (2000) determined the content of phospholipids and lysophospholipids in Grana Cheese samples at different ripening stages. Narváez-Rivas and coworker (2001) developed a HPLC/ELSD method that allowed the quantification of cardiolipin, phosphatidylethanolamine, phosphatidylinositol, phosphatidylserine, phosphatidylcholine, and sphingomyelin in subcutaneous fat from Iberian pig. The limit of quantification were below 0.03 and 0.05 mg kg^{-1} for each phospholipid class. The performance of charged aerosol detection (CAD) was compared to ELSD for the NP-HPLC analysis of *Leishmania* membrane PL (phosphatidic acid, PG, cardiolipin, PI, PE, PS, lysoPE, PC, Sph and lysoPC) classes. The accuracy of the methods ranged from 62.8 to 115.8% and from 58.4 to 110.5% for ELSD and CAD, respectively. With HPLC-ELSD the limits of detection (LODs) were between 71 and 1195 ng and the limits of quantification (LOQs) were

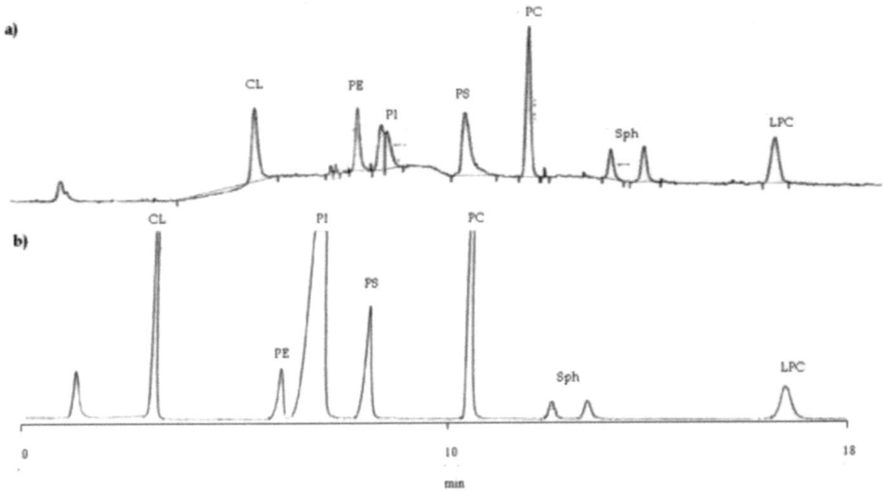

Fig. 3. NP-HPLC-ELSD trace of PL mixture obtained using a silica column (3μ, 150 mm x 4.6 mm) as stationary phase and a) a gradient of solvent A [$CHCl_3$/ CH_3OH /NH_4OH (30%) 70:25:1, v/v] and solvent B [$CHCl_3$/CH_3OH/H_2O/NH_4OH (30%) 60:34:5.5:0.5, v/v], b) a gradient of A [$CHCl_3$/ CH_3OH /NH_4OH (30%) 60:40:1, v/v] and solvent B [$CHCl_3$/CH_3OH/H_2O/NH_4OH (30%) 60:34:5.5:0.5, v/v]. The gas flow of the ELSD was 7.5 L/min and the evaporation temperature was 60 °C.

between 215 and 3622 ng. With HPLC-CAD, the LODs were between 15 and 249 ng whereas the limits of quantification (LOQs) were between 45 and 707 ng (Godoy Ramos et al, 2008). Considering that the ELS detector is really useful for qualitative analysis and that ESI MS detector is really indispensable to discriminate the PL molecular species, some authors suggest the use of the strategy where HPLC apparatus is coupled on-line with both detection systems. Boselli and co-workers (2008) proposed an analytical method for the simultaneous quantification of phospholipid classes and identification of phospholipid molecular species within each class in raw and cooked pork meat. The NP-HPLC apparatus was coupled on-line with two detection systems: an ion trap equipped with electrospray ionization for tandem mass spectrometry and an ELSD. Donato et al (2011) performed the characterization and quantification of phospholipid fraction in cow's and donkey's milk simultaneously coupling HPLC system with hybrid ion trap-time of flight (IT-TOF) mass analyzer and an ELS detector.

5. ES ionization and tandem mass spectrometry (MS/MS)

The ES ionization generates lipid ions by adding a proton $[M+H]^+$, by adding a cation, such as the sodium $[M+Na]^+$ ions or by removing a proton, $[M-H]^-$. Chloride $[M-Cl]^-$ and formiate $[M-HCOO]^-$ adduct ions or multiply-charged ions such as $[M+2H]^{2+}$, are also easily formed. The formation of ion species is strictly affected by the concentration of Na^+ and the pH of the solution, with which the ESI events take place. Dimeric molecular species are often formed during ionization if the PL concentration is high.

The variable formation of adducts and the multiplicity of PL molecular species present in the samples considerably complicate molecular species assignments from such complex spectra. Thus, often, assigning definite molecular identities to individual ion peaks is not possible with single MS. This problem becomes especially important for isobaric molecular species of the same phospholipid class, which have identical mass but differ in combination of acyl chains, or for the identification of oxidized PL. In these cases, experiments of tandem mass spectrometry (MS/MS) are needed as additional approach (Cui & Thomas, 2009; Postle et al., 2007). In MS/MS experiments, the precursor ion created in the first MS undergoes a further fragmentation either by collision-activated dissociation (CAD) or spontaneous dissociation. Larsen et al. (2001) compared the product ion achieved using CAD with an ion-trap mass spectrometry (ITMS) with those obtained with a triple quadrupole. In the fragmentation upon CAD in an ion-trap the main product ions were the *sn*-1 and *sn*-2 lyso-phospholipids whereas CAD in triple quadrupole gave primarily the *sn*-1 and *sn*-2 carboxylate anions. Hsu & Turk (2009) describe the mechanisms underlying the fragmentation processes under low-energy CAD with tandem quadrupole mass spectrometry and with multiple-stage ion-trap mass spectrometry of PL in various ions forms generated by ESI in negative and positive-ion modes ($[M-H]^-$, and $[M-2H+Alk]^-$; $[M+H]^+$, $[M+Alk]^+$, $[M-H+2Alk]^+$; Alk = Na, Li). They noted that the fragment ions leading to structural characterization arising from CAD of the $[M+H]^+$ ions are often of low abundance, and thus are less useful for structural identification. In contrast, fragment ions arising from alkali adduct ions, in particular, the $[M+Li]^+$ ions, as well as from the $[M-H]^-$ ions are abundant and are suitable for unambiguous structural identification. Moreover, for many PL CID of $[M+H]^+$ reveals information about the polar headgroup, while collision induced dissociation of the negative $[M-H]^-$ provides information about the fatty acyl chain.

The discussion of product ions spectra achieved using CAD with ITMS of different PL ions generated by ESI (positive and negative) is reported below.

5.1 Cardiolipin (CL)

The approach using multiple-stage ITMS methods for the characterization of CL as its positive and negative ions was widely studied by several authors. Hsu and co-workers described the characterization of CL molecular species as their $[M-H]^-$, $[M-2H]^{2-}$ and $[M-2H+Na]^-$ ions generated by ESI in negative ions mode (Hsu et al 2005; Hsu & Turk 2006a) and as their sodiated adducts ($[M-2H+3Na]^+$, $[M-H+2Na]^+$, $[M+Na]^+$) formed by positive ES ionization (Hsu & Turk 2006b).

The MS^2-spectra of the $[M - H]^-$ and of the $[M - 2H]^{2-}$ ions contain two sets of prominent fragment ions that comprise a phosphatidic acid (fatty acid/fatty acid-PA), a dehydrated phosphatidylglycerol, and a anion formed by phosphatidic acid and tricyclic glycerophosphate ester (phosphatidic acid + 136). The substantial differences in the abundance of the two distinct phosphatidic anions observed in the MS^2-spectra of the $[M - H]^-$ and $[M - 2H]^{2-}$ and ions lead to the assignment of the phosphatidyl moieties attached to the 1′ or 3′ position of central glycerol. The differences in the intensity of carboxylate anions reflecting the fatty acyl substituent shown, exclusively, in the MS^2-spectra of the $[M - 2H]^{2-}$ provide information to identify the fatty acyl substituents and their position in the glycerol backbone: the carboxylate anion containing the fatty acyl substituent at *sn*-1 (or *sn*-1′) is less abundant than the corresponding ion having that at *sn*-2 (or *sn*-2′). Anyway, the abundance of these two carboxylate ions are reversed in MS^3-spectra of the $[M - H]^-$ ions.

The MS^2-spectra of all sodiated adducts (positive and negative) showed two prominent fragment ion pairs that consist of the phosphatidyl moieties attached to the 1′- and 3′-position of the central glycerol, respectively, resulting from the differential losses of the diacylglycerol moieties containing A and B glycerol, respectively.

The identities of the fatty acyl substituents and their positions on the glycerol backbones (glycerol A and B) are deduced from further degradation (MS^3 experiments) of the above ion pairs that give the fragment ions reflecting the fatty acid substituents at the *sn*-1 (or *sn*-1′) and *sn*-2 (or *sn*-2′) positions. The ions that arise from losses of the fatty acid substituents at *sn*-1 and *sn*-1′, respectively, are prominent.

Different MS^2 spectra of the monosodiated adducts ($[M+Na]^+$) were obtained by Boselli et *al.* (2008). They identified the pork meat CL molecular species and the CAD fragmentation of the *m/z* 1471.6, i.e. $[M+Na]^+$ ($C_{18:1}/C_{18:3}$)/($C_{18:2}/C_{18:2}$)-CL or ($C_{18:2}/C_{18:2}$)/($C_{18:2}/C_{18:2}$)-CL, which resulted the preponderant species in both raw and cooked meat and produced only two complementary fragments due to the cleavage at the phosphatidylglyceridic ester, as reported in **Fig. 4**.

These results point out that only the MS^2-spectra of the $[M-2H]^{2-}$ is sufficient to confirm structural assignment of CL since it contains complementary information. Conversely, the MS^2-spectra of the all CL sodiated adducts are not suitable to elucidate the composition of the isobaric species; thus, the experiments of third order (MS^3) would be needed.

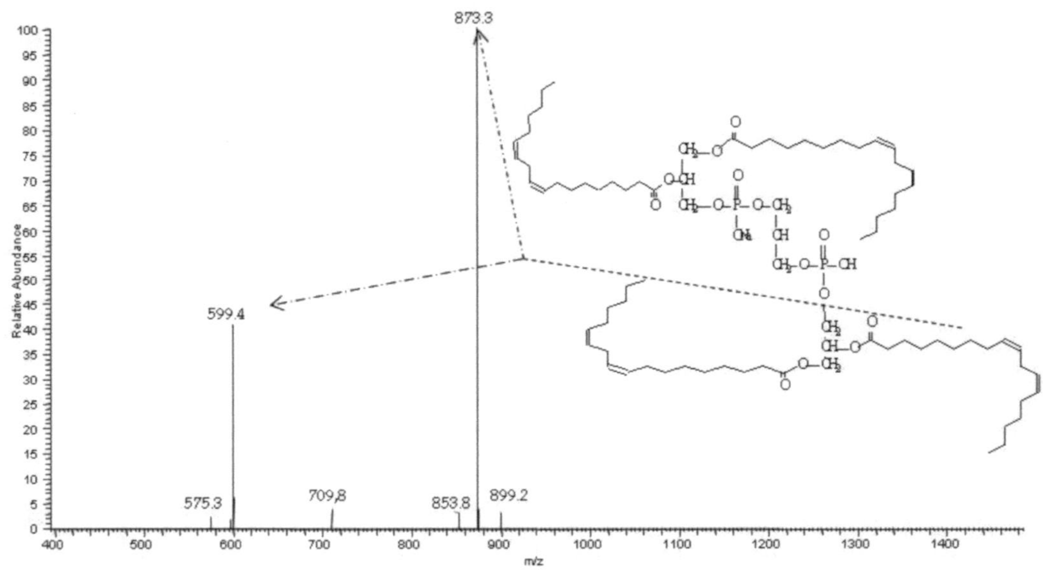

Fig. 4. Positive MS^2-spectra of the $[M+Na]^+$ of CL($C_{18:1}$)/($C_{18:3}$)/($C_{18:2}$)/($C_{18:2}$) or ($C_{18:2}$)/($C_{18:2}$)/($C_{18:2}$)/($C_{18:2}$) at m/z= 1471.6

5.2 Phosphatidic acid (PA)

When being subjected to ESI in negative-ion mode, phosphatidic acid yields a $[M-H]^-$ ion. Its product ion spectra is really suitable to identify the PA species since it shows fragments arising from the loss of both the fatty acyl substituents. In **Fig. 5** is reported a product ion spectrum of *m/z*= 697.5, i.e PA ($C_{18:1}/C_{18:2}$) as $[M-H]^-$ ion which was detected in the almond of the avocado (*Persea americana* Mill) fruit. The spectrum is dominated by the fragments at *m/z* = 415.1 and *m/z* = 417.3 arising from neutral loss of water followed by loss of oleyl moieties ($R_{C18:1}CO$) and linoleyl moieties ($R_{C18:2}CO$), respectively. The fragment ions at *m/z* = 279.3 and *m/z* = 281.3, corresponding to the carboxylate anions, are also visible.

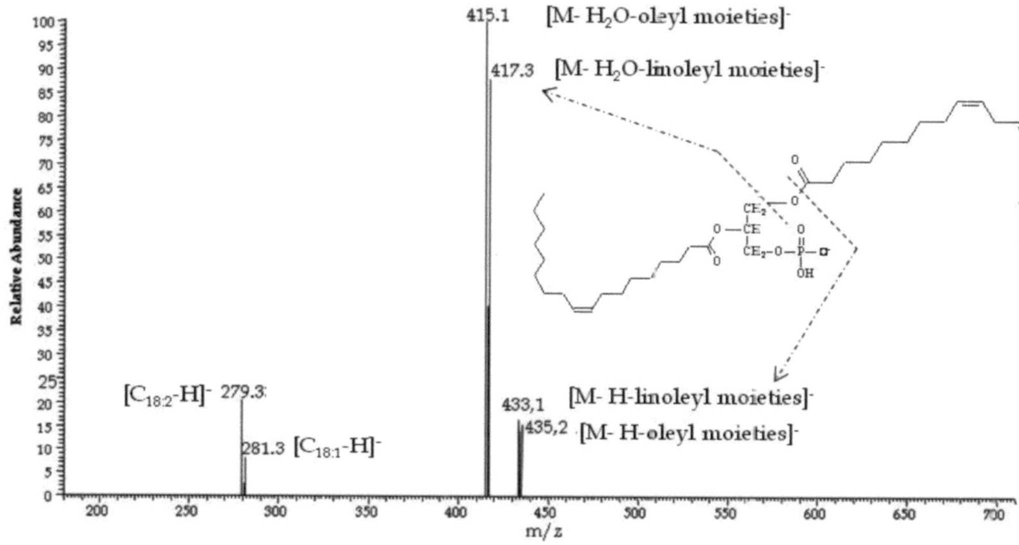

Fig. 5. Negative MS²-spectra of the [M-H]- of PA ($C_{18:2}/C_{18:1}$) at *m/z* 697.5

These observation are consisting with the results obtained by Hsu and Turk (2009) which described the fragmentation processes of [M-H]-, [M+H]+, [M+Li]+ and [M-H+2Li]+ ions of various phospholipids, including PA. Moreover, they found that the sensitivity observed for PA as the [M+Li]+ and [M−H+2Li]+ adduct ions are significantly poorer than those observed as the [M−H]− ions in negative-ion mode. In addition, the fragment ions produced from CAD of the [M+H]+ and of the [M+Li]+ ions are less useful for structural identification than the fragment ions [M-H+2Li]+ and [M-H]-. In fact, the MS²-spectra of [M+H]+ and of [M+Li]+ do not contain the fragments peculiar of individual molecular species, such as fragments having fatty acid moieties. Differently, [M-H+2Li]+ and [M−H]− ions produce MS²-spectra with fragments arising from the losses of the fatty acyl substituents at the *sn*-1 and *sn*-2 positions. In addition, due to the preferential loss of the fatty acid in *sn*-1 position, the fragments containing the fatty acid in *sn*-1 position become more abundant than those containing a fatty acid in *sn*-2 position. For this reasons, characterization of PA as dilithiated adduct and deprotonated ions affords the identification of the fatty acid substituents and their location on the glycerol backbone.

Concerning the ES positive-ion mode ionization, in addition to lithium and protonated adducts, PA can form ammonium adducts. **Fig. 6** shows a positive ESI-MS² product ion spectra of $[M+NH_4]^+$ ions at *m/z*=720.5, i.e. PA ($C_{18:0}/C_{18:1}$) detected in avocado fruit (Pacetti et al 2007).

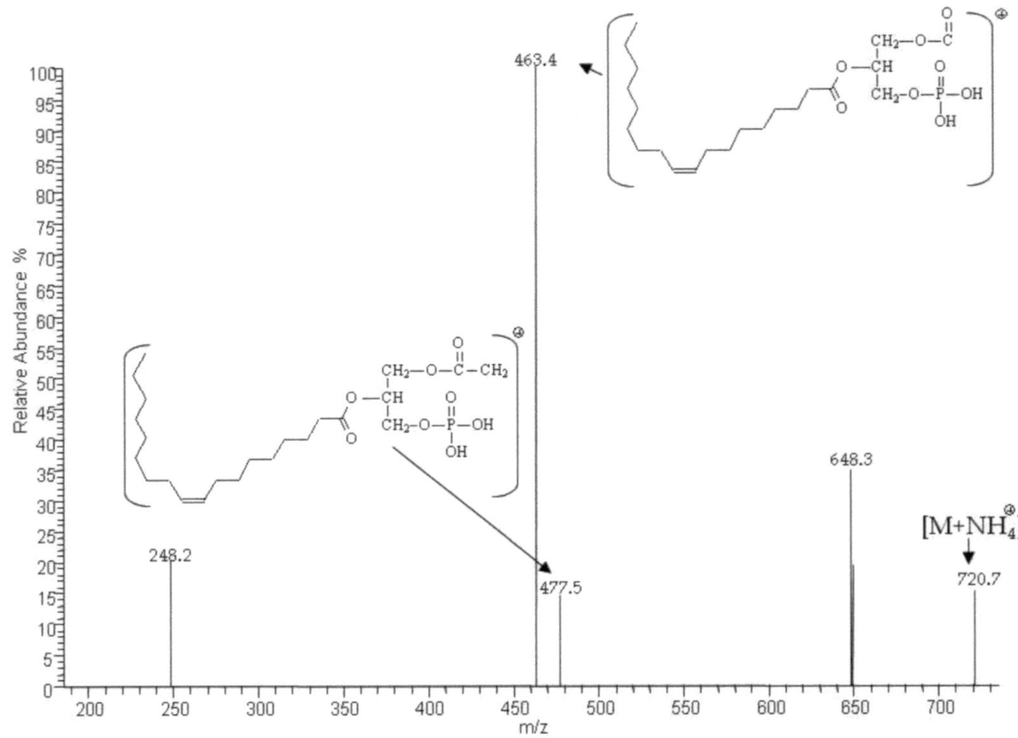

Fig. 6. Positive MS²-spectra of the $[M+NH_4]+$ of PA ($C_{18:0}/C_{18:1}$) at *m/z* 720.5.

The spectrum displayed as major fragment the ion at *m/z*= 463.4 resulting from the loss of the hydrocarbon chain of stearic acid at the carbon C1. Also the ion at *m/z*= 477.5 arising from the chain rupture at the carbon C2 is weakly visible.

5.3 Phosphatidylcholine (PC)

ESI tandem mass spectrometry analysis of PC molecular species can be somewhat complicated because the same molecular species can generate a complex population of ion adducts. The structure of PC is characterized by the presence of a quaternary nitrogen atom having positive charge and by the presence of a negative charge of the phosphate group. Thus, PC is always ionized, independently of the pH. All PC species yield abundant pseudomolecular ions in positive ionization studies, most likely because the phosphate group readily accepts a proton during ESI ($[M+H]^+$). Since ESI solvents contain alkali metal ions, such as Na, K or Li, abundant alkyl adducts species ($[M+Na]^+$, $[M+K]^+$ or $[M+Li]^+$) are observed. The fragment ions arising from the $[M+Li]^+$ adduct ions of PC lipids are abundant and structural identification and distinction among isomers can be achieved (Hsu et al., 1998). In contrast, product-ion spectra from the $[M+Na]^+$ or $[M+K]^+$ ions yield little information about the fatty acids substituent (Hsu and Turk, 2009). Detailed studies of the product-ion spectra arising from the $[M+H]^+$ ions were reported by several authors, such as Pacetti and coworker (2006), Hsu and Turk (2009) and Murphy and Axelsen (2011). The product ion spectra obtained by Pacetti resulted different from the spectra presented by other authors. In detail, according to what reported by Hsu and Turk and by Murphy and Axelsen, the product-ion spectra arising from the $[M+H]^+$ ions is dominated by the *m/z* = 184 ion, representing a phosphocholine ion and by the $[M+H]^+$ ion. Thus, the ions related to the structural information are of low abundance or absent. Differently, the CAD spectra of $[M+H]^+$ showed by Pacetti (**Fig. 7**) are richer in diagnostic fragment ions. The most abundant fragment ion corresponding to the loss of trimethylamine $[M\text{-}N(CH_3)_3]^+$ and the fragments resulting from the loss of one fatty acid $[M+H-R_1CHCOO]^+$, from the loss of an acyl group $[M+H-R_1CHCO]^+$ and from the loss of phosphocholine are clearly detectable (the abundance is higher than 50%).

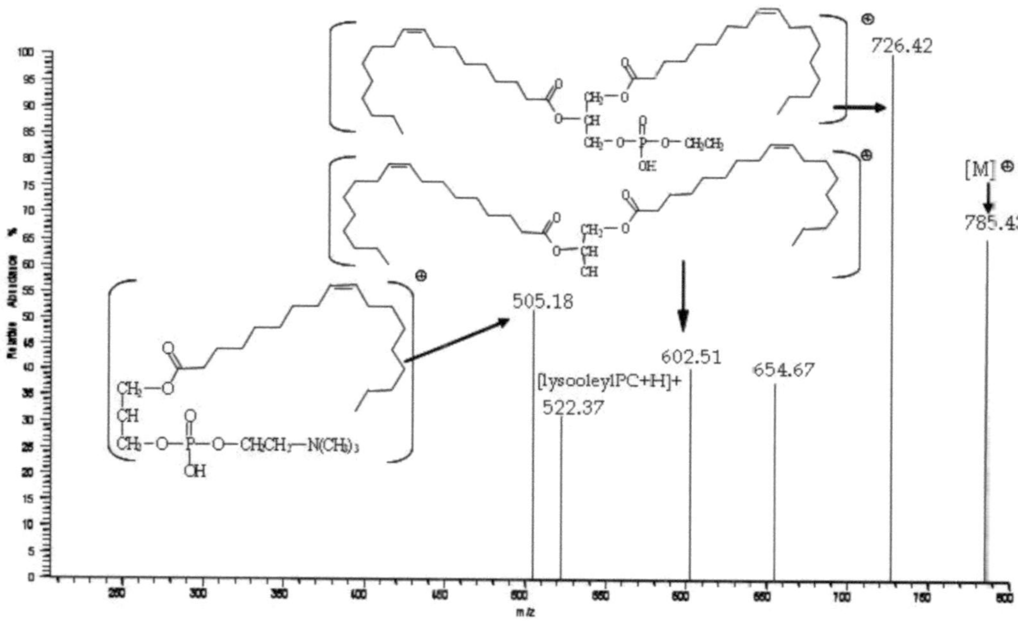

Fig. 7. Positive MS^2-spectra of the $[M+H]^+$ of PC ($C_{18:1}/C_{18:1}$) at *m/z* 786.5.

Studies on the fragmentation of various PC including deuterium-labeled analogs indicated the formation of the ion at *m/z* 184 which involves the participation of mainly the α-hydrogen of the fatty acyl at *sn*-2 (Hsu & Turk, 2003). The studies also revealed that the α-hydrogen of the fatty acyl at *sn*-2 is more labile than that at *sn*-1, resulting in the more favorable formation of the $[M+H-R_2CH\ CO]^+$ ion than the $[M+H-R_1CH\ CO]^+$ ion, arising from the loss of the fatty acyl substituents at *sn*-2 and at *sn*-1 as ketenes, respectively. Therefore, the position of the fatty acyl moieties on the glycerol backbone can be assigned.

The product-ion spectrum from the $[M+Li]^+$ adduct ions show a fragment arising from the loss of trimethylamine moiety as major fragment. The two fragments yielded from the loss of fatty acyl groups are present as minor fragments and have different abundance. According to Hsu and Turk (2009), the α-hydrogen at the fatty acyl chain is involved in the elimination of the adjacent fatty acyl substituent as an acid. Because the α-hydrogens of the fatty acyl at *sn*-2 are more labile than those at *sn*-1, preferential loss of R_1CO_2H to yield the $[M+Li-R_1CO_2H]^+$ ion over the loss of R_2CO_2H that yields $[M+Li-R_2CO_2H]^+$ is observed. The different abundance can be used to distinguish PC positional isomers.

Electrospray ionization also leads to the formation of negative ions derived from PC lipids in spite of the quaternary nitrogen atom with its permanent positive charge. Under negative

ESI condition, PC generate $[M-15]^-$ ions, resulting from the demethylation of the choline moiety and the negative ions by forming adducts with anions, such as chloride $[M+35]^-$, formiate $[M+45]^-$ and acetate $[M+59]^-$. When ESI is performed using a spectrometer with high orifice potentials, the $[M-15]^-$ ions becomes the most important since the high potential leads also to the demethylation of the anion adducts. The MS^2 spectra of $[M-15]^-$ and $[M+59]^-$ ions show anion fragments which are more suitable for identification of the fatty acid substituents and location of their position on the glycerol backbone compared to the product ion spectra obtained from $[M+35]^-$ and from $[M+45]^-$ ions. In detail, the product-ion spectra obtained from $[M-15]^-$ ion are characterized by the presence of both carboxylate anions $[R_1COO]^-$ and $[R_2COO]^-$. The latest fragment results more abundant than $[R_1COO]^-$. Anyway, this discrepancy of abundance is not so visible when the PC molecular species contain a long-chain fatty acid linked to the *sn*-2 position (Berdeaux et al 2010).

Kerwin, Tuininga, & Ericsson (1994) described the formation of an ion cluster with acetate $[M+59]^-$ and the subsequent decomposition of that ion species during collisional activation to yield the $[M-15]^-$ ion that corresponds to the loss of methyl acetate, as well as ions characteristic of the fatty acyl group esterified at the *sn*-1 and *sn*-2 positions.

Pacetti et al. 2005 observed that the product ion spectra obtained from $[M+35]^-$ and from $[M+45]^-$ ions contain only the $[M-CH_3-HCOO]^-$ anion fragment; thus, the negative MS^2-spectrum of the $[M+NH_4]^+$ of PC ($C_{16:0}/C_{18:1}$) at *m/z* 804.4 contains the fragment at 744.4 m/z.

5.4 Lysophosphatidylcholine (LPC)

The behavior of LPC molecular species under ESI is equal to those shown by PC. LPC molecular species can be inferred from their protonated molecules $[M + H]^+$ when positive ESI is performed. Negative ESI yields the demethylated ion ($[M-CH_3]^-$). The presence of metal or anions in the ESI system generates the adduct ions, such as ($[M + Na]^+$, $[M + Li]^+$, $[M+HCOO]^-$, ($[M+ Cl]^-$). As displayed for PC, the product ion mass spectra related to the $[M+H]^+$ ion result more informative than all the other ions.

An example of MS^2 spectra yielded by the $[M+H]^+$ ion at *m/z*= 568.3, i.e *sn*-lyso-docosahexaenoic-phosphatidylcholine [LPC ($C_{22:6}$)], is reported in **Fig. 8**. The spectrum displays two major fragments. One at m/z= 550.2 given by the loss of one water molecule $[M-H_2O]^+$ and the other at *m/z* = 183.9 corresponding to the polar headgroup.

Differently, the product ion spectrum from the ion at *m/z* = 612.3, corresponding to LPC $C_{22:6}$ as a $[M+HCOO]^-$ ion, displays the unique fragment arising from demethylation of choline $[M-CH_3]^+$) with 552.1 *m/z*.

5.5 Phosphatidylethanolamine (PE)

PE can form positive and negative ions. In the ESI mass spectra, $[M+H]^+$ ions are usually detected when working in positive mode, and $[M-H]^-$ ions are displayed in negative mode. The product ion spectra arising from the $[M+H]^+$ are often simple and therefore less applicable, although achievable, for structure characterization. They displayed a most abundant fragment resulting from the loss of the polar headgroup ($[M-NH_3(CH_2)_2OPO_3H]^+$ = $[M+H-141]^+$) and a low intensity fragment resulting from the loss of one acyl group ($[M-R_1CO]^+$). For example, the product-ion spectrum of the $[M+H]^+$ ion of PE ($C_{16:0}/C_{22:6}$) at *m/z* 764.5 (**Fig. 9**) is dominated by the *m/z* 623.4 ($[M+H-141]^+$) ion. The fragments at *m/z* 526.3 and at *m/z* 386.2, corresponding to the loss of palmitoyl acyl moieties and the simultaneous loss of both polar head group and palmitoyl acyl group, respectively, are also visible. Their

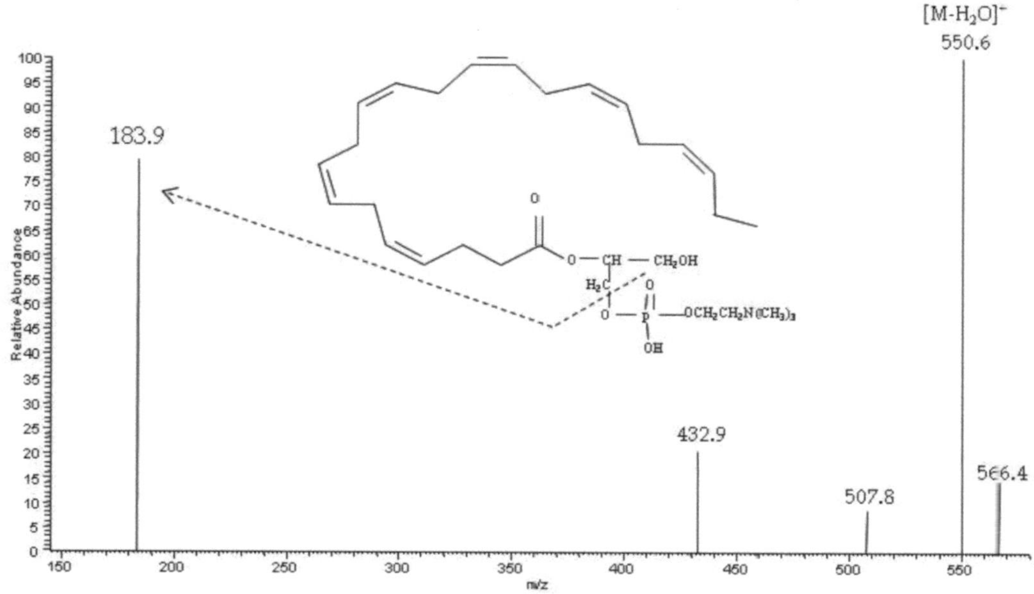

Fig. 8. Positive MS2-spectrum of the $[M+H]^+$ of LPC ($C_{22:6}$) at *m/z* 563.3.

intensity are low as well as those of the fragment at 454.1 *m/z* arising from the loss of docohexanoyl acyl moieties. According to Hsu & Turk (2000), the fragmentation process leading to the $[M+H-141]^+$ ion also involves the participation of the α-hydrogen of the fatty acyl substituent, mainly at *sn*-2. In addition, the observation of $[M+H-141]^+$ rather than a protonated phosphoethanolamine at *m/z* 142 indicates that the phosphoethanolamine (having a zwitterion ion form) is less competitive for proton. For this reason, the response of PE in positive mode is lower than negative mode.

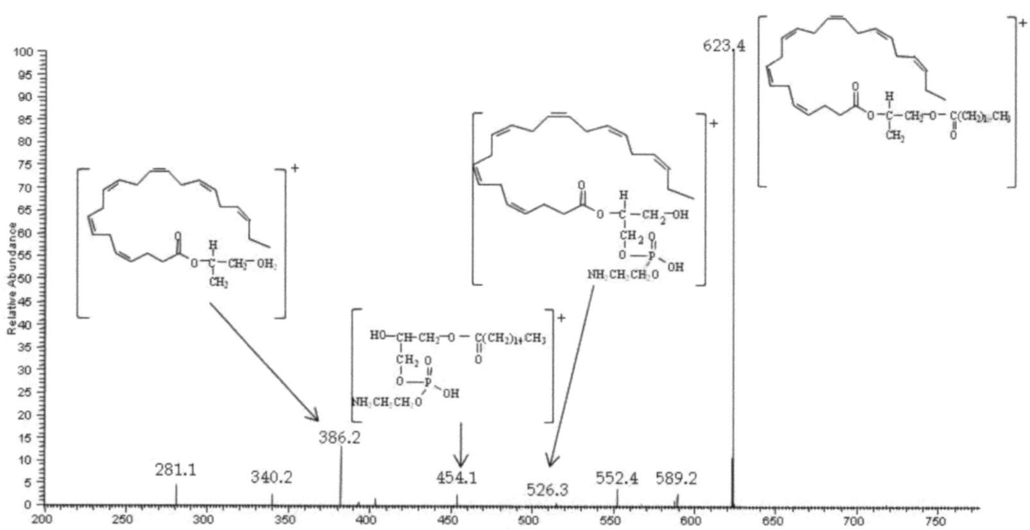

Fig. 9. Positive MS2-spectra of the $[M+H]^+$ of PE ($C_{16:0}/C_{22:6}$) at *m/z* 764.5.

The performance of ESI working in positive mode can be enhanced by inducing the formation of lithium adducts. When ionized in the presence of lithium, PE forms both monolithiated ($[M+Li]^+$) and dilithiated ($[M-H+2Li]^+$) adduct ions. The intensity of the $[M+Li]^+$ and the $[M-H+2Li]^+$ ions are dependent on the concentration of the lithium ion (Hsu & Turk, 2009). The fragmentation of these ions produces the major fragment related to the loss of 147 Da (lithiated polar head) and the minor fragments arising from the loss of aziridine ($[M-CH_2CH_2NH]^+ = [M-43]^+$) and from the loss of one of the fatty acid substituents ($[M-R_1COOH+2Li]^+$ and $[M-R_2COOH+2Li]^+$).
The fragmentation pathway of $[M-H]^-$ leads to the CAD spectra which displayed as major fragments one or two ions resulting from fragmentation at the ester bond ($[LPE-H]^-$) and one or two fatty acid carboxylate anion fragments. As reported elsewhere (Hsu & Turk, 2009; Murphy & Axelsen, 2011), the carboxylate anion and the $[LPE-H]^-$ fragments derived from the fragmentation at the *sn*-2 position are prominent. In Fig. 10, the CAD spectrum of *m/z* = 762.5, i.e. PE ($C_{16:0}/C_{22:6}$) as $[M-H]^-$ ion, shows a minor carboxylate anion at *m/z*= 255.2 ($[C16:0-H]^-$), a major carboxylate anion fragment at *m/z*= 327.12 ($[C22:6-H]^-$), a minor $[LPE-H]^-$ fragment at 524.2 and a major $[LPE-H]^-$ at 452.2, due to the loss of the acyl moieties of the docosahexaenoic acid.

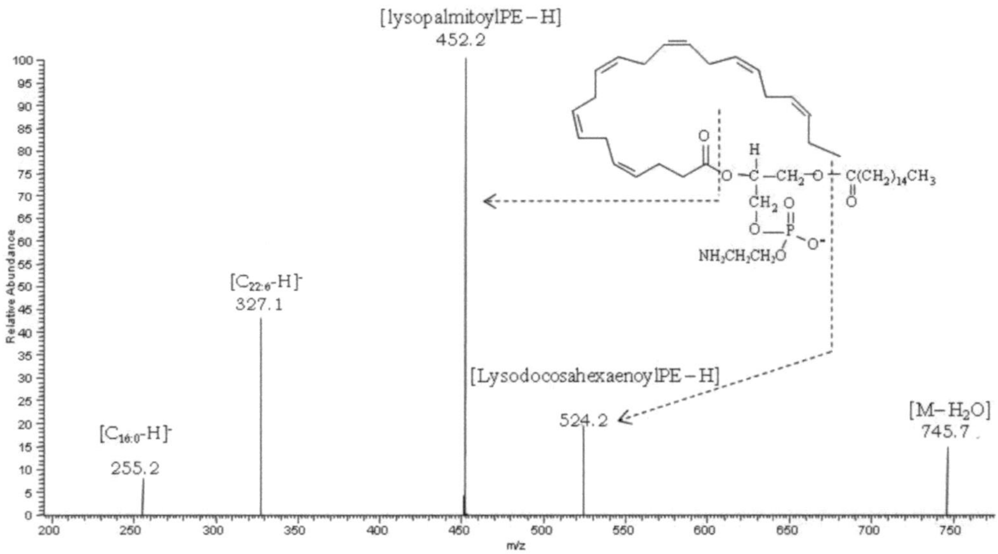

Fig. 10. Negative MS²-spectra of the $[M-H]^-$ of PE ($C_{16:0}/C_{22:6}$) at *m/z* 762.5.

5.6 Phosphatidylethanolamine plasmalogen (pPE)

Phosphatidylethanolamine plasmalogen (pPE) differs from PE because it has a vinyl ether substituent at the *sn*-1 position instead of a fatty acyl moiety.
The use of tandem mass spectrometry to study the positive ion mode collision-induced dissociation of various pPE present in the membrane of human neutrophils and in pork meat was performed by Zemski Berry et al (2004) and by Boselli et al (2008), respectively. They found that the prominent ion detected in positive ion electrospray mass spectrum corresponded to $[M+H]^+$ and showed a product ion spectrum resulting in fragments characteristic of *sn*-1 substituent, in a peculiar fragment of *sn*-2 substituent and in a fragment

ion deriving from the loss of the polar head group ([M-$NH_3(CH_2)_2OPO_3H$]$^+$). As an example, the CAD of *m/z*= 700.4, i.e pPE [M+H]$^+$(p16:0/$C_{18:2}$) (**Fig. 11**) resulted in a major fragment at *m/z* 364.1, containing the phosphatidylethanolamine head group linked with alky-1'-enyl moieties of sn-1 substituent, in a minor fragment at *m/z* 559.5, resulting from the loss of the polar headgroup ([M-$NH_3(CH_2)_2OPO_3H$]$^+$ and in a fragment at *m/z* 337.1, corresponding to the loss of both polar head and 1-O-alky-1'-enyl groups. In contrast, as seen for MS^2 mass spectra of all PE [M+H]$^+$ ions, the fragment resulting from the loss of the polar headgroup is present as minor fragment in the product ion spectrum of pPE [M−H]$^+$. For this reason, the quantification of pPE in a lipid mixture by using constant neutral loss of 141 Da is not suitable as it is for quantification of PE.

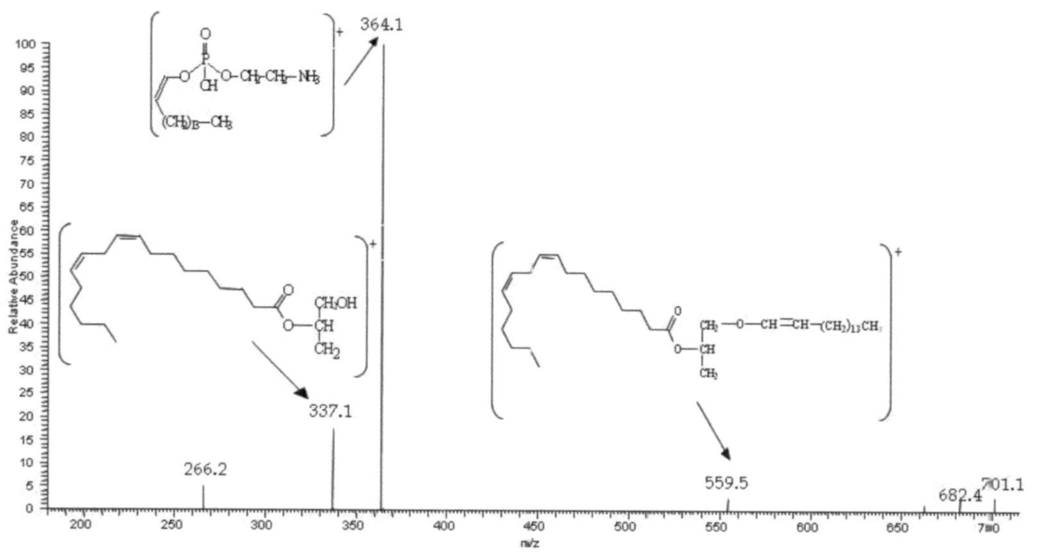

Fig. 11. Positive MS^2-spectra of the [M+H]$^+$ pPE (p16:0/$C_{20:4}$) at *m/z* 700.4.

The mechanism of formation of the predominant fragment ion at *m/z* 364.1 was proposed by Zemski Berry & Murphy (2004). It involves the oxygen atom at the *sn*-1 position attacking the phosphorous atom, which results in the formation of a new O-P bond and concomitant abstraction of the hydrogen from C-2 of the glycerol backbone to form a double bond between C-1 and C-2 of the glycerol structure. Since this fragment is characteristic of the hexadecyl vinyl ether group linked in *sn*-1 position, it is possible to detect all the pPE molecular species containing this substituent by using a precursor ion scan at *m/z*= 364 or all pPE species with an octadecyl vinyl ether by a precursor ion scan of *m/z* = 392.

The characterization of pPE molecular species by means of ESI working in negative ion-mode is reported by Malavolta et al (2004) and by Hsu & Turk 2009. Both groups observed the [M-H]$^-$ ions in negative mode but the product ion spectra obtained by the two authors are different. In detail, Hsu and Turk reported that the CAD spectra of pPE is dominated by the carboxylate anion identifying the fatty acyl moiety of the molecule. Differently, Malavolta et al. showed that the tandem mass spectrum was similar to that obtained with PE. The spectrum has an anion generated from the loss of the fatty acyl moiety ([M-RCO-H]$^-$) as predominat fragment and a carboxylate anion fragment ([RCOO]$^-$) with lower abundance. The fragmentation of the vinyl-ether bond did not occur. The CAD of *m/z*= 746.4, i.e pPE [M-H]- (p16:0/$C_{22:6}$) according to Malavolta et al. is reported in **Fig. 12.**

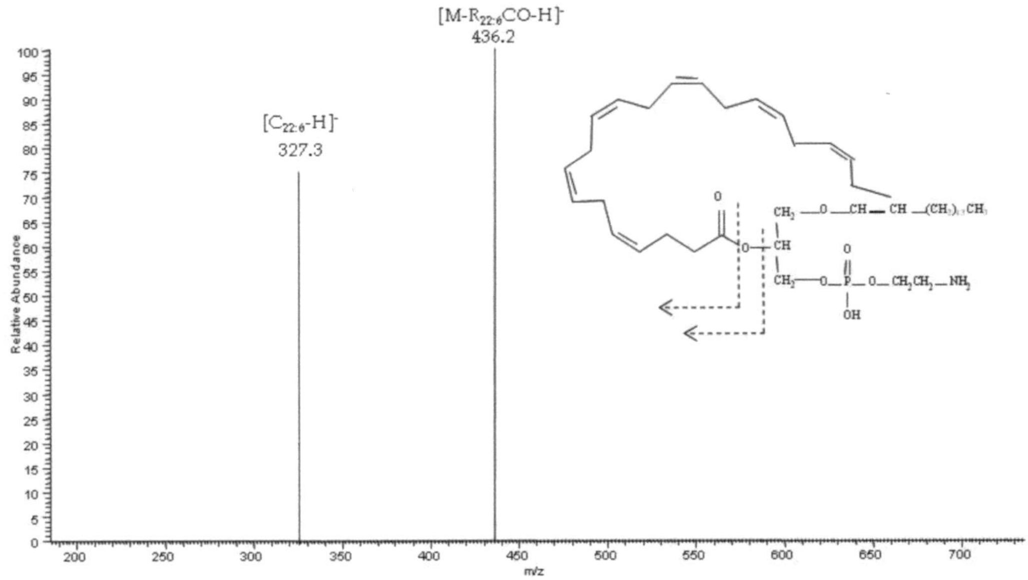

Fig. 12. Negative MS^2-spectra of the [M-H]- of pPE (p16:0/$C_{22:6}$) at *m/z* 746.4.

5.7 Phosphatidylinositol (PI)

PI can be characterized by positive and negative ions. However, negative ions are more abundant and informative because PI contains inositol as a headgroup, which has intrinsic negative charge.

Molecular ions produced by ESI in negative mode corresponded to [M-H]-. Collisional activation of PI negative molecular ion species [M-H]- was reviewed by several authors (Hsu & Turk, 2009; Pacetti et al., 2005; Malavolta et al., 2004). They observed that the MS^2 -spectrum of [M-H]- PI ions is rich of several informative ions, such as carboxylate anions and fragments arising from fragmentation at the ester bond. An example of CAD of *m/z* 909.4, i.e. PI ($C_{18:0}$/$C_{22:6}$) is shown in **Fig. 13**. The spectrum displayed abundant ions resulting from the fragmentation at the ester bond (*m/z* 599.2, [lysoPI-H]$^-$) followed by neutral loss of H_2O (m/z 581.4, [LPI-H-H_2O]$^-$) and inositol (m/z 419.2, [LPI-H-H_2O-$C_6H_{10}O_5$]$^-$). The latest fragment is found for both fatty acids ([LysostearoylPI-H_2O-$C_6H_{10}O_5$]$^-$ and [lysodocosahexaenoyl PI-H_2O-$C_6H_{10}O_5$]$^-$). Only one fatty acid carboxylate anion was detected ([$C_{18:0}$-H]$^-$). These patterns are strongly useful for PI species confirmation.

Molecular ions observed in the positive ion-mode were consistent with [M+H]$^+$ and [M+Na]$^+$. The latest resulted the most abundant ion. Anyway the spectra are complicated by the concomitant presence of both protonated and sodiated adducts. Differently, when the HPLC eluent contains the ammonium ion, the [M+NH_4]$^+$ ion can be formed and it results the only ion yielded (Pacetti et al., 2006). Moreover, PI also forms dilithiated adduct [M-H+2Li]$^+$ ions upon ESI in presence of concentrated Li^+ (Hsu & Turk, 2009). Tandem mass spectrometry of the PI [M+H]$^+$ ion was studied in detail by Murphy and coworkers (2011). Its product ion spectrum showed a major fragment arising from the neutral loss of inositol

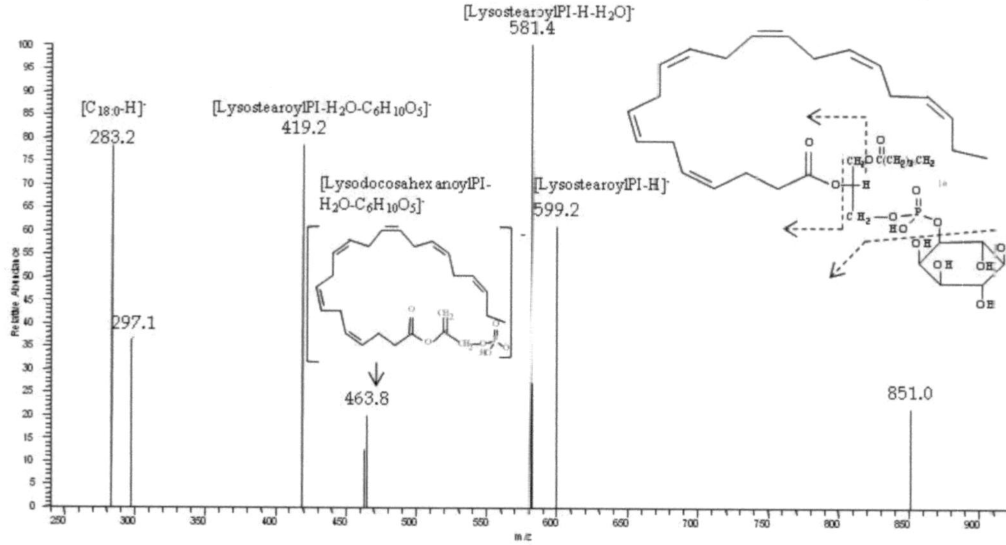

Fig. 13. Negative MS2-spectra of the [M-H]- PI ($C_{18:0}/C_{22:6}$) at *m/z* 909.4.

which corresponds to the PA species [M-259]+. Peculiar fragments containing fatty acid moieties are absent. Also the MS2-spectra of [M-H+2Li]+ did not produce characteristic fragments. It gave rise to a prominent [(LiO)(HO)PO(OX)+Li]+ (where X=inositol−H_2O) at *m/z* = 255. The lithiated diacylglycerol ion was of low abundance. On the contrary, the CAD of $[M+NH_4]^+$ yielded a tipical fragment obtained from the loss of one fatty acid. The fragmentation of the ion $[M+NH_4]^+$ PI ($C_{16:0}/C_{18:2}$) at *m/z* 852.4 is reported in **Fig. 14**. The spectrum is constituted of two fragments: the molecular ion ([M]+, *m/z* 833.6) and the lyso-linoleylPI ion (*m/z* 577.4, $[M-R_{18:2}CO-H_2O]^+$). The molecular ion is the most abundant fragment.

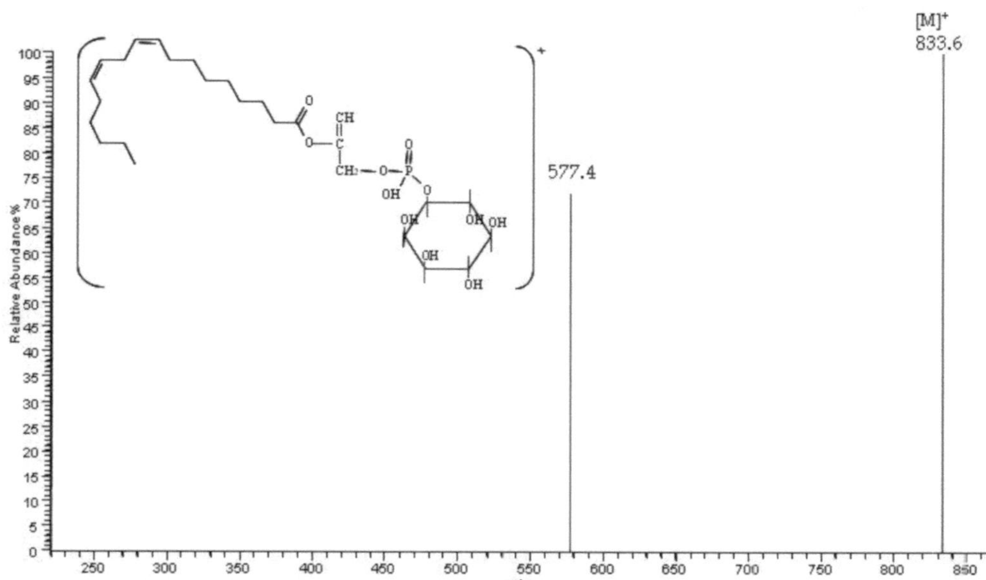

Fig. 14. Positive MS2-spectrum of the $[M-NH_4]^+$ PI ($C_{16:0}/C_{18:2}$) at *m/z* 852.4.

5.8 Phosphatidylserine (PS)

The characterization of PS molecular species by means of ESI ion trap MS yields a positive-ion in form $[M+H]^+$ and negative ions with $[M-H]^-$.

The tandem mass spectrum of the $[M-H]^-$ ion contains complete structural information and offers the utmost sensitivity for structural determination. In contrast, product-ion spectra from the $[M+H]^+$ species are rather simple and are less useful for structural characterization. This statement can be confirmed by comparing the product MS^2 spectra of $[M-H]^-$ ion of PS ($C_{18:0}/C_{22:6}$) at *m/z* = 834.4 (**Fig 15a**) with MS^2 spectra of $[M+H]^+$ of the same PS species at *m/z* 836.5 (**Fig 15b**) detected in the lipid fraction of marine fish. The CAD of $[M-H]^-$ produces an abundant ion resulting from the loss of the head group (*m/z* 747.4, $[PA-H]^-$), two ions resulting from the fragmentation at the ester bond followed by the loss of the serine group $[LPA-H]^-$ and small carboxylate ion fragments. Thus, the two carboxylate anion fragments at *m/z* 283.5 and *m/z* 327.3 and the two $[LPA-H]^-$ ions at 419.4 and at 463.1 identify PS ($C_{18:0}/C_{22:6}$) as the parent ion. Differently, the CAD of $[M+H]^+$ at *m/z* = 836.5 is dominated by *m/z* 651.4 ($[M+H-(HO)_2P(O)O\text{-serine}]^+$) arising from loss of the phosphoserine moiety. Only one of the acylium ions at *m/z* = 329.1 reflecting the fatty acyl sostituents ($C_{22:6}$) is present; anyway, its signal is low.

In the presence of alkali ions, PS molecular species in the form of $[M+Alk]^+$ and $[M-H+2Alk]^+$ (Alk = Li, Na) can also be observed, attributable to the fact that PS possesses two anionic charge sites of which one can attach to an Alk^+ (Hsu & Turk, 2005). Ions informative for structural characterization of PS are of low abundance in the MS^2 spectra of alkali adduct ions. The MS^2-spectrum of the $[M+Alk]^+$ ion contains a unique ion corresponding to the internal loss of a phosphate group probably via the fragmentation processes involving rearrangement steps. The $[M-H+2Alk]^+$ ion of PS yields a major $[M-H+2Alk-87]^+$ ion, which is equivalent to an alkali adduct ion of a monoalkali salt of PA and gives rise to a greater abundance of $[M-H+2Alk-87-R_1CO_2H]^+$ than $[M-H+2Alk-87-R_2CO_2H]^+$.

5.9 Sphingomyelin (Sph)

ESI of sphingomyelin follows closely that of PC and LPC. The presence of the quaternary nitrogen atom, with permanent positive charge, dominates the ESI behaviour of this molecule. Thus, the Sph generates more abundant ions in the positive ion mode, than the negative ion mode. Moreover, in negative ion mode, Sph can be analyzed only if the addition of an anionic reagent (acetate or formiate) in the spray solvent is performed.

In detail, (+)-ESI yields a protonated ion $[M+H]^+$ and a lithium adduct $[M+Li]^+$, whereas, in (-)-ESI, Sph is detected as chlorine $[M+Cl]^-$, formiate $[M+HCOO]^-$ and acetate $[M+CH_3COO]^-$ adducts.

As reported by Murphy and Axelsen (2011) lithiated-sphingomyelin ions produce $[M-59+Li]^+$ ions representing the neutral loss of $(CH_3)_3N$ and $[M-183+Li]^+$ ions representing the neutral loss of phosphocholine. Differently, CAD of $[M+H]^+$ yields the only fragment at *m/z* 184 (phosphocholine group). This evidence contrasts with the results obtained by Boselli et al (2008) where the CAD of $[M+H]^+$ at *m/z* 703.6, i.e. Sph ($C_{16:0}$), yielded fragments deriving from the loss of water (*m/z* 685.2, $[M+H-H_2O]^+$), the loss of the acyl group (*m/z* 464.8), the loss of the amidic moiety (*m/z* 432.2) and a fragment deriving from sphingosin (*m/z* 295.9) (**Fig. 16**).

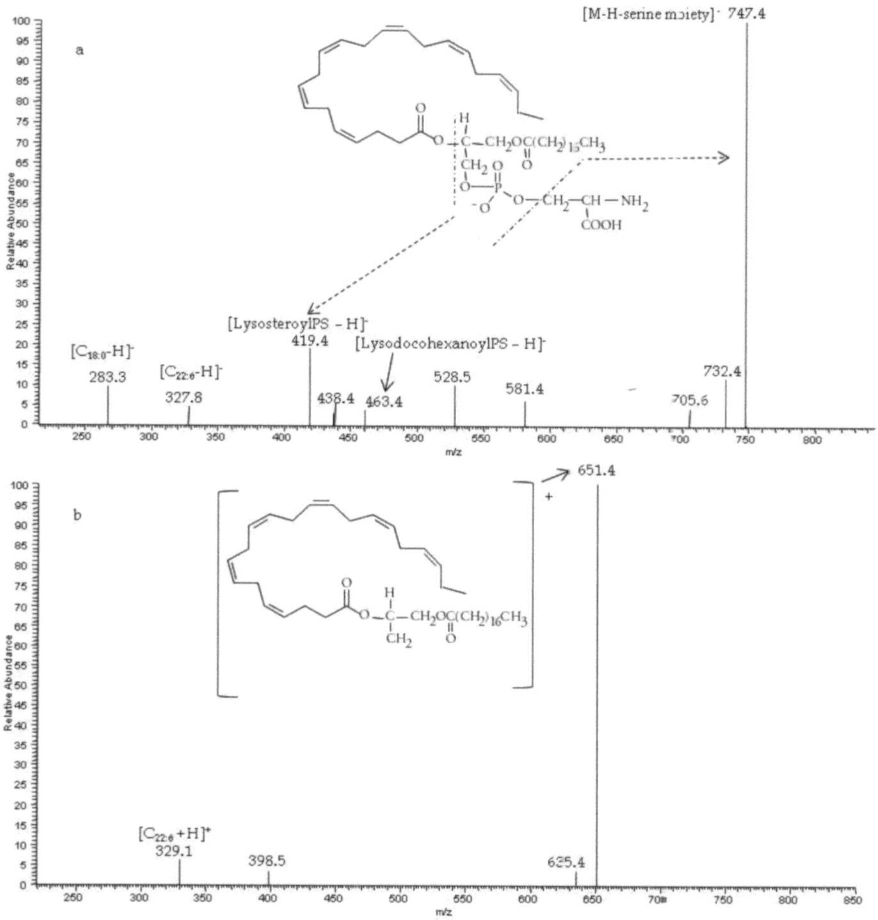

Fig. 15. Positive MS2 spectrum of the PS ($C_{18:0}/C_{22:6}$) ions (a) $[M-H]^-$ ion at *m/z* 834.4 (b) $[M+H]^+$ ion at *m/z* 836.5.

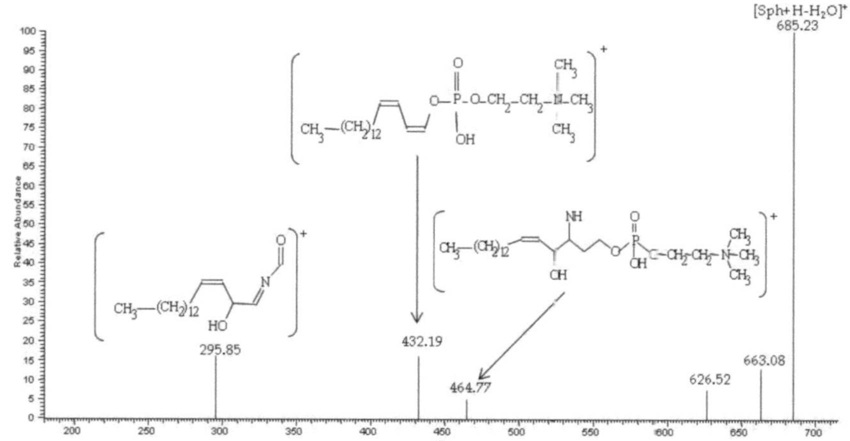

Fig. 16. Positive MS2-spectra of the $[M+H]^+$ Sph ($C_{16:0}$) at *m/z* 703.6.

The product ion spectrum of Sph formiate and acetate adducts ions contain exclusively the $[M-CH_3]^-$ ion fragment, due to the loss of the methyl group of choline (Pacetti et al. 2005; Murphy and Axelsen, 2011). The fragmentation of the chlorine adduct produces only the loss of a methyl group, as well. An example of CAD of ion at *m/z* 747.4, i.e. Sph ($C_{16:0}$) as formiate adduct is reported in **Fig. 17**.

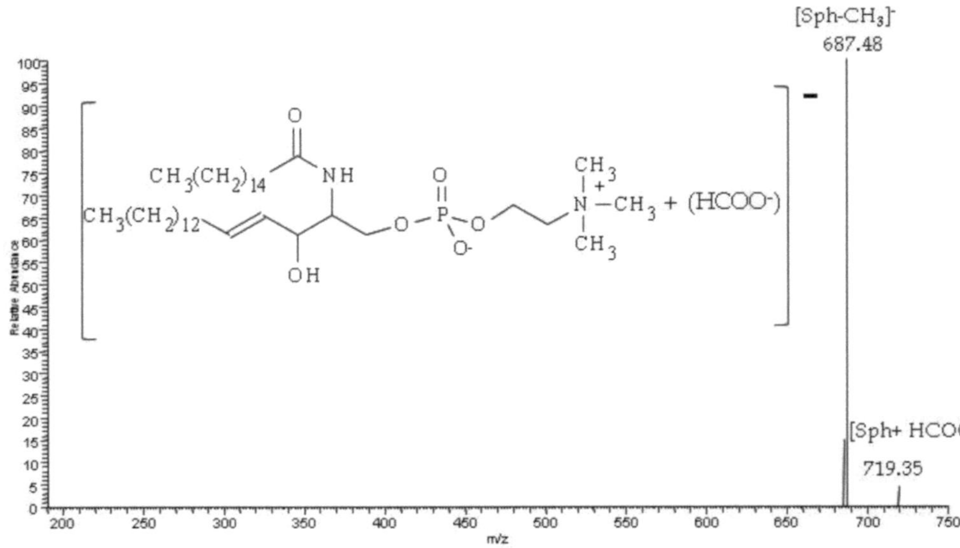

Fig. 17. Negative spectra MS2-spectra of ion at *m/z* 747.5 corresponding to formiate adducts of Sph containing palmitic acid ($C_{16:0}$).

On the basis of these experimental evidences it is possible to establish that the CID with negative ionization is not useful for an unambiguous identification of Sph isobaric molecular species.

6. Conclusion

HPLC-Tandem Mass Spectrometry needs high-value equipment and trained personnel, however it enables researchers to achieve incomparable results with respect to other non-automated techniques. PL molecular species can be determined either in positive or negative ionization depending on the charge of the polar group. Time and chemical consuming clean-up procedures can be avoided in order to fingerprint the (combination of the couple of) fatty acids bound to the phosphoglyceric bone. In certain cases, however, mass spectrometry of the third order would be needed for the identification of isobaric molecular species.

7. References

Abidi, S.L.; & Mounts, T.L. (1992). HPLC separation of molecular species of neutral phospholipids. *Journal of Chromatography* A, Vol.598, 209-218.

Abidi, S.L. & Mounts, T.L. (1996). Reversed-phase high-performance liquid chromatography of molecular species of phospholipid derivatives. *Journal of Chromatography*. A, Vol.741, pp.213-222.

Abidi, S.L.; (1991). HPLC of phosphatidic acids and related polar lipids. *Journal of Chromatography* A, Vol.587, pp. 193-203.

Ahn, E.J.; Kim, H.; Chung, B.C. & Moon, M.H. (2007). Quantitative analysis of phosphatidylcholine in rat liver tissue by nanoflow liquid chromatography/tandem mass spectrometry. *Journal of Separation Science*, Vol.30, No.16, (August 2007), pp. 2598-2604.

Almeida, M.M.M.; Zapata, J.F.F.; Martins, C.B.; & Maia, G.A. (1997) Cholesterol and Phospholipid Levels in Goat Meat As Affected By Dietary Calcium, *Pesquisa Agropecuaria Brasileira*, 32, pp. 555-558.

Annesley, T.M. (2003). Ion Suppression in Mass Spectrometry. *Clinical Chemistry*, Vol.49, pp.1041-1044.

AOCS Official Method Ja 7c-07 Revised (2007) Determination of Lecithin Phospholipids by HPLC-LSD.

Avalli, A. & G. Contarini (2005) Determination of phospholipids in dairy products by SPE/HPLC/ELSD, *Journal of Chromatography A*, 1071 pp. 185–190.

Bang, D.Y.; Ahn, E.J. & Moon, M.H. (2007). Shotgun analysis of phospholipids from mouse liver and brain by nanoflow liquid chromatography/tandem mass spectrometry. *Journal of Chromatography B*, Vol.852, No.1-2, (June 2007), pp. 268–277.

Bang, D.Y.; Kang, D. & Moon, M.H. (2006). Nanoflow liquid chromatography-tandem mass spectrometry for the characterization of intact phosphatidylcholines from soybean, bovine brain, and liver. *Journal of Chromatography A*. Vol. 1104, pp 222–229.

Barroso, B. & Bischoff, R. (2005). LC–MS analysis of phospholipids and lysophospholipids in human bronchoalveolar lavage fluid. *Journal of Chromatography B*, Vol.814, No.1, (January 2005), pp. 21-28.

Becart, J.; Chevalier, C. & Biesse, J.P. (1990). Quantitative analysis of phospholipids by HPLC with light scattering detector: application to raw materials for cosmetic use. *Journal of High Resolution Chromatography*, Vol.13, No.2, February 1990, pp 126-129.

Benfenati, E. & Reginato, R. (1985). A comparison of three methods of soft ionization mass spectrometry of crude phospholipid extracts. *Biological Mass Spectrometry*, Vol.12, No.11, pp. 643–651.

Berdeaux, O.; Juaneda, P.; Martine, L.; Cabaret, S.; Bretillon, L. & Acar, N. (2010) Identification and quantification of phosphatidylcholines containing very-long-chain polyunsaturated fatty acid in bovine and human retina using liquid chromatography/tandem mass spectrometry. *Journal of Chromatography A*, Vol.1217, pp. 7738–7748.

Billah, M.M. & Anthes, J. C. (1990) The regulation and cellular functions of phosphatidylcholine hydrolysis, *Biochem. J.* 269, pp.281-291.

Bligh, E.G. & Dyer, W.J. (1959) A Rapid Method Of Total Lipid Extraction And Purification, *Can. J. Biochem. Physiol.* 37 pp. 911-917.

Boselli, E. & Caboni, M.F. (2000) Supercritical carbon dioxide extraction of phospholipids from dried egg yolk without organic modifier. *Journal of Supercritical Fluids* Vol. 19 pp. 45–50.

Boselli, E.; Pacetti, D.; Curzi, F. & Frega, N.G. (2008). Determination of Phospholipid Molecular Species in Pork Meat by High Performance Liquid Chromatography-Tandem Mass Spectrometry and Evaporative Light Scattering Detection. *Meat Science*, Vol.78, No.3, (March 2008), pp. 305–313.

Boselli, E.; Velazco, V.; Caboni, M.F. & Lercker, G. (2001) Pressurized liquid extraction of lipids for the determination of oxysterols in egg-containing food. *Journal of Chromatography A*, Vol. 917 pp. 239–244.

Bourre J.M. & Dumont O. (2002) The administration of pig brain phospholipids versus soybean phospholipids in the diet during the period of brain development in the rat results in greater increments of brain docosahexaenoic acid, *Neuroscience Letter* 335 pp. 129-133.

Brouwers, J.F.; Versluis, C.; Van Golde, L.M. & Tielens A.G. (1998). 5-Octadecenoic acid: evidence for a novel type of fatty acid modification in schistosomes. *Biochemical Journal*, Vol.334, No.2, (September 1998), pp. 315-319.

Caboni, M. A., Menotta, S., Lercker, G. (1996).Separation and analysis of phospholipids in different foods with a light-scattering detector. J. Am. Oil Chem. Soc.73,1561-1566.

Caboni, M. F.; Boselli, E. & Menotta, S. (2000) HPLC-ELSD determination of Phospholipids and Lysophospholipids in Grana Cheese samples at different ripening stages. *Scienza e Tecnica Lattiero-casearia*, Vol.51, No.1, (February 2000), pp. 7-24.

Cambero, M.I.; de la Hoz, L.; Sanz, B. & Ordonez, J.A. (1991) Lipid and fatty acid composition of rabbit meat: Part 2.--Phospholipids *Meat Science*, 29, pp. 167-176.

Chen, S.S.H. & Kou, A.Y. (1982). Improved procedure for the separation of phospholipids by high-performance liquid chromatography. *Journal of Chromatography B*, Vol.227, No.1, (January 1982), pp. 25–31.

Chicco AJ, & Sparagna GC (2007) Role of cardiolipin alterations in mitochondrial dysfunction and disease. *Am J Physiol Cell Physiol* Vol. 292, pp.33–44.

Christie, W.W. (2011) Complex Glycerolipids, in The lipid library, The American Oil Chemists' Society, (http://lipidlibrary.aocs.org/).

Cockcroft, S. & Garner, K. (2011) Function of the phosphatidylinositol transfer protein gene family: is phosphatidylinositol transfer the mechanism of action? *Critical Reviews in Biochemistry* 46, pp. 89-117.

Cohn, J.S., Wat; E., Kamili, A. & Tandy, S. (2008) Dietary phospholipids, hepatic lipid metabolism and cardiovascular disease, *Current Opinion in Lipidology*, 19, pp. 257-262.

Commission Decision 2000/195/EC authorising phospholipides (85% and 100%) to be placed on the market in the Community, *Official Journal* L 061, 08/03/2000, p. 12.

Commission Decision 2009/752/EC authorising the placing on the market of a lipid extract from Antarctic Krill (*Euphausia superba*) as a novel food ingredient under Regulation (EC) No 258/97 of the European Parliament and of the Council, *Official Journal* L268 of 13 October 2009, p.33.

Commission Regulation (EC) No 2065/2001 of 22 October 2001 laying down detailed rules for the application of Council Regulation (EC) No 104/2000 as regards informing consumers about fishery and aquaculture products (1) *EN Official Journal of the European Communities* 23.10.2001L 278/6.

Cui, Z. & Thomas, M.J. (2009). Phospholipid profiling by tandem mass spectrometry. *Journal of Chromatography B*, Vol.877, No.26, (September 2009), pp. 2709–2715.

Di Marzo, V.; Griinari M.; Carta G.; Murru, E.; Ligresti A.; Cordeddu L.; Giordano E.; Bisogno, T.; Collu M.; Batetta B.; Sanna F.; Uda S.; Berge K. & Banni S. (2010) Dietary krill oil increases docosahexaenoic acid and reduces 2-arachidonoylglycerol but not N-acylethanolamine levels in the brain of obese Zucker rats. *Int Dairy J* 20, pp. 231-235.

Dobson, G & Deighton N. (2001). Analysis of phospholipid molecular species by liquid chromatography-atmospheric pressure chemical ionisation mass spectrometry of diacylglicerols nicotinates. *Chemistry and Physics of Lipids*, Vol.111, No.1, pp. 1-17.

Dodbiba, E.; Xu, C.; Payagala, T.; Wanigasekara, E.; Moon, M.H. & Armstrong, D.W. (2011). Use of ion pairing reagents for sensitive detection and separation of phospholipids in the positive ion mode LC-ESI-MS. *Analyst*, Vol.136, No.8, pp. 1586-1593.

Domingues, M. R. M.; Reis A. & Domingues, P. (2008) Mass spectrometry analysis of oxidized phospholipids, *Chemistry and Physics of Lipids*, Vol. 156, pp. 1-12.

Donato, P.; Cacciola, F.; Cichello, F.; Russo, M.; Dugo, P. & Mondello, L. (2011). Determination of phospholipids in milk samples by means of hydrophilic interaction liquid chromatography coupled to evaporative light scattering and mass spectrometry detection. *Journal of Chromatography A*, Vol.1218, No.37, (September 2011), pp. 6476-6482.

Duckett, S. K.; Wagner, D. G. ; Yates, L. D.; Dolezal H. G. & May, S. G. (1993) Effects of time on feed on beef nutrient composition *J Anim Sci* 71 pp.2079-2088.

Dugan, L.L.; Demediuk, P.; Pendley, C.E. & Horrocks, L.A. (1986). Separation of phospholipids by high-performance liquid chromatography: All major classes, including ethanolamine and choline plasmalogens, and most minor classes, including lysophosphatidylethanolamine. *Journal of Chromatography*, Vol.378, pp. 317–327.

Dumay, Allery, Donnay-Moreno, Barnathan, Jaouen, Carbonneau & Bergé (2009) Optimization of hydrolysis of sardine (Sardina pilchardus) heads with Protamex: enhancement of lipid and phospholipid extraction *Journal of the Science of Food and Agriculture* Vol. 89, pp. 1599–1606.

Enjalbal, C.; Roggero, R.; Cerdan, R.; Martinez, J.; Vial, H. & Aubagnac, J.L. (2004). Automated Monitoring of Phosphatidylcholine Biosyntheses in Plasmodium falciparum by Electrospray Ionization Mass Spectrometry through Stable Isotope Labeling Experiments. *Analytical Chemistry*, Vol.76, No.15, (June 2004), pp. 4515-4521.

Fagan, P. & Wijesundera, C. (2004). Liquid chromatographic analysis of milk phospholipids with on-line pre-concentration. *Journal of Chromatography* A, Vol. 1054, No.1-2, (October 2004), pp. 241-249.

Ferioli, F. & Caboni, M.F. (2010) Composition of phospholipid fraction in raw chicken meat and pre-cooked chicken patties: influence of feeding fat sources and processing technology *Eur Food Res Technol* (2010) 231:117–126.

Folch, J.; Lees, M. & Sloane Stanley, G.H. (1957) A Simple Method For The Isolation And Purification Of Total Lipides From Animal Tissues *J. Biol. Chem.* 226 pp. 497-509.

Fuchs, B.; Süß, R. & Schiller, J. (2010) An update of MALDI-TOF mass spectrometry in lipid research. *Progress in Lipid Research*, Vol.49, pp. 450–475.

Godoy Ramos, R.; Libong, D.; Rakotomanga, M.; Gaudin, K.; Loiseau, P.M. & Chaminade, P. (2008). Comparison between charged aerosol detection and light scattering detection for the analysis of *Leishmania* membrane phospholipids. *Journal of Chromatography A*, Vol. 1209, No.1-2, (October 2008), pp. 88–94.

Graeve, M. & Janssen, D. (2009). Improved separation and quantification of neutral and polar lipid classes by HPLC–ELSD using a monolithic silica phase: Application to exceptional marine lipids. *Journal of Chromatography B*, Vol.877, pp. 1815–1819.

Guan, Z.; Grünler, J.; Piao, S. & Sindelar, P.J. (2001). Separation and Quantitation of Phospholipids and Their Ether Analogues by High-Performance Liquid Chromatography. *Analytical Biochemistry*, Vol.297, No.2 (October 2001), pp. 137–143.

Han, X. & Gross, R.W. (2005). Shotgun lipidomics: Electrospray ionization mass spectrometric analysis and quantitation of cellular lipidomes directly from crude extracts of biological samples. *Mass Spectrometry Reviews,* Vol.24, No.3, pp 367-412.

Hanahan, D. J. (1986) Platelet Activating Factor: A Biologically Active Phosphoglyceride, *Annu. Rev. Biochem.* 55, pp. 483-509.

Hayakawa, J. & Okabayashi, Y.(2004). Simultaneous analysis of phospholipid in rabbit bronchoalveolar lavage fluid by liquid chromatography/mass spectrometry. *Journal of Pharmaceutical and Biomedical Analysis,* Vol.35, No.3, (May 2004), pp. 583-592.

Hemström, P. & Irgum K. (2006). Hydrophilic interaction chromatography. *Journal of Separation Science,* Vol.29, No.12, (August 2006), pp 1784-1821.

Henna Lu, F.S.; Nielsen, N.S.; Timm-Heinrich, M. & Jacobsen, C. (2010) Oxidative Stability of Marine Phospholipids in the Liposomal Form and Their Applications, *Lipids,* 46, pp. 3-23.

Henneberry, A.L.; Lagace, T.A.; Ridgway, N.D.; McMaster, C.R. (2001) Phosphatidylcholine Synthesis Influences the Diacylglycerol Homeostasis Required for Sec14p-dependent Golgi Function and Cell Growth, *Mol. Biol. Cell* Vol. 12, pp. 511-520.

Hermansson, M.; Uphoff, A.; Käkelä, R. & Somerharju, P. (2005). Automated Quantitative Analysis of Complex Lipidomes by Liquid Chromatography/Mass Spectrometry. *Analytical Chemistry,* Vol.77, No.7, (February 2005), pp. 2166-275.

Hoischen, C.; Gura, K.; Luge, C. & Gumpert, J. (1997) Lipid and Fatty Acid Composition of Cytoplasmic Membranes from Streptomyces hygroscopicus and Its Stable Protoplast-Type L Form. *Journal of Bacteriology* Vol. 179, No.11, pp. 3430–3436.

Holinstat, M.; Preininger, A.M.; Milne, S.B.; Hudson, W.J.; Brown, H.A. & Hamm, H.E. (2009) Irreversible Platelet Activation Requires Protease-Activated Receptor 1-Mediated Signaling to Phosphatidylinositol Phosphates, *Mol Pharmacol* Vol. 76 pp. 301–313.

Houjou, T.; Yamatani, K.; Imagawa, M.; Shimizu, T. & Taguchi, R. (2005). A shotgun tandem mass spectrometric analysis of phospholipids with normal-phase and/or reverse-phase liquid chromatography/electrospray ionization mass spectrometry. *Rapid Communication in Mass Spectrometry.* Vol.19, No.5, (February 2005), pp. 654-666.

Houtkooper R.H. & Vaz F.M. (2008) Cardiolipin, the heart of mitochondrial metabolism. *Cell Mol Life Sci* Vol. 65 pp. 2493–2506.

Hsu, F.F.; Bohrer, A. & Turk, J. (1998) Formation of Lithiated Adducts of Glycerophosphocholine Lipids Facilitates their Identification by Electrospray Ionization Tandem Mass Spectrometry. *Journal of the American Society for Mass Spectrometry* Vol. 9, No. 5 pp. 516-526.

Hsu, F.F. & Turk, J. (2000). Charge-remote and charge-driven fragmentation processes in diacyl glycerophosphoethanolamine upon low-energy collisional activation: a mechanistic proposal. *Journal of the American Society for Mass Spectrometry*, Vol.11, No.10 (October 2000) pp. 892-899.

Hsu, F.F. & Turk, J. (2003) Electrospray ionization/tandem quadrupole mass spectrometric studies on phosphatidylcholines: the fragmentation processes. *Journal of the American Society for Mass Spectrometry* Vol. 14, No. 4 pp. 352-363.

Hsu, F.F. & Turk, J. (2006). Characterization of Cardiolipin as the Sodiated Ions by Positive Ion-Electrospray Ionization with Multiple Stage Quadrupole Ion Trap Mass Spectrometry. *Journal of the American Society of Mass Spectrometry*, Vol.17, No.8, (August 2006), pp. 1146-1157.

Hsu, F.F. & Turk, J. (2006). Structural Characterization of Cardiolipin from Escherichia coli by Electrospray Ionization with Multiple Stage Quadrupole Ion-Trap Mass Spectrometric Analysis of $[M-2H+Na]^-$ Ions. *Journal of the American Society of Mass Spectrometry*. Vol.17, No.3, (March 2006), pp. 420–429.

Hsu, F.F. & Turk, J. (2009). Electrospray ionization with low-energy collisionally activated dissociation tandem mass spectrometry of glycerophospholipids: Mechanisms of fragmentation and structural characterization. *Journal of Chromatography B*, Vol.877, No.26, (September 2009), pp. 2673-2695.

Hsu, F.F.; Turk, J.; Rhoades, E.R.; Russell, D.G.; Shi, Y. & Groisman, E.A. (2005) Structural Characterization of Cardiolipin by Tandem Quadrupole and Multiple-Stage Quadrupole Ion-Trap Mass Spectrometry with Electrospray Ionization. *Journal of the American Society of Mass Spectrometry*, Vol.16, No.4, (April 2005), pp. 491–504.

Isaac, G.; Bylund, D.; Mansson, J.E.; Markides, K.E. & Bergquist, J. (2003). Analysis of phosphatidylcholine and sphingomyelin molecular species from brain extracts using capillary liquid chromatography electrospray ionization mass spectrometry. *Journal Neuroscience Methods*, Vol.128, pp. 111–119.

ISO 11701:2009 (2009) Vegetable fats and oils -- Determination of phospholipids content in lecithins by HPLC using a light-scattering detector.

Ito, S.; Nabetani, T.; Shinoda, Y.; Nagatsuka, Y. & Hirabayashi, Y. (2008). Quantitative analysis of a novel glucosylated phospholipid by liquid chromatography–mass spectrometry. *Analytical Biochemistry*, Vol.376, pp. 252–257.

Itoh, K.; Suzuki, A.; Kuroki,Y. & Akino, T. (1985). High performance liquid chromatographic separation of diacylglycerol acetates to quantitate disaturated species of lung phosphatidylcholine. *Lipids*, Vol.20, No.9, (September 1985), pp. 611-616.

Jakab, A. & Forgács, E. (2002). Characterization of plant oils on a monolithic silica column by high-performance liquid chromatography-atmospheric pressure chemical ionization-mass spectrometry. *Chromatographia*, Vol.56, No.1, pp. S69-S73.

Jantscheff, P.; Schlesinger, M.; Fritzsche, J.; Taylor, L.A.; Graeser, R.; Kirfel, G.; Fürst, D.O.; Massing, U.; Bendas, G. (2011) Lysophosphatidylcholine Pretreatment Reduces VLA-4 and P-Selectin–Mediated B16.F10 Melanoma Cell Adhesion In vitro and Inhibits Metastasis-Like Lung Invasion In vivo *Mol. Cancer Ther.* Vol. 10 pp. 186-197.

Joensen, H.; Steingrund, P.; Fjallstein, I.; & Grahl-Nielsen, O. (2000) Discrimination between two reared stocks of cod (*Gadus morhua*) from the Faroe Islands by chemometry of the fatty acid composition in the heart tissue *Marine Biology* Vol. 136, pp. 573-580.

Kakela, R.; Somerharju, P. & Tyynelä, J. (2003). Analysis of phospholipid molecular species in brains from patients with infantile and juvenile neuronal-ceroid lipofuscinosis using liquid chromatography-electrospray ionization mass spectrometry. *Journal of Neurochemistry*, Vol.84, No.3, (March 2003), pp. 1051-1065.

Kerwin, J.L.; Tuininga A.R., Ericsson, L.H. (1994). Identification of molecular species of glycerophospholipids and sphingomyelin using electrospray mass spectrometry. *Journal of Lipid Research*, Vol.35, pp 1102–1114.

Kivini, H.; Järvenpää, E.P.; Aro, H.; Huopalahti, R. & Ryhänen, E.L. (2004). Qualitative and Quantitative Liquid Chromatographic Analysis Methods for the Determination of the Effects of Feed Supplements on Hen Egg Yolk Phospholipids. *Journal of Agricultural and Food Chemistry*, Vol.52, No.13, (June 2004), pp. 4289-4295.

Koc, H.; Mar, M.H.; Ranasinghe, A.; Swenberg, J.A. & Zeisel, S.H. (2002). Quantitation of choline and its metabolites in tissues and foods by liquid chromatography/electrospray ionization-isotope dilution mass spectrometry. *Analytical Chemistry*, Vol.74, No.18, pp. 4734-4740.

Larsen, A.; Uran, S.; Jacobsen, P.B. & Skotland, T. (2001). Collision-induced dissociation of glycero phospholipids using electrospray ion-trap mass spectrometry. *Rapid Communication in Mass Spectrometry*, Vol.15, No.24, pp. 2393-8.

Le Grandois, J.; Marchioni, E.; Zhao, M.; Giuffrida, F.; Ennahar, S. & Bindler, F. (2009) Investigation of Natural Phosphatidylcholine Sources: Separation and Identification by Liquid Chromatography−Electrospray Ionization−Tandem Mass Spectrometry (LC−ESI−MS2) of Molecular Species. *J. Agric. Food Chem.*, Vol. 57 No.14, pp. 6014–6020.

Leseigneur-Meynier, A., & Gandemer, G., (1991) Lipid composition of pork muscle in relation to the metabolic type of the fibres, *Meat Science*, Vol. 29, pp. 229-241.

Lin, J.T.; Ikeda M.D. & McKeon A.T. (2004). Comparison of the Incorporation of Oleate and Ricinoleate into Phosphatidylcholines and Acylglycerols in Soybean Microsomes *Journal Agriculture and Food Chemistry*, Vol.52, No.5, (February 2004), pp. 1152-1156.

Lingwood, D. & Simons, K. (2010) Lipid Rafts As a Membrane-Organizing Principle, *Science* Vol. 327, pp. 46-50.

Ma, Y.C. & Kim, H.Y.(1995). Development of the on-line high-performance liquid chromatography/thermospray mass spectrometry method for the analysis of phospholipid molecular species in rat brain. *Analytical Biochemistry*, Vol.226, No.2, (April 1995), pp. 293-301.

Malavolta, M.; Bocci, F.; Boselli, E. & Frega, N.G. (2004). Normal phase liquid chromatography–electrospray ionization tandem mass spectrometry analysis of phospholipid molecular species in blood mononuclear cells: application to cystic fibrosis. *Journal of Chromatography B*, Vol.810, No.2, (October 2004), pp. 173-186.

Merlin, J.F.; Gresti, J.; Bellenger, S. & Narce, M. (2006). Fast high performance liquid chromatography analysis in lipidomics: Separation of radiolabelled fatty acids and phosphatidylcholine molecular species using a monolithic C18 silica column. *Analytica Chimica Acta*, Vol.565, No.2, (April 2006), pp. 163-167.

Miyakis, S.; Lockshin, M.D.; Atsumi, T.; Branch, D.W.; Brey, R.L.; Cervera, R.; Derksen, R.H.W.M.; De Groot, P.G.; Koike, T.; Meroni, P.L.; Reber, G.; Shoenfeld, Y.; Tincani, A.; Vlachoyiannopoulos, P.G. & Krilis, S.A. (2006) International consensus statement on an update of the classification criteria for definite antiphospholipid syndrome (APS), *Journal of Thrombosis and Haemostasis* Vol. 4 pp.295–306.

Morita, S.; Takeuchi, A. & Kitagawa S. (2010) Functional analysis of two isoforms of phosphatidylethanolamine N-methyltransferase, *Biochem. J.* Vol. 432, pp. 387–398.

Murphy, R.C. & Axelsen , P.H. (2011). Mass spectrometric analysis of long-chain lipids. *Mass Spectrometry Reviews*, Vol.30, No.4, (July-August 2011), pp. 579-599.

Murphy, R.C.; Fiedler J. & Hevko, J. (2001). Analysis of Nonvolatile Lipids by Mass Spectrometry. *Chemical Reviews,* Vol.101, pp. 479-526.

Nagan, N. & Zoeller, R. A. (2001) Plasmalogens: biosynthesis and functions. *Prog. Lipid Res.* Vol. 40, pp. 199–229.

Nakanishi, H.; Iida, Y.; Shimizu, T. & Taguchi, R. (2010). Separation and quantification of sn-1 and sn-2 fatty acid positional isomers in phosphatidylcholine by RPLC-ESIMS/MS. *The Journal of Biochemistry,* Vol.147, No.2, pp. 245-256.

Narváez-Rivas, M.; Gallardo, E.; Ríos, J.J. & León-Camacho, M. (2011). A new high-performance liquid chromatographic method with evaporative light scattering detector for the analysis of phospholipids. Application to Iberian pig subcutaneous fat. *Journal of Chromatography A*, Vol.1218, pp 3453–3458.

Neron, S.; El Amrani, F.; Potus, J. & Nicolas, J. (2004). Separation and quantification by high-performance liquid chromatography with light scattering detection of the main wheat flour phospholipids during dough mixing in the presence of phospholipase. *Journal of Chromatography A.* Vol.1047, No.1, (August 2004), pp. 77-83.

Ogiso, H.; Suzuki, T. & Taguchi, R. (2008). Development of a reverse-phase liquid chromatography electrospray ionization mass spectrometry method for lipidomics, improving detection of phosphatidic acid and phosphatidylserine. *Analytical Biochemistry*, Vol.375, No.1, (April 2008), pp. 124–131.

Pacetti, D.; Hulan, H.W.; Schreiner, M.; Boselli, E. & Frega, N.G. (2005) Positional analysis of egg triacylglycerols and phospholipids from hens fed diets enriched with refined seal blubber oil. *J Sci Food Agric* Vol.85 pp. 1703–1714.

Pacetti, D.; Boselli, E.; Hulan, H.W. & Frega, N.G. (2005). High performance liquid chromatography-tandem mass spectrometry of phospholipid molecular species in eggs from hens fed diets enriched in seal blubber oil. *Journal of Chromatography A,* Vol.1097, No.1-2, (December 2005), pp. 66–73.

Pacetti, D.; Boselli, E.; Lucci, P. & Frega, N.G. (2007). Simultaneous analysis of glycolipids and phospholids molecular species in avocado (Persea americana Mill) fruit. *Journal of Chromatography A*, Vol.1150, No.1-2, (May 2007), pp. 241-251.

Pacetti, D.; Malavolta, M.; Bocci, F.; Boselli, E. & Frega, N.G. (2004). HPLC-ESI-MS-MS Analysis and quantification of phosphatidylcholine molecular species in the serum of cystic fibrosis subjects supplemented with docosahexaenoic acid. *Rapid Communications in Mass Spectrometry*, Vol.18, No.20, (October 2004), pp. 2395-2400.

Pang, L.Q.; Liang, Q.L.; Wang, Y.M.; Ping, L. & Luo, G.A. (2008). Simultaneous determination and quantification of seven major phospholipid classes in human blood using normal-phase liquid chromatography coupled with electrospray mass spectrometry and the application in diabetes nephropathy. *Journal of Chromatography B*, Vol.869, No.1-2, (June 2008), pp. 118–125.

Papahadjopoulos D. (1978) *Liposomes and Their Uses in Biology and Medicine,* Ann NY Acad. Sci., New York Academy of Science, New York, Vol. 308.

Peterson, B.L. & Cummings, B.S. (2006). A review of chromatographic methods for the assessment of phospholipids in biological samples. *Biomedical Chromatography,* Vol.20, No.3, (March 2006), pp. 227-43.

Postle, A.D.; Wilton, D.C.; Hunt, A.N. & Attard G.S. (2007). Probing phospholipid dynamics by electrospray ionisation mass spectrometry. *Progress in Lipid Research,* Vol.46, No.3-4, (May-July 2007), pp. 200–224.

Pulfer, M. & Murphy, R.C. (2003). Electrospray mass spectrometry of phospholipids. *Mass Spectrometry Reviews,* Vol.22, No.5, (August 2003), pp. 332–364.

Qiao, J.; Huang, F.; Naikawadi, R.P.; Kim, K.S.; Said, T.; Lum, H. (2006) Lysophosphatidylcholine impairs endothelial barrier function through the G protein-coupled receptor GPR4, *AJP - Lung Physiol,* Vol. 291, pp. L91-L101.

Rodríguez-Alcalá, L.M. & Fontecha, J. (2010). Major lipid classes separation of buttermilk, and cows, goats and ewes milk by high performance liquid chromatography with an evaporative light scattering detector focused on the phospholipid fraction. *Journal Chromatography A,* Vol.1217, No.18, (April 2010), pp. 3063-3066.

Rombaut R.; Camp, J. V. & Dewettinck, K. (2005) Analysis of Phospho- and Sphingolipids in Dairy Products by a New HPLC Method. *J. Dairy Sci.* Vol. 88 pp. 482–488.

Rombaut, R.; Dewettinck, K. & Van Camp, J. (2007). Phospho and sphingolipid content of selected dairy products as determined by HPLC coupled to an evaporative light scattering detector (HPLC–ELSD). *Journal of Food Composition and Analysis,* Vol.20, No.3-4, (May 2007), pp. 308-312.

Sas, B.; Peys, E. & Helsen, M. (1999). Efficient method for (lyso)phospholipid class separation by high-performance liquid chromatography using an evaporative light-scattering detector. *Journal of Chromatography A,* Vol.864, No.1, pp. 179-182, ISSN 0021-9673

Schiller, J.; Süß, R.; Arnhold, J.; Fuchs, B.; Leßig J.; Müller M.; Petković, M.; Spalteholz, H.; Zschörnig, O. & Arnold, K. (2004). Matrix-assisted laser desorption and ionization time of flight (MALDI-TOF) mass spectrometry in lipid and phospholipid research. *Progress in Lipid Research,* Vol.43, No.5, (September 2004), pp. 449–488.

Schwalbe Herrmann, M.; Willmann, J. & Leibfritz, D. (2010). Separation of phospholipid classes by hydrophilic interaction chromatography detected by electrospray ionization mass spectrometry. *Journal of Chromatography A,* Vol.1217, No.32, (August 2010), pp. 5179–5183.

Seppänen Laakso, T.; Laakso, I.; Vanhanen, H.; Kiviranta, K.; Lehtimäki, T. & Hiltunen, R. (2001). Major human plasma lipid classes determined by quantitative high-performance liquid chromatography, their variation and associations with phospholipid fatty acids. *Journal of Chromatography B,* Vol.754, No.2, (April 2001), pp. 437–445.

Seri, H.I.; Naim, H.Y. & Hassan, T. (2010). A high performance liquid chromatography (HPLC) method with evaporative light scattering detector for quantification of major phospholipids classes of donkey serum. *Veterinarski Arhiv,* Vol.80, No.3, pp. 365-373.

Shirouchi B.; Nagao, K.; Inoue N.; Ohkubo, T.; Hibino, H. & Yanagita T. (2007) Effect of dietary omega 3 phosphatidylcholine on obesity-related disorders in obese Otsuka Long-Evans Tokushima fatty rats. *J Agric Food Chem* Vol. 55 pp.7170-7176.

Silversand, C. & Haux, C. (1997). Improved high-performance liquid chromatographic method for the separation and quantification of lipid classes: application to fish lipids. *Journal of Chromatography B,* Vol. 703, No1-2, (December 1997), pp. 7-14.

Simons K. & Ikonen E. (1997). Functional rafts in cell membranes. *Nature* 387 pp. 569–72.

Smith, R.D. (2006). Future directions for electrospray ionization for biological analysis using mass spectrometry. *Biotechniques,* Vol.41, pp. 147–148.

Sohlenkamp, C.; Lopez-Lara, I.M. & Geiger, O. (2003) Biosynthesis of phosphatidylcholine in bacteria *Progress in Lipid Research* Vol. 42, pp. 115-162.

Standal, I.B.; Axelson, D.E. & Aursand, M. (2010) ^{13}C NMR as a tool for authentication of different gadoid fish species with emphasis on phospholipid profiles. *Food Chemistry* Vol. 121, pp.608-615.

Stith, B.J.; Hall, J.; Ayres, P.; Waggoner, L.; Moore, J.D. & Shaw, W.A. (2000). Quantification of major classes of Xenopus phospholipids by high performance liquid chromatography with evaporative light scattering detection. *Journal of Lipid Research,* Vol.41,(September 2000), pp. 1448–1454.

Sud, M.; Fahy, E.; Cotter, D.; Brown, A.; Dennis, E.A.; Glass, C.K.; Merrill, A.H. Jr.; Murphy, R.C.; Raetz, C.R.; Russell, D.W.; & Subramaniam, S. (2007) LMSD: LIPID MAPS structure database. *Nucleic Acids Research* Vol. 35 pp. D527-32.

Vance, J.E.; Steenbergen, R. (2005). Metabolism and functions of phosphatidylserine. *Progress in Lipid Research* Vol. 44 pp. 207–234.

Vernooij, E.A.; Kettenes van den Bosch, J.; Underberg, W.J. & Crommelin, J.A. (2002). Chemical hydrolysis of DOTAP and DOPE in a liposomal environment. *Journal of Controlled Release,* Vol.79, No.1-3, (February 2002), pp. 299-303.

Villé, H.; Maes G.; De Schrijver, R.; Spincemaille, G.; Rombouts G. & Geers, R. Determination of Phospholipid Content of Intramuscular Fat by Fourier Transform Infrared Spectroscopy. *Meat Science,* Vol. 41, No. 3, pp. 283-291.

Viola, A.; Saywell, V.; Villard, L.; Cozzone, P.J. & Lutz, N.W. (2007) Metabolic Fingerprints of Altered Brain Growth, Osmoregulation and Neurotransmission in a Rett Syndrome Model, *PLoS ONE,* Vol.1, pp. 1-9.

Wang, D.; Xu, W.; Zhou, G.; Zhu, Y.; Li, C. & Yang, M. (2009). Determination of intramuscular phospholipid classes and molecular species in Gaoyou duck. *Food Chemistry,* Vol.112, No.1, (January 2009), pp. 150-155.

Wang, F.; Zhao, Y. & Wang, P. (2003). Separation and Determination of Phospholipids in Biological Samples by High-Performance Liquid Chromatography. *Journal of Chromatographic Science,* Vol.41, No.3, (March 2003), pp. 142–144.

Wang, X.; Pattada Devaiah, S.; Zhang, W. & Welti R. (2006) Signaling functions of phosphatidic acid. *Progress in Lipid Research* Vol. 45, pp. 250-278.

Weihrauch, J.L.; Son, Y.S. (1983) The Phospholipid Content of Foods, *JAOCS,* Vol. 60 pp. 1971-1978.

Winther, B.; Hoem, N.; Berge, K. & Reubsaet, L. (2010). Elucidation of phosphatidylcholine composition in krill oil extracted from *Euphausia superba. Lipids* Vol. 46, pp. 25-36

Xia, Y.Q. & Jemal M. (2009). Phospholipids in liquid chromatography/mass spectrometry bioanalysis: comparison of three tandem mass spectrometric techniques for monitoring plasma phospholipids, the effect of mobile phase composition on phospholipids elution and the association of phospholipids with matrix effects. *Rapid Communication Mass Spectrometry,* Vol.23, pp. 2125–2138.

Xiao, J.; Engel, J.L.; Zhanga, J.; Chen, M.J.; Manning, G. & Dixon, J.E. (2011) Structural and functional analysis of PTPMT1, a phosphatase required for cardiolipin synthesis, *PNAS* Vol.29 pp. 11860-11865.

Yan, K.P.; Zhu, H.L.; Dan, N. & Chen, C. (2010). An Improved Method for the Separation and Quantification of Major Phospholipid Classes by LC-ELSD. *Chromatographia,* Vol.72, No9-10. pp 815-819.

Yandrasitz, J.R.; Berry, G. & Segal, S. (1981). High-performance liquid chromatography of phospholipids with UV detection: optimization of separations on silica. *Journal of Chromatography B*, Vol.225, No.2, (October 1981), pp. 319–328.

Yoon, T.H. & Kim, I.H. (2002). Phosphatidylcholine isolation from egg yolk phospholipids by high-performance liquid chromatography. *Journal of Chromatography A*, Vol.949, No.1-2, (March 2002), pp. 209-216.

Zacarias, A.; Bolanowski, D. & Bhatnagar, A. (2002). Comparative measurements of multicomponent phospholipid mixtures by electrospray mass spectroscopy: relating ion intensity to concentration. *Analytical Biochemistry,* Vol.308, No.1, (September 2002), pp. 152-159.

Zemski Berry, K.A., & Murphy, R.C. (2004). Electrospray Ionization Tandem Mass Spectrometry of Glycerophosphoethanolamine Plasmalogen Phospholipids. *Journal of the American Society of Mass Spectrometry,* Vol.15, No.10, pp. 1499-1508.

Zhao, Y.Y.; Xiong, Y. & Curtis, J.M. (2011). Measurement of phospholipids by hydrophilic interaction liquid chromatography coupled to tandem mass spectrometry: The determination of choline containing compounds in foods. *Journal of Chromatography A*, Vol.1218, No.32, (August 2011), pp. 5470– 5479.

28

Clinical Applications

Joseph Herman[1] and Bori Shushan[2]
[1]*Thermo Fisher Scientific's Application Laboratory at West Chester University of Pennsylvania, West Chester, PA;*
[2]*Clinical Mass Spec Consultants, Toronto, ON,*
[1]*USA*
[2]*Canada*

1. Introduction

The first significant use of mass spectrometry in clinical diagnosis was the determination of inborn errors of metabolism using gas chromatography coupled to mass spectrometry or GC-MS. Biologically significant metabolites such as organic acids, amino acids and others were usually extracted from biological fluids such as urine, and then derivatized to make them volatile, volatility being a prerequisite to separation by gas chromatography. Such elaborate extraction and derivatization schemes have made the use of GC-MS difficult for biologically significant molecules because of their thermal instability. Separations of thermolabile biological molecules by HPLC is a far more straight forward process than by GC and the coupling of HPLC with mass spectrometry, or LC-MS, was a major breakthrough in the application of MS to the measurement of biologically significant molecules such as those measured for clinical diagnosis. Since the late 1970's there has been a great deal of research into coupling HPLC with mass spectrometry. Many different LC-to-MS interfaces had been developed over the last 3 decades including the Moving Belt FAB, Thermospray, Particle Beam, and Flow-FAB, all of which no longer exist today. All these once quite promising and even popular techniques, have been replaced by those involving the use of atmospheric pressure ionization (API) which was first introduced commercially in the late 1980's. The most prominent API LC-MS interfaces are Electrospray (ESI), Nebulizer Assisted Electrospray (a.k.a. IonSpray), Atmospheric Pressure Chemical Ionization (APCI) using a Heated Nebulizer, and Atmospheric Pressure Photo Ionization (APPI) also using a Heated Nebulizer. API facilitated the rapid adoption of LC-MS and a good review of API is available by Thomson [1] and for clinical diagnostics using IonSpray by Henion [2].

As the name implies, in API, ions are created at atmospheric pressure quite apart from the ultra-clean high vacuum analyzer. This means no analyzer contamination and no need to pump away potentially corrosive solvents and buffers using the MS system's expensive vacuum pumps. This alone has made LC-MS coupling exceptionally rugged as well as keeping the operational aspects simple allowing those less skilled in instrumentation but more focused on applications to be successful. This led to a flourishing of bioanalytical applications such as the analysis of a wide variety of biomolecules including biopolymers like polysaccharides, DNA/RNA, proteins/peptides and a plethora of heretofore intractable small molecule analytes such as sphingolipids, phospholipids, acylcarnitines, amino acids, biogenic amines, nucleotides, saccharrides, polar and ionic pharmaceuticals, natural and exogenous metabolites, etc.

API techniques are defined as "soft" ionization methods producing primarily intact molecular or pseudo-molecular ions, i.e. ions are created without fragmentation. This presents a challenge in positively identifying compounds: one must subsequently employ tandem mass spectrometry (MS/MS) or high resolution/high mass accuracies or both to positively identify ionized species. Most clinical diagnostic applications today utilize MS/MS to identify and/or quantify trace analytes in complex biological matrices such as plasma and urine. Excellent review articles on this topic have recently been written by Dooley [3], Vogeser & Seger [4], and Shushan [5].

The most successful and widespread use of MS/MS in clinical diagnosis is in the area of newborn screening for congenital disorders such as amino acidopathies, fatty acid oxidation disorders and organic acidurias employing close to a thousand instruments worldwide [6-9]. Other popular applications in the clinical diagnostic field are: the therapeutic drug monitoring (TDM) of cocktail therapies such as anti-viral treatments [10-11] or immunosuppressents [12-14] or anti-cancer chemotherapies [15]; the analysis of endogenous steroid hormone panels [16]; the determination of peptide-based hormones especially where different isoforms are involved; and the screening and confirmation of drugs-of-abuse and toxicants [17-18]. With respect to steroid analysis there has been an especially rapid adoption of MS/MS since there are now well documented cases of the superiority of LC-MS/MS assays versus immunoassays [19-21]. The evidence of this superiority has led the American Endocrinology Society to issue a statement endorsing the use of LC-MS/MS for the measurement of low levels of endogenous steroids, such as testosterone in children and women, over traditional methods like immunoassays [22]. There is a great deal of well founded interest in the application of MS/MS to endocrinology and the reader is also referred to a recent review article on this subject [23].

In spite of the rapid advances made in the application of LC-MS/MS to clinical assays there are relatively few instruments employed in routine diagnostic labs compared to the traditional clinical analyzer systems which are based upon biochemical- and immuno-assays. The advantages of LC-MS/MS are many including: no costly analyte specific reagents (ASRs); the ability to determine many analytes in a single run with the same low cost of analysis whether one or many analytes are determined; high specificity and sensitivity especially for small-molecule analytes in comparison to immunoassays; and relatively rapid assay development amenable to "homebrew". There are however, significant disadvantages including: there are some classes of compounds, such as proteins, for which LC-MS/MS is not as sensitive as immunoassays; LC-MS/MS systems are complicated pieces of technology which require a great deal of training and skilled operators; the high capital cost of these instruments usually with no "reagent-rental" purchase options (no ASR's); and finally, there is often a significant amount of pre-analytical sample treatment required frequently requiring external robotic liquid-handling systems. The above disadvantages are responsible for the relatively small uptake of LC-MS/MS into routine clinical diagnostic laboratories especially the latter where technicians are more used to simply loading instruments with samples without the requirement for sample pre-treatment.

2. Sample preparation

Perhaps the most important facet of using mass spectrometers for clinical applications is the sample preparation procedure. Dealing with biological matrices presents many unique challenges to performing mass spectrometry and, in particular, using liquid chromatography in conjunction with MS. Consideration has to be made for matrix interferences that produce undesired signals in the channels being monitored as well as dealing with ion suppression effects. Matrix interferences can take the form of isobars that have the same molecular weight and similar (but not identical) structures (e.g. steroids with the same molecular weight), structural isomers of the same compound (i.e. 3-epi-25-hydroxy vitamin D and α-25-hydroxy vitamin D), fragmentation of metabolites back to the starting compound in the source (i.e. glucuronides fragmenting back to the hydroxyl precursor), source fragmentation of substrates to products (i.e. enzymatic profiling), or endogenous background materials that produce ions at the same masses as the compounds of interest. Ion suppression usually is from compounds that behave like a detergent or a surfactant such as endogenous fatty acids or formulation components like PEG or Tween. However, the suppression caused by the large amount of endogenous proteins or phospholipids found in the biological matrix are of primary concern. When analyzing small molecules (>1000 Da) by mass spectrometry the goal of the sample preparation is to remove as much of the proteins and lipids as possible. For large molecule analysis, such as proteins, isolating and purifying the peptide or protein usually must be done during sample preparation, and if not, lengthy chromatography must be used to separate the large number of compounds present in the biological matrix.

2.1 Off-line methods

Sample dilution to reduce the concentration of salts and endogenous materials is the simplest, but least efficient, method of sample clean-up. However, urine and saliva are relatively clean (compared to other biological matrices) and often the concentration of the analyte of interest is high enough (μg/mL) that by simply diluting the sample by a factor of 10 the matrix interferences are minimized enough that no further work-up is necessary.

Protein precipitation is also very straight forward and easy to perform. In this case, an organic solvent or pH change is used that causes the proteins to denature and become insolvent. The proteins precipitate out of solution and can either be filtered out of the "crashed" solutions or are centrifuged and the supernatant is removed for analysis. If an organic protein denaturant is chosen, it must be miscible in water. Protein precipitation is used on all biological matrices including whole blood, plasma and tissue homogenates The key to whether a protein precipitation alone will work is the solubility of the compounds in the organic crashing solution at the pH chosen. Extremely hydrophilic compounds may precipitate due to a lack of solubility. The most common solvents are methanol or acetonitrile. Chilling the solvent will produce a more thorough clean-up. Typically at least a five to one ratio or organic solvent to matrix is required to "crash" the endogenous proteins, but 8 or 10 to 1 is preferred. Concentration of the sample can be done if the supernatant is dried and reconstituted in a smaller volume than the original sample. Another advantage of protein precipitation is that protein binding is destroyed when the proteins denature, so as long as the total faction of the compound of interest is being measured, one does not needed to worry about binding issues. If the free fraction is of interest, a different sample preparation method should be performed. The protein precipitation also destroys the enzymes present in the matrix, so for compounds that are susceptible to degradation due to enzymatic activity in the sample, protein precipitation will help with the overall stability of system.

Often the free or un-bound fraction of an analyte is desired. In this case one wants the compound bound to proteins to remain there (e.g. the analysis of free T4). Equilibrium dialysis or ultrafiltration is normally employed to measure the free fraction of an analyte. However, these methods can also be used to clean-up a sample since only molecules below a certain size can transport through the membranes. The majority of the biological matrix is usually large proteins that can not cross the membrane barriers and therefore the sample is cleaned for MS analysis.

Liquid-liquid extraction (LLE) is based on partitioning of an analyte between two liquids that are immiscible. Typically, an organic solvent that is not miscible in water is chosen where the compound of interest has a high solubility. In this way, the compounds of interest are extracted from the aqueous matrix because they prefer to be in the organic layer of the mixture. The organic layer is then separated from the aqueous layer and can either be directly injected or dried leaving the compound of interest behind. Samples that are dried can then be derivatized or reconstituted as needed for analysis. The samples can also be concentrated if the final reconstituted volume is less then the starting sample volume. LLE is also very good at removing salts since they prefer to stay in the aqueous phase. The disadvantage of LLE is that it is labor intensive and has many steps that can introduce experimental error. The process has been made more palatable with the advent of automated liquid handling systems, but there is a lot of hazardous waste material generated and there is a significant cost associated with all the disposable materials used in the process.

Solid-phase extraction (SPE) works in a similar fashion to LLE but the partitioning is between a solid and a liquid phase. SPE applies the same basic principles used in chromatography. Analytes of interest are absorbed to the solid phase during the clean-up step, which is usually under aqueous conditions. Under the right conditions, most of the matrix components will not be absorbed during the cleaning step and the analyte of interest is thus removed from the matrix. However, compounds with similar chemical properties to the analyte of interest are retained as well. Samples are then released or eluted from the solid phase with an organic solvent. The combination of the right pH and organic content can make the eluted solvent clean of all but a few relatively similar components, of which the analyte of interest is one. The process is completely automatable and has significantly less waste materials and cost then LLE. Once again, the samples can be directly injected or dried. Samples that are dried can then be derivatized or reconstituted as needed for analysis. The samples can also be concentrated if the final reconstituted volume is less then the starting sample volume and the process removes the salts from the sample.

2.2 On-line methods

Solid-phase extraction methods have also been developed on-line. In the majority of cases these methods employ two columns run in tandem, though single column methods exist. For two column approaches, the first column is an SPE column and the second is a normal HPLC column. Matrices are injected onto the HPLC system and samples are cleaned-up on-line because the analyte is retained on the SPE column while the unretained material is washed to waste. Once the analyte of interest has been extracted from the sample matrix, the analyte is eluted to a second analytical column for analysis. The process is done with either disposal cartridges that are changed after each sample injection, or with columns that are cleaned between each injection by the mobile phases. In both cases the on-line methods have the advantage of direct injection and elution of the analyte, which removes the time consuming off-line steps of evaporation, reconstitution, and preparation. Therefore, on-line methods are more efficient, fully automated and require far less consumables then their off-line counterparts. An example of a two column on-line SPE configuration is shown if Figure 1. Isocratic focusing is often used to gain better peak shape with dual column methods. In this case the percent organic used to elute sample from the clean-up column is teed into a aqueous LC flow from a second pump to reduce the amount of organic the analytical column sees during the transfer step. The samples are focused at the head of the analytical column by the isocratic aqueous make up flow from the second pump making hydrophilic compounds easier to capture. Figure 2 illustrates the configuration used for "isocratic focusing".

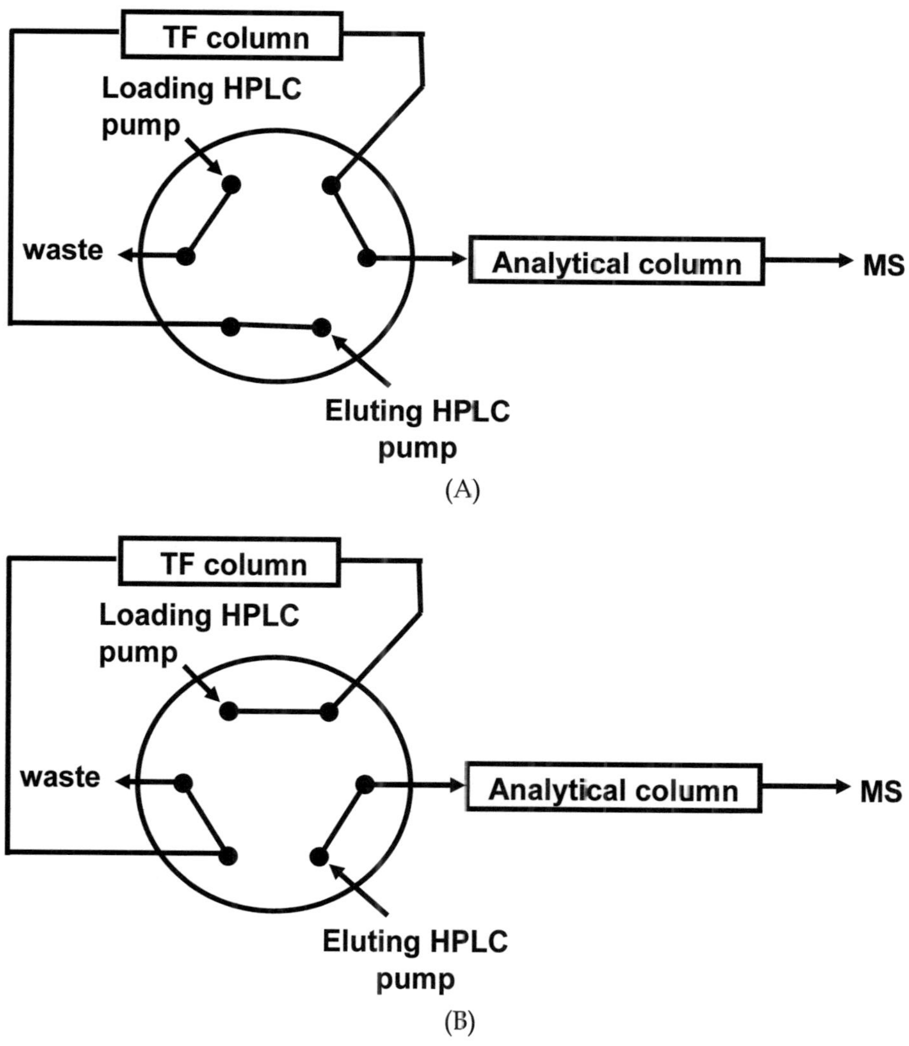

Fig. 1. Dual Column Method Configuration (A) Sample load and Clean-up (B) Sample Elute [25].

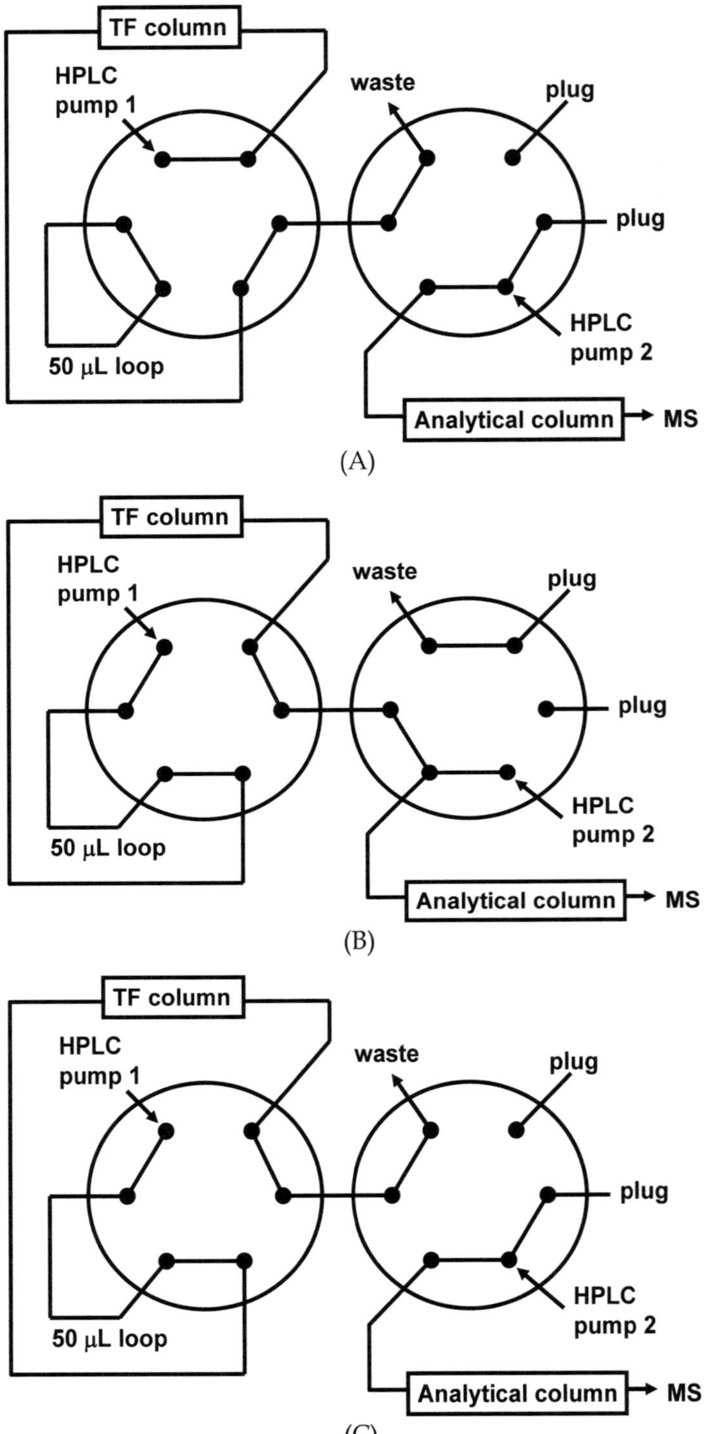

Fig. 2. Isocratic Focusing Method Configuration (A) Sample Clean-up (B) Sample Transfer (C) Sample Elute [25].

Turbulent flow chromatography (TFC) is a second on-line methodology that is performed in a similar manner to SPE. The difference is in how the samples are separated from the matrix. Unlike traditional laminar flow HPLC systems, where the interaction between the stationary phase and the analytes is diffusion controlled, mass transfer is the primary mechanism of separation for turbulent flow. Since the diffusivity of a molecule is inversely proportional to its molecular weight, small molecules have high diffusivity and ready transport into the pores of a packed column under turbulent flow conditions. Large molecules, such as proteins, have low diffusivity and do not have time to enter the pores under turbulent flow conditions. The result is that large molecules are swept away by the turbulent flow while small molecules bind to the stationary phase inside the pores of the packed columns. Traditional it was necessary to use high flow rates to achieve turbulent flow (4-5 mL/min), but 0.5 mm ID TurboFlow® columns are available that reach turbulent velocities at 1-1.5 mL/min. The molecular weight of the molecules excluded from the stationary phase can be adjusted by flow velocity. Figure 3 illustrates the effect of flow rate on the molecular weight exclusion using TFC [24]. Since the primary matrix interferences are due to matrix proteins, separation of proteins in the biological matrix from the analyte of interest is the primary goal of online sample clean-up methods. In fact, when reading the literature on SPE it is often found that much better results are obtained at high flow rates (2-4 mL/min). This author believes that the higher flow rates achieve turbulent flow in the SPE column, even though they were not designed to do so, resulting in a far better mechanism for clean-up then the SPE partitioning alone. A comparison of the effects of ion suppression from protein precipitation, SPE and TFC is shown if Figure 4 [25].

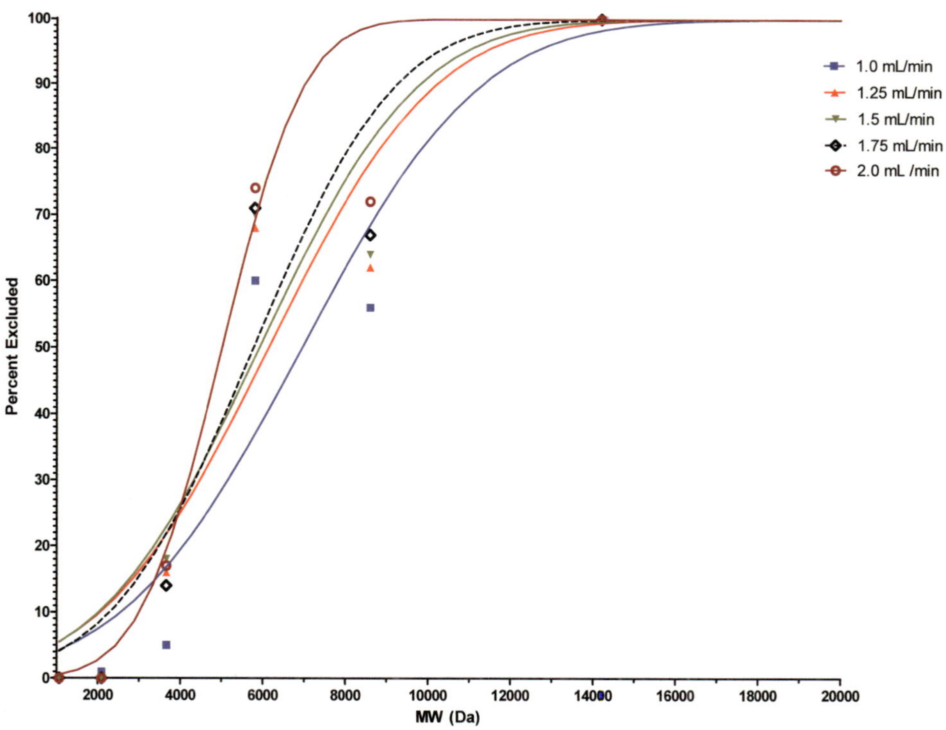

Fig. 3. Comparison of Molecular Weight Exclusion of Proteins by Turbulent Flow Chromatography as a Function of HPLC Flow Rate [24].

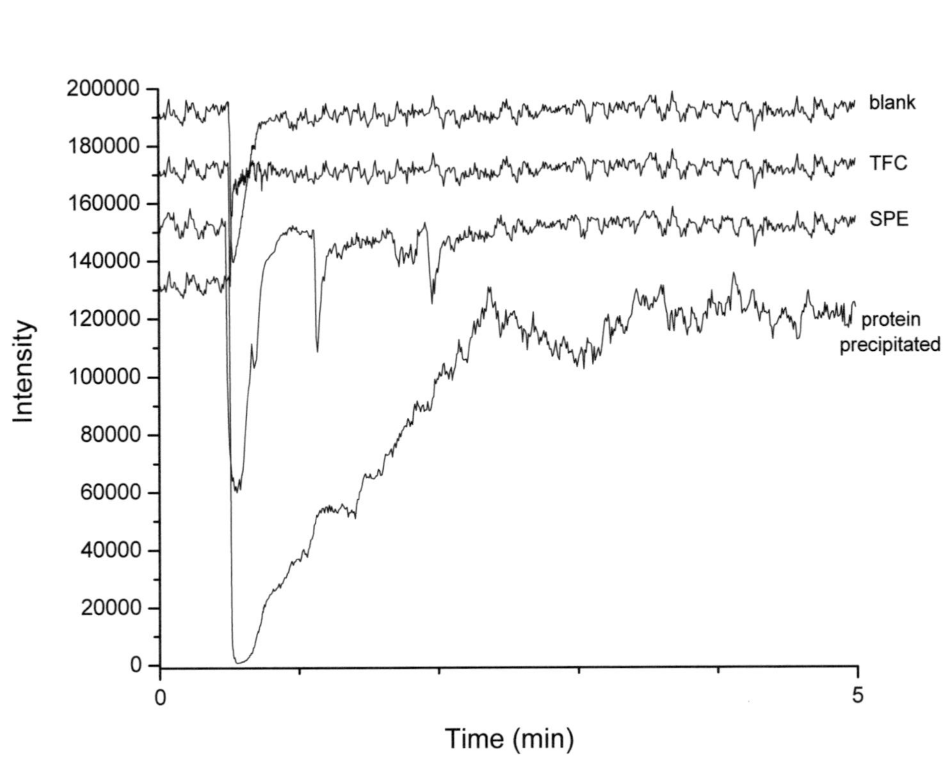

Fig. 4. Ion Suppression Effects from, Blank Injection (50/50 methanol:water), Injection of Protein Precipitated Rat Plasma (4:1 methanol: plasma, centrifuged), Injection of Extracted Rat Plasma in 50/50 methanol water, and TFC of Neat Rat Plasma [25].

Another advantage of the two column methods is that they allow the use of multiplexing, which takes advantage of the time spent cleaning one sample to elute another. Two or more injections are staggered in time and can either have a selection valve that can divert multiple HPLC systems to a single MS, or has two clean-up columns where one is eluting to waste (clean-up) while the other is eluting either directly to the MS or to the MS through an analytical column (analysis). Figure 5 illustrates how multiple injections can be staggered in time to increase sample throughput. Samples are eluted to the mass spectrometer only during the part of the run when the analyte of interest in being eluted. During the HPLC method, when the samples are being cleaned, columns are being washed, or the systems are being equilibrated, flows from other streams are directed to the MS taking advantage of the time needed to perform these functions while not wasting any time on the mass spectrometer acquiring data when there is no analyte of interest eluting.

Restricted access media (RAM) is a third on-line column approach to sample clean-up that can handle neat biological matrices. The particles packed into a RAM column are designed to restrict the access of large macromolecules to the adsorption sites of the stationary phase by coating the normal HPLC packing materials with a second bonded phase that allows small molecules through but repels or excludes large molecules. The cleaning of the sample from the matrix is similar to the mechanism of TFC but accomplished in a different way under laminar flow velocities. However, since the barrier is physical rather then kinetic, RAM columns tend not to last as long as other on-line columns because the restricting layer

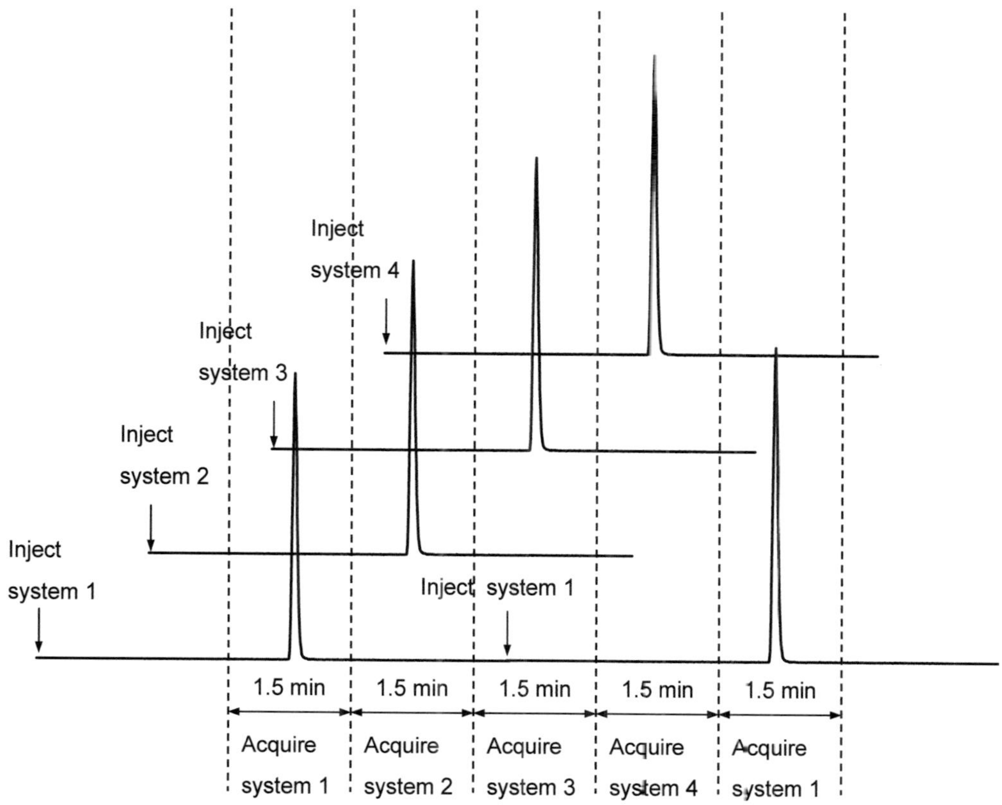

Fig. 5. Staggered multiple inlet methodology (Multiplexing) of several HPLC systems to a single mass spectrometer.

gets clogged over time. A comparison of TFC, SPE and RAM as a function of molecular weight exclusion of proteins is shown if Figure 6 [24].

Immunoaffinity extraction (IAE) uses antibody-antigen interactions to capture analytes with very high selectivity. Antibodies are immobilized onto the stationary phase of an LC column to effect the removal of the sample matrix while leaving the antigen behind. The antigen is then released from the column for analysis. Often these antibodies will interact with a class of compounds rather then only a specific analyte. However, the cross reactivity of similar compounds are much easier to analyze then the original matrix. One of the primary drawbacks to clinical analyzers that do not employ LC/MS/MS is that the detectors can not separate the interferences from the cross-reactivity. The added dimension of separating analytes by their molecular weight allows MS to distinguish compounds within the same class from one another greatly improving the accuracy of the measurement. IAE can also be performed off-line similar to the way SPE is done both off-line and on-line. A problem with IAE columns is that they tend not to have long life times and are quite expensive.

2.3 Derivatization

Often it is desirable to derivatize an analyte in order to improve the sensitivity or assay performance. In GC/MS or LC/MS, derivatization is used primarily to improve the

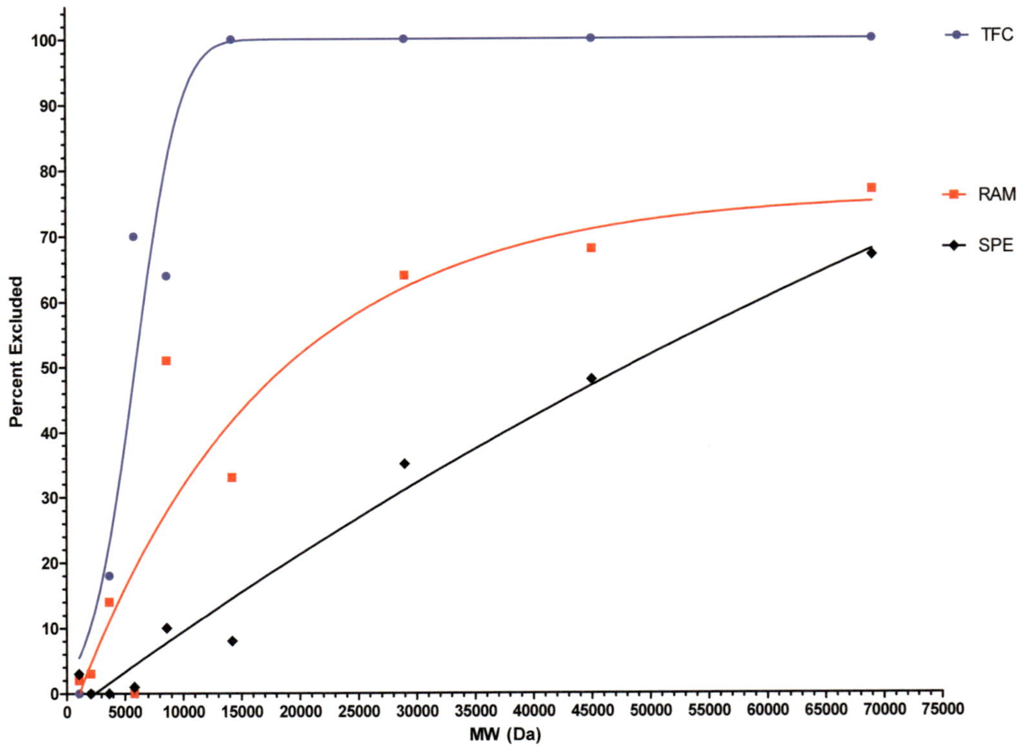

Fig. 6. Comparison of the Molecular Weight Exclusion of Proteins at 1.5 mL/min Flow Rates for Turbulent Flow Chromatography (TFC), Restrcted Access Media (RAM), and Solid Phase Extraction (SPE) Clean-up Columns.

chromatographic separation or increase the ionizability of an analyte. Organic acids tend to give board diffuse peaks by HPLC due to ionic interactions with the stationary phase. Low molecular weight compounds that are very hydrophilic, like amino acids, tend not to be retained on most reverse phased HPLC systems. Converting the acids to an ester not only removes the ionic interactions of the acids but also makes the compounds more hydrophobic resulting in stronger retention. The most common derivatization for both organic acids and amino acids is to form the butyl-ester by reaction with acidic butanol [26-28]. The enhanced chromatography results in better peak shape and better sensitivity. The sensitivity increases because the peaks are narrower, and the analytes elute with more organic in the mobile phase, which promotes desolvation in the source. In other cases, the chromatography is good but the compound just doesn't ionize well. Adding a functional group that promotes ionization can dramatically increase sensitivity. Vitamin D is often derivatized with 4-phenyl-1,2,4-triazoline-3,5-dione (PTAD) to increase sensitivity [29-30]. Drugs of abuse screening is usually preformed by hydrolyzing the samples. The hydrolysis converts all phase two metabolites back to their precursors increasing the likelihood of detecting small concentrations by summing the signals from several sources of the drug into one signal.

The majority of the derivatization preformed in a clinical laboratory is done off-line, is labor intensive, and is time consuming. However, automated liquid handling system can be used to automate the derivatization. If fast kinetics exists, the derivatiztion can be performed on-line. One issue with derivatization is determining how complete the reaction is and what affect that has on what is being measured.

3. Clinical diagnostis

3.1 Vitamin D

Vitamin D deficiency has been linked to several skeletal disease conditions [31-32] prompting an increase in laboratory testing of serum vitamin D levels over the last several years. The Mayo Clinic has reported increases in vitamin D testing of over 80% per year [33]. In fact, Vitamin D analysis is easily the most frequently used LC/MS/MS assay in the clinical laboratory today.

In order to understand vitamin D analysis, it is important to distinguish the various analogs of vitamin D that are analyzed and their relationship and functions, since different laboratories perform the analysis by different methodologies. Figure 7 illustrates the nomenclature for the physiologically relevant vitamin D analogs. Both vitamin D_3 and vitamin D_2 are active, and while both can be absorbed from diet only vitamin D_3 can be made in vitro from exposure to UV light. The active form of vitamin D is the 1,25-hydroxyvitamin D metabolite, but historically it has been difficult to measure. Therefore, the primary measurement in the clinically laboratory has been the 25-hydroxyvitamin D metabolites because they have higher circulating serum levels and much longer half lives.

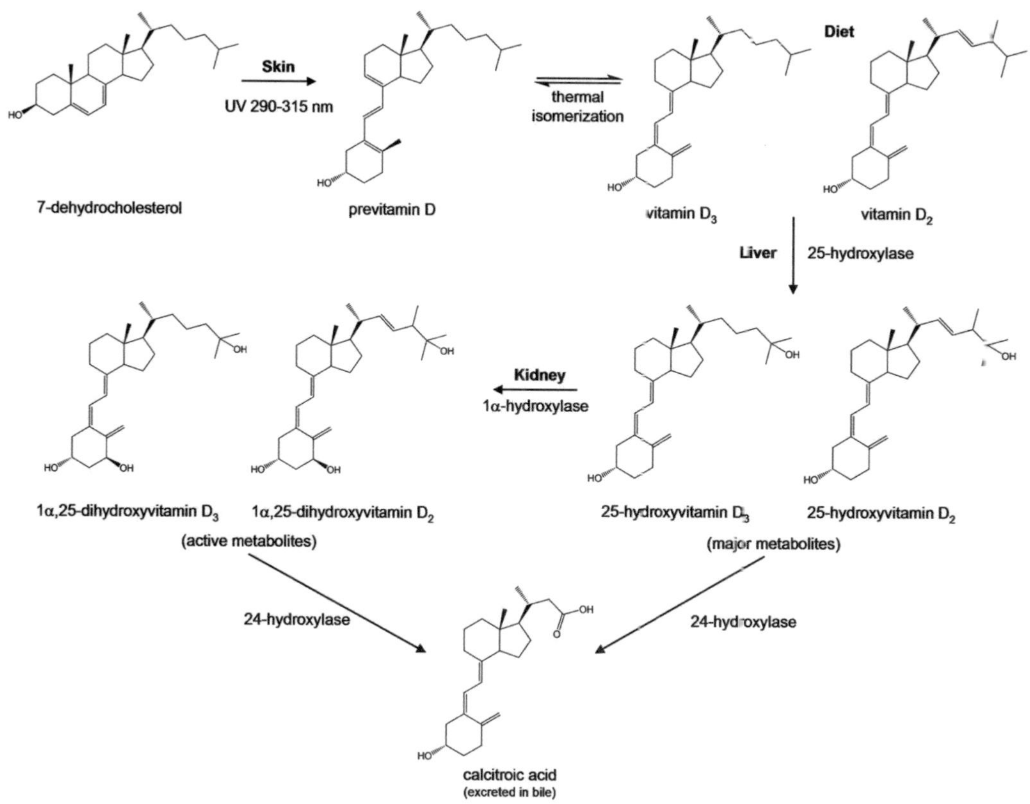

Fig. 7. Vitamin D nomenclature.

Vitamin D_3 (cholecalciferol) is endogenously produced from 7-dehydrocholesterol in the skin from exposure to UV radiation at 290-315 nm. A pre-vitamin D precursor is formed that rapidly isomerizes to Vitamin D_3. Significant variability in the amount of vitamin D3 produced in this manner is observed because of differences in sun exposure due to climate and/or social behavior [34-35]. Both vitamin D_2 (ergocalciferol) and vitamin D_3 are present in food such that diet becomes a significant factor in disease on-set and prevention. Dietary supplements for both vitamin D_2 and D_3 are available as well.

To become physiologically active, both vitamin D_2 and D_3 must be metabolized. While there are several inactive metabolites, it is the 1α,25-dihydroxyvitamin D that is active with the D_3 analog having about twice the activity of the D_2 analog. To form 1α,25-dihydroxyvitamin D, vitamin D is first metabolized to 25-hydroxyvitamin D by the liver. The kidney then metabolizes the 25-hydroxyvitamin D to 1α,25-dihydroxyvitamin D. Regardless of which form of Vitamin D is present, most of the circulating Vitamin D is bound to vitamin D binding protein (DBP).

Clinically, the total amount of 25-hydroxyvitamin D_2 and D_3 is usually reported. There are several reasons both analytically and physiologically for evaluating the 25-hydroxy metabolites. First is that both vitamin D_2 and D_3 have short half-life times ($t_{1/2}$) in circulating blood of 24 hours so their concentrations are greatly influenced by recent sun exposure and diet. Therefore, someone who is vitamin D deficient that just happens to go out in the sun for few hours right before their test would appear to be okay. The 25-hydroxyvitamin Ds have $t_{1/2}$ of 3 weeks and; therefore, are much more representative of the individuals overall vitamin D levels. Several authors have made the case that only the 1,25-hydroxyvitamin D levels are important since that is the active moiety. There is laboratory testing available for 1,25-hydroxyvitamin D but the analysis is more difficult, requires larger sample volumes, the $t_{1/2}$ is only a few hours (though exposure to sun light and diet are no longer regulating factors), and the complexity of the sample preparation takes more time and resources making the assay less cost effective. However, for certain indications the analysis 1,25-hydroxyvitamin D is clearly more useful, such as patients with chronic renal failure, vitamin D dependent rickets, 1,25-hydroxyvitamin D intoxication, lymphoma, and hyper or hypo parathyroidism [33].

Historically 25-hydroxy vitamin D levels were measured using immunoassays. Immunoassays are unable to distinguish between 25-hydroxy vitamin D_2 and D_3 such that only the total 25-hydroxy vitamin level is reported. The LC/MS methodologies are able to measure the levels of 25-hydroxy vitamin D_2 and D_3 separately so the contribution of each to the total can be determined, which is useful in diagnosing the effectiveness of supplemental treatment of vitamin D_2 to patients that are vitamin D deficient. In addition, there are problems with protein binding in the immunoassays that produce results that are lower than the corresponding LC/MS assays [36-37]. There is also some controversy as to whether immunoassays give accurate results due to nonspecific binding of other vitamin D metabolites and or matrix effects [38-40]. Many experts consider LC/MS to be the most accurate method for vitamin D analysis [39, 41] but there is a lot of resistance to changing reference levels that were established by the previous techniques.

Early attempts to run 25(OH) vitamin D by mass spectrometry was challenging due to the fact that it is lipophilic, which makes it's difficult to ionize by API methodologies. Furthermore, fragmentation is also difficult with the loss of water being the only ion formed easily. Other fragment ions are formed at high collision energies; however, under these conditions, multiple fragment ions are formed such that no one particular ion is formed with any reasonable abundance. Early attempts to over come the low ionizability of 25(OH) vitamin Ds involved various types of derivatization. Vreeken et al [42] and others [43-45] used a Diels Alder reaction to enhance sensitivity in API of 25(OH) vitamin D. The method also increased the mass of the 25(OH) vitamin Ds resulting in lower background interferences. Hiagashi et al [43-44] used a Cookson-type reagent to produce a 15 fold increase in the sensitivity over the native form for measuring vitamin D in plasma. However, these derivatization methods are time consuming and cumbersome. Improvements in the sensitivity of mass spectrometers in recent years have made it unnecessary to derivatize vitamin D in order to achieve the desired sensitivity.

Once the mass spectrometers became sensitive enough to measure vitamin D at the concentration required for clinical analysis the main obstacle to LC/MS/MS becoming a routine method for clinic diagnostics was making it high throughput. Vogeser et al [46] reported a rapid LC/MS/MS method in serum employing on-line solid phase extraction (SPE) with a run time of 9 minutes. Chen et al [47] improved upon this method and reduced the total run time to 7 minutes. Knox et al [48] used protein precipitation followed by off-line SPE to run 160 samples per day. The off-line methods produce much faster LC/MS/MS run times but require more preparation in advance. However, these methods can be automated making them an attractive alternate. Hojskov et al [49] used protein precipitation followed by liquid/liquid extraction to get the total run times down to 4 minutes. However, the biggest improvement in sample throughput was realized by combining on-line sample clean-up with multiplexing capabilities. Taylor et al [50] demonstrated the use of on-line clean-up using turbulent flow chromatography with a Cyclone column and a Cohesive Technologies (now part of ThermoFisher Scientific) TX4 multiplexing system to get a sample throughput of 40 samples per hour. Turbulent Flow allows fast and efficient on-line clean-up of biological sample while multiplexing allows coupling of multiple LC systems to a single mass spectrometer, greatly reducing time between injections. Several other authors have reported increased sensitivity in addition to increased sample throughput using TX2 or TX4 on-line multiplexing systems [51]. Vitamin D analyses on these systems are routinely running injection to injection times of one minute (60 samples per hour) Table 1 summarizes the current LC/MS/MS vitamin D assays found in the literature.

3.2 Immunosuppressant drug monitoring

Therapeutic drug monitoring of immunosuppressant drugs (ISDs) is well established [56] The importance of monitoring these drugs is due to their narrow therapeutic window. Elevated dosing of the immunosuppressant drugs can cause significant toxicity while under dosing can result in transplant rejection. Because of this narrow therapeutic window, the immunosuppressant drugs are considered critical dose drugs, requiring individualized drug therapy by measuring the actual drug concentrations in each patient to maximize the therapeutic response and minimize adverse side effects.

Currently there are two main choices in the clinical laboratory for monitoring immunosuppressant drugs; immunoassay or chromatography. However, issues with non-specific binding of the antibody resulting in over estimation with immunoassays [57-60] as well as the long sample run times, complicated sample preparation procedures and lack of sufficient chromophores with HPLC-UV detection [61-63], have made LC/MS/MS methodologies the assay of choice when available. Most of the large-scale Contract Research Organizations (CROs), which operate as central laboratories for clinical diagnostics, analyze immunosuppressents by LC/MS/MS [57, 64].

Reference	Analyte	Matrix	PPT	Sample Prep Extraction	Derivatization	Run time	date
Higashi (43)	25(OH)D2 and 25(OH)D3	human plasma	ACN	LLE (AcOEt)	DMEQTAD	7 min	2001
Higashi (45)	25(OH)D2 and 25(OH)D3	human plasma	ACN	LLE (AcOEt)	NPTAD	7 min	2003
Vogeser (46)	25(OH)D3	human serum	ACN	on-line SPE		9 min	2004
Singh (52)	25(OH)D2 and 25(OH)D3	human serum	ACN	TFC		7 min	2006
Chen (47)	25(OH)D2 and 25(OH)D3	human serum	ACN	on-line SPE		7 min	2008
Knox (48)	25(OH)D2 and 25(OH)D3	human serum/plasma	MeOH	off-line SPE		5 min	2009
Bunch (51)	25(OH)D2 and 25(OH)D3	human serum	ACN	TFC		3 min	2009
Newman (53)	25(OH)D2 and 25(OH)D3	dried blood spots	MeOH	LLE (hexane)		10 min	2009
Eyles (54)	25(OH)D2 and 25(OH)D3	dried blood spots	ACN		PTAD	3 min	2009
Hojskov (49)	25(OH)D2 and 25(OH)D3	human serum	ACN	LLE (heptane)		4 min	2010
Casetta (55)	$1,25(OH)_2D3$	human serium	ACN	on-line SPE		18 min	2010

Table 1. Selected Summary of LC/MS/MS Vitamin D Methods.

The four primary immunosuppressant drugs analyzed today are Sirolimus, Tacrolimus, Everolimus and Cyclosporin A. Clinical Assay kits are available from several vendors for the analysis of these immunosuppressents. These Kits include all the controls, calibrators, sample preparation reagents, internal standards, columns and all the necessary instructions to perform the analysis. Table 2 summarizes the performance of the currently available assay kits for LC/MS/MS analysis of ISD's.

Drug	Linear Range
Cyclosprin A	10-2000
Tacrolimus	1-50
Everolimus	1-50
Sirolimus	1-50

Table 2. A Summary of the Current Analytical Ranges for Immunosuppressant Drugs CV/IVD Kits.

Both electrospray ionization (ESI) and atmospheric chemical ionization (ApCI) are used to analyze the immunosuppressents. Usually the precursor ion is not the protonated molecular ion $(M+H)^+$ but is formed from the ammonium adduct $(M+NH_4)^+$ produced by having ammonium formate or ammonium acetate in the HPLC mobile phases [65]. The use of negative ions has also been reported for cyclosporine A and tacrolimus [66]. ApCI has the advantage of producing fewer matrix effects but sensitivity is reduced due to more fragmentation in the source. However, most mass spectrometers available today have more then enough sensitivity to measure the ISD's using either ionization method. The chromatography of cyclosporine A normally requires elevated temperatures be used on the columns to produce good peak shape. Column temperatures are normally between 60-75°C.

Initial LC/MS/MS methods for the ISDs used off-line liquid-liquid extraction for clean-up or solid phase extraction (SPE) to isolate the drugs from the plasma [67-69]. Off-line methods are rather tedious, involve many manual steps that each can be a source of error, and take time. Automated liquid handling systems are available to reduce the tedium and operator errors but most laboratories have switched to on-line sample preparation processes. The on-line methods usually follow protein precipitation and are of two general types: two-dimensional chromatography using turbulent flow or on-line SPE columns for matrix removal followed by a reverse phase HPLC analytical column to get good chromatographic peak shape [69]; or direct injections that rely on the analytical column to effect suitable chromatographic separation [70-71]. Sample run times using two-dimensional chromatography with as little as 1 minute per sample run times are achievable when combined with multiplexing systems [69]. Whole blood analysis is the most common method used due to the high protein binding of the ISDs. Cell lyses and denaturing of the proteins to release the bound drug is accomplished by using an organic such as methanol or acetonitrile in combination with zinc sulfate [67-72].

The importance of choosing the right internal standards has been demonstrated in the literature [73]. The availability of stable labeled isotope internal standards is critical because the ISDs are known to be more susceptible to matrix effects [73-75] and co-elute with many of the commonly found phospholipids. A study on cyclosporine by Taylor et al [74] clearly demonstrated that the isotope labeled cyclosporine had better analytical performance then any other of the analogs of cyclosporine that were tested.

3.3 Steroids

Steroid analysis is used for diagnosing several endocrine disorders. Traditionally the measurement of steroids was accomplished with immunoassays and radioimmunoassay. However, as stated previously, recent studies have shown the use of immunoassays is problematic [76-79] suffering from a lack of specificity, limited dynamic range and matrix effects. In fact the use of immunoassays has resulted in poor clinical correlation of the test results and substantial disagreement between different manufacturers of the assay kits [22, 76-80].

The first successful use of mass spectrometry to steroid analysis in the clinical laboratory was achieved with GC/MS [80-82]. Significant improvement in steroid measurement was made because of the high resolution separation capabilities of capillary gas chromatography coupled to the high specificity of mass spectrometer. However, in order to get the required sensitivity for steroid analysis, low throughput, labor intensive derivatizations, and other sample preparation requirements resulted in GC/MS methodologies not becoming widely used as a routine technique for steroid measurement in the clinical laboratory.

The development of LC/MS/MS over the last 15 years has made it the technique of choice for analyzing steroids because of the high specificity of the mass spectrometer, the separation capabilities of the liquid chromatography, its wide dynamic range, and the availability of simple sample automated preparation procedures. Steroid analysis is commonly performed from serum, plasma, urine and saliva. The choice of sample preparation procedure is somewhat dependent on whether increasing the concentration of the steroid is necessary for detection. If that is required, off-line LLE or SPE is desirable so that the reconstitution volumes are lower than the amount of starting matrix used; alternatively derivatization can be employed to enhance detection. Both these methods require labor intensive methodologies and significant cost per sample that are not desirable in the clinical setting. All of methods outlined previously are used today including extraction from dried blood spots. Table 3 summarizes the current steroid assays found in the literature.

Aldosterone	113, 81, 119, 120
Androstenedione	81, 83, 96, 88, 117
Androsterone-sulfate	90
Corticosterone	114, 81, 83
Cortisol	108, 112, 114, 81, 116, 83, 88
Cortisone	108, 114, 116, 122
11-deoxycorticosterone	114, 116
11-deoxycortisol	110, 114, 81, 83
21-deoxycortisol	114, 83
DHEA	81, 96, 118
DHEA-sulfate	81, 90
Dihydrotestosterone	83
Epiandrosterone sulfate	90
Estradiol	101, 81, 82, 102, 121
Estriol	82
Estrone	101, 82, 102, 121
17-hydroxyprregnenolone	110, 115
17-hydroxyprogesterone	110, 114, 81, 83, 88, 117
Pregnenolone	110
Progesterone	81, 83
Testosterone	109, 111, 81, 82, 83, 96, 117, 118

Table 3. Steroid assay references.

The biggest problem with analyzing steroids by mass spectrometry is that there are isobaric interferences and considerable cross talk due to the similar fragmentation patterns of other endogenous steroids. Figure 8 illustrates the complexity of the various steroid pathways. By looking at the molecular weights, the loss of water (which easily occurs in the source), and the isotopic distributions, it becomes obvious that there is the potential for a large amount of cross talk between the steroids in the mass spectrometer. Table 4 lists the possible cross talk channels between the steroids. Therefore, the use of chromatography becomes critical to the unambiguous measurement of the various steroids.

Another issue with steroid analysis is protein binding. When the total amount of steroid present is desired simple techniques like protein precipitation easily release the bound steroid. However, if the free fraction measurement is needed, dialysis or ultra-filtration methods are usually used. It is possible to measure the free and total separately if the sensitivity of the method is sufficient to measure the lower levels of the free fraction. Great care must be taken not to release any of the bound protein if the unbound steroid is measured.

Often it is more desirable to measure steroid panels rather then individual steroids. The advantage of steroid panel profiling is not just to simplify the analytical methodology, but monitoring multiple steroid pathways often has clinical relevance [83-87]. Diagnostics for CAH

Steriod	Nominal Mass (Da)	isobars	loss of water interference	loss of two waters interference	loss of actetyl group interference	isotope interference	loss of water isotope interference	loss of two waters isotope interference
cholesterol	386							
cortisol	362					cortisone, aldosterone		
cortisone	360	aldosterone						
aldosterone	360	cortisone						
corticosterone	346	11-deoxycortisol						
11-deoxycortisol	346	corticosterone						
17-OH-pregnenolone	332					17-OH-progesterone, 11-deoxycorticosteone		
11-deoxycorticosterone	330	17-OH-progesterone		cortisol			corticosterone	cortisone, aldosterone
17-OH-progesterone	330	11-deoxycorticosterone		cortisol			corticosterone	cortisone, aldosterone
pregnenolone	316					progesterone	17-OH-pregnenolone	11-deoxycortisol, corticosterone
progesterone	314		17-OH-pregnenolone	11-deoxycortisol, corticosterone			17-OH-progesterone, 11-deoxycorticosteone	
dihydrotestosterone	290				17-OH-pregnenolone	testosterone, estriol, dehydroepiandrosterone		
testosterone	288	estriol, dehydroepiandrosterone			17-OH-progesterone	androstenedione		
estriol	288	testosterone, dehydroepiandrosterone			17-OH-progesterone	androstenedione		
dehydroepiandrosterone	288	testosterone, estriol			17-OH-progesterone	androstenedione		
androstenedione	286					androstanedione		pregnenolone
androstanedione	284			pregnenolone				
estradiol	272				progesterone	estrone	testosterone, estriol, dehydroepiandrosterone	
estrone	270		testosterone, estriol, dehydroepiandrosterone					

Table 4. Mass interferences between the steroids.

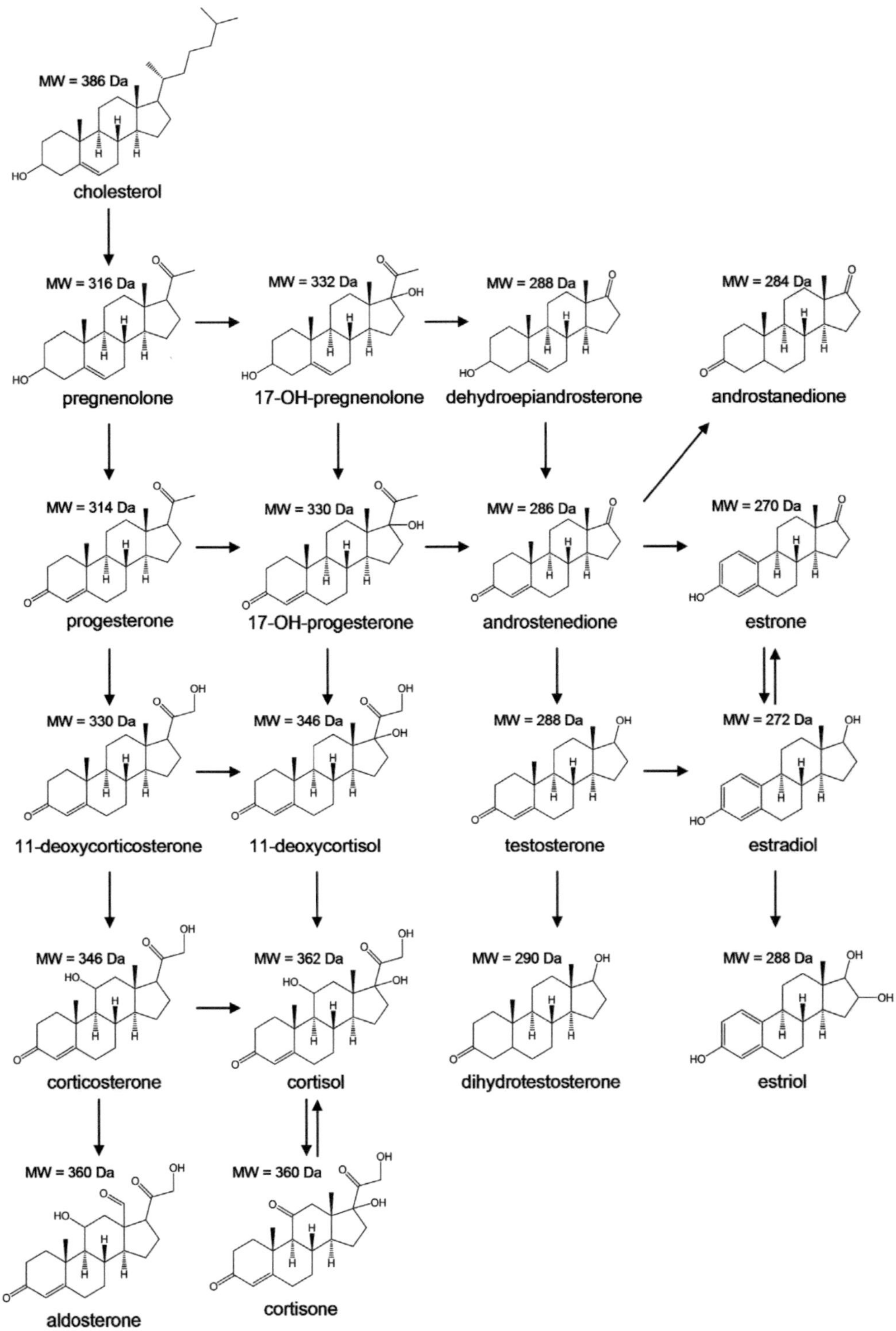

Fig. 8. Complexity of the steroid pathways.

and PCOS, fertility treatment, tumor location, and patient monitoring after gonadal or adrenal surgery are routinely performed by LC/MS/MS [88-90].

Measurement of adrenal steroids is important for the differential diagnosis of CAH and evaluation of adrenal function. The CAH panel includes the following steroids; 17OH-progesterone, androstenedione, cortisol, adrenocorticotropic hormone (ACTH) and pregnenolone. Congenital adrenal hyperplasia is characterized by elevated levels of ACTH, which is the most common disorder associated with the assay. Cross reactivity and poor clinical correlations are observed with immunoassays making agreement with LC/MS/MS methods difficult. Method specificity is also critical to the population being tested and LC/MS/MS has been shown to be much more reliable than other methods especially for newborns, infants and the elderly [83, 91-92]. LC/MS/MS methods for CAH have much lower false positive rates [83, 91-92] than the immunoassays and have been adapted to saliva [93-96] and dried blood spots [91, 97], which are far less invasive sampling techniques then drawing blood.

Measurement of testosterone in men is used to diagnose hypogonadism and to monitor its treatment as well as to monitor androgen suppression therapy during prostate cancer treatment. In women, testosterone levels are markers of alopecia, acne, hirsutism, osteoporosis, tumor screening, late on-set CAH, PCOS, and other endocrine and reproductive diseases [98]. In children, testosterone is analyzed for gender assignment in infants with ambiguous genitalia, delayed onset puberty, and CAH [88-90]. The analysis of testosterone is also complicated by the need to measure both the free and total testosterone levels.

Immunoassays for testosterone work well for healthy normal men but their lack of specificity makes them unreliable for the low concentrations found in women and children [76-78]. Methods for testosterone that utilize LC/MS/MS have demonstrated accurate measurement down to 10 pg/mL [84, 86, 89-100]. Diagnostic methods for others androgens by LC/MS/MS have been reported with similar results including dihydrotestosterone (DHT), androstenedione, and DHEA [84, 86, 89, 100-102]. Methods for the androgens often require derivatization or large sample volumes. However, recent improvements in mass spectrometer sensitivity have allowed analysis without these requirements.

Low concentration of estradiol in females is associated with disturbed puberty, oligoamenorrheam and menopause [88, 98]. Estadiol suffers form the same issues discussed previously for steroids in that there are several endogenous and matrix-related interferences. Problems with cross reactivity with immunoassays is a major problem once again, making LC/MS/MS the method of choice analytically. Like testosterone, estradiol does not ionize well and often derivatization is performed with dansyl chloride or an amine containing sulfonyl halide [103-106]. Two dimensional chromatography methods have been able to measure estradiol without the need for derivatization [85, 105].

Another important steroid that is analyzed by tandem mass spectroscopy is cortisol. Cortisol measurement is used to diagnose adrenal hyperfunction (Cushing"s syndrome) and adrenal insufficiency. Cushing's syndrome results from over expression of cortisol by the adrenal glad and can be caused by pituitary hyperplasia, cancer of the adrenal or pituitary gland, or production of ACTH outside the pituitary gland. Cortisol measurements are done in plasma, serum, salvia and urine depending on the diagnoses.

3.4 Thyroid hormones

Thyroxine (3,3′,5,5′-tetraiodo-L-thyronine or T4) and triiodothyroxine (3,3′,5-triiodo-L-thyronine or T3) are tyrosine-based hormones produced by the thyroid gland that are essential

for regulation of cell metabolism. Both excess and deficiency of thyroxine can cause clinical disorders. Hyperthyroidism is caused by the over production of T3, T4 or both. The most common example is Graves' disease in which both the T3 and T4 levels are elevated. Hypothyroidism is a result of T3 and T4 deficiency and is related to Hashimoto's disease. Both hyper and hypothyriodism can be caused by diet form either not enough or too much intake of iodine. Many thyroids cancers result in the over or under production of T3 and T4 as well.
T4 is produced in the thymus gland and is the primary circulating from of thyroid hormone. The T4 is converted to T3 by tissue deiodinases that remove iodine from the 5′ position of T4. T3 is the active thyroid hormone. Both T3 and T4 are extremely protein bound (>99%), which makes the measurement of the free amounts difficult. The free hormone is considered the clinically relevant concentration because it measures the amount available to the cells.
Free thyroid hormone measurements require separation of the free hormone from the protein bound hormone. It is critical that the separation does not disturb the endogenous equilibrium [107, 123]. Equilibrium dialysis (ED) is the preferred method [124-126] but ultrafiltration (UF) is also used [127-128]. ED is labor intensive, time consuming (overnight, 17-24 hrs), technically demanding, and expensive, which make it unattractive to all but the large, well equipped and well staffed clinical laboratories. UF is much less time consuming (30-40 min), easier to use and has better reproducibility, but is also prone to leakage through the UF-membrane [124].
The use of tandem mass spectrometry following ED or UF has overcome the issues associated with measuring thyroid hormones with immunoassays [129-139]. The first T3 and T4 assays to use tandem mass spectrometry were developed for GC/MS and isotope dilution [129-131, 133, 139-142]. These methods required extensive sample clean-up and derivatization. Development of LC/MS methods eliminated the need for derivatization and allowed the introduction of on-line sample clean-up, greatly increasing the ease of use and removing many sources of experimental error [125].

4. Toxicology

Analytical toxicology is the detection, identification and measurement of drugs, or other foreign compounds (xenobiotics), and their metabolites in biological specimens. Toxicological measurements in the clinical laboratory can be both quantitative and qualitative depending on what question needs to be answered. Therefore, unlike the previous discussions where only quantitative measurements were needed and triple quadrupole MS/MS is the preferred methodology, toxicology screening uses many more types of mass spectrometers on a routine basis. There are several areas of interest for toxicological screening. First is screening of a biological specimen to detect and identify compounds in patients admitted to the hospital with acute intoxication of unknown origin. Intoxication can result from using drugs of abuse, both known and unknown, from accidental expose to hazardous chemicals in the environment, from bacterial or viral infections, or from disease states that produce toxic compounds in vivo. Second is the screening for illegal drug use of known origins where intoxication has not occurred but the substances are regulated, like steroid use in athletes. In the US, there are also programs to monitor pain management for compliance of prescribed drug use. The later is mostly a legal issue to identify patients that take advantage of the medical system and to identify doctors that abuse their privileges and prescribe drugs in manners not consistent with medical doctrine. Third is therapeutic drug monitoring (TDM) and occupational/environmental toxicology. There is considerable overlap between all of the toxicological assays.

One issue with the use of SRM transitions is the presence of isobaric interferences. When only one SRM transition is used there may not be enough specificity in the MS to distinguish between isobars. Chromatographic separation of the isobars is often the easiest approach to resolving isobars. However sometimes it is not possible to separate isobars chromatographically or the time taken to do so is not optimal for laboratory throughput needs or the time to result required is too long for an emergency case. Interference during the analysis of tramadol by LC-MS/MS arising from ingestion of the antidepressant venlafaxine is an example of this problem [143]. In order to overcome isobaric interferences, the use of multiple transitions for a single compound and the ratio between the signal intensities for each transition is used [144-146]. The application of product ion ratios is obviously limited for compounds which do not fragment reproducibly. Interference may also arise from metabolites, or other compounds, which fragment or thermally degrade in-source back to the parent compound. In these cases, chromatographic separation is the only option, but since the metabolites tend to be more hydrophobic then the parent, it is not difficult to achieve. Therefore, to minimize the risk of interference, multiple transitions that avoid non specific transitions such as water loss or low molecular weight fragments should be used whenever possible. For compounds which do not fragment well, or have only one major fragment, interferences must be separated by the chromatography.

In forensic and post-mortem toxicology, systematic toxicological analysis (STA) and general unknown screening (GUS) is the starting point from which further, targeted quantitative analyses follows. Good sensitivity and reliability on as wide a range of compounds as possible is required. For many years GC/MS, despite the problems associated with larger, non-volatile and thermally labile compounds, was considered the best strategy for these analyses. The reproducibility of GC/MS ionization/fragmentation allowed for the development of comprehensive mass spectra libraries for reliable structural identification. There are no equivalent LC/MS spectral libraries due to poor instrument to instrument reproducibility of LC-MS fragmentation. Therefore, in house libraries are needed for LC/MS application of SRM methods [17, 147-157]. SRM methods also suffer from the fact that only known compounds can be searched and that a limited number of transitions can be monitored at one time. For these reasons, a move toward full scanning instruments for screening for both known and unknown compounds is more practical [158-164].

Information or data dependent acquisitions are used to find compounds above a background threshold when acquiring full scan data. Identification is accomplished by triggering MS/MS acquisitions on the precursor ions found above the threshold [1165-167]. Linear ion traps, orbital traps, Qtraps and QTOF are better suited to full scan acquisitions because they acquire full scan data much faster and with no loss of sensitivity compared to triple quadrupoles. However, once the identity of a compound is known, the triple quad is still the best quantitative instrument such that screening is often done by ion traps but conformation and quantification is usually done by triple quads.

An emerging approach to STA analysis is the use of accurate mass (exact mass, or high-resolution) MS. Full-scan MS experiments with 0.1 mDa accuracy are possible that filter the full-scan data and extract analyte chromatograms with very low background noise. Compounds which have the same nominal mass, but different exact masses, can be resolved by the mass spectrometer [168-169]. TOF-HRMS is of interest in the application of empirical formula-based data libraries, with isotope pattern-matching software, and the potential to screen for unknown compounds (and identify their metabolites) by knowledge of elemental composition alone, without the absolute need for reference material [170]. Exact mass

identification of specific metabolites and systematic fragmentation approaches have shown that even structural isomers can be distinguished using accurate mass [169]. Further, retrospective interrogation of full-scan data can be useful to investigate the presence of new compounds/metabolites (such as new 'designer drugs') without re-analysis. Newer Orbitrap®/Exactive™ technology (ThermoFisher Scientific), also capable of HRMS, is finding toxicologically relevant applications [171-173]. "In house" library matching should be carried out for unequivocal compound identification because mass spectral libraries and compound databases may not be completely transferable between instruments. Furthermore, library data matching does not give any information about chromatographic retention time, which is just as important as the mass spectral information [156, 174-177].
Examples of current assays preformed using tandem mass spectrometry for toxicology are listed in Table 5.

4.1 New born screening

Newborn screening programs are designed to identify disease states in infants due to inborn errors in metabolism or genetic defects before they become symptomatic. One problem as the ability to detect more and more disorders in newborns increases is that often there is no treatment for the disease. Researchers need to be cognizant of the implications of knowing a patient has a disease and not be able provide treatment.
Newborn Screening (NBS) for metabolic disorders started in the 1960's with a test for phenylketonuria using bacterial inhibition from dried blood spots [26]. This method was used exclusively for phenylketonuria (PKU) for several decades without many additional aminoacidopathies being added. The beginning of the use of electrospray tandem mass spectrometry changed all that, heralding an explosion of new tests beginning in the mid 1990's where the development of high throughput screening from dried spots using ESI-MS/MS was demonstrated for a large number of inherited metabolic disorders by several authors [26-27, 178-182]. Using ESI-MS/MS, PKU was determined from newborn dried bloodspots (DBS), taken between 24 and 72 hours after birth, by first extracting the amino acids then derivatizing phenylalanine (Phe) and tyrosine (Tyr) to form butylated esters. The butylation reaction was simple and quick making it easily adaptable to high throughput screening. The MS/MS analysis of butylated amino acids was very simple and specific and the use of isotopically labelled internal standards allowed for absolute quantification of the analytes in the DBS sample. Diagnosis of PKU was further improved by taking the ratio of concentrations of Phe to Tyr which is a more sensitive measure of the activity of the enzyme phenylanlanine dehydrogenase. The same extraction/derivatization and ESI-MS/MS methodology was then extended to diagnose other aminoacidopathies such as maple syrup urine disease, by forming and detecting the butylated esters of leucine, isoleucine, alloisoleucine and valine[178]. Once again, absolute quantification was made by the use of isotopically labelled internal standards for each amino acid tested. The same technology was then extended to methionine for diagnoses homocystinuria and hypermethiononemia[179].
In addition to amino acid screens, it was discovered that the butylation reaction was compatible with the simultaneous determination of acylcarnitines from the same DBS sample. The test for medium chain length acylcarnitines was thus developed and used to diagnose medium-chain acyl-CoA dehydrogenase deficiency (MCAD) [27, 180]. The added advantage that the tests for the amino acids and the acylcarnitines could be performed in the same 2 minute assay on the same DBS sample, meant that more then 30 different diseases

could be routinely screened by ESI-MS/MS including aminoacidopathies, organic acidurias and fatty acid oxidation disorders.

One problem with performing amino acid analysis by MS is the inability to distinguish isobars such as Leu and Ile. Iosbars must be separated by the chromatography but that often can extend the run times making the method unsuitable for high throughput screening. Another issue with the use of butyl esterfication derivatives is that it destroys glutamine which is the best marker for ornithine transcarbamylase deficientcy (OTCD). Glutamine is mostly converted to the glutamic acid butyl ester during the derivatiztion and the remaining glutamine butyl ester formed is deaminated in the electrospray source [183].

An alternative derivatization method that has been shown to produce more stabile ions then the corresponding butyl esters is the formation of formamidene butyl esters of amino acids [28, 184]. The method was further optimized to include glutamine and achieved an increase in sensitivity of 50% by forming the isobutyl esters instead of the n-butyl esters. Unfortunately this method can not be used as a substitute in new born screening assays because the acylcarinitines are not fully derivatized at room temperature with this reagent.

In addition to amino acid and acylcaritine analysis, newborn screening is also preformed on free methylmalonic acid and 3-OH propionic acid using the butylated etser derivatization [185]. There are some authors attempting to remove the derivization step from the procedure but no routine screening is being done without it.

Hypothyroidism, which was described previously, is also measured in routine newborn screening; however, the assay is usually performed using dried blood spots and an immunoassay to test for the thyroid stimulating hormone (TSH) [186].

Lysosomal storage disorders (LSD) are a group of rare inherited metabolic diseases that result from deficiency or absence of specific enzymes that breakdown unwanted substances in cells. The resulting build up of undesirable materials leads to disability or death. Recently, treatment for some of the more abundant LSDs has made the need for screening in newborns necessary and is now required in several states and in many European countries. Even though the individual disorders have low incidence their combined abundance is 1 in 3,000 [187-191] warranting their screening in newborns.

Screening for enzymatic deficiencies is done by incubation of samples, in this case dried blood spots, with an added substrate that is known to be converted to a specific product by the enzyme of interest. Normal patient samples will form the product because the enzyme is present. Patients deficient in the enzyme will produce much less or no product from the incubation. Measurement of the product formation over time is then used to diagnose those with enzyme deficiencies.

The initial method for LSDs was designed to analyze Fabry, Gaucher, Krabbe, Neimann-Pick and Pompe disease simultaneously and comprised a 24 hr incubation step prior to mass spectral analysis [187-191]. The method was designed for quick LC/MS/MS analysis to facilitate high throughput and had a sample runtime (injection to injection) of 2 minutes. The draw back of the method was that in order to have fast analysis time on the mass spectrometer no real chromatographic separation was used. The result is that there were interferences from the substrates on the product ion spectra. To reduce the cross talk from the substrate, extensive off-line extractions are conducted to separate the substrates from the products. These extractions used a lot of materials increasing the cost, all the extra manipulation increases the odds of experimental error, and the process took an entire day to perform. More recently, a method that employs turbulent flow chromatography for on-line clean-up to remove matrix effects and UHPLC to separate all the substrates from all the products has been reported by Kasper et al [190]. This method is 4 minutes injection to injection but is multiplexed to reduce the system run time to 2 minutes per sample. Another advantage to the improved method is that the initial method needed to de-tune the ion source conditions to reduce fragmentation in the source to guarantee no signal from the substrates in the product ion channels. In doing this the authors also reduced the signal strength. The UHPLC step employed in the new method separates the substrates and products chromatographically such that no detuning is necessary. The result is that less product formation is needed and reduced incubation times are possible as well. Recently a sixth LSD was added to the LSD panel to screen for MPS 1 disease [191]. Additionally, cost analysis comparing on-line TFC clean-up to off-line extraction methods calculated the total cost per sample for LSD screening to be 0.29 Euro using the on-line method and 0.96 Euro using off-line extraction.

Most of the disorders in the expanded or supplemental newborn screening list used today that are performed by mass spectrometry are shown in Table 10. This expanded screening is not yet universally mandated.

4.2 Proteomics

Barr et.al. [192] were the first to demonstrate the use of proteotypic peptides, usually tryptic peptides of proteins targeted for quantification, as quantifiable surrogate markers for the intact protein. Gerber et.al. [193] employed isotopically labelled proteotypic peptides and LC-SRM. They called these peptides "AQUA" short for Absolute QUAntitation when employing these isotopicallly labeled peptides as internal standards (ISs) for quantifying the surrogate proteotypic peptides. The main advantage of this method of quantification is that the AQUA peptides are chemically identical to the proteotypic peptides making them ideal ISs since they coelute with the analyte of interest. This is important since both IS and analyte peptide are ionized and detected under identical conditions eliminating issues like ion-suppression brought about by co-eluting matrix. The method of detection is SRM or selected reaction monitoring where the precursor molecular ion is transmitted by the first analyzer ("Q1" usually a quadrupole mass filter in a triple quad instrument), fragmented in the quadrupole collision cell (2nd quadrupole) and the appropriate product ion(s) are then monitored by the final quadrupole mass filter ("Q3"). The fragment ions formed are dictated by the proteotypic peptide's amino acid sequence making the SRM process very specific for the analyte and corresponding IS. If more specificity is required, additional fragment ions can be monitored per proteotypic peptide.

The group at the Plasma Proteome Institute improved this process adding an immunoaffinity step called SISCAPA (Stable Isotope Standards and Capture by Anti-Peptide Antibodies), to concentrate the proteotypic and AQUA peptides enhancing their detection by LC-SRM [194]. These SISCAPA techniques have evolved to where they now employ integrated systems incorporating magnetic beads coated with immunoaffinity agents implemented in low-volume trap-wash-elute apparatus suitable for the efficient transfer of highly concentrated analytes to low-flow LC-SRM systems. It is estimated that such systems are capable of enrichments of analytes up to 20,000 times providing detection limits of proteotypic peptides and the proteins they represent down to low ng/mL levels rivaling those of some ELISA assays [195].

One significant advantage of this approach is that the cost of analysis of many analytes by LC-SRM is the same as that of a single analyte which could be especially important when multiple biomarker assays are employed. Anderson and Hunter [196] demonstrated the principle where over 100 SRM's were monitored in a single LC-SRM experiment representing the quantitation of over 50 proteins in a single assay.

Certain immunoassays have poor performance due to a variety of factors including potential interference from endogenous immunoglobulins and imperfect concordance across platforms due to antigen microheterogeneity across patient populations [197]. In such cases the performance of LC-SRM may have significantly better performance as described in the example using SISCAPA to quantify thyroglobulin, a well characterized tumour marker for which interference from endogenous immunoglobulins affects results for 10% to 25% of all patients. Thyroglobulin was successfully quantified down to picomolar levels (3 ng/mL) in non-depleted plasma by SISCAPA LC-SRM. One advantage of the assay is that the trypsinization step digests endogenous immunoglobulins that could potentially interfere with immunoaffinity. LC-SRM overcomes these serious problems exhibited by commercially available immunoassays for thyroglobulin.

Carr [198] recently demonstrated that SISCAPA LC-SRM can be used to quantify a "panel" of protein biomarkers in plasma. In the present example plasma samples were taken at timed intervals during a planned myocardial infarction (PMI) used to treat hypertrophic obstructive cardiomyopathy. It was thought that these samples would be representative of myocardial infarction (MI) and that monitoring well recognized biomarkers of cardiovascular disease such as interleukin-33 (IL-33), for which there is no well-validated immunoassay, and cardiac tropinin I (cTnI) would provide insight into the pathophysiology of MI. The authors suggested that SISCAPA LC-SRM would be better suited to multiple-biomarker assays since there is less likelihood of non-specific binding between the capture and detection antibodies. With SISCAPA each analyte (proteotypic peptide) requires only a capture antibody since the detection is done by LC-SRM. The results of the study indicated a poor correlation between the immunoassay for cTnI and SISCAPA LC-SRM. They speculated that non-specific interactions with endogenous antigens in the immunoassay could account for the poor correlations where LC-SRM would not be as susceptible to these interferences. Even though the SISCAPA LC-SRM exhibited impressive detection limits for IL-33 in spiked serum (low ng/mL), signals for endogenous levels of this analyte in plasma samples were below detection limits. Sensitivity is a common problem for the SISCAPA LC-SRM approach where protein analytes are required to have plasma concentrations above the low ng/mL where many important protein analytes are present at pg/mL levels.

There are some recent examples of the use of LC-SRM to determine endogenous protein biomarkers at higher concentrations. A group at Mayo used LC-SRM of proteotypic peptides to quantify serum Zn-α2 glycoprotein (ZAG), a putative biomarker for prostate cancer [199]. Serum samples were trypsinized without the need for the depletion of abundant proteins and relatively high flow-rate LC-SRM was used instead of nano-flow HPLC. The use of 2 mm ID HPLC columns and higher flow rates greatly enhances the ease-of-use of this method as well as ruggedness and reliability of the assay. Prostate cancer patients showed an average ZAG concentration of 7.6 mg/mL where control subjects had an average ZAG concentration of 3.7 mg/mL demonstrating a clear differentiation. The method itself was validated between 0.32 and 10.2 mg/mL.

Another interesting example involves the determination of human serum albumin (HSA) in urine as a biomarker for renal failure. Renal disease is steadily rising as a complication of type-2 diabetes which is reaching epidemic proportions in North America. Urinary Albumin is a sensitive prognostic and diagnostic biomarker for renal disease. Early diagnosis of the onset of renal disease can prevent much more expensive interventions by giving patients a chance to change diet or life-styles to avoid more serious manifestations of the disease. Accurate and precise methods of rapidly screening for renal disease will become important in population screening to avoid false positives and false negatives; such methods must also be inexpensive to permit their widespread use as disease-screening modalities. A group at the Mayo clinic [200] recently published a method for HSA in serum using LC-SRM and the detection of 2 proteotypic peptides representative of HSA and employing isotopically labeled AQUA peptides as ISs generated in a novel fashion. ISs were introduced into the serum sample as a precisely known amount of intact recombinant ^{15}N-labelled HSA prior to tryptic digestion. During trypsinization both endogenous and labeled HSA forms are degraded into tryptic peptides where the ^{15}N labeled peptides act as ISs for the endogenous HSA. The high flow rate LC-SRM method was validated over the linear range of 3 to 200 mg/mL spanning the clinical requirements. HSA at these levels requires no preconcentration or nano-scale chromatography to be detectable. The run time, however, was 30 minutes making it too long for a population-screening modality. This long analysis time could be significantly shortened using multiplexed sample introduction as described above. The LC-SRM method compared well with immunoturbidity but without the need for sample dilution to cover the full analytical range like immunoturbidity requires.

There is significant interest in the use of SISCAPA LC-SRM for the detection of proteins in biological fluids as evidenced by the following references by Ackerman and others [201-204]. There have been multi-site trials to evaluate the reliability of these types of assays in various laboratories across different LC-SRM platforms [205]. There is controversy as to whether such tests will be transferable to the clinical diagnostic laboratory given the complexity of these tests. This controversy was recently discussed in an article by Diamandis [206] based upon interviews of a panel of experts in the field. The consensus opinion was that MS will best be used in situations where good immunoassays do not exist and that the greatest application for LC-SRM will be in the validation of putative protein biomarkers; while this is not clinical diagnosis, this application cannot be undervalued. Over the past 15 years of intense proteomic research thousands of putative protein biomarkers have been identified in human plasma or serum, yet no significant increase in FDA-approved protein diagnostics has followed [207]. Articles have been written decrying this terrible record and the blame lies with a lack of rigorous validation of putative biomarkers involving clinical studies of sufficiently large cohorts of patient samples. Such studies would require the high throughput quantitative analysis of trace protein biomarkers and there is much hope that MS-based quantitative techniques such as SISCAPA LC-SRM will help in this validation process [184, 208].

One area where MS may also be very useful is for the determination of heterogeneous proteins whose isoforms respond similarly to the immunoassay but whose diagnostic values differ greatly. The sources of protein biomarker microheterogeneity could be genetic, environmental, or a mixture of both. This kind of variability in populations is precisely why large-scale multi-center clinical trials are performed on new diagnostic tests to ensure that this natural protein variability does not confound the interpretation of results. However, what is being ignored here is that the variability itself is likely a source of information about disease and possibly diagnostic of it.

Nelson [209] and Niederkofler [210] pioneered the application of Mass Spectrometry Immuno Assays (MSIA) combining immunoaffinity of intact protein targets with a direct mass spectrometry readout usually using MALDI-TOF to determine the various protein isoforms captured by their proprietary immunoaffinity pipettor tips. This has led to some very informative and thought provoking observations of significant correlations of protein variants and isoforms with disease states. Lopez and Nelson [211] combined MSIA with LC-SRM in the determination of PTH isoforms and their potential correlation with renal disease. The full length form of PTH secreted from the thyroid gland comprises 84 amino acids referred to as PTH(1-84). The measurement of PTH(1-84) and its N-terminal truncated form PTH(7-84) are currently assayed to assist in the diagnosis of hypo/hyperparathyroidism and hypercalcemia and in monitoring for renal osteodystrophy in patients with end-stage renal failure. The PTH(1-84)–to–PTH(7-84) paradigm is a recent example of the biological and clinical utility of microheterogeneity within the PTH protein. These authors began their investigation with an MSIA MALDI-TOF experiment that identified a significant number of PTH variants including the previously unidentified PTH variants PTH(28–84), PTH(34–84), PTH(37– 84), PTH(38–84), PTH(48–84), PTH(34 –77), PTH(37–77), and PTH(38 –77). Quantitative MSIA LC-SRM assays were developed for PTH(1–84), PTH(7–84) and the newly identified variant PTH(34–84), where immunoaffinity captured PTH isoforms were tryptically digested postcapture prior to analysis by LC-SRM using a variety or proteotypic peptides and their corresponding isotopically labeled analogues as internal standards. In all, 32 SRM transitions were simultaneously monitored throughout the analytical run. The peptides monitored by LC-SRM exhibited linear responses (R^2 0.90–0.99) relative to recombinant human PTH concentration, limits of detection for intact PTH of 8 ng/L and limits of quantification of 16–31 ng/L depending on the peptide. Many of these peptides showed good correlations with renal disease and could be used as a basis for differentiating patients with renal disease from healthy controls.

The determination of LSDs, as discussed above in the section on Newborn Screening, is basically the indirect determination of proteins (lysozomal enzymes) by monitoring the conversion of small molecule substrates into products. Many other newborn screening (NBS) methods are designed to assay proteins, usually from dried blood spots (DBS). The most well-known of these is the measurement of Thyroid Stimulating Hormone (TSH) by immunoassay. TSH concentrations from newborn blood spots are significantly below the detection limits of LC-SRM methods; however, there are some recent examples of how mass spectrometry can be employed to screen diseases in newborns employing the measurement of protein biomarkers.

Wilson's Disease (WD) is characterized by reduced ceruloplasmin (CP) levels in blood leading to abnormal transport of copper resulting in dangerous elevations of copper in the body's tissues. The excess copper damages the liver and nervous system and WD affects ca. 1 in 30,000 newborns. Wilson's disease has an autosomal recessive pattern of inheritance characterized by progressive deterioration of the liver and the brain. Consequently liver disease and neuropsychiatric symptoms are the main features that lead to diagnosis. A significant percentage of those diagnosed with WD already require a liver transplant but relatively inexpensive effective treatments are available if diagnosed early [212-213]. The measurement of intact proteins directly is often impractical by LC-MS because of their low concentrations such as TSH. As discussed previously, proteotypic tryptic peptides, which are surrogates for the intact protein, are much easier to detect and quantify by LC-MS/MS. This methodology was employed to determine CP extracted from DBS. The results of the study demonstrated clear differentiation between affected newborns, heterozygous carriers of WD and normal newborns. The levels of CP for the unaffected carrier newborns was ca. 250 mg/L and in a cohort of affected newborns ranged from 10 to 20 mg/L. The LC/MS/MS method was rather lengthy for a newborn screening modality (ca. 7 min);

however, this could be made faster using a multiplexed multi-column system, as suggested by the authors.

Hemoglobinopathies represent a tremendous opportunity for MS/MS. These are rare disorders but the combined incidence can be quite high and every jurisdiction in the USA and many others throughout the world test newborns for these disorders. Outcomes can be significantly improved if treatments are commenced early after birth. Currently used methods to detect hemoglobinopathies include ion exchange chromatography (IEX) and isoelectric focusing (IEF) both of which are time consuming and generally exhibit poor specificity when unusual post-translational modifications occur. Many articles have been published on the use of MS and MS/MS for the determination of hemoglobin (Hb) variants via molecular weight determinations on intact β-globin chains [214-216]. These techniques suffer from situations where the mutation involves a zero (nominal) mass change such as Lys↔Gln or Leu↔Ileu. Most labs use a combination of ESI-MS, IEX and IEF followed by DNA analysis to screen and confirm hemoglobinopathies. MS/MS analysis on its own using proteolytic (tryptic) peptides was suggested as a means of analyzing all mutations [214]. Dooley was the first to suggest that all hemoglobinopathies could be screened and confirmed using only MS/MS of tryptic peptides derived from DBS extracts [217]. Recent work by Dalton and Turner [218-219] demonstrated that MS/MS could be used to screen all the clinically important hemoglobinopathies including HbS, HbC, HbE, HbD^{Punjab}, HbO^{Arab}, Hb^{Lepore}, and β-Thalassemia as well as other clinically important conditions such as delta β-Thalassemia, hereditary persistence of fetal hemoglobin trait (HPFH) and alpha zero thalassemia trait. The important finding of this work was that all these conditions could be screened in a 1 to 2 minute MS/MS experiment using SRM, monitoring 1 SRM per proteotypic peptide and employing proteotypic peptides of wild-type globins as endogenous internal standards. The method employed simple trypsinization of the DBS extract followed by a 1 to 2 minute flow injection analysis (FIA) MS/MS experiment ideal for high throughput screening. These authors demonstrated that by using SRM-triggered full scan MS/MS they were also able to confirm the amino acid sequence of the variant peptide without the need for costly genetic (i.e. DNA) analysis. It was demonstrated that confirmation could easily be built into the screening assay without paying a penalty in time or sensitivity. This work was validated by Boemer, et.al. [220] employing 2 SRM channels per proteotypic peptide and a different brand of MS/MS instrument. Both assays demonstrated 100% sensitivity and specificity using this approach for a wide variety of hemoglobinopathies and even the ability to detect heterozygous genotypes. Dalton and Turner went on to perform a clinical trial [221] on a cohort of 40,000 newborns comparing MS/MS to the predicate IEF method. The results indicated complete concordance between the two methods detecting 199 HbS (including 8 HbS/HbF and 3 HbSC variants), 39 HbC, 52 HbD^{Punjab}, and 48 HbE.

4.3 Metabolomics

As discussed in the previous section, there has been almost 15 years of intensive proteomics research with relatively little to show for it in terms of novel protein diagnostics due mainly to a lack of biological validation of the putative diagnostic targets. Similarly, there is intensive research into metabolomics as a source of new disease biomarkers. Many of the putative metabolomic biomarkers are actually panels of endogenous metabolites for which mass spectrometry has been demonstrated to be very well suited as both a discovery and detection modality.

A recent review by Ellis [222] concluded that of all the "omics" areas of research to discover novel biomarkers, metabolomics is probably the most achievable. There are probably on the order of <5,000 metabolites [223] circulating in blood, where there are about 30,000 expressed genes with hundreds of thousands of single nucleotide polymorphisms (SNPs) and probably well over 1,000,000 proteins considering post-translational modifications and splice variants. With only several thousand metabolites that need consideration, metabolomics is likely to be a more achievable source of new diagnostics than either genetics or proteomics posing much less of a challenge to identify, quantify and validate diagnostic potential. Apart from their potential diagnostic value, metabolite-based biomarkers could also be useful in patient stratification in clinical trials or in disease-treatment where it would be used to monitor whether patients are responding well or poorly to pharmaceuticals or even simple life-style changes. Just as in proteomics research, the majority of recent metabolomic work using mass spectrometry has concentrated on the discovery of novel disease-related biomarkers involving metabolome-wide studies (hundreds or even thousands of metabolites) on cohorts of only relatively few patients and controls. The result has been the creation of panels of many putative metabolomic disease biomarkers without the requisite clinical or biological validation involving broader population studies. To facilitate such studies one group in particular (an Innsbruck-based company called Biocrates, pronounced like "Socrates") has developed CE-IVD approved kits that can be applied in anyone's laboratory to economically perform these rigorously quantitative metabolomic validation studies across relatively large sample sets. Such studies are very important for the translation of metabolomic research into clinical practice.

An interesting example of this application is a collaborative study by systems biology researchers and Biocrates investigating a diabetes mouse model [224] employing a proprietary kit of reagents including all required reactants, standards and protocols to quantitatively determine over 800 endogenous metabolites by LC-MS/MS including: 18 amino acids; 50 reducing mono-, di-, and oligosaccharides, uronic acid, and N-acetylglucosamine; 16 acylcarnitines; 5 hydroxyacylcarnitines; 5 dicarboxylacylcarnitines; free carnitine; and 707 lipids. These lipids were subdivided into 82 different ceramides and glucosylceramides, 110 different sphingomyelins and sphingomyelin derivatives, 95 glycero-phosphatidic acids, 85 glycero-phosphatidylcholines, 103 glycero-phosphatidylethanolamines, 11 glycero-phosphatidylglycerols, 177 glycero-phosphatidylinositols, glycero-phosphatidylinositol-bisphosphates and -triphosphates, and 44 glycero-phosphatidylserines. Glycero-phospholipids were further differentiated with respect to the presence of ester and ether bonds in the glycerol moiety. Using these kits and flow injection analysis (FIA)-MS/MS (SRM) researchers correlated plasma levels of analytes with diabetes treatments; accurate concentrations were derived from the use of isotopically labeled internal standards (ISs) included in the kits. Metabolite concentrations and ratios of concentrations were derived and subjected to unsupervised statistical analysis revealing novel diabetic metabolic phenotypes and the impact medication has on them. Many metabolites were oppositely impacted by treatment which reduced noise in the data to reveal new potential biomarkers of diabetes. This study was subsequently extended to human subjects [225] comparing the Biocrates kit results to other approaches using a cohort of 100 subjects comprising 40 individuals with self-reported diabetes and 60 controls. The other LC-MS approach in this study was a proprietary workflow undertaken by Metabolon as well as an NMR study of the same samples done by Chenomx. In this pilot study a total of 482 metabolites were quantified. The results of this human trial were far less clear than those using the inbred murine subjects of the above earlier study for a variety of reasons including heterogeneity of the human subjects with respect to genetic and environmental factors including varying degrees of glycemic control. The authors concluded that this multi-platform approach to the metabolome-wide analyses of diabetes in a general population was helpful in the identification of a series of known, and also some novel, deregulated metabolites that associate with diabetes under sub-clinical conditions in the general population. However, given the above limitations they observed that a much larger cohort of subjects would be necessary.

In related articles by the same authors [226-227] human genotypes were evaluated against metabolomic data creating what the authors deemed as the first genome-wide association (GWA) study using metabolic traits as phenotypic traits. In these studies 363 serum metabolites from 284 male subjects who were participants in the KORA study [228] were rigorously quantitated using the Biocrates kits and (FIA)-MS/MS (SRM). In parallel, single nucleotide polymorphisms (SNP's) were measured using gene-expression chips and correlated with the metabolomic data. Genetically determined variants in the metabolic phenotype (so called "metabotype") were observed and enhanced using ratios of metabolites with opposite correlations with genotypes. Using this approach, almost 30% of genetic variability correlated with the metabolic data with high statistical significance. The authors conclude that the investigation of the genetically determined metabotypes in their biochemical context may help to better understand the pathogenesis of common diseases and gene-environment interactions. These findings could result in a step toward personalized health care and early nutritional or other life-style interventions based on a combination of genotyping and metabolic characterization.

The above examples of metabolomic applications featured the use of mainly triple quadrupole MS/MS systems, the commonly accepted "workhorse" of quantitative bioanalysis. Over the last 20 years the performance of the triple quad platform has risen steadily due to the incorporation of new technologies and larger, more efficient pumping systems. In fact, like Moore's Law for computing, the performance of the triple quad platform increased by about an order of magnitude every 2 to 3 years more or less in a linear fashion until the last 5 years or so when the triple quad began to asymptotically reach its maximum performance with respect to signal-to-noise and limits of detection.

As the triple quad platform begins to mature technologically, other areas of mass spectrometer performance are beginning to be exploited in the market; specifically there has been a rapid evolution of performance in the area of mass resolving power and mass accuracies. This started with the development of the hybrid MS technologies based upon quadrupoles and time-of-flight systems, the so called Q-ToF instruments, as well as the introduction of linear ion-traps coupled to Fourier transform mass spectrometers (LTQ-FT-MS systems) and its successor the ion-trap coupled to an Orbitrap system (LTQ-Orbitrap systems). These MS/MS systems are capable of high mass accuracies (1-2 ppm) and high mass resolutions (20,000 to 300,000 $m/\Delta m$) which greatly enhances analytical specificity. Generally these hybrid systems are very large, expensive and quite difficult to operate making their incorporation into routine clinical diagnostics quite limited. However, recently lower cost benchtop versions of the Orbitrap instruments are beginning to appear [229-231]. The significance of this new type of high resolution/mass accuracy (HRMA) system is just beginning to be appreciated since the specificity of HRMA of a single MS system could rival that of lower resolution MS/MS systems like the triple quadrupole. The simplicity of a single mass analyzer versus MS/MS could provide the impetus to use benchtop HRMA systems for routine bioanalytical assays instead of MS/MS.

One recent study that exemplifies the use of this technology in metabolomics was recently published by Volmer, et.al. [232] where lipids extracted from plasma samples for a cohort of 10 healthy volunteers were analyzed by UHPLC-MS on a benchtop Orbitrap system. About 240 separate data analyses were undertaken to develop a statistically meaningful data set obtained using 50,000 resolution, and mass accuracies typically well below 2.5 ppm. Like previous lipidomic studies [233] Volmer found that setting m/z ion-extraction windows to ≤2.5 ppm provided unambiguous lipid identification based upon elemental composition. Moreover, the re-constructed ion chromatograms were used for quantitation without significant isobaric interferences using the analyzer constraints of 50,000 resolution and ~1-2 ppm mass accuracy. This large data set was subjected to unsupervised statistical analysis and demonstrated the ability to differentiate subjects' plasma samples and to also pick out those samples enriched with a particular endogenous or exogenous analyte. The study demonstrated that, with high resolving powers and mass accuracies, both targeted and untargeted data analysis can be accomplished on the same data set. Another recent example of the application of this technology was published by Rabinowitz, et.al. [234] involving the UHPLC-MS analysis of anionic metabolites in yeast extract identifying 137 metabolites in a single experiment. This type of platform could be especially useful for metabolomic studies involving human subjects where high numbers of analytes must be rigorously quantitated in a single run but with sufficient flexibility to be able to provide information on unexpected analytes such as exogenous compounds and their metabolites that may be present due to environmental factors (e.g. disease treatment with pharmaceuticals). The translation of the Orbitrap from a research validation tool to a routine instrument for clinical diagnostic measurements is inevitable given the analytical power of this novel technology and the results that have been published to date.

5. Conclusion

The use of tandem mass spectrometry is well documented in the clinical laboratory. In particular, the development of atmospheric pressure ionization techniques to perform LC/MS has allowed the clinical chemist to address new areas of research and drastically improve the quality of clinical assays due to higher specificity and better sensitivity than other approaches. The main draw back to clinical applications of LC/MS/MS is the high initial cost and the need for dedicated highly trained personnel to run and maintain the instruments. The result is that centralized laboratories, either CROs or large medical facilities, where samples are sent for analysis are usually needed in order to take advantage of the emerging technologies. As our understanding of the use of LC/MS/MS and the impact of analyzing biological matrices improves, and mass spectrometers become more and more user friendly, the development of a fully automated clinical analyzer should be possible. However, until an LC/MS/MS clinical analyzer is commercially available, the use of LC/MS/MS in the clinic will remain a specialized tool available by out-sourcing or development core facilitates.

6. References

[1] Thomson BA, "Atmospheric pressure ionization and liquid chromatography/mass spectrometry together at last" J Am Soc Mass Spectrom 1998; 9:187-193
[2] Henion JD, "The origins of ion spray liquid chromatography-tandem mass spectrometry" Clin Chem 2009; 55:1234-1235)

[3] Dooley KC, "Tandem mass spectrometry in the clinical chemistry laboratory" Clin Biochem 2003; 36:471–481)
[4] Vogeser M, Seger C, "Review: A decade of HPLC–MS/MS in the routine clinical laboratory—Goals for further developments" Clin Biochem 2008; 41:649–662
[5] Shushan B, "A Review of Clinical Diagnostic Applications of Liquid Chromatography – Tandem Mass Spectrometry" Mass Spectrometry Reviews 2010; 29: 930– 944
[6] Rashed MS, "Clinical applications of tandem mass spectrometry: ten years of diagnosis and screening for inherited metabolic diseases" J Chromatogr B: Biomed Sci Appl 2001; 758: 27-48
[7] Rashed MS, Bucknall MP, Little D, Awad A, Jacob M, Alamoudi M, Alwattar M, Ozand PT, 1997. 'Screening blood spots for inborn errors of metabolism by electrospray tandem mass spectrometry with a microplate batch process and a computer algorithm for automated flagging of abnormal profiles" 1997 Clin Chem 43:1129–41
[8] Chace DH, Kalas TA, Naylor EW, "Use of tandem mass spectrometry for multianalyte screening of dried blood specimens from newborns" Clin Chem 2003; 49:1797–1817
[9] Chace DH, "Mass spectrometry in newborn and metabolic screening: Historical perspective and future directions" J Mass Spectrom 2009; 44:163–70
[10] Back DJ, Khoo SH, Gibbons SE, Merry C, 2001. "The role of therapeutic drug monitoring in treatment of HIV infection" Br J Clin Pharmacol 2001; 51:301–8
[11] Gu J, Soldin SJ, "Modification of tandem mass spectrometric method to permit simultaneous quantification of 17 anti-HIV drugs which include atazanavir and tipranavir" Clin Chim Acta 2007; 378:222–4
[12] Constanzer M, Chavez C, Matuszewski B, "Development and comparison of high-performance liquid chromatographic methods with tandem mass spectrometric and ultraviolet absorbance detection for the determination of cyclobenzaprine in human plasma and urine" J Chromatogr B 1995; 666:117–126
[13] Taylor PJ, Jones A, Balderson GA, Lynch SV, Norris RLG, Pond SM, "Sensitive, specific quantitative analysis of tacrolimus (FK506) in blood by liquid chromatography-electrospray tandem mass spectrometry" Clin Chem 1996 ; 42: 279–85
[14] Yang Z, Peng Y, Wang S, "Immunosuppressants: Pharmacokinetics, methods of monitoring and role of high performance liquid chromatography/mass spectrometry" Clin Appl Immunol 2005; 5: 405–30)
[15] Lennard L. "Therapeutic drug monitoring of cytotoxic drugs" Br J Clin Pharmacol 2001; 52: 75S–87S
[16] Holst JP, Soldin SJ, Tractenberg RE, Guo T,Kundra P,Verbalis JG, Jonklaas J, "Use of steroid profiles in determining the cause of adrenal insufficiency" Steroids 2007; 72:71–84)
[17] Sauvage FL, Saint-Marcoux F, Duretz B, Deporte D, Lachatre G, Marquet P, "Screening of drugs and toxic compounds with liquid chromatography-linear ion trap tandem mass spectrometry" Clin Chem 2006; 52:1735–42
[18] Mauerer HH, "Current role of liquid chromatography–mass spectrometry in clinical and forensic toxicology" Anal Bioanal Chem 2007; 388:1315–1325

[19] Guo T, Chan M, Soldin SJ, "Steroid profiles using liquid chromatography-tandem mass spectrometry with atmospheric pressure photoionization source" Arch Pathol Lab Med 2004; 128: 469–75

[20] Holst JP, Soldin OP, Guo T, Soldin SJ, "Steroid hormones: Relevance and measurement in the clinical laboratory" Clin Lab Med 2004; 24:105–18

[21] Benton SC, Nuttall M, Nardo L, Laing I, "Measured dehydroepiandrosterone sulfate positively influences testosterone measurement in unextracted female serum: comparison of 2 immunoassays with testosterone measured by LC-MS" Clin Chem 2011; 57: 1074–83

[22] Rosner W, Auchus RJ, Azziz R, Sluss PM, Raff H, "Utility, limitations, and pitfalls in measuring testosterone: An endocrine society position statement" J Clin Endocrinol Metab 207; 92:405–13

[23] Vogeser M, Parhofer KG, "Liquid chromatography tandem-mass spectrometry (LC-MS/MS) - Technique and applications in endocrinology" Exp Clin Endocrinol Diabetes 207; 115: 559–70

[24] Herman JL, Argoti D, Fair SJ, "Molecular weight exclusion of proteins using turbulent flow chromatography" Mass Spectrometry: Applications to the Clinical Lab (MSACL); San Deigo, CA Feb 5-9

[25] Herman JL; Encyclopedia of Mass Spectrometry, Volume 8, "Turbulent Flow LC-MS in Bioanalysis, Principles and Applications," W. Niessen editor, Elsevier, Oxford, UK, (2006)

[26] Chase DH, Millington DS, Terada N, Kahler SG, Roe CR, Hofman LF, "Rapid diagnosis of phenylketonuria by quantitative analysis for phenylalanine and tyrosine in neonatal blood spots by tandem mass spectrometry" Clin Chem 1993; 39: 66-71

[27] Chase DH, Hillman SL, van Hove JLK, Naylor EW, "Rapid diagnosis of MCAD deficiency: quantitative analysis of octanoylcarnitine and other acylcarnitines in newborn blood spots by tandem mass spectrometry" Clin Chem 1997; 43: 2106-13

[28] Johnson DW, "Analysis of amino acids as formamidene butyl esters by electrospray ionization tandem mass spectrometry" Rapid Commun Mass Spectrom 2001; 15:2198-205

[29] Aronov PA, Hall LM, Dettmer K, Stephensen CB, Hammock BD, "Metabolic profiling of major vitamin D metabolites using Diels-Alder derivation and ultra high performance liquid chromatography-tandem mass spectrometry" Anal Bioanal Chem 2008; 392:1917-30

[30] Higashi T, Shibayama Y, Fuji M, Shimada K, "Liquid Chromatography-tandem mass spectrometric method for the determination of salivary 25-hydroxyvitamin D3; a noninvasive tool for the assessment of vitamin D status" Anal Bioanal Chem 2008; 391: 229-38

[31] Holik MF, "Vitamin D deficiency" N Engl J Med 2007:357:266-81

[32] Wang S, "Epidemiology of vitamin D in health and disease" Nutr Res Rev 2009; 22:188-203

[33] Singh RJ, "Are clinical laboratories prepared for accurate testing of 25-hydroxy vitamin D?" Clin Chem 2008; 54:221-3

[34] Chen TC, Chimeh F, Lu Z,Mathieu J, Person KS, Zhang A "Factors that influence the cutaneous synthesis and dietary sources of vitamin D" Arch Biochem Biophys 2007; 460:213-7

[35] Webb AR, "Who, what, where and when-influences on cutaneous vitamin D synthesis" Prog Biophys Mol Biol 2006; 92:17-25

[36] Endres DB, Rude RK; Textbook of Clinical Chemistry and Molecular Diagnostics; "Mineral and Bone Metabolism" CA Burtis and ER Ashwood editors, Elsevier Saunders, 2006:1981-966

[37] Kavathia D, Buckley JD, Rao D, Rybicki B, Burke R, "Elevated 1,25-dihydroxyvitamin D levels are associated with protracted treatment in saccoidosis" Respir Med 2010; 104:564-70

[38] Carter GD, "25-Hydroxyvitamin D assays: the quest for accuracy" Clin Chem 2009; 55:1300-02

[39] Roth HJ, Schidt-Gayk H, Weber H, Niederau C, "Accuracy and clinical implicationsof seven 25-hydroxyvitamin D methods compared with liquid chromatography-tandem mass spectrometry as a reference" Ann Clin Biochem 2008: 45:153-9

[40] Hollis BW, Horst RL, "The assessment of circulating 25(OH)D and 1,25(OH)2D: where we are and where we are going" J Steriod Biochem Mol Biol 2007; 103:473-6

[41] Wallace AM, Gibson S, Hunty Adl, Lamberg-Allardt C, Ashwell M, "Measurement of 25-hydroxyvitamin D in the clinical laboratory:current procedures, performance characteristics and limitations" Steriods 2010; 75:477-88

[42] Vreeken RJ, Honing M, van Baar BL, Ghijsen RT, de Jong GJ, Brinkman UA, "On-line post column Diels-Alder dervitization for the determination of vitamin D3 and it's metabolites by liquid chromatography/thermospray mass spectrometry" Biol Mass Spectrom 1993; 22:621-32

[43] Higashi T, Awada D, Shimada K, "Simultaneous determination of 25-hydroxyvitamin D2 and 25-hydroxyvitamin D3 in human plasma by liquid chromatography-tandem mass spectrometry employing derivatization with a Cookson-type reagent" Biol Pharm Bull 2001; 24:738-43

[44] Higashi T, Homma S, Iwata H, Shimada K, "Characterization of of urinary metabolites of vitamin D3 in man under physiological conditions using liquid chromatography-tandem mass spectrometry" J Parm Biomed Anal 2002; 29:947-55

[45] Higashi T, Yamauchi A, Shimada K, "Application of 4-(4-nitrophenyl)-1, 2, 4-triazoline-3, 5-dione to analysis of 25-hydroxyvitamin D3 in human Plasma by liquid chromatography/electron capture atmospheric pressure chemical ionization-mass spectrometry" Anal Sci 2003: 19:941-3

[46] Vogeser M, Kyriatsoulis A, Huber E, Kobold U, "Candidate reference method for the quantification of circulating 25-hydroxyvitamin D3 by liquid chromatography-tandem mass spectrometry" Clin Chem 2004; 50:1415-7

[47] Chen H, Mccoy LF, Schleicher RL, Pfeiffer CM, "Measurement of 25-hydroxyvitamin D3 (25OHD3) and 25-hydroxyvitamin D2 (25OHD2) in human serum using liquid chromatography-tandem mass spectrometry and its comparison to a radioimmunoassay method" Clin Chim Acta 2008; 391:6-12

[48] Knox S, Harris J, Calton L, Wallace AM, "A simple automated solid-phase extraction procedure for measurement of 25-hydroxyvitamin D3 and D2 by liquid chromatography-tandem mass spectrometry" Anal Clin Biochem 2009; 46:226-30

[49] Hojskov CS, Heickendorff L, Moller HJ, "High-throughput liquid-liquid extraction and LC/MS/MS assay for the determination of circulating 25(OH) vitamin D3 and D2 in the routine clinical laboratory" Clin Chim Acta 2010; 411:114-6

[50] Taylor RL, Grebe SK, Singh RJ, "High throughput analysis of 25-hydroxyvitamins D2 & D3 by LC-MS/MS using an automated on-line extraction" Clin Chem 2005; 51 (Suppl 6); A231-2

[51] Bunch DR, Miller AY, Wang S, "Development and validation of a liquid chromatography-tandem mass spectrometry assay for serum 25-hydroxyvitamin D2/D3 using turbulent flow online extraction technology" Clin Chem Lab Med 2009; 47:1565-72

[52] Singh RJ, Taylor RL, Reddy GS, Grebe SK, "C-3 epimers can account for a significant proportion of total circulating 25-hydroxyvitamin D in infants, complicating accurate measurement and interpretation of vitamin D status" J Clin Endocrinol Metab 2006; 91: 3055-61

[53] Newman MS, Brandon TR, Groves MN, Gregory WL, Kapur S, Zava DT, "A liquid chromatography/tandem mass spectrometry method for determination of 25-hydroxy vitamin D(2) and 25-hydroy vitamin D(3) in dried blood spots: a noninvasive tool for the assessment of vitamin D status" J Diabetes Sci Tecnol 2009; 3: 156-62

[54] Eyles D, Anderson C, Ko P, Jones A, Thomas A, Burne T, "A sensitive LC/MS/MS assay of 25OH vitamin D3 and 25OH vitamin D2 in dreid blood spots" Clin Chim Acta 2009: 403: 145-51

[55] Casetta B, Jans I, Billen J, Vanderschueren D, Bouillon R, "Development of a method for the quantification of 1alpha,25(OH)(2)-vitamin D(3) in serium by liquid chromatography tandem mass spectrometry without derivatization" Eur J Mass Spectrom 2010; 16:81-9

[56] Cox VC, Ensom MH, "Mycophenolate mofetil for solid organ transplantation: does the evidence support the need for clinical pharmacokinetic monitoring?" Ther Drug Monit 2003; 25: 137-57

[57] Korecka M, Shaw LM, "Review of the newest HPLC methods with mass spectrometry detection for determinination od immunosuppressive drugs in clinical practice" Ann Transplant 2009; 14: 61-72

[58] Yang Z, Wang S, "Recent development in application of high performance liquid chromatography-tandem mass spectrometry in therapeutic drug monitoring of immunosuppressents" J Immunol Meth 2008: 336: 98-103

[59] Rostaing L, Cointault O, Marquet P, Josse AG, Lavit M, Saint-Marcoux F, "Falsely elevated whole-blood tacrolimus concentrations in a kidney-transplant patient: potential hazards" Transplt Int 2009

[60] Moscato D, Nonnato A, Adamo R, Vancheri M, Caropreso A, "Therapeutic monitoring of tacrolimus: aberrant results by an immunoassay with automated with automated pretreatment" Clin Chim Acta 2010; 411: 77-80

[61] Khoschsorur G, "Simultaneous measurement of sirolimus and everolimus in whole blood by HPLC with ultraviolet detection" Clin Chem 2005; 51: 1721-4

[62] Baldelli S, Zenoni S, Merlini S, Perico N, Cattaneo D, "Simultaneous determination of everolimus and cyclosporine concentrations by HPLC with ultraviolet detection" Clin Chim Acta 2006; 364: 354-8

[63] Musuamba FT, Di Fazio V, Vanbinst R., Wallemaq P, "A fast ultra-performance liquid chromatography method for simultaneous quantification of mycophenolic acid and it's phenol- and acyl-glucuronides in human plasma" Ther Drug Monit 2009; 31: 110-5

[64] Saint-Marcoux F, Sauvage FL, Marquet P, "Current role of LC-MS in therapeutic drug monitoring" Anal Bioanal Chem 2007; 388: 1327-49

[65] Taylor PJ, "Therapeutic drug monitoring of immunosuppressant drugs by high-performance liquid chromatography-mass spectrometry" Ther Drug Monit 2004; 26: 215-9

[66] Salm P, Taylor PJ, Rooney F, "A high-performance liquid chromatography-mass spectrometry method using a novel atmospheric pressure chemical ionization approach for the rapid simultaneous measurement of tacrolimus and cyclosporine in whole blood" Ther Drug Monit 2008; 30: 292-300

[67] Ansermont N, Fathi M, Veuthey JL, Desmeules J, Rudaz S, Hochstrasser D, "Simultaneous quantification of cyclosporine, tacrolimus, sirolimus and everolimus in whole blood by liquid chromatography-electrospray mass spectrometry" Clin Biochem 2008; 41: 728-35

[68] Seger C, Tentschert K, Stoggl W, Griesmacher A, Ramsay SL, " A rapid HPLC-MS/MS method for simultaneous quantification of cyclosporine A, tacrolimus, sirolimus and everolimus in human blood samples" Nat Protoc 2009; 4: 526-34

[69] Meinitzer A, Gartner G, Pilz S, Stettin M, "Ultra fast liquid chromatography-tandem mass spectrometry routine method for simultaneous determination of cyclosporine A, tacrolimus, sirolimus and everolimus in whole blood using deuterated internal standards for cyclosprin A and everolimus" Ther Drug Monit 2010: 32: 61-6

[70] Bogusz MJ, Enazi EA, Hassan H, Abdel-Jawaad J, Ruwaily JA, Tufil MA, "Simultaneous LC-MS-MS determination of cyclosporine A, tacrolimus, and sirolimus in whole blood as well as mycophenolic acid in plasma using common pretreatment procedure" J Chromatogr B Analyt Technol Biomed Life Sci 2007; 850: 471-80

[71] Koster RA, Dijkers EC, Uges DR, "Robust high-throughput LC-MS-MS method for therapeutic drug monitoring of cyclosporine A, tacrolimus, everolimus and sirolimus in whole blood" Ther Drug Monit 2009; 31: 116-25

[72] Kaier P, Akerboom T, Wood WG, Reinauer H, "A new approach for the determination of immunosuppressive drugs using HPLC-MS/MS and Cs adducts" Ger Med Sci 2006; 4 Doc01

[73] Taylor PJ, "Matrix effects: the Achilles heel of quantitative high-performance liquid chromatography-electrospray-tandem mass spectrometry" Clin Biochem 2005; 38: 328-34

[74] Taylor PJ, Brown SR, Cooper DP, Salm P, Morris MR, Pillans PI, "Evaluation of 3 internal standards for the measurement of cyclosporine by HPLC-mass spectrometry" Clin Chen 2005; 51: 1890-3

[75] Taylor PJ, Franklin ME, Graham KS, Pillans PI, "A HPLC-mass spectrometric method suitable for the therapeutic drug monitoring of everolimus" J Chromatogr B Analyt Technol Biomed Life Sci 2007; 848: 208-14

[76] Taieb J, Mathian B, Millot F, Patricot MC, Mathieu E, Queyrel N, "Testosterone measured by 10 immunoassays and by isotope-dilution gas chromatography-mass spectrometry in sera from 116 men, women and children" Clin Chem 2003; 49: 1381-95

[77] Herold DA, Fitzgerald RL, "Immunoassays for testosterone in women: better than a guess?" Clin Chem 2003; 49: 1250-1

[78] Jones AM, Honour JW, "Unusual results from immunoassays and the role of the clinical endocrinologist" Clin Endocrinol 2006; 64: 234-44

[79] Hoofinagle AN, Werner MH, "The fundamental flaws of immunoassays and potential solutions using tandem mass spectrometry" J Immunol Methods 2009; 347: 3-11

[80] Ismil AA, Walker PL, Cawood ML, Barth JH, "Interference in immunoassays is an underestimated problem" Ann Clin Biochem 2002; 39: 366-73

[81] Siekmann L, "Determination of steroid hormones by the use of isotope dilution-mass spectrometry: a definitive method in clinical chemistry" J Steroid Biochem 1979; 11: 117-23

[82] Shackleton C, "Clinical steroid mass spectrometry: A 45-year history culminating in HPLC-MS/MS becoming an essential tool for patient diagnosis" J Steriod Biochem Mol Biol 2010; 121: 481-90

[83] Kushnir MM, Rockwood AL, Roberts WL, Owen WE, Bunker AM, Meikle AW, "Development and performance evaluation of a tandem mass spectrometry assay for four adrenal steroids" Clin Chem 2006; 52: 1559-67

[84] Guo T, Taylor RL, Singh RJ, Soldin SJ, "Simultaneous determination of 12 steriods by isotope dilution liquid chromatography-photospray ionization tandem mass spectrometry" Clin Chim Acta 2006; 372:76-82

[85] Guo T, Gu J, Soldin OP, Singh RJ, Soldin SJ, "Rapid measurement of estrogens and their metabolites in human serium by liquid chromatography-tandem mass spectrometry without derivatization" Clin Biochem 2008; 41: 736-41

[86] Janzen N, Sander S, Terhardt M, Peter M, Sander J, "Fast and direct quantification of adrenal steroids by tandem mass spectrometry in serum and dried blood spots" J Chromatogr B 2008; 861: 117-22

[87] Soldin OP, Sharma H, Husted L, Soldin SJ, "Periatric reference intervals for aldosterone, 17alpha-hydroxyprogesterone, dehydroepiandrosterone, testosterone and 25-hydroxy vitamin D3 using tamdem mass spectrometry" Clin Biochem 2009; 42: 823-7

[88] Kushnir MM, Naessen T, Kirilovas D, Chaika A, Nosenko J, Mogilevkina I, "Steroid profiles in ovarian follicular fluid from regularly menstruating women and from women after ovarian stimulation" Clin Chem 2009; 55: 519-26

[89] Kulle AE, Riepe FG, Melchior D, Hiort O, Holterhus PM, "A novel ultrapressure liquid chromatography tandem mass spectrometry method for the simultaneous determination of androstenedione, testosterone, and dihydrotestosterone in pediatric blood samples: age- and sex-specific reference data" J Clin Endocrinol Metab 2010; 95: 2399-409

[90] Grumbach MM, Hughes I, Conte FA, "Disorders of sexual differentiation" Williams Textbook of Endocrinology, 10th ed.; Larsen PR, Kronenberg HM, Melmed S, Polonsky KS editors; Saunders NY: 2002, p 842-1002

[91] Lacey JM, Minutti CZ, Magera MJ, Tauscher AL, Casetta B, MsCann M, "Improved specificity of newborn screening for congenital adrenal hyperplasia by second-tier steroid profiling using tandem mass spectrometry" Clin Chem 2004; 50: 621-5

[92] Schwarz E, Liu A, Randall H, Haslip C, Keune F, Murray M, "Use of steroid profiling by UPLC-MS/MS as a second tier test in newborn Screening for congenital adrenal hyperplasis: the Utah experience" Pediatr Res 2009; 66: 230-5

[93] Higashi T, Shibayama Y, Kawasaki I, Shimada K, "Determination of salivary 17-ketosteroid sulfates using liquid chromatography-electrospray ionization-mass spectrometry" J Pharm Biomed Anal 2007; 43: 1782-8

[94] Groshi M, Wagner R, Rauh M, Dorr HG, "Stability of salivary steroids: the influences of storage, food and dental care" Steroids 2001; 66: 737-41

[95] Groshi M, "Current status of salivary hormone analysis" Clin Chem 2008; 58: 1759-69

[96] Walker RF, Hughes IA, Riad-Fahmy D, "Salivary 17 alpha-hydroxyprogesterone in congenital adrenal hyperplasia" Clin Endocrinol 1979; 11: 631-7

[97] Wudy SA, Hartmann M, Svoboda M, "Determination of 17-hydroxyprogesterone in plasma by stable isotope dilution/benchtop liquid chromatography-tandem mass spectrometry" Horm Res 2000; 53: 68-71

[98] Haymond S, Gronowski AM, "Reproductive related disease" Textbook of Clinical Chemistry and Molecular Diagnostics; 4th ed, Ashwood ER, Burtis CA, Bruns DE editors; Saunders, New York: 2005, p 2097-152

[99] Kushnir MM, Blamires T, Roberts WL, Yue B, Rockwood AL, Bunker AM, Meikle AW, "LC-MS/MS assay for androstenedione, dehydroepiandrosterone and testosterone with pediatric and adult reference intervals" Clin Chem 2010; 56: 1138-47

[100] Yue B, Kushnir MM, Meikle AW, Rockwood AL, "High throughput analysis of serum 5aplpha-dihydrotestoserone by 2D-LC/MS/MS" Proceedings of the 57th conference of the American Society of Mass Spectrometry; Philadelphia, PA; 2009

[101] Xu X, Keefer LK, Waterhouse DJ, Saavedra JE, Veenstra TD, Ziegler RG, "Measuring seven endogenous ketoic estrogens simultaneously in human urine by high-performance liquid chromatography-mass spectrometry" Anal Chem 2004; 76: 5829-36

[102] Yamashita K, Miyashiro Y, Maekubo H, Okuyama M, Honm S, Takahashi M, "Development of highly selective quantification method for testosterone and dihydrotestosterone in human serum and prostst tissue by liquid chromatography-electrospray ionization tandem mass spectrometry" Steroids 2009; 74: 920-6

[103] Licea-Perez H, Wang S, Szapacs ME, Yang E, "Development of a highly sensitive and selective UPLC/MS/MS method for the simultaneous determination of

testosterone and 5 alpha-dihydrotestosterone in human serum to support testosterone replacement therapy for hypogonadism" Steroids 2008; 73: 601-10

[104] Kushnir MM, Rockwood AL, Bergquist J, Varshavsky M, Roberts WL, Yue B, "High sensitivity tandem mass spectrometry assay for serum estrone and estradiol" AM J Clin Pathol 2008; 129: 530-9

[105] Nelson Re, Grebe Sk, O'Kane DJ, Singh RJ, "Liquid chromatography-tandem mass spectrometry assay for simultaneous measurement of estradiol and estrone in human plasma" Clin Chem 2004; 50: 373-84

[106] Anari MR, Bakhtir R, Zhu B, Huskey S, Franklin RB, Evans DC, "Derivatization of ethinylestradiol with dansyl chloride to enhance electrospray ionization: application in trace analysis of ethinylestradiol in rhesus monkey plasma" Anal Chem 2002; 74: 4136-44

[107] Midgley JE, "Direct and indirect free thyroxine assay methods: theory and practice" Clin Chem 2001; 47: 1353-63

[108] Kushnir MM, Nelson R, Roberts WL, Rockwood AL, "Cortisol and cortisone analysis in serum and plasma by atomospheric pressure photoionization tandem mass spectemtry" Clin Biochem 2004; 37: 357-62

[109] Kushnir MM, Rockwood AL, Roberts WL, Pattison EG, Bunker AM, Fitzgerald RL, "Performance characterization of a novel tandem mass spectrometry assay for analysis of testosterone in serum" Clin Chem 2006; 52: 120-8

[110] Kushnir MM, Rockwood AL, Roberts WL, Owen WE, Bunker AM, Meikle AW, "Development and performance evaluation of a tandem mass spectrometry assay for four adrenal steroids" Clin Chem 2006; 52: 1559-67

[111] Singh RJ, "Validation of a high throughput method for serum/plasma testosterone using liquid chromatography tandem mass spectrometry (LC-MS/MS)" Steriods 2008; 73: 1339-44

[112] Vogeser M, Briegel J, Jacob K, "Determination of serum cortisol by isotope-dilution liquid-chromatography-electrospray ionization tandem mass spectrometry with on-line extraction" Clin Chem Lab Med 2001: 39: 944-7

[113] Yamashita K, Okuyama M, Nakagawa R, Honma S, Satoh F, Morimoto R, "Development of sensitive derivatization method for aldosterone in liquid chromatography-electrospray ionization tandem mass spectrometry of corticosteroids" J Chromatogr A 2008; 1200: 114-21

[114] Carvalho VM, Nakamura OH, Vieira JG, "Simultaneous quantification of seven endogenous C-21 adrenal steroids by liquid chromatography tandem mass spectrometry in human serum" J Chromatogr B 2008; 872: 154-61

[115] Higashi T, Nishio T, Uchida S, Shimada K, Fukushi M, Maeda M, "Simultaneous determination of 17alpha-hydroxypregnenolone and 17alpha-hydroxyprogesterone in dried blood spots from low birth weight infants using LC-MS/MS" J Pharm Biomed Anal 2008; 48: 177-82

[116] Ionita IA, Fast Dm, Akhlaghi F, "Development of a sensitive and selective method for the quantitive analysis of cortisol, cortisone, prednisolone and prednisone in human plasma" J Chromatogr B 2009; 877: 765-72

[117] Rauh M, Groshl M, Rascher W, Dorr HG, "Automated, fast and sensitive quantification of 17 alpha-hydroxy-progesterone, androstenedione, and testosterone by tandem mass spectrometry with on-line extraction" Steriods 2006; 71: 450-8

[118] Shibayamaa Y, Higashia T, Shimadaa K, Odania A, Mizokamib A, Konakab H, "Simultaneous determination of salivary testosterone and dehydroepiandrosterone using LC-MS/MS: method development and evaluation of applicability for diagnosis and medication for late-onset hypogonadism" J Chromatogr B 2009; 877: 2615-23

[119] Taylor PJ, Cooper DP, Gordon RD, Stowasser M, "Measurement of adlosterone in human plasma by semiautomated HPLC-tandem mass spectrometry" Clin Chem 2009: 55: 115-62

[120] Turpeinen U, Hamalainene E, Stenman UH, "Determination of aldosterone in serum by liquid chromatography-tandem mass spectrometry" J Chromatogr B 2008; 862: 113-8

[121] Yamashita K, Okuyama M, Watanabe Y, Homma S, Kobayashi S, Numazawa M, "Highly sensitive determination of estrone and estradiol in human serum by liquid chromatography-electrospray ionization tandem mass spectrometry" Steriods 2007; 72: 819-27

[122] Kushnir MM, Rockwood AL, Nelson GJ, Terry A, Meikle AW, "Analysis of cortisol in urine by LC-MS/MS" Clin Chem 2003; 49: 965-7

[123] Holm SS, Hansen SH, Faber J, Staun-Olsen P, "Reference methods for the measurement of free thyroid hormones in blood: evaluation of potential reference methods for free thyroxine" Clin Biochem 2004; 37: 85-93

[124] The National Academy of Clinical Biochemistry Laboratory Medicine Practice Guidelines: Laboratory support for the diagnosis and monitoring of thyroid tests for the laboratorian and physician, 13; 202 p 1-30

[125] Yue B, Rockwood AL, Sandrock T, La'ulu SL, Kushnir MM, Meikle AW, "Free thyroid hormones in serum by direct equilibrium dialysis and online solid-phase extraction-liquid chromatography/tandem mass spectrometry" Clin Chem 2008; 54: 642-51

[126] Van Uytfanghe K, Stockl D, Ross HA, Thienpont LM, "Use of frozen sera for ft4 standardization: investigation by equilibrium dialysis combined with isotope dilution-mass spectrometry and immunoassay" Clin Chem 2006; 52: 1817-21

[127] Konno N, Hagiwara K, Taguchi H, Murakami S, Taguchi S, "Measurements of serum-free thyroid hormone concentrations by ultrafiltration - a comparison with equilibrium dialysis and mathematical calculation" Ann Nucl Med 1987; 1: 15-22

[128] Tikanoja SH,Liewendahl BK, "New ultrafiltration method for free thyroxin compared with equilibrium dialysis in patients with thyroid dysfunction and nonthyroidal illness" Clin Chem 1990; 36: 800-4

[129] Gu J, Soldin OP, Soldin SJ, "Simultaneous quantification of free triiodothyronine and free thyroxine by isotope dilution tandem mass spectrometry" Clin Biochem 2007; 40: 1386-91

[130] Soldin SJ, Soukhova N, Janicic N, Jonkaas J, Soldin OP, " The measurement of free thyroxine by isotope dilution tandem mass spectrometry" Clin Chim Acta 2005; 358: 113-8

[131] Soukhova N, Soldin OP, Soldin SJ, "Isotope dilution tandem mass spectrometric method for T4/T3" Clin Chim Acta 2004; 343: 185-90

[132] De Brandere VI, Hou P, Stockl D, Thienpont LM, De Leenherr AP, "Isotope dilution-liquid chromatography/electrospray ionization-tandem mass spectrometry for the determination of serum thyroxine as a potential reference method" Rapid Commun Mass Spectrom 1998; 12: 1099-103

[133] Tai SS, Sniegoski LT, Welch MJ, "Candidate reference method for total thyroxine in human serum: use of isotope-dilution liquid chromatography-mass spectrometry with electrospray ionization" Clin Chem 2002; 48: 637-42

[134] Thienpont LM, Fierens C, De Leenheer AP, Przywara L, "Isotope dilution-gas chromatography/mass spectrometry and liquid chromatography/electrospray ionization-tandem mass spectrometry for the determination of triiodo-thyronine in serum" Rapid Commun Mass Spectrom 1999; 13: 1924-31

[135] Moller B, Falk O, Bjorkhem I, "Isotope dilution-mass spectrometry of thyroxin proposed as a reference method" Clin Chem 1983; 29: 2106-10

[136] Ramsen DB, Farmer MJ, "Development os a gas chromatographic selected ion monitoring assat for thyroxine (t4) in human serum" Biomed Mass Spectom 1984; 11: 421-7

[137] Siekmann L, "Measurement of thyroxine in human serum by isotope dilution mass spectrometry: Definitive methods in clinical chemistry" Biomed Environ Mass Spectrom 1987; 14: 683-8

[138] Hopeley CJ, Stokes P, Webb KS, Bayham M, "The analysis of thyroxine in human serum by an "exact matching" isotope dilution method with liquid chromatography/tandem mass spectrometry" Rapid Commun Mass Spectrom 2004; 18: 1033-8

[139] Tai SS, Bunk DM, Et White, Welch MJ, "Development and evaluation of a reference measurement procedure for the determination of total 3,3',5-triiodothyronine in human serum using isotope-dilution liquid chromatography-tandem mass spectrometry" Anal Chem 2004; 76: 5096-6

[140] Thienport LM, "A major step forward in the routine measurement of serum free thyroid hormones" Anal Chem 2008; 54: 625-6

[141] Thienport LM, Bestall G, Christofides ND, Faix JD, Ieiri T, Jarrige V, "Proposal of a candidate international conventional reference measurement procedure for free thyroxine in serum" Clin Chem Lab Med 2007; 45: 934-6

[142] Gu J, Soldin OP, Soldin SJ, "Development and validation of an isotope dilution tandem mass spectrometry method for the simultaneous quantification of 3-iodothyronamine, thyroxine, triiodothyroxine and 3,3',5-triiodo-l-thyronine in humans" Clin Chem 2010; 56:A60

[143] Allen KR, "Interference by venlafaxine ingestion in the detection of tramadol by liquid chromatography linked to tandem mass spectrometry fir the screening of illict drugs in human urine" Clin Tox 2006; 44: 147-53

[144] Kushnir MM, Rockwood AL, Nelson GJ, Yue B, Urry FM, "Assessing analytical specificity in quantitative analysis using tandem mass spectrometry" Clin Biochem 2005; 38: 319-27

[145] Concheiro M, De Castro A, Quintela O, Lopez-Rivadulla M, "Determination of illict drugs and their metabolites in human urine by liquid chromatography tandem mass spectrometry including relative ion intensity criterion

[146] Savage FL, Gaulier JM, Lachatre G, Marquet P, "Pitfalls and prevention strategies for liquid chromatography-tandem mass spectromrty in the selected reaction-monitoring mode for drug analysis" Clin Chem 2008; 54: 1519-27

[147] Marquet P, Saint-Marcoux F, Gamble TN, LeBlanc JCY, "Comparison of a preliminary procedure for the general unknown screening of drugs and toxic compounds using a quadrupole-linear-ion-trap mass spectrometer with a liquid chromatography-mass spectrometry reference technique" J Chrom B 2003; 789: 9-18

[148] Saint-Marcoux F, Lachatre G, Marquet P, "Evaluation of an improved general unknown screening procedure using liquid-chromatography-electrospray-mass spectrometry by comparison with gas chromatography and high-performance liquid-chromatography-diode array detection" J Am Soc Mass Spectom 2003; 14: 14-22

[149] Mueller CA, Weinman W, Dresen S, Schreiber A, Gergov M, "Development of a multi-target screening analysis for 301 drugs using a Qtrap liquid chromatography/tandem mass spectrometry system and automated library searching" Rapid Commun Mass Spectrom 2005; 19: 1332-38

[150] Dresen S, Kempt J, Weinmann W, "Electrospray-ionization MS/MS library of drugs as database for method development and drug identification" For Sci Int 2006; 161: 86-91

[151] Dresen S, Gergov M, Politi L, Halter C, Weinmann W, "ESI-MS/MS library of 1,253 compounds for application in forensic and clinical toxicology" Anal Bioanal Chem 2009; 395: 2521-6

[152] Dresen S, Ferreiro N, Gnann H, Zimmerman R, Weinmann W, "Detection and identification of 700 drugs by multi-target screening with a 3200 Q Trap® LC-MS/MS system and library searching" Anal Bioanal Chem 2010; 396: 2425-34

[153] Politi L, Morini L, Polettini A, "A direct screening procedure for diuretics in human urine by liquid chromatography-tandem mass spectrometry with information dependent acquisition" Clin Chim Acta 2007; 386: 46-52

[154] Liu HC, Liu RH, Ho HO, Lin DL, "Development of an information-rich LC-MS/MS database for the analysis of drugs in post-mortem specimens" Anal Chem 2009; 81: 9002-11

[155] Liu HC, Liu RH, Lin DL, Ho HO, "Rapid screening and confirmation of drugs and toxic compounds in biological specimens using liquid chromatography/ion trap tandem mass spectrometry and automated library searching" Rapid Commun Mass Spectrom 2010; 24: 75-84

[156] Lynch KL, Breaud AR, Vandenberghe H, Wu AHB, Clarke W, "Performance evaluation of three liquid chromatography mass spectrometry methods for broad spectrum drug screening" Clin Chim Acta 2010; 411: 1474-81

[157] Maurer HH, "Perspectives of liquid chromatography coupled to low- and high-resolution mass spectrometry for screening, identification and quantification of drugs in clinical and forensic toxicology" Thera Drug Monit 2010; 32: 324-7

[158] Gregov M, Boucher B, Ojanpera I, Vuori E, "Toxicological screnning of urine for drugs by liquid chromatography/time-of-flight mass spectrometry with automated target library based on elemental formulas" Rapid Commun Mas Spectom 2001; 15: 521-6

[159] Pelander A, Ojanpera I, Laks S, Rasanen I, Vuori E, "Toxicological screening with formula-based metabolite identification by liquid-chromatography/time-of-flight mass spectrometry" Anal Chem 2003; 75: 5710-8

[160] Pelander A, Ristimaa J, Rasanen I, Vuori E, Ojanpera I, "Screening for basic drug in hair of drug adducts by liquid chromatography/time-of-flight mass spectrometry" Thera Drug Monit 2008; 30: 717-24

[161] Pelander A, Ristimaa J, Ojanpera I, "Vitreous humor as an alternative matrix for comprehensive drug screening in postmortem toxicology by liquid chromatography-time-of flight mass spectrometry" J Anal Tox 2010; 34; 312-8

[162] Ojanpera I, Pelander A, Laks S, Gergov M, Vuori E, Witt M, "Application of accurate mass measurement to urine drug screening" J Anal Tox 2005; 29: 34-40

[163] Polettini A, Gottardo R, Pascali JP, Tagliaro F, "Implemantation and performance evaluation of a database of pharmaco/toxicologically relevant compounds in biological samples using electrospray ionization-time-of-flight mass spectrometry" Anal Chem 2008; 80: 3050-7

[164] Lee HK, Ho CS, Iu YO, Lai PSJ, Shek CC, Lo YC, Klinke HB, Wood M, "Development of a board toxicological screening technique for urine using ultra-performance liquid chromatography and time-of-flight mass spectrometry" Anal Chim Acta 2009; 649: 80-90

[165] Palvic M, Libiseller K, Oberacher H, "Combined use of ESI-QqTOF-MS and ESI-QqTOF-MS/MS with mass-spectral library search for qualitative analysis of drugs" Anal Bioanal Chem 2006; 386: 69-82

[166] Decaestecker TN, Vande Casteele SR, Wallemaq P, Van Petegham CH, Defore DL, Van Bocxlaer JF, "Information-dependent acquisition mediated LC-MS/MS screening procedure with semi-quantitative potential" Anal Chem 2004; 76: 6365-73

[167] Peters RJB, Oosterink JE, Stolker AAM, Georgakopoulos C, Neilen MWF, "Generic sample preparation combined with high-resolution liquid chromatography-time-of-flight mass spectrometry for unification of urine screening in doping-control laboratories" Anal Bioanal Chem 2010; 396: 2583-98

[168] Ojanpera S, Pelander A, Pelzing M, Krebs I, Vuori E, Ojanpera I, "Isotopic pattern and accurate mass determination in urine drug screening by liquid-chromatography/time-of-flight mass spectrometry" Rapid Commun Mass Spectrom 2006; 20: 1161-7

[169] Liotta E, Gottardo R, Bertaso A, Polettini A, "Screening for pharmaco-toxicologically relevant compounds in biosamples using high-resolution mass spectrometry: a "metabolomic" approach to the discrimination between isomers" J Mass Spectrom 2010; 45:261-71

[170] Tyrkko E, Pelander A, Ojanper I, "Differentiation of structural isomers in a target drug database by LC/Q-TOFMS using fragmentation prediction" Drug Test Analysis 2010; 2: 259-70

[171] Thomas A, Guddat S, Kohler M, Krug O, Schanzer W, Petrou M, Thevis M, "Comprehensive plasma-screening for known and unknown substances in doping controls" Rapid Commun Mass Spectrom 2010; 24: 1124-32

[172] Weider ME, Brown PR, Grainger L, Teale P, "Identification of etamiphylline and metabolites in equine plasma and urine by accurate mass and liquid chromatography/tandem mass spectrometry" Dru Test Analysis 2010; 2: 271-7

[173] Johnson K, Kozak M, "Implemantation of ultra-high resolution mass spectrometer Exactive™ for analysis of EtG and EtS in human urine" Toxichem Krimtech 2010; 77: 262

[174] Rivier L, "Criteria for the identification of compounds by liquid chromatography-mass spectrometry and liquid chromatography-multiple mass spectrometry in forensic toxicology and doping analysis" Anal Chim Acta 2003; 492: 69-82

[175] Maralikova B, Weinmann W, "Confirmatory analysis for drugs of abuse in plasma and urine by high-performance liquid chromatography-tandem mass spectrometry with respect to criteria for compound identification" J Chrom B 2004; 811: 21-30

[176] de Zeeuw RA, "Substance identification: the weak link in analytical toxicology" J Chrom B 2004; 811: 3-12

[177] Fox EJ, Twigger S, Allen KR, "Criteria for opiate identification using liquid chromatography linked to tandem mass spectrometry: problems in routine practice" Annals Clin Biochem 2009; 46:50-7

[178] Chase DH, Hillman SL, Millington DS, Kahler CR, Naylor EW, "Rapid diagnosis of maple syrup urine disease in blood spots from newborns by tandem mass spectrometry" Clin Chem 1995; 41:62-8

[179] Chase DH, Hillman SL, Millington DS, Kahler SG, Adam BW, Levy HL, "Rapid diagnosis of homocytinura and other methionemias from newborns' blood spots by tandem mass spectrometry" Clin Chem 1996; 42: 349-55

[180] Rashed MS, Bucknail MP, Little D, Awad A, Jacob M, Alamoudi M, "Screening blood spots for inborn errors of metabolism by electrospray tandem mass spectrometry with a microbatch process and a computer algorithm for automated flagging of abnormal profiles" Clin Chem 1997; 43: 1129-41

[181] Millington DS, Terada N, Chace DH, Chen YT, Ding JH, Kodo N, "The role of tandem mass spectrometry in the diagnosis of fatty acid oxidation disorders" Prog Clin Biol Res 1192; 375: 339-54

[182] Millington DS, Kodo N, Norwood DL, Roe CR, "Tandem mass spectrometry: a new method for acylcaritine profiling with potential for neonatal screening for inborn errors of metabolism" J Inherit Metab Dis 1990; 13: 321-4

[183] Qu J, Chen W, Luo G, Wang Y, Xiao S, Ling Z, Chen G, "Rapid determination of underivatized pyroglutamic acid, glutamic acid, glutamine and other relevant amino acids in fermentation media by LC-MS-MS" Analyst 202; 127: 66-9

[184] Trinh MU, Blake J, Harrison JR, Gerace P, Ranier E, Janice M, Fletcher JM, Johnson DW, "Quantification of glutamine in dried blood spots and plasma by tandem

mass spectrometry for the biochemical diagnosis and monitoring of ornithine transcarbamylase deficiency" Clin Chem 2003; 49: 681-4

[185] La Marca G, Malvagia S, Pasquini E, Innocenti M, Maria A, Donati MA, Zammarchi E, "Rapid 2nd-tier test for measurement of 3-OH-propionic acid and methylmalonic acids on dreid blood spots: reducing the false-positive rate for propionylcarnitine during expanded newborn screening by liquid chromatography-tandem mass spectrometry" Clin Chem 2007; 53: 1364-9

[186] Chace DH, Singleton S, Diperna J, Aiello M, Foley T, "Rapid metabolic and newborn screening of thyroxine (T4) from dried blood spots by MS/MS" Clin Chim Acya 2009; 403: 178-83

[187] Li Y, Brockmann K, Turecek F, Scott CR, Gelb MH, "Tandem mass spectrometry for direct assay of enzymes in dried blood spots: application to newborn screening for Kabbe disease" Clin Chem 2004; 50:638-40

[188] Zhang XK, Elbin CS, Chuang WI, Cooper SK, Marashio CA, Beauregrad C, Keutzer JM, "Multiplex enzyme assay screening of dried blood spots for lysosmal storage disorders by using tandem mass spectrometry" Clin Chem 2008: 54: 1725-8

[189] La Marca G, Casetta B, Malvagia S, Guerrini R, Zammarchi E, "New strategy for the screening of lysosomal storage disorders: the use of the online trapping-and-cleanup liquid chromatography/mass spectrometry" Anal Chem 2009; 81: 6113-21

[190] Kasper DC, Herman JL, De Jesus VR, Mechtler TP, Metz TF, Shushan B, "The application of multiplexed, multi-dimensional ultra-high-performance liquid chromatography/tandem mass spectrometry to the high-throughput screening of lysosomal storage disorders in newborn dried bloodspots" 2010; 15: 986-94

[191] Metz TF, Mechtler TP, Orsini JJ, Martin M, Shushan B, Herman JL, Ratschmann R, Item CB, Streubel B, Herkner KR, Kasper DC, "Simplified Newborn Screening Protocol for Lysosomal Storage Disorders" Clin Chem. 2011; 57: 1286-94

[192] Barr JR, Maggio VL, Patterson DG, Jr., Cooper GR, Henderson LO, Turner WE, Smith SJ, Hannon WH, Needham LL, Sampson EJ, "Isotope dilution-mass spectrometric quantification of specific proteins: Model application with apolipoprotein A-I" Clin Chem 1996: 42:1676–82

[193] Gerber SA, Rush J, Stenman O, Kirschner MW, Gygi SP, "Absolute quantification of proteins and phosphoproteins from cell lysates by tandem MS" Proc Natl Acad Sci USA 2003; 100:6940–45

[194] Anderson NL, Anderson NG, Haines LR, Hardie DB, Olafson RW, Pearson TW "Mass spectrometric quantitation of peptides and proteins using stable isotope standards and capture by anti-peptide antibodies (SISCAPA)" J Proteome Res 2004; 3:235–44

[195] Anderson NL, Jackson A, Smith D, Hardie D, Borchers C, PearsonTW, "SISCAPA peptide enrichment on magnetic beads using an in-line bead trap device" Mol Cell Proteomics 2009; 8: 995–1005

[196] Anderson L, Hunter CL, "Quantitative mass spectrometric multiple reaction monitoring assays for major plasma proteins" Mol Cell Proteomics 2006; 5:573–88

[197] Hoofnagle AN, Becker JO,Wener MH, Heinecke JW, "Quantification of thyroglobulin, a low-abundance serum protein, by immunoaffinity peptide enrichment and tandem mass spectrometry" Clin Chem 2008; 54:1796–804

[198] Carr SA, Anderson L, "Protein quantitation through targeted mass spectrometry: The way out of biomarker purgatory?" Clin Chem 2008; 54: 1749–52

[199] Bondar OP, Barnidge DR, Klee EW, Davis BJ, Klee GG, "LC-MS/MS quantification of Zn-a2 glycoprotein: A potential serum biomarker for prostate cancer" Clin Chem 2007; 53:673–8

[200] Babic N, Larson TS, Grebe SK, Turner ST,Kumar R, Singh RJ, "A liquid chromatography-mass spectrometry method for the quantification of urinary albumin using a novel ^{15}N-isotopically labeled albumin internal standard" Clin Chem 2007; 53:540–542

[201] Ackermann BL, Berna MJ, "Coupling immunoaffinity techniques with MS for quantitative analysis of low-abundance protein biomarkers" Expert Rev Proteomics 2007; 4:175- 86

[202] Berna M, Ackermann B, "Increased throughput for low-abundance protein biomarker verification by liquid chromatography/tandem mass spectrometry" Anal Chem 2009;81:3950–6

[203] Hortin GL, Carr SA, Anderson LN, "Introduction: Advances in Protein Analysis for the Clinical Laboratory" Clinical Chemistry 2010; 56: 149–51

[204] Hoofnagle AN, "Quantitative Clinical Proteomics by Liquid Chromatography-Tandem Mass Spectrometry: Assessing the Platform" Clinical Chemistry 2010; 56: 161–4

[205] Addona TA, Abbatiello SE, Schilling B, Skates SJ, Mani DR, Bunk DM, Spiegelman CH, Zimmerman LJ, Ham AJL, Keshishian H, Hall SC, Allen S, Blackman RK, Borchers CH, Buck C, Cardasis HL, Cusack MP, Dodder NG, Gibson BW, Held JM, Hiltke T, Jackson A, Johansen EB, Kinsinger CR, Jing Li, Mesri M, Neubert TA, Niles RK, Pulsipher TC, Ransohoff D, Rodriguez H, Rudnick PA, Smith D, Tabb DL, Tegeler TJ, Variyath AM, Vega-Montoto LJ, Wahlander A, Waldemarson S, Wang M, Whiteaker JR, Zhao L, Anderson NL, Fisher SJ, Liebler DC, Paulovich AG, Regnier FE, Tempst P, S.A.Carr, "Multi-site assessment of the precision and reproducibility of multiple reaction monitoring-based measurements of proteins in plasma." Nat Biotechnol. 2009; 27: 633–41

[206] Diamandis EP, "Protein quantification by mass spectrometry: Is it ready for prime time?" Clin Chem 2009; 55:1427–30

[207] Anderson NL, "The Clinical Plasma Proteome: A Survey of Clinical Assays for Proteins in Plasma and Serum" Clinical Chemistry 2010; 56: 177–85

[208] Paulovich AG, Whiteaker JR, Hoofnagle AN, Wang P, "The interface between biomarker discovery and clinical validation: The tar pit of the protein biomarker pipeline" Proteomics Clin Appl 2008; 2:1386–1402).

[209] Borges CR, Rehder DS, Jarvis JW, Schaab MR, Oran PE, Nelson RW, "Full-Length Characterization of Proteins in Human Populations" Clin Chem 2010; 56: 202–11

[210] Niederkofler EE, Kiernan UA, O'Rear J, Menon S, Saghir S, Protter AA, "Detection of endogenous B-type natriuretic peptide at very low concentrations in patients with heart failure" Circ Heart Fail 2008, 1: 258 - 64

[211] Lopez MF, Rezai T, Sarracino DA, Prakash A, Krastins B, Athanas M, Singh RJ, Barnidge DR, Oran P, Borges C, Nelson RW, "Design of a Selected Reaction

Monitoring Mass Spectrometric Immunoassay Responsive to Human Parathyroid Hormone (PTH) and Related Variants" Clin Chem 2010; 56: 281–90

[212] Ala A, Walker AP, Ashkan K, Dooley JS, Schilsky ML, "Wilson's disease". Lancet 2007; 369 (9559): 397–408

[213] deWilde A, Sadilkova K, Sadilek M,VastaV, Hahn SH. 2008, "Tryptic peptide analysis of ceruloplasmin in dried blood spots using liquid chromatography–tandem mass spectrometry: Application to newborn screening" Clin Chem 2008; 54:1961–68

[214] Rai DK, Landin B, Alvelius G, Griffiths WJ, "Electrospray tandem mass spectrometry of intact β-chain hemoglobin variants" Anal Chem 2002; 74:2097–102

[215] Bateman RH, Green BN, Morris B, "Electrospray ionization mass spectrometric analysis of the globin chains in hemoglobin heterozygotes can detect the variants HbC, D, and E" Clin Chem 2008; 54:1256–57

[216] Kleinert P, Schmid M, Zurbriggen K, Speer O, Schmugge M, Roschitzki B, Durka SS, Leopold U, Kuster T, Heizmann CW, Frischknecht H, Troxler H, "Mass spectrometry: A tool for enhanced detection of hemoglobin variants" Clin Chem 2008; 54:69–76

[217] Awalt E, Barnard D, Dooley K, "Tandem mass spectrometry—A screening tool for hemoglobinopathies" Clin Chem 2001; 47(6 Suppl): A165

[218] Daniel YA, Turner C, Haynes RM, Hunt BJ, Dalton RN, "Rapid and specific detection of clinically significant haemoglobinopathies using electrospray mass spectrometry-mass spectrometry" Br J Haematol 2005; 130:635–43

[219] Daniel YA, Turner C, Haynes RM, Hunt BJ, Dalton RN, "Quantification of hemoglobin A2 by tandem mass spectrometry" Clin Chem 2007; 53:1448–54

[220] Boemer F, Ketelslegers O, Minon J-M, Bours V, Schoos R, "Newborn screening for sickle cell disease using tandem mass spectrometry" Clin Chem 2008; 54:2036–41

[221] Daniel Y, Turner C, Farrar L, Dalton RN, "A comparison of IEF and MSMS for clinical hemoglobinopathy screening in 40,000 newborns" 50th Annual Meeting of the American Hematology Society, San Francisco, CA, December 2008

[222] Ellis DI, Dunn WB, Griffin JL, Allwood JW, Goodacre R, "Metabolic fingerprinting as a diagnostic tool" Pharmacogenomics 2007; 8:1243–1266

[223] Psychogios N, Hau DD, Peng J, Guo AC, Mandal R, Bouatra S, Sinelnikov I, Krishnamurthy R, Eisner R, Gautam B, Young N, Xia J, Knox C, Dong E, Huang P, Hollander Z, Pedersen TL, Smith SR, Bamforth F, Greiner R, McManus B, Newman JW, Goodfriend T, Wishart DS, "The human serum Metabolome" PLoS One 2011; 6: e16957, 1-23

[224] Altmaier E, Ramsay SL, Graber A, Mewes H-W, Weinberger KM, Suhre K, "Bioinformatics analysis of targeted metabolomics—Uncovering old and new tales of diabetic mice under medication" Endocrinology 2008; 149: 3478–89

[225] Suhre, et.al., "Metabolic Footprint of Diabetes: A Multiplatform Metabolomics Study in an Epidemiological Setting" PLoS One 2101; 5: e13953

[226] Gieger CH, Geistlinger L, Altmaier E, de Angelis H, Kronenberg F, Meitinger T, Mewes H-W, Wichmann H-E, Weinberger KM, Adamski J, Illig T, Suhre K, "Genetics meets metabolomics: A genome-wide association study of metabolite profiles in human serum" PLoS One 2008; Genet 4: e1000282

[227] Illig, T., et.al., "A Genome-wide Perspective of Genetic Variation in Human Metabolism, Nature Genetics" 42(2) 2010, 137-41

[228] Wichmann HE, Gieger C, Illig T, for the KORA Study Group, "KORA-gen. Resource for population genetics, controls and a broad spectrum of disease phenotypes" Gesundheitwesen 2005; 67: 26–30

[229] Makarov A, "Electrostatic axially harmonic orbital trapping: A high performance technique of mass analysis" Anal Chem 200; 72: 1156–62

[230] Bateman KP, Kellmann M, Muenster H, Papp R, Taylor L, "Quantitative-qualitative data acquisition using a bench-top Orbitrap mass spectrometer" J Am Soc Mass Spectrom 2009; 20: 1441–50

[231] Zhang NR, Yu S, Tiller P, Yeh S, Mahan E, Emary WB, "Quantitation of small molecules using high-resolution accurate mass spectrometers—A different approach for analysis of biological samples" Rapid Commun Mass Spectrom 2009; 23: 1085–94

[232] Koulman A,Woffendin G, Narayana VK,Welchman H, Crone C, Volmer DA, "High-resolution extracted ion chromatography, a new tool for metabolomics and lipidomics using a second-generation orbitrap mass spectrometer" Rapid Commun Mass Spectrom 2009; 23:1411–8

[233] Schwudke D, Hannich JT, SurendranathV, GrimardV, Moehring T, Burton L, Kurzchalia T, Shevchenko A, "Top-down lipidomic screens by multivariate analysis of high-resolution survey mass spectra" Anal Chem 2007; 79:4083–93

[234] Rabinowitz JD, Lu W, Clasquin ML, Melamud E, Amador-Noguez D, Caudy AA, "Metabolomic Analysis via Reversed-Phase Ion-Pairing Liquid Chromatography Coupled to a Stand Alone Orbitrap Mass Spectrometer" Anal. Chem. 2010, 82, 3212–21

29

Tandem Mass Spectrometry of Sphingolipids: Application in Metabolic Studies and Diagnosis of Inherited Disorders of Sphingolipid Metabolism

Ladislav Kuchař, Befekadu Asfaw and Jana Ledvincvá
Institute of Inherited Metabolic Disorders, First Faculty of Medicine, Charles University in Prague and General University Hospital in Prague
Czech Republic

1. Introduction

Sphingolipids are an amazingly diverse category of lipids found in all eukaryotes and in some prokaryotes and viruses. They are primarily a component of plasma membranes and of intracellular organelle membranes, including those of the nucleus, mitochondria, endosomes, and lysosomes (Hirabayashi, et al., 2006; Kaushik, et al., 2006; R. Ledeen & Wu, 2011; R. W. Ledeen & Wu, 2008; Prinetti, et al., 2009; van Meer, et al., 2008). Sphingolipids are also an important constituent of plasma lipoprotein classes (Schweppe, et al., 2010; Wiesner, et al., 2009) and of the multilamellar water barrier of the skin (Holleran, et al., 2006). In addition, they are excreted in urine, mostly in the cellular debris of urinary sediment. Urinary sediment analysis, or "indirect biopsy," of kidney cellular elements (Desnick, et al., 1970) can provide information that helps to diagnose certain lysosomal storage diseases (Kitagawa, et al., 2005; Kuchar, et al., 2009; Whitfield, et al., 2001).

Sphingolipids are a heterogenous group. Amide bonds link long-chain fatty acids to aminoalcohols from the sphingoid group, of which sphing-4-enin ({2S,3R,4R}-2-aminooctadec-4-ene-1,3-diol, historically called sphingosine) and its saturated derivative (sphinganine) are the most abundant. Longer or shorter sphingoids, which may be saturated or hydroxylated, also occur in lesser quantities.

The name sphingosine was chosen by German clinician and chemist J. L. W. Tchudichum in 1884 to reflect the enigmatic, "Sphinx-like" properties of the sphingolipid compounds first isolated from the brain. Fatty acid variations include mostly C16-C24 acyl chains, which are often saturated but can also exhibit a degree of unsaturation or hydroxylation (e.g., C24:1, C24:1-OH fatty acids). The general name for N-acylated sphingoids is ceramide. Sphingomyelin and glycosphingolipids have a headgroup in the phosphodiester or glycosyl linkage to the hydroxyl on the carbon-1. The latter compounds are classified as either *neutral glycosphingolipids* with uncharged sugars (glucose, galactose, N-acetylglucosamine, N-acetylglalactosamine and fucose) or *acidic glycosphingolipids* with ionized functional groups (sulfates) or charged sugar moieties (N-acetylneuraminic acid or "sialic" acid).

Two examples of sphingolipid structures are shown in Fig. 1.

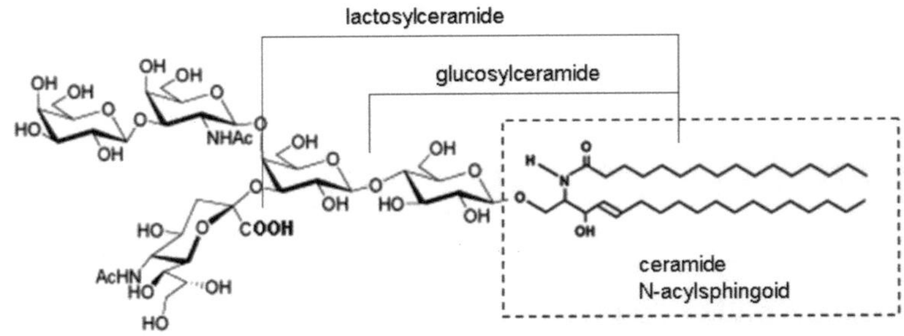

GM1 ganglioside (Galβ1-3GalNAcβ1-4(NeuAcα2-3)Galβ1-4Glcβ-Cer)

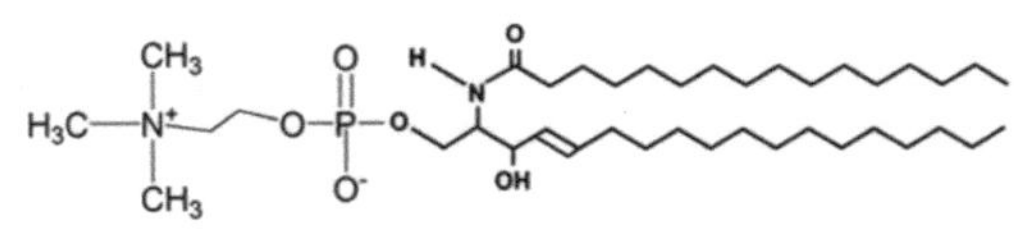

Sphingomyelin (N-acyl-sphing-4-enine-1-phosphocholine)

Fig. 1. GM1 ganglioside - a representative acidic sphingolipid; sphingomyelin - a representative phosphosphingolipid.

The simplest glycosphingolipids are the monohexosylceramides glucosylceramide (GlcCer, Glcβ1-1´Cer) and galactosylceramide (galactocerebroside, GalCer, Galβ1-1´Cer). The latter is less abundant but is specific to neural tissue, where it is also present in a sulfated form (sulfatide). Galactosylceramide also gives rise to small Gala-series of glycosphingolipids (Tab. 1). Glucosylceramide is a key compound in sphingolipid metabolic pathways; more complex glycosphingolipids are derived from the stepwise elongation of the oligosaccharide chain in the Golgi compartment. The addition of β-linked galactose yields lactosylceramide (LacCer, Galβ1-4GlcβCer), whose oligosaccharide chain is the precursor to the different core structures of more complex glycosphingolipids (Tab. 1). These structures are specific to certain tissues: e.g., neolacto-series predominate in leukocytes, lacto-series in secretory organs, globo-series in erythrocytes and ganglio-series in nervous tissue (Schnaar, et al., 2009). This diversity is related to the functional differences between the individual glycosphingolipids.

The catabolism of sphingolipids occurs in acidic cell compartments, i.e. late endosomes and lysosomes. Degraded lipids are embedded in inner membranes rich in negatively charged lipids such as bis(monoacylglycero)phosphate (BMP) (Kolter & Sandhoff, 2010). Sequential degradation steps proceed from the non-reducing end of the oligosaccharide chain catalyzed by soluble lysosomal hydrolases. For lipids with oligosaccharide chains shorter than four sugars, the assistance of small sphingolipid activator proteins (the saposins A, B, C, or D or the GM2 activator protein) is required (Sandhoff, et al., 2001). The sphingoids and fatty acids produced can be degraded in the cytoplasm and processed through the salvage pathway, where they become the building blocks of new membranes (Kitatani, et al., 2008). They can also be used in the regulation systems that control cell function (Kolter & Sandhoff, 2010).

Inherited defects in gene-coding enzymes or proteins involved in sphingolipid degradation result in the accumulation of non-degraded substrates in the lysosomes. "Traffic jams" in the endolysosomal system caused by the accumulation of lipids co-precipitated with other

hydrophobic substances severely impair cell function and lead to lysosomal storage diseases (sphingolipidoses) (Desnick, et al., 2001; Liscum, 2000; Sandhoff, et al., 2001; Schulze & Sandhoff, 2011; von Figura, et al., 2001). Studies of these defects, however, may unveil the complicated mechanisms of cell function and regulation. Recent information learned about the role of NPC1 and NPC2 proteins in the intracellular transport of cholesterol can serve as an example (Infante, et al., 2008; Kwon, et al., 2009; Storch & Xu, 2009; Xu, et al., 2008).
Hitherto, more than 400 structurally distinct sphingolipid variants in mammals have been listed in SphinGOMAP© (http://www.sphingomap.org and http://www.glycoforum.gr.jp) (Aug 2011). A detailed sphingolipid nomenclature is available at http://www.chem.qmul.ac.uk/iupac/lipid/lip1n2.html (Aug 2011) and http://www.chem.qmul.ac.uk/iupac/lipid/lip3n4.html (Aug 2011).

Root names of series	Core structures	Abbreviations
Lacto-	Gal(β1→3)GlcNAc(β1→3)Gal(β1→4)Glc(β1→1')Cer	Lc
Neolacto-	Gal(β1→4)GlcNAc(β1→3)Gal(β1→4)Glc(β1→1')Cer	nLc
Globo-	GalNAc(β1→3)Gal(α1→4)Gal(β1→4)Glc(β1→1')Cer	Gb
Isoglobo-	GalNAc(β1→3)Gal(α1→3)Gal(β1→4)Glc(β1→1')Cer	iGb
Ganglio-	Gal(β1→3)GalNAc(β1→4)Gal(β1→4)Glc(β1→1')Cer	Gg
Gala-	Gal(α1→4)Gal(β1→1')Cer	Ga

Note: key structures characteristic for each series are underlined

Table 1. The major root structures of vertebrate glycosphingolipids.

Physiological sphingolipid function at the cellular level is highly complex. Sphingolipids participate in many cellular events, including cell-cell recognition, the modulation of membrane protein functions, adhesions, intra- and extra-cellular signaling and many still undiscovered processes (Kitatani, et al., 2008). Their rigid, highly saturated character and ability to undergo hydrogen bonding and dipolar interactions predetermines them to cluster into semiordered structures called lipid microdomains - rafts, together with cholesterol and specific set of proteins (Goldschmidt-Arzi, et al., 2011; Helms & Zurzolo, 2004; Holthuis, et al., 2003). They function as important mediators of membrane transport and signaling.
In recent years, investigation of the metabolism and biological functions of sphingolipid biomolecules has increased (Wennekes, et al., 2009). As a result, more accurate methods of analyzing sphingolipids have been developed. A leading method of analysis, tandem mass spectrometry (tandem MS or MS/MS), provides high selectivity and sensitivity of measurement. Indisputable advantage of this technique is a possibility to identify various molecular species of different sphingolipid classes in crude biological samples.
We will focus on the contribution of tandem mass spectrometry to the study of sphingolipids and the usefulness of the technique in diagnosing inherited disorders of sphingolipid degradation. The topics discussed will include the following:

- Tandem mass spectrometry used to analyze sphingolipids in tissues, cells and urine (both the general approach used and its applications in the diagnosis of sphingolipidoses)
- The investigation of the sphingolipid degradation pathway in living cells using stable isotopes or atypical fatty acid labeling
- Sphingolipid isoform profiling, which is a useful tool for diagnosing disorders associated with Gb3Cer and sulfatide storage.

2. Electrospray ionization tandem mass spectrometry of sphingolipids

The equipment used was an AB/MDS SCIEX API 3200 triple quadrupole mass spectrometer. Multiple reaction monitoring was used in positive (neutral glycosphingolipids) and negative (acidic glycosphingolipids) ion mode. Superior sensitivity and selectivity were exhibited when the technique was used in conjunction with normal phase HPLC separation.

2.1 Electrospray ionization

Electrospray ionization abbreviated as ESI is one of the soft ionization methods used in mass spectrometry (Cole, 2010; de Hoffman & Stroobant, 2002). A strong electric field is applied to the analyte solution as it passes through a metal capillary at atmospheric pressure. This field is created using a high voltage (a voltage of up to 6kV) between the capillary tip and the counterelectrode. Droplets of analyte are formed under these conditions on the capillary tip, with ions generated and sorted on the surface. Cations are formed in positive ion mode and anions in negative ion mode. The generation of ions results from the electrochemical processes that occur on the capillary tip and the strong electric field used.

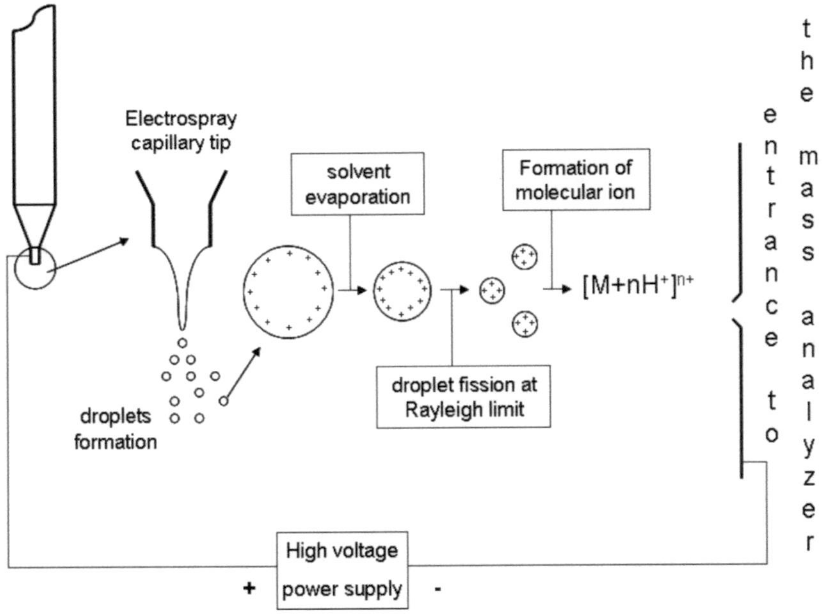

Fig. 2. Scheme of electrospray ionization process.

Droplets flow in the direction of the electric field and shrink in size as the solvent evaporates. When the accumulated surface charge of a droplet exceeds the surface tension force, the droplet breaks into smaller droplets via Rayleigh fission. This process is repeated until the ions on the surface of the droplet are able to overcome the forces holding them, at which point molecular ions are formed (Fig. 2) (Cole, 2010; de Hoffman & Stroobant, 2002; Dulcks & Juraschek, 1999).

The distribution of compounds on the droplet surface results from their relative concentration and solubility. Less soluble compounds tend to be on the surface rather than in the bulk of the solution, which affects their ionization efficiency. The surface of each droplet is limited; thus, the concentration is more important than the total amount of the compound injected in the source (Cole, 2010; de Hoffman & Stroobant, 2002).
Matrix effects occur when additional compound ions are generated on the surface. These compounds can completely mask the analyte (Fig. 3). The salt concentration has a similar effect on ionization; the maximum tolerable concentration of salts is approximately 10^{-3}M (Cole, 2010; de Hoffman & Stroobant, 2002).
The matrix effect can negatively influence electrospray ionization. It is important to choose a sample preparation technique that removes interfering compounds. The use of HPLC or capillary electrophoresis separation prior to electrospray ionization can minimize the matrix effect (Cole, 2010; Micova, et al., 2010); indeed, we saw a 10-fold increase in signal intensity when HPLC was used (data not shown). Established methods of evaluating matrix effects are described in the literature (Taylor, 2005).

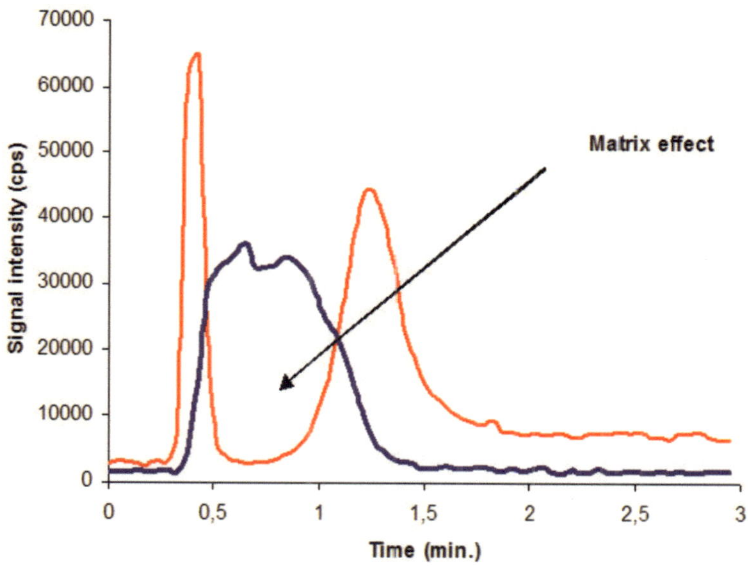

Fig. 3. Matrix effect interference in electrospray ionization conducted on a urinary sample. Total ion current (TIC) of ceramide, ceramide dihexoside and sphingomyelin. Red line: urinary sample after Folch extraction without purification. Visible signal decrease was caused by matrix effect. Blue line: sample after the purification process, which removed the matrix effects.

2.2 Quadrupole mass spectrometer

The quadrupole is a device that uses the stability of ion trajectories in oscillating electric fields to separate ions according to their m/z ratio (de Hoffman & Stroobant, 2002).
The device consists of four parallel circular or hyperbolic metal rods to which radiofrequency voltage (RF) and direct current (DC) are applied. The total electric field is composed of quadrupolar alternative fields superposed on a constant field resulting from

the application of potential on the rods. The result is a mass filter used to separate ions according to their m/z ratio (de Hoffman & Stroobant, 2002; Douglas, 2009).
The mathematical description of the total electric field is based on Mathieu equations (de Hoffman & Stroobant, 2002; Douglas, 2009). The equations can be used to generate a stability diagram for ions in the field. Direct current voltage (U) and radiofrequency voltage amplitude (V) are the only variables in the equations used to define ion position in these diagrams. The other parameters are constants, including the mass and the charge. Thus, changes in U and V determine a given ion's position in the diagram. Only ions in the stable region are able to pass the quadrupole mass filter through the stable trajectories. By continually changing U and V, the quadrupole mass spectrometer is able to scan for ions with different m/z ratios (Fig. 4).

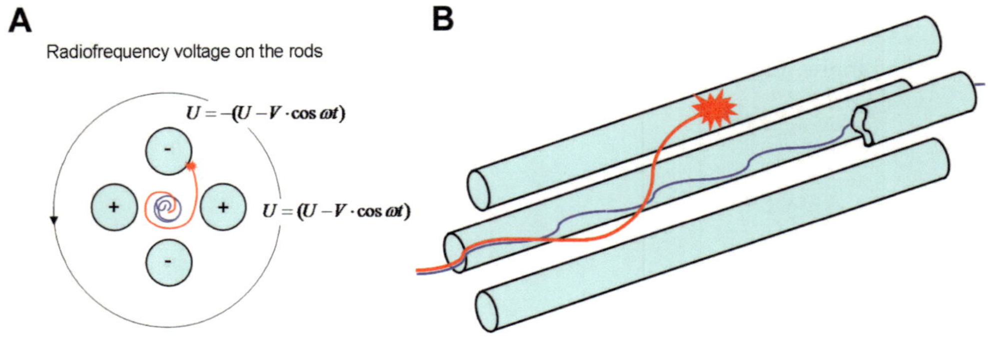

Fig. 4. A quadrupolar mass filter and the trajectories of two ions. Red line: an ion in an unstable region of the stability diagram on an unstable trajectory. Blue line: an ion in a stable region of the stability diagram passing the quadrupole on a stable trajectory. A) Section of the quadrupole illustrating radiofrequency voltage; B) Upper side view of the ion trajectories in the quadrupole.

Practically speaking, quadrupole mass spectrometers are hampered by mass resolution and mass range limitations. The useful level of resolution is one mass unit, and the highest detectable m/z is 4000 (de Hoffman & Stroobant, 2002; McLuckey & Wells, 2001).

2.3 Triple quadrupole tandem mass spectrometry

Triple quadrupole tandem mass spectrometry uses three quadrupoles in a series. The first and third quadrupoles function as mass analyzers, whereas the middle quadrupole, with only the radio frequency voltage used, is employed as a collision cell that does not separate ions according to their m/z ratios. Instead, it works as an ion channel that systematically returns ions to the center of the rods (de Hoffman & Stroobant, 2002; McLuckey & Wells, 2001).
Collision induced dissociation is a process of changing the accelerated ion kinetic energy to internal energy using collisions in a collision cell filled with inert gas (N_2, Ar, or He). Analyte fragments are produced if the volume of accumulated internal energy is greater than the energy of the chemical bonds in the molecule (Cole, 2010; de Hoffman & Stroobant, 2002; McLuckey & Wells, 2001; Sleno & Volmer, 2004).

The static mode (in which only one selected m/z is measured) and the scanning mode (in which a range of m/z ratios is measured) are both used with the first and third quadrupole to attain four specific types of measurement (Cole, 2010; de Hoffman & Stroobant, 2002; McLuckey & Wells, 2001; Sleno & Volmer, 2004):

1. Product ion scans. To attain these measurements, the first quadrupole is used in the static mode, whereas the third quadrupole is used to scan products of collision induced dissociation.
2. Precursor ion scans. To attain these measurements, the first quadrupole is used in scanning mode to determine the mass range of precursor ions which are fragmented in the process of collision induced dissociation. The third quadrupole functions in static mode and is set on the m/z value of the selected fragment which is usually structure specific.
3. Neutral loss scans. The technique used for this purpose monitors the loss of neutral fragments. The first and third quadrupoles are used for scanning with a constant mass offset that represents the neutral fragment loss.
4. Single or multiple reaction monitoring. Here, the first and third quadrupoles work in the static mode and scan the m/z values of the analyte and its selected fragment. A pair of precursor-product ions of this type is called a transition pair. This technique is used extensively in quantitative analysis because it yields the greatest possible measurement sensitivity.

Tandem mass spectrometry exhibits higher selectivity and sensitivity (with a higher signal-to-noise ratio because noise is suppressed) than does simple mass spectrometry. This advantage also makes this technique useful when tandem mass spectrometry is coupled with HPLC separation (Haynes, et al., 2009; Sullards, et al., 2011).

2.4 Tandem mass spectrometry of sphingolipids: ionisation, fragmentation and specificity of sphingolipid analysis

2.4.1 Ionization

The ionization of sphingolipids by electrospray ionization varies for different classes of sphingolipids. This variation can be used in intrasource separation in lipid analyses (Han & Gross, 2005; Haynes, et al., 2009).

Neutral sphingolipids are usually protonated by ammonium formiate, ammonium acetate, formic or acetic acid. Ammonium salts exhibit the highest degree of ionization efficiency and are widely used (Mano, et al., 1997). Neutral salts of Li^+ or Na^+ are also used for ionization and generate ion-lipid adducts (Boscaro, et al., 2002; Olling, et al., 1998).

Acids are generally used for H^+ transfer in solution to basic groups on analyte or to creation of cluster of protonated solvents which later transfer H+ to analyte in the process of charge separation. On the other hand neutral amonium salts are usually added to the solution to facilitate the analysis of polar and neutral analytes by adduct formation and later protonation in the process of ionization through gas-phase reactions. Sodium and lithium aducts are also aded in form of neutral salts; the ionisation process has the character of specific charge separation (Cech & Enke, 2001; Kebarle, 2000).

We found that for positive ion mode measurements, 5 mM ammonium formiate is a more efficient additive for use in sphingolipid ionization than is ammonium acetate. However, when using HPLC combined with mass spectrometry, we prefer to use ammonium acetate and acetic acid because of their better solubility in methanol. The ions generated during the electrospray ionization process were $[M+H]^+$.

Acidic sphingolipids, such as sulfatides or gangliosides, have acidic groups that lose H^+ even in pure methanol. It is also possible to create chloride adducts (Han & Gross, 2005) using halogenated solvents that can also abstract protons (Cech & Enke, 2001).
For measurements in negative ion mode, we used pure methanol solvent to generate $[M-H]^-$ ions.

2.4.2 Fragmentation

Fragmentation studies of sphingolipids revealed characteristic fragments useful for tandem mass spectrometric analysis (Domon & Costello, 1988; Fuller, et al., 2005; Gu, et al., 1997; Hsu, et al., 1998; Hsu & Turk, 2000; Ii, et al., 1995; Kerwin, et al., 1994; Liebisch, et al., 1999; Mano, et al., 1997; Murphy, et al., 2001; Olling, et al., 1998; Whitfield, et al., 2001). The most common fragments used for sphingolipid analysis in positive ion mode are structurally derived from ceramide with C18:1 sphingosine. When the amide bond is broken, followed by the formation of ion derived from sphingoid structure minus one water molecule, the fragment with the m/z value of 282 is produced. If another molecule of water is lost, the fragment with the m/z value of 264 is generated (Fig. 5A) (Gu, et al., 1997; Liebisch, et al., 1999; Murphy, et al., 2001; Olling, et al., 1998). These product ions have different uses. The 264 m/z fragments are used to analyze sphingolipids, whereas the 282 m/z fragments are used to analyze N-deacylated sphingolipids (lysoderivatives) (Gu, et al., 1997; Lieser, et al., 2003; Olling, et al., 1998; Scherer, et al., 2010).
The ceramide fragments mentioned above are commonly used in tandem mass analysis, but other specific sphingolipid structures can also be used for this purpose. Sphingomyelin has characteristic phosphocholines that generate fragments with an m/z value of 184 (Fig. 5B) (Hsu & Turk, 2000; Kerwin, et al., 1994; Murphy, et al., 2001). Sialic acid is another example of a specific sphingolipid structure that is a component of gangliosides. Fragments of sialic acid have m/z values of 290 and 308 (Fig. 5C) (Domon & Costello, 1988; Ii, et al., 1995). A structure that is specific to sulfatides is the sulfate group, which exhibit an m/z value of 97. (Fig. 5D) (Hsu, et al., 1998; Whitfield, et al., 2001).

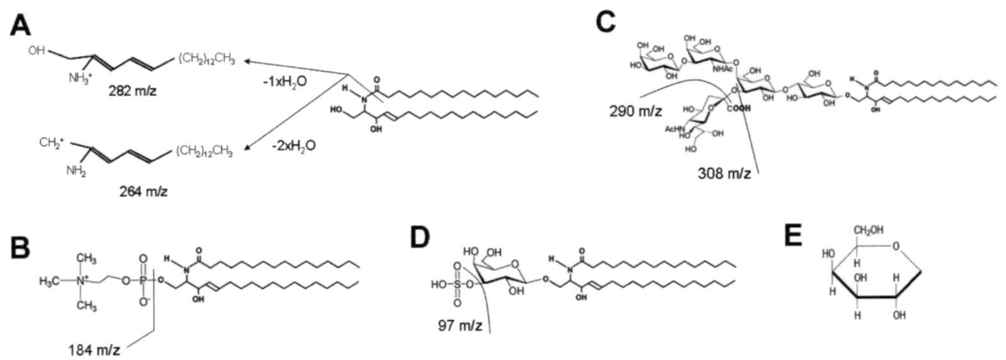

Fig. 5. Fragmentation of ceramide (A) and specific fragments of sphingomyelin (B), GM1 ganglioside (C) and sulphatides (D). Neutral fragment of hexose (galactose) is an example of fragmentation used in a neutral loss scan (E).

Another analytical approach involves scanning for the neutral loss of saccharides from oligosaccharide chains of glycosphingolipids (Fig. 5E and 6) (Boscaro, et al., 2002; Domon &

Costello, 1988; Olling, et al., 1998). In this approach, transition pairs consisting of analyzed glycosphingolipids and their products with shorter or missing oligosaccharides are measured (Fig. 6).

Neutral loss scanning is the technique of choice for glycosphingolipids containing dihydroceramide or sphinganine. Saturating the double bonds of sphing-4-enine reduces the production of sphingoid base fragments (Fig. 5A) in the positive mode, yielding a fragmentation (Fig. 6) efficiency of approximately 2-3%. In contrast, neutral loss scan fragmentation efficiency for the above-mentioned molecules is 10-15%. It is therefore crucial to select the best fragment to conduct a successful tandem mass analysis of sphingolipids.

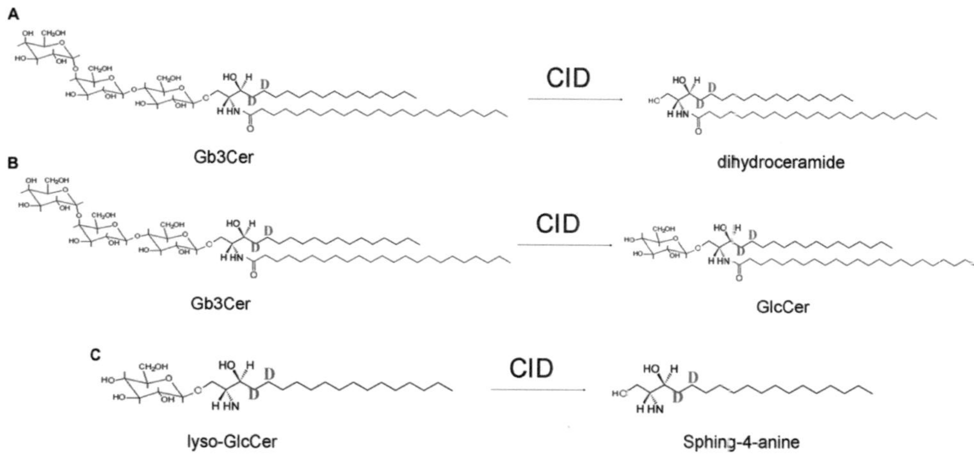

Fig. 6. Principle of neutral loss measurements of sphingolipids with deuterated dihydroceramide and sphing-4-anine in the ceramide region of the molecule. A) Complete loss of Gb3Cer oligosaccharide, B) Shortening of Gb3Cer oligosaccharide C) Neutral loss of saccharide part in lysoglycosphingolipids. CID – collision induced dissociation; Gb3Cer – globotriaosylceramide.

2.4.3 Specificity of mass spectrometry analysis of sphingolipids

Mass spectrometry analyzes molecules according to their m/z values. Different classes of sphingolipids are not represented by one specific molecule; rather, they are a heterogeneous group of molecules with different molecular masses. Their variability is mostly represented by a spectrum of fatty acids that form ceramide structures. Molecular species of individual sphingolipids are called isoforms, and their profiles are usually cell- and tissue-specific (Fig. 7). Therefore, it is important to determine the specific isoform profiles of sphingolipids in biological material before conducting a quantitative analysis.

2.5 Sample preparation for tandem mass spectrometric analysis

A common step in the processing of different biological samples (urine, plasma, cerebrospinal fluid, cells, bioptic or autoptic tissues, etc.) prior to tandem mass spectrometric analysis is the preparation of a lipid extract. Cells and tissues are homogenized, and aliquots of homogenate are usually taken for protein determination. Many extraction procedures have been introduced over the years, most of which have been based on chloroform-methanol mixtures (Bligh & Dyer, 1959; Folch, et al., 1957) or less harmful solvents such as 2-propanol, ethylacetate, hexane or tetrahydrofuran (Heitmann, et al., 1996). The first approach remains the most popular and efficient. Widely used variations on the above-mentioned procedures and recommended methods of removing contaminants from total lipid extract were summarized by Schnaar R (Schnaar, 1994) and van Echten-Deckerd G (van Echten-Deckert, 2000).

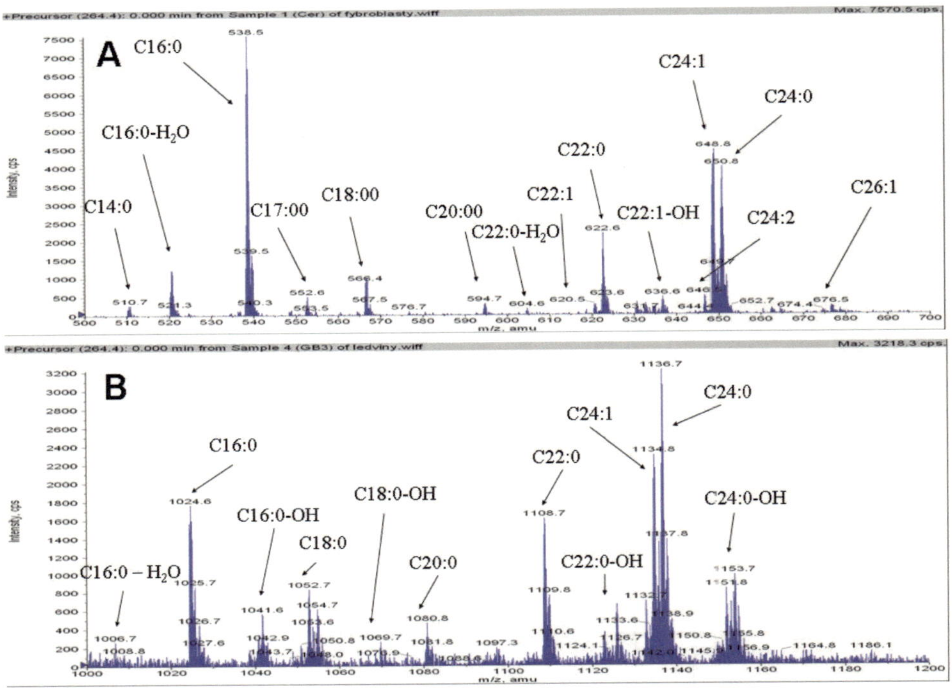

Fig. 7. Isoform profiles measured using precursor ion scans for different lipids and biological materials. Three-milligram protein aliquots of lipid extracts were dissolved in methanol with 5 mM ammonium formiate and measured using 1 min precursor ion scans for ceramide fragments with m/z values of 264 in positive ion mode. Isoforms with different fatty acid chain lengths are identified. A) ceramides in skin fibroblasts, B) Gb3Cer in the kidney.

2.5.1 Preparation of lipid extract from urine

Extracts were prepared as previously published (Kuchar, et al., 2009). First, 150 µl of sonicated urine was extracted with 700 µl of chloroform:methanol (2:1, v:v) containing internal standards in a polypropylene Eppendorf tube. Then, after 15 min of repeated vortexing at 5 min intervals, 150 µl of MilliQ water was added. The vortex mixing procedure was repeated for another 15 min. After a 20 min pause, the samples were centrifuged for 5 min at 14 000 *x* g. The lower organic layer was isolated using a Hamilton syringe and filtered using hydrophilic polytetrafluorethylene (PTFE) syringe filters. The samples were purified after the addition of 500 µl of MilliQ water and 15 min of vortexing. The organic and aqueous phases were separated by 5 min of centrifugation at 14 000 *x* g, and interfering salts and small organic molecules were removed with the upper water layer. The lower organic layer containing the sphingolipids was collected, dried under a stream of nitrogen and stored in a freezer (-20°C).

2.5.2 Preparation of lipid extract from cultured fibroblasts

Fibroblast pellets were extracted by standard procedures using a chloroform:methanol:water mixtures as previously described (Asfaw, et al., 1998). The harvested cells from the 75 cm^2 cultivation flask were homogenized in 250 µl of MilliQ water by sonication. Next, 50 µl of homogenate was used for protein determination (Hartree, 1972). The remaining 200 µl of cell homogenate was mixed with 800 µl of chloroform:methanol (2:1, v/v) in a 15 ml glass tube. The mixture was rigorously vortexed twice for 1 min followed by a 15 min settling time at laboratory temperature. The organic and water layers were then separated during 10 min of centrifugation at 400 *x g*. The upper water and lower organic layers were collected, and the precipitated protein was left in the tube. The protein debris was washed with 500 µl of chloroform:methanol (2:1, v/v), and the organic layer without protein was added to the previously collected phases. The lipid extracts were dried under a stream of nitrogen and redissolved in 400 µl of chloroform:methanol (2:1, v/v). The extracts were then filtered using hydrophilic polytetrafluorethylene syringe pump filters. The filtrates were dried under a stream of nitrogen and stored in a freezer (-20°C) for processing.

2.5.3 Preparation of lipid extract from tissues

The basic procedures used in the tissue extraction process were generally the same as those used by Natomi H (Natomi, et al., 1988), though minor modifications were made. Tissue samples with a wet weight of up to 0.5 g were weighed and homogenized in water or methanol:water (10:1, v/v). Small portions of the homogenate were stored for protein quantification (Hartree, 1972), and the remaining portion was used to prepare the lipid extracts. Chloroform was added to the methanol:water homogenate to get a ratio of chloroform:methanol:water (20:10:1 v/v/v). After vortexing and sonication, the extracted samples were centrifuged and the supernatant collected. The sediment was reextracted with a more polar solvent mixture of chloroform, methanol, and water (10:20:1, v/v/v) and then with chloroform:methanol (1:1, v/v). The total volume of the extraction mixture corresponding to 20 volumes of the original tissue sample was added during every step in the extraction process. The supernatants were combined, filtered and dried under a stream of nitrogen. Dried samples were stored at freezer (-20°C) prior to processing and analysis.

2.5.4 Processing of extracted lipid samples prior to tandem mass analysis

Corresponding internal standards were added to the urinary samples during the extraction process (Kuchar, et al., 2009). The same volume of internal standards as was used for the urine samples was added to the appropriate aliquots of purified lipid extracts: 5 µg of cellular protein or 150 µg of tissue protein. Loading experiments with specifically labeled sphingolipid isoforms required 50 µg protein aliquots.

Finally, the samples were dried under a stream of nitrogen and dissolved in methanol with 5 mM ammonium formiate for measurement in the positive ion mode or in pure methanol for measurement in the negative ion mode.

2.6 Quantitative analysis of sphingolipids by tandem mass spectrometry

We used an AB/MDS SCIEX API 3200 triple quadrupole mass spectrometer equipped with an Agilent 1100 series LC system with an autosampler. The Analyst software version 1.5 was used to operate the hardware and process the measured data. Optimizations of electrospray ionization and tandem mass spectrometry conditions were conducted for each analyzed sphingolipid (Tab. 2). A standard lipid solution with a sphingolipid concentration of 5 µg/ml, was used in the optimization process. For positive ion measurement, 5 mM ammonium formiate in methanol was used to produce $[M+H]^+$ ions. To generate $[M-H]^-$ ions in negative ion mode, pure methanol was used.
We measured lipids using a flow injection analysis of 20 µl sample aliquots samples. We used pure methanol as the mobile phase with a flow rate of 50 µl/min. One lipid was analyzed during one injection to provide the best possible quantitative data. The scan time for a transition pair was usually 100 ms but in some cases was increased to 500 ms for higher sensitivity. The settling time was usually 0 ms. However, it was necessary to increase this parameter to 500-700 ms when the ion optics setting was changed to measure more than one class of sphingolipid in one injection. The resolution was generally set to unit (±1 m/z), but in some cases, we used a high resolution setting for the first quadrupole, as using such a setting can improve mass spectrometer sensitivity.

	CTH	CDH	CMH	lyso-CMH	Cer
Curtain Gas [psi]	10	10	10	10	10
Collision Gas (N_2) [psi]	3	5	5	5	5
Ion Spray Voltage [kV]	5,5	4,5	4,5	4,5	4,5
Temperature [°C]	200	200	200	200	200
Ion Source Gas 1 [psi]	15	20	20	20	20
Ion Source Gas 2 [psi]	20	55	55	55	55
Interface Heater	On	On	On	On	On
Declustering Potential [V]	82,5	65,0	47,0	53,0	60,0
Entrance Potential [V]	8,4	6,0	4,9	4,1	5,0
Collision Energy [V]	77,0	64,0	48,0	31,0	42,0
Collision Cell Exit Potential [V]	10,8	5,6	5,6	9,1	5,7

Table 2. Example of electrospray ionization and ion optics parameters used in tandem mass spectrometry to analyze selected sphingolipids (Sciex API 3200, product ion m/z 264, 5 mM ammonium formiate in methanol). CTH - ceramidetrihexoside; CDH - ceramidedihexoside, CMH and lyso-CMH - ceramidemonohexoside and its N-deacylated derivative.

Procedure of our quantitative analysis was described in a previous study (Kuchar, et al., 2009). Problems associated with the matrix effect were addressed using internal standards, whereas the calibration of the method was based on an external standard. Quantification was performed via single-point calibration using an external calibration point with a standard lipid concentration (an external calibration standard) corrected by the signal ratio toward internal standard (mostly C17:0 isoform which is not naturally abundant) isoform (mostly with C17:0 fatty acid). All standard lipid concentrations were within the broad range of linear response. The internal standard concentration at the external calibration point and in the measured samples was the same. For the quantification procedure, molecular species of sphingolipids with fatty acids of chain lengths from C16 to C26 were selected.

2.7 Preparation of sphingolipids internal standards using enzymatic semi-synthesis

Not all internal standards are commercially available. Thus, we developed a method of enzymatic semi-synthesis using immobilized sphingolipid ceramide N-deacylase (*Pseudomonas sp,* TK4) on porous magnetic cellulose (Kuchar, et al., 2010). Magnetic macroporous bead cellulose was used as carrier for sphingolipid ceramide N-deacylase (SCDase) which was immobilized using a standard procedure (Bilkova, et al., 2005; Korecka, et al., 2005). A 100 µl aliquot of washed settled particles was activated with 0.2 M freshly prepared $NaIO_4$. The activated particles were then washed with 0.1 M phosphate buffer with a pH 7. Binding of 250 mI.U. of sphingolipid ceramide N-deacylase on activated magnetic macroporous bead cellulose was achieved by 10 min incubation in phosphate buffer with a pH 7. The formed Schiff base was stabilized via overnight incubation in a $NaCNBH_3$ solution. The final step consisted of washing particles in phosphate buffer with a pH 7. Immobilized sphingolipid ceramide N-deacylase was stored in phosphate buffer with a pH 7 with 0.1% Triton X-100 at 4°C.

Under specific conditions, this enzyme also catalyzes the reverse reaction (Kita, et al., 2001). Thus, lysoderivates can be reacylated with a specific fatty acid. Using this procedure, we prepared several internal standards with C17:0 fatty acids, e.g. sulfatides, glucosylceramides, and GM1 gangliosides. In this process, we incubated 50 nmol of lyso-glycosphingolipid, 50 nmol of C17:0 fatty acid and immobilized sphingolipid ceramide N-deacylase in 300 µl of pH 7 phosphate buffer with 0.1% Triton X-100 for 20 hrs at 37°C while mixing it on a rotator. The magnetic particles were separated, and the supernatant was transferred and dried under a stream of nitrogen. The quality of the prepared lipids was monitored by HPTLC and tandem mass spectrometry.

2.8 Standardization of quantitative data in urine

The results of sphingolipid quantification in cellular material are often related to protein concentration, a well-established standardization parameter (Liebisch, et al., 1999). Urinary sphingolipids mostly originate from desquamated renal tubular cells. The standardizing parameter commonly used for urinary metabolites is creatinine, but creatinine does not reflect the cellular origin of sphingolipids. Therefore, in urinary samples with a creatinine level lower than 1 mM, the concentration of excreted sphingolipids is artificially inflated (Fig. 8A), which may encourage the incorrect diagnosis of some patients with lysosomal storage diseases (e.g., Fabry disease, prosaposin and saposin B deficiencies, and sulfatidoses). Regarding Fabry disease, this issue has already been pointed out (Forni, et al., 2009). In our experience, such diagnostic errors are especially likely to occur with newborns or children up to six years of age, whose normal concentration of creatinine is generally low (≤ 4 mM). Surprisingly, urinary volume has been found to be much more convenient for use as a standardization parameter (Fig. 8B). It is also possible to use the ratio of the analyzed compound to sphingomyelin (Berna, et al., 1999; Kuchar, et al., 2009) or phosphatidylcholine

(Fuller, et al., 2005; Whitfield, et al., 2001), which are membrane-bound lipids that can be measured simultaneously in the same sample.

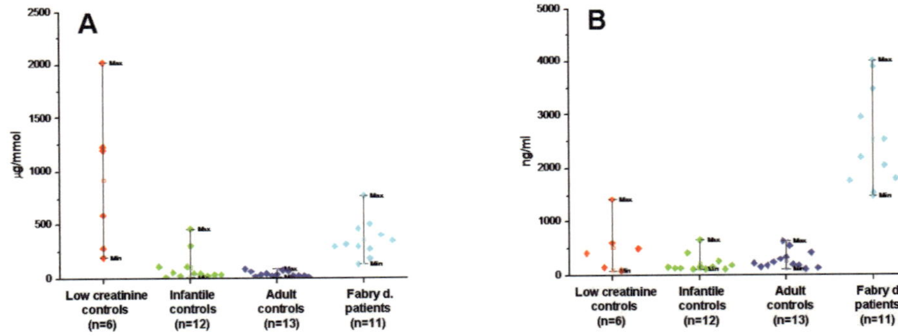

Fig. 8. Comparison of two methods of standardization of urinary Gb3Cer in three groups of samples: in controls with low creatinine (creatinine ≤1 mM), in controls with creatinine within a normal range (creatinine >1 mM-15 mM) and in Fabry patients. It is most critical to appropriately evaluate urine control samples with low creatinine concentrations, which are indistinguishable from samples from patients with Fabry disease (A). The use of urine volume as a standardization parameter makes it possible to differentiate more appropriately between controls and Fabry patients (B).

3. Sphingolipids in lysosomal storage disorders – mass spectrometric data useful for diagnosis and research

Flow injection analysis (FIA) combined with electrospray ionization tandem mass spectrometry makes it possible to determine the concentration of sphingolipids in various biological materials: e.g. urine, cultured fibroblasts and autoptic or bioptic tissue samples from different lysosomal storage disorders.
Examples of these analyses are presented in following paragraphs.

3.1 Sphingolipids in urine

Urine is a non-invasive diagnostic material that is of practical importance in diagnosing lysosomal disorders in which the storage of non-degraded substrate causes pathological processes in the kidneys. These disorders are characterized by the massive excretion of specific sphingolipids, e.g., Gb3Cer in Fabry disease (α-galactosidase A deficiency due to mutations of the *GLA* gene); multiple hydrophobic sphingolipids in complex sphingolipidoses, in which the defect is caused by mutations in the prosaposin gene (sphingolipids with a saccharide chain that is shorter than four monosaccharide units and ceramides are not degraded in prosaposin deficiency and Gb3Cer and sulfatides in saposin B deficiency due to defective activator proteins); sulfatides in metachromatic leukodystrophy (arylsulfatase A deficiency due to mutations of the *ARSA* gene) (Fuller, et al., 2005; Kuchar, et al., 2009; Whitfield, et al., 2001).
We developed a method of tandem mass spectrometry quantification of urinary sphingolipids that can be used in pre-diagnostic screening for the lysosomal disorders mentioned above (Kuchar, et al., 2009). Our data are presented in Table 3.

μg/l	sulfatide	CDH	Gb3Cer	CMH	Cer	SM
pSap-d 44-day-old	**756**	**507**	**2322**	**289**	**184**	1323
SapB-d 50-month-old	**2149**	**513**	**755**	**204**	101	1480
MLD late infantile (n= 6)	**3566 ±1855**	**397** ±220	261 ±188	**191** ±115	113 ±67	**3417 ±2448**
Fabry disease (male) (n=10)	88 ±17	**460 ±199**	**2615 ±915**	45 ±13	62 ±11	1602 ±1081
Controls infantile/juvenile (n=16)	155 ±48	113 ±37	152 ±68	52 ±23	45 ±12	1181 ±437

Table 3. Massive excretion of urinary sphingolipids in the case of saposin-B and prosaposin deficiencies and in patients with Fabry disease and metachromatic leukodystrophy. pSap-d - prosaposin deficiency, SapB-d - saposin B deficiency, MLD - metachromatic leukodystrophy; Cer - ceramide; CTH – ceramidetrihexoside; CDH – ceramidedihexoside, CMH – ceramidemonohexoside SM – sphingomyelin; Gb3Cer - globotriaosylceramide Values are in ng/μg of protein (mean±SD). Non-degraded sphingolipids related to particular lysosomal storage disorders are bolded.

Analyzing non-degraded metabolites can be very helpful in the pre-diagnosis of sphingolipid activator deficiencies in which routine enzymology fails to indicate deficient enzyme activity due to the detergents commonly used in the assays.
Urinary Gb3Cer has been suggested as biomarker for monitoring efficiency of enzyme replacement therapy of Fabry disease. Our tests in a group of Fabry male-patients showed however, that monitoring of this marker is not informative for all treated patients in general but for individual patients only (data not shown). Although excreted Gb3Cer is useful parameter for diagnosis, it is not reliable biomarker for clinical trials as also confirmed by another studies (Schiffmann, et al., 2010). Biological basis of urinary Gb3Cer and its isoforms is still subject of research.

3.2 Sphingolipids in cultured fibroblasts

Although cultured fibroblasts are not typical "storage cells," the concentration of non-degraded lipids increases significantly in some lysosomal storage disorders as documented in Table 4. Investigating sphingolipid profile can help in laboratory diagnosis of these rare diseases, especially among those suspected of having defective activators of lysosomal hydrolases.

3.3 Sphingolipids in tissues: Gb3Cer and lyso-Gb3Cer in Fabry myocardium and kidney

In some cases, a postmortem analysis of autoptic tissue has revealed metabolic defects. Here, we give two examples of tissue analysis that led to a final diagnosis confirmed by DNA analysis later on.

	Cer	CMH	CDH	Gb3Cer	SM
Prosaposin def.	**34,48**	**14,27**	**25,75**	**27,06**	45,70
saposin B def.	6,87	1,44	1,86	**21,36**	107,33
Fabry disease	3,76	1,89	2,10	**35,68**	42,62
Nieman-Pick A	4,15	1,67	5,63	0,51	**195,17**
Control 1	6,21	3,25	5,17	0,52	68,28
Control 2	1,18	2,67	1,37	5,17	21,04

Table 4. Increased concentration of sphingolipids in cultured skin fibroblasts in patients with sphingolipid activator deficiencies (saposin-B and prosaposin deficiencies) and in patients with defective enzyme proteins (in Fabry disease and metachromatic leukodystrophy).Values are in ng/µg of protein. Cer - ceramide; CDH – ceramidedihexoside, CMH – ceramidemonohexoside; SM – sphingomyelin; Gb3Cer - globotriaosylceramide. Non-degraded sphingolipids corresponding to particular lysosomal storage disorders are bolded.

The first example shows the accumulation of Gb3Cer in the kidneys of patients with Fabry disease and in cases of prosaposin deficiency (Tab. 5).

	Cer	CMH	CDH	Gb3Cer	sulfatide	SM
Fabry disease	0,5	1,0	3,7	**115,2**	0,5	24,0
Prosaposin def.	**39,1**	**23,9**	**49,8**	**57,6**	**39,6**	125,8
Control (n=3)	11,2	0,7	3,2	10,2	1,1	57,9

Table 5. Concentration of sphingolipids in the kidneys of Fabry male patient and in a case of prosaposin deficiency. Values are in ng/µg of protein, Control is represented by the mean value. Cer - ceramide; CDH – ceramidedihexoside, CMH – ceramidemonohexoside, SM – shingomyelin; Gb3Cer - globotriaosylceramide. Non-degraded sphingolipids related to particular lysosomal storage disorders are bolded.

Another example demonstrates the storage of Gb3Cer and lyso-Gb3Cer (globotriaosylsphingosine) in the myocardium of Fabry patient (Fig. 9). It is possible that the role of these derivates has been underrated (Dekker, et al., 2011). The role of lyso-Gb3Cer as a molecule that stimulates smooth muscle cell proliferation is now known. These findings indicate the possible role of lyso-Gb3Cer as a signal molecule (Aerts, et al., 2008).

4. The investigation of sphingolipid degradation pathways

4.1 Loading experiments on living cells using lipid substrates labeled with a stable isotope or containing atypical fatty acids

Loading experiments in cell cultures (also called feeding experiments) are frequently used to investigate the metabolic fate of exogenous compounds in living model systems. The main advantage of such experiments is that they assess the entire degradation system, including any nonenzymatic cofactors.

This method can be used to conduct a general analysis of metabolic pathways (Schwarzmann, et al., 1983; Sonderfeld, et al., 1985), intracellular transport or a distribution (Martin & Pagano, 1994) assessment of residual activity in enzyme-deficient cells

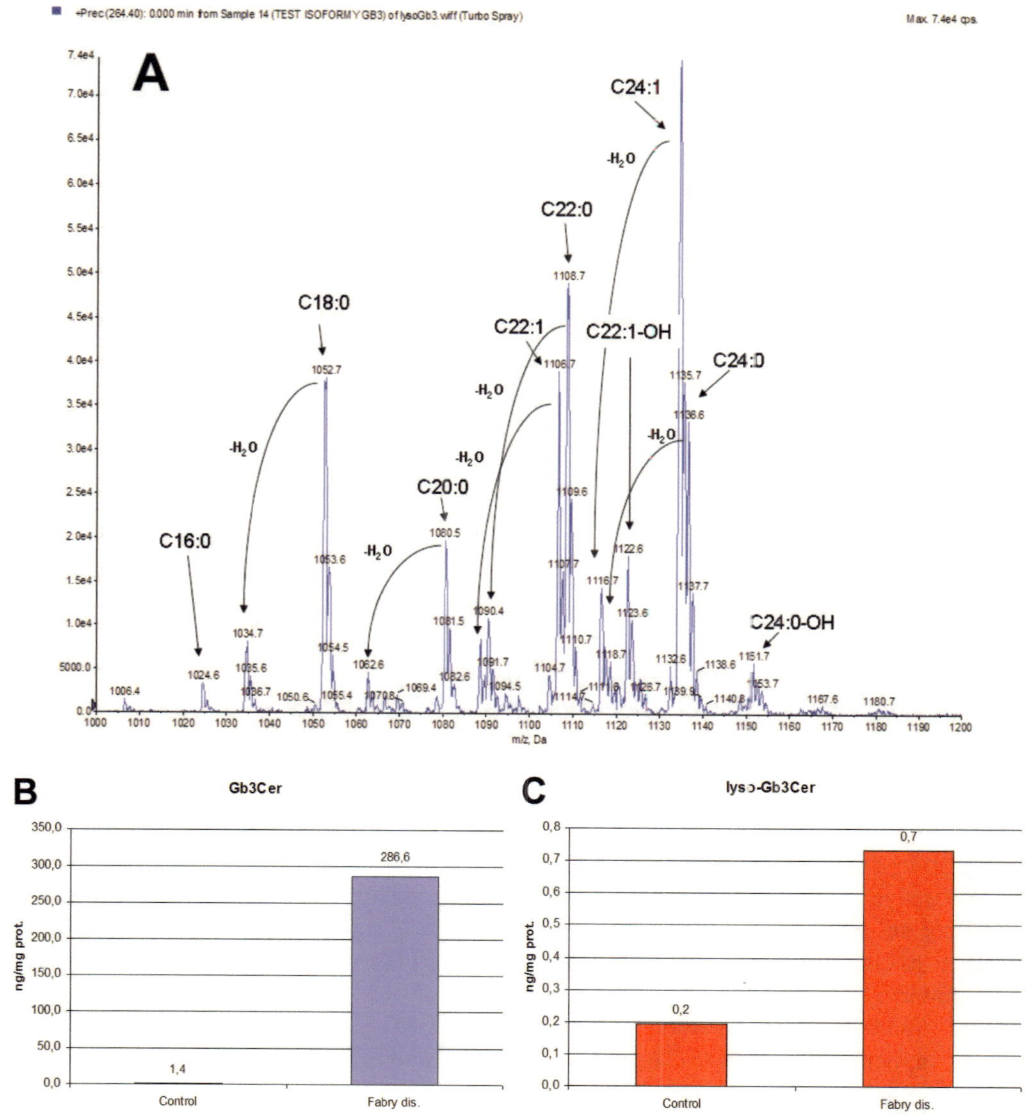

Fig. 9. Gb3Cer and lyso-Gb3Cer (globotriaosylsphingosine) in the myocardium of Fabry patient. A) Precursor ion spectrum of Gb3Cer molecular species in control myocardium. The spectrum was measured by a 1 min scan of a 3 mg protein aliquot of lipid extract dissolved in methanol with 5 mM ammonium formiate. The increased concentration of B) Gb3Cer and C) lyso-Gb3Cer is visible in the autoptic myocardium of Fabry patient in comparison to that of an age-matched control. Quantity was measured by flow injection analysis electrospray ionization tandem mass spectrometry using an multiple reaction monitoring scan. C17:0 Gb3Cer was used as the internal standard for lyso-Gb3Cer quantification similarly as described in (Mills, et al., 2005). Concentrations were measured by method with coefficient of variation - CV <7%.

(Leinekugel, et al., 1992; Porter, et al., 1971). The method can also be used to diagnose storage disorders.
This technique has been used to distinguish between metachromatic leukodystrophy and arylsulfatase A pseudodeficiency (Kihara, et al., 1980) and to identify deficiencies in nonenzymatic protein cofactors of lysosomal glycolipid catabolism (sphingolipid activator proteins) (Klein, et al., 1994; Schepers, et al., 1996; Schmid, et al., 1992; Sonderfeld, et al., 1985; Wrobe, et al., 2000), including prosaposin deficiency (Harzer, et al., 1989; Chatelut, et al., 1997).
In these experiments, the degradation products of labeled exogenous lipid substrates are determined using specific analytical methods. Traditionally, sphingolipids are labeled with radioisotopes, and their degradation products are separated chromatographically and traced using radioactivity assays. Recently, radiolabeled sphingolipids are often replaced by non-radioactive analogues with atypical molecular masses that can be analyzed by tandem mass analysis.
Sphingolipids can be labeled on different parts of the molecule using non-natural fatty acids, creating molecules with atypical m/z values. However, only a few labeled standards are commercially available. Sphingoids can also be deuterium labeled at the double bond to increase mass, but the fragmentation patterns will be altered (see Fragmentation, Fig. 6).
We compared the radioisotope and mass labeling methods in loading experiments involving fibroblast cultures from patients with inherited lysosomal storage diseases such as GM1 gangliosidosis. In this study, genetic variants of the *GLB1* (ß-galactosidase; ß-gal) gene were selected. Both approaches, the use of [^{3}H]GM1 ganglioside and the use of its C18:0-D_3 analogue, clearly showed that the impaired degradation of critical glycosphingolipids resulted from defects in β-gal function, as indicated in Fig. 10.
The experiments conducted with stable isotope-labeled substrates and tandem mass spectrometry facilitated a more accurate quantification analysis of the lipids, and the results were better correlated with the clinical and biochemical phenotypes of the samples. The procedure used to prepare the cellular lipids for tandem mass spectrometry analysis was simple and relatively rapid; unlike radioisotope assays, it did not require separation during the pre-analytical phase (Asfaw, et al., 2002; 1998). However, experiments with radiolabeled glycolipid substrates indicate the entire metabolic pattern (Fig 10A) and can thus make it possible to identify relevant metabolites to be further analyzed via tandem mass spectrometry.
Results similar to those obtained in analyzing the GM1 gangliosidosis were obtained by loading experiments using Gaucher fibroblasts and fibroblasts from patients with prosaposin deficiencies (data not shown).

4.2 Tandem mass determination of *in vitro* acid glycosidase activity

Enzymology, in combination with tandem mass spectrometry, is useful in lysosomal storage disorders screening and in evaluations of enzyme activity (Kasper, et al., 2010; Li, et al., 2004; Spacil, et al., 2011; Turecek, et al., 2007). One practical application of this technique is the analysis of lysosomal β-glucocerebrosidase activity using glucosylceramide with C12:0 fatty acid as the enzyme substrate. Tandem mass spectrometry evaluation techniques can be used with cells and tissue homogenates but also with dried blood spots as the screening material.
We followed Turecek´s method (Turecek, et al., 2007) in measuring lysosomal β-glucocerebrosidase activity in homogenates of cultured skin fibroblasts. The reaction

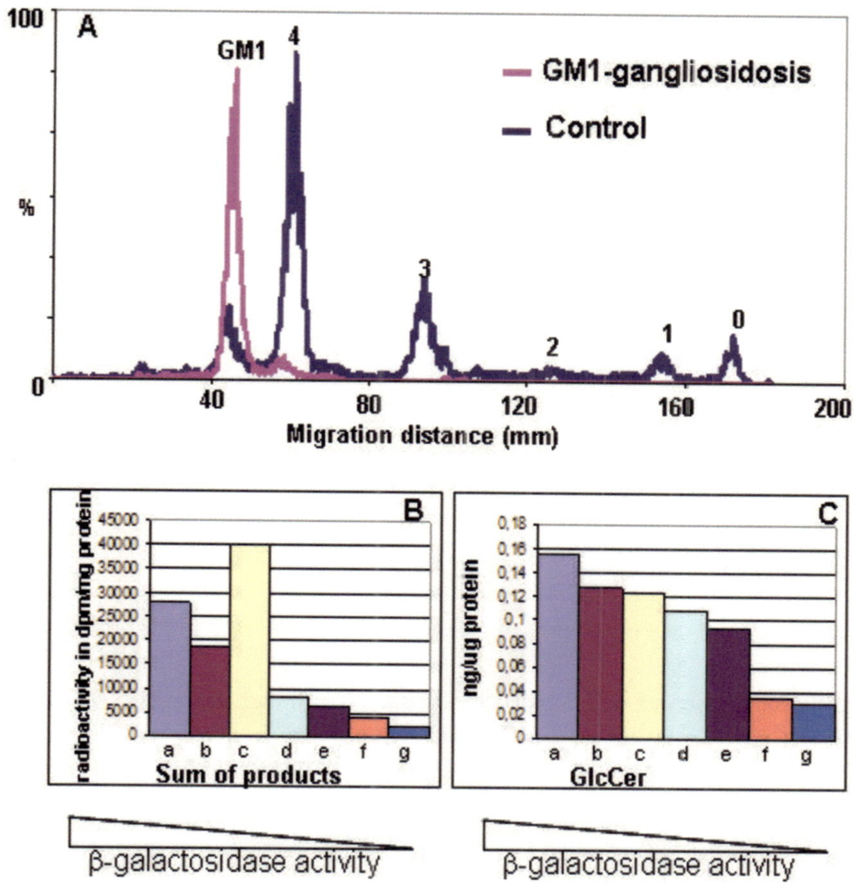

Fig. 10. In situ degradation of GM1 ganglioside by skin fibroblasts from control and β-galactosidase deficient patients. Tritium-labeled glycolipid GM1 ganglioside or a C18:0-D_3 analogue was added to the culture of skin fibroblasts in 25 cm^2 flasks. After 5 days, cells were harvested and lipids extracted. The radio-labeled lipid extracts were separated via TLC and analyzed using a linear scanner (Asfaw, et al., 1998), whereas the GM1 ganglioside with C18:0-D3 fatty acid and its degradation products were extracted and directly analyzed directly in tandem mass spectrometry (details in Materials and Methods). **A:** Degradation pattern of [^{3}H]GM1 ganglioside in control and β-galactosidase-deficient cells. The chromatographic positions of the products are indicated by the number of sugar residues on the glycolipid, which range from 0 (ceramide) to 4 (tetrahexosylceramide). **B:** Quantification of degradation products of [^{3}H]GM1 ganglioside (sum of all products on the TLC plate) in skin fibroblasts from the control and the different β-galactosidase-deficient genetic variants. The cell lines are arranged according to clinical phenotypes of GM1 gangliosidosis: a-control, b-adult GM1 gangliosidosis, c-Morquio B, d-adult GM1 gangliosidosis /Morquio B, e-juvenile GM1 gangliosidosis, f-infantile GM1 gangliosidosis, and g-infantile GM1 gangliosidosis. **C:** Quantification of GlcCer product formed from stable isotope-labeled GM1 ganglioside (C18:0-D_3) in the same cell lines as the radioactive analogue (B). Values are average of two samples.

mixture contained 0.5 μg of sample protein and 0.05% inactivated bovine serum albumin (BSA) to stabilize the enzyme. The mass spectrometry settings were optimized to prevent the artificial conversion of the substrate into the enzyme reaction product. An example of this analysis is presented in Figure 11.

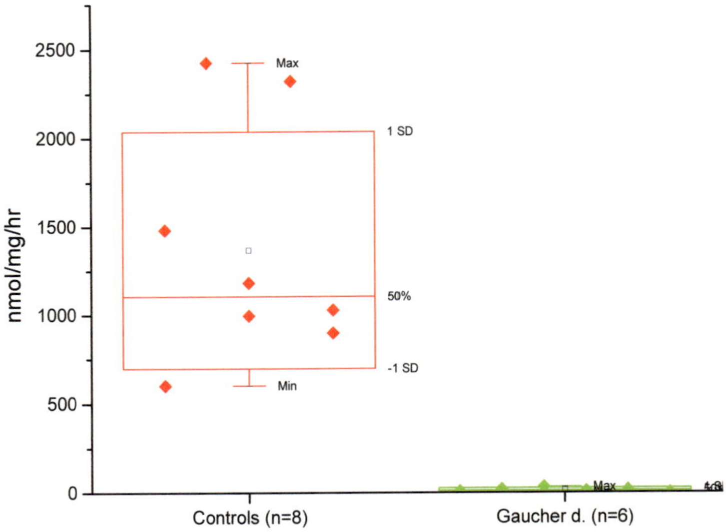

Fig. 11. Activity of lysosomal acid β-glucocerebrosidase in Gaucher and control fibroblasts as measured with natural substrate C12:0 glucosylceramide. The reaction product was analyzed by flow injection analysis electrospray ionization tandem mass spectrometry using a multiple reaction monitoring.

Nowadays, procedures are simplified by skipping laborious process of extraction of reaction mixture after incubation. Instead, HPLC is combined with mass spectrometry. The HPLC step purifies the sample, removing any potential interfering compounds (Kasper, et al., 2010; Spacil, et al., 2011).

5. Sphingolipid isoform profiling – a useful metabolomic approach to disorders involving Gb3Cer and sulfatide storage

Tandem mass spectrometry has advantages over HPLC and other analytical methods in helping to determine the individual molecular species (isoforms) of sphingolipids. Changes in isoform profiles may provide diagnostically important information and indicate specific pathological processes (Fauler, et al., 2005; Paschke, et al., 2011). For example, an analysis of urinary lipid extracts in a case of metachromatic leukodystrophy showed significant differences in sulfatide isoform profiles; such differences were also evident in cases of prosaposin and saposin B deficiency, two other sulfatide storage disorders. We also found changed patterns of globotriaosylceramide species in the urine of Fabry patients and patients with prosaposin gene defects. A shift in the isoform pattern to species with longer chain fatty acids was characteristic of both prosaposin and saposin B deficiencies.

The results presented in Table 6 were evaluated by determining the ratio of the various isoforms to the C18:0 isoform, which is invariable in the profile. The major advantages of this procedure include simple sample preparation without internal standard and a simple data collection process; only a small number of transitions must be measured. The elevation of certain molecular species, particularly those with longer, hydroxylated chains (in the case of the sulfatides), is clearly demonstrated.

A

	Control	n=29	SapB-d	pSap-d	Fabry	n=11
Isoform	Min	Max			Min	Max
16:0	0,64	2,91	2,33	2,14	1,59	5,52
18:0	1,00	1,00	1,00	1,00	1,00	1,00
18:0-OH	0,42	2,23	0,32	0,08	0,03	0,15
20:0	0,46	2,09	1,98	1,69	1,33	2,08
22:0	**0,25**	**3,37**	**6,63**	**5,06**	**4,08**	**7,02**
22:1	0,43	3,71	0,66	0,42	0,34	0,87
22:0-OH	0,18	2,40	0,83	0,37	0,14	0,40
22:1-OH	0,17	5,44	1,10	0,27	0,39	0,56
24:0	**0,00**	**3,23**	**8,33**	**10,50**	**5,66**	**9,52**
24:1	**0,27**	**2,75**	**4,93**	**9,97**	**4,66**	**8,45**
24:2	0,27	4,88	0,65	0,53	0,36	0,88
24:0-OH	0,00	2,02	1,10	0,90	0,53	1,14
24:1-OH	0,47	2,67	0,46	0,33	0,17	0,39
24:2-OH	1,02	10,58	0,27	0,11	0,06	0,21
26:0	0,00	2,18	0,23	0,16	0,08	0,16
26:1	0,23	1,98	0,31	0,24	0,09	0,24
26:2	0,00	2,01	0,13	0,05	0,03	0,11
26:0-OH	0,14	2,35	0,22	0,04	0,02	0,13
26:1-OH	0,00	2,12	0,16	0,02	0,02	0,08
26:2-OH	0,58	6,34	0,10	0,04	0,02	0,10

B

	Control	n=29	SapB-d	pSap-d	MLD	n=15
Isoform	Min	Max			Min	Max
C16:0	0,86	3,02	2,86	3,79	1,92	4,57
C18:0	1,00	1,00	1,00	1,00	1,00	1,00
C18:0-OH	0,58	2,04	0,95	0,62	0,79	1,62
C20:0	0,19	0,80	1,64	1,19	0,67	2,06
C20:0-OH	0,31	1,84	1,30	0,67	1,00	3,20
C22:0	0,34	1,68	**7,18**	**4,25**	**2,65**	**12,11**
C22:1-OH	0,35	1,42	2,06	0,58	1,56	5,95
C22:0-OH	0,30	1,64	**5,21**	**2,28**	**2,85**	**14,29**
C24:1	0,48	1,98	2,59	2,65	1,02	5,21
C24:0	0,45	2,23	**5,22**	**4,24**	**2,27**	**13,28**
C23:0-OH	0,26	1,55	2,86	1,02	2,06	9,45
C24:1-OH	0,21	0,77	**4,18**	**1,53**	**2,33**	**12,56**
C24:0-OH	0,31	1,81	**7,49**	**3,73**	**4,59**	**20,68**
C26:1	0,32	1,84	0,23	0,23	0,18	0,59
C26:0	0,11	0,71	0,55	0,28	0,43	1,31
C26:1-OH	0,09	1,34	0,24	0,17	0,17	0,70
C26:0-OH	0,09	0,62	0,30	0,19	0,15	0,65

Table 6. Changed signal ratios of Gb3Cer isoforms (A) and sulfatide isoforms (B) (to the C18:0 species) in the urine of patients with lysosomal storage disorders. The C18:0 species were selected as the standard invariable parameter in the isoform pattern. Changes in the levels of specific isoforms in patients with lysosomal storage disorders are highlighted. SapB-d - saposin B deficiency; pSap-d - prosaposin deficiency; Fabry - Fabry disease; MLD - metachromatic leukodystrophy.

6. Conclusions

In this chapter, we have introduced a methods of complex sphingolipid analysis, covering sample preparation and final tandem mass spectrometry analysis for various biological materials. This approach has been used in various studies of lysosomal storage disorders and examples showing a range of applications are presented.

Findings in urine are very important in the pre-diagnosis of lysosomal storage disorders and especially in identifying defects in the protein activators of sphingolipid hydrolases. We have also drawn attention to the problems with evaluating urinary sphingolipids using creatinine and have thereby suggested a more reliable approach to standardize sphingolipid excretion.

The tandem mass analysis of sphingolipids in cells and tissues is useful in diagnosing unresolved cases (examples shown in this chapter). This method can also contribute important information to lipidomic studies of the cellular function of these molecules and their bioactive derivatives.

The use of tandem mass spectrometry in loading experiments using labeled sphingolipids can increase quantification accuracy and throughput while eliminating working risk and restrictions by eliminating the need for radioactive analysis. Although tandem mass spectrometry cannot yet fully replace radioisotope methods, using this technique can improve the precision and specificity of the results of metabolic experiments.

The demand for useful screening methods for lysosomal storage disorders has led to the use of tandem mass spectrometry in enzymology. Analyses of enzyme activity using mass spectrometry performed on dried blood spots are highly sensitive and specific, and dried samples are easy to transport. Some methods use natural substrates, which is helpful in research studies of enzyme function and characteristics.

The aforementioned advantage of tandem mass spectrometry is the ability to analyze individual sphingolipid molecules (isoforms). Evaluation of isoform profiles can have diagnostic value in disorders involving storage of Gb3Cer or sulfatides. Metabolomic principles have a tremendous number of research applications, especially in the investigation of various cellular events.

In conclusion, tandem mass spectrometry is robust and sensitive analytical procedure that is still evolving. The method is efficient for determining the composition of endogenous sphingolipid classes in various biological materials and following their metabolic fate. Its ability to establish the metabolomic profiles of sphingolipids under normal and abnormal conditions contributes to a better understanding of the biological significance of sphingolipid molecules.

7. Acknowledgment

This work was supported by the grant project from the Ministry of Education of the Czech Republic MSM 0021620806, the grant No 19509 from the Grant Agency of the Charles University in Prague, the Grant SVV262502 from the Charles University in Prague, Czech Republic and by the grant project from the Ministry of Health of the Czech Republic MZOVFN2005

8. References

Aerts, J. M., Groener, J. E., Kuiper, S., Donker-Koopman, W. E., Strijland, A., Ottenhoff, R., van Roomen, C., Mirzaian, M., Wijburg, F. A., Linthorst, G. E., Vedder, A. C., Rombach, S. M., Cox-Brinkman, J., Somerharju, P., Boot, R. G., Hollak, C. E., Brady, R. O. & Poorthuis, B. J. (2008). Elevated globotriaosylsphingosine is a hallmark of Fabry disease. *Proc Natl Acad Sci U S A*, Vol. 105, No. 8, (Feb 26), pp. (2812-2817), 1091-6490 (Electronic)

Asfaw, B., Ledvinova, J., Dobrovolny, R., Bakker, H. D., Desnick, R. J., van Diggelen, O. P., de Jong, J. G., Kanzaki, T., Chabas, A., Maire, I., Conzelmann, E. & Schindler, D. (2002). Defects in degradation of blood group A and B glycosphingolipids in Schindler and Fabry diseases. *J Lipid Res*, Vol. 43, No. 7, (Jul), pp. (1096-1104), 0022-2275 (Print)

Asfaw, B., Schindler, D., Ledvinova, J., Cerny, B., Smid, F. & Conzelmann, E. (1998). Degradation of blood group A glycolipid A-6-2 by normal and mutant human skin fibroblasts. *J Lipid Res*, Vol. 39, No. 9, (Sep), pp. (1768-1780), 0022-2275 (Print)

Berna, L., Asfaw, B., Conzelmann, E., Cerny, B. & Ledvinova, J. (1999). Determination of urinary sulfatides and other lipids by combination of reversed-phase and thin-layer chromatographies. *Anal Biochem,* Vol. 269, No. 2, (May 1), pp. (304-311), 0003-2697 (Print)

Bilkova, Z., Castagna, A., Zanusso, G., Farinazzo, A., Monaco, S., Damoc, E., Przybylski, M., Benes, M., Lenfeld, J., Viovy, J. L. & Righetti, P. G. (2005). Immunoaffinity reactors for prion protein qualitative analysis. *Proteomics,* Vol. 5, No. 3, (Feb), pp. (639-647), 1615-9853 (Print)

Bligh, E. G. & Dyer, W. J. (1959). A rapid method of total lipid extraction and purification. *Can J Biochem Physiol,* Vol. 37, No. 8, (Aug), pp. (911-917),

Boscaro, F., Pieraccini, G., la Marca, G., Bartolucci, G., Luceri, C., Luceri, F. & Moneti, G. (2002). Rapid quantitation of globotriaosylceramide in human plasma and urine: a potential application for monitoring enzyme replacement therapy in Anderson-Fabry disease. *Rapid Commun Mass Spectrom,* Vol. 16, No. 16, pp. (1507-1514), 0951-4198 (Print)

Cech, N. B. & Enke, C. G. (2001). Practical implications of some recent studies in electrospray ionization fundamentals. *Mass Spectrom Rev,* Vol. 20, No. 6, (Nov-Dec), pp. (362-387), 0277-7037 (Print)

Cole, R. B. (Ed(s).). (2010) *Electrospray and MALDI Mass Spectrometry, Fundamentals, Instrumetation, Practicalities, and Biological Applications,* John Wiley & Sons, Inc., 978-0-471-74107-7, New Jersey

de Hoffman, E. & Stroobant, V. (2002) *Mass Spectrometry, Principles and Applications* (Second), John Wiley and Sons, Ltd, 0-471-48566-7, Chichester

Dekker, N., van Dussen, L., Hollak, C. E., Overkleeft, H., Scheij, S., Ghauharali, K., van Breemen, M. J., Ferraz, M. J., Groener, J. E., Maas, M., Wijburg, F. A., Speijer, D., Tylki-Szymanska, A., Mistry, P. K., Boot, R. G. & Aerts, J. M. (2011). Elevated plasma glucosylsphingosine in Gaucher disease: relation to phenotype, storage cell markers, and therapeutic response. *Blood,* Vol. No. (Aug 25), pp. 1528-0020 (Electronic)

Desnick, R. J., Ioannou, Y. A. & Eng, C. M. (2001). α-Galactosidase A Deficiency: Fabry Disease, In: *The Metabolic and Molecular Bases of Inherited Disease,* Scriver, C. R., Beaudet, A. L., Sly, W. S.&Valle, D., pp. (3733-3774), McGraw-Hill, 0-07-136321-1, New York

Desnick, R. J., Sweeley, C. C. & Krivit, W. (1970). A method for the quantitative determination of neutral glycosphingolipids in urine sediment. *J Lipid Res,* Vol. 11, No. 1, (Jan), pp. (31-37), 0022-2275 (Print)

Domon, B. & Costello, C. E. (1988). Structure elucidation of glycosphingolipids and gangliosides using high-performance tandem mass spectrometry. *Biochemistry,* Vol. 27, No. 5, (Mar 8), pp. (1534-1543), 0006-2960 (Print)

Douglas, D. J. (2009). Linear quadrupoles in mass spectrometry. *Mass Spectrom Rev,* Vol. 28, No. 6, (Nov-Dec), pp. (937-960), 1098-2787 (Electronic)

Dulcks, T. & Juraschek, R. (1999). Electrospray as an ionisation method for mass spectrometry. *Journal of Aerosol Science,* Vol. 30, No. 7, (AUG), pp. (927-943),

Fauler, G., Rechberger, G. N., Devrnja, D., Erwa, W., Plecko, B., Kotanko, P., Breunig, F. & Paschke, E. (2005). Rapid determination of urinary globotriaosylceramide isoform profiles by electrospray ionization mass spectrometry using stearoyl-d35-globotriaosylceramide as internal standard. *Rapid Commun Mass Spectrom,* Vol. 19, No. 11, pp. (1499-1506), 0951-4198 (Print)

Folch, J., Lees, M. & Sloane Stanley, G. H. (1957). A simple method for the isolation and purification of total lipides from animal tissues. *J Biol Chem,* Vol. 226, No. 1, (May), pp. (497-509), 0021-9258 (Print)

Forni, S., Fu, X., Schiffmann, R. & Sweetman, L. (2009). Falsely elevated urinary Gb3 (globotriaosylceramide, CTH, GL3). *Mol Genet Metab,* Vol. 97, No. 1, (May), pp. (91), 1096-7206 (Electronic)

Fuller, M., Sharp, P. C., Rozaklis, T., Whitfield, P. D., Blacklock, D., Hopwood, J. J. & Meikle, P. J. (2005). Urinary lipid profiling for the identification of fabry hemizygotes and heterozygotes. *Clin Chem,* Vol. 51, No. 4, (Apr), pp. (688-694), 0009-9147 (Print)

Goldschmidt-Arzi, M., Shimoni, E., Sabanay, H., Futerman, A. H. & Addadi, L. (2011). Intracellular localization of organized lipid domains of C16-ceramide/cholesterol. *J Struct Biol,* Vol. 175, No. 1, (Jul), pp. (21-30), 1095-8657 (Electronic)

Gu, M., Kerwin, J. L., Watts, J. D. & Aebersold, R. (1997). Ceramide profiling of complex lipid mixtures by electrospray ionization mass spectrometry. *Anal Biochem,* Vol. 244, No. 2, (Jan 15), pp. (347-356), 0003-2697 (Print)

Han, X. & Gross, R. W. (2005). Shotgun lipidomics: electrospray ionization mass spectrometric analysis and quantitation of cellular lipidomes directly from crude extracts of biological samples. *Mass Spectrom Rev,* Vol. 24, No. 3, (May-Jun), pp. (367-412), 0277-7037 (Print)

Hartree, E. F. (1972). Determination of protein: a modification of the Lowry method that gives a linear photometric response. *Anal Biochem,* Vol. 48, No. 2, (Aug), pp. (422-427), 0003-2697 (Print)

Harzer, K., Paton, B. C., Poulos, A., Kustermannkuhn, B., Roggendorf, W., Grisar, T. & Popp, M. (1989). Sphingolipid Activator Protein-Deficiency in a 16-Week-Old Atypical Gaucher Disease Patient and His Fetal Sibling - Biochemical Signs of Combined Sphingolipidoses. *Eur J Pediatr,* Vol. 149, No. 1, (Oct), pp. (31-39), 0340-6199

Haynes, C. A., Allegood, J. C., Park, H. & Sullards, M. C. (2009). Sphingolipidomics: methods for the comprehensive analysis of sphingolipids. *J Chromatogr B Analyt Technol Biomed Life Sci,* Vol. 877, No. 26, (Sep 15), pp. (2696-2708), 1873-376X (Electronic)

Heitmann, D., Lissel, M., Kempken, R. & Muthing, J. (1996). Replacement of chloroform throughout glycosphingolipid isolation. *Biomed Chromatogr,* Vol. 10, No. 5, (Sep-Oct), pp. (245-250), 0269-3879 (Print)

Helms, J. B. & Zurzolo, C. (2004). Lipids as targeting signals: lipid rafts and intracellular trafficking. *Traffic,* Vol. 5, No. 4, (Apr), pp. (247-254), 1398-9219 (Print)

Hirabayashi, Y., Igarashi, Y. & Merrill, A. H., Jr. (2006). Sphingolipids Synthesis, Transport and Cellular Signaling, In: *Sphingolipid Biology,* Hirabayashi, Y., Igarashi, Y.&Merrill, A. H., Jr., pp. (3-22), Springer-Verlag, 4-431-34198-6, Tokyo

Holleran, W. M., Takagi, Y. & Uchida, Y. (2006). Epidermal sphingolipids: metabolism, function, and roles in skin disorders. *FEBS Lett*, Vol. 580, No. 23, (Oct 9), pp. (5456-5466), 0014-5793 (Print)

Holthuis, J. C., van Meer, G. & Huitema, K. (2003). Lipid microdomains, lipid translocation and the organization of intracellular membrane transport (Review). *Mol Membr Biol*, Vol. 20, No. 3, (Jul-Sep), pp. (231-241), 0968-7688 (Print)

Hsu, F. F., Bohrer, A. & Turk, J. (1998). Electrospray ionization tandem mass spectrometric analysis of sulfatide. Determination of fragmentation patterns and characterization of molecular species expressed in brain and in pancreatic islets. *Biochim Biophys Acta*, Vol. 1392, No. 2-3, (Jun 15), pp. (202-216), 0006-3002 (Print)

Hsu, F. F. & Turk, J. (2000). Structural determination of sphingomyelin by tandem mass spectrometry with electrospray ionization. *J Am Soc Mass Spectrom*, Vol. 11, No. 5, (May), pp. (437-449), 1044-0305 (Print)

Chatelut, M., Harzer, K., Christomanou, H., Feunteun, J., Pieraggi, M. T., Paton, B. C., Kishimoto, Y., O'Brien, J. S., Basile, J. P., Thiers, J. C., Salvayre, R. & Levade, T. (1997). Model SV40-transformed fibroblast lines for metabolic studies of human prosaposin and acid ceramidase deficiencies. *Clin Chim Acta*, Vol. 262, No. 1-2, (Jun 27), pp. (61-76), 0009-8981 (Print)

Ii, T., Ohashi, Y. & Nagai, Y. (1995). Structural elucidation of underivatized gangliosides by electrospray-ionization tandem mass spectrometry (ESIMS/MS). *Carbohydr Res*, Vol. 273, No. 1, (Aug 22), pp. (27-40), 0008-6215 (Print)

Infante, R. E., Wang, M. L., Radhakrishnan, A., Kwon, H. J., Brown, M. S. & Goldstein, J. L. (2008). NPC2 facilitates bidirectional transfer of cholesterol between NPC1 and lipid bilayers, a step in cholesterol egress from lysosomes. *Proc Natl Acad Sci U S A*, Vol. 105, No. 40, (Oct 7), pp. (15287-15292), 1091-6490 (Electronic)

Kasper, D. C., Herman, J., De Jesus, V. R., Mechtler, T. P., Metz, T. F. & Shushan, B. (2010). The application of multiplexed, multi-dimensional ultra-high-performance liquid chromatography/tandem mass spectrometry to the high-throughput screening of lysosomal storage disorders in newborn dried bloodspots. *Rapid Commun Mass Spectrom*, Vol. 24, No. 7, (Apr 15), pp. (986-994), 1097-0231 (Electronic)

Kaushik, S., Massey, A. C. & Cuervo, A. M. (2006). Lysosome membrane lipid microdomains: novel regulators of chaperone-mediated autophagy. *EMBO J*, Vol. 25, No. 17, (Sep 6), pp. (3921-3933), 0261-4189 (Print)

Kebarle, P. (2000). A brief overview of the present status of the mechanisms involved in electrospray mass spectrometry. *J Mass Spectrom*, Vol. 35, No. 7, (Jul), pp. (804-817), 1076-5174 (Print)

Kerwin, J. L., Tuininga, A. R. & Ericsson, L. H (1994). Identification of molecular species of glycerophospholipids and sphingomyelin using electrospray mass spectrometry. *J Lipid Res*, Vol. 35, No. 6, (Jun), pp. (1102-1114), 0022-2275 (Print)

Kihara, H., Ho, C. K., Fluharty, A. L., Tsay, K. K. & Hartlage, P. L. (1980). Prenatal diagnosis of metachromatic leukodystrophy in a family with pseudo arylsulfatase A deficiency by the cerebroside sulfate loading test. *Pediatr Res*, Vol. 14, No. 3, (Mar), pp. (224-227), 0031-3998 (Print)

Kita, K., Kurita, T. & Ito, M. (2001). Characterization of the reversible nature of the reaction catalyzed by sphingolipid ceramide N-deacylase. A novel form of reverse hydrolysis reaction. *Eur J Biochem,* Vol. 268, No. 3, (Feb), pp. (592-602), 0014-2956 (Print)

Kitagawa, T., Ishige, N., Suzuki, K., Owada, M., Ohashi, T., Kobayashi, M., Eto, Y., Tanaka, A., Mills, K., Winchester, B. & Keutzer, J. (2005). Non-invasive screening method for Fabry disease by measuring globotriaosylceramide in whole urine samples using tandem mass spectrometry. *Mol Genet Metab,* Vol. 85, No. 3, (Jul), pp. (196-202), 1096-7192 (Print)

Kitatani, K., Idkowiak-Baldys, J. & Hannun, Y. A. (2008). The sphingolipid salvage pathway in ceramide metabolism and signaling. *Cell Signal,* Vol. 20, No. 6, (Jun), pp. (1010-1018), 0898-6568 (Print)

Klein, A., Henseler, M., Klein, C., Suzuki, K., Harzer, K. & Sandhoff, K. (1994). Sphingolipid activator protein D (sap-D) stimulates the lysosomal degradation of ceramide in vivo. *Biochem Biophys Res Commun,* Vol. 200, No. 3, (May 16), pp. (1440-1448), 0006-291X (Print)

Kolter, T. & Sandhoff, K. (2010). Lysosomal degradation of membrane lipids. *FEBS Lett,* Vol. 584, No. 9, (May 3), pp. (1700-1712), 1873-3468 (Electronic)

Korecka, L., Jezova, J., Bilkova, Z., Benes, M., Horak, D., Hradcova, O., Slovakova, M. & Viovy, J. L. (2005). Magnetic enzyme reactors for isolation and study of heterogeneous glycoproteins. *Journal of Magnetism and Magnetic Materials,* Vol. 293, No. 1, (May 1), pp. (349-357), 0304-8853

Kuchar, L., Ledvinova, J., Hrebicek, M., Myskova, H., Dvorakova, L., Berna, L., Chrastina, P., Asfaw, B., Elleder, M., Petermoller, M., Mayrhofer, H., Staudt, M., Krageloh-Mann, I., Paton, B. C. & Harzer, K. (2009). Prosaposin deficiency and saposin B deficiency (activator-deficient metachromatic leukodystrophy): report on two patients detected by analysis of urinary sphingolipids and carrying novel PSAP gene mutations. *Am J Med Genet A,* Vol. 149A, No. 4, (Feb 15), pp. (613-621), 1552-4833 (Electronic)

Kuchar, L., Rotkova, J., Asfaw, B., Lenfeld, J., Horak, D., Korecka, L., Bilkova, Z. & Ledvinova, J. (2010). Semisynthesis of C17:0 isoforms of sulphatide and glucosylceramide using immobilised sphingolipid ceramide N-deacylase for application in analytical mass spectrometry. *Rapid Commun Mass Spectrom,* Vol. 24, No. 16, (Aug 30), pp. (2393-2399), 1097-0231 (Electronic)

Kwon, H. J., Abi-Mosleh, L., Wang, M. L., Deisenhofer, J., Goldstein, J. L., Brown, M. S. & Infante, R. E. (2009). Structure of N-terminal domain of NPC1 reveals distinct subdomains for binding and transfer of cholesterol. *Cell,* Vol. 137, No. 7, (Jun 26), pp. (1213-1224), 1097-4172 (Electronic)

Ledeen, R. & Wu, G. (2011). New findings on nuclear gangliosides: overview on metabolism and function. *J Neurochem,* Vol. 116, No. 5, (Mar), pp. (714-720), 1471-4159 (Electronic)

Ledeen, R. W. & Wu, G. (2008). Nuclear sphingolipids: metabolism and signaling. *J Lipid Res,* Vol. 49, No. 6, (Jun), pp. (1176-1186), 0022-2275 (Print)

Leinekugel, P., Michel, S., Conzelmann, E. & Sandhoff, K. (1992). Quantitative correlation between the residual activity of beta-hexosaminidase A and arylsulfatase A and the severity of the resulting lysosomal storage disease. *Hum Genet,* Vol. 88, No. 5, (Mar), pp. (513-523), 0340-6717 (Print)

Li, Y., Scott, C. R., Chamoles, N. A., Ghavami, A., Pinto, B. M., Turecek, F. & Gelb, M. H. (2004). Direct multiplex assay of lysosomal enzymes in dried blood spots for newborn screening. *Clin Chem,* Vol. 50, No. 10, (Oct), pp. (1785-1796), 0009-9147 (Print)

Liebisch, G., Drobnik, W., Reil, M., Trumbach, B., Arnecke, R., Olgemoller, B., Roscher, A. & Schmitz, G. (1999). Quantitative measurement of different ceramide species from crude cellular extracts by electrospray ionization tandem mass spectrometry (ESI-MS/MS). *J Lipid Res,* Vol. 40, No. 8, (Aug), pp. (1539-1546),

Lieser, B., Liebisch, G., Drobnik, W. & Schmitz, G. (2003). Quantification of sphingosine and sphinganine from crude lipid extracts by HPLC electrospray ionization tandem mass spectrometry. *J Lipid Res,* Vol. 44, No. 11, (Nov), pp. (2209-2216), 0022-2275 (Print)

Liscum, L. (2000). Niemann-Pick type C mutations cause lipid traffic jam. *Traffic,* Vol. 1, No. 3, (Mar), pp. (218-225), 1398-9219 (Print)

Mano, N., Oda, Y., Yamada, K., Asakawa, N. & Katayama, K. (1997). Simultaneous quantitative determination method for sphingolipid metabolites by liquid chromatography/ionspray ionization tandem mass spectrometry. *Anal Biochem,* Vol. 244, No. 2, (Jan 15), pp. (291-300), 0003-2697 (Print)

Martin, O. C. & Pagano, R. E. (1994). Internalization and sorting of a fluorescent analogue of glucosylceramide to the Golgi apparatus of human skin fibroblasts: utilization of endocytic and nonendocytic transport mechanisms. *J Cell Biol,* Vol. 125, No. 4, (May), pp. (769-781), 0021-9525 (Print)

McLuckey, S. A. & Wells, J. M. (2001). Mass analysis at the advent of the 21st century. *Chem Rev,* Vol. 101, No. 2, (Feb), pp. (571-606), 0009-2665 (Print)

Micova, K., Friedecky, D., Faber, E., Polynkova, A. & Adam, T. (2010). Flow injection analysis vs. ultra high performance liquid chromatography coupled with tandem mass spectrometry for determination of imatinib in human plasma. *Clin Chim Acta,* Vol. 411, No. 23-24, (Dec 14), pp. (1957-1962), 1873-3492 (Electronic)

Mills, K., Morris, P., Lee, P., Vellodi, A., Waldek, S., Young, E. & Winchester, B. (2005). Measurement of urinary CDH and CTH by tandem mass spectrometry in patients hemizygous and heterozygous for Fabry disease. *J Inherit Metab Dis,* Vol. 28, No. 1, pp. (35-48), 0141-8955 (Print)

Murphy, R. C., Fiedler, J. & Hevko, J. (2001). Analysis of nonvolatile lipids by mass spectrometry. *Chem Rev,* Vol. 101, No. 2, (Feb), pp. (479-526), 0009-2665 (Print)

Natomi, H., Sugano, K., Iwamori, M., Takaku, F. & Nagai, Y. (1988). Region-specific distribution of glycosphingolipids in the rabbit gastrointestinal tract: preferential enrichment of sulfoglycolipids in the mucosal regions exposed to acid. *Biochim Biophys Acta,* Vol. 961, No. 2, (Jul 22), pp. (213-222), 0006-3002 (Print)

Olling, A., Breimer, M. E., Peltomaa, E., Samuelsson, B. E. & Ghardashkhani, S. (1998). Electrospray ionization and collision-induced dissociation time-of-flight mass spectrometry of neutral glycosphingolipids. *Rapid Commun Mass Spectrom,* Vol. 12, No. 10, pp. (637-645), 0951-4198 (Print)

Paschke, E., Fauler, G., Winkler, H., Schlagenhauf, A., Plecko, B., Erwa, W., Breunig, F., Urban, W., Vujkovac, B., Sunder-Plassmann, G. & Kotanko, P. (2011). Urinary total globotriaosylceramide and isoforms to identify women with Fabry disease: a diagnostic test study. *Am J Kidney Dis,* Vol. 57, No. 5, (May), pp. (673-681), 1523-6838 (Electronic)

Porter, M. T., Fluharty, A. L., Trammell, J. & Kihara, H. (1971). A correlation of intracellular cerebroside sulfatase activity in fibroblasts with latency in metachromatic leukodystrophy. *Biochem Biophys Res Commun,* Vol. 44, No. 3, (Aug 6), pp. (660-666), 0006-291X (Print)

Prinetti, A., Loberto, N., Chigorno, V. & Sonnino, S. (2009). Glycosphingolipid behaviour in complex membranes. *Biochim Biophys Acta,* Vol. 1788, No. 1, (Jan), pp. (184-193), 0006-3002 (Print)

Sandhoff, K., Kolter, T. & Harzer, K. (2001). Sphingolipid Activator Proteins, In: *The Metabolic and Molecular Bases of Inherited Disease,* Scriver, C. R., Beaudet, A. L., Sly, W. S.&Valle, D., pp. (3371-3388), McGraw-Hill, 0-07-136321-1, New York

Schepers, U., Glombitza, G., Lemm, T., Hoffmann, A., Chabas, A., Ozand, P. & Sandhoff, K. (1996). Molecular analysis of a GM2-activator deficiency in two patients with GM2-gangliosidosis AB variant. *Am J Hum Genet,* Vol. 59, No. 5, (Nov), pp. (1048-1056), 0002-9297 (Print)

Scherer, M., Leuthauser-Jaschinski, K., Ecker, J., Schmitz, G. & Liebisch, G. (2010). A rapid and quantitative LC-MS/MS method to profile sphingolipids. *J Lipid Res,* Vol. 51, No. 7, (Jul), pp. (2001-2011), 0022-2275 (Print)

Schiffmann, R., Waldek, S., Benigni, A. & Auray-Blais, C. (2010). Biomarkers of Fabry disease nephropathy. *Clin J Am Soc Nephrol,* Vol. 5, No. 2, (Feb), pp. (360-364), 1555-905X (Electronic)

Schmid, B., Paton, B. C., Sandhoff, K. & Harzer, K. (1992). Metabolism of GM1 ganglioside in cultured skin fibroblasts: anomalies in gangliosidoses, sialidoses, and sphingolipid activator protein (SAP, saposin) 1 and prosaposin deficient disorders. *Hum Genet,* Vol. 89, No. 5, (Jul), pp. (513-518), 0340-6717 (Print)

Schnaar, R. L. (1994). Isolation of glycosphingolipids. *Methods Enzymol,* Vol. 230, No. pp. (348-370), 0076-6879 (Print)

Schnaar, R. L., Suzuki, A. & Stanley, P. (2009). Glycosphingolipids, In: *Essentials of Glycobiology,* Varki, A., Cummings, R. D., Esko, J. D., Freeze, H. H., Stanley, P., Bertozzi, C. R., Hart, G. W.&Etzler, M. E., pp. Cold Spring Harbor Laboratory Press, 978-087969770-9, New York

Schulze, H. & Sandhoff, K. (2011). Lysosomal lipid storage diseases. *Cold Spring Harb Perspect Biol,* Vol. 3, No. 6, pp. 1943-0264 (Electronic)

Schwarzmann, G., Hoffmann-Bleihauer, P., Schubert, J., Sandhoff, K. & Marsh, D. (1983). Incorporation of ganglioside analogues into fibroblast cell membranes. A spin-label study. *Biochemistry,* Vol. 22, No. 21, (Oct 11), pp. (5041-5048), 0006-2960 (Print)

Schweppe, C. H., Hoffmann, P., Nofer, J. R., Pohlentz, G., Mormann, M., Karch, H., Friedrich, A. W. & Muthing, J. (2010). Neutral glycosphingolipids in human blood:

a precise mass spectrometry analysis with special reference to lipoprotein-associated Shiga toxin receptors. *J Lipid Res,* Vol. 51, No. 8, (Aug), pp. (2282-2294), 0022-2275 (Print)

Sleno, L. & Volmer, D. A. (2004). Ion activation methods for tandem mass spectrometry. *J Mass Spectrom,* Vol. 39, No. 10, (Oct), pp. (1091-1112), 1076-5174 (Print)

Sonderfeld, S., Conzelmann, E., Schwarzmann, G., Burg, J., Hinrichs, U. & Sandhoff, K. (1985). Incorporation and metabolism of ganglioside GM2 in skin fibroblasts from normal and GM2 gangliosidosis subjects. *Eur J Biochem,* Vol. 149, No. 2, (Jun 3), pp. (247-255), 0014-2956 (Print)

Spacil, Z., Elliott, S., Reeber, S. L., Gelb, M. H., Scott, C. R. & Turecek, F. (2011). Comparative triplex tandem mass spectrometry assays of lysosomal enzyme activities in dried blood spots using fast liquid chromatography: application to newborn screening of Pompe, Fabry, and Hurler diseases. *Anal Chem,* Vol. 83, No. 12, (Jun 15), pp. (4822-4828), 1520-6882 (Electronic)

Storch, J. & Xu, Z. (2009). Niemann-Pick C2 (NPC2) and intracellular cholesterol trafficking. *Biochim Biophys Acta,* Vol. 1791, No. 7, (Jul), pp. (671-678), 0006-3002 (Print)

Sullards, M. C., Liu, Y., Chen, Y. & Merrill, A. H., Jr. (2011). Analysis of mammalian sphingolipids by liquid chromatography tandem mass spectrometry (LC-MS/MS) and tissue imaging mass spectrometry (TIMS). *Biochim Biophys Acta,* Vol. No. (Jul 1), pp. 0006-3002 (Print)

Taylor, P. J. (2005). Matrix effects: the Achilles heel of quantitative high-performance liquid chromatography-electrospray-tandem mass spectrometry. *Clin Biochem,* Vol. 38, No. 4, (Apr), pp. (328-334), 0009-9120 (Print)

Turecek, F., Scott, C. R. & Gelb, M. H. (2007). Tandem mass spectrometry in the detection of inborn errors of metabolism for newborn screening. *Methods Mol Biol,* Vol. 359, No. pp. (143-157), 1064-3745 (Print)

van Echten-Deckert, G. (2000). Sphingolipid extraction and analysis by thin-layer chromatography. *Methods Enzymol,* Vol. 312, No. pp. (64-79), 0076-6879 (Print)

van Meer, G., Voelker, D. R. & Feigenson, G. W. (2008). Membrane lipids: where they are and how they behave. *Nat Rev Mol Cell Biol,* Vol. 9, No. 2, (Feb), pp. (112-124), 1471-0080 (Electronic)

von Figura, K., Gieselmann, V. & Jaeken, J. (2001). Metachromatic Leukodystrophy, In: *The Metabolic and Molecular Bases of Inherited Disease,* Scriver, C. R., Beaudet, A. L., Sly, W. S.&Valle, D., pp. (3695-3724), McGraw-Hill, 0-07-136321-1, New York

Wennekes, T., van den Berg, R. J., Boot, R. G., van der Marel, G. A., Overkleeft, H. S. & Aerts, J. M. (2009). Glycosphingolipids--nature, function, and pharmacological modulation. *Angew Chem Int Ed Engl,* Vol. 48, No. 47, pp. (8848-8869), 1521-3773 (Electronic)

Whitfield, P. D., Sharp, P. C., Johnson, D. W., Nelson, P. & Meikle, P. J. (2001). Characterization of urinary sulfatides in metachromatic leukodystrophy using electrospray ionization-tandem mass spectrometry. *Mol Genet Metab,* Vol. 73, No. 1, (May), pp. (30-37), 1096-7192 (Print)

Wiesner, P., Leidl, K., Boettcher, A., Schmitz, G. & Liebisch, G. (2009). Lipid profiling of FPLC-separated lipoprotein fractions by electrospray ionization tandem mass spectrometry. *J Lipid Res,* Vol. 50, No. 3, (Mar), pp. (574-585), 0022-2275 (Print)

Wrobe, D., Henseler, M., Huettler, S., Pascual Pascual, S. I., Chabas, A. & Sandhoff, K. (2000). A non-glycosylated and functionally deficient mutant (N215H) of the sphingolipid activator protein B (SAP-B) in a novel case of metachromatic leukodystrophy (MLD). *J Inherit Metab Dis,* Vol. 23, No. 1, (Feb), pp. (63-76), 0141-8955 (Print)

Xu, Z., Farver, W., Kodukula, S. & Storch, J. (2008). Regulation of sterol transport between membranes and NPC2. *Biochemistry,* Vol. 47, No. 42, (Oct 21), pp. (11134-11143), 1520-4995 (Electronic)

30

Newborn Screening by Tandem Mass Spectrometry: Impacts, Implications and Perspectives

François Rousseau, Yves Giguère, Marie-Thérèse Berthier, Dominique Guérette, Jean-Guy Girard and Michel Déry
Service de Biochimie, Dépt. Biologie Médicale, Centre Hospitalier Universitaire de Québec & Université Laval, Canada

1. Introduction

There are over 6000 inherited human diseases described in the McKusick Online Mendelian Inheritance in Man database (OMIM)(McKusick-Nathans Institute of Genetic Medicine, 2011). They show a wide range of variability in frequency, severity, age of onset, diagnostic and treatment approaches. A subset of diseases induce severe metabolic disturbances in the newborn that can lead to irreversible damage and illnesses. For some of those, there is a treatment that prevents such damages as long as it is initiated early after birth. For these diseases, population-based newborn screening has been proposed, developed and implemented in many countries albeit at different rates and based on different criteria. Even if those diseases are rare, the burden that is prevented if treated early has been shown to be cost-effective as it is larger than the cost of systematic screening and treatment of the cases identified (Venditti et al., 2003). Different analytical methods have been deployed since the 1960's to detect abnormal levels of specific metabolites or hormones in the newborn's blood with sufficient reliability and low cost to allow their use as screening methods (see 1.1).
We will describe the tremendous impact that tandem mass spectrometry has had in the field of newborn screening for inherited diseases in the recent decade. After a brief history of newborn screening, we will describe the classical WHO criteria for population-based screening for disease, summarize impacts of tandem mass spectrometry in this field. Then, we will describe the laboratory workflow, the pre-examination, examination, and post-examination aspects of MS/MS-based newborn screening, its advantages and limitations. Other issues will also be reviewed including sample and data management, revision of screening criteria and future perspectives for MS/MS in population-based screening.

1.1 History of newborn screening

The concept of universal population-based newborn screening (NBS) for inherited metabolic diseases was introduced in 1963 when a bacterial inhibition assay for phenylalanine in blood spots collected from newborn babies was proposed (Guthrie & Susi, 1963). This systematic screening identified, within the first days of life, babies with phenylketonuria (PKU). PKU, an autosomal recessive inborn error of metabolism is due to the deficiency of phenylalanine hydroxylase (a cellular enzyme) that leads to the accumulation of phenylalanine in the tissues and blood of affected individuals (20-fold levels as compared to normal) (Scriver, 2007). Phenylanaline at such levels is neurotoxic and impairs the normal development of brain functions after birth, amongst other problems. It affects about one newborn out of 10,000 births in populations of Caucasian origins (1:4500 to 1:18000) (DiLella et al., 1986). If left undetected and untreated, PKU leads to irreversible mental retardation which can be prevented with a low phenylalanine diet initiated rapidly at birth. Newborn screening of PKU has become the classical example of a good screening practice and is accepted as a standard care in all developed countries (Lehotay et al., 2011).

In the following decade, only a few other diseases have been added to NBS, especially congenital hypothyroidism (CH). CH has an estimated incidence of 1:200 to 1:4000 (Rastogi & LaFranchi, 2010). Initially thyroxine (a lipophilic small molecule hormone) was measured as the screening entry test but this was later replaced with measurements of TSH (a peptide hormone) by immunoassay methods. It is estimated that approximately 25% of all newborns worldwide are tested at least for hypothyroidism (Wilcken, 2007). Depending on the incidence of other specific detectable and treatable metabolic diseases some countries have added a few more diseases (e.g. galactosemia, cystic fibrosis (CF), congenital adrenal hyperplasia (CAH) and, tyrosinemia) were added in certain countries (Pitt, 2010).

In 1990, it was proposed that MS/MS could be used to test for multiple analytes simultaneously in dried blood spots from newborn screening programmes (Millington et al., 1990). However, at that time, a significant technical challenge constituted in analyzing hundreds of samples per day injected individually by means of a probe that needed to be cleaned before the next sample (Pollitt, 2006). This laborious process was relieved by the application of electrospray ionization to biomolecules and availability of commercially developed robust and easily cleaned sources of ions, as well as data systems to manage the resultant output (Pollitt, 2006; Johnson et al., 1996; Rashed et al., 1995). The following years revealed that MS/MS, which was already used in the clinical laboratory for quantitation of metabolites and drugs, could also be used as a primary method for NBS of inborn errors of metabolism. This technology was suitable because most of the molecules to measure were small metabolites, it allowed simultaneous analysis of a wide variety of different analytes, it was compatible with high-throughput (2-3 min per sample) and low reagent costs at the same time as being sensitive and specific (Chace, 2009). It has been estimated that, when using the American College of Medical Genetics (ACMG) panel of markers (see below), one affected newborn will be identified by MS/MS for every 3000 births (Rinaldo et al., 2006).

Year	Milestone	Reference
1963	PKU Screening (bacterial growth inhibition)	(Guthrie & Susi, 1963)
1968	WHO Screening Criteria	(Wilson & Junger, 1968)
1973	Addition of Congenital Hypothyroidism Screening	(Dussault & Laberge, 1973)
1975	Tyrosinemia Screening in specific population (Québec)	(Laberge et al., 1975)
1990	Proposal to use MS/MS for newborn screening	(Millington et al., 1990)
1993	Proof of concept for NBS by MS/MS	(Chace et al., 1993)
1995	Electrospray Ionization applied to NBS	(Rashed et al., 1995)
2006	ACMG proposed panel for NBS by MS/MS	(ACMG Newborn Screening Expert Group, 2006)
2008	Revised newborn screening criteria	(Andermann et al., 2008)

Table 1. Some of the major milestones for newborn screening.

1.2 Criteria for population screening

Population screening traditionally has aimed only at diseases that fulfill a set of widely accepted criteria for population-based screening proposed in 1968 by Wilson and Junger, as commissioned by the World Health Organization (WHO)(Wilson & Junger, 1968; Table 2).

Principal Wilson and Jungner Criteria for Screening (WHO, 1968)	In short
• Knowledge of disorder and its natural history if screening did not occur; • Availability of, access to and knowledge of how presymptomatic interventions change natural history of disorder, prevented irreversible harms, improved outcomes and reduced morbidity and mortality; • Screening test is simple with very high sensitivity and specificity and cost benefit ratio; • Resources exist for follow up of individuals with positive screening for confirmatory testing, introduction and monitoring of treatment and for counseling of family in the case of genetic diseases; • These Principles are acceptable to the consumer, the "target population".	• **Knowledge of disorder** • **Presymptomatic intervention improving outcomes** • **Simple and low cost test with excellent performances** • **Resources available for follow-up of positive individuals** • **Acceptability of program**

Table 2. Principal Wilson and Jungner criteria for screening.

These criteria are applicable to systematic or population-based screening (i.e of asymptomatic persons) for any type of disease and have been considered best practice since their publication in 1968. They thus apply not only to inherited diseases but to screening for any disease. They have been used by public health authorities to evaluate candidate tests or programs for population based screening. Since only few tests and diseases have historically met the criteria, population-based screening programs, including newborn screening, have historically targeted very few diseases. The first criteria asks that the natural history of the disorder be very well known, especially if there is no screening offered. This includes the proportion of untreated individuals that will develop severe and irreversible organ damages before they develop symptoms and seek a medical consultation. That is, because population-based screening (pre-symptomatic by definition) will usually not be cost-effective if the progressing disease causes symptoms before producing severe consequences. The second criteria refers to the presence of evidence that, if the disease is detected before it generates symptoms, there is an intervention (or treatment) to propose that will improve the outcomes. If there is no effective intervention with proven beneficial effects that can be offered to prevent the development of the disease after a positive screening test, then there may be no point in performing this test. The third criteria asks for a screening test that is simple, costs little and has good clinical specificity and sensitivity. In a screening context, sensitivity is very important as the very purpose of a screening program is to detect most if not all individuals that will develop the disease, thus the false negative rate (the proportion of disease cases that will test negative) must be as low as possible. However, if the screening test has poor specificity and too many false positive results (normal individuals that will test positive), then too many positive results will need to be confirmed by further testing which will not be cost-effective at all. Indeed, one of the major cost drivers for a screening program is the rate of false positive screening results. This is easily understood, as for example, if a disease has an incidence of 1 in 10000 newborns, and the false positive rate is 5%, then for each true positive case, the screening test will produce 500 false positives that will need to be reclassified as negative after a confirmation test. The fourth criteria refers to the availability of health care and financial resources to provide the true positive cases with appropriate follow-up and treatment, as well as the necessary genetic counseling if the disease screened is an inherited condition. Finally, any population-based screening program also needs to be acceptable to the population to whom it will be offered. This is also an important aspect as social acceptability of a program can significantly affect participation rate and undermine the very objectives of a screening program.

With the advent of multiplex screening methods such as MS/MS, allowing simultaneous detection of several tens of analytes in a single run, these criteria have been challenged and it has been proposed to revise them (Section 4) (Andermann et al., 2008). Countries and health care jurisdictions are still attempting to find a balance between the technological drive to provide all the data available when analyzing a newborn sample by MS/MS and the necessity to fulfill rational criteria before screening entire populations for diseases.

1.3 Impact of MS/MS spectrometry on newborn screening

Tandem mass spectrometry, by its very nature, has the potential to detect and quantify many molecules of similar physicochemical properties at the same time, provided the appropriate sample preparation steps and technical parameters are used (see below). This constitutes a dramatic change as compared to the classical methods used for newborn screening before the advent of MS/MS in this field. This is the major reason for the accepted fact that MS/MS has profoundly changed the paradigm of newborn screening (Chace et al., 2003), and possibly of screening per se. The wide variety of newborn screening panels that are currently in operation in various countries (Table 3) exemplifies the choices that decision makers in each health jurisdiction have to make especially with respect to which diseases to screen for (or not). Not all countries have published the results of their newborn screening programmes but it is interesting to see that the incidence of infants with inherited metabolic disorders identified by MS/MS screening is significant, ranging from 1:1800 to 1:6200 newborn. Also, according to the information publicly available at the present time, the number of metabolic disorders screened by MS/MS in newborns ranges from 0 to over 40. Many jurisdictions have not implemented MS/MS yet, while others are still refining their panel of diseases that are included in the screening assays. The reasons for this wide variety of practices include accessibility to funding for an expensive clinical MS/MS screening infrastructure, but also the absence of uniformly accepted practice guidelines with respect to what diseases should be included in a newborn screening programme. While the American College of Medical Geneticists (ACMG Newborn Screening Expert Group, 2006) recommended a large core panel of 29 inherited metabolic diseases, the UK National screening committee (UK National Screening Committee, 2011) recommends screening for only five diseases (Watson et al., 2006). It is worth mentioning here that, while MS/MS is a powerful diagnostic and measurement technology, some of the analytes to be measured in NBS are not easily amenable to MS/MS (e.g. hemoglobins and thyroid hormones).

Country/ Province/ State	Total number of diseases screened for	Number of diseases screened by MS/MS	Overall prevalence of infants with inherited metabolic disorders identified by MS/MS	Reference
Canada (Québec)	4	1	nd	(Laflamme et al., 2005)
Canada (Ontario)	25	20	nd	(Children's Hospital of Eastern Ontario, 2011)
Canada (Saskashewan)	32	nd	nd	(Lehotay et al., 2011)
USA (NewYork)	47	37	nd	(Wadsworth Center, 2011)
USA (Minnesota)	30	29	1:1816	(Rinaldo et al., 2006)
USA (California)	50	44	nd	(California Department of Public Health, 2011)
UK (recomm.)	5	nd	nd	(UK National Screening Committee, 2011)
Australia (Victoria)	25	23	nd	(Victorian Government Department of Health, 2011)
Spain	nd	10	nd	(Bodamer et al., 2007)
Switzerland	2	0	nd	(Bodamer et al., 2007)
Austria	nd	25	1:2855	(Kasper et al., 2010)
Taiwan	nd	21	1:6219	(Niu et al., 2010)

Table 3. Comparison of some existing NBS programs worldwide (nd=no data).

2. Laboratory considerations

2.1 Laboratory workflow and management

A typical newborn screening laboratory receives between 100 to 1000 dried newborn blood spot samples (See 2.2.1) per working day for analysis. The general target for the turn around time to perform the analytical screening procedure and producing the laboratory report is two to four working days, because for some diseases treatment must be started rapidly to prevent irreversible damage to the infant (See 2.4.2). Confirmatory analyses are needed rapidly after a positive screening test result. Thus, although the same tests are repeated, there are several levels of complexity and management issues to address due to the sheer volume and time pressure facing the NBS laboratory. With such constraints, the NBS laboratory needs to prevent any type of service interruption and have back-up solutions to cover all possible aspects of the process with appropriate redundancy. Also, the clinical newborn screening laboratory should implement a Quality Management System for all laboratory processes as proposed in best practice guidelines such as ISO15189 (International Organization for Standardization, 2003) and NCCLS GP26 (NCCLS, 2004). This must be in place to ensure real-time measurement of appropriate indicators of the system's integrity in all phases of the pre-examination, examination and post-examination processes.
Apart from the laboratory processes themselves (see below), the NBS laboratory will usually be at the center of a complex set of activities surrounding the NBS programme. It is not in the scope of this chapter to describe those aspects in details but here are some examples. There are processes that precede the sampling procedure and for which the NBS laboratory

may be responsible, such as managing and maintaining the stocks of virgin sampling cards and forms in the various hospitals and nurseries participating in the program. Systematic follow-up processes should be in place to ensure that no sample is left unattended and to rapidly refer positive cases to a health care facility that will initiate treatment very rapidly. Laboratory managers may also be in direct contact with the public agency commissioning the NBS programme, but also with health care facilities and professionals upstream and downstream of the sampling-to-reporting process itself. There is wide variability in the organization of the different NBS programmes worldwide.

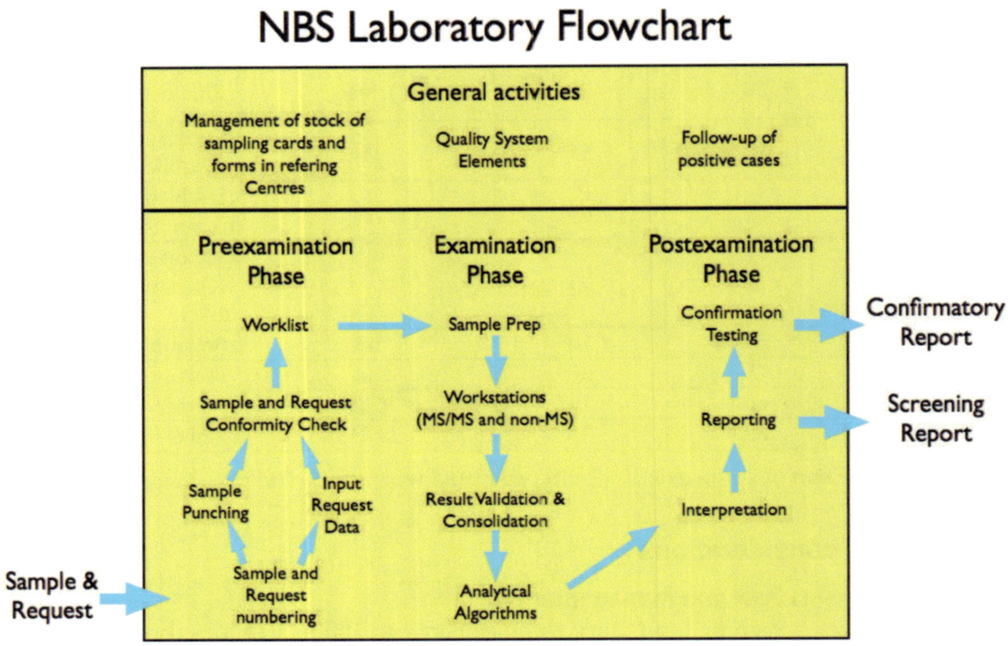

Fig. 1. NBS Laboratory Flowchart.

With respect to the NBS laboratory activities per se (see Figure 1) the three classical pre-examination (or preanalytical), examination and post-examination (or postanalytical) phases include the various stages and processes involved in sample and request forms flows between their arrival in the laboratory and the reporting of results. Data management is described later in this chapter and the details of each phase through which samples and forms go through are also discussed in the following sections.

2.2 Pre-examination considerations

Pre-examination conditions are crucial in order to ensure the performance of specimen analysis. The time of the sampling, the delay between the sampling and reception in the laboratory, the baby's condition (e.g. prematurity, parenteral nutrition, recent transfusion) can influence screening results and thus the life of the baby. Due to these limitations, it is important to note that the screening test result is not diagnostic and requires confirmation with an independent sample and method (Pitt, 2010). It has been estimated that there is a measurement error of ±30% due to the fact of using blood dots on filter paper as sample in part because it is done in many locations and by many people (Lehotay et al., 2011). Of note, with metabolic diseases, several-fold increases of metabolite levels are usually expected. In addition to specimen collection, storage and transport are also potential sources of errors.

2.2.1 Dried blood spots as samples and their management

Specimen collection

Collection of blood onto an absorbent paper card (dried blood spots: DBS) represents the most common type of NBS sample. On a filter paper, a few drops from a heel prick are dried in air for a few hours before being sent to the NBS laboratory. This filter paper is attached with a card that includes specific spaces for writing the family names of the father, the mother, their address and phone number, the date and hour of the newborn's birth, the date and hour of sampling, and the birth hospital (Figure 2). Each NBS card has a unique a number or bar code. In the province of Québec, a multiple identification of the newborn is done by 1) names of the parents, 2) the sampling card number which is unique for the each newborn and provides information of the location of the birth and 3) the date of birth. A complete information card is mandatory for the specimen to be valid and accepted by the NBS laboratory. Otherwise, a new and appropriately identified sample is needed.

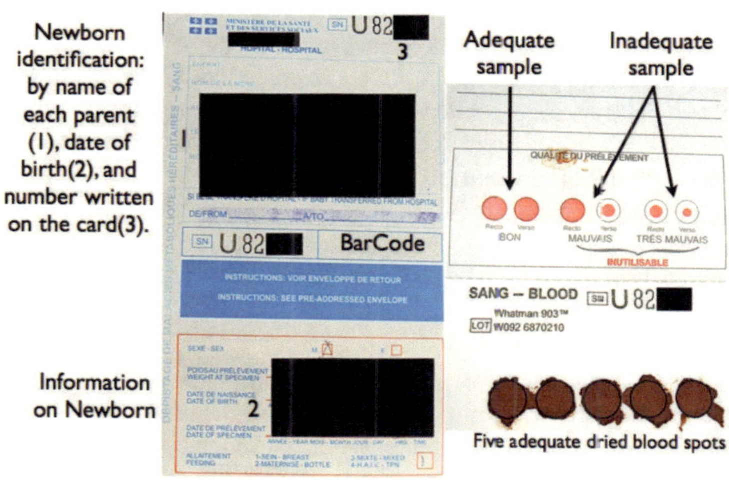

Fig. 2. Example of a newborn screening blood sampling card and its attached filter paper.

The heel puncture must be done on the planar surface of the foot in the safest area (medial to a line drawn posteriorly from the middle of the great toe to the heel or lateral to a line drawn from between the 4th and 5th toe to the heel.) according to the H3A6 CLSI guidelines. No foreign substance around the filter paper must contaminate the sample. Blood must completely fill all printed circles and is applied evenly on one side of the filter (but must also fill completely the minimal sample circle on the other side of the paper), free of layering and clots. It must be correctly dried (4 hours of air drying at room temperature). Laboratory analysis of the sample depends on an assumed amount of blood in the filter paper. Criteria for adequate sampling are summarized by Wadsworth Center (2011) and Kayton (2007).

Timing of collection is important because some metabolite's levels vary with the infant's age (Wilcken & Wiley, 2008). Sampling within 24-72 hours of age is common practice. If it occurs too early (<24 hours of age), a new specimen may be required. Prematurity, birth weight, parenteral nutrition, transfusion, type of feeding, neonatal jaundice and drugs affect metabolite levels (Rolinski et al., 2000) and may influence laboratory results. Such facts should be documented on the NBS card to help interpretation of results (Pitt, 2010).

Transport

Delays in transport can cause degradation of some markers and result in false negative results. Samples should be sent within 24 hours after collection. Batching causes delays in reception and processing of samples and may have serious consequences on the life of an affected newborn (Kayton, 2007). Upon arrival of samples to the NBS laboratory, verification of sampling card information and validation of specimen adequacy is performed. A laboratory identification number is generated for each sample card.

2.2.2 Automation of pre-analytical steps

Given the large volume of samples to analyze in a timely and reproducible way, automation of some pre-analytical steps has been implemented in many NBS laboratories. A micro-volume pipetting station and punchers help to shorten turn around time in the NBS laboratory. The puncher (basic, semi-automatic or automatic) is a convenient tool with minimal user intervention. It punches the sampling cards where a blood spot is located and drops the punch into a microplate. Punchers can read bar codes and some have a light guide to help the technician to optimally position the sample card and verify that the disk falls into the well. More sophisticated punchers are fully automated and accept 300 samples at a time. The micro-volume pipettor can pipet 96 well plates. It is a compact unit, with high precision (1µl / min dispensing, CV 5%) control of dispense rate and calibration values.

2.3 Examination considerations

In the following section, we discuss the particular aspects of tandem mass spectrometry as applied to the analysis of dried blood spot samples for newborn screening (NBS).

2.3.1 Types of sample preparation procedures

In the context of NBS, two types of pre-analytical sample treatments are in use: with or without derivatisation. After reception, verification and labeling of NBS sampling cards, a punch (generally 3.2 mm in diameter) is deposited in a 96 well plate and the position of each specific sample is noted (usually electronically by reading a bar code on the sampling card). Extraction of the analytes from sample card punches is usually performed with a solution comprised mainly of methanol, but including isotope-labeled internal standards for the main amino acids and acylcarnitines that will be measured. Quantification of the metabolites of interest will use these internal standards as a reference. The derivatisation method is more frequently used because of its greater sensitivity. This method will produce butyl derivatives. New methods without derivatisation are becoming more popular with the help of new generations of MS/MS instruments with increased sensitivity. This method simplifies the sample preparation steps by removing the butylation of metabolites with a toxic and corrosive reagent. This reduces significantly sample preparation time. For the majority of quantitative results for acylcarnitines and amino acids, the two methods differ by less than 15% (De Jesus et al., 2010). However, larger differences are observed for dicarboxylic acylcarnitine compounds, especially for short chains such as C5DC (glutaryl-carnitine). Raw counts for the non-derivatised method are much smaller as compared to the derivatisation method, although quantitative results are similar. C5DC is a major marker for glutaric aciduria type I. It is thus important to pay particular attention to this marker. The newer and more sensitive generation of MS/MS instruments may compensate for these lower values. Use of a non-derivatised method can induce a lack of differentiation of isobaric acylcarnitines, as well as increase false-positive rates for certain disorders, and cut-offs should be adjusted accordingly (De Jesus et al., 2010).

2.3.2 Technical configurations of MS/MS for newborn screening

A typical configuration for tandem mass spectrometry for a newborn screening laboratory includes four components (Figure 3). The first one is an HPLC-type of pump ensuring the flow of a mobile phase, but with no liquid chromatography step. The second component injects samples into the mobile phase before it reaches the MS/MS. Pipetting of each sample is performed by an auto-sampler for 96-well plates and hundreds of samples per day.

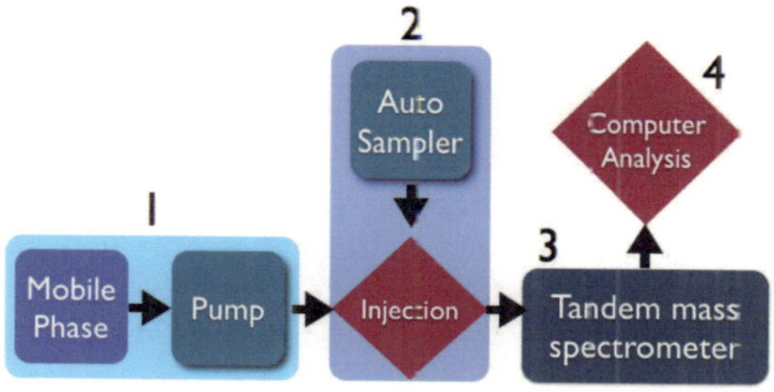

Fig. 3. Schematic configuration for tandem mass spectrometry for newborn screening.

The third component comprises the two mass spectrometers that scan the analytes in each sample, identify and quantify the various signals from metabolites of interest and transfer the raw information to the fourth component, a computer system, which performs the formal quantification of each metabolite and integrate the information into a useful data set.

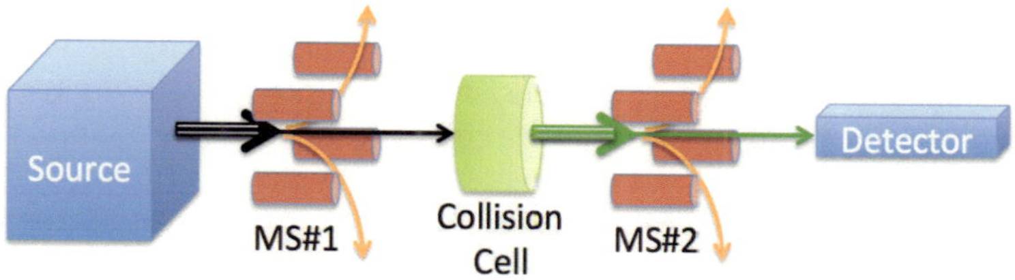

Fig. 4. Schematic representation of the flow of ions in a tandem mass spectrometer. *Adapted from Waters Micromass Quattro Premier XE Mass Spectrometer Operator's Guide.*

The tandem mass spectrometer itself comprises five major elements (Figure 4) : 1- the source where samples in liquid phase are injected, ionized and introduced into the first quadrupole MS; 2- the first quadrupole MS, where a first selection of ions by their m/z ratio is performed (many ions with similar m/z ratios may be selected); 3- the beam is directed into a collision cell where ions collide with Argon molecules which break ions having entered the cell (mother ions) into smaller ions (daughter ions) specific to each mother ion; 4- the second quadrupole MS receives these fragments and filters the daughter ions of interest (green arrow) to identify the presence of molecules of interest, with internal standards to support quantification; 5- an ion detector counts each ion that passes through the second MS at each specific m/z ratio and transfers the information to the computer system for analysis. Many types of sources of ions are used in mass spectrometry, depending on the application. An important aspect of tandem mass spectrometry applied to NBS is electrospray ionization (ESI). ESI is a method of obtaining ions without destroying molecules in the sample. It was developed by John Bennett Fenn who received the Nobel prize of chemistry for this in 2002. Before the development of ESI, it was very difficult to directly inject a liquid phase into a MS, as the interface between a liquid chromatograph and the MS was not easily performed; only gas chromatography was easily interfaced with MS. ESI (Figure 5) operates by sending a liquid sample through an electrically charged capillary tube. A continuous flow of nitrogen along the capillary creates a spray of charged liquid at the end of the capillary. Charged droplets are formed and their evaporation shrinks their size until they burst due to electric repulsion, which leaves individual molecules that are charged and ready to be injected into the MS, after being accelerated according to their charge.

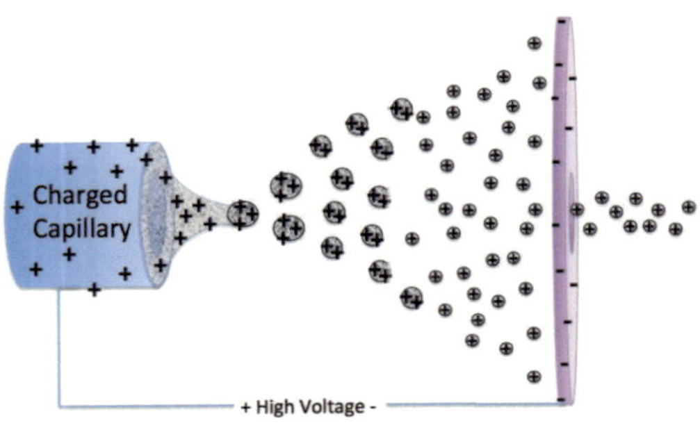

Fig. 5. Schematic representation of electrospray ionization. The liquid sample flows through a charged capillary (left), which produces positively charged droplets attracted by a negatively charged electrode that accelerates charged micro droplets into the MS/MS. *Adapted from Balogh & Stacey (1991).*

Finally, ions are selected and analyzed according to their mass to charge ratio (m/z) by the quadrupoles in tandem. Each quadrupole consists in four parallel rods defining the edges of a rectangular cuboid. Each rod has an opposite charge to its neighbor rod and the same as its opposite rod. The value of the m/z ratio of ions selected by the quadrupole is determined by a radio frequency superimposed with a continuous voltage applied to the quadrupole's electrodes (rods). For a specific radio frequency, there is a cut-off where ions of a specific m/z and larger will pass through. Thus the radio frequency acts as a filter of minimal m/z ratio. Inversely, the combination of continuous potential and the radiofrequency act as a filter of maximal m/z ratio. Thus addition of the appropriate voltage to a radio frequency selects a specific m/z ratio under which ions will not pass through the quadrupole.

2.3.3 Performing the analysis

In the context of newborn screening, the objective of the analysis is to identify selected inherited metabolic disorders of clinical significance.

As for any laboratory analysis, prior to offering a clinical laboratory test, the laboratory must validate (in the case of home-made reagents) or verify (in the case of commercial kits) the analytical method it will be implementing. This will usually include correlation with an established (or other validated) method, intra-run and inter-run reproducibility studies, carryover studies, linearity, recovery, precision, trueness, etc. These studies are needed to ensure the robustness and performance of the analytical method and the overall analytical process in the specific context of a physical laboratory facility. Cut-offs for detection of metabolites of interest should also be established. Accuracy of the assay is important near the cut-off value for each analyte. During the set-up phase for MS/MS assays in the laboratory, up to 2500 to 4000 samples may be needed to establish the appropriate cut-offs for metabolites levels that will be considered as positive upon screening results. Cut-offs are also compared with other screening laboratories and can be adjusted with the collection and analysis of larger number of samples over time. The difficulty is to establish a cutoff value for each metabolite in order to detect diseases and, at the same time, to minimize the occurrence of false negative and false positive screening results.

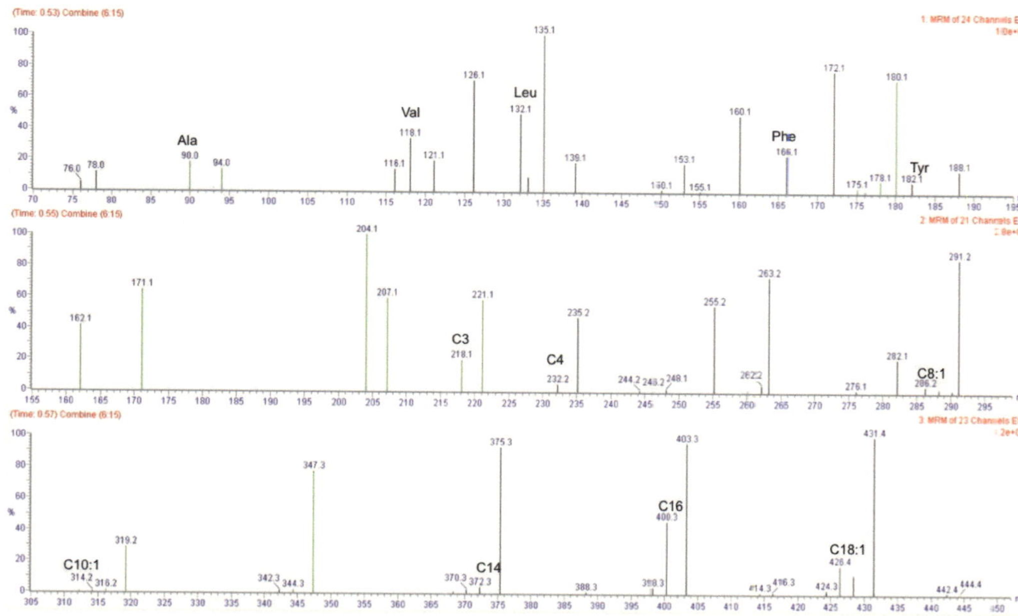

Fig. 6. Typical output for a newborn sample analysed by Multiple Reaction Monitoring (MRM) in MSMS. In this MSMS spectrum, targeted ion masses of interest, corresponding to markers for several inherited metabolic diseases, are divided in three parts : 1 :MRM of 24 aminoacids; 2 : MRM of 21 acyl-carnitines from C0 to C8:1; 3 : MRM of 23 acylcarnitines from C10 to C18OH;. x axis = m/z ; y axis = % intensity which relates to each metabolite concentration. Concentrations are compared with cut-offs for each corresponding analyte, defining an abnormal or normal metabolic profile for this newborn.

Also, prior to launching the test offer, the clinical laboratory must implement a quality assessment program for the analytes that will be measured. A typical output for a newborn sample is shown in Figure 6. Each run is checked by inspection of TIC (total ion current chromatogram) of each sample, internal standard and all quality control (QC) samples. In the laboratory, quality controls (QC) (low and high levels) are run with each batch of newborn samples and contribute to determining the adequacy of the run. Low level controls reflect the normal endogenous metabolites and high level controls reflect concentrations near the cut-off value or an abnormal state for each analyte. For each plate a minimum of one low QC and one high QC must be present. These QCs are "internal quality controls" given that their expected value is known to the lab performing the analysis. It is advised that other controls (such as external controls (of unknown value) or known cases) would also be present in the plate (Lehotay et al., 2011). With proficiency testing (external QC), there are several different dimensions of the screening processes that are examined. The laboratory processes - as if they were real samples - materials of unknown value (to the testing laboratories) provided by a proficiency testing program that then compares the results from several different participating laboratories. The complexity in developing the QC materials for MS/MS in NBS is high. Moreover, participation and comparison of results within collaborative projects such as Region 4S project (Region 4 Genetics Collaborative, 2007), and/or to a quality control program such as the CDC program where performance criteria are collected from several laboratories is relevant and useful to test the performance of the NBS laboratory and essential to insure short and long term quality of results (Chace et al., 1999; Adam et al., 2000; Hannon et al., 1997).

2.3.4 Advantages and limitations of MS/MS

MS/MS technology offers a new approach to NBS by having the ability to screen rapidly for 30 or more metabolic disorders in a single analysis from one small blood sample using a simple protocol (Chace et al., 2003). Indeed, in one run, more than 40 analytes can be quantified. This capacity decreases the NBS laboratory turn around time because many metabolites can be measured in parallel as opposed to sequentially. It is also potentially clinically efficient because screening of metabolic disorders asks for more than one marker to be measured to define the disease. With MS/MS, there is the possibility to produce a rather complete metabolic profile for a patient in a single run. However this great advantage comes with the limitation that some "disorders" detected by MS/MS can be benign or mild. An ethical question is thus raised with respect to how the laboratory, and the physicians, will deal with this information that is of undetermined clinical value (Lehotay et al., 2011). As compared with other methods used in the past, MS/MS showed a high analytical performance for detecting diseases of high frequency, notably MCAD deficiency and PKU. However, the technology's analytical performance is not identical for all analytes measured. (for instance succinyl acetone for the screening of tyrosinemia Type I). Thus, some important diseases of the newborn such as tyrosinemia type I have higher false positive rates by MS/MS than by more conventional methods (Chace et al., 2003). This led to the development of a number of second tier tests requiring a separate testing protocol with a rapid turn around time (Magera et al., 2006; Sander et al., 2006). Another limitation of MS/MS is raised by the diagnostic dilemma when one or several markers are the same for more than one disease. For instance C5OH is a marker of severe disease such as holocarboxylase synthase deficiency as well as a marker for a generally benign condition, namely 3 methylcrotonylCoA Carboxylase deficiency (Dantas et al., 2005; Koeberl et al., 2003). Despite these limitations, implementation of tandem mass spectrometry has generated a significant evolution in newborn screening programs, often referred-to as a paradigm shift.

2.3.5 Metabolites measured

Several metabolites are measured by MS/MS in a newborn screening laboratory. Generally, they arise from protein metabolism (amino acids), fatty acid and organic acid metabolism (acylcarnitines), or from endocrine metabolism (hormones or specific metabolites). To determine those metabolites to be integrated to a newborn screening MS/MS profile, it is important to have an excellent understanding of human metabolism. An enzymatic defect along a metabolic pathway can result in accumulation of the substrate and insufficient levels of the product of this enzyme. The accumulation of substrate - or its by-products -, or the decreased concentration of a normal product - or by-product - can be identified in the MS/MS spectrum and quantified. One challenge in newborn screening testing is that some metabolites are markers for several diseases, that is, they are not specific to a single metabolic disorder. To circumvent the issue of disease specificity, ratios of some markers are usually computed (Region 4 Genetics Collaborative, 2007). Extensive lists of markers that can be analyzed by MS/MS for newborn screening profiles are already published (ACMG, 2006; De Jesus et al., 2010; Lehotay et al., 2011). Some diseases that need to be screened for are not yet readily amenable to MS/MS and thus most NBS laboratories also run other types of assays for these conditions.

Because of pre-examination and examination concerns, screening assays are not equivalent to diagnostic assays. In order to allow for high throughput and multiple metabolite detection, some compromises have to be made on the assay's analytical performances. For [false] positive samples, these are in part compensated by more specific 2nd-tier assays (see 2.4.1). However, negative screening results will not be retested and thus false negative screening results have to be minimized in the context of a screening program, even if somewhat more false positive results will be generated at the initial screening step.

2.4 Post-examination considerations

After the screening examination phase is completed, the laboratory needs to determine which samples need confirmation and also to produce a laboratory report. As the disorders screened in NBS are rare, the majority of results are normal, but test performance will strongly influence delivery of ambiguous or false positive results, with its implications in terms of repeat testing, investigation, and parental stress. The expansion of NBS to a significantly increased number of disorders by MS/MS technology has only amplified and complexified the management of NBS programs during and following sample analysis.

2.4.1 Interpretation of screening results

If the expansion of NBS by MS/MS is a breakthrough in the field, it also represents a major challenge for NBS laboratory personnel who are faced with a significant increase in data generated, as was mentioned in previous sections (see 2.3.3 and 2.3.4).

Performance characteristics of the NBS procedures will depend upon chosen tests cut-offs. If the threshold level for a disorder is too high, a proportion of newborns with a disorder will be missed, while a threshold that is too low will generate many additional positive screen in infants without the disease. The increased number of conditions tested, each of them being rare, led to the implementation of 2nd tier confirmatory tests that are more complex, time-consuming and unsuitable for high throughput mass screening but with performance characteristics of increased sensitivity and specificity. The main aim is to diminish the number of false positive screens. The problem is more acute for primary tests that lack specificity and generate a high rate of false positives (for specific examples, see section 2.4.4). The false positive rate is the proportion of positive tests in newborns that eventually turn out to be normal after follow up evaluation, while the positive predictive value of a test, a function of the sensitivity, the specificity and the prevalence of the condition, is the probability that the infant tested positive has the disease. With about four millions births annually in the USA and each infant screened for about 30 conditions, assuming a specificity of 99.9% of the primary test for each disorder to be screened, this amounts to about 50,000 false positive results per annum in the USA only.

One must emphasize that the introduction of MS/MS in NBS, a very sensitive and specific technology, is not in itself responsible for the increased number of false positive screens, the phenomenon is rather a consequence of the increased number of extremely rare disorders that are included in the NBS panels. This may be unavoidable if NBS programs aim to screen infants for very rare inborn metabolic disorders. As an example, despite a sensitivity of about 100% and a specificity of 99.96%, after confirmatory testing, only eight (8) newborns were true positives out of the 1249 who initially tested positive for maple syrup urine disease (MSUD) in the USA in 2007 (3,364,612 were tested for MSUD overall) (Schulze et al., 2003). Undoubtedly, NBS protocols must include the diagnostic confirmation of a positive screen before any clinical intervention is instituted, otherwise a large number of infants would be given inappropriate treatment. Targets of adequate analytical and post-analytical performance in the era of NBS expansion by MS/MS multiplex testing have been proposed: false positive rate < 0.3%, positive predictive value > 20% (Rinaldo et al., 2006). Interestingly, novel NBS algorithms are being proposed, and 2nd tier testing, characterized by high sensitivity and specificity may also be used to identify rapidly newborns that would have otherwise been missed (i.e. false negatives) by reducing a primary test cut-off followed by 2nd testing on the initial blood spot for the increased number of abnormal results from the primary test, enhancing the performance of NBS (Turgeon et al., 2010).

Example of conditions	**Primary markers**	**Example of informative Ratios**	**Some Other Conditions with same markers**
Argininosuccinic acidemia	citrullin	Citrullin/arginin	Citrullinemia
Citrullinemia types 1 and 2	citrullin	Citrullin/arginin	Citrullinemia
Maple syrup disease	Valine, leucine- isoleucine	Valine/phenylalanine Leu-isoleucine/alanine	ketosis
Phenylketonuria	phenylalanine	Phenylalanine/tyrosine	Diet, prematurity
Type 1 Tyrosinemia	Succinylacetone (the best), tyrosine	Tyrosine/citrulline	Prematurity, types 2-3 tyrosinemias
Types 2 and 3 tyrosinemia	tyrosine	Tyrosine/citrulline	Prematurity, type1-tyrosinemias
Medium chain acyl-coA dehydrogenase deficiency	C6, C8, C10:1, C10	C8/C2; C8/C10	Glutaric type 2
Short chain acyl-coA dehydrogenase deficiency	C4	C4/C2; C4/C3, C4/C8	Glutaric type 2
Glutaric acidemia type1	C5DC	C5DC/C5OH, C5DC/C8, C5DC/C16	Glutaric type 2
Propionic acidemia	C3	C3/C2,C3/C16	Cobalamine deficiency (C, D)
Methylmalonic acidemia (Cobalamin deficiency)	C3	C3/C2,C3/C16, C3/Methionine	Propionic acidemia, Cobalamin deficiency A, B

Table 5. Examples of markers for some conditions, adapted from Lehotay et al. (2011).

Thus, not all abnormal results indicate the presence of an inborn error of metabolism and consequently they must be confirmed by a diagnostic or confirmatory procedure (see 2.4.3). Of note, interpretation of results must distinguish between an ongoing pathologic process and changes in metabolite concentrations due to the maturation process of organ systems. Discrimination between a frankly abnormal result and an undetermined result usually rely on the observed pattern of biochemical abnormality. Most inborn errors of metabolism have specific primary analytes associated with the condition (Table 5).

The presence of secondary markers or analyte ratios can increase the specificity of the screening procedure for the disorder associated with a primary analyte. Other concerns also arise in the presence of a positive screen or ambiguous result. Indeed, some conditions, such as prematurity (delivery < 37 weeks gestation), low birth weight (< 2500 g), some medications, total parenteral nutrition and newborn sickness unrelated to an inborn error of metabolism may cause up to 40% of false positive results (Lehotay et al., 2011). In such circumstances, repeat sampling is usually requested. Recently, the CLSI has issued guidelines and recommendations for repeat testing in these specific circumstances (Miller et al., 2009). Reports of abnormal results must thus include a qualitative interpretation by a NBS professional well versed in metabolic patterns associated with disorders, but also with factors that may mimic the presence of a screened condition.

2.4.2 Reporting of results

Despite recommendations by the American Academy of Pediatrics (AAP) newborn screening task force (Mandl et al., 2002), there exists a wide variety of practices between NBS programs regarding notification of results. While most programs will notify primary health care providers of results from all newborns, whether results are normal or positive, a few states and one Canadian province will issue results only if abnormal. All NBS programs in the USA report abnormal results and monitor follow up activities, but the reporting of abnormal results by NBS programs includes a wide diversity of approaches such as telephone call, letter, certified letter, secured web-based electronic letter and fax. All USA states send a report to the infant's pediatrician, and most report to the hospital of birth, the parents and a geneticist (Kayton, 2007). Considering the importance of follow-up of abnormal results for potentially life threatening conditions, the vast majority of states will follow up until it is confirmed that the screened infant with an abnormal result has a follow up appointment, and about half will track until a diagnosis is made and treatment is initiated. Only a few states however will continue to follow up periodically (Kayton, 2007).

Timely notification of parents of newborns screened positive for disorders with potentially disastrous evolution is critical. Notification of results depends on age at blood sample collection (which varies between NBS programs), dispatch, transfer to the NBS laboratory and processing. The UK Newborn Screening Program Center has recommended targets for reception and processing of samples and notification of results using PKU as the prototype disorder (NHS Newborn Bloodspot Screening Programme, 2008). Overall, 99.5% of samples should be received by 16 days of life, samples should be processed within 4 working days, ideally within 2 working days, and 100% of screened positives should be notified before the infant is 18 days old, with treatment initiation by 21 days of life, if confirmed. Ideally, 80% of screened positives should be notified by 12 days of life, with initiation of therapy before 14 days of life of the infant confirmed positive. As discussed previously, the impact of an expanded NBS panel by MS/MS are significant for both the NBS personnel and health care providers: much more interventions and communications to pediatricians are made by the former, including the reporting of an increased number of false positive and ambiguous results, while the latter are faced with the investigation of infants for disorders for which little is known about the natural history or may be benign conditions. In 2009, the National Academy of Clinical Biochemistry (NACB) issued guidelines for optimal follow up testing for positive newborn screens using MS/MS (Dietzen et al., 2009).

2.4.3 Confirmation testing in a NBS program

Classically, a positive NBS result needs a repeat analysis in duplicate from the initial blood spot before repeat sampling is requested. If the result is confirmed positive with the repeated screening test, the need to perform additional testing depends on the disorders for which the infant is presumably positive and the availability of a second tier test for that disorder. Second tier tests are reflex tests performed on the same dried blood sample as the primary screening test. The relevance to perform a 2nd tier test is justified by the possibility of false positive screens or the presence of an ambiguous result in the presence of poor specificity of the primary screening test for a specific disorder. As mentioned earlier, the expansion of NBS for multiple rare disorders causes a significant increase in the number of false positive results, which increases the burden on laboratory personnel, health care providers, not to mention parental stress. 2nd tier tests allow to confirm results with increased specificity on the initial blood spot. A typical example is NBS screening of congenital adrenal hyperplasia (CAH) using 17-OH progesterone by fluoro-immunoassay. The method lacks specificity, is affected in the presence of prematurity and neonate illnesses, and cross reacts with structurally similar steroids, notably 17-OH pregnenolone. The second tier tests measures 17-OH pregnenolone by LC-MS/MS (Etter et al., 2006), but also cortisol and androstenedione. The ratio of (17 OH-P + androstenedione)/cortisol, increases discrimination since 17-OH-P and androstenedione are increased while cortisol is decreased in CAH (Schwarz et al., 2009; Lacey et al., 2004; Minutti et al., 2004). Another example is the presence of elevated propionylcarnitine (C3), which may suggest the presence of propionic acidemia, but is a non specific marker, also present in methylmalonic acidemia, a disorder of cobalamin metabolism. Further analysis using original blood spots for 3-OH-propionic and methylmalonic acids performed as 2nd tier tests allow to rule in or to rule out propionyl and methylmalonyl acidemia. These 2nd tier tests increase the specificity of the screening procedure, with less than 5% of positive C3 being confirmed as true positives (la Marca et al., 2007). Follow up testing for plasma acylcarnitines, along with other markers, have been developed to determine if the presumptive screened positive are truly positive and if so, for which condition. A number of other inborn errors of metabolism need a 2nd tier test to increase positive predictive value.

2.4.4 Short term follow-up of confirmed positive cases

In the presence of a sample that screened positive for a disorder, the NBS laboratory must report the results for rapid clinical evaluation, diagnosis and appropriate management of the newborn. Actions needed for complete and secure transmission of the information are not only dependent upon the NBS laboratory, but must be coordinated by a structured program. Indeed, a number of services will be needed, such as metabolic, enzymatic and/or molecular diagnostic laboratories, but also specialized health care providers such as geneticists, endocrinologists and metabolic dieticians. Newborns screened positive may be referred to either the primary health care providers or directly to tertiary specialized services, depending to the structure of health care services.

Because of the rapid increase in the number of rare disorders screened with the expansion of NBS by MS/MS, primary health care providers are increasingly faced with the referral of infants with a rare disorder. Results may be presented as analytes measured and/or conditions screened, and should include a clear interpretation, especially in the case of screened positive. Importantly, abnormal results of a screened positive do not mean that the infant has the disease, as additional investigation may be needed before a final diagnosis is reached. Classically, treatment is not initiated before the diagnosis is confirmed and parents have received appropriate counseling. With the increased number of conditions screened, it is thus even more imperative that the reporting of screened positive infant be presented in adequate and comprehensible format to the health care provider. To help physicians structure their approach in the provision of care for these newborn, the American College of Medical Genetics (ACMG) has developed a tool, called the ACT Sheet and confirmatory diagnosis, which provides information for each condition involved in NBS. The tool includes 1) a 1-page ACTion (ACT) sheet that describes short term actions and communication with the family and determining the appropriate follow up steps for the infant that has screened positive, and 2) a 1-page algorithm with an overview of the basic steps involved in determining the final diagnosis in the infant (ACMG, 2011). The ACT sheet and confirmatory algorithm exist for various disorders (endocrine, haematological, genetic and metabolic), and each ACT sheet includes links to informational resources.

There is no doubt that expansion of NBS by MS/MS to a large number of disorders poses organisational challenges. As many rare conditions are screened for, NBS programs and health care professional must deal with the uncertainty about the natural history of these rare conditions. The laboratory is central to the whole NBS process: it must develop a test menu that includes specific 2nd tier testing protocols in order to confirm the biochemical abnormality and identify the underlying metabolic condition (see Table 6 for examples).

Disorder	Example of follow-up marker(s)	Example of follow-up method(s)
Phenylcetonuria	phenylalanine, tyrosine in plasma	ion exchange chromatography (IEC)
	urine organic acids	GCMS
MSUD	valine, leucine, isoleucine alloisoleucine in plasma	IEC
	urine organic acids	GCMS
Argininosuccinic acidemia	argininosuccinate, citrulline	IEC
	urine organic acid	GCMS
Medium chain acyl CoA dehydrogenase deficiency (MCAD)	C6, C8, C10 acylcarnitine in plasma	LCMS/MS
	urinary organic acid	GCMS
glutaric aciduria type1	plasma C5DC acyl carnitine	LCMS/MS
	urine organic acid	GCMS
glutaric aciduria type2	plasma C5DC, C5, C5OH, C6, C8, C10-C16 acyl-carnitines	LCMS/MS
	urine organic acid	GCMS
Additional tests with cells and enzymatic assays can confirm the diagnosis		

Table 6. Examples of follow-up testing for positive cases in newborn screening program (*adapted from Dietzen et al., 2009*).

Highly structured NBS programs will favour the increase in knowledge about these rare conditions, and will ultimately improve provision of care for these infants, communication between parents and health care professional and will reduce parental anxiety related to uncertainty (Deluca et al., 2011). Implementation of electronic health information exchange (HIE) is an opportunity for quality improvement of NBS. With harmonization of standards, coding and terminology and implementation of decision-making and support tools, access to HIE by health care professionals should result in improved effectiveness of short and long term management of infants with an inborn error of metabolism (Downs et al., 2010).

2.4.5 Conservation of dried blood spots

NBS laboratories receive dried blood spots (DBS) specimens from virtually all newborns from their territory. NBS specimens represent an unbiased source of blood that can generate new population-based knowledge, including potential improved children's health, but also new understanding on the background of both child and adult disease. In addition, NBS is probably the only program reaching the entire population, representing, for instance, about 4 millions births/year in the USA. To ensure proper security, confidentiality, privacy and public confidence in NBS practices, each jurisdiction must regulate NBS practices regarding retention and storage of DBS. Owing to the implementation of molecular techniques using DBS, there is a growing interest in the use of these unique samples for purposes other than NBS, such as genomic research. Stored DBS specimens have also been used to establish the presence of congenital cytomegalovirus infection (Barbi et al., 2006; de Vries et al., 2009) and for forensic purposes (Couzin-Frankel, 2009).

Currently, storage and secondary use of DBS is controversial and two distinct approaches are in use, a short-term storage approach (i.e. < 2-3 years) and a long-term approach (more than 5-10 years). While the first approach allows the standard use of DBS for quality insurance and program evaluation, treatment efficacy, test refinement and result verification, the latter one further allows the use of DBS for research purposes, that is for purposes other than those for which they were originally collected. Parents from Minnesota and Texas have charged their respective health departments because DBS specimens had been stored without parents' knowledge or consent. This is not surprising, in the absence of national guidelines and of diverse practices regarding retention and use of specimen in the USA (Lewis et al., 2011) and Canada (Avard et al., 2006). In Texas, as part of the settlement more than 5 millions DBS specimens dating back to 2002 were destroyed (Grody & Howell, 2010; Rollins, 2011). Recently, the Secretary's Advisory Committee on Heritable Disorders in Newborns Children (SACHDNC) published a briefing paper which reviewed the issues facing the state NBS programs related to retention and use of residual DBS specimens and proposed a foundation for developing national guidelines (Secretary's Advisory Committee on Heritable Disorders in Newborns and Children, 2010). This included recommendations for the implementation of a policy on access, disposition, protection of privacy and confidentiality of residual DBS specimens. Biobanking residual DBS for secondary use was also addressed by the President's Council on Bioethics (President's Council on Bioethics, 2008). Noteworthy, in 1993 the Denmark government implemented a national newborn screening biobank addressing most of these issues (Norgaard-Pedersen & Simonsen, 1999). The Organization for Economic Co-operation and Development (OECD) with the International Society for Biological and Environmental Repositories jointly developed best practice guidelines useful to implement such policies in NBS (Baust, 2008; OECD, 2007).

Use of stored DBS specimens requires stability and integrity of samples. Stability will depend on the analyte to be measured. Non-DNA material such as amino acids left at room temperature may degrade within a few months (Therrell et al., 1996), but may be stable for years at -20^{0}C. Long-term storage at ambient temperature results in significant degradation of acylcarnitines, which are hydrolysed into free carnitine and corresponding fatty acids, and aminoacids (Strnadova et al., 2007). The velocity of the decay is logarithmic, depends on chain length of acylcarnitine (Fingerhut et al., 2009), and appropriate correction for storage should be applied. Acylcarnitines are stable at -18^{0}C for at least one year (Fingerhut et al., 2009). On the other hand, DNA quality from stored DBS specimens at room temperature allowed extraction and successful amplification for at least 25 years (Searles Nielsen et al., 2008). Quantitative RNA stability was also shown from stored residual NBS specimens for 20 years at 4^{0}C in controlled relative humidity maintained at 30% (Gauffin et al., 2009). Specimen-to-specimen contamination should be prevented.

3. Data management

Given that each newborn sample is unique and needs to be analyzed with reliability and in a systematic way, data management is an important component of the newborn screening laboratory infrastructure. Typically, the data management system, or laboratory information system (LIS) is expected to manage sample and patient information, interact with the computers onboard both the sample preparation instruments, the analytical instrumentation, manage the various analytical and interpretation algorithms, as well as the production of reports and their transmission to the appropriate health care professional(s). Ideally, the LIS should be able to compute laboratory production statistics, gather and analyze quality control procedures as well as contribute, at least partly, to the laboratory's Quality System indicators. Data storage is usually regulated within each health jurisdiction with respect to its duration, access by various stakeholders and usage. Laboratory data can also be used to evaluate the NBS laboratory's performance. Typically, monitoring of the rate of positive samples (which should remain stable over time), the ratios of various analytical signals within samples, the false positive rate, signal-to-noise ratios can be very useful for laboratory professionals and managers to detect either failures in a laboratory process, or sometimes actual deterioration of the quality of a specific production step or equipment that merit some attention to prevent more serious problems. In the context of laboratories that process all newborn samples from a specific geographic region, particular attention must be placed on preventing a production shutdown which would be very detrimental in the context of diseases that could remain undetected for a longer period of time than acceptable.

4. Revised criteria for population screening

The low reagent cost of MS/MS and its capacity to detect several tens of metabolites in a single run have challenged the paradigm of population screening criteria. As discussed above, different countries have introduced MS/MS technology and expanded the number of disorders in their panel at very different rates after applying different policies and approaches. Two such extremes are the UK, which screen only for a few disorders and the USA where most states now screen for 30 disorders and more. Indeed, if the UK health authorities approved MS/MS in principle, apart from PKU and MCADD, no further expansion is envisioned before more data are produced on evidence of benefits of NBS for other disorders (Pandor et al., 2004). At the other end of the spectrum, most states in the USA now comply to the proposed American College of Medical Genetics (ACMG) core panel of 29 disorders to be screened as primary targets, and an additional 25 disorders (so called secondary targets) that should be identified and reported through full MS/MS profiling (Watson et al., 2006). Many countries, such as Germany, Switzerland and Australia, but also some states in the USA and provinces in Canada, have performed their own evaluation process resulting in expansion to a number of disorders in between both extremes. Clearly, the original screening principles established by Wilson and Junger in 1968 (Wilson & Junger, 1968) - discussed in section 1.2 - have not been applied strictly to NBS. This may be a consequence of the qualitative terms of the Wilson and Junger principles, which make them difficult to use as decision tools. However, the difficulty also arises from the nature of NBS for metabolic diseases which are rare disorders with unclear natural history and benefits of preventive measures/treatment due to desperately lacking data. Consequently, as more disorders with very low prevalence and heterogeneous clinical characteristics are added to the NBS panel, it becomes difficult to establish the balance between potential benefits and harms, as well as to evaluate cost-effectiveness. Interestingly, a recent report showed that parents of children affected with very rare disorders are much in favor of NBS, even for untreatable disorders, mainly because NBS may eliminate a painful diagnostic odyssey (Plass et al., 2010). On the other hand, the recent report from the US Chair on Bioethics unequivocally rejected the technological imperative as a argument for NBS (President's Council on Bioethics, 2008). It is beyond the scope of this chapter to detail the ethical stakes of NBS expansion, but as pointed out by Botkin et al (2006), when neither benefits nor harms are well characterized, a more cautious approach may be warranted.

There is no consensus on whether Wilson and Junger screening principles might or might not be applicable for NBS and the ACMG panel shows a number of flaws (Moyer et al., 2008). Randomized controlled trials may be warranted in NBS, and there is a need to develop a rigorous process to assess available evidence and develop decision tools. Recently, a proposal for reviewed Wilson and Junger screening principles was published in the WHO-bulletin (Andermann et al., 2008). In addition the UK National Screening Committee (NSC) further expanded the 10 Wilson and Junger screening principles into 19 more specific criteria and 87 items of information under 35 general headings (UK National Screening Committee, 2011). These attempts at better defining screening criteria and their evaluation may help public health authorities to better determine which disorders should be implemented into universal public health NBS programs as innovations and new knowledge is reported. Also, cost-effectiveness studies, when available, will inform decision makers about screening options and the impact of various parameters within specific health jurisdictions and population contexts (Venditti et al., 2003, Pandor et al., 2004).

5. Perspectives for tandem mass spectrometry in population screening

Tandem mass spectrometry has become a standard method in routine clinical laboratories. The field of newborn screening for inherited metabolic diseases has shown to be a natural fit for this technology that is both sensitive, precise and allows multiplex analyses of several tens of analytes from very small samples. Since year 2000, tens of thousands infants and their families have already benefited from the early identification and treatment of their inherited metabolic disease that has been allowed by this technology. Industry has also rapidly adapted and now proposes turnkey solutions for large panels of relevant (and perhaps less relevant) NBS metabolites (Lehotay et al., 2011). New methods are proposed on a regular basis and, in the near future, it would not be surprising to witness the availability of MS/MS approaches for the more complex molecules such as the peptides measured in congenital hypothyroidism and hemoglobinopathies. Progresses at automation of many steps of MS/MS in the clinical laboratory (Vogeser & Kirchhoff, 2011), as well as miniaturization of instruments (e.g. by the use of nanosprays and microfluidics) also open new avenues for further improvement of the efficacy of these techniques.

The existence of unresolved questions is highlighted by the great variety in the size of testing panels offered in different jurisdictions, as well as significant discrepancies between existing guidelines for NBS programs with respect to the diseases to be screened. Future applications of this powerful technology will likely include the detection and measurement of peptides and complex molecules with potential applications in NBS. Although other analytical methods will continue to be needed either to screen for specific molecules, or for confirmatory studies of samples positive on screening, tandem mass spectrometry in the field of newborn screening is already a pillar technology and is promised to a great future.

6. Acknowledgments

FR holds a Research Chair in health technology assessment and evidence-based laboratory medicine (Fonds de Recherche en Santé du Québec (FRSQ)/Québec Ministry of Health and Social Services/Centre Hospitalier Universitaire de Québec). YG holds a Senior clinician-scientist award from FRSQ. This work having led to this chapter was in part supported by the APOGÉE-Net/CanGèneTest Research and Knowledge Network in Genetic health Services, funded by the Canadian Institutes for Health Research (www.cangenetest.org).

7. References

Adam BW, Alexander JR, Smith SJ, Chace DH, Loeber JG, Elvers LH, et al. (2000). Recoveries of phenylalanine from two sets of dried-blood-spot reference materials: prediction from hematocrit, spot volume, and paper matrix. *Clin Chem*, vol.46, No.1, (Jan), pp.(126-8), 0009-9147 (Print), 0009-9147 (Linking).

American College of Medical Genetics. (2011). Available from: <http://www.acmg.net/AM/Template.cfm?Section=act_sheets_and_confirmatory_algorithms>.

American College of Medical Genetics Newborn Screening Expert Group. (2006). Newborn screening: toward a uniform screening panel and system--executive summary. *Pediatrics*, vol.117, No.5 Pt 2, (May), pp.(S296-307), 1098-4275 (Electronic), 0031-4005 (Linking).

Andermann A, Blancquaert I, Beauchamp S & Dery V. (2008). Revisiting Wilson and Jungner in the genomic age: a review of screening criteria over the past 40 years. *Bull World Health Organ*, vol.86, No.4, (Apr), pp.(317-9), 1564-0604 (Electronic), 0042-9686 (Linking).

Avard D, Vallance H, Greenberg C, Laberge C, Kharaboyan L & Plant M. (2006), Variability in the Storage and Use of Newborn Dried Bloodspots in Canada: Is it Time for National Standards?, *Genomics, Society and Policy*, vol.2, No.3, pp.(80-95), 1746-5354

Balogh MP & Stacey CC. (1991). Chromatographic developments in low flow delivery for liquid chromatography-mass spectrometry. *J Chromatogr*, vol.562, No.1-2, (Jan 2), pp.(73-9), 0021-9673 (Print), 0021-9673 (Linking).

Barbi M, Binda S, Caroppo S & Primache V. (2006). Neonatal screening for congenital cytomegalovirus infection and hearing loss. *J Clin Virol*, vol.35, No.2, (Feb), pp.(206-9), 1386-6532 (Print), 1386-6532 (Linking).

Baust JG. (2008). ISBER: Best Practices for Repositories and Trends at the Institute for Problems of Cryobiology and Medicine. *Cell Preservation Technology*, vol.6, No.1, pp.(1).

Bodamer OA, Hoffmann GF & Lindner M. (2007). Expanded newborn screening in Europe 2007. *J Inherit Metab Dis*, vol.30, No.4, (Aug), pp.(439-44), 1573-2665 (Electronic), 0141-8955 (Linking).

Botkin JR, Clayton EW, Fost NC, Burke W, Murray TH, Baily MA, et al. (2006). Newborn screening technology: proceed with caution. *Pediatrics*, vol.117, No.5, (May), pp.(1793-9), 1098-4275 (Electronic), 0031-4005 (Linking).

Chace DH, Millington DS, Terada N, Kahler SG, Roe CR & Hofman LF. (1993). Rapid diagnosis of phenylketonuria by quantitative analysis for phenylalanine and tyrosine in neonatal blood spots by tandem mass spectrometry. *Clin Chem*, vol.39, No.1, (Jan), pp.(66-71), 0009-9147 (Print), 0009-9147 (Linking).

Chace DH, Adam BW, Smith SJ, Alexander JR, Hillman SL & Hannon WH. (1999). Validation of accuracy-based amino acid reference materials in dried-blood spots by tandem mass spectrometry for newborn screening assays. *Clin Chem*, vol.45, No.8 Pt 1, (Aug), pp.(1269-77), 0009-9147 (Print), 0009-9147 (Linking).

Chace DH, Kalas TA & Naylor EW. (2003). Use of tandem mass spectrometry for multianalyte screening of dried blood specimens from newborns. *Clin Chem*, vol.49, No.11, (Nov), pp.(1797-817), 0009-9147 (Print), 0009-9147 (Linking).

Chace DH. (2009). Mass spectrometry in newborn and metabolic screening: historical perspective and future directions. *J Mass Spectrom*, vol.44, No.2, (Feb), pp.(163-70), 1076-5174 (Print), 1076-5174 (Linking).

Couzin-Frankel J. (2009). Newborn blood collections. Science gold mine, ethical minefield. *Science*, vol.324, No.5924, (Apr 10), pp.(166-8), 1095-9203 (Electronic), 0036-8075 (Linking).

Dantas MF, Suormala T, Randolph A, Coelho D, Fowler B, Valle D, et al. (2005). 3-Methylcrotonyl-CoA carboxylase deficiency: mutation analysis in 28 probands, 9 symptomatic and 19 detected by newborn screening. *Hum Mutat*, vol.26, No.2, (Aug), pp.(164), 1098-1004 (Electronic), 1059-7794 (Linking).

De Jesus VR, Chace DH, Lim TH, Mei JV & Hannon WH. (2010). Comparison of amino acids and acylcarnitines assay methods used in newborn screening assays by tandem mass spectrometry. *Clin Chim Acta*, vol.411, No.9-10, (May 2), pp.(684-9), 1873-3492 (Electronic), 0009-8981 (Linking).

de Vries JJ, Claas EC, Kroes AC & Vossen AC. (2009). Evaluation of DNA extraction methods for dried blood spots in the diagnosis of congenital cytomegalovirus infection. *J Clin Virol*, vol.46 Suppl 4, (Dec), pp.(S37-42), 1873-5967 (Electronic), 1386-6532 (Linking).

Deluca JM, Kearney MH, Norton SA & Arnold GL. (2011). Parents' experiences of expanded newborn screening evaluations. *Pediatrics*, vol.128, No.1, (Jul), pp.(53-61), 1098-4275 (Electronic), 0031-4005 (Linking).

Dietzen DJ, Rinaldo P, Whitley RJ, Rhead WJ, Hannon WH, Garg UC, et al. (2009). National academy of clinical biochemistry laboratory medicine practice guidelines: follow-up testing for metabolic disease identified by expanded newborn screening using tandem mass spectrometry; executive summary. *Clin Chem*, vol.55, No.9, (Sep), pp.(1615-26), 1530-8561 (Electronic), 0009-9147 (Linking).

DiLella AG, Kwok SC, Ledley FD, Marvit J & Woo SL. (1986). Molecular structure and polymorphic map of the human phenylalanine hydroxylase gene. *Biochemistry*, vol.25, No.4, (Feb 25), pp.(743-9), 0006-2960 (Print), 0006-2960 (Linking).

Downs SM, van Dyck PC, Rinaldo P, McDonald C, Howell RR, Zuckerman A, et al. (2010). Improving newborn screening laboratory test ordering and result reporting using health information exchange. *J Am Med Inform Assoc*, vol.17, No.1, (Jan-Feb), pp.(13-8), 1527-974X (Electronic), 1067-5027 (Linking).

Dussault JH & Laberge C. (1973). [Thyroxine (T4) determination by radioimmunological method in dried blood eluate: new diagnostic method of neonatal hypothyroidism?]. *Union Med Can*, vol.102, No.10, (Oct), pp.(2062-4), 0041-6959 (Print), 0041-6959 (Linking).

Etter ML, Eichhorst J & Lehotay DC. (2006). Clinical determination of 17-hydroxyprogesterone in serum by LC-MS/MS: comparison to Coat-A-Count RIA method. *J Chromatogr B Analyt Technol Biomed Life Sci*, vol.840, No.1, (Aug 7), pp.(69-74), 1570-0232 (Print), 1570-0232 (Linking).

Fingerhut R, Ensenauer R, Roschinger W, Arnecke R, Olgemoller B & Roscher AA. (2009). Stability of acylcarnitines and free carnitine in dried blood samples: implications for retrospective diagnosis of inborn errors of metabolism and neonatal screening for carnitine transporter deficiency. *Anal Chem*, vol.81, No.9, (May 1), pp.(3571-5), 1520-6882 (Electronic), 0003-2700 (Linking).

Gauffin F, Nordgren A, Barbany G, Gustafsson B & Karlsson H. (2009). Quantitation of RNA decay in dried blood spots during 20 years of storage. *Clin Chem Lab Med*, vol.47, No.12, pp.(1467-9), 1434-6621 (Print), 1434-6621 (Linking).

Grody WW & Howell RR. (2010). The fate of newborn screening blood spots. *Pediatr Res*, vol.67, No.3, (Mar), pp.(237), 1530-0447 (Electronic), 0031-3998 (Linking).

Guthrie R & Susi A. (1963). A Simple Phenylalanine Method for Detecting Phenylketonuria in Large Populations of Newborn Infants. *Pediatrics*, vol.32, (Sep), pp.(338-43), 0031-4005 (Print), 0031-4005 (Linking).

Hannon WH, Boyle J, Davin B, Marsden A, McCabe ER, Schwartz M, et al. (1997). *Blood collection on filter paper for neonatal screening programs: approved standard*, Third Edition, NCCLS document LA4-A3, NCCLS, 940 West Valley Road, Suite 1400, Wayne (PA) 19087-1898 USA

International Organization for Standardization. (2003). *ISO 15189:2003. Medical laboratories - Particular requirements for quality and competence*, International Organization for Standardization, Geneva, Switzerland

Johnson AW, Mills K & Clayton PT. (1996). The use of automated electrospray ionization tandem MS for the diagnosis of inborn errors of metabolism from dried blood spots. *Biochem Soc Trans*, vol.24, No.3, (Aug), pp.(932-8), 0300-5127 (Print), 0300-5127 (Linking).

Kasper DC, Ratschmann R, Metz TF, Mechtler TP, Moslinger D, Konstantopoulou V, et al. (2010). The national Austrian newborn screening program - eight years experience with mass spectrometry. past, present, and future goals. *Wien Klin Wochenschr*, vol.122, No.21-22, (Nov), pp.(607-13), 1613-7671 (Electronic), 0043-5325 (Linking).

Kayton A. (2007). Newborn screening: a literature review. *Neonatal Netw*, vol.26, No.2, (Mar-Apr), pp.(85-95), 0730-0832 (Print), 0730-0832 (Linking).

Koeberl DD, Millington DS, Smith WE, Weavil SD, Muenzer J, McCandless SE, et al. (2003). Evaluation of 3-methylcrotonyl-CoA carboxylase deficiency detected by tandem mass spectrometry newborn screening. *J Inherit Metab Dis*, vol.26, No.1, pp.(25-35), 0141-8955 (Print), 0141-8955 (Linking).

la Marca G, Malvagia S, Pasquini E, Innocenti M, Donati MA & Zammarchi E. (2007). Rapid 2nd-tier test for measurement of 3-OH-propionic and methylmalonic acids on dried blood spots: reducing the false-positive rate for propionylcarnitine during expanded newborn screening by liquid chromatography-tandem mass spectrometry. *Clin Chem*, vol.53, No.7, (Jul), pp.(1364-9), 0009-9147 (Print), 0009-9147 (Linking).

Laberge C, Scriver CR, Clow CL & Dufour D. (1975). [The genetic medicine network in Quebec: An integrated program for diagnosis, counseling and treatment of hereditary metabolic diseases]. *Union Med Can*, vol.104, No.3, (Mar), pp.(428-32), 0041-6959 (Print), 0041-6959 (Linking).

Lacey JM, Minutti CZ, Magera MJ, Tauscher AL, Casetta B, McCann M, et al. (2004). Improved specificity of newborn screening for congenital adrenal hyperplasia by second-tier steroid profiling using tandem mass spectrometry. *Clin Chem*, vol.50, No.3, (Mar), pp.(621-5), 0009-9147 (Print), 0009-9147 (Linking).

Laflamme N, Fortier M, Lindsay C & Turgeon J. (2005). *Rapport d'évaluation du programme Québécois de dépistage sanguin des maladies génétiques chez le nouveau-né*, ISBN 2-550-47125-3, Institut National de Santé Publique du Québec, Québec

Lehotay DC, Hall P, Lepage J, Eichhorst JC, Etter ML & Greenberg CR. (2011). LC-MS/MS progress in newborn screening. *Clin Biochem*, vol.44, No.1, (Jan), pp.(21-31), 1873-2933 (Electronic), 0009-9120 (Linking).

Lewis MH, Goldenberg A, Anderson R, Rothwell E & Botkin J. (2011). State laws regarding the retention and use of residual newborn screening blood samples. *Pediatrics*, vol.127, No.4, (Apr), pp.(703-12), 1098-4275 (Electronic), 0031-4005 (Linking).

Magera MJ, Gunawardena ND, Hahn SH, Tortorelli S, Mitchell GA, Goodman SI, et al. (2006). Quantitative determination of succinylacetone in dried blood spots for newborn screening of tyrosinemia type I. *Mol Genet Metab*, vol.88, No.1, (May), pp.(16-21), 1096-7192 (Print), 1096-7192 (Linking).

Mandl KD, Feit S, Larson C & Kohane IS. (2002). Newborn screening program practices in the United States: notification, research, and consent. *Pediatrics*, vol.109, No.2, (Feb), pp.(269-73), 1098-4275 (Electronic), 0031-4005 (Linking).

McKusick-Nathans Institute of Genetic Medicine; Johns Hopkins University. (2011). *Online Mendelian Inheritance in Man, OMIM,* Available from: <http://omim.org/>.

Miller J, Tuerck MS, Awad K, Chace DH, Copeland S, Rasmussen SA, et al. (2009). *Newborn Screening for Preterm, Low Birth Wieght, and Sick Newborns: Approved Guideline*, CLSI document I/LA31-A, ISBN 1-56238-710-3, CLSI, 940 West Valley Road, Suite 1400, Wayne (PA) 19087-1898 USA

Millington DS, Kodo N, Norwood DL & Roe CR. (1990). Tandem mass spectrometry: a new method for acylcarnitine profiling with potential for neonatal screening for inborn errors of metabolism. *J Inherit Metab Dis*, vol.13, No.3, pp.(321-4), 0141-8955 (Print), 0141-8955 (Linking).

Minutti CZ, Lacey JM, Magera MJ, Hahn SH, McCann M, Schulze A, et al. (2004). Steroid profiling by tandem mass spectrometry improves the positive predictive value of newborn screening for congenital adrenal hyperplasia. *J Clin Endocrinol Metab*, vol.89, No.8, (Aug), pp.(3687-93), 0021-972X (Print), 0021-972X (Linking).

Moyer VA, Calonge N, Teutsch SM & Botkin JR. (2008). Expanding newborn screening: process, policy, and priorities. *Hastings Cent Rep*, vol.38, No.3, (May-Jun), pp.(32-9), 0093-0334 (Print), 0093-0334 (Linking).

NCCLS. (2004). *Application of a Quality Management System Model for Laboratory Services: Approved Guideline*, Third Edition, NCCLS document GP26-A3, ISBN 1-56238-553-4, NCCLS, 940 West Valley Road, Suite 1400, Wayne, Pensylvania 19087-1898 USA

NHS Newborn Bllodspot Screening Programme. (2008). Available from: <http://newbornbloodspot.screening.nhs.uk/>.

Niu DM, Chien YH, Chiang CC, Ho HC, Hwu WL, Kao SM, et al. (2010). Nationwide survey of extended newborn screening by tandem mass spectrometry in Taiwan. *J Inherit Metab Dis*, vol.33, No.Suppl 2, (Oct), pp.(S295-305), 1573-2665 (Electronic), 0141-8955 (Linking).

Norgaard-Pedersen B & Simonsen H. (1999). Biological specimen banks in neonatal screening. *Acta Paediatr Suppl*, vol.88, No.432, (Dec), pp.(106-9), 0803-5326 (Print), 0803-5326 (Linking).

Organisation for Economic Cooperation and Development (OECD). (2007). *OECD Best Practice Guidelines for Biological Resource Centres*, OCDE,

Pandor A, Eastham J, Beverley C, Chilcott J & Paisley S. (2004). Clinical effectiveness and cost-effectiveness of neonatal screening for inborn errors of metabolism using tandem mass spectrometry: a systematic review. *Health Technol Assess*, vol.8, No.12, (Mar), pp.(iii, 1-121), 1366-5278 (Print), 1366-5278 (Linking).

Pitt JJ. (2010). Newborn screening. *Clin Biochem Rev*, vol.31, No.2, (May), pp.(57-68), 0159-8090 (Print), 0159-8090 (Linking).

Plass AM, van El CG, Pieters T & Cornel MC. (2010). Neonatal screening for treatable and untreatable disorders: prospective parents' opinions. *Pediatrics*, vol.125, No.1, (Jan), pp.(e99-106), 1098-4275 (Electronic), 0031-4005 (Linking).

Pollitt RJ. (2006). International perspectives on newborn screening. *J Inherit Metab Dis*, vol.29, No.2-3, (Apr-Jun), pp.(390-6), 1573-2665 (Electronic), 0141-8955 (Linking).

President's Council on Bioethics. (2008). *The Changing Moral Focus of Newborn Screening*, President's Council on Bioethics, Washington, DC

Rashed MS, Ozand PT, Bucknall MP & Little D. (1995). Diagnosis of inborn errors of metabolism from blood spots by acylcarnitines and amino acids profiling using automated electrospray tandem mass spectrometry. *Pediatr Res*, vol.38, No.3, (Sep), pp.(324-31), 0031-3998 (Print), 0031-3998 (Linking).

Rastogi MV & LaFranchi SH. (2010). Congenital hypothyroidism. *Orphanet J Rare Dis*, vol.5, pp.(17), 1750-1172 (Electronic), 1750-1172 (Linking).

Region 4 Genetics Collaborative. (2007). Available from: <www.region4genetics.org/>.

Rinaldo P, Zafari S, Tortorelli S & Matern D. (2006). Making the case for objective performance metrics in newborn screening by tandem mass spectrometry. *Ment Retard Dev Disabil Res Rev*, vol.12, No.4, pp.(255-61), 1080-4013 (Print), 1080-4013 (Linking).

Rolinski B, Arnecke R, Dame T, Kreischer J, Olgemoller B, Wolf E, et al. (2000). The biochemical metabolite screen in the Munich ENU Mouse Mutagenesis Project: determination of amino acids and acylcarnitines by tandem mass spectrometry. *Mamm Genome*, vol.11, No.7, (Jul), pp.(547-51), 0938-8990 (Print), 0938-8990 (Linking).

Rollins G. (2011), Who Owns Newborn Screening Samples? Murky State Laws Threaten to Hinder Research using Residual Blood Spots, *Clinical Laboratory News*, vol.37, No.6,

Sander J, Janzen N, Peter M, Sander S, Steuerwald U, Holtkamp U, et al. (2006). Newborn screening for hepatorenal tyrosinemia: Tandem mass spectrometric quantification of succinylacetone. *Clin Chem*, vol.52, No.3, (Mar), pp.(482-7), 0009-9147 (Print), 0009-9147 (Linking).

Schulze A, Lindner M, Kohlmuller D, Olgemoller K, Mayatepek E & Hoffmann GF. (2003). Expanded newborn screening for inborn errors of metabolism by electrospray ionization-tandem mass spectrometry: results, outcome, and implications. *Pediatrics*, vol.111, No.6 Pt 1, (Jun), pp.(1399-406), 1098-4275 (Electronic), 0031-4005 (Linking).

Schwarz E, Liu A, Randall H, Haslip C, Keune F, Murray M, et al. (2009). Use of steroid profiling by UPLC-MS/MS as a second tier test in newborn screening for congenital adrenal hyperplasia: the Utah experience. *Pediatr Res*, vol.66, No.2, (Aug), pp.(230-5), 1530-0447 (Electronic), 0031-3998 (Linking).

Scriver CR. (2007). The PAH gene, phenylketonuria, and a paradigm shift. *Hum Mutat*, vol.28, No.9, (Sep), pp.(831-45), 1098-1004 (Electronic), 1059-7794 (Linking).

Searles Nielsen S, Mueller BA, De Roos AJ & Checkoway H. (2008). Newborn screening archives as a specimen source for epidemiologic studies: feasibility and potential for bias. *Ann Epidemiol*, vol.18, No.1, (Jan), pp.(58-64), 1047-2797 (Print), 1047-2797 (Linking).

Secretary's Advisory Committee on Heritable Disorders in Newborns and Children. (2010). *Considerations and Recommendations for National Guidance Regarding the Retention and Use of Residual Dried Blood Spot Specimens after Newborn Screening*, Briefing paper,

Strnadova KA, Holub M, Muhl A, Heinze G, Ratschmann R, Mascher H, et al. (2007). Long-term stability of amino acids and acylcarnitines in dried blood spots. *Clin Chem*, vol.53, No.4, (Apr), pp.(717-22), 0009-9147 (Print), 0009-9147 (Linking).

Therrell BL, Hannon WH, Pass KA, Lorey F, Brokopp C, Eckman J, et al. (1996). Guidelines for the retention, storage, and use of residual dried blood spot samples after newborn screening analysis: statement of the Council of Regional Networks for Genetic Services. *Biochem Mol Med*, vol.57, No.2, (Apr), pp.(116-24), 1077-3150 (Print), 1077-3150 (Linking).

Turgeon CT, Magera MJ, Cuthbert CD, Loken PR, Gavrilov DK, Tortorelli S, et al. (2010). Determination of total homocysteine, methylmalonic acid, and 2-methylcitric acid in dried blood spots by tandem mass spectrometry. *Clin Chem*, vol.56, No.11, (Nov), pp.(1686-95), 1530-8561 (Electronic), 0009-9147 (Linking).

UK National Screening Committee. (2011). Available from: <http://www.nsc.nhs.uk/>.

Venditti LN, Venditti CP, Berry GT, Kaplan PB, Kaye EM, Glick H, et al. (2003). Newborn screening by tandem mass spectrometry for medium-chain Acyl-CoA dehydrogenase deficiency: a cost-effectiveness analysis. *Pediatrics*, vol.112, No.5, (Nov), pp.(1005-15), 1098-4275 (Electronic), 0031-4005 (Linking).

Vogeser M & Kirchhoff F. (2011). Progress in automation of LC-MS in laboratory medicine. *Clin Biochem*, vol.44, No.1, (Jan), pp.(4-13), 1873-2933 (Electronic), 0009-9120 (Linking).

Wadsworth Center. (n.d.). Available from: <http:www.wadsworth.org/>.

Watson MS, Lloyd-Puryear MA, Mann MY, Rinaldo P & Howell RR. (2006). Newborn screening: Toward a uniform screening panel and system. *Genet Med*, vol.8, No.5 (suppl.), pp.(12S-252S).

Wilcken B. (2007). Recent advances in newborn screening. *J Inherit Metab Dis*, vol.30, No.2, (Apr), pp.(129-33), 1573-2665 (Electronic), 0141-8955 (Linking).

Wilcken B & Wiley V. (2008). Newborn screening. *Pathology*, vol.40, No.2, (Feb), pp.(104-15), 0031-3025 (Print), 0031-3025 (Linking).

Wilson JM & Junger G. (1968). *Principles and Practice of Screening for Disease*, World Health Organization, ISBN 9789241300346.